Formulas and Equations

$$m = \frac{y_2 - y_1}{x_2 - x_1}$$ Slope of a line

$ax + b = 0$ Linear equation

$y = mx + b$ Slope–intercept form

$y = m(x - x_1) + y_1$ Point–slope form

$x = h$ Vertical line

$y = b$ Horizontal line

$d = rt$ Distance, rate, and time

$s = 16t^2$ Distance for a falling object

$a^2 - b^2 = (a - b)(a + b)$ Difference of two squares

$(a + b)^2 = a^2 + 2ab + b^2$
$(a - b)^2 = a^2 - 2ab + b^2$ Square of a binomial

$x = -\dfrac{b}{2a}$ Vertex formula (x-coordinate)

$y = a(x - h)^2 + k$ Vertex form

$ax^2 + bx + c = 0$ Quadratic equation

$x = \dfrac{-b \pm \sqrt{b^2 - 4ac}}{2a}$ Quadratic formula

$(x - h)^2 + (y - k)^2 = r^2$ Equation of a circle

$d = \sqrt{(x_2 - x_1)^2 + (y_2 - y_1)^2}$ Distance between two points

Geometry

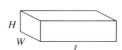

Rectangle
$A = LW$
$P = 2L + 2W$

Triangle
$A = \frac{1}{2}bh$
$P = a + b + c$

Pythagorean Theorem
$c^2 = a^2 + b^2$

Circle
$C = 2\pi r$
$A = \pi r^2$

Rectangular (Parallelepiped) Box
$V = LWH$
$S = 2LW + 2LH + 2WH$

Cylinder
$V = \pi r^2 h$
$S = 2\pi rh + 2\pi r^2$

Sphere
$V = \frac{4}{3}\pi r^3$
$S = 4\pi r^2$

Cone
$V = \frac{1}{3}\pi r^2 h$
$S = \pi r^2 + \pi r\sqrt{r^2 + h^2}$

Beginning and Intermediate Algebra

with Applications and Visualization

Gary K. Rockswold

Minnesota State University, Mankato

Terry A. Krieger

Winona State University

PEARSON

Addison
Wesley

Boston • San Francisco • New York • London • Toronto
Sydney • Tokyo • Singapore • Madrid • Mexico City
Munich • Paris • Cape Town • Hong Kong • Montreal

Publisher	Greg Tobin
Editor in Chief	Maureen O'Connor
Acquisitions Editor	Jennifer Crum
Project Editor	Lauren Morse
Editorial Assistant	Marcia Emerson
Managing Editor	Ron Hampton
Production Supervisor	Kathleen A. Manley
Production Services	Kathy Diamond
Photo Researcher	Beth Anderson
Compositor	Nesbitt Graphics, Inc.
Media Producer	Lynne Blaszak
Software Development	Chris Tragasz and Kathleen Bowler
Marketing Manager	Dona Kenly
Marketing Coordinator	Lindsay Skay
Prepress Supervisor	Caroline Fell
Manufacturing Buyer	Evelyn Beaton
Design Supervisor	Dennis Schaefer
Cover Designer	Andrea Menza
Cover Photograph	Comstock IMAGES

Photo Credits

p. 1: Digital Vision; p. 66: NASA/JPL; p. 83: PhotoDisc Blue; p. 146: Matt BenDaniel, Starmatt Astrophotography; p. 236: Brand X Pictures; pp. 289 and 319, NASA; p. 352: JohnFarmer, Cordaiy Photo Library Ltd./Corbis; p. 402: PhotoDisc; p. 454: Glen Allison/Stone; p. 476: Beth Anderson; p. 552: Corbis RF; p. 593: PhotoDisc RF; p. 625: Sara Anderson; pp. 661, 731, and 801: Digital Vision; p. 839: PhotoDisc

Library of Congress Cataloging-in-Publication Data
Rockswold, Gary K.
   Beginning and intermediate algebra with applications and visualization / Gary K. Rockswold, Terry A. Krieger.
      p. cm.
   Includes bibliographical references and index.
   ISBN 0-321-15891-1
   1. Algebra. I. Krieger, Terry A. II. Title.

QA152.3.R595 2005
512.9—dc22   2003070660

6 7 8 9 10—RRDW—08 07

To

Mel for Friendship, Mona for Sun Moon, and

Deb and Sen loo Chow for *Simple Truths*

CONTENTS

7   Rational Expressions   402

8   Introduction to Functions   476

LIST OF APPLICATIONS

PREFACE

Beginning and Intermediate Algebra with Applications and Visualization offers an innovative approach to the beginning and intermediate algebra curriculum that allows students to gain both skills and understanding. This textbook not only demonstrates the relevance of mathematics, but it also prepares students for future courses. The early introduction of graphs allows instructors to use applications and visualization to present mathematical concepts. Real data, graphs, and tables play an important role in the course, giving meaning to the numbers and equations that students encounter. This approach increases student interest, motivation, and the likelihood for success.

This textbook covers both beginning and intermediate algebra topics without the repetition of instruction that often occurs when using separate textbooks. It is one of three textbooks in an algebra series which also includes *Beginning Algebra with Applications and Visualization* and *Intermediate Algebra with Applications and Visualization,* Second Edition.

 ## Approach

In *Beginning and Intermediate Algebra with Applications and Visualization*, mathematical concepts are often introduced by moving from the concrete to the abstract. Relevant applications underscore mathematical concepts. This text includes a diverse collection of unique, up-to-date applications that answer the commonly asked question: "When will I ever use this?" Applications, visualization, and the Rule of Four (verbal, graphical, numerical, and symbolic methods) allow greater access to mathematics for students having different learning styles. However, the primary purpose of this text is to teach mathematical concepts and skills. Building skills is an objective of every section. Problem-solving skills are given a strong emphasis throughout, which often helps students succeed in more advanced mathematics courses. Whenever possible, problem solving is combined with current applications to increase student interest. Standard mathematical definitions, theorems, symbolism, and rigor are maintained.

A comprehensive curriculum is presented with a balanced and flexible approach that is essential for today's algebra courses. Instructors have the flexibility to strike their own balance with regard to emphasis on skills, Rule of Four, applications, and graphing calculator technology. This text contains numerous practical applications, including modeling of real-world data. Instructors have the freedom to integrate graphing calculators in this course; however, graphing calculator use is *optional*.

Organization

This text consists of 14 chapters and 75 sections with beginning algebra topics typically occuring in the first six or seven chapters. Functions are introduced in Chapter 8 and used throughout the second half of the book. Systems of linear equations in two variables are covered in Chapter 4, with more advanced topics about systems of linear equations covered in Chapter 9.

Chapter 1: Introduction to Algebra
This chapter provides a basic introduction to real numbers, the real number line, exponents, and properties of real numbers. Students are also introduced to variables, algebraic expressions, and equations.

Chapter 2: Linear Equations and Inequalities
This chapter introduces basic concepts related to equations and inequalities. It concentrates on linear equations and linear inequalities and integrates many applications to give meaning to the equations and inequalities that students encounter. A four-step process for problem solving is introduced, which is used throughout the textbook.

Chapter 3: Graphing Equations
Graphing equations in the rectangular coordinate system is discussed in this chapter. Linear equations in two variables are graphed, including lines in slope–intercept form and point–slope form. Slope as a rate of change is emphasized, and students are often asked to interpret slope in applications. The last section in this chapter provides a basic introduction to modeling data with linear equations.

Chapter 4: Systems of Linear Equations in Two Variables
This chapter follows naturally after Chapter 3. After graphing a single linear equation in two variables, students graph two linear equations in two variables. Both substitution and elimination are discussed as methods for solving systems of linear equations. Graphs and tables of values are also used to solve systems of linear equations. In the last section, systems of linear inequalities are solved by shading the solution set in the xy-plane.

Chapter 5: Polynomials and Exponents
This chapter begins with a review of basic concepts about exponents. Integer exponents are introduced later in the chapter. Addition, subtraction, multiplication, and division of polynomials are covered, along with special products.

Chapter 6: Factoring Polynomials and Solving Equations
This chapter discusses factoring polynomials and then uses factoring to solve polynomial equations. Factoring trinomials is presented in separate sections, first by factoring $x^2 + bx + c$ and then by factoring $ax^2 + bx + c$. Factoring is used to solve both basic quadratic equations (Section 6.5) and higher degree equations (Section 6.6).

Chapter 7: Rational Expressions
Rational expressions with many real-life applications are included. Addition, subtraction, multiplication, division, and simplification of rational expressions are covered. Separate sections focus on addition of rational expressions with like and unlike denominators. Complex fractions and rational equations with applications are discussed. The last section covers proportions and variation.

Chapter 8: Introduction to Functions
This chapter provides a basic introduction to functions, including their representations (Rule of Four), domain, and range. After this chapter, functions continue to be used throughout the textbook whenever appropriate. Linear functions are emphasized, but absolute value, polynomial, and rational (optional) functions are also introduced. Interval notation is used to identify a function's domain and range. The chapter concludes with a discussion of absolute value

equations and inequalities, using both symbolic and graphical methods that are based on a function approach.

Chapter 9: Matrices and Systems of Linear Equations

This chapter covers more advanced topics about systems of linear equations, including matrix solutions and determinants.

Chapter 10: Radical Expressions and Functions

Radical notation and some basic functions, such as the square root and cube root functions, are discussed. Presentation of properties of rational exponents is delayed until this chapter so that students can immediately apply them. Material on simplifying radical expressions is included. The chapter concludes with a section on complex numbers.

Chapter 11: Quadratic Functions and Equations

Quadratic functions and their graphs are discussed early in this chapter so that students may use these concepts to understand solutions to quadratic equations. The connection between *x*-intercepts, zeros of a quadratic function, and solutions to a quadratic equation is made. Methods for solving quadratic equations are discussed extensively.

Chapter 12: Exponential and Logarithmic Functions

This chapter begins with a section on composite and inverse functions. Logarithms, properties of logarithms, exponential and logarithmic functions, and exponential and logarithmic equations are covered in this chapter. Linear growth and exponential growth are compared, enabling students to grasp the fundamental difference between these two types of growth. The chapter contains many applications.

Chapter 13: Conic Sections

This chapter introduces parabolas, circles, ellipses, and hyperbolas. It concludes with a section on solving nonlinear systems of equations and inequalities.

Chapter 14: Sequences and Series

This chapter introduces the basic concepts of both sequences and series, concentrating on arithmetic and geometric sequences and series. Both graphical and symbolic representations of sequences are discussed, along with applications and models. The chapter concludes with the binomial theorem.

 ## Features

Applications and Models

Interesting, straightforward applications are a strength of this textbook, helping students become more effective problem solvers. Applications are intuitive and not overly technical so that they can be introduced in a minimum of class time. Current data are utilized to create meaningful mathematical models, exposing students to a wealth of actual uses of mathematics. A unique feature of this text is that the applications are woven into both the discussions and the exercises. Students can more easily learn how to solve applications when they are discussed within the text. (See pages 204, 274, and 406.)

Putting It All Together
This helpful feature occurs at the end of each section to summarize techniques and reinforce the mathematical concepts presented in the section. It is given in an easy-to-follow grid format. (See pages 90, 119, and 183-184.)

Section Exercise Sets
The exercise sets are the heart of any mathematics text, and this textbook includes a wide variety of exercises that are instructive for student learning. Each exercise set contains exercises involving basic concepts, skill-building, and applications. In addition, many exercises ask students to read and interpret graphs. Writing About Mathematics exercises are also included at the end of every exercise set. The exercise sets are carefully graded and categorized by topic, making it easier for instructors to select appropriate assignments. (See pages 172, 265, and 409.)

Checking Basic Concepts
After every two sections, this feature presents a brief set of exercises that students can use for review purposes or group activities. These exercises require 10–15 minutes to complete and can be used during class if time permits. (See pages 24, 124, and 210.)

Making Connections
This feature occurs throughout the text and helps students relate previously learned concepts to new concepts. (See pages 14, 87, and 161.)

Critical Thinking
One or more Critical Thinking exercises are included in most sections. They pose questions that can be used for either classroom discussion or homework assignments. These exercises typically ask students to extend a mathematical concept beyond what has already been discussed. (See pages 99, 183, and 273.)

Technology Notes
Occurring throughout the text, Technology Notes offer students guidance, suggestions, and cautions on the use of the graphing calculator. (See pages 95, 151, and 270.)

Group Activities: Working with Real Data
This feature occurs after selected sections (1 or 2 per chapter) and provides an opportunity for students to work collaboratively on a problem that involves real-world data. Most activities can be completed with limited use of class time. (See pages 175, 316, and 335.)

Chapter and Section Introductions
Many algebra students have little or no understanding of what mathematics is about. Chapter and section introductions present and explain some of the reasons for studying mathematics. They provide insights into the relevance of mathematics to many aspects of real life. (See pages 155, 236, and 402.)

Chapter Summaries
Chapter summaries are presented in an easy-to-read grid format for students to use in reviewing the important topics in the chapter. (See pages 73, 137, and 219.)

Chapter Review Exercises

Chapter Review Exercises contain both skill-building exercises, which are keyed to the appropriate sections within the chapter, and application exercises, which stress the practical relevance of mathematical concepts. The Chapter Review Exercises stress techniques for solving problems and provide students with the review necessary to pass a chapter test successfully. (See pages 141, 225, and 283.)

Chapter Tests

A test is provided in the end-of-chapter review material of every chapter so that students can apply their knowledge and practice their skills. (See pages 144, 230, and 287.)

Extended and Discovery Exercises

These exercises occur at the end of each chapter and are usually more complex than the Review Exercises, requiring extension or discovery of a topic presented in the chapter. They can be utilized for either collaborative learning or extra homework assignments. (See pages 82, 231, and 288.)

Cumulative Review Exercises

This feature appears after every three chapters and after Chapter 14. These exercise sets allow students to review skills and concepts related to more than one chapter. (See pages 232, 397, and 589.)

Graphing Calculator Icons and Calculator Helps

The icon 🖩 is used to denote an exercise that requires students to have access to *graphing* calculators. This feature allows instructors to easily make assignments for students who do not have graphing calculators. However, exercises without an icon can often be completed with or without the aid of technology. Calculator Helps are found in the margin, and refer students to Appendix A, Using the Graphing Calculator, which gives actual keystrokes for the TI-83, TI-83 Plus, and TI-84 Plus graphing calculators. (See pages 160, 242, and 383.)

Sources

For the numerous applications that appear throughout the text, genuine sources are cited to help establish the practical applications of mathematics. (See pages 2, 209, and 402.) In addition, a comprehensive bibliography appears at the end of the text.

Supplements

For a comprehensive list of the supplements and study aids that accompany *Beginning and Intermediate Algebra with Applications and Visualization*, see pages xxiii and xxiv.

Acknowledgments

Many individuals contributed to the development of this textbook. We thank the following reviewers, whose comments and suggestions were invaluable in preparing *Beginning and Intermediate Algebra with Applications and Visualization*.

Mary Lou Baker, *Columbia State Community College*
Debra Bryant, *Tennessee Technological University*
Pat C. Cook, *Weatherford College*

Stephan DeLong, *Tidewater Community College—Virginia Beach*
Laura Ferguson, *Weatherford College*
Margret Hathaway, *Kansas City Community College*
Kamal P. Hennayake, *Chesapeake College*
Betty Jo Major, *Fayetteville Technical Community College*
Kim Nunn, *Northeast State Technical Community College*
Larry Pontaski, *Pueblo Community College*
R. B. Pruitt, *South Plains College*
Pat Riley, *Hopkinsville Community College*
Don Rose, *College of the Sequoias*
Rebecca Schantz, *Prairie State College*
Jack C. Sharp, *Floyd College*
Paul Swatzel, *California Southern University—San Bernardino*
Charlene Tintera, *Texas A&M University—Corpus Christi*

Elina Niemelä and Janis Cimperman deserve special credit for their help with accuracy checking. Without the excellent cooperation from the professional staff at Addison-Wesley Publishing Company, this project would have been impossible. Thanks go to Greg Tobin and Maureen O'Connor for giving their support. Particular recognition is due Jennifer Crum and Lauren Morse, who gave essential advice and assistance. The outstanding contributions of Kathy Manley, Joe Vetere, Dennis Schaefer, Dona Kenly, Lindsay Skay, Marcia Emerson, and Lynne Blaszak are greatly appreciated. Special thanks go to Kathy Diamond, who was instrumental in the success of this project.

A deep sense of gratitude goes to Wendy Rockswold and Carrie Krieger, who not only proofread the manuscript, but also gave invaluable encouragement and suggestions. Their support and love made this endeavor possible.

Please feel free to send us your comments and questions at either of the following e-mail addresses: *gary.rockswold@mnsu.edu* or *tkrieger@mcleodusa.net*. Your opinion is important to us.

Gary K. Rockswold
Terry A. Krieger

STUDENT SUPPLEMENTS

Student's Solutions Manual
- By co-author Terry Krieger
- Contains solutions for the odd-numbered section-level exercises (excluding Writing About Mathematics and Group Activity exercises), and solutions to all Checking Basic Concepts exercises, Chapter Review Exercises, Chapter Test Questions, and Cumulative Review Exercises

  ISBN: 0-321-20601-0

Videos
- Created specifically to accompany *Beginning and Intermediate Algebra with Applications and Visualization*
- Cover every section of every chapter; the lecturers present examples that are taken directly from the textbook
- Include a "stop the tape" feature that encourages students to stop the videotape, work through an example, and resume play to watch the video instructor go over the solution

  ISBN: 0-321-21308-4

Digital Video Tutor
- The entire set of videotapes for this textbook in digital format on CD-ROM
- Easy and convenient for students to watch video segments displayed on a computer, at home or on campus
- Available for student purchase with the text at minimal cost
- Ideal for distance learning or supplemental instruction

  ISBN: 0-321-22783-2

Addison-Wesley Math Tutor Center
- The Addison-Wesley Math Tutor Center is staffed by qualified mathematics instructors who provide students with tutoring on examples and odd-numbered exercises from the textbook. Tutoring is available via toll-free telephone, toll-free fax, e-mail, or the Internet. White Board technology allows tutors and students to actually see problems worked while they "talk" in real time over the Internet during tutoring sessions.

TI Rebate Coupon
- Can be packaged with the text for qualified adoptions at no extra charge.
  $15 value

INSTRUCTOR SUPPLEMENTS

Annotated Instructor's Edition
- Contains Teaching Tips and provides answers to every exercise in the textbook except the Writing About Mathematics exercises
- Answers that do not fit on the same page as the exercises themselves are supplied in the Instructor's Answers at the back of the textbook.

  ISBN: 0-321-15894-6

Instructor's Solutions Manual
- By co-author Terry Krieger
- Provides solutions to all section-level exercises (excluding Writing About Mathematics and Group Activity exercises), and solutions to all Checking Basic Concepts exercises, Chapter Review Exercises, Chapter Test Questions, and Cumulative Review Exercises

  ISBN:0-321-20602-9

Adjunct Support Manual
- Includes resources designed to help both new and adjunct faculty with course preparation and classroom management
- Offers helpful teaching tips and additional exercises for selected content

  ISBN: 0-321-26902-0

Printed Test Bank/Instructor's Resource Guide
- The Test Bank contains: three free-response test forms per chapter, one of which (Form C) places stronger emphasis on applications and graphing calculator technology; one multiple-choice test form per chapter; and one free-response and one multiple-choice final exam.
- The Resource Guide contains: five sets of cumulative review exercises that cover Chapters 1–3, 1–6, 1–9, 1–12, and 1–14; notes for presenting graphing calculator topics, as well as supplemental activities; transparency masters consisting of tables figures, and examples from the text.

  ISBN: 0-321-22788-3

TestGen
- Enables instructors to build, edit, print, and administer tests
- Features a computerized bank of questions developed to cover all text objectives
- Instructors can modify questions or add new questions by using the built-in question editor, which allows users to create graphs, import graphics, insert math notation, and insert variable numbers or text.
- Tests can be printed or administered online via the Web or other network.
- Available on a dual-platform Windows/Macintosh CD-ROM

  ISBN:0-321-21312-2

MathXL® Tutorials on CD

This interactive tutorial CD-ROM provides algorithmically generated practice exercises that are correlated at the objective level to the content of the text. Every exercise is accompanied by an example and a guided solution designed to involve students in the solution process. Selected exercises may also include a video clip to help students visualize concepts. The software recognizes common student errors and provides appropriate feedback; it also tracks student activity and scores, and can generate printed summaries of students' progress.

ISBN 0-321-21310-6

MathXL

MathXL is an online homework, tutorial, and assessment system that accompanies your Addison-Wesley textbook in mathematics or statistics. With MathXL, instructors can create, edit, and assign online homework and tests using algorithmically generated exercises correlated to your textbook. All student work is tracked in MathXL's online gradebook. Students can take chapter tests in MathXL and receive personalized study plans based on their test results. The study plan diagnoses weaknesses and links students directly to tutorial exercises for the objectives they need to study and retest. Students can also access supplemental video clips directly from selected exercises.

www.mathxl.com

MyMathLab

MyMathLab is a series of text-specific, easily customizable online courses for Addison-Wesley textbooks in mathematics and statistics. MyMathLab is powered by CourseCompass—Pearson Education's online teaching and learning environment—and by MathXL—our online homework, tutorial, and assessment system. MyMathLab gives instructors the tools they need to deliver all or a portion of their course online, whether students are in a lab setting or working from home. MyMathLab provides a rich and flexible set of course materials, featuring free-response exercises that are algorithmically generated for unlimited practice and mastery. Students can also use online tools, such as video lectures, animations, and a multimedia textbook, to independently improve their understanding and performance. Instructors can use MyMathLab's homework and test managers to select and assign online exercises correlated directly to the textbook, and they can import TestGen tests into MyMathLab for added flexibility. MyMathLab's online gradebook—designed specifically for mathematics and statistics—automatically tracks students' homework and test results and gives the instructor control over how to calculate final grades. Instructors can also add offline (paper-and-pencil) grades to the MathXL gradebook. MyMathLab is available to qualified adopters. For more information, visit our Web site at www.mymathlab.com or contact your Addison-Wesley sales representative.

WALKTHROUGH

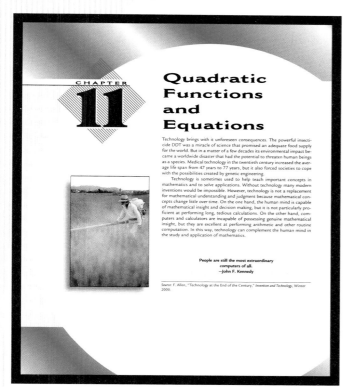

Source: F. Allen, "Technology at the End of the Century," *Invention and Technology*, Winter 2000.

Chapter Openers

Each chapter opener describes an application to motivate students by giving them insight into the relevance of that chapter's central mathematical concepts.

Rule of Four

Throughout the text, concepts are presented by means of verbal, graphical, numerical, and symbolic representations to support multiple learning styles and methods of problem solving. This textbook provides a flexible approach to the Rule of Four that lets instructors easily emphasize whichever method(s) they prefer.

Walkthrough

Applications and Models

Both the exposition and the exercises contain unique applications that model real-world data. Examples often begin with concrete applications that are used to derive the abstract mathematical concepts—an approach that motivates students by illustrating the relevance of the math from the very start. Application headings in the exercise sets call out the real-world topics presented, and sources of data are cited throughout.

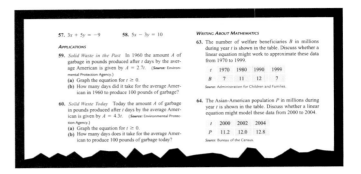

Graphing Calculator Technology

Graphing calculator use is optional, but for instructors who want to integrate graphing calculators, this textbook thoroughly and seamlessly integrates graphing technology without sacrificing traditional algebraic skills. Students can solve problems using multiple methods, learning to evaluate graphing calculator techniques against other problem-solving methods.

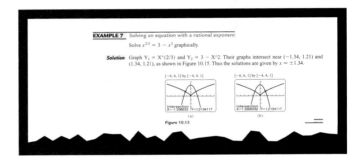

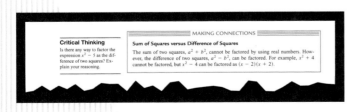

 ## Making Connections

This feature occurs throughout and helps students see how topics are interrelated. It also helps motivate students by calling out connections between mathematical concepts.

 ## Critical Thinking

The Critical Thinking feature appears in most sections and poses a question that can be used for either classroom discussion or homework. Critical Thinking questions typically ask students to extend a mathematical concept beyond what has already been discussed in the text.

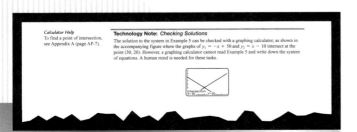

Technology Notes

Occurring throughout, Technology Notes offer students guidance and suggestions for using the graphing calculator to solve problems and explore concepts.

Walkthrough

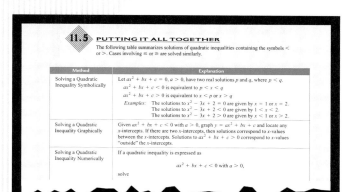

Putting It All Together

This helpful feature occurs at the end of each section to summarize techniques and reinforce concepts. These unique summaries give students a consistent, visual study aid for every section of the textbook.

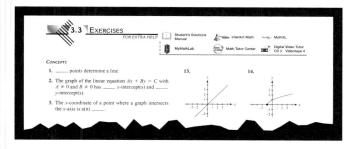

Section Exercise Sets

Each exercise set contains a wide variety of problems involving basic concepts, skill-building, and applications, and many exercises ask students to read and interpret graphs. Exercise sets are carefully graded and categorized according to the topics within sections, making it easier for instructors to select appropriate assignments. Each set concludes with Writing About Mathematics exercises to help students verbalize and synthesize concepts.

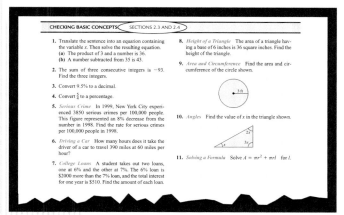

Checking Basic Concepts

Provided after every two sections, Checking Basic Concepts can be used for review or for group activities. These brief exercise sets require 10–15 minutes to complete and can be used during class if time permits.

Group Activities: Working with Real Data

Group Activities (1–2 per chapter) ask students to work collaboratively to solve a problem involving real data. Most activities can be completed with limited use of class time as needed.

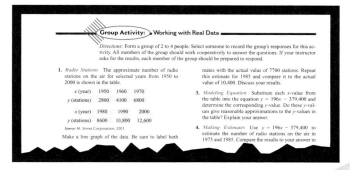

Walkthrough

For Extra Help

Found at the beginning of each exercise set, these boxes direct students to the various supplementary resources available to them for extra help and practice.

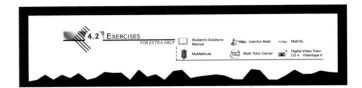

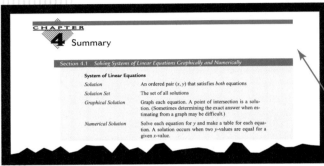

End-of-Chapter Material

The end-of-chapter material includes the following features to ensure that students have ample opportunities to practice skills, synthesize concepts, and explore material in greater depth:

- A *Chapter Summary* to present the key concepts from each section

- Extensive *Chapter Review Exercises* that include both skill-building exercises and applications, and stress different problem-solving techniques

- A *Chapter Test* to give students a chance to apply their knowledge and practice their skills

- *Extended and Discovery Exercises,* which are ideal for collaborative learning or extra-credit homework assignments, allowing students to extend their knowledge of a particular topic

- *Cumulative Review Exercises* presented after every three chapters and Chapter 14 that stress both skills and concepts

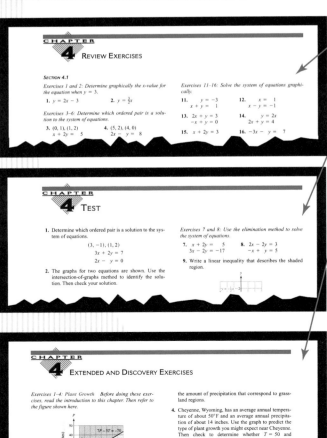

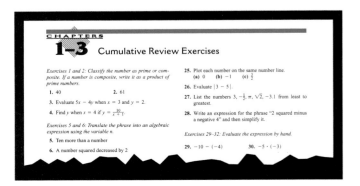

Introduction to Algebra

"Why do I need to learn math?" This question is commonly asked by students throughout the country. There are many reasons for studying mathematics. A good starting point for answering this question is to look at the past.

Centuries ago, many people were not only illiterate, but they were also unable to do even basic arithmetic. This inability was not due to a lack of intelligence, but a lack of education. In about A.D. 400, Saint Augustine declared that people who could add and subtract were in conspiracy with the devil. One hundred years ago many Americans were educated in rural schools where it was difficult to find teachers who could multiply and divide. In the 1940s and 1950s, many mathematics teachers had no training beyond algebra. Today, courses such as algebra, trigonometry, and even calculus are essential tools for many occupations.

Mathematics is important because it opens doors in our society. Without basic mathematical skills, opportunities are lost. Most vocations and professions require a higher level of mathematical understanding than in the past. Seldom are employers' mathematical expectations of employees lowered, as the workplace becomes more technical—not less. Mathematics is the *language of technology*. The Department of Education cites mathematics as a key to achievement in our society.

Decisions made by students in high school or college to avoid mathematics classes can have lifelong ramifications that affect vocations, incomes, and lifestyles. Switching majors to escape taking mathematics may prevent students from pursuing their real dreams. Like reading and writing, mathematics is a necessary component of success.

What skills will be needed in society during the next 50 years? Although no one can answer this question with certainty, history tells us that the use of mathematics will not diminish but likely flourish.

**It is not enough to have a good mind; the
main thing is to use it well.**
—René Descartes

Sources: A. Toffler and H. Toffler, *Creating a New Civilization; USA Today.*

1.1   NUMBERS, VARIABLES, AND EXPRESSIONS

Natural Numbers and Whole Numbers · Prime Numbers and Composite Numbers · Variables, Algebraic Expressions, and Equations · Translating Words to Expressions

INTRODUCTION

The need for numbers has existed in nearly every society. Numbers first occurred in the measurement of time, currency, goods, and land. As the complexity of a society increased, so did its numbers. One tribe that lived near New Guinea counted only to 6. Any number higher than 6 was referred to as "many." This number system met the needs of that society. However, a highly technical society could not function with so few numbers. We begin this section by discussing two sets of numbers that are vital to our technological society: natural numbers and whole numbers.   (**Source:** *Historical Topics for the Mathematics Classroom, Thirty-first Yearbook, NCTM.*)

NATURAL NUMBERS AND WHOLE NUMBERS

One important set of numbers found in most societies is the set of **natural numbers**. These numbers comprise the *counting numbers* and may be expressed as follows.

$$1, 2, 3, 4, 5, 6, \ldots$$

Because there are infinitely many natural numbers, three dots show that the list continues in the same pattern without end. A second set of numbers is called the **whole numbers**, given by

$$0, 1, 2, 3, 4, 5, \ldots.$$

Whole numbers include the natural numbers and the number 0.

> **Critical Thinking**
>
> Give an example from everyday life of natural number or whole number use.

Natural numbers and whole numbers can be used when data are not broken into fractional parts. For example, Table 1.1 lists the number of bachelor degrees awarded during selected academic years. Note that both natural numbers and whole numbers are appropriate to describe these data because a fraction of a degree is not awarded.

TABLE 1.1   Bachelor Degrees Awarded

Year	1969–1970	1979–1980	1989–1990	1999–2000
Degrees	792,317	929,417	1,051,344	1,185,000

Source: Department of Education.

PRIME NUMBERS AND COMPOSITE NUMBERS

When two natural numbers are multiplied, the result is another natural number. For example, if the natural numbers 3 and 4 are multiplied, the result is 12, a natural number. We say that the **product** of 3 and 4 is 12 and express this product as $3 \times 4 = 12$. The numbers 3 and 4 are **factors** of 12. Note that other natural numbers also are factors of 12, such as 2 and 6 or 1 and 12.

A natural number greater than 1 that has *only* itself and 1 as natural number factors is a **prime number**. The number 7 is prime because the only natural number factors of 7 are 1 and 7. The following is a partial list of prime numbers. There are infinitely many prime numbers.

$$2, \quad 3, \quad 5, \quad 7, \quad 11, \quad 13, \quad 17, \quad 19, \quad 23, \quad 29$$

A natural number greater than 1 that is not prime is a **composite number**. The natural number 15 is a composite number because $3 \times 5 = 15$, so 15 has factors other than itself and 1. Any composite number can be written as a product of prime numbers. For example, the natural number 120 is composite because 10 and 12 are natural number factors of 120. We can use the tree diagram shown in Figure 1.1(a) to factor 120 into prime numbers. We start by showing the factors 10 and 12 below 120 and then continue downward, factoring **10** and **12**. The bottom of the tree shows

$$120 = 2 \times 5 \times 3 \times 2 \times 2$$
$$= 2 \times 2 \times 2 \times 3 \times 5.$$

The tree is not unique because 120 could be factored as 3×40 instead of 10×12, as illustrated in Figure 1.1(b). However, the **prime factorization** is *always* unique.

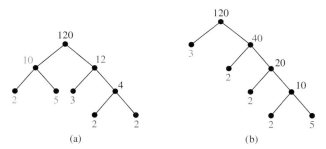

Figure 1.1 Prime Factorization of 120

─────────────── MAKING CONNECTIONS ───────────────

Prime Numbers and the Internet

Have you ever ordered something on the Internet by credit card? To ensure the privacy of your credit card number, security codes are used. Large prime numbers often play an essential role in the development of these codes.

EXAMPLE 1   Classifying numbers as prime or composite

Classify each natural number as prime or composite, if possible. If a number is composite, write it as a product of prime numbers.
(a) 31   **(b)** 1   **(c)** 35   **(d)** 200

Solution   **(a)** The only factors of 31 are itself and 1, so the natural number 31 is prime.
(b) The number 1 is neither prime nor composite because prime and composite numbers must be greater than 1.
(c) The number 35 is composite because 5 and 7 are factors. It can be written as a product of prime numbers as $35 = 5 \times 7$.

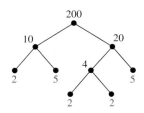

Figure 1.2 Prime Factorization of 200

Critical Thinking

Suppose that you draw a tree diagram for the prime factorization of a prime number. Describe what the tree will look like.

(d) The number 200 is composite because 10 and 20 are factors. A tree diagram can be used to write 200 as a product of prime numbers, as shown in Figure 1.2. The tree diagram reveals that 200 can be factored as $200 = 2 \times 5 \times 2 \times 2 \times 5$, or it can be factored as $200 = 2 \times 2 \times 2 \times 5 \times 5$.

VARIABLES, ALGEBRAIC EXPRESSIONS, AND EQUATIONS

There are 12 inches in 1 foot, so in 5 feet there must be $5 \times 12 = 60$ inches. Similarly, in 3 feet there are $3 \times 12 = 36$ inches. If we need to convert feet to inches frequently, Table 1.2 might help.

TABLE 1.2   Converting Feet to Inches

Feet	1	2	3	4	5	6	7
Inches	12	24	36	48	60	72	84

One difficulty with this table is that a needed value might not occur in it. For example, the table is not helpful in converting 11 feet to inches. We could expand Table 1.2 into Table 1.3 for $11 \times 12 = 132$. However, expanding the table to accommodate every possible value for feet would be impossible.

TABLE 1.3   Converting Feet to Inches

Feet	1	2	3	4	5	6	7	11
Inches	12	24	36	48	60	72	84	132

Variables are often used in mathematics when tables of numbers are inadequate. A **variable** is a symbol, typically a letter such as x, y, or z, used to represent an unknown quantity. In the preceding example the number of feet to be converted to inches could be represented by the variable F. The corresponding number of inches could be represented by the variable I. The number of inches in F feet is given by the *algebraic expression* $12 \times F$. That is, to calculate the number of inches in F feet, multiply F by **12**. The relationship between feet F and inches I is shown by using the *equation* or *formula*

$$I = \mathbf{12} \times F.$$

Sometimes a dot ($\cdot$) is used to indicate multiplication because a multiplication sign ($\times$) can be confused with the variable x. Other times the multiplication sign is omitted altogether. Thus all four formulas

$$I = 12 \times F, \quad I = 12 \cdot F, \quad I = 12F, \quad \text{and} \quad I = 12(F)$$

represent the same relationship between feet and inches. If we let $F = \mathbf{10}$ in the second formula,

$$I = 12 \cdot \mathbf{10} = 120.$$

That is, there are 120 inches in 10 feet.

More formally, an **algebraic expression** consists of numbers, variables, operation symbols, such as $+$, $-$, $\times$, and $\div$, and grouping symbols, such as parentheses. An **equation** is a mathematical statement that two algebraic expressions are equal. Equations *always* contain an equals sign. A **formula** is a special type of equation used to calculate one quantity from given values of other quantities. A formula frequently establishes a relationship between two or more quantities. The formula $I = 12F$ shows that there are 12 inches in each foot.

EXAMPLE 2 Evaluating algebraic expressions with one variable

Evaluate each algebraic expression for $x = 4$.

(a) $x + 5$ (b) $5x$ (c) $15 - x$ (d) $\dfrac{x}{(x - 2)}$

Solution (a) Replace x with 4 in the expression $x + 5$ to obtain $4 + 5 = 9$.
(b) The expression $5x$ indicates multiplication of 5 and x. Thus $5x = 5 \cdot 4 = 20$.
(c) $15 - x = 15 - 4 = 11$
(d) Perform all arithmetic operations inside parentheses first.

$$\frac{x}{(x - 2)} = \frac{4}{(4 - 2)} = \frac{4}{2} = 2$$

Some algebraic expressions contain more than one variable. For example, if a car travels 120 miles on 6 gallons of gasoline, then the car's *mileage* is $\frac{120}{6} = 20$ miles per gallon. In general, if a car travels M miles on G gallons of gasoline, then its mileage is given by the expression $\frac{M}{G}$. Note that $\frac{M}{G}$ contains two variables, M and G, whereas $12F$ contains only one variable, F.

EXAMPLE 3 Evaluating algebraic expressions with two variables

Evaluate each algebraic expression for $y = 2$ and $z = 8$.

(a) $3yz$ (b) $z - y$ (c) $\dfrac{z}{y}$

Solution (a) Replace y with 2 and z with 8 to obtain $3yz = 3 \cdot 2 \cdot 8 = 48$.
(b) $z - y = 8 - 2 = 6$

(c) $\dfrac{z}{y} = \dfrac{8}{2} = 4$

EXAMPLE 4 Evaluating formulas

Find the value of y for $x = 15$ and $z = 10$.

(a) $y = x - 3$ (b) $y = \dfrac{x}{5}$ (c) $y = 8xz$

Solution (a) Substitute 15 for x and then evaluate the right side of the formula to find y.

$$
\begin{aligned}
y &= x - 3 &&\text{Given formula}\\
&= 15 - 3 &&\text{Let } x = 15.\\
&= 12 &&\text{Subtract.}
\end{aligned}
$$

(b) $y = \dfrac{x}{5} = \dfrac{15}{5} = 3$

(c) $y = 8xz = 8 \cdot 15 \cdot 10 = 1200$

TRANSLATING WORDS TO EXPRESSIONS

Many times in mathematics, algebraic expressions are not given; rather, we must write our own expressions. To accomplish this task, we often translate words to symbols.

EXAMPLE 5   Translating words to expressions

Translate each phrase to an algebraic expression. Specify what each variable represents.
(a) Twice the cost of a CD
(b) Four more than a number
(c) Ten less than the president's age
(d) A number plus 10, all divided by a different number

Solution (a) If the cost were \$15, then twice this cost would be $2 \cdot 15 = \$30$. In general, if we let C be the cost of a CD, then twice the cost of a CD would be $2 \cdot C$, or $2C$.
(b) If the number were 20, then four more than the number would be $20 + 4 = 24$. If we let n represent the number, then four more than the number would be $n + 4$.
(c) If the president's age were 55, then ten less than the president's age would be $55 - 10 = 45$. If we let A represent the president's age, then ten less would be $A - 10$.
(d) Let x be the first number and y be the other number. Then the expression can be written as $(x + 10) \div y$. Note that parentheses are used around $x + 10$ because it is "all" divided by y. This expression could also be written as $\frac{x + 10}{y}$.

Cities are made up of large amounts of concrete and asphalt that heat up in the daytime from sunlight but do not cool off completely at night. As a result, urban areas tend to be warmer than surrounding rural areas. This effect is called the *urban heat island* and has been documented in cities throughout the world. The next example discusses the impact of this effect in Phoenix, Arizona.

EXAMPLE 6   Translating words to a formula

For each year after 1950, the average nighttime temperature in Phoenix has increased by about 0.1°C.
(a) What was the increase in the nighttime temperature after 20 years, or in 1970?
(b) Write a formula (or equation) that gives the increase T in average nighttime temperature x years after 1950.
(c) Use your formula to estimate the increase in nighttime temperature in 1980.

Solution (a) The average nighttime temperature has increased by 0.1°C per year, so after 20 years the temperature increase would be $0.1 \times 20 = 2.0°C$.
(b) To calculate the nighttime increase in temperature multiply the number of years past 1950 by 0.1. Thus $T = 0.1x$, where x represents the number of years after 1950.
(c) Because $1980 - 1950 = 30$, let $x = 30$ in the formula $T = 0.1x$ to get

$$T = 0.1(30) = 3.$$

The average nighttime temperature increased by $3°C$.

Another use of translating words to formulas is in finding the areas of various shapes.

EXAMPLE 7 Finding the area of a rectangle

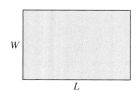

The area A of a rectangle equals its length L times its width W, as illustrated in the accompanying figure.
(a) Write a formula that shows the relationship between these three quantities.
(b) Find the area of a standard sheet of paper that is 8.5 inches wide and 11 inches long.

Solution **(a)** $A = LW$
(b) $A = 11 \cdot 8.5 = 93.5$ square inches

1.1 PUTTING IT ALL TOGETHER

The following table summarizes some of the topics discussed in this section.

Concept	Comments	Examples
Natural Numbers	Sometimes referred to as the *counting numbers*	1, 2, 3, 4, 5, . . .
Whole Numbers	Includes the natural numbers and 0	0, 1, 2, 3, 4, . . .
Prime Number	A natural number greater than 1 whose only factors are itself and 1; there are infinitely many prime numbers.	2, 3, 5, 7, 11, 13, and 17
Composite Number	A natural number greater than 1 that is *not* a prime number; there are infinitely many composite numbers.	4, 6, 8, 9, 10, 12, and 14
Variable	Represents an unknown quantity	x, y, z, A, F, and T
Algebraic Expression	May consist of variables, numbers, operation symbols, such as $+, -, \times$, and $\div$, and grouping symbols, such as parentheses.	$x + 3, \frac{x}{y}, 2z + 5, 12F, x + y + z,$ and $x(y + 5)$
Equation	An equation is a statement that two algebraic expressions are equal. An equation always includes an equals sign.	$2 + 3 = 5, x + 5 = 7, I = 12F,$ and $y = 0.1x$
Formula	A formula is a special type of equation that can be used to calculate one quantity for given values of other quantities. Formulas show relationships between two or more quantities.	$I = 12F, y = 0.1x, A = lw,$ and $F = 3Y$

 **1.1 EXERCISES**

FOR EXTRA HELP

| | Student's Solutions Manual | InterAct Math | *MathXL* MathXL |
| | MyMathLab | Math Tutor Center | Digital Video Tutor CD 1 Videotape 1 |

CONCEPTS

1. The natural numbers are also called the _____ numbers.

2. A whole number is either a natural number or the number _____.

3. The only factors of a prime number are itself and _____.

4. If a natural number is greater than 1 and not a prime number, then it is called a(n) _____ number.

5. The number 11 is a (prime/composite) number.

6. The number 9 is a (prime/composite) number.

7. Because $3 \times 6 = 18$, the numbers 3 and 6 are _____ of 18.

8. If two numbers are multiplied, the result is called the _____ of the two numbers.

9. A symbol or letter used to represent an unknown quantity is called a(n) _____.

10. Equations always contain a(n) _____.

PRIME NUMBERS AND COMPOSITE NUMBERS

Exercises 11–22: Classify the number as prime, composite, or neither. If the number is composite, write it as a product of prime numbers.

11. 4   **12.** 36

13. 1   **14.** 0

15. 29   **16.** 13

17. 92   **18.** 69

19. 225   **20.** 900

21. 149   **22.** 101

Exercises 23–32: Write the composite number as a product of prime numbers.

23. 6   **24.** 8

25. 12   **26.** 20

27. 32   **28.** 100

29. 294   **30.** 175

31. 300   **32.** 455

Exercises 33–40: State whether the given quantity could accurately be described by the whole numbers.

33. The population of a country

34. The cost of a candy bar in dollars

35. A student's grade point average

36. The Fahrenheit temperature in Antarctica

37. The number of bytes stored on a computer's hard drive

38. The exact winning time in a marathon, measured in hours

39. The number of students in a class

40. The number of bald eagles living in the United States

ALGEBRAIC EXPRESSIONS, FORMULAS, AND EQUATIONS

Exercises 41–50: Evaluate the expression for the given value of x.

41. $2x$ $x = 5$ **42.** $x + 8$ $x = 7$

43. $8 - x$ $x = 1$ **44.** $7x$ $x = 0$

45. $\dfrac{x}{8}$ $x = 32$ **46.** $\dfrac{5}{(x - 3)}$ $x = 8$

47. $3(x + 1)$ $x = 5$ **48.** $7(6 - x)$ $x = 3$

49. $\left(\dfrac{x}{2}\right) + 1$ $x = 6$ **50.** $3 - \left(\dfrac{6}{x}\right)$ $x = 2$

Exercises 51–56: Evaluate the expression for the given values of x and y.

51. $x + y$ $x = 5, \quad y = 4$

52. $5xy$ $x = 2, \quad y = 3$

53. $4 \cdot \dfrac{x}{y}$ $x = 8, \quad y = 4$

54. $y - x$ $x = 8,\quad y = 11$

55. $y(x - 3)$ $x = 5,\quad y = 7$

56. $(x + y) - 5$ $x = 6,\quad y = 3$

Exercises 57–60: Find the value of y for the given value of x.

57. $y = x + 1$ $x = 0$

58. $y = x \cdot x$ $x = 5$

59. $y = 4x$ $x = 7$

60. $y = 2(x - 3)$ $x = 3$

Exercises 61–64: Find the value of F for the given value of z.

61. $F = z - 5$ $z = 12$

62. $F = \dfrac{z}{4}$ $z = 40$

63. $F = \dfrac{30}{z}$ $z = 6$

64. $F = z \cdot z \cdot z$ $z = 5$

Exercises 65–68: Find the value of y for the given values of x and z.

65. $y = x + z$ $x = 3,\quad z = 15$

66. $y = 3xz$ $x = 2,\quad z = 0$

67. $y = \dfrac{x}{z}$ $x = 9,\quad z = 3$

68. $y = x - z$ $x = 9,\quad z = 1$

TRANSLATING WORDS TO SYMBOLS

Exercises 69–80: Translate the phrase to an algebraic expression. State what each variable represents.

69. Three times the cost of a soda

70. Twice the cost of a gallon of gasoline

71. Five more than a number

72. Five less than a number

73. Triple a number

74. A number plus two

75. Two hundred less than the population of a town

76. The total number of dogs and cats in a city

77. A number divided by six

78. A number divided by another number

79. A number plus seven, all divided by a different number

80. One-fourth of a number plus one-tenth of a different number

APPLICATIONS

81. *Yards to Feet*   Make a table of values that converts y yards to F feet. Let $y = 1, 2, 3, \ldots, 7$. Write a formula that converts y yards to F feet.

82. *Gallons to Quarts*   Make a table of values that converts g gallons to Q quarts. Let $g = 1, 2, 3, \ldots, 6$. Write a formula that converts g gallons to Q quarts.

83. *Dollars to Pennies*   Write a formula that converts D dollars to P pennies.

84. *Quarters to Nickels*   Write a formula that converts Q quarters to N nickels.

85. *Distance Traveled*   A car travels at a constant speed of 50 miles per hour. Write a formula that gives the miles M that the car travels in H hours. How far does the car travel in 3 hours?

86. *Distance Traveled*   A train travels at a constant speed of 75 miles per hour. Write a formula that gives the miles M that the train travels in H hours. How far does the train travel in 5 hours?

87. *Heart Beat*   The resting heart beat of a person is 70 beats per minute. Write a formula that gives the number of beats B that occur in x minutes. How many beats are there in an hour?

88. *Water Pressure*   Water pressure increases by about one-half pound per square inch for each foot of water depth. Write a formula that gives the pressure on the eardrum of a person who dives to a depth of x feet. What is the pressure at a depth of 30 feet?

89. *Cost of a CD*   The table lists the cost C of buying x compact discs. Write an equation that relates C and x.

Number of CDs (x)	1	2	3	4
Cost (C)	$12	$24	$36	$48

90. *Gallons of Water*   The table lists the gallons G of water coming from a garden hose after m minutes. Write an equation that relates G and m.

Minutes (m)	1	2	3	4
Gallons (G)	4	8	12	16

91. *Area of a Rectangle*   The area of a rectangle equals its length times its width. Find the area of the rectangle shown in the figure.

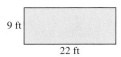

9 ft

22 ft

92. *Area of a Triangle*   The area A of a triangle is given by the formula $A = \frac{1}{2}bh$, where b is the base and h is the height. See the accompanying figure. Find the area of a triangle that has a base of 6 inches and a height of 5 inches.

93. Give an example where the whole numbers are not sufficient to describe a quantity in real life. Explain your reasoning.

94. Explain what a prime number is. How can you determine whether a number is prime?

95. Explain what a composite number is. How can you determine whether a number is composite?

96. When are variables used? Give an example.

1.2   FRACTIONS

Basic Concepts · Reducing Fractions to Lowest Terms · Multiplication and Division of Fractions · Addition and Subtraction of Fractions · Applications

INTRODUCTION

Historically, natural and whole numbers have not been sufficient for most societies. Early on, the concept of splitting a quantity into parts was common, and as a result, fractions were developed. For instance, pouring a quart of milk into four cups gives rise to the concept of a fourth of a quart. In this section we discuss fractions and how to add, subtract, multiply, and divide them.

BASIC CONCEPTS

If we divide a circular pie into 6 equal slices, as shown in Figure 1.3, then each piece represents one-sixth of the pie, and can be represented by the fraction $\frac{1}{6}$. Five slices of the pie would represent five-sixths of the pie and can be represented by the fraction $\frac{5}{6}$.

Figure 1.3

The parts of a fraction are named as follows.

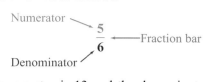

In the fraction $\frac{13}{17}$, the numerator is 13 and the denominator is 17. Sometimes we can represent a general fraction by using variables. The fraction $\frac{a}{b}$ can represent any fraction with numerator a and denominator b. However, the value of b cannot equal 0, which is denoted $b \neq 0$. (The symbol $\neq$ means "not equal to.")

Note: The fraction bar represents division. For example, the fraction $\frac{1}{2}$ represents the result when 1 is divided by 2, which is 0.5. We discuss this concept further in Section 1.4.

EXAMPLE 1   Identifying numerators and denominators

Give the numerator and denominator of each fraction.

(a) $\dfrac{6}{13}$ **(b)** $\dfrac{ac}{b}$ **(c)** $\dfrac{x-5}{y+z}$

Solution   **(a)** The numerator is 6, and the denominator is 13.
(b) The numerator is ac, and the denominator is b.
(c) The numerator is $x - 5$, and the denominator is $y + z$.

REDUCING FRACTIONS TO LOWEST TERMS

Consider the amount of pizza shown in each of the three pies in Figure 1.4. The first pie was cut into sixths and there are 3 pieces remaining, the second pie was cut into fourths and there are two pieces remaining, and the third pie was cut into only two pieces with one piece remaining. In all three cases half a pizza remains.

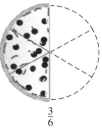
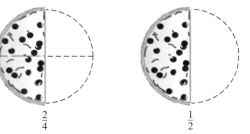

Figure 1.4

Figure 1.4 illustrates that the fractions $\frac{3}{6}, \frac{2}{4}$, and $\frac{1}{2}$ are equal. The fraction $\frac{1}{2}$ is in **lowest terms** because its numerator and denominator have no factors in common, whereas the fractions $\frac{3}{6}$ and $\frac{2}{4}$ are not in lowest terms. In the fraction $\frac{3}{6}$, the numerator and denominator have a common factor of 3, and in the fraction $\frac{2}{4}$, the numerator and denominator have a common factor of 2. The fraction $\frac{3}{6}$ can be reduced as follows.

$$\frac{3}{6} = \frac{1 \cdot 3}{2 \cdot 3} \qquad \text{Factor out 3.}$$

$$= \frac{1}{2} \qquad \frac{a \cdot c}{b \cdot c} = \frac{a}{b}$$

To reduce $\frac{3}{6}$, we used the *basic principle of fractions*: The value of a fraction is unchanged if the numerator and denominator of the fraction are multiplied (or divided) by the same nonzero number. We can also reduce the fraction $\frac{2}{4}$ to $\frac{1}{2}$ by using the basic principle of fractions.

$$\frac{2}{4} = \frac{1 \cdot 2}{2 \cdot 2} = \frac{1}{2}$$

When reducing fractions, we usually factor out the *greatest common factor* (GCF) for the numerator and the denominator. For example, to reduce $\frac{27}{36}$ we first find the greatest common factor for 27 and 36. The **greatest common factor** is the largest factor that is common to both 27 and 36. Because $27 = 9 \times 3$ and $36 = 9 \times 4$, their greatest common factor is 9. Note that 3 is a factor of 27 and 36, but it is not the *greatest* common factor. Greatest common factors can also be found by determining the largest number that divides evenly into each number. Because 9 is the largest number that divides evenly into 27 and 36, it is the greatest common factor. Thus, reduced to lowest terms, $\frac{27}{36} = \frac{3 \cdot 9}{4 \cdot 9} = \frac{3}{4}$.

REDUCING FRACTIONS

To reduce a fraction to lowest terms, begin by factoring out the greatest common factor c in the numerator and in the denominator. Then apply the **basic principle of fractions:**

$$\frac{a \cdot c}{b \cdot c} = \frac{a}{b}.$$

Note: This principle is true because multiplying a fraction by $\frac{c}{c}$ or 1 does not change its value.

EXAMPLE 2   Finding the greatest common factor

Find the greatest common factor (GCF) for each pair of numbers.
(a) 8, 12   **(b)** 24, 60

Solution **(a)** Because $8 = 4 \cdot 2$ and $12 = 4 \cdot 3$, the number 4 is the largest factor that is common to both 8 and 12. Thus the GCF of 8 and 12 is 4. (Note that 2 is a *common* factor of 8 and 12 because $8 = 2 \cdot 4$ and $12 = 2 \cdot 6$, but 2 is not the *greatest* common factor.)

(b) When we are working with larger numbers, one way to determine the greatest common factor is to find the prime factorization of each number, as in

$$24 = 4 \cdot 6 = \mathbf{2 \cdot 2 \cdot 2 \cdot 3} \quad \text{and}$$
$$60 = 6 \cdot 10 = 2 \cdot 3 \cdot 2 \cdot 5 = \mathbf{2 \cdot 2 \cdot 3 \cdot 5}.$$

Each prime factorization has two 2s and one 3 in common. Thus the GCF of 24 and 60 is $2 \cdot 2 \cdot 3 = 12$.

EXAMPLE 3 Reducing fractions to lowest terms

Reduce each fraction to lowest terms.

(a) $\frac{8}{12}$ (b) $\frac{24}{60}$ (c) $\frac{42}{105}$

Solution (a) From Example 2(a), the GCF of 8 and 12 is 4. Thus

$$\frac{8}{12} = \frac{2 \cdot 4}{3 \cdot 4} = \frac{2}{3}.$$

(b) From Example 2(b), the GCF of 24 and 60 is 12. Thus

$$\frac{24}{60} = \frac{2 \cdot 12}{5 \cdot 12} = \frac{2}{5}.$$

(c) The prime factorizations of 42 and 105 are

$$42 = 6 \cdot 7 = 2 \cdot 3 \cdot 7 \quad \text{and} \quad 105 = 5 \cdot 21 = 5 \cdot 3 \cdot 7.$$

The GCF of 42 and 105 is $3 \cdot 7 = 21$. Thus

$$\frac{42}{105} = \frac{2 \cdot 21}{5 \cdot 21} = \frac{2}{5}.$$

Figure 1.5

MULTIPLICATION AND DIVISION OF FRACTIONS

Suppose we cut *half* an apple into *thirds*, as illustrated in Figure 1.5. Then each piece represents one-sixth of the original apple. One-third of one-half is described by the product

$$\frac{1}{3} \cdot \frac{1}{2} = \frac{1}{6}.$$

(Note that the word "of" in mathematics often indicates multiplication.)

This example demonstrates that the numerator of the product of two fractions is found by multiplying the numerators of the two fractions. Similarly, the denominator of the product of two fractions is found by multiplying the denominators of the two fractions. For example, the product of $\frac{2}{3}$ and $\frac{5}{7}$ is

Multiply numerators.

$$\frac{2}{3} \cdot \frac{5}{7} = \frac{2 \cdot 5}{3 \cdot 7} = \frac{10}{21}.$$

Multiply denominators.

MULTIPLICATION OF FRACTIONS

The product of $\frac{a}{b}$ and $\frac{c}{d}$ is given by

$$\frac{a}{b} \cdot \frac{c}{d} = \frac{ac}{bd},$$

where b and d are not 0.

EXAMPLE 4 Multiplying fractions

Multiply each expression and reduce the result when appropriate.

(a) $\dfrac{4}{5} \cdot \dfrac{6}{7}$ (b) $\dfrac{8}{9} \cdot \dfrac{3}{4}$ (c) $3 \cdot \dfrac{5}{9}$ (d) $\dfrac{x}{y} \cdot \dfrac{z}{3}$

Solution (a) $\dfrac{4}{5} \cdot \dfrac{6}{7} = \dfrac{4 \cdot 6}{5 \cdot 7} = \dfrac{24}{35}$

(b) $\dfrac{8}{9} \cdot \dfrac{3}{4} = \dfrac{8 \cdot 3}{9 \cdot 4} = \dfrac{24}{36}$; the GCF of 24 and 36 is 12, so

$$\frac{24}{36} = \frac{2 \cdot 12}{3 \cdot 12} = \frac{2}{3}.$$

(c) Start by writing 3 as $\frac{3}{1}$.

$$3 \cdot \frac{5}{9} = \frac{3}{1} \cdot \frac{5}{9} = \frac{3 \cdot 5}{1 \cdot 9} = \frac{15}{9}$$

The GCF of 15 and 9 is 3, so

$$\frac{15}{9} = \frac{5 \cdot 3}{3 \cdot 3} = \frac{5}{3}.$$

(d) $\dfrac{x}{y} \cdot \dfrac{z}{3} = \dfrac{x \cdot z}{y \cdot 3} = \dfrac{xz}{3y}$

When we write the product of a variable and a number, such as $y \cdot 3$, we typically write the number first, followed by the variable. That is, $y \cdot 3 = 3y$.

≡ MAKING CONNECTIONS ≡

Multiplying and Reducing Fractions

When multiplying fractions, sometimes it is possible to use the commutative property to rewrite the product so that it is easier to reduce. In Example 4(b) the product could be written as

$$\frac{8}{9} \cdot \frac{3}{4} = \frac{3 \cdot 8}{9 \cdot 4} = \frac{3}{9} \cdot \frac{8}{4} = \frac{1}{3} \cdot \frac{2}{1} = \frac{2}{3}.$$

Instead of reducing $\frac{24}{36}$, which contains larger numbers, the simpler fractions of $\frac{3}{9}$ and $\frac{8}{4}$ were reduced first.

EXAMPLE 5 Finding fractional parts

Find each fractional part.
(a) One-fifth of two-thirds   (b) Four-fifths of three-sevenths   (c) Three-fifths of ten

Solution (a) The word "of" indicates multiplication. The fractional part is

$$\frac{1}{5} \cdot \frac{2}{3} = \frac{1 \cdot 2}{5 \cdot 3} = \frac{2}{15}.$$

(b) $\dfrac{4}{5} \cdot \dfrac{3}{7} = \dfrac{4 \cdot 3}{5 \cdot 7} = \dfrac{12}{35}$

(c) $\dfrac{3}{5} \cdot 10 = \dfrac{3}{5} \cdot \dfrac{10}{1} = \dfrac{30}{5} = 6$

In the next example we use fractions in an application.

EXAMPLE 6   Estimating college completion rates

About four-fifths of the U.S. population over the age of 25 has a high school diploma. About one-third of those people have gone on to complete 4 or more years of college. What fraction of the U.S. population over the age of 25 has completed 4 or more years of college?

Solution   One-third of four-fifths is

$$\frac{1}{3} \cdot \frac{4}{5} = \frac{1 \cdot 4}{3 \cdot 5} = \frac{4}{15}.$$

About four-fifteenths of the U.S. population over the age of 25 has completed 4 or more years of college.

TABLE 1.4   Numbers and Their Reciprocals

Number	Reciprocal
3	$\frac{1}{3}$
$\frac{1}{4}$	4
$\frac{3}{2}$	$\frac{2}{3}$
$\frac{21}{37}$	$\frac{37}{21}$

The **multiplicative inverse** or **reciprocal** of a nonzero number a is $\frac{1}{a}$. Table 1.4 lists several numbers and their reciprocals. Note that the product of a number and its reciprocal is always 1. For example, the reciprocal of 2 is $\frac{1}{2}$, and their product is $2 \cdot \frac{1}{2} = 1$.

Suppose that a group of children wants to buy gum from a gum ball machine that costs a half dollar for each gum ball. If the caregiver for the children has 4 dollars, then the number of gum balls that can be bought equals the number of half dollars that there are in 4 dollars. Thus 8 gum balls can be bought.

This calculation is given by $4 \div \frac{1}{2}$. To divide a number by a fraction, multiply the number by the reciprocal of the fraction. That is, "invert and multiply."

$$4 \div \frac{1}{2} = 4 \cdot \frac{2}{1} \qquad \text{Multiply by the reciprocal of } \tfrac{1}{2}.$$

$$= \frac{4}{1} \cdot \frac{2}{1} \qquad \text{Write 4 as } \tfrac{4}{1}.$$

$$= \frac{8}{1} \qquad \text{Multiply the fractions.}$$

$$= 8 \qquad \tfrac{a}{1} = a \text{ for all values of } a.$$

Justification for multiplying by the reciprocal when dividing two fraction is

$$\frac{a}{b} \div \frac{c}{d} = \frac{\dfrac{a}{b}}{\dfrac{c}{d}} = \frac{\dfrac{a}{b} \cdot \dfrac{d}{c}}{\dfrac{c}{d} \cdot \dfrac{d}{c}} = \frac{\dfrac{a}{b} \cdot \dfrac{d}{c}}{1} = \frac{a}{b} \cdot \frac{d}{c}.$$

These results are summarized as follows.

DIVISION OF FRACTIONS

For real numbers a, b, c, and d, with b, c, and d not equal to 0,

$$\frac{a}{b} \div \frac{c}{d} = \frac{a}{b} \cdot \frac{d}{c}.$$

EXAMPLE 7 Dividing fractions

Divide each expression.

(a) $\dfrac{1}{3} \div \dfrac{3}{5}$ (b) $\dfrac{4}{5} \div \dfrac{4}{5}$ (c) $5 \div \dfrac{10}{3}$ (d) $\dfrac{x}{2} \div \dfrac{y}{z}$

Solution (a) To divide $\frac{1}{3}$ by $\frac{3}{5}$, multiply $\frac{1}{3}$ by $\frac{5}{3}$, which is the reciprocal of $\frac{3}{5}$.

$$\frac{1}{3} \div \frac{3}{5} = \frac{1}{3} \cdot \frac{5}{3} = \frac{1 \cdot 5}{3 \cdot 3} = \frac{5}{9}$$

(b) $\frac{4}{5} \div \frac{4}{5} = \frac{4}{5} \cdot \frac{5}{4} = \frac{4 \cdot 5}{5 \cdot 4} = \frac{20}{20} = 1$. Note that when we divide any nonzero number by itself, the result is 1.

(c) $5 \div \frac{10}{3} = \frac{5}{1} \cdot \frac{3}{10} = \frac{15}{10} = \frac{3}{2}$

(d) $\dfrac{x}{2} \div \dfrac{y}{z} = \dfrac{x}{2} \cdot \dfrac{z}{y} = \dfrac{xz}{2y}$

Note: In Example 5(c) the answer was left as $\frac{3}{2}$. In arithmetic, $\frac{3}{2}$ is sometimes called an **improper fraction** and is often written as the mixed number $1\frac{1}{2}$. However, in algebra there is nothing "improper" about $\frac{3}{2}$, and fractions are often left as improper fractions.

Technology Note

When entering a mixed number into a calculator, it is usually easiest first to convert the mixed number to an improper fraction. For example, enter $2\frac{2}{3}$ as $\frac{8}{3}$. Otherwise, enter $2\frac{2}{3}$ as $2 + \frac{2}{3}$. See the accompanying figure.

```
8/3
        2.666666667
2+2/3
        2.666666667
```

EXAMPLE 8 Writing a problem

Describe a problem for which the solution could be found by dividing 5 by $\frac{1}{6}$.

Solution If five pies are each cut into sixths, how many pieces of pie are there? (Note that one way to answer this question is to determine how many one-sixths there are in 5, which is the equivalent of multiplying 5 by 6.)

ADDITION AND SUBTRACTION OF FRACTIONS

Suppose that a member of a group cuts a sheet of paper into eighths. If one member picks up two pieces and another member picks up three pieces, then together they have

$$\frac{2}{8} + \frac{3}{8} = \frac{5}{8}$$

of a sheet of paper, as illustrated in Figure 1.6. When the denominator of one fraction is the same as the denominator of a second fraction, the sum of the two fractions can be found by adding their numerators and keeping the common denominator.

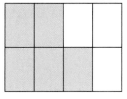

Figure 1.6

Similarly, if someone picks up 5 pieces of paper and gives 2 away, then that person has

$$\frac{5}{8} - \frac{2}{8} = \frac{3}{8}$$

of a sheet of paper. To subtract two fractions with common denominators, subtract their numerators and keep the common denominator.

> ### ADDITION AND SUBTRACTION OF FRACTIONS
>
> To add or subtract fractions with a common denominator d, use the equations
>
> $$\frac{a}{d} + \frac{b}{d} = \frac{a + b}{d} \quad \text{and} \quad \frac{a}{d} - \frac{b}{d} = \frac{a - b}{d},$$
>
> where d is not 0.

EXAMPLE 9 Adding and subtracting fractions with common denominators

Add or subtract as indicated. Reduce your answer to lowest terms when appropriate.
(a) $\frac{5}{13} + \frac{12}{13}$ **(b)** $\frac{11}{8} - \frac{5}{8}$

Solution **(a)** Because the fractions have a common denominator, add the numerators and keep the common denominator.

$$\frac{5}{13} + \frac{12}{13} = \frac{5 + 12}{13} = \frac{17}{13}$$

(b) Because the fractions have a common denominator, subtract the numerators and keep the common denominator.

$$\frac{11}{8} - \frac{5}{8} = \frac{11 - 5}{8} = \frac{6}{8}$$

The fraction $\frac{6}{8}$ can be reduced to $\frac{3}{4}$.

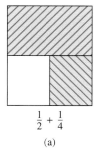

$\frac{1}{2} + \frac{1}{4}$

(a)

$\frac{2}{4} + \frac{1}{4} = \frac{3}{4}$

(b)

Figure 1.7

Suppose that one person mows half a large lawn while another person mows a fourth of the lawn. To determine how much they mowed together we need to find the sum $\frac{1}{2} + \frac{1}{4}$. See Figure 1.7(a). Before we can add fractions with unlike denominators, we must write each fraction with a common denominator. The least common denominator of 2 and 4 is 4. Thus we need to write $\frac{1}{2}$ as $\frac{?}{4}$ by multiplying the numerator and denominator by the *same nonzero number*.

$$\frac{1}{2} = \frac{1}{2} \cdot \frac{2}{2} \qquad \text{Multiply by 1.}$$

$$= \frac{2}{4} \qquad \text{Multiply fractions.}$$

Now we can find the needed sum.

$$\frac{1}{2} + \frac{1}{4} = \frac{2}{4} + \frac{1}{4} = \frac{3}{4}$$

Together the two people mow three-fourths of the lawn, as illustrated in Figure 1.7(b).

To add fractions with unlike denominators, we first determine a common denominator. Generally we find the *least* common denominator (LCD).

FINDING THE LEAST COMMON DENOMINATOR (LCD)

STEP 1: Find the prime factorization for each denominator.

STEP 2: List each factor that appears in one or more of the factorizations. If a factor is repeated in any of the factorizations, list this factor the maximum number of times that it is repeated.

STEP 3: The product of this list of factors is the LCD.

EXAMPLE 10   Finding the least common denominator

Find the LCD for each set of fractions.

(a) $\frac{5}{6}, \frac{3}{4}$ (b) $\frac{2}{27}, \frac{5}{12}$ (c) $\frac{1}{4}, \frac{2}{5}, \frac{7}{10}$

Solution   (a) **STEP 1:** The prime factorizations are $6 = 2 \cdot 3$ and $4 = 2 \cdot 2$.
   STEP 2: List the factors: **2, 2, 3**. Note that, because the factor 2 appears a maximum of two times, it is listed twice.
   STEP 3: The LCD is the product of this list, or $2 \cdot 2 \cdot 3 = 12$.

Note: Finding an LCD is equivalent to finding the smallest number that each denominator divides into evenly. Both 6 and 4 divide into 12 evenly, and 12 is the smallest such number. Thus 12 is the LCD for $\frac{5}{6}$ and $\frac{3}{4}$.

   (b) **STEP 1:** The prime factorizations are $27 = 3 \cdot 3 \cdot 3$ and $12 = \mathbf{2 \cdot 2} \cdot 3$.
   STEP 2: List the factors: **3, 3, 3, 2, 2**.
   STEP 3: The LCD is $3 \cdot 3 \cdot 3 \cdot 2 \cdot 2 = 108$.
   (c) **STEP 1:** The prime factorizations are $4 = 2 \cdot 2, 5 = 5$, and $10 = 2 \cdot 5$.
   STEP 2: List the factors: **2, 2, 5**.
   STEP 3: The LCD is $2 \cdot 2 \cdot 5 = 20$.

In the next example we apply this procedure in adding and subtracting fractions.

EXAMPLE 11   Adding and subtracting fractions with unlike denominators

Add or subtract as indicated. Reduce your answer to lowest terms.

(a) $\frac{5}{6} + \frac{3}{4}$ (b) $\frac{5}{12} - \frac{2}{27}$ (c) $\frac{1}{4} + \frac{2}{5} + \frac{7}{10}$

Solution   (a) From Example 10(a) the LCD is 12. Begin by writing each fraction with a denominator of 12.

$$\frac{5}{6} + \frac{3}{4} = \frac{5}{6} \cdot \frac{2}{2} + \frac{3}{4} \cdot \frac{3}{3} \qquad \text{Change to LCD of 12.}$$

$$= \frac{10}{12} + \frac{9}{12} \qquad \text{Multiply the fractions.}$$

$$= \frac{10 + 9}{12} \qquad \text{Add the numerators.}$$

$$= \frac{19}{12} \qquad \text{Simplify.}$$

(b) From Example 10(b) the LCD is 108.

$$\frac{5}{12} - \frac{2}{27} = \frac{5}{12} \cdot \frac{9}{9} - \frac{2}{27} \cdot \frac{4}{4}$$ Change to LCD of 108.

$$= \frac{45}{108} - \frac{8}{108}$$ Multiply the fractions.

$$= \frac{45 - 8}{108}$$ Subtract the numerators.

$$= \frac{37}{108}$$ Simplify.

(c) From Example 10(c) the LCD is 20.

$$\frac{1}{4} + \frac{2}{5} + \frac{7}{10} = \frac{1}{4} \cdot \frac{5}{5} + \frac{2}{5} \cdot \frac{4}{4} + \frac{7}{10} \cdot \frac{2}{2}$$ Change to LCD of 20.

$$= \frac{5}{20} + \frac{8}{20} + \frac{14}{20}$$ Multiply the fractions.

$$= \frac{5 + 8 + 14}{20}$$ Add the numerators.

$$= \frac{27}{20}$$ Simplify.

APPLICATIONS

Fractions occur in a variety of applications. We illustrate two here.

EXAMPLE 12   Applying fractions to carpentry

A board measures $35\frac{3}{4}$ inches and needs to be cut into four equal parts, as depicted in Figure 1.8. Find the length of each piece.

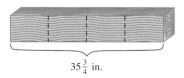

$35\frac{3}{4}$ in.

Figure 1.8

Solution   Begin by writing $35\frac{3}{4}$ as the improper fraction $\frac{143}{4}$. As the board is to be cut into four equal parts, the length of each piece should be

$$\frac{143}{4} \div 4 = \frac{143}{4} \cdot \frac{1}{4} = \frac{143}{16}, \quad \text{or} \quad 8\frac{15}{16} \text{ inches.}$$

EXAMPLE 13 Applying fractions to cooking

A recipe requires $2\frac{1}{4}$ cups of flour and makes enough for 9 people. How much flour should be used if the chef wants to make enough for 15 people?

Solution   The amount of flour in the recipe needs to be increased by the factor $\frac{15}{9} = \frac{5}{3}$. Because the original recipe requires $2\frac{1}{4} = \frac{9}{4}$ cups of flour, the chef should use

$$\frac{9}{4} \cdot \frac{5}{3} = \frac{45}{12} = 3\frac{9}{12} = 3\frac{3}{4}$$

cups of flour.

Critical Thinking

Think of a situation in every-day life in which fractions would be needed.

1.2   PUTTING IT ALL TOGETHER

The following table summarizes some important concepts related to fractions.

Topic	Comments	Examples
Fraction	The fraction $\frac{a}{b}$ has numerator a and denominator b.	The fraction $\frac{xy}{2}$ has numerator xy and denominator 2.
Greatest Common Factor (GCF)	The GCF of two numbers equals the largest number that divides into both evenly.	The GCF of 12 and 18 is 6 because 6 is the largest number that divides into 12 and 18 evenly.
Reducing Fractions	Use the principle $$\frac{a \cdot c}{b \cdot c} = \frac{a}{b}$$ to reduce fractions, where c is the GCF of the numerator and denominator.	The GCF of 24 and 32 is 8, so $$\frac{24}{32} = \frac{3 \cdot 8}{4 \cdot 8} = \frac{3}{4}.$$ The GCF of 20 and 8 is 4, so $$\frac{20}{8} = \frac{5 \cdot 4}{2 \cdot 4} = \frac{5}{2}.$$
Multiplicative Inverse or Reciprocal	The reciprocal of $\frac{a}{b}$ is $\frac{b}{a}$, where a and b are not zero.	The reciprocals of 5 and $\frac{3}{4}$ are $\frac{1}{5}$ and $\frac{4}{3}$, respectively. The product of a number and its reciprocal is 1.
Multiplication and Division of Fractions	$$\frac{a}{b} \cdot \frac{c}{d} = \frac{ac}{bd}$$ $$\frac{a}{b} \div \frac{c}{d} = \frac{a}{b} \cdot \frac{d}{c}$$	$$\frac{3}{5} \cdot \frac{4}{9} = \frac{12}{45} = \frac{4}{15}$$ $$\frac{3}{2} \div \frac{6}{5} = \frac{3}{2} \cdot \frac{5}{6} = \frac{15}{12} = \frac{5}{4}$$

Topic	Comments	Examples
Addition and Subtraction of Fractions with Like Denominators	$\dfrac{a}{d} + \dfrac{c}{d} = \dfrac{a+c}{d}$ and $\dfrac{a}{d} - \dfrac{c}{d} = \dfrac{a-c}{d}$	$\dfrac{3}{5} + \dfrac{4}{5} = \dfrac{3+4}{5} = \dfrac{7}{5}$ and $\dfrac{17}{12} - \dfrac{11}{12} = \dfrac{17-11}{12} = \dfrac{6}{12} = \dfrac{1}{2}$
Least Common Denominator (LCD)	The LCD of two fractions equals the smallest number that both denominators divide into evenly.	The LCD of $\dfrac{5}{12}$ and $\dfrac{7}{18}$ is 36 because 36 is the smallest number that both 12 and 18 divide into evenly.
Addition and Subtraction of Fractions with Unlike Denominators	First write each fraction with the least common denominator. Then add or subtract the numerators.	The LCD of $\dfrac{3}{4}$ and $\dfrac{7}{10}$ is 20. $\dfrac{3}{4} + \dfrac{7}{10} = \dfrac{3}{4} \cdot \dfrac{5}{5} + \dfrac{7}{10} \cdot \dfrac{2}{2}$ $= \dfrac{15}{20} + \dfrac{14}{20}$ $= \dfrac{29}{20}$

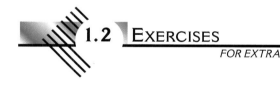

1.2 EXERCISES

FOR EXTRA HELP

Student's Solutions Manual

InterAct Math

MathXL

MyMathLab

Math Tutor Center

Digital Video Tutor
CD 1   Videotape 1

CONCEPTS

1. A small pie is cut into 4 pieces. If someone eats 3 of the pieces, what fraction of the pie did the person eat? What fraction of the pie remains?

2. In the fraction $\dfrac{11}{21}$ the numerator is _____ and the denominator is _____.

3. In the fraction $\dfrac{a}{b}$, the variable b cannot equal _____.

4. The fraction $\dfrac{a}{a}$ with $a \neq 0$ equals _____.

5. The numerator of the product of two fractions is found by multiplying the _____ of the two fractions.

6. The denominator of the product of two fractions is found by multiplying the _____ of the two fractions.

7. $\dfrac{a}{b} \cdot \dfrac{c}{d} =$ _____

8. $\dfrac{ac}{bc} =$ _____

9. What fractional part is half of a half?

10. What is the reciprocal of a, provided $a \neq 0$?

11. $\dfrac{a}{b} \div \dfrac{c}{d} =$ _____

12. $\dfrac{a}{b} + \dfrac{c}{b} =$ _____

13. $\dfrac{a}{b} - \dfrac{c}{b} =$ _____

14. The number 6 can always be written as the fraction $\dfrac{?}{1}$.

15. The greatest common factor of 4 and 6 is _____.

16. The least common denominator for the fractions $\dfrac{1}{4}$ and $\dfrac{5}{6}$ is _____.

LOWEST TERMS

Exercises 17–22: Find the greatest common factor.

17. 4, 12

18. 3, 27

19. 50, 75

20. 45, 105

21. 100, 60, 70

22. 36, 48, 72

Exercises 23–26: Use the basic principle of fractions to reduce the expression.

23. $\frac{3 \cdot 4}{5 \cdot 4}$

24. $\frac{2 \cdot 7}{9 \cdot 7}$

25. $\frac{3 \cdot 8}{8 \cdot 5}$

26. $\frac{7 \cdot 16}{16 \cdot 3}$

Exercises 27–38: Reduce the fraction to lowest terms.

27. $\frac{4}{8}$

28. $\frac{4}{12}$

29. $\frac{5}{15}$

30. $\frac{3}{27}$

31. $\frac{10}{25}$

32. $\frac{5}{20}$

33. $\frac{12}{36}$

34. $\frac{16}{24}$

35. $\frac{12}{30}$

36. $\frac{60}{105}$

37. $\frac{19}{76}$

38. $\frac{17}{51}$

MULTIPLICATION AND DIVISION OF FRACTIONS

Exercises 39–54: Multiply and reduce to lowest terms when appropriate.

39. $\frac{1}{2} \cdot \frac{1}{3}$

40. $\frac{2}{3} \cdot \frac{7}{5}$

41. $\frac{3}{4} \cdot \frac{1}{5}$

42. $\frac{3}{2} \cdot \frac{5}{8}$

43. $\frac{5}{3} \cdot \frac{3}{5}$

44. $\frac{21}{32} \cdot \frac{32}{21}$

45. $\frac{5}{6} \cdot \frac{18}{25}$

46. $\frac{7}{9} \cdot \frac{3}{14}$

47. $4 \cdot \frac{3}{5}$

48. $5 \cdot \frac{7}{10}$

49. $2 \cdot \frac{3}{8}$

50. $10 \cdot \frac{1}{100}$

51. $\frac{x}{y} \cdot \frac{y}{x}$

52. $\frac{x}{y} \cdot \frac{y}{z}$

53. $\frac{a}{b} \cdot \frac{3}{2}$

54. $\frac{5}{8} \cdot \frac{4x}{5y}$

Exercises 55–60: Find the fractional part.

55. One-fourth of three-fourths

56. Three-sevenths of nine-sixteenths

57. Two-thirds of six

58. Three-fourths of seven

59. One-half of two-thirds

60. Five-elevenths of nine-eighths

Exercises 61–64: Give the reciprocal of each number.

61. (a) 5   (b) 7   (c) $\frac{4}{7}$ (d) $\frac{9}{8}$

62. (a) 3   (b) 2   (c) $\frac{6}{5}$ (d) $\frac{3}{8}$

63. (a) $\frac{1}{2}$ (b) $\frac{1}{9}$ (c) $\frac{12}{101}$ (d) $\frac{31}{17}$

64. (a) $\frac{1}{5}$ (b) $\frac{7}{3}$ (c) $\frac{23}{64}$ (d) $\frac{63}{29}$

Exercises 65–82: Divide and reduce to lowest terms when appropriate.

65. $\frac{1}{2} \div \frac{1}{3}$

66. $\frac{3}{4} \div \frac{1}{5}$

67. $\frac{3}{4} \div \frac{1}{2}$

68. $\frac{6}{7} \div \frac{3}{14}$

69. $\frac{4}{3} \div \frac{1}{6}$

70. $\frac{12}{21} \div \frac{4}{7}$

71. $\frac{32}{27} \div \frac{8}{9}$

72. $\frac{8}{15} \div \frac{2}{25}$

73. $\frac{9}{1} \div \frac{8}{7}$

74. $2 \div \frac{3}{4}$

75. $10 \div \frac{5}{6}$

76. $8 \div \frac{4}{3}$

77. $\frac{9}{10} \div 3$

78. $\frac{32}{27} \div 16$

79. $\frac{a}{b} \div \frac{2}{b}$

80. $\frac{3a}{b} \div \frac{3}{c}$

81. $\frac{x}{y} \div \frac{x}{y}$

82. $\frac{x}{3y} \div \frac{x}{3}$

ADDITION AND SUBTRACTION OF FRACTIONS

Exercises 83–88: Add or subtract. Write each answer in lowest terms.

83. (a) $\frac{2}{3} + \frac{1}{3}$ (b) $\frac{2}{3} - \frac{1}{3}$

84. (a) $\frac{5}{12} + \frac{1}{12}$ (b) $\frac{5}{12} - \frac{1}{12}$

85. (a) $\frac{3}{2} + \frac{1}{2}$ (b) $\frac{3}{2} - \frac{1}{2}$

86. (a) $\frac{18}{29} + \frac{7}{29}$ (b) $\frac{18}{29} - \frac{7}{29}$

87. (a) $\frac{5}{33} + \frac{2}{33}$ (b) $\frac{5}{33} - \frac{2}{33}$

88. (a) $\frac{91}{104} + \frac{17}{104}$ (b) $\frac{91}{104} - \frac{17}{104}$

Exercises 89–100: Find the least common denominator for the fractions.

89. $\frac{1}{5}, \frac{3}{10}$

90. $\frac{5}{12}, \frac{1}{18}$

91. $\frac{4}{9}, \frac{2}{15}$

92. $\frac{1}{11}, \frac{1}{2}$

93. $\frac{2}{5}, \frac{3}{15}$

94. $\frac{9}{21}, \frac{3}{7}$

95. $\frac{1}{6}, \frac{5}{8}$

96. $\frac{1}{9}, \frac{5}{12}$

97. $\frac{1}{2}, \frac{1}{3}, \frac{1}{4}$

98. $\frac{2}{5}, \frac{2}{3}, \frac{1}{6}$

99. $\frac{1}{4}, \frac{3}{8}, \frac{1}{12}$

100. $\frac{2}{15}, \frac{7}{20}, \frac{1}{30}$

Exercises 101–116: Add or subtract. Write your answer in lowest terms.

101. $\frac{1}{2} + \frac{1}{3}$

102. $\frac{2}{3} + \frac{1}{4}$

103. $\frac{5}{8} + \frac{3}{16}$

104. $\frac{1}{9} + \frac{2}{15}$

105. $\frac{1}{2} - \frac{1}{4}$

106. $\frac{2}{3} - \frac{2}{9}$

107. $\frac{25}{24} - \frac{7}{8}$

108. $\frac{4}{5} - \frac{1}{4}$

109. $\frac{11}{14} + \frac{2}{35}$

110. $\frac{7}{8} + \frac{4}{15}$

111. $\frac{5}{12} - \frac{1}{18}$

112. $\frac{9}{20} - \frac{7}{30}$

113. $\frac{3}{100} + \frac{1}{300} - \frac{1}{200}$

114. $\frac{45}{36} + \frac{4}{9} + \frac{1}{4}$

115. $\frac{7}{8} - \frac{1}{6} + \frac{5}{12}$

116. $\frac{9}{40} - \frac{3}{50} - \frac{1}{100}$

APPLICATIONS

117. *Accidental Deaths* For the age group 15 to 24, motor vehicle accidents account for $\frac{31}{42}$ of all accidental deaths, and firearms account for $\frac{31}{1260}$ of all accidental deaths. What fraction of all accidental deaths do vehicle accidents *and* firearms account for? (*Source:* National Safety Council.)

118. *Illicit Drug Use* For the age group 18 to 25, the fraction of people who used illicit drugs during their lifetime was $\frac{12}{25}$, whereas the fraction who used illicit drugs during the past year was $\frac{27}{100}$. What fraction of this population has used illicit drugs but not during the past year? (*Source:* Department of Health and Human Services.)

119. *Carpentry* A board measures $64\frac{5}{8}$ inches and needs to be cut in half. Find the length of each half.

120. *Carpentry* A rope measures $15\frac{1}{2}$ feet and needs to be cut in four equal parts. Find the length of each piece.

121. *Cooking* A recipe requires $3\frac{1}{2}$ cups of flour and makes enough for 8 people. How much flour should be used to make enough for 12 people?

122. *Cooking* A recipe requires $1\frac{3}{4}$ cups of sugar and makes enough for 6 people. How much sugar should be used to make enough for 15 people?

123. *Geometry* Find the area of the triangle shown with base $1\frac{2}{3}$ yards and height $\frac{3}{4}$ yard. (*Hint:* The area of a triangle equals half the product of its base and height.)

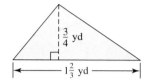

124. *Geometry* Find the area of the rectangle shown.

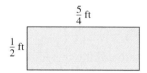

125. *Distance* Use the map to find the distance between Smalltown and Bigtown by traveling through Middletown.

126. *Distance* An athlete jogs $1\frac{3}{8}$ miles, $5\frac{3}{4}$ miles, and $3\frac{5}{8}$ miles. In all, how far did the athlete jog?

127. *Adult Smokers* About $\frac{1}{4}$ of the adult population in the United States smoked in 1995. Within the part of the population that smoked, $\frac{7}{25}$ of them were aged 18 to 24. What fraction of the adult population in the United States was aged 18 to 24 and smoked? (*Source:* Department of Health and Human Services.)

128. *Smokeless Tobacco Use* About $\frac{2}{25}$ of the teenage population in the United States used smokeless tobacco in 1997. Within the part of the teenage population that used smokeless tobacco, $\frac{29}{31}$ were males. What fraction of the teenage population was male and used smokeless tobacco? (*Source:* Department of Health and Human Services.)

WRITING ABOUT MATHEMATICS

129. Describe a problem in real life for which the solution could be found by multiplying 30 by $\frac{1}{4}$.

130. Describe a problem in real life for which the solution could be found by dividing $2\frac{1}{2}$ by $\frac{1}{3}$.

CHECKING BASIC CONCEPTS   SECTIONS 1.1 AND 1.2

1. Classify each number as prime, composite, or neither. If a number is composite, write it as a product of primes.
 (a) 19   **(b)** 28   **(c)** 1   **(d)** 180

2. Evaluate $\frac{10}{(x+2)}$ for $x = 3$.

3. Find y for $x = 5$ if $y = 6x$.

4. Translate the phrase "a number plus five" into an algebraic expression.

5. Find the area of the rectangle shown.

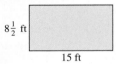

$8\frac{1}{2}$ ft

15 ft

6. Reduce each fraction to lowest terms.
 (a) $\frac{25}{35}$ **(b)** $\frac{26}{39}$

7. Give the reciprocal of $\frac{4}{3}$.

8. Evaluate each expression. Write each answer in lowest terms.
 (a) $\frac{2}{3} \cdot \frac{3}{4}$ **(b)** $\frac{5}{6} \div \frac{10}{3}$
 (c) $\frac{3}{10} + \frac{1}{10}$ **(d)** $\frac{3}{4} - \frac{1}{6}$

9. A recipe calls for $1\frac{2}{3}$ cups of flour. How much flour should be used if the recipe is doubled?

1.3   EXPONENTS AND ORDER OF OPERATIONS

**Natural Number Exponents · Order of Operations ·
Translating Words to Expressions**

INTRODUCTION

In elementary school you learned addition. Later, you learned that multiplication is a fast way to add. For example, rather than adding

$$3 + 3 + 3 + 3, \quad \longleftarrow \text{4 terms}$$

we can multiply $3 \cdot 4$. Similarly, exponents represent a fast way to multiply. Rather than multiplying

$$3 \cdot 3 \cdot 3 \cdot 3, \quad \longleftarrow \text{4 factors}$$

we can evaluate the *exponential expression* 3^4. We begin this section by discussing natural numbers as exponents.

NATURAL NUMBER EXPONENTS

The area of a square equals the length of one of its sides times itself. If the square is 5 inches on a side, then its area is

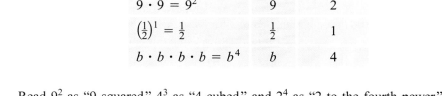

$$5 \cdot 5 = 5^2 = 25 \text{ square inches.}$$

The expression 5^2 is an **exponential expression** with *base* 5 and *exponent* 2. Exponential expressions occur in a variety of applications. For example, suppose an investment doubles 3 times. Then, its final value is

$$2 \cdot 2 \cdot 2 = 2^3 = 8$$

times larger. Table 1.5 contains examples of exponential expressions.

TABLE 1.5

Expression	Base	Exponent
$2 \cdot 2 \cdot 2 \cdot 2 = 2^4$	2	4
$4 \cdot 4 \cdot 4 = 4^3$	4	3
$9 \cdot 9 = 9^2$	9	2
$\left(\frac{1}{2}\right)^1 = \frac{1}{2}$	$\frac{1}{2}$	1
$b \cdot b \cdot b \cdot b = b^4$	b	4

Read 9^2 as "9 squared," 4^3 as "4 cubed," and 2^4 as "2 to the fourth power." The terms *squared* and *cubed* come from geometry. If the length of a side of a square is 3, then its area is

$$3 \cdot 3 = 3^2 = 9$$

square units, as illustrated in Figure 1.9. Similarly, if the length of an edge of a cube is 3, then its volume is

$$3 \cdot 3 \cdot 3 = 3^3 = 27$$

cubic units, as shown in Figure 1.10.

Note: The expressions 3^2 and 3^3 can also be read as "3 to the second power" and "3 to the third power," respectively. In general, the expression x^n is read as "x to the nth power" or "the nth power of x."

Figure 1.9
3 Squared

Figure 1.10
3 Cubed

EXPONENTIAL NOTATION

The expression b^n, where n is a natural number, means

$$b^n = \underbrace{b \cdot b \cdot b \cdot \cdots \cdot b}_{n \text{ factors}}.$$

The **base** is b and the **exponent** is n.

EXAMPLE 1   Evaluating exponential notation

Evaluate each expression.
(a) 3^4 **(b)** 10^3 **(c)** 23^1 **(d)** $\left(\frac{3}{4}\right)^2$

Solution   **(a)** The exponential expression 3^4 indicates that 3 is to be multiplied times itself 4 times.

$$3^4 = \underbrace{3 \cdot 3 \cdot 3 \cdot 3}_{4 \text{ factors}} = 9 \cdot 9 = 81$$

(b) $10^3 = 10 \cdot 10 \cdot 10 = 1000$
(c) $23^1 = 23$
(d) $\left(\frac{3}{4}\right)^2 = \frac{3}{4} \cdot \frac{3}{4} = \frac{9}{16}$

EXAMPLE 2   Writing numbers in exponential notation

Use the given base to write each number as an exponential expression. Check your results with a calculator, if one is available.
(a) 100 (base 10)   **(b)** 16 (base 2)   **(c)** 27 (base 3)

Solution   **(a)** $100 = 10 \cdot 10 = 10^2$
(b) $16 = 4 \cdot 4 = 2 \cdot 2 \cdot 2 \cdot 2 = 2^4$
(c) $27 = 3 \cdot 9 = 3 \cdot 3 \cdot 3 = 3^3$

These values are supported in Figure 1.11, where exponential expressions are evaluated with a calculator by using the "^" key. (Note that some calculators may have a different key for evaluating exponential expressions.)

```
10^2
                100
2^4
                 16
3^3
                 27
```

Figure 1.11

AN APPLICATION OF EXPONENTS   Computer memory is often measured in bytes, with each *byte* capable of storing one letter of the alphabet. For example, it takes four bytes to store the word "math" in a computer. Bytes of computer memory are often manufactured in amounts equal to powers of 2, as illustrated in the next example.

EXAMPLE 3   Analyzing computer memory

In computer technology 1K (kilobyte) of memory equals 2^{10} bytes, and 1MB (megabyte) of memory equals 2^{20} bytes. Determine whether 1K of memory equals one thousand bytes and whether 1MB equals one million bytes.   (*Source*: D. Horn, *Basic Electronics Theory*.)

Solution   Figure 1.12 shows that $2^{10} = 1024$ and $2^{20} = 1,048,576$. Thus 1K represents slightly more than one thousand bytes, and 1MB represents more than one million bytes.

```
2^10
               1024
2^20
            1048576
```

Figure 1.12

Critical Thinking

One gigabyte is often referred to as 1 billion bytes. Use Example 3 to write an expression that gives the number of bytes in a gigabyte. If you have a calculator available, determine whether 1 gigabyte is exactly 1 billion bytes.

ORDER OF OPERATIONS

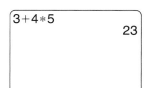

Figure 1.13

When the expression $3 + 4 \cdot 5$ is evaluated, is the result

$$7 \cdot 5 = 35 \quad \text{or} \quad 3 + 20 = 23?$$

Figure 1.13 shows that a calculator gives a result of 23. The reason is that multiplication is performed before addition.

Because it is important to evaluate arithmetic expressions consistently, the following rules are used.

ORDER OF OPERATIONS

Use the following order of operations. First perform all calculations within parentheses and absolute values, or above and below the fraction bar.
1. Evaluate all exponential expressions.
2. Do all multiplication and division from *left to right*.
3. Do all addition and subtraction from *left to right*.

EXAMPLE 4　Evaluating arithmetic expressions

Evaluate each expression by hand.
(a) $10 - 4 - 3$　　**(b)** $10 - (4 - 3)$　　**(c)** $5 + \frac{12}{3}$　　**(d)** $\frac{4}{2 + 6}$

Solution　**(a)** There are no parentheses, so we evaluate subtraction from *left to right*.

$$10 - 4 - 3 = 6 - 3 = 3$$

(b) Note the similarity between this part and part (a). The difference is the parentheses, so subtraction inside the parentheses must be performed first.

$$10 - (4 - 3) = 10 - 1 = 9$$

(c) In this part we perform division before addition.

$$5 + \frac{12}{3} = 5 + 4 = 9$$

(d) The implication is that both the numerator and the denominator of a fraction have parentheses around them.

$$\frac{4}{2 + 6} = \frac{4}{(2 + 6)} = \frac{4}{8} = \frac{1}{2}$$

Note that the given expression does *not* equal $\frac{4}{2} + 6 = 8$.

EXAMPLE 5　Evaluating arithmetic expressions

Evaluate each expression.
(a) $25 - 4 \cdot 6$　　**(b)** $6 + 7 \cdot 2 - (4 - 1)$　　**(c)** $\frac{3 + 3^2}{14 - 2}$　　**(d)** $5 \cdot 2^3 - (3 + 2)$

Solution　**(a)** Because multiplication is performed before subtraction, we evaluate the expression as follows.

$$25 - 4 \cdot 6 = 25 - 24 \qquad \text{Multiply.}$$
$$= 1 \qquad \text{Subtract.}$$

(b) Start by performing the subtraction within the parentheses first and then perform the multiplication. Finally perform the addition and subtraction from left to right.

$$6 + 7 \cdot 2 - (4 - 1) = 6 + 7 \cdot 2 - 3 \qquad \text{Subtract.}$$
$$= 6 + 14 - 3 \qquad \text{Multiply.}$$
$$= 20 - 3 \qquad \text{Add.}$$
$$= 17 \qquad \text{Subtract.}$$

(c) First note that parentheses are implied around the numerator and denominator.

$$\frac{3 + 3^2}{14 - 2} = \frac{(3 + 3^2)}{(14 - 2)} \qquad \text{Insert parentheses.}$$
$$= \frac{(3 + 9)}{(14 - 2)} \qquad \text{Do the exponent first.}$$
$$= \frac{12}{12} \qquad \text{Add and subtract.}$$
$$= 1 \qquad \text{Reduce.}$$

(d) Begin by evaluating the expression inside parentheses.

$$5 \cdot 2^3 - (3 + 2) = 5 \cdot 2^3 - 5 \qquad \text{Add.}$$
$$= 5 \cdot 8 - 5 \qquad \text{Evaluate the exponent.}$$
$$= 40 - 5 \qquad \text{Multiply.}$$
$$= 35 \qquad \text{Subtract.}$$

TRANSLATING WORDS TO EXPRESSIONS

Sometimes before we can solve a problem we must translate words into mathematical expressions. For example, if the gasoline tank in your car holds 20 gallons and there are 11 gallons in it, then "twenty minus eleven," or $20 - 11 = 9$, equals the number of gallons of gasoline needed to fill the tank.

EXAMPLE 6 Writing and evaluating expressions

Write each expression and then evaluate it.
(a) Two to the fourth power plus ten
(b) Twenty less five times three
(c) Ten cubed divided by five squared
(d) Sixty divided by the quantity ten minus six

Solution **(a)** $2^4 + 10 = 2 \cdot 2 \cdot 2 \cdot 2 + 10 = 16 + 10 = 26$
(b) $20 - 5 \cdot 3 = 20 - 15 = 5$
(c) $\frac{10^3}{5^2} = \frac{1000}{25} = 40$

(d) Here, the word "quantity" indicates that parentheses should be used.

$$60 \div (10 - 6) = \frac{60}{10 - 6} = \frac{60}{4} = 15$$

PUTTING IT ALL TOGETHER

The following table summarizes some topics related to exponential expressions. Remember that exponential expressions are a fast way to multiply.

Topic	Comments	Examples
Exponential Expression	If n is a natural number, then b^n equals $$\underbrace{b \cdot b \cdot b \cdot \cdots \cdot b}_{n \text{ factors}}$$ and is read "b to the nth power."	$5^1 = 5, 7^2 = 7 \cdot 7 = 49,$ $4^3 = 4 \cdot 4 \cdot 4 = 64,$ and $k^4 = k \cdot k \cdot k \cdot k$
Base and Exponent	The base in b^n is b and the exponent is n.	7^4 has base 7 and exponent 4 and x^3 has base x and exponent 3

Without rules for the order of operations, the value of many arithmetic expressions would be ambiguous. For example, without rules of precedence, one person might evaluate $10 - 4 \cdot 2$ to be $6 \cdot 2 = 12$, whereas another person might evaluate it to be $10 - 8 = 2$. However, because multiplication is performed before subtraction, 2 is the correct answer.

Use the following order of operations. First perform all calculations within parentheses and absolute values, or above and below the fraction bar.

1. Evaluate all exponential expressions.
2. Do all multiplication and division from *left to right*.
3. Do all addition and subtraction from *left to right*.

1.3 EXERCISES

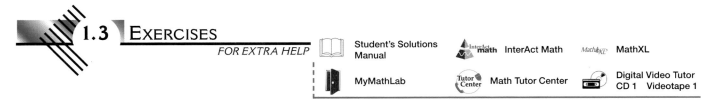

FOR EXTRA HELP

Student's Solutions Manual   InterAct Math   MathXL
MyMathLab   Math Tutor Center   Digital Video Tutor CD 1   Videotape 1

CONCEPTS

1. Multiplication is a fast way to _____.

2. Exponential expressions represent a fast way to _____.

3. The expression $a \cdot a \cdot a \cdot a \cdot a \cdot a$ can be written as _____.

4. In the expression 5^3, the number 5 is called the _____ and the number 3 is called the _____.

5. Use symbols to write "6 squared."

6. Use symbols to write "8 cubed."

7. When evaluating the expression $5 + 6 \cdot 2$, the result is _____ because _____ is performed before _____.

8. When evaluating the expression $10 - 2^3$, the result is _____ because _____ are evaluated before _____ is performed.

9. The expression $10 - 4 - 2$ equals _____ because subtraction is performed from _____ to _____.

10. The expression $16 \div 4 \div 2$ equals _____ because division is performed from _____ to _____.

11. Are the expressions 2^3 and 3^2 equal? Explain.

12. Is 5^2 equal to $5 \cdot 5$ or $5 \cdot 2$?

NATURAL NUMBER EXPONENTS

Exercises 13–20: Write the expression as an exponential expression.

13. $2 \cdot 2 \cdot 2 \cdot 2 \cdot 2$ **14.** $4 \cdot 4 \cdot 4$

15. $3 \cdot 3 \cdot 3 \cdot 3$ **16.** $10 \cdot 10$

17. $\frac{1}{2} \cdot \frac{1}{2} \cdot \frac{1}{2} \cdot \frac{1}{2}$ **18.** $\frac{5}{7} \cdot \frac{5}{7} \cdot \frac{5}{7} \cdot \frac{5}{7} \cdot \frac{5}{7}$

19. $a \cdot a \cdot a \cdot a \cdot a$ **20.** $b \cdot b \cdot b \cdot b$

Exercises 21–30: Use multiplication to rewrite the expression, and then evaluate the result.

21. (a) 2^4 **(b)** 4^2 **22. (a)** 3^2 **(b)** 5^3

23. (a) 6^1 **(b)** 1^6 **24. (a)** 17^1 **(b)** 1^{17}

25. (a) 2^5 **(b)** 10^3 **26. (a)** 10^5 **(b)** 3^4

27. (a) $\left(\frac{2}{3}\right)^2$ **(b)** $\left(\frac{1}{2}\right)^5$ **28. (a)** $\left(\frac{1}{10}\right)^3$ **(b)** $\left(\frac{4}{3}\right)^1$

29. (a) $\left(\frac{2}{5}\right)^3$ **(b)** $\left(\frac{9}{7}\right)^2$ **30. (a)** $\left(\frac{3}{10}\right)^4$ **(b)** $\left(\frac{3}{4}\right)^4$

Exercises 31–40: (Refer to Example 2.) Use the given base to write the number as an exponential expression. Check your result if you have a calculator available.

31. 8   (base 2)   **32.** 9   (base 3)

33. 25   (base 5)   **34.** 32   (base 2)

35. 49   (base 7)   **36.** 81   (base 3)

37. 1000   (base 10)   **38.** 256   (base 4)

39. $\frac{1}{16}$ $\left(\text{base } \frac{1}{2}\right)$ **40.** $\frac{9}{25}$ $\left(\text{base } \frac{3}{5}\right)$

ORDER OF OPERATIONS

Exercises 41–64: Evaluate the expression by hand.

41. $5 + 4 \cdot 6$ **42.** $6 \cdot 7 - 8$

43. $6 \div 3 + 2$ **44.** $20 - 10 \div 5$

45. $100 - \frac{50}{5}$ **46.** $\frac{200}{100} + 6$

47. $10 - 6 - 1$ **48.** $30 - 9 - 5$

49. $20 \div 5 \div 2$ **50.** $500 \div 100 \div 5$

51. $3 + 2^4$ **52.** $10 - 3^2 + 1$

53. $4 \cdot 2^3$ **54.** $100 - 2 \cdot 3^3$

55. $(3 + 2)^3$ **56.** $5 \cdot (3 - 2)^8 - 5$

57. $\frac{4 + 8}{1 + 3}$ **58.** $5 - \frac{3 + 1}{3 - 1}$

59. $\frac{2^3}{4 - 2}$ **60.** $\frac{10 - 3^2}{2 \cdot 4^2}$

61. $10^2 - (30 - 2 \cdot 5)$ **62.** $5^2 + 3 \cdot 5 \div 3 - 1$

63. $\left(\frac{1}{2}\right)^4 + \frac{5 + 4}{3}$ **64.** $\left(\frac{7}{9}\right)^2 - \frac{6 - 5}{3}$

TRANSLATING WORDS TO SYMBOLS

Exercises 65–74: Use symbols to write the expression and then evaluate it.

65. Two cubed minus eight

66. Five squared plus nine

67. Thirty less four times three

68. One hundred plus five times six

69. Four squared divided by two cubed

70. Three cubed times two squared

71. Forty divided by ten, plus two

72. Thirty times ten, minus three

73. One hundred times the quantity two plus three

74. Fifty divided by the quantity eight plus two

APPLICATIONS

75. *Computer Memory* (Refer to Example 3.) Determine the number of bytes on a 1.44 MB floppy disk.

76. *Computer Memory* Determine the number of bytes on a 10 GB hard drive. (*Hint:* One gigabyte equals 2^{30} bytes.)

77. *GPS Clocks* The Global Positioning System (GPS) is made up of 24 satellites that allow individuals with GPS receivers to pinpoint their positions on Earth. Every 1024 weeks the clocks in the GPS satellites

reset to zero. The first time they reset was on August 21, 1999.   (*Source:* Associated Press.)
(a) Find an exponent k so that $2^k = 1024$.
(b) Estimate the number of years in 1024 weeks.

78. *Doubling Effect*   Suppose that a savings account containing $1000 doubles its value every 7 years. How much money will be in the account after 28 years?

WRITING ABOUT MATHEMATICS

79. Explain how exponential expressions are related to multiplication. Give an example.

80. Explain why agreement on the order of operations is necessary.

Group Activity: Working with Real Data

Directions: Form a group of 2 to 4 people. Select someone to record the group's responses for this activity. All members of the group should work cooperatively to answer the questions. If your instructor asks for your results, each member of the group should be prepared to respond.

Converting Temperatures   To convert Celsius degrees C to Fahrenheit degrees F use the formula $F = 32 + \frac{9}{5}C$. This exercise illustrates the importance of understanding the order of operations.
(a) Complete the following table by evaluating the formula in the two ways shown.

Celsius	$F = \left(32 + \frac{9}{5}\right)C$	$F = 32 + \left(\frac{9}{5}C\right)$
$-40°\text{C}$		
$0°\text{C}$		
$5°\text{C}$	$169°\text{F}$	$41°\text{F}$
$20°\text{C}$		
$30°\text{C}$		
$100°\text{C}$		

(b) At what Celsius temperature does water freeze? At what Fahrenheit temperature does water freeze?
(c) Which column gives the correct Fahrenheit temperatures? Why?
(d) Explain why having an agreed order for operations in mathematics is necessary.

1.4   REAL NUMBERS AND THE NUMBER LINE

Signed Numbers · Integers and Rational Numbers · Square Roots · Real and Irrational Numbers · The Number Line · Absolute Value · Inequality

INTRODUCTION

Fractions were invented before decimals and negative numbers. The concept of splitting something into parts was readily understood by most people. However, performing arithmetic with fractions was quite cumbersome, particularly if the fractions had unlike denominators. For example, to find the sum $\frac{1}{2} + \frac{1}{4}$, a common denominator needed to be found.

With the invention of decimals, this addition problem became much simpler and could be expressed as $0.50 + 0.25 = 0.75$. One important reason for the invention of decimal numbers is that arithmetic with decimals is much easier to perform than arithmetic with fractions. Even the stock market recently switched from fractions to decimals. In this section we discuss decimal numbers.

SIGNED NUMBERS

The idea that numbers could be negative was a difficult concept for many mathematicians. As late as the eighteenth century negative numbers were not readily accepted by everyone. After all, how could a person have -5 oranges?

However, negative numbers make more sense when someone is working with money. If you owe someone 100 dollars, this amount can be thought of as -100, whereas if you have a balance of 100 dollars in your checking account, this amount can be thought of as $+100$. (The positive sign is usually omitted.) The number 0 is neither positive nor negative.

The **opposite**, or **additive inverse**, of a number a is $-a$. For example, the opposite of 25 is -25, the opposite of -5 is 5, and the opposite of 0 is 0. The following double negative rule is helpful in simplifying expressions containing negative signs.

DOUBLE NEGATIVE RULE

Let a be any number. Then $-(-a) = a$.

Thus $-(-8) = 8$ and $-\left(-(-10)\right) = -10$.

EXAMPLE 1 Finding opposites (or additive inverses)

Find the opposite of each expression.

(a) 13   **(b)** $-\frac{4}{7}$ **(c)** $5 - \frac{4}{2}$ **(d)** $-(-7)$

Solution **(a)** The opposite of 13 is -13.

(b) The opposite of $-\frac{4}{7}$ is $\frac{4}{7}$.

(c) $5 - \frac{4}{2} = 5 - 2 = 3$, so the opposite of $5 - \frac{4}{2}$ is -3.

(d) $-(-7) = 7$, so the opposite of $-(-7)$ is -7.

Note: To find the opposite of an exponential expression evaluate the exponent first. For example, the opposite of 2^4 is

$$-2^4 = -(2 \cdot 2 \cdot 2 \cdot 2) = -16.$$

EXAMPLE 2 Finding an additive inverse (or opposite)

Find the additive inverse of $-t$, if $t = -\frac{2}{3}$.

Solution The additive inverse of $-t$ is $t = -\frac{2}{3}$ because $-(-t) = t$ by the double negative rule.

INTEGERS AND RATIONAL NUMBERS

The *integers* are given by

$$\ldots, \quad -3, \quad -2, \quad -1, \quad 0, \quad 1, \quad 2, \quad 3, \quad \ldots.$$

A **rational number** is any number that can be expressed as the ratio of two integers $\frac{p}{q}$, where $q \neq 0$. Rational numbers can be written as fractions, and they include all integers. Rational numbers may be positive, negative, or zero. Some examples of rational numbers are

$$\frac{2}{3}, \quad -\frac{3}{5}, \quad \frac{-7}{2}, \quad 1.2, \quad \text{and} \quad 3.$$

The numbers 1.2 and 3 are both rational numbers because they can be written as $\frac{12}{10}$ and $\frac{3}{1}$.

Note: $\frac{-7}{2} = \frac{7}{-2} = -\frac{7}{2}$; the position of the negative sign does not affect the value of the fraction.

The fraction bar can be thought of as a division symbol. As a result, rational numbers have decimal equivalents. For example, $\frac{1}{2}$ is equivalent to $1 \div 2$. Because

$$\frac{0.5}{2 \overline{)1.0}},$$

$\frac{1}{2} = 0.5$. In general, a rational number may be expressed in a decimal form that either *repeats* or *terminates*. The fraction $\frac{1}{3}$ may be expressed as $0.\overline{3}$, a repeating decimal, and the fraction $\frac{1}{4}$ may be expressed as 0.25, a terminating decimal. The overbar indicates that $0.\overline{3} = 0.3333333\ldots.$

Integers and rational numbers are used to describe quantities such as temperature. Table 1.6 lists equivalent temperatures in both degrees Fahrenheit and degrees Celsius. Note that both positive and negative numbers are used to describe temperature.

TABLE 1.6   Fahrenheit and Celsius Temperature

°F	°C	Observation
−89	$-67.\overline{2}$	Alcohol freezes
−40	−40	Mercury freezes
0	$-17.\overline{7}$	A salt and snow mixture freezes
32	0	Water freezes
100	$37.\overline{7}$	A very warm day
212	100	Water boils

EXAMPLE 3   **Classifying numbers**

Classify each number as one or more of the following: natural number, whole number, integer, or rational number.

(a) $\frac{6}{3}$ **(b)** -1 **(c)** 0 **(d)** $-\frac{11}{3}$

Solution   **(a)** Because $\frac{6}{3} = 2$, the number $\frac{6}{3}$ is a natural number, whole number, integer, and rational number.

(b) The number -1 is an integer and rational number but not a natural number or a whole number.

(c) The number 0 is a whole number, integer, and rational number but not a natural number.

(d) The fraction $-\frac{11}{3}$ is a rational number because it is the ratio of two integers. However, it is not a natural number, a whole number, or an integer.

SQUARE ROOTS

Square roots are frequently used in algebra. The number b is a **square root** of a number a if $b \cdot b = a$. Every positive number has two square roots. For example, one square root of 9 is 3 because $3 \cdot 3 = 9$, and a second square root of 9 is -3 because $-3 \cdot (-3) = 9$. (We discuss multiplication of negative numbers in Section 1.6.) The square root of a that is nonnegative (or not negative) is called the **principal square root**, denoted $\sqrt{a}$. For example, $\sqrt{25} = 5$ because $5 \cdot 5 = 25$ and 5 is nonnegative.

EXAMPLE 4   Calculating principal square roots

Evaluate each square root. Approximate your answer to three decimal places when appropriate.

(a) $\sqrt{36}$ **(b)** $\sqrt{100}$ **(c)** $\sqrt{5}$

Solution   **(a)** $\sqrt{36} = 6$ because $6 \cdot 6 = 36$ and 6 is nonnegative.

(b) $\sqrt{100} = 10$ because $10 \cdot 10 = 100$ and 10 is nonnegative.

(c) $\sqrt{5}$ is a number between 2 and 3 because $2 \cdot 2 = 4$ and $3 \cdot 3 = 9$. We can estimate the value of $\sqrt{5}$ with a calculator. Figure 1.14, reveals that $\sqrt{5}$ *approximately* equals 2.236. However, 2.236 does not exactly equal $\sqrt{5}$ because $2.236 \times 2.236 = 4.999696$, which does not equal 5.

```
√(5)
        2.236067977
2.236*2.236
        4.999696
```

Figure 1.14

Calculator Help
To evaluate a square root, see Appendix A (page AP-1).

REAL AND IRRATIONAL NUMBERS

Real numbers can be represented by decimal numbers. That is, if a number can be represented by a decimal number, then it is a real number. Every fraction has a decimal form, so real numbers include rational numbers. However, some real numbers cannot be expressed by fractions. They are called **irrational numbers**. The numbers $\sqrt{2}$, $\sqrt{15}$, and π are examples of irrational numbers. They can be represented by decimals, but not by decimals that either repeat or terminate. Examples of real numbers include

$$-10, \quad 151\frac{1}{4}, \quad -131.37, \quad \frac{1}{3}, \quad -\sqrt{5}, \quad \text{and} \quad \sqrt{11}.$$

Any real number may be approximated by a terminating decimal. The symbol $\approx$ represents **approximately equal**. Each of the following real numbers has been approximated to two *decimal places*.

$$\pi \approx 3.14, \quad \frac{2}{3} \approx 0.67, \quad \text{and} \quad \sqrt{200} \approx 14.14$$

If you have a calculator, verify these results.

Figure 1.15 shows the relationship between the different sets of numbers. Note that each real number is either a rational number or an irrational number but not both. The natural numbers, whole numbers, and integers are rational numbers.

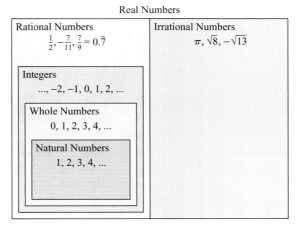

Figure 1.15 The Real Numbers

=== MAKING CONNECTIONS ===

Rational and Irrational Numbers

Both rational and irrational numbers can be written as decimals. However, rational numbers can be represented by either terminating or repeating decimals. For example, $\frac{1}{2} = 0.5$ is a terminating decimal and $\frac{1}{3} = 0.333\ldots$ is a repeating decimal. Irrational numbers are represented by decimals that neither terminate nor repeat.

EXAMPLE 5 Classifying numbers

Identify the natural numbers, integers, rational numbers, and irrational numbers in the following list.

$$5, \quad -1.2, \quad \frac{13}{7}, \quad -\sqrt{7}, \quad -12, \quad \text{and} \quad \sqrt{16}$$

Solution Natural numbers: 5 and $\sqrt{16} = 4$

Integers: 5, -12, and $\sqrt{16} = 4$

Rational numbers: 5, -1.2, $\frac{13}{7}$, -12, and $\sqrt{16} = 4$

Irrational number: $-\sqrt{7}$

Even though a data set may comprise only integers, decimals are often needed to describe it. One common way to describe data is to find their **average**. To find the average of a set of numbers we add the numbers and divide by how many numbers there are in the set. For example, the average of 5, 9, and 16 equals 10 because

$$\frac{5 + 9 + 16}{3} = \frac{30}{3} = 10.$$

EXAMPLE 6   Analyzing data

Table 1.7 lists the number of higher education institutions, such as colleges and universities, for various years. Find the average number of institutions during this 4-year period. Is the result a natural number, a rational number, or an irrational number?

TABLE 1.7   Higher Education Institutions

Year	1995	1996	1997	1998
Institutions	3706	4009	4064	4070

Source: National Center for Education Statistics.

Solution The average number of institutions was

$$\frac{3706 + 4009 + 4064 + 4070}{4} = \frac{15{,}849}{4} = 3962.25.$$

Critical Thinking

Think of an example in which the sum of two irrational numbers is a rational number.

The average of these four natural numbers is an integer divided by an integer, which is a rational number. However, it is neither a natural number nor an irrational number.

THE NUMBER LINE

The real numbers can be represented visually by using a number line, as shown in Figure 1.16. Each real number corresponds to a unique point on the number line. The point associated with the real number 0 is called the **origin**. The positive integers are equally spaced to the right of the origin, and the negative integers are equally spaced to the left of the origin. The number line extends indefinitely both left and right. Other real numbers can also be located on the number line. For example, the number $\frac{1}{2}$ can be identified by placing a dot halfway between the integers 0 and 1. The numbers $-\sqrt{2} \approx -1.41$ and $\frac{5}{4} = 1.25$ can also be placed (approximately) on this number line.

Figure 1.16   The Number Line

EXAMPLE 7   Plotting numbers on a number line

Plot each real number on a number line.
(a) $-\frac{3}{2}$ **(b)** $\sqrt{3}$ **(c)** π

Solution **(a)** $-\frac{3}{2} = -1.5$. Place a dot halfway between -2 and -1, as shown in Figure 1.17.
(b) A calculator gives $\sqrt{3} \approx 1.73$. Place a dot between 1 and 2 so that it is about three-fourths of the way toward 2, as shown in Figure 1.17.
(c) $\pi \approx 3.14$. Place a dot just past the integer 3, as shown in Figure 1.17.

Figure 1.17   Plotting Real Numbers

ABSOLUTE VALUE

The absolute value of a real number equals its distance on the number line from the origin. Because distance is never negative, the absolute value of a real number is *never negative.* The absolute value of a real number a is denoted $|a|$ and is read "the absolute value of a." Figure 1.18 shows that the absolute values of -2 and 2 equal 2 because both have distance 2 from the origin. That is, $|-2| = 2$ and $|2| = 2$.

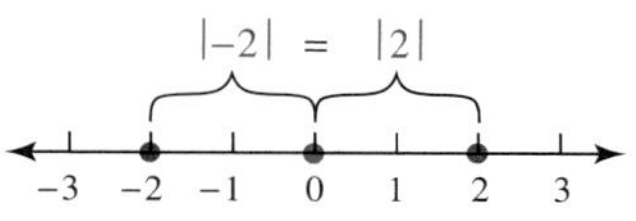

Figure 1.18

EXAMPLE 8   Finding the absolute value of a real number

Write the expression without the absolute value sign.
(a) $|-7|$ **(b)** $|0|$
(c) $|-a|$, if a is a positive number   **(d)** $|a|$, if a is a negative number

Solution **(a)** $|-7| = 7$ because 7 is a positive number.
(b) $|0| = 0$ because the distance is 0 between the origin and 0.
(c) If a is positive, then $-a$ is negative. Thus $|-a| = a$.
(d) If a is negative, then $-a$ is positive. Thus $|a| = -a$. For example, if $a = -5$, then $|-5| = -(-5) = 5$.

Our results about absolute value can be summarized as follows.

$$|a| = a, \quad \text{if } a \text{ is positive or } 0.$$
$$|a| = -a, \quad \text{if } a \text{ is negative.}$$

INEQUALITY

If a real number a is located to the left of a real number b on the number line, we say that a is **less than** b and write $a < b$. Similarly, if a real number b is located to the right of a real number a, we say that b is **greater than** a and write $b > a$. Thus $-3 < 2$ because -3 is located to the left of 2, and $2 > -3$ because 2 is located to the right of -3. See Figure 1.19. In general, any negative number will always be less than any positive number, and any positive number will always be greater than any negative number.

We say that a is **less than or equal to** b, denoted $a \le b$, if either $a < b$ or $a = b$ is true. Similarly, a is **greater than or equal to** b, denoted $a \ge b$, if either $a > b$ or $a = b$ is true.

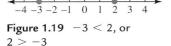
Figure 1.19 $-3 < 2$, or $2 > -3$

EXAMPLE 9   Ordering real numbers

List the following numbers from least to greatest. Then plot these numbers on a number line.
$$-2, \quad -\pi, \quad \sqrt{2}, \quad 0, \quad \text{and} \quad 2.5$$

Solution First note that $-\pi \approx -3.14 < -2$. The two negative numbers are less than 0, and the two positive numbers are greater than 0. Also, $\sqrt{2} \approx 1.41$, so $\sqrt{2} < 2.5$. Listing the numbers from least to greatest results in
$$-\pi, \quad -2, \quad 0, \quad \sqrt{2}, \quad \text{and} \quad 2.5.$$

These numbers are plotted on the number line shown in Figure 1.20. Note that these numbers increase from left to right on the number line.

Figure 1.20

1.4 PUTTING IT ALL TOGETHER

Because of the variety of information produced and used by our society, different sets of numbers had to be developed. Without numbers, information can only be described qualitatively, not quantitatively. For example, we might say that it rained a lot last night, but we

continued on next page

continued from previous page

would not be able to give an actual measurement of the rainfall. The following table summarizes several sets of numbers.

Set of Numbers	Comments	Examples
Integers	Includes the natural numbers, their opposites, and 0	$\ldots, -2, -1, 0, 1, 2, \ldots$
Rational Numbers	Includes integers, all fractions $\frac{p}{q}$, where p and $q \neq 0$ are integers, and all repeating and terminating decimals	$\frac{1}{2}, -3, \frac{128}{6}, -0.335, 0, 0.25 = \frac{1}{4}$, and $0.\overline{3} = \frac{1}{3}$
Irrational Numbers	Any decimal number that neither terminates nor repeats; a real number that is not rational	$\pi, \sqrt{3},$ and $\sqrt{15}$
Real Numbers	Any number that can be expressed in decimal form; includes the rational and irrational numbers	$\pi, \sqrt{3}, -\frac{4}{7}, 0, -10, 0.\overline{6} = \frac{2}{3}, 1000,$ and $\sqrt{15}$

A number line can be used to visualize the real number system, as illustrated in the figure. The point associated with the number 0 is called the origin.

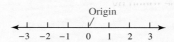

The absolute value of a number a equals its distance on the number line from the origin. If a number a is located to the left of a number b, then a is less than b (written $a < b$). If a is located to the right of b, then a is greater than b (written $a > b$).

Topic	Notation	Examples
Absolute Value	If $a \geq 0$, then $$\lvert a \rvert = a.$$ If $a < 0$, then $$\lvert a \rvert = -a.$$ $\lvert a \rvert$ is *never* negative.	$\lvert 17 \rvert = 17, \lvert -12 \rvert = 12,$ and $\lvert 0 \rvert = 0$
Inequality	Symbols of inequality include: $<, >, \leq, \geq,$ and $\neq$.	$-3 < -2$ less than $6 > 4$ greater than $-5 \leq -5$ less than or equal $18 \geq 0$ greater than or equal $7 \neq 8$ not equal

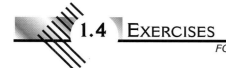

1.4 EXERCISES

FOR EXTRA HELP

 Student's Solutions Manual

 InterAct Math

MathXL MathXL

 MyMathLab

 Math Tutor Center

 Digital Video Tutor CD 1   Videotape 1

CONCEPTS

1. The opposite of the number b is _____.

2. The additive inverse of -7 is _____.

3. $-(-b) =$ _____.

4. The integers include the natural numbers, zero, and the additive inverses of the _____ numbers.

5. A number that can be written as $\frac{p}{q}$, where p and q are integers with $q \neq 0$, is a(n) _____ number.

6. If a number can be written in decimal form, then it is a(n) _____ number.

7. If a real number is not a rational number, then it is a(n) _____ number.

8. The decimal equivalent for $\frac{1}{4}$ can be found by dividing _____ by _____.

9. Give an example of an irrational number.

10. The symbol $\neq$ is used to indicate that two numbers are _____.

11. The symbol $\approx$ is used to indicate that two numbers are _____.

12. To calculate the average of three numbers, add the three numbers and divide the sum by _____.

13. The origin on the number line corresponds to the number _____.

14. The negative numbers are to the (left/right) of the origin on the number line.

15. The absolute value of a number a gives its distance on the number line from the _____.

16. If $b > 0$, then $|-b| =$ _____.

Exercises 17–20: Insert the symbol $<$, $=$, or $>$ to make the statement true.

17. If a number a is located to the right of a number b on the number line, then a __ b.

18. If $b > 0$ and $a < 0$, then b __ a.

19. If $a \geq b$, then either $a > b$ or a __ b.

20. If $a \geq b$ and $b \geq a$, then a __ b.

SIGNED NUMBERS

Exercises 21–28: Find the opposite of each expression.

21. (a) 9 (b) -9

22. (a) -6 (b) 6

23. (a) $\frac{2}{3}$ (b) $-\frac{2}{3}$

24. (a) $-\left(\frac{-4}{5}\right)$ (b) $-\left(\frac{-4}{-5}\right)$

25. (a) $-(-8)$ (b) $-\left(-(-8)\right)$

26. (a) $-\left(-(-2)\right)$ (b) $-(-2)$

27. (a) a (b) $-a$

28. (a) $-b$ (b) $-(-b)$

29. Find the additive inverse of t, if $-t = 6$.

30. Find the additive inverse of $-t$, if $t = -\frac{4}{5}$.

31. Find the additive inverse of $-b$, if $b = \frac{1}{2}$.

32. Find the additive inverse of b, if $-b = \frac{5}{-6}$.

NUMBERS AND THE NUMBER LINE

Exercises 33–44: Find the decimal equivalent for the rational number.

33. $\frac{1}{4}$ 34. $\frac{3}{5}$

35. $\frac{7}{8}$ 36. $\frac{3}{10}$

37. $\frac{3}{2}$ 38. $\frac{3}{50}$

39. $\frac{1}{20}$ 40. $\frac{3}{16}$

41. $\frac{2}{3}$ 42. $\frac{2}{9}$

43. $\frac{7}{9}$ 44. $\frac{5}{11}$

Exercises 45–50: Classify the number as one or more of the following: natural number, whole number, integer, or rational number.

45. 8 46. -8

47. $\frac{16}{4}$

48. $\frac{5}{7}$

49. 0

50. $-\frac{15}{31}$

Exercises 51–58: Classify the number as one or more of the following: natural number, integer, rational number, or irrational number.

51. -4.5

52. π

53. $\sqrt{11}$

54. $\sqrt{25}$

55. $\frac{8}{4}$

56. -5

57. $-\sqrt{3}$

58. $3.\overline{3}$

Exercises 59–68: Plot each number on a number line.

59. (a) 0 **(b)** -2 **(c)** 3

60. (a) -1 **(b)** -3 **(c)** 4

61. (a) $\frac{1}{2}$ **(b)** $-\frac{1}{2}$ **(c)** 2

62. (a) $-\frac{3}{2}$ **(b)** $\frac{3}{2}$ **(c)** 0

63. (a) 1.3 **(b)** -2.5 **(c)** 0.7

64. (a) 3.2 **(b)** -2.8 **(c)** 0.5

65. (a) -10 **(b)** -20 **(c)** 30

66. (a) 5 **(b)** 10 **(c)** -10

67. (a) π **(b)** $\sqrt{2}$ **(c)** $-\sqrt{3}$

68. (a) $\sqrt{11}$ **(b)** $-\sqrt{5}$ **(c)** $\sqrt{4}$

ABSOLUTE VALUE

Exercises 69–78: Evaluate the expression.

69. $|5.23|$

70. $|\pi|$

71. $|-7|$

72. $|-\sqrt{2}|$

73. $\left|-\frac{1}{2}\right|$

74. $\left|\frac{2}{3} - \frac{1}{3}\right|$

75. $|\pi - 3|$

76. $|9 - 4|$

77. $|b|$, if b is negative

78. $|-b|$, if b is positive

INEQUALITY

Exercises 79–86: Insert the symbol $>$ or $<$ to make the statement true.

79. $5 \underline{\quad} 7$

80. $-5 \underline{\quad} 7$

81. $-5 \underline{\quad} -7$

82. $\frac{3}{5} \underline{\quad} \frac{2}{5}$

83. $-\frac{1}{3} \underline{\quad} -\frac{2}{3}$

84. $-\frac{1}{10} \underline{\quad} 0$

85. $-1.9 \underline{\quad} -1.3$

86. $5.1 \underline{\quad} -6.2$

Exercises 87–92: List the given numbers from least to greatest.

87. $-3, 0, 1, -9, -2^3$

88. $4, -2^3, \frac{1}{2}, -\frac{3}{2}, \frac{3}{2}$

89. $-2, \pi, \frac{1}{3}, -\frac{3}{2}, \sqrt{5}$

90. $9, 14, -\frac{1}{12}, -\frac{3}{16}, \sqrt{7}$

91. $-\frac{4}{7}, -\frac{17}{28}, -4^2, \sqrt{7}, \sqrt{2}$

92. $-3.1, 2^3, -3^3, 6, 9.4$

APPLICATIONS

93. *Higher Education* The table lists the total enrollment, in millions of students, in higher education.

Year	1997	1998	1999	2000
Enrollment	14.5	14.6	14.9	15.1

Source: Department of Education.

  (a) What was the enrollment in 1998?
  (b) Mentally estimate the average enrollment for this 4-year period.
  (c) Calculate the average enrollment. Is your mental estimate in reasonable agreement with your calculated result?

94. *Federal Budget* The table shows the budget in trillions of dollars of the federal government for selected years.

Year	1997	1998	1999	2000	2001
Budget	1.60	1.65	1.70	1.79	1.86

Source: Office of Management and Budget.

  (a) What was the budget in 1999?
  (b) Mentally estimate the average budget for this 5-year period.
  (c) Calculate the average budget. Is your mental estimate in reasonable agreement with your calculated result?

WRITING ABOUT MATHEMATICS

95. What is a rational number? Is every integer a rational number? Why or why not?

96. Explain why $\frac{3}{7} > \frac{1}{3}$. Now explain in general how to determine whether $\frac{a}{b} > \frac{c}{d}$. Assume that a, b, c, and d are natural numbers.

CHECKING BASIC CONCEPTS   SECTIONS 1.3 AND 1.4

1. Evaluate each expression.
   (a) 2^3 (b) 10^4 (c) $\left(\frac{2}{3}\right)^3$ (d) -3^4

2. Evaluate each expression without a calculator.
   (a) $6 + 5 \cdot 4$ (b) $6 + 6 \div 2$
   (c) $5 - 2 - 1$ (d) $\frac{6 - 3}{2 + 4}$
   (e) $12 \div (6 \div 2)$ (f) $2^3 - 2\left(2 + \frac{4}{2}\right)$

3. Translate the phrase "five cubed divided by three" to an algebraic expression.

4. Find the opposite of each expression.
   (a) -17 (b) a

5. Classify each number as one or more of the following: natural number, integer, rational number, or irrational number.
   (a) $\frac{10}{2}$ (b) -5 (c) $\sqrt{5}$ (d) $-\frac{5}{6}$

6. Plot each number on the same number line.
   (a) 0 (b) -3 (c) 2
   (d) $\frac{3}{4}$ (e) $-\sqrt{2}$

7. Evaluate each expression.
   (a) $|-12|$ (b) $|-a|$, if $a > 0$

8. List the following numbers from least to greatest.

$$\sqrt{3},\quad -7,\quad 0,\quad \frac{1}{3},\quad -1.6,\quad 3^2$$

1.5   ADDITION AND SUBTRACTION OF REAL NUMBERS

Addition of Real Numbers · Subtraction of Real Numbers · Applications

INTRODUCTION

There are four arithmetic operations: addition, subtraction, multiplication, and division. You probably learned to use arithmetic operations on whole numbers in elementary school. In this section we extend those concepts of addition and subtraction of real numbers to include addition and subtraction of positive *and* negative numbers. Both addition and subtraction require two numbers to calculate an answer. As a result, these operations are called **binary operations**. Multiplication and division are also binary operations. Some operations, such as finding the opposite (negation), are called **unary operations** because they require only one number.

ADDITION OF REAL NUMBERS

In an addition problem the two numbers added are called **addends**, and the answer is called the **sum**. In the addition problem $3 + 5 = 8$, the numbers 3 and 5 are the addends and the number 8 is the sum.

The *opposite* (or *additive inverse*) of a real number a is $-a$. When we add opposites, the result is 0. That is, $a + (-a) = 0$ for every real number a.

EXAMPLE 1   Adding opposites

Find the opposite of each number and calculate the sum of the number and its opposite.
   (a) 45 (b) $\sqrt{2}$ (c) $-\frac{1}{2}$

Solution   (a) The opposite of 45 is -45. Their sum is $45 + (-45) = 0$.
   (b) The opposite of $\sqrt{2}$ is $-\sqrt{2}$. Their sum is $\sqrt{2} + (-\sqrt{2}) = 0$.
   (c) The opposite of $-\frac{1}{2}$ is $\frac{1}{2}$. Their sum is $-\frac{1}{2} + \frac{1}{2} = 0$.

When adding real numbers, it may be helpful to think of money. A positive number represents income, and a negative number indicates a debt. The sum

$$8 + (-6) = 2$$

would represent being paid $8 and owing $6. In this case $2 would be left over. Similarly,

$$-2 + (-3) = -5$$

would represent debts of $2 and $3, resulting in a total debt of $5. To add two real numbers we can use the following rules.

ADDITION OF REAL NUMBERS

To add two numbers that are either *both positive* or *both negative*, add their absolute values. Their sum has the same sign as the two numbers.

To add two numbers with *opposite signs*, find the absolute value of each number. Subtract the smaller absolute value from the larger. The sum has the same sign as the sign of the number with the largest absolute value. If the two numbers are opposites, their sum is 0.

EXAMPLE 2   Adding real numbers

Evaluate each expression.
(a) $-3 + (-5)$ **(b)** $-\frac{1}{10} + \frac{2}{5}$ **(c)** $8.4 + (-9.5)$

Solution
(a) The numbers are both negative, so we add the absolute values $|-3|$ and $|-5|$ to obtain 8. The signs of the addends are both negative, so the answer is -8. That is, $-3 + (-5) = -8$. If we owe $3 and then owe an additional $5, the total amount owed is $8.

(b) The numbers have opposite signs, so we subtract their absolute values to obtain

$$\frac{2}{5} - \frac{1}{10} = \frac{4}{10} - \frac{1}{10} = \frac{3}{10}.$$

The sum is positive because $\left|\frac{2}{5}\right|$ is greater than $\left|-\frac{1}{10}\right|$. That is, $-\frac{1}{10} + \frac{2}{5} = \frac{3}{10}$. If we spend $0.10 and receive $0.40, then we have $0.30 left.

(c) $8.4 + (-9.5) = -1.1$ because $|-9.5|$ is 1.1 more than $|8.4|$. If we have $8.40 and owe $9.50, we are short $1.10.

One way to add real numbers visually is to use a number line. To add $4 + (-3)$ start at 0 (the origin) and draw an arrow to the right 4 units long from 0 to 4. The number -3 is a negative number, so draw an arrow 3 units long to the left, starting at the tip of the first arrow. See Figure 1.21. The tip of the second arrow is at 1, which equals the sum of 4 and -3.

To find the sum $-2 + (-3)$, draw an arrow 2 units long to the left, starting at the origin. Then draw an arrow 3 units long to the left, starting at the tip of the first arrow, which is located at -2. See Figure 1.22. Because the tip of the second arrow coincides with -5 on the number line, the sum of -2 and -3 is -5.

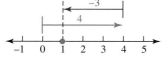

Figure 1.21 $4 + (-3) = 1$

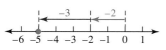

Figure 1.22 $-2 + (-3) = -5$

SUBTRACTION OF REAL NUMBERS

The answer to a subtraction problem is called the **difference**. Addition and subtraction of real numbers can occur in everyday life at grocery stores, where the cost of various items are added to the total and discounts from coupons are subtracted. When subtracting two real numbers, changing the problem to an addition problem may be helpful.

SUBTRACTION OF REAL NUMBERS

For any real numbers a and b,

$$a - b = a + (-b).$$

To subtract b from a, add a and the opposite of b.

EXAMPLE 3 Subtracting real numbers

Evaluate each expression by hand.
(a) $10 - 20$ **(b)** $-5 - 2$ **(c)** $-2.1 - (-3.2)$ **(d)** $\frac{1}{2} - \left(-\frac{3}{4}\right)$

Solution **(a)** $10 - 20 = 10 + (-20) = -10$
(b) $-5 - 2 = -5 + (-2) = -7$
(c) $-2.1 - (-3.2) = -2.1 + 3.2 = 1.1$
(d) $\frac{1}{2} - \left(-\frac{3}{4}\right) = \frac{1}{2} + \frac{3}{4} = \frac{2}{4} + \frac{3}{4} = \frac{5}{4}$

In the next example, we show how to add and subtract groups of numbers. Remember that when there are no parentheses, addition and subtraction are performed from *left* to *right*.

EXAMPLE 4 Adding and subtracting real numbers

Evaluate each expression.
(a) $5 - 4 - (-6) + 1$ **(b)** $\frac{1}{2} - \frac{3}{4} + \frac{1}{3}$ **(c)** $-6.1 + 5.6 - 10.1$

Solution **(a)** Rewrite the expression in terms of addition only, and then find the sum.

$$5 - 4 - (-6) + 1 = 5 + (-4) + 6 + 1$$
$$= 1 + 6 + 1$$
$$= 8$$

(b) Begin by rewriting the fractions with the LCD of 12.

$$\frac{1}{2} - \frac{3}{4} + \frac{1}{3} = \frac{6}{12} - \frac{9}{12} + \frac{4}{12}$$
$$= \frac{6}{12} + \left(-\frac{9}{12}\right) + \frac{4}{12}$$
$$= -\frac{3}{12} + \frac{4}{12}$$
$$= \frac{1}{12}$$

Critical Thinking

Explain how subtraction of real numbers can be performed on a number line.

(c) The expression can be evaluated by changing subtraction to addition.

$$-6.1 + 5.6 - 10.1 = -6.1 + 5.6 + (-10.1)$$
$$= -0.5 + (-10.1)$$
$$= -10.6$$

APPLICATIONS

In word problems, certain phrases often indicate that we should add two numbers. For example, if a person's age is 10 more than 15, then we add 10 and 15 to obtain 25. The phrase *more than* often indicates addition. Other words that indicate addition are *sum*, *greater*, *greater than*, *plus*, and *increased by*.

Similarly, if the outside temperature is 10°F less than 32°F, then the outside temperature is $32 - 10 = 22°F$. The phrase *less than* often indicates subtraction. Other words that indicate subtraction are *difference*, *less*, *minus*, and *decreased by*.

EXAMPLE 5   Calculating temperature differences

The hottest outdoor temperature ever recorded in the shade was 136°F in the Sahara desert, and the coldest outside temperature ever recorded was −129°F in Antarctica. Find the temperature difference between these two temperatures.   (**Source:** *Guiness Book of Records*, 1995.)

Solution   The word *difference* indicates subtraction. We must subtract the two temperatures.

$$136 - (-129) = 136 + 129 = 265°F.$$

Addition of positive and negative numbers occurs at banks if we let positive numbers represent deposits and negative numbers represent withdrawals.

EXAMPLE 6   Balancing a checking account

The initial balance in a checking account is $285. Find the final balance if the following represents a list of withdrawals and deposits: −$15, −$20, $500, and −$100.

Solution   Find the sum of the five numbers.

$$285 + (-15) + (-20) + 500 + (-100) = 270 + (-20) + 500 + (-100)$$
$$= 250 + 500 + (-100)$$
$$= 750 + (-100)$$
$$= 650$$

The final balance is $650. This result may be supported by evaluating the expression with a calculator, as illustrated in Figure 1.23. Note that the expression has been evaluated two different ways.

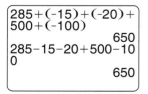

```
285+(-15)+(-20)+
500+(-100)
                650
285-15-20+500-10
0
                650
```

Figure 1.23

Technology Note: *Subtraction and Negation*

Calculators typically have *different* keys to represent subtraction and negation. Be sure to use the correct key.

1.5 PUTTING IT ALL TOGETHER

The following table outlines how to add and subtract real numbers.

Operation	Comments	Examples
Addition	To add real numbers, use a number line or follow the rules found in the box on page 42.	$-2 + 8 = 6$ $0.8 + (-0.3) = 0.5$ $-\frac{1}{7} + \left(-\frac{3}{7}\right) = \frac{-1 + (-3)}{7} = -\frac{4}{7}$ $-4 + 4 = 0$ $-3 + 4 + (-2) = -1$
Subtraction	To subtract real numbers, transform the problem to an addition problem by adding the opposite. $a - b = a + (-b)$	$6 - 8 = 6 + (-8) = -2$ $-3 - 4 = -3 + (-4) = -7$ $-\frac{1}{2} - \left(-\frac{3}{2}\right) = -\frac{1}{2} + \frac{3}{2} = \frac{2}{2} = 1$ $-5 - (-5) = -5 + 5 = 0$ $9.4 - (-1.2) = 9.4 + 1.2 = 10.6$

1.5 EXERCISES

CONCEPTS

1. $-16 + 16 =$ _____.

2. When you add opposites, the result always equals _____.

3. In an addition problem, the two numbers added are called _____.

4. The solution to an addition problem is called the _____.

5. The solution to a subtraction problem is called the _____.

6. If two positive numbers are added, the sum is always a _____ number.

7. If two negative numbers are added, the sum is always a _____ number.

8. To subtract b from a, add the _____ of b to a. That is, $a - b = a +$ _____.

9. The words *sum*, *more*, and *plus* indicate that _____ should be performed.

10. The words *difference*, *less than*, and *minus* indicate that _____ should be performed.

ADDITION AND SUBTRACTION OF REAL NUMBERS

Exercises 11–16: Find the opposite of the number and then calculate the sum of the number and its opposite.

11. 25

12. $-\frac{1}{2}$

13. $-\sqrt{21}$

14. $-\pi$

15. 5.63

16. -5^2

Exercises 17–28: Use a number line to find the sum.

17. $1 + 3$ **18.** $3 + 1$

19. $4 + (-2)$ **20.** $-4 + 6$

21. $-1 + (-2)$ **22.** $-2 + (-2)$

23. $-1 + 3$ **24.** $3 + (-1)$

25. $-10 + 20$ **26.** $15 + (-5)$

27. $-50 + (-100)$ **28.** $-100 + 100$

Exercises 29–46: Find the sum.

29. $5 + (-4)$ **30.** $-9 + 7$

31. $-1 + (-6)$ **32.** $-10 + (-23)$

33. $\frac{3}{4} + \left(-\frac{1}{2}\right)$ **34.** $-\frac{5}{12} + \left(-\frac{1}{6}\right)$

35. $-\frac{6}{7} + \frac{3}{14}$ **36.** $-\frac{1}{5} + \frac{2}{5}$

37. $-\frac{1}{2} + \left(-\frac{3}{4}\right)$ **38.** $-\frac{2}{9} + \left(-\frac{1}{12}\right)$

39. $0.6 + (-1.7)$ **40.** $4.3 + (-2.4)$

41. $-52 + 86$ **42.** $-103 + (-134)$

43. $8 + (-7) + (-2)$ **44.** $-10 + 20 + (-10)$

45. $\frac{1}{2} + \frac{3}{4} + \left(-\frac{1}{2}\right) + \left(-\frac{3}{4}\right)$

46. $0.1 + (-0.4) + (-1.5) + 0.9$

Exercises 47–62: Find the difference.

47. $5 - 8$ **48.** $3 - 5$

49. $-2 - (-9)$ **50.** $-10 - (-19)$

51. $\frac{1}{3} - \left(-\frac{2}{3}\right)$ **52.** $\frac{3}{4} - \frac{5}{4}$

53. $\frac{6}{7} - \frac{13}{14}$ **54.** $-\frac{5}{6} - \frac{1}{6}$

55. $-\frac{1}{10} - \left(-\frac{3}{5}\right)$ **56.** $-\frac{2}{11} - \left(-\frac{5}{11}\right)$

57. $0.8 - (-2.1)$ **58.** $-9.6 - (-5.7)$

59. $-73 - 91$ **60.** $201 - 502$

61. $-7 - (-6) - 10$ **62.** $-20 - 30 - (-40)$

Exercises 63–76: Evaluate the expression.

63. $10 - 19$ **64.** $5 + (-9)$

65. $19 - (-22) + 1$ **66.** $53 + (-43) - 10$

67. $-3 + 4 - 6$ **68.** $-11 + 8 - 10$

69. $100 - 200 + 100 - (-50)$

70. $-50 - (-40) + (-60) + 80$

71. $1.5 - 2.3 + 9.6$

72. $10.5 - (-5.5) + (-1.5)$

73. $-\frac{1}{2} + \frac{1}{4} - \left(-\frac{3}{4}\right)$ **74.** $\frac{1}{4} - \left(-\frac{2}{5}\right) + \left(-\frac{3}{20}\right)$

75. $|4 - 9| - |1 - 7|$

76. $|-5 - (-3)| - |-6 + 8|$

Exercises 77–86: Write an arithmetic expression for the given phrase and then simplify it.

77. The sum of two and negative five

78. Subtract ten from negative six

79. Negative five increased by seven

80. Negative twenty decreased by eight

81. The <u>additive inverse</u> of the quantity two cubed

82. Five minus the quantity two cubed

83. The difference between negative six and seven (*Hint:* Write the numbers for the subtraction problem in the order given.)

84. The difference between one-half and three-fourths

85. Six plus negative ten minus five

86. Ten minus seven plus negative twenty

APPLICATIONS

87. *Checking Account* The initial balance in a checking account is \$358. Find the final balance resulting from the following withdrawals and deposits: $-\$45$, \$37, \$120, and $-\$240$.

88. *Savings Account* A savings account has \$1245 in it. Find the final balance resulting from the following withdrawals and deposits: $-\$189$, \$975, $-\$226$, and $-\$876$.

89. *Deepest and Highest* The deepest point in the ocean is the Mariana Trench, which is 35,839 feet below sea level. The highest point on Earth is Mount Everest, which is 29,029 feet above sea level. What is the difference in height between Mount Everest and the Mariana Trench?   (***Source:*** *The Guinness Book of Records.*)

90. *Greatest Temperature Ranges*   The greatest temperature range on Earth occurs in Siberia, where the temperature can vary between 98°F in the summer and −90°F in the winter. Find the difference between these two temperatures.   (***Source:*** *The Guinness Book of Records.*)

WRITING ABOUT MATHEMATICS

91. Explain how to add two negative numbers. Give an example.

92. Explain how to subtract a negative number from a positive number. Give an example.

1.6   MULTIPLICATION AND DIVISION OF REAL NUMBERS

Multiplication of Real Numbers · **Division of Real Numbers** · **Data and Number Sense**

INTRODUCTION

In some businesses, such as auto-parts stores, sometimes it is necessary to convert fractions to decimal numbers and decimals to fractions. For example, it might be necessary to convert the size of a $2\frac{3}{8}$-inch bolt to a decimal number before it can be ordered. This conversion makes use of division of real numbers. In this section we discuss both multiplication and division of real numbers.   (***Source:*** NAPA.)

MULTIPLICATION OF REAL NUMBERS

In a multiplication problem, the numbers multiplied are called the *factors*, and the answer is called the *product*. In the problem $3 \cdot 5 = 15$, the numbers 3 and 5 are factors and 15 is the product. Multiplication is a fast way to perform addition. For example, $5 \cdot 2 = 10$ is equivalent to finding the sum of five 2s, or

$$2 + 2 + 2 + 2 + 2 = 10.$$

Similarly, the product $5 \cdot (-2)$ is equivalent to finding the sum of five −2s, or

$$(-2) + (-2) + (-2) + (-2) + (-2) = -10.$$

Thus $5 \cdot (-2) = -10$. In general, the product of a positive number and a negative number is a negative number.

What sign should the product of two negative numbers have? To answer this question, consider the following patterns.

$$
\begin{array}{ll}
3 \cdot 2 = 6 & \quad 3 \cdot (-2) = -6 \\
2 \cdot 2 = 4 & \quad 2 \cdot (-2) = -4 \\
1 \cdot 2 = 2 & \quad 1 \cdot (-2) = -2 \\
0 \cdot 2 = 0 & \quad 0 \cdot (-2) = 0 \\
-1 \cdot 2 = -2 & \quad -1 \cdot (-2) = ? \\
-2 \cdot 2 = -4 & \quad -2 \cdot (-2) = ?
\end{array}
$$

In the first column, each time the first factor is reduced by 1, the product is found by *subtracting* 2 from the product in the previous row. Note that because multiplication is a fast way to perform addition, the number of 2s being added is one less as we move down the rows in the first column.

In the second column, each time the first factor is reduced by 1, the product is found by *subtracting* -2 from the previous product, which is equivalent to *adding* 2. If the pattern in the second column is to continue, then

$$-1 \cdot (-2) = 2$$
$$-2 \cdot (-2) = 4$$
$$-3 \cdot (-2) = 6,$$

and so on. Note that, in general, the product of two negative numbers is a positive number.

 ## SIGNS OF PRODUCTS

The product of two numbers with *like* signs is positive. The product of two numbers with *unlike* signs is negative.

As a result of these two rules, $4 \cdot 5 = 20$, $4 \cdot (-5) = -20$, $-4 \cdot 5 = -20$, and $-4 \cdot (-5) = 20$.

EXAMPLE 1   Multiplying real numbers

Evaluate each expression by hand.
(a) $-11 \cdot 8$ **(b)** $\frac{3}{5} \cdot \frac{4}{7}$ **(c)** $-1.2(-10)$ **(d)** $(1.2)(5)(-7)$

Solution   **(a)** The product is negative because the factors -11 and 8 have unlike signs. Thus $-11 \cdot 8 = -88$.
(b) The product is positive because both factors are positive.

$$\frac{3}{5} \cdot \frac{4}{7} = \frac{3 \cdot 4}{5 \cdot 7} = \frac{12}{35}$$

(c) As both factors are negative, the product is positive. Thus $-1.2(-10) = 12$.
(d) $(1.2)(5)(-7) = 6 \cdot (-7) = -42$

DIVISION OF REAL NUMBERS

In the division problem $20 \div 4 = 5$, the number 20 is the **dividend**, 4 is the **divisor**, and 5 is the **quotient**. This division problem can also be written in fraction form as $\frac{20}{4} = 5$. Division of real numbers can be defined in terms of multiplication and reciprocals. The reciprocal, or multiplicative inverse, of a real number a is $\frac{1}{a}$. The number 0 has *no reciprocal*.

Technology Note

Try dividing 5 by 0 with a calculator. On most calculators dividing a number by 0 results in an error message. Does it on your calculator?

 ## DIVIDING REAL NUMBERS

For real numbers a and b with $b \neq 0$,

$$\frac{a}{b} = a \cdot \frac{1}{b}.$$

That is, to divide a by b, multiply a by the reciprocal of b.

Note: Division by 0 is undefined because 0 has no reciprocal.

EXAMPLE 2   Dividing real numbers

Evaluate each expression by hand.

(a) $-12 \div \frac{1}{2}$ (b) $\dfrac{\frac{2}{3}}{-7}$ (c) $\frac{-4}{-24}$ (d) $6 \div 0$

Solution   (a) $-12 \div \frac{1}{2} = \frac{-12}{1} \cdot \frac{2}{1} = \frac{-24}{1} = -24$

(b) $\dfrac{\frac{2}{3}}{-7} = \frac{2}{3} \div (-7) = \frac{2}{3} \cdot \left(-\frac{1}{7}\right) = -\frac{2}{21}$

(c) $\frac{-4}{-24} = -4 \cdot \left(-\frac{1}{24}\right) = \frac{4}{24} = \frac{1}{6}$

(d) $6 \div 0 = \frac{6}{0}$ is undefined. The number 0 has no reciprocal.

When determining the sign of a quotient, the following rules may be helpful.

> ### SIGNS OF QUOTIENTS
>
> The quotient of two numbers with *like* signs is positive. The quotient of two numbers with *unlike* signs is negative.

As a result of these two rules, $\frac{20}{4} = 5$, $\frac{-20}{4} = -5$, $\frac{20}{-4} = -5$, and $\frac{-20}{-4} = 5$. You may wonder why a negative number divided by a negative number is a positive number. Remember that division is a fast way to perform subtraction. For example, because

$$20 - 4 - 4 - 4 - 4 - 4 = 0,$$

there are five 4s in 20 and so $\frac{20}{4} = 5$. To check this answer, multiply 5 and 4 to obtain $5 \cdot 4 = 20$. Similarly,

$$-20 - (-4) - (-4) - (-4) - (-4) - (-4) = 0.$$

Thus there are five -4s in -20 and so $\frac{-20}{-4} = 5$. To check this answer, multiply 5 and -4 to obtain $5 \cdot (-4) = -20$.

In business, employees may need to convert fractions and mixed numbers to decimal numbers. The next example illustrates this process.

EXAMPLE 3   Converting fractions to decimals

Convert each measurement to a decimal number.

(a) $2\frac{3}{8}$-inch bolt   (b) $\frac{15}{16}$-inch diameter   (c) $1\frac{1}{3}$-cup flour

Solution   (a) Because $\frac{3}{8}$ indicates the division problem $3 \div 8$, begin by dividing 3 by 8.

$$
\begin{array}{r}
0.375 \\
8\overline{)3.000} \\
\underline{24} \\
60 \\
\underline{56} \\
40 \\
\underline{40} \\
0
\end{array}
$$

Thus the mixed number $2\frac{3}{8}$ is equivalent to the decimal number 2.375.

(b) Start by dividing 15 by 16.

$$
\begin{array}{r}
0.9375 \\
16\overline{)15.0000} \\
\underline{144} \\
60 \\
\underline{48} \\
120 \\
\underline{112} \\
80 \\
\underline{80} \\
0
\end{array}
$$

Thus the fraction $\frac{15}{16}$ is equivalent to the decimal number 0.9375.

(c) Start by dividing 1 by 3.

$$
\begin{array}{r}
0.333\ldots \\
3\overline{)1.000} \\
\underline{9} \\
10 \\
\underline{9} \\
10 \\
\underline{9} \\
1
\end{array}
$$

Thus the mixed number $1\frac{1}{3}$ is equivalent to the decimal number $1.333\ldots$, or $1.\overline{3}$.

In the next example, numbers expressed as terminating decimals are converted to fractions.

EXAMPLE 4   Converting decimals to fractions

Convert each decimal number to a fraction.
(a) 0.06   **(b)** 0.375   **(c)** 0.0025

Solution   **(a)** The decimal 0.06 equals six hundredths, or $\frac{6}{100}$. Reducing this fraction gives

$$
\frac{6}{100} = \frac{3 \cdot 2}{50 \cdot 2} = \frac{3}{50}.
$$

(b) The decimal 0.375 equals three hundred seventy-five thousandths, or $\frac{375}{1000}$. Reducing this fraction gives

$$
\frac{375}{1000} = \frac{3 \cdot 125}{8 \cdot 125} = \frac{3}{8}.
$$

(c) The decimal 0.0025 equals twenty-five ten thousandths, or $\frac{25}{10,000}$. Reducing this fraction gives

$$
\frac{25}{10,000} = \frac{1 \cdot 25}{400 \cdot 25} = \frac{1}{400}.
$$

═══════════ MAKING CONNECTIONS ═══════════

The Four Arithmetic Operations

If you know how to add real numbers, then you also know how to subtract real numbers because subtraction is defined in terms of addition. That is,

$$a - b = a + (-b).$$

If you know how to multiply real numbers, then you also know how to divide real numbers because division is defined in terms of multiplication. That is,

$$\frac{a}{b} = a \cdot \frac{1}{b}.$$

FRACTIONS AND CALCULATORS (OPTIONAL) Many calculators have the capability to perform arithmetic on fractions and express the answer as either a decimal or a fraction. The next example illustrates this capability.

EXAMPLE 5 Performing arithmetic operations with technology

Use a calculator to evaluate each expression. Express your answer as a decimal and as a fraction.

(a) $\frac{1}{3} + \frac{2}{5} - \frac{4}{9}$ **(b)** $\left(\frac{4}{9} \cdot \frac{3}{8}\right) \div \frac{2}{3}$

Solution   **(a)** Figure 1.24(a) shows that

$$\frac{1}{3} + \frac{2}{5} - \frac{4}{9} = 0.2\overline{8}, \quad \text{or} \quad \frac{13}{45}.$$

Note: Generally it is a good idea to put parentheses around fractions when you are using a calculator.

(b) Figure 1.24(b) shows that $\left(\frac{4}{9} \cdot \frac{3}{8}\right) \div \frac{2}{3} = 0.25$, or $\frac{1}{4}$.

Calculator Help
To convert answers to fractions, see Appendix A (page AP-2).

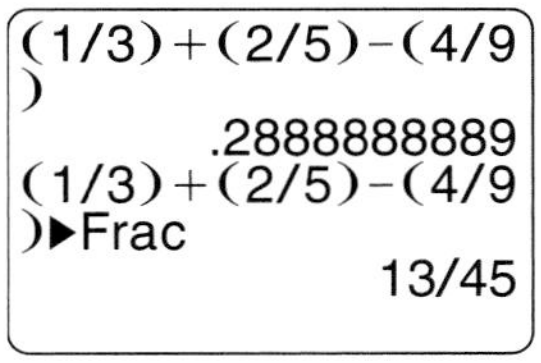

```
(1/3)+(2/5)-(4/9
)
          .2888888889
(1/3)+(2/5)-(4/9
)▶Frac
               13/45
```

```
((4/9)*(3/8))/(2
/3)
              .25
((4/9)*(3/8))/(2
/3)▶Frac
             1/4
```

(a) 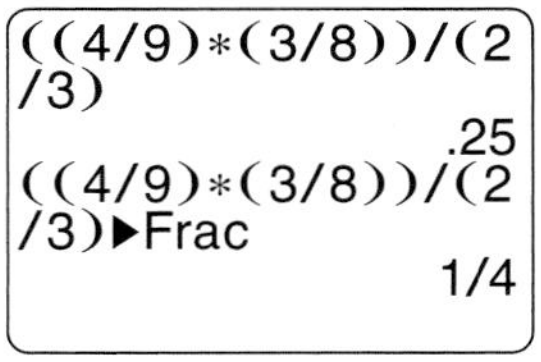   (b)

Figure 1.24

DATA AND NUMBER SENSE

Approximations are commonly used in everyday life. When making an estimate, we often use arithmetic, and we should always ask ourselves if a result seems reasonable.

EXAMPLE 6 Determining a reasonable answer

Table 1.8 lists the number of reported endangered species in the United States for selected years.

TABLE 1.8 Endangered Species of Animals in the United States

Year	1997	1998	1999	2000
Species	343	357	358	368

Source: Fish and Wildlife Service.

Determine mentally which of the following values approximates the average number of endangered species from 1997 to 2000: 339, 357, or 368. Explain your reasoning. Then calculate the true average.

Solution The average value would lie between the maximum and minimum numbers of endangered species. The value of 339 is less than the minimum value of 343 in Table 1.8, and 368 is the maximum value in the table. The only reasonable choice is 357 because it is located between the maximum and minimum of the four data items. To find the actual average add the four values and then divide by 4.

$$\frac{343 + 357 + 358 + 368}{4} = \frac{1426}{4} = 356.5.$$

EXAMPLE 7 Determining a reasonable answer

It is 2823 miles from New York to Los Angeles. Determine mentally which of the following would best estimate the actual driving time for this distance in a car: 50, 100, or 120 hours.

Solution Average speed equals distance divided by time. Dividing by 100 is easy, so start by dividing 2800 miles by 100 hours. The average speed would be about $\frac{2800}{100} = 28$ miles per hour, which is too slow for most drivers. A more reasonable choice is 50 hours because then the average speed would double to about 56 miles per hour.

EXAMPLE 8 Estimating a numeric value

Table 1.9 lists the number of cellular phone subscribers in millions for selected years. Estimate the number of subscribers in 1998 by assuming that this trend continued.

TABLE 1.9 Cellular Phone Subscribers

Year	1994	1995	1996	1997	1998
Subscribers	24.1	33.8	44.0	55.3	?

Source: Cellular Telecommunications Industry Association.

Solution The data show that the number of subscribers has increased each year by about 10 or 11 million. A reasonable estimate might be $55 + 11 = 66$ million, but estimates may vary slightly. (Note that the actual value in 1998 was 69.2 million, which was somewhat higher than the estimate.)

Critical Thinking

Suppose that you know the average cost of public college tuition in 1990 and in 2000. Using only these two values, do you think it would be more accurate to estimate the tuition in 1995 or in 2005? Explain your reasoning.

1.6 PUTTING IT ALL TOGETHER

The following table summarizes some of the information presented in this section.

Topic	Comments	Examples	
Multiplication	The product of two numbers with like signs is positive, and the product of two numbers with unlike signs is negative.	$6 \cdot 7 = 42$	Like signs
		$6 \cdot (-7) = -42$	Unlike signs
		$-6 \cdot 7 = -42$	Unlike signs
		$-6 \cdot (-7) = 42$	Like signs
Division	For real numbers a and b, with $b \neq 0$, $$\frac{a}{b} = a \cdot \frac{1}{b}.$$ The quotient of two numbers with like signs is positive, and the quotient of two numbers with unlike signs is negative.	$\dfrac{42}{6} = 7$	Like signs
		$\dfrac{-42}{6} = -7$	Unlike signs
		$\dfrac{42}{-6} = -7$	Unlike signs
		$\dfrac{-42}{-6} = 7$	Like signs
		$\dfrac{\frac{3}{2}}{-6} = \dfrac{3}{2} \cdot \left(-\dfrac{1}{6}\right)$	
		$= -\dfrac{3}{12}$	Unlike signs
		$= -\dfrac{1}{4}$	

continued on next page

continued from previous page

Topic	Comments	Examples
Converting Fractions to Decimals	The fraction $\frac{a}{b}$ is equivalent to $a \div b$.	The fraction $\frac{2}{9}$ is equivalent to $0.222\ldots$, or $0.\overline{2}$, because the division problem $$\begin{array}{r} 0.222\ldots \\ 9\overline{)2.00000} \\ \underline{18} \\ 20 \\ \underline{18} \\ 20 \\ \underline{18} \\ 2 \end{array}$$ results in a repeating decimal.
Converting Terminating Decimals to Fractions	Write the decimal as a fraction with a denominator equal to a power of 10 and then reduce this fraction.	$$0.55 = \frac{55}{100}$$ $$= \frac{11 \cdot 5}{20 \cdot 5}$$ $$= \frac{11}{20}$$

1.6 EXERCISES

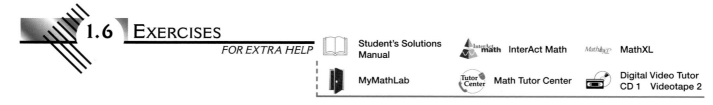

FOR EXTRA HELP

Student's Solutions Manual   InterAct Math   MathXL

MyMathLab   Math Tutor Center   Digital Video Tutor CD 1   Videotape 2

CONCEPTS

1. In a multiplication problem, the numbers multiplied are called _____.

2. The solution to a multiplication problem is called the _____.

3. The product of a positive number and a negative number is a _____ number.

4. The product of two negative numbers is a _____ number.

5. The solution to a division problem is called the _____.

6. In the problem $15 \div 5$, the number 15 is the _____ and the number 5 is the _____.

7. The reciprocal of a nonzero number a is _____.

8. The reciprocal of $-\frac{3}{4}$ is _____.

9. To divide a by b, multiply a by the _____ of b.

10. $\dfrac{a}{b} = a \cdot$ _____

11. A negative number divided by a negative number is a _____ number.

12. A negative number divided by a positive number is a _____ number.

13. $\frac{-4}{-2} = $ _____

14. To convert $\frac{5}{8}$ to a decimal, divide _____ by _____.

MULTIPLICATION AND DIVISION OF REAL NUMBERS

Exercises 15–36: Multiply.

15. $-3 \cdot 4$

16. $-5 \cdot 7$

17. $6 \cdot (-3)$

18. $2 \cdot (-1)$

19. $0 \cdot (-2.13)$

20. $-2 \cdot (-7)$

21. $-6 \cdot (-10)$

22. $-3 \cdot (-1.7) \cdot 0$

23. $-\frac{1}{2} \cdot \left(-\frac{2}{4}\right)$

24. $-\frac{3}{4} \cdot \left(-\frac{5}{12}\right)$

25. $-\frac{3}{7} \cdot \frac{7}{3}$

26. $\frac{5}{8} \cdot \left(-\frac{4}{15}\right)$

27. $-10 \cdot (-20)$

28. $1000 \cdot (-70)$

29. $-50 \cdot 100$

30. $-0.5 \cdot (-0.3)$

31. $-2 \cdot 3 \cdot (-4) \cdot 5$

32. $-3 \cdot (-5) \cdot (-2) \cdot 10$

33. $-6 \cdot \frac{1}{6} \cdot \frac{7}{9} \cdot \left(-\frac{9}{7}\right) \cdot \left(-\frac{3}{2}\right)$

34. $-\frac{8}{5} \cdot \frac{1}{8} \cdot \left(-\frac{5}{7}\right) \cdot -7$

35. $(-1) \cdot (-1) \cdot (-1) \cdot (-1)$

36. $5 \cdot (-2) \cdot (-2) \cdot (-2)$

Exercises 37–58: Divide.

37. $-10 \div 5$

38. $-8 \div 4$

39. $-20 \div (-2)$

40. $-15 \div (-3)$

41. $\frac{-12}{3}$

42. $\frac{-25}{-5}$

43. $\frac{39}{-13}$

44. $-\frac{100}{-20}$

45. $-16 \div \frac{1}{2}$

46. $10 \div \left(-\frac{1}{3}\right)$

47. $\frac{1}{2} \div (-11)$

48. $-\frac{3}{4} \div (-6)$

49. $\frac{-\frac{4}{5}}{-3}$

50. $\frac{\frac{7}{8}}{-7}$

51. $\frac{5}{6} \div \left(-\frac{8}{9}\right)$

52. $-\frac{11}{12} \div \left(-\frac{11}{4}\right)$

53. $-\frac{1}{2} \div 0$

54. $-9 \div 0$

55. $-0.5 \div \frac{1}{2}$

56. $-0.25 \div \left(-\frac{3}{4}\right)$

57. $-\frac{2}{3} \div 0.5$

58. $\frac{1}{6} \div 1.5$

CONVERTING BETWEEN FRACTIONS AND DECIMALS

Exercises 59–70: Write the fraction or mixed number as a decimal.

59. $\frac{1}{2}$

60. $\frac{3}{4}$

61. $\frac{3}{16}$

62. $\frac{1}{9}$

63. $3\frac{1}{2}$

64. $2\frac{1}{4}$

65. $5\frac{2}{3}$

66. $6\frac{7}{9}$

67. $1\frac{7}{16}$

68. $\frac{11}{16}$

69. $\frac{7}{8}$

70. $6\frac{1}{12}$

Exercises 71–78: Write the decimal number as a fraction in lowest terms.

71. 0.25

72. 0.8

73. 0.16

74. 0.35

75. 0.625

76. 0.0125

77. 0.6875

78. 0.21875

APPLICATIONS

79. *Top Tourist Destinations* The table lists the number of tourists in millions visiting selected countries during 2001.

Country	China	Italy	USA	Spain	France
Tourists	33.2	39.1	44.5	49.5	76.5

Source: World Tourism Organization.

(a) How many tourists visited Spain in 2001?
(b) Calculate the average number of tourists for the five countries listed.

80. *Buying a Computer* An advertisement for a computer states that it costs \$202 down and \$98.99 per month for 24 months. Mentally estimate which of the following best estimates the cost of the computer: \$2600, \$3200, or \$3800. Explain your reasoning.

81. *Leasing a Car* To lease a car costs \$1497 down and \$249 per month for 36 months. Mentally estimate the cost of the lease. Explain your reasoning.

82. *Amway Sales* The table lists worldwide retail sales in billions of dollars for Amway products.

Year	1988	1990	1992	1994	1996	1998
Sales	1.8	2.1	3.9	5.3	6.8	5.7

Source: Amway.

(a) What were Amway sales in 1994?

(b) Estimate sales in 1993. Explain how you arrived at your estimate and compare it to the actual value of $4.5 billion.

(c) Estimate sales in 1997. Actual sales were $7.0 billion in 1997. Discuss the difficulty with obtaining an accurate estimate if you use only the data in the table.

83. *Cable Modems* Cable modems provided by the cable TV industry give high-speed access to the Internet. The table lists the number of cable modem subscribers in millions for selected years.

Year	1999	2000	2001	2002
Subscribers	1.0	2.0	3.0	4.3

Source: Yankee Group.

(a) Discuss any trends in cable modem subscribers from 1999 to 2002.

(b) Estimate the number of subscribers in 2003. Explain your reasoning.

84. *Digital Subscriber Lines* Local phone companies provide digital subscriber lines (DSL) for high-speed Internet access. This technology competes directly with cable modem service provided by cable TV companies. The table lists the number of digital subscriber lines in millions for selected years.

Year	1999	2000	2001	2002
Subscribers	0.3	0.7	1.5	2.7

Source: Yankee Group.

(a) Discuss any trends in DSL from 1999 to 2002.

(b) Estimate the number of subscribers in 2003. Explain your reasoning.

WRITING ABOUT MATHEMATICS

85. Division is a fast way to subtract. Consider the division problem $\frac{-6}{-2}$, whose quotient represents the number of -2s in -6. Using this idea, explain why the answer is a positive number.

86. Explain how to determine whether the product of three signed numbers is positive or negative.

CHECKING BASIC CONCEPTS   SECTIONS 1.5 AND 1.6

1. Find each sum.
  (a) $-4 + 4$ (b) $-10 + (-12) + 3$

2. Evaluate each expression.
  (a) $\frac{2}{3} - \left(-\frac{2}{9}\right)$ (b) $-1.2 - 5.1 + 3.1$

3. The hottest temperature ever recorded at International Falls, Minnesota, was $98°F$, and the coldest temperature ever recorded was $-46°F$. What is the difference between these two temperatures?

4. Find each product.
  (a) $-5 \cdot (-7)$ (b) $-\frac{1}{2} \cdot \frac{2}{3} \cdot \left(-\frac{4}{5}\right)$

5. Evaluate each expression.
  (a) $-5 \div \frac{2}{3}$ (b) $-\frac{5}{8} \div \left(-\frac{4}{3}\right)$

6. What is the reciprocal of $-\frac{7}{6}$?

7. Simplify each expression.
  (a) $\frac{-10}{2}$ (b) $\frac{10}{-2}$ (c) $-\frac{10}{2}$ (d) $\frac{-10}{-2}$

8. Convert each fraction or mixed number to a decimal number.
  (a) $\frac{3}{5}$ (b) $3\frac{7}{8}$

9. The table lists the number of prisoners per 100,000 resident population, or incarceration rate (IR), for the United States during selected years.

Year	1980	1985	1990	1995	2000
IR	139	201	292	411	477

Source: Bureau of Justice Statistics Bulletin.

(a) Discuss any trend in the incarceration rate between 1980 and 2000.

(b) Assuming that the current trend continues, estimate the incarceration rate in 2010. Explain your reasoning.

1.7   PROPERTIES OF REAL NUMBERS

Commutative Properties · Associative Properties · Distributive Properties · Identity and Inverse Properties · Mental Calculations

INTRODUCTION

The order in which you perform actions is often important. For example, putting on your socks and then your shoes is not the same as putting on your shoes and then your socks. In mathematical terms, these two actions are not *commutative*. However, the actions of tying your shoes and putting on a sweatshirt probably are *commutative*. In mathematics some operations are commutative and others are not. In this section we discuss several properties of real numbers.

COMMUTATIVE PROPERTIES

The **commutative property for addition** states that two numbers, a and b, can be added in any order and the result will be the same. That is, $a + b = b + a$. For example, if a person is paid $5 and then $7 or paid $7 and then $5, the result is the same. Either way the person is paid a total of

$$5 + 7 = 7 + 5 = 12 \text{ dollars.}$$

There is also a **commutative property for multiplication**. It states that two numbers, a and b, can be multiplied in any order and the result will be the same. That is, $a \cdot b = b \cdot a$. For example, 3 groups of 5 people or 5 groups of 3 people both contain

$$3 \cdot 5 = 5 \cdot 3 = 15 \text{ people.}$$

We can summarize these results as follows.

COMMUTATIVE PROPERTIES

For any real numbers a and b,

$$a + b = b + a \qquad \text{Addition}$$

and

$$a \cdot b = b \cdot a. \qquad \text{Multiplication}$$

EXAMPLE 1   Applying the commutative properties

Use the commutative properties to rewrite each expression.
(a) $15 + 100$ (b) $a \cdot 8$ (c) $x(y + z)$

Solution   (a) By the commutative property for addition $15 + 100$ can be written as $100 + 15$.
(b) By the commutative property for multiplication $a \cdot 8$ can be written as $8 \cdot a$ or $8a$.
(c) We can apply both commutative properties.

$$
\begin{aligned}
x(y + z) &= (y + z)x \qquad && \text{Commutative property for multiplication} \\
&= (z + y)x \qquad && \text{Commutative property for addition}
\end{aligned}
$$

In the next example we determine whether the other arithmetic operations are commutative.

EXAMPLE 2 Determining whether subtraction or division is commutative

Determine whether there is a commutative property for subtraction or division.

Solution The arithmetic operation of subtraction is *not* commutative. For example, $5 - 3 = 2$ but $3 - 5 = -2$. Similarly, division is *not* commutative because $4 \div 2 = 2$ but $2 \div 4 = \frac{1}{2}$.

ASSOCIATIVE PROPERTIES

The commutative properties allow us to interchange the order of two numbers when we add or multiply. The *associative properties* allow us to change how numbers are grouped. For example, if a person earns $5, $4, and $6, then the total amount earned can be calculated either as

$$(5 + 4) + 6 = 9 + 6 = 15 \quad \text{or as}$$
$$5 + (4 + 6) = 5 + 10 = 15.$$

In either case we obtain the same answer, $15, which is the result of the **associative property for addition**. We did not change the order of the numbers; we only changed how the numbers were grouped. There is also an **associative property for multiplication**, which is illustrated by

$$(5 \cdot 4) \cdot 6 = 20 \cdot 6 = 120 \quad \text{and}$$
$$5 \cdot (4 \cdot 6) = 5 \cdot 24 = 120.$$

We can summarize these results as follows.

ASSOCIATIVE PROPERTIES

For any real numbers a, b, and c,

$$(a + b) + c = a + (b + c) \qquad \text{Addition}$$

and

$$(a \cdot b) \cdot c = a \cdot (b \cdot c). \qquad \text{Multiplication}$$

Note: Sometimes we omit the multiplication dot. Thus $a \cdot b = ab$ and $5 \cdot x \cdot y = 5xy$.

EXAMPLE 3 Applying the associative properties

Use the associative property to rewrite each expression.
(a) $(5 + 6) + 7$ **(b)** $x(yz)$

Solution **(a)** The given expression is equivalent to $5 + (6 + 7)$.
(b) The given expression is equivalent to $(xy)z$.

EXAMPLE 4   Identifying properties of real numbers

State the property that each equation illustrates.

(a) $5 \cdot (8y) = (5 \cdot 8)y$ **(b)** $3 \cdot 7 = 7 \cdot 3$ **(c)** $x + yz = yz + x$

Solution   **(a)** This equation illustrates the associative property for multiplication because the grouping of the numbers has been changed.

(b) This equation illustrates the commutative property for multiplication because the order of the numbers 3 and 7 has been changed.

(c) This equation illustrates the commutative property for addition because the order of the terms x and yz has been changed.

In the next example we determine whether the other arithmetic operations are associative.

EXAMPLE 5   Determining whether subtraction or division is associative

Determine whether there is an associative property for subtraction or division.

Solution   Subtraction is *not* associative because

$$(5 - 3) - 2 = 2 - 2 = 0 \quad \text{but} \quad 5 - (3 - 2) = 5 - 1 = 4.$$

Division is *not* associative because

$$(24 \div 4) \div 2 = 6 \div 2 = 3 \quad \text{but} \quad 24 \div (4 \div 2) = 24 \div 2 = 12.$$

≡ MAKING CONNECTIONS ≡

Commutative and Associative Properties

Both the commutative and associative properties work for addition and multiplication. However, neither property works for subtraction or division.

DISTRIBUTIVE PROPERTIES

The distributive properties are used frequently in algebra to simplify expressions. An example is

$$4(2 + 3) = 4 \cdot 2 + 4 \cdot 3.$$

The 4 must be multiplied by *both* the 2 and the 3—not just the 2. We illustrate a distributive property geometrically in Figure 1.25. Note that the area of one rectangle that is 4 squares by 5 squares is the same as the area of two rectangles: one that is 4 squares by 2 squares and another that is 4 squares by 3 squares. In either case the total area is 20 square units.

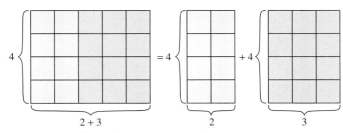

Figure 1.25 $4(2 + 3) = 4 \cdot 2 + 4 \cdot 3$

The distributive property remains valid when addition is replaced with subtraction. For example,

$$4(2 - 3) = 4 \cdot 2 - 4 \cdot 3$$

is a true statement. We can summarize these results about the distributive properties as follows.

DISTRIBUTIVE PROPERTIES

For any real numbers a, b, and c,

$$a(b + c) = ab + ac \quad \text{and} \quad a(b - c) = ab - ac.$$

Note: Because multiplication is commutative, the distributive properties can also be written as

$$(b + c)a = ba + ca \quad \text{and} \quad (b - c)a = ba - ca.$$

EXAMPLE 6   Applying the distributive properties

Apply a distributive property to each expression.
(a) $5a + 2a$ **(b)** $-6(a - 2)$ **(c)** $20 - (x + 7)$

Solution   **(a)** A distributive property can be used to factor out an a.

$$5a + 2a = (5 + 2)a = 7a$$

(b) $-6(a - 2) = -6 \cdot a - (-6) \cdot 2 = -6a + 12$

(c) $20 - (x + 7) = 20 + (-1)(x + 7)$ Change subtraction to addition.

$$= 20 + (-1) \cdot x + (-1) \cdot 7 \qquad \text{Distributive property}$$

$$= 20 - x - 7 \qquad\qquad\qquad \text{Multiply.}$$

$$= 13 - x \qquad\qquad\qquad\quad \text{Subtract.}$$

Note: To simplify the expression $20 - (x + 7)$, we subtract *both* the x and the 7. Thus we can quickly simplify the given expression to

$$20 - (x + 7) = 20 - x - 7.$$

EXAMPLE 7   Identifying properties of real numbers

State the property or properties illustrated by each equation.
(a) $4(5 - x) = 20 - 4x$ **(b)** $(4 + x) + 5 = x + 9$
(c) $5z + 7z = 12z$ **(d)** $x(y + z) = zx + yx$

Solution   **(a)** This equation illustrates use of the distributive property with subtraction.

$$4(5 - x) = 4 \cdot 5 - 4 \cdot x = 20 - 4x$$

(b) This equation illustrates use of both the commutative and associative properties for addition.

$$(4 + x) + 5 = (x + 4) + 5 \qquad \text{Commutative property for addition}$$

$$= x + (4 + 5) \qquad \text{Associative property for addition}$$

$$= x + 9 \qquad\qquad\quad \text{Simplify}$$

(c) This equation illustrates use of the distributive property with addition.

$$5z + 7z = (5 + 7)z = 12z$$

(d) This equation illustrates use of both commutative properties and a distributive property.

$$x(y + z) = xy + xz \qquad \text{Distributive property}$$
$$= xz + xy \qquad \text{Commutative property for addition}$$
$$= zx + yx \qquad \text{Commutative property for multiplication}$$

IDENTITY AND INVERSE PROPERTIES

The **identity property of 0** states that if 0 is added to any real number a, the result is a. The number 0 is called the **additive identity**. Examples include

$$-3 + 0 = -3 \quad \text{and} \quad 0 + 18 = 18.$$

The **identity property of 1** states that if any number a is multiplied by 1, the result is a. The number 1 is called the **multiplicative identity**. Examples include

$$-7 \cdot 1 = -7 \quad \text{and} \quad 1 \cdot 9 = 9.$$

We can summarize these results as follows.

IDENTITY PROPERTIES

For any real number a,

$$a + 0 = 0 + a = a \qquad \text{Additive identity}$$

and

$$a \cdot 1 = 1 \cdot a = a. \qquad \text{Multiplicative identity}$$

The *additive inverse* or *opposite* of a number a is $-a$. The number 0 is its own opposite. The opposite of -5 is $-(-5) = 5$ and the opposite of x is $-x$. The sum of a number a and its additive inverse equals the additive identity 0. Thus $-5 + 5 = 0$ and $x + (-x) = 0$.

The *multiplicative inverse* or *reciprocal* of a nonzero number a is $\frac{1}{a}$. The number 0 has no multiplicative inverse. The multiplicative inverse of $-\frac{5}{4}$ is $-\frac{4}{5}$. The product of a number and its multiplicative inverse equals the multiplicative identity 1. Thus $-\frac{5}{4} \cdot \left(-\frac{4}{5}\right) = 1$.

INVERSE PROPERTIES

For any real number a,

$$a + (-a) = 0 \quad \text{and} \quad -a + a = 0. \qquad \text{Additive inverse}$$

For any *nonzero* real number a,

$$a \cdot \frac{1}{a} = 1 \quad \text{and} \quad \frac{1}{a} \cdot a = 1. \qquad \text{Multiplicative inverse}$$

EXAMPLE 8 Identifying identity and inverse properties

State the property or properties illustrated by each equation.
(a) $0 + xy = xy$ **(b)** $\frac{36}{30} = \frac{6}{5} \cdot \frac{6}{6} = \frac{6}{5}$
(c) $x + (-x) + 5 = 0 + 5 = 5$ **(d)** $\frac{1}{9} \cdot 9y = 1 \cdot y = y$

Solution **(a)** This equation illustrates use of the identity property for 0.
(b) Because $\frac{6}{6} = 1$, these equations illustrate how a fraction can be reduced by using the identity property for 1.
(c) These equations illustrate use of the additive inverse property and the identity property for 0.
(d) These equations illustrate use of the multiplicative inverse property and the identity property of 1.

MENTAL CALCULATIONS

Properties of numbers can be used to simplify calculations. For example, to find the sum

$$4 + 7 + 6 + 3$$

we might apply the commutative and associative properties for addition to obtain

$$(4 + 6) + (7 + 3) = 10 + 10 = 20.$$

Similarly, we might calculate

$$14 + 7 + 16 + 33 \quad \text{as}$$

$$(14 + 16) + (7 + 33) = 30 + 40 = 70.$$

Suppose that we are to add $128 + 19$ mentally. One way is to add 20 to 128 and then subtract 1.

$$
\begin{aligned}
128 + 19 &= 128 + (20 - 1) & 19 = 20 - 1 \\
&= (128 + 20) - 1 & \text{Associative property} \\
&= 148 - 1 & \text{Add.} \\
&= 147. & \text{Subtract.}
\end{aligned}
$$

Similarly, we could find the sum $128 + 98$ by first adding 100 to 128 to obtain 228 and then subtracting 2, resulting in 226.

The distributive property can be helpful when we are multiplying mentally. For example, to estimate the number of people in a marching band with 7 columns and 23 rows we need to find the product $7 \cdot 23$. To evaluate the product mentally, think of 23 as $20 + 3$.

$$
\begin{aligned}
7(20 + 3) &= 7 \cdot 20 + 7 \cdot 3 & \text{Distributive property} \\
&= 140 + 21 & \text{Multiply.} \\
&= 161 & \text{Add.}
\end{aligned}
$$

EXAMPLE 9 Performing calculations mentally

Use properties of real numbers to calculate each expression mentally.
(a) $21 + 15 + 9 + 5$ **(b)** $\frac{1}{2} \cdot \frac{2}{3} \cdot 2 \cdot \frac{3}{2}$
(c) $523 + 199$ **(d)** $6 \cdot 55$

Solution   **(a)** Use the commutative and associative properties to group numbers that sum to a multiple of 10.

$$21 + 15 + 9 + 5 = (21 + 9) + (15 + 5) = 30 + 20 = 50$$

(b) Use the commutative and associative properties to group numbers with their reciprocals.

$$\frac{1}{2} \cdot \frac{2}{3} \cdot 2 \cdot \frac{3}{2} = \left(\frac{1}{2} \cdot 2\right) \cdot \left(\frac{2}{3} \cdot \frac{3}{2}\right) = 1 \cdot 1 = 1$$

Critical Thinking

How could you quickly calculate $5283 - 198$ without a calculator?

(c) Instead of adding 199, add 200 and then subtract 1.

$$523 + 200 - 1 = 723 - 1 = 722$$

(d) Think of 55 as $50 + 5$ and then apply the distributive property.

$$6 \cdot (50 + 5) = 300 + 30 = 330$$

1.7 PUTTING IT ALL TOGETHER

The following table summarizes some properties of real numbers.

Property	Definition	Examples
Commutative	For any real numbers a and b, $$a + b = b + a \quad \text{and}$$ $$a \cdot b = b \cdot a.$$	$4 + 6 = 6 + 4$ and $4 \cdot 6 = 6 \cdot 4$
Associative	For any real numbers a, b, and c, $$(a + b) + c = a + (b + c) \quad \text{and}$$ $$(a \cdot b) \cdot c = a \cdot (b \cdot c).$$	$(3 + 4) + 5 = 3 + (4 + 5)$ and $(3 \cdot 4) \cdot 5 = 3 \cdot (4 \cdot 5)$
Distributive	For any real numbers a, b, and c, $$a(b + c) = ab + ac \quad \text{and}$$ $$a(b - c) = ab - ac.$$	$5(x + 2) = 5x + 10$ and $5(x - 2) = 5x - 10$
Identity (0 and 1)	The identity for addition is 0, and the identity for multiplication is 1. For any real number a, $a + 0 = a$ and $a \cdot 1 = a.$	$5 + 0 = 5$ and $5 \cdot 1 = 5$
Inverse	The additive inverse of a is $-a$, and $a + (-a) = 0$. The multiplicative inverse of a nonzero number a is $\frac{1}{a}$, and $a \cdot \frac{1}{a} = 1.$	$8 + (-8) = 0$ and $\frac{2}{3} \cdot \frac{3}{2} = 1$

1.7 EXERCISES

FOR EXTRA HELP

CONCEPTS

1. $a + b = b + a$ illustrates the _____ property for _____.

2. $a \cdot b = b \cdot a$ illustrates the _____ property for _____.

3. $(a + b) + c = a + (b + c)$ illustrates the _____ property for _____.

4. $(a \cdot b) \cdot c = a \cdot (b \cdot c)$ illustrates the _____ property for _____.

5. $a(b + c) = ab + ac$ illustrates the _____ property.

6. $a(b - c) = ab - ac$ illustrates the _____ property.

7. $a + 0 = 0 + a = a$ illustrates the _____ property for _____.

8. $a \cdot 1 = 1 \cdot a = a$ illustrates the _____ property for _____.

9. The additive inverse of a is _____.

10. The multiplicative inverse or reciprocal of a nonzero number a is _____.

PROPERTIES OF REAL NUMBERS

Exercises 11–18: Use a commutative property to rewrite the expression. Do not simplify.

11. $-6 + 10$

12. $23 + 7$

13. $-5 \cdot 6$

14. $25 \cdot (-46)$

15. $a + 10$

16. $b + c$

17. $b \cdot 7$

18. $a \cdot 23$

Exercises 19–26: Use an associative property to rewrite the expression by changing the parentheses. Do not simplify.

19. $(1 + 2) + 3$

20. $-7 + (5 + 15)$

21. $2 \cdot (3 \cdot 4)$

22. $(9 \cdot (-4)) \cdot 5$

23. $(a + 5) + c$

24. $(10 + b) + a$

25. $(x \cdot 3) \cdot 4$

26. $5 \cdot (x \cdot y)$

Exercises 27–38: Use a distributive property to rewrite the expression. Then simplify the expression.

27. $4(3 + 2)$

28. $5(6 - 9)$

29. $a(b - 8)$

30. $3(x + y)$

31. $-1(t + z)$

32. $-1(a + 6)$

33. $-(5 - a)$

34. $12 - (4u - b)$

35. $(a + 5)3$

36. $(x + y)7$

37. $(6 - z)(-3)$

38. $4x - 2(3y - 5)$

Exercises 39–52: State the property or properties that the equation illustrates.

39. $x \cdot 5 = 5x$

40. $7 + a = a + 7$

41. $(a + 5) + 7 = a + 12$

42. $(9 + a) + 8 = a + 17$

43. $4(5 + x) = 20 + 4x$

44. $3(5 + x) = 3x + 15$

45. $x(3 - y) = 3x - xy$

46. $-(u - v) = -u + v$

47. $6x + 9x = 15x$

48. $9x - 11x = -2x$

49. $3 \cdot (4 \cdot a) = 12a$

50. $(x \cdot 3) \cdot 5 = 15x$

51. $-(t - 7) = -t + 7$

52. $z \cdot 5 - y \cdot 6 = 5z - 6y$

IDENTITY AND INVERSE PROPERTIES

Exercises 53–62: State the property or properties illustrated.

53. $0 + x = x$

54. $5x + 0 = 5x$

55. $1 \cdot a = a$

56. $\frac{1}{7} \cdot (7 \cdot a) = a$

57. $\frac{25}{15} = \frac{5}{3} \cdot \frac{5}{5} = \frac{5}{3}$

58. $\frac{50}{40} = \frac{5}{4} \cdot \frac{10}{10} = \frac{5}{4}$

59. $\frac{1}{xy} \cdot xy = 1$

60. $\frac{1}{a + b} \cdot (a + b) = 1$

61. $-xyz + xyz = 0$

62. $\frac{1}{y} + \left(-\frac{1}{y}\right) = 0$

MENTAL CALCULATIONS

Exercises 63–80: Use properties of real numbers to calculate the expression mentally.

63. $4 + 2 + 9 + 8 + 1 + 6$

64. $21 + 32 + 19 + 8$

65. $45 + 43 + 5 + 7$

66. $5 + 7 + 12 + 13 + 8$

67. $129 + 49$ 68. $87 + 99$

69. $379 + 98$ 70. $4570 + 998$

71. $178 - 99$ 72. $500 - 101$

73. $6 \cdot 15$ 74. $4 \cdot 56$

75. $8 \cdot 102$ 76. $5 \cdot 999$

77. $\frac{1}{2} \cdot \frac{1}{2} \cdot \frac{1}{2} \cdot 2 \cdot 2 \cdot 2$ 78. $\frac{1}{2} \cdot \frac{4}{5} \cdot \frac{7}{3} \cdot 2 \cdot \frac{5}{4}$

79. $\frac{7}{6} \cdot \frac{1}{2} \cdot \frac{1}{2} \cdot \frac{1}{2} \cdot \frac{8}{7}$ 80. $\frac{4}{11} \cdot \frac{11}{6} \cdot \frac{6}{7} \cdot \frac{7}{4}$

MULTIPLYING AND DIVIDING BY POWERS OF 10 MENTALLY

81. *Multiplying by 10*   Multiplying by 10 is easy in the decimal system. To multiply an integer by 10, append one 0 to the number. For example, $10 \times 23 = 230$. Simplify each expression mentally.
    (a) 10×41 (b) 10×997
    (c) -630×10 (d) $-14,000 \times 10$

82. *Multiplying by 10*   To multiply a decimal number by 10, move the decimal point one place to the right. For example, $10 \times 23.45 = 234.5$. Simplify each expression mentally.
    (a) 10×101.68 (b) $10 \times (-1.235)$
    (c) -113.4×10 (d) 0.567×10
    (e) 10×0.0045 (f) -0.05×10

83. *Multiplying by Powers of 10*   To multiply an integer by a power of 10 in the form
$$10^k = 1\underbrace{00 \ldots 0}_{k \text{ zeros}},$$
append k zeros to the number. Some examples of this are $100 \times 45 = 4500$, $1000 \times 235 = 235,000$, and $10,000 \times 12 = 120,000$. Simplify each expression mentally.
    (a) 1000×19 (b) $100 \times (-451)$
    (c) $10,000 \times 6$ (d) $-79 \times 100,000$

84. *Multiplying by Powers of 10*   To multiply a decimal number by a power of 10 in the form
$$10^k = 1\underbrace{00 \ldots 0}_{k \text{ zeros}},$$
move the decimal point k places to the right. For example, $100 \times 1.234 = 123.4$. Simplify each expression mentally.
    (a) 1000×1.2345 (b) $100 \times (-5.1)$
    (c) 45.67×1000 (d) $0.567 \times 10,000$
    (e) 100×0.0005 (f) $-0.05 \times 100,000$

85. *Dividing by 10*   To divide a number by 10, move the decimal point one place to the left. For example, $78.9 \div 10 = 7.89$. Simplify each expression mentally.
    (a) $12.56 \div 10$ (b) $9.6 \div 10$
    (c) $0.987 \div 10$ (d) $-0.056 \div 10$
    (e) $1200 \div 10$ (f) $4578 \div 10$

86. *Dividing by Powers of 10*   To divide a decimal number by a power of 10 in the form
$$10^k = 1\underbrace{00 \ldots 0}_{k \text{ zeros}},$$
move the decimal point k places to the left. For example, $123.4 \div 100 = 1.234$. Simplify each expression mentally.
    (a) $78.89 \div 100$ (b) $0.05 \div 1000$
    (c) $5678 \div 10,000$ (d) $-9.8 \div 1000$
    (e) $-101 \div 100,000$ (f) $7.8 \div 100$

APPLICATIONS

87. *Earnings*   Earning $100 one day and $75 the next day is equivalent to earning $75 the first day and $100 the second day. What property of real numbers does this example illustrate?

88. *Leasing a Car*   An advertisement for a lease on a new car states that it costs $2480 down and $201 per month for 20 months. Mentally calculate the cost of the lease. Explain your reasoning.

89. *Gasoline Mileage*   A car travels 198 miles on 10 gallons of gasoline. Mentally calculate the number of miles that the car travels on 1 gallon of gasoline.

90. *Gallons of Water*   A wading pool is 50 feet long, 20 feet wide, and 1 foot deep.
   (a) Mentally determine the number of cubic feet in the pool. (*Hint:* Volume equals length times width times height.)
   (b) One cubic foot equals about 7.5 gallons. Mentally calculate the number of gallons of water in the pool.

91. *Digital Images of Io*   Satellites take digital pictures and transmit them back to Earth. The accompanying picture of Jupiter's moon Io is a digital picture, created by using a rectangular pattern of small pixels. This image is 500 pixels wide and 400 pixels high, so the total number of pixels in it is $500 \times 400 = 200{,}000$ pixels.   (*Source:* NASA.)

   (a) Find the total number of pixels in an image 400 pixels wide and 500 pixels high.
   (b) Suppose that a picture is x pixels wide and y pixels high. What property states that it has the same number of pixels as a picture y pixels wide and x pixels high?

92. *Dimensions of a Pool*   A small pool of water is 13 feet long, 5 feet wide, and 2 feet deep. The volume V of the pool in cubic feet is found by multiplying 13 by 5 by 2.
   (a) If you did this calculation mentally would you multiply $(13 \times 5) \times 2$ or $13 \times (5 \times 2)$? Why?
   (b) What property allows you to do either calculation and still obtain the correct answer?

WRITING ABOUT MATHEMATICS

93. To determine the cost of tuition, a student tries to compute $16 \times \$96$ with a calculator and gets $\$15{,}360$. What would you tell this person?

94. A student performs the following computation by hand.

$$20 - 6 - 2 + 8 \div 4 \div 2 \overset{?}{=} 20 - 4 + 8 \div 2$$
$$\overset{?}{=} 20 - 4 + 4$$
$$\overset{?}{=} 20$$

Find any incorrect steps and explain what is wrong. What is the correct answer?

95. The computation $3 + 10^{20} - 10^{20}$ performed on a calculator gives a result of 0. (Try it.) What is the correct answer? Why is it important to know properties of real numbers even though you have a calculator?

96. Does the "distributive property"
$$a + (b \cdot c) \overset{?}{=} (a + b) \cdot (a + c)$$
hold for any real numbers a, b, and c? Explain your reasoning.

Group Activity: Working with Real Data

Directions: Form a group of 2 to 4 people. Select someone to record the group's responses for this activity. All members of the group should work cooperatively to answer the questions. If your instructor asks for your results, each member of the group should be prepared to respond.

1. *Walk to the Moon*   The distance to the moon is about 237,000 miles. Walking at 4 miles per hour, estimate how many hours it would take to travel this distance. How many years is this?

2. *Salary*   Suppose that for full-time work a person earns 1¢ for the first week, 2¢ for the second week,

4¢ for the third week, 8¢ for the fourth week, and so on for 1 year.
   (a) Discuss whether you think this would be a good deal. (*Hint:* There are 52 weeks in a year.)
   (b) Estimate how much this person would make the last week of the year.

1.8 SIMPLIFYING AND WRITING ALGEBRAIC EXPRESSIONS

Terms · Combining Like Terms · Simplifying Expressions · Writing Expressions

INTRODUCTION

In arithmetic it is common to simplify expressions such as

$$\frac{1}{2} + \frac{1}{4} + \frac{1}{4}.$$

This expression equals 1, and it is easier to perform calculations with the number 1 than with these fractions. However, simplifying expressions requires mathematical skills. In this section we introduce some of these basic skills.

TERMS

One way to simplify expressions is to combine like terms. A **term** is a number, a variable, or a *product* of numbers and variables raised to powers. Examples of terms include

$$4, \quad z, \quad 5x, \quad \frac{2}{5}z, \quad -4xy, \quad -x^2, \quad \text{and} \quad 6x^3y^4.$$

Terms do not contain addition or subtraction signs, but they can contain negative signs. The **coefficient** of a term is the number that appears in the term. If no number appears, the coefficient is either -1 or 1. The coefficient of $\frac{1}{2}xy$ is $\frac{1}{2}$, the coefficient of -15 is -15, and the coefficient of $-y^2$ is -1.

EXAMPLE 1   Identifying terms

Determine whether each expression is a term. If the expression is a term, identify its coefficient.
(a) 51   **(b)** $5a$ **(c)** $2x + 3y$ **(d)** $-3x^2$

Solution **(a)** A number is a term. Its coefficient is 51.
(b) The product of a number and a variable is a term. Its coefficient is 5
(c) The sum (or difference) of two terms is not a term.
(d) The product of a number and a variable with an exponent is a term. Its coefficient is -3.

> ═══ MAKING CONNECTIONS ═══
>
> **Factors and Terms**
>
> When variables and numbers are multiplied, they are called *factors*. For example, the expression $4xy$ has factors of 4, x, and y. When variables and numbers are added or subtracted, they are called *terms*. For example, the expression $x - 5xy + 1$ has terms of x, $5xy$, and 1.

Combining Like Terms

Suppose that we have two boards with lengths $2x$ and $3x$, where the value of x could be any length such as 2 feet. See Figure 1.26. Because $2x$ and $3x$ are *like terms* we can find the total length of the two boards by *adding like terms*.

$$2x + 3x = (2 + 3)x = 5x$$

The combined length of the two boards is $5x$ units.

Figure 1.26 $2x + 3x = 5x$

We can also determine how much longer the second board is than the first board by subtracting terms.

$$3x - 2x = (3 - 2)x = 1x = x$$

Thus the second board is x units longer than the first board.

If two terms contain the same variables raised to the same powers, we call them **like terms**. We can add or subtract like terms, but not *unlike* terms. If one board has length $2x$ and the other board has length $3y$, then we cannot determine the total length other than to say that it is $2x + 3y$. See Figure 1.27. The reason is that the lengths of x and y might not be equal. The terms $2x$ and $3y$ are unlike terms and *cannot* be combined.

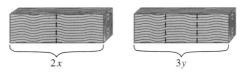

Figure 1.27 $2x + 3y$

EXAMPLE 2   Identifying like terms

Determine whether each pair of terms are like or unlike.
(a) $-4m, 7m$ (b) $8x^2, 8y^2$ (c) $-x^2, 3x^2$ (d) $\frac{1}{2}z, -3z^2$ (e) $5, -4n$

Solution   (a) The variable in both terms is m, so they are like terms.
(b) The variables are different, so they are unlike terms.
(c) Both terms have the same variable raised to the same power (x^2), so they are like terms.
(d) The second term contains $z^2 = z \cdot z$, whereas the first term contains only z. Thus they are unlike terms.
(e) The first term has no variable, whereas the second term contains the variable n. They are unlike terms.

EXAMPLE 3 Combining like terms

Combine terms in each expression, if possible.
(a) $-3x + 5x$ (b) $8y - y$ (c) $-x^2 + 5x^2$ (d) $\frac{1}{2}y - 3y^3$

Solution (a) Combine terms by applying a distributive property.
$$-3x + 5x = (-3 + 5)x = 2x$$

(b) Note that y can be written as $1y$. They are like terms and can be combined.
$$8y - y = (8 - 1)y = 7y$$

(c) Note that $-x^2$ can be written as $-1x^2$. They are like terms and can be combined.
$$-x^2 + 5x^2 = (-1 + 5)x^2 = 4x^2$$

(d) They are unlike terms, so they cannot be combined.

SIMPLIFYING EXPRESSIONS

Critical Thinking

Use rectangles to explain how to add $xy + 2xy$.

The area of a rectangle equals length times width. In Figure 1.28 the area of the first rectangle is $3x$, the area of the second rectangle is $2x$, and the area of the third rectangle is x. The area of the last rectangle equals the total area of the three smaller rectangles. That is,
$$3x + 2x + x = (3 + 2 + 1)x = 6x.$$

The expression $3x + 2x + x$ can be *simplified* to $6x$.

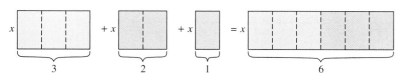

Figure 1.28 $3x + 2x + x = 6x$

EXAMPLE 4 Simplifying expressions

Simplify each expression.
(a) $2 + x - 6 + 5x$ (b) $2y - (y + 3)$ (c) $\dfrac{-1.5x}{-1.5}$ (d) $\dfrac{5x}{3x}$

Solution (a) Combine like terms by applying the properties of real numbers.

$2 + x - 6 + 5x = 2 + (-6) + x + 5x$	Commutative property
$= 2 + (-6) + (1 + 5)x$	Distributive property
$= -4 + 6x$	Add.

(b)
$2y - 1(y + 3) = 2y - 1y - 3$	Distributive property
$= (2 - 1)y - 3$	Distributive property
$= y - 3$	Subtract.

(c)
$\dfrac{-1.5x}{-1.5} = \dfrac{-1.5}{-1.5} \cdot \dfrac{x}{1}$	Multiplication of fractions
$= 1 \cdot x$	Reduce the fractions.
$= x$	Multiplicative identity

(d)
$$\frac{5x}{3x} = \frac{5}{3} \cdot \frac{x}{x}$$ Property of fractions

$$= \frac{5}{3} \cdot 1$$ Reduce the fractions.

$$= \frac{5}{3}$$ Multiplicative identity ▬

Note: The expressions in Example 4(c) and 4(d) can be simplified directly by using the basic principle of fractions: $\frac{ac}{bc} = \frac{a}{b}$.

WRITING EXPRESSIONS

In real-life situations, we often have to translate words to symbols. For example, to calculate federal and state income tax we might have to multiply taxable income by 0.15 and by 0.05 and then find the sum. If we let x represent taxable income, then the total federal and state income tax is $0.15x + 0.05x$. This expression can be simplified with a distributive property.

$$0.15x + 0.05x = (0.15 + 0.05)x$$
$$= 0.20x$$

Thus the total income tax on a taxable income of $x = \$20,000$ would be

$$0.20(20,000) = \$4000.$$

EXAMPLE 5   **Writing and simplifying an expression**

A sidewalk has a constant width and comprises several short sections with lengths 12, 6, and 5 feet, as illustrated in Figure 1.29.
(a) Write and simplify an expression that gives the number of square feet of sidewalk.

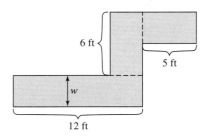

Figure 1.29

(b) Find the area of the sidewalk if its width is 3 feet.

Solution   **(a)** Let w be the constant width of the sidewalk in feet. The area of each sidewalk section equals its length times its width. The total area of the sidewalk is

$$12w + 6w + 5w = (12 + 6 + 5)w = 23w.$$

(b) When $w = 3$, the area is $23w = 23 \cdot 3 = 69$ square feet. ▬

1.8 PUTTING IT ALL TOGETHER

The following table summarizes some of the concepts presented in this section.

Concept	Comments	Examples
Term	A term is a number, variable, or product of numbers and variables raised to powers	$12, -10, x, -y$ $-3x, 5z, xy, 6x^2$ $10y^3, \frac{3}{4}x, 25xyz^2$
Coefficient of a Term	The coefficient of a term is the number that appears in the term. If no number appears, then the coefficient is either 1 or -1.	12 Coefficient is 12 x Coefficient is 1 $-xy$ Coefficient is -1 $-4x$ Coefficient is -4
Like Terms	Like terms have the same variables raised to the same powers.	$5m$ and $-6m$ x^2 and $-74x^2$ $2xy$ and $-xy$
Combining Like Terms	Like terms can be combined by using the distributive property.	$5x + 2x = (5 + 2)x = 7x$ $y - 3y = (1 - 3)y = -2y$ $8x^2 + x^2 = (8 + 1)x^2 = 9x^2$

1.8 EXERCISES

FOR EXTRA HELP

 Student's Solutions Manual

 InterAct Math

 MathXL

MyMathLab

Math Tutor Center

Digital Video Tutor CD 1 Videotape 2

CONCEPTS

1. A _____ is a number, a variable, or a product of numbers and variables raised to powers.

2. The expression $a + b$ (is/is not) a term, whereas the expression ab (is/is not) a term.

3. The number 7 in the term $7x^2y$ is called the _____ of the term.

4. If two terms contain the same variables raised to the same powers, we call them (like/unlike) terms.

5. The terms $2x$ and $5x$ are (like/unlike) terms, whereas $9x$ and $9z$ are (like/unlike) terms.

6. We can add or subtract (like/unlike) terms.

LIKE TERMS

Exercises 7–18: Determine whether the expression is a term. If the expression is a term, identify its coefficient.

7. 91

8. -12

9. $-6b$

10. $9z$

11. $x + 10$

12. $20 - 2y$

13. x^2

14. $4x^3$

15. $4x - 5$

16. $5z + 6x$

17. $-9xyz$

18. $-a^2b^2$

Exercises 19–28: Determine whether the terms are like or unlike.

19. $6, -8$

20. $2x, 19$

21. $5x, -22x$

22. $19y, -y$

23. $18x, 18y$

24. $-6a, -6b$

25. $x^2, -15x^2, 6x^2$

26. $xyz, 19xyz, -xyz$

27. $xy, xz, 2xy$

28. $-8x^2y, 2x^2z, x^2y$

Exercises 29–40: Combine terms, if possible.

29. $3x + 5x$

30. $6x - 8x$

31. $19y - 5y$

32. $22z + z$

33. $5x - 7y$

34. $3y + 3z$

35. $5 + 5y$

36. $x + x^2$

37. $5x^2 - 2x^2$

38. $25z^3 - 10z^3$

39. $8xy - 10xy + xy$

40. $100xy^2 + 25xy^2 - 5xy^2$

SIMPLIFYING AND WRITING EXPRESSIONS

Exercises 41–68: Simplify the expression.

41. $5 + x - 3 + 2x$

42. $x - 5 - 5x + 7$

43. $-\frac{3}{4} + z - 3z + \frac{5}{4}$

44. $\frac{4}{3}z - 100 + 200 - \frac{1}{3}z$

45. $4y - y + 8y$

46. $14z - 15z - z$

47. $-3 + 6z + 2 - 2z$

48. $19a - 12a + 5 - 6$

49. $-2(3z - 6y) - z$

50. $6\left(\frac{1}{2}a - \frac{1}{6}b\right) - 3b$

51. $2 - \frac{3}{4}(4x + 8)$

52. $-5 - (5x - 6)$

53. $-x - (5x + 1)$

54. $2x - 4(x + 2)$

55. $1 - \frac{1}{3}(x + 1)$

56. $-3 - 3(4 - x)$

57. $\frac{3}{5}(x + y) - \frac{1}{5}(x - 1)$

58. $-5(a + b) - (a + b)$

59. $0.2x^2 + 0.3x^2 - 0.1x^2$

60. $32z^3 - 52z^3 + 20z^3$

61. $2x^2 - 3x + 5x^2 - 4x$

62. $\frac{5}{6}y^2 - 4 + \frac{1}{12}y^2 + 3$

63. $\frac{8x}{8}$

64. $\frac{-0.1y}{-0.1}$

65. $\frac{-3y}{-y}$

66. $\frac{2x}{7x}$

67. $\frac{-108z}{-108}$

68. $\frac{3xy}{-6xy}$

Exercises 69–74: Translate the phrase into a mathematical expression and then simplify. Use the variable x.

69. The sum of five times a number and six times the same number

70. The sum of a number and three times the same number

71. The sum of a number squared and twice the same number squared

72. One-half of a number minus three-fourths of the same number

73. Six times a number minus four times the same number

74. Two cubed times a number minus three squared times the same number

APPLICATIONS

75. *Street Dimensions* (Refer to Example 5.) A street has a constant width w and comprises several straight sections having lengths 400, 350, 220, and 600 feet.
 (a) Write and simplify an expression that gives the square footage of the street.
 (b) Find the area of the street if its width is 42 feet.

76. *Sidewalk Dimensions* A sidewalk has a constant width w and comprises several short sections having lengths 12, 14, and 10 feet.
 (a) Write and simplify an expression that gives the number of square feet of sidewalk.
 (b) Find the area of the sidewalk if its width is 5 feet.

77. *Snowblowers* Two snowblowers are being used to clear a driveway. The first blower can remove 20 cubic feet per minute, and the second blower can remove 30 cubic feet per minute.
 (a) Write and simplify an expression that gives the number of cubic feet of snow removed in x minutes.
 (b) Find the total number of cubic feet of snow removed in 48 minutes.

(c) How many minutes would it take to remove the snow from a driveway that measures 30 feet by 20 feet, if the snow is 2 feet deep?

78. *Winding Rope* Two motors are winding up rope. The first motor can wind up rope at 2 feet per second, and the second motor can wind up rope at 5 feet per second.
(a) Write and simplify an expression that gives the length of rope wound by both motors in x seconds.
(b) Find the total length of rope wound up in 3 minutes.
(c) How many minutes would it take to wind up 2100 feet of rope by using both motors?

WRITING ABOUT MATHEMATICS

79. Explain how to add like terms. What property of real numbers is used?

80. A student simplified the following expression *incorrectly*.

$$3(x - 5) + 5x \overset{?}{=} 3x - 5 + 5x$$
$$\overset{?}{=} 3x + 5x - 5$$
$$\overset{?}{=} (3 + 5)x - 5$$
$$\overset{?}{=} 8x - 5$$

Find the error and explain what went wrong. What should the final answer be?

CHECKING BASIC CONCEPTS SECTIONS 1.7 AND 1.8

1. Use the commutative property to rewrite each expression.
   (a) $y \cdot 18$ (b) $10 + x$

2. Use the associative and commutative properties to simplify $5 \cdot (y \cdot 4)$.

3. Simplify each expression.
   (a) $10 - (5 + x)$ (b) $5(x - 7)$

4. State the property that the equation $5x + 3x = 8x$ illustrates.

5. Simplify $-4xy + 4xy$.

6. Mentally evaluate each expression.
   (a) $32 + 17 + 8 + 3$ (b) $\frac{5}{6} \cdot \frac{7}{8} \cdot \frac{6}{5} \cdot 8$
   (c) $567 - 199$

7. Combine like terms in each expression.
   (a) $5z + 9z$ (b) $5y - 4 - 8y + 7$

8. Simplify each expression.
   (a) $2y - (5y + 3)$
   (b) $-4(x + 3y) + 2(2x - y)$
   (c) $\dfrac{20x}{20}$ (d) $\dfrac{35x^2}{x^2}$

CHAPTER

1 Summary

Section 1.1 *Numbers, Variables, and Expressions*

Sets of Numbers
Natural Numbers   1, 2, 3, 4, . . .

Whole Numbers   0, 1, 2, 3, . . .

Prime and Composite Numbers

Prime Number	Only natural number factors are itself and 1; must be a natural number greater than 1

        Examples: 2, 3, 5, 7, 11, 13, and 17

Composite Number	A natural number greater than 1 that is not prime; a composite number can be written as a product of two or more prime numbers.

        Examples: $24 = 2 \times 2 \times 2 \times 3$ and
                      $18 = 2 \times 3 \times 3$

Important Terms

Multiplicative Inverse or Reciprocal	The reciprocal of a nonzero number a is $\frac{1}{a}$.

        Examples: The reciprocal of -5 is $-\frac{1}{5}$, and the reciprocal of $\frac{2x}{y}$ is $\frac{y}{2x}$, provided $x \neq 0$ and $y \neq 0$.

Variable	A symbol or letter that represents an unknown quantity

        Examples: a, b, F, x, and y

Algebraic Expression	Consists of numbers, variables, arithmetic symbols, and grouping symbols

        Examples: $3x + 1$ and $5(x + 2) - y$

Equation	A statement that two algebraic expressions are equal; an equation always contains an equals sign.

        Examples: $1 + 2 = 3$ and $z - 7 = 8$

Section 1.2   *Fractions*

Parts of a Fraction

$$\text{Numerator} \diagdown \; \frac{N}{D} \; \text{— Fraction bar}$$
$$\text{Denominator} \diagup$$

Reducing Fractions $\dfrac{a \cdot c}{b \cdot c} = \dfrac{a}{b}$

Example: $\dfrac{20}{35} = \dfrac{4 \cdot 5}{7 \cdot 5} = \dfrac{4}{7}$

Multiplication and Division

$$\frac{a}{b} \cdot \frac{c}{d} = \frac{ac}{bd} \qquad b \text{ and } d \text{ are nonzero.}$$

$$\frac{a}{b} \div \frac{c}{d} = \frac{a}{b} \cdot \frac{d}{c} \qquad b, c, \text{ and } d \text{ are nonzero.}$$

Examples: $\frac{3}{4} \cdot \frac{7}{5} = \frac{21}{20}$ and $\frac{3}{4} \div \frac{7}{5} = \frac{3}{4} \cdot \frac{5}{7} = \frac{3 \cdot 5}{4 \cdot 7} = \frac{15}{28}$

Addition and Subtraction with a Common Denominator

$$\frac{a}{b} + \frac{c}{b} = \frac{a + c}{b} \quad \text{and} \quad \frac{a}{b} - \frac{c}{b} = \frac{a - c}{b}$$

Examples: $\frac{3}{5} + \frac{1}{5} = \frac{3 + 1}{5} = \frac{4}{5}$ and $\frac{3}{5} - \frac{1}{5} = \frac{3 - 1}{5} = \frac{2}{5}$

Addition and Subtraction with Unlike Denominators Write the expressions with an LCD. Then add or subtract the numerators, keeping the denominator unchanged.

Examples: $\frac{2}{9} + \frac{1}{6} = \frac{2}{9} \cdot \frac{2}{2} + \frac{1}{6} \cdot \frac{3}{3} = \frac{4}{18} + \frac{3}{18} = \frac{7}{18}$ LCD is 18.

$\quad\quad\quad\quad \frac{2}{9} - \frac{1}{6} = \frac{2}{9} \cdot \frac{2}{2} - \frac{1}{6} \cdot \frac{3}{3} = \frac{4}{18} - \frac{3}{18} = \frac{1}{18}$

Section 1.3 Exponents and Order of Operations

Exponential Expression

$$\text{Base} \longrightarrow 5^3 \longleftarrow \text{Exponent}$$

Examples: $3^4 = 3 \cdot 3 \cdot 3 \cdot 3 = 81$

Order of Operations
Use the following order of operations. First perform all calculations within parentheses and absolute values, or above and below the fraction bar.

1. Evaluate all exponential expressions.

2. Do all multiplication and division from *left to right*.

3. Do all addition and subtraction from *left to right*.

Note: Negative signs are evaluated after exponents so $-2^4 = -16$.

Examples: $100 - 5^2 \cdot 2 = 100 - 25 \cdot 2$ and $\frac{4 + 2}{5 - 3} \cdot 4 = \frac{6}{2} \cdot 4$

$\quad\quad\quad\quad\quad\quad\quad\quad = 100 - 50 \quad\quad\quad\quad\quad\quad = 3 \cdot 4$

$\quad\quad\quad\quad\quad\quad\quad\quad = 50 \quad\quad\quad\quad\quad\quad\quad\quad = 12$

Section 1.4 Real Numbers and the Number Line

Opposite or Additive Inverse The opposite of the number a is $-a$.

Example: The opposite of 5 is -5, and the opposite of -8 is $-(-8) = 8$.

Sets of Numbers

Integers $\quad\quad\quad\quad\quad\quad\quad \ldots, -3, -2, -1, 0, 1, 2, 3, \ldots$

Rational Numbers $\quad\quad\quad\quad \frac{p}{q}$, where p and $q \neq 0$ are integers; rational numbers can be written as decimal numbers that either repeat or terminate.

$\quad\quad\quad\quad\quad\quad\quad\quad\quad\quad$ *Examples:* $-\frac{3}{4}, 5, -7.6, \frac{6}{3},$ and $\frac{1}{3}$

Real Numbers $\quad\quad\quad\quad\quad\quad$ Numbers that can be written as decimal numbers

$\quad\quad\quad\quad\quad\quad\quad\quad\quad\quad$ *Examples:* $-\frac{3}{4}, 5, -7.6, \frac{6}{3}, \pi, \sqrt{3},$ and $-\sqrt{7}$

Irrational Numbers   Real numbers that are not rational

Examples: π, $\sqrt{3}$, and $-\sqrt{7}$

Average   To calculate the average of a set of numbers, find the sum of the numbers and then divide the sum by how many numbers there are in the set.

Example: The average of 4, 5, 20, and 11 is

$$\frac{4 + 5 + 20 + 11}{4} = \frac{40}{4} = 10.$$

We divide by 4 because we are finding the average of 4 numbers.

The Number Line

The origin corresponds to the number 0.

Absolute Value   If a is positive or 0, then $|a| = a$, and if a is negative, then $|a| = -a$. The absolute value of a number is *never* negative.

Examples: $|5| = 5$, $|-5| = 5$,   and $|0| = 0$

Inequality   If a is located to the left of b on the number line, then $a < b$. If a is located to the right of b on the number line, then $a > b$.

Examples: $-3 < 6$ and $-1 > -2$

Section 1.5   *Addition and Subtraction of Real Numbers*

Addition of Opposites

$$a + (-a) = 0$$

Examples: $3 + (-3) = 0$ and $-\frac{5}{7} + \frac{5}{7} = 0$

Addition of Real Numbers

To add two numbers that are either *both positive* or *both negative*, add their absolute values. Their sum has the same sign as the two numbers.

To add two numbers with *opposite signs*, find the absolute value of each number. Subtract the smaller absolute value from the larger. The sum has the same sign as the sign of the number with the largest absolute value. If the two numbers are opposites, their sum is 0.

Examples: $-4 + 5 = 1$, $3 + (-7) = -4$, $-4 + (-2) = -6$,   and
$8 + 2 = 10$

Subtraction of Real Numbers   For any real numbers a and b, $a - b = a + (-b)$

Examples: $5 - 9 = 5 + (-9) = -4$ and $-4 - (-3) = -4 + 3 = -1$

Section 1.6   *Multiplication and Division of Real Numbers*

Important Terms

Factors	Numbers multiplied in a multiplication problem

        Example: 5 and 4 are factors of 20 because $5 \cdot 4 = 20$.

Product   The answer to a multiplication problem

        Example: The product of 5 and 4 is 20.

Dividend, Divisor,   If $\frac{a}{b} = c$, then a is the dividend, b is the divisor, and c is the
and Quotient   quotient.

        Example: In the division problem $30 \div 5 = \frac{30}{5} = 6$, 30 is
                   the dividend, 5 is the divisor, and 6 is the quotient.

Signs of Products or Quotients   The product or quotient of two numbers with like signs is positive. The product or quotient of two numbers with unlike signs is negative.

Examples: $-4 \cdot 6 = -24$, $-2 \cdot (-5) = 10$, $\frac{-18}{6} = -3$,   and $-4 \div -2 = 2$

Writing Fractions as Decimals   To write the fraction $\frac{a}{b}$ as a decimal, divide b into a.

Example: $\frac{4}{9} = 0.\overline{4}$ because division of 4 by 9 gives the repeating decimal 0.4444. . . .

$$
\begin{array}{r}
0.444\ldots \\
9)\overline{4.000} \\
\underline{36} \\
40 \\
\underline{36} \\
40 \\
\underline{36} \\
4
\end{array}
$$

Section 1.7   *Properties of Real Numbers*

Important Properties

Commutative $a + b = b + a$ and $a \cdot b = b \cdot a$

        Examples: $3 + 4 = 4 + 3$ and $-6 \cdot 3 = 3 \cdot (-6)$

Associative $(a + b) + c = a + (b + c)$ and $(a \cdot b) \cdot c = a \cdot (b \cdot c)$

        Examples: $(2 + 3) + 4 = 2 + (3 + 4)$
                     $(2 \cdot 3) \cdot 4 = 2 \cdot (3 \cdot 4)$

Distributive $a(b + c) = ab + ac$ and $a(b - c) = ab - ac$

        Examples: $3(x + 5) = 3x + 15$
                     $4(5 - 2) = 4 \cdot 5 - 4 \cdot 2$

Identity $a + 0 = 0 + a = a$ Additive identity is 0.
        $a \cdot 1 = 1 \cdot a = a$ Multiplicative identity is 1.

        Examples: $5 + 0 = 0 + 5 = 5$ and $1 \cdot (-4) = -4 \cdot 1 = -4$

Inverse	$a + (-a) = 0$ and $-a + a = 0$
	$a \cdot \frac{1}{a} = 1$ and $\frac{1}{a} \cdot a = 1$
	Examples: $5 + (-5) = 0$ and $\frac{1}{2} \cdot 2 = 1$

Note: The commutative and associative properties apply to addition and multiplication but *not* to subtraction and division.

Section 1.8   *Simplifying and Writing Algebraic Expressions*

Important Terms

Term	A term is a number, a variable, or a product of numbers and variables raised to powers.
	Examples: 5, $-10x$, $3xy$, and x^2
Coefficient	The number portion of a term
	Examples: The coefficients for the terms $3xy$, $-x^2$, and -7 are 3, -1, and -7, respectively.
Like Terms	Two terms containing the same variables raised to the same powers; their coefficients may be different.
	Examples: The following pairs are like terms:
	$5x$ and $-x$; $6x^2$ and $-2x^2$; $3xy$ and $-\frac{1}{2}xy$.

Combining Like Terms   To add or subtract like terms, apply the distributive property.

Examples: $4x + 5x = (4 + 5)x = 9x$ and $5xy - 7xy = (5 - 7)xy = -2xy$

CHAPTER

1 Review Exercises

SECTION 1.1

Exercises 1–4: Classify the number as prime or composite. If the number is composite, write it as a product of prime numbers.

1. 29

2. 27

3. 108

4. 91

Exercises 5–8: Evaluate the expression for the given values of x and y.

5. $2x - 5$ $x = 4$

6. $7 - \dfrac{10}{x}$ $x = 5$

7. $9x - 2y$ $x = 2,$ $y = 3$

8. $\dfrac{2x}{x - y}$ $x = 6,$ $y = 4$

Exercises 9 and 10: Find the value of y for the given values of x and z.

9. $y = x - 5$ $x = 12$

10. $y = xz + 1$ $x = 2,$ $z = 3$

Exercises 11–14: Translate the phrase into an algebraic expression. State precisely what each variable represents when appropriate.

11. Five times the cost of a CD

12. Five less than a number

13. Three squared increased by five

14. Two cubed divided by the quantity three plus one

SECTION 1.2

15. Use the basic principle of fractions to reduce each expression.
   (a) $\frac{5 \cdot 7}{8 \cdot 7}$ **(b)** $\frac{3a}{4a}$

16. Reduce each fraction to lowest terms.
   (a) $\frac{9}{12}$ **(b)** $\frac{36}{60}$

Exercises 17–20: Multiply and then reduce the result to lowest terms when appropriate.

17. $\frac{3}{4} \cdot \frac{5}{6}$

18. $\frac{1}{2} \cdot \frac{4}{9}$

19. $\frac{2}{3} \cdot \frac{5}{11} \cdot \frac{9}{10}$

20. $\frac{12}{11} \cdot \frac{22}{23} \cdot \frac{1}{2}$

21. Find the fractional part: one-fifth of three-sevenths.

22. Find the reciprocal of each number.
   (a) 8   **(b)** 1   **(c)** $\frac{5}{19}$ **(d)** $\frac{3}{2}$

Exercises 23–26: Divide and then reduce to lowest terms when appropriate.

23. $\frac{3}{2} \div \frac{1}{6}$

24. $\frac{9}{10} \div \frac{7}{5}$

25. $8 \div \frac{2}{3}$

26. $\frac{3}{4} \div 6$

Exercises 27 and 28: Find the least common denominator for the fractions.

27. $\frac{1}{8}, \frac{5}{12}$

28. $\frac{2}{14}, \frac{1}{21}$

Exercises 29–34: Add or subtract and then reduce to lowest terms when appropriate.

29. $\frac{2}{15} + \frac{3}{15}$

30. $\frac{5}{4} - \frac{3}{4}$

31. $\frac{11}{12} - \frac{1}{8}$

32. $\frac{6}{11} - \frac{3}{22}$

33. $\frac{2}{3} - \frac{1}{2} + \frac{1}{4}$

34. $\frac{1}{6} + \frac{2}{3} - \frac{1}{9}$

SECTION 1.3

Exercises 35–38: Write the expression as an exponential expression.

35. $5 \cdot 5 \cdot 5 \cdot 5 \cdot 5 \cdot 5$

36. $\frac{7}{6} \cdot \frac{7}{6} \cdot \frac{7}{6}$

37. $3 \cdot 3 \cdot 3 \cdot 3$

38. $x \cdot x \cdot x \cdot x \cdot x$

39. Use multiplication to rewrite each expression, and then evaluate the result.
   (a) 4^3 **(b)** 7^2 **(c)** 8^1

40. Find a natural number n such that $2^n = 32$.

Exercises 41–52: Evaluate the expression by hand.

41. $7 + 3 \cdot 6$

42. $15 - 5 - 3$

43. $24 \div 4 \div 2$

44. $30 - 15 \div 3$

45. $18 \div 6 - 2$

46. $\frac{18}{4 + 5}$

47. $9 - 3^2$

48. $2^3 - 8$

49. $2^4 - 8 + \frac{4}{2}$

50. $3^2 - 4(5 - 3)$

51. $7 - \frac{4 + 6}{2 + 3}$

52. $3^3 - 2^3$

SECTION 1.4

Exercises 53–58: Classify the number as one or more of the following: natural number, whole number, integer, rational number, or irrational number.

53. 0

54. $-\frac{5}{6}$

55. -7

56. $\sqrt{17}$

57. π

58. 3.4

59. Plot each number on the same number line.
   (a) 0   **(b)** -2 **(c)** $\frac{5}{4}$

60. Evaluate each expression.
   (a) $|-5|$ **(b)** $|-\pi|$ **(c)** $|\sqrt{2} - 1|$

61. Insert the symbol $>$ or $<$ to make each statement true.
   (a) $-5 \underline{\quad} 4$ **(b)** $-\frac{1}{2} \underline{\quad} -\frac{5}{2}$

62. List the numbers $\sqrt{3}, -3, 3, -\frac{2}{3},$ and $\pi - 1$ from least to greatest.

SECTIONS 1.5 AND 1.6

Exercises 63 and 64: Use a number line to find the sum.

63. $-1 + 2$

64. $-2 + (-3)$

Exercises 65–76: Evaluate the expression.

65. $5 + (-4)$

66. $-9 - (-7)$

67. $11 \cdot (-4)$

68. $-8 \cdot (-5)$

69. $11 \div (-4)$

70. $-4 \div \frac{4}{7}$

71. $-\frac{1}{2} + \left(-\frac{3}{4}\right)$

72. $-\frac{5}{9} - \left(-\frac{1}{3}\right)$

73. $-\frac{1}{3} \cdot \left(-\frac{6}{7}\right)$

74. $\dfrac{\frac{4}{5}}{-7}$

75. $-\frac{3}{2} \div \left(-\frac{3}{8}\right)$

76. $\frac{3}{8} \div (-0.5)$

Exercises 77 and 78: Write an arithmetic expression for the given phrase and then simplify it.

77. Three plus negative five

78. Subtract negative four from two

Exercises 79 and 80: Write the fraction or mixed number as a decimal.

79. $\frac{7}{9}$

80. $2\frac{1}{5}$

Exercises 81 and 82: Write the decimal number as a fraction in lowest terms.

81. 0.6

82. 0.375

SECTION 1.7

Exercises 83–92: State the property that the equation illustrates.

83. $z \cdot 3 = 3z$

84. $6 + (7 + 5x) = (6 + 7) + 5x$

85. $2(5x - 2) = 10x - 4$

86. $5 + x + 3 = 5 + 3 + x$

87. $1 \cdot a = a$

88. $3 \cdot (5x) = (3 \cdot 5)x$

89. $12 - (x + 7) = 12 - x - 7$

90. $a + 0 = a$

91. $-5x + 5x = 0$

92. $-5 \cdot \left(-\frac{1}{5}\right) = 1$

Exercises 93–98: Use properties of real numbers to evaluate the expression mentally.

93. $7 + 9 + 12 + 8 + 1 + 3$

94. $500 - 199$

95. $25 \cdot 99$

96. $4581 + 1999$

97. 54.98×10

98. $4356 \div 100$

SECTION 1.8

Exercises 99–102: Determine whether the expression is a term. If the expression is a term, identify its coefficient.

99. $55x$

100. $-xy$

101. $9xy + 2z$

102. $x - 7$

Exercises 103–110: Simplify the expression.

103. $-10x + 4x$

104. $19z - 4z$

105. $3x^2 + x^2$

106. $7 + 2x - 6 + x$

107. $-\frac{1}{2} + \frac{3}{2}z - z + \frac{5}{2}$

108. $5(x - 3) - (4x + 3)$

109. $\dfrac{35a}{7a}$

110. $\dfrac{0.5c}{0.5}$

APPLICATIONS

111. *Painting a Wall*   Two people are painting a large wall. The first person paints 3 square feet per minute, while the second person paints 4 square feet per minute.
  (a) Write and simplify an expression that gives the total number of square feet the two people can paint in x minutes.
  (b) Find the number of square feet painted in 1 hour.
  (c) How many minutes would it take for them to paint a wall 8 feet tall and 21 feet wide?

112. *Area of a Triangle*   Find the area of the triangle shown.

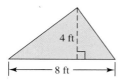

113. *Gallons to Pints*   There are 8 pints in 1 gallon. Make a table of values that converts G gallons to P pints. Let $G = 1, 2, 3, \ldots, 6$. Write a formula that converts G gallons to P pints.

114. *Blank CDs*   The table lists the cost C of buying x blank compact discs. Write an equation that relates C and x.

Blank CDs (x)	1	2	3	4
Cost (C)	$0.25	$0.50	$0.75	$1.00

115. *Aging in the United States*   In 2050, about $\frac{1}{5}$ of the population will be aged 65 or over and about $\frac{1}{20}$ of the population will be aged 85 or over. Estimate the fraction of the population that will be between the ages of 65 and 85 in 2050.   (*Source:* Bureau of the Census.)

116. *Carpentry*   A board measures $5\frac{3}{4}$ feet and needs to be cut in five equal pieces. Find the length of each piece.

117. *Distance*   Over four days an athlete jogs $3\frac{1}{8}$ miles, $4\frac{3}{8}$ miles, $6\frac{1}{4}$ miles, and $1\frac{5}{8}$ miles. How far did the athlete jog in all?

118. *Federal Employees*   The table lists the number of federal employees in thousands for selected years.

Year	1900	1920	1940
Employees	235	655	1042

Year	1960	1980	2000
Employees	2399	2876	2790

Source: Office of Management and Budget.

(a) How many federal employees were there in 1940?
(b) Mentally estimate the number of federal employees in 1950. Explain how you made your estimate. (The actual value is 1961 thousand, or 1,961,000.)

119. *Checking Account*   The initial balance in a checking account is $1652. Find the final balance resulting from the following sequence of withdrawals and deposits: $-$78$, $-$91$, 256, and $-$638$.

120. *Temperature Range*   The highest temperature ever recorded in Amarillo, Texas, was $108°F$ and the coldest was $-16°F$. Find the difference between these two temperatures.   (*Source: The Weather Almanac, 1995.*)

121. *Average Annual Rainfall*   The table lists monthly average rainfall in inches at Daytona Beach, Florida.

Month	Jan	Feb	Mar	Apr	May	June
Rainfall	2.5	3.0	3.2	2.5	3.1	5.7

Month	July	Aug	Sept	Oct	Nov	Dec
Rainfall	5.5	6.3	6.7	4.6	2.6	2.4

Source: The Weather Almanac, 1995.

(a) What was the average rainfall in March?
(b) Estimate the *yearly* average rainfall in Daytona Beach.

122. *Buying a Car*   A used car costs $2800 down and $310 per month for 20 months. Mentally calculate how much the car will cost. Explain your reasoning.

CHAPTER 1 Test

1. Classify the number 56 as prime or composite. If 56 is composite, write it as a product of prime numbers.

2. Evaluate the expression $\dfrac{5x}{2x-1}$ for $x = -3$.

3. Translate the phrase "four squared decreased by three" to an algebraic expression. Then find the value of the expression.

4. Reduce $\frac{24}{32}$ to lowest terms.

5. Evaluate each expression. Write your answer in lowest terms.

    (a) $\frac{5}{8} + \frac{1}{8}$ **(b)** $\frac{5}{9} - \frac{3}{15}$

    (c) $\frac{3}{5} \cdot \frac{10}{21}$ **(d)** $6 \div \frac{8}{5}$

6. Evaluate each expression.

    (a) $6 + 10 \div 5$ **(b)** $4^3 - (3 - 5 \cdot 2)$

    (c) $-6^2 - 6 + \frac{4}{2}$

7. State the property or properties that each equation illustrates.

    (a) $6x - 2x = 4x$ **(b)** $12 \cdot (3x) = 36x$

    (c) $4 + x + 8 = 12 + x$

8. Simplify each expression.

    (a) $5 - 5z + 7 + z$ **(b)** $12x - (6 - 3x)$

    (c) $5 - 4(x + 6) + \dfrac{15x}{3}$

9. *Mowing a Lawn* Two people are mowing a lawn. The first person has a riding mower and can mow $\frac{4}{3}$ acres per hour; the second person has a push mower and can mow $\frac{1}{4}$ acre per hour.

    (a) Write and simplify an expression that gives the total number of acres that the two people mow in x hours.

    (b) Find the total acreage that they can mow in an 8-hour work day.

10. *Cost Equation* The table lists the cost C of buying x tickets to a hockey game.

Tickets (x)	3	4	5	6
Cost (C)	$39	$52	$65	$78

    (a) Find an equation that relates C and x.

    (b) What is the cost of 17 tickets?

11. A wire $7\frac{4}{5}$ feet long is to be cut in 3 equal parts. How long should each part be?

12. The initial balance for a savings account is $892. Find the final balance resulting from withdrawals and deposits of $-\$57$, $\$150$, and $-\$345$.

CHAPTER 1
Extended and Discovery Exercises

1. *Arithmetic Operations* Insert $+$, $-$, $\times$, or $\div$ between the numbers to obtain the given answer. Do not use any parentheses.

$$2 \quad 2 \quad 2 \quad 2 = 0$$
$$3 \quad 3 \quad 3 \quad 3 = 10$$
$$4 \quad 4 \quad 4 \quad 4 = 1$$
$$6 \quad 6 \quad 6 \quad 6 = 36$$
$$7 \quad 7 \quad 7 \quad 7 = 63$$

2. *Magic Squares* The following square is called a "magic square" because the numbers in each row, column, and diagonal sum to 15.

8	3	4
1	5	9
6	7	2

Complete the following magic square having 4 rows and 4 columns by arranging the numbers 1 through 16 so that each row, column, and diagonal sums to 34. The four corners will also sum to 34.

	2		13
5			
	6		
4			1

Linear Equations and Inequalities

Mathematics is a unique subject because it does not depend on people making experiments, but it is an essential tool to be used in describing, or modeling, events in the real world. For example, sunbathing is a popular pastime, but ultraviolet light from the sun is responsible for both tanning and burning exposed skin. Mathematics lets us use numbers to describe the intensity of ultraviolet light. The table shows the maximum ultraviolet intensity measured in milliwatts per square meter for various latitudes and dates.

Latitude	Mar. 21	June 21	Sept. 21	Dec. 21
0°	325	254	325	272
10°	311	275	280	220
20°	249	292	256	143
30°	179	248	182	80
40°	99	199	127	34
50°	57	143	75	13

If a student from Chicago, located at a latitude of 42°, spends spring break in Hawaii with a latitude of 20°, the sun's ultraviolet rays in Hawaii will be approximately $\frac{249}{99} \approx 2.5$ times more intense than in Chicago. Suppose that you travel to the equator for spring break. How much more intense will the sun be at the equator than where you presently live?

**Education is not the filling of a pail,
but the lighting of a fire.
—William Butler Yeats**

Source: J. Williams, *The USA Today Weather Almanac 1995*.

2.1   INTRODUCTION TO EQUATIONS

Basic Concepts   ·   The Addition Property of Equality   ·
The Multiplication Property of Equality

INTRODUCTION

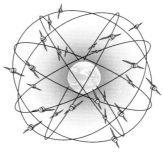

Figure 2.1 GPS Satellites

A new and exciting technology is the Global Positioning System (GPS), which consists of 24 satellites that travel around Earth in nearly circular orbits, as illustrated in Figure 2.1. The GPS can be used to determine locations and velocities of cars, airplanes, and hikers with a high degree of accuracy. New cars often come equipped with the GPS, and their drivers can determine their cars' locations to within a few feet.   (***Source:*** J. Van Sickle, *GPS for Land Surveyors*.)

   To create the GPS thousands of equations were solved, and mathematics was an essential tool in their solution. In this section we discuss many of the basic concepts needed to solve equations.

BASIC CONCEPTS

Suppose that during a storm it rains 2 inches before noon and 1 inch per hour thereafter until 5 P.M. Table 2.1 lists the total rainfall R after various elapsed times x, where $x = 0$ corresponds to noon.

TABLE 2.1

Elapsed Time: x (hours)	0	1	2	3	4	5
Total Rainfall: R (inches)	2	3	4	5	6	7

   The data suggest that the total rainfall R in inches is 2 more than the elapsed time x. A formula that *models* or *describes* the rainfall x hours past noon is given by

$$R = x + 2.$$

For example, 3 hours past noon, or at 3 P.M.,

$$R = 3 + 2 = 5$$

inches of rain had fallen. Even though $x = 4.5$ does not appear in the table, we can calculate the amount of rainfall at 4:30 P.M. with the formula as

$$R = 4.5 + 2 = 6.5 \text{ inches.}$$

The advantage that a formula has over a table of values is that a formula can be used to calculate the rainfall at any time x, not just at the times listed in the table.

   At what time had 6 inches of rain fallen? From Table 2.1 the *solution* is 4, or 4 P.M. To find this solution without the table, we can *solve* the equation

$$x + 2 = 6.$$

   An equation can be either true or false. For example, the equation $1 + 2 = 3$ is true, whereas the equation $1 + 2 = 4$ is false. When an equation contains a variable, the equation may be true for some values of the variable and false for other values of the variable.

Each value of the variable that makes the equation true is called a **solution** to the equation, and the *set of all solutions* is called the **solution set.** *Solving an equation* means finding all its solutions. Because $4 + 2 = 6$, the solution to the equation

$$x + 2 = 6$$

is 4, and the solution set is $\{4\}$. Note that **braces** $\{\}$ are used to denote a set. Sometimes an equation can have more than one solution. For example, the equation $x^2 = 4$ has two solutions, -2 and **2**, because

$$(-2)^2 = -2 \cdot (-2) = 4 \quad \text{and} \quad 2^2 = 2 \cdot 2 = 4.$$

For the equation $x^2 = 4$, the solution set is $\{-2, 2\}$.

Many times we cannot solve an equation simply by looking at it. In these situations we must use a step-by-step procedure. During each step an equation is transformed into a different but equivalent equation. **Equivalent equations** are equations that have the same solution set. For example, the equations

$$x + 2 = 5 \quad \text{and} \quad x = 3$$

are equivalent equations because the solution set for both equations is $\{3\}$.

Two important principles used to solve equations are the *addition property of equality* and the *multiplication property of equality.*

THE ADDITION PROPERTY OF EQUALITY

When solving an equation, we have to apply the same operation to each side of the equation. For example, one way to solve the equation

$$x + 2 = 5$$

is to add -2 to each side. This step results in isolating the x on one side of the equation.

$x + 2 = 5$	Given equation
$x + 2 + (-2) = 5 + (-2)$	Add -2 to each side.
$x + 0 = 3$	Addition of real numbers
$x = 3$	Additive identity

These four equations are equivalent, but the solution is most apparent in the last equation. To transform one equation to the next, a step-by-step procedure based on mathematical properties is used. The reasons that justify each step are written at the right in blue. When -2 is added to each side of the equation, the addition property of equality is used.

ADDITION PROPERTY OF EQUALITY

If a, b, and c are real numbers, then

$$a = b \quad \text{is equivalent to} \quad a + c = b + c.$$

Generally, the addition property of equality is based on the concept that "if equals are added to equals, the results are equal."

Note: Because any subtraction problem can be changed to an addition problem, the addition property of equality also works for subtraction. That is, if the same number is subtracted from each side of an equation, the result is an equivalent equation.

EXAMPLE 1   Using the addition property of equality

Solve each equation.
(a) $x + 10 = 7$ **(b)** $t - 4 = 3$ **(c)** $\frac{1}{2} = -\frac{3}{4} + y$

Solution   **(a)** When solving an equation, we try to isolate the variable on one side of the equation. If we add -10 to (or subtract 10 from) each side of the equation, the value of x becomes apparent.

$$x + 10 = 7 \qquad \text{Given equation}$$
$$x + 10 + (-10) = 7 + (-10) \qquad \text{Add } -10 \text{ to each side.}$$
$$x + 0 = -3 \qquad \text{Addition of real numbers}$$
$$x = -3 \qquad \text{Additive identity}$$

The solution is -3.

(b) To isolate the variable t, add 4 to each side.

$$t - 4 = 3 \qquad \text{Given equation}$$
$$t - 4 + 4 = 3 + 4 \qquad \text{Add 4 to each side.}$$
$$t = 7 \qquad \text{Addition of real numbers}$$

The solution is 7.

(c) To isolate the variable y, add $\frac{3}{4}$ to each side.

$$\frac{1}{2} = -\frac{3}{4} + y \qquad \text{Given equation}$$
$$\frac{1}{2} + \frac{3}{4} = -\frac{3}{4} + \frac{3}{4} + y \qquad \text{Add } \frac{3}{4} \text{ to each side.}$$
$$\frac{5}{4} = y \qquad \text{Addition of real numbers}$$

The solution is $\frac{5}{4}$.

CHECKING A SOLUTION   To check a solution, substitute it in the given equation to find out if a true statement results. To check the solution for Example 1(c), substitute $\frac{5}{4}$ for y in the given equation.

$$\frac{1}{2} = -\frac{3}{4} + y \qquad \text{Given equation}$$
$$\frac{1}{2} \stackrel{?}{=} -\frac{3}{4} + \frac{5}{4} \qquad \text{Let } y = \frac{5}{4}.$$
$$\frac{1}{2} \stackrel{?}{=} \frac{2}{4} \qquad \text{Add fractions.}$$
$$\frac{1}{2} = \frac{1}{2} \qquad \text{The answer checks.}$$

The answer of $\frac{5}{4}$ checks because the left side of the equation equals the right side of the equation.

Critical Thinking

When you are checking a solution, why do you substitute your answer in the *given* equation?

EXAMPLE 2   Solving and checking a solution

Solve the equation $-5 + y = 3$. Write the solution set and then check it.

Solution   Isolate y by adding 5 to each side.

$$-5 + y = 3 \qquad \text{Given equation}$$
$$5 + (-5) + y = 5 + 3 \qquad \text{Add 5 to each side.}$$
$$0 + y = 8 \qquad \text{Add.}$$
$$y = 8 \qquad \text{Additive identity}$$

The solution set is $\{8\}$. To check this answer substitute **8** for y in the given equation.

$$-5 + y = 3 \qquad \text{Given equation}$$
$$-5 + 8 \stackrel{?}{=} 3 \qquad \text{Let } y = 8.$$
$$3 = 3 \qquad \text{The answer checks.}$$

═══ MAKING CONNECTIONS ═══

Equations and Scales

Think of an equation as an old-fashioned scale, where two pans must balance, as illustrated in the accompanying figure. If two weights initially balance the pans, then adding an equal amount of weight to each pan results in the pans remaining balanced.

THE MULTIPLICATION PROPERTY OF EQUALITY

The multiplication property of equality is another important property used to solve equations. We can illustrate this property by considering a formula that converts yards to feet. Because there are 3 feet in 1 yard, the formula $F = 3Y$ computes F, the number of feet in Y yards. For example, if $Y = 5$ yards, then $F = 3 \cdot 5 = 15$ feet.

Now consider the reverse, converting 27 feet to yards. The answer to this conversion corresponds to the solution to

$$27 = 3Y.$$

To find the solution, multiply each side of the equation by the reciprocal of 3, or $\frac{1}{3}$.

$$27 = 3Y \qquad \text{Given equation}$$
$$\frac{1}{3} \cdot 27 = \frac{1}{3} \cdot 3 \cdot Y \qquad \text{Multiply each side by } \frac{1}{3}.$$
$$9 = 1 \cdot Y \qquad \text{Multiplication of fractions}$$
$$9 = Y \qquad \text{Multiplicative identity}$$

Thus 27 feet is equivalent to 9 yards.

When each side of an equation is multiplied by the same nonzero number, the multiplication property of equality is being applied.

MULTIPLICATION PROPERTY OF EQUALITY

If a, b, and c are real numbers with $c \neq 0$, then

$$a = b \quad \text{is equivalent to} \quad ac = bc.$$

Generally, the multiplication property of equality is based on the concept that "if equals are multiplied by equals, the results are equal."

Note: Because any division problem can be changed to a multiplication problem, the multiplication property of equality also works for division. That is, if each side of an equation is divided by the same nonzero number, the result is an equivalent equation.

EXAMPLE 3   **Using the multiplication property of equality**

Solve each equation.
(a) $\frac{1}{3}x = 4$ **(b)** $-4y = 8$ **(c)** $5 = \frac{3}{4}z$

Solution   **(a)** The coefficient of the x-term is $\frac{1}{3}$, and we want it to be 1. We start by multiplying each side of the equation by **3**, the reciprocal of $\frac{1}{3}$.

$$\frac{1}{3}x = 4 \qquad \text{Given equation}$$

$$\mathbf{3} \cdot \frac{1}{3}x = \mathbf{3} \cdot \mathbf{4} \qquad \text{Multiply each side by 3.}$$

$$1 \cdot x = 12 \qquad \text{Multiplicative inverses}$$

$$x = 12 \qquad \text{Multiplicative identity}$$

The solution is 12.

(b) The coefficient of the y-term is -4, so we can either multiply each side of the equation by $-\frac{1}{4}$ or divide each side by -4. This step will make the coefficient of y equal to 1.

$$-4y = 8 \qquad \text{Given equation}$$

$$\frac{-4y}{-4} = \frac{8}{-4} \qquad \text{Divide each side by } -4.$$

$$y = -2 \qquad \text{Reduce fractions.}$$

The solution is -2.

(c) To change the coefficient of z from $\frac{3}{4}$ to 1, multiply each side of the equation by $\frac{4}{3}$, the reciprocal of $\frac{3}{4}$.

$$5 = \frac{3}{4}z \qquad \text{Given equation}$$

$$\frac{4}{3} \cdot 5 = \frac{4}{3} \cdot \frac{3}{4}z \qquad \text{Multiply each side by } \frac{4}{3}.$$

$$\frac{20}{3} = z \qquad \text{Multiply.}$$

The solution is $\frac{20}{3}$.

EXAMPLE 4 Solving and checking a solution

Solve the equation $\frac{3}{4} = -\frac{3}{7}t$. Write the solution set and then check it.

Solution Multiply each side of the equation by $-\frac{7}{3}$, the reciprocal of $-\frac{3}{7}$.

$$\frac{3}{4} = -\frac{3}{7}t \qquad \text{Given equation}$$

$$-\frac{7}{3} \cdot \frac{3}{4} = -\frac{7}{3} \cdot \left(-\frac{3}{7}\right)t \qquad \text{Multiply each side by } -\frac{7}{3}.$$

$$-\frac{7}{4} = 1 \cdot t \qquad \text{Multiply and reduce fractions.}$$

$$-\frac{7}{4} = t \qquad \text{Multiplicative identity}$$

The solution set is $\left\{-\frac{7}{4}\right\}$. To check this answer, substitute $-\frac{7}{4}$ for t in the given equation.

$$\frac{3}{4} = -\frac{3}{7}t \qquad \text{Given equation}$$

$$\frac{3}{4} \overset{?}{=} -\frac{3}{7} \cdot \left(-\frac{7}{4}\right) \qquad \text{Let } t = -\frac{7}{4}.$$

$$\frac{3}{4} = \frac{3}{4} \qquad \text{The answer checks.}$$

AN APPLICATION Alaska's major glaciers are melting at an alarming rate. From the mid 1950s to the mid 1990s, these glaciers melted at a rate of 13 cubic miles per year. Recently, Alaska's glaciers have been melting at a rate of 24 cubic miles per year, almost double the earlier rate. Ecologists are concerned that this phenomenon is a sign of global warming. (**Source:** *USA Today*.)

EXAMPLE 5 Estimating glacial melting in Alaska

The glaciers in Alaska are currently melting at a rate of 24 cubic miles per year.
(a) Write a formula that gives the cubic miles of ice I that will melt in x years.
(b) At this rate, determine how long it will take for 300 cubic miles of ice to melt.

Solution **(a)** In 1 year $24 \cdot 1 = 24$ cubic miles will melt, in 2 years $24 \cdot 2 = 48$ cubic miles will melt, and in x years $24 \cdot x = 24x$ cubic miles will melt. Thus $I = 24x$, where x is in years and I is in cubic miles.

(b) To determine how long it will take for 300 cubic miles to melt let $I = 300$ in the formula.

$$I = 24x \qquad \text{Formula from part (a)}$$

$$300 = 24x \qquad \text{Let } I = 300.$$

$$\frac{300}{24} = \frac{24x}{24} \qquad \text{Divide each side by 24.}$$

$$12.5 = x \qquad \text{Simplify.}$$

At current rates, it will take 12.5 years for 300 cubic miles of ice to melt.

2.1 PUTTING IT ALL TOGETHER

The following table summarizes some of the topics discussed in this section.

Concept	Comments	Examples
Solution	A value for a variable that makes an equation a true statement	The solution to $x + 5 = 20$ is 15, and the solutions to $x^2 = 9$ are -3 and 3.
Solution Set	The set of all solutions to an equation	The solution set to $x + 5 = 20$ is $\{15\}$, and the solution set to $x^2 = 9$ is $\{-3, 3\}$.
Equivalent Equations	Two equations are equivalent if they have the same solution set.	The equations $$2x = 14 \quad \text{and} \quad x = 7$$ are equivalent because the solution set to both equations is $\{7\}$.
Addition Property of Equality	The equations $a = b$ and $a + c = b + c$ are equivalent. This property is used to solve equations.	To solve $x - 3 = 8$ add 3 to each side of the equation. $$x - 3 + 3 = 8 + 3$$ $$x = 11$$ The solution is 11.
Multiplication Property of Equality	The equations $a = b$ and $a \cdot c = b \cdot c$ with $c \neq 0$ are equivalent. This property is used to solve equations.	To solve $\frac{1}{5}x = 10$ multiply each side of the equation by 5. $$5 \cdot \frac{1}{5}x = 5 \cdot 10$$ $$x = 50$$ The solution is 50.
Checking a Solution	Substitute the solution in the given equation and simplify each side to check it.	The solution to $$x + 12 = 20$$ is 8 because $$8 + 12 \stackrel{?}{=} 20$$ $$20 = 20.$$

 **2.1 EXERCISES**

FOR EXTRA HELP

📖 Student's Solutions Manual	InterAct Math
🚪 MyMathLab	Tutor Center Math Tutor Center

 MathXL

Digital Video Tutor
CD 1 Videotape 3

CONCEPTS

1. Every equation contains a(n)_____sign.

2. Each value of a variable that makes an equation true is called a(n) _____.

3. The set of all solutions to an equation is called the _____.

4. _____ equations have the same solution sets.

5. To solve an equation, you must find all _____.

6. To check an answer, substitute it in the _____ equation.

7. If $a = b$, then $a + c =$ _____.

8. If $a = b$ and $c \neq 0$, then $ac =$ _____.

9. The solution to $x - 1 = 0$ is _____.

10. The solution to $3x = 6$ is _____.

THE ADDITION PROPERTY OF EQUALITY

11. To solve $x - 22 = 4$, add _____ to each side of the equation.

12. To solve $\frac{5}{6} = \frac{1}{6} + x$, add _____ to each side of the equation.

Exercises 13–26: Solve the equation. Check your answer.

13. $x + 5 = 0$

14. $x + 3 = 7$

15. $x - 7 = 1$

16. $x - 23 = 0$

17. $9 = y - 8$

18. $97 = -23 + y$

19. $\frac{1}{2} = z - \frac{3}{2}$

20. $\frac{3}{4} + z = -\frac{1}{2}$

21. $t - 0.8 = 4.2$

22. $4 = -9 + t$

23. $25 + x = 10$

24. $85 = x - 20$

25. $1989 = 26 + y$

26. $y - 1.23 = -0.02$

THE MULTIPLICATION PROPERTY OF EQUALITY

27. To solve $5x = 4$ for x, divide each side by _____.

28. To solve $\frac{4}{3}y = 8$ for y, multiply each side by _____.

Exercises 29–42: Solve the equation. Check your answer.

29. $5x = 15$

30. $-2x = 8$

31. $-7x = 0$

32. $25x = 0$

33. $\frac{1}{2}x = \frac{3}{2}$

34. $\frac{3}{4}x = \frac{5}{8}$

35. $\frac{1}{2} = \frac{2}{5}z$

36. $-\frac{3}{4} = -\frac{1}{8}z$

37. $25 = 5z$

38. $-10 = -4z$

39. $0.5t = 3.5$

40. $2.2t = -9.9$

41. $\frac{3}{8} = \frac{1}{4}y$

42. $1.2 = 0.3y$

APPLICATIONS

43. *Rainfall* On a stormy day it rains 3 inches before noon and $\frac{1}{2}$ inch per hour thereafter until 6 P.M.
    (a) Make a table that shows the total rainfall R in inches, x hours past noon, ending at 6 P.M.
    (b) Write a formula that calculates R.
    (c) Use your formula to calculate the total rainfall at 3 P.M. Does the answer agree with the value in your table from part (a)?
    (d) How much rain had fallen by 2:15 P.M.?

44. *Cold Weather* A furnace is turned on at midnight when the temperature inside a cabin is $0°$F. The cabin warms at a rate of $10°$F per hour until 7 A.M.
    (a) Make a table that shows the cabin temperature T in degrees Fahrenheit, x hours past midnight, ending at 7 A.M.
    (b) Write a formula that calculates T.
    (c) Use your formula to calculate the temperature at 5 A.M. Does the answer agree with the value in your table from part (a)?
    (d) Find the cabin temperature at 2:45 A.M.

45. *Football Field*   A football field is 300 feet long.
   (a) Write a formula that gives the length L of x football fields in feet.
   (b) Use your formula to write an equation whose solution gives the number of football fields in 870 feet.
   (c) Solve your equation from part (b).

46. *Acreage*   An acre equals 43,560 square feet.
   (a) Write a formula that converts A acres to S square feet.
   (b) Use your formula to write an equation whose solution gives the number of acres in 871,200 square feet.
   (c) Solve your equation from part (b).

47. *Glacial Melting in Alaska*   (Refer to Example 5.) The Alaskan glaciers are melting at a rate of 24 cubic miles per year. At this rate, how many years will it take for 420 cubic miles of the glacier to melt? (*Source: USA Today.*)

48. *Arctic Ice Cap*   The Arctic ice cap contains 680,000 cubic miles of ice. If this ice cap melted at a rate of 50 cubic miles per year, how many years would it take for the entire ice cap to melt? (*Source:* Department of the Interior, Geological Survey.)

49. *Cost of a Car*   When the cost of a car is multiplied by 0.07 the result is $1750. Find the cost of the car.

50. *Raise in Salary*   If an employee's salary is multiplied by 1.06, which corresponds to a 6% raise, the result is $58,300. Find the employee's current salary.

WRITING ABOUT MATHEMATICS

51. A student solves an equation as follows.

$$x + 30 = 64$$
$$x \overset{?}{=} 64 + 30$$
$$x \overset{?}{=} 94$$

Identify the student's mistake. What is the solution?

52. What is a good first step for solving the equation $\frac{a}{b}x = 1$, where a and b are natural numbers? What is the solution? Explain your answers.

2.2   LINEAR EQUATIONS

Basic Concepts · Solving Linear Equations · Applying the Distributive Property · Clearing Fractions and Decimals · Equations with Zero or Infinitely Many Solutions

INTRODUCTION

Solving equations is an important concern in mathematics. Billions of dollars are spent each year to solve equations that lead to the creation of better products. If our society could not solve equations, we would not have TV sets, DVD players, satellites, fiber optics, CAT scans, computers, or accurate weather forecasts. In this section we discuss linear equations and some of their applications. One of the simplest types of equations, linear equations, can always be solved by hand.

BASIC CONCEPTS

Suppose that a bicyclist is 5 miles from home, riding *away* from home at 10 miles per hour. The distance between the bicyclist and home for various elapsed times is shown in Table 2.2.

TABLE 2.2 Distance from Home

Elapsed Time (hours)	0	1	2	3
Distance (miles)	5	15	25	35

10 miles   10 miles   10 miles

The bicyclist is moving at a constant speed, so the distance increases by 10 miles every hour. The data suggest that the distance D from home after x hours could be calculated by the formula

$$D = 10x + 5.$$

For example, after **2** hours the distance is

$$D = 10(2) + 5 = 25 \text{ miles.}$$

Table 2.2 verifies that the bicyclist is 25 miles from home after 2 hours. However, the table is less helpful if we want to find the elapsed time when the bicyclist is 18 miles from home. To answer this question, we could substitute **18** for D and then solve the equation

$$18 = 10x + 5.$$

Subtracting 18 from each side allows us to write an equivalent equation.

$$18 - 18 = 10x + 5 - 18 \qquad \text{Subtract 18 from each side.}$$
$$0 = 10x - 13 \qquad \text{Simplify.}$$
$$10x - 13 = 0 \qquad \text{Rewrite the equation.}$$

The equation $10x - 13 = 0$ is an example of a *linear equation* and is solved in Example 3. Linear equations often model applications in which things either move or change at a constant rate.

LINEAR EQUATION IN ONE VARIABLE

A **linear equation** in one variable is an equation that can be written in the form

$$ax + b = 0,$$

where $a \neq 0$.

Note: The equation $18 = 10x + 5$ is linear because it can be written as $10x - 13 = 0$.

Table 2.3 gives examples of linear equations and values for a and b.

TABLE 2.3 Linear Equations

$ax + b = 0$	a	b
$x - 1 = 0$	1	-1
$-5x + 1 = 0$	-5	1
$2.5x = 0$	2.5	0

EXAMPLE 1   Determining whether an equation is linear

Determine whether the equation is linear. If the equation is linear, give values for a and b.
(a) $4x + 5 = 0$ **(b)** $5 = -\frac{3}{4}x$ **(c)** $4x^2 + 6 = 0$

Solution   **(a)** The equation is linear because it is in the form $ax + b = 0$ with $a = 4$ and $b = 5$.
(b) The equation can be rewritten as follows.

$$5 = -\frac{3}{4}x \qquad \text{Given equation}$$

$$\frac{3}{4}x + 5 = \frac{3}{4}x + \left(-\frac{3}{4}x\right) \qquad \text{Add } \frac{3}{4}x \text{ to each side.}$$

$$\frac{3}{4}x + 5 = 0 \qquad \text{Rewrite the equation.}$$

This equation is linear because it is in the form $ax + b = 0$ with $a = \frac{3}{4}$ and $b = 5$.

Note: If 5 had been subtracted from each side, the result would be $0 = -\frac{3}{4}x - 5$, which is an equivalent linear equation with $a = -\frac{3}{4}$ and $b = -5$.

(c) The equation is *not* linear because it cannot be written in the form $ax + b = 0$. It has a nonzero term containing x^2.

SOLVING LINEAR EQUATIONS

One way to solve a linear equation is to make a table.

EXAMPLE 2   Using a table to solve an equation

Complete Table 2.4 for the given values of x. Then solve the equation $2x - 5 = -7$.

TABLE 2.4

x	-3	-2	-1	0	1	2	3
$2x - 5$	-11						

Solution   To complete the table, substitute $x = -2, -1, 0, 1, 2,$ and 3 into the expression $2x - 5$. For example, if $x = -2$, then $2x - 5 = 2(-2) - 5 = -9$. The other values shown in Table 2.5 can be found similarly.

Calculator Help
To create a table of values, see
Appendix A (page AP-3).

TABLE 2.5

x	-3	-2	-1	0	1	2	3
$2x - 5$	-11	-9	-7	-5	-3	-1	1

From the table, $2x - 5$ equals -7 when $x = -1$. Thus the solution to $2x - 5 = -7$ is -1.

A more efficient method for solving linear equations than making a table is to use the addition and multiplication properties of equality.

EXAMPLE 3   Solving linear equations

Solve each linear equation. Check the answer for part (b).
(a) $10x - 13 = 0$ **(b)** $\frac{1}{2}x + 3 = 6$ **(c)** $5x + 7 = 2x + 3$

Solution **(a)** First, isolate the x-term on the left side of the equation by adding 13 to each side.

Technology Note
Graphing Calculators and Tables

Many graphing calculators have the capability to make tables. Table 2.5 is shown in the accompanying figure.

X	Y₁
-3	-11
-2	-9
-1	-7
0	-5
1	-3
2	-1
3	1

Y₁ ▉ 2X − 5

$$10x - 13 = 0 \qquad \text{Given equation}$$
$$10x - 13 + \mathbf{13} = 0 + \mathbf{13} \qquad \text{Add 13 to each side.}$$
$$10x = 13 \qquad \text{Add the real numbers.}$$

To obtain a coefficient of 1 on the x-term, divide each side by 10.

$$\frac{10x}{\mathbf{10}} = \frac{13}{\mathbf{10}} \qquad \text{Divide each side by 10.}$$
$$x = \frac{13}{10} \qquad \text{Reduce.}$$

The solution is $\frac{13}{10}$.

Note: In the second equation of the solution the addition property is used, and in the fourth equation of the solution the multiplication property is used.

(b) Start by subtracting 3 from each side.

$$\frac{1}{2}x + 3 = 6 \qquad \text{Given equation}$$

$$\frac{1}{2}x + 3 - \mathbf{3} = 6 - \mathbf{3} \qquad \text{Subtract 3 from each side.}$$

$$\frac{1}{2}x = 3 \qquad \text{Subtract the real numbers.}$$

$$\mathbf{2} \cdot \frac{1}{2}x = \mathbf{2} \cdot 3 \qquad \text{Multiply each side by 2.}$$

$$x = 6 \qquad \text{Multiply the real numbers.}$$

The solution is 6. To check it substitute **6** for x in the given equation.

$$\frac{1}{2} \cdot \mathbf{6} + 3 \overset{?}{=} 6 \qquad \text{Let } x = 6.$$

$$3 + 3 \overset{?}{=} 6 \qquad \text{Multiply.}$$

$$6 = 6 \qquad \text{The answer checks.}$$

(c) In this equation there are two x-terms. Generally it is a good idea to move all x-terms to one side of the equation and all real numbers to the other side. To accomplish this task, begin by subtracting $2x$ from each side.

$$5x + 7 = 2x + 3 \qquad \text{Given equation}$$
$$5x - \mathbf{2x} + 7 = 2x - \mathbf{2x} + 3 \qquad \text{Subtract } 2x \text{ from each side.}$$
$$3x + 7 = 3 \qquad \text{Combine like terms.}$$
$$3x + 7 - \mathbf{7} = 3 - \mathbf{7} \qquad \text{Subtract 7 from each side.}$$
$$3x = -4 \qquad \text{Simplify.}$$
$$\frac{3x}{\mathbf{3}} = \frac{-4}{\mathbf{3}} \qquad \text{Divide each side by 3.}$$
$$x = -\frac{4}{3} \qquad \text{Reduce the fractions.}$$

The solution is $-\frac{4}{3}$.

```
═══════════════ MAKING CONNECTIONS ═══════════════
```

Linear Equations and Solutions

A linear equation has *exactly one* solution because $ax + b = 0$, with $a \neq 0$, implies that $x = -\frac{b}{a}$.

The next example involves an application that requires us to solve a linear equation.

EXAMPLE 4 Estimating Internet users

The number of Internet users I in millions during year x can be approximated by the formula

$$I = \frac{44}{3}x - 29{,}219,$$

where $x \geq 1996$. Estimate the year when there were 100 million Internet users. (***Source:*** The Internet Society.)

Solution Let $I = 100$ in the formula and solve for x.

$$100 = \frac{44}{3}x - 29{,}219 \qquad \text{Equation to be solved}$$

$$29{,}319 = \frac{44}{3}x \qquad \text{Add 29,219 to each side.}$$

$$\frac{3}{44}(29{,}319) = \frac{3}{44} \cdot \left(\frac{44}{3}x\right) \qquad \text{Multiply each side by } \frac{3}{44}.$$

$$\frac{87{,}957}{44} = x \qquad \text{Simplify.}$$

$$x \approx 1999 \qquad \text{Approximate with a calculator.}$$

During 1999 the number of Internet users reached 100 million.

APPLYING THE DISTRIBUTIVE PROPERTY

Sometimes the distributive property is helpful in solving linear equations. The next example demonstrates how to apply the distributive property in such situations.

EXAMPLE 5 Applying the distributive property

Solve each linear equation. Check the answer for part (a).
(a) $4(x - 3) + x = 0$
(b) $2(3z - 4) + 1 = 3(z + 1)$

Solution (a) Begin by applying the distributive property.

$$4(x - 3) + x = 0 \qquad \text{Given equation}$$

$$4x - 12 + x = 0 \qquad \text{Distributive property}$$

$$5x - 12 = 0 \qquad \text{Combine like terms.}$$

$$5x - 12 + 12 = 0 + 12 \qquad \text{Add 12 to each side.}$$

$$5x = 12 \qquad \text{Add the real numbers.}$$

$$\frac{5x}{5} = \frac{12}{5} \qquad \text{Divide each side by 5.}$$

$$x = \frac{12}{5} \qquad \text{Reduce.}$$

To check whether $\frac{12}{5}$ is the solution, substitute $\frac{12}{5}$ for x in the given equation.

$$4\left(\frac{12}{5} - 3\right) + \frac{12}{5} \stackrel{?}{=} 0 \qquad \text{Let } x = \frac{12}{5}.$$

$$\frac{48}{5} - 12 + \frac{12}{5} \stackrel{?}{=} 0 \qquad \text{Distributive property}$$

$$12 - 12 \stackrel{?}{=} 0 \qquad \text{Add the fractions and reduce.}$$

$$0 = 0 \qquad \text{The answer checks.}$$

(b) Begin by applying the distributive property to each side of the equation. Then move z-terms to the left side and terms containing only real numbers to the right side.

$$2(3z - 4) + 1 = 3(z + 1) \qquad \text{Given equation}$$

$$6z - 8 + 1 = 3z + 3 \qquad \text{Distributive property}$$

$$6z - 7 = 3z + 3 \qquad \text{Add the real numbers.}$$

$$6z = 3z + 10 \qquad \text{Add 7 to each side.}$$

$$3z = 10 \qquad \text{Subtract } 3z \text{ from each side.}$$

$$\frac{3z}{3} = \frac{10}{3} \qquad \text{Divide each side by 3.}$$

$$z = \frac{10}{3} \qquad \text{Reduce.}$$

CLEARING FRACTIONS AND DECIMALS

Because most people find it easier to do hand calculations without fractions or decimals, clearing an equation of fractions or decimals before solving for the variable is often helpful. To clear fractions or decimals we often multiply each side of the equation by the least common denominator (LCD).

EXAMPLE 6   Clearing fractions from linear equations

Solve each linear equation.

(a) $\frac{1}{7}x - \frac{5}{7} = \frac{3}{7}$ (b) $\frac{2}{3}x - \frac{1}{6} = x$

Solution **(a)** Multiply each side of the equation by the LCD 7 to clear (remove) fractions from the equation.

$$\frac{1}{7}x - \frac{5}{7} = \frac{3}{7}$$ Given equation

$$7\left(\frac{1}{7}x - \frac{5}{7}\right) = 7 \cdot \frac{3}{7}$$ Multiply each side by 7.

$$x - 5 = 3$$ Distributive property

$$x = 8$$ Add 5 to each side.

The solution is 8.

(b) The LCD for 3 and 6 is 6. Multiply each side of the equation by 6.

$$\frac{2}{3}x - \frac{1}{6} = x$$ Given equation

$$6\left(\frac{2}{3}x - \frac{1}{6}\right) = 6 \cdot x$$ Multiply each side by 6.

$$4x - 1 = 6x$$ Distributive property.

$$-1 = 2x$$ Subtract 4x from each side.

$$-\frac{1}{2} = x$$ Divide each side by 2.

The solution is $-\frac{1}{2}$.

EXAMPLE 7 **Clearing decimals from a linear equation**

Solve each linear equation.
(a) $0.2x - 0.7 = 0.4$ **(b)** $0.01x - 0.42 = -0.2x$

Solution **(a)** The least common denominator for 0.2, 0.7, and 0.4 $\left(\text{or } \frac{2}{10}, \frac{7}{10}, \text{ and } \frac{4}{10}\right)$ is 10. Multiply each side by 10. When multiplying by 10, move the decimal point 1 place to the right.

$$0.2x - 0.7 = 0.4$$ Given equation

$$10(0.2x - 0.7) = 10(0.4)$$ Multiply each side by 10.

$$2x - 7 = 4$$ Distributive property

$$2x = 11$$ Add 7 to each side.

$$x = \frac{11}{2}$$ Divide each side by 2.

The solution is $\frac{11}{2}$, or 5.5.

(b) The least common denominator for 0.01, 0.2, and 0.42 $\left(\text{or } \frac{1}{100}, \frac{2}{10}, \text{ and } \frac{42}{100}\right)$ is 100. Multiply each side by 100. When multiplying by 100, move the decimal point 2 places to the right.

$$0.01x - 0.42 = -0.2x$$ Given equation

$$100(0.01x - 0.42) = 100(-0.2x)$$ Multiply each side by 100.

$$x - 42 = -20x$$ Distributive property

$$x - 42 + 20x + 42 = -20x + 20x + 42$$ Add 20x and 42.

$$21x = 42$$ Combine like terms.

$$x = 2$$ Divide each side by 21.

The solution is 2.

EQUATIONS WITH ZERO OR INFINITELY MANY SOLUTIONS

Some equations that appear to be linear are not because when they are written in the form $ax + b = 0$ the value of a equals 0 and so no x-term appears. This type of equation can have no solutions or infinitely many solutions. An example of an equation that has no solutions is

$$x + 1 = x$$

because a number x cannot equal itself plus 1. If we attempt to solve this equation by subtracting x from each side, we obtain the equation $1 = 0$, which is always false, indicating there are no solutions. This equation has no x-term.

An example of an equation with infinitely many solutions is

$$5x = 2x + 3x,$$

because the equation simplifies to

$$5x = 5x,$$

which is true for any real number x. If $5x$ is subtracted from each side the result is $0 = 0$, which has no x-term and is always true.

EXAMPLE 8   Determining numbers of solutions

Determine whether the equation has zero, one, or infinitely many solutions.
(a) $3x = 2(x + 1) + x$ (b) $2x - (x + 1) = x - 1$ (c) $5x = 2(x - 4)$

Solution   (a) Start by applying the distributive property.

$$
\begin{array}{ll}
3x = 2(x + 1) + x & \text{Given equation} \\
3x = 2x + 2 + x & \text{Distributive property} \\
3x = 3x + 2 & \text{Combine like terms.} \\
0 = 2 & \text{Subtract } 3x.
\end{array}
$$

Because the equation $0 = 2$ is always false, there are no solutions.

(b) Start by applying the distributive property.

$$
\begin{array}{ll}
2x - (x + 1) = x - 1 & \text{Given equation} \\
2x - x - 1 = x - 1 & \text{Distributive property} \\
x - 1 = x - 1 & \text{Combine like terms.} \\
x = x & \text{Add 1 to each side.} \\
0 = 0 & \text{Subtract } x \text{ from each side.}
\end{array}
$$

Because the equation $0 = 0$ is always true, there are infinitely many solutions. Note that the solution set contains all real numbers.

(c) Start by applying the distributive property.

$$
\begin{array}{ll}
5x = 2(x - 4) & \text{Given equation} \\
5x = 2x - 8 & \text{Distributive property} \\
3x = -8 & \text{Subtract } 2x \text{ from each side.} \\
x = -\dfrac{8}{3} & \text{Divide each side by 3.}
\end{array}
$$

Thus there is one solution.

Critical Thinking

What must be true about b and d for the equation

$$bx - 2 = dx + 7$$

to have no solutions? What must be true about b and d for this equation to have exactly one solution?

2.2 PUTTING IT ALL TOGETHER

The following table summarizes some of the topics discussed in this section.

Concept	Comments	Examples
Linear Equation	Can be written as $$ax + b = 0,$$ where $a \neq 0$; has one solution.	The equation $5x - 8 = 0$ is linear, with $a = 5$ and $b = -8$.
Solving Linear Equations	Use the addition and multiplication properties of equality.	$5x - 8 = 0$ — Given equation $5x = 8$ — Add 8 to each side. $x = \dfrac{8}{5}$ — Divide each side by 5.
Equations with No Solutions	Some equations that appear to be linear have no solutions.	The equation $$x = x + 5$$ has no solutions because a number cannot equal itself plus 5.
Equations with Infinitely Many Solutions	Some equations that appear to be linear have infinitely many solutions.	The equation $$2x = x + x$$ has infinitely many solutions because the equation is true for all values of x.

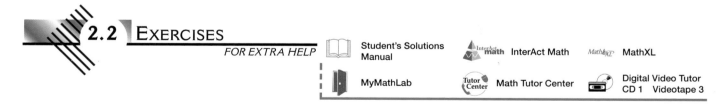

2.2 EXERCISES

FOR EXTRA HELP

Student's Solutions Manual   InterAct Math   MathXL

MyMathLab   Math Tutor Center   Digital Video Tutor CD 1   Videotape 3

CONCEPTS

1. A linear equation can be written in the form _____ with $a \neq 0$.

2. The equation $x^2 - 1 = 0$ (is/is not) a linear equation.

3. What two properties of equality are frequently used to solve linear equations?

4. What property justifies that $4(x - 3) = 4x - 12$ is true for any value of x?

5. To clear fractions from an equation, multiply each side by the _____ .

6. To clear decimals from $0.3x + 1.2 = 0.01$, multiply each side by _____ .

7. How many solutions does the equation $3x = 2x + x$ have?

8. How many solutions does the equation $x = x + 10$ have?

IDENTIFYING LINEAR EQUATIONS

Exercises 9–20: (Refer to Example 1.) Determine whether the equation is linear. If the equation is linear, give values for a and b so that it can be written in the form $ax + b = 0$.

9. $3x - 7 = 0$

10. $-2x + 1 = 4$

11. $\frac{1}{2}x = 0$

12. $-\frac{3}{4}x = 0$

13. $4x^2 - 6 = 11$

14. $-2x^2 + x = 4$

15. $1.1x = 0.9$

16. $-5.7x = -3.4$

17. $2(x - 3) = 0$

18. $\frac{1}{2}(x + 4) = 0$

19. $6x - x^2 = 0$

20. $5 = 4x^3$

SOLVING LINEAR EQUATIONS

Exercises 21–24: Evaluate the expression for each value of x in the table. Then use the table to solve the given equation.

21. $-4x + 8 = 0$

x	1	2	3	4	5
$-4x + 8$	4				

22. $3x + 2 = 5$

x	-2	-1	0	1	2
$3x + 2$	-4				

23. $4 - 2x = 6$

x	-2	-1	0	1	2
$4 - 2x$	8				

24. $9 - (x + 3) = 4$

x	-2	-1	0	1	2
$9 - (x + 3)$	8				

Exercises 25–54: Solve the equation and check the solution.

25. $11x = 3$

26. $-5x = 15$

27. $x - 18 = 5$

28. $8 = 5 + 3x$

29. $2x - 1 = 13$

30. $4x + 3 = 39$

31. $5x + 5 = -6$

32. $-7x - 4 = 31$

33. $3z + 2 = z - 5$

34. $z - 5 = 5z - 3$

35. $12y - 6 = 33 - y$

36. $-13y + 2 = 22 - 3y$

37. $4(x - 1) = 5$

38. $-2(2x + 7) = 1$

39. $1 - (3x + 1) = 5 - x$

40. $6 + 2(x - 7) = 10 - 3(x - 3)$

41. $(5t - 6) + 2(t + 1) = 0$

42. $-2(t - 7) - (t + 5) = 5$

43. $3(4z - 1) - 2(z + 2) = 2(z + 1)$

44. $-(z + 4) + (3z + 1) = -2(z + 1)$

45. $7.3x - 1.7 = 5.6$

46. $5.5x + 3x = 51$

47. $-9.5x - 0.05 = 10.5x + 1.05$

48. $0.04x + 0.03 = 0.02x - 0.1$

49. $\frac{1}{2}x - \frac{3}{2} = \frac{5}{2}$

50. $-\frac{1}{4}x + \frac{5}{4} = \frac{3}{4}$

51. $-\frac{3}{8}x + \frac{1}{4} = \frac{1}{2}x + \frac{1}{8}$

52. $\frac{1}{3}x + \frac{1}{4} = \frac{1}{6} - x$

53. $4y - 2(y + 1) = 0$

54. $(15y + 20) - 5y = 5 - 10y$

Exercises 55–64: Determine whether the equation has zero, one, or infinitely many solutions.

55. $5x = 5x + 1$

56. $2(x - 3) = 2x - 6$

57. $8x = 0$

58. $9x = x + 1$

59. $4(x + 2) - 2(2x + 3) = 10$

60. $5(2x + 7) - (10x + 5) = 30$

61. $4x = 5(x + 3) - x$

62. $x - (3x + 2) = 15 - 2x$

63. $2x - (x + 5) = x - 5$

64. $5x = 15 - 2(x + 7)$

APPLICATIONS

65. *Distance Traveled* A bicyclist is 4 miles from home, riding away from home at 8 miles per hour.
   (a) Make a table that shows the bicyclist's distance D from home after 1, 2, 3, and 4 hours.
   (b) Write a formula that calculates D after x hours.
   (c) Use your formula to determine D when $x = 3$ hours. Does your answer agree with the value found in your table?
   (d) Find x when $D = 22$ miles. Interpret the result.

66. *Distance Traveled*   An athlete is 16 miles from home, running *toward* home at 6 miles per hour.
   (a) Write a formula that calculates the distance D that the athlete is from home after x hours.
   (b) Use your formula to determine D when $x = 1.5$ hours.
   (c) Find x when $D = 5.5$ miles. Interpret the result.

67. *HIV Infections*   The cumulative number of HIV infections N in thousands for the United States in year x can be approximated by the formula

$$N = 40x - 79{,}065,$$

where $x \geq 1990$. Estimate the year when this number reached 815 thousand.   (*Source:* Centers for Disease Control and Prevention.)

68. *Internet Users*   (Refer to Example 4.) The number of Internet users I in millions during year x, where $x \geq 1996$, can be approximated by the formula

$$I = \frac{44}{3}x - 29{,}219.$$

Estimate the year in which there were 85 million Internet users for the first time.

69. *State and Federal Inmates*   From 1988 to 1995 the number N of state and federal inmates in thousands during year x can be approximated by the formula $N = 70x - 138{,}532$. Determine the year in which there were 908 thousand inmates.   (*Source:* Bureau of Justice.)

70. *Government Costs*   From 1960 to 2000 the cost C (in billions of 1992 dollars) to regulate social and economic programs could be approximated by the formula $C = 0.35x - 684$ during year x. Estimate the year in which the cost reached $6.6 billion.   (*Source:* Center for the Study of American Business.)

WRITING ABOUT MATHEMATICS

71. A student says that the equation $4x - 1 = 1 - x$ is not a linear equation because it is not in the form $ax + b = 0$. Is the student correct? Explain.

72. A student solves a linear equation as follows.

$$4(x + 3) = 5 - (x + 3)$$
$$4x + 3 \overset{?}{=} 5 - x + 3$$
$$4x + 3 \overset{?}{=} 8 - x$$
$$5x \overset{?}{=} 5$$
$$x \overset{?}{=} 1$$

Identify and explain the errors that the student made. What is the correct answer?

CHECKING BASIC CONCEPTS   SECTIONS 2.1 AND 2.2

1. Evaluate $4x - 3$ for each value of x in the table. Then use the table to solve $4x - 3 = 13$.

x	3	3.5	4	4.5	5
$4x - 3$	9				17

2. Solve each equation and check your answer.
   (a) $x - 12 = 6$
   (b) $\frac{3}{4}z = \frac{1}{8}$
   (c) $0.6t + 0.4 = 2$
   (d) $5 - 2(x - 2) = 3(4 - x)$

3. Determine whether each equation has zero, one, or infinitely many solutions.
   (a) $x - 5 = 6x$
   (b) $-2(x - 5) = 10 - 2x$
   (c) $-(x - 1) = -x - 1$

4. *Distance Traveled* A driver is 300 miles from home and is traveling toward home on a freeway at a constant speed of 75 miles per hour.
   (a) Write a formula to calculate the distance D that the driver is from home after x hours.
   (b) Write an equation whose solution gives the hours needed for the driver to reach home.
   (c) Solve the equation from part (b).

2.3 INTRODUCTION TO PROBLEM SOLVING

Steps for Solving a Problem · **Percent Problems** · **Distance Problems** · **Other Types of Problems**

INTRODUCTION

One important characteristic of human beings is their ability to solve problems. For example, throughout history people have wanted to predict the weather but have found it difficult to do. After all, how could anyone predict the future? Today, with the help of mathematics, science, computers, and years of data collection, meteorologists are providing increasingly accurate weather forecasts. Solving problems requires problem-solving skills, which we introduce in this section.

STEPS FOR SOLVING A PROBLEM

Word problems are challenging because formulas and equations are not usually given. To solve such problems we need a strategy. The following steps are based on George Polya's (1888–1985) four-step process for problem solving.

STEPS FOR SOLVING A PROBLEM

STEP 1: Read the problem carefully and be sure that you understand it. (You may need to read the problem more than once.). Assign a variable to what you are being asked to find. If necessary, write other quantities in terms of this variable.

STEP 2: Write an equation that relates the quantities described in the problem. You may need to sketch a diagram or refer to known formulas.

STEP 3: Solve the equation and determine the solution.

STEP 4: Look back and check your solution. Does it seem reasonable?

In the next example we apply these steps to a word problem involving unknown numbers.

EXAMPLE 1   Solving a number problem

The sum of three consecutive natural numbers is 81. Find the three numbers.

Solution   **STEP 1:** Start by assigning a variable n to an unknown quantity.

n: smallest of the three natural numbers

Next, write the other two natural numbers in terms of n.

$n + 1$: next consecutive natural number

$n + 2$: largest of the three consecutive natural numbers

STEP 2: Write an equation that relates these unknown quantities. The sum of the three consecutive natural numbers is 81, so the needed equation is

$$n + (n + 1) + (n + 2) = 81.$$

STEP 3: Solve the equation in STEP 2.

$$n + (n + 1) + (n + 2) = 81 \qquad \text{Equation to be solved.}$$
$$(n + n + n) + (1 + 2) = 81 \qquad \text{Commutative and associative properties}$$
$$3n + 3 = 81 \qquad \text{Combine like terms.}$$
$$3n = 78 \qquad \text{Subtract 3 from each side.}$$
$$n = 26 \qquad \text{Divide each side by 3.}$$

The smallest of the three numbers is 26, so the three numbers are 26, 27, and 28.

STEP 4: To check this solution we can add the three numbers to find out if their sum is 81.

$$26 + 27 + 28 = 81$$

The solution checks.

PERCENT PROBLEMS

Problems involving percentages often occur in everyday life. Wage increases, sales tax, and government data all make use of percentages. For example, it is estimated that 10% of the world's population will be older than 65 in 2025. (*Source:* World Health Organization.) As 1% (one percent) represents one part in one hundred, 10 of every 100 people in 2025 will be older than 65. Percent notation can be changed to either fraction or decimal notation.

PERCENT NOTATION

The expression $x\%$ represents the fraction $\frac{x}{100}$ or the decimal number $x \times 0.01$.

Note: To write $x\%$ as a decimal number, move the decimal point two places to the *left*.

EXAMPLE 2 Converting percent notation

Convert each percentage to fraction and decimal notation.
(a) 23% (b) 5.2% (c) 0.3%

Solution
(a) *Fraction Notation*: To convert to fraction notation divide 23 by 100. Thus $23\% = \frac{23}{100}$.
Decimal Notation: Move the decimal point two places to the left: $23\% = 0.23$.
(b) *Fraction Notation*: $5.2\% = \frac{5.2}{100} = \frac{52}{1000} = \frac{13 \cdot 4}{250 \cdot 4} = \frac{13}{250}$
Decimal Notation: Move the decimal point two places to the left: $5.2\% = 0.052$.
(c) *Fraction Notation*: $0.3\% = \frac{0.3}{100} = \frac{3}{1000}$
Decimal Notation: Move the decimal point two places to the left: $0.3\% = 0.003$.

EXAMPLE 3 Converting to percent notation

Convert each real number to a percentage.
(a) 0.234 (b) $\frac{1}{4}$ (c) 2.7

Solution
(a) Move the decimal point two places to the *right* to obtain $0.234 = 23.4\%$.
(b) $\frac{1}{4} = 0.25$, so $\frac{1}{4} = 25\%$.
(c) Move the decimal point two places to the right to obtain $2.7 = 270\%$. Note that percentages can be greater than 100%.

PERCENT CHANGE Suppose that the cost of tuition increases from T_1 dollars to T_2 dollars. Then the **percent change** in tuition is given by

$$\frac{T_2 - T_1}{T_1} \times 100.$$

The reason for multiplying by 100 is to change the decimal representation to a percentage.

EXAMPLE 4 Calculating percent increase

From 1980 to 2000 consumer credit increased from $350 billion to $1530 billion. Calculate the percent increase in consumer credit from 1980 to 2000.   (*Source:* Federal Reserve Bulletin.)

Solution Let $T_1 = 350$ and $T_2 = 1530$. The percent increase in consumer credit is

$$\frac{T_2 - T_1}{T_1} \times 100 = \frac{1530 - 350}{350} \times 100 \approx 337\%.$$

As we previously indicated, percentages frequently occur in applications.

EXAMPLE 5 Solving a percent problem

During the 1998–1999 academic year, tuition and fees at public colleges and universities was $3283, on average, and increased by 2.2% during the next year. Find the average cost of tuition and fees during the 1999–2000 academic year.

Solution **STEP 1:** Let T represent the tuition and fees during 1999–2000.

STEP 2: T equals $3283 plus 2.2% of $3283 and 2.2% = 0.022, so

$$T = 3283 + (0.022)3283.$$

Critical Thinking

If your salary increased 200%, by what factor did it increase?

Note: The word "of" in percent problems often indicates multiplication.

STEP 3: To find T, evaluate the expression on the right side to get

$$T = 3283 + (0.022)3283 \approx 3355.$$

Tuition and fees were $3355.

STEP 4: To check the answer, calculate the percent increase. Tuition and fees went up by

$$\frac{3355 - 3283}{3283} \times 100 \approx 2.2\%,$$

so the answer checks.

EXAMPLE 6 Solving a percent problem

In 2025, 10% of the world's population, or 800 million people, will be older than 65. Find the estimated population of the world in 2025.

Solution **STEP 1:** Let P represent the world's population in 2025.

STEP 2: 10% of P equals 800 million. As 10% = 0.10, this information is described by

$$0.10P = 800.$$

STEP 3: To solve the equation in STEP 2, multiply each side by 10.

$$0.10P = 800 \qquad \text{Equation to be solved}$$

$$P = 8000 \qquad \text{Multiply by 10.}$$

In 2025 the world population could reach 8000 million, or 8 billion.

STEP 4: To check the answer, determine whether 10% of 8 billion is 800 million.

$$(0.10)8{,}000{,}000{,}000 = 800{,}000{,}000$$

The answer checks.

DISTANCE PROBLEMS

If a person drives on an interstate highway at 70 miles per hour for 3 hours, then the total distance traveled is $70 \cdot 3 = 210$ miles. In general, $d = rt$, where d is the distance traveled, r is the rate (or speed), and t is time.

EXAMPLE 7   Solving a distance problem

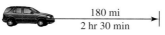
180 mi
2 hr 30 min

Figure 2.2

A person drives for 2 hours and 30 minutes at a constant speed and travels 180 miles. See Figure 2.2. Find the speed of the car in miles per hour.

Solution   **STEP 1:** Let r represent the car's rate, or speed, in miles per hour.

STEP 2: The rate is to be given in miles per hour, so change 2 hours and 30 minutes to 2.5 or $\frac{5}{2}$ hours. Because $d = 180$ and $r = \frac{5}{2}$, the equation $d = rt$ becomes

$$180 = r \cdot \frac{5}{2}.$$

STEP 3: Solve the equation in STEP 2 for r by multiplying each side of the equation by $\frac{2}{5}$, which is the reciprocal of $\frac{5}{2}$.

$$180 = \frac{5}{2} \cdot r \qquad \text{Equation to be solved}$$

$$\frac{2}{5} \cdot 180 = \frac{2}{5} \cdot \frac{5}{2} \cdot r \qquad \text{Multiply each side by } \frac{2}{5}.$$

$$72 = r \qquad \text{Simplify.}$$

The speed of the car is 72 miles per hour.

STEP 4: Traveling at 72 miles per hour for $\frac{5}{2}$ hours results in a distance of

$$d = rt = 72 \cdot \frac{5}{2} = 180 \text{ miles.}$$

The answer checks.

EXAMPLE 8   Solving a distance problem

An athlete jogs at two speeds, covering a distance of 7 miles in $\frac{3}{4}$ hour. If the athlete ran $\frac{1}{4}$ hour at 8 miles per hour, find the second speed.

Solution   **STEP 1:** Let r represent the second speed of the jogger.

STEP 2: The total time spent jogging is $\frac{3}{4}$ hour, so the time spent jogging at the second speed must be $\frac{3}{4} - \frac{1}{4} = \frac{1}{2}$ hour. The total distance of 7 miles is the result of jogging at 8 miles per hour for $\frac{1}{4}$ hour and at r miles per hour for $\frac{1}{2}$ hour. Thus

$$8 \cdot \frac{1}{4} + r \cdot \frac{1}{2} = 7.$$

STEP 3: Solve the equation in STEP 2 for r.

$$8 \cdot \frac{1}{4} + r \cdot \frac{1}{2} = 7 \qquad \text{Equation to be solved}$$

$$2 + \frac{r}{2} = 7 \qquad \text{Simplify the equation.}$$

$$\frac{r}{2} = 5 \qquad \text{Subtract 2 from each side.}$$

$$r = 10 \qquad \text{Multiply each side by 2.}$$

The athlete's second speed is 10 miles per hour.

STEP 4: Substituting $r = 10$ in the equation yields

$$8 \cdot \frac{1}{4} + 10 \cdot \frac{1}{2} = 2 + 5 = 7,$$

So the answer checks.

OTHER TYPES OF PROBLEMS

Many applied problems involve linear equations. For example, right after people have their wisdom teeth pulled, they may need to rinse their mouth with salt water. In the next example, we use linear equations to determine how much water must be added to dilute a concentrated saline solution.

EXAMPLE 9 Diluting a saline solution

A solution contains 4% salt. How much pure water should be added to 30 ounces of the solution to dilute it to a 1.5% solution?

Solution **STEP 1:** Assign a variable x as follows.

x: ounces of pure water (0% salt solution)

30: ounces of 4% salt solution

$x + 30$: ounces of 1.5% salt solution

In Figure 2.3, three beakers illustrate this situation.

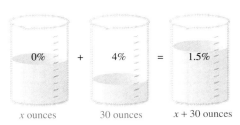

Figure 2.3 Mixing a Saline Solution

STEP 2: Note that the amount of salt in the first two beakers must equal the amount of salt in the third beaker. We use Table 2.6 to organize our calculations. The amount of salt in a solution equals the concentration times the solution amount.

TABLE 2.6 Mixing a Saline Solution

Solution Type	Concentration (as a decimal)	Solution Amount (ounces)	Salt (ounces)
Pure Water	0.00 (0%)	x	$0.00x$
Initial Solution	0.04 (4%)	30	0.04(30)
Final Solution	0.015 (1.5%)	$x + 30$	$0.015(x + 30)$

The amount of salt in the pure water and the 4% solution is

$$0.00x + 0.04(30) = 0 + 1.2 = 1.2 \text{ ounces.}$$

The amount of salt in the final solution, after x ounces of water are added, is

$$0.015(x + 30) \text{ ounces.}$$

Because the amounts of salt in the solutions before and after mixing must be equal, the following equation must hold.

$$0.015(x + 30) = 1.2$$

STEP 3: Solve the equation in STEP 2.

$$0.015(x + 30) = 1.2 \qquad \text{Equation to be solved}$$
$$(x + 30) = \frac{1.2}{0.015} \qquad \text{Divide both sides by 0.015.}$$
$$x + 30 = 80 \qquad \text{Simplify.}$$
$$x = 50 \qquad \text{Subtract 30.}$$

Fifty ounces of water should be added.

STEP 4: Adding 50 ounces of water will yield $50 + 30 = 80$ ounces of water containing $0.04(30) = 1.2$ ounces of salt. The concentration is $\frac{1.2}{80} = 0.015$ or 1.5%, so the answer checks.

Many times interest rates for student loans vary. In the next example, we present a situation in which a student has to borrow money at two different interest rates.

EXAMPLE 10 Calculating interest on college loans

A student takes out a loan for a limited amount of money at 5% interest and then must pay 7% for any additional money. If the student borrows $2000 more at 7% than at 5%, then the total interest for one year is $440. How much did the student borrow at each rate?

Solution **STEP 1:** Assign a variable x as follows.

$$x: \text{loan amount at 5\% interest}$$
$$x + 2000: \text{loan amount at 7\%.}$$

STEP 2: The amount of interest paid for the 5% loan is 5% of x, or $0.05x$. The amount of interest paid for the 7% loan is 7% of $x + 2000$, or $0.07(x + 2000)$. The total interest equals $440, so we solve the equation

$$0.05x + 0.07(x + 2000) = 440.$$

STEP 3: Solve the equation in STEP 2 for x.

$0.05x + 0.07(x + 2000) = 440$	Equation to be solved
$5x + 7(x + 2000) = 44{,}000$	Multiply by 100 to clear decimals.
$5x + 7x + 14{,}000 = 44{,}000$	Distributive property
$12x + 14{,}000 = 44{,}000$	Combine like terms.
$12x = 30{,}000$	Subtract 14,000.
$x = 2500$	Divide each side by 12.

The student borrowed $2500 at 5% and $4500 at 7%.

STEP 4: To check this solution add 5% of $2500 and 7% of $4500.

$$0.05(2500) + 0.07(4500) = 440$$

The interest is $440, so the answer checks.

2.3 PUTTING IT ALL TOGETHER

In this section we presented a four-step approach to problem solving. However, because no approach works in every situation, solving mathematical problems takes time, effort, and creativity.

Concept	Comments	Examples
Percent Notation	x% represents either the fraction $\frac{x}{100}$ or the decimal $x \times 0.01$.	$17\% = \frac{17}{100} = 0.17$ $1.5\% = \frac{1.5}{100} = \frac{15}{1000} = 0.015$ $234\% = \frac{234}{100} = 2.34$
Converting Fractions to Percent Notation	Divide the numerator by the denominator and multiply by 100.	$\frac{1}{2} = 0.5 = 50\%$ $\frac{2}{3} = 0.\overline{6} \approx 66.7\%$ $\frac{8}{5} = 1.6 = 160\%$
Percent Change	If the quantity changes from T_1 to T_2, then the percent change is $$\frac{T_2 - T_1}{T_1} \times 100.$$	If the price of gas changes from $1.50 to $1.35, the percent change is $$\frac{1.35 - 1.50}{1.50} \times 100 = -10\%.$$ The price decreases by 10%.
Distance Problems	Distance d equals rate r times time t, or $$d = r \cdot t.$$	If a car travels 60 mph for 3 hours, then the distance d is $$d = r \cdot t = 60 \cdot 3 = 180 \text{ miles.}$$

2.3 EXERCISES

FOR EXTRA HELP

CONCEPTS

1. When you are solving a word problem, what is the last step?

2. Given an integer n, what are the next two consecutive integers?

3. The expression $x\%$ equals the fraction _____ .

4. The expression $x\%$ equals the decimal $x \times$ _____ .

5. The number 0.5 equals _____ %.

6. The number $\frac{3}{4}$ equals _____ %.

7. If a price changes from P_1 to P_2, then the percent change equals _____ .

8. If a car travels at speed r for time t, then distance is given by $d =$ _____ .

NUMBER PROBLEMS

Exercises 9–16: Translate the sentence into an equation using the variable x. Then solve the resulting equation.

9. The sum of 2 and a number is 12.

10. Twice a number plus 7 equals 9.

11. A number divided by 5 equals the number decreased by 24.

12. 25 times a number is 125.

13. If a number is increased by 5 and then divided by 2, the result is 7.

14. A number subtracted from 8 is 5.

15. The quotient of a number and 2 is 17.

16. The product of 5 and a number equals 95.

Exercises 17–26: Find the number or numbers.

17. The sum of three consecutive natural numbers is 96.

18. The sum of three consecutive integers is -123.

19. Three times a number equals 102.

20. A number plus 18 equals twice the number.

21. Five times a number is 24 more than twice the number.

22. Three times a number is 18 less than the number.

23. Six times a number divided by 7 equals 18.

24. Two less than twice a number divided by 5 equals 4.

25. Four times the sum of a number and 5 equals 64.

26. The opposite of the sum of a number and -5 equals 24.

PERCENT PROBLEMS

Exercises 27–34: Convert the percentage to fraction and decimal notation.

27. 37%

28. 52%

29. 148%

30. 252%

31. 6.9%

32. 8.1%

33. 0.05%

34. 0.12%

Exercises 35–44: Convert the number to a percentage.

35. 0.45

36. 0.08

37. 1.8

38. 2.97

39. $\frac{2}{5}$

40. $\frac{1}{3}$

41. $\frac{3}{4}$

42. $\frac{7}{20}$

43. $\frac{5}{6}$

44. $\frac{53}{50}$

45. *Voter Turnout* In the 1980 election for president there were 86.5 million voters, whereas in 2000 there were 105.4 million voters. Find the percent change in the number of voters.

46. *College Degrees* In 1960, about 392,000 people received a bachelor's degree, and by 1998 this number had increased to 1,184,000. Find the percent change

in the number of bachelor's degrees received over this time period. (**Source:** The College Board.)

47. *Wages*   A part-time instructor is receiving $950 per credit taught. If the instructor receives a 4% increase, how much will the new per credit compensation be?

48. *Tuition Increase*   Tuition is currently $125 per credit. There are plans to raise tuition by 8% for next year. What will the new tuition be per credit?

49. *Income Tax Returns*   In 1999, the Internal Revenue Service processed 125 million individual income tax returns, up 1.62% from the preceding year. How many individual income tax returns were processed in 1998?   (**Source:** Internal Revenue Service.)

50. *Medicare Enrollment*   In 2000, 39.3 million people were enrolled in Medicare, an increase of 6.2% from 1995. How many people were enrolled in Medicare in 1995?   (**Source:** Health Care Financing Administration.)

51. *AIDS Deaths*   There were 17,047 AIDS deaths in 1998, or 34.5% of the 1995 AIDS deaths. Determine the number of AIDS deaths in 1995.   (**Source:** Centers for Disease Control and Prevention.)

52. *Gasoline Prices*   Calculate the percent change if the price of a gallon of gasoline increases from $1.20 to $1.50. Now calculate the percent change if the price of a gallon of gasoline decreases from $1.50 to $1.20.

53. *Meeting a Future Mate*   About 32% of adult Americans believe that people will meet their mates on the Internet. In a recent survey, 480 people stated this belief. Determine the number of people participating in the survey.   (**Source:** Men's Health.)

54. *Union Membership*   In 1998, 6.9 million, or 37.5%, of all government workers were unionized. How many government workers were there in 1998? (**Source:** Department of Labor.)

DISTANCE PROBLEMS

Exercises 55–60: Use the formula $d = rt$ to find the value of the missing variable.

55. $r = 4$ mph, $t = 2$ hours

56. $r = 70$ mph, $t = 2.5$ hours

57. $d = 1000$ feet, $t = 50$ seconds

58. $d = 1250$ miles, $t = 5$ days

59. $d = 200$ miles, $r = 40$ mph

60. $d = 1700$ feet, $r = 10$ feet per second

61. *Driving a Car*   A person drives a car at a constant speed for 4 hours and 15 minutes, traveling 255 miles. Find the speed of the car in miles per hour.

62. *Flying an Airplane*   A pilot flies a plane at a constant speed for 5 hours and 30 minutes, traveling 715 miles. Find the speed of the plane in miles per hour.

63. *Jogging Speeds*   One runner passes another runner on a hiking trail. The faster runner is jogging 2 miles per hour faster than the slower runner. Determine how long it will be before the faster runner is $\frac{3}{4}$ mile ahead of the slower runner.

64. *Distance Running*   An athlete runs 8 miles, first at a slower speed and then at a faster speed. The total time spent running is 1 hour. If the athlete runs $\frac{1}{3}$ hour at 6 miles per hour, find the second speed.

65. *Jogging Speeds*   At first an athlete jogs at 6 miles per hour and then jogs at 5 miles per hour, traveling 7 miles in 1.3 hours. How long did the athlete jog at each speed?

66. *Distance and Time*   A train is 100 miles west of St. Louis, Missouri, traveling east at 60 miles per hour. How long will it take for the train to be 410 miles east of St. Louis?

67. *Distance and Time*   A plane is 300 miles west of Chicago, Illinois, and is flying west at 500 miles per hour. How long will it take for the plane to be 2175 miles west of Chicago?

68. *Finding Speeds*   Two cars pass on a straight highway while traveling in opposite directions. One car is traveling 6 miles per hour faster than the other car. After 1.5 hours the two cars are 171 miles apart. Find the speed of each car.

OTHER TYPES OF PROBLEMS

69. *Saline Solution*   (Refer to Example 9.) A solution contains 3% salt. How much water should be added to 20 ounces of this solution to dilute it to a 1.2% solution?

70. *Acid Solution*   A solution contains 15% hydrochloric acid. How much water should be added to 50 milliliters of this solution to dilute it to a 2% solution?

71. *College Loans*   (Refer to Example 10.) A student takes out two loans, one at 5% interest and the other at 6% interest. The 5% loan is $1000 more than the 6% loan, and the total interest for 1 year is $215. How much is each loan?

72. *Bank Loans*   Two bank loans, one for $5000 and the other for $3000, cost a total of $550 in interest for one year. The $5000 loan has an interest rate 3% lower than the interest rate for the $3000 loan. Find the interest rate for each loan.

73. *Mixing Antifreeze*   How many gallons of 70% antifreeze should be mixed with 10 gallons of 30% antifreeze to obtain a 45% antifreeze mixture?

74. *Mixing Antifreeze*   How many gallons of 65% antifreeze and how many gallons of 20% antifreeze should be mixed to obtain 50 gallons of a 56% mixture of antifreeze?

WRITING ABOUT MATHEMATICS

75. State the four steps for solving a word problem.

76. The cost of living has increased about 600% during the past 50 years. Does this percent change correspond to a cost of living increase of 6 times? Explain.

Group Activity: Working with Real Data

Directions: Form a group of 2 to 4 people. Select someone to record the group's responses for this activity. All members of the group should work cooperatively to answer the questions. If your instructor asks for your results, each member of the group should be prepared to respond.

Exercises 1–5: In this set of exercises you are to use your mathematical problem-solving skills to find the thickness of a piece of aluminum foil without measuring it directly.

1. *Area of a Rectangle*   The area of a rectangle equals length times width. Find the area of a rectangle with length 12 centimeters and width 11 centimeters.

2. *Volume of a Box*   The volume of a box equals length times width times height. Find the volume of the box shown, which is 12 centimeters long, 11 centimeters wide, and 5 centimeters high.

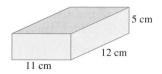

3. *Height of a Box*   Suppose that a box has a volume of 100 cubic centimeters and that the area of the bottom of the box is 50 square centimeters. Find the height of the box.

4. *Volume of Aluminum Foil*   One cubic centimeter of aluminum weighs 2.7 grams. If a piece of aluminum foil weighs 5.4 grams, find the volume of the aluminum foil.

5. *Thickness of Aluminum Foil*   A rectangular sheet of aluminum foil is 50 centimeters long and 20 centimeters wide, and weighs 5.4 grams. Find the thickness of the aluminum foil in centimeters.

2.4 FORMULAS

Basic Concepts · Formulas from Geometry · Solving for a Variable · Other Formulas

INTRODUCTION

Have you ever wondered how the registrar's office calculates your grade point average (GPA)? A formula is used that involves the number of credits earned at each possible grade. Once a formula has been derived, it can be used over and over by any number of people. This fact makes formulas a convenient and easy way to solve certain types of recurring problems. In this section we discuss several types of formulas.

BASIC CONCEPTS

A *formula* is an equation that can be used to calculate an unknown quantity by using known values of other quantities. Formulas also establish a relationship between different quantities. For example, to calculate the area A of a rectangular room with length L and width W, the formula

$$A = LW$$

can be used. If a dormitory room is 16 feet by 10 feet, then its area is

$$A = 16 \cdot 10 = 160 \text{ square feet.}$$

The formula $A = LW$ calculates area by using the quantities of length and width.

Another useful formula is $M = \frac{D}{G}$, which can be used to calculate a car's gas mileage M after it has traveled D miles on G gallons of gasoline.

EXAMPLE 1   Calculating mileage of a trip

A tourist starts a trip with a full tank of gas and an odometer that reads 45,682 miles. At the end of the trip, it takes 9.7 gallons of gas to fill the tank, and the odometer reads 45,903 miles. Find the gas mileage for the car.

Solution   The distance traveled is $D = 45{,}903 - 45{,}682 = \mathbf{221}$ miles and the number of gallons used is $G = \mathbf{9.7}$. Thus

$$M = \frac{D}{G} = \frac{221}{9.7} \approx 22.8 \text{ miles per gallon.} \qquad \overline{}$$

FORMULAS FROM GEOMETRY

Formulas from geometry are frequently used in various fields, including surveying and construction. In this subsection we discuss several important formulas from geometry.

TRIANGLES   If a triangle has base b and height h, then its area is $A = \frac{1}{2}bh$, as shown in Figure 2.4(a) on the next page. If a triangle has a base of $\mathbf{4}$ feet and a height of 18 inches, then its area is

$$A = \frac{1}{2}bh = \frac{1}{2}(\mathbf{4})(\mathbf{1.5}) = 3 \text{ square feet,}$$

as illustrated in Figure 2.4(b). Note that 18 inches needs to be converted to **1.5** feet so that all the units of length are the same. If the base had been converted to 48 inches, then the area would have been

$$A = \frac{1}{2}bh = \frac{1}{2}(48)(18) = 432 \text{ square } \textit{inches}.$$

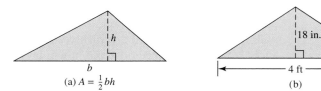

(a) $A = \frac{1}{2}bh$ (b)

Figure 2.4 Triangles

EXAMPLE 2 Calculating area of a region

A residential lot is shown in Figure 2.5. It comprises a rectangular region and an adjacent triangular region.
(a) Find the area of this lot.
(b) An acre contains 43,560 square feet. How many acres are there in this lot?

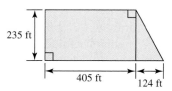

Figure 2.5

Solution **(a)** The rectangular portion of the lot has length **405** feet and width **235** feet. Its area A_R is

$$A_R = LW = 405 \cdot 235 = 95{,}175 \text{ square feet.}$$

The triangular region has base **124** feet and height **235** feet. Its area A_T is

$$A_T = \frac{1}{2}bh = \frac{1}{2} \cdot 124 \cdot 235 = 14{,}570 \text{ square feet.}$$

The total area A of the lot equals the sum of A_R and A_T.

$$A = 95{,}175 + 14{,}570 = 109{,}745 \text{ square feet}$$

(b) Each acre equals 43,560 square feet, so divide 109,745 by 43,560 to calculate the number of acres.

$$\frac{109{,}745}{43{,}560} \approx 2.5 \text{ acres}$$

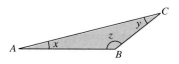

Figure 2.6

ANGLES Angles are often measured in degrees. A **degree** (°) is $\frac{1}{360}$ of a revolution, so there are 360° in one complete revolution. In any triangle, the sum of the measures of the angles equals 180°. In Figure 2.6, triangle ABC has angles with measures x, y, and z. Therefore

$$x + y + z = 180°.$$

EXAMPLE 3   Finding angles in a triangle

In a triangle the two smaller angles are equal in measure and are half the measure of the largest angle. Find the measure of each angle.

Solution   Let x represent the measure of each of the two smaller angles, as illustrated in Figure 2.7. Then the measure of the largest angle is $2x$, and the sum of the measures of the three angles is given by

$$x + x + 2x = 180°.$$

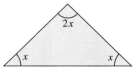

Figure 2.7

This equation can be solved as follows.

$x + x + 2x = 180°$	Equation to be solved
$4x = 180°$	Combine like terms.
$\dfrac{4x}{4} = \dfrac{180°}{4}$	Divide each side by 4.
$x = 45°$	Divide the real numbers.

The measure of the largest angle is $2x = 2 \cdot 45° = 90°$. Thus the measures of the three angles are $45°$, $45°$, and $90°$.

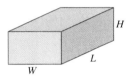

Figure 2.8
Rectangular Box

BOXES   The box in Figure 2.8 has length L, width W, and height H. Its volume V is given by

$$V = LWH.$$

The surface of the box comprises six rectangular regions: top and bottom, front and back, and left and right sides. The total surface area S of the box is given by

$$S = LW + LW + WH + WH + LH + LH.$$
(top + bottom + front + back + left side + right side)

When we combine like terms, this expression simplifies to

$$S = 2LW + 2WH + 2LH.$$

EXAMPLE 4   Finding the volume and surface area of a box

Find the volume and surface area of the box shown in Figure 2.9.

Solution   Figure 2.9 shows that the box has length $L = 10$ inches, width $W = 8$ inches, and height $H = 5$ inches. The volume of the box is

$$V = LWH = 10 \cdot 8 \cdot 5 = 400 \text{ cubic inches.}$$

The surface area of the box is

$$S = 2LW + 2WH + 2LH$$
$$= 2(10)(8) + 2(8)(5) + 2(10)(5)$$
$$= 160 + 80 + 100$$
$$= 340 \text{ square inches.}$$

Figure 2.9

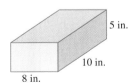

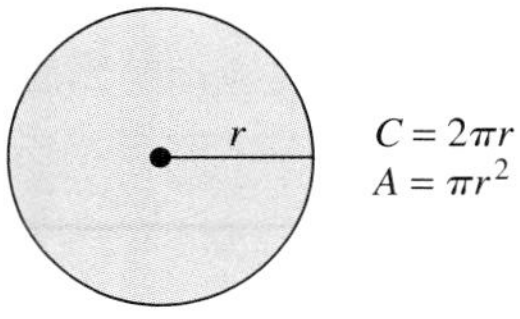

Figure 2.10 Circle

CIRCLES   The perimeter of a circle is called its **circumference** C and is given by $C = 2\pi r$, where r is the radius of the circle. The area A of a circle is given by $A = \pi r^2$. See Figure 2.10. (Recall that $\pi \approx 3.14$.)

EXAMPLE 5 Finding the circumference and area of a circle

A circle has a diameter of 25 inches. Approximate its circumference and area.

Solution The radius is half the diameter, or 12.5 inches.

$$\text{Circumference: } C = 2\pi r = 2\pi(12.5) = 25\pi \approx 78.5 \text{ inches.}$$

$$\text{Area: } A = \pi r^2 = \pi(12.5)^2 = 156.25\pi \approx 491 \text{ square inches.}$$

Figure 2.11
Cylinder

Critical Thinking

Write a formula for the circumference and area of a circle having diameter d.

CYLINDERS A soup can is usually made in the shape of a cylinder. The volume of a cylinder having radius r and height h is $V = \pi r^2 h$. See Figure 2.11.

EXAMPLE 6 Calculating the volume of a soda can

A cylindrical soda can has a radius of $1\frac{1}{4}$ inches and a height of $4\frac{3}{8}$ inches.
(a) Find the volume of the can.
(b) If 1 cubic inch equals 0.554 fluid ounces, find the number of fluid ounces in the can.

Solution (a) Changing mixed numbers to improper fractions gives the radius as $r = \frac{5}{4}$ inches and the height as $h = \frac{35}{8}$ inches.

$$V = \pi r^2 h \qquad \text{Volume of the soda can}$$

$$= \pi\left(\frac{5}{4}\right)^2\left(\frac{35}{8}\right) \qquad \text{Substitute.}$$

$$= \pi\left(\frac{875}{128}\right) \qquad \text{Multiply the fractions.}$$

$$\approx 21.48 \text{ cubic inches} \qquad \text{Approximate.}$$

(b) To calculate the number of fluid ounces in 21.48 cubic inches, multiply by 0.554.

$$21.48(0.554) \approx 11.9 \text{ fluid ounces}$$

Note that a typical aluminum soda can holds 12 fluid ounces.

SOLVING FOR A VARIABLE

A formula establishes a relationship between two or more variables (or quantities). Sometimes a formula is not solved for the needed variable. For example, if the area A and the width W of a rectangular region is given, then its length L can be found by solving the formula $A = LW$ for L.

$$A = LW \qquad \text{Area formula}$$

$$\frac{A}{W} = \frac{LW}{W} \qquad \text{Divide each side by } W.$$

$$\frac{A}{W} = L \qquad \text{Reduce the fraction.}$$

$$L = \frac{A}{W} \qquad \text{Rewrite the equation.}$$

If the area A of a rectangle is 400 square inches and its width W is 16 inches, then rectangle's length L is

$$L = \frac{A}{W} = \frac{400}{16} = 25 \text{ inches.}$$

There is a useful formula for calculating how far away a bolt of lightning is. It is based on the fact that sometimes we see a flash of lighting before we hear the thunder. This phenomenon happens because light travels much faster than sound. The farther away the lightning, the greater is the delay between our seeing the flash and hearing the thunder. In general, if the delay is x seconds, then the lightning is $D = \frac{x}{5}$ miles away. For instance, if there is a 10-second delay, then the lightning is $D = \frac{10}{5} = 2$ miles away.

EXAMPLE 7   Calculating the delay between lightning and thunder

Doppler radar shows an electrical storm 3.5 miles away. If you see lightning from this storm, how long will it be before you hear the thunder?

Solution   Solve the formula $D = \frac{x}{5}$ for x.

$$D = \frac{x}{5} \qquad \text{Given formula}$$

$$5D = x \qquad \text{Multiply each side by 5.}$$

If $D = 3.5$ miles, then the delay is

$$x = 5D = 5(3.5) = 17.5 \text{ seconds.}$$

In the next example we apply the area formula for a trapezoid.

EXAMPLE 8   Finding the base of a trapezoid

The area of the trapezoid shown in Figure 2.12 is given by

$$A = \frac{1}{2}(a + b)h,$$

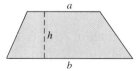

Figure 2.12   Trapezoid

where a and b are the bases of the trapezoid and h is the height.
(a) Solve the formula for b.
(b) A trapezoid has area $A = 36$ square inches, height $h = 4$ inches, and base $a = 8$ inches. Find b.

Solution   **(a)** To clear the equation of the fraction, multiply each side by 2.

$$A = \frac{1}{2}(a + b)h \qquad \text{Area formula}$$

$$2A = (a + b)h \qquad \text{Multiply each side by 2.}$$

$$\frac{2A}{h} = a + b \qquad \text{Divide each side by } h.$$

$$\frac{2A}{h} - a = b \qquad \text{Subtract } a \text{ from each side.}$$

$$b = \frac{2A}{h} - a \qquad \text{Rewrite the formula.}$$

(b) Let $A = 36$, $h = 4$, and $a = 8$ in $b = \frac{2A}{h} - a$. Then

$$b = \frac{2(36)}{4} - 8 = 18 - 8 = 10 \text{ inches.}$$

EXAMPLE 9   Solving for a variable

Solve each equation for the indicated variable.

(a) $c = \frac{a + b}{2}$ for b **(b)** $ab - bc = ac$ for c

Solution **(a)** To clear the equation of the fraction, multiply each side by 2.

$$c = \frac{a + b}{2} \qquad \text{Given formula}$$

$$2c = a + b \qquad \text{Multiply each side by 2.}$$

$$2c - a = b \qquad \text{Subtract } a.$$

The formula solved for b is $b = 2c - a$.

(b) Begin by moving the term on the left side containing c to the right side of the equation.

$$ab - bc = ac \qquad \text{Given formula}$$

$$ab - bc + bc = ac + bc \qquad \text{Add } bc \text{ to each side.}$$

$$ab = (a + b)c \qquad \text{Combine terms; distributive property}$$

$$\frac{ab}{a + b} = \frac{(a + b)c}{(a + b)} \qquad \text{Divide each side by } (a + b).$$

$$\frac{ab}{a + b} = c \qquad \text{Reduce the fraction.}$$

The formula solved for c is $c = \frac{ab}{a + b}$.

Critical Thinking

Are the formulas $c = \frac{1}{a - b}$ and $c = \frac{-1}{b - a}$ equivalent? Why?

OTHER FORMULAS

To calculate a student's GPA, the number of credits earned with a grade of A, B, C, D, and F must be known. If a, b, c, d, and f represent these credit counts respectively, then

$$\text{GPA} = \frac{4a + 3b + 2c + d}{a + b + c + d + f}.$$

This formula is based on the assumption that a 4.0 GPA is an A, a 3.0 GPA is a B, and so on.

EXAMPLE 10   Calculating a student's GPA

A student has earned 16 credits of A, 32 credits of B, 12 credits of C, 2 credits of D, and 5 credits of F. Calculate the student's GPA to the nearest hundredth.

Solution Let $a = 16$, $b = 32$, $c = 12$, $d = 2$, and $f = 5$. Then

$$\text{GPA} = \frac{4 \cdot 16 + 3 \cdot 32 + 2 \cdot 12 + 2}{16 + 32 + 12 + 2 + 5} = \frac{186}{67} \approx 2.78.$$

The student's GPA is 2.78.

EXAMPLE 11 Converting temperature scales

In the United States, temperature is measured with either the Fahrenheit or the Celsius temperature scales. To convert Fahrenheit degrees F to Celsius degrees C, the formula $C = \frac{5}{9}(F - 32)$ can be used.

(a) Solve the formula for F to find a formula that converts Celsius degrees to Fahrenheit degrees.

(b) If the outside temperature is $20°C$, find the equivalent Fahrenheit temperature.

Solution **(a)** The reciprocal of $\frac{5}{9}$ is $\frac{9}{5}$, so multiply each side by $\frac{9}{5}$.

$$C = \frac{5}{9}(F - 32) \qquad \text{Given equation}$$

$$\frac{9}{5}C = \frac{9}{5} \cdot \frac{5}{9}(F - 32) \qquad \text{Multiply each side by } \frac{9}{5}.$$

$$\frac{9}{5}C = F - 32 \qquad \text{Multiplicative inverses}$$

$$\frac{9}{5}C + 32 = F \qquad \text{Add 32 to each side.}$$

The required formula is $F = \frac{9}{5}C + 32$.

(b) If $C = 20°C$, then $F = \frac{9}{5}(20) + 32 = 36 + 32 = 68°F$.

2.4 PUTTING IT ALL TOGETHER

In this section we discussed formulas and how to solve for a variable. The following table summarizes some of these formulas.

Concept	Formula	Examples
Area of a Rectangle	$A = LW$, where L is the length and W is the width.	If $L = 10$ feet and $W = 5$ feet, then the area is $A = 5 \cdot 10 = 50$ square feet.
Gas Mileage	$M = \frac{D}{G}$, where D is the distance and G is the gasoline used.	If a car travels 100 miles on 5 gallons of gasoline, then its mileage is $M = \frac{100}{5} = 20$ miles per gallon.

continued on next page

continued from previous page

Concept	Formula	Examples
Area of a Triangle	$A = \frac{1}{2}bh$, where b is the base and h is the height.	If $b = 5$ inches and $h = 6$ inches, then the area is $$A = \frac{1}{2}(5)(6) = 15 \text{ square inches.}$$
Angle Measure in a Triangle	$x + y + z = 180°$, where x, y, and z are the angle measures.	If $x = 40°$ and $y = 60°$, then $z = 80°$ because $$40° + 60° + 80° = 180°$$
Volume and Surface Area of a Box	If a box has length L, width W, and height H, then its volume is $$V = LWH$$ and its surface area is $$S = 2LW + 2WH + 2LH.$$	If a box has dimensions $L = 4$ feet, $W = 3$ feet, and $H = 2$ feet, then $$V = 4(3)(2) = 24 \text{ cubic feet}$$ and $$S = 2(4)(3) + 2(3)(2) + 2(4)(2)$$ $$= 52 \text{ square feet.}$$
Volume of a Cylinder	$V = \pi r^2 h$, where r is the radius and h is the height.	If $r = 5$ inches and $h = 20$ inches, then the volume is $$V = \pi(5^2)(20) = 500\pi \text{ square inches.}$$
Distance from Lightning	$D = \frac{x}{5}$, where x is the seconds between seeing the flash and hearing the thunder.	If $x = 15$ seconds, then the lightning is $D = \frac{15}{5} = 3$ miles away.

Concept	Formula	Examples
Area of a Trapezoid	$A = \frac{1}{2}(a + b)h$, where a and b are the bases and h is the height.	If $a = 4$, $b = 6$, and $h = 3$, then the area is $$A = \frac{1}{2}(4 + 6)(3) = 15 \text{ square units.}$$
Converting Between Fahrenheit and Celsius Degrees	$$C = \frac{5}{9}(F - 32)$$ $$F = \frac{9}{5}C + 32$$	212°F is equivalent to $$C = \frac{5}{9}(212 - 32) = 100°C.$$ 100°C is equivalent to $$F = \frac{9}{5}(100) + 32 = 212°F.$$
Calculating Grade Point Average (GPA)	GPA is calculated by $$\frac{4a + 3b + 2c + d}{a + b + c + d + f},$$ where a, b, c, d, and f represent the credits earned with grades of A, B, C, D, and F, respectively.	10 credits of A, 8 credits of B, 6 credits of C, 12 credits of D, and 8 credits of F results in a GPA of $$\frac{4(10) + 3(8) + 2(6) + 12}{10 + 8 + 6 + 12 + 8} = 2.0.$$

 **2.4 EXERCISES**

FOR EXTRA HELP

 Student's Solutions Manual
 InterAct Math
 MathXL

MyMathLab
Math Tutor Center
 Digital Video Tutor CD 1 Videotape 3

CONCEPTS

1. A(n) _____ can be used to calculate one quantity by using known values of other quantities.

2. The area A of a rectangle with length L and width W is $A = $ _____.

3. The area A of a triangle with base b and height h is $A = $ _____.

4. One degree equals _____ of a revolution.

5. There are _____ degrees in one revolution.

6. The sum of the measures of the angles in a triangle equals _____ degrees.

7. The circumference C of a circle with radius r is $C = $ _____.

8. The area A of a circle with radius r is $A = $ _____.

9. The volume V of a box with length L, width W, and height H is $V = $ _____.

10. The area A of a trapezoid with height h and bases a and b is $A =$ _____ .

FORMULAS FROM GEOMETRY

Exercises 11–18: Find the area of the region shown.

11.

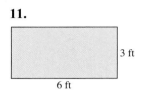

3 ft

6 ft

12.

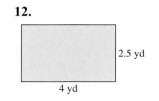

2.5 yd

4 yd

13.

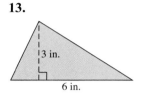

3 in.

6 in.

14.

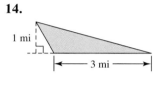

1 mi

3 mi

15.

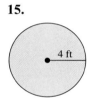

4 ft

16.

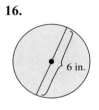

6 in.

17.

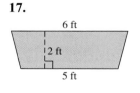

6 ft

2 ft

5 ft

18.

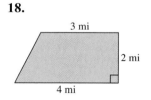

3 mi

2 mi

4 mi

19. Find the area of a rectangle having a 7-inch width and a 13-inch length.

20. Find the area of a rectangle having a 5-foot width and a 7-yard length.

21. Find the area of a triangle having a 12-inch base and a 6-inch height.

22. Find the area of a triangle having a 9-foot base and a 72-inch height.

23. Find the circumference of a circle having an 8-inch diameter.

24. Find the area of a circle having a 9-foot radius.

25. *Area of a Lot*   Find the area of the lot shown at the top of the next column, which consists of a square and a triangle.

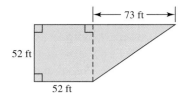

73 ft

52 ft

52 ft

26. *Area of a Lot*   Find the area of the lot shown, which consists of a rectangle and two triangles.

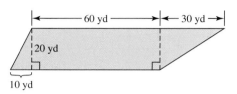

60 yd

30 yd

20 yd

10 yd

Exercises 27 and 28: Angle Measure   Find the measure of the third angle in the triangle.

27.

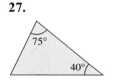

75°

40°

28.

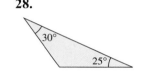

30°

25°

29. A triangle contains two angles having measures of 23° and 76°. Find the measure of the third angle.

30. The measures of the angles in an *equilateral triangle* are equal. Find their measure.

31. The measures of the angles in a triangle are x, $2x$, and $3x$. Find the value of x.

32. The measures of the angles in a triangle are $3x$, $4x$, and $11x$. Find the value of x.

33. In a triangle the two smaller angles are equal in measure and are each one third of the measure of the larger angle. Find the measure of each angle.

34. In a triangle the two larger angles differ by 10°. The smaller angle is 50° less than the largest angle. Find the measure of each angle.

35. The diameter of a circle is 12 inches. Find its circumference and area.

36. The radius of a circle is $\frac{5}{4}$ feet. Find its circumference and area.

37. The circumference of a circle is 2π inches. Find its radius and area.

38. The circumference of a circle is 13π feet. Find its radius and area.

Exercises 39–42: A box with a top has length L, width W, and height H. Find the volume and surface area of the box.

39. $L = 22$ inches, $W = 12$ inches, $H = 10$ inches

40. $L = 5$ feet, $W = 3$ feet, $H = 6$ feet

41. $L = \frac{2}{3}$ yard, $W = \frac{2}{3}$ foot, $H = \frac{3}{2}$ feet

42. $L = 1.2$ meters, $W = 0.8$ meter, $H = 0.6$ meter

Exercises 43–46: Use the formula $V = \pi r^2 h$ to find the volume of a cylindrical container for the given r and h. Leave your answer in terms of π.

43. $r = 2$ inches, $h = 5$ inches

44. $r = \frac{1}{2}$ inch, $h = \frac{3}{2}$ inches

45. $r = 5$ inches, $h = 2$ feet

46. $r = 2.5$ feet, $h = 1.5$ yards

47. *Volume of a Barrel* (Refer to Example 6.) A cylindrical barrel has a diameter of $1\frac{3}{4}$ feet and a height of 3 feet. Find the volume of the barrel.

48. *Volume of a Can* (Refer to Example 6.)
  (a) Find the volume of an aluminum can with a radius of $\frac{3}{4}$ inch and a height of $2\frac{1}{2}$ inches.
  (b) Find the number of fluid ounces in the can if one cubic inch equals 0.554 fluid ounces.

SOLVING FOR A VARIABLE

Exercises 49–62: Solve the formula for the given variable.

49. $A = LW$ for W

50. $A = \frac{1}{2}bh$ for b

51. $V = \pi r^2 h$ for h

52. $V = \frac{1}{3}\pi r^2 h$ for h

53. $A = \frac{1}{2}(a + b)h$ for a

54. $C = 2\pi r$ for r

55. $V = LWH$ for W

56. $P = 2x + 2y$ for y

57. $s = \dfrac{a + b + c}{2}$ for b

58. $t = \dfrac{x - y}{3}$ for x

59. $\dfrac{a}{b} - \dfrac{c}{b} = 1$ for b

60. $\dfrac{x}{y} + \dfrac{z}{y} = 5$ for z

61. $ab = cd + ad$ for a

62. $S = 2LW + 2LH + 2WH$ for W

63. *Perimeter of a Rectangle* The perimeter of a rectangle equals the sum of the lengths of its four sides. If the width of a rectangle is 5 inches and its perimeter is 40 inches, find the length of the rectangle.

64. *Perimeter of a Triangle* Two sides of a triangle have lengths of 5 feet and 7 feet. If the perimeter of the triangle is 21 feet, what is the length of the third side?

OTHER FORMULAS AND APPLICATIONS

Exercises 65–68: (Refer to Example 10.) Let a represent the number of credits with a grade of A, b the number of credits with a grade of B, and so on. Calculate the corresponding grade point average (GPA). Round your answer to the nearest hundredth.

65. $a = 30, b = 45, c = 12, d = 4, f = 4$

66. $a = 70, b = 35, c = 5, d = 0, f = 0$

67. $a = 0, b = 60, c = 80, d = 10, f = 6$

68. $a = 3, b = 5, c = 8, d = 0, f = 22$

Exercises 69–72: (Refer to Example 11.) Convert the Celsius temperature to an equivalent Fahrenheit temperature.

69. $25°C$

70. $100°C$

71. $-40°C$

72. $0°C$

Exercises 73–76: (Refer to Example 11.) Convert the Fahrenheit temperature to an equivalent Celsius temperature.

73. $23°F$

74. $98.6°F$

75. $-4°F$

76. $-31°F$

77. *Lightning* (Refer to Example 7.) The time delay between a flash of lightning and the sound of thunder is 12 seconds. How far away was the lightning?

78. *Lightning* (Refer to Example 7.) Doppler radar shows an electrical storm 2.5 miles away. If you see the lightning from this storm, how long will it be before you hear the thunder?

WRITING ABOUT MATHEMATICS

79. A student solves the formula $A = \frac{1}{2}bh$ for h and obtains the formula $h = \frac{1}{2}bA$. Explain the error that the student is making. What is the correct answer?

80. Give an example of a formula that you have used and explain how you used it.

---

1. Translate the sentence into an equation containing the variable x. Then solve the resulting equation.
   (a) The product of 3 and a number is 36.
   (b) A number subtracted from 35 is 43.

2. The sum of three consecutive integers is -93. Find the three integers.

3. Convert 9.5% to a decimal.

4. Convert $\frac{5}{4}$ to a percentage.

5. *Serious Crime* In 1999, New York City experienced 3850 serious crimes per 100,000 people. This figure represented an 8% decrease from the number in 1998. Find the rate for serious crimes per 100,000 people in 1998.

6. *Driving a Car* How many hours does it take the driver of a car to travel 390 miles at 60 miles per hour?

7. *College Loans* A student takes out two loans, one at 6% and the other at 7%. The 6% loan is $2000 more than the 7% loan, and the total interest for one year is $510. Find the amount of each loan.

8. *Height of a Triangle* The area of a triangle having a base of 6 inches is 36 square inches. Find the height of the triangle.

9. *Area and Circumference* Find the area and circumference of the circle shown.

10. *Angles* Find the value of x in the triangle shown.

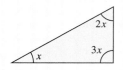

11. *Solving a Formula* Solve $A = \pi r^2 + \pi r l$ for l.

2.5 LINEAR INEQUALITIES

Solutions and Number Line Graphs · **The Addition Property of Inequalities** · **The Multiplication Property of Inequalities** · **Applications**

INTRODUCTION

On a freeway, the speed limit might be 70 miles per hour. A driver traveling x miles per hour is obeying the speed limit if $x \leq 70$ and breaking the speed limit if $x > 70$. A speed of $x = 70$ represents the boundary between obeying the speed limit and breaking it. A posted speed limit, or *boundary*, allows drivers to easily determine whether they are speeding.

Solving linear inequalities is closely related to solving linear equations because equality is the boundary between *greater than* and *less than*. In this section we discuss techniques used to solve linear inequalities.

SOLUTIONS AND NUMBER LINE GRAPHS

A **linear inequality** results whenever the equals sign in a linear equation is replaced with any one of the symbols $<$, $\leq$, $>$, or $\geq$. Examples of linear equations include

$$x = 5, \quad 2x + 1 = 0, \quad 1 - x = 6, \quad \text{and} \quad 5x + 1 = 3 - 2x.$$

Therefore examples of linear inequalities include

$$x > 5, \quad 2x + 1 < 0, \quad 1 - x \geq 6, \quad \text{and} \quad 5x + 1 \leq 3 - 2x.$$

A **solution** to an inequality is a value of the variable that makes the statement true. The set of all solutions is called the **solution set.** Two inequalities are *equivalent* if they have the same solution set. Inequalities frequently have infinitely many solutions. For example, the solution set for the inequality $x > 5$ includes all real numbers greater than 5.

A number line can be used to graph the solution set for an inequality. The graph of all real numbers satisfying $x < 2$ is shown in Figure 2.13(a), and the graph of all real numbers satisfying $x \leq 2$ is shown in Figure 2.13(b). (The symbol $\leq$ is read "less than or equal to." Similarly, the symbol $\geq$ is read "greater than or equal to.") A parenthesis ")" is used to show that $x = 2$ is not included in Figure 2.13(a), and a bracket "]" is used to show that $x = 2$ is included in Figure 2.13(b).

Figure 2.13

EXAMPLE 1   Graphing inequalities on a number line

Use a number line to graph the solution set to each inequality.
(a) $x > 0$ **(b)** $x \geq 0$ **(c)** $x \leq -1$ **(d)** $x < 3$

Solution **(a)** First locate $x = 0$ (or the origin) on a number line. Numbers greater than 0 are located to the right of the origin, so shade the number line to the right of the origin. Because $x > 0$, the number 0 is not included, so place a parenthesis "(" at 0, as shown in Figure 2.14(a).
(b) Figure 2.14(b) is similar to the graph in part (a) except that a bracket "[" is placed at the origin because $x = 0$ is included in the solution set.
(c) First locate $x = -1$ on the number line. Numbers less than -1 are located to the left of -1. Because -1 is included, a bracket "]" is placed at -1, as shown in Figure 2.14(c).
(d) Real numbers less than 3 are graphed in Figure 2.14(d).

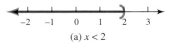
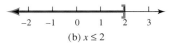

Figure 2.14

We can check possible solutions to an inequality in the same way that we checked possible solutions to an equation. For example, to check whether 5 is a solution to $2x + 3 = 13$, we substitute $x = 5$ in the equation.

$$2(5) + 3 \stackrel{?}{=} 13 \qquad \text{Let } x = 5.$$
$$13 = 13 \qquad \text{A true statement}$$

Thus 5 is a solution to this equation. Similarly, to check whether 7 is a solution to $2x + 3 > 13$, we substitute $x = 7$ in the inequality.

$$2(7) + 3 \overset{?}{>} 13 \qquad \text{Let } x = 7.$$

$$17 > 13 \qquad \text{A true statement}$$

Thus 7 is a solution to the inequality.

EXAMPLE 2   Checking possible solutions

Check to find out whether each given value of x is a solution to each inequality.
(a) $3x - 4 < 10, \quad x = 6$ **(b)** $4 - 2x \leq 8, \quad x = -2$

Solution   **(a)** Substitute 6 for x and simplify.

$$3(6) - 4 \overset{?}{<} 10 \qquad \text{Let } x = 6.$$

$$14 \overset{?}{<} 10 \qquad \text{A false statement}$$

Thus 6 is *not* a solution to the inequality.
(b) Let $x = -2$ and simplify.

$$4 - 2(-2) \overset{?}{\leq} 8 \qquad \text{Let } x = -2.$$

$$8 \leq 8 \qquad \text{A true statement}$$

Thus -2 is a solution to the inequality.

EXAMPLE 3   Finding solutions to equations and inequalities

In Table 2.7 the expression $2x - 6$ has been evaluated for each x. Use the table to determine any solutions to each equation or inequality.
(a) $2x - 6 = 0$ **(b)** $2x - 6 > 0$ **(c)** $2x - 6 \geq 0$ **(d)** $2x - 6 < 0$

TABLE 2.7

x	0	1	2	3	4	5	6
$2x - 6$	-6	-4	-2	0	2	4	6

Solution   **(a)** From Table 2.7, $2x - 6$ equals 0 when $x = 3$.
(b) From Table 2.7, $2x - 6$ is greater than 0 when $x > 3$. Solutions to this inequality satisfy $x > 3$.
(c) The expression $2x - 6$ is greater than or equal to 0 when $x \geq 3$.
(d) The expression $2x - 6$ is less than 0 when $x < 3$.

THE ADDITION PROPERTY OF INEQUALITIES

Suppose that the speed limit on a country road is 55 miles per hour, which is 25 miles per hour faster than the speed limit in town. If x represents lawful speeds in town, then x satisfies the inequality

$$x + 25 \leq 55.$$

To solve this inequality we add -25 to each side of the inequality.

$$x + 25 + (-25) \leq 55 + (-25) \qquad \text{Add } -25 \text{ to each side.}$$
$$x \leq 30 \qquad \text{Add the real numbers.}$$

Thus drivers are obeying the speed limit in town when they travel at 30 miles per hour or less. To solve this inequality the addition property of inequalities was used.

ADDITION PROPERTY OF INEQUALITIES

Let a, b, and c be expressions that represent real numbers. The inequalities

$$a < b \quad \text{and} \quad a + c < b + c$$

are equivalent. That is, the same number may be added to (or subtracted from) each side of an inequality. Similar properties exist for $>$, $\leq$, and $\geq$.

To solve some inequalities, we apply the addition property of inequities to obtain a simpler, equivalent inequality.

EXAMPLE 4   **Applying the addition property of inequalities**

Solve each inequality.
(a) $x - 1 > 4$ **(b)** $3 + 2x \leq 5 + x$

Solution   **(a)** Begin by adding 1 to each side of the inequality.

$$x - 1 > 4 \qquad \text{Given inequality}$$
$$x - 1 + 1 > 4 + 1 \qquad \text{Add 1 to each side.}$$
$$x > 5 \qquad \text{Simplify.}$$

The solution set is given by $x > 5$.

(b) Begin by subtracting 3 from (or adding -3 to) each side of the inequality.

$$3 + 2x \leq 5 + x \qquad \text{Given inequality}$$
$$3 + 2x - 3 \leq 5 + x - 3 \qquad \text{Subtract 3 from each side.}$$
$$2x \leq 2 + x \qquad \text{Subtract the real numbers.}$$
$$2x - x \leq 2 + x - x \qquad \text{Subtract } x \text{ from each side.}$$
$$x \leq 2 \qquad \text{Combine like terms.}$$

The solution set is given by $x \leq 2$.

Solution sets can be written in **set-builder notation**. For example, the solution set consisting of "all real numbers x such that x is less than or equal to 2" can be written as $\{x \mid x \leq 2\}$. The vertical line segment "$\mid$" is read "such that."

═══════════════ MAKING CONNECTIONS ═══════════════

The addition property of inequalities can be illustrated by an old-fashioned pan balance. In the figure on the left, the weight on the right pan is heavier than the weight on the left because the right pan rests lower than the left pan. If we add the same amount of weight to both pans, the right pan still weighs more than the left pan by the same amount, as illustrated in the figure on the right.

EXAMPLE 5   Applying the addition property of inequalities

Solve $5 + \frac{1}{2}x \le 3 - \frac{1}{2}x$. Write the solution set in set-builder notation.

Solution   Begin by subtracting 5 from each side of the inequality.

$$5 + \frac{1}{2}x \le 3 - \frac{1}{2}x \qquad \text{Given inequality}$$

$$\frac{1}{2}x \le -\frac{1}{2}x - 2 \qquad \text{Subtract 5 from each side.}$$

$$\frac{1}{2}x + \frac{1}{2}x \le -\frac{1}{2}x + \frac{1}{2}x - 2 \qquad \text{Add } \frac{1}{2}x \text{ to each side.}$$

$$x \le -2 \qquad \text{Combine like terms.}$$

Written in set-builder notation, the solution set is $\{x \mid x \le -2\}$.   ─────

THE MULTIPLICATION PROPERTY OF INEQUALITIES

Although the addition property of inequalities can be used to solve the inequality $3 + x \le 15$, it cannot be used to solve $3x \le 15$. The multiplication property of inequalities is needed to solve this second inequality.

▌▐▌▐▌▐   **MULTIPLICATION PROPERTY OF INEQUALITIES**

Let a, b, and c be expressions that represent real numbers.

1. If $c > 0$, then the inequalities $a < b$ and $ac < bc$ are equivalent. That is, each side of an inequality may be multiplied (or divided) by the same positive number.
2. If $c < 0$, then the inequalities $a < b$ and $ac > bc$ are equivalent. That is, each side of an inequality may be multiplied (or divided) by the same negative number, provided the inequality symbol is reversed.

Note that similar properties exist for $\le$ and $\ge$.

Note: It is important to reverse the inequality symbol when either multiplying or dividing by a negative number. For instance, consider multiplying each side of the inequality $-3 < 2$ by -4.

$$-3 < 2$$
$$-3 \cdot (-4) > 2 \cdot (-4)$$
$$12 > -8$$

If the inequality symbol had not been reversed, we would have obtained $12 < -8$, which is a false statement.

EXAMPLE 6   Applying the multiplication property of inequalities

Solve each inequality. Write the solution set in set-builder notation.
(a) $3x < 18$ **(b)** $-7 \le -\frac{1}{2}x$

Solution   **(a)** To solve for x, divide each side by 3.

$$3x < 18 \qquad \text{Given inequality}$$
$$\frac{3x}{3} < \frac{18}{3} \qquad \text{Divide each side by 3.}$$
$$x < 6 \qquad \text{Reduce fractions.}$$

The solution set is $\{x \mid x < 6\}$.
(b) To isolate x, multiply each side by -2. Remember to reverse the inequality symbol.

$$-7 \le -\frac{1}{2}x \qquad \text{Given inequality}$$
$$-2(-7) \ge -2\left(-\frac{1}{2}\right)x \qquad \text{Multiply by } -2; \text{ reverse the inequality.}$$
$$14 \ge 1 \cdot x \qquad \text{Multiply the real numbers.}$$
$$x \le 14 \qquad \text{Rewrite the inequality.}$$

The solution set is $\{x \mid x \le 14\}$.

EXAMPLE 7   Applying both properties of inequalities

Solve each inequality. Write the solution set in set-builder notation.
(a) $4x - 7 \ge -6$ **(b)** $-8 + 3x \le 5x + 3$ **(c)** $0.4(2x - 5) < 1.1x + 2$

Solution   **(a)** Start by adding 7 to each side.

$$4x - 7 \ge -6 \qquad \text{Given inequality}$$
$$4x \ge 1 \qquad \text{Add 7 to each side.}$$
$$x \ge \frac{1}{4} \qquad \text{Divide each side by 4.}$$

The solution set is $\left\{x \mid x \ge \frac{1}{4}\right\}$.

(b) Begin by adding 8 to each side.

$$-8 + 3x \leq 5x + 3 \qquad \text{Given inequality}$$
$$3x \leq 5x + 11 \qquad \text{Add 8 to each side.}$$
$$3x - 5x \leq 5x + 11 - 5x \qquad \text{Subtract } 5x \text{ from each side.}$$
$$-2x \leq 11 \qquad \text{Combine like terms.}$$
$$\frac{-2x}{-2} \geq \frac{11}{-2} \qquad \text{Divide by } -2; \text{ reverse the inequality.}$$
$$x \geq -\frac{11}{2} \qquad \text{Reduce fractions.}$$

The solution set is $\left\{ x \mid x \geq -\frac{11}{2} \right\}$.

(c) Begin by multiplying each side by 10 to eliminate decimals.

$$0.4(2x - 5) < 1.1x + 2 \qquad \text{Given inequality}$$
$$4(2x - 5) < 11x + 20 \qquad \text{Multiply each side by 10.}$$
$$8x - 20 < 11x + 20 \qquad \text{Distributive property}$$
$$8x < 11x + 40 \qquad \text{Add 20 to each side.}$$
$$-3x < 40 \qquad \text{Subtract } 11x \text{ from each side.}$$
$$x > -\frac{40}{3} \qquad \text{Divide by } -3; \text{ reverse the inequality.}$$

The solution set is $\left\{ x \mid x > -\frac{40}{3} \right\}$.

Critical Thinking

Solve $-5 - 3x > -2x + 7$ without having to reverse the inequality symbol.

APPLICATIONS

To work applications involving inequalities, we often have to translate words to mathematical statements. For example, the phrase "at least" can be translated to "greater than or equal to."

EXAMPLE 8 Translating words to inequalities

Translate each phrase to an inequality. Let the variable be x.
(a) A number that is more than 30
(b) An age that is at least 18
(c) A grade point average that is at most 3.25

Solution **(a)** The inequality $x > 30$ represents a number x that is more than 30.
(b) The inequality $x \geq 18$ represents an age x that is at least 18.
(c) The inequality $x \leq 3.25$ represents a grade point average x that is at most 3.25.

In the lower atmosphere the air temperature generally becomes colder as the altitude increases. One mile above Earth's surface the temperature is about $29°F$ colder than the ground-level temperature. As the air cools, there is an increased chance of clouds forming. In the next example we estimate the altitudes where clouds may form. (*Source:* A. Miller and R. Anthes, *Meteorology.*)

EXAMPLE 9   Finding the altitude of clouds

If the ground temperature is $90°F$, then the temperature T above Earth's surface is given by the formula $T = 90 - 29x$, where x is the altitude in miles. Suppose that clouds form only where the temperature is $3°F$ or colder. Determine the heights at which clouds may form.

Solution   Clouds may form at altitudes at which the temperature T is less than or equal to $3°F$. Thus we must solve the inequality $90 - 29x \le 3$.

$$90 - 29x \le 3 \qquad \text{Inequality to be solved}$$

$$-29x \le -87 \qquad \text{Subtract 90 from each side.}$$

$$\frac{-29x}{-29} \ge \frac{-87}{-29} \qquad \text{Divide by } -29; \text{ reverse the inequality.}$$

$$x \ge 3 \qquad \text{Reduce the fraction.}$$

Clouds may form at 3 miles or higher.

EXAMPLE 10   Modeling AIDS Research

AIDS research funding F in billions of dollars increased from 1994 to 2000 and can be modeled by the formula $F = 0.083x - 164.2$, where x is the year. Estimate when the funding F was greater than or equal to $1.55 billion. (*Source:* National Institute of Health.)

Solution   We must solve the inequality $F \ge 1.55$.

$$0.083x - 164.2 \ge 1.55 \qquad \text{Inequality to be solved}$$

$$0.083x \ge 165.75 \qquad \text{Add 164.2 to each side.}$$

$$x \ge \frac{165.75}{0.083} \qquad \text{Divide each side by 0.083.}$$

Because $\frac{165.75}{0.083} \approx 1997$, funding for AIDS was greater than or equal to $1.55 billion during 1997 and after.

EXAMPLE 11   Calculating revenue, cost, and profit

For a computer company, the cost to produce one laptop computer is $1320 plus a one-time fixed cost of $200,000 for research and development. The revenue received from selling one laptop computer is $1850.
(a) Write a formula that gives the cost C of producing x laptop computers.
(b) Write a formula that gives the revenue R from selling x laptop computers.
(c) Profit equals revenue minus cost. Write a formula that calculates the profit P from selling x laptop computers.
(d) How many computers need to be sold to yield a positive profit?

Solution   **(a)** The cost of producing the first laptop is

$$1320 \times 1 + 200,000 = \$201,320.$$

The cost of producing the second laptop is

$$1320 \times 2 + 200,000 = \$202,640.$$

And, in general, the cost of producing x laptops is

$$1320 \times x + 200{,}000 = 1320x + 200{,}000.$$

Thus $C = 1320x + 200{,}000$.

(b) Because the company receives \$1850 for each laptop, the revenue for x laptops is $R = 1850x$.

(c) Profit equals revenue minus cost, so

$$
\begin{aligned}
P &= R - C \\
&= 1850x - (1320x + 200{,}000) \\
&= 530x - 200{,}000.
\end{aligned}
$$

Thus $P = 530x - 200{,}000$.

(d) To determine how many laptops need to be sold to yield a positive profit, we must solve the inequality $P > 0$.

$$530x - 200{,}000 > 0 \qquad \text{Inequality to be solved}$$
$$530x > 200{,}000 \qquad \text{Add 200,000.}$$
$$x > \frac{200{,}000}{530} \qquad \text{Divide by 530.}$$

Because $\frac{200{,}000}{530} \approx 377.4$, the company must sell at least 378 laptops. Note that the company cannot sell a fraction of a laptop.

2.5   PUTTING IT ALL TOGETHER

In this section we discussed linear inequalities and how to solve them. A linear inequality has infinitely many solutions and can be solved by using the addition and multiplication properties of inequalities. When multiplying or dividing an inequality by a negative number, you must reverse the inequality symbol. The following table summarizes some of the concepts presented in this section.

Concept	Comments	Examples
Linear Inequality	If the equals sign in a linear equation is replaced with $<, >, \le,$ or $\ge$, a linear inequality results.	*Linear Equation* *Linear Inequality* $4x - 1 = 0$ $4x - 1 > 0$ $2 - x = 3x$ $2 - x \le 3x$ $4(x + 3) = 1 - x$ $4(x + 3) < 1 - x$ $-6x + 3 = 5$ $-6x + 3 \ge 5$
Solution to an Inequality	A value for a variable that makes the inequality a true statement	5 is a solution to $2x > 5$ because $2(5) > 5$ is a true statement.
Set-Builder Notation	A notation that can be used to identify the solution set to an inequality	The solution set for $x - 2 < 5$ can be written as $\{x \mid x < 7\}$ and is read "the set of real numbers x such that x is less than 7."
Solution Set	The set of all solutions to an inequality	The solution set to $x + 1 > 5$ is $\{x \mid x > 4\}$.

Concept	Comments	Examples
Number Line Graphs	The solutions to an inequality can be graphed on a number line.	$x < 2$ is graphed as follows. $x \geq -1$ is graphed as follows.
Addition Property of Inequalities	$a < b$ is equivalent to $a + c < b + c,$ where a, b, and c represent real number expressions.	$x - 5 \geq 6$ Given inequality $x \geq 11$ Add 5. $3x > 5 + 2x$ Given inequality $x > 5$ Subtract $2x$.
Multiplication Property of Inequalities	$a < b$ is equivalent to $ac < bc,$ when $c > 0$ and is equivalent to $ac > bc$ when $c < 0$.	$\frac{1}{2}x \geq 6$ Given inequality $x \geq 12$ Multiply by 2. $-3x > 5$ Given inequality $x < -\dfrac{5}{3}$ Divide by -3; reverse the inequality symbol.

2.5 EXERCISES

FOR EXTRA HELP

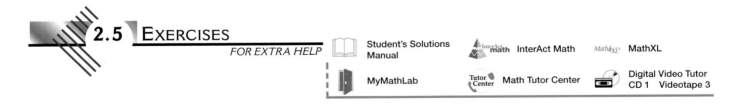

Student's Solutions Manual   InterAct Math   MathXL

MyMathLab   Math Tutor Center   Digital Video Tutor   CD 1   Videotape 3

CONCEPTS

1. A linear inequality results whenever the _____ in a linear equation is replaced by any one of the symbols _____, _____, _____, or _____.

2. Two linear inequalities are _____ if they have the same solution set.

3. The solution set to a linear equation contains (one/infinitely many) solution(s).

4. The solution set to a linear inequality contains (one/infinitely many) solution(s).

5. The solution set to a linear inequality can be graphed by using a _____.

6. The value of 5 (is/is not) a solution to the inequality $3x < 10$.

7. The addition property of inequalities states that, if $a > b$, then $a + c$ _____ $b + c$.

8. The multiplication property of inequalities states that, if $a < b$ and $c > 0$, then ac _____ bc.

9. The multiplication property of inequalities states that, if $a < b$ and $c < 0$, then ac _____ bc.

10. Are $-4x < 8$ and $x < -2$ equivalent inequalities? Explain.

SOLUTIONS AND NUMBER LINE GRAPHS

Exercises 11–18: Use a number line to graph the solution set to the inequality.

11. $x < 0$

12. $x > -2$

13. $x > 1$

14. $x < -\frac{5}{2}$

15. $x \leq 1.5$

16. $x \geq -3$

17. $z \geq -2$

18. $z \leq -\pi$

Exercises 19–24: Express the set of real numbers graphed on the number line with an inequality.

19.   **20.**

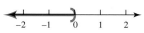

21.   **22.**

23.   **24.**

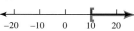 

Exercises 25–34: Determine whether the given value of the variable is a solution to the inequality.

25. $x + 5 > 5$ $x = 4$

26. $x - 7 < 0$ $x = 6$

27. $5x \geq 25$ $x = 5$

28. $-3x \leq -8$ $x = -2$

29. $4y - 3 \leq 5$ $y = -3$

30. $3y + 5 \geq -8$ $y = -3$

31. $5(z + 1) < 3z - 7$ $z = -7$

32. $-(z + 7) > 3(6 - z)$ $z = 2$

33. $\frac{3}{2}t - \frac{1}{2} \geq 1 - t$ $t = -2$

34. $2t - 3 > 5t - (2t + 1)$ $t = 5$

TABLES AND LINEAR INEQUALITIES

Exercises 35–38: Use the table to solve the inequality.

35. $3x + 6 > 0$

x	-4	-3	-2	-1	0
$3x + 6$	-6	-3	0	3	6

36. $6 - 3x \leq 0$

x	1	2	3	4	5
$6 - 3x$	3	0	-3	-6	-9

37. $-2x + 7 > 5$

x	-1	0	1	2	3
$-2x + 7$	9	7	5	3	1

38. $5(x - 3) \leq 4$

x	3.2	3.4	3.6	3.8	4
$5(x - 3)$	1	2	3	4	5

Exercises 39–42: Complete the table. Then use the table to solve the inequality.

39. $-2x + 6 \leq 0$

x	1	2	3	4	5
$-2x + 6$	4				-4

40. $3x - 1 < 8$

x	0	1	2	3	4
$3x - 1$	-1				

41. $5 - x > x + 7$

x	-3	-2	-1	0	1
$5 - x$	8				4
$x + 7$	4				8

42. $2(3 - x) \geq -3(x - 2)$

x	-2	-1	0	1	2
$2(3 - x)$					
$-3(x - 2)$					

SOLVING LINEAR INEQUALITIES

Exercises 43–50: Use the addition property of inequalities to solve the inequality. Then graph the solution set.

43. $x - 3 > 0$

44. $x + 6 < 3$

45. $3 - y \leq 5$

46. $8 - y \geq 10$

47. $12 < 4 + z$

48. $2z \leq z + 17$

49. $5 - 2t \geq 10 - t$

50. $-2t > -3t + 1$

Exercises 51–58: Use the multiplication property of inequalities to solve the inequality. Then graph the solution set.

51. $2x < 10$

52. $3x > 9$

53. $-\frac{1}{2}t \geq 1$

54. $-5t \leq -6$

55. $\frac{3}{4} > -5y$

56. $10 \geq -\frac{1}{7}y$

57. $-\frac{2}{3} \leq \frac{1}{7}z$

58. $-\frac{3}{10}z < 11$

Exercises 59–86: Solve the linear inequality. Write the solution set in set-builder notation.

59. $3x + 1 < 22$

60. $4 + 5x \leq 9$

61. $5 - \frac{3}{4}x \geq 6$

62. $10 - \frac{2}{5}x > 0$

63. $45 > 6 - 2x$

64. $69 \geq 3 - 11x$

65. $5x - 2 \leq 3x + 1$

66. $12x + 1 < 25 - 3x$

67. $-x + 24 < x + 23$

68. $6 - 4x \leq x + 1$

69. $-(x + 1) \geq 3(x - 2)$

70. $5(x + 2) > -2(x - 3)$

71. $3(2x + 1) > -(5 - 3x)$

72. $4x \geq -3(7 - 2x) + 1$

73. $-(7x + 5) + 1 \geq 3x - 1$

74. $3(2 - x) - 5 > -4(5 - x)$

75. $1.6x + 0.4 \leq 0.4x$

76. $-5.1x + 1.1 < 0.1 - 0.1x$

77. $0.8x - 0.5 < x + 1 - 0.5x$

78. $0.1(x + 1) - 0.1 \leq 0.2x - 0.5$

79. $-\frac{1}{2}\left(\frac{2}{3}x + 4\right) \geq x$

80. $-5x > \frac{4}{5}\left(\frac{10}{3}x + 10\right)$

81. $\frac{3}{7}x + \frac{2}{7} > -\frac{1}{7}x - \frac{5}{14}$

82. $\frac{5}{6} - \frac{1}{3}x \geq -\frac{1}{3}\left(\frac{5}{6}x - 1\right)$

83. $\frac{x}{3} + \frac{5x}{6} \leq \frac{2}{3}$

84. $\frac{3x}{4} - \frac{x}{2} < 1$

85. $\frac{6x}{7} < \frac{1}{3}x + 1$

86. $\frac{5x}{8} - \frac{3x}{4} \leq 8$

TRANSLATING INEQUALITIES

Exercises 87–94: Translate the statement to an inequality. Let x be the variable.

87. A speed that is greater than 60 miles per hour

88. A speed that is at most 60 miles per hour

89. An age that is at least 21 years old

90. An age that is less than 21 years old

91. A salary that is more than $40,000

92. A salary that is less than or equal to $40,000

93. A speed that does not exceed 70 miles per hour

94. A speed that is not less than 70 miles per hour

APPLICATIONS

95. *Geometry*   Find values for x so that the perimeter of the rectangle is less than 50 feet.

96. *Geometry*   A rectangle is twice as long as it is wide. If the rectangle is to have a perimeter of at least 36 inches, what values for the width are possible?

97. *Geometry*   A triangle with height 12 inches is to have area less than 120 square inches. What must be true about the base of the triangle?

98. *Geometry*   A trapezoid with height 6 inches is to have area not more than 120 square inches. What must be true about the sum of the bases of the trapezoid? (*Hint:* $A = \frac{1}{2}h(a + b)$, where a and b are the bases of the trapezoid.)

99. *Grade Average*   A student scores 74 out of 100 on a test. If the maximum score on the next test is also 100 points, what score does the student need to maintain at least an average of 80?

100. *Grade Average*   A student scores 65 and 82 on two different 100-point tests. If the maximum score on the next test is also 100 points, what score does the student need to maintain at least an average of 70?

101. *Parking Rates*   Parking in a student lot costs $2 for the first half hour and $1.25 for each hour thereafter. A partial hour is charged the same as a full hour. What is the longest time that a student can park in this lot for $8?

102. *Parking Rates*   Parking in a student lot costs $2.50 for the first hour and $1 for each hour thereafter. Parking at a nearby lot costs $1.25 for each hour. In both lots a partial hour is charged as a full hour. In which lot can a student park the longest for $5? For $11?

103. *Car Rental*   A rental car costs $25 per day plus $0.20 per mile. If someone has $200 to spend and needs to drive the car 90 miles each day, for how many days can that person rent the car? Assume that the car cannot be rented for part of a day.

104. *Car Rental*   One car rental agency charges $20 per day plus $0.25 per mile. A different agency charges $37 per day with unlimited mileage. For what mileages is the second rental agency a better deal?

105. *Revenue and Cost*   (Refer to Example 11.) The cost to produce one compact disc is $1.50 plus a one-time fixed cost of $2000. The revenue received from selling one compact disc is $12.
  (a) Write a formula that gives the cost C of producing x compact discs. Be sure to include the fixed cost.
  (b) Write a formula that gives the revenue R from selling x compact discs.
  (c) Profit equals revenue minus cost. Write a formula that calculates the profit P from selling x compact discs.
  (d) What numbers of compact discs need to be sold to yield a positive profit?

106. *Revenue and Cost*   The cost to produce one laptop computer is $890 plus a one-time fixed cost of $100,000 for research and development. The revenue received from selling one laptop computer is $1520.
  (a) Write a formula that gives the cost C of producing x laptop computers.
  (b) Write a formula that gives the revenue R from selling x laptop computers.
  (c) Profit equals revenue minus cost. Write a formula that calculates the profit P from selling x laptop computers.

  (d) How many computers need to be sold to yield a positive profit?

107. *Distance and Time*   Two cars are traveling in the same direction along a freeway. After x hours the first car's distance from a rest stop is given by $70x$ and the second car's distance is given by $60x + 35$.
  (a) When are the cars the same distance from the rest stop?
  (b) When is the first car farther from the rest stop than the second?

108. *Sales of CDs and LP Records*   From 1985 to 1990, U.S. sales of CDs in millions can be modeled by

$$C = 51.6(x - 1985) + 9.1,$$

and sales of vinyl LP records in millions can be modeled by

$$L = -31.9(x - 1985) + 167.7.$$

In both expressions x represents the year.   (*Source: Recording Industry Association of America.*)
  (a) Estimate the years in which CD sales were greater than or equal to LP record sales.
  (b) What happened to sales of CDs and LP records after 1990?

109. *Altitude and Temperature*   (Refer to Example 9.) If the temperature on the ground is $60°$F, then the air temperature x miles high is given by $T = 60 - 29x$. Determine the altitudes at which the air temperature is greater than $2°$F.   (*Source: A. Miller.*)

110. *Altitude and Dew Point*   If the dew point on the ground is $70°$F, then the dew point x miles high is given by $D = 70 - 5.8x$. Determine the altitudes at which the dew point is greater than $41°$F.   (*Source: A. Miller.*)

111. *Hepatitis-C Research*   The hepatitis-C virus (HCV) can live in a person for years without the person experiencing any symptoms. Some 4 million Americans have HCV, and 10,000 people die from it each year. Research funding in millions of dollars from 1994 to 2000 for hepatitis-C can be approximated by

$$F = 4.4(x - 1994) + 7,$$

where x is the year. Determine when funding was less than or equal to $29 million.   (*Source: National Institutes of Health.*)

112. *Size and Weight of a Fish* If the length of a bass is between 20 and 25 inches, its weight W in pounds can be estimated by the formula $W = 0.96x - 14.4$, where x is the length of the fish. (*Source:* Minnesota Department of Natural Resources.)

(a) What length of bass is likely to weigh 7.2 pounds?

(b) What lengths of bass are likely to weigh less than 7.2 pounds?

WRITING ABOUT MATHEMATICS

113. Explain each of the terms and give an example.
(a) Linear equation
(b) Linear inequality

114. Suppose that a student says that a linear equation and a linear inequality can be solved the same way. How would you respond?

CHECKING BASIC CONCEPTS SECTION 2.5

1. Use a number line to graph the solution set to $x + 1 \geq -1$.

2. Express the set of real numbers graphed on the number line by using an inequality.

3. Complete the table. Then use the table to solve $5 - 2x \leq 7$.

x	-2	-1	0	1	2
$5 - 2x$				1	

4. Solve each inequality. Write the solution set in set-builder notation.
(a) $x + 5 > 8$
(b) $-\frac{5}{7}x \leq 25$
(c) $3x \geq -2(1 - 2x) + 3$

5. *Geometry* The length of a rectangle is 5 inches longer than twice its width. If the perimeter of the rectangle is more than 88 inches, find possible widths for the rectangle.

CHAPTER
2 Summary

Section 2.1 *Introduction to Equations*

Equations Every equation contains an equals sign. An equation can either be true or false.

Important Terms

Solution	A value for a variable that makes the equation true
Solution Set	The set of all solutions
Equivalent Equations	Equations that have the same solution set

Checking a Solution	Substituting the result in the given equation to find out if the equation is true

> *Example:* 3 is the solution to $4x - 2 = 10$ because $4(3) - 2 = 10$ is a true statement.

Properties of Equality

Addition Property	$a = b$ is equivalent to $a + c = b + c$.

> *Example:* $x - 3 = 0$ and $x - 3 + 3 = 0 + 3$ are equivalent equations.

Multiplication Property	$a = b$ is equivalent to $ac = bc$, provided $c \neq 0$.

> *Example:* $2x = 5$ and $\frac{1}{2} \cdot 2x = \frac{1}{2} \cdot 5$ are equivalent equations.

Section 2.2   *Linear Equations*

Linear Equation   Can be written in the form $ax + b = 0$, where $a \neq 0$

Examples: $3x - 5 = 0$ is linear, whereas $5x^2 + 2x = 0$ is *not* linear.

Solving Linear Equations   To solve a linear equation use the addition and multiplication properties of equality.

Example:

$$\frac{1}{2}x - 6 = 4 \qquad \text{Given equation}$$

$$\frac{1}{2}x = 10 \qquad \text{Add 6 to each side.}$$

$$x = 20 \qquad \text{Multiply each side by 2.}$$

Distributive Properties $a(b + c) = ab + ac$ or $a(b - c) = ab - ac$

Examples: $5(2x + 3) = 10x + 15$ and $5(2x - 3) = 10x - 15$

Clearing Fractions and Decimals   When fractions or decimals appear in an equation, multiplying each side by the least common denominator can be helpful.

Examples: Multiply $\frac{1}{3}x - \frac{1}{6} = \frac{2}{3}$ by 6 to obtain $2x - 1 = 4$.

Multiply $0.04x + 0.1 = 0.07$ by 100 to obtain $4x + 10 = 7$.

Number of Solutions   Equations that can be written in the form $ax + b = 0$, where a and b are *any* real number, can have zero, one, or infinitely many solutions.

Examples: $x + 3 = x$ is equivalent to $3 = 0$.   (Zero solutions)

$2y + 1 = 9$ is equivalent to $y = 4$.   (One solution)

$z + z = 2z$ is equivalent to $0 = 0$.   (Infinitely many solutions)

Section 2.3   *Introduction to Problem Solving*

Steps for Solving a Problem   The following steps can be used as a guide for solving word problems.

STEP 1: Read the problem carefully and be sure that you understand it. (You may need to read the problem more than once.) Assign a variable to what you are being asked to find. If necessary, write other quantities in terms of this variable.

STEP 2: Write an equation that relates the quantities described in the problem. You may need to sketch a diagram or refer to known formulas.

STEP 3: Solve the equation and determine the solution.

STEP 4: Look back and check your solution. Does it seem reasonable?

Percent Problems

The expression x%   Represents the fraction $\frac{x}{100}$ or the decimal number given by $x \times 0.01$.

> *Examples:* $45\% = \dfrac{45}{100} = 0.45$
>
> $7.1\% = \dfrac{7.1}{100} = \dfrac{71}{1000} = 0.071$

Percent Change   If a quantity changes from T_1 to T_2, then the percent change equals $\dfrac{T_2 - T_1}{T_1} \times 100$.

> *Example:*   If a price increases from \$2 to \$3, then the percent change equals $\dfrac{3 - 2}{2} \times 100 = 50\%$.

Distance Problems   If an object travels at speed (rate) r for time t, then the distance d traveled is calculated by $d = rt$.

Example: A car moving at 65 mph for 2 hours travels

$$d = rt = 65 \cdot 2 = 130 \text{ miles}.$$

Section 2.4   *Formulas*

Formula   A formula is an equation that can be used to calculate a quantity by using known values of other quantities.

Example: The formula $M = \frac{D}{G}$ can be used to calculate the gas mileage M obtained by a car traveling D miles on G gallons of gasoline.

Formulas from Geometry

Area of a Rectangle $A = LW$, where L is the length and W is the width

Area of a Triangle $A = \frac{1}{2}bh$, where b is the base and h is the height

Degree Measure   There are $360°$ in one complete revolution.

Angle Measure   The sum of the angles in a triangle equals $180°$.

Circumference	$C = 2\pi r$, where r is the radius
Area of a Circle	$A = \pi r^2$, where r is the radius
Volume of a Cylinder	$V = \pi r^2 h$, where r is the radius and h is the height
Area of a Trapezoid	$A = \frac{1}{2}(a + b)h$, where h is the height and a and b are the bases of the trapezoid.

Other Formulas

GPA	$\text{GPA} = \frac{4a + 3b + 2c + d}{a + b + c + d + f}$, where a represents the number of A credits earned, b the number of B credits earned, and so on.
Temperature Scales	$F = \frac{9}{5}C + 32$ and $C = \frac{5}{9}(F - 32)$, where F is the Fahrenheit temperature and C is the Celsius temperature.

See Putting It All Together in Section 2.4 for examples.

Section 2.5   *Linear Inequalities*

Linear Inequality   A linear inequality results when the equals sign in a linear equation is replaced with any one of the symbols $<, \leq, >,$ or $\geq$.

Examples: $x > 0$, $6 - \frac{2}{3}x \leq 7$,   and $4(x - 1) < 3x - 1$

Number Line Graphs   A number line can be used to graph the solution set to a linear inequality.

Example: The graph of $x \leq 1$ is shown in the figure.

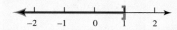

Properties of Inequality

Addition Property	$a < b$ is equivalent to $a + c < b + c$.
	Example: $x - 3 < 0$ and $x - 3 + 3 < 0 + 3$ are equivalent inequalities.
Multiplication Property	When $c > 0$, $a < b$ is equivalent to $ac < bc$. When $c < 0$, $a < b$ is equivalent to $ac > bc$.
	Examples: $2x < 6$ is equivalent to $2x\left(\frac{1}{2}\right) < 6\left(\frac{1}{2}\right)$ or $x < 3$. $-2x < 6$ is equivalent to $-2x\left(-\frac{1}{2}\right) > 6\left(-\frac{1}{2}\right)$ or $x > -3$.

CHAPTER 2 Review Exercises

SECTION 2.1

Exercises 1–8: Solve the equation. Check your solution.

1. $x + 9 = 3$

2. $x - 4 = -2$

3. $x - \frac{3}{4} = \frac{3}{2}$

4. $x + 0.5 = 0$

5. $4x = 12$

6. $3x = -7$

7. $-0.5x = 1.25$

8. $-\frac{1}{3}x = \frac{7}{6}$

SECTION 2.2

Exercises 9–12: Decide whether the equation is linear. If the equation is linear, give values for a and b so that it can be written in the form ax + b = 0.

9. $5x - 3 = 0$

10. $-4x + 3 = 2$

11. $0.55x = 0.05$

12. $\frac{3}{8}x^2 - x = \frac{1}{4}$

Exercises 13–20: Solve the equation and check the solution.

13. $4x - 5 = 3$

14. $7 - \frac{1}{2}x = -4$

15. $5(x - 3) = 12$

16. $1 - (x - 3) = 6 + 2x$

17. $3.4x - 4 = 5 - 0.6x$

18. $-\frac{1}{3}(3 - 6x) = -(x + 2) + 1$

19. $\frac{2}{3}x - \frac{1}{6} = \frac{5}{12}$

20. $2y - 3(2 - y) = 5 + y$

Exercises 21–24: Determine whether the equation has zero, one, or infinitely many solutions.

21. $4(3x - 2) = 2(6x + 5)$

22. $5(3x - 1) = 15x - 5$

23. $8x = 5x + 3x$

24. $9x - 2 = 8x - 2$

Exercises 25 and 26: Complete the table. Then use the table to solve the given equation.

25. $-2x + 3 = 0$

x	0.5	1.0	1.5	2.0	2.5
$-2x + 3$	2				

26. $-(x + 1) + 3 = 2$

x	-2	-1	0	1	2
$-(x + 1) + 3$				0	

SECTION 2.3

Exercises 27–30: Translate the sentence into an equation by using the variable x. Then solve the resulting equation.

27. The product of a number and 6 is 72.

28. The sum of a number and 18 is -23.

29. Twice a number minus 5 equals the number plus 4.

30. The sum of a number and 4 equals the product of the number and 3.

Exercises 31 and 32: Find the number or numbers.

31. The sum of four consecutive natural numbers is 70.

32. The sum of three consecutive integers is -153.

Exercises 33–36: Convert the percentage to fraction and decimal notation.

33. 85%

34. 5.6%

35. 0.03%

36. 342%

Exercises 37–40: Convert the number to a percentage.

37. 0.89

38. 0.005

39. 2.3

40. 1

SECTION 2.4

Exercises 41–44: Use the formula d = rt to find the value of the missing variable.

41. $r = 8$ mph, $t = 3$ hours

42. $r = 70$ feet per second, $t = 55$ seconds

43. $d = 500$ yards, $t = 20$ seconds

44. $d = 125$ miles, $r = 15$ miles per hour

Exercises 45 and 46: Find the area of the region shown.

45. **46.**

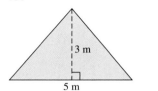

47. Find the area of a rectangle having a 24-inch width and a 3-foot length.

48. Find the area of a triangle having a 13-inch base and a 7-inch height.

49. Find the circumference of a circle having an 18-foot diameter.

50. Find the area of a circle having a 5-inch radius.

51. Find the measure of the third angle in the triangle.

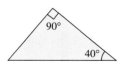

52. The angles in a triangle have measures x, $3x$, and $4x$. Find the value of x.

53. If a cylinder has radius 5 inches and height 25 inches, find its volume. (*Hint:* $V = \pi r^2 h$.)

54. Find the area of a trapezoid with height 5 feet and bases 3 feet and 18 inches. (*Hint:* $A = \frac{1}{2}(a + b)h$.)

Exercises 55–60: Solve the formula for the specified variable.

55. $a = x + y$ for x

56. $P = 2x + 2y$ for x

57. $z = 2xy$ for y

58. $S = \dfrac{a + b + c}{3}$ for b

59. $T = \dfrac{a}{3} + \dfrac{b}{4}$ for b

60. $cd = ab + bc$ for c

SECTION 2.5

Exercises 61–64: Use a number line to graph the solution set to the inequality.

61. $x < 2$

62. $x > -1$

63. $y \geq -\frac{3}{2}$

64. $y \leq 2.5$

Exercises 65 and 66: Express the set of real numbers graphed on the number line with an inequality.

65. **66.**

Exercises 67–70: Determine whether the given value of x is a solution to the inequality.

67. $2x + 1 \leq 5$ $x = -3$

68. $5 - \frac{1}{2}x > -1$ $x = 4$

69. $1 - (x + 3) \geq x$ $x = -2$

70. $4(x + 1) < -(5 - x)$ $x = -1$

Exercises 71 and 72: Complete the table and then use the table to solve the inequality.

71. $5 - x > 3$

x	0	1	2	3	4
$5 - x$	5				

72. $2x - 5 \leq 0$

x	1		1.5	2	2.5	3
$2x - 5$	-3					

Exercises 73–78: Solve the inequality. Write the solution set in set-builder notation.

73. $x - 3 > 0$

74. $-2x \leq 10$

75. $5 - 2x \geq 7$

76. $3(x - 1) < 20$

77. $5x \leq 3 - (4x + 2)$

78. $3x - 2(4 - x) \geq x + 1$

Exercises 79–82: Translate the statement to an inequality. Let x be the variable.

79. A speed that is less than 50 miles per hour

80. A salary that is at most $45,000

81. An age that is at least 16 years old

82. A year before 1995

APPLICATIONS

83. *Rainfall* On a stormy day 2 inches of rain fall before noon and $\frac{3}{4}$ inch per hour fall thereafter until 5 P.M.
  (a) Make a table that shows the total rainfall at each hour starting at noon and ending at 5 P.M.
  (b) Write a formula that calculates the rainfall R in inches, x hours past noon.
  (c) Use your formula to calculate the total rainfall at 5 P.M. Does your answer agree with the value in your table from part (a)?
  (d) How much rain had fallen at 3:45 P.M.?

84. *Cost of a Laptop* A 5% sales tax on a laptop computer amounted to $106.25. Find the cost of the laptop.

85. *Distance Traveled* At noon a bicyclist is 50 miles from home, riding toward home at 10 miles per hour.
  (a) Make a table that shows the bicyclist's distance D from home after 1, 2, 3, 4, and 5 hours.
  (b) Write a formula that calculates the distance D from home after x hours.
  (c) Use your formula to determine D when $x = 3$ hours. Does your answer agree with the value shown in your table?
  (d) For what times was the bicyclist at least 20 miles from home? Assume that $x \geq 0$.

86. *High School Graduates* The number of high school graduates N in millions during year x, $x \geq 1995$, can be approximated by the formula

$$N = \frac{1}{15}x - 130.4.$$

Estimate the year during which this number reached 2.8 million.   (*Source:* Department of Education.)

87. *Master's Degree* In 1971, about 230,500 people received a master's degree, and in 1997, 419,401 did. Find the percent change in the number of master's degrees received between 1971 and 1997.

88. *Car Speeds* One car passes another car on a freeway. The faster car is traveling 12 miles per hour faster than the slower car. Determine how long it will be before the faster car is 2 miles ahead of the slower car.

89. *Saline Solution* A saline solution contains 3% salt. How much water should be added to 100 milliliters of this solution to dilute it to a 2% solution?

90. *Investment Money* A student invests two sums of money, $500 and $800, at different interest rates, receiving a total of $55 in interest after one year. The $500 investment received an interest rate 2% lower than the interest rate for the $800 investment. Find the interest rate for each investment.

91. *Dimensions of a Rectangle* The width of a rectangle is 10 inches less than its length. If the perimeter of the rectangle is 112 inches, find the dimensions of the rectangle.

92. *Geometry* A triangle with height 8 inches is to have an area that is not more than 100 square inches. What lengths are possible for the base of the triangle?

93. *Grade Average* A student scores 75 and 91 on two different tests of 100 points. If the maximum score on the next test is also 100 points, what score does the student need to maintain an average of at least 80?

94. *Parking Rates* Parking in a lot costs $2.25 for the first hour and $1.25 for each hour thereafter. A partial hour is charged the same as a full hour. What is the longest time that someone can park for $9?

95. *Profit* The cost to produce one DVD player is $85 plus a one-time fixed cost of $150,000. The revenue received from selling one DVD player is $225.
  (a) Write a formula that gives the cost C of producing x DVD players.
  (b) Write a formula that gives the revenue R from selling x DVD players.
  (c) Profit equals revenue minus cost. Write a formula that calculates the profit P from selling x DVD players.
  (d) What numbers of DVD players sold will result in a loss? (*Hint:* A loss corresponds to a negative profit.)

CHAPTER 2 Test

Exercises 1–4: Solve the equation. Check your solution.

1. $9 = 3 - x$ **2.** $4x - 3 = 7$

3. $4x - (2 - x) = -3(2x + 6)$

4. $\frac{1}{12}x - \frac{2}{3} = \frac{1}{2}\left(\frac{3}{4} - \frac{1}{3}x\right)$

5. Determine the number of solutions to the equation
$$6(2x - 1) = -4(3 - 3x).$$

6. Complete the table. Then use the table to solve $6 - 2x = 0$.

x	0	1	2	3	4
$6 - 2x$	6				

Exercises 7 and 8: Translate the sentence into an equation, using the variable x. Then solve the resulting equation.

7. The sum of a number and -7 is 6.

8. Twice a number plus 6 equals the number minus 7.

9. The sum of three consecutive natural numbers is 336. Find the three numbers.

10. Convert 5.6% to fraction and decimal notation.

11. Convert 0.345 to a percentage.

12. Find 7.5% of $500.

13. *Speed of Sound* Sound can travel one mile (5280 feet) in 5 seconds. Find the speed of sound in feet per second.

14. *Area* Find the area of the triangle shown.

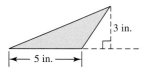

15. Find the circumference and area of a circle with a 30-inch diameter.

16. The measures of the angles in a triangle are x, $2x$, and $3x$. Find the value of x.

17. Solve $z = y - 3xy$ for the variable x.

18. Solve $3(6 - 5x) < 20 - x$. Write the solution set in set-builder notation.

19. *Snowfall* On a stormy day 5 inches of snow fall before noon and 2 inches per hour fall thereafter until 10 P.M.
(a) Write a formula that calculates the snowfall S in inches, x hours past noon.
(b) Use your formula to calculate the total snowfall at 8 P.M.
(c) How much snow had fallen at 6:15 P.M.?

20. *Mixing Acid* A solution is 45% hydrochloric acid. How much water should be added to 1000 milliliters of this solution to dilute it to a 15% solution?

21. *Medical Malpractice* Malpractice insurance premiums at Thomas Jefferson University Hospital in Philadelphia increased from $8 million in 1998 to $32 million in 2003. Calculate the percent change in insurance premiums over this 5-year period. (*Source: The New York Times.*)

2 Extended and Discovery Exercises

Exercises 1–4: Average Speed If someone travels a distance d in time t, then the person's average speed is $\frac{d}{t}$. Use this fact to solve the problem.

1. A driver travels at 50 mph for the first hour and then travels at 70 mph for the second hour. What was the average speed of the car?

2. A bicyclist rides 1 mile uphill at 5 mph and then rides 1 mile downhill at 10 miles per hour. Find the average speed of the bicyclist. Does your answer agree with what you expected?

3. At a 3-mile cross-country race an athlete runs 2 miles at 8 mph and 1 mile at 10 mph. What was the athlete's average speed?

4. A pilot flies an airplane between two cities and travels half the distance at 200 mph and the other half at 100 mph. Find the average speed of the airplane.

5. *A Puzzle About Coins* Suppose that seven coins look exactly alike but that one coin weighs less than any of the other six coins. If you have only a balance with two pans, devise a plan to find the lighter coin. What is the minimum number of weighings necessary? Explain your answer.

6. *Global Warming* If the global climate were to warm significantly as a result of the greenhouse effect or other climatic change, the Arctic ice cap would start to melt. This ice cap contains the equivalent of some 680,000 cubic miles of water. More than 200 million people currently live on land that is less than 3 feet above sea level. In the United States several large cities have low average elevations. Three examples are Boston (14 feet), New Orleans (4 feet), and San Diego (13 feet). In this exercise you are to estimate the rise in sea level if the Arctic ice cap were to melt and to determine whether this event would have a significant impact on people living in coastal areas.

    (a) The surface area of a sphere is given by the formula $4\pi r^2$, where r is its radius. Although the shape of Earth is not exactly spherical, it has an average radius of 3960 miles. Estimate the surface area of Earth.

    (b) Oceans cover approximately 71% of the total surface area of Earth. How many square miles of Earth's surface are covered by oceans?

    (c) Approximate the potential rise in sea level by dividing the total volume of the water from the ice cap by the surface area of the oceans. Convert your answer from miles to feet.

    (d) Discuss the implications of your calculation. How would cities such as Boston, New Orleans, and San Diego be affected?

    (e) The Antarctic ice cap contains some 6,300,000 cubic miles of water. Estimate how much the sea level would rise if this ice cap melted. (*Source: Department of the Interior, Geological Survey.*)

Graphing Equations

Global warming has been in the news for years. Scientists are beginning to document that both the arctic and antarctic ice packs are melting. This phenomenon is an indication that global temperatures may be rising. The amount of light from the sun that Earth reflects into space is an important factor. If more light is reflected, then Earth stays cooler. Astronomer Phil Goode from the New Jersey Institute of Technology has developed a simple way to determine how reflective Earth is by measuring the amount of earth-shine illuminating the moon. The area of the moon illuminated by earth-shine is the darker portion of the moon that is visible but not as visible as the portion illuminated by the sun. Goode found that the percent change in earthshine (from the average) varied with the month. An excellent way to display these data is with the *line graph* shown in the accompanying figure. We discuss line graphs and other types of graphs in this chapter.

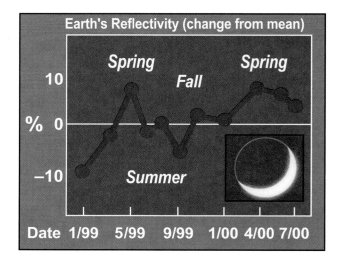

Originality is the essence of true scholarship.
—Nnamdi Azikiwe

Source: Hannah Hoag, *Discover*, November 2002, p. 11. Graphic by Matt Zang. (Reprinted with permission.)

3.1 INTRODUCTION TO GRAPHING

The Rectangular Coordinate System · Scatterplots and Line Graphs

INTRODUCTION

TABLE 3.1
Per Capita Income for the United States

Year	Amount
1970	$4072
1980	$10,029
1990	$19,142
2000	$29,676

Source: Bureau of Economic Analysis.

Table 3.1 lists the per capita (per person) personal income for the United States for selected years. These income amounts have *not* been adjusted for inflation. When a table contains only four data values, as in Table 3.1, we can easily see that income increased significantly from 1970 to 2000. If a table listed income for every year from 1970 to 2000, seeing trends in the data would be more difficult. And if a table contained 1000 data values, determining trends in the data would be extremely difficult. In mathematical problems, there are frequently infinitely many data points!

Rather than always using tables to display data, presenting data on a graph is often more useful. For example, the data in Table 3.1 are graphed in Figure 3.1. This bar graph is more visual than the table and shows the trend at a glance.

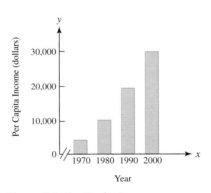

Figure 3.1 Per Capita Income

THE RECTANGULAR COORDINATE SYSTEM

One common way to graph data is to use the **rectangular coordinate system**, or *xy*-plane. In the *xy*-plane the horizontal axis is the *x*-axis, and the vertical axis is the *y*-axis. The axes intersect at the **origin** and divide the *xy*-plane into four regions called **quadrants**. They are numbered I, II, III, and IV counterclockwise, as shown in Figure 3.2.

Before we can plot data, we must first understand the concept of an **ordered pair** (x, y). In Table 3.1 we can let *x*-values correspond to the year and *y*-values correspond to the per capita income. Then the fact that the per capita income in 1970 was $4072 can be summarized by the ordered pair (1970, 4072). Similarly, the ordered pair (1980, 10029) indicates that the per capita income was $10,029 in 1980. *Order is important in an ordered pair*. The ordered pairs (1950, 2010) and (2010, 1950) are different. The first ordered pair indicates that the per capita income in 1950 was $2010, whereas the second ordered pair indicates that the per capita income in 2010 will be $1950.

To plot the ordered pair $(-2, 3)$ in the *xy*-plane, begin by locating $x = -2$ on the *x*-axis. Then move upward until a height of $y = 3$ is reached. Thus the point $(-2, 3)$ is located 2 units left of the origin and 3 units above the origin. In Figure 3.3, the point $(-2, 3)$ is plotted in quadrant II.

Note: A point that lies on one of the axes is not located in a quadrant.

Figure 3.2 The *xy*-plane

Figure 3.3 Plotting a Point

EXAMPLE 1   Plotting points

Plot the following ordered pairs on the same *xy*-plane. State the quadrant in which each point is located, if possible.

(a) $(3, 2)$ **(b)** $(-2, -3)$ **(c)** $(-3, 0)$

Solution   **(a)** The point $(3, 2)$ is located in quadrant I, 3 units to the right of the origin and 2 units above the origin. See Figure 3.4.

(b) The point $(-2, -3)$ is located in quadrant III, 2 units to the left of the origin and 3 units below the origin. See Figure 3.4.

(c) The point $(-3, 0)$ is not in any quadrant because it is located 3 units left of the origin on the *x*-axis. See Figure 3.4.

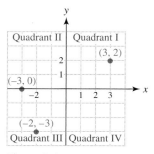

Figure 3.4

EXAMPLE 2   Reading a graph

During the past decade, frozen pizza makers have improved their pizzas to taste more like homemade. As a result, their sales have increased. Figure 3.5 shows U.S. retail sales of frozen pizzas in billions of dollars. Estimate frozen pizza sales in 1994 and in 2000. (***Source:*** Business Trend Analyst.)

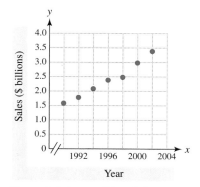

Figure 3.5 Frozen Pizza Sales

Solution   To estimate frozen pizza sales in 1994, first locate 1994 on the *x*-axis. Then move upward to the data point and approximate its *y*-coordinate. Figure 3.6(a) shows that there were about

$2.1 billion in frozen pizza sales in 1994. Similarly, frozen pizza sales in 2000 were about $3 billion, as shown in Figure 3.6(b).

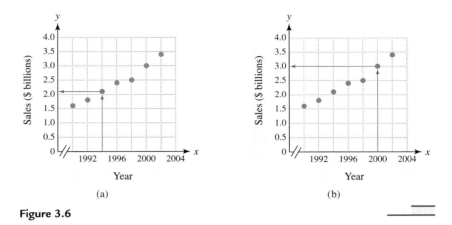

(a)   (b)

Figure 3.6

SCATTERPLOTS AND LINE GRAPHS

If distinct points are plotted in the *xy*-plane, then the resulting graph is called a **scatterplot**. Figure 3.5 is an example of a scatterplot that shows frozen pizza sales. A different scatterplot is shown in Figure 3.7, in which the points $(1, 3)$, $(2, 2)$, $(3, 1)$, $(4, 4)$, and $(5, 1)$ are plotted.

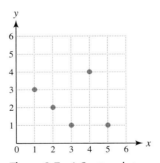

Figure 3.7   A Scatterplot

The next example illustrates how to make a scatterplot by using real data.

EXAMPLE 3   Making a scatterplot of gasoline prices

Table 3.2 lists the average price of a gallon of gasoline for selected years. Make a scatterplot of the data.

TABLE 3.2   Average Price of Gasoline

Year	1950	1960	1970	1980	1990	2000
Cost (per gal)	27¢	31¢	36¢	119¢	115¢	156¢

Source: Department of Energy.

Solution Plot the six data points (1950, 27), (1960, 31), (1970, 36), (1980, 119), (1990, 115), and (2000, 156) in the *xy*-plane. The *x*-values vary from 1950 to 2000, so label the *x*-axis from 1950 to 2000 every 10 years. The *y*-values vary from 27 to 156, so label the *y*-axis from 0 to 175 every 25¢. Note that the labels on the *xy*-axes may vary. However, the *x*- and *y*-scales must be large enough to accommodate every data point. Figure 3.8 shows the scatterplot.

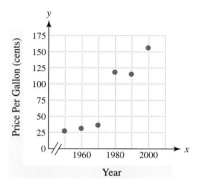

Figure 3.8 Price of Gasoline

Sometimes it is helpful to connect consecutive data points in a scatterplot with straight-line segments. This type of graph visually emphasizes changes in the data and is called a **line graph**.

EXAMPLE 4 Making a line graph

Use the data in Table 3.3 to make a line graph.

TABLE 3.3

x	-2	-1	0	1	2
y	1	2	-2	-1	1

Solution The data in Table 3.3 are represented by the five ordered pairs $(-2, 1)$, $(-1, 2)$, $(0, -2)$, $(1, -1)$, and $(2, 1)$. Plot these points and then connect consecutive points with a straight-line segment, as shown in Figure 3.9.

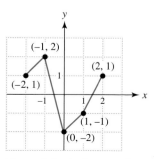

Figure 3.9 A Line Graph

Technology Note: *Scatterplots and Line Graphs*

Graphing calculators are capable of creating both line graphs and scatterplots. The line graph in Figure 3.9 is shown to the left and the corresponding scatterplot is shown to the right.

Calculator Help

To make a scatterplot or a line graph, see Appendix A (page AP-4).

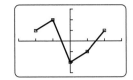
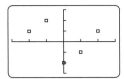

EXAMPLE 5   Analyzing a line graph

The line graph in Figure 3.10 shows the per capita energy consumption in the United States. Units are in millions of Btu, where 1 Btu equals the amount of heat necessary to raise 1 pound of water 1°F.   (***Source:*** Department of Energy.)

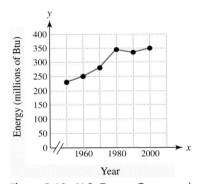

Figure 3.10   U.S. Energy Consumption

(a) Did energy consumption ever decrease during this time period? Explain.
(b) Estimate the energy consumption in 1960 and in 2000.
(c) Estimate the percent change in energy consumption from 1960 to 2000.

Solution   **(a)** Yes, energy consumption decreased slightly between 1980 and 1990.
(b) From the graph, per capita energy consumption in 1960 was about 250 million Btu. In 2000 it was about 350 million Btu.

Note: When different people read a graph, values they obtain may vary slightly.

(c) The percent change from 1960 to 2000 was

$$\frac{350 - 250}{250} \times 100 = 40,$$

so the increase was 40%.

Critical Thinking

When analyzing data, do you prefer a table of values, a scatterplot, or a line graph? Explain your answer.

3.1   PUTTING IT ALL TOGETHER

The rectangular coordinate plane, or xy-plane, can be used to plot ordered pairs in the form (x, y). Application data can frequently be represented by ordered pairs. For example, the ordered pair $(12, 45)$ might indicate that the average high temperature in the twelfth month (December) is $45°$F.

Concept	Explanation	Examples
Ordered Pair	Has the form (x, y), where the order of x and y is important	$(1, 2)$, $(-2, 3)$, $(2, 1)$ and $(-4, -2)$ are distinct ordered pairs.
Rectangular Coordinate System, or xy-plane	Consists of a horizontal x-axis and a vertical y-axis Can be used to graph ordered pairs	The points $(1, 2)$, $(0, -2)$, and $(-1, 1)$ are plotted in the graph. 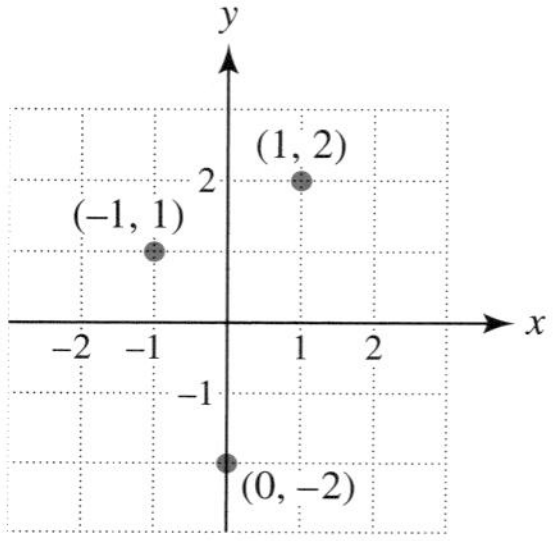
Scatterplot	Individual points are plotted in the xy-plane.	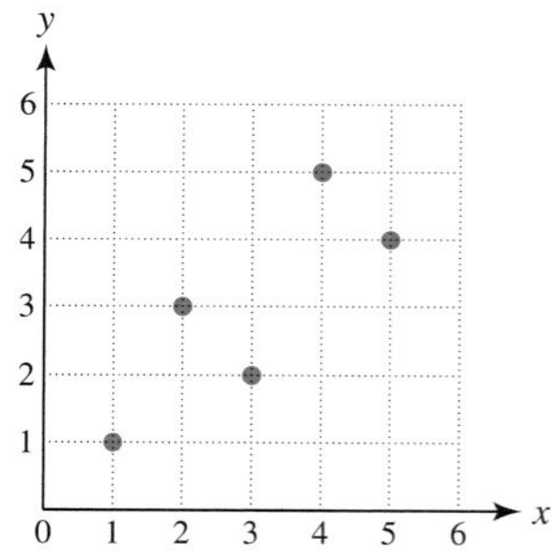
Line Graph	Similar to a scatterplot except that line segments are drawn between consecutive data points	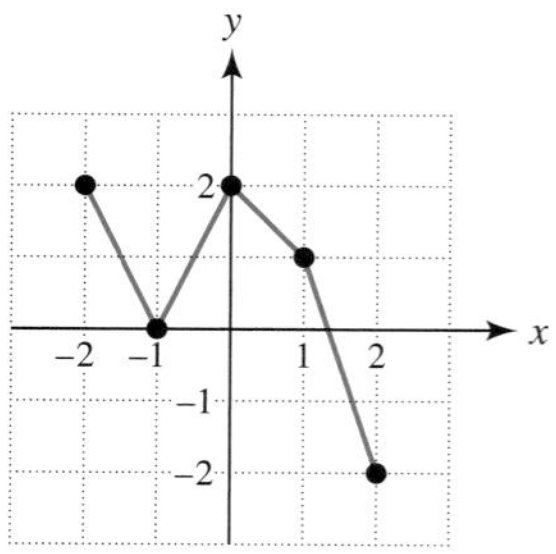

3.1   EXERCISES

FOR EXTRA HELP

 Student's Solutions Manual

 InterAct Math

 MathXL

MyMathLab

 Math Tutor Center

Digital Video Tutor
CD 2   Videotape 4

CONCEPTS

1. Another name for the rectangular coordinate system is the _____ .

2. The origin corresponds to the point _____ in the xy-plane.

3. How many quadrants are there in the xy-plane?

4. The point $(-2, 3)$ is located in quadrant _____ .

5. If both x and y are negative, then the point (x, y) is located in quadrant _____ .

6. A point that lies on one of the _____ is not located in a quadrant.

7. If distinct points are plotted in the xy-plane, the resulting graph is called a _____ .

8. If the consecutive points in a scatterplot are connected with line segments, the resulting graph is called a _____ graph.

CARTESIAN COORDINATE PLANE

Exercises 9–12: Identify the coordinates of each point in the graph.

9.

10.

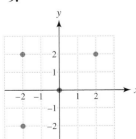

11.

12.

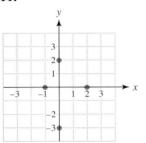

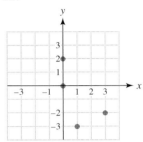

Exercises 13–18: State the quadrant in which each point lies.

13. **(a)** $(1, 4)$ **(b)** $(-1, -4)$

14. **(a)** $(-2, 3)$ **(b)** $(2, -3)$

15. **(a)** $(7, 0)$ **(b)** $(0.1, 7)$

16. **(a)** $(100, -3)$ **(b)** $(-100, -3)$

17. **(a)** $\left(-\frac{1}{2}, \frac{3}{4}\right)$ **(b)** $\left(\frac{3}{4}, -\frac{1}{2}\right)$

18. **(a)** $(1.2, 0)$ **(b)** $(0, -1.2)$

Exercises 19–28: Make a scatterplot by plotting the given points. Be sure to label each axis.

19. $(0, 0), (1, 2), (-3, 2), (-1, -2)$

20. $(0, -3), (-2, 1), (2, 2), (-4, -4)$

21. $(-1, 0), (4, -3), (0, -1), (3, 4)$

22. $(1, 1), (-2, 2), (-3, -3), (4, -4)$

23. $(2, 4), (-4, 4), (0, -4), (-6, 2)$

24. $(4, 8), (8, 4), (-8, -4), (-4, 0)$

25. $(5, 0), (5, -5), (-10, -20), (10, -10)$

26. $(10, 30), (-20, 10), (40, 0), (-30, -10)$

27. $(0, 0.1), (0.2, -0.3), (-0.1, 0.4)$

28. $(1.5, 2.5), (-1, -1.5), (-2.5, 0)$

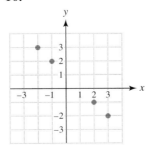

Exercises 29–34: Use the table of xy-values to make a line graph.

29.

x	−2	−1	0	1	2
y	2	1	0	−1	−2

30.

x	−4	−2	0	2	4
y	4	−2	3	−1	2

31.

x	−10	−5	0	5	10
y	20	−10	10	0	−20

32.

x	1	2	3	4	5
y	2	3	1	5	4

33.

x	−5	5	−10	10	0
y	10	20	30	20	40

34.

x	3	−2	2	1	−3
y	4	3	3	−2	−3

Exercises 35 and 36: Identify the coordinates of each point in the graph. Then explain what the coordinates of the first point indicate.

35. Billions of dollars spent on military personnel in the United States. (***Source:*** Department of Defense.)

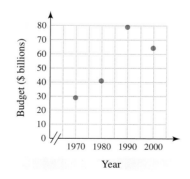

36. Cases of Tetanus in the United States. (***Source:*** Department of Health and Human Services.)

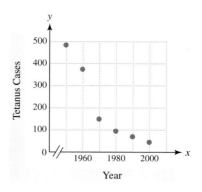

GRAPHING REAL DATA

Exercises 37–42: Graphing Real Data   The table contains real data.

   (a) Make a line graph of the data. Be sure to label the axes.

   (b) Comment on any trends in the data.

37. Head Start participation *H* in thousands during year *t*

t	1970	1980	1990	2000
H	480	380	540	858

Source: Department of Health and Human Services.

38. Percent *P* of total music sales that were CDs during year *t*

t	1996	1997	1998	1999	2000
P	68.4%	70.2%	74.8%	83.2%	89.3%

Source: Recording Industry Association of America.

39. Welfare beneficiaries *B* in millions during year *t*

t	1970	1980	1990	1999
B	7	11	12	7

Source: Administration for Children and Families.

40. Total Medicaid recipients *R* in millions during year *x*

x	1975	1981	1990	1996	1998
R	3.6	3.4	3.2	4.3	4.0

Source: Health Care Financing Administration.

41. Asian-American population y in millions during year x

x	1998	2000	2002	2004
y	10.5	11.2	12.0	12.8

Source: Bureau of the Census.

42. U.S. Internet users y in millions during year x

x	1998	1999	2000	2001	2002
y	25	30	35	42	48

Source: Department of Commerce.

WRITING ABOUT MATHEMATICS

43. Explain how to identify the quadrant that a point lies in if it has coordinates (x, y).

44. Explain the difference between a scatterplot and a line graph. Give an example of each.

3.2   LINEAR EQUATIONS IN TWO VARIABLES

Basic Concepts · Tables of Solutions · Graphing Linear Equations in Two Variables

INTRODUCTION

Figure 3.11 shows a scatterplot of average college tuition and fees at public colleges and universities, together with a line that models the data. If we could find an equation for this line, then we could use it to estimate tuition and fees for years without data points. In this section we discuss linear equations, whose graphs are lines. Linear equations and lines are often used to approximate data.

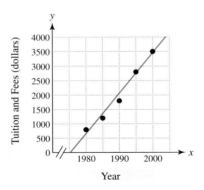

Figure 3.11   Public Institutions

BASIC CONCEPTS

Equations can have any number of variables. In Chapter 2 we solved equations having one variable. The following are examples of equations with two variables.

$$y = 2x, \quad 3x + 2y = 4, \quad z = t^2, \quad \text{and} \quad a - b = 1$$

A solution to an equation with one variable is one number that makes the statement true. For example, the solution to $x - 1 = 3$ is 4 because $4 - 1 = 3$ is a true statement. A solution to an equation with two variables consists of two numbers, one for each variable, which can be expressed as an ordered pair. For example, one solution to the equation $y = 5x$ is given by $x = 1$ and $y = 5$ because $5 = 5(1)$ is a true statement. This solution can be expressed as the ordered pair $(1, 5)$. Another solution to $y = 5x$ is $(2, 10)$, because $10 = 5(2)$ is also a true statement. In fact the equation $y = 5x$ has infinitely many solutions because, for every value of x that we choose, there is a corresponding value of y that makes the equation true.

EXAMPLE 1   Testing solutions to equations

Determine whether each ordered pair is a solution to each equation.

(a) $y = x + 3$, $(1, 4)$ **(b)** $2x - y = 5$, $\left(\frac{1}{2}, -4\right)$ **(c)** $-4x + 5y = 20$, $(-5, 1)$

Solution   **(a)** Let $x = 1$ and $y = 4$ in the given equation.

$$y = x + 3 \qquad \text{Given equation}$$
$$4 \stackrel{?}{=} 1 + 3 \qquad x = 1, y = 4$$
$$4 = 4 \qquad \text{A true statement}$$

The ordered pair $(1, 4)$ is a solution.

(b) Let $x = \frac{1}{2}$ and $y = -4$ in the given equation.

$$2x - y = 5 \qquad \text{Given equation}$$
$$2\left(\frac{1}{2}\right) - (-4) \stackrel{?}{=} 5 \qquad x = \frac{1}{2}, y = -4$$
$$1 + 4 \stackrel{?}{=} 5 \qquad \text{Simplify the left side.}$$
$$5 = 5 \qquad \text{A true statement}$$

The ordered pair $\left(\frac{1}{2}, -4\right)$ is a solution.

(c) Let $x = -5$ and $y = 1$ in the given equation.

$$-4x + 5y = 20 \qquad \text{Given equation}$$
$$-4(-5) + 5(1) \stackrel{?}{=} 20 \qquad x = -5, y = 1$$
$$20 + 5 \stackrel{?}{=} 20 \qquad \text{Simplify the left side.}$$
$$25 \stackrel{?}{=} 20 \qquad \text{A \textit{false} statement}$$

The ordered pair $(-5, 1)$ is *not* a solution.

TABLES OF SOLUTIONS

A table can be used to list solutions to an equation. For example, Table 3.4 lists solutions to $x + y = 5$, where the sum of each xy-pair equals 5.

TABLE 3.4	$x + y = 5$				
x	-2	-1	0	1	2
y	7	6	5	4	3

Most equations in two variables have infinitely many solutions, so it is impossible to list all solutions in a table. However, when you are graphing an equation, having a table that lists a few solutions to the equation is often helpful. The next two examples demonstrate how to complete a table for a given equation.

EXAMPLE 2 Completing a table of solutions

Complete the table for the equation $y = 2x - 3$.

x	−4	−2	0	2
y				

Solution Start by determining the corresponding y-value for each x-value in the table. For example, when $x = -2$, the equation $y = 2x - 3$ implies that $y = 2(-2) - 3 = -7$. Filling in the y-values results in Table 3.5.

TABLE 3.5

x	−4	−2	0	2
y	−11	−7	−3	1

EXAMPLE 3 Making a table of solutions

Use $y = 0, 5, 10,$ and 15 to make a table of solutions to $5x + 2y = 10$.

Solution Begin by listing the required y-values in the table. Next determine the corresponding x-values for each y-value by using the equation $5x + 2y = 10$. For example, when $y = 0$, the equation $5x + 2(0) = 10$ implies that $5x = 10$, or $x = 2$. Filling in the x-values results in Table 3.6.

TABLE 3.6

x	2	0	−2	−4
y	0	5	10	15

Formulas can be difficult to work with. As a result, newspapers, magazines, and books often list numbers in a table rather than presenting a formula for the reader to use. The next example illustrates a situation in which a table might be preferable to a formula.

EXAMPLE 4 Calculating appropriate lengths of crutches

People who sustain leg injuries often require crutches. An appropriate crutch length L in inches for an injured person who is t inches tall is estimated by $L = 0.72t + 2$.
(*Source: Journal of the American Physical Therapy Association.*)

(a) Complete the table. Round values to the nearest inch.

t	60	65	70	75	80
L					

(b) Use the table to determine the appropriate crutch length for a person 5 feet 10 inches tall.

Solution **(a)** If $t = 60$, then $L = 0.72(60) + 2 = 43.2 + 2 = 45.2$, or 45 inches rounded to the nearest inch. Other values in Table 3.7 are found similarly. (You may want to use a calculator to help complete the table.)

(b) A person who is 5 feet 10 inches tall is $5 \cdot 12 + 10 = 70$ inches tall. Table 3.7 reveals that a person 70 inches tall needs crutches that are about 52 inches long.

TABLE 3.7
Crutch Lengths

t	L
60	45
65	49
70	52
75	56
80	60

GRAPHING LINEAR EQUATIONS IN TWO VARIABLES

In the preceding section we showed that a line graph visually displays data. Many times graphs are used in mathematics to make concepts easier to understand.

EXAMPLE 5   Graphing an equation with two variables

Make a table of values for the equation $y = 2x$, and then use the table to graph this equation.

Solution   Start by selecting a few convenient values for x, such as $x = -1, 0, 1$, and 2. Then complete the table.

Note: A table of values can be either horizontal or vertical. Table 3.8 is presented in a vertical format.

TABLE 3.8 $y = 2x$

x	y
-1	-2
0	0
1	2
2	4

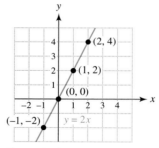
Figure 3.12

To graph the equation $y = 2x$, start by plotting the points $(-1, -2)$, $(0, 0)$, $(1, 2)$, and $(2, 4)$ given in Table 3.8, as shown in Figure 3.12. Note that all four points appear to lie on the same (straight) line. Because there are infinitely many points that satisfy the equation, we can draw a line through these points.

An equation whose graph is a line is called a *linear equation*. Linear equations with two variables can be written in the following **standard form**.

LINEAR EQUATION IN TWO VARIABLES

A **linear equation in two variables** can be written as

$$Ax + By = C,$$

where A, B, and C are fixed numbers and A and B are not both equal to 0. The graph of a linear equation in two variables is a line.

Note: In mathematics a line is always *straight*.

In Example 5, the equation $y = 2x$ is a linear equation because it can be written in standard form by adding $-2x$ to each side.

$$y = 2x \qquad \text{Given equation}$$
$$-2x + y = -2x + 2x \qquad \text{Add } -2x \text{ to each side.}$$
$$-2x + y = 0 \qquad \text{Additive inverses.}$$

The equation $-2x + y = 0$ is a linear equation in two variables because it is in the form $Ax + By = C$ with $A = -2$, $B = 1$, and $C = 0$.

EXAMPLE 6   Graphing linear equations

Graph each linear equation.
(a) $y = \frac{1}{2}x - 1$ (b) $x + y = 4$

Solution   (a) Two points determine a line. However, it is a good idea to plot three points to be sure that the line is graphed correctly. Start by choosing three values for x and then calculate the corresponding y-values, as shown in Table 3.9. In Figure 3.13, the points $(-2, -2)$, $(0, -1)$, and $(2, 0)$ are plotted with the line passing through each one.

TABLE 3.9 $\quad y = \frac{1}{2}x - 1$

x	y
-2	-2
0	-1
2	0

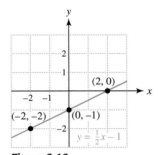

Figure 3.13

(b) If an ordered pair (x, y) is a solution to the given equation, then the sum of x and y is 4. Table 3.10 shows three examples. In Figure 3.14, the points $(0, 4)$, $(2, 2)$, and $(4, 0)$ are plotted with the line passing through each one.

TABLE 3.10 $\quad x + y = 4$

x	y
0	4
2	2
4	0

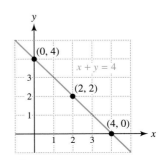

Figure 3.14

Technology Note: *Graphing Equations*

Calculator Help

To graph an equation, see
Appendix A (page AP-5).

Graphing calculators can be used to graph equations. Before graphing the equation $x + y = 4$, solve the equation for y to obtain $y = 4 - x$. A calculator graph of Figure 3.14 is shown in the figure, except that the three points have not been plotted.

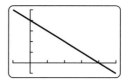

EXAMPLE 7 **Solving for y and then graphing**

Graph each linear equation by solving for y first.
(a) $4x - 3y = 12$ **(b)** $-2x + 4y = 8$

Solution **(a)** First solve the given equation for y.

$$4x - 3y = 12 \qquad \text{Given equation}$$

$$-3y = -4x + 12 \qquad \text{Subtract } 4x \text{ from each side.}$$

$$y = \frac{4}{3}x - 4 \qquad \text{Divide each side by } -3.$$

Table 3.11 lists the solutions $(0, -4)$, $(3, 0)$, and $(6, 4)$, which are plotted in Figure 3.15 with the line passing through each one.

TABLE 3.11	$y = \frac{4}{3}x - 4$
x	y
0	-4
3	0
6	4

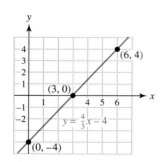

Figure 3.15

(b) First solve the given equation for y.

$$-2x + 4y = 8 \qquad \text{Given equation}$$

$$4y = 2x + 8 \qquad \text{Add } 2x \text{ to each side.}$$

$$y = \frac{1}{2}x + 2 \qquad \text{Divide each side by } 4.$$

Table 3.12 lists the solutions $(-2, 1)$, $(0, 2)$, and $(2, 3)$, which are plotted in Figure 3.16 with the line passing through each one.

TABLE 3.12 $y = \frac{1}{2}x + 2$

x	y
-2	1
0	2
2	3

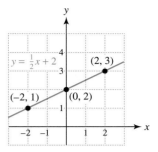

Figure 3.16

═══ MAKING CONNECTIONS ═══

Graphs and Solution Sets

A graph visually depicts the set of solutions to an equation. Each point on the graph represents one solution to the equation.

Note: In a linear equation, each x-value determines a unique y-value. Because there are infinitely many x-values, there are infinitely many points located on the graph of a linear equation. Thus the graph of a linear equation is a *continuous* line with no breaks.

 3.2   PUTTING IT ALL TOGETHER

The following table summarizes several important topics covered in this section.

Concept	Explanation	Examples
Equation in Two Variables	An equation that has two variables	$y = 4x + 5$, $4x - 5y = 20$, and $u - v = 100$
Solution to an Equation in Two Variables	An ordered pair (x, y) whose x- and y-values satisfy the equation	$(1, 3)$ is a solution to $3x + y = 6$ because $3(1) + 3 = 6$ is a true statement.
	Has infinitely many solutions	The equation $y = 2x$ has infinitely many solutions, such as $(1, 2)$, $(2, 4)$, $(3, 6)$, and so on.
Linear Equation in Two Variables	Can be written as $$Ax + By = C,$$ where A, B, and C are fixed numbers and A and B are not both equal to 0	$3x + 4y = 5$ $y = 4 - 3x$ (or $3x + y = 4$) $x = 2y + 1$ (or $x - 2y = 1$) The graph of each equation is a line.

3.2 EXERCISES

FOR EXTRA HELP

 Student's Solutions Manual

 InterAct Math InterAct Math

 MathXL MathXL

MyMathLab

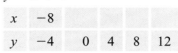 Math Tutor Center

 Digital Video Tutor
CD 2 Videotape 4

CONCEPTS

1. $4x + 6y = 24$ is an example of an equation in _____ variables.

2. The ordered pair $(1, 3)$ is a(n) _____ to the equation $y = 3x$.

3. An equation that can be written in the form $Ax + By = C$ is called a(n) _____ equation in two variables.

4. The equation $4x - 3y = 10$ (is/is not) a linear equation in two variables.

5. A _____ of an equation visually depicts its solution set.

6. The graph of a linear equation in two variables is a(n) _____.

SOLUTIONS TO EQUATIONS

Exercises 7–16: Determine whether the ordered pair is a solution for the given equation.

7. $y = x + 1, (5, 6)$ 8. $y = 4 - x, (6, 2)$

9. $y = 4x + 7, (2, 13)$ 10. $y = -3x + 2, (-2, 8)$

11. $4x - y = -13, (-2, 3)$

12. $3y + 2x = 0, (-2, 3)$

13. $y - 6x = -1, \left(\frac{1}{2}, 2\right)$ 14. $\frac{1}{2}x + \frac{3}{2}y = 0, \left(-\frac{3}{2}, \frac{1}{2}\right)$

15. $0.31x - 0.42y = -9, (100, 100)$

16. $0.5x - 0.6y = 4, (20, 10)$

Exercises 17–22: Complete the table for the given equation.

17. $y = 4x$

x	-2	-1	0	1	2
y	-8				

18. $y = \frac{1}{2}x - 1$

x	0	1	2	3	4
y	-1				

19. $y = x + 4$

x	-8				
y	-4	0	4	8	12

20. $2x - y = 1$

x	-1				
y	-3	-1	0	1	3

21. $3y + 2x = 6$

x					
y	-2	0	2	4	8

22. $3x - 5y = 30$

x	-5	0	5	10	15
y					

Exercises 23–30: Use the given values of the variable to make a table of solutions for the equation.

23. $y = 3x$ $x = -3, 0, 3, 6$

24. $y = 1 - 2x$ $x = 0, 1, 2, 3$

25. $y = \dfrac{x + 4}{2}$ $x = -8, -4, 0, 4$

26. $y = \dfrac{x}{3} - 1$ $x = 0, 2, 4, 6$

27. $x + y = 6$ $y = -2, 0, 2, 4$

28. $2x - 3y = 9$ $y = -3, 0, 1, 2$

29. $y - 4x = 0$ $y = -2, -1, 0, 1$

30. $-4x = 6y - 4$ $y = -1, 0, 1, 2$

GRAPHING EQUATIONS

Exercises 31–58: Graph the equation.

31. $y = x$ 32. $y = \frac{1}{2}x$

33. $y = \frac{1}{3}x$

34. $y = -2x$

35. $y = x + 3$

36. $y = x - 2$

37. $y = x - 4$

38. $y = x + 2$

39. $y = 2x + 1$

40. $y = \frac{1}{2}x - 1$

41. $y = 4 - 2x$

42. $y = 2 - 3x$

43. $y = 7 + x$

44. $y = 2 + 2x$

45. $y = -\frac{1}{2}x + \frac{1}{2}$

46. $y = -\frac{3}{4}x + 2$

47. $2x + 3y = 6$

48. $3x + 2y = 6$

49. $x + 4y = 4$

50. $4x + y = -4$

51. $-x + 2y = 8$

52. $-2x + 6y = 12$

53. $y - 2x = 7$

54. $3y - x = 2$

55. $5x - 4y = 20$

56. $4x - 5y = -20$

57. $3x + 5y = -9$

58. $5x - 3y = 10$

APPLICATIONS

59. *Solid Waste in the Past*   In 1960 the amount A of garbage in pounds produced after t days by the average American is given by $A = 2.7t$.   (*Source:* Environmental Protection Agency.)
   (a) Graph the equation for $t \geq 0$.
   (b) How many days did it take for the average American in 1960 to produce 100 pounds of garbage?

60. *Solid Waste Today*   Today the amount A of garbage in pounds produced after t days by the average American is given by $A = 4.3t$.   (*Source:* Environmental Protection Agency.)
   (a) Graph the equation for $t \geq 0$.
   (b) How many days does it take for the average American to produce 100 pounds of garbage today?

61. *Compact Disc Sales*   From 1996 to 2000 the percent P of total music sales with a compact disc format is modeled by $P = 5.5t - 10{,}911$, where t is the year. (*Source:* Recording Industry Association of America.)
   (a) Evaluate P for $t = 1996$ and for $t = 2000$.
   (b) Use your results from part (a) to graph the equation from 1996 to 2000.
   (c) In what year was $P = 83.5\%$?

62. *HIV Infections in the United States*   From 1990 to 2000 the cumulative number of HIV infections I in thousands during year t is modeled by the equation $I = 40t - 79{,}065$.   (*Source:* Department of Health and Human Services.)
   (a) Evaluate I for $t = 1990$ and for $t = 2000$.
   (b) Use your results from part (a) to graph the equation from 1990 to 2000.
   (c) In what year was $I = 775$?

WRITING ABOUT MATHEMATICS

63. The number of welfare beneficiaries B in millions during year t is shown in the table. Discuss whether a linear equation might work to approximate these data from 1970 to 1999.

t	1970	1980	1990	1999
B	7	11	12	7

Source: Administration for Children and Families.

64. The Asian-American population P in millions during year t is shown in the table. Discuss whether a linear equation might model these data from 2000 to 2004.

t	2000	2002	2004
P	11.2	12.0	12.8

Source: Bureau of the Census.

CHECKING BASIC CONCEPTS   SECTIONS 3.1 AND 3.2

1. Identify the coordinates of the four points in the graph. State the quadrant, if any, in which each point lies.

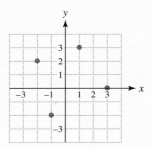

2. Make a scatterplot of the five points $(-2, -2)$, $(-1, -3)$, $(0, 0)$, $(1, 2)$, and $(2, 3)$.

3. Complete the table for the equation $y = -2x + 1$.

x	-2	-1	0	1	2
y	5				

4. Find the value of y in each equation for the given value of x.
   (a) $y = 3 - x$ $x = -2$
   (b) $y = \dfrac{4 + x}{3}$ $x = 8$
   (c) $2x - 3y = 6$ $x = 0$

5. Graph the equation.
   (a) $y = \frac{1}{2}x$ (b) $4x + 6y = 12$

6. *World Population*   The line graph shows the world population from 1800 to 2000. Estimate the coordinates of each data point. Then explain the meaning of the coordinates of the last point in the line graph.   (*Source:* Census Bureau, *World Population Profile*.)

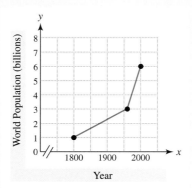

7. *Total Federal Receipts*   The total amount of money A in $ trillions collected by the federal government in year t from 1995 to 2001 can be approximated by $A = 0.115t - 228$.   (*Source:* Budget of the U.S. Government, Fiscal Year 2002.)
   (a) Find A when $t = 1995$ and $t = 2001$. Interpret each result.
   (b) Use your results from part (a) to graph the equation from 1995 to 2001. Be sure to label the axes.
   (c) In what year were the total receipts $1.55 trillion?

3.3   MORE GRAPHING OF LINES

Finding Intercepts · Horizontal Lines · Vertical Lines

INTRODUCTION

The graph of a linear equation is a line, which can be used to model data. Two points, such as the x- and y-intercepts, can be used to sketch such a line. We begin this section by discussing intercepts and their significance in applications and then consider horizontal and vertical lines.

FINDING INTERCEPTS

Suppose that someone leaves a rest stop on an Interstate highway and drives home at a constant speed of 50 miles per hour. The graph shown in Figure 3.17 reflects the distance of the driver from home at various times. The graph intersects the y-axis at 200 miles, which is called the *y-intercept*. In this situation the y-intercept represents the initial distance (when $x = 0$) between the driver and home. The graph also intersects the x-axis at 4 hours, which is called the *x-intercept*. This intercept represents the elapsed time when the distance of the driver from home was 0 miles.

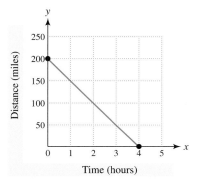

Figure 3.17

▌▌▌▌▌	**FINDING x- AND y-INTERCEPTS**

The x-coordinate of a point where a graph intersects the x-axis is an **x-intercept**. To find an x-intercept, let $y = 0$ in the equation and solve for x.

The y-coordinate of a point where a graph intersects the y-axis is a **y-intercept**. To find a y-intercept, let $x = 0$ in the equation and solve for y.

The graph of the linear equation $3x + 2y = 6$ is shown in Figure 3.18. The x-intercept is 2 and the y-intercept is 3. These two intercepts can be found without a graph. To find the x-intercept let $y = 0$ in the equation and solve for x.

$$3x + 2(0) = 6 \qquad \text{Let } y = 0.$$
$$3x = 6 \qquad \text{Simplify.}$$
$$x = 2 \qquad \text{Divide each side by 3.}$$

To find the y-intercept let $x = 0$ in the equation and solve for y.

$$3(0) + 2y = 6 \qquad \text{Let } x = 0.$$
$$2y = 6 \qquad \text{Simplify.}$$
$$y = 3 \qquad \text{Divide each side by 2.}$$

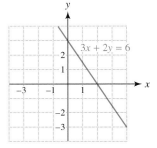

Figure 3.18

EXAMPLE 1   Using intercepts to graph a line

Use intercepts to graph $2x - 6y = 12$.

Solution   The x-intercept is found by letting $y = 0$.

$$2x - 6(0) = 12 \qquad \text{Let } y = 0.$$
$$x = 6 \qquad \text{Solve for } x.$$

The y-intercept is found by letting $x = 0$.

$$2(0) - 6y = 12 \qquad \text{Let } y = 0.$$
$$y = -2 \qquad \text{Solve for } y.$$

Therefore the graph passes through the points $(6, 0)$ and $(0, -2)$, as shown in Figure 3.19.

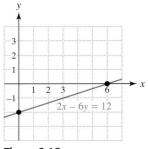

Figure 3.19

EXAMPLE 2   Using intercepts to graph a line

Complete the following for the linear equation $2(y + 2) = 3x - 5$.
(a) Write the equation in the form $Ax + By = C$. Identify A, B, and C.
(b) Find the intercepts.
(c) Graph the equation.

Solution   **(a)** Start by applying the distributive property.

$$2(y + 2) = 3x - 5 \qquad \text{Given equation}$$
$$2y + 4 = 3x - 5 \qquad \text{Distributive property}$$
$$2y = 3x - 9 \qquad \text{Subtract 4 from each side.}$$
$$-3x + 2y = -9 \qquad \text{Subtract } 3x \text{ from each side.}$$

One form is $-3x + 2y = -9$ with $A = -3$, $B = 2$, and $C = -9$. Note that other forms, such as $3x - 2y = 9$, are also possible.

(b) The x-intercept is found by letting $y = 0$.

$$-3x + 2(0) = -9 \qquad \text{Let } y = 0.$$
$$-3x = -9 \qquad \text{Simplify.}$$
$$x = 3 \qquad \text{Divide each side by } -3.$$

The y-intercept is found by letting $x = 0$.

$$-3(0) + 2y = -9 \qquad \text{Let } x = 0.$$
$$2y = -9 \qquad \text{Simplify.}$$
$$y = -\frac{9}{2} \qquad \text{Divide each side by 2.}$$

The x-intercept is 3, and the y-intercept is $-\frac{9}{2}$.

(c) The graph is a line that passes through the points $(3, 0)$ and $\left(0, -\frac{9}{2}\right)$. See Figure 3.20.

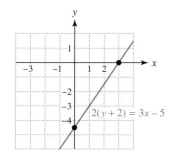

Figure 3.20

EXAMPLE 3 Modeling the velocity of a toy rocket

A toy rocket is shot vertically into the air. Its velocity v in feet per second after t seconds is given by $v = 160 - 32t$. Assume that $t \geq 0$ and $t \leq 5$.

(a) Graph the equation by finding the intercepts. Let t correspond to the horizontal axis (x-axis) and v correspond to the vertical axis (y-axis).

(b) Interpret each intercept.

Solution (a) To find the t-intercept let $v = 0$.

$$0 = 160 - 32t \qquad \text{Let } v = 0.$$
$$32t = 160 \qquad \text{Add } 32t \text{ to each side.}$$
$$t = 5 \qquad \text{Divide each side by 32.}$$

To find the v-intercept let $t = 0$.

$$v = 160 - 32(0) \qquad \text{Let } t = 0.$$
$$v = 160 \qquad \text{Simplify.}$$

Therefore the graph passes through $(5, 0)$ and $(0, 160)$, as shown in Figure 3.21.

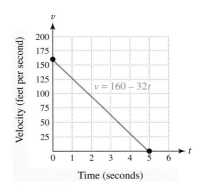

Figure 3.21

(b) The t-intercept indicates that the rocket had a velocity of 0 feet per second after 5 seconds. The v-intercept indicates that the rocket's initial velocity was 160 feet per second.

TABLE 3.13
Speed of a Car

t	y
1	70
2	70
3	70
4	70
5	70

HORIZONTAL LINES

Suppose that someone drives a car on a freeway at a constant speed of 70 miles per hour. Table 3.13 shows the speed y after t hours.

We can make a scatterplot of the data by plotting the five points $(1, 70)$, $(2, 70)$, $(3, 70)$, $(4, 70)$, and $(5, 70)$, as shown in Figure 3.22(a) on the following page. Note that, regardless of the value of time t, the speed is always 70 miles per hour. Thus the graph of the car's speed is a horizontal line, as shown in Figure 3.22(b). The equation of this line is $y = 70$ with y-intercept 70. There are no x-intercepts.

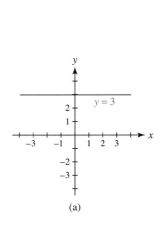

(a)

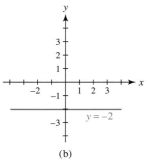

(b)

Figure 3.23

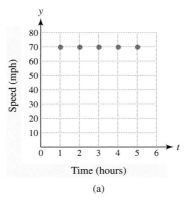

(a)

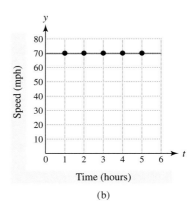

(b)

Figure 3.22

In general, the equation of a horizontal line is $y = b$, where b is a constant. Examples of horizontal lines are shown in Figure 3.23. Note that every point on the graph of $y = 3$ in Figure 3.23(a) has a y-coordinate of 3, and that every point on the graph of $y = -2$ in Figure 3.23(b) has a y-coordinate of -2.

HORIZONTAL LINE

The equation of a horizontal line with y-intercept b is $y = b$.

The equation $y = b$ is an example of a linear equation in the form $Ax + By = C$ with $A = 0$, $B = 1$, and $C = b$. (Note that in general B and b do not represent the same number.)

EXAMPLE 4   Graphing a horizontal line

Graph the equation $y = -1$ and identify its y-intercept.

Solution   The graph of $y = -1$ is a horizontal line passing through the point $(0, -1)$, as shown in Figure 3.24. Its y-intercept is -1.

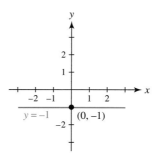

Figure 3.24

Critical Thinking

What is the equation of the x-axis? Explain your reasoning.

VERTICAL LINES

TABLE 3.14
Falling Object

t (seconds)	d (feet)
1	16
1	15
1	18
1	17

A scientist is experimentally determining the distance d that an object falls in 1 second. Because it is an experiment, the distances vary slightly on each trial. Table 3.14 shows the results.

We can make a scatterplot of the data by plotting the points $(1, 16)$, $(1, 15)$, $(1, 18)$, and $(1, 17)$, as shown in Figure 3.25(a). In each case the time is always 1 second and each point lies on the graph of a vertical line, as shown in Figure 3.25(b). This vertical line has the equation $x = 1$ because each point on the line has an x-coordinate of 1 and there are no restrictions on the y-coordinate. This line has x-intercept 1 but no y-intercept.

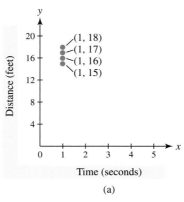 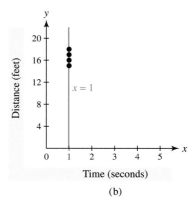

Figure 3.25

In general, the graph of a vertical line is $x = k$, where k is a constant. Examples of vertical lines are shown in Figure 3.26. Note that every point on the graph of $x = 3$ in Figure 3.26(a) has an x-coordinate of 3 and that every point on the graph of $x = -2$ shown in Figure 3.26(b) has an x-coordinate of -2.

Calculator Help
To graph a vertical line see
Appendix A (page AP-6).

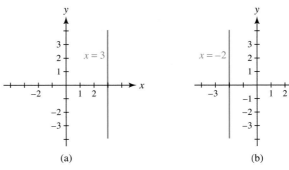

Figure 3.26

VERTICAL LINE

The equation of a vertical line with x-intercept k is $x = k$.

The equation $x = k$ is an example of a linear equation in the form $Ax + By = C$ with $A = 1$, $B = 0$, and $C = k$.

EXAMPLE 5   Graphing a vertical line

Graph the equation $x = -3$, and identify its x-intercept.

Solution   The graph of $x = -3$ is a vertical line passing through the point $(-3, 0)$, as shown in Figure 3.27. Its x-intercept is -3.

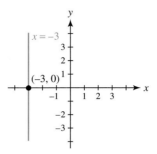

Critical Thinking

What is the equation of the y-axis? Explain your reasoning.

Figure 3.27

EXAMPLE 6   Writing equations of lines

Write the equation of the line shown in each graph.

(a)

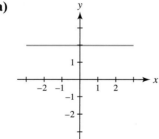

(b)

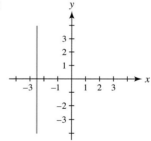

Solution   **(a)** The graph is a horizontal line with y-intercept 2. Its equation is $y = 2$.
(b) The graph is a vertical line with x-intercept -2.5. Its equation is $x = -2.5$.

═══════════════ MAKING CONNECTIONS ═══════════════

Lines and Linear Equations

The equation of any line can be written in the standard form $Ax + By = C$.

1. If $A = 0$ and $B \neq 0$, the line is horizontal.
2. If $A \neq 0$ and $B = 0$, the line is vertical.
3. If $A \neq 0$ and $B \neq 0$, the line is neither horizontal nor vertical.

3.3 PUTTING IT ALL TOGETHER

The following table summarizes several important topics presented in this section.

Concept	Explanation	Examples
x- and y-Intercepts	The x-coordinate of a point at which a graph intersects the x-axis is called an x-intercept. The y-coordinate of a point at which a graph intersects the y-axis is called a y-intercept.	x-intercept, -3 y-intercept, 2
Finding Intercepts	To find x-intercepts let $y = 0$ in the equation and solve for x. To find y-intercepts let $x = 0$ in the equation and solve for y.	Let $4x - 5y = 20$. x-intercept: $4x - 5(0) = 20$ $\qquad\qquad\qquad\qquad x = 5$ y-intercept: $4(0) - 5y = 20$ $\qquad\qquad\qquad\qquad y = -4$ The x-intercept is 5, and the y-intercept is -4.
Horizontal Line	Has equation $y = b$, where b is a constant Has y-intercept b and no x-intercept if $b \neq 0$	y-intercept, b
Vertical Line	Has equation $x = k$, where k is a constant Has x-intercept k and no y-intercept if $k \neq 0$	x-intercept, k

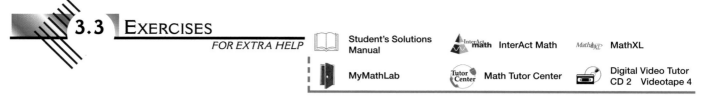

3.3 EXERCISES

FOR EXTRA HELP

Student's Solutions Manual

InterAct Math   InterAct Math

MathXL   MathXL

MyMathLab

Tutor Center   Math Tutor Center

Digital Video Tutor
CD 2   Videotape 4

CONCEPTS

1. _____ points determine a line.

2. The graph of the linear equation $Ax + By = C$ with $A \neq 0$ and $B \neq 0$ has _____ x-intercept(s) and _____ y-intercept(s).

3. The x-coordinate of a point where a graph intersects the x-axis is a(n) _____ .

4. To find an x-intercept, let $y =$ _____ and solve for x.

5. The y-coordinate of a point at which a graph intersects the y-axis is a(n) _____ .

6. To find a y-intercept, let $x =$ _____ and solve for y.

7. The graph of $y = 3$ is a _____ line with y-intercept _____ .

8. The equation for a horizontal line with y-intercept b is _____ .

9. The graph of $x = 3$ is a _____ line with x-intercept _____ .

10. The equation for a vertical line with x-intercept k is _____ .

FINDING INTERCEPTS

Exercises 11–18: Identify any x-intercepts and y-intercepts in the graph.

11.

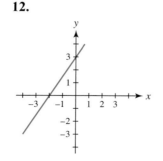

12.

13.

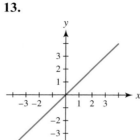

14.

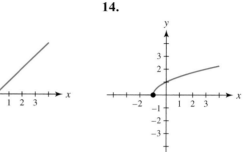

15.

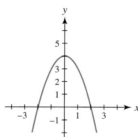

16.

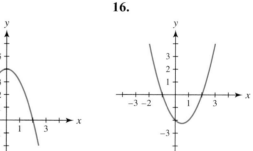

17.

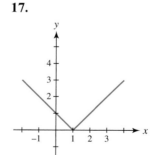

18.

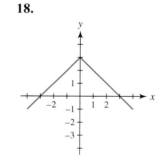

Exercises 19–22: Complete the table. Then determine the x-intercept and the y-intercept for the graph of the equation.

19. $y = x + 2$

x	−2	−1	0	1	2
y					

20. $y = 2x - 4$

x	-2	-1	0	1	2
y					

21. $-x + y = -2$

x	-4	-2	0	2	4
y					

22. $x + y = 1$

x	-2	-1	0	1	2
y					

Exercises 23–42: Find any intercepts. Then graph the linear equation.

23. $-2x + 3y = -6$ **24.** $4x + 3y = 12$

25. $3x - 5y = 15$ **26.** $2x - 4y = -4$

27. $x - 3y = 6$ **28.** $5x + y = -5$

29. $6x - y = -6$ **30.** $5x + 7y = -35$

31. $3x + 7y = 21$ **32.** $-3x + 8y = 24$

33. $40y - 30x = -120$ **34.** $10y - 20x = 40$

35. $\frac{1}{2}x - y = 2$ **36.** $x - \frac{1}{2}y = 4$

37. $-\dfrac{x}{4} + \dfrac{y}{3} = 1$ **38.** $\dfrac{x}{3} - \dfrac{y}{4} = 1$

39. $\dfrac{x}{3} + \dfrac{y}{2} = 1$ **40.** $\dfrac{x}{5} - \dfrac{y}{4} = 1$

41. $0.6y - 1.5x = 3$ **42.** $0.5y - 0.4x = 2$

HORIZONTAL AND VERTICAL LINES

Exercises 43–50: Graph each equation.

43. (a) $y = 2$ (b) $x = 2$

44. (a) $y = -2$ (b) $x = -2$

45. (a) $y = -4$ (b) $x = -4$

46. (a) $y = 0$ (b) $x = 0$

47. (a) $y = -1$ (b) $x = -1$

48. (a) $y = -\frac{1}{2}$ (b) $x = -\frac{1}{2}$

49. (a) $y = \frac{3}{2}$ (b) $x = \frac{3}{2}$

50. (a) $y = -1.5$ (b) $x = 2.5$

Exercises 51–58: Write an equation for the line shown in the graph.

51.
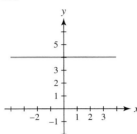

52.
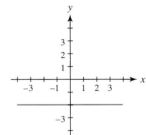

53.
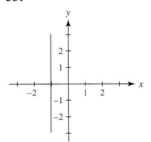

54.
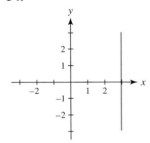

55.
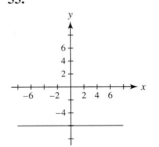

56.
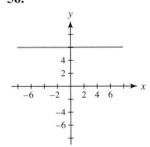

57.
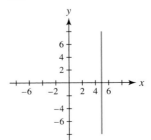

58.
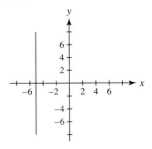

Exercises 59–62: Write an equation for the line that passes through the points shown in the table.

59.

x	−2	−1	0	1	2
y	1	1	1	1	1

60.

x	0	1	2	3	4
y	−10	−10	−10	−10	−10

61.

x	−6	−6	−6	−6	−6
y	5	4	3	2	1

62.

x	20	20	20	20	20
y	−2	−1	0	1	2

Exercises 63–68: Write the equations of a horizontal line and a vertical line that pass through the given point. (Hint: Make a sketch.)

63. $(1, 2)$

64. $(−3, 4)$

65. $(20, −45)$

66. $(−5, 12)$

67. $(0, 5)$

68. $(−3, 0)$

Exercises 69–76: Find an equation for a line satisfying the following conditions. (Hint: Make a sketch.)

69. Vertical, passing through $(−1, 6)$

70. Vertical, passing through $(2, −7)$

71. Horizontal, passing through $\left(\frac{3}{4}, -\frac{5}{6}\right)$

72. Horizontal, passing through $(5.1, 6.2)$

73. Perpendicular to $y = \frac{1}{2}$, passing through $(4, −9)$

74. Perpendicular to $x = 2$, passing through $(3, 4)$

75. Parallel to $x = 4$, passing through $\left(-\frac{2}{3}, \frac{1}{2}\right)$

76. Parallel to $y = −2.1$, passing through $(7.6, 3.5)$

APPLICATIONS

Exercises 77 and 78: **Distance**   *The distance of a driver from home is illustrated in the graph.*
  (a) *Find the intercepts.*
  (b) *Interpret each intercept.*

77.

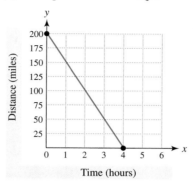

78.

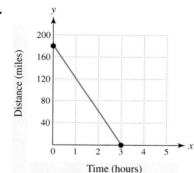

Exercises 79 and 80: **Water in a Pool**   *The amount of water in a swimming pool is depicted in the graph.*
  (a) *Find the intercepts.*
  (b) *Interpret each intercept.*

79.

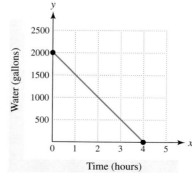

80.

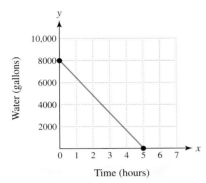

Exercises 81 and 82: Modeling a Toy Rocket (*Refer to Example 3.*) *The velocity v of a toy rocket in feet per second after t seconds of flight is given, where $t \geq 0$.*

    (a) *Find the intercepts and then graph the equation.*

    (b) *Interpret each intercept.*

81. $v = 128 - 32t$ **82.** $v = 96 - 32t$

WRITING ABOUT MATHEMATICS

83. Given an equation, explain how to find an x-intercept and a y-intercept.

84. The form $\frac{x}{a} + \frac{y}{b} = 1$ is called the **intercept form** of a linear equation. Explain how you can use this equation to find the intercepts. (*Hint:* Graph $\frac{x}{2} + \frac{y}{3} = 1$ and find its intercepts.)

Group Activity: Working with Real Data

Directions: Form a group of 2 to 4 people. Select someone to record the group's responses for this activity. All members of the group should work cooperatively to answer the questions. If your instructor asks for the results, each member of the group should be prepared to respond.

1. *Radio Stations* The approximate number of radio stations on the air for selected years from 1950 to 2000 is shown in the table.

x (year)	1950	1960	1970
y (stations)	2800	4100	6800

x (year)	1980	1990	2000
y (stations)	8600	10,800	12,600

Source: M. Street Corporation, 2001.

Make a line graph of the data. Be sure to label both axes.

2. *Estimation* Discuss ways to estimate the number of radio stations on the air in 1975. Compare your esti-

mates with the actual value of 7700 stations. Repeat this estimate for 1985 and compare it to the actual value of 10,400. Discuss your results.

3. *Modeling Equation* Substitute each x-value from the table into the equation $y = 196x - 379,400$ and determine the corresponding y-value. Do these y-values give reasonable approximations to the y-values in the table? Explain your answer.

4. *Making Estimates* Use $y = 196x - 379,400$ to estimate the number of radio stations on the air in 1975 and 1985. Compare the results to your answer in Exercise 2.

3.4   SLOPE AND RATES OF CHANGE

Finding Slopes of Lines · **Slope as a Rate of Change**

INTRODUCTION

Figure 3.28 shows some graphs of lines, where the horizontal axis represents time.

Which graph might represent the distance traveled by you if you are walking?

Which graph might represent the temperature in your freezer?

Which graph might represent the amount of gasoline in your car's tank while you are driving?

To be able to answer these questions, you probably used the concept of slope. In mathematics, slope is a real number that measures the "tilt" or "angle" of a line. In this section we discuss slope and how it is used in applications.

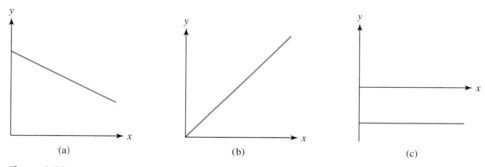

Figure 3.28

FINDING SLOPES OF LINES

The graph shown in Figure 3.29 illustrates the cost of parking for x hours. The graph tilts upward from left to right, which indicates that the cost increases as the number of hours increases. Note that, for each hour of parking, the cost increases by \$2. The graph *rises* 2 units for every unit of *run*, and the ratio $\frac{\text{rise}}{\text{run}}$ equals the *slope* of the line. The slope m of this line is 2, which indicates that the cost of parking is \$2 per hour.

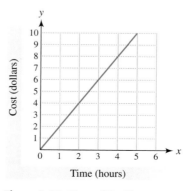

Figure 3.29 Cost of Parking

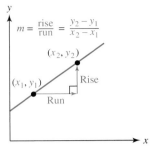

Figure 3.30 Slope *m* of a Line

A more general case of the slope of a line is shown in Figure 3.30 where a line passes through the points (x_1, y_1) and (x_2, y_2). The **rise**, or *change in y*, is $y_2 - y_1$, and the **run**, or *change in x*, is $x_2 - x_1$. Slope *m* is given by $m = \frac{y_2 - y_1}{x_2 - x_1}$.

Note: The symbol x_1 has a **subscript** of 1 and is read "*x* sub one" or "*x* one". Thus x_1 and x_2 are used to denote two different *x*-values. Similar comments apply to y_1 and y_2.

SLOPE

The **slope** *m* of the line passing through the points (x_1, y_1) and (x_2, y_2) is

$$m = \frac{\text{rise}}{\text{run}} = \frac{y_2 - y_1}{x_2 - x_1}$$

where $x_1 \neq x_2$. That is, slope equals rise over run.

EXAMPLE 1   Calculating the slope of a line

Use the two points labeled in Figure 3.31(a) to find the slope of the line. What are the rise and run between these two points? Interpret the slope in terms of rise and run.

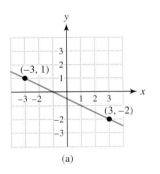
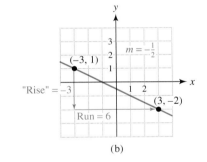

(a)   (b)

Figure 3.31

Solution   The line passes through the points $(-3, 1)$ and $(3, -2)$, so let $(x_1, y_1) = (-3, 1)$ and $(x_2, y_2) = (3, -2)$. The slope is

$$m = \frac{y_2 - y_1}{x_2 - x_1} = \frac{-2 - 1}{3 - (-3)} = \frac{-3}{6} = -\frac{1}{2}.$$

Between the points $(-3, 1)$ and $(3, -2)$ the "rise" is -3 units and the run is 6 units. See Figure 3.31(b). The ratio $\frac{\text{rise}}{\text{run}}$ is $\frac{-3}{6}$, or $-\frac{1}{2}$. Thus the graph falls 1 unit for each 2 units of run. A negative slope indicates that the line falls from left to right.

Note: In Example 1 the same slope would result if we let $(x_1, y_1) = (3, -2)$ and $(x_2, y_2) = (-3, 1)$. In this case the calculation would be

$$m = \frac{y_2 - y_1}{x_2 - x_1} = \frac{1 - (-2)}{-3 - 3} = \frac{3}{-6} = -\frac{1}{2}.$$

If a line has **positive slope**, it *rises from left to right*, as shown in Figure 3.32. If a line has **negative slope**, it *falls from left to right*, as shown in Figure 3.33. Slope 0 indicates that a line is horizontal, which is shown in Figure 3.34. Any two points on a vertical line have the same *x*-coordinate so the run always equals 0. Thus the slope is undefined for a vertical line, which is shown in Figure 3.35.

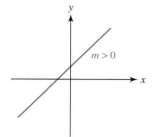

Figure 3.32
Positive Slope

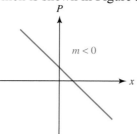

Figure 3.33
Negative Slope

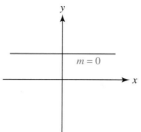

Figure 3.34
Zero Slope

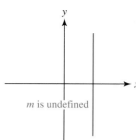

Figure 3.35
Undefined Slope

EXAMPLE 2   Finding slope from a graph

Find the slope of each line.

(a)

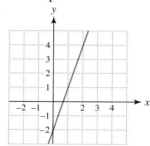

(b)

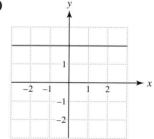

Solution   **(a)** The graph rises 3 units for each unit of run. For example, the graph passes through $(1, 1)$ and $(2, 4)$, so the line rises $4 - 1 = 3$ units with a $2 - 1 = 1$ unit run, as shown in Figure 3.36. Therefore the slope is

$$m = \frac{\text{rise}}{\text{run}} = \frac{3}{1} = 3.$$

(b) The line is horizontal, so the slope is 0.

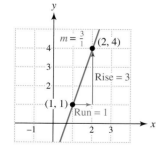

Figure 3.36

SLOPE OF A LINE

1. A line that rises *from left to right* has positive slope.
2. A line that falls *from left to right* has negative slope.
3. A horizontal line has slope 0.
4. A vertical line has undefined slope.

Two points determine a line and its slope, as demonstrated in the next example.

EXAMPLE 3   Calculating the slope of a line

Calculate the slope of the line passing through each pair of points. Graph the line.
(a) $(-2, 3), (2, 1)$ **(b)** $(-1, 3), (2, 3)$ **(c)** $(-3, 3), (-3, -2)$

Solution   **(a)** $m = \dfrac{y_2 - y_1}{x_2 - x_1} = \dfrac{1 - 3}{2 - (-2)} = \dfrac{-2}{4} = -\dfrac{1}{2}$. This slope indicates that the line falls 1 unit for every 2 units of horizontal run, as shown in Figure 3.37(a).

(b) $m = \dfrac{y_2 - y_1}{x_2 - x_1} = \dfrac{3 - 3}{2 - (-1)} = \dfrac{0}{3} = 0$. The line is horizontal, as shown in Figure 3.37(b).

(c) $m = \dfrac{y_2 - y_1}{x_2 - x_1} = \dfrac{-2 - 3}{-3 - (-3)} = \dfrac{-5}{0}$, which is an undefined expression. The line has undefined slope and is vertical, as shown in Figure 3.37(c).

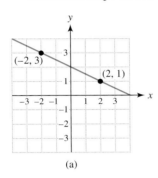
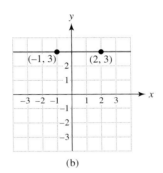
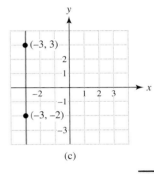

(a)   (b)   (c)

Figure 3.37

A point and a slope also determine a line, as illustrated in the next example.

EXAMPLE 4   Sketching a line with a given slope

Sketch a line passing through the point $(1, 4)$ and having slope $-\dfrac{2}{3}$.

Solution   Start by plotting the point $(1, 4)$. A slope of $-\dfrac{2}{3}$ indicates that the y-values *decrease* 2 units each time the x-values increase by 3 units. That is, the line *falls* 2 units for every 3-unit increase in the run. Because the line passes through $(1, 4)$, a **2**-unit decrease in y and a **3**-unit increase in x results in the line passing through the point $(1 + 3, 4 - 2)$ or $(4, 2)$. See Figure 3.38.

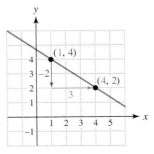

Figure 3.38 Slope $m = -\dfrac{2}{3}$

EXAMPLE 5   Sketching a line with a given *y*-intercept

Sketch a line with slope -2 and *y*-intercept 3.

Solution   For the *y*-intercept of 3, plot the point $(0, 3)$. The slope is -2, so the *y*-values decrease 2 units for each unit increase in *x*. Therefore the line must also pass through $(0 + 1, 3 - 2)$ or $(1, 1)$. Plot $(0, 3)$ and $(1, 1)$ and then sketch the line, as shown in Figure 3.39. ⎯⎯

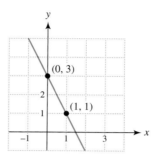

Figure 3.39 Slope: -2;
y-intercept: 3

If we know the slope of a line and a point on the line, we can complete a table of points, as demonstrated in the next example.

EXAMPLE 6   Completing a table of values

A line has slope 2 and passes through the first point listed in the table. Complete the table so that each point lies on the line.

x	-2	-1	0	1
y	1			

Solution   Consecutive *x*-values in the table increase by 1 unit. Because the slope is 2, consecutive *y*-values increase by 2 units for each unit increase in *x*, as shown in Table 3.15

TABLE 3.15

x	-2	-1	0	1
y	1	3	5	7

Slope as a Rate of Change

When lines are used to model physical quantities in applications, their slopes provide important information. Slope measures the **rate of change** in a quantity. We illustrated this concept in the next four examples.

EXAMPLE 7   Interpreting slope

The distance y in miles that an athlete training for a marathon is from home after x hours is shown in Figure 3.40.
(a) Find the y-intercept. What does the y-intercept represent?
(b) The graph passes through the point $(1, 10)$. Discuss the meaning of this point.
(c) Find the slope of this line. Interpret the slope as a rate of change.

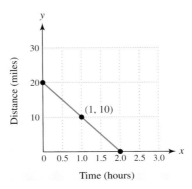

Figure 3.40   Distance from Home

Solution   **(a)** On the graph the y-intercept is 20, so the athlete is initially 20 miles from home.
(b) The point $(1, 10)$ means that after 1 hour the athlete is 10 miles from home.
(c) The line passes through the points $(0, 20)$ and $(1, 10)$. Its slope is

$$m = \frac{10 - 20}{1 - 0} = -10.$$

Slope -10 means that the athlete is running *toward* home at 10 miles per hour. A *negative* slope indicates that the distance between the runner and home is decreasing.

EXAMPLE 8   Interpreting slope

When a company manufactures 2000 compact disc players, its profit is $10,000, and when it manufactures 4500 compact disc players, its profit is $35,000.
(a) Find the slope of the line passing through $(2000, 10000)$ and $(4500, 35000)$.
(b) Interpret the slope as a rate of change.

Solution   **(a)** $m = \frac{35,000 - 10,000}{4500 - 2000} = \frac{25,000}{2500} = 10$
(b) Profit increases, *on average*, by $10 for each additional compact disc player made.

EXAMPLE 9 Analyzing tetanus cases

Table 3.16 lists numbers of reported cases of tetanus in the United States for selected years.

TABLE 3.16

Year	1950	1960	1970	1980	1990	2000
Cases of Tetanus	486	368	148	95	64	45

Source: Department of Health and Human Services.

(a) Make a line graph of the data.
(b) Find the slope of each line segment.
(c) Interpret each slope as a rate of change.

Solution **(a)** A line graph connecting the points (1950, 486), (1960, 368), (1970, 148), (1980, 95), (1990, 64), and (2000, 45) is shown in Figure 3.41.

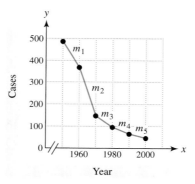

Figure 3.41 Tetanus Cases

(b) The slope of each line segment may be calculated as follows.

$$m_1 = \frac{368 - 486}{1960 - 1950} = -11.8$$

$$m_2 = \frac{148 - 368}{1970 - 1960} = -22.0$$

$$m_3 = \frac{95 - 148}{1980 - 1970} = -5.3$$

$$m_4 = \frac{64 - 95}{1990 - 1980} = -3.1$$

$$m_5 = \frac{45 - 64}{2000 - 1990} = -1.9$$

(c) Slope $m_1 = -11.8$ indicates that, *on average*, the number of tetanus cases *decreased* by 11.8 cases per year between 1950 and 1960. The other four slopes can be interpreted similarly.

Note: The number of tetanus cases did not decrease by *exactly* 11.8 cases per year between 1950 and 1960. Why? However, the yearly *average* decrease is 11.8 cases.

EXAMPLE 10 Sketching a Model

During a storm, rain falls at the rate of 2 inches per hour from 1 A.M. to 3 A.M., 1 inch per hour from 3 A.M. to 4 A.M., and $\frac{1}{2}$ inch per hour from 4 A.M. to 6 A.M.
(a) Sketch a graph that shows the total accumulation of rainfall from 1 A.M. to 6 A.M.
(b) What does the slope of each line segment represent?

Solution **(a)** At 1 A.M. the accumulated rainfall is 0, so place a point at $(1, 0)$. Rain falls at a constant rate of 2 inches per hour for the next 2 hours, so at 3 A.M. the total rainfall is 4 inches. Place a point at $(3, 4)$. Because the rainfall is constant, sketch a line segment from $(1, 0)$ to $(3, 4)$, as shown in Figure 3.42. Similarly, during the next hour 1 inch of rain falls, so draw a line segment from $(3, 4)$ to $(4, 5)$. Finally, 1 inch of rain falls from 4 A.M. to 6 A.M., so draw a line segment from $(4, 5)$ to $(6, 6)$.

Critical Thinking

An athlete runs 10 miles per hour for 30 minutes away from home and then jogs back home at 5 miles per hour. Sketch a graph that shows the distance between the athlete and home. What does the slope of each line segment represent?

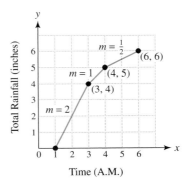

Figure 3.42 Total Rainfall

(b) The slope of each line segment represents the rate at which rain was falling. For example, the first segment has slope 2 because rain fell at a rate of 2 inches per hour during that period of time.

3.4 PUTTING IT ALL TOGETHER

The "tilt" of a line is called the slope and equals rise over run. A positive slope indicates that the line *rises* from left to right, whereas a negative slope indicates that the line *falls* from left to right. A horizontal line has slope 0, and a vertical line has undefined slope. When a quantity can be approximated with a line, its slope indicates a rate of change in the quantity. The following table summarizes some basic concepts about slope.

continued on next page

continued from previous page

Concept	Comments	Example
Rise, Run, and Slope	Rise is a vertical change in a line, and run is a horizontal change in a line. The ratio $\frac{rise}{run}$ is the slope m.	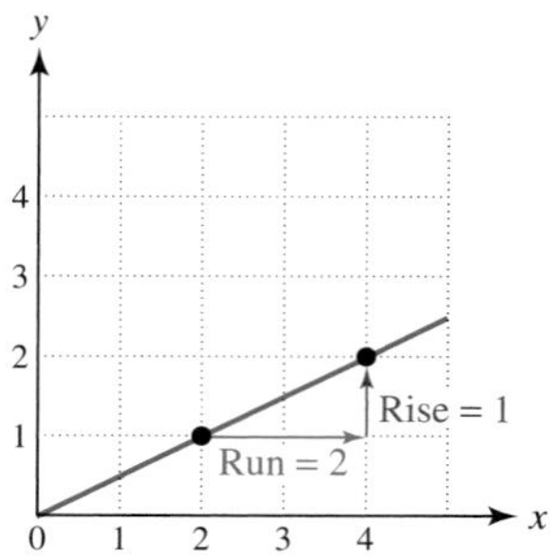 $$m = \frac{\text{rise}}{\text{run}} = \frac{1}{2}$$
Calculating Slope	For any two points (x_1, y_1) and (x_2, y_2) slope m is $$m = \frac{y_2 - y_1}{x_2 - x_1}.$$	The slope of the line passing through $(-2, 3)$ and $(1, 5)$ is $$m = \frac{5 - 3}{1 - (-2)} = \frac{2}{3}.$$ The line rises 2 units for every 3 units of run along the x-axis.
Slope as a Rate of Change	Slope indicates how fast the graph of a line is changing.	The graph shows that water is being pumped from a swimming pool. Slope $m = -66\frac{2}{3}$ indicates that water is leaving the tank at the rate of $66\frac{2}{3}$ gallons per hour. 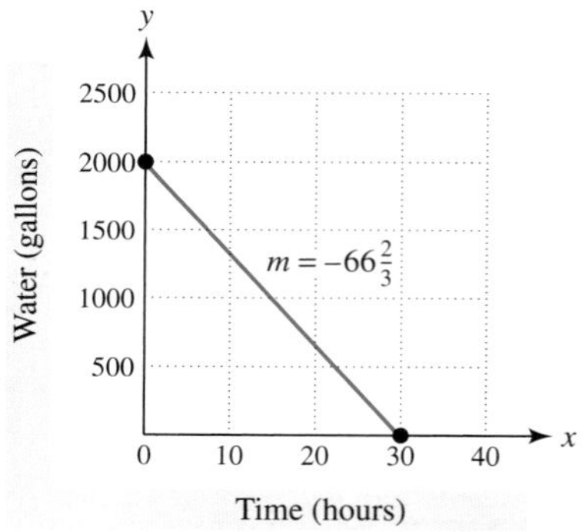

 **3.4** EXERCISES

FOR EXTRA HELP

 Student's Solutions Manual

 MyMathLab

 InterAct Math

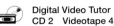 MathXL

Tutor Center   Math Tutor Center

Digital Video Tutor
CD 2   Videotape 4

CONCEPTS

1. Run is the change in the (horizontal/vertical) distance along a line.

2. Rise is the change in the (horizontal/vertical) distance along a line.

3. Slope m of a line is _____ over _____.

4. Slope 0 indicates that a line is _____.

5. Undefined slope indicates that a line is _____.

6. If a line passes through (x_1, y_1) and (x_2, y_2), then $m =$ _____.

Exercises 7–12: State whether the slope of the line is positive, negative, zero, or undefined.

7.

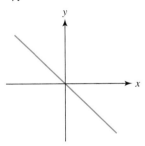

8.

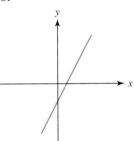

9.

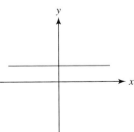

10.

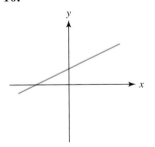

11.

12.

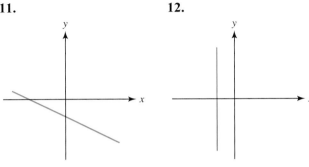

FINDING SLOPES OF LINES

Exercises 13–24: Find the slope of the line. Interpret the slope in terms of rise and run.

13.

14.

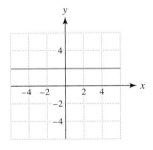

15.

16.

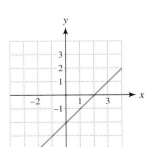
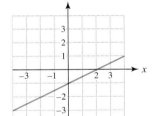
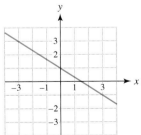

17.

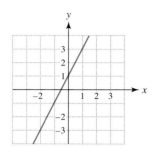

18.

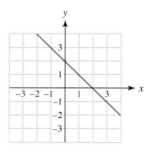

19.

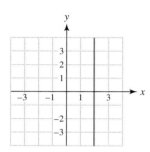

20.

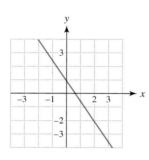

21.

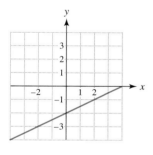

22.

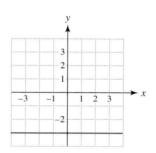

23.

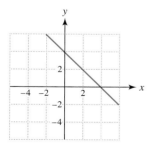

24.

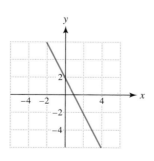

Exercises 25–28: Find the slope of the line passing through the two points. Graph the line.

25. $(1, 2), (2, 4)$

26. $(-3, 2), (2, -3)$

27. $(2, 1), (-1, 3)$

28. $(0, -1), (-2, 4)$

Exercises 29–44: Find the slope of the line passing through the two points.

29. $(4, -2), (-3, -7)$

30. $(15, -3), (20, 9)$

31. $(-3, 4), (4, -2)$

32. $(1, -3), (3, -5)$

33. $(-3, 6), (4, 6)$

34. $(-3, 5), (-5, 5)$

35. $(-1, 6), (-1, -4)$

36. $\left(\frac{1}{2}, -\frac{2}{7}\right), \left(\frac{1}{2}, \frac{13}{17}\right)$

37. $(1980, 5), (2000, 18)$

38. $(1989, 10), (1999, 16)$

39. $(1950, 6.1), (2000, 10.6)$

40. $(1900, 10), (1950, 35)$

41. $\left(\frac{1}{3}, -\frac{2}{7}\right), \left(-\frac{2}{3}, \frac{3}{7}\right)$

42. $(-1.3, 5.6), (-2.6, -2.5)$

43. $(12, -34), (14, 64)$

44. $(-25, 105), (60, 55)$

Exercises 45–52: (Refer to Example 4.) Sketch a line passing through the point with slope m.

45. $(0, 2), m = -1$

46. $(0, -1), m = 2$

47. $(1, 1), m = 3$

48. $(1, -1), m = -2$

49. $(-2, 3), m = -\frac{1}{2}$

50. $(-1, -2), m = \frac{3}{4}$

51. $(-3, 1), m = \frac{1}{2}$

52. $(-2, 2), m = -3$

Exercises 53–56: The table lists points located on a line. Find the slope, x-intercept, and y-intercept of the line.

53.

x	0	1	2	3
y	-2	0	2	4

54.

x	-1	0	1	2
y	0	5	10	15

55.

x	-2	-1	0	1
y	9	0	-9	-18

56.

x	-4	-2	0	2
y	6	3	0	-3

Exercises 57–62: (Refer to Example 6.) A line has the given slope m and passes through the first point listed in the table. Complete the table so that each point in the table lies on the line.

57. $m = 2$

x	0	1	2	3
y	−4			

58. $m = -\frac{1}{2}$

x	0	1	2	3
y	2			

59. $m = -3$

x	1	2	3	4
y	4			

60. $m = -1$

x	−1	0	1	2
y	10			

61. $m = \frac{3}{2}$

x	−4	−2	0	2
y	0			

62. $m = 3$

x	−2	0	2	4
y	−4			

Exercises 63–70: Do the following.
  (a) Graph the equation.
  (b) Find the slope of the line.

63. $y = 2x - 1$

64. $y = x - 2$

65. $3x + y = 2$

66. $-\frac{1}{2}x + y = -1$

67. $-x + 3y = 0$

68. $2x + y = 0$

69. $y = 2$

70. $y = -3$

SLOPE AS A RATE OF CHANGE

Exercises 71–74: Modeling   Choose the graph (a.–d.) that models the situation best.

71. Money a person earns after x hours, working for $10 per hour

72. Total acres of rain forests in the world during the past 20 years

73. World population from 1980 to 2000

74. Square miles of land in Nevada from 1950 to 2000

a.

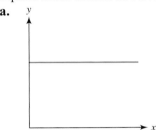

b.

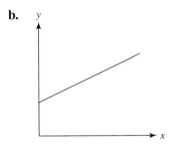

c.

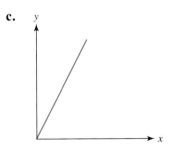

d.

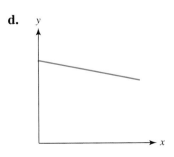

Exercises 75 and 76: Modeling   *The line graph represents the gallons of water in a small swimming pool after x hours. Assume that there is a pump that can either add water to or remove water from the pool.*

(a) *Estimate the slope of each line segment.*
(b) *Interpret each slope as a rate of change.*
(c) *Describe what happened to the amount of water in the pool.*

75.

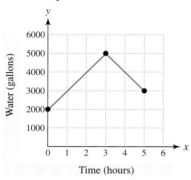

76.

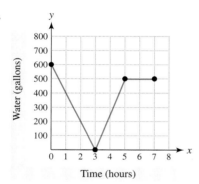

Exercises 77 and 78: Modeling   *An individual is driving a car along a straight road. The graph shows the distance that the driver is from home after x hours.*

(a) *Find the slope of each line segment in the graph.*
(b) *Interpret each slope as a rate of change.*
(c) *Describe both the motion and location of the car.*

77.

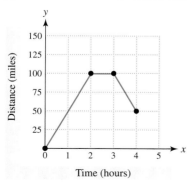

78.

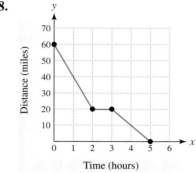

Exercises 79–82: Sketching a Model   *Sketch a graph that models the given situation.*

79. The distance that a bicycle rider is from home if the rider is initially 20 miles from home and arrives home after riding at a constant speed for 2 hours

80. The distance that an athlete is from home if the athlete starts at home, runs away from home at 8 miles per hour for 30 minutes, and then walks back home at 4 miles per hour

81. The distance that a person is from home if this individual drives to a mall, stays 2 hours, and then drives home. Assume that the distance to the mall is 20 miles and that the trip takes 30 minutes.

82. The amount of water in a 10,000-gallon swimming pool that is filled at the rate of 1000 gallons per hour, left full for 10 hours, and then drained at the rate of 2000 gallons per hour

APPLICATIONS

83. *Older Mothers*   The number of children born to mothers 40 years old or older in 1990 was 50 thousand and in 2000 it was 70 thousand.   (*Source:* National Center for Health Statistics.)
(a) Calculate the slope of the line passing through (1990, 50) and (2000, 70).
(b) Interpret the slope as a rate of change.

84. *Profit from Laptops*   When a company manufactures 500 laptop computers, its profit is $100,000, and when it manufactures 1500 laptop computers, its profit is $400,000.
(a) Find the slope of the line passing through the points (500, 100000) and (1500, 400000).
(b) Interpret the slope as a rate of change.

85. *Revenue*   The graph shows revenue received from selling x screwdrivers.

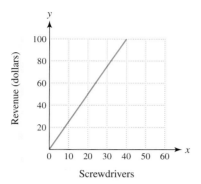

Screwdrivers

(a) Find the slope of the line shown.
(b) Interpret the slope as a rate of change.

86. *Electricity*   The graph shows how voltage is related to amperage in an electrical circuit. The slope corresponds to the resistance in ohms. Find the resistance in this electrical circuit.

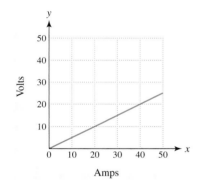

Amps

87. *Two-Cycle Engines*   Two-cycle engines used in snowmobiles, jet skis, chain saws, and outboard motors require a mixture of gas and oil to run properly. For certain engines the gallons of oil O that should be added to G gallons of gasoline is given by $O = \frac{1}{50}G$. (*Source:* Johnson Outboard Motor Company.)
(a) How many gallons of oil should be added to 50 gallons of gasoline and to 100 gallons of gasoline?
(b) Find the slope of the graph of the equation.
(c) Interpret the slope as a rate of change.

88. *Plant Growth*   The graph shows the seasonal growth of a type of grass for various amounts of rainfall.

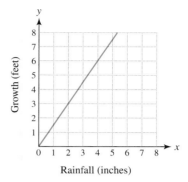

Rainfall (inches)

(a) Find the slope of the line shown.
(b) Interpret the slope as a rate of change.

89. *Walking for Charities*   The table lists the amount of money M in dollars raised for walking various distances in miles for a charity.

x	0	5	10	15
M	0	100	250	450

(a) Make a line graph of the data.
(b) Calculate the slope of each line segment.
(c) Interpret each slope as a rate of change.

90. *Insect Population*   The table lists the number N of black flies in thousands per acre after x weeks.

x	0	2	4	6
N	3	4	10	18

(a) Make a line graph of the data.
(b) Calculate the slope of each line segment.
(c) Interpret each slope as a rate of change.

91. *Median Household Income*   In 1980, median family income was about $18,000, and in 2000 it was about $42,000.   (Department of the Treasury.)
(a) Find the slope of the line passing through the points (1980, 18000) and (2000, 42000).
(b) Interpret the slope as a rate of change.
(c) If this trend continues, estimate the median family income in 2005.

92. *Minimum Wage*   In 1980, the minimum wage was about $3.10 per hour, and in 2000 it was $5.15. (*Source:* Department of Labor.)

   (a) Find the slope of the line passing through the points (1980, 3.1) and (2000, 5.15).

   (b) Interpret the slope as a rate of change.

   (c) If this trend continues, estimate the minimum wage in 2005.

WRITING ABOUT MATHEMATICS

93. If you have two points and the slope formula

$$m = \frac{y_2 - y_1}{x_2 - x_1},$$

does it matter which point is (x_1, y_1) and which point is (x_2, y_2)? Explain.

94. Suppose that a line approximates the distance y in miles that a person drives in x hours. What does the slope of the line represent? Give an example.

95. Describe the information that the slope m gives about a line. Be as complete as possible.

96. Could one line have two different slopes? Explain your answer.

CHECKING BASIC CONCEPTS   SECTIONS 3.3 AND 3.4

1. Identify the x- and y-intercepts in the graph.

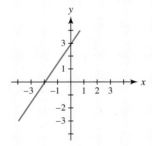

2. Complete the table for the equation $2x - y = 2$. Then determine the x- and y-intercepts.

x	-2	-1	0	1	2
y	-6				

3. Find any intercepts. Then graph the linear equation.
   (a) $x - 2y = 6$ **(b)** $y = 2$ **(c)** $x = -1$

4. Write the equations of a horizontal line and a vertical line that pass through the point $(-2, 4)$.

5. If possible, find the slope of the line passing through each pair of points.
   (a) $(-2, 3), (2, 6)$ **(b)** $(-5, 3), (0, 3)$
   (c) $(1, 5), (1, 8)$

6. Find the slope of the line shown.

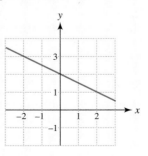

7. Sketch a line passing through the point $(-3, 1)$ and having slope 2.

8. *Modeling*   The line graph shows the depth of water in a small pond before and after a rain storm.
   (a) Estimate the slope of each line segment.
   (b) Interpret each slope as a rate of change.
   (c) Describe what happened to the amount of water in the pond.

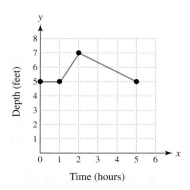

Time (hours)

9. *Minnesota County Population*   Dodge County has an area of 450 square miles with a population of 18,100. Meeker County has an area of 600 square miles with a population of 22,300. (***Source:*** Bureau of the Census.)
   (a) Find the slope of the line passing through the points (450, 18100) and (600, 22300).
   (b) Interpret the slope as a rate of change.
   (c) Is this rate of change a valid predictor of the population of a county with an area of 700 square miles? Explain your reasoning.

3.5   SLOPE–INTERCEPT FORM

Finding Slope–Intercept Form · Parallel and Perpendicular Lines

INTRODUCTION

For any two points in the xy-plane, we can draw a unique line passing through them, as illustrated in Figure 3.43(a). Another way to determine a unique line is to know the y-intercept and the slope. For example, if a line has y-intercept 2 and slope $m = 1$, then the resulting line is shown in Figure 3.43(b). In this section we discuss how to find the equation of a line given its slope and y-intercept.

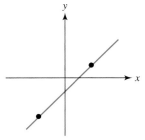

(a) Two points determine a line.

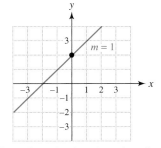

(b) One point and a slope determine a line.

Figure 3.43

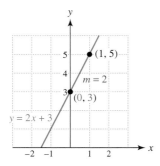

Figure 3.44 Slope 2,
y-intercept 3

FINDING SLOPE–INTERCEPT FORM

The graph of $y = 2x + 3$ passes through $(0, 3)$ and $(1, 5)$, as shown in Figure 3.44. The slope of this line is

$$m = \frac{5 - 3}{1 - 0} = 2.$$

If $x = 0$ in $y = 2x + 3$, then $y = 2(0) + 3 = 3$. Thus the graph of $y = 2x + 3$ has slope 2 and y-intercept 3. In general, the graph of $y = mx + b$ has slope m and y-intercept b. The form $y = mx + b$ is called the *slope–intercept form*.

SLOPE–INTERCEPT FORM

The line with slope m and y-intercept b is given by

$$y = mx + b,$$

the **slope–intercept form** of a line.

The graph of the equation $y = 4x - 3$ has slope 4 and y-intercept -3, and the graph of the equation $y = -\frac{1}{2}x + 8$ has slope $-\frac{1}{2}$ and y-intercept 8.

EXAMPLE 1 Sketching a line

Sketch a line with slope $-\frac{1}{2}$ and y-intercept 1. Write its slope–intercept form.

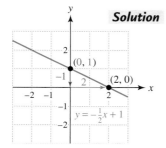

Figure 3.45

Solution For the y-intercept of 1 plot the point $(0, 1)$. Slope $-\frac{1}{2}$ indicates that the graph falls **1** unit for each **2**-unit increase in x. Thus the line passes through the point $(0 + 2, 1 - 1)$, or $(2, 0)$, as shown in Figure 3.45. The slope–intercept form of this line is $y = -\frac{1}{2}x + 1$.

EXAMPLE 2 Using a graph to write the slope–intercept form

For each graph write the slope–intercept form of each line.

(a)

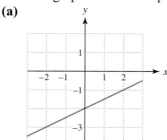

(b)

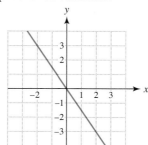

Solution **(a)** The graph intersects the y-axis at -2, so the y-intercept is -2. Because the graph rises **1** unit for each **2**-unit increase in x, the slope is $\frac{1}{2}$. The slope–intercept form of the line is $y = \frac{1}{2}x - 2$.

(b) The graph intersects the y-axis at 0, so the y-intercept is 0. Because the graph falls 3 units for each 2-unit increase in x, the slope is $-\frac{3}{2}$. The slope–intercept form of the line is $y = -\frac{3}{2}x + 0$, or $y = -\frac{3}{2}x$.

EXAMPLE 3   **Writing an equation in slope–intercept form**

Write each equation in slope–intercept form. Then give the slope and y-intercept of the line.
(a) $2x + 3y = 12$ **(b)** $x = 2y + 4$

Solution **(a)** To write the equation in slope–intercept form, solve for y.

$$2x + 3y = 12 \qquad \text{Given equation}$$
$$3y = -2x + 12 \qquad \text{Subtract } -2x \text{ from each side.}$$
$$y = -\frac{2}{3}x + 4 \qquad \text{Divide each side by 3.}$$

The slope of the line is $-\frac{2}{3}$, and the y-intercept is **4**.

(b) This equation is *not* in slope–intercept form because it is solved for x, not y.

$$x = 2y + 4 \qquad \text{Given equation}$$
$$x - 4 = 2y \qquad \text{Subtract 4 from each side.}$$
$$\frac{1}{2}x - 2 = y \qquad \text{Divide each side by 2.}$$
$$y = \frac{1}{2}x - 2 \qquad \text{Rewrite the equation.}$$

The slope of the line is $\frac{1}{2}$, and the y-intercept is -2 .

EXAMPLE 4   **Graphing an equation in slope–intercept form**

Write the equation $y = 3 - 2x$ in slope–intercept form and then graph it.

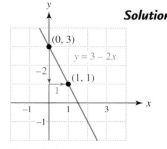

Figure 3.46

Solution First write the given equation in slope–intercept form.

$$y = 3 - 2x \qquad \text{Given equation}$$
$$y = 3 + (-2x) \qquad \text{Change subtraction to addition.}$$
$$y = -2x + 3 \qquad \text{Commutative property}$$

The slope–intercept form is $y = -2x + 3$, with slope -2 and y-intercept 3. To graph this equation plot the point $(0, 3)$. The line falls 2 units for each unit increase in x, so plot the point $(0 + 1, 3 - 2) = (1, 1)$. Sketch a line passing through $(0, 3)$ and $(1, 1)$, as shown in Figure 3.46.

EXAMPLE 5   Modeling cell phone costs

Roaming with a cell phone costs $5 for the initial connection and $0.50 per minute.
(a) If someone talks for 23 minutes, what is the charge?
(b) Write the slope–intercept form that gives the cost of talking for x minutes.
(c) If the charge is $8.50, how long did the person talk?

Solution **(a)** The charge for 23 minutes at $0.50 per minute plus $5 would be

$$0.50 \times 23 + 5 = \$16.50.$$

(b) The rate of increase is $0.50 per minute with an initial cost of $5. Let $y = 0.5x + 5$, where the slope or rate of change is **0.5** and the y-intercept is **5**.

(c) To determine how long a person can talk for $8.50, we can solve the following equation.

$0.5x + 5 = 8.5$	Equation to solve.
$0.5x = 3.5$	Subtract 5 from each side.
$x = \dfrac{3.5}{0.5}$	Divide each side by 0.5
$x = 7$	Simplify.

The person talked for 7 minutes. Note that this solution is based on the assumption that the phone company did not round up a fraction of a minute.

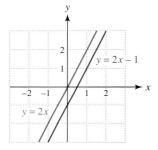

Figure 3.47

PARALLEL AND PERPENDICULAR LINES

Slope is an important concept for determining whether two lines are parallel. If two lines have the same slope, they are parallel. For example, the lines $y = 2x$ and $y = 2x - 1$ are parallel because they both have slope **2**, as shown in Figure 3.47.

PARALLEL LINES
Two lines with the same slope are parallel.
Two nonvertical parallel lines have the same slope.

EXAMPLE 6   Finding parallel lines

Find the slope–intercept form of a line parallel to $y = -2x + 3$ and passing through the point $(-2, 3)$. Sketch a graph of each line.

Solution Because the line $y = -2x + 3$ has slope -2, any parallel line also has slope -2 with slope–intercept form $y = -2x + b$ for some b. The value of b can be found by substituting the point $(-2, 3)$ into the slope–intercept form.

$y = -2x + b$	Slope–intercept form
$3 = -2(-2) + b$	Let $x = -2$ and $y = 3$.
$3 = 4 + b$	Multiply the real numbers.
$-1 = b$	Subtract 4 from each side.

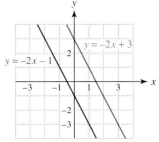

Figure 3.48

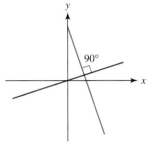

Figure 3.49 Perpendicular Lines

The y-intercept is -1, and so the slope–intercept form is $y = -2x - 1$. The graphs of the equations $y = -2x + 3$ and $y = -2x - 1$ are shown in Figure 3.48. Note that they are parallel lines, both with slope -2 but with different y-intercepts, 3 and -1 .

The lines shown in Figure 3.49 are perpendicular because they intersect at a $90°$ angle. Rather than measure the angle between two intersecting lines, we can determine whether two lines are perpendicular from their slopes. The slopes of perpendicular lines satisfy the following properties.

PERPENDICULAR LINES

If two perpendicular lines have nonzero slopes m_1 and m_2, then $m_1 \cdot m_2 = -1$.

If two lines have slopes m_1 and m_2 such that $m_1 \cdot m_2 = -1$, then they are perpendicular lines.

Table 3.17 shows examples of slopes m_1 and m_2 that result in perpendicular lines. Note that $m_1 \cdot m_2 = -1$ and that $m_2 = -\frac{1}{m_1}$. That is, the product of the two slopes is -1, and the two slopes are negative reciprocals of each other.

TABLE 3.17 Slopes of Perpendicular Lines

m_1	1	$-\frac{1}{2}$	-4	$\frac{2}{3}$	$\frac{3}{4}$
m_2	-1	2	$\frac{1}{4}$	$-\frac{3}{2}$	$-\frac{4}{3}$

We can use these concepts to find equations of perpendicular lines, as illustrated in the next two examples.

EXAMPLE 7 Finding perpendicular lines

Find the slope–intercept form of a line passing through the origin that is perpendicular to each line.
(a) $y = 3x$ **(b)** $y = -\frac{2}{5}x + 5$ **(c)** $-3x + 4y = 24$

Solution **(a)** If a line passes through the origin, then its y-intercept is 0 with slope–intercept form $y = mx$. The given line $y = 3x$ has slope $m_1 = 3$, so a line perpendicular to it has slope

$$m_2 = -\frac{1}{m_1} = -\frac{1}{3}.$$

The required slope–intercept form is $y = -\frac{1}{3}x$.
(b) The given line $y = -\frac{2}{5}x + 5$ has slope $m_1 = -\frac{2}{5}$, so a line perpendicular to it has slope $m_2 = \frac{5}{2}$. The required slope–intercept form is $y = \frac{5}{2}x$.

(c) To determine the slope of the given line, first write the equation in slope–intercept form.

$$-3x + 4y = 24 \qquad \text{Given equation}$$
$$4y = 3x + 24 \qquad \text{Add } 3x \text{ to each side.}$$
$$y = \frac{3}{4}x + 6 \qquad \text{Divide each side by 4.}$$

The slope of the given line is $m_1 = \frac{3}{4}$, so a line perpendicular to it has slope $m_2 = -\frac{4}{3}$. The required slope–intercept form is $y = -\frac{4}{3}x$.

EXAMPLE 8   Finding a perpendicular line

Find the slope–intercept form of the line perpendicular to $y = -\frac{1}{2}x + 1$ and passing through the point $(1, -1)$. Sketch each line in the same xy-plane.

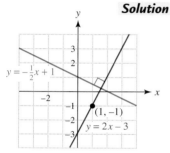

Figure 3.50

Solution   The line $y = -\frac{1}{2}x + 1$ has slope $m_1 = -\frac{1}{2}$. Any line perpendicular to it has slope $m_2 = 2$ with slope–intercept form $y = 2x + b$ for some b. The value of b can be found by substituting the point $(1, -1)$ in the slope–intercept form.

$$y = 2x + b \qquad \text{Slope–intercept form}$$
$$-1 = 2(1) + b \qquad \text{Let } x = 1 \text{ and } y = -1.$$
$$-1 = 2 + b \qquad \text{Multiply the real numbers.}$$
$$-3 = b \qquad \text{Subtract 2 from each side.}$$

The slope–intercept form is $y = 2x - 3$. The graphs of $y = -\frac{1}{2}x + 1$ and $y = 2x - 3$ are shown in Figure 3.50. Note that the point $(1, -1)$ lies on the graph of $y = 2x - 3$.

3.5   PUTTING IT ALL TOGETHER

The following table shows important forms of an equation of a line.

Concept	Comments	Example
Slope–Intercept Form $y = mx + b$	A unique equation for a line, determined by the slope m and the y-intercept b	An equation of the line with slope $m = 3$ and y-intercept $b = -5$ is $y = 3x - 5$.
Parallel Lines	$y = m_1x + b_1$ and $y = m_2x + b_2$, where $m_1 = m_2$ Nonvertical parallel lines have the same slope.	The lines $y = 2x - 1$ and $y = 2x + 2$ are parallel because they both have slope 2.

Concept	Comments	Example
Perpendicular Lines	$y = m_1x + b_1$ and $y = m_2x + b_2$, where $m_1m_2 = -1$ Perpendicular lines which are neither vertical nor horizontal have slopes whose product equals -1.	The lines $y = 3x - 1$ and $y = -\frac{1}{3}x + 2$ are perpendicular because $$m_1m_2 = 3\left(-\frac{1}{3}\right) = -1.$$ 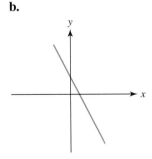

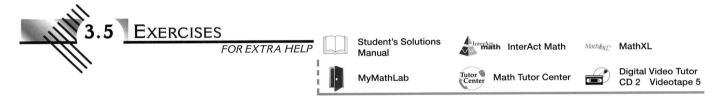

3.5 EXERCISES

FOR EXTRA HELP

Student's Solutions Manual

MyMathLab

InterAct Math

Math Tutor Center

MathXL

Digital Video Tutor
CD 2   Videotape 5

CONCEPTS

1. The slope–intercept form of a line is _____.

2. In the slope–intercept form of a line, m represents the _____ of the line.

3. In the slope–intercept form of a line, b represents the _____ of the line.

4. If $m = 0$ in the slope–intercept form of a line, then its graph is a _____ line.

5. If $b = 0$ in the slope–intercept form of a line, then its graph passes through the _____.

6. If a line passes through $(0, 5)$ with slope 2, then its slope–intercept form is _____.

Exercises 7–12: Match the description with its graph (a.–f.) on this page and the following page.

7. A line with positive slope and negative y-intercept

8. A line with positive slope and positive y-intercept

9. A line with negative slope and y-intercept 0

10. A line with negative slope and nonzero y-intercept

11. A line with no x-intercept

12. A line with no y-intercept

a.

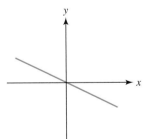

b.

c.

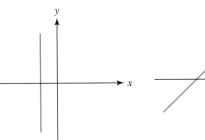

d.

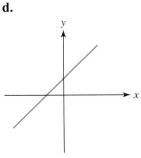

e.

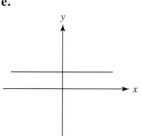

f.

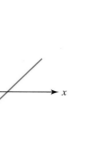

19.

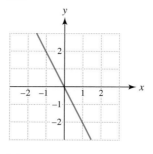

20.

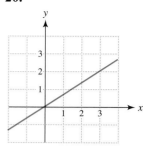

SLOPE–INTERCEPT FORM

Exercises 13–22: Write the slope–intercept form for the line shown.

13.

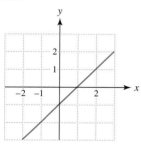

14.

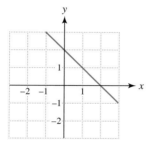

21.

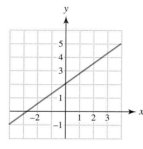

22.

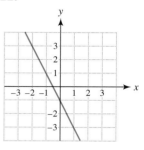

15.

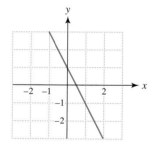

16.

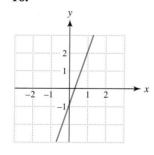

Exercises 23–32: Sketch a line with slope m and y-intercept b. Write its slope–intercept form.

23. $m = 1, b = 2$

24. $m = -1, b = 3$

25. $m = 2, b = -1$

26. $m = -3, b = 2$

27. $m = -\frac{1}{2}, b = -2$

28. $m = -\frac{2}{3}, b = 0$

29. $m = \frac{1}{3}, b = 0$

30. $m = 3, b = -3$

31. $m = -2, b = 1$

32. $m = \frac{1}{2}, b = -1$

17.

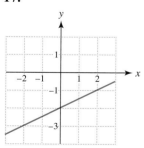

18.

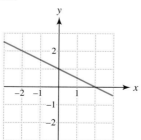

Exercises 33–44: Do the following.

    (a) Write the equation in slope–intercept form.

    (b) Give the slope and y-intercept of the line.

33. $x + y = 4$

34. $x - y = 6$

35. $2x + y = 4$

36. $-4x + y = 8$

37. $x - 2y = -4$

38. $x + 3y = -9$

39. $2x - 3y = 6$

40. $4x + 5y = 20$

41. $x = 4y - 6$

42. $x = -3y + 2$

43. $\frac{1}{2}x + \frac{3}{2}y = 1$

44. $-\frac{3}{4}x + \frac{1}{2}y = \frac{1}{2}$

Exercises 45–56: Graph the equation.

45. $y = -3x + 2$

46. $y = \frac{1}{2}x - 1$

47. $y = \frac{1}{3}x$

48. $y = -2x$

49. $y = 2$

50. $y = -3$

51. $y = -x + 3$

52. $y = \frac{2}{3}x - 2$

53. $y = -\frac{1}{2}x + 1$

54. $y = -2x + 2$

55. $y = 2 - x$

56. $y = 2 - 3x$

Exercises 57–60: The table shows points that all lie on the same line. Find the slope–intercept form for the line.

57.

x	0	1	2
y	2	4	6

58.

x	−1	0	1
y	4	8	12

59.

x	−2	0	2
y	−4	−2	0

60.

x	0	2	4
y	6	3	0

PARALLEL AND PERPENDICULAR LINES

Exercises 61–70: Find the slope–intercept form of the line satisfying the given conditions.

61. Slope $\frac{4}{7}$, y-intercept 3

62. Slope $-\frac{1}{2}$, y-intercept -7

63. Parallel to $y = 3x + 1$, passing through $(0, 0)$

64. Parallel to $y = -2x$, passing through $(0, 1)$

65. Parallel to $2x + 4y = 5$, passing through $(1, 2)$

66. Parallel to $-x - 3y = 9$, passing through $(-3, 1)$

67. Perpendicular to $y = -\frac{1}{2}x - 3$, passing through $(0, -4)$

68. Perpendicular to $y = \frac{3}{4}x - \frac{1}{2}$, passing through $(3, -2)$

69. Perpendicular to $x = -\frac{1}{3}y$, passing through $(-1, 0)$

70. Perpendicular to $6x - 3y = 18$, passing through $(4, -3)$

APPLICATIONS

71. *Rental Cars*   Driving a rental car x miles costs $y = 0.25x + 25$ dollars.
   (a) How much would it cost to rent the car but not drive it?
   (b) How much does it cost to drive the car 1 *additional* mile?
   (c) What is the y-intercept of $y = 0.25x + 25$? What does it represent?
   (d) What is the slope of $y = 0.25x + 25$? What does it represent?

72. *Calculating Rainfall*   The total rainfall y in inches that fell x hours past noon is given by $y = \frac{1}{2}x + 3$.
   (a) How much rainfall was there at noon?
   (b) At what rate was rain falling in the afternoon?
   (c) What is the y-intercept of $y = \frac{1}{2}x + 3$? What does it represent?
   (d) What is the slope of $y = \frac{1}{2}x + 3$? What does it represent?

73. *Long-Distance Phone Service*   Long-distance phone service costs $3.95 per month plus $0.07 per minute. (Assume that a partial minute is not rounded up.)
   (a) During July, a person talks a total of 50 minutes. What is the charge?
   (b) Write an equation in slope–intercept form that gives the monthly cost C of talking long distance for x minutes.
   (c) If the long-distance charge for one month is $8.64, how much time did the person spend talking on the phone?

74. *Electrical Rates*   Electrical service costs $8 per month plus $0.10 per kilowatt-hour of electricity used. (Assume that a partial kilowatt-hour is not rounded up.)
   (a) If the resident of an apartment uses 650 kilowatt-hours in 1 month, what is the charge?
   (b) Write an equation in slope–intercept form that gives the cost C of using x kilowatt-hours in 1 month.
   (c) If the monthly electrical bill for the apartment's resident is $43, how many kilowatt-hours were used?

75. *Cost of Driving*   The cost of driving a car includes both fixed costs and mileage costs. Assume that it costs $189.20 per month for insurance and car payments and $0.30 per mile for gasoline, oil, and routine maintenance.
   (a) Find values for m and b so that $y = mx + b$ models the monthly cost of driving the car x miles.
   (b) What does the value of b represent?

76. *Antarctic Ozone Layer*   The ozone layer occurs in Earth's atmosphere between altitudes of 12 and 18 miles and is an important filter of ultraviolet light from the sun. The thickness of the ozone layer is frequently measured in Dobson units. An average value is 300 Dobson units. In 1991, the reported minimum in the antarctic *ozone hole* was about 110 Dobson units. (*Source:* R. Huffman, *Atmospheric Ultraviolet Remote Sensing.*)

   (a) The equation $T = 0.01D$ describes the thickness T in millimeters of an ozone layer that is D Dobson units. How many millimeters thick was the ozone layer over the antarctic in 1991?

   (b) What is the average thickness of the ozone layer in millimeters?

WRITING ABOUT MATHEMATICS

77. Explain how the values of m and b can be used to graph the equation $y = mx + b$.

78. Explain how to find the value of b in the equation $y = 2x + b$ if the point $(3, 4)$ lies on the line.

Group Activity: Working with Real Data

Directions: Form a group of 2 to 4 people. Select someone to record the group's responses for this activity. All members of the group should work cooperatively to answer the questions. If your instructor asks for the results, each member of the group should be prepared to respond.

Exercises 1–5: In this set of exercises you are to use your knowledge of equations of lines to model the average annual cost of tuition and fees.

1. *Cost of Tuition*   In 1980, the average cost of tuition and fees at *private* four-year colleges was $3600, and in 2000 it was $16,300. Sketch a line that passes through the points $(1980, 3600)$ and $(2000, 16300)$. (*Source:* The College Board.)

2. *Rate of Change in Tuition*   Calculate the slope of the line in your graph. Interpret this slope as a rate of change.

3. *Modeling Tuition*   Find the slope–intercept form of the line in your sketch. What is the y-intercept and does it have meaning in this situation?

4. *Predicting Tuition*   Use your equation to estimate tuition and fees in 1998 and compare it to the known value of $14,700. Estimate tuition and fees in 2005.

5. *Public Tuition*   In 1980, the average cost of tuition and fees at *public* four-year colleges was $800, and in 2000 it was $3500. Repeat Exercises 1–4 for these data. Note that the known value for 1998 is $3250. (*Source:* The College Board.)

3.6   POINT–SLOPE FORM

Derivation of Point–Slope Form · Finding Point–Slope Form · Applications

INTRODUCTION

In 1995, there were 690 female officers in the Marine Corps, and by 2000 this number had increased to 932. This growth is illustrated in Figure 3.51, where the line passes through the points $(1995, 690)$ and $(2000, 932)$. Because two points determine a unique line, we can find the equation of this line and use it to *estimate* the number of female officers in other years. In this section we discuss how to find this equation by using the *point–slope form*, rather than the slope–intercept form.   (***Source:*** Department of Defense.)

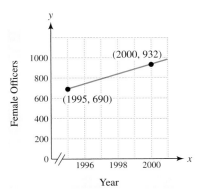

Figure 3.51 Female Officers in the Marines

DERIVATION OF POINT–SLOPE FORM

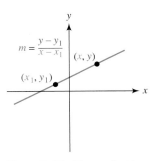

Figure 3.52 Slope of a Line

If we know the slope and y-intercept of a line, we can write its slope–intercept form, $y = mx + b$, which is an example of an **equation of a line**. The point–slope form is a different type of equation of a line.

Suppose that a (nonvertical) line with slope m passes through the point (x_1, y_1). If (x, y) is a different point on this line, then $m = \frac{y - y_1}{x - x_1}$. See Figure 3.52. Using this slope formula, we can find the point–slope form.

$$m = \frac{y - y_1}{x - x_1}$$ Slope formula

$$m \cdot (x - x_1) = \frac{y - y_1}{x - x_1} \cdot (x - x_1)$$ Multiply each side by $(x - x_1)$.

$$m(x - x_1) = y - y_1$$ Simplify.

$$y - y_1 = m(x - x_1)$$ Rewrite the equation.

$$y = m(x - x_1) + y_1$$ Add y_1 to each side.

The equation $y - y_1 = m(x - x_1)$ is traditionally called the *point–slope form*. An equivalent form that is helpful when graphing is $y = m(x - x_1) + y_1$. Both equations are in *point–slope form*.

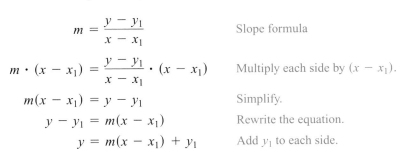

POINT–SLOPE FORM

The line with slope m passing through the point (x_1, y_1) is given by

$$y - y_1 = m(x - x_1), \text{ or equivalently,}$$

$$y = m(x - x_1) + y_1,$$

the **point–slope form** of a line.

FINDING POINT–SLOPE FORM

In the next example we find a point–slope form for a line. Note that *any* point that lies on the line can be used in its point–slope form.

EXAMPLE 1   Finding a point–slope form

Find a point–slope form for a line passing through the point $(-2, 3)$ with slope $-\frac{1}{2}$. Does the point $(2, 1)$ lie on this line?

Solution   Let $m = -\frac{1}{2}$ and $(x_1, y_1) = (-2, 3)$ in the point–slope form.

$$y - y_1 = m(x - x_1) \qquad \text{Point–slope form}$$

$$y - 3 = -\frac{1}{2}\big(x - (-2)\big) \qquad x_1 = -2, y_1 = 3, \text{ and } m = -\frac{1}{2}$$

$$y - 3 = -\frac{1}{2}(x + 2) \qquad \text{Simplify.}$$

To determine whether the point $(2, 1)$ lies on the line, substitute $x = 2$ and $y = 1$ in the equation.

$$1 - 3 \overset{?}{=} -\frac{1}{2}(2 + 2) \qquad \text{Let } x = 2 \text{ and } y = 1.$$

$$-2 \overset{?}{=} -\frac{1}{2}(4) \qquad \text{Simplify.}$$

$$-2 = -2 \qquad \text{A true statement}$$

The point $(2, 1)$ lies on the line because it satisfies the point–slope form.

═══════════════ MAKING CONNECTIONS ═══════════════

Slope–Intercept and Point–Slope Forms

The slope–intercept form, $y = mx + b$, is unique because any nonvertical line has one slope m and one y-intercept b. The point–slope form, $y - y_1 = m(x - x_1)$, is *not* unique because (x_1, y_1) can be any point that lies on the line. However, any point–slope form can be simplified to a unique slope–intercept form.

EXAMPLE 2   Finding a point–slope form

Use the labeled point in Figure 3.53 to write a point–slope form for the line and then simplify it to the slope–intercept form.

Solution   The graph rises 2 units for each 3 units of horizontal run, so the slope is $\frac{2}{3}$. Let $m = \frac{2}{3}$ and $(x_1, y_1) = (1, 1)$ in the point–slope form.

$$y - y_1 = m(x - x_1) \qquad \text{Point–slope form}$$

$$y - 1 = \frac{2}{3}(x - 1) \qquad \text{Let } m = \frac{2}{3}, x_1 = 1, \text{ and } y_1 = 1.$$

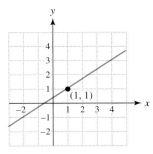

Figure 3.53

To simplify to the slope–intercept form, apply the distributive property.

$$y - 1 = \frac{2}{3}x - \frac{2}{3} \qquad \text{Distributive property}$$

$$y = \frac{2}{3}x + \frac{1}{3} \qquad \text{Add 1, or } \tfrac{3}{3}, \text{ to each side.}$$

The slope–intercept form is $y = \frac{2}{3}x + \frac{1}{3}$.

In the next example we use the point–slope form to find the equation of a line passing through two points.

EXAMPLE 3 Finding the equation of a line

Use the point–slope form to find an equation of the line passing through the points $(1, -4)$ and $(-2, 5)$.

Solution Before we can apply the point–slope form, we must find the slope.

$$m = \frac{y_2 - y_1}{x_2 - x_1} \qquad \text{Slope formula}$$

$$= \frac{5 - (-4)}{-2 - 1} \qquad x_1 = 1, y_1 = -4, x_2 = -2, \text{ and } y_2 = 5$$

$$= -3 \qquad \text{Simplify.}$$

We can let either the point $(1, -4)$ or the point $(-2, 5)$ be (x_1, y_1) in the point–slope form. If $(x_1, y_1) = (1, -4)$, then the equation of the line becomes the following.

$$y - y_1 = m(x - x_1) \qquad \text{Point–slope form}$$

$$y - (-4) = -3(x - 1) \qquad x_1 = 1, y_1 = -4, \text{ and } m = -3$$

$$y + 4 = -3(x - 1) \qquad \text{Simplify.}$$

If we let $(x_1, y_1) = (-2, 5)$, the point–slope form becomes

$$y - 5 = -3(x + 2).$$

Although the two point–slope forms in Example 3 might appear to be different, they actually are equivalent because they simplify to the same slope–intercept form.

$y + 4 = -3(x - 1)$	$y - 5 = -3(x + 2)$	Point–slope forms
$y = -3(x - 1) - 4$	$y = -3(x + 2) + 5$	Addition property
$y = -3x + 3 - 4$	$y = -3x - 6 + 5$	Distributive property
$y = -3x - 1$	$y = -3x - 1$	Identical slope–intercept forms

EXAMPLE 4 Finding equations of lines

Find the slope–intercept form for the line that satisfies the conditions.
(a) Slope $\frac{1}{2}$, passing through $(-2, 4)$
(b) x-intercept -3, y-intercept 2
(c) Perpendicular to $y = -\frac{2}{3}x$, passing through $\left(\frac{2}{3}, 3\right)$

Solution **(a)** Substitute $m = \frac{1}{2}$, $x_1 = -2$, and $y_1 = 4$ in the point–slope form.

$$y - y_1 = m(x - x_1) \qquad \text{Point–slope form}$$

$$y - 4 = \frac{1}{2}(x + 2) \qquad \text{Substitute.}$$

$$y - 4 = \frac{1}{2}x + 1 \qquad \text{Distributive property}$$

$$y = \frac{1}{2}x + 5 \qquad \text{Add 4 to each side.}$$

(b) The line passes through the points $(-3, 0)$ and $(0, 2)$. The slope of the line is

$$m = \frac{2 - 0}{0 - (-3)} = \frac{2}{3}.$$

Because the line has slope $\frac{2}{3}$ and y-intercept 2, the slope–intercept form is

$$y = \frac{2}{3}x + 2.$$

(c) The slope of the given line is $m_1 = -\frac{2}{3}$, so the slope of the line perpendicular to it is $\frac{3}{2}$. Let $m = \frac{3}{2}$, $x_1 = \frac{2}{3}$, and $y_1 = 3$ in the point–slope form.

$$y - y_1 = m(x - x_1) \qquad \text{Point–slope form}$$

$$y - 3 = \frac{3}{2}\left(x - \frac{2}{3}\right) \qquad \text{Substitute.}$$

$$y - 3 = \frac{3}{2}x - 1 \qquad \text{Distributive property}$$

$$y = \frac{3}{2}x + 2 \qquad \text{Add 3 to each side.}$$

APPLICATIONS

In the next example we find the equation of the line that models the Marine Corps data presented in the introduction to this section.

EXAMPLE 5 Modeling numbers of female officers

In 1995, there were 690 female officers in the Marine Corps, and by 2000 this number had increased to 932. See Figure 3.51.
(a) Use the point $(1995, 690)$ to find a point–slope form of the line shown in Figure 3.51.
(b) Interpret the slope as a rate of change.
(c) Use Figure 3.51 to estimate the number of female officers in 1997. Then use your equation from part (a) to approximate this number. How do your answers compare?

Solution **(a)** The slope of the line passing through $(1995, 690)$ and $(2000, 932)$ is

$$m = \frac{932 - 690}{2000 - 1995} = 48.4.$$

If we let $x_1 = 1995$ and $y_1 = 690$, then the point–slope form becomes

$$y - 690 = 48.4(x - 1995) \quad \text{or}$$
$$y = 48.4(x - 1995) + 690.$$

(b) Slope $m = 48.4$ indicates that the number of female officers increased, *on average,* by about 48.4 officers per year.

(c) From Figure 3.51, it appears that the number of female officers in 1997 was about 790. See Figure 3.54. To estimate this value let $x = 1997$ in the equation of the point–slope form from part (a).

$$y = 48.4(1997 - 1995) + 690 \approx 786.8$$

Although the graphical estimate and calculated answers are not exactly equal, they are approximately equal. Estimations made from a graph usually are not exact.

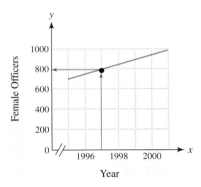

Figure 3.54

In the next example we review several concepts of lines.

EXAMPLE 6 Modeling water in a pool

A small swimming pool is being emptied by a pump that removes water at a constant rate. After 1 hour the pool contains 5000 gallons, and after 3 hours it contains 3000 gallons.

(a) How fast is the pump removing water?

(b) Find the slope–intercept form of a line that models the amount of water in the pool. Interpret the slope.

(c) Find the y-intercept and the x-intercept. Interpret each.

(d) Sketch a graph of the amount of water in the pool during the first 6 hours.

(e) The point $(2, 4000)$ lies on the graph. Explain its meaning.

Solution **(a)** The pump removes $5000 - 3000 = 2000$ gallons of water in 2 hours, or 1000 gallons per hour.

(b) The line passes through the points $(1, 5000)$ and $(3, 3000)$, so the slope is

$$m = \frac{3000 - 5000}{3 - 1} = -1000.$$

One way to find the slope–intercept form is to use the point–slope form.

$y - y_1 = m(x - x_1)$	Point–slope form
$y - 5000 = -1000(x - 1)$	$m = -1000, x_1 = 1,$ and $y_1 = 5000$
$y - 5000 = -1000x + 1000$	Distributive property
$y = -1000x + 6000$	Add 5000 to each side.

The slope -1000 means that the pump is *removing* 1000 gallons per hour.

(c) The y-intercept is 6000 and indicates that the pool initially contained 6000 gallons. To find the x-intercept let $y = 0$ in the slope–intercept form.

$0 = -1000x + 6000$	Let $y = 0$.
$1000x = 6000$	Add $1000x$ to each side.
$x = \dfrac{6000}{1000}$	Divide by 1000.
$x = 6$	Reduce.

An x-intercept of 6 indicates that the pool is empty after 6 hours.

(d) The x-intercept is **6**, and the y-intercept is **6000**. Sketch a line passing through $(6, 0)$ and $(0, 6000)$, as shown in Figure 3.55.

(e) The point $(2, 4000)$ indicates that after 2 hours the pool contains 4000 gallons of water.

Critical Thinking

Suppose that a line models the amount of water in a swimming pool. What does a positive slope indicate? What does a negative slope indicate?

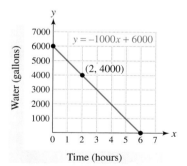

Figure 3.55 Water in a Pool

3.6 PUTTING IT ALL TOGETHER

The following table summarizes the point–slope form of a line. Note that the slope–intercept form is unique for a given line, whereas the point–slope form depends on the point used for (x_1, y_1).

Concept	Comments	Example
Point–Slope Form $$y - y_1 = m(x - x_1) \quad \text{or}$$ $$y = m(x - x_1) + y_1$$	Used to find the equation of a line, given two points or one point and the slope Can always be simplified to slope–intercept form	For two points $(1, 2)$ and $(3, 5)$, first compute $m = \frac{5 - 2}{3 - 1} = \frac{3}{2}$. An equation of this line is $$y - 2 = \frac{3}{2}(x - 1) \quad \text{or}$$ $$y = \frac{3}{2}(x - 1) + 2.$$ 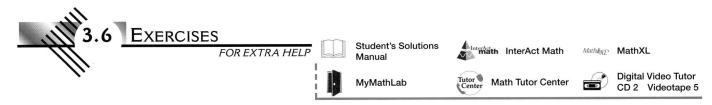

3.6 EXERCISES

FOR EXTRA HELP

CONCEPTS

1. How many lines are determined by two distinct points?

2. How many lines are determined by a point and a slope?

3. Give the slope–intercept form of a line.

4. Give the point–slope form of a line.

5. Is the slope–intercept form of a line unique? Explain.

6. Is the point–slope form of a line unique? Explain.

POINT–SLOPE FORM

Exercises 7–12: Determine whether the given point lies on the line.

7. $(-3, 3)$ $y - 1 = -\frac{2}{3}x$

8. $(4, 0)$ $y + 1 = \frac{1}{4}x$

9. $(1, 4)$ $y - 3 = -(x - 1)$

10. $(3, -11)$ $y + 1 = -2(x + 3)$

11. $(0, 4)$ $y = \frac{1}{2}(x + 4) + 2$

12. $(2, -8)$ $y = 3(x - 5) - 1$

Exercises 13–16: Use the labeled point to write the point–slope form for the line.

13.
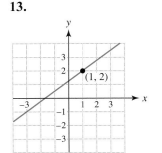

14.
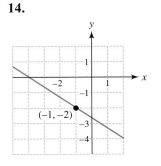

15.   **16.**

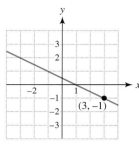

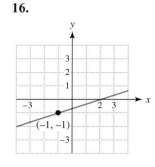

Exercises 17–26: Find a point–slope form for the line sat-
isfying the stated conditions. When two points are given,
use the first point in the point–slope form.

17. Slope -3, passing through $(1, -2)$

18. Slope $-\frac{2}{3}$, passing through $(-3, 6)$

19. Slope 1.5, passing through $(2000, 30)$

20. Slope -10, passing through $(2004, 100)$

21. Passing through $(2, 4)$ and $(-1, -3)$

22. Passing through $(3, -1)$ and $(1, -4)$

23. Passing through $(5, 0)$ and $(0, -3)$

24. Passing through $(-2, 0)$ and $(0, -1)$

25. Passing through $(1990, 15)$ and $(2000, 65)$

26. Passing through $(1999, 5)$ and $(2004, 30)$

Exercises 27–34: Write the point–slope form in slope–
intercept form.

27. $y - 4 = 3(x - 2)$ **28.** $y - 3 = -2(x + 1)$

29. $y + 2 = \frac{1}{3}(x + 6)$ **30.** $y + \frac{2}{3} = -\frac{1}{6}(x - 2)$

31. $y = -2(x - 2) + 5$ **32.** $y = -\frac{1}{2}(x + 4) - 6$

33. $y = -16(x + 1.5) + 5$

34. $y = -15(x - 1) + 100$

Exercises 35–44: Find the slope–intercept form for the
line satisfying the conditions.

35. Slope -2, passing through $(4, -3)$

36. Slope $\frac{1}{5}$, passing through $(-2, 5)$

37. Passing through $(3, -2)$ and $(2, -1)$

38. Passing through $(8, 3)$ and $(-7, 3)$

39. x-intercept 3, y-intercept $\frac{1}{3}$

40. x-intercept 2, y-intercept -3

41. Parallel to $y = 2x - 1$, passing through $(2, -3)$

42. Parallel to $y = -\frac{3}{2}x$, passing through $(0, 20)$

43. Perpendicular to $y = -\frac{1}{2}x + 3$, passing through $(6, -3)$

44. Perpendicular to $y = \frac{3}{5}(x + 1) + 3$, passing through $(1, -2)$

Exercises 45–48: The points in the table lie on a line. Find
the slope–intercept form of the line.

45.

x	1	2	3	4
y	-3	-5	-7	-9

46.

x	2	3	4	5
y	5	8	11	14

47.

x	-1	1	3	5
y	-3	-2	-1	0

48.

x	-1	5	11	17
y	1	-3	-7	-11

GRAPHICAL INTERPRETATION

49. *Distance and Speed*   A person is driving a car along a straight road. The graph shows the distance y in miles that the driver is from home after x hours.

    (a) Is the person traveling toward or away from home?

    (b) The graph passes through $(1, 250)$ and $(4, 100)$. Discuss the meaning of these points.

    (c) How fast is the driver traveling?

    (d) Find the slope–intercept form of the line. Interpret the slope as a rate of change.

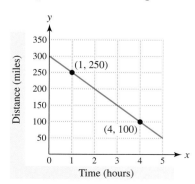

50. *Distance and Speed*   A person rides a bicycle at 10 miles per hour, first away from home for 1 hour and then toward home for 1 hour. Sketch a graph that shows the distance *d* between the bicyclist and home after *x* hours.

51. *Water and Flow*   The graph shows the amount of water *y* in a 500-gallon tank after *x* minutes have elapsed.
   (a) Is water entering or leaving the tank? How much water is in the tank after 4 minutes?
   (b) Find the *y*-intercept. Explain its meaning.
   (c) Find the slope–intercept form of the line. Interpret the slope as a rate of change.
   (d) After how many minutes will the tank be full?

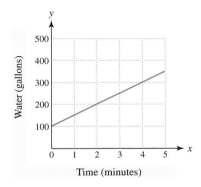

Time (minutes)

52. *Water and Flow*   A hose is used to fill a 100-gallon barrel. If the hose delivers 5 gallons of water per minute, sketch a graph of the amount *A* of water in the barrel during the first 20 minutes.

53. *Change in Temperature*   The outside temperature was 40°F at 1 A.M. and 15°F at 6 A.M. Assume that the temperature changed at a constant rate.
   (a) At what rate did the temperature change?
   (b) Find the slope–intercept form of a line that models the temperature *T* at *x* A.M. Interpret the slope as a rate of change.
   (c) Assuming that your equation is valid for times after 6 A.M., find and interpret the *x*-intercept.
   (d) Sketch a graph that shows the temperature from 1 A.M. to 9 A.M.
   (e) The point (4, 25) lies on the graph. Explain its meaning.

54. *Cost of Fuel*   The cost of buying 5 gallons of fuel oil is $5.25 and the cost of buying 15 gallons of fuel oil is $15.75.
   (a) What is the cost of a gallon of fuel oil?

   (b) Find the slope–intercept form of a line that models the cost of buying *x* gallons of fuel oil. Interpret the slope as a rate of change.
   (c) Find and interpret the *x*-intercept.
   (d) Sketch a graph that shows the cost of buying 20 gallons or less of fuel oil.
   (e) The point (11, 11.55) lies on the graph. Explain its meaning.

APPLICATIONS

55. *State and Federal Inmates*   From 1988 to 2000, the number of state and federal inmates *y* in thousands can be modeled by $y = 64(x - 1988) + 628$, where *x* is the year.
   (*Source:* Bureau of Justice.)
   (a) Find the number of inmates in 1988 and 2000.
   (b) What is the slope of the graph of *y*? Interpret the slope as a rate of change.

56. *Municipal Waste*   From 1960 to 2000, municipal solid waste *y* in millions of tons can be modeled by $y = 3.6(x - 1960) + 87.8$, where *x* is the year.
   (*Source:* Environmental Protection Agency.)
   (a) Find the tons of waste in 1999.
   (b) What is the slope of the graph of *y*? Interpret the slope as a rate of change.

57. *Tuition and Fees*   The graph models average tuition and fees, in dollars, at public four-year colleges from 1982 to 1996. (*Source:* The College Board.)
   (a) The line passes through the points (1984, 1225) and (1987, 1621). Explain the meaning of each point.
   (b) Find a point–slope form for this line. Interpret the slope as a rate of change.
   (c) Use the graph to estimate tuition and fees in 1990. Then use your equation from part (b) to estimate tuition and fees in 1990.

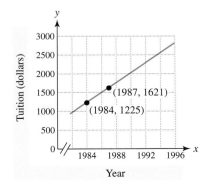

Year

58. *Western Population*   In 1950 the western region of the United States had a population of 20.2 million, and in 1960 it was 28.4 million. (*Source:* Bureau of the Census.)

   (a) Find the slope–intercept form of a line passing through (1950, 20.2) and (1960, 28.4).

   (b) Use your equation from part (a) to estimate the western population in 1967.

59. *HIV Infection Rates*   In 1996, there were an estimated 22 million HIV infections worldwide with an annual infection rate of 3.1 million. (*Source:* U.S. Centers for Disease Control and Prevention.)

   (a) Find the slope–intercept form of a line that approximates the cumulative number of HIV infections in millions during year x, where $x \geq 1996$.

   (b) Estimate the number of HIV infections in 2003.

60. *Farm Pollution*   In 1979, the number of farm pollution incidents reported in England and Wales was about 1500, and in 1988 it was about 4000. (*Source:* C. Mason, *Biology of Freshwater Pollution.*)

   (a) Find a slope–intercept form of a line passing through (1979, 1500) and (1988, 4000).

   (b) Use your equation from part (a) to estimate the number of incidents in 1986.

WRITING ABOUT MATHEMATICS

61. Explain how to find the equation of a line passing through two points with coordinates (x_1, y_1) and (x_2, y_2). Give an example.

62. Explain how slope is related to rate of change. Give an example.

CHECKING BASIC CONCEPTS   SECTIONS 3.5 AND 3.6

1. Write the slope–intercept form for the line shown in the graph.

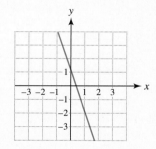

2. Write the equation $4x - 5y = 20$ in slope–intercept form. Give the slope and y-intercept.

3. Graph $y = \frac{1}{2}x - 3$.

4. Write the slope–intercept form of a line that satisfies the following.

   (a) Slope 3, passing through $(0, -2)$

   (b) Perpendicular to $y = \frac{2}{3}x$, passing through $(-2, 3)$

   (c) Passing through $(1, -4)$ and $(-2, 3)$

5. Find the slope–intercept form of the line passing through the points in the table.

x	-2	0	1
y	-1	3	5

6. *Distance and Speed*   A bicyclist is riding at a constant speed and is 36 miles from home at 1 P.M. Two hours later the bicyclist is 12 miles from home.

   (a) Find the slope–intercept form of a line passing through (1, 36) and (3, 12).

   (b) How fast is the bicyclist traveling?

   (c) When will the bicyclist arrive home?

   (d) How far was the bicyclist from home at noon?

7. *Snowfall*   The total amount of snowfall S in inches t hours past noon is given by $S = 2t + 5$.

   (a) How many inches of snow fell by noon?

   (b) At what rate did snow fall in the afternoon?

   (c) What is the y-intercept for the graph of this equation? What does it represent?

   (d) What is the slope for the graph of this equation? What does it represent?

3.7 INTRODUCTION TO MODELING

Basic Concepts · Modeling Linear Data

INTRODUCTION

For centuries people have tried to understand the world around them by creating models. For example, a weather forecast is based on a model. Mathematics is used to create these weather models, which often contain thousands of equations.

A model is an *abstraction* of something that people observed. Not only should a good model describe *known* data, but it should also be able to predict *future* data. In this section we discuss linear models. Linear models are used to describe data that have a constant rate of change.

BASIC CONCEPTS

Figure 3.56(a) shows a scatterplot of the number of inmates in the federal prison system from 1997 to 2000. The four points in the graph appear to be "nearly" collinear. That is, they appear almost to lie on the same line. Using mathematical modeling, we can find an equation for such a line. Once we have found it, we can use it to make estimates about the federal inmate population. An example of such a line is shown in Figure 3.56(b). (See Exercise 49 at the end of this section.)

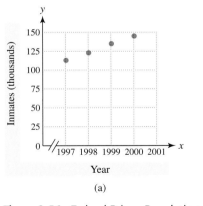
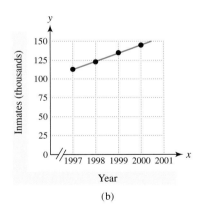

(a)   (b)

Figure 3.56 Federal Prison Population

Generally, mathematical models are not exact representations of data. Data that might appear to be linear may not be, as is the case in Figure 3.56(b), where the line does not pass through all four points. Figure 3.57(a) on the following page shows data modeled *exactly* by a line, whereas Figure 3.57(b) shows data modeled *approximately* by a line. In applications a model is more apt to be approximate than exact.

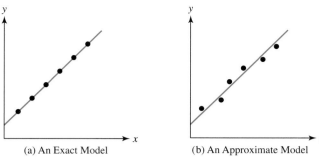

(a) An Exact Model   (b) An Approximate Model

Figure 3.57

EXAMPLE 1   Determining whether a model is exact

A person can vote in the United States at age 18 or over. Table 3.18 shows the voting-age population P in thousands for selected years x. Does the equation $P = 1.5x - 2797$ model the data exactly? Explain.

TABLE 3.18   Voting-Age Population

x	1996	1998	2000
P	197	198	203

Source: Bureau of the Census.

Solution   To determine whether the equation models the data exactly, let $x = 1996, 1998$, and 2000 in the given equation.

$$x = 1996: \qquad P = 1.5(1996) - 2797 = 197$$
$$x = 1998: \qquad P = 1.5(1998) - 2797 = 200$$
$$x = 2000: \qquad P = 1.5(2000) - 2797 = 203$$

The model is *not exact* because it does not predict a voting-age population of 198 thousand in 1998.

MODELING LINEAR DATA

Linear data can be modeled with a line. In the next example we use a line to model gas mileage.

EXAMPLE 2   Determining gas mileage

TABLE 3.19

x	2	4	6	8
y	30	60	90	120

Table 3.19 shows the number of miles y traveled by an SUV on x gallons of gasoline.
(a) Plot the data in the xy-plane. Be sure to label each axis.
(b) Sketch a line that models the data. (You may want to use a ruler.)
(c) Find the equation of the line and interpret the slope of the line.
(d) How far could this SUV travel on 11 gallons of gasoline?

Solution (a) Plot the points (2, 30), (4, 60), (6, 90), and (8, 120), as shown Figure 3.58(a).

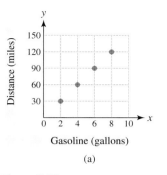
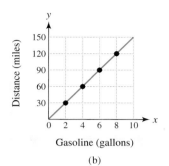

(a)   (b)

Figure 3.58

(b) Sketch a line similar to the one shown in Figure 3.58(b). This particular line passes through each data point.

(c) First find the slope m of the line by choosing two points that the line passes through, such as (2, 30) and (8, 120).

$$m = \frac{120 - 30}{8 - 2} = \frac{90}{6} = 15$$

Now find the equation of the line passing through (2, 30) with slope 15.

$$y - y_1 = m(x - x_1) \qquad \text{Point–slope form}$$
$$y - 30 = 15(x - 2) \qquad x_1 = 2, y_1 = 30, \text{ and } m = 15$$
$$y - 30 = 15x - 30 \qquad \text{Distributive property}$$
$$y = 15x \qquad \text{Add 30 to each side.}$$

The data are modeled by the equation $y = 15x$. Slope 15 indicates that the mileage of this SUV is 15 miles per gallon.

(d) On 11 gallons of gasoline the SUV could go $y = 15(11) = 165$ miles. _____

EXAMPLE 3   Modeling linear data

Table 3.20 contains ordered pairs that can be modeled approximately by a line.
(a) Plot the data. Could a line pass through all five points?
(b) Sketch a line that models the data and then determine its equation.

TABLE 3.20

x	1	2	3	4	5
y	3	5	6	10	11

Solution (a) Plot the ordered pairs (1, 3), (2, 5), (3, 6), (4, 10), and (5, 11), as shown in Figure 3.59(a) on the following page. The points are not collinear, so it is impossible to sketch a line that passes through all five points.

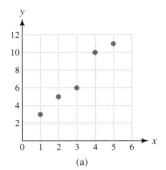

(a)

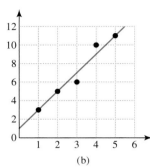

(b)

Figure 3.59

(b) One possibility for a line is shown in Figure 3.59(b). This line passes through three of the five points, and is above one point and below another point. To determine the equation of this line, pick two points that the line passes through. For example, the points (1, 3) and (5, 11) lie on the line. The slope of this line is

$$m = \frac{11 - 3}{5 - 1} = \frac{8}{4} = 2.$$

The equation of the line passing through (1, 3) with slope 2 can be found as follows.

$$
\begin{aligned}
y - y_1 &= m(x - x_1) && \text{Point–slope form} \\
y - 3 &= 2(x - 1) && x_1 = 1, y_1 = 3, \text{ and } m = 2 \\
y - 3 &= 2x - 2 && \text{Distributive property} \\
y &= 2x + 1 && \text{Add 3 to each side.}
\end{aligned}
$$

The equation of this line is $y = 2x + 1$.

When a quantity increases at a constant rate, it can be modeled with the linear equation $y = mx + b$. This concept is illustrated in the next example.

EXAMPLE 4   Modeling worldwide HIV cases in children

At the beginning of 2000, a total of 5.1 million children (under age 15) had been infected with HIV. The rate of new infections was 0.6 million per year.
(a) Write a linear equation $C = mx + b$ that models the total number of children C in millions that had been infected with HIV, x years after January 1, 2000.
(b) Estimate C at the beginning of 2005.

Solution   **(a)** In the equation $C = mx + b$, the rate of change in HIV infections corresponds to the slope m, and the initial number of infections at the beginning of 2000 corresponds to b. Therefore the equation $C = 0.6x + 5.1$ models the data.
(b) The beginning of 2005 is 5 years after January 1, 2000, so let $x = 5$.

$$C = 0.6(5) + 5.1 = 8.1 \text{ million}$$

Note: There were a total of 5.1 million infected children at the beginning of 2000 with an infection rate of 0.6 million per year. After 5 years there would be an additional $0.6(5) = 3$ million infected children, raising the total number to $5.1 + 3 = 8.1$ million.

EXAMPLE 5   Modeling with linear equations

Find a linear equation in the form $y = mx + b$ that models the quantity y after x days.
(a) A quantity y is initially 500 and increases at a rate of 6 per day.
(b) A quantity y is initially 1800 and decreases at a rate of 25 per day.
(c) A quantity y is initially 10,000 and remains constant.

Solution   **(a)** In the equation $y = mx + b$, b represents the initial amount and m represents the rate of change. Therefore $y = 6x + 500$.

(b) The quantity y is decreasing at the rate of 25 per day with an initial amount of 1800, so $y = -25x + 1800$.

(c) The quantity is constant, so $m = 0$ and $y = 10,000$.

These concepts about modeling are summarized as follows.

MODELING WITH A LINEAR EQUATION

To model a quantity y that has a constant rate of change, use the equation

$$y = mx + b,$$

where

$$m = \text{(constant rate of change)} \quad \text{and} \quad b = \text{(initial amount)}.$$

3.7 PUTTING IT ALL TOGETHER

The following table summarizes important concepts related to linear modeling.

Concept	Comments	Example
Linear Model	Used to model a quantity that has a constant rate of change.	If a total of 2 inches of rain fell before noon, and if rain fell at the rate of $\frac{1}{2}$ inch per hour, then $y = \frac{1}{2}x + 2$ models the total rainfall x hours past noon.
Modeling Linear Data with a Line	1. Plot the data. 2. Sketch a line that either passes through or nearly through the points. 3. Pick two points on the line and find the equation of the line.	To model $(0, 4)$, $(1, 3)$, and $(2, 2)$ plot the points and sketch a line as shown in the accompanying figure. Many times one line cannot pass through all the points. The equation of the line is $y = -x + 4$.

3.7 EXERCISES

FOR EXTRA HELP

 Student's Solutions Manual

 MyMathLab

 InterAct Math

 MathXL

Tutor Center Math Tutor Center

Digital Video Tutor CD 2 Videotape 5

CONCEPTS

1. A model is a(n) _____ of something that people observe.

2. A good model should be able to _____ future data.

3. Linear data are modeled by a(n) _____ equation.

4. If a line passes through all the data points, it is a(n) _____ model.

5. If a line passes near, but not through each data point, it is a(n) _____ model.

6. If a quantity is modeled by the equation $y = mx + b$, then $m =$ _____ and $b =$ _____.

Exercises 7–12: Modeling Match the situation to the graph (a.–f.) that models it best. In each case the x-axis represents time.

7. College tuition from 1980 to 2000

8. Yearly average temperature in degrees Celsius at the North Pole

9. Profit from selling boxes of candy if it costs $200 to make the candy

10. Height of Mount Hood in Oregon

11. Total amount of water delivered by a garden hose if it flows at a constant rate

12. Sales of 8-track music tapes from 1970 to 1980 (*Hint:* 8-track tapes are obsolete.)

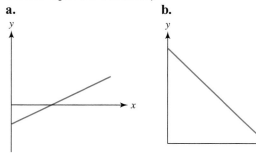

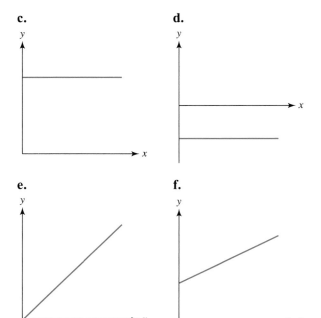

MODELING LINEAR DATA

Exercises 13–18: State whether the ordered pairs shown in the table are modeled exactly by the linear equation.

13. $y = 2x + 2$

x	0	1	2
y	2	4	6

14. $y = -2x + 5$

x	0	1	2
y	5	3	0

15. $y = -4x$

x	−1	0	1
y	4	0	−8

16. $y = 5 - x$

x	−2	1	4
y	7	4	1

17. $y = 1.4x - 4$

x	0	5	10
y	−4	3	9

18. $y = -\frac{4}{3}x - \frac{13}{3}$

x	-7	-4	-1
y	5	1	-3

Exercises 19–24: State whether the linear model shown in the graph is exact or approximate. Then find the equation of the line.

19.

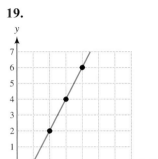

20.

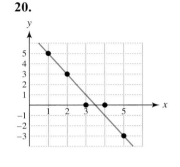

21.

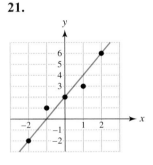

22.

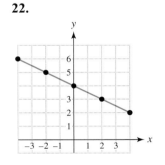

23.

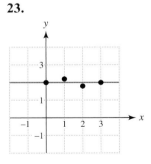

24.

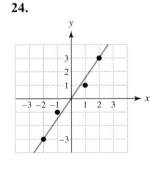

Exercises 25–30: Find an equation $y = mx + b$ that models the quantity y after x units of time.

25. A quantity y is initially 40 and increases at a rate of 5 per minute.

26. A quantity y is initially -60 and increases at a rate of 1.7 per minute.

27. A quantity y is initially -5 and decreases at a rate of 20 per day.

28. A quantity y is initially 5000 and decreases at a rate of 35 per day.

29. A quantity y is initially 8 and remains constant.

30. A quantity y is initially -45 and remains constant.

Exercises 31–38: Modeling Data   The table contains ordered pairs.
    (a) Plot the data. Could a line pass through all five points?
    (b) Sketch a line that models the data.
    (c) Determine an equation of the line. Answers may vary.

31.

x	0	1	2	3	4
y	4	2	0	-2	-4

32.

x	0	1	2	2	3
y	7	6	5	4	3

33.

x	-2	-1	0	1	2
y	4	1	0	-1	-4

34.

x	-2	-1	0	1	2
y	7	5	1	-3	-5

35.

x	-6	-4	-2	0	2
y	1	0	-1	-2	-3

36.

x	-4	-2	0	2	4
y	1	2	3	4	5

37.

x	-6	-3	0	3	6
y	-3	-2	-0.5	0	0.5

38.

x	-3	-2	-1	0	1
y	1	2	3	4	5

APPLICATIONS

Exercises 39–46: Write an equation that models the described quantity. Specify what each variable represents.

39. A barrel contains 200 gallons of water and is being filled at a rate of 5 gallons per minute.

40. A barrel contains 40 gallons of gasoline and is being drained at a rate of 3 gallons per minute.

41. An athlete has run 5 miles and is jogging at 6 miles per hour.

42. A new car has 141 miles on its odometer and is traveling at 70 miles per hour.

43. A worker has already earned $200 and is being paid $8 per hour.

44. A gambler has lost $500 and is losing money at a rate of $150 per hour.

45. A carpenter has already shingled 5 garage roofs and is shingling roofs at a rate of 1 per day.

46. A vinyl record has been playing for 3 minutes and continues to rotate at $33\frac{1}{3}$ revolutions per minute.

47. *Kilimanjaro Glacier*   Mount Kilimanjaro is located in Tanzania, Africa, and has an elevation of 19,340 feet. In 1912, the glacier on this peak covered 5 acres. By 2002 this glacier had melted to only 1 acre. (*Source: NBC News.*)
   (a) Assume that this glacier melted at a constant rate each year. Find this *yearly* rate.
   (b) Use your answer in part (a) to write a linear equation that gives the acreage A of this glacier t years past 1912.

48. *World Population*   In 1975, the world's population reached 4 billion people, and in 2000 the world's population reached 6 billion people.
   (*Source: The World Almanac 2001.*)
   (a) Find the average yearly increase in the world's population from 1975 to 2000.
   (b) Write a linear equation that estimates the world's population P in billions x years after 1975.

49. *Prison Population*   The data points in Figure 3.56 are (1997, 113), (1998, 123), (1999, 135), and (2000, 145).
   (a) Use the first and last data points to determine a line that models the data. Write the equation in slope–intercept form.

(b) Use the line to estimate the prison population in 2003.

50. *Niagara Falls*   The average flow of water over Niagara Falls is 212,000 cubic feet of water per second.
   (a) Write an equation that gives the number of cubic feet of water F that flow over the falls in x seconds.
   (b) How much water flows over Niagara Falls in 1 minute?

51. *Gas Mileage*   (Refer to Example 2.) The table shows the number of miles y traveled by a car on x gallons of gasoline.

x (gallons)	3	6	9	12
y (miles)	60	120	180	240

   (a) Plot the data in the xy-plane. Be sure to label each axis.
   (b) Sketch a line that models this data. (You may want to use a ruler.)
   (c) Calculate the slope of the line. Interpret the slope.
   (d) Find an equation of the line.
   (e) How far could this car travel on 7 gallons of gasoline?

52. *Air Temperature*   Generally, the air temperature becomes colder as the altitude above the ground increases. The table lists typical air temperatures x miles high when the ground temperature is $80°$ F.

x (miles)	0	1	2	3
y (°F)	80	62	44	26

   (a) Plot the data in the xy-plane. Be sure to label each axis.
   (b) Sketch a line that models these data. (You may want to use a ruler.)
   (c) Calculate the slope of the line. Interpret the slope.
   (d) Find the slope–intercept form of the line.
   (e) Estimate the air temperature 5 miles high.

WRITING ABOUT MATHEMATICS

53. In Example 2 the gas mileage of an SUV is modeled with a linear equation. Explain why it is reasonable to use a linear equation to model this situation. (*Hint:* Compare the amounts of gasoline that the SUV uses traveling 30 miles and 60 miles.)

54. Explain the steps for finding the equation of a line that models a table of data points.

CHECKING BASIC CONCEPTS — SECTION 3.7

1. State whether the ordered pairs shown in the table are modeled exactly by $y = -5x + 10$.

x	-2	-1	0	1
y	20	15	10	5

2. State whether the linear model shown in the graph is exact or approximate. Then find the equation of the line.

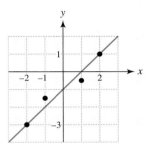

3. Find an equation, $y = mx + b$, that models the quantity y after x units of time.

(a) A quantity y is initially 50 pounds and increases at a rate of 10 pounds per day.

(b) A quantity y is initially $200°\text{F}$ and decreases at a rate of $2°\text{F}$ per minute.

4. The table contains ordered pairs.

(a) Plot the data. Could a line pass through all four points?

(b) Sketch a line that models the data.

(c) Determine the equation of the line.

x	-2	0	2	4
y	2	1	0	-1

5. *Global Warming* The average annual temperature in Antarctica increased by $0.05°\text{F}$ per year between 1959 and 1988. (*Source: USA Today.*)

(a) Write an equation that models the average temperature *increase T*, x years after 1959.

(b) Use your equation to calculate the temperature increase between 1959 and 1975.

CHAPTER

3 Summary

Section 3.1 Introduction to Graphing

The Rectangular Coordinate System (*xy*-plane)

Points	Plotted as (x, y) ordered pairs
Four Quadrants	I, II, III, and IV; the axes do not lie in a quadrant.

Note: The point $(1, 0)$ does not lie in a quadrant.

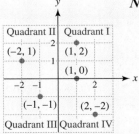

Types of Data Graphs

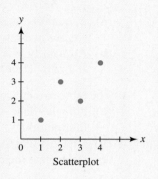

Scatterplot

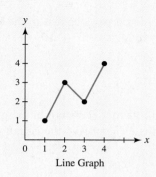

Line Graph

Section 3.2   *Linear Equations in Two Variables*

Equations in Two Variables   An equation with two variables and possibly some constants

Examples: $y = 3x + 7$ and $x + y = 100$

Solution to an Equation in Two Variables   The solution to an equation in two variables is an ordered pair that makes the equation a true statement.

Example: $(1, 2)$ is a solution to $2x + y = 4$ because $2(1) + 2 = 4$ is true.

Graphing a Linear Equation in Two Variables

Linear Equation $y = mx + b$ or $Ax + By = C$

Graphing   Plot at least three points and connect them with a line.

Example: $y = 2x - 1$

x	y
0	−1
1	1
2	3

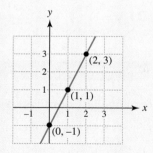

Section 3.3   *More Graphing of Lines*

Intercepts

x-Intercept   The x-coordinate of a point at which a graph intersects the x-axis; to find an x-intercept, let $y = 0$ in the equation and solve for x.

y-Intercept   The y-coordinate of a point at which a graph intersects the y-axis; to find a y-intercept, let $x = 0$ in the equation and solve for y.

Example: $x + 3y = 3$

x-intercept: Solve $x + 3(0) = 3$ to find the *x*-intercept of 3.

y-intercept: Solve $0 + 3y = 3$ to find the *y*-intercept of 1.

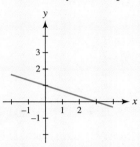

x-intercept: 3, *y*-intercept: 1

Horizontal and Vertical Lines

The equation of a horizontal line with *y*-intercept *b* is $y = b$.

The equation of a vertical line with *x*-intercept *k* is $x = k$.

Example: The horizontal line $y = -1$ has *y*-intercept -1 and no *x*-intercepts. The vertical line $x = 2$ has *x*-intercept 2 and no *y*-intercepts.

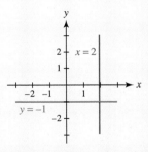

Section 3.4 *Slope and Rates of Change*

Slope   The ratio $\frac{\text{rise}}{\text{run}}$, or $\frac{\text{change in } y}{\text{change in } x}$, is the slope *m* of a line. A positive slope indicates that a line rises from left to right, and a negative slope indicates that a line falls from left to right.

Example: Slope $\frac{2}{3}$ indicates that a line rises 2 units for every 3 units of run along the *x*-axis. The line shown in the graph has slope $\frac{2}{3}$.

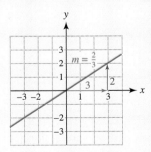

Calculating Slope	A line passing through (x_1, y_1) and (x_2, y_2) has slope

$$m = \frac{y_2 - y_1}{x_2 - x_1}.$$

Example: The line through $(-2, 2)$ and $(3, 4)$ has slope

$$m = \frac{4 - 2}{3 - (-2)} = \frac{2}{5}.$$

Horizontal Line	Has slope 0
Vertical Line	Has undefined slope

Slope as a Rate of Change   Slope indicates how fast a line is rising or falling. In applications, slope can indicate how fast a quantity is changing.

Example: The line shown in the graph has slope -2 and depicts an initial outside temperature of $6°$F. Slope -2 indicates that the temperature is *decreasing* at a rate of $2°$F per hour.

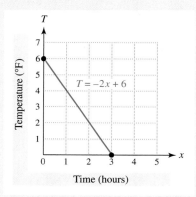

Section 3.5   *Slope–Intercept Form*

Slope–Intercept Form

$y = mx + b$ Slope m, y-intercept b

Example: $y = -\frac{1}{2}x + 2$ has slope $-\frac{1}{2}$ and y-intercept 2, as shown in the graph.

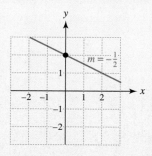

Parallel Lines   Lines with the same slope are parallel; nonvertical parallel lines have the same slope.

Example: The equations $y = -2x + 1$ and $y = -2x$ determine parallel lines because $m_1 = m_2 = -2$.

Perpendicular Lines

If two perpendicular lines have nonzero slopes m_1 and m_2, then $m_1 \cdot m_2 = -1$.

If two lines have nonzero slopes satisfying $m_1 \cdot m_2 = -1$, then they are perpendicular.

Example: The equations $y = -\frac{1}{2}x$ and $y = 2x - 2$ determine perpendicular lines because $m_1 \cdot m_2 = -\frac{1}{2} \cdot 2 = -1$.

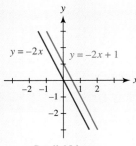

Parallel Lines   Perpendicular Lines

Section 3.6   *Point–Slope Form*

Point–Slope Form   The equation of the line passing through (x_1, y_1) with slope m is

$$y - y_1 = m(x - x_1) \quad \text{or} \quad y = m(x - x_1) + y_1.$$

Example: If $m = -2$ and $(x_1, y_1) = (-2, 3)$, then the point–slope form is either

$$y - 3 = -2(x + 2) \quad \text{or} \quad y = -2(x + 2) + 3.$$

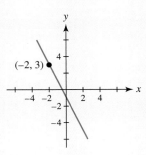

Example: The slope of the line passing through $(-2, 5)$ and $(4, 2)$ is

$$m = \frac{2 - 5}{4 - (-2)} = \frac{-3}{6} = -\frac{1}{2}.$$

Either $(-2, 5)$ or $(4, 2)$ may be used in the point–slope form. The point $(4, 2)$ results in the equation

$$y - 2 = -\frac{1}{2}(x - 4) \quad \text{or} \quad y = -\frac{1}{2}(x - 4) + 2.$$

Section 3.7   *Introduction to Modeling*

Mathematical Modeling   Mathematics is used to describe or approximate the behavior of real-life phenomena.

Exact Model   The equation describes the data precisely without error.

Example: $y = 3x$ models the data exactly.

x	0	1	2	3
y	0	3	6	9

Approximate Model   The equation describes the data approximately. An approximate model occurs most often in applications.

Example: The line in the graph models the data approximately.

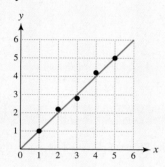

Modeling with a Linear Equation   To model a quantity y that has a constant rate of change, use the equation

$$y = mx + b,$$

where

$$m = \text{(constant rate of change)} \quad \text{and} \quad b = \text{(initial amount)}.$$

Example: If the temperature is initially $100°F$ and cools at $5°F$ per hour, then

$$T = -5x + 100$$

models the temperature T after x hours.

CHAPTER

3 Review Exercises

1. Identify the coordinates of each point in the graph. Identify the quadrant, if any, in which each point lies.

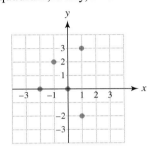

2. Make a scatterplot by plotting the following four points: $(-2, 3)$, $(-1, -1)$, $(0, 3)$, and $(2, -1)$.

Exercises 3 and 4: Use the table of xy-values to make a line graph.

3.

x	-2	-1	0	1	2
y	-3	2	-1	-2	3

4.

x	-10	-5	0	5	10
y	5	-10	10	-5	0

SECTION 3.2

Exercises 5–8: Determine whether the ordered pair is a solution for the given equation.

5. $y = x - 3$ $(6, 3)$

6. $y = 5 - 2x$ $(-2, 1)$

7. $3x - y = 3$ $(-1, 6)$

8. $\frac{1}{2}x + 2y = -8$ $(-4, -3)$

Exercises 9 and 10: Complete the table for the given equation.

9. $y = -3x$

x	-2	-1	0	1	2
y					

10. $2x + y = 5$

x					
y	-3	-1	0	1	3

Exercises 11–14: Use the given values of the variable to find solutions for the equation. Put them in a table.

11. $y = 3x + 2$ $x = -2, 0, 2, 4$

12. $y = 7 - x$ $x = 1, 2, 3, 4$

13. $y - 2x = 0$ $y = -1, 0, 1, 2$

14. $2y + x = 1$ $y = 1, 2, 3, 4$

Exercises 15–22: Graph the equation.

15. $y = 2x$

16. $y = x + 1$

17. $y = \frac{1}{2}x - 1$

18. $y = -3x + 2$

19. $x + y = 2$

20. $3x - 2y = 6$

21. $-4x + y = 8$

22. $2x + 3y = 12$

SECTION 3.3

Exercises 23 and 24: Identify the x- and y-intercepts.

23.

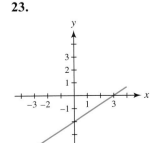

24.

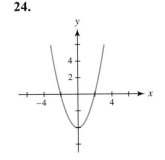

Exercises 25 and 26: Complete the table. Then determine the x- and y-intercepts for the graph of the equation.

25. $y = 2 - x$

x	-2	-1	0	1	2
y					

26. $x - 2y = 4$

x	-4	-2	0	2	4
y					

Exercises 27–30: Find any intercepts. Then graph the linear equation.

27. $2x - 3y = 6$ **28.** $5x - y = 5$

29. $0.1x - 0.2y = 0.4$ **30.** $\dfrac{x}{2} + \dfrac{y}{3} = 1$

31. Graph each equation.
   (a) $y = 1$ **(b)** $x = -3$

32. Write an equation for each line shown in the graph.

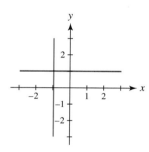

Exercises 33 and 34: Write an equation for the line that passes through the points shown in the table.

33.

x	-2	-1	0	1	2
y	1	1	1	1	1

34.

x	3	3	3	3	3
y	-2	-1	0	1	2

35. Write the equations of a horizontal line and a vertical line that pass through the point $(-2, 3)$.

36. *Distance* The distance from driver to home is illustrated in the graph at the top of the next column.
   (a) Find the intercepts.
   (b) Interpret each intercept.

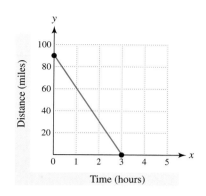

SECTION 3.4

Exercises 37–40: Find the slope, if possible, of the line passing through the two points.

37. $(2, 3), (4, 7)$ **38.** $(-3, 1), (2, -1)$

39. $(2, 1), (5, 1)$ **40.** $(-5, 6), (-5, 10)$

Exercises 41 and 42: Find the slope of the line shown in the graph.

41.

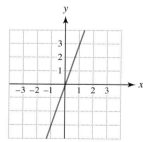

42.

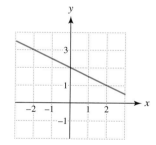

Exercises 43–46: Do the following.
   (a) Graph the linear equation.
   (b) What are the slope and y-intercept of the line?

43. $y = -2x$ **44.** $y = x - 1$

45. $x + 2y = 4$ **46.** $2x - 3y = -6$

Exercises 47–50: Sketch a line passing through the given point (x, y) and having slope m.

47. $(0, -3), m = 2$ **48.** $(0, 1), m = -\dfrac{1}{2}$

49. $(-1, 1), m = -\dfrac{2}{3}$ **50.** $(2, 2), m = 1$

51. The table lists points located on a line. Find the slope and intercepts of the line.

x	-1	0	1	2
y	-4	-2	0	2

52. A line with slope $\frac{1}{2}$ passes through the first point shown in the table. Complete the table so that each point in the table lies on the line.

x	0	1	2	3
y	1			

SECTION 3.5

Exercises 53 and 54: Write the slope–intercept form for the line.

53.

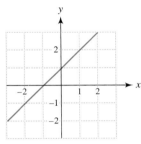

54.

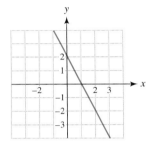

Exercises 55 and 56: Sketch a line with slope m and y-intercept b. Write its slope–intercept form.

55. $m = 2, b = -2$

56. $m = -\frac{3}{4}, b = 3$

Exercises 57–60: Do the following.

   (a) *Write the equation in slope–intercept form.*

   (b) *Give the slope and y-intercept of the line.*

57. $x + y = 3$

58. $-3x + 2y = -6$

59. $20x - 10y = 200$

60. $5x - 6y = 30$

Exercises 61–68: Graph the equation.

61. $y = \frac{1}{2}x + 1$

62. $y = 3x - 2$

63. $y = -\frac{1}{3}x$

64. $y = 3x$

65. $y = 2$

66. $y = -1$

67. $y = 4 - x$

68. $y = 2 - \frac{2}{3}x$

Exercises 69 and 70: All the points shown in the table lie on the same line. Find the slope–intercept form for the line.

69.

x	0	1	2
y	-5	0	5

70.

x	-1	0	1
y	2	0	-2

Exercises 71–74: Find the slope–intercept form for the line satisfying the given conditions.

71. Slope $-\frac{5}{6}$, y-intercept 2

72. Parallel to $y = -2x + 1$, passing through $(1, -1)$

73. Perpendicular to $y = -\frac{3}{2}x$, passing through $(3, 0)$

74. Perpendicular to $y = 5x - 3$, passing through $(0, -2)$

SECTION 3.6

Exercises 75 and 76: Determine whether the given point lies on the line.

75. $(-3, 1)$ $y - 1 = 2(x + 3)$

76. $(3, -8)$ $y = -3(x - 1) + 2$

Exercises 77–84: Find the slope–intercept form for the line that satisfies the conditions given.

77. Slope 5, passing through $(1, 2)$

78. Slope 20, passing through $(3, -5)$

79. Passing through $(-2, 1)$ and $(1, -1)$

80. Passing through $(20, -30)$ and $(40, 30)$

81. x-intercept 3, y-intercept -4

82. x-intercept $\frac{1}{2}$, y-intercept -1

83. Parallel to $y = 2x$, passing through $(5, 7)$

84. Perpendicular to $y - 4 = \frac{3}{2}(x + 1)$, passing through $(-1, 0)$

85. State whether the ordered pairs shown in the table are modeled exactly by $y = -x + 4$.

x	0	1	2	2
y	4	3	2	1

86. State whether the linear model shown in the graph is exact or approximate. Then find the equation of the line.

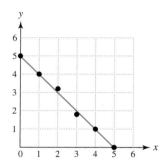

Exercises 87–90: Find an equation $y = mx + b$ that models y after x units of time.

87. y is initially 40 pounds and decreases at a rate of 2 pounds per minute.

88. y is initially 200 gallons and increases at 20 gallons per hour.

89. y is initially 50 and remains constant.

90. y is initially 20 feet *below* sea level and rises at 5 feet per second.

Exercises 91 and 92: The table contains ordered pairs.
  (a) Plot the data. Could a line pass through all five points?
  (b) Sketch a line that models the data.
  (c) Determine an equation of the line. Answers may vary.

91.

x	0	1	2	3	4
y	10	6	2	-2	-6

92.

x	-4	-2	0	2	4
y	1	2.1	3	3.9	5

APPLICATIONS

93. *Graphing Real Data*   The table contains real data on divorces D in millions during year t.
  (a) Make a line graph of the data. Be sure to label the axes.
  (b) Comment on any trends in the data.

t	1950	1960	1970	1980	1990
D	0.4	0.4	0.7	1.2	1.2

Source: National Center for Health Statistics.

94. *Water Usage*   The average American uses 100 gallons of water each day.
  (a) Write an equation that gives the gallons G of water that a person uses in t days.
  (b) Graph the equation for $t \geq 0$.
  (c) How many days does it take for the average American to use 5000 gallons of water?

95. *Modeling a Toy Rocket*   The velocity v of a toy rocket in feet per second after t seconds of flight is given by $v = 160 - 32t$, where $t \geq 0$.
  (a) Graph the equation.
  (b) Interpret each intercept.

96. *Modeling*   The accompanying line graph represents the insect population on 1 acre of land after x weeks. During this time a farmer sprayed pesticides on the land.
  (a) Estimate the slope of each line segment.
  (b) Interpret each slope as a rate of change.
  (c) Describe what happened to the insect population.

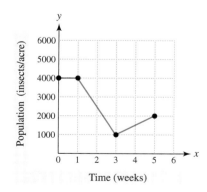

97. *Sketching a Graph*   An athlete jogs 4 miles away from home at a constant rate of 8 miles per hour and then turns around and jogs back home at the same speed. Sketch a graph that shows the distance that the athlete is from home. Be sure to label each axis.

98. *Nursing Homes*   In 1985, there were 19,100 nursing homes, and in 1997, there were 17,000. (*Source:* National Center for Health Statistics.)

(a) Calculate the slope of the line passing through (1985, 19100) and (1997, 17000).

(b) Interpret the slope as a rate of change.

99. *School Enrollment*   In 1990, school enrollment at all levels was 60 million students, and in 2010, it is projected to be 70 million. (*Source:* National Center for Education Statistics.)

(a) Find the slope of the line passing through the points (1990, 60) and (2010, 70).

(b) Interpret the slope as a rate of change.

(c) If this trend continues, estimate the school enrollment in 2030.

100. *Rental Cars*   The cost C in dollars for driving a rental car x miles is $C = 0.2x + 35$.

(a) How much would it cost to rent the car but not drive it?

(b) How much does it cost to drive the car one *additional* mile?

(c) What is the y-intercept of the graph of $C = 0.2x + 35$? What does it represent?

(d) What is the slope of the graph of $C = 0.2x + 35$? What does it represent?

101. *Distance and Speed*   A person is driving a car along a straight road. The accompanying graph shows the distance y in miles that the driver is from home after x hours.

(a) Is the person traveling toward or away from home? Why?

(b) The graph passes through (1, 200) and (3, 100). Discuss the meaning of these points.

(c) Find the slope–intercept form of the line. Interpret the slope as a rate of change.

(d) Use the graph to estimate the distance traveled after 2 hours. Then check your answer by using your equation from part (c).

102. *Deaths from Pneumonia*   From 1990 to 1998, deaths D in thousands from pneumonia can be modeled by $D = 1.59(x - 1990) + 77.4$, where x is the year. (*Source:* National Center for Health Statistics.)

(a) Find the number of deaths from pneumonia in 1990 and 1998.

(b) What is the slope of the graph of D? Interpret the slope as a rate of change.

103. *Arctic Glaciers*   Owing to global warming the arctic ice cap has been breaking up. As a result there has been an increase in the number of icebergs floating in the Arctic Ocean. The table gives an iceberg count I for various years t. (*Source:* NBC News.)

t	1970	1980	2000
I	400	600	1000

Source: National Broadcasting Company.

(a) Make a scatterplot of the data.

(b) Find the slope–intercept form of a line that models the number of icebergs I in year t. Interpret the slope of this line.

(c) Is the line you found in part (b) an exact model for the data in the table?

(d) If current trends continue, what will be the iceberg count in 2005?

104. *Gas Mileage*   The table shows the number of miles y traveled by a car on x gallons of gasoline.

x	2	4	8	10
y	40	79	161	200

(a) Plot the data in the xy-plane. Be sure to label each axis.

(b) Sketch a line that models these data. (You may want to use a ruler.)

(c) Calculate the slope of the line. Interpret the slope of this line.

(d) Find the equation of the line. Is your line an *exact* model?

(e) How far could this car travel on 9 gallons of gasoline?

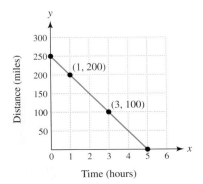

CHAPTER

3 TEST

1. Identify the coordinates of each point in the graph. State the quadrant, if any, in which each point lies.

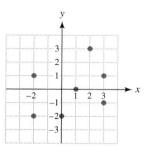

2. Make a scatterplot by plotting the four points $(0, 0)$, $(-2, -2)$, $(3, 0)$, and $(3, -2)$.

3. Complete the table for the equation $y = 2x - 4$. Then determine the x- and y-intercepts for the graph of the equation.

x	-2	-1	0	1	2
y					

4. Sketch a line passing through the point $(2, 1)$ and having slope $-\frac{1}{2}$.

Exercises 5–8: Graph the equation.

5. $y = 2$

6. $x = -3$

7. $y = -3x + 3$

8. $4x - 3y = 12$

9. Write the slope–intercept form for the line shown in the graph.

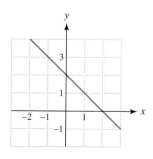

10. Write the equation $-4x + 2y = 1$ in slope–intercept form. Give the slope and the y-intercept.

Exercises 11–14: Find the slope–intercept form for the line that satisfies the given conditions.

11. Slope $-\frac{4}{3}$, y-intercept -5

12. Parallel to $y = 3x - 1$, passing through $(2, -5)$

13. Perpendicular to $y = \frac{1}{3}x$, passing through $(1, 2)$

14. Passing through $(-4, 2)$ and $(2, -1)$

15. State whether the linear model shown in the graph is exact or approximate. Then find the equation of the line.

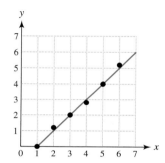

16. *Modeling* The line graph represents the total fish population P in a small lake after x years. One winter the lake almost froze completely.
    (a) Estimate the slope of each line segment.
    (b) Interpret each slope as a rate of change.
    (c) Describe what happened to the fish population.

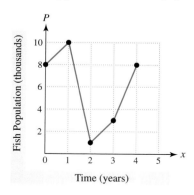

17. *Sketching a Graph* A cyclist rides a bicycle at 10 miles per hour for 2 hours and then at 8 miles per hour for 1 hour. Sketch a graph that shows the total distance d traveled after x hours. Be sure to label each axis.

18. *Delinquency Cases* In 1988, there were 47 delinquency cases per 1000 youths at risk. In 1998, there were 60 delinquency cases per 1000 youths at risk.
(*Source*: National Center for Juvenile Justice.)
(a) Find the slope of the line passing through the points (1988, 47) and (1998, 60).
(b) Interpret the slope as a rate of change.

(c) If this trend continues, estimate the number of delinquency cases in 2004.

19. *Modeling Insects* Write an equation in slope–intercept form that models the number of insects N after x days if there are initially 2000 insects and they increase at a rate of 100 insects per day.

CHAPTER 3 Extended and Discovery Exercises

Exercises 1 and 2: The table of real data can be modeled by a line. However, the line may not be an exact model, so answers may vary.

1. *U.S. Population* The population P of the United States in millions is shown in the table during year x.

x	1960	1970	1980	1990	2000
P	179	203	227	249	281

Source: Bureau of the Census.

(a) Make a scatterplot of the data.
(b) Find a point–slope form of a line that models these data.
(c) Use your equation to estimate the U.S. population in 2005.

2. *Women in Politics* The table lists percentages P of women in state legislatures during year x.

x	1981	1983	1985	1987
P	12.1	13.3	14.8	15.7
x	1989	1991	1993	1995
P	17.0	18.3	20.5	20.7

Source: National Women's Political Caucus.

(a) Make a scatterplot of the data.
(b) Find a point–slope form of a line that models these data.
(c) Use your equation to estimate the percentage of women in state legislatures in 2000.

CREATING GEOMETRIC SHAPES

Exercises 3–6: Many geometric shapes can be created with intersecting lines. For example, a triangle is formed when three lines intersect at three distinct points called vertices. Find equations of lines that satisfy the characteristics described. Sketching a graph may be helpful.

3. *Triangle* A triangle has vertices (0, 0), (2, 3), and (3, 6).
(a) Find slope–intercept forms for three lines that pass through each pair of points.
(b) Graph the three lines. Is a triangle formed by the line segments connecting the three points?

4. *Parallelogram* A parallelogram has four sides, with opposite sides parallel. Three sides of a parallelogram are given by the equations $y = 2x + 2$, $y = 2x - 1$, and $y = -x - 2$.
(a) If the fourth side of the parallelogram passes through the point $(-2, 3)$, find its equation.
(b) Graph all four lines. Is a parallelogram formed?

5. *Rectangle* The two vertices (0, 0) and (4, 2) determine one side of a rectangle. The side parallel to this side passes through the point (0, 3).
(a) Find slope–intercept forms for the four lines that determine the sides of the rectangle.
(b) Graph all four lines. Is a rectangle formed? (*Hint*: If you use a graphing calculator be sure to set a square window.)

6. *Square* Three vertices of a square are (1, 2), (4, 2), and (4, 5).
(a) Find the fourth vertex.
(b) Find equations of lines that correspond to the four sides of the square.
(c) Graph all four lines. Is a square formed?

CHAPTERS 1-3   Cumulative Review Exercises

Exercises 1 and 2: Classify the number as prime or composite. If a number is composite, write it as a product of prime numbers.

1. 40

2. 61

3. Evaluate $5x - 4y$ when $x = 3$ and $y = 2$.

4. Find y when $x = 4$ if $y = \frac{10}{x + 1}$.

Exercises 5 and 6: Translate the phrase into an algebraic expression using the variable n.

5. Ten more than a number

6. A number squared decreased by 2

7. Find the greatest common factor of 32 and 24.

8. Reduce $\frac{32}{40}$ to lowest terms.

9. Find the fractional part: one-fourth of two-fifths.

10. Find the reciprocal of $\frac{7}{3}$.

Exercises 11–14: Evaluate by hand and then reduce to lowest terms.

11. $\frac{4}{3} \cdot \frac{21}{24}$

12. $\frac{3}{4} \div \frac{9}{8}$

13. $\frac{2}{3} + \frac{4}{3}$

14. $\frac{7}{10} - \frac{2}{15}$

15. Write $4 \cdot 4 \cdot 4$ as an exponential expression.

16. Use multiplication to rewrite 2^4 and then evaluate the result.

Exercises 17–22: Evaluate the expression by hand.

17. $20 - 2 \cdot 3$

18. $14 - 5 - 2$

19. $\frac{1 + 4}{1 + 2}$

20. -3^2

21. $10 \div 2 \cdot 5$

22. $10 - 2^3$

Exercises 23 and 24: Classify the number as one or more of the following: natural number, whole number, integer, rational number, or irrational number.

23. $-\frac{4}{5}$

24. $\sqrt{3}$

25. Plot each number on the same number line.
(a) 0   (b) -1 (c) $\frac{3}{2}$

26. Evaluate $|3 - 5|$.

27. List the numbers $3, -\frac{1}{2}, \pi, \sqrt{2}, -3.1$ from least to greatest.

28. Write an expression for the phrase "2 squared minus a negative 4" and then simplify it.

Exercises 29–32: Evaluate the expression by hand.

29. $-10 - (-4)$

30. $-5 \cdot (-3)$

31. $-12 \div \left(-\frac{2}{3}\right)$

32. $-\frac{2x}{5y} \div \left(\frac{x}{10y}\right)$

33. State the properties of real numbers that the equation $3 \cdot (x \cdot 5) = 15x$ illustrates.

34. Use properties of real numbers to evaluate the expression $11 + 22 + 8 + 9$ mentally.

Exercises 35 and 36: Simplify the expression.

35. $3 + 4x - 2 + 3x$

36. $2(x - 1) - (x + 2)$

Exercises 37–40: Solve the equation. Check your solution.

37. $x + 5 = 2$

38. $\frac{1}{3}z = 7$

39. $3t - 5 = 1$

40. $\frac{1}{2}(x - 2) = 3 - (x + 4)$

41. Complete the table. Then use the table to solve the equation $6 - 2x = 4$.

x	-2	-1	0	1	2
$6 - 2x$					

42. Translate the sentence "Twice a number increased by 2 equals the number decreased by 5" to an equation, using the variable n. Then solve the equation.

43. The sum of four consecutive integers is -98. Find the integers.

44. Convert 9.8% to fraction and decimal notation.

45. Convert 0.234 to a percentage.

46. If $r = 10$ mph and $d = 80$ miles, use the formula $d = rt$ to find t.

Exercises 47 and 48: Find the area of the region shown.

47.
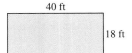
40 ft
18 ft

48.

5 ft
8 ft

49. Find the area of a circle with a 4-foot diameter.

50. Two angles in a triangle are 23° and 95°. Find the measure of the third angle.

Exercises 51 and 52: Solve the formula for the specified variable.

51. $A = \frac{1}{2}bh$ for b

52. $P = 2W + 2L$ for L

53. Use an inequality to express the set of real numbers graphed.

-2 -1 0 1 2 3

54. Use a number line to graph the solution set to $x \geq -1$.

55. Determine whether 2 is a solution to the linear inequality $3x - 2(1 + x) \geq 0$.

56. Complete the table. Then use the table to solve the linear inequality $2x - 3 \leq 1$.

x	0	1	2	3	4
$2x - 3$					

Exercises 57 and 58: Solve the inequality. Write the solution set in set-builder notation.

57. $3 - 6x < 3$

58. $2x \leq 1 - (2x - 1)$

59. Identify the coordinates of each point in the graph. State the quadrant, if any, in which each point lies.

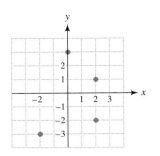

60. Make a scatterplot and a line graph with the points $(-2, 1)$, $(-1, 3)$, $(0, 0)$, $(1, -2)$, and $(2, 3)$.

61. Determine whether $(5, -1)$ is a solution to the equation $x - 3y = 8$.

62. Complete the table for the equation $x + 2y = 4$.

x	-2	-1	0	1	2
y					

Exercises 63–70: Graph the equation.

63. $y = 3x$

64. $y = x - 2$

65. $y = -\frac{1}{2}x + 1$

66. $x + y = 3$

67. $2x + 3y = -6$

68. $5x - 2y = 10$

69. $x = 3$

70. $y = -\frac{3}{2}$

Exercises 71 and 72: Identify the x- and y-intercepts. Then write the slope–intercept form of the line.

71.
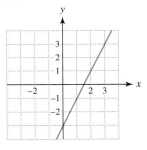

72.
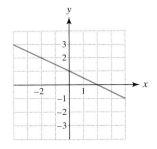

73. Determine the intercepts for the graph of the equation $-4x + 5y = 40$.

74. Write an equation for each line shown in the graph.

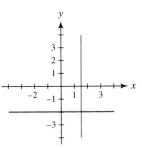

75. Sketch a line with slope $m = -2$, passing through the point $(-1, 2)$.

76. Sketch a line with undefined slope, passing through the point $(2, -1)$.

77. The table lists points located on a line. Write the slope–intercept form of the line.

x	-1	0	1	2
y	-6	-3	0	3

78. Write the equation $3x - 5y = 15$ in slope–intercept form. Graph the equation.

Exercises 79–82: Find the slope–intercept form for the line satisfying the given conditions.

79. Parallel to $y = -\frac{1}{3}x - 5$, passing through $(-3, 8)$

80. Perpendicular to $3x - 2y = 6$, passing through the point $(0, -3)$

81. Passing through $(-1, 3)$ and $(2, -3)$

82. x-intercept -2, y-intercept $\frac{1}{2}$

83. State whether the linear model shown in the graph is exact or approximate. Then find the equation of the line.

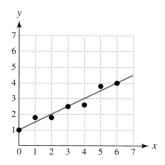

84. Let the initial value of y be 100 pounds, increasing at 5 pounds per hour. Find an equation in slope–intercept form that models y after x hours.

85. An insect population is initially 20,000 and increasing at 5000 per day. Write an equation that gives the number of insects I after x days.

86.

x	0	1	2	3
y	5	3	1	-1

    (a) Plot the data in the table. Could a line pass through all four points?
    (b) Sketch a line that models the data.
    (c) Determine the equation of the line.

APPLICATIONS

87. *Mowing Grass*   Two people are mowing a large lawn. The first person cuts 50 square yards per

minute while the second person cuts 30 square yards per minute.
    (a) Write and simplify an expression that gives the total number of square yards the two people can mow in x minutes.
    (b) How many minutes would it take for them to mow a lawn that is 20,000 square yards?

88. *Pizzas*   The table lists the cost C of buying x large pepperoni pizzas. Write an equation that relates C and x.

x	2	3	4
C	$16	$24	$32

89. *U.S. Postal Service*   In 2000, about $\frac{11}{20}$ of the mail consisted of first-class mail and periodicals. For every periodical there were 9 pieces of first-class mail. Estimate the fraction of the mail that was first-class mail.   (*Source:* U.S. Postal Service, *2000 Annual Report.*)

90. *Distance*   A person walks the following distances over three days: $2\frac{1}{2}$ miles, $4\frac{3}{4}$ miles, and $3\frac{1}{4}$ miles. How far did the person walk?

91. *Cost of a Car*   A 6% sales tax on a used car amounted to $330. Find the cost of the car.

92. *U.S. Per Capita Income*   The per capita income I in dollars from 1970 to 2000 can be modeled by

$$I = 807.4x - 1,587,300,$$

where x is the year.   (*Source:* Bureau of the Census.)
    (a) What has been the yearly average increase in per capita income?
    (b) Estimate the year when this income reached $19,400.

93. *Income Tax Returns*   In 1990, about 112 million income tax returns were filed. By 2000, this number had increased to about 127 million. Find the percent change in the number of income tax returns filed between 1990 and 2000.   (*Source:* Internal Revenue Service.)

94. *Investment Money*   A student invests two sums of money at 3% and 4% interest, receiving a total of $110 in interest after 1 year. Twice as much money is invested at 4% than at 3%. Find the amount invested at each interest rate.

95. *Dimensions of a Rectangle*   The width of a rectangle is 5 inches less than its length. If the perimeter of the rectangle is greater than 4550 inches, what values are possible for its width?

96. *Profit*   The cost to produce one music CD is $0.25 plus a one-time fixed cost of $200. The revenue received from selling one CD is $10.
(a) Write a formula that gives the cost C of producing x music CDs.
(b) Write a formula that gives the revenue R from selling x music CDs.
(c) Profit equals revenue minus cost. Write a formula that calculates the profit P from selling x music CDs.
(d) How many music CDs need to be sold to yield a (positive) profit?

97. *Modeling*   In an experiment a sample of fish is fed a minimum amount of food plus additional amounts of food. The accompanying line graph shows the average increase in the length of a fish.
(a) Find the slope of each line segment.
(b) Interpret each slope as a rate of change.
(c) Why does the line graph "level off" as the amount of additional food increases?

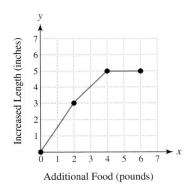

98. *School Lunch Program*   In 1970, 22.4 million students participated in the national school lunch program. By 2000, this number increased to 27.2 million students.   (*Source:* Department of Agriculture.)
(a) Find the slope of the line passing through the points (1970, 22.4) and (2000, 27.2).
(b) Interpret the slope as a rate of change.
(c) If this trend continues, estimate the number of participants in 2005.

99. *Rental Cars*   The cost C in dollars for driving a rental car x miles is $C = 0.3x + 25$.
(a) How much does it cost to drive the car 200 miles?
(b) How much does it cost to rent the car but not drive it?
(c) How much does it cost to drive the car 1 *additional* mile?

100. *Sales of Baseball Hats*   The revenue generated from selling various numbers of hats at a baseball park is shown in the accompanying graph.
(a) The line passes through (20, 240) and (50, 600). Discuss the meaning of these points.
(b) Find the slope–intercept form of the line. Interpret the slope as a rate of change.

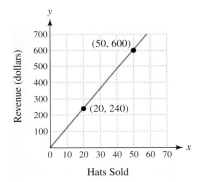

Systems of Linear Equations in Two Variables

Two factors that have a critical effect on plant growth are temperature and precipitation. If a region has too little precipitation, it will be a desert. Forests often grow in regions where trees can exist at relatively low temperatures and there is sufficient rainfall. At other levels of temperature and precipitation, grasslands may prevail. The accompanying figure illustrates the relationships among forests, grasslands, and deserts suggested by average annual temperature T in degrees Fahrenheit and precipitation P in inches. For example, Cheyenne, Wyoming, has an average annual temperature of about 50°F and average annual precipitation of about 14 inches. What does the graph predict about the plant growth near Cheyenne?

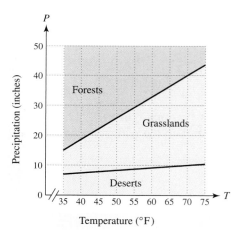

This graph illustrates systems of linear inequalities, which we discuss in this chapter. Mathematics is essential to creating these types of graphs and solving application problems.

**We can do anything we want to do
if we stick to it long enough.**
—Helen Keller

Source: A. Miller and J. Thompson, *Elements of Meteorology.*

4.1 SOLVING SYSTEMS OF LINEAR EQUATIONS GRAPHICALLY AND NUMERICALLY

Basic Concepts · Solutions to Systems of Equations

INTRODUCTION

Many equations from everyday life involve more than one variable. To calculate the area of a rectangle we need to know both its width and its length. To calculate wind chill we need to know both the outside temperature and the wind velocity. In this section we consider linear equations in two variables. The concepts discussed in this section are used in many applications.

BASIC CONCEPTS

In Chapter 3 we showed that the graph of $y = mx + b$ is a line with slope m and y-intercept b, as illustrated in Figure 4.1. Each point on this line represents a solution to the equation $y = mx + b$.

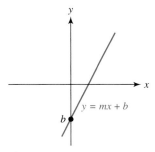

Figure 4.1

Now consider the following application of a line. If renting a moving truck for one day costs $25 plus $0.50 per mile, then the equation $C = 0.5x + 25$ represents the cost C in dollars of driving the rental truck x miles. The graph of this line is shown in Figure 4.2(a) for $x \geq 0$.

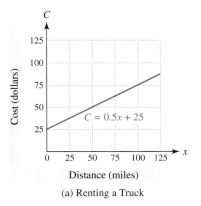

(a) Renting a Truck

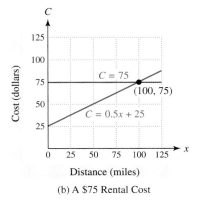

(b) A $75 Rental Cost

Figure 4.2

Suppose that we want to determine the number of miles that the truck is driven when the rental cost is $75. One way to solve this problem *graphically* is to graph both $C = 0.5x + 25$ and $C = 75$ in the same *xy*-plane, as shown in Figure 4.2(b) on the previous page. The lines intersect at the point (100, 75), which is a solution to $C = 0.5x + 25$ and to $C = 75$. That is, if the rental cost is $75, then the mileage must be 100 miles. This graphical technique of solving two equations is sometimes called the **intersection-of-graphs method**. To find a solution with this method, we locate a point where two graphs intersect.

EXAMPLE 1   Solving an equation graphically

The equation $P = 10x$ calculates an employee's pay for working x hours at $10 per hour. Use the intersection-of-graphs method to find the number of hours that the employee worked if the amount paid is $40.

Solution   Begin by graphing the equations $P = 10x$ and $P = 40$, as illustrated in Figure 4.3.

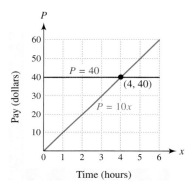

Figure 4.3

The graphs intersect at the point (**4**, **40**), which indicates that the employee must work **4** hours to earn **$40**.

Equations can be solved in more than one way. In Example 1 we determined graphically that $P = 10x$ is equal to 40 when $x = 4$. We could also solve this problem by making a table of values, as illustrated by Table 4.1. Note that, when $x = 4$, $P = \$40$. A table of values provides a *numerical solution.*

TABLE 4.1   Wages Earned at $10 per hour

x (hours)	0	1	2	3	4	5	6
P (pay)	$0	$10	$20	$30	$40	$50	$60

Note: Although graphical and numerical methods are different, both methods should give the same solution. Exceptions might occur, for example, when reading a graph precisely is difficult, or when a needed value does not appear in a table.

Technology Note: *Intersection of Graphs and Table of Values*

A graphing calculator can be used to find the intersection of the two graphs shown in Figure 4.3. It can also be used to create Table 4.1. The accompanying figures illustrate how a calculator can be used to determine that $y_1 = 10x$ equals $y_2 = 40$ when $x = 4$.

Calculator Help

To find a point of intersection see Appendix A (page AP-7). To make a table, see Appendix A (page AP-3).

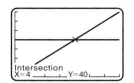

X	Y₁	Y₂
0	0	40
1	10	40
2	20	40
3	30	40
4	40	40
5	50	40
6	60	40

X=4

EXAMPLE 2 **Solving an equation graphically**

Use a graph to find the *x*-value when $y = 3$.
(a) $y = 2x - 1$ **(b)** $-3x + 2y = 12$

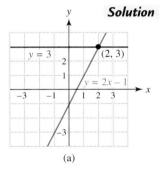

(a)

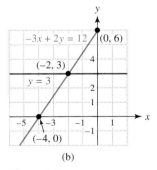

(b)

Figure 4.4

Solution **(a)** Begin by graphing the equations $y = 2x - 1$ and $y = 3$. The graph of $y = 2x - 1$ is a line with slope 2 and *y*-intercept -1. The graph of $y = 3$ is a horizontal line with *y*-intercept 3. In Figure 4.4(a) their graphs intersect at the point $(2, 3)$. Therefore an *x*-value of 2 corresponds to a *y*-value of 3.

(b) One way to graph $-3x + 2y = 12$ is to write this equation in slope–intercept form.

$$-3x + 2y = 12 \qquad \text{Given equation}$$
$$2y = 3x + 12 \qquad \text{Add } 3x \text{ to each side.}$$
$$y = \frac{3}{2}x + 6 \qquad \text{Divide each side by 2.}$$

The line has slope $\frac{3}{2}$ and *y*-intercept 6. Its graph and the graph of $y = 3$ are shown in Figure 4.4(b). Their graphs intersect at $(-2, 3)$. Therefore an *x*-value of -2 corresponds to a *y*-value of 3.

═══ MAKING CONNECTIONS ═══

Different Ways to Graph the Same Line

In Example 2(b) the line $-3x + 2y = 12$ was graphed by finding its slope–intercept form. A second way to graph this line is to find the *x*- and *y*-intercepts of the line. If $y = 0$, then $x = -4$ makes this equation true, and if $x = 0$, then $y = 6$ makes this equation true. Thus the *x*-intercept is -4, and the *y*-intercept is 6. Note that the (slanted) line in Figure 4.4(b) passes through the points $(-4, 0)$ and $(0, 6)$, which could be used to graph the line.

SOLUTIONS TO SYSTEMS OF EQUATIONS

In Example 2(b) we determined the x-value when $y = 3$ in the equation $-3x + 2y = 12$. This problem can be thought of as solving the following *system of two equations in two variables*.

$$-3x + 2y = 12$$
$$y = 3$$

The solution is the ordered pair $(-2, 3)$, which indicates that, when $x = -2$ and $y = 3$, each equation is a true statement.

$$-3(-2) + 2(3) = 12 \qquad \text{A true statement}$$
$$3 = 3 \qquad \text{A true statement}$$

Suppose that the sum of two numbers is 10 and that their difference is 4. If we let x and y represent the two numbers, then the equations

$$x + y = 10 \qquad \text{Sum is 10.}$$
$$x - y = 4 \qquad \text{Difference is 4.}$$

describe this situation. Each equation is a linear equation in two variables, so we call these equations a **system of linear equations in two variables**. Its graph typically consists of two lines. A **solution to a system** of two equations is an ordered pair (x, y) that makes *both* equations true. This ordered pair gives the coordinates of a point where the two lines intersect.

EXAMPLE 3   Testing for solutions

Determine whether $(4, 6)$ or $(7, 3)$ is a solution to

$$x + y = 10$$
$$x - y = 4.$$

Solution   To determine whether $(4, 6)$ is a solution, substitute $x = 4$ and $y = 6$ into each equation.

$x + y = 10$	$x - y = 4$	Given equations
$4 + 6 \stackrel{?}{=} 10$	$4 - 6 \stackrel{?}{=} 4$	Let $x = 4, y = 6$.
$10 = 10$ (True)	$-2 = 4$ (False)	Second equation is false.

Because $(4, 6)$ does not satisfy both equations, it is not a solution for the system of equations. Next let $x = 7$ and $y = 3$ to determine whether $(7, 3)$ is a solution.

$x + y = 10$	$x - y = 4$	Given equations
$7 + 3 \stackrel{?}{=} 10$	$7 - 3 \stackrel{?}{=} 4$	Let $x = 7, y = 3$.
$10 = 10$ (True)	$4 = 4$ (True)	Both are true.

Because $(7, 3)$ makes *both* equations true, it is a solution for the system of equations.

In the next example we find the solution to a system of equations graphically and numerically.

EXAMPLE 4   Solving a system graphically and numerically

Solve

$$x + 2y = 4$$
$$2x - y = 3$$

with a graph and with a table of values.

Solution   *Graphically*   Begin by writing each equation in slope–intercept form.

$x + 2y = 4$	First equation	$2x - y = 3$	Second equation
$2y = -x + 4$	Subtract x.	$-y = -2x + 3$	Subtract $2x$.
$y = -\dfrac{1}{2}x + 2$	Divide by 2.	$y = 2x - 3$	Multiply by -1.

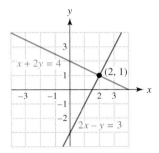

Figure 4.5

The graphs of $y = -\frac{1}{2}x + 2$ and $y = 2x - 3$ are shown in Figure 4.5. Their graphs intersect at the point $(2, 1)$.

Numerically   Table 4.2 shows the equations $y = -\frac{1}{2}x + 2$ and $y = 2x - 3$ evaluated for various values of x. Note that, when $x = 2$, both equations have a y-value of 1. Thus $(2, 1)$ is a solution.

Critical Thinking

If the graph of a system of linear equations consists of two parallel lines, how many solutions are there?

TABLE 4.2   A Numerical Solution

x	-1	0	1	2	3
$y = -\frac{1}{2}x + 2$	2.5	2	1.5	1	0.5
$y = 2x - 3$	-5	-3	-1	1	3

In the next example, we solve an application involving a system of equations.

EXAMPLE 5   Traveling to watch sports

In 2002, about 50 million Americans traveled to watch either football or basketball. About 10 million more people traveled to watch football than basketball. How many Americans traveled to watch each sport?   (*Source: Sports Travel Magazine.*)

Solution   **STEP 1** *Identify each variable.*

x: millions of Americans who traveled to watch football

y: millions of Americans who traveled to watch basketball

STEP 2 *Write a system of equations.* The total number of Americans who watched either sport is 50 million, so we know that $x + y = 50$. Because 10 million more people watched football than basketball, we also know that $x - y = 10$. Thus a system of equations representing this situation is

$$x + y = 50$$
$$x - y = 10.$$

STEP 3   *Solve the system of equations.* To solve this system graphically, write each equation in slope–intercept form.

$$y = -x + 50$$
$$y = x - 10$$

Their graphs intersect at the point (30, 20), as shown in Figure 4.6. Thus about 30 million Americans traveled to watch football and 20 million Americans traveled to watch basketball.

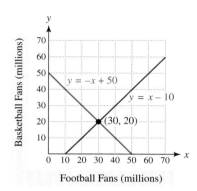

Figure 4.6

STEP 4   *Check your solution.* Note that $30 + 20 = 50$ million Americans traveled to watch either football or basketball and that $30 - 20 = 10$ million more Americans watched football than basketball.

Calculator Help
To find a point of intersection, see Appendix A (page AP-7).

Technology Note: *Checking Solutions*

The solution to the system in Example 5 can be checked with a graphing calculator, as shown in the accompanying figure where the graphs of $y_1 = -x + 50$ and $y_2 = x - 10$ intersect at the point (30, 20). However, a graphing calculator cannot read Example 5 and write down the system of equations. A human mind is needed for these tasks.

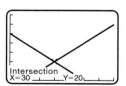

4.1   PUTTING IT ALL TOGETHER

Some types of situations are described by using two equations in two variables. For example, suppose that a person walks a total of 8 miles in 2 days by walking 2 miles farther on the first day than on the second day. If x represents the number of miles walked on the first day and if y represents the number of miles walked on the second day, then the system of equations

$$x + y = 8$$
$$x - y = 2$$

describes this situation. The methods discussed in this section reveal that the solution to this system is (5, 3), which indicates that the person walked 5 miles on the first day and 3 miles on the second day. The following table summarizes some important concepts of systems of equations.

Concept	Explanation	Example
System of Linear Equations in Two Variables	Can be written as $$Ax + By = C$$ $$Dx + Ey = F$$	$$x + y = 8$$ $$x - y = 2$$
Solution to a System of Equations	An ordered pair (x, y) that satisfies *both* equations	The solution to the preceding system is (5, 3) because, when $x = 5$ and $y = 3$ are substituted, both equations are true. $$5 + 3 \overset{?}{=} 8 \qquad \text{True}$$ $$5 - 3 \overset{?}{=} 2 \qquad \text{True}$$
Graphical Solution to a System of Equations	Graph each equation. A point of intersection represents a solution.	The graphs of $y = -x + 8$ and $y = x - 2$ intersect at (5, 3).
Numerical Solution to a System of Equations	Make a table for each equation. A solution occurs when one x-value gives the same y-values in both equations.	Make tables for $y = -x + 8$ and $y = x - 2$. When $x = 5$, $y = 3$ in both equations, so (5, 3) is a solution. <table>

x	4	5	6
$y = -x + 8$	4	3	2
$y = x - 2$	2	3	4

4.1 EXERCISES

FOR EXTRA HELP

CONCEPTS

1. The solution to a system of two equations in two variables is a(n) _____ pair.

2. A graphical technique for solving a system of two equations in two variables is the _____ method.

3. The graphs of
$$x + y = 20$$
$$x - y = 2$$
intersect at (11, 9). What is the solution to the system?

4. Is the ordered pair (0, 0) a solution to the system
$$2x + 3y = 8$$
$$5x - 4y = -3?$$

5. To find a numerical solution, start by creating a _____ of values for the equations.

6. If a graphical method and a numerical method (table of values) are used to solve the same system of equations, then the two solutions should be (the same/different).

SOLVING SYSTEMS OF EQUATIONS

Exercises 7–14: Determine graphically the x-value when $y = 2$ in the given equation.

7. $y = 2x$

8. $y = \frac{1}{3}x$

9. $y = 4 - x$

10. $y = -2 - x$

11. $y = -\frac{1}{2}x + 1$

12. $y = 3x - 1$

13. $2x + y = 6$

14. $-3x + 4y = 11$

Exercises 15–20: Determine which ordered pair is a solution to the system of equations.

15. (0, 0), (1, 1)
$$x + y = 2$$
$$x - y = 0$$

16. (−1, 2), (1, −2)
$$2x + y = 0$$
$$x - 2y = -5$$

17. (−1, −1), (2, −3)
$$2x + 3y = -5$$
$$4x - 5y = 23$$

18. (2, −1), (−2, −2)
$$-x + 4y = -6$$
$$6x - 7y = 19$$

19. (2, 0), (−1, −3)
$$-5x + 5y = -10$$
$$4x + 9y = 8$$

20. $\left(\frac{1}{2}, \frac{3}{2}\right), \left(\frac{3}{4}, \frac{5}{4}\right)$
$$x + y = 2$$
$$3x - y = 0$$

Exercises 21–26: The graphs of two equations are shown. Use the intersection-of-graphs method to identify the solution to both equations. Then check your answer.

21.

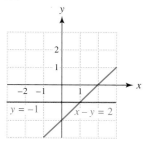

22.

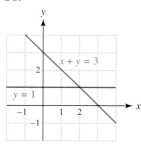

23.

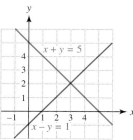

24.

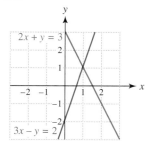

25.

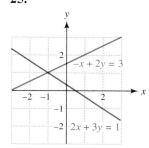

26.

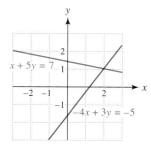

Exercises 27–30: A table for two equations is given. Identify the solution to both equations.

27.

x	1	2	3	4
$y = 2x$	2	4	6	8
$y = 4$	4	4	4	4

28.

x	2	3	4	5
$y = 6 - x$	4	3	2	1
$y = x - 2$	0	1	2	3

29.

x	1	2	3	4
$y = 4 - x$	3	2	1	0
$y = x - 2$	-1	0	1	2

30.

x	1	2	3	4
$y = 6 - 3x$	3	0	-3	-6
$y = 2 - x$	1	0	-1	-2

Exercises 31 and 32: Complete the table for each equation. Then identify the solution to both equations.

31.

x	0	1	2	3
$y = x + 2$				
$y = 4 - x$				

32.

x	-5	-4	-3	-2
$y = 2x + 1$				
$y = x - 1$				

Exercises 33–38: Use the specified method to solve the system of equations.

    (a) Graphically
    (b) Numerically (table of values)

33. $y = 2x + 3$
    $y = 1$

34. $y = 2 - x$
    $y = 0$

35. $y = 4 - x$
    $y = x - 2$

36. $y = 2x$
    $y = -\frac{1}{2}x$

37. $y = 3x$
    $y = x + 2$

38. $y = 2x - 3$
    $y = -x + 3$

Exercises 39–50: Solve the system of equations graphically.

39. $x + y = -3$
    $x - y = 1$

40. $x - y = 3$
    $2x + y = 3$

41. $2x - y = 3$
    $3x + y = 2$

42. $x + 2y = 6$
    $-x + 3y = 4$

43. $-4x + 2y = 0$
    $x - y = -1$

44. $4x - y = 2$
    $y = 2x$

45. $2x - y = 4$
    $x + 2y = 7$

46. $x - y = 2$
    $\frac{1}{2}x + y = 4$

47. $x = -y + 4$
    $x = 3y$

48. $x = 2y$
    $y = -\frac{1}{2}x$

49. $x + y = 3$
    $x = \frac{1}{2}y$

50. $2x - 4y = 8$
    $\frac{1}{2}x + y = -4$

APPLICATIONS

51. *Renting a Truck* A rental truck costs $50 plus $0.50 per mile.
    (a) Write an equation that gives the cost of driving the truck x miles.
    (b) Use the intersection-of-graphs method to determine the number of miles that the truck was driven if the rental cost is $80.
    (c) Solve part (b) numerically with a table of values.

52. *Renting a Car* A rental car costs $25 plus $0.25 per mile.
    (a) Write an equation that gives the cost of driving the car x miles.
    (b) Use the intersection-of-graphs method to determine the number of miles that the car was driven if the rental cost is $100.
    (c) Solve part (b) numerically with a table of values.

53. *Recorded Music* In 2000, rock and country music accounted for 35% of all music sales. Rock music sales were 2.5 times greater than country music sales. (*Source:* Recording Industry Association of America.)
    (a) Let x be the percentage of sales due to rock music and let y be the percentage of music sales due to country music. Write a system of two equations that describes the given information.
    (b) Solve your system graphically.

54. *Sales of Radios*   During 1999 and 2000, about 40 thousand radios were sold, excluding car radios. The 2000 sales exceeded the 1999 sales by 2 thousand radios.   (***Source:*** M. Street Corporation.)
   (a) Let y be the radio sales in thousands during 1999 and x be the radio sales in thousands during 2000. Write a system of two equations that describes the given information.
   (b) Solve your system graphically.

55. *Dimensions of a Rectangle*   A rectangle is 4 inches longer than it is wide. Its perimeter is 28 inches.
   (a) Write a system of two equations in two variables that describes this information. Be sure to specify what each variable means.
   (b) Solve your system graphically. Interpret your results.

56. *Dimensions of a Triangle*   An isosceles triangle has a perimeter of 17 inches with its two shorter sides equal in length. The longest side measures 2 inches more than either of the shorter sides.

   (a) Write a system of two equations in two variables that describes this information. Be sure to specify what each variable means.
   (b) Solve your system graphically. Explain what your results mean.

WRITING ABOUT MATHEMATICS

57. Use the intersection-of-graphs method to help explain why you typically expect a linear system in two variables to have one solution.

58. Could a system of two linear equations in two variables have exactly two solutions? Explain your reasoning.

59. Give one disadvantage of using a table to solve a system of equations. Explain your answer.

60. Do the equations $y = 2x + 1$ and $y = 2x - 1$ have a common solution? Explain your answer.

4.2   SOLVING SYSTEMS OF LINEAR EQUATIONS BY SUBSTITUTION

The Method of Substitution · Types of Systems of Linear Equations · Applications

INTRODUCTION

In Section 4.1 we solved systems of linear equations by using graphs and tables. A disadvantage of a graph is that reading the graph precisely can be difficult. A disadvantage of using a table is that locating the solution can be difficult when it is either a fraction or a large number. In this section we introduce the method of substitution, which involves the use of symbols only to solve systems of equations. The advantage of this method is that the exact solution can always be found (provided it exists.)

THE METHOD OF SUBSTITUTION

If you and a friend earned $120 together, then there is no way to determine how much each one earned. You might have earned $70 while your friend earned $50, or vice versa. If x represents how much your friend earned and y represents how much you earned, then the equation $x + y = 120$ describes this situation. However, if we know that you earned twice as much as your friend, then we can include a second equation, $y = 2x$. The amount that each of you earned can now be determined by *substituting* $2x$ for y in the first equation.

$$
\begin{array}{ll}
x + y = 120 & \text{First equation} \\
x + 2x = 120 & \text{Substitute } 2x \text{ for } y. \\
3x = 120 & \text{Combine like terms.} \\
x = 40 & \text{Divide by 3.}
\end{array}
$$

Thus your friend earned $40, and you earned twice as much, or $80.

This technique of substituting an expression for a variable and solving the resulting equation is called the **method of substitution**.

EXAMPLE 1   Using the method of substitution

Solve each system of equations.
(a) $2x + y = 10$ **(b)** $-2x + 3y = -8$
$y = 3x$ $x = 3y + 1$

Solution   **(a)** From the second equation, substitute $3x$ for y in the first equation.

$$
\begin{array}{ll}
2x + y = 10 & \text{First equation} \\
2x + 3x = 10 & \text{Substitute } 3x \text{ for } y. \\
5x = 10 & \text{Combine like terms.} \\
x = 2 & \text{Divide by 5.}
\end{array}
$$

The solution to this system is an *ordered pair*, so we must also find y. Because $y = 3x$ and $x = 2$, it follows that $y = 3(2) = 6$. The solution is $(2, 6)$. (Check it.)
(b) The second equation, $x = 3y + 1$, is solved for x. Substitute $(3y + 1)$ for x in the first equation. Be sure to include parentheses around the expression $3y + 1$.

$$
\begin{array}{ll}
-2x + 3y = -8 & \text{First equation} \\
-2(3y + 1) + 3y = -8 & \text{Substitute } (3y + 1) \text{ for } x. \\
-6y - 2 + 3y = -8 & \text{Distributive property} \\
-3y - 2 = -8 & \text{Combine like terms.} \\
-3y = -6 & \text{Add 2 to each side.} \\
y = 2 & \text{Divide each side by } -3.
\end{array}
$$

To find x, substitute $y = 2$ in $x = 3y + 1$ to obtain $x = 7$. The solution is $(7, 2)$.

Note: When an expression contains two or more terms, it is usually best to place parentheses around it before substituting in an equation. In Example 1(b), the distributive property would not have been applied correctly without the parentheses.

Sometimes it is necessary to solve for a variable before substitution can be used, as demonstrated in the next example.

EXAMPLE 2   Using the method of substitution

Solve each system of equations.
(a) $x + y = 8$ **(b)** $3a - 2b = 2$
$2x - 3y = 6$ $a + 4b = 3$

Solution **(a)** Neither equation is solved for a variable, but we can easily solve the first equation for y.

$$x + y = 8 \qquad \text{First equation}$$
$$y = 8 - x \qquad \text{Subtract } x \text{ from each side.}$$

Now we can substitute $(8 - x)$ for y in the second equation.

$$2x - 3y = 6 \qquad \text{Second equation}$$
$$2x - 3(8 - x) = 6 \qquad \text{Substitute } (8 - x) \text{ for } y.$$
$$2x - 24 + 3x = 6 \qquad \text{Distributive property}$$
$$5x = 30 \qquad \text{Combine terms; add 24.}$$
$$x = 6 \qquad \text{Divide each side by 5.}$$

Because $y = 8 - x$ and $x = 6$, $y = 8 - 6 = 2$. The solution is $(6, 2)$.
(b) Although we could solve either equation for either variable, solving the second equation for a is easiest because its coefficient is 1.

$$a + 4b = 3 \qquad \text{Second equation}$$
$$a = 3 - 4b \qquad \text{Subtract } 4b \text{ from each side.}$$

Now substitute $(3 - 4b)$ for a in the first equation.

$$3a - 2b = 2 \qquad \text{First equation}$$
$$3(3 - 4b) - 2b = 2 \qquad \text{Substitute } (3 - 4b) \text{ for } a.$$
$$9 - 12b - 2b = 2 \qquad \text{Distributive property}$$
$$-14b = -7 \qquad \text{Combine terms; subtract 9.}$$
$$b = \frac{1}{2} \qquad \text{Divide each side by } -14.$$

To find a, substitute $b = \frac{1}{2}$ in $a = 3 - 4b$ to obtain $a = 1$. The solution is $\left(1, \frac{1}{2}\right)$.

Note: When a system of equations contains variables other than x and y, we will list them alphabetically in an ordered pair.

TYPES OF SYSTEMS OF LINEAR EQUATIONS

A system of linear equations typically has one solution. However, in the next example we solve two systems of equations that do not have one solution.

EXAMPLE 3   Solving other types of systems

If possible, use substitution to solve the system of equations. Then use graphing to help explain the result.
(a) $3x + y = 4$ **(b)** $x + y = 2$
$6x + 2y = 2$ $2x + 2y = 4$

Solution   **(a)** Solve the first equation for y to obtain $y = 4 - 3x$. Next substitute $(4 - 3x)$ for y in the second equation.

$$6x + 2y = 2 \qquad \text{Second equation}$$
$$6x + 2(4 - 3x) = 2 \qquad \text{Substitute } (4 - 3x) \text{ for } y.$$
$$6x + 8 - 6x = 2 \qquad \text{Distributive property}$$
$$8 = 2 \ (\text{False}) \qquad \text{Combine terms.}$$

The equation $8 = 2$ is *always false*, which indicates that there are *no solutions*. One way to graph each equation is to write the equations in slope–intercept form.

$3x + y = 4$	First equation	$6x + 2y = 2$	Second equation
$y = -3x + 4$	Subtract $3x$.	$y = -3x + 1$	Subtract $6x$, divide by 2.

The graphs of these equations are parallel lines with slope -3, as shown in Figure 4.7(a). Because the lines *do not intersect* there are *no solutions* to the system of equations.

(b) Solve the first equation for y to obtain $y = 2 - x$. Now substitute $(2 - x)$ for y in the second equation.

$$2x + 2y = 4 \qquad \text{Second equation}$$
$$2x + 2(2 - x) = 4 \qquad \text{Substitute } (2 - x) \text{ for } y.$$
$$2x + 4 - 2x = 4 \qquad \text{Distributive property}$$
$$4 = 4 \ (\text{True}) \qquad \text{Combine terms.}$$

The equation $4 = 4$ is *always true*, which means that there are *infinitely many solutions*. One way to graph these equations is to write them in slope–intercept form first.

$x + y = 2$	First equation	$2x + 2y = 4$	Second equation
$y = -x + 2$	Subtract x.	$y = -x + 2$	Subtract $2x$, divide by 2.

Because the equations have the same slope–intercept form, their graphs are identical, as shown in Figure 4.7(b). *Every point* on this line *represents a solution* to the system of equations, so there are infinitely many solutions.

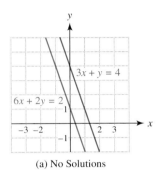

(a) No Solutions

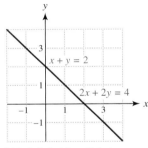

(b) Infinitely Many Solutions

Figure 4.7

┃┃┃┃┃┃ TYPES OF EQUATIONS AND NUMBER OF SOLUTIONS

A system of linear equations can have zero, one, or infinitely many solutions. It cannot have any other number of solutions. If a system has

1. no solutions, it is an **inconsistent system**. A graphical solution results in parallel lines.
2. one solution, it is a **consistent system**, and its equations are **independent equations**. A graphical solution results in two lines that intersect at one point.
3. infinitely many solutions, it is a **consistent system**, and its equations are **dependent equations**. A graphical solution results in identical lines. ┃┃┃┃┃┃

EXAMPLE 4   **Identifying types of equations**

Graphs of two equations are shown. State the number of solutions to each system of equations. Then state whether the system is consistent or inconsistent. If it is consistent, state whether the equations are dependent or independent.

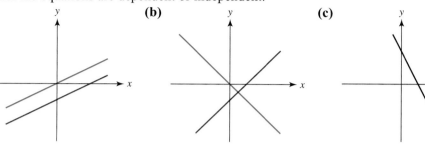

(a)   (b)   (c)

Solution   (a)   The lines are parallel, so there are no solutions. The system is inconsistent.

(b)   The lines intersect at one point, so there is one solution. The system is consistent, and the equations are independent.

(c)   There is only one line, which indicates that the lines are identical, or coincide, so there are infinitely many solutions. The system is consistent and the equations are dependent.

APPLICATIONS

In the next two examples we use the method of substitution to solve applications.

EXAMPLE 5   **Determining pizza sales**

In 2002, combined sales of frozen and ready-to-eat pizza reached $28.9 billion. Ready-to-eat pizza sales were 7.5 times more than frozen pizza sales. Find the amount of sales for each type of pizza.   (*Source: Business Trend Analyst.*)

Solution   **STEP 1**   *Identify each variable.* Clearly identify what each variable represents.

x: sales of ready-to-eat pizza, in billions of dollars

y: sales of frozen pizza, in billions of dollars

STEP 2   *Write a system of equations.*

$$x + y = 28.9 \qquad \text{Sales total \$28.9 billion.}$$

$$x = 7.5y \qquad \begin{array}{l}\text{Ready-to-eat pizza sales } x \text{ are}\\ \text{7.5 times frozen pizza sales } y.\end{array}$$

STEP 3   *Solve the system of linear equations.* Substitute $7.5y$ for x in the first equation.

$$x + y = 28.9 \qquad \text{First equation}$$

$$\mathbf{7.5y} + y = 28.9 \qquad \text{Substitute } 7.5y \text{ for } x.$$

$$8.5y = 28.9 \qquad \text{Combine like terms.}$$

$$y = \frac{28.9}{8.5} \qquad \text{Divide each side by 8.5.}$$

$$= 3.4 \qquad \text{Simplify.}$$

Because $x = 7.5y$, it follows that $x = 7.5(3.4) = 25.5$. Thus frozen pizza sales were \$3.4 billion and ready-to-eat pizza sales were \$25.5 billion in 2002.

STEP 4 *Check the solution.* The sum of these sales is $25.5 + 3.4 = \$28.9$ billion, and ready-to-eat pizza sales exceeded frozen pizza sales by $\frac{25.5}{3.4} = 7.5$ times. The answer checks.

EXAMPLE 6   Determining airplane speed and wind speed

An airplane flies 2400 miles into (or against) the wind in 8 hours. The return trip takes 6 hours. Find the speed of the airplane with no wind and the speed of the wind.

Solution   **STEP 1** *Identify each variable.*

x: the speed of the airplane without wind

y: the speed of the wind

STEP 2 *Write a system of equations.* The speed of the airplane against the wind is $\frac{2400}{8} = 300$ miles per hour, because it traveled 2400 miles in 8 hours. The wind slowed the plane, so $x - y = 300$. Similarly, the airplane flew $\frac{2400}{6} = 400$ miles per hour with the wind because it traveled 2400 miles in 6 hours. The wind made the plane fly faster, so $x + y = 400$.

$$x - y = 300 \qquad \text{Speed against the wind}$$
$$x + y = 400 \qquad \text{Speed with the wind}$$

Critical Thinking

A boat travels 10 miles per hour upstream and 16 miles per hour downstream. How fast is the current?

STEP 3 *Solve the system of linear equations.* Solve the first equation for x to obtain $x = y + 300$. Substitute $(y + 300)$ for x in the second equation.

$$x + y = 400 \qquad \text{Second equation}$$
$$(y + \mathbf{300}) + y = 400 \qquad \text{Substitute } (y + 300) \text{ for } x.$$
$$2y = 100 \qquad \text{Combine like terms; subtract 300.}$$
$$y = 50 \qquad \text{Divide by 2.}$$

Because $x = y + 300$, it follows that $x = 50 + 300 = 350$. Thus the airplane can fly 350 miles per hour with no wind, and the wind speed is 50 miles per hour.

STEP 4 *Check the solution.* The plane flies $350 - 50 = 300$ miles per hour into the wind, taking $\frac{2400}{300} = 8$ hours. The plane flies $350 + 50 = 400$ miles per hour with the wind, taking $\frac{2400}{400} = 6$ hours. The answers check.

4.2   PUTTING IT ALL TOGETHER

A system of linear equations can be solved symbolically by using the method of substitution. This method always provides the exact solution (provided one exists.) A system of linear equations can have zero, one, or infinitely many solutions. These concepts are summarized in the following table.

continued on next page

continued from previous page

Concept	Explanation	Example
Method of Substitution	Can be used to solve a system of equations **STEP 1** Solve one equation for one variable. **STEP 2** Substitute the result in the other equation and solve. **STEP 3** Find the other variable. **STEP 4** Check the solution.	$x + y = 5$ Sum is 5. $x - y = 1$ Difference is 1. **STEP 1** Solve for x in the second equation. $$x = y + 1$$ **STEP 2** Substitute $(y + 1)$ in the first equation for x. $$(y + 1) + y = 5$$ $$2y + 1 = 5$$ $$2y = 4$$ $$y = 2$$ **STEP 3** Because $x = y + 1$, $$x = 2 + 1 = 3.$$ **STEP 4** $3 + 2 = 5$ $3 - 2 = 1$ $(3, 2)$ checks.
Inconsistent System	Linear equations with no solutions Graphs result in parallel lines.	 No Solutions
Consistent System with Independent Equations	Linear equations with 1 solution Graphs result in intersecting lines.	 One Solution

Concept	Explanation	Example
Consistent System with Dependent Equations	Linear equations with infinitely many solutions Graphs result in identical lines, or lines that coincide.	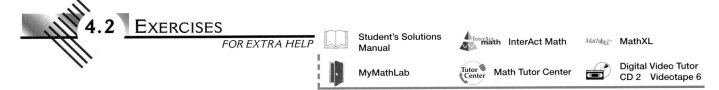 Infinitely Many Solutions

4.2 EXERCISES

FOR EXTRA HELP

Student's Solutions Manual   InterAct Math   MathXL

MyMathLab   Math Tutor Center   Digital Video Tutor   CD 2   Videotape 6

CONCEPTS

1. The technique of substituting an expression for a variable in an equation is called the method of _____.

2. A system of linear equations can have _____, _____, or _____ solutions.

3. Suppose that the method of substitution results in the equation $1 = 1$. What does this indicate about the number of solutions to the system of equations?

4. Suppose that the method of substitution results in the equation $0 = 1$. What does this indicate about the number of solutions to the system of equations?

5. If a system of linear equations has at least one solution, then it is a(n) (consistent/inconsistent) system.

6. If a system of linear equations has no solutions, then it is a(n) (consistent/inconsistent) system.

7. If a system of linear equations has exactly one solution, then the equations are (dependent/independent).

8. If a system of linear equations has infinitely many solutions, then the equations are (dependent/independent).

SOLVING SYSTEMS OF EQUATIONS

Exercises 9–36: Use the method of substitution to solve the system of linear equations.

9. $x + y = 9$
   $y = 2x$

10. $x + y = -12$
    $y = -3x$

11. $x + 2y = 4$
    $x = 2y$

12. $-x + 3y = -12$
    $x = 5y$

13. $2x + y = -2$
    $y = x + 1$

14. $-3x + y = -10$
    $y = x - 2$

15. $x + 3y = 3$
    $x = y + 3$

16. $x - 2y = -5$
    $x = 4 - y$

17. $3x + 2y = \frac{3}{2}$
    $y = 2x - 1$

18. $-3x + 5y = 4$
    $y = 2 - 3x$

19. $2x - 3y = -12$
$x = 2 - \frac{1}{2}y$

20. $\frac{3}{4}x + \frac{1}{4}y = -\frac{7}{4}$
$x = 1 - 2y$

21. $2x - 3y = -4$
$3x - y = 1$

22. $\frac{1}{2}x - y = -1$
$2x - \frac{1}{2}y = \frac{13}{2}$

23. $x - 5y = 26$
$2x + 6y = -12$

24. $4x - 3y = -4$
$x + 7y = -63$

25. $\frac{1}{2}y - z = 5$
$y - 3z = 13$

26. $3y - 7z = -2$
$5y - z = 2$

27. $10r - 20t = 20$
$r + 60t = -29$

28. $-r + 10t = 22$
$-10r + 5t = 30$

29. $3x + 2y = 9$
$2x - 3y = -7$

30. $5x - 2y = -5$
$2x - 5y = 19$

31. $2a - 3b = 6$
$-5a + 4b = -8$

32. $-5a + 7b = -1$
$3a + 2b = 13$

33. $-\frac{1}{2}x + 3y = 5$
$2x - \frac{1}{2}y = 3$

34. $3x - \frac{1}{2}y = 2$
$-\frac{1}{2}x + 5y = \frac{19}{2}$

35. $3a + 5b = 16$
$-8a + 2b = 34$

36. $5a - 10b = 20$
$10a + 5b = 15$

Exercises 37–42: The graphs of two equations are shown. State the number of solutions to each system of equations. Then state whether the system is consistent or inconsistent. If it is consistent, state whether the equations are dependent or independent.

37.

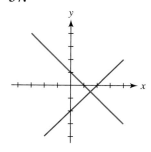

38.

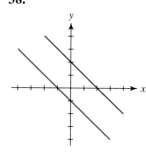

39.

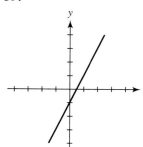

40.

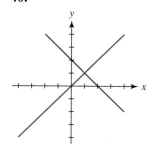

41.

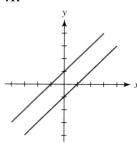

42.

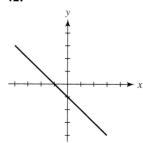

Exercises 43–50: Use the method of substitution and the intersection-of-graphs method to solve the system of linear equations. Then state whether the system is consistent or inconsistent. If it is consistent, state whether the equations are dependent or independent.

43. $x + y = 4$
$x + y = 2$

44. $x + y = 10$
$x - y = 2$

45. $2x - y = 3$
$x - 2y = 0$

46. $-2x + y = 0$
$y = 2x + 1$

47. $x - y = 1$
$2x - 2y = 2$

48. $x + 2y = 4$
$-x - 2y = -4$

49. $x - 2y = 4$
$x = 2y$

50. $-3x + y = 4$
$y = 3x + 4$

Exercises 51–68: Use the method of substitution to solve the system of linear equations. These systems may have zero, one, or infinitely many solutions.

51. $x + y = 9$
$x + y = 7$

52. $x - y = 8$
$x - y = 4$

53. $x - y = 4$
$2x - 2y = 8$

54. $2x + y = 5$
$4x + 2y = 10$

55. $x + y = 4$
$x - y = 2$

56. $x - y = 3$
$2x - y = 7$

57. $x - y = 7$
$-x + y = -7$

58. $2u - v = 6$
$-4u + 2v = -12$

59. $u - 2v = 5$
$2u - 4v = -2$

60. $3r + 3t = 9$
$2r + 2t = 4$

61. $2r + 3t = 1$
$r - 3t = -5$

62. $5x - y = -1$
$2x - 7y = 7$

63. $y = 5x$
$y = -3x$

64. $a = b + 1$
$a = b - 1$

65. $5a = 4 - b$
$5a = 3 - b$

66. $3y = x$
$3y = 2x$

67. $2x + 4y = 0$
$3x + 6y = 5$

68. $-5x + 10y = 3$
$\frac{1}{2}x - y = 1$

APPLICATIONS

69. *Rectangle* A rectangular garden is 10 feet longer than it is wide. Its perimeter is 72 feet.
(a) Let W be the width of the garden and L be the length. Write a system of linear equations whose solution gives the width and length of the garden.
(b) Use the method of substitution to solve the system. Check your answer.

70. *Isosceles Triangle* The measures of the two smaller angles in a triangle are equal and their sum equals the largest angle.
(a) Let x be the measure of each of the two smaller angles and y be the measure of the largest angle. Write a system of linear equations whose solution gives the measures of these angles.
(b) Use the method of substitution to solve the system. Check your answer.

71. *Complementary Angles* The smaller of two complementary angles is half the measure of the larger angle.
(a) Let x be the measure of the smaller angle and y be the measure of the larger angle. Write a system of linear equations whose solution gives the measures of these angles.
(b) Use the method of substitution to solve your system.
(c) Use graphing to solve the system.

72. *Supplementary Angles* The smaller of two supplementary angles is one-fourth the measure of the larger angle.
(a) Let x be the measure of the smaller angle and y be the measure of the larger angle. Write a system of linear equations whose solution gives the measures of these angles.
(b) Use the method of substitution to solve the system.

73. *Average Room Prices* In 2002, the average room price for a hotel chain was $1.68 less than in 2001. The 2002 room price was 98% of the 2001 room price. (*Source:* Smith Travel Research.)
(a) Let x be the average room price in 2001 and y be this price in 2002. Write a system of linear equations whose solution gives the average room prices for each year.
(b) Use the method of substitution to solve the system.

74. *Ticket Prices* Two hundred tickets were sold for a baseball game, which amounted to $840. Student tickets cost $3, and adult tickets cost $5.
(a) Let x be the number of student tickets sold and y be the number of adult tickets sold. Write a system of linear equations whose solution gives the number of each type of ticket sold.
(b) Use the method of substitution to solve the system.

75. *NBA Basketball Court* An official NBA basketball court is 44 feet longer than it is wide. If its perimeter is 288 feet, find its dimensions.

76. *Football Field* A U.S. football field is 139.5 feet longer than it is wide. If its perimeter is 921 feet, find its dimensions.

77. *Number Problem* The sum of two numbers is 70. The larger number is two more than three times the smaller number. Find the two numbers.

78. *Number Problem* The difference of two numbers is 12. The larger number is one less than twice the smaller number. Find the two numbers.

79. *Mixture Problem* A chemist has 20% and 50% solutions of acid available. How many liters of each solution should be mixed to obtain 10 liters of a 40% acid solution?

80. *Mixture Problem* A mechanic needs a radiator to have a 40% antifreeze solution. The radiator currently is filled with 4 gallons of a 25% antifreeze solution. How much of the antifreeze mixture should be drained from the car if the mechanic replaces it with pure antifreeze?

81. *Speed Problem* A tugboat goes 120 miles upstream in 15 hours. The return trip downstream takes 10 hours. Find the speed of the tugboat without a current and the speed of the current.

82. *Speed Problem* An airplane flies 1200 miles into the wind in 3 hours. The return trip takes 2 hours. Find the speed of the airplane without a wind and the speed of the wind.

83. *Great Lakes* Together, Lake Superior and Lake Michigan cover 54 thousand square miles. Lake Superior is approximately 10 thousand square miles larger than Lake Michigan. Find the size of each lake. (*Source:* National Oceanic and Atmospheric Administration.)

84. *Longest Rivers*   The two longest rivers in the world are the Nile and the Amazon. Together, they are 8145 miles with the Amazon being 145 miles shorter than the Nile. Find the length of each river.   (*Source:* National Oceanic and Atmospheric Administration.)

WRITING ABOUT MATHEMATICS

85. State one advantage that the method of substitution has over the intersection-of-graphs method. Explain your answer.

86. When applying the method of substitution, how do you know that there are no solutions?

87. When applying the method of substitution, how do you know that there are infinitely many solutions?

88. When applying the intersection-of-graphs method, how do you know that there are no solutions?

CHECKING BASIC CONCEPTS   SECTIONS 4.1 AND 4.2

1. Determine graphically the x-value in each equation when $y = 2$.
   (a) $y = 1 - \frac{1}{2}x$ **(b)** $2x - 3y = 6$

2. Determine whether $(-1, 0)$ or $(4, 2)$ is a solution to

$$2x - 5y = -2$$
$$3x + 2y = 16.$$

3. Solve the system of equations graphically. Check your answer.

$$x - y = 1$$
$$2x + y = 5$$

4. Use the method of substitution to solve each system of equations. How many solutions does each have?

   (a) $x + y = -1$
   $y = 2 - x$
   (b) $4x - y = 5$
   $-x + y = -2$
   (c) $x + 2y = 3$
   $-x - 2y = -3$

5. *Room Prices*   A hotel rents single and double rooms for $50 and $60, respectively. The hotel receives $17,000 for renting 300 rooms.
   (a) Let x be the number of single rooms rented and let y be the number of double rooms rented. Write a system of linear equations whose solution gives the values of x and y.
   (b) Use the method of substitution to solve the system. Check your answer.

4.3   SOLVING SYSTEMS OF LINEAR EQUATIONS BY ELIMINATION

The Elimination Method · Recognizing Other Types of Systems · Applications

INTRODUCTION

Two important methods for solving systems of linear equations are the intersection-of-graphs method and the substitution method. The intersection-of-graphs method is a graphical method because it involves the use of graphs to find the solution. The substitution method is a symbolic method because it involves the use of only symbols and algebra to find the solution. In this section we introduce a second symbolic method called the *elimination method*. This method is very efficient for solving some types of systems of linear equations.

THE ELIMINATION METHOD

The elimination method is based on the addition property of equality. Simply put, it is based on the concept that "if equals are added to equals, the results are equal." That is, if

$$a = b \quad \text{and} \quad c = d,$$

then

$$a + c = b + d.$$

For example, if the sum of two numbers is 20 and their difference is 4, then the system of equations

$$x + y = 20$$
$$x - y = 4$$

describes these two numbers. By the addition property of equality, the sum of the left sides of these equations equals the sum of their right sides.

$$(x + y) + (x - y) = 20 + 4 \qquad \text{Add left sides of these equations.}$$
$$\text{Add right sides of these equations.}$$
$$2x = 24 \qquad \text{Combine terms.}$$

Note that the y-variable is eliminated by adding the left sides. The resulting equation, $2x = 24$, simplifies to $x = 12$. Because $x + y = 20$ and $x = 12$, it follows that $y = 8$. To organize the elimination method better, we can carry out the addition in a vertical format.

$$
\begin{array}{r}
x + y = 20 \\
\underline{x - y = 4} \\
2x + 0y = 24
\end{array}
\qquad \text{Add left sides and right sides.}
$$

EXAMPLE 1   Applying the elimination method

Solve each system of equations. Check each solution.
(a) $2x + y = 1$ (b) $-2a + b = -3$
    $3x - y = 9$ $2a + 3b = 7$

Solution   (a) Adding these two equations eliminates the y-variable.

$$
\begin{array}{rl}
2x + y = 1 & \qquad \text{First equation} \\
\underline{3x - y = 9} & \qquad \text{Second equation} \\
5x = 10, \quad \text{or} \quad x = 2 & \qquad \text{Add and solve for } x.
\end{array}
$$

To find y, substitute 2 for x into either of the *given* equations.

$$
\begin{array}{rl}
2(2) + y = 1 & \qquad \text{Let } x = 2 \text{ in first equation.} \\
y = -3 & \qquad \text{Subtract 4 from each side.}
\end{array}
$$

The solution is the *ordered pair* $(2, -3)$, which can be checked by substituting 2 for x and -3 for y in the given equations.

$$2x + y = 1 \qquad 3x - y = 9 \qquad \text{Given equations}$$
$$2(2) + (-3) \overset{?}{=} 1 \qquad 3(2) - (-3) \overset{?}{=} 9 \qquad \text{Let } x = 2 \text{ and } y = -3.$$
$$4 - 3 \overset{?}{=} 1 \qquad 6 + 3 \overset{?}{=} 9 \qquad \text{Simplify.}$$
$$1 = 1 \qquad 9 = 9 \qquad \text{The solution checks.}$$

(b) Adding these two equations eliminates the a-variable.

$$\begin{aligned}-2a + b &= -3 \\ 2a + 3b &= 7 \\ \hline 4b &= 4, \quad \text{or} \quad b = 1 \qquad \text{Add and solve for } b.\end{aligned}$$

To find a, substitute 1 for b in either of the *given* equations.

$$2a + 3(1) = 7 \qquad \text{Let } b = 1 \text{ in second equation.}$$
$$2a = 4 \qquad \text{Subtract 3 from each side.}$$
$$a = 2 \qquad \text{Divide each side by 2.}$$

The solution is the *ordered pair* $(2, 1)$, which can be checked by substituting 2 for a and 1 for b in the given equations.

$$-2a + b = -3 \qquad 2a + 3b = 7 \qquad \text{Given equations}$$
$$-2(2) + 1 \overset{?}{=} -3 \qquad 2(2) + 3(1) \overset{?}{=} 7 \qquad \text{Let } a = 2 \text{ and } b = 1.$$
$$-4 + 1 \overset{?}{=} -3 \qquad 4 + 3 \overset{?}{=} 7 \qquad \text{Simplify.}$$
$$-3 = -3 \qquad 7 = 7 \qquad \text{The solution checks.}$$

Adding two equations does not always eliminate a variable. For example, adding the following equations eliminates neither variable.

$$\begin{aligned}3x - 2y &= 11 \qquad \text{First equation} \\ 4x + y &= 11 \qquad \text{Second equation} \\ \hline 7x - y &= 22 \qquad \text{Add the equations.}\end{aligned}$$

However, by the multiplication property of equality, we can multiply the second equation by 2. Then adding the equations eliminates the y-variable.

$$\begin{aligned}3x - 2y &= 11 \\ 8x + 2y &= 22 \qquad \text{Multiply the second equation by 2.} \\ \hline 11x &= 33, \quad \text{or} \quad x = 3 \qquad \text{Add and solve for } x.\end{aligned}$$

In the next two examples, we use the multiplication property of equality.

EXAMPLE 2 Multiplying before applying elimination

Solve each system of equations.

(a) $5x - y = -11$
$ 2x + 3y = -1$

(b) $3x + 2y = 1$
$ 2x - 3y = 5$

Solution (a) We multiply the first equation by 3 and then add to eliminate the y-variable.

$$15x - 3y = -33 \qquad \text{Multiply first equation by 3.}$$
$$\underline{2x + 3y = -1}$$
$$17x = -34, \quad \text{or} \quad x = -2 \qquad \text{Add and solve for } x.$$

We can find y by substituting -2 for x in the second equation.

$$2(-2) + 3y = -1 \qquad \text{Let } x = -2 \text{ in second equation.}$$
$$3y = 3 \qquad \text{Add 4 to each side.}$$
$$y = 1 \qquad \text{Divide each side by 3.}$$

The solution is $(-2, 1)$.

(b) For this system we must apply the multiplication property to both equations. If we multiply the first equation by 3 and the second equation by 2, then the coefficients of the y-variables will be opposites. Adding eliminates the y-variable.

$$9x + 6y = 3 \qquad \text{Multiply by 3.}$$
$$\underline{4x - 6y = 10} \qquad \text{Multiply by 2.}$$
$$13x = 13, \quad \text{or} \quad x = 1 \qquad \text{Add and solve for } x.$$

To find y, substitute 1 for x in the first *given* equation.

$$3(1) + 2y = 1 \qquad \text{Let } x = 1 \text{ in first equation.}$$
$$2y = -2 \qquad \text{Subtract 3 from each side.}$$
$$y = -1 \qquad \text{Divide each side by 2.}$$

The solution is $(1, -1)$.

EXAMPLE 3 Multiplying before applying elimination

Solve each system of equations.

(a) $4x + 5y = 11$ (b) $2y = -6 - 5x$
     $2x - 7y = -23$ $2x = -5y + 6$

Solution (a) Multiply the second equation by -2 and then add the equations to eliminate the x-variable.

$$4x + 5y = 11$$
$$\underline{-4x + 14y = 46} \qquad \text{Multiply by } -2.$$
$$19y = 57, \quad \text{or} \quad y = 3 \qquad \text{Add and solve for } y.$$

We can find x by substituting 3 for y in the second equation, $2x - 7y = -23$.

$$2x - 7(3) = -23 \qquad \text{Substitute 3 for } y.$$
$$2x = -2 \qquad \text{Add 21 to each side.}$$
$$x = -1 \qquad \text{Divide each side by 2.}$$

The solution is $(-1, 3)$.

(b) It is best to write each equation in the standard form: $Ax + By = C$.

$$5x + 2y = -6 \qquad \text{First equation in standard form}$$
$$2x + 5y = 6 \qquad \text{Second equation in standard form}$$

If we multiply the top equation by -2 and the bottom equation by 5, then we can eliminate the x-variable by adding.

$$
\begin{array}{ll}
-10x - 4y = 12 & \text{Multiply by } -2. \\
\underline{10x + 25y = 30} & \text{Multiply by } 5. \\
\qquad\quad 21y = 42, \quad \text{or} \quad y = 2 & \text{Add and solve for } y.
\end{array}
$$

To find x, substitute 2 for y in the first given equation, $2y = -6 - 5x$.

$$
\begin{array}{ll}
2(2) = -6 - 5x & \text{Let } y = 2 \text{ in first equation.} \\
10 = -5x & \text{Add 6 to each side.} \\
-2 = x & \text{Divide by } -5.
\end{array}
$$

The solution is $(-2, 2)$.

In the next example we use three different methods to solve a system of equations.

EXAMPLE 4 Solving a system with different methods

Solve the system of equations symbolically, graphically, and numerically. Do your answers agree?

$$
\begin{array}{l}
x + y = 2 \\
x - 3y = 6
\end{array}
$$

Solution **Symbolic Solution** The elimination method is a symbolic method. We can solve the system by multiplying the second equation by -1 and adding to eliminate the x-variable.

$$
\begin{array}{ll}
x + y = 2 & \\
\underline{-x + 3y = -6} & \text{Multiply by } -1. \\
\qquad 4y = -4, \quad \text{or} \quad y = -1 & \text{Add and solve for } y.
\end{array}
$$

We can find x by substituting -1 for y in the first equation, $x + y = 2$.

$$
\begin{array}{ll}
x + (-1) = 2 & \text{Let } y = -1 \text{ in first equation.} \\
x = 3 & \text{Add 1 to each side.}
\end{array}
$$

The solution is $(3, -1)$.

Graphical Solution For a graphical solution, we solve each equation for y to obtain the slope–intercept form.

$$
\begin{array}{llll}
x + y = 2 & \text{First equation} & x - 3y = 6 & \text{Second equation} \\
y = -x + 2 & \text{Subtract } x. & -3y = -x + 6 & \text{Subtract } x. \\
& & y = \dfrac{1}{3}x - 2 & \text{Divide by } -3.
\end{array}
$$

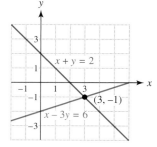

Figure 4.8

The graphs of $y = -x + 2$ and $y = \frac{1}{3}x - 2$ are shown in Figure 4.8. They intersect at $(3, -1)$. (These graphs could also be obtained by finding the x- and y-intercepts for each equation.)

Numerical Solution   A numerical solution consists of a table of values, as shown in Table 4.3. Note that, when $x = 3$, both y-values equal -1. Therefore the solution is $(3, -1)$.

TABLE 4.3

x	0	1	2	3	4
$y = -x + 2$	2	1	0	-1	-2
$y = \frac{1}{3}x - 2$	-2	$-\frac{5}{3}$	$-\frac{4}{3}$	-1	$-\frac{2}{3}$

RECOGNIZING OTHER TYPES OF SYSTEMS

In Section 4.2 we discussed how a system of linear equations can have zero, one, or infinitely many solutions. Elimination can also be used on systems that have no solutions or infinitely many solutions.

EXAMPLE 5   Solving other types of systems

Solve each system of equations by using the elimination method. Then graph the system.
(a) $x - 2y = 4$ (b) $3x + 3y = 6$
       $-2x + 4y = -8$ $x + y = 1$

Solution   (a) We multiply the first equation by 2 and then add, which eliminates both variables.

$$
\begin{aligned}
2x - 4y &= 8 \qquad &\text{Multiply by 2.}\\
-2x + 4y &= -8 \\
\hline
0 &= 0 \qquad &\text{Add.}
\end{aligned}
$$

The equation $0 = 0$ is *always true*, which indicates that the system has *infinitely many solutions*. A graph of the two equations is shown in Figure 4.9. Note that the two lines are identical so there actually is only one line, and *every point on this line represents a solution*.

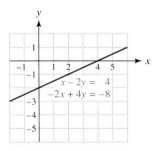

Figure 4.9

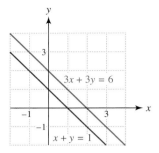

Figure 4.10

(b) We multiply the second equation, $x + y = 1$, by -3 and add, eliminating both variables.

$$3x + 3y = 6$$
$$\underline{-3x - 3y = -3} \qquad \text{Multiply by 3.}$$
$$ 0 = 3 \quad \text{(False)} \qquad \text{Add.}$$

The equation $0 = 3$ is *always false*, which indicates that the system has *no solutions*. A graph of the two equations is shown in Figure 4.10. Note that the two lines are parallel and thus do not intersect.

═══════════════════ MAKING CONNECTIONS ═══════════════════

Numerical and Graphical Solutions

The left-hand figure shows a numerical solution for Example 5(a), where each equation is solved for y to obtain

$$Y_1 = (4 - X)/(-2) \quad \text{and} \quad Y_2 = (-8 + 2X)/4.$$

Note that $y_1 = y_2$ for each value of x, indicating that the graphs of y_1 and y_2 are the same line. Similarly, the right-hand figure shows a numerical solution for Example 5(b), where

$$Y_1 = (6 - 3X)/3 \quad \text{and} \quad Y_2 = 1 - X.$$

Note that $y_1 \neq y_2$ and $y_1 - y_2 = 1$ for every value of x, indicating that the graphs of y_1 and y_2 are parallel lines that do not intersect.

X	Y₁	Y₂
0	-2	-2
1	-1.5	-1.5
2	-1	-1
3	-.5	-.5
4	0	0
5	.5	.5
6	1	1

$Y_1 \blacksquare (4 - X)/(-2)$

X	Y₁	Y₂
-1	3	2
0	2	1
1	1	0
2	0	-1
3	-1	-2
4	-2	-3
5	-3	-4

$Y_1 \blacksquare (6 - 3X)/3$

Calculator Help
To make a table, see Appendix A (page AP-3).

APPLICATIONS

In the next two examples we use elimination to solve applications relating to new cancer cases and burning calories during exercise.

EXAMPLE 6   Determining new cancer cases

In 2001, there were 1,268,000 new cancer cases. Men accounted for 18,000 more new cases than women. How many new cases of cancer were there for each gender?   (*Source:* American Cancer Society.)

Solution   **STEP 1** *Identify each variable.*

x: new cancer cases for men in 2001

y: new cancer cases for women in 2001

STEP 2 *Write a system of equations.*

$$x + y = 1{,}268{,}000$$
$$x - y = \phantom{1{,}26}18{,}000$$

STEP 3 *Solve the system of linear equations.* Add the two equations to eliminate the y-variable.

$$x + y = 1{,}268{,}000$$
$$\underline{x - y = 18{,}000}$$
$$2x = 1{,}286{,}000, \quad \text{or} \quad x = 643{,}000$$

There were 18,000 fewer new cases of cancer for women, so it follows from the second equation that $y = 643{,}000 - 18{,}000 = 625{,}000$.

STEP 4 *Check the solution.* The total number of cases was

$$643{,}000 + 625{,}000 = 1{,}268{,}000.$$

The number of new cases for men exceeded the number of new cases for women by

$$643{,}000 - 625{,}000 = 18{,}000.$$

The answer checks.

EXAMPLE 7   **Burning calories during exercise**

During strenuous exercise, an athlete can burn 10 calories per minute on a rowing machine and 11.5 calories per minute on a stair climber. If an athlete burns 433 calories in a 40-minute workout, how many minutes did the athlete spend on each machine? (*Source: Runner's World.*)

Solution   **STEP 1** *Identify each variable.*

x: number of minutes spent on a rowing machine

y: number of minutes spent on a stair climber

STEP 2 *Write a system of equations.* The total workout took 40 minutes, so $x + y = 40$. The athlete burned $10x$ calories on the rowing machine and $11.5y$ calories on the stair climber. Because the total number of calories equals 433, it follows that $10x + 11.5y = 433$.

$$x + y = 40 \qquad \text{Workout is 40 minutes.}$$
$$10x + 11.5y = 433 \qquad \text{Total calories burned is 433.}$$

STEP 3 *Solve the system of linear equations.* Multiply the first equation by -10 and add.

$$-10x - 10y = -400 \qquad\qquad \text{Multiply by } -10.$$
$$\underline{10x + 11.5y = 433}$$
$$ 1.5y = 33, \quad \text{or} \quad y = \frac{33}{1.5} = 22 \qquad \text{Add and solve for } y.$$

Because $x + y = 40$ and $y = 22$, it follows that $x = 18$. Thus the athlete spent 18 minutes on the rowing machine and 22 minutes on the stair climber.

STEP 4 *Check your answer.* Because $18 + 22 = 40$, the athlete worked out for 40 minutes. Also,

$$10(18) + 11.5(22) = 433,$$

so the athlete burned 433 calories. The answer checks.

4.3 PUTTING IT ALL TOGETHER

The substitution and elimination methods are two symbolic techniques that can be used to solve systems of linear equations. The elimination method makes use of the addition and multiplication properties of equality and is based on the idea that "if equals are added to equals, the results are equal." The following table highlights the important aspects of elimination.

Concept	Explanation	Example
Elimination Method	If $a = b$ and $c = d$, then $$a + c = b + d.$$ May be used to solve systems of equations	$$\begin{aligned} x + y &= 5 \\ \underline{x - y} &= \underline{-1} \\ 2x &= 4, \quad \text{or} \quad x = 2 \quad \text{Add.} \end{aligned}$$ Because $x + y = 5$ and $x = 2$, it follows that $y = 3$. The solution is $(2, 3)$.
Other Types of Systems	Elimination can be used to recognize systems having no solutions or infinitely many solutions.	$$\begin{aligned} -x - y &= -4 \\ \underline{x + y} &= \underline{2} \\ 0 &= -2 \quad \text{Add.} \end{aligned}$$ Because $0 = -2$ is always false, there are no solutions. Given $$\begin{aligned} x + y &= 4 \\ 2x + 2y &= 8, \end{aligned}$$ multiply the first equation by -2. $$\begin{aligned} -2x - 2y &= -8 \\ \underline{2x + 2y} &= \underline{8} \\ 0 &= 0 \quad \text{Add.} \end{aligned}$$ Because $0 = 0$ is always true, there are infinitely many solutions.

4.3 EXERCISES

FOR EXTRA HELP

| | Student's Solutions Manual | | InterAct Math | | MathXL |
| | MyMathLab | | Math Tutor Center | | Digital Video Tutor CD 2 Videotape 6 |

CONCEPTS

1. Name two symbolic methods for solving a system of linear equations.

2. The elimination method is based on the _____ property of equality.

3. The addition property of equality states that if $a = b$ and $c = d$, then $a + c$ ___ $b + d$.

4. The multiplication property of equality states that if $a = b$, then ca ___ cb.

5. When you are using elimination to solve
$$2x + y = 6$$
$$x - y = 2,$$
what is a good first step?

6. When you are using elimination to solve
$$x + 2y = 8$$
$$3x - 5y = 2,$$
what is a good first step?

USING ELIMINATION

Exercises 7–14: If possible, use the graph to solve the system of equations. Then use the elimination method to verify your answer.

7. $x + y = 2$
   $x - y = 0$

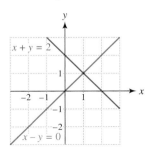

8. $x + y = 6$
   $2x - y = 3$

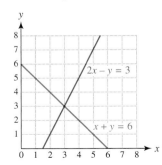

9. $2x + 3y = -1$
   $2x - 3y = -7$

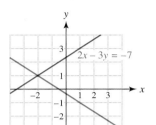

10. $-2x + y = -3$
    $4x - 3y = 7$

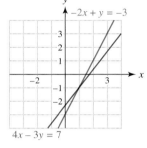

11. $x + y = 3$
    $x + y = -1$

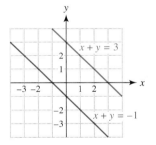

12. $2x - y = 4$
    $-2x + y = -4$

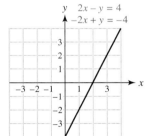

13. $2x + 2y = 6$
    $x + y = 3$

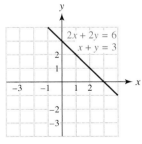

14. $x - 3y = 3$
    $-x + 3y = 4$

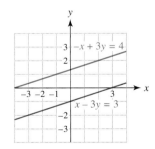

Exercises 15–36: Use the elimination method to solve the system of equations.

15. $x + y = 7$
    $x - y = 5$

16. $x - y = 8$
    $x + y = 4$

17. $-x + y = 5$
$x + y = 3$

18. $x - y = 10$
$-x - y = 20$

19. $2x + y = 8$
$3x - y = 2$

20. $-x + 2y = 3$
$x + 6y = 5$

21. $-2x + y = -3$
$2x - 4y = 0$

22. $2x + 6y = -5$
$7x - 6y = -4$

23. $a + 6b = 2$
$a + 3b = -1$

24. $5a - 6b = -2$
$5a + 5b = 9$

25. $3r - t = 7$
$2r - t = 2$

26. $-r + 2t = 0$
$3r + 2t = 8$

27. $3u + 2v = -16$
$2u + v = -9$

28. $5u - v = 0$
$3u + 3v = -18$

29. $2x + 7y = 6$
$4x - 3y = -22$

30. $5x - 7y = 5$
$-2x + 2y = -2$

31. $5x - 3y = 4$
$x + 4y = 10$

32. $-3x - 8y = 1$
$2x + 5y = 0$

33. $\frac{1}{2}x - y = 3$
$\frac{3}{2}x + y = 5$

34. $x - \frac{1}{4}y = 4$
$-4x + \frac{1}{4}y = -9$

35. $-5x - 10y = -22$
$10x + 15y = 35$

36. $-15x + 4y = -20$
$5x + 7y = 90$

Exercises 37–40: A table of values is given for two linear equations. Use the table to solve this system.

37.

x	0	1	2	3	4
$y = -x + 5$	5	4	3	2	1
$y = 2x - 4$	-4	-2	0	2	4

38.

x	-3	-2	-1	0	1
$y = x + 1$	-2	-1	0	1	2
$y = -x - 3$	0	-1	-2	-3	-4

39.

x	-2	-1	0	1	2
$y = 3x + 1$	-5	-2	1	4	7
$y = -x + 1$	3	2	1	0	-1

40.

x	-2	-1	0	1	2
$y = 2x$	-4	-2	0	2	4
$y = -x$	2	1	0	-1	-2

USING MORE THAN ONE METHOD

Exercises 41–46: Solve the system of equations
(a) symbolically,
(b) graphically, and
(c) numerically.

41. $2x + y = 5$
$x - y = 1$

42. $-x + y = 2$
$3x + y = -2$

43. $2x + y = 5$
$x + y = 1$

44. $-x + y = 2$
$3x - y = -2$

45. $6x + 3y = 6$
$-2x + 2y = -2$

46. $-x + 2y = 5$
$2x + 2y = 8$

ELIMINATION AND OTHER TYPES OF SYSTEMS

Exercises 47–56: Use elimination to determine whether the system of equations has zero, one, or infinitely many solutions. Then graph the system.

47. $2x - 2y = 4$
$-x + y = -2$

48. $-2x + y = 4$
$4x - 2y = -8$

49. $x - y = 0$
$x + y = 0$

50. $x - y = 2$
$x + y = 2$

51. $x - y = 4$
$x - y = 1$

52. $-2x + 3y = 5$
$4x - 6y = 10$

53. $x - y = 5$
$2x - y = 4$

54. $6x + 9y = 18$
$4x + 6y = 12$

55. $4x - 8y = 24$
$6x - 12y = 36$

56. $x - 3y = 2$
$-x + 3y = 4$

APPLICATIONS

57. *Skin Cancer*   In 2001, there were 56,400 new cases of skin cancer in the United States. Men represented 7000 more cases than women. How many new cases of skin cancer were there for men and for women? (*Source:* American Cancer Society.)

58. *Health Care Expenses*   In 2002, out-of-pocket expenses for an elderly person in poor health were $3353 more than for an elderly person in good health. Combined out-of-pocket expenses for one person in poor health and one person in good health were $6213. Find the out-of-pocket expenses for each type of person. (*Source:* "Trends in Medicare-Choice Benefits and Premiums, 1999–2002." *Commonwealth Fund.*)

59. *Burning Calories*   During strenuous exercise an athlete can burn 9 calories per minute on a stationary bicycle and 11.5 calories per minute on a stair climber. In a 30-minute workout an athlete burns 300 calories. How many minutes did the athlete spend on each type of exercise equipment? (*Source: Runner's World.*)

60. *Distance Running* An athlete runs at 9 mph and then at 12 mph, covering 10 miles in 1 hour. How long did the athlete run at each speed?

61. *River Current* A riverboat takes 8 hours to travel 64 miles downstream and 16 hours for the return trip. What is the speed of the current and the speed of the riverboat in still water?

62. *Airplane Speed* An airplane travels 3000 miles with the wind in 5 hours and takes 6 hours for the return trip into the wind. What is the speed of the wind and the speed of the airplane without any wind?

63. *Investments* A total of $5000 is invested at 3% and 5% annual interest. After 1 year the total interest equals $210. How much money is invested at each interest rate?

64. *Mixing Antifreeze* A car radiator holds 2 gallons of fluid and initially is empty. If a mixture of water and antifreeze contains 70% antifreeze and another mixture contains 15% antifreeze, how much of each should be combined to fill the radiator with a 50% antifreeze mixture?

65. *Number Problem* The sum of two integers is -17, and their difference is 69. Find the two integers.

66. *Supplementary Angles* The measures of two supplementary angles differ by 74°. Find the two angles.

67. *Picture Dimensions* The figure shows a red graph that gives possible dimensions for a rectangular picture frame with perimeter 120 inches. The blue graph shows possible dimensions for a rectangular frame whose length L is twice its width W.

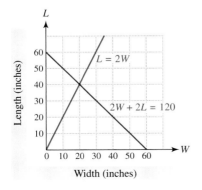

(a) Use the figure to determine the dimensions of a frame with a perimeter of 120 inches and a length that is twice the width.

(b) Solve this problem symbolically.

68. *Sales of CDs and Tapes* A company sells compact discs d and cassette tapes t. The figure shows a red graph of $d + t = 2000$. The blue graph shows a revenue of $15,000 received from selling d compact discs at $12 each and t tapes at $6 each.

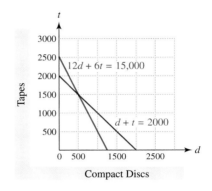

Compact Discs

(a) If the total number of discs and tapes sold is 2000, determine how many of each were sold to obtain a revenue of $15,000.

(b) Solve this problem symbolically.

WRITING ABOUT MATHEMATICS

69. Suppose that a system of linear equations is solved symbolically, numerically, and graphically. How do the solutions from each method compare? Explain your answer.

70. When you are solving a system of linear equations by elimination, how can you recognize that the system has no solutions?

Group Activity: Working with Real Data

Directions: Form a group of 2 to 4 people. Select someone to record the group's responses for this activity. All members of the group should work cooperatively to answer the questions. If your instructor asks for your results, each member of the group should be prepared to respond.

Exercises 1–4: Per Capita Income   *In 2000, the average of the per capita (per person) incomes for Massachusetts and Maine was $32,000. The per capita income in Massachusetts exceeded the per capita income in Maine by $12,000.*

1. Set up a system of equations whose solution gives the per capita income in each state. Identify what each variable represents.

2. Use substitution to solve this system. Interpret the result.

3. Use elimination to solve this system.

4. Solve this system graphically. Do all your answers agree?

4.4   SYSTEMS OF LINEAR INEQUALITIES

Basic Concepts · Solutions to One Inequality · Solutions to Systems of Inequalities · Applications

INTRODUCTION

Although there is no *ideal* weight for a person, government agencies and insurance companies sometimes recommend a *range* of weights for various heights. *Inequalities* are used with these recommendations. One example is shown in Figure 4.11, where the blue region contains ordered pairs (w, h) that give recommended weight–height combinations. Describing this region mathematically requires systems of linear inequalities, which we discuss in this section.   (***Source:*** Department of Agriculture.)

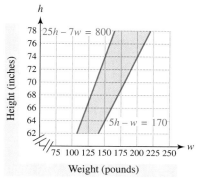

Figure 4.11

BASIC CONCEPTS

Suppose that a student works both at the library and at a department store. The library pays $10 per hour, and the department store pays $8 per hour. The equation $A = 10L + 8D$ calculates the amount of money earned from working L hours at the library and D hours at the department store. If the cost of one college credit is $80, then solutions to the equation

$$10L + 8D = 80$$

are ordered pairs (L, D) that result in the student earning enough to pay for one credit. Its graph is the line shown in Figure 4.12(a). The point $(4, 5)$ lies on this line, which indicates that, if the student works 4 hours at the library and 5 hours at the department store, then the pay is $80.

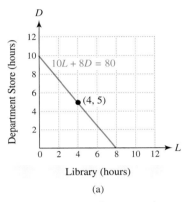
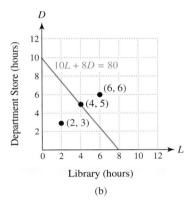

Figure 4.12 A Student's Earnings

There are many situations in which the student can make more than $80. For example, the **test point** $(6, 6)$ lies above the line in Figure 4.12(b), indicating that, if the student works 6 hours at both the library and the department store, then the pay is more than $80. In fact, any point *above* the line results in pay *greater than* $80. The region **above** the line is described by the inequality

$$10L + 8D > 80.$$

The test point $(6, 6)$ represents earnings of $108 and *satisfies* this inequality because

$$10(6) + 8(6) > 80$$

is a true statement. Similarly, any point *below* the line gives an ordered pair (L, D) that results in earnings *less than* $80. The point $(2, 3)$ in Figure 4.12(b) lies below the line and represents earnings of $44. That is,

$$10(2) + 8(3) < 80.$$

The region **below** the line is described by the inequality

$$10L + 8D < 80.$$

SOLUTIONS TO ONE INEQUALITY

Any linear equation in two variables can be written in standard form as

$$Ax + By = C,$$

where A, B, and C are constants. When the equals sign is replaced with $<$, $>$, $\leq$, or $\geq$, a **linear inequality in two variables** results. Examples of linear *equations* in two variables include

$$2x + 3y = 10 \quad \text{and} \quad y = \frac{1}{2}x - 5,$$

and so examples of linear *inequalities* in two variables include

$$2x + 3y < 10 \quad \text{and} \quad y \geq \frac{1}{2}x - 5.$$

A *solution* to a linear inequality in two variables is an ordered pair (x, y) that makes the inequality a true statement. The *solution set* is the set of all solutions to the inequality. The solution set to an inequality in two variables is typically a region in the xy-plane, which means that there are infinitely many solutions.

EXAMPLE 1   Writing a linear inequality

Write a linear inequality that describes each shaded region.

(a)

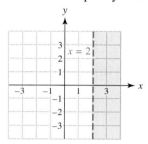

(b)

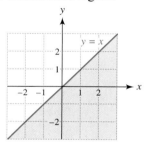

Solution   **(a)** The shaded region is bounded by the line $x = 2$. The *dashed* line indicates that the line is not included in the solution set. Only points with x-coordinates greater than 2 are shaded. Thus every point in the shaded region satisfies $x > 2$.

(b) The solution set includes all points that are on or below the line $y = x$. An inequality that describes this region is $y \leq x$, which can also be written as $-x + y \leq 0$. _____

Technology Note: *Shading an Inequality*

Graphing calculators can be used to shade a solution set to an inequality. The left-hand screen shows how to enter the equation from Example 1(b), and the right-hand screen shows the resulting graph.

Calculator Help
To shade an inequality, see Appendix A (page AP-7).

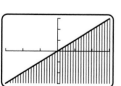

EXAMPLE 2   Graphing a linear inequality

Shade the solution set for each inequality.
(a) $y \leq 1$ **(b)** $x + y < 3$ **(c)** $-x + 2y \geq 2$

Solution (a) First graph the horizontal line $y = 1$, where a solid line indicates that the line is included in the solution set. Next decide whether to shade above or below this line. The inequality $y \le 1$ indicates that a solution (x, y) must have a y-coordinate less than or equal to 1. There are no restrictions on the x-coordinate. The shaded region below the line $y = 1$ in Figure 4.13(a) depicts all ordered pairs (x, y) satisfying $y \le 1$. The horizontal line $y = 1$ is solid because it is included in the solution set.

(b) Graph the line $x + y = 3$, as shown in Figure 4.13(b). Because the inequality is $<$ and not $\le$, the line is not included and is dashed rather than solid. To decide whether to shade above or below this dashed line, select a *test point* in either region. For example, the point $(0, 0)$ satisfies the inequality $x + y < 3$, because $0 + 0 < 3$ is a true statement. Therefore shade the region that does contain $(0, 0)$, which is *below* the line.

(c) Graph $-x + 2y = 2$ as a solid line, as shown in Figure 4.13(c). To decide whether to shade above or below the solid line, use the test point $(0, 0)$ again. (Note that other test points can be used.) The point $(0, 0)$ does not satisfy the inequality $-x + 2y \ge 2$ because $-0 + 2(0) \ge 2$ is a false statement. Therefore shade the region that does not contain $(0, 0)$, which is *above* the line.

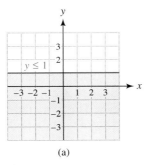   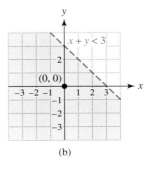   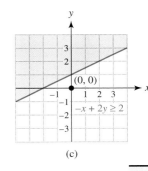

(a)   (b)   (c)

Figure 4.13

GRAPHING A LINEAR INEQUALITY

1. Replace the inequality symbol with an equals sign and graph the resulting line. If the inequality is $<$ or $>$, use a dashed line, and if it is $\le$ or $\ge$, use a solid line.
2. Pick a test point that does *not* lie on the line. Substitute this point in the given inequality. Determine whether the resulting statement is true or false.
3. If the statement is true, shade the region containing the test point. If the statement is false, shade the region not containing the test point.

SOLUTIONS TO SYSTEMS OF INEQUALITIES

Sometimes a solution set must satisfy two inequalities. The point $(2, 1)$ satisfies both of the inequalities $x > 1$ and $y < 2$. We show how to shade the solution set to the system of inequalities

$$x > 1$$
$$y < 2$$

in Figure 4.14 on the next page. Figure 4.14(a) shows the solution set to $x > 1$ in blue, and Figure 4.14(b) shows the solution set to $y < 2$ in red. The solution set to the system of equations includes all points in *both* the red and blue regions. It is shaded purple in Figure 4.14(c).

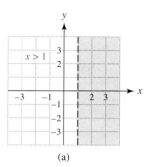

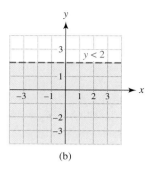

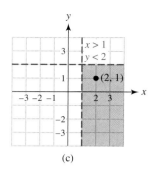

(a)   (b)   (c)

Figure 4.14

EXAMPLE 3   Graphing systems of linear inequalities

Shade the solution set to the system of inequalities.

$$x > -1$$
$$x + y \le 1$$

Solution   In this example, we start by graphing the solution set to each inequality. The solution set to $x > -1$ is the blue region that lies to the right of the vertical line $x = -1$, as shown in Figure 4.15(a). The solution set to $x + y \le 1$ includes the line $x + y = 1$ and the red region that lies below it, as shown in Figure 4.15(b). For a point to satisfy the *system* of inequalities it must satisfy *both* inequalities. Therefore the solution set is the *intersection* of the blue and red regions. This region is shown as the purple region in Figure 4.15(c). Note that the test point $(0, 0)$ satisfies both inequalities and that it is located in the shaded region.

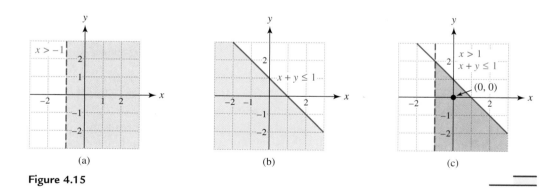

(a)   (b)   (c)

Figure 4.15

EXAMPLE 4   Graphing systems of linear inequalities

Shade the solution set to the system of inequalities.

$$x + 2y < -2$$
$$2x + y \ge 2$$

Solution   In this example we use test points to determine the solution set. We start by graphing the dashed line $x + 2y = -2$ and the solid line $2x + y = 2$, as shown in Figure 4.16(a). Note that these two lines divide the xy-plane into 4 regions, numbered 1, 2, 3, and 4. If we let $(0, 0)$ be a test point, it does not satisfy either inequality. Therefore we do not shade region 2, which contains $(0, 0)$. However, there are still 3 possible regions. If we try the test point $(4, -4)$ in region 4, it satisfies both the given inequalities.

$$4 + 2(-4) < -2 \qquad \text{A true statement}$$
$$2(4) + (-4) \geq 2 \qquad \text{A true statement}$$

Thus we shade region 4, as shown in Figure 4.16(b).

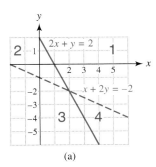

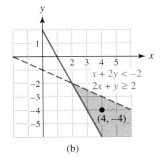

(a)   (b)

Figure 4.16

Critical Thinking

Does the solution set in Figure 4.16(b) include the point of intersection, $(2, -2)$? Explain your reasoning.

=== MAKING CONNECTIONS ===

Solving for y and shading inequalities

Another way to solve the system of inequalities in Example 4 that does not involve test points is to solve each inequality for y to obtain

$$y < -\frac{1}{2}x - 1$$
$$y \geq -2x + 2.$$

The solution set is the region in Figure 4.16(b) that lies *below* the line $y = -\frac{1}{2}x - 2$ and *above and including* the line $y = -2x + 2$.

APPLICATIONS

The next example illustrates an application of inequalities.

EXAMPLE 5 **Manufacturing radios and CD players**

A business manufactures radios and CD players. Because every CD player contains a radio, it must produce at least as many radios as CD players. In addition, the total number of radios and CD players cannot exceed 50 because of limited resources. Shade the region that shows numbers of radios R and CD players P that can be produced within these restrictions. Label the horizontal axis R and the vertical axis P.

Solution Because the company must produce *at least* as many radios R as CD players P, we have $R \geq P$, which can also be written as $P \leq R$. The total number of radios and CD players *cannot exceed* 50 so $R + P \leq 50$. To shade the solution set for

$$P \leq R$$
$$R + P \leq 50,$$

we first graph the lines $P = R$ and $R + P = 50$, as shown in Figure 4.17(a). Because the number of radios and CD players cannot be negative, the graph includes only quadrant I. These lines divide this quadrant into four regions, and we can determine the correct region to shade by selecting one test point from each region. The region containing the test point satisfying both inequalities is the one to be shaded. For example, the test point (**20**, **10**) with $R = 20$, $P = 10$ satisfies both inequalities.

$$10 \leq 20 \qquad \text{A true statement; } P \leq R$$

$$20 + 10 \leq 50 \qquad \text{A true statement; } R + P \leq 50$$

The solution set is shaded in Figure 4.17(b).

Note: An alternative solution is to write the inequalities as $P \leq R$ and $P \leq -R + 50$. Then the solution set lies *below both lines*. This region is shaded in Figure 4.17(b).

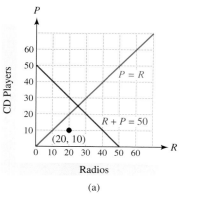

(a)

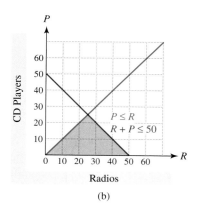

(b)

Figure 4.17

The next example discusses the application from the introduction to this section.

EXAMPLE 6   Finding weight–height combinations

Figure 4.18 shows a shaded region containing recommended weights w for heights h.

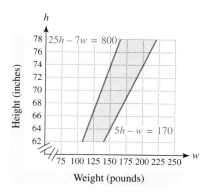

Figure 4.18

(a) What does this graph indicate about someone who is 68 inches tall and 150 pounds?

(b) The shaded region in Figure 4.18 is determined by the following system of inequalities.

$$25h - 7w \leq 800$$
$$5h - w \geq 170$$

Verify that $h = 68$ and $w = 150$ satisfies the system of inequalities.

(c) What do ordered pairs (w, h) to the left of the shaded region indicate?

Solution   **(a)** The point $(150, 68)$ lies in the shaded region. Therefore someone who is 68 inches tall and weighs 150 pounds falls within the recommended guidelines.

Critical Thinking

In Example 6 what do ordered pairs (w, h) to the right of the shaded region represent? Explain your answer.

(b) Both inequalities are satisfied by $h = 68$, $w = 150$.

$$25(68) - 7(150) = 650 \leq 800$$
$$5(68) - 150 = 190 \geq 170$$

(c) To the left of the shaded region are ordered pairs (w, h) that represent smaller weights and taller heights. This region corresponds to people who weigh less than recommended.

4.4 PUTTING IT ALL TOGETHER

The following table summarizes important concepts related to linear inequalities in two variables.

Concept	Explanation	Examples
Linear Inequality in Two Variables	An inequality that can be written as $$Ax + By < C,$$ where $<$ can also be $\leq$, $>$, or $\geq$.	$3x + y \geq 10$, $-x + 3y < 5$, $y \leq 5 - x$, and $x > 5$
Solution	A solution (x, y) makes the inequality a true statement.	The point $(0, 0)$ satisfies $$2x - y < 2,$$ so it is a solution to the inequality.
Solution Set	The set of all solutions Usually a region in the xy-plane	The solution set to $x + y > 2$ is all points above the line $x + y = 2$.

continued on next page

continued from previous page

Concept	Explanation	Examples
System of Linear Inequalities in Two Variables	Solutions to systems must satisfy both inequalities. Usually includes infinitely many solutions	The point $(0, 0)$ is a solution to $$x + y \leq 2$$ $$2x - y > -4,$$ because both inequalities are true when $x = 0$ and $y = 0$. 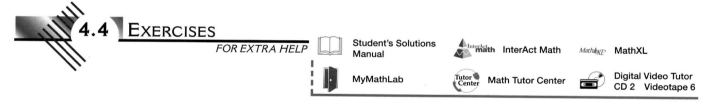

4.4 EXERCISES

FOR EXTRA HELP

Student's Solutions Manual   InterAct Math   MathXL
MyMathLab   Math Tutor Center   Digital Video Tutor CD 2   Videotape 6

CONCEPTS

1. Describe the graph of the solution set to $y \leq k$ for some number k.

2. Describe the graph of the solution set to $x > k$ for some number k.

3. Describe the graph of the solution set to $y \geq x$.

4. When graphing the solution set to a linear inequality, one way to determine which region to shade is to use a _____ point.

5. When graphing a linear inequality containing either $<$ or $>$, use a _____ line.

6. When graphing a linear inequality containing either $\leq$ or $\geq$, use a _____ line.

7. When graphing the linear inequality $Ax + By < C$, a first step is to graph the line _____.

8. A solution to a system of two inequalities must make (both inequalities/one inequality) true.

SOLUTIONS TO LINEAR INEQUALITIES

Exercises 9–20: Determine whether the test point is a solution to the linear inequality.

9. $(3, 1)$, $x > 2$

10. $(-3, 4)$, $x \leq -3$

11. $(0, 0)$, $y \geq 2$

12. $(0, 0)$, $y < -3$

13. $(5, 4)$, $y \geq x$

14. $(-1, 2)$, $y < x$

15. $(3, 0)$, $y < x - 1$

16. $(0, 5)$, $y > 2x + 4$

17. $(-2, 6)$, $x + y \leq 4$

18. $(2, -4)$, $x - y \geq 7$

19. $(-1, -1)$, $2x + y \geq -1$

20. $(0, 1)$, $-x - 5y \geq -1$

Exercises 21–24: Use the graph to identify one solution to the inequality. Then shade the solution set.

21. $x > 2$

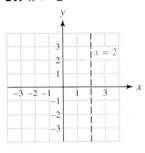

22. $y \leq 1 - x$

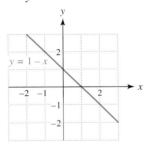

23. $\frac{1}{2}x + y \leq 0$

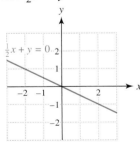

24. $3x - 2y > 6$

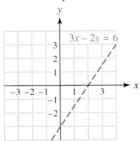

Exercises 25–30: Determine if the test point is a solution to the system of linear inequalities.

25. $(3, 1)$
$x - y < 3$
$x + y > 3$

26. $(0, 0)$
$x - 2y < 1$
$2x - y > -1$

27. $(-2, 3)$
$3x - 2y \geq 1$
$-x + 3y > 3$

28. $(1, 2)$
$2x - 2y < 5$
$x - y > -1$

29. $(4, -2)$
$x - 2y \geq 8$
$-2x - 5y > 0$

30. $(-1, -2)$
$x + y < 0$
$-2x - 3y \leq -1$

Exercises 31–34: The graphs of two equations are shown with four test points labeled. Use these points to decide which region should be shaded to solve the given system of inequalities.

31. $x \leq 2$
       $x + y \geq 2$

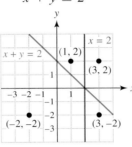

32. $y \geq 1$
         $2x - y \geq 3$

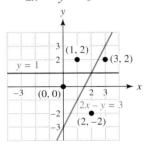

33. $x + y \leq 3$
       $y \leq 2x$

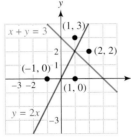

34. $y \leq x$
       $y \geq -x$

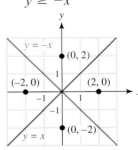

Exercises 35–42: Write a linear inequality that describes the shaded region.

35.

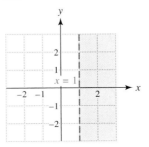

36.

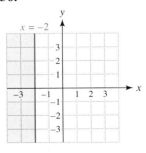

37.

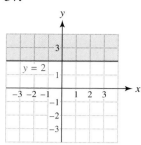

38.

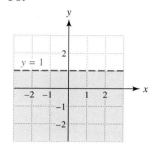

39.

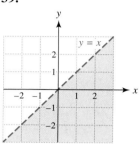

40.

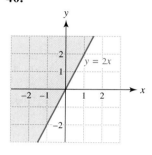

41.

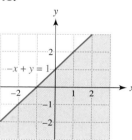

42.

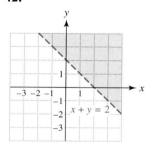

Exercises 43–54: Shade the solution set to the inequality.

43. $x \leq -1$ **44.** $x > 3$

45. $y < -2$ **46.** $y \geq 0$

47. $y > x$ **48.** $y \leq x$

49. $y \geq 3x$ **50.** $y < -2x$

51. $x + y \leq 1$ **52.** $x + y \geq -2$

53. $2x - y > 2$ **54.** $-x - y < 1$

Exercises 55–72: Shade the solution set to the system of inequalities.

55. $x > 2$ **56.** $x \leq -1$
    $y < 3$ $y \geq 3$

57. $x \leq -2$ **58.** $y > 2$
    $y < 2x$ $y \geq -x$

59. $y \leq x$ **60.** $y \leq \frac{1}{2}x$
    $y > -x$ $y \geq -2x$

61. $x + y \leq 3$ **62.** $x + y > 2$
    $-x + y \leq 1$ $x - y < 2$

63. $2x + y > -3$ **64.** $-x + y \geq 3$
    $x + y \leq -1$ $2x - y \geq -2$

65. $2x + y \geq -3$ **66.** $-x + y \geq 2$
    $x + y > -1$ $3x - y \geq -2$

67. $y > -2$ **68.** $x \geq 2$
    $x + 2y \leq -4$ $3y < x - 3$

69. $x + 2y > -4$ **70.** $x + 3y \geq 3$
    $2x + y \leq 3$ $3x - 2y \geq 6$

71. $3x + 4y \leq 12$ **72.** $-2x + y \geq 6$
    $5x + 3y \geq 15$ $x - 2y \geq -4$

APPLICATIONS

73. ***Working on Two Projects***   An employee is required to spend more time on project X than on project Y. The employee can work at most 40 hours on these two projects. Shade the region in the *xy*-plane that represents the number of hours that the employee can spend on each project.

74. ***Radios and CDs***   (Refer to Example 5.) A business manufactures at least two radios for each CD player. The total number of radios and CD players must be less than 90. Shade the region that represents the number of radios *R* and CD players *P* that can be produced within these restrictions. Put *P* on the horizontal axis.

75. ***Maximum Heart Rate***   When exercising, people often try to maintain target heart rates that are a percentage of their maximum heart rate. A person's maximum heart rate *R* is $R = 220 - A$, where *A* is the person's age and *R* is the heart rate in beats per minute.
   (a) What is *R* for a person 20 years old? 70 years old?
   (b) Sketch a graph of $R \leq 220 - A$. Assume that *A* is between 20 and 70 and put *A* on the horizontal axis of your graph.
   (c) Interpret your graph.

76. *Target Heart Rate* (Refer to the preceding exercise.) A target heart rate T that is half a person's maximum heart rate is given by $T = 110 - \frac{1}{2}A$, where A is a person's age.

  (a) What is T for a person 30 years old? 50 years old?

  (b) Sketch a graph of the system of inequalities.

$$T \geq 110 - \frac{1}{2}A$$

$$T \leq 220 - A$$

    Assume that A is between 20 and 60.

  (c) Interpret your graph.

77. *Height and Weight* Use Figure 4.18 in Example 6 to determine the range of recommended weights for a person who is 74 inches tall.

78. *Height and Weight* Use Figure 4.18 in Example 6 to determine the range of recommended heights for a person who weighs 150 pounds.

WRITING ABOUT MATHEMATICS

79. What is the solution set to the following system of inequalities? Explain your reasoning.

$$y > x$$

$$y < x - 1$$

80. Write down a system of linear inequalities whose solution set is the entire xy-plane. Explain your reasoning.

CHECKING BASIC CONCEPTS   SECTIONS 4.3 AND 4.4

1. Use elimination to solve the system of equations.

$$2x + 3y = 5$$

$$x - 7y = -6$$

2. Use elimination to solve each system of equations. How many solutions are there in each case?

  (a) $x + y = -1$

       $x - 2y = 2$

  (b) $5x - 6y = 4$

       $-5x + 6y = 1$

  (c) $x - 3y = 0$

      $2x - 6y = 0$

3. Solve the system of equations symbolically, graphically, and numerically.

$$-2x + y = 0$$

$$y = 2x$$

4. Shade the solution set to each inequality.

  (a) $y < -1$ **(b)** $x + y < 1$

5. Shade the solution set to the system of inequalities.

$$x \leq -1$$

$$-2x + y > -3$$

6. *Large Cities in the United States* The combined population of New York and Chicago was 11 million people in 2000. The population of New York exceeded the population of Chicago by 5 million people.

  (a) Let x be the population of New York and y be the population of Chicago. Write a system of equations whose solution gives the population of each city in 2000.

  (b) Solve the system of equations.

CHAPTER 4 Summary

Section 4.1   *Solving Systems of Linear Equations Graphically and Numerically*

System of Linear Equations

Solution	An ordered pair (x, y) that satisfies *both* equations
Solution Set	The set of all solutions
Graphical Solution	Graph each equation. A point of intersection is a solution. (Sometimes determining the exact answer when estimating from a graph may be difficult.)
Numerical Solution	Solve each equation for y and make a table for each equation. A solution occurs when two y-values are equal for a given x-value.

Example: The ordered pair $(3, 1)$ is the solution to the following system.

$$x + y = 4 \qquad 3 + 1 = 4 \text{ is a true statement.}$$
$$x - y = 2 \qquad 3 - 1 = 2 \text{ is a true statement.}$$

A Graphical Solution

The point of intersection, $(3, 1)$, is the solution to the system of equations.

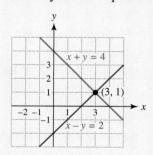

A Numerical Solution

The ordered pair $(\mathbf{3}, \mathbf{1})$ is the solution. When $x = 3$, both y-values equal 1.

x	1	2	3	4
$y = 4 - x$	3	2	1	0
$y = x - 2$	-1	0	1	2

Section 4.2   *Solving Systems of Linear Equations by Substitution*

Method of Substitution   This method can be used to solve a system of equations symbolically and always gives the exact answer.

Example: $-2x + y = -3$
$x + y = 3$

STEP 1   Solve one of the equations for a convenient variable.

$$x + y = 3 \quad \text{becomes} \quad y = 3 - x.$$

STEP 2   Substitute this result in the other equation and then solve.

$$-2x + (3 - x) = -3 \qquad \text{Substitute } (3 - x) \text{ for } y.$$

$$-3x = -6 \qquad \text{Combine like terms; subtract 3.}$$

$$x = 2 \qquad \text{Divide each side by } -3.$$

STEP 3   Find the value of the other variable. Because $y = 3 - x$ and $x = 2$, it follows that $y = 3 - 2 = 1$.

STEP 4   Check to determine that $(2, 1)$ is the solution.

$$-2(2) + (1) \overset{?}{=} -3 \qquad \text{A true statement}$$

$$2 + 1 \overset{?}{=} 3 \qquad \text{A true statement}$$

The solution $(2, 1)$ checks.

Types of Systems of Linear Equations   Can have zero, one, or infinitely many solutions.

No solutions	Inconsistent (parallel lines)
One solution	Consistent (independent equations)
Infinitely many solutions	Consistent (dependent equations)

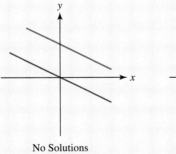
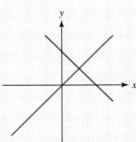
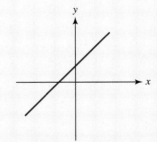

No Solutions
parallel lines

One Solution
intersecting lines (1 point)

Infinitely Many Solutions
identical (or coincident) lines

Section 4.3   *Solving Systems of Linear Equations by Elimination*

Method of Elimination   This method can be used to solve a system of linear equations symbolically and always gives the exact answer.

Example:
$$x + 3y = 1$$
$$\underline{-x + y = 3}$$
$$4y = 4, \quad \text{or} \quad y = 1 \qquad \text{Add and solve for } y.$$

Substitute $y = 1$ into first (or second) equation. $x + 3(1) = 1$ implies that $x = -2$, so $(-2, 1)$ is the solution.

Note: To eliminate a variable, it may be necessary to multiply one or both equations by a constant before adding.

Recognizing Types of Systems

No solutions	Final equation is always false, such as $0 = 1$.
One solution	Final equation has one solution, such as $x = 1$.
Infinitely many solutions	Final equation is always true, such as $0 = 0$.

Section 4.4   *Systems of Linear Inequalities*

Graphing a Linear Inequality

1. Replace the inequality symbol with an equals sign and graph the resulting line. If the inequality is $<$ or $>$ use a dashed line, and if it is $\leq$ or $\geq$ use a solid line.
2. Pick a *test point* that does *not* lie on the line. Substitute this point in the given inequality. Determine whether the resulting statement is true or false.
3. If the statement is true, shade the region containing the test point. If the statement is false, shade the region on the other side of the line.

Example: $x \leq 1$
$$x + y \leq 2$$

Graph the lines $x = 1$ and $x + y = 2$. Then pick a test point, such as $(0, 0)$, and substitute it in each inequality.

$$0 \leq 1 \qquad \text{A true statement}$$
$$0 + 0 \leq 2 \qquad \text{A true statement}$$

Because $(0, 0)$ satisfies *both* inequalities, shade the region containing $(0, 0)$.

Note: When shading the solution set to a *system* of inequalities, you may need to try more than one test point.

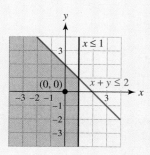

Note: An alternative way to determine the solution set is to shade the region to the *left* of the line $x = 1$ (because $x \leq 1$) and *below* the line $y = -x + 2$ (because $y \leq -x + 2$.)

CHAPTER 4 REVIEW EXERCISES

SECTION 4.1

Exercises 1 and 2: Determine graphically the x-value for the equation when $y = 3$.

1. $y = 2x - 3$

2. $y = \frac{3}{2}x$

Exercises 3–6: Determine which ordered pair is a solution to the system of equations.

3. $(0, 1), (1, 2)$
$$x + 2y = 5$$
$$x - y = -1$$

4. $(5, 2), (4, 0)$
$$2x - y = 8$$
$$x + 3y = 11$$

5. $(2, 2), (4, 3)$
$$\tfrac{1}{2}x = y - 1$$
$$2x = 3y - 1$$

6. $(2, -4), (-1, 2)$
$$5x - 2y = 18$$
$$y = -2x$$

Exercises 7 and 8: The graphs for two equations are shown. Use the intersection-of-graphs method to identify the solution to both equations. Then check your result.

7.

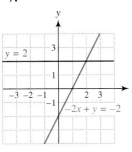

8.

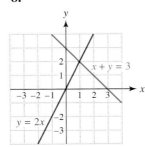

Exercises 9 and 10: A table for two equations is given. Identify the solution to both equations.

9.

x	1	2	3	4
$y = 3x$	3	6	9	12
$y = 6$	6	6	6	6

10.

x	-1	0	1	2
$y = 2x - 1$	-3	-1	1	3
$y = 2 - x$	3	2	1	0

Exercises 11–16: Solve the system of equations graphically.

11.
$$y = -3$$
$$x + y = 1$$

12.
$$x = 1$$
$$x - y = -1$$

13.
$$2x + y = 3$$
$$-x + y = 0$$

14.
$$y = 2x$$
$$2x + y = 4$$

15.
$$x + 2y = 3$$
$$2x + y = 3$$

16.
$$-3x - y = 7$$
$$2x + 3y = -7$$

SECTION 4.2

Exercises 17–22: Use the method of substitution to solve the system of linear equations.

17. $x + y = 8$
$$y = 3x$$

18. $x - 2y = 22$
$$y = -5x$$

19.
$$2x + y = 5$$
$$-3x + y = 0$$

20. $3x - y = 5$
$$x - y = -5$$

21.
$$x + 3y = 1$$
$$-2x + 2y = 6$$

22. $3x - 2y = -4$
$$2x - y = -4$$

Exercises 23–26: The graphs of two equations are shown.

(a) State the number of solutions to the system of equations.

(b) Is the system consistent or inconsistent? If the system is consistent, state whether the equations are dependent or independent.

23.

24.

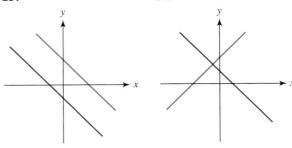

25.

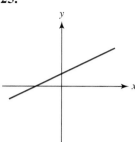

26.

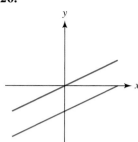

Exercises 27–30: *Use the method of substitution to solve the system of linear equations. Then solve the system graphically. Note that these systems may have zero, one, or infinitely many solutions.*

27. $x + y = 2$
$\quad\quad y = -x$

28. $x + y = -2$
$\quad\quad x + y = 3$

29. $-x + 2y = 2$
$\quad\quad x - 2y = -2$

30. $-x - y = -2$
$\quad\quad 2x - y = 1$

SECTION 4.3

Exercises 31 and 32: *Use the graph to solve the system of equations. Then use the elimination method to verify your answer.*

31. $x + y = 3$
$\quad\quad x - y = 1$

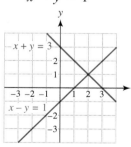

32. $2x + 3y = 4$
$\quad\quad x - 2y = -5$

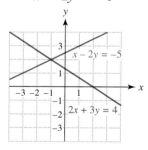

Exercises 33–40: *Use the elimination method to solve the system of equations.*

33. $x + y = 10$
$\quad\quad x - y = 12$

34. $2x - y = 2$
$\quad\quad 3x + y = 3$

35. $-2x + 2y = -1$
$\quad\quadx - 3y = -3$

36. $2x - 5y = 0$
$\quad\quad 2x + 4y = 9$

37. $2a + b = 3$
$\quad -3a - 2b = -1$

38. $a - 3b = 2$
$\quad 3a + b = 26$

39. $5r + 3t = -1$
$\quad -2r - 5t = -11$

40. $5r + 2t = 5$
$\quad\quad 3r - 7t = 3$

Exercises 41 and 42: *Solve the following system of equations (a) symbolically, (b) graphically, and (c) numerically.*

41. $3x + y = 6$
$\quad\quad x - y = -2$

42. $2x + y = 3$
$\quad -x + 2y = -4$

Exercises 43–46: *Use elimination to determine whether the system of equations has zero, one, or infinitely many solutions.*

43. $x - y = 5$
$\quad -x + y = -5$

44. $3x - 3y = 0$
$\quad -x + y = 0$

45. $-2x + y = 3$
$\quad2x - y = 3$

46. $-2x + y = 2$
$\quad3x - y = 3$

SECTION 4.4

Exercises 47–50: *Determine whether the test point is a solution to the linear inequality.*

47. $(5, -3)\quad y \le 2$

48. $(-1, 3)\quad x > -1$

49. $(1, 2)\quad x + y < -2$

50. $(1, -4)\quad 2x - 3y \ge 2$

Exercises 51 and 52: *Use the graph to identify one solution to the inequality. Answers may vary. Then shade the solution set.*

51. $y \le -1$

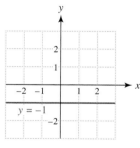

52. $2x + y > 1$

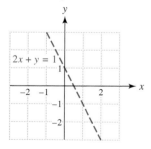

Exercises 53 and 54: *Determine whether the test point is a solution to the system of linear inequalities.*

53. $(1, -2)$
$\quad\quad x - 2y > 3$
$\quad\quad 2x + y < 3$

54. $(4, -3)$
$\quad\quad x - y \ge 1$
$\quad\quad 4x + 3y \le 4$

Exercises 55 and 56: The graphs of two equations are shown with four test points labeled. Use these points to decide which region should be shaded to solve the system of inequalities.

55. $y \le 1$
$2x + y \ge -1$

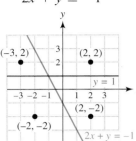

56. $y \ge x$
$x + y \ge 2$

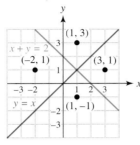

Exercises 57 and 58: Write a linear inequality that describes the shaded region.

57.

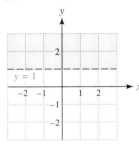

58.

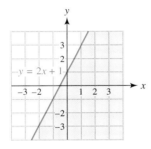

Exercises 59–64: Shade the solution set for the inequality.

59. $x \le 1$

60. $y > 2$

61. $y > 3x$

62. $x \ge 2y$

63. $y < x + 1$

64. $2x + y \ge -2$

Exercises 65–70: Shade the solution set for the system of inequalities.

65. $x > -1$
$y < -2$

66. $y \le x$
$y \ge -2x$

67. $x + y \le 3$
$y \ge -x$

68. $2x + y < 3$
$y > x$

69. $\frac{1}{2}x + y \ge 2$
$x - 2y \le 0$

70. $2x - y < 3$
$4x + 2y > -6$

APPLICATIONS

71. *Motor Vehicle Fatalities*   The number of motor vehicle deaths increased by 13.3 times from 1912 to 1999. There were 38,130 more deaths in 1999 than in 1912. Find the number of motor vehicle deaths in each of the two years. Note that the number of motor vehicles on the road increased from 1 million to 218 million between 1912 and 1999.   (*Source:* Department of Health and Human Services.)

72. *Lung Cancer*   In 2001, 185,000 new cases of lung cancer were reported. There were 20,000 more new cases for men than for women. How many new cases of lung cancer were there for men and for women? (*Source:* American Cancer Society.)

73. *Renting a Car*   A rental car costs $40 plus $0.20 per mile.
  (a) Write an equation that gives the cost C of driving the car x miles.
  (b) Use the intersection-of-graphs method to determine the number of miles that the car is driven if the rental cost is $90.
  (c) Solve part (b) numerically with a table of values.

74. *Dimensions of a Garden*   A rectangular garden has 88 feet of fencing around it. The garden is 4 feet longer than it is wide. Find the dimensions of the garden.

75. *Triangle*   In an isosceles triangle, the measures of the two smaller angles are equal and their sum is 40° more than the larger angle.
  (a) Let x be the measure of each of the two smaller angles and y be the measure of the larger angle. Write a system of linear equations whose solution gives the measures of these angles.
  (b) Use the method of substitution to solve the system.
  (c) Use the method of elimination to solve the system.

76. *Supplementary Angles*   The smaller of two supplementary angles is 30° less than the measure of the larger angle. Find each angle.

77. *Room Prices*   Ten rooms are rented at rates of $80 and $120 per night. The total collected for the 10 rooms is $920.
  (a) Write a system of linear equations whose solution gives the number of each type of room rented. Be sure to state what each variable represents.
  (b) Solve the system of equations.

78. *Mixture Problem*   One type of candy sells for $2 per pound, and another type sells for $3 per pound. An order for 18 pounds of candy costs $47. How much of each type of candy was bought?

79. *Burning Calories*   An athlete burns 9 calories per minute on a stationary bicycle and 11 calories per minute on a stair climber. In a 60-minute workout the athlete burns 590 calories. How many minutes did the athlete spend on each type of exercise equipment? (**Source:** *Runner's World.*)

80. *River Current*   A riverboat travels 140 miles downstream in 10 hours, and the return trip takes 14 hours. What is the speed of the current?

81. *Garage Dimensions*   The blue graph shown in the figure gives possible dimensions for a rectangular garage with perimeter 80 feet. The red graph shows possible dimensions for a garage that has width W two-thirds of its length L.

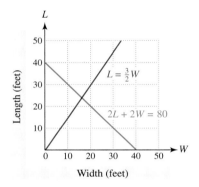

(a) Use the graph to estimate the dimensions of a garage with perimeter 80 feet and width two-thirds its length.

(b) Solve this problem symbolically.

82. *Wheels and Trailers*   A business manufactures at least two wheels for each trailer it makes. The total number of trailers and wheels manufactured cannot exceed 30 per week. Shade the region that represents numbers of wheels W and trailers T that can be produced each week within these restrictions. Label the horizontal axis W and the vertical axis T.

83. *Target Heart Rate*   A target heart rate T that is 70% of a person's maximum heart rate is approximated by $T = 150 - 0.7A$, where A is a person's age.

(a) What is T for a person 20 years old? 60 years old?

(b) Sketch a graph of $T \geq 150 - 0.7A$. Assume that A is between 20 and 60.

(c) Interpret this graph.

4 TEST

1. Determine which ordered pair is a solution to the system of equations.

$$(3, -1), (1, 2)$$
$$3x + 2y = 7$$
$$2x - y = 0$$

2. The graphs for two equations are shown. Use the intersection-of-graphs method to identify the solution. Then check your solution.

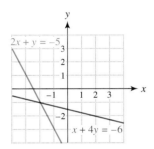

3. A table for two equations is given. Identify the solution to both equations.

x	-2	-1	0	1
$y = 2x$	-4	-2	0	2
$y = 3x + 1$	-5	-2	1	4

4. Solve the system of equations graphically.

$$x + 2y = 4$$
$$x + y = 1$$

5. Use the method of substitution to solve the system of linear equations.

$$3x + 2y = 9$$
$$y = 3x$$

6. Use the method of substitution to solve the system of linear equations. How many solutions are there? Is the system consistent or inconsistent?

(a) $x + 3y = 5$ **(b)** $-x + \frac{1}{2}y = 12$
     $3x - 2y = 4$ $2x - y = -4$

Exercises 7 and 8: Use the elimination method to solve the system of equations.

7. $x + 2y = 5$ **8.** $2x - 2y = 3$
    $3x - 2y = -17$ $-x + y = 5$

9. Write a linear inequality that describes the shaded region.

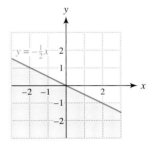

Exercises 10 and 11: Shade the solution set for the inequality.

10. $x \le 4$ **11.** $x + y > 2$

Exercises 12 and 13: Shade the solution set for the system of inequalities.

12. $x > 2$ **13.** $2x + y \le 3$
     $y < 2x$ $x - y \ge 0$

14. *IRS Collections* In 1998 and 1999, the IRS collected a total of $3.7 trillion in taxes. The IRS collected $0.1 trillion more in 1999 than in 1998. How much did the IRS collect in each year? (**Source:** Internal Revenue Service.)

15. *Jogging Speed* An athlete jogs at 6 miles per hour and at 9 miles per hour for a total time of 1 hour, covering a distance of 7 miles. How long did the athlete jog at each speed?

CHAPTER 4 EXTENDED AND DISCOVERY EXERCISES

Exercises 1–4: Plant Growth   *Before doing these exercises, read the introduction to this chapter. Then refer to the figure shown here.*

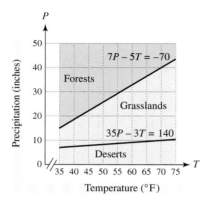

1. The equation of the line that separates grasslands and forests is

$$7P - 5T = -70.$$

Write an inequality that describes temperatures and the amount of precipitation that correspond to forested regions. (Include the line.)

2. The equation of the line that separates grasslands and deserts is

$$35P - 3T = 140.$$

Write an inequality that describes temperatures and the amount of precipitation that correspond to desert regions. (Include the line.)

3. Using the information from Exercises 1 and 2, write a system of inequalities that describes temperatures and the amount of precipitation that correspond to grassland regions.

4. Cheyenne, Wyoming, has an average annual temperature of about 50°F and an average annual precipitation of about 14 inches. Use the graph to predict the type of plant growth you might expect near Cheyenne. Then check to determine whether $T = 50$ and $P = 14$ satisfy the proper inequalities.

Exercises 5–8: Numerical Solutions   *If a solution to a linear equation does not appear in a table of values, can you still find it? Use only the table to solve the equation. Then explain how you got your answer.*

5. $2x + 1 = 0$

x	-2	-1	0	1	2
$y = 2x + 1$	-3	-1	1	3	5

6. $4x + 3 = 5$

x	-2	-1	0	1	2
$y = 4x + 3$	-5	-1	3	7	11

7. $\frac{1}{2}x + 3 = 3.75$

x	-2	-1	0	1	2
$y = \frac{1}{2}x + 3$	2	2.5	3	3.5	4

8. $3x - 1 = 0$

x	-2	-1	0	1	2
$y = 3x - 1$	-7	-4	-1	2	5

Polynomials and Exponents

Digital images were first sent between New York and London by cable in the early 1920s. Unfortunately the transmission time was 3 hours and the quality was poor. Digital photography was developed further by NASA in the 1960s because ordinary pictures were subject to interference when transmitted through space. Digital pictures remain crystal clear even if they travel millions of miles. The following digital picture shows Mars, a planet in our solar system.

Digital images comprise tiny units called pixels, which are represented in a computer or camera by numbers. As a result, mathematics plays an important role in digital images. In this chapter we illustrate some of the ways mathematics is used to describe digital pictures.

If you want to do something, do it!
—Plautus

Source: NASA. (Photograph reprinted with permission.)

5.1 RULES FOR EXPONENTS

Review of Bases and Exponents · Zero Exponents · The Product Rule · Power Rules

INTRODUCTION

Exponents occur throughout mathematics. Regardless of how many mathematics courses you plan to take, exponents are sure to play an important role in each. Because exponents are so important, this section is essential for your becoming successful in mathematics. It takes practice, so set aside some extra time.

REVIEW OF BASES AND EXPONENTS

The expression 5^3 is an exponential expression with *base* 5 and *exponent* 3. Its value is

$$5 \cdot 5 \cdot 5 = 125.$$

In general, b^n is an exponential expression with base b and exponent n. If n is a natural number, then

$$\text{Exponent} \longrightarrow \atop b^n = \underbrace{b \cdot b \cdot b \cdots \cdot b}_{n \text{ times}}.$$
$$\text{Base} \longrightarrow$$

When evaluating expressions, evaluate exponents before performing addition, subtraction, multiplication, division, or negation.

EXAMPLE 1   Evaluating exponential expressions

Evaluate each expression.

(a) $1 + \dfrac{2^4}{4}$ **(b)** $3\left(\frac{1}{3}\right)^2$ **(c)** -2^4 **(d)** $(-2)^4$

Solution   **(a)** Evaluate the exponent first.

$$1 + \frac{2^4}{4} = 1 + \frac{2 \cdot 2 \cdot 2 \cdot 2}{4} = 1 + \frac{16}{4} = 1 + 4 = 5$$

(b) $3\left(\frac{1}{3}\right)^2 = 3\left(\frac{1}{3} \cdot \frac{1}{3}\right) = 3 \cdot \frac{1}{9} = \frac{3}{9} = \frac{1}{3}$

(c) Because exponents are evaluated before negation is performed,

$$-2^4 = -(2 \cdot 2 \cdot 2 \cdot 2) = -16.$$

(d) $(-2)^4 = \overbrace{(-2)(-2)(-2)(-2)}^{4 \text{ factors}} = 16$. Note how the expressions in parts (c) and (d) differ.

Technology Note: *Evaluating Exponents*

Exponents can often be evaluated on calculators by using the ^ key. The four expressions from Example 1 are evaluated with a calculator and the results are shown in the two figures.

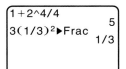

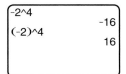

ZERO EXPONENTS

So far we have discussed the meaning of natural number exponents. What if an exponent is equal to 0? What does 2^0 equal? To answer these questions consider Table 5.1, which shows values for decreasing powers of 2. Note that each time the power of 2 decreases by 1, the resulting value is divided by 2. For this pattern to continue, we need to define 2^0 to be 1 because if we divide 2 by 2, the result is 1.

This discussion suggests that $2^0 = 1$, and is generalized as follows.

TABLE 5.1 Powers of 2

Power of 2	Value
2^3	8
2^2	4
2^1	2
2^0	?

ZERO EXPONENT

For any nonzero real number b,

$$b^0 = 1.$$

The expression 0^0 is undefined.

EXAMPLE 2 Evaluating zero exponents

Evaluate each expression. Assume that all variables represent nonzero numbers.

(a) 7^0 **(b)** $3\left(\frac{4}{9}\right)^0$ **(c)** $\left(\frac{x^2 y^5}{3z}\right)^0$

Solution
(a) $7^0 = 1$
(b) $3\left(\frac{4}{9}\right)^0 = 3(1) = 3$. (Note that the exponent 0 does not apply to 3.)
(c) All variables are nonzero, so the expression inside the parentheses is also nonzero. Thus $\left(\frac{x^2 y^5}{3z}\right)^0 = 1$.

THE PRODUCT RULE

We can calculate products of exponential expressions *provided their bases are the same*. For example,

$$4^3 \cdot 4^2 = (4 \cdot 4 \cdot 4) \cdot (4 \cdot 4) = 4^5.$$

The expression $4^3 \cdot 4^2$ has a total of $3 + 2 = 5$ factors of 4, so the result is $4^{3+2} = 4^5$. To multiply exponential expressions with the *same* base, we add exponents and the base does not change. We generalize this discussion as the following rule.

THE PRODUCT RULE

For any real number a and natural numbers m and n,

$$a^m \cdot a^n = a^{m+n}.$$

Note: The product $2^4 \cdot 3^2$ cannot be simplified by using the product rule because the exponential expressions have different bases: 2 and 3.

EXAMPLE 3   Using the product rule

Multiply and simplify.
(a) $2^3 \cdot 2^2$ **(b)** $x^4 x^5$ **(c)** $2x^2 \cdot 5x^6$ **(d)** $x^3(2x + 3x^2)$

Solution   **(a)** $2^3 \cdot 2^2 = 2^{3+2} = 2^5 = 32$
(b) $x^4 x^5 = x^{4+5} = x^9$
(c) $2x^2 \cdot 5x^6 = 2 \cdot 5 \cdot x^2 \cdot x^6 = 10x^{2+6} = 10x^8$
(d) To simplify this expression, first apply a distributive property.

$$x^3(2x + 3x^2) = x^3 \cdot 2x + x^3 \cdot 3x^2 = 2x^4 + 3x^5$$
Exponent is 1.

If an exponent does not appear on an expression, it is assumed to be 1. For example, x can be written as x^1 and $(x + y)$ can be written as $(x + y)^1$.

EXAMPLE 4   Applying the product rule

Multiply $(a + b)(a + b)^4$.

Solution   First write $(a + b)$ as $(a + b)^1$. Then

$$(a + b)^1 \cdot (a + b)^4 = (a + b)^{1+4} = (a + b)^5.$$

VISUALIZING EXPONENTS   We can visualize exponents by stacking wooden blocks having different lengths. For example, the product $2 \cdot 2 \cdot 2$ can be thought of as three blocks with length 2, as illustrated in Figure 5.1 Because they all have the same length (same base), we can stack them and the result represents 2^3.

Figure 5.1 $2 \cdot 2 \cdot 2 = 2^3$

In Figure 5.2 the expression $3^2 \cdot 3^3$ can be simplified by stacking blocks of length 3. Note that the stack is 5 blocks high, so the result represents 3^5.

Figure 5.2 $3^2 \cdot 3^3 = 3^5$

However, Figure 5.3 reveals that $2^2 \cdot 3^3$ cannot be stacked properly because the blocks have different lengths (bases).

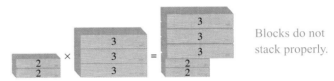

Blocks do not stack properly.

Figure 5.3 $2^2 \cdot 3^3 = ?$

POWER RULES

How should $(4^3)^2$ be evaluated? To answer this question consider

$$(4^3)^2 = \underbrace{4^3 \cdot 4^3}_{\text{2 factors}} = 4^{3+3} = 4^6.$$

Similarly,

$$(a^5)^3 = \underbrace{a^5 \cdot a^5 \cdot a^5}_{\text{3 factors}} = a^{5+5+5} = a^{15}.$$

This discussion suggests that, to raise a power to a power, we multiply the exponents, which we generalize as follows.

RAISING POWERS TO POWERS

For any real number a and natural numbers m and n,

$$(a^m)^n = a^{mn}.$$

EXAMPLE 5   Raising powers to powers

Simplify and write the expression by using exponents.
(a) $(3^2)^4$ **(b)** $(5^6)^3$ **(c)** $(x^2)^3$ **(d)** $(a^3)^2$

Solution   **(a)** $(3^2)^4 = 3^{2 \cdot 4} = 3^8$ **(b)** $(5^6)^3 = 5^{6 \cdot 3} = 5^{18}$
(c) $(x^2)^3 = x^{2 \cdot 3} = x^6$ **(d)** $(a^3)^2 = a^{3 \cdot 2} = a^6$

To decide how to simplify the expression $(2x)^3$, consider

$$(2x)^3 = \underbrace{2x \cdot 2x \cdot 2x}_{3 \text{ factors}} = \underbrace{(2 \cdot 2 \cdot 2)}_{3 \text{ factors}} \cdot \underbrace{(x \cdot x \cdot x)}_{3 \text{ factors}} = 2^3 x^3.$$

To raise a product to a power, we raise each factor to the power, which we generalize as follows.

RAISING A PRODUCT TO A POWER

For any real numbers a and b and natural number n,

$$(ab)^n = a^n b^n.$$

EXAMPLE 6   Raising a product to a power

Simplify the expression.
(a) $(3z)^2$ (b) $(-2x^2)^3$ (c) $4(x^2y^3)^5$ (d) $(-2^2a^5)^3$

Solution   (a) $(3z)^2 = 3^2 z^2 = 9z^2$
(b) $(-2x^2)^3 = (-2)^3(x^2)^3 = -8x^6$
(c) $4(x^2y^3)^5 = 4(x^2)^5(y^3)^5 = 4x^{10}y^{15}$
(d) $(-2^2a^5)^3 = (-4a^5)^3 = (-4)^3(a^5)^3 = -64a^{15}$

The following equation illustrates a third power rule.

$$\left(\frac{2}{3}\right)^4 = \frac{2}{3} \cdot \frac{2}{3} \cdot \frac{2}{3} \cdot \frac{2}{3} = \frac{2^4}{3^4}$$

That is, to raise a quotient to a power, raise both the numerator and the denominator to the power, which we generalize as follows.

RAISING A QUOTIENT TO A POWER

For any real numbers a and b and natural number n,

$$\left(\frac{a}{b}\right)^n = \frac{a^n}{b^n}. \qquad b \neq 0$$

EXAMPLE 7   Raising a quotient to a power

Simplify the expression.

(a) $\left(\dfrac{2}{3}\right)^3$ (b) $\left(\dfrac{a}{b}\right)^9$ (c) $\left(\dfrac{(a+b)}{5}\right)^2$

Solution   (a) $\left(\dfrac{2}{3}\right)^3 = \dfrac{2^3}{3^3} = \dfrac{8}{27}$ (b) $\left(\dfrac{a}{b}\right)^9 = \dfrac{a^9}{b^9}$

(c) $\left(\dfrac{(a+b)}{5}\right)^2 = \dfrac{(a+b)^2}{5^2} = \dfrac{(a+b)^2}{25}$

Simplification of some types of expressions requires application of more than one rule of exponents.

EXAMPLE 8   Combining rules for exponents

Simplify the expression.

(a) $(2a)^2(3a)^3$ **(b)** $\left(\dfrac{a^2b^3}{c}\right)^4$ **(c)** $(2x^3y)^2(-4x^2y^3)^3$

Solution   **(a)** $(2a)^2(3a)^3 = 2^2a^2 \cdot 3^3a^3$ Raising a product to a power
$= 4 \cdot 27 \cdot a^2 \cdot a^3$ Evaluate powers; commutative property
$= 108a^5$ Product rule

(b) $\left(\dfrac{a^2b^3}{c}\right)^4 = \dfrac{(a^2)^4(b^3)^4}{c^4}$ Raising a quotient to a power; raising a product to a power

$= \dfrac{a^8b^{12}}{c^4}$ Raising a power to a power

(c) $(2x^3y)^2(-4x^2y^3)^3 = 2^2(x^3)^2y^2(-4)^3(x^2)^3(y^3)^3$ Raising a product to a power
$= 4x^6y^2(-64)x^6y^9$ Raising a power to a power
$= 4(-64)x^6x^6y^2y^9$ Commutative property
$= -256x^{12}y^{11}$ Product rule

EXAMPLE 9   Calculating growth of an investment

If a piece of property increases in value by about 11% each year for 20 years, then its value will double three times.
(a) Write an exponential expression that represents "doubling three times."
(b) If the property is initially worth $25,000, how much will it be worth if it doubles 3 times?

Solution   **(a)** Doubling three times is represented by 2^3.
(b) $2^3(25,000) = 8(25,000) = \$200,000$

 5.1   **PUTTING IT ALL TOGETHER**

The following table summarizes properties of exponents.

Concept	Explanation	Examples
Bases and Exponents	In the expression b^n, b is the base and n is the exponent. $$b^n = \underbrace{b \cdot b \cdots \cdot b}_{n \text{ times}}$$	2^3 has base 2 and exponent 3. $9^1 = 9$, $3^2 = 3 \cdot 3 = 9$, $4^3 = 4 \cdot 4 \cdot 4 = 64$, and $-6^2 = -(6 \cdot 6) = -36$

continued on next page

continued from previous page

Concept	Explanation	Examples
Zero Exponents	$b^0 = 1$ for any nonzero number b.	$5^0 = 1$, $x^0 = 1$, and $(xy^3)^0 = 1$
The Product Rule	$a^m \cdot a^n = a^{m+n}$, where m and n are natural numbers.	$2^4 \cdot 2^3 = 2^{4+3} = 2^7$, $x \cdot x^2 \cdot x^6 = x^{1+2+6} = x^9$, and $(x+1) \cdot (x+1)^2 = (x+1)^3$
Raising Powers to Powers	$(a^m)^n = a^{mn}$, where m and n are natural numbers.	$(2^4)^2 = 2^{4\cdot2} = 2^8$, $(x^2)^5 = x^{2\cdot5} = x^{10}$, and $(a^4)^3 = a^{4\cdot3} = a^{12}$
Raising Products to Powers	$(ab)^n = a^n b^n$, where n is a natural number.	$(3x)^3 = 3^3 x^3 = 27x^3$, $(x^2 y)^4 = (x^2)^4 y^4 = x^8 y^4$, and $(-xy)^6 = (-x)^6 y^6 = x^6 y^6$
Raising Quotients to Powers	$\left(\dfrac{a}{b}\right)^n = \dfrac{a^n}{b^n},$ where $b \neq 0$ and n is a natural number.	$\left(\dfrac{x}{y}\right)^5 = \dfrac{x^5}{y^5}$ and $\left(\dfrac{a^2 b}{d^3}\right)^4 = \dfrac{(a^2)^4 b^4}{(d^3)^4} = \dfrac{a^8 b^4}{d^{12}}$

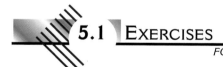

5.1 EXERCISES

FOR EXTRA HELP

Student's Solutions Manual

InterAct Math

MathXL

MyMathLab

Tutor Center   Math Tutor Center

Digital Video Tutor
CD 2   Videotape 7

CONCEPTS

1. In the expression b^n, b is the _____ and n is the _____.

2. The expression $b^0 = $ _____ for any nonzero number b.

3. Write $\frac{1}{2} \cdot \frac{1}{2} \cdot \frac{1}{2}$ by using exponents.

4. Write $x \cdot x \cdot x \cdot x$ by using exponents.

5. $a^m \cdot a^n = $ _____.   6. $(a^m)^n = $ _____.

7. $(ab)^n = $ _____.   8. $\left(\dfrac{a}{b}\right)^n = $ _____.

PROPERTIES OF EXPONENTS

Exercises 9–24: Evaluate the expression.

9. 8^2

10. 4^3

11. $(-2)^3$

12. $(-3)^4$

13. -2^3

14. -3^4

15. 6^0

16. $(-0.5)^0$

17. $2 \cdot 4^2$

18. $-3 \cdot 2^4$

19. $1 + 5^2$

20. $5^2 - 4^2$

21. $\dfrac{4^2}{2}$

22. $\left(\dfrac{-4}{2}\right)^2$

23. $4 \cdot \dfrac{1}{2^3}$

24. $\dfrac{2}{2^3} - \dfrac{1}{2^3}$

Exercises 25–72: Simplify the expression.

25. $4^2 \cdot 4^6$

26. $5^3 \cdot 5^3$

27. $2^3 \cdot 2^2$

28. $10^4 \cdot 10^3$

29. $x^3 \cdot x^6$

30. $a^5 \cdot a^2$

31. $z^0 z^4$

32. $a^2 a^3 a^1$

33. $x^2 x^2 x^2$

34. $y^7 y^3 y^0$

35. $4x^2 \cdot 5x^5$

36. $-2y^6 \cdot 5y^2$

37. $3(-xy^3)(x^2 y)$

38. $(a^2 b^3)(-ab^2)$

39. $(2^3)^2$

40. $(10^3)^4$

41. $(n^3)^4$

42. $(z^7)^3$

43. $x(x^3)^2$

44. $(z^3)^2(5z^5)$

45. $(-7b)^2$

46. $(-4z)^3$

47. $(ab)^3$

48. $(xy)^8$

49. $(2x^2)^2$

50. $(3a^2)^4$

51. $(-4b^2)^3$

52. $(-3r^4 t^3)^2$

53. $(x^2 y^3)^7$

54. $(rt^2)^5$

55. $(y^3)^2(x^4 y)^3$

56. $(ab^3)^2(ab)^3$

57. $\left(\dfrac{2}{3}\right)^3$

58. 1

59. $\left(\dfrac{a}{b}\right)^5$

60. $\left(\dfrac{x}{2}\right)^4$

61. $\left(\dfrac{2x}{5}\right)^3$

62. $\left(\dfrac{3y}{2}\right)^4$

63. $\left(\dfrac{3x^2}{5y^4}\right)^3$

64. $\left(\dfrac{a^2 b^3}{3}\right)^5$

65. $(a + b)^2(a + b)^3$

66. $(x - y)^5(x - y)^4$

67. $6(x^4 y^6)^0$

68. $\left(\dfrac{xy}{z^2}\right)^0$

69. $a(a^2 + 2b^2)$

70. $x^3(3x - 5y^4)$

71. $(r + t)(rt)$

72. $(x - y)(x^2 y^3)$

APPLICATIONS

73. *Area of a Rectangle*   Find the area of the rectangle.

74. *Area of a Square*   Find the area of the square whose sides have length $2ab$.

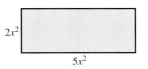

75. *Volume of a Box*   A box has the dimensions shown. Write an expression for its volume.

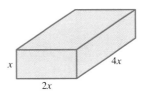

76. *Volume of a Cube*   Each side of a cube is $3a^2$. Write an expression for its volume.

77. *Area of a Circle*   Write an expression for the area of the circle.

78. *Volume of a Sphere*   The volume V of a sphere with radius r is given by $V = \frac{4}{3}\pi r^3$. Find an expression for the volume of a sphere with radius $2b$.

79. *Compound Interest* If P dollars are deposited in an account that pays 5% annual interest, then the amount of money in the account after 3 years is $P(1 + 0.05)^3$. Find the amount when $P = \$1000$.

80. *Compound Interest* If P dollars are deposited in an account that pays 9% annual interest, then the amount of money in the account after 4 years is $P(1 + 0.09)^4$. Find the amount when $P = \$500$.

Exercises 81–84: (Refer to the discussion about visualizing exponents found on pages 292–293.) Make a sketch of blocks that represents the given exponential equation.

81. $4^2 \cdot 4^1 = 4^3$

82. $2^3 \cdot 2^3 = 2^6$

83. $2^3 \cdot 3^2 \cdot 3^1 \cdot 2^2 = 2^5 \cdot 3^3$

84. $4^2 \cdot 2^2 \cdot 4^2 = 4^5$ (*Hint:* Write 2^2 with a different base.)

WRITING ABOUT MATHEMATICS

85. Are the expressions $(4x)^2$ and $4x^2$ equal? Explain your answer.

86. Are the expressions $3^3 \cdot 2^3$ and 6^6 equal? Explain your answer.

5.2 ADDITION AND SUBTRACTION OF POLYNOMIALS

Monomials and Polynomials · **Addition of Polynomials** · **Subtraction of Polynomials** · **Evaluating Polynomial Expressions**

INTRODUCTION

If you have ever exercised strenuously and then taken your pulse immediately afterward, you may have discovered that your pulse slowed quickly at first and then gradually leveled off. A typical scatterplot is shown in Figure 5.4(a). These data points cannot be modeled accurately with a line, so a new expression, called a *polynomial,* is needed to model them. A graph of this new expression is shown in Figure 5.4(b) and discussed in Exercise 83. (*Source:* V. Thomas, *Science and Sport.*)

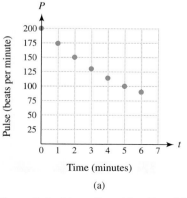

(a)

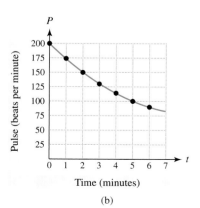

(b)

Figure 5.4 Heart Rate After Exercising

MONOMIALS AND POLYNOMIALS

A **monomial** is a number, a variable, or a product of numbers and variables raised to natural number powers. Examples of monomials include

$$-3, \quad xy^2, \quad 5a^2, \quad -z^3, \quad \text{and} \quad -\frac{1}{2}xy^3.$$

A monomial may contain more than one variable, but monomials do not contain division by variables. For example, the expression $\frac{3}{z}$ is not a monomial. If an expression contains addition or subtraction signs, it is *not* a monomial.

The **degree of a monomial** is the sum of the exponents of the variables. If the monomial has only one variable, its degree is the exponent of that variable. A nonzero number has degree 0, and the number 0 has *undefined* degree. The number in a monomial is called the **coefficient of the monomial**. Table 5.2 contains the degree and coefficient of several monomials.

TABLE 5.2   Properties of Monomials

Monomial	-5	$6a^3b$	$-xy$	$7y^3$
Degree	0	4	2	3
Coefficient	-5	6	-1	7

A **polynomial** is the sum of one or more monomials. Each monomial is called a *term* of the polynomial. Addition or subtraction signs separate terms. The expression $2x^2 - 3x + 5$ is a **polynomial in one variable** with three terms. Examples of polynomials in one variable include

$$-2x, \quad 3x + 1, \quad 4y^2 - y + 7, \quad \text{and} \quad x^5 - 3x^3 + x - 7.$$

These polynomials have, 1, 2, 3, and 4 terms, respectively. A polynomial with *two terms* is called a **binomial**, and a polynomial with *three terms* is called a **trinomial**.

A polynomial can have more than one variable, as in

$$x^2y^2, \quad 2xy^2 + 5x^2y - 1, \quad \text{and} \quad a^2 + 2ab + b^2.$$

Note that all variables in a polynomial are raised to natural number powers. The **degree of a polynomial** is the degree of the term (or monomial) with highest degree.

EXAMPLE 1   Identifying properties of polynomials

Determine whether the expression is a polynomial. If it is, state how many terms and variables the polynomial contains and its degree.

(a) $7x^2 - 3x + 1$ **(b)** $5x^3 - 3x^2y^3 + xy^2 - 2y^3$ **(c)** $4x^2 + \dfrac{5}{x + 1}$

Solution   **(a)** The expression $7x^2 - 3x + 1$ is a polynomial with three terms and one variable. The term with highest degree is $7x^2$, so the polynomial has degree 2.

   (b) The expression $5x^3 - 3x^2y^3 + xy^2 - 2y^3$ is a polynomial with four terms and two variables. The term with highest degree is $-3x^2y^3$, so the polynomial has degree $2 + 3 = 5$.

   (c) The expression $4x^2 + \dfrac{5}{x + 1}$ is a not a polynomial because it contains division by $x + 1$.

Monomials occur in applications involving geometry, as illustrated in the next example.

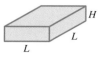

EXAMPLE 2 Writing monomials

The box shown in Figure 5.5 has a square bottom having length L and sides having height H. Write a monomial that gives the following quantities.

(a) Area of the bottom   **(b)** Volume of the box   **(c)** Volume of 6 identical boxes

Figure 5.5 Square-Bottomed Box

Solution **(a)** The area of a rectangle equals length times width. The bottom is a square, so its area is

$$L \cdot L, \quad \text{or} \quad L^2.$$

Critical Thinking

Write an expression that gives the volume of six identical cubes having sides of length L.

(b) The volume of this box equals length times width times height. Thus its volume is

$$L \cdot L \cdot H, \quad \text{or} \quad L^2 H.$$

(c) The volume of 6 boxes is 6 times the volume of one box, or $6L^2 H$.

ADDITION OF POLYNOMIALS

Suppose that we have 2 identical rectangles with length L and width W, as illustrated in Figure 5.6. Then the area of one rectangle is LW and the total area is

$$LW + LW.$$

This area is equivalent to 2 times LW, which can be expressed as $2LW$, or

$$LW + LW = 2LW.$$

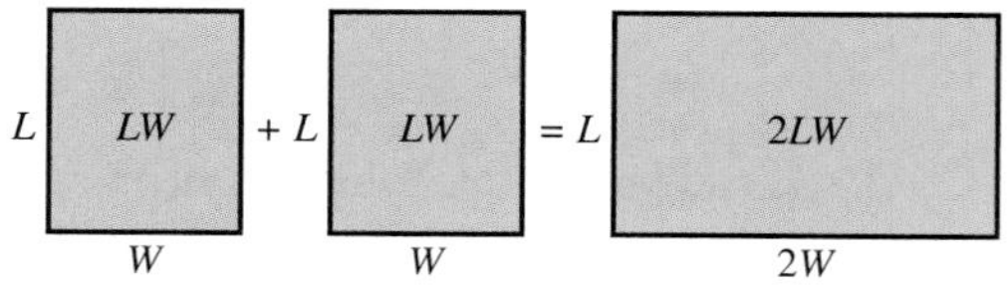

$$L \boxed{LW} + L \boxed{LW} = L \boxed{2LW}$$
$$\quad W \qquad\qquad W \qquad\qquad 2W$$

Figure 5.6 Adding $LW + LW$

If two monomials contain the same variables raised to the same powers, we call them **like terms**. We can add or subtract *like* terms but not *unlike* terms. The terms LW and $2LW$ are like terms and can be combined geometrically, as shown in Figure 5.7. If we joined one of the small rectangles with area LW and a larger rectangle with area $2LW$, then the total area is $3LW$.

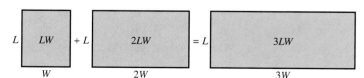

Figure 5.7 Adding $LW + 2LW$

The *distributive property* justifies combining like terms.

$$1LW + 2LW = (1 + 2)LW = 3LW$$

The rectangles shown in Figure 5.8 have areas of ab and xy. Their combined area is their sum, $ab + xy$. However, because these monomials are unlike terms, they cannot be combined into one term.

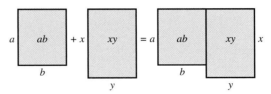

Figure 5.8 Unlike terms: $ab + xy$

EXAMPLE 3 Adding like terms

State whether each pair of expressions contains like terms or unlike terms. If they are like terms, add them.
(a) $5x^2, -x^2$ **(b)** $7a^2b, 10ab^2$ **(c)** $4rt^2, \frac{1}{2}rt^2$

Solution **(a)** The terms $5x^2$ and $-x^2$ have the same variable raised to the same power, so they are like terms. To add like terms add their coefficients. Note that the coefficient of $-x^2$ is -1.

$$5x^2 + (-x^2) = \big(5 + (-1)\big)x^2 \qquad \text{Distributive property}$$
$$= 4x^2 \qquad \text{Add.}$$

(b) The terms $7a^2b$ and $10ab^2$ have the same variables, but these variables are not raised to the same power. They are unlike terms and therefore cannot be added.

(c) The terms $4rt^2$ and $\frac{1}{2}rt^2$ have the same variables raised to the same powers, so they are like terms.

$$4rt^2 + \frac{1}{2}rt^2 = \left(4 + \frac{1}{2}\right)rt^2 \qquad \text{Distributive property}$$

$$= \frac{9}{2}rt^2 \qquad \text{Add.}$$

To add two polynomials, add like terms, as illustrated in the next example.

EXAMPLE 4 Adding polynomials

Add each pair of polynomials by combining like terms.
(a) $(3x + 4) + (-4x + 2)$
(b) $(y^2 - 2y + 1) + (3y^2 + y + 11)$
(c) $(3a^3 - 4a + 1) + (4a^3 + a^2 - 5)$

Solution **(a)** $(3x + 4) + (-4x + 2) = 3x + (-4x) + 4 + 2$
$$= (3 - 4)x + (4 + 2)$$
$$= -x + 6$$

(b) $(y^2 - 2y + 1) + (3y^2 + y + 11) = y^2 + 3y^2 - 2y + y + 1 + 11$
$$= (1 + 3)y^2 + (-2 + 1)y + (1 + 11)$$
$$= 4y^2 - y + 12$$

Note: With practice the first two steps can be done mentally.

(c) $(3a^3 - 4a + 1) + (4a^3 + a^2 - 5) = 3a^3 + 4a^3 + a^2 - 4a + 1 - 5$
$$= (3 + 4)a^3 + a^2 - 4a - 4$$
$$= 7a^3 + a^2 - 4a - 4$$

Polynomials can also be added vertically, as demonstrated in the next example.

EXAMPLE 5   Adding polynomials vertically

Simplify $(3x^2 - 3x + 5) + (-x^2 + x - 6)$.

Solution   Write the polynomials in a vertical format and then add each column of like terms.

$$\begin{array}{r} 3x^2 - 3x + 5 \\ \underline{-x^2 + x - 6} \\ 2x^2 - 2x - 1 \end{array} \quad \text{Add.}$$

Regardless of the method used, the same answer should be obtained. However, adding vertically requires that *like terms be placed in the same column.*

SUBTRACTION OF POLYNOMIALS

To subtract two integers add the first integer with the *additive inverse* or *opposite* of the second integer. Fox example, $3 - 5$ is evaluated as follows.

$$3 - 5 = 3 + (-5) \quad \text{Add the opposite.}$$
$$= -2 \quad \text{Simplify.}$$

Similarly, to subtract two polynomials add the first polynomial and the *opposite* of the second polynomial. To find the opposite of a polynomial, simply negate each term. Table 5.3 lists some polynomials and their opposites.

Critical Thinking

What is the result when a polynomial and its opposite are added?

TABLE 5.3   Opposites of Polynomials

Polynomial	Opposite
$2x - 4$	$-2x + 4$
$-x^2 - 2x + 9$	$x^2 + 2x - 9$
$6x^3 - 12$	$-6x^3 + 12$
$-3x^4 - 2x^2 - 8x + 3$	$3x^4 + 2x^2 + 8x - 3$

EXAMPLE 6   Subtracting polynomials

Simplify each expression.
(a) $(3x - 4) - (5x + 1)$
(b) $(5x^2 + 2x - 3) - (6x^2 - 7x + 9)$
(c) $(6x^3 + x^2) - (-3x^3 - 9)$

Solution **(a)** The opposite of $(5x + 1)$ is $(-5x - 1)$.

$$(3x - 4) - (5x + 1) = (3x - 4) + (-5x - 1)$$
$$= (3 - 5)x + (-4 - 1)$$
$$= -2x - 5$$

(b) The opposite of $(6x^2 - 7x + 9)$ is $(-6x^2 + 7x - 9)$.

$$(5x^2 + 2x - 3) - (6x^2 - 7x + 9) = (5x^2 + 2x - 3) + (-6x^2 + 7x - 9)$$
$$= (5 - 6)x^2 + (2 + 7)x + (-3 - 9)$$
$$= -x^2 + 9x - 12$$

(c) The opposite of $(-3x^3 - 9)$ is $(3x^3 + 9)$.

$$(6x^3 + x^2) - (-3x^3 - 9) = (6x^3 + x^2) + (3x^3 + 9)$$
$$= (6 + 3)x^3 + x^2 + 9$$
$$= 9x^3 + x^2 + 9$$

EXAMPLE 7   Subtracting polynomials vertically

Simplify $(5x^2 - 2x + 7) - (-3x^2 + 3)$.

Solution To subtract two polynomials vertically simply add the first polynomial and the opposite of the second polynomial. No x-term occurs in the second polynomial, so insert $0x$.

$$5x^2 - 2x + 7$$
$$\underline{3x^2 + 0x - 3}$$ Opposite of $-3x^2 + 3$ is $3x^2 - 3$ or $3x^2 + 0x - 3$.
$$8x^2 - 2x + 4$$ Add like terms in each column.

EVALUATING POLYNOMIAL EXPRESSIONS

Frequently, monomials and polynomials represent formulas that may be evaluated. We illustrate such applications in the next two examples.

EXAMPLE 8   Writing and evaluating a monomial

The monomial (or polynomial) x^2y represents the volume of the box having a square bottom, as shown in Figure 5.9. Find the volume of the box if $x = 3$ feet and $y = 2$ feet.

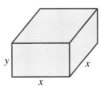

Figure 5.9   Volume $= x^2y$

Solution To calculate the volume let $x = 3$ and $y = 2$ in the polynomial x^2y.

$$x^2y = 3^2 \cdot 2 = 9 \cdot 2 = 18 \text{ cubic feet}$$

EXAMPLE 9   Modeling sales of personal computers

Worldwide sales of personal computers have increased dramatically in recent years, as illustrated in Figure 5.10. The polynomial

$$0.7868x^2 + 12x + 79.5$$

approximates the number of computers sold in millions during year x, where $x = 0$ corresponds to 1997, $x = 1$ to 1998, and so on. Estimate the number of personal computers sold in 2002 by using both the graph and the polynomial.   (**Source:** International Data Corporation.)

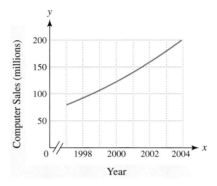

Figure 5.10

Solution   From the graph shown in Figure 5.11, it appears that personal computer sales were slightly more than 150 million, or about 160 million, in 2002. The year 2002 corresponds to $x = 5$ in the given polynomial, so substitute 5 for x and evaluate the resulting expression.

$$0.7868x^2 + 12x + 79.5 = 0.7868(5)^2 + 12(5) + 79.5$$
$$\approx 159 \text{ million}$$

The graph and the polynomial give similiar results.

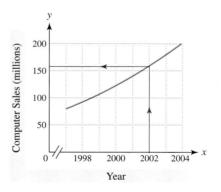

Figure 5.11

5.2   PUTTING IT ALL TOGETHER

In this section we discussed monomials and polynomials, including how to add, subtract, and evaluate them. The following table summarizes several important concepts related to these topics.

Concept	Explanation	Examples
Monomial	A number, variable, or product of numbers and variables raised to natural number powers Degree is the sum of the exponents. Coefficient is the number in a monomial.	$4x^2y$ Degree: 3, coefficient: 4 $-6x^2$ Degree: 2, coefficient: -6 $-a^4$ Degree: 4, coefficient: -1 x Degree: 1, coefficient: 1 -8 Degree: 0, coefficient: -8
Polynomial	A sum of one or more monomials	$4x^2 + 8xy^2 + 3y^2$ Trinomial $5x^2 - 2x + 8$ Trinomial $-9x^4 + 100$ Binomial
Like Terms	Monomials containing the same variables raised to the same powers	$10x$ and $-2x$ $4x^2$ and $3x^2$ $5ab^2$ and $-ab^2$ $5z$ and $\frac{1}{2}z$
Addition of Polynomials	To add polynomials combine like terms by applying the distributive property.	$(x^2 + 3x + 1) + (2x^2 - 2x + 7)$ $= (1 + 2)x^2 + (3 - 2)x + (1 + 7)$ $= 3x^2 + x + 8$ $3x^2y + 5x^2y = (3 + 5)x^2y$ $= 8x^2y$
Opposite of a Polynomial	To obtain the opposite of a polynomial negate each term.	*Polynomial* *Opposite* $-2x^2 + x - 6$ $2x^2 - x + 6$ $a^2 - b^2$ $-a^2 + b^2$ $-3x - 18$ $3x + 18$
Subtraction of Polynomials	To subtract polynomials add the first polynomial to the opposite of the second polynomial.	$(x^2 + 3x) - (2x^2 - 5x)$ $= (x^2 + 3x) + (-2x^2 + 5x)$ $= (1 - 2)x^2 + (3 + 5)x$ $= -x^2 + 8x$
Evaluating a Polynomial	To evaluate a polynomial in x substitute a value for x in the expression and simplify.	To evaluate the polynomial $3x^2 - 2x + 1$ when $x = 2$, substitute 2 for x and simplify. $3(2)^2 - 2(2) + 1 = 9$

5.2 EXERCISES

FOR EXTRA HELP   Student's Solutions Manual   InterAct Math   MathXL

MyMathLab   Math Tutor Center   Digital Video Tutor CD 2 Videotape 7

CONCEPTS

1. A _____ is a number, a variable, or a product of numbers and variables raised to a natural number power.

2. Is $\frac{3}{x}$ an example of a monomial?

3. A _____ is a monomial or a sum of monomials.

4. The _____ of a monomial is the sum of the exponents of the variables.

5. A polynomial with two terms is called a _____.

6. A polynomial with three terms is called a _____.

7. The polynomial $4x^3 + x^2$ has _____ terms and its degree is _____.

8. A box with length L, width W, and height H has a volume given by _____.

9. To add two polynomials, combine _____ terms.

10. To obtain the opposite of a polynomial _____ each term.

11. To subtract two polynomials, add the first polynomial to the _____ of the second polynomial.

12. Polynomials can be added either horizontally or _____.

PROPERTIES OF POLYNOMIALS

Exercises 13–20: Identify the degree and coefficient of the monomial.

13. $3x^2$

14. y

15. $-ab$

16. $2xy$

17. $-5rt$

18. $8x^2y^5$

19. -6

20. $\frac{1}{2}$

Exercises 21–28: Determine whether the expression is a polynomial. If it is, state how many terms and variables the polynomial contains. Then state its degree.

21. $-x$

22. $7z$

23. $4x^2 - 5x + 9$

24. $x^3 - 9$

25. $x + \dfrac{1}{x}$

26. $\dfrac{5}{xy + 1}$

27. $3x^2y - xy^3$

28. $a^3 + 3a^2b + 3ab^2 + b^3$

Exercises 29–38: State whether the given pair of expressions are like terms. If they are like terms, add them.

29. $5x, -4x$

30. $x^2, 8x^2$

31. $x^3, -6x^3$

32. $4xy, -9xy$

33. $9x, -xy$

34. $5x^2y, -3xy^2$

35. ab, ba

36. $rt^2, -2t^2r$

37. $7xy^2, -3xy^2$

38. a, b

ADDITION OF POLYNOMIALS

Exercises 39–50: Add the polynomials.

39. $(3x + 5) + (-4x + 4)$

40. $(-x + 5) + (2x - 5)$

41. $(3x^2 + 4x + 1) + (x^2 + 4x - 6)$

42. $(-x^2 - x + 7) + (2x^2 + 3x - 1)$

43. $(a^3 - 6) + (4a^3 + 7)$

44. $(2b^4 - 3b^2) + (-3b^4 + b^2)$

45. $(y^3 + 3y^2 - 5) + (3y^3 + 4y - 4)$

46. $(4z^4 + z^2 - 10) + (-z^4 + 4z - 5)$

47. $(-xy + 5) + (5xy - 4)$

48. $(2a^2 + b^2) + (3a^2 - 5b^2)$

49. $(a^3b^2 + a^2b^3) + (a^2b^3 - a^3b^2)$

50. $(a^2 + ab + b^2) + (a^2 - ab + b^2)$

Exercises 51–54: Add the polynomials vertically.

51. $4x^2 - 2x + 1$
$5x^2 + 3x - 7$

52. $8x^2 + 3x + 5$
$-x^2 - 3x - 9$

53. $-x^2 + x$
$2x^2 - 8x - 1$

54. $a^3 - 3a^2b + 3ab^2 - b^3$
$a^3 + 3a^2b + 3ab^2 + b^3$

SUBTRACTION OF POLYNOMIALS

Exercises 55–64: Write the opposite of the polynomial.

55. $5x^2$

56. $17x + 12$

57. $3a^2 - a + 4$

58. $-b^3 + 3b$

59. $-2t^2 - 3t + 4$

60. $7t^2 + t - 10$

61. $-3x^3 - x + 5$

62. $-x^3 + 4x^2 - 5x$

63. $9xy - x^2$

64. $a^3 + ab - 2b^2$

Exercises 65–76: Subtract the polynomials.

65. $(3x + 1) - (-x + 3)$

66. $(-2x + 5) - (x + 7)$

67. $(-x^2 + 6x + 8) - (2x^2 + x - 2)$

68. $(2y^2 + 3y - 2) - (y^2 - y - 4)$

69. $(a^2 - 2a) - (4a^2 + 3a)$

70. $(7b^3 + 3b) - (-3b^3 - b)$

71. $(z^3 - 2z^2 - z) - (4z^2 + 5z + 1)$

72. $(3z^4 - z) - (-z^4 + 4z^2 - 5)$

73. $(4xy + x^2y^2) - (xy - x^2y^2)$

74. $(a^2 + b^2) - (-a^2 + b^2)$

75. $(ab^2) - (ab^2 + a^3b)$

76. $(x^2 + 3xy + 4y^2) - (x^2 - xy + 4y^2)$

Exercises 77–80: (Refer to Example 7.) Subtract the polynomials vertically.

77. $(x^2 + 2x - 3) - (2x^2 + 7x + 1)$

78. $(5x^2 - 9x - 1) - (x^2 - x + 3)$

79. $(3x^3 - 2x) - (5x^3 + 4x + 2)$

80. $(a^2 + 3ab + 2b^2) - (a^2 - 3ab + 2b^2)$

APPLICATIONS

81. *Volume of a Box*   The volume of a cube having sides of length x is given by x^3. Find the volume of a cube with sides of length 3 inches.

82. *Volume of a Box*   (Refer to Example 8.) A box has a square base of length 10 inches and height 6 inches. Use the formula x^2y to find its volume.

83. *Exercise and Heart Rate*   The polynomial given by $1.6t^2 - 28t + 200$ calculates the heart rate shown in Figure 5.4(b) where t represents the elapsed time in minutes since exercise stopped.
  (a) What was the heart rate when the athlete first stopped exercising?
  (b) What was the heart rate after 5 minutes?
  (c) Describe what happens to the heart rate after exercise stopped.

84. *Cellular Phone Subscribers*   In the early years of cellular phone technology—from 1986 through 1991—the number of subscribers in millions could be modeled by the polynomial $0.163x^2 - 0.146x + 0.205$, where $x = 1$ corresponds to 1986, $x = 2$ to 1987, and so on. The graph illustrates this growth.

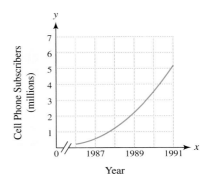

  (a) Use the graph to estimate the number of cellular phone subscribers in 1990.
  (b) Use the polynomial to estimate the number of cellular phone subscribers in 1990.
  (c) Do your answers from parts (a) and (b) agree?

85. *Areas of Squares*   Write a monomial that equals the sum of the areas of the squares. Then calculate this sum for $z = 10$ inches.

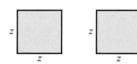

86. *Areas of Rectangles*   Find a monomial that equals the sum of the areas of the three rectangles. Find this sum for $a = 5$ yards and $b = 3$ yards.

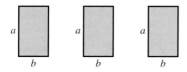

87. *Area of a Figure*   Find a polynomial that equals the area of the figure. Calculate its area for $x = 6$ feet.

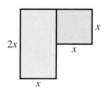

88. *Area of a Rectangle*   Write a polynomial that gives the area of the rectangle. Calculate its area for $x = 3$ feet.

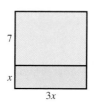

89. *Areas of Circles*   Write a polynomial that gives the sum of the areas of two circles, one with radius x and the other with radius y. Find this sum for $x = 2$ feet and $y = 3$ feet. Leave your answer in terms of π.

90. *Squares and Circles*   Write a polynomial that gives the sum of the areas of a square having sides of length x and a circle having diameter x. Approximate this sum to the nearest hundredth of a square foot for $x = 6$ feet.

91. *World Population*   The table lists actual and projected world population P in billions for selected years t.

t	1974	1987	1999	2012
P	4	5	6	7

Source: Bureau of the Census.

   (a) Find the slope of the line segment connecting consecutive data points in the table. Can these data be modeled with a line? Explain.
   (b) Does the polynomial $0.077t - 148$ give good estimates for the world population in year t? Explain how you decided.

92. *Price of a Stamp*   The table lists the price P of a first-class postage stamp for selected years t.

t	1963	1975	1987	2002
P	5¢	13¢	25¢	37¢

Source: U.S. Postal Service.

   (a) Does the polynomial $0.84t - 1645$ model the data in the table exactly?
   (b) Does it give approximations that are within 1.5¢ of the actual values?

WRITING ABOUT MATHEMATICS

93. Explain what the terms monomial, binomial, trinomial, and polynomial mean. Give an example of each.

94. Explain how to determine the degree of a polynomial having one variable. Give an example.

95. Explain how to obtain the opposite of a polynomial. Give an example.

96. Explain how to subtract two polynomials. Give an example.

1. Evaluate each expression.
   (a) -5^2 (b) $3^2 - 2^3$

2. Simplify each expression.
   (a) $10^3 \cdot 10^5$ (b) $(3x^2)(-4x^5)$

   (c) $(a^3b)^2$ (d) $\left(\dfrac{x}{z^3}\right)^4$

3. State the number of terms and variables in the polynomial $5x^3y - 2x^2y + 5$. What is its degree?

4. A box has a rectangular bottom twice as long as it is wide.

   (a) If the bottom has width W and the box has height H, write a monomial that gives the volume of the box.

   (b) Find the volume of the box for $W = 12$ inches and $H = 10$ inches.

5. Simplify each expression.
   (a) $(2a^2 + 3a - 1) + (a^2 - 3a + 7)$
   (b) $(4z^3 + 5z) - (2z^3 - 2z + 8)$
   (c) $(x^2 + 2xy + y^2) - (x^2 - 2xy + y^2)$

6. Evaluate $5x^2 - 7x$ for $x = -2$.

5.3 MULTIPLICATION OF POLYNOMIALS

Multiplying Monomials · Review of the Distributive Properties · Multiplying Monomials and Polynomials · Multiplying Polynomials

INTRODUCTION

The study of polynomials dates back to Babylonian civilization in about 1800–1600 B.C. Many eighteenth-century mathematicians devoted their entire careers to the study of polynomials. Polynomials still play an important role in mathematics, often being used to approximate unknown quantities. In this section we discuss the basics of multiplying polynomials.   (**Source:** *Historical Topics for the Mathematics Classroom, Thirty-first Yearbook,* NCTM.)

MULTIPLYING MONOMIALS

A monomial is a number, a variable, or a product of numbers and variables raised to natural number powers. To multiply monomials, we often use the product rule.

EXAMPLE 1   Multiplying monomials

Multiply.
(a) $-5x^2 \cdot 4x^3$ (b) $(7xy^4)(x^3y^2)$

Solution   (a)
$$
\begin{aligned}
-5x^2 \cdot 4x^3 &= (-5)(4)x^2x^3 && \text{Commutative property} \\
&= -20x^{2+3} && \text{The product rule} \\
&= -20x^5 && \text{Simplify.}
\end{aligned}
$$

(b)
$$
\begin{aligned}
(7xy^4)(x^3y^2) &= 7xx^3y^4y^2 && \text{Commutative property} \\
&= 7x^{1+3}y^{4+2} && \text{The product rule} \\
&= 7x^4y^6 && \text{Simplify.}
\end{aligned}
$$

REVIEW OF THE DISTRIBUTIVE PROPERTIES

Distributive properties are used frequently for multiplying monomials and polynomials. For all real numbers a, b, and c

$$a(b + c) = ab + ac \quad \text{and}$$
$$a(b - c) = ab - ac.$$

The distributive property can be visualized geometrically. For example, the equation

$$3(x + 2) = 3x + 6$$

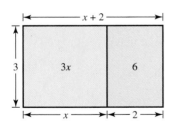

Figure 5.12   Area: $3x + 6$

is illustrated in Figure 5.12. The dimensions of the large rectangle are 3 by $x + 2$, and its area is $3(x + 2)$. The areas of the two small rectangles, $3x$ and 6, equal the area of the large rectangle. Therefore $3(x + 2) = 3x + 6$.

In the next example we use the distributive properties to multiply expressions.

EXAMPLE 2   Using distributive properties

Multiply.
(a) $2(3x + 4)$ **(b)** $(3x^2 + 4)5$ **(c)** $-x(3x - 6)$

Solution   **(a)** $2(3x + 4) = 2 \cdot 3x + 2 \cdot 4 = 6x + 8$

(b) $(3x^2 + 4)5 = 3x^2 \cdot 5 + 4 \cdot 5 = 15x^2 + 20$

(c) $-x(3x - 6) = -x \cdot 3x + x \cdot 6 = -3x^2 + 6x$

MULTIPLYING MONOMIALS AND POLYNOMIALS

A monomial consists of one term, whereas a polynomial consists of one or more terms separated by $+$ or $-$ signs. To multiply a monomial by a polynomial, we apply the distributive properties and the product rule.

EXAMPLE 3   Multiplying monomials and polynomials

Multiply.
(a) $9x(2x^2 - 3)$ **(b)** $(5x - 8)x^2$
(c) $-7(2x^2 - 4x + 6)$ **(d)** $4x^3(x^4 + 9x^2 - 8)$

Solution   **(a)** $9x(2x^2 - 3) = 9x \cdot 2x^2 - 9x \cdot 3$ Distributive property
$= 18x^3 - 27x$ The product rule

(b) $(5x - 8)x^2 = 5x \cdot x^2 - 8 \cdot x^2$ Distributive property
$= 5x^3 - 8x^2$ The product rule

(c) $-7(2x^2 - 4x + 6) = -7 \cdot 2x^2 + 7 \cdot 4x - 7 \cdot 6$ Distributive property
$= -14x^2 + 28x - 42$ Simplify.

(d) $4x^3(x^4 + 9x^2 - 8) = 4x^3 \cdot x^4 + 4x^3 \cdot 9x^2 - 4x^3 \cdot 8$ Distributive property
$= 4x^7 + 36x^5 - 32x^3$ The product rule

We can also multiply monomials and polynomials that contain more than one variable.

EXAMPLE 4   Multiplying monomials and polynomials

Multiply.
(a) $2xy(7x^2y^3 - 1)$ **(b)** $-ab(a^2 - b^2)$

Solution   **(a)** $2xy(7x^2y^3 - 1) = 2xy \cdot 7x^2y^3 - 2xy \cdot 1$ Distributive property
$= 14xx^2yy^3 - 2xy$ Commutative property
$= 14x^3y^4 - 2xy$ The product rule

(b) $-ab(a^2 - b^2) = -ab \cdot a^2 + ab \cdot b^2$ Distributive property
$= -aa^2b + abb^2$ Commutative property
$= -a^3b + ab^3$ The product rule

MULTIPLYING POLYNOMIALS

Monomials, binomials, and trinomials are examples of polynomials. Recall that a monomial has one term, a binomial has two terms, and a trinomial has three terms. In the next example we multiply two binomials, using both geometric and symbolic techniques.

EXAMPLE 5   Multiplying binomials

Multiply $(x + 4)(x + 2)$
(a) geometrically and   **(b)** symbolically.

Solution   **(a)** To multiply $(x + 4)(x + 2)$ geometrically draw a rectangle $x + 4$ long and $x + 2$ wide, as shown in Figure 5.13(a). The area of this rectangle equals length times width, or $(x + 4)(x + 2)$. The large rectangle can be divided into four smaller rectangles, which have areas of x^2, $4x$, $2x$, and 8, as shown in Figure 5.13(b). Thus

$$(x + 4)(x + 2) = x^2 + 4x + 2x + 8$$
$$= x^2 + 6x + 8.$$

(b) To multiply $(x + 4)(x + 2)$ symbolically apply the distributive property.

$$(x + 4)(x + 2) = (x + 4)(x) + (x + 4)(2)$$
$$= x \cdot x + 4 \cdot x + x \cdot 2 + 4 \cdot 2$$
$$= x^2 + 4x + 2x + 8$$
$$= x^2 + 6x + 8$$

To multiply $(x + 4)$ by $(x + 2)$ we multiplied every term in $x + 4$ by every term in $x + 2$.

$$(x + 4)(x + 2) = x^2 + 2x + 4x + 8$$
$$= x^2 + 6x + 8$$

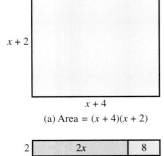

(a) Area = $(x + 4)(x + 2)$

(b) Area = $x^2 + 4x + 2x + 8$

Figure 5.13

Note: This process of multiplying binomials is sometimes called *FOIL*. This acronym may be used to remind us to multiply the first terms (*F*), outside terms (*O*), inside terms (*I*), and last terms (*L*).

Multiply the *First terms* to obtain x^2. $(x + 4)(x + 2)$

Multiply the *Outside terms* to obtain $2x$. $(x + 4)(x + 2)$

Multiply the *Inside terms* to obtain $4x$. $(x + 4)(x + 2)$

Multiply the *Last terms* to obtain 8. $(x + 4)(x + 2)$

The following statement summarizes how to multiply two polynomials in general.

MULTIPLYING POLYNOMIALS

The product of two polynomials may be found by multiplying every term in the first polynomial by every term in the second polynomial and then combining like terms.

EXAMPLE 6   Multiplying binomials

Multiply. Draw arrows to show how each term is found.
  (a) $(3x + 2)(x + 1)$ **(b)** $(1 - x)(1 + 2x)$ **(c)** $(4x - 3)(x^2 - 2x)$

Solution **(a)** $(3x + 2)(x + 1) = 3x \cdot x + 3x \cdot 1 + 2 \cdot x + 2 \cdot 1$
$$= 3x^2 + 3x + 2x + 2$$
$$= 3x^2 + 5x + 2$$

  (b) $(1 - x)(1 + 2x) = 1 \cdot 1 + 1 \cdot 2x - x \cdot 1 - x \cdot 2x$
$$= 1 + 2x - x - 2x^2$$
$$= 1 + x - 2x^2$$

  (c) $(4x - 3)(x^2 - 2x) = 4x \cdot x^2 - 4x \cdot 2x - 3 \cdot x^2 + 3 \cdot 2x$
$$= 4x^3 - 8x^2 - 3x^2 + 6x$$
$$= 4x^3 - 11x^2 + 6x$$

EXAMPLE 7   Multiplying polynomials

Multiply.
  (a) $(2x + 3)(x^2 + x - 1)$ **(b)** $(a - b)(a^2 + ab + b^2)$
  (c) $(x^4 + 2x^2 - 5)(x^2 + 1)$

Solution **(a)** Multiply every term in $(2x + 3)$ by every term in $(x^2 + x - 1)$.

$(2x + 3)(x^2 + x - 1) = 2x \cdot x^2 + 2x \cdot x - 2x \cdot 1 + 3 \cdot x^2 + 3 \cdot x - 3 \cdot 1$
$$= 2x^3 + 2x^2 - 2x + 3x^2 + 3x - 3$$
$$= 2x^3 + 5x^2 + x - 3$$

  (b) $(a - b)(a^2 + ab + b^2) = a \cdot a^2 + a \cdot ab + a \cdot b^2 - b \cdot a^2 - b \cdot ab - b \cdot b^2$
$$= a^3 + a^2b + ab^2 - a^2b - ab^2 - b^3$$
$$= a^3 - b^3$$

  (c) $(x^4 + 2x^2 - 5)(x^2 + 1) = x^4 \cdot x^2 + x^4 \cdot 1 + 2x^2 \cdot x^2 + 2x^2 \cdot 1 - 5 \cdot x^2 - 5 \cdot 1$
$$= x^6 + x^4 + 2x^4 + 2x^2 - 5x^2 - 5$$
$$= x^6 + 3x^4 - 3x^2 - 5$$

Polynomials can be multiplied vertically in a manner similar to multiplication of real numbers. For example, multiplication of 123 times 12 is performed as follows.

$$
\begin{array}{r}
1\;2\;3 \\
\times\;1\;2 \\
\hline
2\;4\;6 \\
1\;2\;3 \\
\hline
1\;4\;7\;6
\end{array}
$$

A similar method can be used to multiply polynomials vertically.

EXAMPLE 8   Multiplying polynomials vertically

Multiply $2x^2 - 4x + 1$ and $x + 3$.

Solution   Write the polynomials vertically. Then multiply every term in the second polynomial times every term in the first polynomial. Arrange the results so that *like terms are in the same column.*

$$
\begin{array}{r}
2x^2 - 4x + 1 \\
x + 3 \\
\hline
6x^2 - 12x + 3 \\
2x^3 - 4x^2 + x \\
\hline
2x^3 + 2x^2 - 11x + 3
\end{array}
$$

Multiply top row by 3.

Multiply top row by x.

Add each column.

$\equiv$ MAKING CONNECTIONS $\equiv$

Vertical and Horizontal Formats

Whether you decide to add, subtract, or multiply polynomials vertically or horizontally, remember that the same answer is obtained either way.

EXAMPLE 9   Finding the volume of a box

A box has a width 3 inches less than its height and a length 4 inches more than its height.
(a) If h represents the height of the box, write a polynomial that represents the volume of the box.
(b) Use this polynomial to calculate the volume of the box if $h = 10$ inches.

Solution   **(a)** If h is the height, then $h - 3$ is the width and $h + 4$ is the length, as illustrated in Figure 5.14. Its volume equals the product of these three expressions.

$$
\begin{aligned}
h(h - 3)(h + 4) &= (h^2 - 3h)(h + 4) \\
&= h^2 \cdot h + h^2 \cdot 4 - 3h \cdot h - 3h \cdot 4 \\
&= h^3 + 4h^2 - 3h^2 - 12h \\
&= h^3 + h^2 - 12h
\end{aligned}
$$

(b) If $h = 10$, then the volume is

$$10^3 + 10^2 - 12(10) = 1000 + 100 - 120 = 980 \text{ cubic inches.}$$

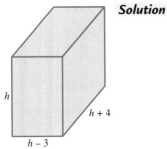

Figure 5.14

5.3   PUTTING IT ALL TOGETHER

The following table summarizes multiplication of polynomials.

Concept	Explanation	Examples
Distributive Properties	For all real numbers a, b, and c, $$a(b + c) = ab + ac \quad \text{and}$$ $$a(b - c) = ab - ac.$$	$5(x + 3) = 5x + 15$, $3(x - 6) = 3x - 18$, and $-2x(3 - 5x^3) = -6x + 10x^4$
Multiplying Polynomials	The product of two polynomials may be found by multiplying every term in the first polynomial by every term in the second polynomial.	$3x(5x^2 + 2x - 7)$ $\quad = 3x \cdot 5x^2 + 3x \cdot 2x - 3x \cdot 7$ $\quad = 15x^3 + 6x^2 - 21x$ $(x + 2)(7x - 3)$ $\quad = x \cdot 7x - x \cdot 3 + 2 \cdot 7x - 2 \cdot 3$ $\quad = 7x^2 - 3x + 14x - 6$ $\quad = 7x^2 + 11x - 6$

5.3   EXERCISES

FOR EXTRA HELP

 Student's Solutions Manual

 MyMathLab

 InterAct Math

 Math Tutor Center

*Math*XL MathXL

Digital Video Tutor
CD 3   Videotape 7

CONCEPTS

1. The equation $x^2 \cdot x^3 = x^5$ illustrates what rule of exponents?

2. The equation $3(x - 2) = 3x - 6$ illustrates what property?

3. Give an example of a binomial.

4. Give an example of a trinomial.

5. The product of two polynomials may be found by multiplying every ____ in the first polynomial by every ____ in the second polynomial and then combining like terms.

6. Are $(x + y)(x + y)$ and $x^2 + y^2$ equivalent expressions?

MULTIPLICATION OF MONOMIALS

Exercises 7–16: Multiply.

7. $x^2 \cdot x^5$

8. $-a \cdot a^5$

9. $-3a \cdot 4a$

10. $7x \cdot 5x$

11. $4x^3 \cdot 5x^2$

12. $6b^6 \cdot 3b^5$

13. $xy^2 \cdot 4xy$

14. $3ab \cdot ab^2$

15. $(-3xy^2)(4x^2y)$

16. $(-r^2t^2)(-r^3t)$

MULTIPLICATION OF MONOMIALS AND POLYNOMIALS

Exercises 17–40: Multiply and simplify the expression.

17. $3(x + 4)$

18. $-7(4x - 1)$

19. $-5(9x + 1)$

20. $10(1 - 6x)$

21. $(4 - z)z$

22. $3z(1 - 5z)$

23. $-y(5 + 3y)$

24. $(2y - 8)2y$

25. $3x(5x^2 - 4)$

26. $-6x(2x^3 + 1)$

27. $(6x - 6)x^2$

28. $(1 - 2x^2)3x^2$

29. $-8(4t^2 + t + 1)$

30. $7(3t^2 - 2t - 5)$

31. $4m(-2m^2 + 7m - 6)$

32. $-9m(-m^2 - 5m + 10)$

33. $n^2(-5n^2 + n - 2)$

34. $6n^3(2 - 4n + n^2)$

35. $xy(x + y)$

36. $ab(2a - 3b)$

37. $x^2(x^2y - xy^2)$

38. $2y^2(xy - 5)$

39. $-ab(a^3 - 2b^3)$

40. $5rt(r^2 + 2rt + t^2)$

MULTIPLICATION OF POLYNOMIALS

Exercises 41–46: (Refer to Example 5.) Multiply the expression **(a)** *geometrically and* **(b)** *symbolically.*

41. $x(x + 3)$

42. $2x(x + 5)$

43. $(x + 2)(x + 2)$

44. $(x + 1)(x + 3)$

45. $(x + 3)(x + 6)$

46. $(x + 5)(x + 2)$

Exercises 47–74: Multiply and simplify the expression.

47. $(x + 3)(x + 5)$

48. $(x - 4)(x - 7)$

49. $(x - 8)(x - 9)$

50. $(x + 10)(x + 10)$

51. $(3z - 2)(2z - 5)$

52. $(z + 6)(2z - 1)$

53. $(8b - 1)(8b + 1)$

54. $(3t + 2)(3t - 2)$

55. $(10y + 7)(y - 1)$

56. $(y + 6)(2y + 7)$

57. $(5 - 3a)(1 - 2a)$

58. $(4 - a)(5 + 3a)$

59. $(1 - 3x)(1 + 3x)$

60. $(10 - x)(5 - 2x)$

61. $(x - 1)(x^2 + 1)$

62. $(x + 2)(x^2 - x)$

63. $(x^2 + 4)(4x - 3)$

64. $(3x^2 - 1)(3x^2 + 1)$

65. $(2n + 1)(n^2 + 3)$

66. $(2 - n^2)(1 + n^2)$

67. $(m + 1)(m^2 + 3m + 1)$

68. $(m - 2)(m^2 - m + 5)$

69. $(3x - 2)(2x^2 - x + 4)$

70. $(5x + 4)(x^2 - 3x + 2)$

71. $(x + 1)(x^2 - x + 1)$

72. $(x - 2)(x^2 + 4x + 4)$

73. $(4b^2 + 3b + 7)(b^2 + 3)$

74. $(-3a^2 - 2a + 1)(3a^2 - 3)$

APPLICATIONS

75. *Volume of a Box* (Refer to Example 9.) A box has a width 4 inches less than its height and a length 2 inches more than its height.
    (a) If h is the height of the box, write a polynomial that represents the volume of the box.
    (b) Use this polynomial to calculate the volume for $h = 25$ inches.

76. *Surface Area of a Box* Use the drawing of the box to do the following.

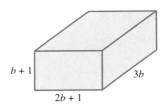

    (a) Write a polynomial that represents the area of its bottom.
    (b) Write a polynomial that represents the area of its front.
    (c) Write a polynomial that represents the area of its right side.
    (d) Write a polynomial that represents the total area of its six sides.

77. *Surface Area of a Cube* Write a polynomial that represents the total area of the six sides of the cube having sides of length $x + 1$.

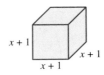

78. *Surface Area of a Sphere* The surface area of a sphere with radius r is $4\pi r^2$. Write a polynomial that gives the surface area of a sphere with radius $x + 2$. Leave your answer in terms of π.

79. *Toy Rocket* A toy rocket is shot straight up into the air. Its height h in feet above the ground after t seconds is represented by the expression $t(64 - 16t)$.
(a) Multiply this expression.
(b) Evaluate the expression obtained in part (a) and the given expression for $t = 2$.
(c) Are your answers in part (b) the same? Should they be the same?

80. *Toy Rocket on the Moon* (Refer to the preceding exercise.) If the same toy rocket were flown on the moon, then its height h in feet after t seconds would be $t(64 - \frac{5}{2}t)$.
(a) Multiply this expression.
(b) Evaluate the expression obtained in part (a) and the given expression for $t = 2$. Did the rocket go higher on the moon?

81. *Perimeter of a Pen* A rectangular pen for a pet has a perimeter of 100 feet. If one side of the pen has length x, then its area is given by $x(50 - x)$.
(a) Multiply this expression.
(b) Evaluate the expression obtained in part (a) for $x = 25$.

82. *Rectangular Garden* A rectangular garden has a perimeter of 500 feet.
(a) If one side of the garden has length x, then write a polynomial expression that gives its area. Multiply this expression completely.
(b) Evaluate the expression obtained for $x = 50$ and interpret your answer.

WRITING ABOUT MATHEMATICS

83. Explain how the acronym FOIL relates to multiplying two binomials, such as $x + 3$ and $2x + 1$.

84. Does the FOIL method work for multiplying a binomial and a trinomial? Explain.

85. Explain in words how to multiply any two polynomials. Give an example.

86. Give two properties of real numbers that are used for multiplying $3x(5x^2 - 3x + 2)$. Explain your answer.

Group Activity: Working with Real Data

Directions: Form a group of 2 to 4 people. Select someone to record the group's responses for this activity. All members of the group should work cooperatively to answer the questions. If your instructor asks for your results, each member of the group should be prepared to respond.

Biology Some types of worms have a remarkable ability to live without moisture. The following table from one study shows the number of worms W surviving after x days.

x (days)	0	20	40	80	120	160
W (worms)	50	48	45	36	20	3

Source: D. Brown and P. Rothery, *Models in Biology.*

(a) Use the equation $W = -0.0014x^2 - 0.076x + 50$ to find W for each x-value in the table.
(b) Discuss how well this equation approximates the data.
(c) Use this equation to estimate the number of worms on day 60 and on day 180. Which answer is most accurate? Explain.

5.4 SPECIAL PRODUCTS

Product of a Sum and Difference · **Squaring Binomials** · **Cubing Binomials**

INTRODUCTION

Polynomials are often used to approximate real-world phenomena in applications. Polynomials have played an important role in the development of everyday products such as computers, cell phones, and automobiles. Even digital images in computers and interest calculations at a bank make use of polynomials. In this section we discuss how to multiply some special types of binomials.

PRODUCT OF A SUM AND DIFFERENCE

Products of the form $(a + b)(a - b)$ occur frequently in mathematics. Other examples include

$$(x + y)(x - y) \quad \text{and} \quad (2r + 3t)(2r - 3t).$$

These products can always be multiplied by using the techniques discussed in Section 5.3. However, there is a faster way to multiply these special products.

$$(a + b)(a - b) = a \cdot a - a \cdot b + b \cdot a - b \cdot b$$
$$= a^2 - ab + ba - b^2$$
$$= a^2 - b^2$$

In words, the product of a sum of two numbers and their difference equals the difference of their squares. We generalize this method as follows.

PRODUCT OF A SUM AND DIFFERENCE

For any real numbers a and b,

$$(a + b)(a - b) = a^2 - b^2.$$

EXAMPLE 1 Finding products of sums and differences

Multiply.
(a) $(x + y)(x - y)$ **(b)** $(z - 2)(z + 2)$
(c) $(2r + 3t)(2r - 3t)$ **(d)** $(5m^2 - 4n^2)(5m^2 + 4n^2)$

Solution **(a)** If we let $a = x$ and $b = y$, then we can apply the rule

$$(a + b)(a - b) = a^2 - b^2.$$

Thus

$$(x + y)(x - y) = (x)^2 - (y)^2$$
$$= x^2 - y^2.$$

(b) Because the expressions $(z + 2)(z - 2)$ and $(z - 2)(z + 2)$ are equal by the commutative property, we can apply the formula for the product of a sum and difference.

$$(z - 2)(z + 2) = (z)^2 - (2)^2$$
$$= z^2 - 4$$

(c) Let $a = 2r$ and $b = 3t$. Then the product can be evaluated as follows.

$$(2r + 3t)(2r - 3t) = (2r)^2 - (3t)^2$$
$$= 4r^2 - 9t^2$$

(d) $(5m^2 - 4n^2)(5m^2 + 4n^2) = (5m^2)^2 - (4n^2)^2 = 25m^4 - 16n^4$

The next example demonstrates how to multiply mentally some products of numbers.

EXAMPLE 2   Finding a product

Use the product of a sum and difference to find $22 \cdot 18$.

Solution   Rewrite and evaluate $22 \cdot 18$ as follows.

$$22 \cdot 18 = (20 + 2)(20 - 2) \qquad \text{Product of a sum and difference}$$
$$= 20^2 - 2^2 \qquad \qquad (a + b)(a - b) = a^2 - b^2$$
$$= 400 - 4 \qquad \qquad \text{Evaluate exponents.}$$
$$= 396 \qquad \qquad \text{Subtract.}$$

SQUARING BINOMIALS

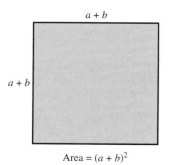

$a + b$

$a + b$

Area $= (a + b)^2$

Figure 5.15

Because each side of the square shown in Figure 5.15 has length $(a + b)$, its area equals

$$(a + b)(a + b),$$

which can be written as $(a + b)^2$. We can multiply this expression as follows.

$$(a + b)^2 = (a + b)(a + b)$$
$$= a^2 + ab + ba + b^2$$
$$= a^2 + 2ab + b^2$$

This result is illustrated geometrically in Figure 5.16, where the area of the large square equals $(a + b)^2$. This area can also be calculated by adding the areas of the four small rectangles.

$$a^2 + ab + ba + b^2 = a^2 + 2ab + b^2$$

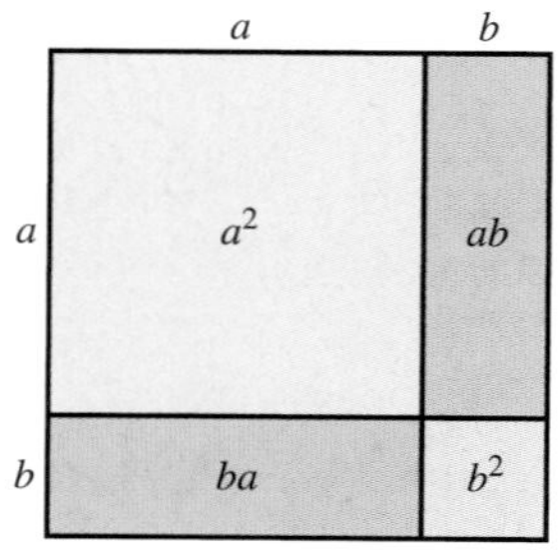

The geometric and symbolic results are the same. Note that to obtain the middle term, $2ab$, we can multiply the two terms in the binomial and *double* the result.

A similar product that is also the square of a binomial can be calculated as

$$(a - b)^2 = (a - b)(a - b)$$
$$= a^2 - ab - ba + b^2$$
$$= a^2 - 2ab + b^2.$$

The results are summarized as follows.

a b

a a^2 ab

b ba b^2

$(a + b)^2 = a^2 + 2ab + b^2$

Figure 5.16

SQUARING A BINOMIAL

For any real numbers a and b,

$$(a + b)^2 = a^2 + 2ab + b^2 \quad \text{and}$$
$$(a - b)^2 = a^2 - 2ab + b^2.$$

That is, the square of a binomial equals the square of the first term, plus (or minus) twice the product of the two terms, plus the square of the last term.

Note: $(a + b)^2 \neq a^2 + b^2$. Do not forget the middle term when squaring a binomial.

EXAMPLE 3   Squaring a binomial

Multiply.
(a) $(x + 3)^2$ **(b)** $(2x - 5)^2$ **(c)** $(1 - 5y)^2$ **(d)** $(7a^2 + 3b)^2$

Solution   **(a)** If we let $a = x$ and $b = 3$, then we can apply the formula

$$(a + b)^2 = a^2 + 2ab + b^2.$$

Thus

$$(x + 3)^2 = (x)^2 + 2(x)(3) + (3)^2$$
$$= x^2 + 6x + 9.$$

(b) Applying the formula $(a - b)^2 = a^2 - 2ab + b^2$ with $a = 2x$ and $b = 5$ gives

$$(2x - 5)^2 = (2x)^2 - 2(2x)(5) + (5)^2$$
$$= 4x^2 - 20x + 25.$$

(c) $(1 - 5y)^2 = (1)^2 - 2(1)(5y) + (5y)^2 = 1 - 10y + 25y^2$
(d) $(7a^2 + 3b)^2 = (7a^2)^2 + 2(7a^2)(3b) + (3b)^2 = 49a^4 + 42a^2b + 9b^2$

═══ MAKING CONNECTIONS ═══

Multiplying Binomial and Special Products

If you forget these special products, you can still multiply polynomials by using earlier techniques. For example, the binomial in Example 3(b) can be multiplied as

$$(2x - 5)^2 = (2x - 5)(2x - 5)$$
$$= 2x \cdot 2x - 2x \cdot 5 - 5 \cdot 2x + 5 \cdot 5$$
$$= 4x^2 - 10x - 10x + 25$$
$$= 4x^2 - 20x + 25.$$

Figure 5.17 Digital Picture

AN APPLICATION   NASA first developed digital pictures because they were easy to transmit through space and because they provided clear images. A digital image supplied by NASA is shown in Figure 5.17.

Today, digital cameras are readily available, and the Internet uses digital images exclusively. The next example shows how polynomials relate to digital pictures.

EXAMPLE 4 Calculating the size of a digital picture

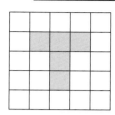

A digital picture comprises tiny square units called *pixels*. Shading individual pixels creates a picture. A simplified version of a digital picture of the letter T is shown in Figure 5.18. The dimensions of this picture are 3 pixels by 3 pixels with a 1-pixel border.

(a) Suppose that a square digital picture, including its border, is $x + 2$ pixels by $x + 2$ pixels. Find a polynomial that gives the total number of pixels in the picture, including the border.

(b) Let $x = 3$ and evaluate the polynomial. Does it agree with Figure 5.18?

Figure 5.18

Solution **(a)** The total number of pixels equals $(x + 2)$ times $(x + 2)$, or $(x + 2)^2$.

$$(x + 2)^2 = x^2 + 4x + 4$$

(b) For $x = 3$, the polynomial evaluates to $3^2 + 4 \cdot 3 + 4 = 25$, the total number of pixels. This result agrees with Figure 5.18, which has a total of $5 \cdot 5 = 25$ pixels with a 3×3–pixel picture of the letter T inside.

CUBING BINOMIALS

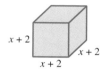

$x + 2$
$x + 2$
$x + 2$

Figure 5.19
Volume $= (x + 2)^3$

To calculate the volume of the cube shown in Figure 5.19, we find the product of its length, width, and height. Because all sides have the same length, its volume is $(x + 2)^3$. That is, the volume equals the **cube** of $x + 2$.

To multiply the expression $(x + 2)^3$, we proceed as follows.

$(x + 2)^3 = (x + 2)(x + 2)^2$ $a^3 = a \cdot a^2$

$= (x + 2)(x^2 + 4x + 4)$ Square the binomial.

$= x \cdot x^2 + x \cdot 4x + x \cdot 4 + 2 \cdot x^2 + 2 \cdot 4x + 2 \cdot 4$ Multiply the polynomials.

$= x^3 + 4x^2 + 4x + 2x^2 + 8x + 8$ Simplify terms.

$= x^3 + 6x^2 + 12x + 8$ Add like terms.

EXAMPLE 5 Cubing a binomial

Multiply $(2z - 3)^3$.

Solution $(2z - 3)^3 = (2z - 3)(2z - 3)^2$ $a^3 = a \cdot a^2$

$= (2z - 3)(4z^2 - 12z + 9)$ Square the binomial.

$= 8z^3 - 24z^2 + 18z - 12z^2 + 36z - 27$ Multiply the polynomials.

$= 8z^3 - 36z^2 + 54z - 27$ Add like terms.

Note: $(2z - 3)^3 \neq (2z)^3 - (3)^3 = 8z^3 - 27.$

Critical Thinking

Suppose that a student is convinced that the expressions

$(x + y)^3$ and $x^3 + y^3$

are equal. How could you convince the student otherwise?

EXAMPLE 6 Calculating interest

If a savings account pays x percent annual interest, where x is expressed as a decimal, then after 3 years a sum of money will grow by a factor of $(1 + x)^3$.
(a) Multiply this expression.
(b) Evaluate the expression for $x = 0.10$ (or 10%), and interpret the result.

Solution **(a)**

$$
\begin{aligned}
(1 + x)^3 &= (1 + x)(1 + x)^2 && a^3 = a \cdot a^2 \\
&= (1 + x)(1 + 2x + x^2) && \text{Square the binomial.} \\
&= 1 + 2x + x^2 + x + 2x^2 + x^3 && \text{Multiply the binomials.} \\
&= 1 + 3x + 3x^2 + x^3 && \text{Add like terms.}
\end{aligned}
$$

(b) Let $x = 0.1$ in the expression. Then

$$1 + 3(0.1) + 3(0.1)^2 + (0.1)^3 = 1.331.$$

The sum of money will increase by a factor of 1.331. For example, if $1000 are deposited in this account, it will grow to $1331 after 3 years.

5.4 PUTTING IT ALL TOGETHER

The following table summarizes some special products of polynomials.

Concept	Explanation	Examples
Product of a Sum and Difference	For any real numbers x and y, $$(x + y)(x - y) = x^2 - y^2.$$	$(x + 6)(x - 6) = x^2 - 36$, $(2x - 3)(2x + 3) = 4x^2 - 9$, and $(x^2 + y^2)(x^2 - y^2) = x^4 - y^4$
Squaring a Binomial	For all real numbers x and y, $$(x + y)^2 = x^2 + 2xy + y^2$$ and $$(x - y)^2 = x^2 - 2xy + y^2.$$	$(x + 4)^2 = x^2 + 8x + 16$, $(5x - 2)^2 = 25x^2 - 20x + 4$, $(1 - 7x)^2 = 1 - 14x + 49x^2$, and $(x^2 + y^2)^2 = x^4 + 2x^2y^2 + y^4$
Cubing a Binomial	Multiply the binomial by its square.	$$\begin{aligned}(x + 3)^3 &= (x + 3)(x + 3)^2 \\ &= (x + 3)(x^2 + 6x + 9) \\ &= x^3 + 6x^2 + 9x + 3x^2 + 18x + 27 \\ &= x^3 + 9x^2 + 27x + 27\end{aligned}$$

5.4 EXERCISES

FOR EXTRA HELP

Student's Solutions Manual

 InterAct Math

MathXL

MyMathLab

Math Tutor Center

Digital Video Tutor
CD 3   Videotape 7

CONCEPTS

1. For any real numbers a and b, $(a + b)(a - b) =$ _____ .

2. Are the expressions $(3z - 1)(3z + 1)$ and $9z^2 - 1$ equal for all real numbers z? If not, choose a value for z that supports your answer.

3. For any real numbers a and b, $(a + b)^2 =$ _____ .

4. For any real numbers a and b, $(a - b)^2 =$ _____ .

5. Are the expressions $(x + y)^2$ and $x^2 + y^2$ equal for all real numbers x and y? If not, choose values for x and y that support your answer.

6. Are the expressions $(r - t)^2$ and $r^2 - t^2$ equal for all real numbers r and t? If not, choose values for r and t that support your answer.

7. Are the expressions $(z + 5)^3$ and $z^3 + 5^3$ equal for all real numbers z? If not, choose a value for z that supports your answer.

8. $(a + b)^3 = (a + b) \cdot$ _____ .

PRODUCT OF A SUM AND DIFFERENCE

Exercises 9–24: Multiply.

9. $(x + 1)(x - 1)$

10. $(x - 7)(x + 7)$

11. $(4x - 1)(4x + 1)$

12. $(10x + 3)(10x - 3)$

13. $(1 + 2a)(1 - 2a)$

14. $(4 - 9b)(4 + 9b)$

15. $(2x + 3y)(2x - 3y)$

16. $(5r - 6t)(5r + 6t)$

17. $(ab - 5)(ab + 5)$

18. $(2xy + 7)(2xy - 7)$

19. $(ab + 4)(ab - 4)$

20. $(2 - 3yz)(2 + 3yz)$

21. $(a^2 - b^2)(a^2 + b^2)$

22. $(3x^2 + y^2)(3x^2 - y^2)$

23. $(x^3 - y^3)(x^3 + y^3)$

24. $(2a^4 + b^4)(2a^4 - b^4)$

Exercises 25–30: (Refer to Example 2.) Use the product of a sum and difference to evaluate the expression.

25. $101 \cdot 99$

26. $52 \cdot 48$

27. $23 \cdot 17$

28. $29 \cdot 31$

29. $90 \cdot 110$

30. $38 \cdot 42$

SQUARING BINOMIALS

Exercises 31–46: Multiply.

31. $(x + 1)^2$

32. $(a + 3)^2$

33. $(a - 2)^2$

34. $(x - 7)^2$

35. $(2x + 3)^2$

36. $(7x - 2)^2$

37. $(3b + 5)^2$

38. $(7t + 10)^2$

39. $\left(\frac{3}{4}a - 4\right)^2$

40. $\left(\frac{1}{5}a + 1\right)^2$

41. $(1 - b)^2$

42. $(1 - 4a)^2$

43. $(5 + y^3)^2$

44. $(9 - 5x^2)^2$

45. $(a^2 + b)^2$

46. $(x^3 - y^3)^2$

CUBING BINOMIALS

Exercises 47–56: Multiply.

47. $(a + 1)^3$

48. $(b + 4)^3$

49. $(x - 2)^3$

50. $(y - 7)^3$

51. $(2x + 1)^3$

52. $(4z + 3)^3$

53. $(6u - 1)^3$

54. $(5v + 3)^3$

55. $t(t + 2)^3$

56. $a(2a - b)^3$

MULTIPLICATION OF POLYNOMIALS

Exercises 57–70: Multiply, using any appropriate method.

57. $4(5x + 9)$

58. $(2x + 1)(3x - 5)$

59. $(x - 5)(x + 7)$

60. $(x + 10)(x + 10)$

61. $(3x - 5)^2$

62. $(x - 3)(x + 9)$

63. $(5x + 3)(5x + 4)$

64. $-x^3(x^2 - x + 1)$

65. $(4b - 5)(4b + 5)$

66. $(x + 5)^3$

67. $-5x(4x^2 - 7x + 2)$

68. $(4x^2 - 5)(4x^2 + 5)$

69. $(4 - a)^3$

70. $(x - y)(x + y)(x^2 + y^2)$

APPLICATIONS

Exercises 71–74: Do each part and verify that your answers are the same.
  (a) Find the area of the large square by multiplying its length and width.
  (b) Find the sum of the areas of the smaller rectangles inside the large square.

71.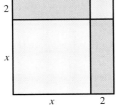

72.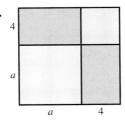

73.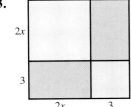

74.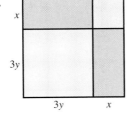

Exercises 75 and 76: Find a polynomial that represents the following.
  (a) Outside surface area of the six sides of the cube
  (b) Volume of the cube

75.

76.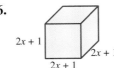

77. *Compound Interest* (Refer to Example 6.) If a sum of money is deposited in a savings account paying x percent

$$x^3 + 15x^2 + 75x + 125$$

annual interest (expressed as a decimal), then this sum of money increases by a factor of $(1 + x)^2$ after 2 years.
(a) Multiply this expression.
(b) Evaluate the polynomial in part (a) for $x = 0.10$ and interpret the answer.

78. *Compound Interest* If a sum of money is deposited in a savings account paying x percent annual interest, then this sum of money increases by a factor of $\left(1 + \frac{1}{100}x\right)^3$ after 3 years.
(a) Multiply this expression.
(b) Evaluate the polynomial in part (a) for $x = 8$ and interpret the answer.

79. *Probability* If there is an x percent chance of rain on each of two consecutive days, then there is a $(1 - x)^2$ percent chance of no rain either day. Assume that all percentages are expressed as decimals.
(a) Multiply this expression.
(b) Evaluate the expression in part (a) for $x = 0.50$ and interpret the answer.

80. *Probability* If there is an x percent chance of rolling a 6 with one die, then there is a $(1 - x)^3$ percent chance of not rolling a 6 with three dice. Assume that all percentages are expressed as decimals or fractions.
(a) Multiply this expression.
(b) Evaluate the expression in part (a) for $x = \frac{1}{6}$ and interpret the answer.

81. *Swimming Pool* A square swimming pool has an 8-foot–wide sidewalk around it.
(a) If the sides of the pool have length z, find a polynomial that gives the area of the sidewalk.
(b) Evaluate the polynomial in part (a) for $z = 60$ and interpret the answer.

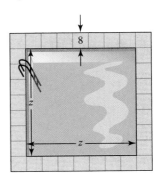

82. *Digital Picture*   (Refer to Example 4.) Suppose that a digital picture, including its border, is $x + 2$ pixels by $x + 2$ pixels and that the actual picture inside the border is $x - 2$ pixels by $x - 2$ pixels.

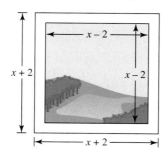

(a) Find a polynomial that gives the number of pixels in the border.

(b) Evaluate the polynomial in part (a) for $x = 5$.

(c) Sketch a digital picture of the letter H with $x = 5$. Does the picture agree with the answer in part (b)?

(d) Digital pictures typically have large values for x. If a picture has $x = 500$, find the total number of pixels in its border.

WRITING ABOUT MATHEMATICS

83. Explain why $(a + b)^2$ does not equal $a^2 + b^2$ for all real numbers a and b.

84. Explain how to find the cube of a binomial.

CHECKING BASIC CONCEPTS   SECTIONS 5.3 AND 5.4

1. Multiply each expression.
   (a) $(-3xy^4)(5x^2y)$ **(b)** $-x(6 - 4x)$
   (c) $3ab(a^2 - 2ab + b^2)$

2. Multiply each expression.
   (a) $(x + 3)(4x - 3)$
   (b) $(x^2 - 1)(2x^2 + 2)$
   (c) $(x + y)(x^2 - xy + y^2)$

3. Multiply each expression.
   (a) $(5x + 2)(5x - 2)$ **(b)** $(x + 3)^2$
   (c) $(2 - 7x)^2$ **(d)** $(t + 2)^3$

4. *Surface Area*   A sphere with radius r has surface area $A = 4\pi r^2$.
   (a) Find an expression for the surface area for $r = a + 7$.
   (b) Find the surface area for $a = 8$. Leave your answer in terms of π.

5. Complete each part and verify that your answers are the same.
   (a) Find the area of the large square by squaring the length of one of its sides.
   (b) Find the sum of the areas of the smaller rectangles inside the large square.

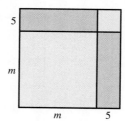

5.5 INTEGER EXPONENTS AND THE QUOTIENT RULE

Negative Integers as Exponents · The Quotient Rule · Other Rules for Exponents · Scientific Notation

INTRODUCTION

Technology has brought with it the need for both very small and very large numbers. The size of an average virus is 5 millionths of a centimeter, whereas the distance to the nearest star, Alpha Centauri, is 25 trillion miles. Exponents are often used to represent such numbers. In this section we discuss properties of integer exponents and some of their applications. (**Source:** C. Ronan, *The Natural History of the Universe.*)

NEGATIVE INTEGERS AS EXPONENTS

So far we have defined whole numbers as exponents. For example,

$$5^0 = 1 \quad \text{and} \quad 2^3 = 2 \cdot 2 \cdot 2 = 8.$$

We can also define negative integers as exponents as follows.

NEGATIVE INTEGER EXPONENTS

Let a be a nonzero real number and n be a positive integer. Then

$$a^{-n} = \frac{1}{a^n}.$$

That is, a^{-n} is the reciprocal of a^n.

In the next example we evaluate negative exponents.

EXAMPLE 1 Evaluating negative exponents

Simplify each expression.
(a) 2^{-3} **(b)** 7^{-1} **(c)** x^{-2} **(d)** $(x + y)^{-8}$

Solution **(a)** Because $a^{-n} = \frac{1}{a^n}$, $2^{-3} = \frac{1}{2^3} = \frac{1}{2 \cdot 2 \cdot 2} = \frac{1}{8}$.

(b) $7^{-1} = \frac{1}{7^1} = \frac{1}{7}$

(c) $x^{-2} = \frac{1}{x^2}$

(d) $(x + y)^{-8} = \frac{1}{(x + y)^8}$

Calculator Help
To use the fraction feature
(Frac), see Appendix A
(page AP-2).

Technology Note: *Negative Exponents*

Calculators can be used to evaluate negative exponents. The figure
shows how a graphing calculator evaluates the expressions in parts
(a) and (b) of Example 1.

```
2^(-3)►Frac
          1/8
7^(-1)►Frac
          1/7
```

The rules for exponents discussed so far also apply to expressions having negative exponents. For example, we can apply the product rule, $a^m \cdot a^n = a^{m+n}$, as follows.

$$2^{-3} \cdot 2^2 \overset{\text{Add}}{=} 2^{-3+2} = 2^{-1} = \frac{1}{2}$$

We can check this result by evaluating the expression without using the product rule.

$$2^{-3} \cdot 2^2 = \frac{1}{2^3} \cdot 2^2 = \frac{1}{8} \cdot 4 = \frac{4}{8} = \frac{1}{2}$$

EXAMPLE 2 Using the product rule with negative exponents

Evaluate each expression.
(a) $5^2 \cdot 5^{-4}$ **(b)** $3^{-2} \cdot 3^{-1}$

Solution **(a)** $5^2 \cdot 5^{-4} = 5^{2+(-4)} = 5^{-2} = \frac{1}{5^2} = \frac{1}{25}$

(b) $3^{-2} \cdot 3^{-1} = 3^{-2+(-1)} = 3^{-3} = \frac{1}{3^3} = \frac{1}{27}$

EXAMPLE 3 Using the rules of exponents

Simplify and write each expression, using positive exponents.
(a) $x^2 \cdot x^{-5}$ **(b)** $(y^3)^{-4}$ **(c)** $(rt)^{-5}$ **(d)** $(ab)^{-3}(a^{-2}b)^3$

Solution **(a)** Using the product rule, $a^m \cdot a^n = a^{m+n}$, gives

$$x^2 \cdot x^{-5} = x^{2+(-5)} = x^{-3} = \frac{1}{x^3}.$$

(b) Using the power rule, $(a^m)^n = a^{mn}$, gives

$$(y^3)^{-4} = y^{3\cdot(-4)} = y^{-12} = \frac{1}{y^{12}}.$$

(c) Using the power rule, $(ab)^n = a^n b^n$, gives

$$(rt)^{-5} = r^{-5}t^{-5} = \frac{1}{r^5} \cdot \frac{1}{t^5} = \frac{1}{r^5 t^5}.$$

This expression could also be simplified as follows.

$$(rt)^{-5} = \frac{1}{(rt)^5} = \frac{1}{r^5 t^5}$$

(d)
$$(ab)^{-3}(a^{-2}b)^3 = a^{-3}b^{-3}a^{-6}b^3$$
$$= a^{-3+(-6)}b^{-3+3}$$
$$= a^{-9}b^0$$
$$= \frac{1}{a^9}(1)$$
$$= \frac{1}{a^9}$$

THE QUOTIENT RULE

Consider the division problem

$$\frac{3^4}{3^2} = \frac{3 \cdot 3 \cdot 3 \cdot 3}{3 \cdot 3} = \frac{3}{3} \cdot \frac{3}{3} \cdot 3 \cdot 3 = 1 \cdot 1 \cdot 3^2 = 3^2.$$

Because there are two more 3s in the numerator than in the denominator, the result is $3^{4-2} = 3^2$. That is, to divide exponential expressions having the *same base*, subtract the exponent of the denominator from the exponent of the numerator and keep the same base. This rule is called the *quotient rule,* which we express in symbols as follows.

THE QUOTIENT RULE

For any nonzero number a and integers m and n,

$$\frac{a^m}{a^n} = a^{m-n}.$$

EXAMPLE 4   Using the quotient rule

Simplify and write each expression, using positive exponents.

(a) $\dfrac{4^3}{4^5}$ **(b)** $\dfrac{6a^7}{3a^4}$ **(c)** $\dfrac{xy^7}{x^2y^5}$

Solution

(a) $\dfrac{4^3}{4^5} = 4^{3-5} = 4^{-2} = \dfrac{1}{4^2} = \dfrac{1}{16}$ $\quad$ ⌐ Subtract

(b) $\dfrac{6a^7}{3a^4} = \dfrac{6}{3} \cdot \dfrac{a^7}{a^4} = 2a^{7-4} = 2a^3$

(c) $\dfrac{xy^7}{x^2y^5} = \dfrac{x^1}{x^2} \cdot \dfrac{y^7}{y^5} = x^{1-2}y^{7-5} = x^{-1}y^2 = \dfrac{y^2}{x}$

===== MAKING CONNECTIONS =====

The Quotient Rule and Simplifying Quotients

Some quotients can be simplified mentally. Because

$$\frac{x^5}{x^3} = \frac{x \cdot x \cdot x \cdot x \cdot x}{x \cdot x \cdot x},$$

the quotient $\frac{x^5}{x^3}$ has five factors of x in the numerator and three factors of x in the denominator. There are two more factors of x in the numerator than in the denominator, $5 - 3 = 2$, so this expression simplifies to x^2. Similarly,

$$\frac{x^3}{x^5} = \frac{x \cdot x \cdot x}{x \cdot x \cdot x \cdot x \cdot x}$$

has two more factors of x in the denominator than in the numerator. This expression simplifies to $\frac{1}{x^2}$. Use this technique to simplify the expressions

$$\frac{z^7}{z^4}, \quad \frac{a^5}{a^8}, \quad \text{and} \quad \frac{x^6 y^2}{x^3 y^7}.$$

OTHER RULES FOR EXPONENTS

Other rules can be used to simplify expressions with negative exponents.

QUOTIENTS AND NEGATIVE EXPONENTS

The following three rules hold for any nonzero real numbers a and b and positive integers m and n.

1. $\dfrac{1}{a^{-n}} = a^n$ **2.** $\dfrac{a^{-n}}{b^{-m}} = \dfrac{b^m}{a^n}$ **3.** $\left(\dfrac{a}{b}\right)^{-n} = \left(\dfrac{b}{a}\right)^n$

We demonstrate the validity of these rules as follows.

1. $\dfrac{1}{a^{-n}} = \dfrac{1}{\frac{1}{a^n}} = 1 \cdot \dfrac{a^n}{1} = a^n$

2. $\dfrac{a^{-n}}{b^{-m}} = \dfrac{\frac{1}{a^n}}{\frac{1}{b^m}} = \dfrac{1}{a^n} \cdot \dfrac{b^m}{1} = \dfrac{b^m}{a^n}$

3. $\left(\dfrac{a}{b}\right)^{-n} = \dfrac{a^{-n}}{b^{-n}} = \dfrac{\frac{1}{a^n}}{\frac{1}{b^n}} = \dfrac{1}{a^n} \cdot \dfrac{b^n}{1} = \dfrac{b^n}{a^n} = \left(\dfrac{b}{a}\right)^n$

In the next example we apply these rules.

EXAMPLE 5   Working with quotients and negative exponents

Simplify and write each expression, using positive exponents.

(a) $\dfrac{1}{2^{-5}}$ (b) $\dfrac{3^{-3}}{4^{-2}}$ (c) $\dfrac{5x^{-4}y^2}{10x^2y^{-4}}$ (d) $\left(\dfrac{2}{z^2}\right)^{-4}$

Solution   (a) $\dfrac{1}{2^{-5}} = 2^5 = 2 \cdot 2 \cdot 2 \cdot 2 \cdot 2 = 32$

(b) $\dfrac{3^{-3}}{4^{-2}} = \dfrac{4^2}{3^3} = \dfrac{16}{27}$

(c) $\dfrac{5x^{-4}y^2}{10x^2y^{-4}} = \dfrac{y^2y^4}{2x^2x^4} = \dfrac{y^6}{2x^6}$

(d) $\left(\dfrac{2}{z^2}\right)^{-4} = \left(\dfrac{z^2}{2}\right)^4 = \dfrac{z^8}{2^4} = \dfrac{z^8}{16}$

Technology Note:
Powers of 10

Calculators make use of scientific notation, as illustrated in the accompanying figure. The letter E denotes a power of 10. That is,

$2.5\text{E}13 = 2.5 \times 10^{13}$ and
$5\text{E}{-6} = 5 \times 10^{-6}$.

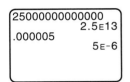

25000000000000
                2.5E13
.000005
                5E-6

Note: The calculator has been set in *scientific mode.*

SCIENTIFIC NOTATION

Powers of 10 are important because they are used in science to express numbers that are either very small or very large in absolute value. Table 5.4 lists some powers of 10. Note that, if the power of 10 decreases by 1, the result decreases by a factor of $\frac{1}{10}$, or equivalently, the decimal point is moved one place to the left. Table 5.5 shows the values of some important powers of 10.

TABLE 5.4   Powers of 10

Power of 10	Value
10^3	1000
10^2	100
10^1	10
10^0	1
10^{-1}	$\frac{1}{10} = 0.1$
10^{-2}	$\frac{1}{100} = 0.01$
10^{-3}	$\frac{1}{1000} = 0.001$

TABLE 5.5   Important Powers of 10

Number	Value
10^3	Thousand
10^6	Million
10^9	Billion
10^{12}	Trillion
10^{-1}	Tenth
10^{-2}	Hundredth
10^{-3}	Thousandth
10^{-6}	Millionth

Recall that numbers written in decimal notation are sometimes said to be in *standard form. Scientific notation* can be used to express decimal numbers that are either very large or very small in absolute value.

The distance to the nearest star is 25 trillion miles. This distance can be written in scientific notation as

$$25,\!000,\!000,\!000,\!000 = 2.5 \times 10^{13}.$$
13 decimal places

The 10^{13} indicates that the decimal point in 2.5 should be moved 13 places to the right.

Calculator Help
To set a calculator in scientific mode, see Appendix A (page AP-2).

A typical virus is about 5 millionths of a centimeter in diameter, which can be written in scientific notation as

$$0.\underbrace{000005}_{\text{6 decimal places}} = 5 \times 10^{-6} \text{ cm.}$$

The 10^{-6} indicates that the decimal point in 5 should be moved 6 places to the left. The following definition provides a more complete explanation of scientific notation.

SCIENTIFIC NOTATION

A real number a is in **scientific notation** when a is written as $b \times 10^n$, where $1 \le |b| < 10$ and n is an integer.

EXAMPLE 6   Converting scientific notation to standard form

Write each number in standard form.
(a) 5.23×10^4 **(b)** 8.1×10^{-3} **(c)** 6×10^{-2}

Solution   **(a)** The positive exponent 4 indicates that the decimal point in 5.23 is to be moved 4 places to the *right*.

Calculator Help
To enter numbers in scientific notation, see Appendix A (page AP-3).

$$5.23 \times 10^4 = 5.\underset{1\ 2\ 3\ 4}{2\ 3\ 0\ 0.} = 52{,}300$$

(b) A negative exponent -3 indicates that the decimal point in 8.1 is to be moved 3 places to the *left*.

$$8.1 \times 10^{-3} = 0.\underset{1\ 2\ 3}{0\ 0\ 8}.1 = 0.0081$$

(c) $6 \times 10^{-2} = 0.\underset{1\ 2}{0\ 6}. = 0.06$

The following steps can be used for writing a positive number a in scientific notation.

WRITING A POSITIVE NUMBER a IN SCIENTIFIC NOTATION

1. Move the decimal point in a number a until it becomes a number b such that $1 \le b < 10$.
2. Count the number of places that the decimal point was moved. Let this positive integer be n.
3. If the decimal point was moved to the *left*, then $a = b \times 10^n$. If the decimal point was moved to the *right*, then $a = b \times 10^{-n}$.

Note: The scientific notation for a negative number a is the opposite of the scientific notation of $|a|$. For example, $450 = 4.5 \times 10^2$ and $-450 = -4.5 \times 10^2$.

EXAMPLE 7 Writing a number in scientific notation

Write each number in scientific notation.
(a) 281,000,000   (U.S. population in 2000)
(b) 0.001   (Approximate time in seconds for sound to travel one foot)

Solution (a) Move the assumed decimal point in 281,000,000 eight places to the *left* to obtain 2.81.

$$2.\underbrace{8\ 1\ 0\ 0\ 0\ 0\ 0\ 0.}_{1\ 2\ 3\ 4\ 5\ 6\ 7\ 8}$$

The scientific notation for 281,000,000 is 2.81×10^8.
(b) Move the decimal point in 0.001 three places to the *right* to obtain 1.

$$0.\underbrace{0\ 0\ 1}_{1\ 2\ 3}.$$

The scientific notation for 0.001 is 1×10^{-3}.

Numbers in scientific notation can be multiplied by applying properties of real numbers and properties of exponents.

$$(6 \times 10^4) \cdot (3 \times 10^3) = (6 \cdot 3) \times (10^4 \cdot 10^3) \quad \text{Properties of real numbers}$$
$$= 18 \times 10^7 \quad \text{Product rule}$$
$$= 1.8 \times 10^8 \quad \text{Scientific notation}$$

Division can also be performed with scientific notation.

$$\frac{6 \times 10^4}{3 \times 10^3} = \frac{6}{3} \times \frac{10^4}{10^3} \quad \text{Property of fractions}$$
$$= 2 \times 10^1 \quad \text{Quotient rule}$$

These results are supported in Figure 5.20, where the calculator is set in scientific mode.

In the next example we show how to use scientific notation in an application.

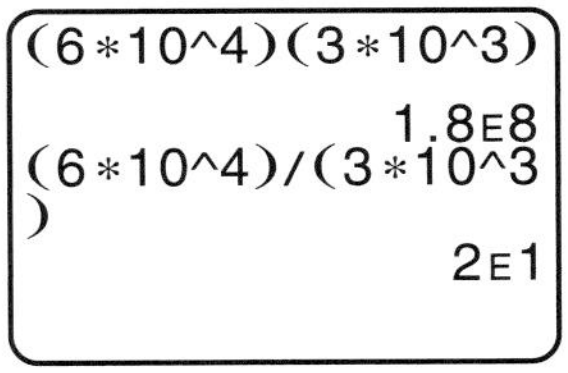

```
(6*10^4)(3*10^3)
            1.8E8
(6*10^4)/(3*10^3
)
              2E1
```

Figure 5.20

EXAMPLE 8 Analyzing the cost of advertising

In 2000, a total of 2.34×10^{11} was spent on advertising in the United States. At that time the population of the United States was 2.81×10^8. Determine how much was spent per person on advertising.

Solution To determine the amount spent per person, divide 2.34×10^{11} by 2.81×10^8.

$$\frac{2.34 \times 10^{11}}{2.81 \times 10^8} = \frac{2.34}{2.81} \times 10^{11-8} \approx 0.833 \times 10^3$$

In 2000, about $833 was spent on advertising for every person in the United States.

Critical Thinking

Estimate the number of seconds that you have been alive. Write your answer in scientific notation.

5.5   PUTTING IT ALL TOGETHER

In this section we discussed negative integer exponents and scientific notation. The following table summarizes several important concepts related to these topics. Assume that a and b are nonzero real numbers and that m and n are integers.

Concept	Explanation	Examples
Negative Exponents	$a^{-n} = \dfrac{1}{a^n}$	$2^{-4} = \dfrac{1}{2^4} = \dfrac{1}{16}$, $a^{-8} = \dfrac{1}{a^8}$, and $(xy)^{-2} = \dfrac{1}{(xy)^2} = \dfrac{1}{x^2 y^2}$
Quotient Rule	$\dfrac{a^m}{a^n} = a^{m-n}$	$\dfrac{7^2}{7^4} = 7^{2-4} = 7^{-2} = \dfrac{1}{7^2} = \dfrac{1}{49}$ and $\dfrac{x^6}{x^3} = x^{6-3} = x^3$
Quotients and Negative Exponents	1. $\dfrac{1}{a^{-n}} = a^n$ 2. $\dfrac{a^{-n}}{b^{-m}} = \dfrac{b^m}{a^n}$ 3. $\left(\dfrac{a}{b}\right)^{-n} = \left(\dfrac{b}{a}\right)^n$	$\dfrac{1}{5^{-2}} = 5^2 = 25$ $\dfrac{x^{-4}}{y^{-2}} = \dfrac{y^2}{x^4}$ $\left(\dfrac{2}{3}\right)^{-3} = \left(\dfrac{3}{2}\right)^3 = \dfrac{3^3}{2^3} = \dfrac{27}{8}$
Scientific Notation	Write a as $b \times 10^n$, where $1 \le |b| < 10$ and n is an integer.	$23{,}500 = 2.35 \times 10^4$, $0.0056 = 5.6 \times 10^{-3}$, and $1000 = 1 \times 10^3$

5.5   EXERCISES

FOR EXTRA HELP

Student's Solutions Manual

 InterAct Math

 MathXL

 MyMathLab

 Math Tutor Center

 Digital Video Tutor
CD 3   Videotape 7

CONCEPTS

1. $a^{-n} = $ _____

2. $2^{-3} = $ _____

3. $\dfrac{1}{a^{-n}} = $ _____

4. $\dfrac{1}{2^{-3}} = $ _____

5. $\dfrac{a^m}{a^n} = $ _____

6. $\dfrac{2^5}{2^2} = $ _____

7. $\left(\dfrac{a}{b}\right)^{-n} = $ _____.

8. $\left(\dfrac{3}{5}\right)^{-2} = $ _____.

9. To write a number a in scientific notation as $b \times 10^n$, the number b must satisfy _____.

10. Are 0.005 and 5×10^3 equivalent numbers?

NEGATIVE EXPONENTS

Exercises 11–34: Simplify the expression.

11. 4^{-1}

12. 6^{-2}

13. $\left(\dfrac{1}{3}\right)^{-2}$

14. 2.5^{-1}

15. $2^3 \cdot 2^{-2}$

16. $3^{-4} \cdot 3^2$

17. $10^4 \cdot 10^{-2}$

18. $10^{-1} \cdot 10^{-2}$

19. $3^{-2} \cdot 3^{-1} \cdot 3^{-1}$

20. $2^{-3} \cdot 2^5 \cdot 2^{-4}$

21. $(2^3)^{-1}$

22. $(3^{-2})^{-2}$

23. $(3^2 4^3)^{-1}$

24. $(2^{-2} 3^2)^{-2}$

25. $\dfrac{4^5}{4^2}$

26. $\dfrac{5^5}{5^3}$

27. $\dfrac{1^9}{1^7}$

28. $\dfrac{-6^4}{6}$

29. $\dfrac{1}{4^{-3}}$

30. $\dfrac{1}{6^{-2}}$

31. $\dfrac{5^{-2}}{5^{-4}}$

32. $\dfrac{7^{-3}}{7^{-1}}$

33. $\left(\dfrac{2}{7}\right)^{-2}$

34. $\left(\dfrac{3}{4}\right)^{-3}$

Exercises 35–82: Simplify and write the expression using positive exponents.

35. x^{-1}

36. y^{-2}

37. a^{-4}

38. z^{-7}

39. $x^{-2} \cdot x^{-1} \cdot x$

40. $y^{-3} \cdot y^4 \cdot y^{-5}$

41. $a^{-5} \cdot a^{-2} \cdot a^{-1}$

42. $b^5 \cdot b^{-3} \cdot b^{-6}$

43. $x^2 y^{-3} x^{-5} y^6$

44. $a^{-2} b^{-6} b^3 a^{-1}$

45. $(xy)^{-3}$

46. $(ab)^{-1}$

47. $(2t)^{-4}$

48. $(8c)^{-2}$

49. $(x + 1)^{-7}$

50. $(a + b)^{-9}$

51. $(a^{-2})^{-4}$

52. $(4x^3)^{-3}$

53. $(rt^3)^{-2}$

54. $(xy^{-3})^{-2}$

55. $(ab)^2 (a^2)^{-3}$

56. $(x^3)^{-2}(xy)^4 y^{-5}$

57. $\dfrac{x^4}{x^2}$

58. $\dfrac{y^9}{y^5}$

59. $\dfrac{a^{10}}{a^{-3}}$

60. $\dfrac{b^5}{b^{-2}}$

61. $\dfrac{4z}{2z^4}$

62. $\dfrac{12x^2}{24x^7}$

63. $\dfrac{-4xy^5}{6x^3 y^2}$

64. $\dfrac{12a^6 b^2}{8ab^3}$

65. $\dfrac{x^{-4}}{x^{-1}}$

66. $\dfrac{y^{-2}}{y^{-7}}$

67. $\dfrac{10b^{-4}}{5b^{-5}}$

68. $\dfrac{8a^{-2}}{2a^{-3}}$

69. $\left(\dfrac{a}{b}\right)^3$

70. $\left(\dfrac{2x}{y}\right)^5$

71. $\dfrac{1}{y^{-5}}$

72. $\dfrac{1}{z^{-6}}$

73. $\dfrac{4}{2t^{-3}}$

74. $\dfrac{5}{10b^{-5}}$

75. $\dfrac{1}{(xy)^{-2}}$

76. $\dfrac{1}{(ab)^{-1}}$

77. $\dfrac{1}{(a^2 b)^{-3}}$

78. $\dfrac{1}{(rt^4)^{-2}}$

79. $\left(\dfrac{a}{b}\right)^{-2}$

80. $\left(\dfrac{2x}{y}\right)^{-3}$

81. $\left(\dfrac{u}{4v}\right)^{-1}$

82. $\left(\dfrac{5u}{3v}\right)^{-2}$

SCIENTIFIC NOTATION

Exercises 83–88: (Refer to Table 5.5.) Write the value of the power of 10 in words.

83. 10^3

84. 10^6

85. 10^9

86. 10^{-1}

87. 10^{-2}

88. 10^{-6}

Exercises 89–100: Write the expression in standard form.

89. 2×10^3

90. 5×10^2

91. 4.5×10^4

92. 7.1×10^6

93. 8×10^{-3}

94. 9×10^{-1}

95. 4.56×10^{-4}

96. 9.4×10^{-2}

97. 3.9×10^{7}

98. 5.27×10^{6}

99. -5×10^{5}

100. -9.5×10^{3}

Exercises 101–112: Write the number in scientific notation.

101. 2000

102. 11,000

103. 567

104. 9300

105. 12,000,000

106. 600,000

107. 0.004

108. 0.0008

109. 0.000895

110. 0.0123

111. -0.05

112. -0.934

Exercises 113–120: Evaluate the expression. Write the answer in standard form.

113. $(5 \times 10^{3})(3 \times 10^{2})$

114. $(2.1 \times 10^{2})(2 \times 10^{4})$

115. $(-3 \times 10^{-3})(5 \times 10^{2})$

116. $(4 \times 10^{2})(1 \times 10^{3})(5 \times 10^{-4})$

117. $\dfrac{4 \times 10^{5}}{2 \times 10^{2}}$

118. $\dfrac{9 \times 10^{2}}{3 \times 10^{6}}$

119. $\dfrac{8 \times 10^{-6}}{4 \times 10^{-3}}$

120. $\dfrac{6.3 \times 10^{2}}{2 \times 10^{-3}}$

APPLICATIONS

121. *Light-year* The distance that light travels in 1 year is called a *light-year*. Light travels at 1.86×10^{5} miles per second, and there are about 3.15×10^{7} seconds in 1 year.
(a) Estimate the number of miles in 1 light-year.
(b) Except for the sun, Alpha Centauri is the nearest star, and its distance is 4.27 light-years from Earth. Estimate its distance in miles. Write your answer in scientific notation.

122. *Milky Way* It takes 2×10^{8} years for the sun to make one orbit around the Milky Way galaxy. Write this number in standard form.

123. *Speed of the Sun* (Refer to Exercises 121 and 122.) Assume that the sun's orbit in the Milky Way galaxy is circular with a diameter of 10^{5} light-years. Estimate how many miles the sun travels in 1 year.

124. *Distance to the Moon* The moon is about 240,000 miles from Earth.
(a) Write this number in scientific notation.
(b) If a rocket traveled at 4×10^{4} miles per hour, how long would it take for the rocket to reach the moon?

125. *Gross Domestic Product* The gross domestic product (GDP) is the total national output of goods and services valued at market prices *within* the United States. The GDP of the United States in 2000 was \$9,963,000,000,000.   (*Source:* Bureau of Economic Analysis.)
(a) Write this number in scientific notation.
(b) In 2000, the U.S. population was 2.81×10^{8}. On average, how many dollars of goods and services were produced by each individual?

126. *Average Family Net Worth* A family refers to a group of two or more people related by birth, marriage, or adoption who reside together. In 2000, the average family net worth was \$280,000, and there were about 7.2×10^{7} families. Calculate the total family net worth in the United States in 2000. (*Source:* Bureau of the Census.)

WRITING ABOUT MATHEMATICS

127. Explain what a negative exponent represents and how it is different from a positive exponent. Give an example.

128. Explain why scientific notation is helpful for writing some numbers.

Group Activity: Working with Real Data

Directions: Form a group of 2 to 4 people. Select a person to record the group's responses for this activity. All members of the group should work cooperatively to answer the questions. If your instructor asks for your results, each member of the group should be prepared to respond.

Water in a Lake   East Battle Lake in Minnesota covers an area of about 1950 acres or 8.5×10^7 square feet, and its average depth is about 3.2×10^1 feet.
(a) Estimate the cubic feet of water in the lake. (*Hint:* volume = area $\times$ average depth.)

(b) One cubic foot of water equals about 7.5 gallons. How many gallons of water are in this lake?
(c) The population of the United States is about 2.81×10^8, and the average American uses 10^2 gallons of water per day. Could this lake supply the American population with water for 1 day?

5.6   DIVISION OF POLYNOMIALS

Division by a Monomial · Division by a Polynomial

INTRODUCTION

The study of polynomials has occupied the minds of mathematicians for centuries. During the sixteenth century, Italian mathematicians discovered how to solve higher degree polynomial equations. In this section we demonstrate how to divide polynomials. Division is often needed to factor polynomials and to solve polynomial equations.   (*Source:* H. Eves, *An Introduction to the History of Mathematics.*)

DIVISION BY A MONOMIAL

To add two fractions with like denominators, we use the property

$$\frac{a}{d} + \frac{b}{d} = \frac{a + b}{d}.$$

For example, $\frac{1}{7} + \frac{3}{7} = \frac{1 + 3}{7} = \frac{4}{7}$.

To divide a polynomial by a monomial we use the same property, only in reverse. That is,

$$\frac{a + b}{d} = \frac{a}{d} + \frac{b}{d}.$$

EXAMPLE 1 Dividing a polynomial by a monomial

Divide.

(a) $\dfrac{a^5 + a^3}{a^2}$ **(b)** $\dfrac{5x^4 + 10x}{10x}$ **(c)** $\dfrac{3y^2 + 2y - 12}{6y}$

Solution **(a)** $\dfrac{a^5 + a^3}{a^2} = \dfrac{a^5}{a^2} + \dfrac{a^3}{a^2} = a^3 + a$

(b) $\dfrac{5x^4 + 10x}{10x} = \dfrac{5x^4}{10x} + \dfrac{10x}{10x} = \dfrac{1}{2}x^3 + 1$

(c) $\dfrac{3y^2 + 2y - 12}{6y} = \dfrac{3y^2}{6y} + \dfrac{2y}{6y} - \dfrac{12}{6y} = \dfrac{1}{2}y + \dfrac{1}{3} - \dfrac{2}{y}$

≡ MAKING CONNECTIONS ≡

Division and Simplification

A common mistake made when dividing expressions is to "cancel" incorrectly. Note in Example 1(b) that

$$\frac{5x^4 + 10x}{10x} \neq 5x^4 + \frac{10x}{10x}.$$

The monomial must be divided into *every* term in the numerator.

When dividing two natural numbers, we can check our work by multiplying. For example, $\frac{10}{5} = 2$, and we can check this result by finding the product $5 \cdot 2 = 10$. Similarly, to check

$$\frac{a^5 + a^3}{a^2} = a^3 + a$$

we can multiply a^2 and $a^3 + a$.

$$a^2(a^3 + a) = a^2 \cdot a^3 + a^2 \cdot a \qquad \text{Distributive property}$$
$$= a^5 + a^3 \qquad \text{It checks.}$$

EXAMPLE 2 Dividing and checking

Divide the expression $\frac{8x^3 - 4x^2 + 6x}{2x^2}$ and then check the result.

Solution Be sure to divide $2x^2$ into *every* term in the numerator.

$$\frac{8x^3 - 4x^2 + 6x}{2x^2} = \frac{8x^3}{2x^2} - \frac{4x^2}{2x^2} + \frac{6x}{2x^2} = 4x - 2 + \frac{3}{x}$$

Check:

$$2x^2\left(4x - 2 + \frac{3}{x}\right) = 2x^2 \cdot 4x - 2x^2 \cdot 2 + 2x^2 \cdot \frac{3}{x}$$

$$= 8x^3 - 4x^2 + 6x$$

EXAMPLE 3 Finding the length of a rectangle

The rectangle in Figure 5.21 has an area $A = x^2 + 2x$ and width x. Find an expression for its length L.

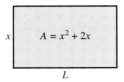

Figure 5.21

Solution The area A of a rectangle equals length L times width W, or $A = LW$. Solving for L gives

$$L = \frac{A}{W}.$$

Thus to find the length of the given rectangle, divide the area by its width.

$$L = \frac{x^2 + 2x}{x} = \frac{x^2}{x} + \frac{2x}{x} = x + 2$$

The length of the rectangle is $x + 2$. The answer checks because $x(x + 2) = x^2 + 2x$.

DIVISION BY A POLYNOMIAL

To enable you to understand division by a polynomial better, we first need to review some terminology related to long division of natural numbers.

$$
\begin{array}{r}
\text{Quotient} \longrightarrow 67 \text{ R } 3 \longleftarrow \text{Remainder} \\
\text{Divisor} \longrightarrow 4\overline{)271} \longleftarrow \text{Dividend} \\
\underline{24} \\
31 \\
\underline{28} \\
3
\end{array}
$$

To check this result, we find the product of the quotient and divisor and then add the remainder. Because $67 \cdot 4 + 3 = 271$, the answer checks. The quotient and remainder can also be expressed as $67\frac{3}{4}$. Division of polynomials is similar to long division of natural numbers.

EXAMPLE 4 Dividing polynomials

Divide $\frac{6x^2 + 13x + 3}{3x + 2}$ and check.

Solution Begin by dividing the first term of $3x + 2$ into the first term of $6x^2 + 13x + 2$. That is, divide $3x$ into $6x^2$ to obtain $2x$. Then take the product of $2x$ and $3x + 2$, place it below $6x^2 + 13x + 2$, and subtract. Bring down the 3.

$$
\begin{array}{r}
2x \\
3x + 2\overline{)6x^2 + 13x + 3} \\
\underline{6x^2 + 4x} \\
9x + 3
\end{array}
$$

$\dfrac{6x^2}{3x} = 2x$

$2x(3x + 2) = 6x^2 + 4x$

Subtract: $13x - 4x = 9x$.

Bring down the 3.

In the next step, divide $3x$ into the first term of $9x + 3$ to obtain 3. Then take the product of 3 and $3x + 2$, place it below $9x + 3$, and subtract.

$$
\begin{array}{r}
2x + 3 \\
3x + 2\overline{)6x^2 + 13x + 3} \\
\underline{6x^2 + 4x} \\
9x + 3 \\
\underline{9x + 6} \\
-3
\end{array}
\qquad
\begin{array}{l}
\dfrac{9x}{3x} = 3 \\[2ex]
\\
\\
3(3x + 2) = 9x + 6 \\
\text{Subtract: } 3 - 6 = -3.
\end{array}
$$

The quotient is $2x + 3$ with remainder -3. This result can also be written as

$$
2x + 3 + \frac{-3}{3x + 2}, \qquad \text{Quotient} + \frac{\text{Remainder}}{\text{Divisor}}
$$

in the same manner that 67 R 3 can be written as $67\frac{3}{4}$.

Check polynomial division by adding the remainder to the product of the divisor and the quotient. That is,

$$
(\text{Divisor})(\text{Quotient}) + \text{Remainder} = \text{Dividend}.
$$

For this example, the equation becomes

$$
\begin{aligned}
(3x + 2)(2x + 3) + (-3) &= 3x \cdot 2x + 3x \cdot 3 + 2 \cdot 2x + 2 \cdot 3 - 3 \\
&= 6x^2 + 9x + 4x + 6 - 3 \\
&= 6x^2 + 13x + 3. \qquad\qquad \text{It checks.}
\end{aligned}
$$

EXAMPLE 5   Dividing polynomials

Simplify $(3x^3 + 2x - 4) \div (x - 2)$.

Solution   Because the dividend does not have an x^2-term, insert $0x^2$ as a "place holder." Then begin by dividing x into $3x^3$ to obtain $3x^2$.

$$
\begin{array}{r}
3x^2 \\
x - 2\overline{)3x^3 + 0x^2 + 2x - 4} \\
\underline{3x^3 - 6x^2} \\
6x^2 + 2x
\end{array}
\qquad
\begin{array}{l}
\dfrac{3x^3}{x} = 3x^2 \\[2ex]
\\
3x^2(x - 2) = 3x^3 - 6x^2 \\
\text{Subtract: } 0x^2 - (-6x^2) = 6x^2. \\
\text{Bring down } 2x.
\end{array}
$$

In the next step, divide x into $6x^2$.

$$
\begin{array}{r}
3x^2 + 6x \\
x - 2\overline{)3x^3 + 0x^2 + 2x - 4} \\
\underline{3x^3 - 6x^2} \\
6x^2 + 2x \\
\underline{6x^2 - 12x} \\
14x - 4
\end{array}
\qquad
\begin{array}{l}
\dfrac{6x^2}{x} = 6x \\[2ex]
\\
\\
6x(x - 2) = 6x^2 - 12x \\
\text{Subtract: } 2x - (-12x) = 14x. \\
\text{Bring down } -4.
\end{array}
$$

Now divide x into $14x$.

$$\begin{array}{r} 3x^2 + 6x + 14 \\ x - 2\overline{)3x^3 + 0x^2 + 2x - 4} \\ \underline{3x^3 - 6x^2} \\ 6x^2 + 2x \\ \underline{6x^2 - 12x} \\ 14x - 4 \\ \underline{14x - 28} \\ 24 \end{array}$$

$\dfrac{14x}{x} = 14$

$14(x - 2) = 14x - 28$

Subtract: $-4 - (-28) = 24.$

The quotient is $3x^2 + 6x + 14$ with remainder 24. This result can also be written as

$$3x^2 + 6x + 14 + \frac{24}{x - 2}.$$

___

EXAMPLE 6 Dividing with a quadratic divisor

Divide $x^3 - 3x^2 + 3x + 2$ by $x^2 + 1$.

Solution Begin by writing $x^2 + 1$ as $x^2 + 0x + 1$.

$$\begin{array}{r} x - 3 \\ x^2 + 0x + 1\overline{)x^3 - 3x^2 + 3x + 2} \\ \underline{x^3 + 0x^2 + x} \\ -3x^2 + 2x + 2 \\ \underline{-3x^2 + 0x - 3} \\ 2x + 5 \end{array}$$

Critical Thinking

Two polynomials are divided and the remainder is 0. The divisor has degree m, the quotient has degree n, and the dividend has degree p. What is the relationship among m, n, and p?

The quotient is $x - 3$ with remainder $2x + 5$. This result can also be written as

$$x - 3 + \frac{2x + 5}{x^2 + 1}.$$

5.6 PUTTING IT ALL TOGETHER

In this section we discussed division of polynomials by monomials and other polynomials. The following table summarizes several important concepts related to division of polynomials.

Concept	Explanation	Examples
Division by a Monomial	Use the property $$\frac{a + b}{d} = \frac{a}{d} + \frac{b}{d}.$$ Be sure to divide the denominator into every term in the numerator.	$\dfrac{2x^3 + 4x}{2x^2} = \dfrac{2x^3}{2x^2} + \dfrac{4x}{2x^2} = x + \dfrac{2}{x}$ and $\dfrac{a^2 - 2a}{4a} = \dfrac{a^2}{4a} - \dfrac{2a}{4a} = \dfrac{a}{4} - \dfrac{1}{2}$

continued on next page

continued from previous page

Concept	Explanation	Examples
Division by a Polynomial	Is done similarly to how long division of natural numbers is performed	Divide $x^2 + 3x + 3$ by $x + 1$. $$\begin{array}{r} x + 2 \\ x + 1 \overline{)x^2 + 3x + 3} \\ \underline{x^2 + x} \\ 2x + 3 \\ \underline{2x + 2} \\ 1 \end{array}$$ The quotient is $x + 2$ with remainder 1, which can be expressed as $$x + 2 + \frac{1}{x + 1}.$$
Checking a Result	Dividend $=$ (Divisor)(Quotient) $+$ Remainder	When $x^2 + 3x + 3$ is divided by $x + 1$, the quotient is $x + 2$ with remainder 1. Thus $$(x + 1)(x + 2) + 1 = x^2 + 3x + 3,$$ and the answer checks.

5.6 EXERCISES

FOR EXTRA HELP

📖 Student's Solutions Manual

🚪 MyMathLab

InterAct Math

Tutor Center Math Tutor Center

MathXL

Digital Video Tutor
CD 3   Videotape 7

CONCEPTS

1. $\frac{a + b}{d} =$ _____

2. $\frac{a + b - c}{d} =$ _____

3. Are the expressions $\frac{5x^2 + 2x}{2x}$ and $5x^2 + 1$ equal?

4. Are the expressions $\frac{5x^2 + 2x}{2x}$ and $\frac{5x^2}{2x}$ equal?

5. Because $\frac{37}{9} = 4$ with remainder 1, it follows that $37 =$ _____ $\cdot$ _____ $+$ _____.

6. Because $2x^3 - x + 5$ divided by $x + 1$ equals $2x^2 - 2x + 1$ with remainder 4, it follows that $2x^3 - x + 5 =$ _____ $\cdot$ _____ $+$ _____.

DIVISION BY A MONOMIAL

Exercises 7–12: Divide and check.

7. $\frac{6x^2}{3x}$

8. $\frac{-5x^2}{10x^4}$

9. $\frac{z^4 + z^3}{z}$

10. $\frac{t^3 - t}{t}$

11. $\frac{a^5 - 6a^3}{2a^3}$

12. $\frac{b^4 - 4b}{4b^2}$

Exercises 13–20: Divide.

13. $\dfrac{4x - 7x^4}{x^2}$

14. $\dfrac{1 + 6x^4}{3x^3}$

15. $\dfrac{9x^4 - 3x + 6}{3x}$

16. $\dfrac{y^3 - 4y + 6}{y}$

17. $\dfrac{12y^4 - 3y^2 + 6y}{3y^2}$

18. $\dfrac{2x^2 - 6x + 9}{12x}$

19. $\dfrac{15m^4 - 10m^3 + 20m^2}{5m^2}$

20. $\dfrac{n^8 - 8n^6 + 4n^4}{2n^5}$

Exercises 21–26: Divide and check.

21. $\dfrac{2x^2 - 3x + 1}{x - 2}$

22. $\dfrac{4x^2 - x + 3}{x + 2}$

23. $\dfrac{x^2 + 2x + 1}{x + 1}$

24. $\dfrac{4x^2 - 4x + 1}{2x - 1}$

25. $\dfrac{x^3 - x^2 + x - 2}{x - 1}$

26. $\dfrac{2x^3 + 3x^2 + 3x - 1}{2x + 1}$

Exercises 27–38: Divide.

27. $\dfrac{4x^3 - 3x^2 + 7x + 3}{4x + 1}$

28. $\dfrac{10x^3 - x^2 - 17x - 7}{5x + 2}$

29. $\dfrac{x^3 - x + 2}{x - 2}$

30. $\dfrac{6x^3 + 8x^2 + 4}{3x + 4}$

31. $(3x^3 + 2) \div (x - 1)$

32. $(-3x^3 + 8x^2 + x) \div (3x + 4)$

33. $(x^3 + 3x^2 + 1) \div (x^2 + 1)$

34. $(x^4 - x^3 + x^2 - x + 1) \div (x^2 - 1)$

35. $\dfrac{x^3 + 1}{x^2 - x + 1}$

36. $\dfrac{4x^3 + 3x + 2}{2x^2 - x + 1}$

37. $\dfrac{x^3 + 8}{x + 2}$

38. $\dfrac{x^4 - 16}{x - 2}$

APPLICATIONS

39. *Area of a Rectangle*   The area A of a rectangle is $8x^2$, and one of its sides has length $2x$. Find an expression for the length L of the other side.

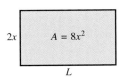

40. *Area of a Rectangle*   The area A of a rectangle is $x^2 - 1$, and one of its sides has length $x + 1$. Find the width W of the other side.

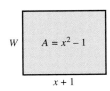

41. *Volume of a Box*   The volume V of a box is $2x^3 + 4x^2$, and the area of its bottom is $2x^2$. Find the height of the box in terms of x. Make a possible sketch of the box, and label the length of each side.

42. *Area of a Triangle*   A triangle has height h and area $A = 2h^2 - 4h$. Find its base b in terms of h. Make a possible sketch of the triangle, and label the height and base. (*Hint:* $A = \frac{1}{2}bh$.)

WRITING ABOUT MATHEMATICS

43. Suppose that one polynomial is divided into another polynomial and the remainder is 0. What does the product of the divisor and quotient equal? Explain.

44. A student simplifies the expression $\frac{4x^3 - 1}{4x^2}$ to $x - 1$. Explain the student's error.

CHECKING BASIC CONCEPTS ⟨ SECTIONS 5.5 AND 5.6 ⟩

---

1. Simplify each expression. Write the result with positive exponents.

   (a) 9^{-2} **(b)** $\dfrac{3x^{-3}}{6x^4}$ **(c)** $(4ab^{-4})^{-2}$

2. Simplify each expression. Write the result with positive exponents.

   (a) $\dfrac{1}{z^{-5}}$ **(b)** $\dfrac{x^{-3}}{y^{-6}}$ **(c)** $\left(\dfrac{3}{x^2}\right)^{-3}$

3. Write each expression in scientific notation.

   (a) 45,000 **(b)** 0.000234 **(c)** 0.01

4. Write each expression in standard form.

   (a) 4.71×10^4 **(b)** 6×10^{-3}

5. Simplify $\dfrac{25a^4 - 15a^3}{5a^3}$.

6. Divide $3x^2 - x - 4$ by $x - 1$. State the quotient and remainder.

7. *Distance to the Sun* The distance to the sun is approximately 93 million miles.
   (a) Write this distance in scientific notation.
   (b) Light travels at 1.86×10^5 miles per second. How long does it take for the sun's light to reach Earth?

CHAPTER

5 Summary

Section 5.1   *Rules for Exponents*

Bases and Exponents The expression b^n has base b, exponent n, and equals the expression $\underbrace{b \cdot b \cdot b \cdot \cdots \cdot b}_{n \text{ times}}$.

Example: 2^3 has base 2, exponent 3, and equals $2 \cdot 2 \cdot 2 = 8$.

Evaluating Expressions   1. Evaluate exponents.
2. Perform negation.
3. Do multiplication and division from left to right.
4. Do addition and subtraction from left to right.

Example: $-3^2 + 3 \cdot 4 = -9 + 3 \cdot 4 = -9 + 12 = 3$

Zero Exponents   For any nonzero number b, $b^0 = 1$. Note that 0^0 is undefined.

Examples: $5^0 = 1$ and $\left(\dfrac{x}{y}\right)^0 = 1$, where x and y are nonzero.

Product Rule   For any real number a and natural numbers m and n,

$$a^m \cdot a^n = a^{m+n}.$$

Examples: $3^4 \cdot 3^2 = 3^6$ and $x^3 x^2 x^4 = x^9$

Power Rules   For any real numbers a and b and natural numbers m and n,

$$(a^m)^n = a^{mn}, \quad (ab)^n = a^n b^n, \quad \text{and} \quad \left(\frac{a}{b}\right)^n = \frac{a^n}{b^n}, b \neq 0.$$

Examples: $(x^2)^3 = x^6, \quad (3x)^4 = 3^4 x^4 = 81x^4, \quad \text{and} \quad \left(\frac{2}{y}\right)^3 = \frac{2^3}{y^3} = \frac{8}{y^3}$

Section 5.2   *Addition and Subtraction of Polynomials*

Terms Related to Polynomials

Monomial	A number, variable, or product of numbers and variables raised to natural number powers
Degree of a Monomial	Sum of the exponents of the variables
Coefficient of a Monomial	The number in a monomial

Example: The monomial $-3x^2 y^3$ has degree 5 and coefficient -3.

Polynomial	A monomial or a sum of monomials
Term of a Polynomial	Each monomial is a term of the polynomial
Binomial	A polynomial with two terms
Trinomial	A polynomial with three terms
Degree of a Polynomial	The degree of the term with highest degree
Opposite of a Polynomial	The opposite is found by negating each term.

Example: $2x^3 - 4x + 5$ is a trinomial with degree 3. Its opposite is $-2x^3 + 4x - 5$.

Like Terms	Two monomials with the same variables raised to the same powers

Examples: $3xy^2$ and $-xy^2$ are like terms.

$5x^3$ and $3x^3$ are like terms.

$5x^2$ and $5x$ are unlike terms.

Addition of Polynomials   Combine like terms, using the distributive property.

Example: $(2x^2 - 4x) + (-x^2 - x) = (2 - 1)x^2 + (-4 - 1)x$
$= x^2 - 5x$

Subtraction of Polynomials   Add the first polynomial to the opposite of the second polynomial.

Example: $(4x^4 - 5x) - (7x^4 + 6x) = (4x^4 - 5x) + (-7x^4 - 6x)$
$= (4 - 7)x^4 + (-5 - 6)x$
$= -3x^4 - 11x$

Section 5.3   *Multiplication of Polynomials*

Multiplication of Monomials   Use the commutative property and the product rule.

Examples: $-2x^3 \cdot 3x^2 = -2 \cdot 3 \cdot x^3 \cdot x^2 = -6x^5$

$(2xy^2)(3x^2y^3) = 2 \cdot 3 \cdot x \cdot x^2 \cdot y^2 \cdot y^3 = 6x^3y^5$

⌐——— Assumed exponent of 1

Distributive Properties

$$a(b + c) = ab + ac \quad \text{and} \quad a(b - c) = ab - ac$$

Examples: $4x(3x + 6) = 4x \cdot 3x + 4x \cdot 6 = 12x^2 + 24x$

$ab(a^2 - b^2) = ab \cdot a^2 - ab \cdot b^2 = a^3b - ab^3$

Multiplication of Monomials and Polynomials   Apply the distributive properties. Be sure to multiply every term in the polynomial by the monomial.

Example: $-2x^2(4x^2 - 5x - 3) = -8x^4 + 10x^3 + 6x^2$

Multiplication of Polynomials   The product of two polynomials may be found by multiplying every term in the first polynomial by every term in the second polynomial. Be sure to combine like terms.

Examples: $(x + 3)(2x - 5) = 2x^2 - 5x + 6x - 15$

$\qquad\qquad = 2x^2 + x - 15$

$(2x + 1)(x^2 - 5x + 2) = 2x^3 - 10x^2 + 4x + x^2 - 5x + 2$

$\qquad\qquad\qquad\qquad = 2x^3 - 9x^2 - x + 2$

Section 5.4   *Special Products*

Product of a Sum and Difference
$$(a + b)(a - b) = a^2 - b^2$$

Examples: $(x + 4)(x - 4) = x^2 - 16$

$(2r - 3t)(2r + 3t) = (2r)^2 - (3t)^2 = 4r^2 - 9t^2$

Squaring Binomials

$$(a + b)^2 = a^2 + 2ab + b^2 \quad \text{and} \quad (a - b)^2 = a^2 - 2ab + b^2$$

Examples: $(2x + 1)^2 = (2x)^2 + 2(2x)1 + 1^2 = 4x^2 + 4x + 1$

$(z^2 - 2)^2 = (z^2)^2 - 2z^2(2) + 2^2 = z^4 - 4z^2 + 4$

Cubing Binomials   To multiply $(a + b)^3$ write it as $(a + b)(a + b)^2$ and then expand $(a + b)^2$.

Example: $(x + 4)^3 = (x + 4)(x + 4)^2$

$\qquad\qquad = (x + 4)(x^2 + 8x + 16)$ Expand the binomial.

$\qquad\qquad = x^3 + 8x^2 + 16x + 4x^2 + 32x + 64$ Distributive property

$\qquad\qquad = x^3 + 12x^2 + 48x + 64$ Combine like terms.

Section 5.5   *Integer Exponents and the Quotient Rule*

Negative Integers as Exponents   For any nonzero real number a and positive integer n,

$$a^{-n} = \frac{1}{a^n}.$$

Example: $5^{-2} = \frac{1}{5^2} = \frac{1}{25}$ and $x^{-4} = \frac{1}{x^4}$

The Quotient Rule   For any nonzero real number a and integers m and n,

$$\frac{a^m}{a^n} = a^{m-n}.$$

Examples: $\frac{6^4}{6^2} = 6^2 = 36$ and $\frac{xy^3}{x^4y^2} = x^{-3}y^1 = \frac{y}{x^3}$

Other Rules   For any nonzero real numbers a and b and positive integers m and n,

$$\frac{1}{a^{-n}} = a^n, \quad \frac{a^{-n}}{b^{-m}} = \frac{b^m}{a^n}, \quad \text{and} \quad \left(\frac{a}{b}\right)^{-n} = \left(\frac{b}{a}\right)^n.$$

Examples: $\frac{1}{4^{-3}} = 4^3$, $\frac{x^{-3}}{y^{-2}} = \frac{y^2}{x^3}$,   and $\left(\frac{4}{5}\right)^{-2} = \left(\frac{5}{4}\right)^2$

Scientific Notation   A real number a written as $b \times 10^n$, where $1 \le |b| < 10$ and n is an integer.

Examples: $2.34 \times 10^3 = 2340$ Move decimal 3 places to the right.
$2.34 \times 10^{-3} = 0.00234$ Move decimal 3 places to the left.

Section 5.6   *Division of Polynomials*

Division of a Polynomial by a Monomial   Divide the monomial into *every* term of the polynomial.

Example: $\dfrac{5x^3 - 10x^2 + 15x}{5x} = \dfrac{5x^3}{5x} - \dfrac{10x^2}{5x} + \dfrac{15x}{5x} = x^2 - 2x + 3$

Division of a Polynomial by a Polynomial   Division of polynomials is performed similarly to long division of natural numbers.

Example: Divide $2x^3 + 4x^2 - 3x + 1$ by $x + 1$.

$$
\begin{array}{r}
2x^2 + 2x - 5 \\
x + 1 \overline{\smash{)}\,2x^3 + 4x^2 - 3x + 1} \\
\underline{2x^3 + 2x^2 } \\
2x^2 - 3x \\
\underline{2x^2 + 2x } \\
-5x + 1 \\
\underline{-5x - 5} \\
6
\end{array}
$$

The quotient is $2x^2 + 2x - 5$ with remainder 6, which can be written as

$$2x^2 + 2x - 5 + \frac{6}{x + 1}.$$

Review Exercises

SECTION 5.1

Exercises 1–6: Evaluate the expression.

1. 5^3
2. -3^4
3. $4(-2)^0$
4. $3 + 3^2 - 3^0$
5. $\dfrac{-5^2}{5}$
6. $\left(\dfrac{-5}{5}\right)^2$

Exercises 7–20: Simplify the expression.

7. $6^2 \cdot 6^3$
8. $10^5 \cdot 10^7$
9. $z^4 \cdot z^5$
10. $y^2 \cdot y \cdot y^3$
11. $5x^2 \cdot 6x^7$
12. $(ab^3)(a^3b)$
13. $(2^5)^2$
14. $(m^4)^5$
15. $(ab)^3$
16. $(x^2y^3)^4$
17. $(xy)^3(x^2y^4)^2$
18. $(a^2b^9)^0$
19. $(r - t)^4(r - t)^5$
20. $(a + b)^2(a + b)^4$

SECTION 5.2

Exercises 21 and 22: Identify the degree and coefficient of the monomial.

21. $6x^7$
22. $-x^2y^3$

Exercises 23–26: Determine whether the expression is a polynomial. If it is, state how many terms and variables the polynomial contains. Then state its degree.

23. $8y$
24. $8x^3 - 3x^2 + x - 5$
25. $a^2 + 2ab + b^2$
26. $\dfrac{1}{xy}$

27. Add the polynomials vertically.
$$3x^2 + 4x + 8$$
$$\underline{2x^2 - 5x - 5}$$

28. Write the opposite of $6x^2 - 3x - 7$.

Exercises 29–34: Simplify.

29. $(4x - 3) + (-x + 7)$
30. $(3x^2 - 1) - (5x^2 + 12)$
31. $(x^2 + 5x + 6) - (3x^2 - 4x + 1)$
32. $(a^3 + 4a^2) + (a^3 - 5a^2 + 7a)$
33. $(xy + y^2) + (4y^2 - 4xy)$
34. $(7x^2 + 2xy + y^2) - (7x^2 - 2xy + y^2)$

SECTION 5.3

Exercises 35–48: Multiply and simplify.

35. $-x^2 \cdot x^3$
36. $-(r^2t^3)(rt)$
37. $-3(2t - 5)$
38. $2y(1 - 6y)$
39. $6x^3(3x^2 + 5x)$
40. $-x(x^2 - 2x + 9)$
41. $-ab(a^2 - 2ab + b^2)$
42. $(a - 2)(a + 5)$
43. $(8x - 3)(x + 2)$
44. $(2x - 1)(1 - x)$
45. $(y^2 + 1)(2y + 1)$
46. $(y^2 - 1)(2y^2 + 1)$
47. $(z + 1)(z^2 - z + 1)$
48. $(4z - 3)(z^2 - 3z + 1)$

Exercises 49–50: Multiply the expression
    (a) *geometrically and*
    (b) *symbolically.*

49. $z(z + 1)$
50. $2x(x + 2)$

SECTION 5.4

Exercises 51–66: Multiply.

51. $(z + 2)(z - 2)$
52. $(5z - 9)(5z + 9)$
53. $(1 - 3y)(1 + 3y)$
54. $(5x + 4y)(5x - 4y)$
55. $(rt + 1)(rt - 1)$
56. $(2m^2 - n^2)(2m^2 + n^2)$

57. $(x + 1)^2$

58. $(4x + 3)^2$

59. $(y - 3)^2$

60. $(2y - 5)^2$

61. $(4 + a)^2$

62. $(4 - a)^2$

63. $(x^2 + y^2)^2$

64. $(xy - 2)^2$

65. $(z + 5)^3$

66. $(2z - 1)^3$

SECTION 5.5

Exercises 67–72: Simplify the expression.

67. 9^{-1}

68. 3^{-2}

69. $4^3 \cdot 4^{-2}$

70. $10^{-6} \cdot 10^3$

71. $\dfrac{1}{6^{-2}}$

72. $\dfrac{5^7}{5^9}$

Exercises 73–88: Simplify and write the expression with positive exponents.

73. z^{-2}

74. y^{-4}

75. $a^{-4} \cdot a^2$

76. $x^2 \cdot x^{-5} \cdot x$

77. $(2t)^{-2}$

78. $(ab^2)^{-3}$

79. $(xy)^{-2}(x^{-2}y)^{-1}$

80. $\dfrac{x^6}{x^2}$

81. $\dfrac{4x}{2x^4}$

82. $\dfrac{20x^5y^3}{30xy^6}$

83. $\left(\dfrac{a}{b}\right)^5$

84. $\dfrac{4}{t^{-4}}$

85. $\left(\dfrac{x}{3}\right)^{-3}$

86. $\dfrac{2}{(ab)^{-1}}$

87. $\left(\dfrac{x}{y}\right)^{-2}$

88. $\left(\dfrac{3u}{2v}\right)^{-1}$

Exercises 89–92: Write the expression in standard form.

89. 6×10^2

90. 5.24×10^4

91. 3.7×10^{-3}

92. 6.234×10^{-2}

Exercises 93–96: Write the number in scientific notation.

93. $10,000$

94. $56,100,000$

95. 0.000054

96. 0.001

Exercises 97 and 98: Evaluate the expression. Write the result in standard form.

97. $(4 \times 10^2)(6 \times 10^4)$

98. $\dfrac{8 \times 10^3}{4 \times 10^4}$

SECTION 5.6

Exercises 99–106: Divide and check.

99. $\dfrac{5x^2 + 3x}{3x}$

100. $\dfrac{6b^4 - 4b^2 + 2}{2b^2}$

101. $\dfrac{3x^2 - x + 2}{x - 1}$

102. $\dfrac{9x^2 - 6x - 2}{3x + 2}$

103. $\dfrac{4x^3 - 11x^2 - 7x - 1}{4x + 1}$

104. $\dfrac{2x^3 - x^2 - 1}{2x - 1}$

105. $\dfrac{x^3 - x^2 - x + 1}{x^2 + 1}$

106. $\dfrac{x^4 + 3x^3 + 8x^2 + 7x + 5}{x^2 + x + 1}$

APPLICATIONS

107. *Heart Rate*   An athlete starts running and continues for 10 seconds. The polynomial $t^2 + 60$ calculates the heart rate of the athlete in beats per minute t seconds after beginning to run.

   (a) What is the athlete's heart rate when the athlete first starts to run?

   (b) What is the athlete's heart rate after 10 seconds?

   (c) What happens to the athlete's heart rate while the athlete is running?

108. *Areas of Rectangles*   Find a monomial equal to the sum of the areas of the rectangles. Calculate this sum for $x = 3$ feet and $y = 4$ feet.

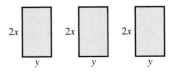

109. *Area of a Rectangle* Write a polynomial that gives the area of the rectangle. Calculate its area for $z = 6$ inches.

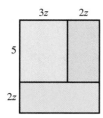

110. *Area of a Square* Find the area of the square whose sides have length x^2y.

111. *Compound Interest* If P dollars are deposited in an account that pays 6% annual interest, then the amount of money after 3 years is given by $P(1 + 0.06)^3$. Find this amount when $P = \$700$.

112. *Surface Area of a Box* Use the drawing of the box to answer the following.

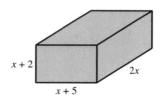

(a) Write a polynomial that represents the area of the bottom.
(b) Write a polynomial that represents the area of the front.
(c) Write a polynomial that represents the area of the right side.
(d) Write a polynomial that represents the area of the six sides of the box.

113. *Volume of a Sphere* The expression for the volume of a sphere with radius r is $\frac{4}{3}\pi r^3$. Find a polynomial that gives the volume of a sphere with radius $x + 2$. Leave your answer in terms of π.

114. *Height of a Baseball* A baseball is hit straight up. Its height h in feet above the ground after t seconds is given by $t(96 - 16t)$.
(a) Multiply this expression.
(b) Evaluate both the expression in part (a) and the given expression for $t = 2$. Interpret the result.

115. *Rectangular Building* A rectangular building has a perimeter of 1200 feet.
(a) If one side of the building has length L, write a polynomial expression that gives its area. (Be sure to multiply your expression.)
(b) Evaluate the expression in part (a) for $L = 50$ and interpret the answer.

116. *Geometry* Complete each part and verify that your answers are equal.
(a) Find the area of the large square by multiplying its length and width.
(b) Find the sum of the areas of the smaller rectangles inside the large square.

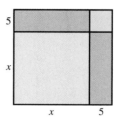

117. *Digital Picture* A digital picture, including its border, is $x + 4$ pixels by $x + 4$ pixels, and the actual picture inside the border is $x - 4$ pixels by $x - 4$ pixels.

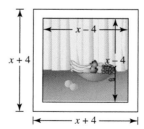

(a) Find a polynomial that gives the number of pixels in the border.
(b) Let $x = 100$ and evaluate the polynomial.

118. *World Population* If current trends continue, world population P in billions may be modeled by the equation $P = 6(1.014)^x$, where x represents the number of years after 2000. Estimate the world population in 2005. (***Source:*** United Nations Population Fund.)

119. *Federal Debt*   In 1990, the federal debt held by the public was \$2.19 trillion, and the population of the United States was 249 million. Use scientific notation to approximate the national debt per person. (*Source:* Department of the Treasury.)

120. *Alcohol Consumption*   In 1994, about 211 million people in the United States were aged 14 or older. They consumed, on average, 2.21 gallons of alcohol per person. Use scientific notation to estimate the total number of gallons of alcohol consumed by this age group.   (*Source:* Department of Health and Human Services.)

CHAPTER 5 Test

1. Is $5x^2 - 3xy - 7y^3$ a polynomial? If it is, state how many terms and variables the polynomial contains. Then state its degree.

2. Write the opposite of $-x^3 + 4x - 8$.

Exercises 3–5: Simplify.

3. $(-3x + 4) + (7x + 2)$

4. $(5x^2 - x + 3) - (4x^2 - 2x + 10)$

5. $(a^3 + 5ab) + (3a^3 - 3ab)$

6. Evaluate each expression by hand.
   (a) $-4^2 + 10$ **(b)** 8^{-2} **(c)** $\dfrac{1}{2^{-3}}$
   (d) $-3x^0$

Exercises 7–11: Write the given expression with positive exponents.

7. $x^7 \cdot x^{-3}$

8. $(a^{-1}b^2)^{-3}$

9. $ab(a^2 - b^2)$

10. $\left(\dfrac{3a^2}{2b^{-3}}\right)^{-2}$

11. $\dfrac{12xy^4}{6x^2y}$

Exercises 12–16: Multiply and simplify.

12. $3x^2(4x^3 - 6x + 1)$ **13.** $(z - 3)(2z + 4)$

14. $(7y^2 - 3)(7y^2 + 3)$ **15.** $(3x - 2)^2$

16. $(m + 3)^3$

17. Write 6.1×10^{-3} in standard form.

18. Write 5410 in scientific notation.

Exercises 19–20: Divide.

19. $\dfrac{9x^3 - 6x^2 + 3x}{3x^2}$ **20.** $\dfrac{x^3 + x^2 - x + 1}{x + 2}$

21. *Concert Tickets*   Tickets for a concert are sold for \$20 each.
   (a) Write a polynomial that gives the revenue from selling t tickets.
   (b) Putting on the concert costs management \$2000 to hire the band plus \$2 for each ticket sold. What is the total cost of the concert if t tickets are sold?
   (c) Subtract the polynomial that you found in part (b) from the polynomial that you found in part (a). What does this polynomial represent?

22. *Areas of Rectangles*   Find a polynomial representing the sum of the areas of two identical rectangles that have width $2x$ and length $3x$. Calculate this sum for $x = 10$ feet.

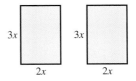

23. *Volume of a Box*   Write a polynomial that represents the volume of the box. Be sure to multiply your answer completely.

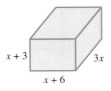

24. *Height of a Baseball* A baseball is hit straight up. Its height in feet above the ground after t seconds is given by $t(88 - 16t)$.

(a) Multiply this expression.

(b) Evaluate the expression in part (a) for $t = 3$. Interpret the result.

CHAPTER 5

Extended and Discovery Exercises

Exercises 1–6: Arithmetic and Scientific Notation *The product* $(4 \times 10^3) \times (2 \times 10^2)$ *can be evaluated as*

$$(4 \times 2) \times (10^3 \times 10^2) = 8 \times 10^5,$$

and the quotient $(4 \times 10^3) \div (2 \times 10^2)$ *can be evaluated as*

$$\frac{4 \times 10^3}{2 \times 10^2} = \frac{4}{2} \times \frac{10^3}{10^2} = 2 \times 10^1.$$

How would you evaluate the sum $(4 \times 10^3) + (2 \times 10^2)$? *How would you evaluate the difference* $(4 \times 10^3) - (2 \times 10^2)$? *Make a conjecture as to how numbers in scientific notation should be added and subtracted. Try your method on these problems and then check your answers with a calculator set in scientific mode. Does your method work?*

1. $(4 \times 10^3) + (3 \times 10^3)$

2. $(5 \times 10^{-2}) - (2 \times 10^{-2})$

3. $(1.2 \times 10^4) - (3 \times 10^3)$

4. $(2 \times 10^2) + (6 \times 10^1)$

5. $(2 \times 10^{-1}) + (4 \times 10^{-2})$

6. $(2 \times 10^{-3}) - (5 \times 10^{-2})$

Exercises 7–8: Constructing a Box *A box is constructed from a rectangular piece of metal by cutting squares from the corners and folding up the sides. The square, cutout corners are x inches by x inches.*

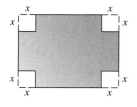 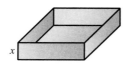

7. Suppose that the dimensions of the metal piece are 20 inches by 30 inches.

(a) Write a polynomial that gives the volume of the box.

(b) Find the volume of the the box for $x = 4$ inches.

8. Suppose that the metal piece is square with sides of length 25 inches.

(a) Write a polynomial expression that gives the outside surface area of the box. (Assume that the box does not have a top.)

(b) Find this area for $x = 3$ inches.

 Exercises 9–12: Calculators and Polynomials *A graphing calculator can be used to help determine whether two polynomial expressions in one variable are equal. For example, suppose that a student believes that $(x + 2)^2$ and $x^2 + 4$ are equal. Then the first two calculator tables shown demonstrate that the two expressions are not equal except for x = 0.*

X	Y1
-3	1
-2	0
-1	1
0	4
1	9
2	16
3	25
Y1■(X+2)2	

X	Y1
-3	13
-2	8
-1	5
0	4
1	5
2	8
3	13
Y1■X2+4	

The next two calculator tables support the fact that $(x + 1)^2$ and $x^2 + 2x + 1$ are equal for all x.

X	Y1
-3	4
-2	1
-1	0
0	1
1	4
2	9
3	16
Y1■(X+1)2	

X	Y1
-3	4
-2	1
-1	0
0	1
1	4
2	9
3	16
Y1■X2+2X+1	

Use a graphing calculator to determine whether the first expression is equal to the second expression. If the expressions are not equal, multiply the first expression and simplify it.

9. $3x(4 - 5x)$, $12x - 5x$

10. $(x - 1)^2$, $x^2 - 1$

11. $(x - 1)(x^2 + x + 1)$, $x^3 - 1$

12. $(x - 2)^3$, $x^3 - 8$

CHAPTER 6

Factoring Polynomials and Solving Equations

Have you ever noticed that cars are designed so that their exteriors are curved and smooth? This characteristic is especially true for solar cars. In fact, a side view of a solar car often resembles the cross section of an airplane wing, as illustrated in the accompanying figure. The reason for this design is to reduce drag from air resistance. Reducing drag increases fuel efficiency and gives the car a smoother ride in the wind.

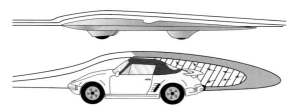

Views of a Solar Car and a Standard Car

Mathematics plays an important role in the design of cars. Cubic polynomials called *cubic splines* are used extensively by engineers to obtain a smooth shape for new cars. Although the topic of cubic splines is covered in advanced mathematics and engineering courses, in this chapter we introduce many of the concepts necessary for understanding how cars are designed.

What we see depends mainly on what we look for.
—John Lubbock

Source: R. Burden and J. Faires. *Numerical Analysis.*

6.1 INTRODUCTION TO FACTORING

Common Factors · Factoring by Grouping

INTRODUCTION

Polynomials are frequently used in applications to approximate data. As a result, scientists and mathematicians commonly solve equations involving polynomials. One way to solve these equations is to use **factoring**. When factoring a polynomial, we usually write it as a *product* of lower degree polynomials. For example, because $x(x - 1) = x^2 - x$, we say that the polynomial $x^2 - x$ can be *factored* as $x(x - 1)$. Note that $x^2 - x$ is a polynomial of degree 2, whereas x and $x - 1$ are both degree 1 polynomials. In this section we introduce two basic methods of factoring polynomials.

COMMON FACTORS

When factoring a polynomial, we first look for factors that are common to each term. By applying a distributive property we can often write a polynomial as a product. For example, each term in the polynomial $8x^2 + 6x$ has a factor of $2x$ because

$$8x^2 = 2x \cdot 4x \quad \text{and} \quad 6x = 2x \cdot 3.$$

Therefore by the distributive property,

$$8x^2 + 6x = 2x(4x + 3).$$

Thus the product $2x(4x + 3)$ equals $8x^2 + 6x$. We check this result by multiplying.

$$2x(4x + 3) = 2x \cdot 4x + 2x \cdot 3$$
$$= 8x^2 + 6x \qquad \text{It checks.}$$

This factorization is shown visually in Figure 6.1, where possible dimensions for a rectangle with an area of $8x^2 + 6x$ are $2x$ by $4x + 3$.

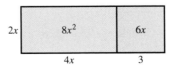

Figure 6.1 $8x^2 + 6x = 2x(4x + 3)$

The expressions $8x^2 + 6x$ and $2x(4x + 3)$ are equal for all values of x. Figure 6.2 also illustrates this fact with a table of values for each expression.

Calculator Help
To make a table of values, see Appendix A (page AP-3).

X	Y₁
-3	54
-2	20
-1	2
0	0
1	14
2	44
3	90

Y₁≣8X^2+6X

(a)

X	Y₁
-3	54
-2	20
-1	2
0	0
1	14
2	44
3	90

Y₁≣2X(4X+3)

(b)

Figure 6.2

EXAMPLE 1 **Finding common factors**

Factor.

(a) $15x^2 + 10x$ **(b)** $6y^3 - 2y^2$ **(c)** $3z^3 + 9z^2 - 6z$ **(d)** $2x^2y^2 + 4xy^3$

Solution **(a)** In the expression $15x^2 + 10x$, the terms $15x^2$ and $10x$ both contain a common factor of $5x$ because

$$15x^2 = 5x \cdot 3x \quad \text{and} \quad 10x = 5x \cdot 2.$$

Therefore this polynomial can be factored as

$$15x^2 + 10x = 5x(3x + 2).$$

(b) In the expression $6y^3 - 2y^2$, the terms $6y^3$ and $2y^2$ both contain a common factor of $2y^2$ because

$$6y^3 = 2y^2 \cdot 3y \quad \text{and} \quad 2y^2 = 2y^2 \cdot 1.$$

Therefore this polynomial can be factored as

$$6y^3 - 2y^2 = 2y^2(3y - 1).$$

(c) In the expression $3z^3 + 9z^2 - 6z$, the terms $3z^3$, $9z^2$, and $6z$ all contain a common factor of $3z$ because

$$3z^3 = 3z \cdot z^2, \quad 9z^2 = 3z \cdot 3z, \quad \text{and} \quad 6z = 3z \cdot 2.$$

Therefore this polynomial can be factored as

$$3z^3 + 9z^2 - 6z = 3z(z^2 + 3z - 2).$$

(d) In the expression $2x^2y^2 + 4xy^3$, the terms $2x^2y^2$ and $4xy^3$ both contain a common factor of $2xy^2$ because

$$2x^2y^2 = 2xy^2 \cdot x \quad \text{and} \quad 4xy^3 = 2xy^2 \cdot 2y.$$

Thus $2x^2y^2 + 4xy^3 = 2xy^2(x + 2y)$.

═══════════ MAKING CONNECTIONS ═══════════

Checking Common Factors with Multiplication

When factoring we can check our work by multiplying. For example, if we are uncertain whether the equation

$$6y^3 - 2y^2 = 2y^2(3y - 1)$$

is correct, we can apply the distributive property to the right side to obtain

$$2y^2(3y - 1) = 2y^2 \cdot 3y - 2y^2 \cdot 1$$
$$= 6y^3 - 2y^2. \qquad \text{It checks.}$$

In most situations we factor out the *greatest common factor* (GCF). For example, the polynomial $12b^3 + 8b^2$ has a common factor of $2b$. We could factor this polynomial as

$$12b^3 + 8b^2 = 2b(6b^2 + 4b).$$

However, we can factor out $4b^2$ instead.

$$12b^3 + 8b^2 = 4b^2(3b + 2)$$

Because $4b^2$ is the common factor with the greatest coefficient and highest degree, we say that $4b^2$ is the **greatest common factor** of $12b^3 + 8b^2$. In Example 1 we factored out the greatest common factor for each expression.

EXAMPLE 2 Finding the greatest common factor

Find the greatest common factor for each expression.
  (a) $9x^2 + 6x$ **(b)** $4z^4 + 8z^2$ **(c)** $8a^2b^3 - 16a^3b^2$

Solution **(a)** Because

$$9x^2 = 3 \cdot 3 \cdot x \cdot x \quad \text{and}$$
$$6x = 3 \cdot 2 \cdot x,$$

both terms have common factors of **3** and x. The GCF is the product of these two factors, or $3 \cdot x = 3x$.

(b) Because

$$4z^4 = 2 \cdot 2 \cdot z \cdot z \cdot z \cdot z \quad \text{and}$$
$$8z^2 = 2 \cdot 2 \cdot 2 \cdot z \cdot z,$$

both terms have common factors of **2**, **2**, z, and z. The GCF is the product of these four factors, or $2 \cdot 2 \cdot z \cdot z = 4z^2$.

(c) Because

$$8a^2b^3 = 2 \cdot 2 \cdot 2 \cdot a \cdot a \cdot b \cdot b \cdot b \quad \text{and}$$
$$16a^3b^2 = 2 \cdot 2 \cdot 2 \cdot 2 \cdot a \cdot a \cdot a \cdot b \cdot b,$$

both terms have common factors of **2**, **2**, **2**, a, a, b, and b. The GCF is the product of these seven factors, or $2 \cdot 2 \cdot 2 \cdot a \cdot a \cdot b \cdot b = 8a^2b^2$.

Note: With practice, you may find that you can determine the GCF mentally without factoring each term as was done in Example 2.

In the next example, we factor an expression that occurs in a scientific application.

EXAMPLE 3 Modeling the flight of a golf ball

If a golf ball is hit upward at 66 feet per second (45 miles per hour), then its height in feet after t seconds is approximated by $66t - 16t^2$. Factor this expression.

Solution Both $66t$ and $16t^2$ contain a common factor of $2t$ because

$$66t = 2t \cdot 33 \quad \text{and} \quad 16t^2 = 2t \cdot 8t.$$

Therefore this polynomial can be factored as

$$66t - 16t^2 = 2t(33 - 8t).$$

FACTORING BY GROUPING

Factoring by grouping is a technique that makes use of the associative and distributive properties. The next example illustrates one step in this factoring technique.

EXAMPLE 4 Factoring out binomials

Factor.
(a) $5x(x + 3) + 6(x + 3)$ **(b)** $x^2(2x - 5) - 4x(2x - 5)$

Solution **(a)** Each term in the expression $5x(x + 3) + 6(x + 3)$ contains the binomial $(x + 3)$. Therefore the distributive property can be used to factor this expression.

$$5x(x + 3) + 6(x + 3) = (5x + 6)(x + 3)$$

(b) Each term in the expression $x^2(2x - 5) - 4x(2x - 5)$ contains the binomial $(2x - 5)$. Therefore the distributive property can be used to factor this expression.

$$x^2(2x - 5) - 4x(2x - 5) = (x^2 - 4x)(2x - 5)$$
$$= x(x - 4)(2x - 5)$$

Now consider the polynomial

$$x^3 + x^2 + 2x + 2.$$

We can factor this polynomial by first grouping it into two binomials.

$(x^3 + x^2) + (2x + 2)$ Associative property
$x^2(x + 1) + 2(x + 1)$ Factor out common factors.
$(x^2 + 2)(x + 1)$ Factor out $(x + 1)$.

When factoring by grouping, we factor out a common factor more than once.

EXAMPLE 5 Factoring by grouping

Factor each polynomial
(a) $2t^3 - 4t^2 + 3t - 6$ **(b)** $3x + 3y + ax + ay$
(c) $3y^3 - y^2 - 9y + 3$ **(d)** $z^3 + 4z^2 - 5z - 20$

Solution **(a)** $2t^3 - 4t^2 + 3t - 6 = (2t^3 - 4t^2) + (3t - 6)$ Associative property
$= 2t^2(t - 2) + 3(t - 2)$ Factor out common factors.
$= (2t^2 + 3)(t - 2)$ Factor out $(t - 2)$.
(b) $3x + 3y + ax + ay = (3x + 3y) + (ax + ay)$ Associative property
$= 3(x + y) + a(x + y)$ Factor out common factors.
$= (3 + a)(x + y)$ Factor out $(x + y)$.
(c) $3y^3 - y^2 - 9y + 3 = (3y^3 - y^2) + (-9y + 3)$ Associative property
$= y^2(3y - 1) - 3(3y - 1)$ Factor out common factors.
$= (y^2 - 3)(3y - 1)$ Factor out $(3y - 1)$.
(d) $z^3 + 4z^2 - 5z - 20 = (z^3 + 4z^2) + (-5z - 20)$ Associative property
$= z^2(z + 4) - 5(z + 4)$ Factor out common factors.
$= (z^2 - 5)(z + 4)$ Factor out $(z + 4)$.

6.1   PUTTING IT ALL TOGETHER

In this section we introduced basic concepts used to factor polynomials. They are summarized in the following table.

Concept	Explanation	Examples
Common Factor	Factor out a monomial common to each term in a polynomial.	$6z^2 - 6z = 6z(z - 1)$ $4y^3 - 6y^2 = 2y^2(2y - 3)$ $5x^3 - 10x^2 + 15x = 5x(x^2 - 2x + 3)$ $2a^3b^3 - 4a^2b^3 = 2a^2b^3(a - 2)$
Greatest Common Factor (GCF)	The common factor with the greatest coefficient and highest degree	The GCF of $10x^4 + 15x^2$ is $5x^2$. Common factors include 1, 5, x, $5x$, x^2, and $5x^2$. However, $5x^2$ is *the greatest* common factor.
Factoring by Grouping	Factoring by grouping is a method that can be used to factor *four terms* into a product of two binomials. It makes use of the associative and distributive properties.	$2x^3 + 3x^2 + 2x + 3$ $\quad = (2x^3 + 3x^2) + (2x + 3)$ $\quad = x^2(2x + 3) + 1(2x + 3)$ $\quad = (x^2 + 1)(2x + 3)$ $4x^3 - 24x^2 - 3x + 18$ $\quad = (4x^3 - 24x^2) + (-3x + 18)$ $\quad = 4x^2(x - 6) - 3(x - 6)$ $\quad = (4x^2 - 3)(x - 6)$

6.1   EXERCISES

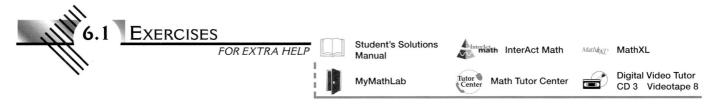

FOR EXTRA HELP

Student's Solutions Manual

MyMathLab

InterAct Math

Math Tutor Center

MathXL

Digital Video Tutor
CD 3   Videotape 8

CONCEPTS

1. When you write a polynomial as a product of two or more polynomials, it is called _____.

2. A common factor in the expression $ab + ac$ is _____.

3. The _____ of a polynomial is the common factor with the greatest coefficient and highest degree.

4. Factoring by _____ is a method that can be used to factor four terms into a product of two binomials by using the associative and distributive properties.

5. Identify four common factors of $2x^2 + 4x$.

6. Identify the greatest common factor (GCF) of the expression $2x^2 + 4x$.

COMMON FACTORS

Exercises 7–10: Factor the expression. Then make a sketch of a rectangle that illustrates this factorization.

7. $2x + 4$ 8. $6 + 3x$

9. $z^2 + 4z$ **10.** $2z^2 + 10z$

Exercises 11–28: Identify the greatest common factor. Then factor the expression.

11. $6x - 18x^2$ **12.** $16x^2 - 24x^3$

13. $8y^3 - 12y^2$ **14.** $12y^3 - 8y^2 + 4y$

15. $6z^3 + 3z^2 + 9z$ **16.** $16z^3 - 24z^2 - 36z$

17. $x^4 - 5x^3 - 4x^2$ **18.** $2x^4 + 8x^2$

19. $5y^5 + 10y^4 - 15y^3 + 10y^2$

20. $7y^4 - 14y^3 - 21y^2 + 7y$

21. $xy + xz$ **22.** $ab - bc$

23. $ab^2 - a^2b$ **24.** $4x^2y + 6xy^2$

25. $5x^2y^4 + 10x^3y^3$ **26.** $3r^3t^3 - 6r^4t^2$

27. $a^2b + ab^2 + ab$ **28.** $6ab^2 - 9ab + 12b^2$

FACTORING BY GROUPING

Exercises 29–36: Factor.

29. $x(x + 1) - 2(x + 1)$

30. $5x(3x - 2) + 2(3x - 2)$

31. $(z + 5)z + (z + 5)4$

32. $(4z + 3)2z^2 - (4z + 3)6z$

33. $y^2(y + 7) - 4y(y + 7)$

34. $3y^2(y - 2) + 5(y - 2)$

35. $4x^3(x - 5) + (x - 5)$

36. $8x^2(x + 3) + (x + 3)$

Exercises 37–56: Factor by grouping.

37. $x^3 + 2x^2 + 3x + 6$ **38.** $x^3 + 6x^2 + x + 6$

39. $2y^3 + y^2 + 2y + 1$ **40.** $4y^3 + 10y^2 + 2y + 5$

41. $2z^3 - 6z^2 + 5z - 15$ **42.** $15z^3 - 5z^2 + 6z - 2$

43. $4t^3 - 20t^2 + 3t - 15$ **44.** $4t^3 - 12t^2 + 3t - 9$

45. $9r^3 + 6r^2 - 6r - 4$ **46.** $3r^3 + 12r^2 - 2r - 8$

47. $7x^3 + 21x^2 - 2x - 6$ **48.** $6x^3 + 3x^2 - 10x - 5$

49. $2y^3 - 7y^2 - 4y + 14$ **50.** $y^3 - 5y^2 - 3y + 15$

51. $z^3 - 4z^2 - 7z + 28$

52. $12z^3 - 18z^2 - 10z + 15$

53. $2x^4 - 3x^3 + 4x - 6$ **54.** $x^4 + x^3 + 5x + 5$

55. $ax + bx + ay + by$ **56.** $ax - bx + ay - by$

APPLICATIONS

57. *Flight of a Golf Ball*   (Refer to Example 3.) The height of a golf ball in feet after t seconds is given by $80t - 16t^2$.
(a) Identify the greatest common factor.
(b) Factor this expression.

58. *Flight of a Golf Ball*   Repeat Exercise 57 if the height of a golf ball in feet after t seconds is given by $128t - 16t^2$.

59. *Volume of a Box*   A box is constructed by cutting out square corners of a rectangular piece of cardboard and folding up the sides. If the cutout corners have sides with length x, then the volume of the box is given by the polynomial $4x^3 - 60x^2 + 200x$.

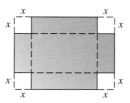 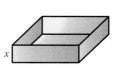

(a) Find the volume of the box when $x = 3$ inches.
(b) Factor out the greatest common factor for this expression.

60. *Volume of a Box*   (Refer to the preceding exercise.) A box is constructed from a square piece of metal that is 20 inches on a side.
(a) If the square corners of length x are cut out, write a polynomial that gives the volume of the box.
(b) Evaluate the polynomial when $x = 4$ inches.
(c) Factor out the greatest common factor for this polynomial expression.

WRITING ABOUT MATHEMATICS

61. Use an example to explain the difference between a common factor and the greatest common factor.

62. Use an example to explain how to factor a polynomial by grouping. What two properties of real numbers did you use?

6.2 FACTORING TRINOMIALS I ($x^2 + bx + c$)

Review of the FOIL Method · Factoring Trinomials Having a Leading Coefficient of 1

INTRODUCTION

In Section 6.1 we discussed two types of factoring: common factors and grouping. In this section we introduce methods for factoring certain types of trinomials. These techniques are frequently used in mathematics classes that you might take in the future. We use factoring later in this chapter to solve equations.

REVIEW OF THE FOIL METHOD

A **trinomial** is a polynomial that has three terms. We begin by reviewing products of binomials that result in trinomials.

$$(x + 2)(x + 3) = x \cdot x + x \cdot 3 + 2 \cdot x + 2 \cdot 3$$
$$= x^2 + 5x + 6$$

Note that the first term, x^2, in the trinomial results from multiplying the *first* terms of each binomial. The middle term, $5x$, results from adding the product of the *outside* terms and the product of the *inside* terms. Finally the last term, 6, results from multiplying the *last* terms of each binomial. We discussed this method of multiplying binomials, called FOIL, in Section 5.3 and illustrate it as follows.

$$(x + 2)(x + 3) = x^2 + 5x + 6$$

$$\begin{aligned} 2x \\ +3x \\ \overline{5x} \end{aligned}$$ The middle term checks.

FACTORING TRINOMIALS HAVING A LEADING COEFFICIENT OF 1

Any trinomial of degree 2 in the variable x can be written in *standard form* as $ax^2 + bx + c$, where a, b, and c are constants. The constant a is called the **leading coefficient** of the trinomial. In this section we focus on trinomials where $a = 1$.

Recall that the binomials $(x + m)$ and $(x + n)$ are multiplied as follows.

$$(x + m)(x + n) = x^2 + nx + mx + mn$$
$$= x^2 + (m + n)x + mn$$

Note that the coefficient of the x-term is the sum of m and n and that the constant (or third) term is the product of m and n. Thus to factor a trinomial in the form $x^2 + bx + c$, we start by finding two numbers, m and n, such that $m \cdot n = c$ and $m + n = b$. We illustrate this statement with $x^2 + 6x + 8$.

Standard Form	*Example*
$ax^2 + bx + c$	$x^2 + 6x + 8$
$m \cdot n = c$	$m \cdot n = 8$
$m + n = b$	$m + n = 6$

To determine possible values for m and n, we list factor pairs for 8 and search for a pair whose sum is 6, as in Table 6.1.

TABLE 6.1 Factor Pairs for 8

Factors	1, 8	2, 4
Sum	9	6

Because $2 \cdot 4 = 8$ and $2 + 4 = 6$, we let $m = 2$ and $n = 4$. We then factor the trinomial as

$$x^2 + 6x + 8 = (x + 2)(x + 4).$$

Note that, if you can find this factor pair mentally, making a table is not necessary.
We check the result by multiplying the two binomials.

$$(x + 2)\ (x + 4) = x^2 + 6x + 8$$

$2x$

$+4x$

$6x$ ——— The middle term checks.

FACTORING $x^2 + bx + c$

To factor the trinomial $x^2 + bx + c$, find two numbers m and n that satisfy

$$m \cdot n = c \quad \text{and} \quad m + n = b.$$

Then $x^2 + bx + c = (x + m)(x + n)$.

EXAMPLE 1 **Factoring a trinomial having only positive coefficients**

Factor each trinomial.
(a) $x^2 + 7x + 12$ **(b)** $x^2 + 13x + 30$ **(c)** $z^2 + 9z + 20$

Solution **(a)** To factor $x^2 + 7x + 12$ we need to find a factor pair for 12 whose sum is 7. To do so we make Table 6.2.

TABLE 6.2 Factor Pairs for 12

Factors	1, 12	2, 6	3, 4
Sum	13	8	7

The required factor pair is **3** and **4** because $3 \cdot 4 = 12$ and $3 + 4 = 7$. Therefore the given trinomial can be factored as

$$x^2 + 7x + 12 = (x + 3)(x + 4).$$

(b) To factor $x^2 + 13x + 30$ we need to find a factor pair for 30 whose sum is 13. The required pair is 3 and 10. Thus

$$x^2 + 13x + 30 = (x + 3)(x + 10).$$

(c) To factor $z^2 + 9z + 20$ we need to find a factor pair for 20 whose sum is 9. The required pair is 4 and 5. Thus

$$z^2 + 9z + 20 = (z + 4)(z + 5).$$

In the next example, the coefficients of the middle terms are negative.

EXAMPLE 2 Factoring trinomials having a negative middle coefficient

Factor each trinomial.
(a) $x^2 - 7x + 10$ (b) $x^2 - 8x + 15$ (c) $y^2 - 9y + 18$

Solution (a) To factor $x^2 - 7x + 10$ we need to find a factor pair for 10 whose sum equals -7. To have a positive product *and* a negative sum, *both* numbers must be negative, as shown in Table 6.3.

TABLE 6.3   Factor Pairs for 10

Factors	$-1, -10$	$-2, -5$
Sum	-11	-7

The required pair is -2 and -5 because $-2 \cdot (-5) = 10$ and $-2 + (-5) = -7$. Therefore the given trinomial can be factored as

$$x^2 - 7x + 12 = (x - 2)(x - 5).$$

(b) To factor $x^2 - 8x + 15$ we need to find a factor pair for 15 whose sum is -8. The required pair is -3 and -5. Thus

$$x^2 - 8x + 15 = (x - 3)(x - 5).$$

(c) To factor $y^2 - 9y + 18$ we need to find a factor pair for 18 whose sum is -9. The required pair is -3 and -6. Thus

$$y^2 - 9y + 18 = (y - 3)(y - 6).$$

In Examples 1 and 2 the coefficient of the last term was always positive. In the next example, this coefficient is negative and the coefficient of the middle term is either positive or negative.

EXAMPLE 3 Factoring trinomials having a negative constant term

Factor each trinomial.
(a) $x^2 - 3x - 4$ (b) $x^2 + 7x - 8$ (c) $t^2 - 2t - 24$

Solution (a) To factor $x^2 - 3x - 4$ we need to find a factor pair for -4 whose sum is -3. To have a negative product one factor must be positive and the other factor must be negative, as shown in Table 6.4.

TABLE 6.4   Factor Pairs for -4

Factors	$-1, 4$	$1, -4$	$-2, 2$
Sum	3	-3	0

The required pair is 1 and -4 because $1 \cdot (-4) = -4$ and $1 + (-4) = -3$. Therefore the given trinomial can be factored as

$$x^2 - 3x - 4 = (x + 1)(x - 4),$$

which can be checked by multiplying $(x + 1)(x - 4)$.

(b) To factor $x^2 + 7x - 8$ we need to find a factor pair for -8 whose sum is 7. The required pair is -1 and 8. Thus

$$x^2 + 7x - 8 = (x - 1)(x + 8).$$

(c) To factor $t^2 - 2t - 24$ we need to find a factor pair for -24 whose sum is -2. The required pair is -6 and 4. Thus

$$t^2 - 2t - 24 = (t - 6)(t + 4).$$

Critical Thinking

A cube has a surface area of $6x^2 + 24x + 24$. What is the length of each side?

EXAMPLE 4   **Finding the dimensions of a rectangle**

Find one possibility for the dimensions of a rectangle that has an area of $x^2 + 6x + 5$.

Solution   The area of a rectangle equals length times width. If we can factor $x^2 + 6x + 5$, then the factors can represent its length and width. Because

$$x^2 + 6x + 5 = (x + 1)(x + 5),$$

one possibility for the rectangle's dimensions is width $x + 1$ and length $x + 5$, as illustrated in Figure 6.3.

Figure 6.3   Area $= x^2 + 6x + 5$

6.2   PUTTING IT ALL TOGETHER

In this section we discussed factoring trinomials of the form $x^2 + bx + c$. The following table summarizes these techniques.

Concept	Explanation	Examples
Factoring Trinomials of the Form $x^2 + bx + c$	Find two numbers m and n that satisfy $mn = c$ and $m + n = b$. Then $$x^2 + bx + c = (x + m)(x + n)$$	$x^2 + 9x + 20 = (x + 4)(x + 5)$ because $4 \cdot 5 = 20$ and $4 + 5 = 9$. $x^2 + x - 6 = (x - 2)(x + 3)$ because $-2 \cdot 3 = -6$ and $-2 + 3 = 1$. $x^2 - 8x + 12 = (x - 6)(x - 2)$ because $-6 \cdot (-2) = 12$ and $-6 + (-2) = -8$.

6.2 EXERCISES

FOR EXTRA HELP

 Student's Solutions Manual

 InterAct Math

 MathXL

MyMathLab

 Math Tutor Center

Digital Video Tutor
CD 3   Videotape 8

CONCEPTS

1. Explain what each letter represents in the word FOIL.

2. Multiply $(x + m)(x + n)$. What is the coefficient of the x-term? What is the constant term?

3. To factor $x^2 + bx + c$, start by finding two numbers m and n that satisfy _____ = c and _____ = b.

4. Find two integers m and n such that $mn = 24$ and $m + n = 11$.

5. List all positive integer pairs that have a product of 12.

6. List all negative integer pairs that have a product of 30.

Exercises 7–14: Find the integer pair that has the given product and sum.

7. Product: 28   Sum: 11

8. Product: 35   Sum: 12

9. Product: 30   Sum: 13

10. Product: 100   Sum: 29

11. Product: -50 Sum: 5

12. Product: -15 Sum: -2

13. Product: 28   Sum: -11

14. Product: 80   Sum: -42

FACTORING TRINOMIALS

Exercises 15–54: Factor each trinomial.

15. $x^2 + 3x + 2$

16. $x^2 + 5x + 4$

17. $y^2 + 4y + 4$

18. $y^2 + 8y + 7$

19. $z^2 + 6z + 9$

20. $z^2 + 6z + 5$

21. $x^2 + 8x + 15$

22. $x^2 + 9x + 14$

23. $m^2 + 13m + 36$

24. $m^2 + 15m + 36$

25. $n^2 + 20n + 100$

26. $n^2 + 52n + 100$

27. $x^2 - 6x + 5$

28. $x^2 - 6x + 8$

29. $y^2 - 7y + 12$

30. $y^2 - 12y + 27$

31. $z^2 - 13z + 40$

32. $z^2 - 15z + 54$

33. $a^2 - 16a + 63$

34. $a^2 - 82a + 81$

35. $b^2 - 30b + 125$

36. $b^2 - 19b + 90$

37. $x^2 + 13x - 90$

38. $x^2 + 15x - 100$

39. $m^2 + 4m - 45$

40. $m^2 + 4m - 60$

41. $n^2 + 10n - 200$

42. $n^2 + 2n - 120$

43. $x^2 + 22x - 23$

44. $x^2 + 18x - 19$

45. $a^2 + 4a - 32$

46. $a^2 + 9a - 36$

47. $b^2 - b - 20$

48. $b^2 - b - 12$

49. $x^2 - x - 72$

50. $x^2 - 2x - 80$

51. $y^2 - 15y - 34$

52. $y^2 - 10y - 39$

53. $z^2 - 5z - 66$

54. $z^2 - 6z - 55$

Exercises 55–62: Factor each trinomial.

55. $5 + 6x + x^2$
    (*Hint:* Write the expression in standard form.)

56. $8 + 6x + x^2$

57. $3 - 4x + x^2$

58. $10 - 7x + x^2$

59. $12 + 4x - x^2$
    (*Hint:* Write $(m - x)(n + x)$ and find m and n.)

60. $28 + 3x - x^2$

61. $32 - 4x - x^2$

62. $40 - 3x - x^2$

GEOMETRY

63. A square has an area of $x^2 + 2x + 1$. Find the length of a side. Make a sketch of the square.

64. A square has an area of $x^2 + 6x + 9$. Find the length of a side. Make a sketch of the square.

65. A rectangle has an area of $x^2 + 3x + 2$. Find one possibility for its width and length. Make a sketch of the rectangle.

66. A rectangle has an area of $x^2 + 9x + 8$. Find one possibility for its width and length. Make a sketch of the rectangle.

67. A cube has a surface area of $6x^2 + 12x + 6$. Find the length of a side. (*Hint:* First factor out the GCF.)

68. A cube has a surface area of $6x^2 + 36x + 54$. Find the length of a side.

69. Write a polynomial in factored form that represents the total area of the figure.

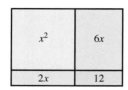

70. Write a polynomial in factored form that represents the total area of the figure.

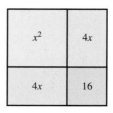

WRITING ABOUT MATHEMATICS

71. Explain how to determine whether a trinomial has been factored correctly. Give an example.

72. Factoring $x^2 + bx + c$ involves finding two integers m and n such that $mn = c$ and $m + n = b$. When you are finding m and n, is it better first to determine values of m and n so that the product is c or the sum is b? Explain your reasoning.

CHECKING BASIC CONCEPTS   SECTIONS 6.1 AND 6.2

1. What is the greatest common factor for the expression $8x^3 - 12x^2 + 24x$?

2. Factor $12z^3 - 18z^2$.

3. Factor each expression by grouping.
   (a) $6y(y - 2) + 5(y - 2)$
   (b) $2x^3 + x^2 + 10x + 5$

4. Factor each trinomial.
   (a) $x^2 + 6x + 8$ **(b)** $x^2 - x - 42$

5. Write a polynomial in factored form that represents the total area of the figure.

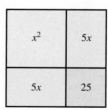

6.3   FACTORING TRINOMIALS II $(ax^2 + bx + c)$

Factoring Trinomials by Grouping · Factoring with FOIL

INTRODUCTION

The sum of the areas of the four small rectangles shown in Figure 6.4 is $2x^2 + 5x + 2$. Note that the length of the large rectangle is $2x + 1$ and that its width is $x + 2$. One way to determine these dimensions is to factor $2x^2 + 5x + 2$. However, this trinomial has a lead-

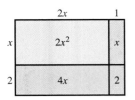

$2x^2 + 5x + 2 = (2x + 1)(x + 2)$

Figure 6.4

ing coefficient of 2. Thus far we have factored only trinomials with a leading coefficient of 1. In this section we discuss two methods used to factor trinomials in the form $ax^2 + bx + c$, where $a \neq 0$ or 1. In Example 1(a), we demonstrate how to factor the trinomial $2x^2 + 5x + 2$ as $(2x + 1)(x + 2)$.

FACTORING TRINOMIALS BY GROUPING

To factor the polynomial given by $x^2 + 6x + 5$ we find two numbers, m and n, such that $mn = 5$ and $m + n = 6$. For this trinomial we let $m = 1$ and $n = 5$, which gives

$$x^2 + 6x + 5 = (x + 1)(x + 5).$$

To factor the polynomial $2x^2 + 7x + 3$, which has a leading coefficient of 2, we must find two numbers m and n such that $mn = 2 \cdot 3 = 6$ and $m + n = 7$. One solution is $m = 1$ and $n = 6$. Using grouping, we can now factor this trinomial by writing $7x$ as $1x + 6x$.

$$\begin{aligned}
2x^2 + 7x + 3 &= 2x^2 + x + 6x + 3 && \text{Write } 7x \text{ as } x + 6x.\\
&= (2x^2 + x) + (6x + 3) && \text{Associative property}\\
&= x(2x + 1) + 3(2x + 1) && \text{Distributive property}\\
&= (x + 3)(2x + 1) && \text{Factor by grouping.}
\end{aligned}$$

This technique of factoring trinomials by grouping is summarized as follows.

FACTORING $ax^2 + bx + c$ BY GROUPING

To factor $ax^2 + bx + c$ perform the following steps. (Assume that a, b, and c have no common factors.)

1. Find two numbers, m and n, such that $mn = ac$ and $m + n = b$.
2. Write the trinomial as $ax^2 + mx + nx + c$.
3. Use grouping to factor this expression into two binomials.

EXAMPLE 1 Factoring $ax^2 + bx + c$ by grouping

Factor each trinomial.
(a) $2x^2 + 5x + 2$ **(b)** $3z^2 + z - 2$ **(c)** $10t^2 - 11t + 3$

Solution **(a)** To factor $2x^2 + 5x + 2$ we need to find m and n that satisfy $mn = 2 \cdot 2 = 4$ and $m + n = 5$. Two such numbers are $m = 1$ and $n = 4$.

$$\begin{aligned}
2x^2 + 5x + 2 &= 2x^2 + x + 4x + 2 && \text{Write } 5x \text{ as } x + 4x.\\
&= (2x^2 + x) + (4x + 2) && \text{Associative property}\\
&= x(2x + 1) + 2(2x + 1) && \text{Distributive property}\\
&= (x + 2)(2x + 1) && \text{Factor by grouping.}
\end{aligned}$$

(b) To factor $3z^2 + 1z - 2$ we need to find m and n that satisfy $mn = 3 \cdot (-2) = -6$ and $m + n = 1$. Two such numbers are $m = 3$ and $n = -2$.

$$3z^2 + z - 2 = 3z^2 + 3z - 2z - 2 \qquad \text{Write } z \text{ as } 3z - 2z.$$
$$= (3z^2 + 3z) + (-2z - 2) \qquad \text{Associative property}$$
$$= 3z(z + 1) - 2(z + 1) \qquad \text{Distributive property}$$
$$= (3z - 2)(z + 1) \qquad \text{Factor by grouping.}$$

(c) To factor $10t^2 - 11t + 3$ we need to find m and n that satisfy $mn = 10 \cdot 3 = 30$ and $m + n = -11$. Two such numbers are $m = -5$ and $n = -6$.

$$10t^2 - 11t + 3 = 10t^2 - 5t - 6t + 3 \qquad \text{Write } -11t \text{ as } -5t - 6t.$$
$$= (10t^2 - 5t) + (-6t + 3) \qquad \text{Associative property}$$
$$= 5t(2t - 1) - 3(2t - 1) \qquad \text{Distributive property}$$
$$= (5t - 3)(2t - 1) \qquad \text{Factor by grouping.}$$

=== MAKING CONNECTIONS ===

Different Ways to Factor by Grouping

In Example 1(c) we could have written $-11t$ as $-6t - 5t$, rather than $-5t - 6t$. Then the factoring could have been written as

$$10t^2 - 11t + 3 = 10t^2 - 6t - 5t + 3$$
$$= (10t^2 - 6t) + (-5t + 3)$$
$$= 2t(5t - 3) - 1(5t - 3)$$
$$= (2t - 1)(5t - 3),$$

which gives the same result.

FACTORING WITH FOIL

Rather than factoring a trinomial by grouping, we can use FOIL in reverse to determine its binomial factors. For example, the factors of $2x^2 + 5x + 2$ are two binomials:

$$2x^2 + 5x + 2 = (\_\_ + \_\_)(\_\_ + \_\_),$$

where the expressions to be placed in the four blanks are yet to be found. By the FOIL method, we know that the product of the first terms of these two binomials is $2x^2$. Because $2x^2 = 2x \cdot x$, we can write

$$2x^2 + 5x + 2 = (2x + \_\_)(x + \_\_).$$

By FOIL, the product of the last terms in each binomial must be 2. Because $2 = 1 \cdot 2$, we could put 1 and 2 in the blanks. However, we must be sure to place them correctly so that the product of the outside terms plus the product of the inside terms is $5x$.

$$(2x + 1)(x + 2) = 2x^2 + 5x + 2$$

$1x$

$+4x$

$5x \longleftarrow$ Middle term checks.

If we had mistakenly interchanged the 1 and 2, we would have obtained an incorrect result.

$$(2x + 2)(x + 1) = 2x^2 + 4x + 2$$

Middle term is *not* 5x.

In the next example we factor expressions of the form $ax^2 + bx + c$, where $a \neq 1$. We use FOIL in reverse and *trial and error* to find the correct factors.

EXAMPLE 2 Factoring the form $ax^2 + bx + c$

Factor each trinomial.
(a) $3x^2 + 5x + 2$ **(b)** $6x^2 + 7x - 3$ **(c)** $6 + 4y^2 - 11y$

Solution **(a)** To factor $3x^2 + 5x + 2$ we start by finding factors of $3x^2$, which include $3x$ and x, so we write

$$3x^2 + 5x + 2 = (3x + \underline{\quad})(x + \underline{\quad}).$$

The factors of the last term 2 are **1** and **2**. (Because the middle term is positive, we do not consider -1 and -2.) If we place values of 1 and 2 in the binomial as follows, the middle term becomes $7x$ rather than $5x$.

$$(3x + 1)(x + 2) = 3x^2 + 7x + 2$$

Middle term is *not* 5x.

By reversing the positions of 1 and 2, we obtain a correct factorization.

$$(3x + 2)(x + 1) = 3x^2 + 5x + 2$$

Middle term checks.

(b) To factor $6x^2 + 7x - 3$, we start by finding factors of $6x^2$, which include $2x$ and $3x$ or $6x$ and x. Factors of the last term, -3, include -1 and 3 or 1 and -3. The following two factorizations give incorrect results.

$$(2x + 1)(3x - 3) = 6x^2 - 3x - 3$$

Middle term is *not* 7x.

$$(6x + 1)(x - 3) = 6x^2 - 17x - 3$$

Middle term is *not* 7x.

To obtain a middle term of $7x$ we use the following arrangement.

$$(3x - 1)(2x + 3) = 6x^2 + 7x - 3$$

Middle term checks.

To find the correct factorization we often need to try more than once.

(c) To factor $6 + 4y^2 - 11y$, we start by writing the trinomial in standard form as $4y^2 - 11y + 6$. Then we find possible factors of the first term, $4y^2$, which include $2y$ and $2y$ or $4y$ and y. Factors of the last term, 6, include either -1 and -6 or -2 and -3.

Critical Thinking

Is it possible to factor every trinomial with the methods discussed in this section? Try to factor the following polynomials and then make a conjecture.

$x^2 + x + 5$, $x^2 - 2x + 6$,

and $2x^2 + 3x + 2$

(Because the middle term is negative, we do not use the positive factors of 1 and 6 or 2 and 3.) To obtain a middle term of $-11y$ we use the following arrangement.

$$(4y - 3)\ (y - 2) = 4y^2 - 11y + 6$$

$$-3y$$

$$-8y$$

$$-11y \longleftarrow \text{Middle term checks.}$$

In the next example we demonstrate how to factor trinomials having a negative leading coefficient. This task is sometimes accomplished by first factoring out -1.

EXAMPLE 3   Factoring trinomials having a negative leading coefficient

Factor each trinomial.

(a) $-6x^2 + 17x - 5$ (b) $1 - x - 2x^2$

Solution   (a) One way to factor $-6x^2 + 17x - 5$ is to start by factoring out a -1. Then we can apply the FOIL method in reverse.

$$
\begin{aligned}
-6x^2 + 17x - 5 &= -1(6x^2 - 17x + 5) &&\text{Factor out } -1. \\
&= -1(3x - 1)(2x - 5) &&\text{Factor the trinomial.} \\
&= -(3x - 1)(2x - 5) &&\text{Rewrite.}
\end{aligned}
$$

(b) We write $1 - x - 2x^2$ in standard form, factor out -1, and then apply FOIL in reverse.

$$
\begin{aligned}
1 - x - 2x^2 &= -2x^2 - x + 1 &&\text{Standard form} \\
&= -1(2x^2 + x - 1) &&\text{Factor out } -1. \\
&= -(2x - 1)(x + 1) &&\text{Factor the trinomial.}
\end{aligned}
$$

6.3   PUTTING IT ALL TOGETHER

In this section we discussed factoring trinomials in the form $ax^2 + bx + c$ by using grouping and FOIL in reverse. The following table summarizes these two methods.

Concept	Explanation	Examples
Factoring Trinomials by Grouping	To factor $ax^2 + bx + c$, find two numbers, m and n, such that $mn = ac$ and $m + n = b$. Then write $$(ax^2 + mx) + (nx + c).$$ Use grouping to factor this expression into two binomials.	For $3x^2 + 10x + 8$ let $m = 6$ and $n = 4$ because $mn = 3 \cdot 8 = 24$ and $m + n = 10$. $$\begin{aligned} 3x^2 + 10x + 8 &= 3x^2 + 6x + 4x + 8 \\ &= (3x^2 + 6x) + (4x + 8) \\ &= 3x(x + 2) + 4(x + 2) \\ &= (3x + 4)(x + 2) \end{aligned}$$

Concept	Explanation	Examples
Factoring Trinomials by Using FOIL in Reverse	To factor $ax^2 + bx + c$ find factors of ax^2 and of c. Choose and arrange these factors in two binomials so that the middle term is bx.	For $3x^2 + 10x + 8$ the factors of $3x^2$ are $3x$ and x. The positive factors of 8 are either 2 and 4, or 1 and 8. A middle term of $10x$ can be obtained as follows. $(3x + 4)(x + 2) = 3x^2 + 10x + 8$ $\lfloor\ 4x \rfloor$ $+6x$ $10x$ ← The middle term checks.
Choosing Signs When Factoring $ax^2 + bx + c$	For $ax^2 + bx + c$ with $a > 0$, **1.** If $c > 0$ and $b > 0$, use $(\_ + \_)(\_ + \_)$. **2.** If $c > 0$ and $b < 0$, use $(\_ - \_)(\_ - \_)$. **3.** If $c < 0$, use $(\_ - \_)(\_ + \_)$ or $(\_ + \_)(\_ - \_)$.	**1.** $2x^2 + 7x + 3 = (2x + 1)(x + 3)$ $(c = 3, b = 7)$ **2.** $2x^2 - 7x + 3 = (2x - 1)(x - 3)$ $(c = 3, b = -7)$ **3.** $2x^2 + 5x - 3 = (2x - 1)(x + 3)$ $(c = -3, b = 5)$ $2x^2 - 5x - 3 = (2x + 1)(x - 3)$ $(c = -3, b = -5)$

6.3 EXERCISES

FOR EXTRA HELP

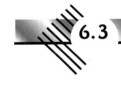 Student's Solutions Manual   InterAct Math   MathXL

MyMathLab   Math Tutor Center   Digital Video Tutor   CD 3   Videotape 8

CONCEPTS

1. To factor the polynomial $ax^2 + bx + c$ by grouping, you first find two numbers, m and n, such that $mn =$ _____ and $m + n =$ _____.

2. To factor the polynomial $ax^2 + bx + c$ with FOIL, you first find possible factors for _____ and for _____.

3. If $3x^2 + 5x + 2$ is factored, the result is the product $(3x \_ 2)(x \_ 1)$.

4. If $3x^2 - x - 2$ is factored, the result is the product $(3x \_ 2)(x \_ 1)$.

5. If $3x^2 - 5x + 2$ is factored, the result is the product $(3x \_ 2)(x \_ 1)$.

6. If $3x^2 + x - 2$ is factored, the result is the product $(3x \_ 2)(x \_ 1)$.

7. If $4x^2 + 11x + 6$ is factored, the result is the product $(4x + \_)(\_ + 2)$.

8. If $4x^2 - 5x - 6$ is factored, the result is the product $(x - \_)(\_ + 3)$.

9. If $4x^2 + 4x - 3$ is factored, the result is the product $(2x - \_)(\_ + 3)$.

10. If $4x^2 - 8x + 3$ is factored, the result is the product $(2x - \_)(\_ - 3)$.

FACTORING TRINOMIALS

Exercises 11–42: Factor.

11. $2x^2 + 7x + 3$

12. $2x^2 + 3x + 1$

13. $3x^2 + 4x + 1$

14. $3x^2 + 10x + 3$

15. $6x^2 + 11x + 3$

16. $6x^2 + 17x + 5$

17. $5x^2 - 11x + 2$

18. $7x^2 - 8x + 1$

19. $2y^2 - 7y + 5$

20. $2y^2 - 11y + 12$

21. $7z^2 - 37z + 10$

22. $3z^2 - 11z + 6$

23. $3t^2 - 7t - 6$

24. $8t^2 - 6t - 9$

25. $15r^2 + r - 6$

26. $12r^2 + r - 6$

27. $24m^2 - 23m - 12$

28. $24m^2 + 29m - 4$

29. $25x^2 + 5x - 2$

30. $30x^2 + 7x - 2$

31. $6x^2 + 11x - 2$

32. $12x^2 + 28x - 5$

33. $21n^2 + 4n - 1$

34. $21n^2 + 10n + 1$

35. $14y^2 + 23y + 3$

36. $28y^2 + 25y + 3$

37. $28z^2 - 25z + 3$

38. $15z^2 - 19z + 6$

39. $30x^2 - 29x + 6$

40. $50x^2 - 55x + 12$

41. $18t^2 + 23t - 6$

42. $33t^2 + 7t - 10$

Exercises 43–52: Factor.

43. $2 + 15x + 7x^2$

44. $3 + 16x + 5x^2$

45. $2 - 5x + 2x^2$

46. $5 - 6x + x^2$

47. $3 - 2x - 8x^2$

48. $5 - 3x - 2x^2$

49. $-2x^2 - 7x + 15$

50. $-5x^2 - 19x + 4$

51. $-5x^2 + 14x + 3$

52. $-6x^2 + 17x + 14$

53. A rectangle has an area of $6x^2 + 7x + 2$. Find possible dimensions for this rectangle. Make a sketch of the rectangle.

54. A rectangle has an area of $2x^2 + 5x + 3$. Find possible dimensions for the rectangle. Make a sketch of the rectangle.

55. Write a polynomial in factored form that represents the total area of the figure.

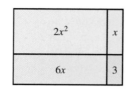

56. Write a polynomial in factored form that represents the total area of the figure.

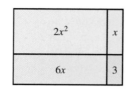

WRITING ABOUT MATHEMATICS

57. Explain how the sign of the third term in a trinomial affects how it is factored.

58. Explain the steps to be used to factor $ax^2 + bx + c$ by grouping.

6.4   SPECIAL TYPES OF FACTORING

Difference of Two Squares · Perfect Square Trinomials ·
Sum and Difference of Two Cubes

INTRODUCTION

Some polynomials can be factored by using special methods. These methods are used to solve equations and simplify expressions. In this section we discuss some of these methods.

DIFFERENCE OF TWO SQUARES

In Section 5.4 we showed that

$$(a - b)(a + b) = a^2 - b^2.$$

We can use this equation to factor a difference of two squares. For example, to factor the expression $x^2 - 25$ we can write it in the form $a^2 - b^2$, where $a = x$ and $b = 5$. Then the equation

$$a^2 - b^2 = (a - b)(a + b)$$

becomes

$$x^2 - 25 = (x - 5)(x + 5).$$

DIFFERENCE OF TWO SQUARES

For any real numbers a and b,

$$a^2 - b^2 = (a - b)(a + b).$$

In the next example we apply this method to other expressions.

EXAMPLE 1   Factoring the difference of two squares

Factor each difference of two squares.
(a) $x^2 - 36$ **(b)** $4x^2 - 9$ **(c)** $100 - 16t^2$ **(d)** $49y^2 - 64z^2$

Solution   **(a)** Substitute x for a and **6** for b. Then the equation

$$a^2 - b^2 = (a - b)(a + b)$$

becomes

$$x^2 - 6^2 = (x - 6)(x + 6).$$

(b) The expression $4x^2 - 9$ can be written as $(2x)^2 - 3^2$. Thus

$$4x^2 - 9 = (2x - 3)(2x + 3).$$

(c) The expression $100 - 16t^2$ can be written as $(10)^2 - (4t)^2$. Thus

$$100 - 16t^2 = (10 - 4t)(10 + 4t).$$

(d) The expression $49y^2 - 64z^2$ can be written as $(7y)^2 - (8z)^2$. Thus

$$49y^2 - 64z^2 = (7y - 8z)(7y + 8z).$$

Critical Thinking

Is there any way to factor the expression $x^2 - 5$ as the difference of two squares? Explain your reasoning.

MAKING CONNECTIONS

Sum of Squares versus Difference of Squares

The sum of two squares, $a^2 + b^2$, cannot be factored by using real numbers. However, the difference of two squares, $a^2 - b^2$, can be factored. For example, $x^2 + 4$ cannot be factored, but $x^2 - 4$ can be factored as $(x - 2)(x + 2)$.

PERFECT SQUARE TRINOMIALS

In Section 5.4 we also showed how to multiply $(a + b)^2$ and $(a - b)^2$:

$$(a + b)^2 = a^2 + 2ab + b^2 \quad \text{and}$$
$$(a - b)^2 = a^2 - 2ab + b^2.$$

The expressions $a^2 + 2ab + b^2$ and $a^2 - 2ab + b^2$ are called **perfect square trinomials.** If we can recognize a perfect square trinomial, we can use the following formulas to factor it.

PERFECT SQUARE TRINOMIALS

For any real numbers a and b,

$$a^2 + 2ab + b^2 = (a + b)^2 \quad \text{and}$$
$$a^2 - 2ab + b^2 = (a - b)^2.$$

We demonstrate this method in the next example. When factoring a trinomial as a perfect square trinomial, we must first verify that the middle term is correct. This technique is also demonstrated in the next example.

EXAMPLE 2   Factoring perfect square trinomials

If possible, factor each trinomial.
(a) $x^2 + 10x + 25$ **(b)** $4x^2 - 4x + 1$
(c) $9z^2 + 18z + 4$ **(d)** $x^2 - 4xy + 4y^2$

Solution   **(a)** To factor $x^2 + 10x + 25$ we let $a^2 = x^2$ and $b^2 = 5^2$ so that $a = x$ and $b = 5$. To be a perfect square trinomial, the middle term must be $2ab$. That is, the middle term must be *twice* the product of x and 5. So

$$2ab = 2 \cdot x \cdot 5 = 10x,$$

which is the middle term in the given expression. Thus

$$a^2 + 2ab + b^2 = (a + b)^2$$

becomes

$$x^2 + 10x + 25 = (x + 5)^2.$$

(b) To factor $4x^2 - 4x + 1$ we let $a^2 = (2x)^2$ and $b^2 = 1^2$ so that $a = 2x$ and $b = 1$. The middle term must be $2ab$. So

$$2ab = 2 \cdot 2x \cdot 1 = 4x,$$

which is the middle term in the given expression, except for the subtraction sign. Thus

$$a^2 - 2ab + b^2 = (a - b)^2$$

becomes

$$4x^2 - 4x + 1 = (2x - 1)^2.$$

(c) To factor $9z^2 + 18z + 4$ we let $a^2 = (3z)^2$ and $b^2 = 2^2$ so that $a = 3z$ and $b = 2$. The middle term must be twice the product of $3z$ and 2. So

$$2ab = 2 \cdot 3z \cdot 2 = 12z,$$

which is *not* the middle term of $18z$. We cannot factor this expression as a perfect square trinomial.

(d) To factor $x^2 - 4xy + 4y^2$ we let $a^2 = x^2$ and $b^2 = (2y)^2$ so that $a = x$ and $b = 2y$. The middle term must be twice the product of x and $2y$. So

$$2ab = 2 \cdot x \cdot 2y = 4xy,$$

which is the middle term, except for the subtraction sign. Thus

$$a^2 - 2ab + b^2 = (a - b)^2$$

becomes

$$x^2 - 4xy + 4y^2 = (x - 2y)^2.$$

MAKING CONNECTIONS

Special Factoring and General Techniques

If you do not recognize a polynomial as the difference of two squares or a perfect square trinomial, you can still factor the polynomial by using the methods discussed in earlier sections.

SUM AND DIFFERENCE OF TWO CUBES

The sum or difference of two cubes may be factored—a result of the two equations

$$(a + b)(a^2 - ab + b^2) = a^3 + b^3 \quad \text{and}$$
$$(a - b)(a^2 + ab + b^2) = a^3 - b^3.$$

These equations can be verified by multiplying the left side to obtain the right side. For example,

$$
\begin{aligned}
(a + b)(a^2 - ab + b^2) &= a \cdot a^2 - a \cdot ab + a \cdot b^2 + b \cdot a^2 - b \cdot ab + b \cdot b^2 \\
&= a^3 - a^2 b + ab^2 + a^2 b - ab^2 + b^3 \\
&= a^3 + b^3
\end{aligned}
$$

SUM AND DIFFERENCE OF TWO CUBES

For any real numbers a and b,

Opposite Signs

$$a^3 + b^3 = (a + b)(a^2 - ab + b^2) \quad \text{and}$$
$$a^3 - b^3 = (a - b)(a^2 + ab + b^2).$$

Opposite Signs

EXAMPLE 3 Factoring the sum and difference of two cubes

Factor each polynomial.
(a) $z^3 + 8$ (b) $x^3 - 27$ (c) $8x^3 - 1$ (d) $16y^3 + 54z^3$

Solution (a) To factor $z^3 + 8$ we let $a^3 = z^3$ and $b^3 = 2^3$ so that $a = z$ and $b = 2$. Then

$$a^3 + b^3 = (a + b)(a^2 - ab + b^2)$$

becomes

$$z^3 + 2^3 = (z + 2)(z^2 - z \cdot 2 + 2^2)$$
$$= (z + 2)(z^2 - 2z + 4).$$

(b) To factor $x^3 - 27$ we let $a^3 = x^3$ and $b^3 = 3^3$ so that $a = x$ and $b = 3$. Then

$$a^3 - b^3 = (a - b)(a^2 + ab + b^2)$$

becomes

$$x^3 - 3^3 = (x - 3)(x^2 + x \cdot 3 + 3^2)$$
$$= (x - 3)(x^2 + 3x + 9).$$

(c) To factor $8x^3 - 1$ we let $a^3 = (2x)^3$ and $b^3 = 1^3$ so that $a = 2x$ and $b = 1$. Then

$$(2x)^3 - 1^3 = (2x - 1)((2x)^2 + 2x \cdot 1 + 1^2)$$
$$= (2x - 1)(4x^2 + 2x + 1).$$

(d) Always start by checking for a common factor. In this example, we factor out the common factor of 2 in $16y^3 + 54z^3$. Then we can factor the resulting sum of two cubes by letting $a = 2y$ and $b = 3z$.

$$16y^3 + 54z^3 = 2(8y^3 + 27z^3)$$
$$= 2((2y)^3 + (3z)^3)$$
$$= 2((2y + 3z)((2y)^2 - 2y \cdot 3z + (3z)^2))$$
$$= 2(2y + 3z)(4y^2 - 6yz + 9z^2).$$

6.4   PUTTING IT ALL TOGETHER

In this section we discussed some special types of factoring, which are summarized in the following table.

Factoring	Explanation	Examples
Difference of Two Squares	$a^2 - b^2 = (a - b)(a + b)$ **Note:** The *sum* of two squares, $a^2 + b^2$, cannot be factored by using real numbers.	$x^2 - 49 = (x - 7)(x + 7)$ $81 - z^2 = (9 - z)(9 + z)$ $4r^2 - 25t^2 = (2r - 5t)(2r + 5t)$ $16a^2 + b^2$ cannot be factored.

Factoring	Explanation	Examples
Perfect Square Trinomial	$a^2 + 2ab + b^2 = (a + b)^2$ $a^2 - 2ab + b^2 = (a - b)^2$ Be sure to verify that the given middle term equals $2ab$ before factoring.	$m^2 + 2m + 1 = (m + 1)^2$ $25y^2 - 30y + 9 = (5y - 3)^2$ $36r^2 + 12rt + t^2 = (6r + t)^2$ $x^2 + 5x + 4$ is *not* a perfect square trinomial because $2ab = 2 \cdot x \cdot 2 = 4x \neq 5x.$
Sum and Difference of Two Cubes	$a^3 + b^3 = (a + b)(a^2 - ab + b^2)$ $a^3 - b^3 = (a - b)(a^2 + ab + b^2)$	$y^3 + 27 = (y + 3)(y^2 - y \cdot 3 + 3^2)$ $\quad\quad = (y + 3)(y^2 - 3y + 9)$ $27r^3 - 64t^3$ $\quad = (3r - 4t)((3r)^2 + 3r \cdot 4t + (4t)^2)$ $\quad = (3r - 4t)(9r^2 + 12rt + 16t^2)$

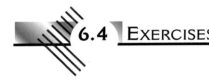

6.4 EXERCISES

FOR EXTRA HELP

CONCEPTS

1. $a^2 - b^2 = $ _____

2. The expression $a^2 + b^2$ (can/cannot) be factored by using real numbers.

3. If the expression $36x^2 - 49y^2$ is written in the form $a^2 - b^2$, then $a = $ _____ and $b = $ _____.

4. $a^2 + 2ab + b^2 = $ _____

5. $a^2 - 2ab + b^2 = $ _____

6. $x^2 + $ _____ $+ 9$ is a perfect square trinomial.

7. $4r^2 - $ _____ $+ 25t^2$ is a perfect square trinomial.

8. $a^3 + b^3 = $ _____

9. $a^3 - b^3 = $ _____

10. If the expression $8x^3 + 27y^3$ is written in the form $a^3 + b^3$, then $a = $ _____ and $b = $ _____.

11. If $y^3 - 8$ is factored, the result is the product $(y \_\_ 2)(y^2 \_\_ 2y + 4)$.

12. If $64z^3 + 27$ is factored, the result is the product $(4z \_\_ 3)(16z^2 \_\_ 12z + 9)$.

FACTORING THE DIFFERENCE OF TWO SQUARES

Exercises 13–30: Factor.

13. $x^2 - 1$

14. $x^2 - 16$

15. $z^2 - 100$

16. $z^2 - 81$

17. $4y^2 - 1$

18. $9y^2 - 16$

19. $36z^2 - 25$

20. $49z^2 - 64$

21. $9 - x^2$

22. $25 - x^2$

23. $1 - 9y^2$

24. $49 - 16y^2$

25. $4a^2 - 9b^2$

26. $16a^2 - b^2$

27. $36m^2 - 25n^2$

28. $49m^2 - 100n^2$

29. $81r^2 - 49t^2$

30. $625r^2 - 121t^2$

FACTORING PERFECT SQUARE TRINOMIALS

Exercises 31–52: Factor as a perfect square trinomial whenever possible.

31. $x^2 + 8x + 16$

32. $x^2 + 4x + 4$

33. $z^2 + 12z + 25$

34. $z^2 - 18z + 36$

35. $x^2 - 6x + 9$

36. $x^2 - 10x + 25$

37. $9y^2 + 6y + 1$

38. $16y^2 + 8y + 1$

39. $4z^2 - 4z + 1$

40. $25z^2 - 12z + 1$

41. $9t^2 + 16t + 4$

42. $4t^2 + 12t + 9$

43. $9x^2 + 30x + 25$

44. $25x^2 + 60x + 36$

45. $4a^2 - 36a + 81$

46. $9a^2 - 60a + 100$

47. $x^2 + 2xy + y^2$

48. $x^2 - 6xy + 9y^2$

49. $r^2 - 10rt + 25t^2$

50. $15r^2 + 10rt + t^2$

51. $4y^2 - 10yz + 9z^2$

52. $25y^2 - 20yz + 4z^2$

FACTORING SUMS AND DIFFERENCES OF TWO CUBES

Exercises 53–70: Factor.

53. $z^3 + 1$

54. $z^3 + 8$

55. $x^3 + 64$

56. $x^3 + 125$

57. $y^3 - 8$

58. $y^3 - 27$

59. $n^3 - 1$

60. $n^3 - 64$

61. $8x^3 + 1$

62. $27x^3 - 1$

63. $m^3 - 64n^3$

64. $m^3 + 8n^3$

65. $16a^3 + 2b^3$

66. $3a^3 + 81b^3$

67. $8x^3 + 125y^3$

68. $27x^3 + 64y^3$

69. $500r^3 - 32t^3$

70. $8r^3 - 64t^3$

GENERAL FACTORING

Exercises 71–86: Use any method to factor the polynomial.

71. $9x^4 - 3x^3$

72. $4x^3 - 6x^2$

73. $4y^2 - 9$

74. $25 - y^2$

75. $25z^2 - 80z + 64$

76. $9z^2 + 48z + 64$

77. $2x^2 - 15x + 7$

78. $3x^2 + 11x - 4$

79. $2n^2 + 21n + 27$

80. $n^2 - 8n - 20$

81. $8x^3 - 27$

82. $216x^3 + y^3$ (*Hint:* $6^3 = 216$.)

83. $x^3 + 2x^2 - 5x - 10$

84. $2x^3 + 5x^2 + 8x + 20$

85. $a^2 + 18ab + 81b^2$

86. $25a^2 + 10ab + b^2$

GEOMETRY

87. A square has an area of $4x^2 + 12x + 9$. Find the length of a side. Make a sketch of the square.

88. A square has an area of $9x^2 + 30x + 25$. Find the length of a side. Make a sketch of the square.

WRITING ABOUT MATHEMATICS

89. Explain how factoring $x^3 + y^3$ is different from factoring $x^3 - y^3$.

90. Using the techniques discussed in this section, can you factor the expression $4x^2 + 9y^2$ into two binomials? Explain your reasoning.

CHECKING BASIC CONCEPTS   SECTIONS 6.3 AND 6.4

1. Factor each trinomial.
   (a) $2x^2 - 5x - 12$ (b) $6x^2 + 17x - 14$

2. Write a polynomial in factored form that represents the total area of the figure.

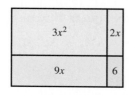

3. Factor the polynomials.
   (a) $z^2 - 64$ (b) $9r^2 - 4t^2$
   (c) $x^2 + 12x + 36$ (d) $9a^2 - 12ab + 4b^2$

4. Factor.
   (a) $m^3 - 27$ (b) $125n^3 + 27$

6.5 SOLVING EQUATIONS BY FACTORING I (QUADRATICS)

The Zero-Product Property · Solving Quadratic Equations · Applications

INTRODUCTION

If a golf ball is hit upward at 132 feet per second, or 90 miles per hour, then its height h in feet above the ground after t seconds is given by $h = 132t - 16t^2$. The expression $132t - 16t^2$ is an example of a *quadratic polynomial*. To determine the elapsed time between when the ball is hit and when it strikes the ground (or when $h = 0$) we solve the *quadratic equation*

$$132t - 16t^2 = 0.$$

One method for solving this equation is by factoring. (See Example 4.) In this section we discuss how to use factoring to solve a variety of equations.

THE ZERO-PRODUCT PROPERTY

To solve equations we often use the **zero-product property**, which states that, if the product of two numbers is 0, then at least one of the numbers must be 0.

 ZERO-PRODUCT PROPERTY

For all real numbers a and b, if $ab = 0$, then $a = 0$ or $b = 0$ (or both).

Note: The zero-product property works only for 0. If $ab = 1$, then it does *not* follow that $a = 1$ or $b = 1$. For example, $a = \frac{1}{3}$ and $b = 3$ satisfy the equation $ab = 1$.

After factoring an expression, we can use the zero-product property to solve an equation. The left side of the equation

$$3t^2 - 9t = 0$$

may be factored to obtain

$$3t(t - 3) = 0.$$

Note that the product of $3t$ and $t - 3$ is 0. By the zero-product property, either

$$3t = 0 \quad \text{or} \quad t - 3 = 0.$$

Solving each equation for t results in

$$t = 0 \quad \text{or} \quad t = 3.$$

These values can be checked by substituting them into the given equation $3t^2 - 9t = 0$.

$$3(0)^2 - 9(0) = 0 \qquad \text{Let } t = 0. \text{ It checks.}$$
$$3(3)^2 - 9(3) = 0 \qquad \text{Let } t = 3. \text{ It checks.}$$

The t–values of 0 and 3 are called **zeros** of the polynomial $3t^2 - 9t$, because when either is substituted in this polynomial, the result is 0.

EXAMPLE 1 Applying the zero-product property

Solve each equation.
(a) $x(x - 1) = 0$ (b) $2z^2 = 0$
(c) $(t + 3)(t + 2) = 0$ (d) $x(x - 2)(2x + 1) = 0$

Solution   (a) By the zero-product property, $x(x - 1) = 0$ when $x = 0$ or $x - 1 = 0$. The solutions are 0 and 1.
(b) $2z^2 = 2 \cdot z \cdot z$ and $2 \neq 0$, so $2z^2 = 0$ when $z = 0$.
(c) $(t + 3)(t + 2) = 0$ implies that $t + 3 = 0$ or $t + 2 = 0$. The solutions to the equation are -3 and -2.
(d) We apply the zero-product property to $x(x - 2)(2x + 1) = 0$. Thus $x = 0$ or $x - 2 = 0$ or $2x + 1 = 0$. The solutions are $-\frac{1}{2}$, 0, and 2.

SOLVING QUADRATIC EQUATIONS

Any **quadratic polynomial** in the variable x can be written as $ax^2 + bx + c$ with $a \neq 0$. Any **quadratic equation** in the variable x can be written as $ax^2 + bx + c = 0$ with $a \neq 0$. This form of quadratic equation is called the **standard form** of a quadratic equation. For example, $x^2 + 2x - 3$ is a quadratic polynomial and $x^2 + 2x - 3 = 0$ is a quadratic equation written in standard form.

To solve a quadratic equation we often use factoring and the zero-product property. This method is summarized by the following steps. Although it is not necessary to label each step in the solution to a quadratic equation, it is important to keep these steps in mind.

SOLVING QUADRATIC EQUATIONS

To solve a quadratic equation by factoring, follow these steps.

STEP 1: If necessary, write the equation in standard form as $ax^2 + bx + c = 0$.

STEP 2: Factor the left side of the equation using any method.

STEP 3: Apply the zero-product property.

STEP 4: Solve each of the resulting equations.

EXAMPLE 2 Solving equations by factoring

Solve each quadratic equation. Check your answers.
(a) $x^2 + 2x = 0$ (b) $y^2 = 16$ (c) $z^2 - 3z + 2 = 0$ (d) $2x^2 = 5 - 9x$

Solution   (a) Because $x^2 + 2x = 0$ is in standard form, we begin by factoring out the GCF of x.

$x^2 + 2x = 0$	Given equation
$x(x + 2) = 0$	Factor out x. (STEP 2)
$x = 0$ or $x + 2 = 0$	Zero-product property (STEP 3)
$x = 0$ or $x = -2$	Solve for x. (STEP 4)

To verify these values, substitute -2 and 0 for x in the given equation.

$$(-2)^2 + 2(-2) \stackrel{?}{=} 0 \qquad (0)^2 + 2(0) \stackrel{?}{=} 0 \qquad \text{Substitute } -2 \text{ and } 0.$$
$$0 = 0 \qquad\qquad\quad 0 = 0 \qquad\qquad \text{Both answers check.}$$

Therefore the solution are -2 and 0.

(b) To write $y^2 = 16$ in standard form we begin by subtracting 16 from each side to obtain 0 on the right side.

$$
\begin{array}{ll}
y^2 = 16 & \text{Given equation} \\
y^2 - 16 = 0 & \text{Subtract 16. (STEP 1)} \\
(y - 4)(y + 4) = 0 & \text{Difference of squares (STEP 2)} \\
y - 4 = 0 \quad \text{or} \quad y + 4 = 0 & \text{Zero-product property (STEP 3)} \\
y = 4 \quad \text{or} \quad y = -4 & \text{Solve for } y. \text{ (STEP 4)}
\end{array}
$$

To verify these values, substitute -4 and 4 for y in the given equation.

$$(-4)^2 \stackrel{?}{=} 16 \qquad (4)^2 \stackrel{?}{=} 16 \qquad \text{Substitute } -4 \text{ and } 4.$$
$$16 = 16 \qquad\quad 16 = 16 \qquad \text{Both answers check.}$$

The solutions are -4 and 4.

(c) We begin by factoring the left side of the equation, $z^2 - 3z + 2$.

$$
\begin{array}{ll}
z^2 - 3z + 2 = 0 & \text{Given equation} \\
(z - 1)(z - 2) = 0 & \text{Factor. (STEP 2)} \\
z - 1 = 0 \quad \text{or} \quad z - 2 = 0 & \text{Zero-product property (STEP 3)} \\
z = 1 \quad \text{or} \quad z = 2 & \text{Solve for } z. \text{ (STEP 4)}
\end{array}
$$

To verify these values, substitute 1 and 2 for z in the given equation.

$$1^2 - 3(1) + 2 \stackrel{?}{=} 0 \qquad 2^2 - 3(2) + 2 \stackrel{?}{=} 0 \qquad \text{Substitute 1 and 2.}$$
$$0 = 0 \qquad\qquad\quad 0 = 0 \qquad\qquad \text{Both answers check.}$$

The solutions are 1 and 2.

(d) We write $2x^2 = 5 - 9x$ in standard form by adding -5 and $9x$ to each side.

$$
\begin{array}{ll}
2x^2 = 5 - 9x & \text{Given equation} \\
2x^2 + 9x - 5 = 0 & \text{Add } -5 \text{ and } 9x. \text{ (STEP 1)} \\
(2x - 1)(x + 5) = 0 & \text{Factor. (STEP 2)} \\
2x - 1 = 0 \quad \text{or} \quad x + 5 = 0 & \text{Zero-product property (STEP 3)} \\
x = \dfrac{1}{2} \quad \text{or} \quad x = -5 & \text{Solve for } x. \text{ (STEP 4)}
\end{array}
$$

To verify these values, substitute -5 and $\frac{1}{2}$ for x in the given equation.

$$2(-5)^2 \stackrel{?}{=} 5 - 9(-5) \qquad 2\left(\frac{1}{2}\right)^2 \stackrel{?}{=} 5 - 9\left(\frac{1}{2}\right) \qquad \text{Substitute 1 and 2.}$$
$$50 = 50 \qquad\qquad\qquad \frac{1}{2} = \frac{1}{2} \qquad\qquad\quad \text{Both answers check.}$$

The solutions are -5 and $\frac{1}{2}$.

EXAMPLE 3 Solving an equation by factoring

Solve $6x^2 - x = 12$.

Solution We cannot solve the equation $6x^2 - x = 12$ by factoring out the common factor of x in $6x^2 - x$ and setting each factor equal to 12. Instead we apply the zero-product property by first writing the given equation in the standard form: $ax^2 + bx + c = 0$.

$$6x^2 - x = 12 \qquad \text{Given equation}$$
$$6x^2 - x - 12 = 0 \qquad \text{Subtract 12. (STEP 1)}$$
$$(2x - 3)(3x + 4) = 0 \qquad \text{Factor. (STEP 2)}$$
$$2x - 3 = 0 \quad \text{or} \quad 3x + 4 = 0 \qquad \text{Zero-product property (STEP 3)}$$
$$x = \frac{3}{2} \quad \text{or} \quad x = -\frac{4}{3} \qquad \text{Solve for } x. \text{ (STEP 4)}$$

The solutions are $-\frac{4}{3}$ and $\frac{3}{2}$.

APPLICATIONS

To solve applications problems we often need to solve equations. The next example illustrates how to solve the application presented in the introduction to this section.

EXAMPLE 4 Modeling the flight of a golf ball

If a golf ball is hit upward at 132 feet per second, or 90 miles per hour, then its height h in feet after t seconds is $h = 132t - 16t^2$. After how long does the golf ball strike the ground?

Solution The golf ball strikes the ground when its height is 0.

$$132t - 16t^2 = 0 \qquad \text{Let } h = 0.$$
$$4t(33 - 4t) = 0 \qquad \text{Factor out } 4t.$$
$$4t = 0 \quad \text{or} \quad 33 - 4t = 0 \qquad \text{Zero-product property}$$
$$t = 0 \quad \text{or} \quad -4t = -33 \qquad \text{Divide by 4; subtract 33.}$$
$$t = 0 \quad \text{or} \quad t = \frac{33}{4} \qquad \text{Solve for } t.$$

The ball strikes the ground after $\frac{33}{4} = 8.25$ seconds. The solution of 0 is not used in this problem because it corresponds to the time when the ball is hit.

When you try to stop a car, the greater the speed the greater is the stopping distance. In fact, if you drive twice as fast, the braking distance increases by about four times. And if you drive three times faster, the braking distance increases by about nine times.

EXAMPLE 5 Modeling braking distance

The braking distance D in feet required for a car traveling at x miles per hour to stop on dry, level pavement can be approximated by $D = \frac{1}{11}x^2$.
(a) Calculate the braking distance for a car traveling 70 miles per hour.
(b) If the braking distance is 44 feet, calculate the speed of the car.

(c) If you have a calculator available, use it to solve part (b) numerically with a table of values.

Solution (a) If $x = 70$, then $D = \frac{1}{11}(70)^2 = \frac{4900}{11} \approx 445$ feet.

(b) ***Symbolic Solution***   Let $D = 44$ in the given equation and solve.

$$\frac{1}{11}x^2 = 44 \qquad \text{Let } D = 44.$$

$$\frac{1}{11}x^2 - 44 = 0 \qquad \text{Subtract 44.}$$

$$x^2 - 484 = 0 \qquad \text{Multiply by 11.}$$

$$(x - 22)(x + 22) = 0 \qquad \text{Difference of two squares}$$
$$\text{Note: } 22^2 = 484$$

$$x - 22 = 0 \quad \text{or} \quad x + 22 = 0 \qquad \text{Zero-product property}$$

$$x = 22 \quad \text{or} \quad x = -22 \qquad \text{Solve for } x.$$

The car is traveling at (approximately) 22 miles per hour. (Note that $x = -22$ has no physical meaning in this problem.)

(c) ***Numerical Solution***   Let $Y_1 = X^2/11$ and make a table of values. Scroll through the table, as shown in Figure 6.5, to where $x = 22$ when $y_1 = 44$. Thus the solution is 22 miles per hour.

X	Y₁
20	36.364
21	40.091
22	**44**
23	48.091
24	52.364
25	56.818
26	61.455

X=22

Figure 6.5

Calculator Help

To make a table of values, see Appendix A (page AP-3).

Quadratic equations sometimes are used in applications involving rectangular shapes. The next application involves finding the dimensions of a digital photograph.

EXAMPLE 6   **Finding the dimensions of a digital photograph**

A small digital photograph is 20 pixels longer than it is wide, as illustrated in Figure 6.6. It has a total of 2400 pixels. Find the dimensions of this photograph.

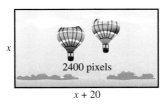

x

2400 pixels

$x + 20$

Figure 6.6 Dimensions of a Photograph

Solution   From Figure 6.6 the rectangular photograph has an area of 2400 pixels.

$$x(x + 20) = 2400 \qquad \text{Area} = \text{width} \times \text{length}$$

$$x^2 + 20x = 2400 \qquad \text{Multiply.}$$

$$x^2 + 20x - 2400 = 0 \qquad \text{Subtract 2400.}$$

$$(x - 40)(x + 60) = 0 \qquad \text{Factor.}$$

$$x - 40 = 0 \quad \text{or} \quad x + 60 = 0 \qquad \text{Zero-product property}$$

$$x = 40 \quad \text{or} \quad x = -60 \qquad \text{Solve for } x.$$

Critical Thinking

Are the two solutions to $x(2x + 1) = 1$ found by letting $x = 1$ or $2x + 1 = 1$? Explain your reasoning. What are the solutions to this equation?

The valid solution is 40. Thus the dimensions of the photograph are 40 pixels by $40 + 20 = 60$ pixels.

6.5 PUTTING IT ALL TOGETHER

In this section we discussed solving quadratic equations by using the factoring and zero-product property. These topics are summarized in the following table.

Concept	Explanation	Examples
Zero-Product Property	If the product of two or more expressions is 0, then at least one of the expressions must equal 0.	$ab = 0$ implies that $a = 0$ or $b = 0$. $x(x + 1)$ implies that $x = 0$ or $x + 1 = 0$. $z(z - 1)(z + 2) = 0$ implies that $z = 0$ or $z - 1 = 0$ or $z + 2 = 0$.
Solving Quadratic Equations by Factoring	1. Write the equation as $$ax^2 + bx + c = 0.$$ 2. Factor the left side of this equation. 3. Apply the zero-product property. 4. Solve each resulting equation.	$2x^2 + 11x = 6$ $2x^2 + 11x - 6 = 0$ STEP 1 $(2x - 1)(x + 6) = 0$ STEP 2 $2x - 1 = 0$ or $x + 6 = 0$ STEP 3 $x = \dfrac{1}{2}$ or $x = -6$ STEP 4

6.5 EXERCISES

FOR EXTRA HELP

Student's Solutions Manual

InterAct Math

MathXL

MyMathLab

Math Tutor Center

Digital Video Tutor
CD 3   Videotape 8

CONCEPTS

1. If $ab = 0$, then either $a =$ _____ or $b =$ _____.

2. Does the zero-product property state that, if $(x - 1)(x - 2) = 3$, then either $x - 1 = 3$ or $x - 2 = 3$? Explain your answer.

3. If $2x(x + 6) = 0$, then either _____ or _____.

4. What is a good first step when you are solving the equation $4x^2 + 1 = 4x$ by factoring?

5. What is the next step when you are solving the equation $(x + 5)(x - 4) = 0$?

6. Factoring is an important tool in _____ equations.

7. Any quadratic equation in the variable x can be written in standard form as _____.

8. Standard form for $x^2 + 1 = 6x$ is _____.

ZERO-PRODUCT PROPERTY

Exercises 9–20: Solve the equation.

9. $xy = 0$

10. $m^2 = 0$

11. $2x(x + 8) = 0$

12. $x(x + 10) = 0$

13. $(y - 1)(y - 2) = 0$

14. $(y + 4)(y - 3) = 0$

15. $(2z - 1)(4z - 3) = 0$

16. $(6z + 5)(z - 7) = 0$

17. $(1 - 3n)(3 - 7n) = 0$ **18.** $(5 - n)(5 + n) = 0$

19. $x(x - 5)(x - 8) = 0$ **20.** $x(x + 1)(x - 6) = 0$

SOLVING QUADRATIC EQUATIONS

Exercises 21–52: Solve and check.

21. $x^2 - x = 0$ **22.** $2x^2 + 4x = 0$

23. $z^2 - 5z = 0$ **24.** $6z^2 - 3z = 0$

25. $10y^2 + 15y = 0$ **26.** $2y^2 + 3y = 0$

27. $x^2 - 1 = 0$ **28.** $x^2 - 9 = 0$

29. $4n^2 - 1 = 0$ **30.** $9n^2 - 4 = 0$

31. $z^2 + 3z + 2 = 0$ **32.** $z^2 - 2z - 3 = 0$

33. $x^2 - 12x + 35 = 0$ **34.** $x^2 - x - 20 = 0$

35. $2b^2 + 3b - 2 = 0$ **36.** $3b^2 + b - 2 = 0$

37. $6y^2 + 19y + 10 = 0$ **38.** $4y^2 - 25y - 21 = 0$

39. $x^2 = 25$ **40.** $x^2 = 81$

41. $t^2 = 5t$ **42.** $10t^2 = -5t$

43. $3m^2 = -9m$ **44.** $4m^2 = 9$

45. $x^2 = 5x + 6$ **46.** $2x^2 + 3x = 14$

47. $12z^2 + 11z = 15$ **48.** $12z^2 = 5 - 4z$

49. $t(t + 1) = 2$ **50.** $t(t - 7) = -12$

51. $x(2x + 5) = 3$ **52.** $x(3x + 2) = 5$

GEOMETRY

53. *Dimensions of a Square*   A square has an area of 144 square feet. Find the length of a side.

54. *Dimensions of a Cube*   A cube has a surface area of 96 square feet. Find the length of a side.

55. *Radius of a Circle*   The numerical difference between the area and the circumference of a circle is 8π inches. Find the radius of the circle. (*Hint:* First factor out π in your equation.)

56. *Dimensions of a Rectangle*   A rectangle is 5 feet longer than it is wide and has an area of 126 square feet. What are its dimensions?

Exercises 57 and 58: **Pythagorean Theorem**   *Suppose that a right triangle has legs a and b and hypotenuse c, as illustrated in the figure. Then these values satisfy* $a^2 + b^2 = c^2$.

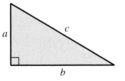

Find the value of x in the figure.

57.   **58.**

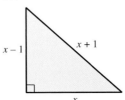   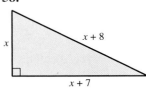

APPLICATIONS

59. *Flight of a Golf Ball*   (Refer to Example 4.) The height h in feet of a golf ball after t seconds is given by $h = 96t - 16t^2$.
   (a) How long did it take for the ball to hit the ground?
   (b) Make a table of h for $t = 0, 1, 2, \ldots, 6$. After how many seconds did the golf ball reach its maximum height?

60. *Flight of a Baseball*   The height h in feet of a baseball after t seconds is given by $h = -16t^2 + 88t + 3$. At what values of t is the height of the baseball 75 feet?

61. *Braking Distance*   (Refer to Example 5.) The braking distance D in feet required for a car traveling x miles per hour to stop on dry, level pavement can be approximated by $D = \frac{1}{11}x^2$.
   (a) Calculate the braking distance for 30 miles per hour and 60 miles per hour. How do your answers compare?
   (b) If the braking distance is 33 feet, estimate the speed of the car.
   (c) If you have a calculator, use it to solve part (b) numerically. Do your answers agree?

62. *Braking Distance* The braking distance D in feet required for a car traveling x miles per hour to stop on wet, level pavement is approximated by $D = \frac{1}{9}x^2$.
   (a) Calculate the braking distance for 36 miles per hour and 72 miles per hour. How do your answers compare?
   (b) If the braking distance is 49 feet, estimate the speed of the car.
   (c) If you have a calculator, use it to solve part (b) numerically. Do your answers agree?

63. *Women in the Workforce* The number of women W in the workforce in millions can be estimated by the equation $W = \frac{19}{3125}x^2 + \frac{11}{2}$, where $x = 0$ corresponds to 1900, $x = 10$ to 1910, and so on until $x = 100$ corresponds to 2000.
   (a) How many women were in the workforce in 1930 and in 2000?
   (b) Use a table of values to estimate the year in which 45 million women were in the workforce.

64. *AIDS Deaths* The cumulative number of AIDS deaths D from 1984 to 1994 can be modeled by

$$D = 2375x^2 + 5134x + 5020,$$

where $x = 1$ corresponds to 1984, $x = 2$ to 1985, and so on until $x = 11$ corresponds to 1994.

   (a) Estimate the cumulative number of AIDS deaths in 1993.
   (b) Use a table of values to estimate the year in which the cumulative number of AIDS deaths reached 90,000.

65. *Digital Photographs* (Refer to Example 6.) A digital photograph is 10 pixels longer than it is wide and has a total area of 2000 pixels. Find the dimensions of this rectangular photograph.

66. *Dimensions of a Building* The rectangular floor of a shed has a length 4 feet longer than its width, and its area is 140 square feet. Let x be the width of the floor.
   (a) Write a quadratic equation whose solution gives the width of the floor.
   (b) Solve this equation.

WRITING ABOUT MATHEMATICS

67. List four steps for solving a quadratic equation by factoring.

68. Explain why factoring is important.

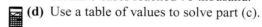

Group Activity: Working with Real Data

Directions: Form a group of 2 to 4 people. Select someone to record the group's responses for this activity. All members of the group should work cooperatively to answer the questions. If your instructor asks for your results, each member of the group should be prepared to respond.

AIDS Cases From 1984 to 1994 the cumulative number N of AIDS cases in thousands can be approximated by $N = 3x^2 + 14x + 7$, where $x = 0$ corresponds to 1984.

Year	1984	1986	1988	1990	1992	1994
Cases	11	42	106	197	329	442

Source: Department of Health and Human Services.

   (a) Use the equation to find N for each year in the table.
   (b) Discuss how well this equation approximates the data.
   (c) Use the equation to find the year in which the number of AIDS cases reached 76 thousand.
   (d) Use a table of values to solve part (c).

6.6 SOLVING EQUATIONS BY FACTORING II (HIGHER DEGREE)

Polynomials Having Common Factors · Special Types of Polynomials

INTRODUCTION

In this section we discuss factoring polynomials having higher degree. Polynomials of degree 2 or higher are often used in applications. For example, the polynomial

$$0.0013x^3 - 0.085x^2 + 1.6x + 12$$

models natural gas consumption in the United States (in trillions of cubic feet) during year x, where $x = 0$ corresponds to 1960, $x = 10$ to 1970, and so on until $x = 40$ corresponds to 2000, as shown in Figure 6.7. In Exercises 73–76 we discuss further the consumption of natural gas. (**Source:** Department of Energy.)

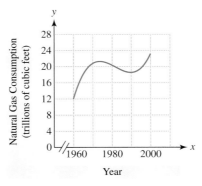

Figure 6.7 Natural Gas Consumption

POLYNOMIALS HAVING COMMON FACTORS

The first step in factoring a polynomial is to factor out the greatest common factor (GCF). For example, to factor $x^3 - x$ we start by factoring out the GCF, which is x. The resulting expression can be factored as the difference of two squares.

$$x^3 - x = x(x^2 - 1) \qquad \text{Factor out } x.$$
$$= x(x - 1)(x + 1) \qquad \text{Difference of two squares}$$

EXAMPLE 1   **Factoring trinomials with common factors**

Factor each trinomial completely.
(a) $-4x^2 + 28x - 40$ **(b)** $10x^3 + 28x^2 - 6x$

Solution   **(a)** Each term in $-4x^2 + 28x - 40$ has a factor of -4, so we start by factoring out -4. (In this example, we factor out the *negative* of the GCF to obtain a positive leading coefficient.)

$$-4x^2 + 28x - 40 = -4(x^2 - 7x + 10) \qquad \text{Factor out } -4.$$
$$= -4(x - 5)(x - 2) \qquad \text{Factor the trinomial.}$$

(b) We start by factoring out the GCF for $10x^3 + 28x^2 - 6x$, which is **2x**.

$$10x^3 + 28x^2 - 6x = 2x(5x^2 + 14x - 3) \qquad \text{Factor out } 2x.$$
$$= 2x(5x - 1)(x + 3) \qquad \text{Factor the trinomial.}$$

EXAMPLE 2 Solving polynomial equations

Solve each equation.
(a) $x^3 - x^2 - 6x = 0$ **(b)** $4x^4 + 10x^3 = 6x^2$

Solution **(a)** We start by factoring out the GCF, which is x.

$$x^3 - x^2 - 6x = 0 \qquad \text{Given equation}$$
$$x(x^2 - x - 6) = 0 \qquad \text{Factor out } x.$$
$$x(x - 3)(x + 2) = 0 \qquad \text{Factor the trinomial.}$$
$$x = 0 \quad \text{or} \quad x - 3 = 0 \quad \text{or} \quad x + 2 = 0 \qquad \text{Zero-product property}$$
$$x = 0 \quad \text{or} \quad x = 3 \quad \text{or} \quad x = -2 \qquad \text{Solve for } x.$$

The solutions are -2, 0, and 3.
(b) We start by subtracting $6x^2$ from each side to obtain 0 on one side of the equation.

$$4x^4 + 10x^3 = 6x^2 \qquad \text{Given equation}$$
$$4x^4 + 10x^3 - 6x^2 = 0 \qquad \text{Subtract } 6x^2.$$
$$2x^2(2x^2 + 5x - 3) = 0 \qquad \text{Factor out the GCF, } 2x^2.$$
$$2x^2(2x - 1)(x + 3) = 0 \qquad \text{Factor the trinomial.}$$
$$2x \cdot x = 0 \quad \text{or} \quad 2x - 1 = 0 \quad \text{or} \quad x + 3 = 0 \qquad \text{Zero-product property}$$
$$x = 0 \quad \text{or} \quad x = \frac{1}{2} \quad \text{or} \quad x = -3 \qquad \text{Solve for } x.$$

The solutions are -3, 0, and $\frac{1}{2}$.

The corners of a square piece of metal are cut out to form a box, as shown in Figure 6.8. This square piece of metal has sides with length 10 inches, and the cutout corners are squares with length x. The outside surface area A of this box, including the bottom and the sides but *not* the top, is $A = 100 - 4x^2$. (See Critical Thinking in the margin.)

Critical Thinking

Calculate the outside surface area A of the box shown in Figure 6.8 two different ways.

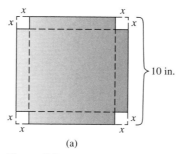

(a)

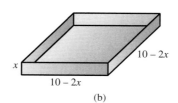

(b)

Figure 6.8

EXAMPLE 3   Finding a dimension of a box

Find the value of x in Figure 6.8 if its outside surface area A is 84 square inches.

Solution   Solve the equation $A = 84$ for x.

$$100 - 4x^2 = 84 \qquad \text{\small $A = 84$}$$
$$16 - 4x^2 = 0 \qquad \text{\small Subtract 84.}$$
$$4(4 - x^2) = 0 \qquad \text{\small Factor out 4.}$$
$$4(2 - x)(2 + x) = 0 \qquad \text{\small Difference of squares}$$
$$2 - x = 0 \quad \text{or} \quad 2 + x = 0 \qquad \text{\small Zero-product property}$$
$$x = 2 \quad \text{or} \quad x = -2 \qquad \text{\small Solve for x.}$$

If squares measuring 2 inches on a side are cut out of each corner, the surface area of the box is 84 square inches.

SPECIAL TYPES OF POLYNOMIALS

Some types of polynomials of higher degree can be factored by using methods that we have already presented.

EXAMPLE 4   Factoring higher degree polynomials

Factor each polynomial completely.
(a) $x^4 - 16$ **(b)** $y^4 + 5y^2 + 4$ **(c)** $x^4 + 2x^2y^2 + y^4$ **(d)** $r^4 - t^4$

Solution   **(a)** We view this polynomial as the difference of two squares, where $x^4 = (x^2)^2$. Then we factor twice.

$$x^4 - 16 = (x^2)^2 - 4^2 \qquad \text{\small Rewrite.}$$
$$= (x^2 - 4)(x^2 + 4) \qquad \text{\small Difference of squares}$$
$$= (x - 2)(x + 2)(x^2 + 4) \qquad \text{\small Difference of squares}$$

Note that $x^2 + 4$ does not factor.
(b) Because $x^2 + 5x + 4 = (x + 1)(x + 4)$,

$$y^4 + 5y^2 + 4 = (y^2 + 1)(y^2 + 4).$$

Note that neither $y^2 + 1$ nor $y^2 + 4$ are factored further.
(c) Because $a^2 + 2ab + b^2 = (a + b)^2$, we let $a = x^2$ and $b = y^2$ and then factor the given trinomial.

$$x^4 + 2x^2y^2 + y^4 = (x^2 + y^2)^2$$

(d) Note that $a^2 - b^2 = (a - b)(a + b)$. We let $a = r^2$ and $b = t^2$ and factor.

$$r^4 - t^4 = (r^2 - t^2)(r^2 + t^2)$$
$$= (r - t)(r + t)(r^2 + t^2)$$

Note that $r^2 + t^2$ cannot be factored.

EXAMPLE 5   **Solving an equation**

Solve $x^5 - 81x = 0$.

Solution   We start by factoring out the common factor of x.

$$x^5 - 81x = 0 \qquad \text{Given equation}$$
$$x(x^4 - 81) = 0 \qquad \text{Factor out } x.$$
$$x(x^2 - 9)(x^2 + 9) = 0 \qquad \text{Difference of two squares}$$
$$x(x - 3)(x + 3)(x^2 + 9) = 0 \qquad \text{Difference of two squares}$$
$$x = 0 \ \text{ or } \ x - 3 = 0 \ \text{ or } \ x + 3 = 0 \ \text{ or } \ x^2 + 9 = 0 \qquad \text{Zero-product property}$$
$$x = 0 \ \text{ or } \ x = 3 \ \text{ or } \ x = -3 \qquad \text{Solve for } x.$$

Note that $x^2 + 9 = 0$ has no real number solutions because the square of a number plus 9 is never 0. The solutions are -3, 0, and 3.

6.6   PUTTING IT ALL TOGETHER

The following table summarizes factoring higher degree polynomials and solving higher degree polynomial equations.

Concept	Explanation	Examples
Common Factors	A first step when factoring polynomials is to factor out the GCF.	$x^3 - 4x^2 - 5x = x(x^2 - 4x - 5)$ $= x(x - 5)(x + 1)$ $4x^4 - 16x^2 = 4x^2(x^2 - 4)$ $= 4x^2(x - 2)(x + 2)$
Factoring Trinomials	Some higher degree trinomials can be factored by using the same methods that we used to factor trinomials. **Note:** $a^2 + b^2$ cannot be factored.	$x^4 + 6x^2 + 5 = (x^2 + 5)(x^2 + 1)$ $4y^4 - 25 = (2y^2 - 5)(2y^2 + 5)$ $2z^4 - 32 = 2(z^4 - 16)$ $= 2(z^2 - 4)(z^2 + 4)$ $= 2(z - 2)(z + 2)(z^2 + 4)$
Solving Equations by Factoring	Use factoring and the zero-product property to solve polynomial equations.	$3x^3 - 12x^2 + 9x = 0$ $3x(x^2 - 4x + 3) = 0$ $3x(x - 3)(x - 1) = 0$ $3x = 0 \ \text{ or } \ x - 3 = 0 \ \text{ or } \ x - 1 = 0$ $x = 0 \ \text{ or } \ x = 3 \ \text{ or } \ x = 1$ The solutions are 0, 1, and 3.

6.6 EXERCISES

FOR EXTRA HELP

 Student's Solutions Manual

 InterAct Math

 MathXL

 MyMathLab

Tutor Center Math Tutor Center

Digital Video Tutor CD 3 Videotape 8

CONCEPTS

1. When you are factoring polynomials, a good first step is to factor out the _____.

2. When you are solving an equation by factoring, the _____ property is used.

3. Because $x^2 + 3x + 2 = (x + 1)(x + 2)$, it follows that $z^4 + 3z^2 + 2 = $ _____.

4. Is $x^4 - 1 = (x^2 - 1)(x^2 + 1)$ factored completely? Explain.

5. When you are solving the equation $x^3 = x$, what is a good first step?

6. When you are solving $x^4 - x^2 = 0$, what is a good first step?

FACTORING POLYNOMIALS

Exercises 7–18: Factor the polynomial completely.

7. $5x^2 - 5x - 30$

8. $3x^2 - 15x + 12$

9. $-4y^2 - 32y - 48$

10. $-7y^2 + 14y + 21$

11. $-20z^2 - 110z - 50$

12. $-12z^2 - 54z + 30$

13. $60 - 64t - 28t^2$

14. $18 - 45t - 27t^2$

15. $r^3 - r$

16. $r^3 + 2r^2 - 3r$

17. $3x^3 + 3x^2 - 18x$

18. $6x^3 - 26x^2 - 20x$

Exercises 19–44: Factor completely.

19. $72z^3 + 12z^2 - 24z$

20. $6z^3 - 4z^2 - 42z$

21. $x^4 - 4x^2$

22. $4x^4 - 36x^2$

23. $t^4 + t^3 - 2t^2$

24. $t^4 + 5t^3 - 24t^2$

25. $x^4 - 5x^2 + 6$

26. $x^4 - 3x^2 - 10$

27. $2x^4 + 7x^2 + 3$

28. $3x^4 - 8x^2 + 5$

29. $y^4 + 6y^2 + 9$

30. $y^4 - 10y^2 + 25$

31. $x^4 - 9$

32. $x^4 - 25$

33. $x^4 - 81$

34. $4x^4 - 64$

35. $z^5 + 2z^4 + z^3$

36. $6z^5 - 47z^4 + 35z^3$

37. $2x^2 + xy - y^2$

38. $2x^2 + 5xy + 2y^2$

39. $a^4 - 2a^2b^2 + b^4$

40. $a^3 + 2a^2b + ab^2$

41. $x^3 - xy^2$

42. $2x^2y - 2y^3$

43. $4x^3 + 4x^2y + xy^2$

44. $x^2y - 6xy^2 + 9y^3$

SOLVING EQUATIONS

Exercises 45–48: Do the following.

45. (a) Factor $x^3 - 4x$.
    (b) Solve $x^3 - 4x = 0$.

46. (a) Factor $4x^3 - 16x$.
    (b) Solve $4x^3 - 16x = 0$.

47. (a) Factor $2y^3 - 6y^2 - 36y$.
    (b) Solve $2y^3 - 6y^2 - 36y = 0$.

48. (a) Factor $z^4 - 13z^2 + 36$.
    (b) Solve $z^4 - 13z^2 + 36 = 0$.

Exercises 49–70: Solve.

49. $3x^2 + 33x + 72 = 0$

50. $4x^2 - 16x - 20 = 0$

51. $25x^2 = 50x + 75$

52. $10x^2 = 20x + 80$

53. $y^3 - 3y^2 - 4y = 0$

54. $y^3 - 3y^2 + 2y = 0$

55. $3z^3 + 6z^2 = 72z$

56. $4z^3 = 4z^2 + 24z$

57. $x^4 - 36x^2 = 0$

58. $4x^4 = 100x^2$

59. $r^4 + 6r^3 = 7r^2$

60. $r^4 - 11r^3 + 30r^2 = 0$

61. $x^4 - 13x^2 + 36 = 0$

62. $x^4 - 17x^2 + 16 = 0$

63. $x^4 + 1 = 2x^2$

64. $x^4 - 8x^2 + 16 = 0$

65. $a^4 = 81$

66. $b^3 = -8$

67. $x^3 - 2x^2 - x + 2 = 0$ (*Hint:* Use grouping.)

68. $x^3 - x^2 + 4x - 4 = 0$

69. $x^3 - 5x^2 + x - 5 = 0$

70. $3x^3 + 2x^2 - 27x - 18 = 0$

APPLICATIONS

71. *Dimensions of a Box*   (Refer to Example 3.) A box is made from a rectangular piece of metal with length 20 inches and width 15 inches. The box has no top.
   (a) Are there any limitations on the size of x? Explain.
   (b) Write an expression that gives the outside surface area of the box. (*Hint:* Consider the size of the metal sheet and how much was cut out.)
   (c) If the outside surface area of the box is 275 square inches, find x.

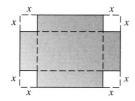 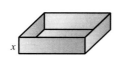

72. *Dimensions of a Box*   (Refer to Exercise 71.)
   (a) Find a polynomial that gives the volume of the box for a given x.
   (b) Factor your polynomial completely.
   (c) What are the zeros of your polynomial? What do they represent in this problem?

Exercises 73–76: U.S. Natural Gas Consumption   (*Refer to the introduction to this section.*) *The polynomial*

$$0.0013x^3 - 0.085x^2 + 1.6x + 12$$

models natural gas consumption in trillions of cubic feet during year x, where $x = 0$ corresponds to 1960, $x = 1$ to 1961, and so on.

73. How much natural gas was consumed in 1990?

74. In which year (between 1970 and 1990) was natural gas consumption about 20.4 trillion cubic feet?

75. Explain any difficulties encountered when you try to solve the equation

$$0.0013x^3 - 0.085x^2 + 1.6x + 12 = 23.2.$$

76. How might you solve this equation without factoring? If you were to find the solution to this equation, what would it represent?

WRITING ABOUT MATHEMATICS

77. Compare factoring the polynomial $x^2 + 6x + 5$ with factoring the polynomial $z^4 + 6z^2 + 5$.

78. Suppose that a polynomial can be factored. Explain how its factors can be used to find the zeros of the polynomial. Give an example.

CHECKING BASIC CONCEPTS   SECTIONS 6.5 AND 6.6

1. Solve each quadratic equation.
   (a) $4y^2 - 6y = 0$ **(b)** $5z^2 + 2z = 3$

2. Solve $x^2 + 2x - 3 = 0$ symbolically and numerically with a table of values.

3. If a golf ball is hit upward at 60 miles per hour, then its height in feet after t seconds is given by $88t - 16t^2$. Use factoring to determine when the golf ball strikes the ground.

4. Factor each polynomial.
   (a) $x^4 - 8x^2 + 16$ **(b)** $2y^3 + 17y^2 - 30y$
   (c) $x^4 - 16y^4$

5. Solve $t^4 + t^3 = 12t^2$.

Summary

Terms Related to Factoring Polynomials

Factoring	Writing a polynomial as a product, usually of lower degree polynomials
Common Factor	An expression that is a factor of each term in a polynomial

> **Example:**   Examples of common factors of $4x^4 + 8x^2$ are $2x$, x^2, and $4x^2$.

Greatest Common Factor (GCF)	The common factor with greatest degree and coefficient that can be factored out of a polynomial

> **Example:**   The GCF of $4x^4 + 8x^2$ is $4x^2$.
>
> $$4x^4 + 8x^2 = 4x^2(x^2 + 2)$$

Factoring by Grouping	Used to factor a four-term polynomial into two binomials

> **Example:** $x^3 + 5x^2 + 3x + 15 = x^2(x + 5) + 3(x + 5)$
> $$= (x^2 + 3)(x + 5)$$

Review of the FOIL Method   A method used for multiplying two binomials

First Terms	$(2x + 3)(5x + 4)$:	$2x \cdot 5x = 10x^2$
Outside Terms	$(2x + 3)(5x + 4)$:	$2x \cdot 4 = 8x$
Inside Terms	$(2x + 3)(5x + 4)$:	$3 \cdot 5x = 15x$
Last Terms	$(2x + 3)(5x + 4)$:	$3 \cdot 4 = 12$

The product is the sum of these four terms:

$$(2x + 3)(5x + 4) = 10x^2 + 8x + 15x + 12 = 10x^2 + 23x + 12.$$

Factoring Trinomials Having a Leading Coefficient of 1   To factor the trinomial $x^2 + bx + c$, find two numbers, m and n, that satisfy

$$m \cdot n = c \quad \text{and} \quad m + n = b.$$

Then $x^2 + bx + c = (x + m)(x + n)$.

Example: Because $-3 \cdot 5 = -15$ and $-3 + 5 = 2$,

$$x^2 + 2x - 15 = (x - 3)(x + 5).$$

Section 6.3   *Factoring Trinomials II* $(ax^2 + bx + c)$

Factoring Trinomials by Grouping   To factor $ax^2 + bx + c$, perform the following steps. (Assume that a, b, and c have no factor in common.)

1. Find two numbers, m and n, such that $mn = ac$ and $m + n = b$.
2. Write the trinomial as $ax^2 + mx + nx + c$.
3. Use grouping to factor this expression into two binomials.

Example: To factor $3x^2 + 10x - 8$, find two numbers whose product is -24 and whose sum is 10. These two numbers are $m = 12$ and $n = -2$, so write $10x$ as $12x - 2x$.

$$
\begin{aligned}
3x^2 + 10x - 8 &= 3x^2 + 12x - 2x - 8 \\
&= (3x^2 + 12x) + (-2x - 8) \\
&= 3x(x + 4) - 2(x + 4) \\
&= (3x - 2)(x + 4)
\end{aligned}
$$

Factoring with FOIL   Use trial and error and FOIL in reverse to find the factors of a trinomial.

Example: To factor $3x^2 + 10x - 8$, first find factors of $3x^2$.

$$(3x + \underline{\quad})(x + \underline{\quad})$$

Then place factors of -8 so that the resulting middle term is $10x$.

$$(3x + \underline{-2}) \cdot (x + \underline{4})$$

$-2x$

$12x$

$10x$ ⟵ The middle term checks.

Section 6.4   *Special Types of Factoring*

Difference of Two Squares

$$a^2 - b^2 = (a - b)(a + b)$$

Examples: $x^2 - 16 = (x - 4)(x + 4)$ $(a = x, b = 4)$
$4r^2 - 9t^2 = (2r - 3t)(2r + 3t)$ $(a = 2r, b = 3t)$

Perfect Square Trinomials

$$a^2 + 2ab + b^2 = (a + b)^2 \quad \text{and} \quad a^2 - 2ab + b^2 = (a - b)^2$$

Examples: $4x^2 + 4x + 1 = (2x)^2 + 2(2x)1 + 1^2 = (2x + 1)^2$ $(a = 2x, b = 1)$

$x^2 - 10x + 25 = x^2 - 2 \cdot x \cdot 5 + 5^2 = (x - 5)^2$ $(a = x, b = 5)$

Sums and Differences of Two Cubes

$$a^3 + b^3 = (a + b)(a^2 - ab + b^2) \quad \text{and} \quad a^3 - b^3 = (a - b)(a^2 + ab + b^2)$$

Examples: $x^3 + 8 = (x + 2)(x^2 - 2x + 4)$ $(a = x, b = 2)$
$27x^3 - 1 = (3x - 1)(9x^2 + 3x + 1)$ $(a = 3x, b = 1)$

Section 6.5 *Solving Equations by Factoring I*

Zero-Product Property

For any real numbers a and b, if $ab = 0$, then $a = 0$ or $b = 0$ (or both). The zero-product property is used to solve equations.

Examples: $xy = 0$ implies that $x = 0$ or $y = 0$.

$(x + 5)(x - 3) = 0$ implies $x + 5 = 0$ or $x - 3 = 0$.

Zero of a Polynomial
A number a is a zero of a polynomial if the result is 0 when a is substituted in that polynomial.

Example: The number -2 is a zero of $x^2 - 4$ because $(-2)^2 - 4 = 0$.

Solving Quadratic Equations by Factoring
To solve a quadratic equation by factoring, follow these steps.

STEP 1: If necessary, use algebra to write the equation as $ax^2 + bx + c = 0$.

STEP 2: Factor the left side of the equation using any method.

STEP 3: Apply the zero-product property.

STEP 4: Solve each of the resulting equations.

Example:

$$
\begin{aligned}
x^2 + 7x &= 8 && \text{Given equation} \\
x^2 + 7x - 8 &= 0 && \text{STEP 1} \\
(x + 8)(x - 1) &= 0 && \text{STEP 2} \\
x + 8 = 0 \quad \text{or} \quad x - 1 &= 0 && \text{STEP 3} \\
x = -8 \quad \text{or} \quad x &= 1 && \text{STEP 4}
\end{aligned}
$$

Section 6.6 *Solving Equations by Factoring II*

Factoring Polynomials of Higher Degree
The distributive property and the techniques for factoring quadratic polynomials can also be applied to polynomials of higher degree.

Examples: $10r^3 + 15r = 5r(2r^2 + 3)$ (To check, multiply the right side.)

Because $2x^2 + x - 1 = (x + 1)(2x - 1)$,
it follows that $2z^4 + z^2 - 1 = (z^2 + 1)(2z^2 - 1)$.

Solving Equations by Factoring
Use algebra to obtain 0 on one side of the equation. Factor the other side and apply the zero-product property.

Example:

$$
\begin{aligned}
x^3 &= 4x \\
x^3 - 4x &= 0 && \text{Subtract } 4x. \\
x(x^2 - 4) &= 0 && \text{Factor out the GCF, } x. \\
x(x - 2)(x + 2) &= 0 && \text{Difference of squares} \\
x = 0 \quad \text{or} \quad x - 2 = 0 \quad \text{or} \quad x + 2 &= 0 && \text{Zero-product property} \\
x = 0 \quad \text{or} \quad x = 2 \quad \text{or} \quad x &= -2 && \text{Solve for } x.
\end{aligned}
$$

Review Exercises

SECTION 6.1

Exercises 1–4: Identify the greatest common factor for the expression and then factor the expression.

1. $8z^3 - 4z^2$

2. $6x^4 + 3x^3 - 12x^2$

3. $9xy + 15yz^2$

4. $a^2b^3 + a^3b^2$

Exercises 5–10: Factor by grouping.

5. $x(x + 2) - 3(x + 2)$

6. $y^2(x - 5) + 3y(x - 5)$ **7.** $z^3 - 2z^2 + 5z - 10$

8. $t^3 + t^2 + 8t + 8$ **9.** $x^3 - 3x^2 + 6x - 18$

10. $ax + bx - ay - by$

SECTION 6.2

Exercises 11–14: Find an integer pair that has the given product and sum.

11. Product: 20   Sum: 9

12. Product: -21 Sum: 4

13. Product: 36   Sum: -13

14. Product: -100 Sum: -21

Exercises 15–22: Factor the trinomial

15. $x^2 - x - 12$ **16.** $x^2 + 10x + 24$

17. $x^2 + 6x - 16$ **18.** $x^2 - x - 42$

19. $x^2 + 2x - 3$ **20.** $x^2 + 22x + 120$

21. $10 - 7x + x^2$ **22.** $24 + 2x - x^2$

SECTION 6.3

Exercises 23–30: Factor the trinomial.

23. $9x^2 + 3x - 2$ **24.** $2x^2 + 3x - 5$

25. $3x^2 + 14x + 15$ **26.** $35x^2 - 2x - 1$

27. $24x^2 - 7x - 5$ **28.** $4x^2 + 33x - 27$

29. $12 - 5x - 2x^2$ **30.** $1 + 3x - 10x^2$

SECTION 6.4

Exercises 31–40: Factor.

31. $z^2 - 4$ **32.** $9z^2 - 64$

33. $36 - y^2$ **34.** $100a^2 - 81b^2$

35. $x^2 + 14x + 49$ **36.** $x^2 - 10x + 25$

37. $4x^2 - 12x + 9$ **38.** $9x^2 + 48x + 64$

39. $8t^3 - 1$ **40.** $27r^3 + 8t^3$

SECTION 6.5

Exercises 41–52: Solve the equation.

41. $mn = 0$ **42.** $y^2 = 0$

43. $(4x - 3)(x + 9) = 0$

44. $(1 - 4x)(6 + 5x) = 0$

45. $z(z - 1)(z - 2) = 0$ **46.** $z^2 - 7z = 0$

47. $y^2 - 64 = 0$ **48.** $y^2 + 9y + 14 = 0$

49. $x^2 = x + 6$ **50.** $10x^2 + 11x = 6$

51. $t(t - 14) = 72$ **52.** $t(2t - 1) = 10$

SECTION 6.6

Exercises 53–62: Factor completely.

53. $5x^2 - 15x - 50$ **54.** $-3x^2 - 6x + 45$

55. $y^3 - 4y$ **56.** $3y^3 + 6y^2 - 9y$

57. $2z^4 + 14z^3 + 20z^2$ **58.** $8z^4 - 32z^2$

59. $x^4 - 6x^2 + 9$ **60.** $2x^4 - 15x^2 - 27$

61. $a^2 + 10ab + 25b^2$ **62.** $x^3 - xy^2$

Exercises 63–70: Solve.

63. $16x^2 - 72x - 40 = 0$

64. $2x^3 - 11x^2 + 15x = 0$ **65.** $t^3 = 25t$

66. $t^4 - 7t^3 + 12t^2 = 0$ **67.** $z^4 + 16 = 8z^2$

68. $z^4 - 256 = 0$ **69.** $y^3 = -64$

70. $y^3 - y^2 - y + 1 = 0$

APPLICATIONS

71. A square has area $9x^2 + 42x + 49$. Find the length of a side. Make a sketch of the square.

72. A rectangle has area $x^2 + 6x + 5$. Find possible dimensions for the rectangle. Make a sketch of the rectangle.

73. A cube has surface area $6x^2 + 12x + 6$. Find the length of a side.

74. Write a polynomial in factored form that represents the total area of the rectangle.

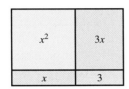

75. Write a polynomial in factored form that represents the total area of the rectangle.

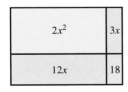

76. *Radius of a Circle*   The area and the circumference of a circle are numerically equal. Find the radius of the circle.

77. *Dimensions of a Shed*   The floor of a rectangular shed is 7 feet longer than it is wide and has an area of 120 square feet. What are its dimensions?

78. *Flight of a Ball*   A ball is hit upward. Its height h in feet after t seconds is given by $h = -16t^2 + 80t + 4$. At what times is the ball 100 feet in the air?

79. *Stopping Distance*   The stopping distance D in feet that it takes for a car traveling x miles per hour to stop on wet, level pavement can be approximated by the formula $D = \frac{1}{9}x^2 + \frac{11}{3}x$.
   (a) Estimate the distance required for the car to stop when it is traveling 45 miles per hour.

(b) If the stopping distance is 80 feet, what is the speed of the car?

 (c) If you have a calculator, use it to solve part (b) numerically with a table of values. Do your answers agree?

80. *Revenue*   A company makes tops for the boxes of pickup trucks. The total revenue R in dollars from selling the tops for p dollars each is given by $R = p(200 - p)$, where $p \le 200$.
   (a) Find R when $p = \$100$.
   (b) Find p when $R = \$7500$.
 (c) If you have a calculator, use it to solve part (b) numerically with a table of values. Do your answers agree?

81. *Airline Passengers*   The number N of worldwide airline passengers in millions during year y from 1950 to 1990 is approximated by

$$N = 0.68y^2 + 3.8y + 24,$$

where $y = 0$ corresponds to 1950, $y = 1$ to 1951, and so on until $y = 40$ corresponds to 1990.
   (a) Estimate the number of airline passengers in 1970.
 (b) Use a table of values to estimate the year in which the number of airline passengers reached 544 million.

82. *Digital Photographs*   A digital photograph is 30 pixels longer than it is wide and has a total area of 4000 pixels. Find the dimensions of this photograph.

83. *Dimensions of a Box*   A box is made from a rectangular piece of metal with length 50 inches and width 40 inches by cutting out square corners and folding up the sides.
   (a) Write an expression that gives the surface area of the inside of the box.
   (b) If the surface area of the box is 1900 square inches, find x.

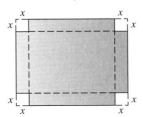 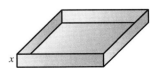

CHAPTER 6 Test

Exercises 1 and 2: Identify the greatest common factor for the expression. Then factor the expression.

1. $4x^2y - 20xy^2 + 12xy$ **2.** $9a^3b^2 + 3a^2b^2$

Exercises 3 and 4: Factor by grouping.

3. $ay + by + az + bz$ **4.** $3x^3 + x^2 - 15x - 5$

Exercises 5–8: Factor the trinomial.

5. $y^2 + 4y - 12$ **6.** $4x^2 + 20x + 25$

7. $4z^2 - 19z + 12$ **8.** $21 - 17t + 2t^2$

Exercises 9–12: Solve the equation.

9. $x^2 - 16 = 0$ **10.** $y^2 = y + 20$

11. $9z^2 + 16 = 24z$ **12.** $x(x - 5) = 66$

Exercises 13 and 14: Factor completely.

13. $6x^3 + 3x^2 - 3x$ **14.** $2z^4 - 12z^2 - 54$

Exercises 15 and 16: Solve the equation.

15. $y^3 = 9y$ **16.** $x^4 - 5x^2 + 4 = 0$

17. A square has area $9x^2 + 30x + 25$. Find the length of a side in terms of x.

18. Write a polynomial in factored form that represents the total area of the rectangle.

x^2	$3x$
$2x$	6

19. *Braking Distance*   The braking distance D in feet required for a car traveling at x miles per hour to stop on dry, level pavement can be modeled by $D = \frac{1}{11}x^2$.
   (a) Calculate the distance required for the car to stop when it is traveling 55 miles per hour.
   (b) If the braking distance is 99 feet, estimate the speed of the car.

20. *Flight of a Ball*   A ball is thrown upward. Its height h in feet after t seconds is given by $h = -16t^2 + 48t + 4$. At what times is the ball 36 feet in the air?

CHAPTER 6

Extended and Discovery Exercises

Exercises 1–6: *Difference of Two Squares*   The difference of two squares can be factored by using

$$a^2 - b^2 = (a - b)(a + b).$$

This equation can also be used in some situations where an expression may not appear to be the difference of two squares. For example, because $(\sqrt{3})^2 = 3$, $x^2 - 3$ can be written and then factored as

$$x^2 - 3 = x^2 - (\sqrt{3})^2$$
$$= (x - \sqrt{3})(x + \sqrt{3}).$$

Use this concept to factor the following expressions as the difference of two squares.

1. $x^2 - 5$

2. $y^2 - 7$

3. $3z^2 - 25$

4. $7t^2 - 11$

5. $x - 4$ for $x \ge 0$ (*Hint:* $(\sqrt{x})^2 = x$.)

6. $x - 7$ for $x \ge 0$

Exercises 7–12: *Solving Equations*   (Refer to Exercises 1–6.) Solve the equation by factoring it as the difference of two squares.

7. $x^2 - 3 = 0$

8. $y^2 - 7 = 0$

9. $3x^2 - 25 = 0$

10. $7x^2 - 11 = 0$

11. $x^4 - 9 = 0$

12. $x^4 - 25 = 0$

CHAPTERS 1–6

Cumulative Review Exercises

1. Write 144 as a product of prime numbers.

2. Evaluate $-2x + 3y$ for $x = -2$ and $y = 4$.

3. Translate the phrase "Five more than twice a number" to an algebraic expression containing the variable n.

4. Find the reciprocal of $-\frac{8}{5}$.

Exercises 5–8: Evaluate by hand and then reduce to lowest terms.

5. $\frac{3}{5} \cdot \frac{15}{21}$

6. $\frac{1}{2} \div \frac{5}{4}$

7. $\frac{5}{8} + \frac{1}{8}$

8. $\frac{4}{5} - \frac{1}{10}$

Exercises 9 and 10: Evaluate by hand.

9. $26 - 3 \cdot 6 \div 2$

10. $-2^2 + \frac{3 + 2}{8 + 2}$

11. Plot each number on the same number line.
   (a) -2 **(b)** 3 **(c)** $-\frac{1}{2}$

12. Simplify $\dfrac{4x}{9y} \div \dfrac{6x}{3y}$.

Exercises 13 and 14: Simplify the expression.

13. $-2 + 7x + 4 - 5x$

14. $-4(4 - y) + (5 - 3y)$

15. Solve $4t - 7 = 25$ and check.

16. Complete the table. Then use the table to solve the equation $2x + 3 = 5$.

x	-2	-1	0	1	2
$2x + 3$					

17. Translate the sentence "Triple a number decreased by 5 equals the number decreased by 7" into an equation using the variable n. Then solve the equation.

18. Convert 5.7% to fraction and decimal notation.

19. Convert 0.123 to a percentage.

20. If $r = 60$ mph and $d = 150$ miles, use the formula $d = rt$ to find t.

21. Find the area of a circle with a 5-foot diameter. Leave your answer in terms of π.

22. Solve $P = 2W + 2L$ for W.

23. Use a number line to graph the solution set for the inequality $x + 2 \geq -1$.

24. Solve $5 - 3z < -1$.

25. Make a scatterplot having the following five points: $(-2, 3), (-1, 2), (0, -1), (1, 1),$ and $(2, 2)$.

Exercises 26–29: Graph the given equation. Determine any intercepts.

26. $y = 3x - 2$ **27.** $2x - 3y = 6$

28. $x = 1$ **29.** $y = -2$

30. Identify the x-intercept and the y-intercept. Then write the slope–intercept form of the line.

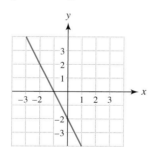

31. Sketch a line with slope $m = 3$ passing through $(-2, -1)$. Write its slope–intercept form.

32. The table lists points located on a line. Write the slope–intercept form of the line.

x	-2	-1	0	1
y	-5	-3	-1	1

Exercises 33–35: Find the slope–intercept form for the line satisfying the given conditions.

33. Parallel to $y = -\frac{2}{3}x + 1$ and passing through the point $(2, -1)$

34. Perpendicular to $2x - 3y = -6$ and passing through the point $(1, 2)$

35. Passing through the points $(-2, 1)$ and $(1, 5)$

36. A fish population is initially 2000 and is increasing by 200 per month. Write an equation that gives the number N of fish after x months.

37. Determine which ordered pair is a solution to the system of equations.

$$(2, -4), (1, -2)$$
$$4x + y = 2$$
$$x - 4y = 9$$

38. The graphs of two equations are shown. Use the graphs to identify the solution to the system of equations. Then check your answer.

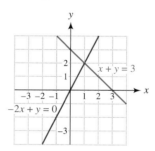

39. A table for two equations is given. Identify the solution to both equations.

x	-1	0	1	2
$y = 2x$	-2	0	2	4
$y = 3 - x$	4	3	2	1

Exercises 40–42: Solve the system of equations.

40. $y = -1$
$2x + y = 1$

41. $5x + y = -5$
$-x + 2y = 12$

42. $-3r - t = 2$
$2r + t = -4$

Exercises 43 and 44: The graphs of two linear equations are shown.
    (a) State the number of solutions to the system of equations.
    (b) Is the system consistent or inconsistent? If the system is consistent, state whether the equations are dependent or independent.

43.   **44.**

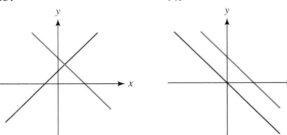

Exercises 45 and 46: Solve the system of linear equations. Note that these systems may have zero, one, or infinitely many solutions.

45. $\;\;2x - y = \;\;5$
    $-2x + y = -5$

46. $\;\;4x - 6y = 12$
    $-6x + 9y = 18$

47. Solve the system of equations symbolically, graphically, and numerically.

$$-2x + y = 0$$
$$x + y = 3$$

48. Use the graph to identify one solution to $2x + y < 1$. Then shade the solution set.

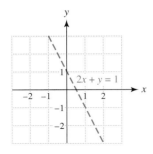

Exercises 49 and 50: Shade the solution set to the inequality.

49. $x \le 2$ **50.** $2x + 3y \ge 6$

Exercises 51 and 52: Shade the solution set to the system of inequalities.

51. $x > -1$ **52.** $2x - \;y \le 4$
    $y < x$ $x + 2y \ge 2$

Exercises 53–68: Simplify the expression.

53. $(5x^2 - 3) + (-x^2 + 4)$

54. $(2xy + x^2) - (2x^2 - 5xy)$

55. -2^4 **56.** $(xy)^0$

57. $3z^2 \cdot 5z^6$ **58.** $(a^2b^4)(ab^2)$

59. $(ab)^3$ **60.** $(xy)^4(x^3y^{-4})^2$

61. $7x^3(-2x^2 + 3x)$

62. $(a - b)(a^2 + ab + b^2)$

63. $(7x - 2)(3x + 5)$ **64.** $(2y^2 - 3)(5y^2 + 2)$

65. $(2rt + 3)(2rt - 3)$ **66.** $(2t + 5)^2$

67. $(x - 7)^2$ **68.** $(x^2 - y^2)^2$

Exercises 69–75: Simplify and write the expression using positive exponents.

69. $2^{-4} \cdot 2^5$ **70.** $\dfrac{1}{3^{-2}}$

71. $a^{-4} \cdot a^2$ **72.** $(2t^3)^{-2}$

73. $(xy)^{-3}(x^{-1}y^2)^{-1}$ **74.** $\dfrac{4x^2}{2x^4}$

75. $\left(\dfrac{2x}{y^{-2}}\right)^5$

Exercises 76 and 77: Write the expression in standard form.

76. 8.3×10^4 **77.** 6.23×10^{-3}

Exercises 78 and 79: Write the number in scientific notation.

78. 543,000   **79.** 0.00123

Exercises 80–82: Divide and check.

80. $\dfrac{6x^3 + 12x^2}{3x}$ **81.** $\dfrac{2x^2 - x + 3}{x - 1}$

82. $\dfrac{3x^3 - x + 1}{x^2 + 1}$

Exercises 83 and 84: Identify the greatest common factor for the expression. Then factor the expression.

83. $20z^3 - 15z^2$

84. $12x^2y + 15xy^2$

Exercises 85 and 86: Factor by grouping.

85. $2y^2(x + 2) - 5(x + 2)$

86. $t^3 + 6t^2 + t + 6$

Exercises 87–100: Factor completely.

87. $x^2 + 3x - 28$

88. $6y^2 + y - 12$

89. $6 + 13x - 5x^2$

90. $9z^2 - 4$

91. $25x^2 - 4y^2$

92. $t^2 + 16t + 64$

93. $64x^2 - 16x + 1$

94. $27t^3 - 8$

95. $-4x^2 + 4x + 24$

96. $x^3 - 16x$

97. $x^3 + 2x^2 - 99x$

98. $x^4 - 12x^2 + 27$

99. $a^2 + 6ab + 9b^2$

100. $x^3y - x^2y^2$

Exercises 101–107: Solve the equation.

101. $(2x - 7)(x + 5) = 0$

102. $2x^2 - 4x = 0$

103. $y^2 + 5y - 14 = 0$ **104.** $y^4 = 25y^2$

105. $8z^2 + 8z - 16 = 0$ **106.** $4z^3 = 49z$

107. $x^4 - 18x^2 + 81 = 0$

APPLICATIONS

108. *Shoveling the Driveway*   Two people are shoveling snow from a driveway. The first person shovels 10 square feet per minute, while the second person shovels 8 square feet per minute.
   (a) Write and simplify an expression that gives the total square feet that the two people shovel in x minutes.
   (b) How many minutes would it take for them to clear a driveway with an area of 900 square feet?

109. *Running Distance*   A person runs the following distances over three days: $1\frac{3}{4}$ miles, $2\frac{1}{2}$ miles, and $2\frac{2}{3}$ miles. How far did the person run altogether?

110. *Cost of a Car*   A 7% sales tax on a new car amounted to $1470. Find the cost of the car.

111. *Burning Calories*   An athlete can burn 12 calories per minute while cross-country skiing and 9 calories per minute while running at 5 miles per hour. If the athlete burns 615 calories in 60 minutes, how long is spent on each activity?

112. *Head Start*   In 1980, 376 thousand children participated in the federal Head Start program. By 2000 this number had increased to 858 thousand students. (*Source*: Department of Health and Human Services.)
   (a) Find the slope of the line passing through the points (1980, 376) and (2000, 858).
   (b) Interpret the slope as a rate of change.
   (c) If trends continue, estimate the number of participants in 2010.

113. *Renting a Car*   A rental car costs $20 plus $0.25 per mile.
   (a) Write an equation that gives the cost C of driving the car x miles.
   (b) Determine the number of miles that the car is driven if the rental cost is $100.

114. *Triangle*   In an isosceles triangle, the measures of the two smaller angles are equal and their sum is 20° more than the larger angle.
   (a) Let x be the measure of one of the two smaller angles and y be the measure of the larger angle. Write a system of linear equations whose solution gives the measures of these angles.
   (b) Solve your system.

115. *Radios and CD Players*   A business manufactures at least three radios for each CD player. The total number of radios and CD players cannot exceed 90 per week. Shade the region that represents numbers of radios R and CD players P that can be produced each week within these restrictions.

116. *Energy*   The U.S. consumption of natural gas G in trillions of cubic feet from 1990 to 2000 can be approximated by

$$G = -0.036x^2 + 0.76x + 18.7,$$

where $x = 0$ corresponds to 1990, $x = 1$ to 1991, and so on.   (*Source*: Department of Energy.)
   (a) Evaluate this polynomial for $x = 5$ and interpret the answer.
   (b) In which year was natural gas consumption 22.7 trillion cubic feet?

117. *Areas of Rectangles*   Find a monomial equal to the sum of the areas of the rectangles. Calculate this sum for $x = 2$ yards and $y = 3$ yards.

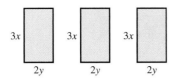

118. *Volume of a Cube*   Find an expression for the volume of a cube whose sides have length $2xy^2$.

119. *Area of a Square*   Complete each part and verify that your answers are the same.
   (a) Find the area of the large square by multiplying its length and width
   (b) Find the sum of the areas of the four smaller rectangles inside the large square.

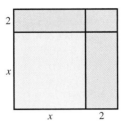

120. *Area of a Square*   A square has area $x^2 + 12x + 36$. Find the length of a side.

121. *Flight of a Golf Ball*   A golf ball is hit upward. Its height h in feet after t seconds is given by the formula $h = 64t - 16t^2$.
   (a) How long does it take for the ball to hit the ground?
   (b) At what times is the ball 48 feet in the air?

7

Rational Expressions

One of the most significant problems facing the U.S. transportation system is chronic highway congestion. Traffic congestion costs motorists more than $72 billion annually in wasted time and fuel. From 1982 to 1997, the amount of wasted time caused by traffic congestion increased from 1.9 to 4.3 billion hours. Some 6.6 billion gallons of fuel were wasted by vehicles of all kinds idling in traffic. Although people drive 143 percent more miles today than they did 30 years ago, there has been only a 5 percent increase in additional roads.

Traffic congestion is subject to a nonlinear effect. If the amount of traffic doubles on a highway, the wait in traffic may more than double. At times, only a slight increase in the traffic rate can result in a dramatic increase in the time spent waiting. Mathematics can describe this effect by using rational expressions, which are used frequently to diagnose and solve traffic and highway engineering problems. In this chapter we introduce rational expressions and some of their applications.

**There is nothing wrong with making mistakes.
Just don't respond with encores.
—Anonymous**

Sources: Philip J. Longman, "American Gridlock." *U.S. News & World Report*, May 28, 2001; U.S. Department of Transportation.

7.1 INTRODUCTION TO RATIONAL EXPRESSIONS

Basic Concepts · Simplifying Rational Expressions · Applications

INTRODUCTION

Have you ever been moving smoothly in traffic and noticed how it can come to a halt all of a sudden? Mathematics shows that, if the number of cars on a road increases even slightly, then the movement of traffic can slow dramatically. To understand this process better, we need to discuss rational expressions.

BASIC CONCEPTS

In Chapters 5 and 6 we discussed polynomials. Examples of polynomials include

$$3, \quad 2x, \quad x^2 + 4, \quad \text{and} \quad x^3 - 1.$$

In this chapter we discuss *rational expressions*. Rational expressions can be written as quotients (fractions) of two polynomials. Examples of rational expressions include

$$\frac{3}{2x}, \quad \frac{2x}{x^2 + 4}, \quad \frac{x^2 + 4}{3}, \quad \text{and} \quad \frac{x^3 - 1}{x^2 + 4}.$$

RATIONAL EXPRESSION

A rational expression can be written as $\frac{P}{Q}$, where P and Q are polynomials. A rational expression is defined whenever $Q \neq 0$.

We can evaluate polynomials for different values of a variable. For example, for $x = 2$ the polynomial $x^2 - 3x + 1$ evaluates to

$$(2)^2 - 3(2) + 1 = -1.$$

Rational expressions can be evaluated similarly.

EXAMPLE 1   Evaluating rational expressions

If possible, evaluate each expression for the given value of the variable.

(a) $\dfrac{1}{x + 1}$ $\quad x = 2$ $\quad$ (b) $\dfrac{y^2}{2y - 1}$ $\quad y = -4$

(c) $\dfrac{5z + 8}{z^2 - 2z + 1}$ $\quad z = 1$ $\quad$ (d) $\dfrac{2 - x}{x - 2}$ $\quad x = -3$

Solution (a) If $x = 2$, then $\frac{1}{x + 1}$ becomes $\frac{1}{2 + 1} = \frac{1}{3}$.

(b) If $y = -4$, then $\frac{y^2}{2y - 1}$ becomes $\frac{(-4)^2}{2(-4) - 1} = -\frac{16}{9}$.

(c) If $z = 1$, then $\frac{5z + 8}{z^2 - 2z + 1}$ becomes $\frac{5(1) + 8}{1^2 - 2(1) + 1} = \frac{13}{0}$, which is undefined because division by 0 is not possible.

(d) If $x = -3$, then $\frac{2 - x}{x - 2}$ becomes $\frac{2 - (-3)}{-3 - 2} = \frac{5}{-5} = -1$.

Rational expressions are different from polynomials because they are undefined whenever their denominators are 0. For example, the expression in Example 1(c) is undefined when $z = 1$.

EXAMPLE 2 **Determining when a rational expression is undefined**

Find all values of the variable for which each expression is undefined.

(a) $\dfrac{1}{x}$ **(b)** $\dfrac{4t}{t - 3}$ **(c)** $\dfrac{1 - 6r}{r^2 - 4}$ **(d)** $\dfrac{4}{x^2 + 1}$

Solution **(a)** A rational expression is undefined when its denominator is 0. Thus $\frac{1}{x}$ is undefined when $x = 0$.

(b) The expression $\frac{4t}{t - 3}$ is undefined when its denominator, $t - 3$, is 0, or when $t = 3$.

(c) The expression $\frac{1 - 6r}{r^2 - 4}$ is undefined when its denominator, $r^2 - 4$, is 0. Here

$$r^2 - 4 = (r - 2)(r + 2) = 0$$

implies that the denominator is 0 when $r = -2$, or $r = 2$.

(d) In the expression $\frac{4}{x^2 + 1}$ the denominator, $x^2 + 1$, is never 0 because any real number squared plus 1 is always greater than or equal to 1. Thus this rational expression is defined for all real numbers x.

SIMPLIFYING RATIONAL EXPRESSIONS

In Chapter 1, we used the *basic principle of fractions*,

$$\frac{a \cdot c}{b \cdot c} = \frac{a}{b}.$$

When we use this basic principle, the fraction $\frac{8}{12}$, for example, reduces to

$$\frac{8}{12} = \frac{2 \cdot 4}{3 \cdot 4} = \frac{2}{3}.$$

We can also apply this basic principle to rational expressions. For example,

$$\frac{x(x - 1)}{4(x - 1)} = \frac{x}{4},$$

provided that $x \neq 1$. Note that this simplification is not valid when $x = 1$. When simplifying rational expressions, we assume that values of the variable making a rational expression undefined are excluded.

||||||| **BASIC PRINCIPLE OF RATIONAL EXPRESSIONS**

The following property can be used to simplify rational expressions, where P, Q, and R are polynomials.

$$\frac{P \cdot R}{Q \cdot R} = \frac{P}{Q} \qquad Q \text{ and } R \text{ are nonzero.}$$

|||||||

Note: $\frac{P \cdot R}{Q \cdot R} = \frac{P}{Q} \cdot \frac{R}{R} = \frac{P}{Q} \cdot 1 = \frac{P}{Q}$, provided that $Q \neq 0$ and $R \neq 0$.

Like fractions, rational expressions can be written in *lowest terms*. For example, the rational expression $\frac{x^2 - 1}{x^2 + 2x + 1}$ can be written in lowest terms by factoring the numerator and the denominator and then applying the basic principle of rational expressions.

$$\frac{x^2 - 1}{x^2 + 2x + 1} = \frac{(x - 1)(x + 1)}{(x + 1)(x + 1)} \qquad \text{Factor the numerator and the denominator.}$$

$$= \frac{x - 1}{x + 1} \qquad \text{Apply } \frac{PR}{QR} = \frac{P}{Q} \text{ with } R = x + 1.$$

Because the basic principle of rational expressions cannot be applied further to $\frac{x-1}{x+1}$, we say that this expression is written in lowest terms.

EXAMPLE 3   Simplifying rational expressions

Simplify each expression and write it in lowest terms.

(a) $\frac{8y}{4y^2}$ **(b)** $\frac{2x + 6}{3x + 9}$ **(c)** $\frac{(z + 1)(z - 5)}{(z - 5)(z + 3)}$ **(d)** $\frac{x^2 - 9}{2x^2 + 7x + 3}$

Solution   **(a)** Factor out the greatest common factor, $4y$, in the numerator and the denominator. Then apply the basic principle of rational expressions.

$$\frac{8y}{4y^2} = \frac{2 \cdot 4y}{y \cdot 4y} = \frac{2}{y}$$

(b) Use the distributive property and then the basic principle of rational fractions.

$$\frac{2x + 6}{3x + 9} = \frac{2(x + 3)}{3(x + 3)} = \frac{2}{3}$$

(c) Use the commutative property to rewrite the denominator of the expression.

$$\frac{(z + 1)(z - 5)}{(z - 5)(z + 3)} = \frac{(z + 1)(z - 5)}{(z + 3)(z - 5)} = \frac{z + 1}{z + 3}$$

(d) Start by factoring the numerator and the denominator.

$$\frac{x^2 - 9}{2x^2 + 7x + 3} = \frac{(x - 3)(x + 3)}{(2x + 1)(x + 3)} = \frac{x - 3}{2x + 1} \qquad \underline{\qquad}$$

Recall that a negative sign can be placed in a fraction in a number of different ways. For example,

$$-\frac{5}{7} = \frac{-5}{7} = \frac{5}{-7}$$

illustrates three fractions that are equal. This property can also be applied to rational expressions, as demonstrated in the next example.

EXAMPLE 4   Distributing a negative sign

Simplify each expression.

(a) $-\frac{5 - x}{x - 5}$ **(b)** $\frac{-x - 6}{2x + 12}$ **(c)** $\frac{10 - z}{z - 10}$

Solution (a) Rewrite the expression with the negative sign in the numerator and then apply the distributive property. Be sure to include parentheses around the numerator.

$$-\frac{5 - x}{x - 5} = \frac{-(5 - x)}{x - 5} = \frac{-5 + x}{x - 5} = \frac{x - 5}{x - 5} = 1$$

The same answer can be obtained by distributing the negative sign over the denominator.

$$-\frac{5 - x}{x - 5} = \frac{5 - x}{-(x - 5)} = \frac{5 - x}{-x + 5} = \frac{5 - x}{5 - x} = 1$$

(b) Factor -1 out of the numerator and 2 out of the denominator.

$$\frac{-x - 6}{2x + 12} = \frac{-1(x + 6)}{2(x + 6)} = -\frac{1}{2}$$

(c) Start by factoring -1 out of the numerator.

$$\frac{10 - z}{z - 10} = \frac{-1(-10 + z)}{z - 10} = \frac{-1(z - 10)}{z - 10} = -1$$

═══ MAKING CONNECTIONS ═══

Negative Signs and Rational Expression

In general, $(b - a)$ equals $-1(a - b)$. As a result, if $a \neq b$, then

$$\frac{b - a}{a - b} = -1.$$

See Example 4(c).

APPLICATIONS

The next example is based on the introduction to this section and illustrates modeling of traffic flow with a rational expression.

EXAMPLE 5 Modeling traffic flow

Suppose that 10 cars per minute can pass through a construction zone. If traffic arrives *randomly* at an average rate of x cars per minute, the average time T in minutes spent waiting in line is given by

$$T = \frac{1}{10 - x},$$

where $x < 10$. (***Source:*** N. Garber and L. Hoel, *Traffic and Highway Engineering.*)

(a) Complete Table 7.1 by finding T for each value of x.

TABLE 7.1 Waiting in Traffic

x (cars/minute)	5	7	9	9.5	9.9	9.99
T (minutes)						

(b) Interpret the results.

Solution (a) When $x = 5$ cars per minute, then $T = \frac{1}{10-5} = \frac{1}{5}$ minute. Other values are found similarly, as shown in Table 7.2.

TABLE 7.2   Waiting in Traffic

x (cars/minute)	5	7	9	9.5	9.9	9.99
T (minutes)	$\frac{1}{5}$	$\frac{1}{3}$	1	2	10	100

(b) Note that, as the average traffic rate increases from 9 cars per minute to 9.9 cars per minute, the time needed to pass through the construction zone increases from 1 minute to 10 minutes. As x nears 10 cars per minute, a small increase in x increases the waiting time dramatically.

This nonlinear effect for traffic congestion in Example 5 is shown in Figure 7.1, where points from Table 7.2 have been plotted and a curve passing through them has been sketched. A vertical dashed line was also sketched at $x = 10$. This dashed line is called a **vertical asymptote** and indicates that the rational expression is undefined at this value of x. Near the vertical asymptote, the waiting time T increases dramatically for small increases in x. Note that the graph of T does not intersect or cross this vertical asymptote.

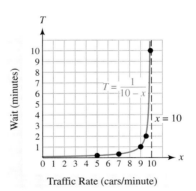

Figure 7.1

Technology Note: *Making Tables*

Table 7.2 can also be created with a graphing calculator by using the Ask feature, as illustrated in the following displays.

Calculator Help
To make a table of values, see Appendix A (page AP-3).

```
TABLE SETUP
 TblStart=5
 ΔTbl=1
 Indpnt:  Auto  Ask
 Depend:  Auto  Ask
```

X	Y₁
5	.2
7	.33333
9	1
9.5	2
9.9	10
9.99	100

Y₁≣1/(10−X)

AN APPLICATION INVOLVING PROBABILITY   If 10 marbles, one blue and nine red, are placed in a jar, then the *probability,* or *likelihood,* of picking the blue marble at random is 1 *chance* in 10, or $\frac{1}{10}$. The probability of drawing a red marble at random is 9 chances in 10, or $\frac{9}{10}$. **Probability** is a real number between 0 and 1. A probability of 0 indicates that an event is impossible, whereas a probability of 1 indicates that an event is certain.

EXAMPLE 6   Calculating probability

Suppose that n balls, numbered 1 to n, are placed in a container and only one ball has the winning number.
(a) What is the probability of drawing the winning ball at random?
(b) Calculate this probability for $n = 100$, 1000, and 10,000.
(c) What happens to the probability of drawing the winning ball as the number of balls increases?

Solution   **(a)** There is 1 chance in n of drawing the winning ball, so the probability is $\frac{1}{n}$.
(b) For $n = 100$, 1000, and 10,000, the probabilities are $\frac{1}{100}$, $\frac{1}{1000}$, and $\frac{1}{10,000}$.
(c) As the number of balls increases, the probability of picking the winning ball decreases.

Critical Thinking

What is the probability of *not* drawing a winning number if there are 100 balls and n balls have a winning number, where $n \leq 100$?

7.1   PUTTING IT ALL TOGETHER

In this section we discussed some basic concepts of rational expressions. They are summarized in the following table.

Concept	Explanation	Examples
Rational Expression	An expression of the form $\frac{P}{Q}$, where P and Q are polynomials with $Q \neq 0$.	$\dfrac{1}{x}, \dfrac{x-3}{2x^2-1}, \dfrac{2x+9}{5x}$, and $\dfrac{x^2+3x-5}{1}$
Undefined Values for a Rational Expression	If the *denominator* of a rational expression is 0, then the rational expression is undefined for that value of the variable.	$\dfrac{1}{x-3}$ is undefined when $x = 3$. $\dfrac{5y}{y^2-1}$ is undefined when $y = 1$ or when $y = -1$.
Basic Principle of Rational Expressions	Factor the numerator and the denominator. Then apply $$\frac{P \cdot R}{Q \cdot R} = \frac{P}{Q}.$$	$\dfrac{4xy^2}{6xy^3} = \dfrac{2(2xy^2)}{3y(2xy^2)} = \dfrac{2}{3y}$ $\dfrac{4x(x-4)}{(x+1)(x-4)} = \dfrac{4x}{x+1}$

7.1 EXERCISES

FOR EXTRA HELP

 Student's Solutions Manual

MyMathLab

InterAct Math

Math Tutor Center

MathXL

Digital Video Tutor CD 3 Videotape 9

CONCEPTS

1. A rational expression can be written as ____, where P and Q are ____ with $Q \neq 0$.

2. Is $\frac{x}{2x^2 + 1}$ a rational expression? Why or why not?

3. A rational expression is undefined whenever ____.

4. The rational expression $\frac{1}{x - a}$ is undefined whenever $x =$ ____.

5. The basic principle of fractions states that a fraction can be reduced by using $\frac{a \cdot c}{b \cdot c} =$ ____.

6. The basic principle of rational expressions can be used to simplify ____ expressions.

7. The expression $\frac{x - a}{a - x}$ simplifies to ____.

8. The expression $-\frac{x - 1}{x + 1}$ simplifies to $\frac{?}{x + 1}$.

EVALUATING RATIONAL EXPRESSIONS

Exercises 9–20: If possible, evaluate the expression at the given value of the variable.

9. $\dfrac{3}{x}$ $\qquad x = -7$

10. $\dfrac{3}{x + 3}$ $\qquad x = 0$

11. $-\dfrac{x}{x - 5}$ $\qquad x = -4$

12. $-\dfrac{4x}{5x + 1}$ $\qquad x = 1$

13. $\dfrac{y + 1}{y^2}$ $\qquad y = -2$

14. $\dfrac{3y - 1}{y^2 + 1}$ $\qquad y = -1$

15. $\dfrac{7z}{z^2 - 4}$ $\qquad z = -2$

16. $\dfrac{5}{z^2 - 3z + 2}$ $\qquad z = -1$

17. $\dfrac{5}{3t + 6}$ $\qquad t = -2$

18. $\dfrac{4t}{2t + 5}$ $\qquad t = -\dfrac{5}{2}$

19. $-\dfrac{6 - x}{x - 6}$ $\qquad x = 0$

20. $\dfrac{8 - 2x}{2x - 8}$ $\qquad x = -5$

Exercises 21–24: Complete the table for the given rational expression. If a value is undefined, place a dash in the table.

21.

x	-2	-1	0	1	2
$\dfrac{x}{x + 1}$					

22.

x	-2	-1	0	1	2
$\dfrac{2x}{3x - 1}$					

23.

x	-2	-1	0	1	2
$\dfrac{3x}{2x^2 + 1}$					

24.

x	-2	-1	0	1	2
$\dfrac{2x - 1}{x^2 - 1}$					

Exercises 25–38: Find any values of the variable that make the expression undefined.

25. $-\dfrac{8}{x}$

26. $\dfrac{7}{x + 1}$

27. $\dfrac{4}{z - 3}$

28. $\dfrac{7 - z}{z - 7}$

29. $\dfrac{4y}{5y + 4}$

30. $\dfrac{3 + y}{3y - 7}$

31. $\dfrac{5t + 2}{t^2 + 1}$

32. $\dfrac{8t}{t^2 + 25}$

33. $\dfrac{8x}{x^2 - 25}$

34. $\dfrac{x + 4}{x^2 - 36}$

35. $\dfrac{x^2 + 3x + 2}{x^2 + 5x + 6}$

36. $\dfrac{2x - 1}{x^2 - 7x + 10}$

37. $\dfrac{8z^2 + z + 1}{2z^2 - 7z + 5}$

38. $\dfrac{4n^2 + 17n - 15}{3n^2 - 8n + 4}$

SIMPLIFYING RATIONAL EXPRESSIONS

Exercises 39–46: Reduce the fraction to lowest terms.

39. $\frac{12}{18}$

40. $\frac{24}{32}$

41. $\frac{24}{48}$

42. $-\frac{22}{33}$

43. $-\frac{6}{15}$

44. $\frac{8}{22}$

45. $-\frac{25}{75}$

46. $-\frac{36}{42}$

Exercises 47–78: Simplify the expression and write it in lowest terms.

47. $\dfrac{5x^4}{10x^6}$

48. $\dfrac{6y^2}{9y}$

49. $\dfrac{8xy^3}{6x^2y^2}$

50. $\dfrac{36x^2y^5}{6x^5y}$

51. $\dfrac{x + 4}{2x + 8}$

52. $\dfrac{5x - 10}{x - 2}$

53. $\dfrac{3z - 9}{5z - 15}$

54. $\dfrac{4z + 8}{10 + 5z}$

55. $\dfrac{(x + 1)(x - 1)}{(x + 6)(x - 1)}$

56. $\dfrac{(2x + 1)(x + 9)}{(4x - 3)(x + 9)}$

57. $\dfrac{(5y + 3)(2y - 1)}{(2y - 1)(y + 2)}$

58. $\dfrac{(4y - 1)(5y + 7)}{(5y + 7)(1 - 4y)}$

59. $\dfrac{x - 7}{7 - x}$

60. $\dfrac{5 - x}{x - 5}$

61. $\dfrac{a - b}{b - a}$

62. $\dfrac{2t - 3r}{3r - 2t}$

63. $\dfrac{(3x + 5)(x - 1)}{(3x - 5)(1 - x)}$

64. $\dfrac{(2 - x)(x - 2)}{(x - 2)(2 - x)}$

65. $\dfrac{n^2 - n}{n^2 - 5n}$

66. $\dfrac{3n^2 - 4n}{n^2 + 4n}$

67. $\dfrac{x^2 - 3x}{6x - 18}$

68. $\dfrac{4x^2 + 16x}{5x^2 + 20x}$

69. $\dfrac{z^2 - 3z + 2}{z^2 - 4z + 3}$

70. $\dfrac{z^2 - 3z - 10}{z^2 - 2z - 8}$

71. $\dfrac{2x^2 + 7x - 4}{6x^2 + x - 2}$

72. $\dfrac{5x^2 + 3x - 2}{5x^2 + 13x - 6}$

73. $\dfrac{x - 3}{3x^2 - 11x + 6}$

74. $\dfrac{2x - 1}{4x^2 + 6x - 4}$

75. $-\dfrac{a - 9}{9 - a}$

76. $-\dfrac{-b - 6}{b + 6}$

77. $\dfrac{-2x - 1}{4x + 2}$

78. $\dfrac{4x + 3}{-8x - 6}$

APPLICATIONS

79. *Modeling Traffic Flow* (Refer to Example 5.) Five vehicles per minute can pass through a construction zone. If the traffic arrives randomly at an average rate of x vehicles per minute, the average time T in minutes spent waiting in line is given by $T = \frac{1}{5 - x}$ for $x < 5$. (*Source:* N. Garber.)
   (a) Evaluate the expression for $x = 3$ and interpret the result.
   (b) Complete the table and interpret the results.

x	2	4	4.5	4.9	4.99
T					

80. *Standing in Line* A checker in a grocery store can serve 20 customers per hour. If people arrive randomly at an average rate of x per hour, then the average number of customers N waiting in line is given by $N = \frac{x^2}{400 - 20x}$ for $x < 20$. (*Source:* N. Garber.)
   (a) Complete the table.

x	5	10	18	19	20
N					

(b) Compare the number of customers waiting in line if the average rate increases from 18 to 19 customers per hour.

81. *Probability* Suppose that a coin is flipped. What is the probability that a head appears?

82. *Probability* A die shows the numbers 1, 2, 3, 4, 5, and 6. If each number has an equal chance of appearing on any given roll, what is the probability that a 2 or 4 appears?

83. *Probability* (Refer to Example 6.) Suppose that there are *n* balls in a container and that three balls have a winning number. If a ball is drawn randomly, do each of the following.
   (a) Write a rational expression that gives the probability of drawing a winning ball.
   (b) Write a rational expression that gives the probability of not drawing a winning ball. Evaluate your expression for $n = 100$ and interpret the result.

84. *Surface Area of a Cylinder* If a cylindrical container has a volume of π cubic feet, then its surface area *S* in square feet (excluding the top and bottom) is given by $S = \frac{2\pi}{r}$, where *r* is the radius of the cylinder.

   (a) Calculate *S* when $r = \frac{1}{2}$ foot.
   (b) What happens to this surface area when *r* becomes large? Sketch this situation.
   (c) What happens to the surface area when *r* becomes small (nearly 0)? Sketch this situation.

85. *Distance and Time* A car is traveling at 60 miles per hour.
   (a) How long does it take the car to travel 360 miles?
   (b) Write a rational expression that gives the time that it takes the car to travel *M* miles.

86. *Distance and Time* A bicyclist rides uphill at 10 miles per hour for 5 miles and then rides downhill at 20 miles per hour for 5 miles. What was the bicyclist's average speed? (*Hint:* Average speed equals distance divided by time.)

Exercises 87 and 88: Traffic Flow (*Refer to Example 5.*) *The figure shows a graph of the waiting time T in minutes at a construction zone when cars are arriving randomly at an average rate of x cars per minute.*
   (a) *Give the equation of the vertical asymptote.*
   (b) *Explain what the graph indicates about the traffic flow.*

87.

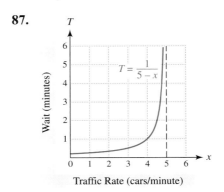

$$T = \frac{1}{5 - x}$$

Traffic Rate (cars/minute)

88.

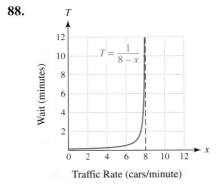

$$T = \frac{1}{8 - x}$$

Traffic Rate (cars/minute)

WRITING ABOUT MATHEMATICS

89. Explain what a rational expression is. When is a rational expression undefined?

90. Does the rational expression $\frac{5x + 2}{10x + 4}$ equal $\frac{5x}{10x} + \frac{2}{4}$? Explain your answer.

=== Group Activity: Working with Real Data ===

Directions: Form a group of 2 to 4 people. Select someone to record the group's responses for this activity. All members of the group should work cooperatively to answer the questions. If your instructor asks for your results, each member of the group should be prepared to respond.

Students Per Computer The following table lists students per computer in U.S. public schools for selected years.

Year	1983	1985	1987	1989
Students/Computer	125	50	32	22

Year	1991	1993	1995	1997
Students/Computer	18	14	10	6

Source: Quality Education Data, Inc.

(a) Make a scatterplot of the data. Would a straight line model the data accurately? Explain.

(b) Discuss how well the formula

$$S = \frac{125}{1 + 0.7(y - 1983)}$$

models these data, where S represents the students per computer and y represents the year.

(c) In what year does the formula reveal that there were about 17 students per computer?

7.2 MULTIPLICATION AND DIVISION OF RATIONAL EXPRESSIONS

Review of Multiplication and Division of Fractions ·
Multiplication of Rational Expressions · Division of Rational Expressions

INTRODUCTION

In previous chapters we reviewed how to add, subtract, multiply, and divide real numbers and polynomials. In this section we show how to multiply and divide rational expressions; in the next section we discuss addition and subtraction of rational expressions.

REVIEW OF MULTIPLICATION AND DIVISION OF FRACTIONS

To multiply two fractions we use the property

$$\frac{a}{b} \cdot \frac{c}{d} = \frac{ac}{bd}.$$

In the next example we review multiplication of fractions. (See Section 1.2.)

EXAMPLE 1   Multiplying fractions

Multiply and reduce your answers to lowest terms.
(a) $\frac{3}{7} \cdot \frac{4}{5}$ (b) $2 \cdot \frac{3}{4}$ (c) $\frac{4}{21} \cdot \frac{7}{8}$

Solution (a) $\frac{3}{7} \cdot \frac{4}{5} = \frac{12}{35}$ (b) $2 \cdot \frac{3}{4} = \frac{2}{1} \cdot \frac{3}{4} = \frac{6}{4} = \frac{3}{2}$

(c) $\frac{4}{21} \cdot \frac{7}{8} = \frac{7 \cdot 4}{21 \cdot 8} = \frac{1}{3} \cdot \frac{1}{2} = \frac{1}{6}$

To divide two fractions we "invert and multiply." That is, we change a division problem to a multiplication problem by using

$$\frac{a}{b} \div \frac{c}{d} = \frac{a}{b} \cdot \frac{d}{c}.$$

EXAMPLE 2 Dividing fractions

Divide and reduce your answers to lowest terms.

(a) $\frac{1}{3} \div \frac{5}{7}$ (b) $\frac{4}{5} \div 8$ (c) $\frac{8}{9} \div \frac{10}{3}$

Solution (a) $\frac{1}{3} \div \frac{5}{7} = \frac{1}{3} \cdot \frac{7}{5} = \frac{7}{15}$

(b) $\frac{4}{5} \div 8 = \frac{4}{5} \cdot \frac{1}{8} = \frac{4}{40} = \frac{1}{10}$

(c) $\frac{8}{9} \div \frac{10}{3} = \frac{8}{9} \cdot \frac{3}{10} = \frac{24}{90} = \frac{4 \cdot 6}{15 \cdot 6} = \frac{4}{15}$

MULTIPLICATION OF RATIONAL EXPRESSIONS

Multiplying rational expressions is similar to multiplying fractions.

PRODUCTS OF RATIONAL EXPRESSIONS

To multiply two rational expressions multiply the numerators and multiply the denominators. That is,

$$\frac{A}{B} \cdot \frac{C}{D} = \frac{AC}{BD},$$

where B and D are nonzero.

EXAMPLE 3 Multiplying rational expressions

Multiply and reduce to lowest terms. Leave your answers in factored form.

(a) $\frac{3}{x} \cdot \frac{2x - 5}{x - 1}$ (b) $\frac{x - 1}{4x} \cdot \frac{x + 3}{x - 1}$

(c) $\frac{x^2 - 4}{x + 3} \cdot \frac{x + 3}{x + 2}$ (d) $\frac{4}{x^2 + 3x + 2} \cdot \frac{x^2 + 2x + 1}{8}$

Solution (a) $\frac{3}{x} \cdot \frac{2x - 5}{x - 1} = \frac{3(2x - 5)}{x(x - 1)}$ Multiply the numerators and the denominators.

(b) $\frac{x - 1}{4x} \cdot \frac{x + 3}{x - 1} = \frac{(x - 1)(x + 3)}{4x(x - 1)}$ Multiply the numerators and the denominators.

$= \frac{(x + 3)(x - 1)}{4x(x - 1)}$ Commutative property

$= \frac{x + 3}{4x}$ Reduce. Note that $\frac{(x - 1)}{(x - 1)} = 1$.

(c) $\dfrac{x^2 - 4}{x + 3} \cdot \dfrac{x + 3}{x + 2} = \dfrac{(x - 2)(x + 2)}{x + 3} \cdot \dfrac{x + 3}{x + 2}$ Factor.

$\qquad\qquad\qquad = \dfrac{(x - 2)(x + 2)(x + 3)}{(x + 3)(x + 2)}$ Multiply the numerators and the denominators.

$\qquad\qquad\qquad = \dfrac{(x - 2)(x + 2)(x + 3)}{(x + 2)(x + 3)}$ Commutative property

$\qquad\qquad\qquad = x - 2$ Reduce. Note that $\dfrac{(x + 2)(x + 3)}{(x + 2)(x + 3)} = 1$.

(d) $\dfrac{4}{x^2 + 3x + 2} \cdot \dfrac{x^2 + 2x + 1}{8} = \dfrac{4(x^2 + 2x + 1)}{8(x^2 + 3x + 2)}$ Multiply the numerators and the denominators.

$\qquad\qquad\qquad = \dfrac{4(x + 1)(x + 1)}{8(x + 2)(x + 1)}$ Factor.

$\qquad\qquad\qquad = \dfrac{x + 1}{2(x + 2)}$ Reduce. Note that $\dfrac{(x + 1)}{(x + 1)} = 1$.

DIVISION OF RATIONAL EXPRESSIONS

Dividing rational expressions is similar to dividing fractions.

QUOTIENTS OF RATIONAL EXPRESSIONS

To divide two rational expressions multiply by the reciprocal of the divisor. That is,

$$\frac{A}{B} \div \frac{C}{D} = \frac{A}{B} \cdot \frac{D}{C},$$

where B, C, and D are nonzero.

EXAMPLE 4   Dividing rational expressions

Divide and reduce to lowest terms.

(a) $\dfrac{5}{2x} \div \dfrac{10}{x - 4}$

(b) $\dfrac{x^2 - 9}{x^2 + 4} \div (x - 3)$

(c) $\dfrac{x^2 - x}{x^2 - x - 2} \div \dfrac{x}{x - 2}$

Solution **(a)** $\dfrac{5}{2x} \div \dfrac{10}{x - 4} = \dfrac{5}{2x} \cdot \dfrac{x - 4}{10}$ Invert and multiply.

$\qquad\qquad\qquad = \dfrac{5(x - 4)}{20x}$ Multiply the numerators and the denominators.

$\qquad\qquad\qquad = \dfrac{x - 4}{4x}$ Reduce. Note that $\frac{5}{20} = \frac{1}{4}$.

(b) $\dfrac{x^2 - 9}{x^2 + 4} \div (x - 3) = \dfrac{x^2 - 9}{x^2 + 4} \cdot \dfrac{1}{x - 3}$ Invert and multiply.

$\qquad\qquad\qquad\qquad = \dfrac{x^2 - 9}{(x^2 + 4)(x - 3)}$ Multiply the numerators and the denominators.

$\qquad\qquad\qquad\qquad = \dfrac{(x + 3)(x - 3)}{(x^2 + 4)(x - 3)}$ Factor the numerator.

$\qquad\qquad\qquad\qquad = \dfrac{x + 3}{x^2 + 4}$ Reduce. Note that $\dfrac{(x - 3)}{(x - 3)} = 1$.

(c) $\dfrac{x^2 - x}{x^2 - x - 2} \div \dfrac{x}{x - 2} = \dfrac{x^2 - x}{x^2 - x - 2} \cdot \dfrac{x - 2}{x}$ Invert and multiply.

$\qquad\qquad\qquad\qquad = \dfrac{(x^2 - x)(x - 2)}{(x^2 - x - 2)x}$ Multiply the numerators and the denominators.

$\qquad\qquad\qquad\qquad = \dfrac{x(x - 1)(x - 2)}{x(x + 1)(x - 2)}$ Factor numerator and denominator.

$\qquad\qquad\qquad\qquad = \dfrac{x - 1}{x + 1}$ Reduce. Note that $\dfrac{x(x - 2)}{x(x - 2)} = 1$.

Critical Thinking

Simplify the expression to lowest terms mentally.

$\dfrac{x^2 - 1}{x + 5} \div \dfrac{x - 1}{x + 5} \div (x + 1)$

PUTTING IT ALL TOGETHER

In this section we showed how to multiply and divide rational expressions. The following table summarizes these methods.

Concept	Explanation	Examples
Basic Principle of Rational Expressions	$\dfrac{PR}{QR} = \dfrac{P}{Q}$, where Q and R are nonzero polynomials. Apply this principle to reduce rational expressions to lowest terms.	$\dfrac{(x + 2)(2x - 3)}{(x - 1)(2x - 3)} = \dfrac{x + 2}{x - 1}$
Multiplication of Rational Expressions	Multiply the numerators and multiply the denominators: $$\dfrac{A}{B} \cdot \dfrac{C}{D} = \dfrac{AC}{BD}.$$ Then reduce the result to lowest terms.	$\dfrac{4x}{x - 1} \cdot \dfrac{x - 1}{x + 1} = \dfrac{4x(x - 1)}{(x - 1)(x + 1)}$ $$= \dfrac{4x}{x + 1}$$
Division of Rational Expressions	Multiply the first expression by the reciprocal of the second expression: $$\dfrac{A}{B} \div \dfrac{C}{D} = \dfrac{A}{B} \cdot \dfrac{D}{C}.$$ Then reduce the result to lowest terms.	$\dfrac{x + 1}{x - 3} \div \dfrac{x + 1}{x - 5} = \dfrac{x + 1}{x - 3} \cdot \dfrac{x - 5}{x + 1}$ $$= \dfrac{(x - 5)(x + 1)}{(x - 3)(x + 1)}$$ $$= \dfrac{x - 5}{x - 3}$$

7.2 EXERCISES

FOR EXTRA HELP

 Student's Solutions Manual

 MyMathLab

 InterAct Math

Tutor Center   Math Tutor Center

 MathXL

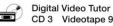 Digital Video Tutor   CD 3   Videotape 9

CONCEPTS

1. $\dfrac{A}{B} \cdot \dfrac{C}{D} =$ _____.

2. $\dfrac{A}{B} \div \dfrac{C}{D} =$ _____.

3. Simplify $\dfrac{(x+7)(x+2)}{(x+1)(x+2)}$.

4. $\dfrac{AC}{BC} =$ _____.

5. Does $\dfrac{x+2}{x}$ equal $\dfrac{x}{x} + 2$?

6. To divide $\frac{1}{3}$ by $\frac{4}{5}$ multiply _____ by _____.

REVIEW OF FRACTIONS

Exercises 7–18: Multiply and reduce to lowest terms.

7. $\frac{1}{2} \cdot \frac{4}{5}$

8. $\frac{6}{7} \cdot \frac{7}{18}$

9. $\frac{3}{7} \cdot 4$

10. $5 \cdot \frac{4}{5}$

11. $\frac{1}{3} \cdot \frac{2}{3} \cdot \frac{9}{11}$

12. $\frac{2}{5} \cdot \frac{10}{11} \cdot \frac{1}{4}$

13. $\frac{2}{3} \div \frac{1}{6}$

14. $\frac{5}{7} \div \frac{5}{8}$

15. $\frac{8}{9} \div \frac{5}{3}$

16. $\frac{7}{3} \div 6$

17. $8 \div \frac{4}{5}$

18. $\frac{1}{2} \div \frac{5}{4} \div \frac{2}{5}$

MULTIPLYING RATIONAL EXPRESSIONS

Exercises 19–26: Reduce the expression to lowest terms.

19. $\dfrac{x+5}{x+5}$

20. $\dfrac{2x-3}{2x-3}$

21. $\dfrac{(z+1)(z+2)}{(z+4)(z+2)}$

22. $\dfrac{(2z-7)(z+5)}{(3z+5)(z+5)}$

23. $\dfrac{8y(y+7)}{12y(y+7)}$

24. $\dfrac{6(y+1)}{12(y+1)}$

25. $\dfrac{x(x+2)(x+3)}{x(x-2)(x+3)}$

26. $\dfrac{2(x+1)(x-1)}{4(x+1)(x-1)}$

Exercises 27–46: Multiply and reduce to lowest terms. Leave your answers in factored form.

27. $\dfrac{8}{x} \cdot \dfrac{x+1}{x}$

28. $\dfrac{7}{2x} \cdot \dfrac{x}{x-1}$

29. $\dfrac{8+x}{x} \cdot \dfrac{x-3}{x+8}$

30. $\dfrac{5x^2+x}{2x-1} \cdot \dfrac{1}{x}$

31. $\dfrac{z+3}{z+4} \cdot \dfrac{z+4}{z-7}$

32. $\dfrac{2z+1}{3z} \cdot \dfrac{3z}{z+2}$

33. $\dfrac{5x+1}{3x+2} \cdot \dfrac{3x+2}{5x+1}$

34. $\dfrac{x+1}{x+3} \cdot \dfrac{x+3}{x+1}$

35. $\dfrac{(t+1)^2}{t+2} \cdot \dfrac{(t+2)^2}{t+1}$

36. $\dfrac{(t-1)^2}{(t+5)^2} \cdot \dfrac{t+5}{t-1}$

37. $\dfrac{x^2}{x^2+4} \cdot \dfrac{x+4}{x}$

38. $\dfrac{x-1}{x^2} \cdot \dfrac{x^2}{x^2+1}$

39. $\dfrac{z^2-1}{z^2-4} \cdot \dfrac{z-2}{z+1}$

40. $\dfrac{z^2-9}{z-5} \cdot \dfrac{z-5}{z+3}$

41. $\dfrac{y^2-2y}{y^2-1} \cdot \dfrac{y+1}{y-2}$

42. $\dfrac{y^2-4y}{y+1} \cdot \dfrac{y+1}{y-4}$

43. $\dfrac{2x^2-x-3}{3x^2-8x-3} \cdot \dfrac{3x+1}{2x-3}$

44. $\dfrac{6x^2+11x-2}{3x^2+11x-4} \cdot \dfrac{3x-1}{6x-1}$

45. $\dfrac{(x-3)^3}{x^2-2x+1} \cdot \dfrac{x-1}{(x-3)^2}$

46. $\dfrac{x^2+4x+4}{x^2-2x+1} \cdot \dfrac{(x-1)^2}{(x+2)^2}$

Exercises 47–68: Divide and reduce to lowest terms. Leave your answers in factored form.

47. $\dfrac{2}{x} \div \dfrac{2x+3}{x}$

48. $\dfrac{6}{2x} \div \dfrac{x+2}{2x}$

49. $\dfrac{x-2}{3x} \div \dfrac{2-x}{6x}$

50. $\dfrac{x+1}{2x-1} \div \dfrac{x+1}{x}$

51. $\dfrac{z+2}{z+1} \div \dfrac{z+2}{z-1}$

52. $\dfrac{z+7}{z-4} \div \dfrac{z+7}{z-4}$

53. $\dfrac{3y+4}{2y+1} \div \dfrac{3y+4}{y+2}$

54. $\dfrac{y+5}{y-2} \div \dfrac{y}{y+3}$

55. $\dfrac{t^2-1}{t^2+1} \div \dfrac{t+1}{4}$

56. $\dfrac{4}{2t^3} \div \dfrac{8}{t^2}$

57. $\dfrac{y^2 - 9}{y^2 - 25} \div \dfrac{y + 3}{y + 5}$ **58.** $\dfrac{y + 1}{y - 4} \div \dfrac{y^2 - 1}{y^2 - 16}$

59. $\dfrac{2x^2 - 4x}{2x - 1} \div \dfrac{x - 2}{2x - 1}$ **60.** $\dfrac{x - 4}{x^2 + x} \div \dfrac{5}{x + 1}$

61. $\dfrac{2z^2 - 5z - 3}{z^2 + z - 20} \div \dfrac{z - 3}{z - 4}$

62. $\dfrac{z^2 + 12z + 27}{z^2 - 5z - 14} \div \dfrac{z + 3}{z + 2}$

63. $\dfrac{t^2 - 1}{t^2 + 5t - 6} \div (t + 1)$

64. $\dfrac{t^2 - 2t - 3}{t^2 - 5t - 6} \div (t - 3)$

65. $\dfrac{a - b}{a + b} \div \dfrac{a - b}{2a + 3b}$

66. $\dfrac{x^3 - y^3}{x^2 - y^2} \div \dfrac{x^2 + xy + y^2}{x - y}$

67. $\dfrac{x - y}{x^2 + 2xy + y^2} \div \dfrac{1}{(x + y)^2}$

68. $\dfrac{a^2 - b^2}{4a^2 - 9b^2} \div \dfrac{a - b}{2a + 3b}$

APPLICATIONS

69. *Probability* Suppose that one jar holds n balls and that a second jar holds $n + 1$ balls. Each jar contains one winning ball.
  (a) The probability, or chance, of drawing the winning ball from the first jar and *not* drawing it from the second jar is
$$\frac{1}{n} \cdot \frac{n}{n + 1}.$$
  Simplify this expression.
  (b) Find this probability for $n = 99$.

70. *U.S. AIDS Cases* The cumulative number of AIDS cases C in the United States from 1982 to 1994 can be modeled by $C = 3200x^2 + 1586$, and the cumulative number of AIDS deaths D from 1982 to 1994 can be modeled by $D = 1900x^2 + 619$. In these equations $x = 0$ corresponds to 1982, $x = 1$ to 1983, and so on until $x = 12$ corresponds to 1994.
  (a) Write the rational expression $\frac{D}{C}$ in terms of x.
  (b) Evaluate your expression for $x = 4, 7$, and 10. Round your answers to the nearest thousandth. Interpret the results.
  (c) Explain what the rational expression $\frac{D}{C}$ represents.

WRITING ABOUT MATHEMATICS

71. Explain how to multiply two rational expressions.

72. Explain how to divide two rational expressions.

CHECKING BASIC CONCEPTS   SECTIONS 7.1 AND 7.2

1. If possible, evaluate the expression $\dfrac{3}{x^2 - 1}$ for $x = -1$ and $x = 3$.

2. Reduce each expression to lowest terms.
  (a) $\dfrac{6x^3y^2}{15x^2y^3}$ **(b)** $\dfrac{5x - 15}{x - 3}$ **(c)** $\dfrac{x^2 - x - 6}{x^2 + x - 12}$

3. Multiply and reduce each expression to lowest terms.
  (a) $\dfrac{4}{3x} \cdot \dfrac{2x}{6}$ **(b)** $\dfrac{2x + 4}{x^2 - 1} \cdot \dfrac{x + 1}{x + 2}$

4. Divide and reduce each expression to lowest terms.
  (a) $\dfrac{7}{3z^2} \div \dfrac{14}{5z^3}$ **(b)** $\dfrac{x^2 + x}{x - 3} \div \dfrac{x}{x - 3}$

5. *Waiting in Line* Customers are waiting in line at a department store. They arrive randomly at an average rate of x per minute. If the clerk can wait on 2 customers per minute, then the average time in minutes spent waiting in line is given by $T = \dfrac{1}{2 - x}$ for $x < 2$. (*Source:* N. Garber, *Traffic and Highway Engineering.*)
  (a) Complete the table.

x	0.5	1.0	1.5	1.9
T				

  (b) What happens to the waiting time as x increases but remains less than 2?

7.3 ADDITION AND SUBTRACTION WITH LIKE DENOMINATORS

Review of Addition and Subtraction of Fractions ·
Rational Expressions Having Like Denominators

INTRODUCTION

Addition of rational expressions is frequently used in applications involving probability. For example, if two dice are rolled, there are 2 chances in 36 that the sum will be 3 and 5 chances in 36 that the sum will be 6. Thus the probability that the sum will be either 3 or 6 is

$$\frac{2}{36} + \frac{5}{36} = \frac{7}{36},$$

or 7 chances in 36.

In this section we discuss methods for adding and subtracting rational expressions having like denominators. These methods are similar to the ones used to add and subtract fractions having like denominators.

REVIEW OF ADDITION AND SUBTRACTION OF FRACTIONS

In Section 1.2 we demonstrated how the property

$$\frac{a}{c} + \frac{b}{c} = \frac{a + b}{c}$$

can be used to add fractions having like denominators. For example,

$$\frac{3}{7} + \frac{2}{7} = \frac{3 + 2}{7} = \frac{5}{7}.$$

To subtract two fractions having like denominators the property

$$\frac{a}{c} - \frac{b}{c} = \frac{a - b}{c}$$

is used. For example,

$$\frac{2}{5} - \frac{4}{5} = \frac{2 - 4}{5} = -\frac{2}{5}.$$

EXAMPLE 1   Adding and subtracting fractions having like denominators

Simplify each expression and reduce to lowest terms.

(a) $\frac{3}{8} + \frac{4}{8}$ **(b)** $\frac{5}{9} + \frac{1}{9}$ **(c)** $\frac{12}{5} - \frac{7}{5}$ **(d)** $\frac{23}{20} - \frac{13}{20}$

Solution   **(a)** $\frac{3}{8} + \frac{4}{8} = \frac{3 + 4}{8} = \frac{7}{8}$

(b) $\frac{5}{9} + \frac{1}{9} = \frac{5 + 1}{9} = \frac{6}{9} = \frac{2}{3}$

(c) $\frac{12}{5} - \frac{7}{5} = \frac{12 - 7}{5} = \frac{5}{5} = 1$

(d) $\frac{23}{20} - \frac{13}{20} = \frac{23 - 13}{20} = \frac{10}{20} = \frac{1}{2}$

Technology Note: *Arithmetic of Fractions*

Many calculators have the capability to perform addition and subtraction of fractions, as illustrated in the following figures.

Calculator Help

To express a result as a fraction see Appendix A (page AP-2).

```
(3/8)+(4/8)▶Frac
                7/8
(5/9)+(1/9)▶Frac
                2/3
```

```
(12/5)-(7/5)▶Fra
c
                  1
(23/20)-(13/20)▶
Frac
                1/2
```

RATIONAL EXPRESSIONS HAVING LIKE DENOMINATORS

Addition and subtraction of rational expressions having like denominators are similar to addition and subtraction of fractions. The following property can be used to add two rational expressions having like denominators.

SUMS OF RATIONAL EXPRESSIONS

To add two rational expressions having like denominators, add their numerators. Keep the same denominator.

$$\frac{A}{C} + \frac{B}{C} = \frac{A + B}{C} \qquad C \text{ is nonzero.}$$

EXAMPLE 2   Adding rational expressions having like denominators

Add and reduce to lowest terms.

(a) $\dfrac{3}{b} + \dfrac{2}{b}$ (b) $\dfrac{z}{z + 2} + \dfrac{2}{z + 2}$

(c) $\dfrac{x - 1}{x^2 + x} + \dfrac{1}{x^2 + x}$ (d) $\dfrac{t^2 + t}{t - 1} + \dfrac{1 - 3t}{t - 1}$

Solution (a) $\dfrac{3}{b} + \dfrac{2}{b} = \dfrac{3 + 2}{b} = \dfrac{5}{b}$ Add the numerators.

(b) $\dfrac{z}{z + 2} + \dfrac{2}{z + 2} = \dfrac{z + 2}{z + 2}$ Add the numerators.

$= 1$ Reduce.

(c) $\dfrac{x - 1}{x^2 + x} + \dfrac{1}{x^2 + x} = \dfrac{x - 1 + 1}{x^2 + x}$ Add the numerators.

$= \dfrac{x}{x(x + 1)}$ Factor the denominator.

$= \dfrac{1}{x + 1}$ Reduce. Note that $\dfrac{x}{x} = 1$.

(d) $\dfrac{t^2 + t}{t - 1} + \dfrac{1 - 3t}{t - 1} = \dfrac{t^2 + t + 1 - 3t}{t - 1}$ Add the numerators.

$\qquad\qquad\qquad = \dfrac{t^2 - 2t + 1}{t - 1}$ Simplify the numerator.

$\qquad\qquad\qquad = \dfrac{(t - 1)(t - 1)}{t - 1}$ Factor the numerator.

$\qquad\qquad\qquad = t - 1$ Reduce. Note that $\frac{t - 1}{t - 1} = 1$.

EXAMPLE 3   Adding rational expressions having two variables

Add and reduce to lowest terms.

(a) $\dfrac{4}{xy} + \dfrac{5}{xy}$ **(b)** $\dfrac{a}{a^2 - b^2} + \dfrac{b}{a^2 - b^2}$ **(c)** $\dfrac{1}{x - y} + \dfrac{-1}{y - x}$

Solution   **(a)** $\dfrac{4}{xy} + \dfrac{5}{xy} = \dfrac{4 + 5}{xy} = \dfrac{9}{xy}$ Add the numerators.

(b) $\dfrac{a}{a^2 - b^2} + \dfrac{b}{a^2 - b^2} = \dfrac{a + b}{a^2 - b^2}$ Add the numerators.

$\qquad\qquad\qquad\qquad = \dfrac{a + b}{(a - b)(a + b)}$ Factor the denominator.

$\qquad\qquad\qquad\qquad = \dfrac{1}{a - b}$ Reduce. Note that $\dfrac{a + b}{a + b} = 1$.

(c) First write $\frac{1}{x - y} + \frac{-1}{y - x}$ with a common denominator. Note that, if we multiply the second term by 1, written in the form $\frac{-1}{-1}$, it becomes

$$\frac{-1}{y - x} \cdot \frac{-1}{-1} = \frac{(-1)(-1)}{(y - x)(-1)} = \frac{1}{-y + x} = \frac{1}{x - y}.$$

Thus the given sum can be simplified as follows.

$$\frac{1}{x - y} + \frac{-1}{y - x} = \frac{1}{x - y} + \frac{1}{x - y}$$ Rewrite the second term.

$$= \frac{2}{x - y}$$ Add the numerators.

Next we consider subtraction of rational expressions having like denominators.

DIFFERENCES OF RATIONAL EXPRESSIONS

To subtract two rational expressions having like denominators, subtract their numerators. Keep the same denominator.

$$\frac{A}{C} - \frac{B}{C} = \frac{A - B}{C} \qquad C \text{ is nonzero.}$$

EXAMPLE 4 **Subtracting rational expressions having like denominators**

Subtract and reduce to lowest terms.

(a) $\dfrac{a+1}{a} - \dfrac{1}{a}$ (b) $\dfrac{2y}{3y-1} - \dfrac{3y}{3y-1}$ (c) $\dfrac{1+x}{2x^2+5x-3} - \dfrac{-2}{2x^2+5x-3}$

Solution (a) $\dfrac{a+1}{a} - \dfrac{1}{a} = \dfrac{a+1-1}{a} = \dfrac{a}{a} = 1$

(b) $\dfrac{2y}{3y-1} - \dfrac{3y}{3y-1} = \dfrac{2y-3y}{3y-1} = \dfrac{-y}{3y-1}$ or $-\dfrac{y}{3y-1}$

(c) $\dfrac{1+x}{2x^2+5x-3} - \dfrac{-2}{2x^2+5x-3} = \dfrac{1+x-(-2)}{2x^2+5x-3}$

$= \dfrac{x+3}{(2x-1)(x+3)}$

$= \dfrac{1}{2x-1}$

EXAMPLE 5 **Subtracting rational expressions having like denominators**

Subtract and reduce to lowest terms.

(a) $\dfrac{2x}{x+1} - \dfrac{x-1}{x+1}$ (b) $\dfrac{x+y}{3y} - \dfrac{x-y}{3y}$

Solution (a) $\dfrac{2x}{x+1} - \dfrac{x-1}{x+1} = \dfrac{2x-(x-1)}{x+1}$ Subtract the numerators.

$= \dfrac{2x-x+1}{x+1}$ Distribute the minus sign.

$= \dfrac{x+1}{x+1}$ Simplify the numerator.

$= 1$ Reduce.

Note: When subtracting $x - 1$ from $2x$ in the numerator, *you must place parentheses around $x - 1$.*

(b) $\dfrac{x+y}{3y} - \dfrac{x-y}{3y} = \dfrac{x+y-(x-y)}{3y}$ Subtract the numerators.

$= \dfrac{x+y-x+y}{3y}$ Distribute the minus sign.

$= \dfrac{2y}{3y}$ Simplify the numerator.

$= \dfrac{2}{3}$ Reduce.

EXAMPLE 6   Analyzing quality control

A container holds n batteries of two different sizes: A and AA. There are 2 defective A batteries and 4 defective AA batteries. If a battery is picked at random by a quality control inspector, then the probability, or chance, of one of the defective batteries being chosen is given by the expression $\frac{2}{n} + \frac{4}{n}$.
(a) Simplify this expression.
(b) Interpret the result.

Solution   (a) Because the denominators are the same, we simply add the numerators.

$$\frac{2}{n} + \frac{4}{n} = \frac{2 + 4}{n} = \frac{6}{n}$$

(b) There are 6 in n chances that a defective battery is chosen.

7.3   PUTTING IT ALL TOGETHER

In this section we discussed how to add and subtract rational expressions having like denominators. After adding or subtracting rational expressions, be sure to factor the numerator and the denominator completely. Then reduce the result to lowest terms by applying the basic principle of rational expressions. The following table summarizes this discussion, where A, B, and C represent polynomials and C is nonzero.

Concept	Explanation	Examples
Addition of Rational Expressions	$\dfrac{A}{C} + \dfrac{B}{C} = \dfrac{A + B}{C}$	$\dfrac{x}{x + 1} + \dfrac{1 - x}{x + 1} = \dfrac{x + 1 - x}{x + 1} = \dfrac{1}{x + 1}$ $\dfrac{2x}{x^2 - 1} + \dfrac{x}{x^2 - 1} = \dfrac{2x + x}{x^2 - 1} = \dfrac{3x}{x^2 - 1}$
Subtraction of Rational Expressions If B consists of more than one term, put parentheses around B and apply the distributive property.	$\dfrac{A}{C} - \dfrac{B}{C} = \dfrac{A - B}{C}$	$\dfrac{2x}{x^2 - 4} - \dfrac{x + 2}{x^2 - 4} = \dfrac{2x - (x + 2)}{x^2 - 4}$ $= \dfrac{2x - x - 2}{x^2 - 4}$ $= \dfrac{x - 2}{(x + 2)(x - 2)}$ $= \dfrac{1}{x + 2}$

7.3 EXERCISES

FOR EXTRA HELP

 Student's Solutions Manual

 InterAct Math

 MathXL

MyMathLab

 Math Tutor Center

Digital Video Tutor
CD 3   Videotape 9

CONCEPTS

1. To add two rational expressions having like denominators, _____ their _____. The _____ do not change.

2. To subtract two rational expressions having like denominators, _____ their _____. The _____ do not change.

3. $\dfrac{A}{C} + \dfrac{B}{C} = $ _____

4. $\dfrac{A}{C} - \dfrac{B}{C} = $ _____

5. $5 - (x + 1) = $ _____

6. $\dfrac{5}{x - 1} - \dfrac{x + 1}{x - 1} = $ _____

7. $3x - (2x - 5) = $ _____

8. $\dfrac{3x}{7} - \dfrac{2x - 5}{7} = $ _____

ADDITION AND SUBTRACTION OF FRACTIONS

Exercises 9–22: Simplify and reduce to lowest terms.

9. $\frac{1}{2} + \frac{1}{2}$

10. $\frac{3}{7} + \frac{2}{7}$

11. $\frac{4}{5} + \frac{2}{5}$

12. $\frac{3}{11} + \frac{5}{11}$

13. $\frac{1}{6} + \frac{5}{6}$

14. $\frac{3}{10} + \frac{5}{10}$

15. $\frac{4}{7} - \frac{1}{7}$

16. $\frac{5}{13} - \frac{7}{13}$

17. $\frac{7}{8} - \frac{3}{8}$

18. $\frac{9}{16} - \frac{5}{16}$

19. $\frac{11}{12} - \frac{5}{12}$

20. $\frac{7}{24} - \frac{3}{24}$

21. $\frac{7}{15} + \frac{4}{15} - \frac{1}{15}$

22. $\frac{11}{36} - \frac{5}{36} + \frac{1}{36}$

ADDITION AND SUBTRACTION OF RATIONAL EXPRESSIONS

Exercises 23–54: Simplify and reduce to lowest terms.

23. $\dfrac{2}{x} + \dfrac{1}{x}$

24. $\dfrac{9}{x} - \dfrac{7}{x}$

25. $\dfrac{7 + 2x}{4x} - \dfrac{7}{4x}$

26. $\dfrac{x - 1}{5x} + \dfrac{2x + 1}{5x}$

27. $\dfrac{y + 3}{y - 3} + \dfrac{2y - 12}{y - 3}$

28. $\dfrac{5 - y}{y + 2} + \dfrac{y}{y + 2}$

29. $\dfrac{5z}{4z + 3} - \dfrac{z}{4z + 3}$

30. $\dfrac{z}{2z + 1} - \dfrac{1 - z}{2z + 1}$

31. $\dfrac{t + 5}{t + 6} + \dfrac{t + 7}{t + 6}$

32. $\dfrac{t + 1}{t - 4} + \dfrac{1 - t}{t - 4}$

33. $\dfrac{5x}{2x + 3} - \dfrac{3x - 3}{2x + 3}$

34. $\dfrac{x}{5 - x} - \dfrac{2x - 5}{5 - x}$

35. $\dfrac{x^2 + 4x - 1}{4x + 2} - \dfrac{x^2 - 4x - 5}{4x + 2}$

36. $\dfrac{2x^2 - x + 5}{x^2 - 9} - \dfrac{x^2 - x + 14}{x^2 - 9}$

37. $\dfrac{3y}{5} + \dfrac{2y - 5}{5}$

38. $\dfrac{3y - 22}{11} + \dfrac{8y}{11}$

39. $\dfrac{x + y}{4} + \dfrac{x - y}{4}$

40. $\dfrac{x + y}{4} - \dfrac{x - y}{4}$

41. $\dfrac{z^2 + 4}{z - 2} - \dfrac{4z}{z - 2}$

42. $\dfrac{z^2 + 2z}{z + 1} + \dfrac{1}{z + 1}$

43. $\dfrac{2x^2 - 5x}{2x + 1} - \dfrac{3}{2x + 1}$

44. $\dfrac{2x^2}{x + 2} + \dfrac{9x + 10}{x + 2}$

45. $\dfrac{3n}{2n^2 - n + 5} + \dfrac{4n}{2n^2 - n + 5}$

46. $\dfrac{n}{n^2 + n + 1} - \dfrac{1}{n^2 + n + 1}$

47. $\dfrac{1}{x + 3} + \dfrac{2}{x + 3} + \dfrac{3}{x + 3}$

48. $\dfrac{x}{2x - 5} - \dfrac{1}{2x - 5} + \dfrac{2x + 1}{2x - 5}$

49. $\dfrac{x}{x+y} + \dfrac{y}{x+y}$

50. $\dfrac{x}{x^2-y^2} - \dfrac{y}{x^2-y^2}$

51. $\dfrac{a^2}{a+b} - \dfrac{b^2}{a+b}$

52. $\dfrac{a^2}{a+b} + \dfrac{2ab+b^2}{a+b}$

53. $\dfrac{4x^2}{2x+3y} - \dfrac{9y^2}{2x+3y}$

54. $\dfrac{x^3}{x^2+xy+y^2} - \dfrac{y^3}{x^2+xy+y^2}$

APPLICATIONS

55. *Quality Control*   (Refer to Example 6.) A container holds $n+1$ batteries of three different sizes: B, C, and D. There are 6 defective B batteries, 5 defective C batteries, and 3 defective D batteries. If a battery is chosen at random by a quality control inspector, the probability, or chance, of one of the defective batteries being chosen is

$$\frac{6}{n+1} + \frac{5}{n+1} + \frac{3}{n+1}.$$

(a) Simplify this expression.

(b) Evaluate the simplified expression for $n = 99$ and interpret the result.

56. *Intensity of a Light Bulb*   The farther a person is from a light bulb, the less intense its light is. The equation $I = \frac{19}{4d^2}$ approximates the intensity of light from a 60-watt light bulb at a distance of d meters, where I is measured in watts per square meter.

(a) Find I for $d = 2$ meters and interpret the result.

(b) The intensity of light from a 100-watt light bulb is about $I = \frac{32}{4d^2}$. Find an expression for the sum of the intensities of light from a 100-watt bulb and a 60-watt bulb.

WRITING ABOUT MATHEMATICS

57. Explain how to add two rational expressions having like denominators. Give an example.

58. Explain how to subtract two rational expressions having like denominators. Give an example.

7.4   ADDITION AND SUBTRACTION WITH UNLIKE DENOMINATORS

Finding Least Common Multiples · Review of Fractions Having Unlike Denominators · Rational Expressions Having Unlike Denominators

INTRODUCTION

In Section 7.3, we added and subtracted rational expressions having like denominators. Although the denominators of rational expressions are often unlike, rational expressions can still be added and subtracted after a common denominator is found. One way to find the least common denominator for a sum or difference of rational expressions is to find the least common multiple of the denominators. In the following subsection we show how to find the least common multiple.

FINDING LEAST COMMON MULTIPLES

Two friends work part-time at a store. The first person works every sixth day, and the second person works every eighth day. If they both work today, how many days will pass before they work on the same day again?

We can answer this question by listing the days that each person works.

First person: 6,   12,   18,   **24**,   30,   36,   42,   **48**,   54

Second person: 8,   16,   **24**,   32,   40,   **48**,   56,   64

After 24 days, the two friends work on the same day. The next time is after 48 days. The numbers 24 and 48 are *common multiples* of 6 and 8. (Find another.) However, 24 is the *least* common multiple (LCM) of 6 and 8.

Another way to find the least common multiple of 6 and 8 is to factor each number into prime numbers.

$$6 = 2 \cdot 3 \quad \text{and} \quad 8 = 2 \cdot 2 \cdot 2$$

To find the least common multiple, first list each factor the greatest number of times that it occurs in either factorization. Then find the product of these numbers. For this example, the factor 2 occurs three times in the factorization of 8 and only once in the factorization of 6, so list 2 three times. The factor 3 appears only once in the factorization of 6 so list it once:

2, 2, 2, 3.

The least common multiple is their product: $2 \cdot 2 \cdot 2 \cdot 3 = 24$.

This same procedure can also be used to find the least common multiple for two or more polynomials.

FINDING THE LEAST COMMON MULTIPLE

The least common multiple (LCM) of two or more polynomials can be found as follows.

STEP 1 Factor each polynomial completely.

STEP 2 List each factor the greatest number of times that it occurs in any factorization.

STEP 3 Find the product of this list of factors. The result is the LCM.

EXAMPLE 1   Finding least common multiples

Find the least common multiple of each pair of expressions.
(a) $2x, 5x^2$ **(b)** $x^2 - x, x - 1$
(c) $x + 2, x - 3$ **(d)** $x^2 + 2x + 1, x^2 + 3x + 2$

Solution **(a) STEP 1** Factor $2x$ and $5x^2$ completely.

$$2x = 2 \cdot x \quad \text{and} \quad 5x^2 = 5 \cdot x \cdot x$$

STEP 2 In either factorization, the factor 2 occurs at most once, the factor 5 occurs at most once, and the factor x occurs at most twice. The list of factors is: **2, 5,** x, x.

STEP 3 The LCM equals the product

$$2 \cdot 5 \cdot x \cdot x = 10x^2.$$

(b) STEP 1 Factor $x^2 - x$ and $x - 1$ completely. Note that $x - 1$ cannot be factored.

$$x^2 - x = x(x - 1) \quad \text{and} \quad x - 1 = x - 1$$

STEP 2 Both factors of x and $x - 1$ occur at most once in either factorization. The list of factors is $x, (x - 1)$.

STEP 3 The LCM is the product $x(x - 1)$, or $x^2 - x$.

(c) STEP 1 Neither $x + 2$ nor $x - 3$ can be factored.

STEP 2 The list of factors is $(x + 2), (x - 3)$.

STEP 3 The LCM is the product $(x + 2)(x - 3)$, or $x^2 - x - 6$.

(d) STEP 1 Factor $x^2 + 2x + 1$ and $x^2 + 3x + 2$ completely.

$$x^2 + 2x + 1 = (x + 1)(x + 1) \quad \text{and} \quad x^2 + 3x + 2 = (x + 1)(x + 2)$$

STEP 2 In either factorization, the factor $(x + 1)$ occurs at most twice and the factor $(x + 2)$ occurs at most once. The list is $(x + 1), (x + 1), (x + 2)$.

STEP 3 The LCM is the product $(x + 1)^2(x + 2)$, which is left in factored form.

REVIEW OF FRACTIONS HAVING UNLIKE DENOMINATORS

Before we can find the sum $\frac{1}{2} + \frac{1}{3}$ by hand, we need to rewrite these fractions by using their least common denominator. The least common denominator (LCD) of $\frac{1}{2}$ and $\frac{1}{3}$ corresponds to the least common multiple of 2 and 3, which is 6. As a result, we rewrite these fractions as

$$\frac{1}{2} \cdot \frac{3}{3} = \frac{3}{6} \quad \text{and} \quad \frac{1}{3} \cdot \frac{2}{2} = \frac{2}{6}.$$

Their sum is

$$\frac{1}{2} + \frac{1}{3} = \frac{3}{6} + \frac{2}{6} = \frac{5}{6}.$$

EXAMPLE 2 Adding and subtracting fractions having unlike denominators

Simplify each expression.

(a) $\frac{3}{10} + \frac{4}{15}$ **(b)** $\frac{7}{8} - \frac{1}{6}$

Solution **(a)** The LCD for $\frac{3}{10}$ and $\frac{4}{15}$ equals the LCM of 10 and 15, which is 30. We rewrite these fractions as

$$\frac{3}{10} \cdot \frac{3}{3} = \frac{9}{30} \quad \text{and} \quad \frac{4}{15} \cdot \frac{2}{2} = \frac{8}{30}.$$

Their sum is

$$\frac{3}{10} + \frac{4}{15} = \frac{9}{30} + \frac{8}{30} = \frac{17}{30}.$$

(b) The LCD for $\frac{7}{8}$ and $\frac{1}{6}$ equals the LCM of 8 and 6, which is 24. We rewrite these fractions as

$$\frac{7}{8} \cdot \frac{3}{3} = \frac{21}{24} \quad \text{and} \quad \frac{1}{6} \cdot \frac{4}{4} = \frac{4}{24}.$$

Their difference is

$$\frac{7}{8} - \frac{1}{6} = \frac{21}{24} - \frac{4}{24} = \frac{17}{24}.$$

RATIONAL EXPRESSIONS HAVING UNLIKE DENOMINATORS

The first step in adding or subtracting rational expressions having unlike denominators is to rewrite each expression by using their least common denominator. Then the sum or difference can be found by using the techniques discussed in Section 7.3.

Note: The LCD of two or more rational expressions equals the LCM of their denominators.

EXAMPLE 3   Adding rational expressions having unlike denominators

Find each sum and leave your answer in factored form.

(a) $\dfrac{5}{8y} + \dfrac{7}{4y^2}$ **(b)** $\dfrac{1}{x-1} + \dfrac{1}{x+1}$ **(c)** $\dfrac{x}{x^2 + 2x + 1} + \dfrac{1}{x+1}$

Solution   **(a)**   First find the LCM for $8y$ and $4y^2$.

$$8y = 2 \cdot 2 \cdot 2 \cdot y \quad \text{and} \quad 4y^2 = 2 \cdot 2 \cdot y \cdot y$$

Thus the LCM is $2 \cdot 2 \cdot 2 \cdot y \cdot y = 8y^2$.

$$\frac{5}{8y} + \frac{7}{4y^2} = \frac{5}{8y} \cdot \frac{y}{y} + \frac{7}{4y^2} \cdot \frac{2}{2} \qquad \text{Rewrite by using the LCD.}$$

$$= \frac{5y}{8y^2} + \frac{14}{8y^2} \qquad \text{Multiply the fractions.}$$

$$= \frac{5y + 14}{8y^2} \qquad \text{Add the numerators.}$$

(b)   The LCM for $x - 1$ and $x + 1$ is their product, $(x - 1)(x + 1)$.

$$\frac{1}{x-1} + \frac{1}{x+1} = \frac{1}{x-1} \cdot \frac{x+1}{x+1} + \frac{1}{x+1} \cdot \frac{x-1}{x-1} \qquad \text{Rewrite by using the LCD.}$$

$$= \frac{x+1}{(x-1)(x+1)} + \frac{x-1}{(x+1)(x-1)} \qquad \text{Multiply the fractions.}$$

$$= \frac{x+1+x-1}{(x-1)(x+1)} \qquad \text{Add the numerators.}$$

$$= \frac{2x}{(x-1)(x+1)} \qquad \text{Simplify the numerator.}$$

(c) First find the LCM for $x^2 + 2x + 1$ and $x + 1$. Because
$$x^2 + 2x + 1 = (x + 1)(x + 1),$$
their LCM is $(x + 1)(x + 1) = (x + 1)^2$.

$$\frac{x}{x^2 + 2x + 1} + \frac{1}{x + 1} = \frac{x}{(x + 1)^2} + \frac{1}{x + 1} \cdot \frac{x + 1}{x + 1} \qquad \text{Rewrite by using the LCD.}$$

$$= \frac{x}{(x + 1)^2} + \frac{x + 1}{(x + 1)^2} \qquad \text{Multiply the fractions.}$$

$$= \frac{2x + 1}{(x + 1)^2} \qquad \text{Add the numerators.}$$

EXAMPLE 4   **Subtracting rational expressions having unlike denominators**

Simplify each expression. Write your answer in lowest terms and leave it in factored form.

(a) $\dfrac{5}{z} - \dfrac{z}{z - 1}$ 　　　　　**(b)** $\dfrac{5}{x + 1} - \dfrac{1}{x^2 - 1}$

(c) $\dfrac{x}{x^2 - 2x} - \dfrac{1}{x^2 + 2x}$ 　　　**(d)** $\dfrac{3}{x - 1} - \dfrac{3}{x^2 - x} + \dfrac{5}{x}$

Solution　**(a)** The LCD is $z(z - 1)$.

$$\frac{5}{z} - \frac{z}{z - 1} = \frac{5}{z} \cdot \frac{z - 1}{z - 1} - \frac{z}{z - 1} \cdot \frac{z}{z} \qquad \text{Rewrite by using the LCD.}$$

$$= \frac{5(z - 1)}{z(z - 1)} - \frac{z^2}{z(z - 1)} \qquad \text{Multiply the fractions.}$$

$$= \frac{5(z - 1) - z^2}{z(z - 1)} \qquad \text{Add the numerators.}$$

$$= \frac{-z^2 + 5z - 5}{z(z - 1)} \qquad \text{Simplify the numerator.}$$

(b) The LCD is $(x - 1)(x + 1)$. Note that $x^2 - 1 = (x - 1)(x + 1)$.

$$\frac{5}{x + 1} - \frac{1}{x^2 - 1} = \frac{5}{x + 1} \cdot \frac{x - 1}{x - 1} - \frac{1}{(x - 1)(x + 1)} \qquad \text{Rewrite by using the LCD.}$$

$$= \frac{5(x - 1) - 1}{(x - 1)(x + 1)} \qquad \text{Subtract the numerators.}$$

$$= \frac{5x - 5 - 1}{(x - 1)(x + 1)} \qquad \text{Distributive property}$$

$$= \frac{5x - 6}{(x - 1)(x + 1)} \qquad \text{Simplify the numerator.}$$

(c) Start by factoring each denominator. Because
$$x^2 - 2x = x(x - 2) \quad \text{and} \quad x^2 + 2x = x(x + 2),$$
the LCD is $x(x - 2)(x + 2)$.

$$\frac{x}{x^2 - 2x} - \frac{1}{x^2 + 2x} = \frac{x}{x(x-2)} \cdot \frac{x+2}{x+2} - \frac{1}{x(x+2)} \cdot \frac{x-2}{x-2}$$ Rewrite by using the LCD.

$$= \frac{x(x+2)}{x(x-2)(x+2)} - \frac{x-2}{x(x-2)(x+2)}$$ Multiply the rational expressions.

$$= \frac{x(x+2) - (x-2)}{x(x-2)(x+2)}$$ Subtract the numerators.

$$= \frac{x^2 + 2x - x + 2}{x(x-2)(x+2)}$$ Distributive property

$$= \frac{x^2 + x + 2}{x(x-2)(x+2)}$$ Simplify the numerator.

(d) The given expression contains three rational expressions. Begin by finding the LCM of the three denominators: $x - 1, x^2 - x$, and x. Because $x^2 - x = x(x - 1)$, the LCM is $x(x - 1)$.

$$\frac{3}{x-1} - \frac{3}{x^2 - x} + \frac{5}{x} = \frac{3}{x-1} \cdot \frac{x}{x} - \frac{3}{x(x-1)} + \frac{5}{x} \cdot \frac{x-1}{x-1}$$ Rewrite by using the LCD.

$$= \frac{3x}{x(x-1)} - \frac{3}{x(x-1)} + \frac{5(x-1)}{x(x-1)}$$ Multiply the expressions.

$$= \frac{3x - 3 + 5x - 5}{x(x-1)}$$ Combine the expressions.

$$= \frac{8x - 8}{x(x-1)}$$ Simplify the numerator.

$$= \frac{8(x-1)}{x(x-1)}$$ Factor the numerator.

$$= \frac{8}{x}$$ Reduce to lowest terms.

Critical Thinking

Find the reciprocal of the sum $x + \frac{1}{x}$.

AN APPLICATION INVOLVING ELECTRICITY Sums of rational expressions occur in applications involving electricity. The flow of electricity through a wire can be compared to the flow of water through a hose. Voltage is the force "pushing" the electricity and corresponds to water pressure in a hose. Resistance is the opposition to the flow of electricity and corresponds to the diameter of a hose. More resistance results in less flow of electricity. An ordinary light bulb is an example of a resistor in an electrical circuit. Resistance is often measured in units called *ohms*. For example, a standard 60-watt light bulb has a resistance of about 200 ohms.

Suppose that two light bulbs are wired in parallel so that electricity can flow through either light bulb, as illustrated in Figure 7.2. If the individual resistances of the light bulbs are R and S, then the total resistance of the circuit is given by the *reciprocal* of the sum

$$\frac{1}{R} + \frac{1}{S}.$$

(*Source:* R. Weidner and R. Sells, *Elementary Classical Physics, Vol. 2.*)

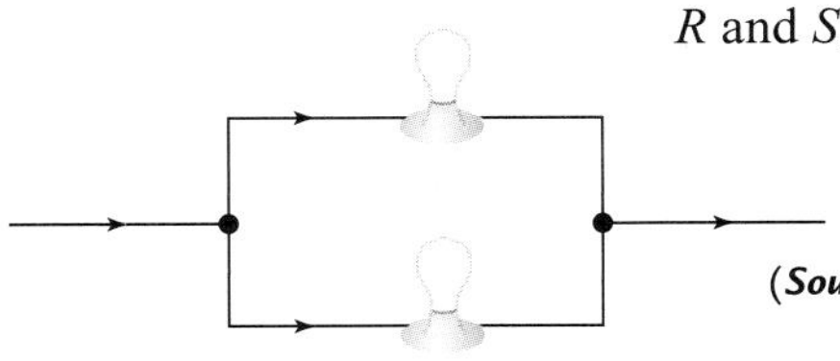

Figure 7.2

EXAMPLE 5   Modeling electrical resistance

Add $\frac{1}{R} + \frac{1}{S}$, and then find the reciprocal of the result.

Solution   The LCD for $\frac{1}{R}$ and $\frac{1}{S}$ is RS.

$$\frac{1}{R} + \frac{1}{S} = \frac{1}{R} \cdot \frac{S}{S} + \frac{1}{S} \cdot \frac{R}{R} \qquad \text{Rewrite by using the LCD.}$$

$$= \frac{S}{RS} + \frac{R}{RS} \qquad \text{Multiply the fractions.}$$

$$= \frac{S + R}{RS} \qquad \text{Add the numerators.}$$

In general, the reciprocal of $\frac{a}{b}$ is $\frac{b}{a}$, so the reciprocal of $\frac{S + R}{RS}$ is $\frac{RS}{S + R}$. This final expression can be used to find the total resistance of the circuit.

7.4   PUTTING IT ALL TOGETHER

In this section we discussed how to add and subtract rational expressions having unlike denominators. The following table summarizes this discussion.

Concept	Explanation	Examples
Least Common Multiple (LCM)	1. Factor each polynomial completely. 2. List each factor the greatest number of times that it occurs in any factorization. 3. The LCM is the product of this list.	1. $x^2 - 2x = x(x - 2)$ $x^2 - 6x + 8 = (x - 2)(x - 4)$ 2. $x, (x - 2), (x - 4)$ 3. LCM $= x(x - 2)(x - 4)$
Least Common Denominator (LCD)	The LCD of two or more rational expressions equals the LCM of their denominators.	The LCD of $\frac{2}{x^2 - 2x}$ and $\frac{3}{x^2 - 6x + 8}$ is $x(x - 2)(x - 4)$ because the LCM of $x^2 - 2x$ and $x^2 - 6x + 8$ is $x(x - 2)(x - 4)$, as shown in the preceding numbered list.
Addition and Subtraction of Rational Expressions Having Unlike Denominators	First rewrite each expression by using the LCD. Then add or subtract the expressions. Finally write your answer in lowest terms.	The LCM of x and $x + 2$ is $x(x + 2)$. $\frac{2}{x} + \frac{5}{x + 2} = \frac{2}{x} \cdot \frac{x + 2}{x + 2} + \frac{5}{x + 2} \cdot \frac{x}{x}$ $= \frac{2x + 4}{x(x + 2)} + \frac{5x}{x(x + 2)}$ $= \frac{7x + 4}{x(x + 2)}$

7.4 EXERCISES

FOR EXTRA HELP

 Student's Solutions Manual

 InterAct Math

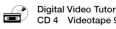 MathXL

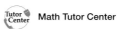 MyMathLab

 Math Tutor Center

Digital Video Tutor
CD 4 Videotape 9

CONCEPTS

1. Give a common multiple of 6 and 9 that is not the *least* common multiple.

2. The LCM of x and y is _____.

3. To rewrite $\frac{3}{4}$ having denominator 12, multiply $\frac{3}{4}$ by the fraction _____.

4. To rewrite $\frac{4}{x-1}$ having denominator $x^2 - 1$, multiply $\frac{4}{x-1}$ by the rational expression _____.

5. What is the LCD for $\frac{1}{4}$ and $\frac{5}{6}$?

6. What is the LCD for $\frac{1}{2x}$ and $\frac{5}{6x^2}$?

Exercises 7–14: Find the least common multiple.

7. 4, 6

8. 6, 9

9. 2, 3

10. 5, 4

11. 10, 15

12. 8, 12

13. 24, 36

14. 32, 40

Exercises 15–38: Find the least common multiple. Leave your answer in factored form.

15. $4x, 6x$

16. $6x, 9x$

17. $5x, 10x^2$

18. $4x^2, 12x$

19. $x, x + 1$

20. $4x, x - 1$

21. $2x + 1, x + 3$

22. $5x + 3, x + 9$

23. $4x^2, 9x^3$

24. $12x^3, 15x^5$

25. $x^2 - x, x^2 + x$

26. $x^2 + 2x, x^2$

27. $(x + 1)^2, x + 1$

28. $(x - 8)^2, (x - 8)(x + 1)$

29. $(2x - 1)^3, (2x - 1)(x + 3)$

30. $(x - 4)(x + 4), (x - 4)(x + 3)$

31. $4x^2 - 1, 2x + 1$

32. $x^2 + 4x + 3, x + 3$

33. $x^2 - 1, x + 1$

34. $x^2 - 4, x - 2$

35. $x^2 + 4, 4x$

36. $4x^2 + x, x$

37. $2x^2 + 7x + 6, x^2 + 5x + 6$

38. $x^2 - 3x + 2, x^2 + 2x - 3$

ADDITION AND SUBTRACTION OF RATIONAL EXPRESSIONS

Exercises 39–50: Rewrite the rational expression so it has the given denominator D. For example, if the fraction $\frac{1}{4}$ is written with denominator $D = 8$, it becomes $\frac{2}{8}$.

39. $\frac{1}{3}, D = 9$

40. $\frac{3}{4}, D = 24$

41. $\frac{5}{7}, D = 21$

42. $\frac{4}{5}, D = 30$

43. $\frac{1}{4x}, D = 8x^3$

44. $\frac{5}{3x}, D = 9x^2$

45. $\frac{1}{x + 2}, D = x^2 - 4$

46. $\frac{3}{x - 3}, D = x^2 - 9$

47. $\frac{1}{x + 1}, D = x^2 + x$

48. $\frac{3}{x - 3}, D = x^2 - 3x$

49. $\frac{2x}{x + 1}, D = x^2 + 2x + 1$

50. $\frac{x}{2x - 1}, D = 2x^2 + 11x - 6$

Exercises 51–92: Simplify the expression. Write your answer in lowest terms and leave it in factored form.

51. $\frac{4}{5} + \frac{1}{2}$

52. $\frac{3}{8} + \frac{1}{4}$

53. $\frac{5}{9} - \frac{1}{3}$

54. $\frac{7}{10} - \frac{3}{15}$

55. $\frac{1}{3x} + \frac{3}{4x}$

56. $\frac{4}{2x^2} + \frac{7}{3x}$

57. $\frac{5}{z^2} - \frac{7}{z^3}$

58. $\frac{8}{z} - \frac{3}{2z}$

59. $\frac{1}{x} - \frac{1}{y}$

60. $\frac{1}{xy} - \frac{4}{y}$

61. $\dfrac{a}{b} + \dfrac{b}{a}$

62. $\dfrac{3}{x} - \dfrac{4}{y}$

63. $\dfrac{1}{2x + 4} + \dfrac{3}{x + 2}$

64. $\dfrac{1}{5x - 10} - \dfrac{x}{x - 2}$

65. $\dfrac{2}{t - 2} - \dfrac{1}{t}$

66. $\dfrac{7}{2t} + \dfrac{1}{t + 5}$

67. $\dfrac{5}{n - 1} + \dfrac{n}{n + 1}$

68. $\dfrac{4n}{3n - 2} + \dfrac{n}{n + 1}$

69. $\dfrac{3}{x - 3} + \dfrac{6}{3 - x}$

70. $\dfrac{x}{x - 8} + \dfrac{x}{8 - x}$

71. $\dfrac{1}{5k - 1} + \dfrac{1}{1 - 5k}$

72. $\dfrac{4}{4 - 3k} + \dfrac{3k}{3k - 4}$

73. $\dfrac{2x}{(x - 1)^2} + \dfrac{4}{x - 1}$

74. $\dfrac{5}{(x + 5)} - \dfrac{x}{(x + 5)^2}$

75. $\dfrac{2y}{y(2y - 1)} + \dfrac{1}{2y - 1}$

76. $\dfrac{5y}{y(y + 1)} - \dfrac{5}{y + 1}$

77. $\dfrac{1}{x + 2} - \dfrac{1}{x^2 + 2x}$

78. $\dfrac{1}{x - 3} - \dfrac{2}{x^2 - 3x}$

79. $\dfrac{x}{x^2 + 4x + 4} + \dfrac{1}{x + 2}$

80. $\dfrac{1}{x^2 - 3x - 4} - \dfrac{1}{x + 1}$

81. $\dfrac{x}{(x + 1)(x + 2)} - \dfrac{1}{(x + 2)(x + 3)}$

82. $\dfrac{2x}{(x - 1)(x - 2)} - \dfrac{5}{x - 2}$

83. $\dfrac{1}{a + b} - \dfrac{1}{a - b}$

84. $\dfrac{x}{x^2 - y^2} - \dfrac{1}{x + y}$

85. $\dfrac{r}{r - t} + \dfrac{t}{t - r} - 1$

86. $\dfrac{1}{x} + \dfrac{2}{x^2 - 2x} + \dfrac{5}{x - 2}$

87. $\dfrac{1}{2a} + \dfrac{1}{3a} + \dfrac{1}{4a}$

88. $\dfrac{1}{b} + \dfrac{1}{b + 1} + \dfrac{1}{b + 2}$

89. $\dfrac{2}{x - y} + \dfrac{3}{y - x} + \dfrac{1}{x - y}$

90. $\dfrac{a}{a - b} + \dfrac{b}{b - a} + \dfrac{3}{b}$

91. $\dfrac{3}{x - 3} - \dfrac{3}{x^2 - 3x} - \dfrac{6}{x(x - 3)}$

92. $\dfrac{3}{2a - 4} + \dfrac{5}{2a} - \dfrac{3}{a^2 - 2a}$

APPLICATIONS

93. *Electricity* In Example 5, we showed that the expressions

$$\frac{1}{R} + \frac{1}{S} \quad \text{and} \quad \frac{S + R}{RS}$$

are equivalent. Evaluate both expressions by using $R = 120$ and $S = 200$. Are your answers the same?

94. *Intensity of a Light Bulb* The formula $I = \frac{32}{4d^2}$ approximates the intensity of light from a 100-watt light bulb at a distance of d meters, where I is in watts per square meter. For light from a 40-watt bulb the equation for its intensity becomes $I = \frac{16}{5d^2}$.
  (a) Find an expression for the sum of the intensities of light from the two light bulbs.
  (b) Find the combined intensity of their light at $d = 5$ meters.

95. *Photography* A lens in a camera has a focal length, which is important for focusing the camera. If an object is at a distance D from the lens that has a focal length F, then to be in focus the distance S between the lens and the film should satisfy the equation

$$\frac{1}{S} = \frac{1}{F} - \frac{1}{D},$$

as illustrated in the figure. Write the difference $\frac{1}{F} - \frac{1}{D}$ as one term.

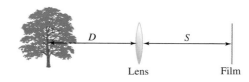

Lens   Film

96. *Geometry*   Find the sum of the areas of the two rectangles shown in the figure. Write your answer in factored form.

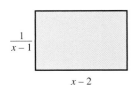

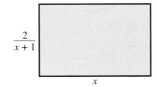

97. Explain how to find the least common multiple of two polynomials.

98. Explain how to subtract two rational expressions having unlike denominators.

CHECKING BASIC CONCEPTS   SECTIONS 7.3 AND 7.4

1. Simplify each expression and reduce to lowest terms.

(a) $\dfrac{x}{x + 2} + \dfrac{2}{x + 2}$ **(b)** $\dfrac{2}{3x} - \dfrac{x}{3x}$

(c) $\dfrac{z^2 + z}{z + 2} + \dfrac{z}{z + 2}$

2. Find the least common multiple of each pair of expressions.

(a) $3x, 5x$ **(b)** $4x, x^2 + x$

(c) $x + 1, x - 1$

3. Simplify each expression and reduce to lowest terms.

(a) $\dfrac{1}{x + 1} + \dfrac{5}{x}$ **(b)** $\dfrac{5}{x - 3} + \dfrac{1}{3 - x}$

(c) $\dfrac{-4}{4x + 2} - \dfrac{x + 2}{2x + 1}$

4. Simplify the expression $\dfrac{a}{a - b} - \dfrac{b}{a + b}$.

7.5   COMPLEX FRACTIONS

Basic Concepts · Simplifying Complex Fractions

INTRODUCTION

If two and a half pizzas are cut so that each piece equals one-fourth of a pizza, then there are 10 pieces of pizza. This problem can be written as

$$\frac{2\frac{1}{2}}{\frac{1}{4}} \quad \text{or} \quad \frac{2 + \frac{1}{2}}{\frac{1}{4}}.$$

These expressions are examples of *complex* fractions. Typically, we want to rewrite a complex fraction as a standard fraction in the form $\frac{a}{b}$. In this section we show how to simplify complex fractions.

BASIC CONCEPTS

A **complex fraction** is a rational expression that contains fractions in its numerator, denominator, or both. Examples of complex fractions include

$$\frac{1 + \dfrac{1}{2}}{1 - \dfrac{1}{3}}, \qquad \frac{5z}{\dfrac{7}{z} - \dfrac{4}{2z}}, \qquad \text{and} \qquad \frac{\dfrac{2x}{3} + \dfrac{1}{x}}{x - \dfrac{1}{x+1}}.$$

When simplifying a complex fraction, we can sometimes multiply its numerator by the reciprocal of its denominator, the "invert and multiply" strategy.

SIMPLIFYING COMPLEX FRACTIONS

For any real numbers a, b, c, and d,

$$\frac{\dfrac{a}{b}}{\dfrac{c}{d}} = \frac{a}{b} \cdot \frac{d}{c},$$

where b, c, and d are nonzero.

Note: The expression $\dfrac{\frac{a}{b}}{\frac{c}{d}}$ is equivalent to $\dfrac{a}{b} \div \dfrac{c}{d}$.

EXAMPLE 1 Simplifying basic complex fractions

Simplify each complex fraction.

(a) $\dfrac{\frac{3}{4}}{\frac{9}{8}}$ (b) $\dfrac{3\frac{3}{4}}{1\frac{1}{2}}$ (c) $\dfrac{\frac{x}{4}}{\frac{3}{2y}}$ (d) $\dfrac{\frac{(x-1)^2}{4}}{\frac{x-1}{8}}$

Solution (a) To simplify $\frac{3}{4} \div \frac{9}{8}$, multiply $\frac{3}{4}$ by the reciprocal of $\frac{9}{8}$, or $\frac{8}{9}$.

$$\frac{\dfrac{3}{4}}{\dfrac{9}{8}} = \frac{3}{4} \cdot \frac{8}{9} \qquad \text{Invert and multiply.}$$

$$= \frac{24}{36} \qquad \text{Multiply the fractions.}$$

$$= \frac{2}{3} \qquad \text{Reduce.}$$

(b) Start by writing $3\frac{3}{4}$ and $1\frac{1}{2}$ as improper fractions.

$$\frac{3\frac{3}{4}}{1\frac{1}{2}} = \frac{\frac{15}{4}}{\frac{3}{2}} \qquad \text{Write as improper fractions.}$$

$$= \frac{15}{4} \cdot \frac{2}{3} \qquad \text{Invert and multiply.}$$

$$= \frac{5}{2} \qquad \text{Multiply and reduce.}$$

(c) To simplify $\frac{x}{4} \div \frac{3}{2y}$, multiply $\frac{x}{4}$ by the reciprocal of $\frac{3}{2y}$, or $\frac{2y}{3}$.

$$\frac{\frac{x}{4}}{\frac{3}{2y}} = \frac{x}{4} \cdot \frac{2y}{3} \qquad \text{Invert and multiply.}$$

$$= \frac{2xy}{12} \qquad \text{Multiply the fractions.}$$

$$= \frac{xy}{6} \qquad \text{Reduce.}$$

(d) To simplify $\frac{(x-1)^2}{4} \div \frac{x-1}{8}$, multiply $\frac{(x-1)^2}{4}$ by the reciprocal of $\frac{x-1}{8}$, or $\frac{8}{x-1}$.

$$\frac{\frac{(x-1)^2}{4}}{\frac{x-1}{8}} = \frac{(x-1)^2}{4} \cdot \frac{8}{x-1} \qquad \text{Invert and multiply.}$$

$$= \frac{8(x-1)(x-1)}{4(x-1)} \qquad \text{Multiply the fractions.}$$

$$= 2(x-1) \qquad \text{Reduce.}$$

SIMPLIFYING COMPLEX FRACTIONS

There are two basic methods of simplifying a complex fraction. The first is to simplify both the numerator and the denominator and then divide the resulting two fractions. The second is to multiply the numerator and denominator by the least common denominator of all fractions appearing in the complex fraction.

SIMPLIFYING THE NUMERATOR AND DENOMINATOR The next example illustrates use of the first method whereby both the numerator and the denominator are simplified.

EXAMPLE 2   Simplifying complex fractions

Simplify. Write your answer in lowest terms.

(a) $\dfrac{\dfrac{1}{a} + \dfrac{1}{b}}{\dfrac{1}{a} - \dfrac{1}{b}}$ (b) $\dfrac{x - \dfrac{4}{x}}{x + \dfrac{4}{x}}$ (c) $\dfrac{\dfrac{1}{x} + \dfrac{2}{x-1}}{\dfrac{2}{x} - \dfrac{1}{x-1}}$ (d) $\dfrac{\dfrac{1}{x} + \dfrac{1}{y}}{\dfrac{1}{x^2} - \dfrac{1}{y^2}}$

Solution   (a) First, write the numerator as one fraction by using the LCD, ab.

$$\frac{1}{a} + \frac{1}{b} = \frac{b}{ab} + \frac{a}{ab} = \frac{b + a}{ab}$$

Second, write the denominator as one fraction by using the LCD, ab.

$$\frac{1}{a} - \frac{1}{b} = \frac{b}{ab} - \frac{a}{ab} = \frac{b - a}{ab}$$

Finally, use these results to simplify the given complex fraction.

$$\frac{\dfrac{1}{a} + \dfrac{1}{b}}{\dfrac{1}{a} - \dfrac{1}{b}} = \frac{\dfrac{b + a}{ab}}{\dfrac{b - a}{ab}}$$ Simplify the numerator and the denominator.

$$= \frac{b + a}{ab} \cdot \frac{ab}{b - a}$$ Invert and multiply.

$$= \frac{ab(b + a)}{ab(b - a)}$$ Multiply.

$$= \frac{b + a}{b - a}$$ Reduce.

(b) The LCD for both the numerator and the denominator is x.

$$\frac{x - \dfrac{4}{x}}{x + \dfrac{4}{x}} = \frac{\dfrac{x^2}{x} - \dfrac{4}{x}}{\dfrac{x^2}{x} + \dfrac{4}{x}}$$ Write by using the LCD.

$$= \frac{\dfrac{x^2 - 4}{x}}{\dfrac{x^2 + 4}{x}}$$ Combine terms.

$$= \frac{x^2 - 4}{x} \cdot \frac{x}{x^2 + 4}$$ Invert and multiply.

$$= \frac{x(x^2 - 4)}{x(x^2 + 4)}$$ Multiply.

$$= \frac{x^2 - 4}{x^2 + 4}$$ Reduce.

(c) The LCD for the numerator and the denominator is $x(x - 1)$.

$$\frac{\dfrac{1}{x} + \dfrac{2}{x - 1}}{\dfrac{2}{x} - \dfrac{1}{x - 1}} = \frac{\dfrac{x - 1}{x(x - 1)} + \dfrac{2x}{x(x - 1)}}{\dfrac{2(x - 1)}{x(x - 1)} - \dfrac{x}{x(x - 1)}}$$ Write by using the LCD.

$$= \frac{\dfrac{x - 1 + 2x}{x(x - 1)}}{\dfrac{2(x - 1) - x}{x(x - 1)}}$$ Combine terms.

$$= \frac{\dfrac{3x - 1}{x(x - 1)}}{\dfrac{x - 2}{x(x - 1)}}$$ Simplify.

$$= \frac{3x - 1}{x(x - 1)} \cdot \frac{x(x - 1)}{x - 2}$$ Invert and multiply.

$$= \frac{3x - 1}{x - 2}$$ Multiply and reduce.

(d) The LCD for the numerator is xy, and the LCD for the denominator is x^2y^2.

$$\frac{\dfrac{1}{x} + \dfrac{1}{y}}{\dfrac{1}{x^2} - \dfrac{1}{y^2}} = \frac{\dfrac{y}{xy} + \dfrac{x}{xy}}{\dfrac{y^2}{x^2y^2} - \dfrac{x^2}{x^2y^2}}$$ Write by using the LCD.

$$= \frac{\dfrac{y + x}{xy}}{\dfrac{y^2 - x^2}{x^2y^2}}$$ Combine terms.

$$= \frac{y + x}{xy} \cdot \frac{x^2y^2}{y^2 - x^2}$$ Invert and multiply.

$$= \frac{x^2y^2(y + x)}{xy(y^2 - x^2)}$$ Multiply.

$$= \frac{x^2y^2(y + x)}{xy(y - x)(y + x)}$$ Factor the denominator.

$$= \frac{xy}{y - x}$$ Reduce.

MULTIPLYING BY THE LCD   The second method of simplifying a complex fraction is to multiply the numerator and denominator by the least common denominator of all the fractions in *both* the numerator and the denominator. We can use this method to simplify the complex fraction in Example 2(a). Because the LCD for the numerator and the denominator is ab, we multiply the complex fraction by 1, expressed in the form $\frac{ab}{ab}$. This method is equivalent to multiplying the numerator and the denominator by ab.

$$\frac{\dfrac{1}{a} + \dfrac{1}{b}}{\dfrac{1}{a} - \dfrac{1}{b}} = \frac{\left(\dfrac{1}{a} + \dfrac{1}{b}\right)ab}{\left(\dfrac{1}{a} - \dfrac{1}{b}\right)ab} \qquad \text{Multiply by } \frac{ab}{ab} = 1.$$

$$= \frac{\dfrac{ab}{a} + \dfrac{ab}{b}}{\dfrac{ab}{a} - \dfrac{ab}{b}} \qquad \text{Distributive property}$$

$$= \frac{b + a}{b - a} \qquad \text{Reduce.}$$

EXAMPLE 3   **Simplifying complex fractions**

Simplify.

(a) $\dfrac{2z}{\dfrac{4}{z} + \dfrac{3}{z}}$ **(b)** $\dfrac{\dfrac{1}{x - 3}}{\dfrac{1}{x} + \dfrac{3}{x - 3}}$ **(c)** $\dfrac{\dfrac{1}{a} - \dfrac{1}{b}}{\dfrac{1}{2b^2} - \dfrac{1}{2a^2}}$

Solution   **(a)** The LCD for the numerator *and* the denominator is z, so multiply each by z.

$$\frac{2z}{\left(\dfrac{4}{z} + \dfrac{3}{z}\right)} = \frac{(2z)z}{\left(\dfrac{4}{z} + \dfrac{3}{z}\right)z} \qquad \text{Multiply by } \frac{z}{z}.$$

$$= \frac{2z^2}{\dfrac{4z}{z} + \dfrac{3z}{z}} \qquad \text{Distributive property}$$

$$= \frac{2z^2}{4 + 3} \qquad \text{Reduce.}$$

$$= \frac{2z^2}{7} \qquad \text{Simplify.}$$

(b) The LCD is the product $x(x - 3)$.

$$\frac{\dfrac{1}{x - 3}}{\left(\dfrac{1}{x} + \dfrac{3}{x - 3}\right)} = \frac{\dfrac{1}{x - 3}}{\left(\dfrac{1}{x} + \dfrac{3}{x - 3}\right)} \cdot \frac{x(x - 3)}{x(x - 3)} \qquad \text{Multiply by 1.}$$

$$= \frac{\dfrac{x(x - 3)}{x - 3}}{\dfrac{x(x - 3)}{x} + \dfrac{3x(x - 3)}{x - 3}} \qquad \text{Distributive property}$$

$$= \frac{x}{(x-3)+3x}$$ Reduce the fractions.

$$= \frac{x}{4x-3}$$ Simplify the denominator.

(c) The LCD for the numerator *and* the denominator is $2a^2b^2$.

$$\frac{\dfrac{1}{a}-\dfrac{1}{b}}{\dfrac{1}{2b^2}-\dfrac{1}{2a^2}} = \frac{\dfrac{1}{a}-\dfrac{1}{b}}{\dfrac{1}{2b^2}-\dfrac{1}{2a^2}}\cdot\frac{2a^2b^2}{2a^2b^2}$$ Multiply by 1.

Critical Thinking

Are the expressions $\dfrac{\frac{a}{b}}{\frac{a}{b}+1}$ and

$\dfrac{1}{1+1}$ equal? Explain.

Are the expressions $\dfrac{\frac{a}{b}+1}{\frac{a}{b}}$ and

$1+\dfrac{b}{a}$ equal? Explain.

$$= \frac{\left(\dfrac{1}{a}-\dfrac{1}{b}\right)2a^2b^2}{\left(\dfrac{1}{2b^2}-\dfrac{1}{2a^2}\right)2a^2b^2}$$ Multiply the fractions.

$$= \frac{\dfrac{2a^2b^2}{a}-\dfrac{2a^2b^2}{b}}{\dfrac{2a^2b^2}{2b^2}-\dfrac{2a^2b^2}{2a^2}}$$ Distributive property

$$= \frac{2ab^2-2a^2b}{a^2-b^2}$$ Reduce the fractions.

$$= \frac{2ab(b-a)}{(a-b)(a+b)}$$ Factor.

$$= -\frac{2ab}{a+b}$$ Reduce.

7.5   PUTTING IT ALL TOGETHER

In this section we discussed how to simplify complex fractions to the form $\frac{a}{b}$. The following table summarizes this discussion.

Concept	Explanation	Examples
Complex Fraction	A rational expression that contains fractions in its numerator, denominator, or both	$\dfrac{3+\dfrac{1}{x+1}}{3-\dfrac{1}{x+1}}$ and $\dfrac{\dfrac{x}{y}-\dfrac{y}{x}}{\dfrac{x}{y}+\dfrac{y}{x}}$
Simplifying Complex Fractions	$\dfrac{\dfrac{a}{b}}{\dfrac{c}{d}}=\dfrac{a}{b}\cdot\dfrac{d}{c}$	$\dfrac{\dfrac{2}{x}}{\dfrac{4}{x-1}}=\dfrac{2}{x}\cdot\dfrac{x-1}{4}=\dfrac{x-1}{2x}$

continued on next page

continued from previous page

Concept	Explanation	Examples
Method I: Simplifying the Numerator and Denominator First	Combine the terms in the numerator, combine the terms in the denominator, and then invert and multiply.	$\dfrac{\dfrac{1}{x}+\dfrac{3}{x}}{\dfrac{5}{y}-\dfrac{4}{y}}=\dfrac{\dfrac{4}{x}}{\dfrac{1}{y}}=\dfrac{4}{x}\cdot\dfrac{y}{1}=\dfrac{4y}{x}$
Method II: Multiplying the Numerator and Denominator by the LCD	Multiply the numerator *and* the denominator by the LCD of *all* fractions appearing in the expression.	$\dfrac{\dfrac{2}{x}+\dfrac{1}{y}}{\dfrac{4}{y}-\dfrac{1}{x}}=\dfrac{\left(\dfrac{2}{x}+\dfrac{1}{y}\right)xy}{\left(\dfrac{4}{y}-\dfrac{1}{x}\right)xy}$ $=\dfrac{2y+x}{4x-y}$ Note that the LCD is xy.

7.5 EXERCISES

CONCEPTS

1. $\dfrac{\dfrac{1}{2}}{\dfrac{3}{4}}=$ _____

2. $\dfrac{\dfrac{a}{b}}{\dfrac{c}{d}}=$ _____

3. A complex fraction is a rational expression that contains _____ in its numerator, denominator, or both.

4. Write the phrase "the quantity x plus one half divided by the quantity x minus one half" as a complex fraction.

5. What operation does the fraction bar represent?

6. Write the expression $\dfrac{x}{2}\div\dfrac{1}{x-1}$ as a complex fraction.

7. Write the expression $\dfrac{a}{b}\div\dfrac{c}{d}$ as a complex fraction.

8. What is the LCD for $\dfrac{1}{x+2}$ and $\dfrac{1}{x}$?

SIMPLIFYING COMPLEX FRACTIONS

Exercises 9–14: For the complex fraction, determine the LCD of all the fractions appearing in both the numerator and the denominator.

9. $\dfrac{\dfrac{x}{5}-\dfrac{1}{6}}{\dfrac{2}{15}-3x}$

10. $\dfrac{\dfrac{1}{2}-\dfrac{1}{x}}{\dfrac{1}{2}+\dfrac{1}{x}}$

11. $\dfrac{\dfrac{2}{x+1}-x}{\dfrac{2}{x-1}+x}$

12. $\dfrac{\dfrac{1}{4x}-\dfrac{4}{x}}{\dfrac{1}{2x}+\dfrac{1}{3x}}$

13. $\dfrac{\dfrac{1}{2x-1}-\dfrac{1}{2x+1}}{\dfrac{x+1}{x}}$

14. $\dfrac{\dfrac{1}{4x^2}-\dfrac{1}{2x^3}}{\dfrac{1}{x-1}+\dfrac{1}{x-1}}$

Exercises 15–46: Simplify the complex fraction.

15. $\dfrac{\frac{2}{3}}{\frac{5}{6}}$

16. $\dfrac{\frac{8}{9}}{\frac{5}{4}}$

17. $\dfrac{\frac{r}{t}}{\frac{2r}{t}}$

18. $\dfrac{\frac{8}{p}}{\frac{4}{p}}$

19. $\dfrac{\frac{6}{x}}{\frac{2}{y}}$

20. $\dfrac{\frac{3}{14x}}{\frac{6}{7x}}$

21. $\dfrac{\frac{6}{m-2}}{\frac{2}{m-2}}$

22. $\dfrac{\frac{3}{n+1}}{\frac{6}{n+1}}$

23. $\dfrac{\frac{p+1}{p}}{\frac{p+2}{p}}$

24. $\dfrac{\frac{2p}{2p+5}}{\frac{1}{4p+10}}$

25. $\dfrac{\frac{5}{z^2-1}}{\frac{z}{z^2-1}}$

26. $\dfrac{\frac{z}{z-2}}{\frac{z}{z-2}}$

27. $\dfrac{\frac{y}{y^2-9}}{\frac{1}{y+3}}$

28. $\dfrac{\frac{2y}{2y-1}}{\frac{1}{4y^2-1}}$

29. $\dfrac{\frac{3}{x+1}}{\frac{4}{x+1}-\frac{1}{x+1}}$

30. $\dfrac{\frac{5}{2x-3}-\frac{4}{2x-3}}{\frac{7}{2x-3}+\frac{8}{2x-3}}$

31. $\dfrac{\frac{1}{m^2n}+\frac{1}{mn^2}}{\frac{1}{m^2n}-\frac{1}{mn^2}}$

32. $\dfrac{\frac{3}{x-1}-\frac{2}{x}}{\frac{3}{x-1}+\frac{2}{x}}$

33. $\dfrac{\frac{1}{2x}+\frac{1}{y}}{\frac{1}{y}-\frac{1}{2x}}$

34. $\dfrac{\frac{3}{x}-\frac{2}{y}}{\frac{3}{x}+\frac{2}{y}}$

35. $\dfrac{\frac{1}{ab}+\frac{1}{a}}{\frac{1}{ab}-\frac{1}{b}}$

36. $\dfrac{\frac{1}{a}+\frac{2}{3b}}{\frac{1}{a}-\frac{5}{2b}}$

37. $\dfrac{\frac{2}{q}-\frac{1}{q+1}}{\frac{1}{q+1}}$

38. $\dfrac{\frac{5}{p}+\frac{4}{p-5}}{\frac{5}{p}-\frac{5}{p-5}}$

39. $\dfrac{\frac{1}{x+1}+\frac{1}{x+2}}{\frac{1}{x+1}-\frac{1}{x+2}}$

40. $\dfrac{\frac{1}{x-3}-\frac{1}{x+3}}{1-\frac{1}{x^2-9}}$

41. $\dfrac{\frac{1}{2x-1}-\frac{1}{2x+1}}{\frac{x+1}{x}}$

42. $\dfrac{\frac{1}{4x^2}-\frac{1}{x^3}}{\frac{1}{x-1}+\frac{1}{x-1}}$

43. $\dfrac{\frac{1}{ab^2}-\frac{1}{a^2b}}{\frac{1}{b}-\frac{1}{a}}$

44. $\dfrac{\frac{1}{x^2}-\frac{1}{y^2}}{\frac{1}{x}-\frac{1}{y}}$

45. $\dfrac{1}{a^{-1}+b^{-1}}$

46. $\dfrac{a^2-b^2}{a^{-2}-b^{-2}}$

APPLICATIONS

47. *Annuity* If P dollars are deposited every 2 weeks in an account paying an annual interest rate r expressed as a decimal, then the amount A in the account after 2 years can be approximated by

$$\left(P\left(1+\frac{r}{26}\right)^{52}-P\right)\div\frac{r}{26}.$$

Write this expression as a complex fraction.

48. *Annuity* (Continuation of the preceding exercise) Use a calculator to evaluate the expression when $r=0.026$ (2.6%) and $P=\$100$. Interpret the result.

49. *Resistance in Electricity* Light bulbs are often wired so that electricity can flow through either bulb, as illustrated in the accompanying figure.

In this way, if one bulb burns out, the other bulb still works. If two light bulbs have resistances T and S, then their combined resistance R is given by the complex fraction.

$$R = \frac{1}{\dfrac{1}{T} + \dfrac{1}{S}}$$

Simplify this formula.

50. *Resistance in Electricity* (Refer to the preceding exercise.) Evaluate the formula

$$R = \frac{1}{\dfrac{1}{T} + \dfrac{1}{S}}$$

when $T = 100$ and $S = 200$.

WRITING ABOUT MATHEMATICS

51. A student simplifies a complex fraction as shown below. Explain the student's mistake and how you would simplify the complex fraction correctly.

$$\frac{\dfrac{1}{x} + 1}{\dfrac{1}{x}} \stackrel{?}{=} \frac{\dfrac{1}{x}}{\dfrac{1}{x}} + 1 = 2$$

52. Explain one method for simplifying a complex fraction.

7.6 RATIONAL EQUATIONS AND FORMULAS

Solving Rational Equations · Graphical and Numerical Solutions · Solving a Formula for a Variable · Applications

INTRODUCTION

In Section 7.1 we demonstrated that, if cars arrive randomly at a construction site at an average rate of x cars per minute and if 10 cars per minute can pass through the site, then the average time T spent waiting in line by a car is

$$T = \frac{1}{10 - x},$$

where T is in minutes. If the highway department wants to limit the average wait for a car to $\frac{1}{2}$ minute or less, then mathematics can be used to determine the corresponding value of x. However, before solving this problem, we discuss rational equations.   (*Source:* N. Garber and L. Hoel, *Traffic and Highway Engineering.*)

SOLVING RATIONAL EQUATIONS

If an equation contains one or more rational expressions, it is called a **rational equation.** Rational equations occur in mathematics whenever a rational expression is set equal to a constant. For example, the formula

$$T = \frac{1}{10 - x}$$

from the introduction can be used to estimate the average time that cars wait to get through a construction site. If the wait is $\frac{1}{2}$ minute, we determine x by solving the *rational equation*

$$\frac{1}{10 - x} = \frac{1}{2}.$$

To solve this equation we multiply each side by the LCD: $2(10 - x)$.

$$\frac{2(10 - x)}{10 - x} = \frac{2(10 - x)}{2} \qquad \text{Multiply by the LCD.}$$

$$2 = 10 - x \qquad \text{Reduce.}$$

$$x = 8 \qquad \text{Add } x; \text{ subtract 2}$$

Thus the average wait is $\frac{1}{2}$ minute when cars arrive randomly at an average rate of 8 cars per minute.

In general, if a rational equation is in the form

$$\frac{a}{b} = \frac{c}{d},$$

we can multiply each side of this equation by the least common denominator bd to obtain

$$\frac{a(bd)}{b} = \frac{c(bd)}{d},$$

which simplifies to $ad = cb$. This technique can be used to solve some types of basic rational equations.

SOLVING BASIC RATIONAL EQUATIONS

The equations

$$\frac{a}{b} \bowtie \frac{c}{d} \qquad \text{and} \qquad ad = bc$$

are equivalent, provided that b and d are nonzero. Note that converting the first equation to the second equation is sometimes called *cross multiplying*.

EXAMPLE 1 Solving rational equations

Solve each equation.

(a) $\dfrac{5}{3} = \dfrac{4}{x}$ **(b)** $\dfrac{x + 1}{5} = \dfrac{3x}{2}$ **(c)** $\dfrac{4}{3x - 4} = x$ **(d)** $\dfrac{1}{x} + \dfrac{2}{x} = \dfrac{3}{7}$

Solution **(a)**

$$\frac{5}{3} = \frac{4}{x} \qquad \text{Given equation}$$

$$5x = 12 \qquad \text{Cross multiply.}$$

$$x = \frac{12}{5} \qquad \text{Divide by 5.}$$

The solution is $\frac{12}{5}$.

(b)

$$\frac{x+1}{5} = \frac{3x}{2}$$ Given equation

$$2(x+1) = 15x$$ Cross multiply.

$$2x + 2 = 15x$$ Distributive property

$$-13x = -2$$ Subtract 2 and $15x$.

$$x = \frac{2}{13}$$ Divide by -13.

The solution is $\frac{2}{13}$.

(c)

$$\frac{4}{3x-4} = \frac{x}{1}$$ Write x as $\frac{x}{1}$.

$$x(3x-4) = 4 \cdot 1$$ Cross multiply.

$$3x^2 - 4x = 4$$ Distributive property

$$3x^2 - 4x - 4 = 0$$ Subtract 4.

$$(3x+2)(x-2) = 0$$ Factor.

$$3x + 2 = 0 \quad \text{or} \quad x - 2 = 0$$ Zero-product property

$$x = -\frac{2}{3} \quad \text{or} \quad x = 2$$ Solve each equation.

The solutions are $-\frac{2}{3}$ and 2.

(d)

$$\frac{1}{x} + \frac{2}{x} = \frac{3}{7}$$ Given equation

$$\frac{3}{x} = \frac{3}{7}$$ Add the rational expressions.

$$3x = 21$$ Cross multiply.

$$x = 7$$ Divide by 3.

The solution is 7.

Another technique for solving rational equations is to multiply each side by the least common denominator.

EXAMPLE 2   Multiplying by the LCD

Solve each equation. Check your answer.

(a) $\dfrac{1}{x-1} - \dfrac{1}{x} = \dfrac{1}{9x}$ **(b)** $\dfrac{1}{x-1} + \dfrac{1}{x+1} = \dfrac{12}{x^2-1}$

Solution   **(a)** Start by multiplying each term by the LCD, $9x(x-1)$.

$$\frac{1}{x-1} - \frac{1}{x} = \frac{1}{9x}$$ Given equation

$$\frac{9x(x-1)}{x-1} - \frac{9x(x-1)}{x} = \frac{9x(x-1)}{9x}$$ Multiply each term by the LCD.

$$9x - 9(x-1) = x - 1$$ Reduce each rational expression.

$$9 = x - 1$$ Distributive property

$$x = 10$$ Add 1 and rewrite.

Check:

$$\frac{1}{10 - 1} - \frac{1}{10} \overset{?}{=} \frac{1}{9(10)}$$

Let $x = 10$.

$$\frac{10}{90} - \frac{9}{90} \overset{?}{=} \frac{1}{90}$$

LCD $= 90$.

$$\frac{1}{90} = \frac{1}{90}$$

The answer checks.

(b) Start by multiplying each term by the LCD, $x^2 - 1 = (x - 1)(x + 1)$.

$$\frac{1}{x - 1} + \frac{1}{x + 1} = \frac{12}{x^2 - 1}$$

Given equation

$$\frac{(x - 1)(x + 1)}{x - 1} + \frac{(x - 1)(x + 1)}{x + 1} = \frac{12(x - 1)(x + 1)}{(x - 1)(x + 1)}$$

Multiply each term by the LCD.

$$(x + 1) + (x - 1) = 12$$

Reduce each rational expression.

$$2x = 12$$

Combine like terms.

$$x = 6$$

Divide by 2.

Check:

$$\frac{1}{6 - 1} + \frac{1}{6 + 1} \overset{?}{=} \frac{12}{6^2 - 1}$$

Let $x = 6$.

$$\frac{1}{5} + \frac{1}{7} \overset{?}{=} \frac{12}{35}$$

Simplify.

$$\frac{7}{35} + \frac{5}{35} \overset{?}{=} \frac{12}{35}$$

The LCD is 35.

$$\frac{12}{35} = \frac{12}{35}$$

The answer checks.

When solving a rational equation, be sure to check your answer, as shown in Example 3.

EXAMPLE 3   Solving an equation with an extraneous solution

If possible, solve $\dfrac{1}{x - 2} + \dfrac{1}{x + 2} = \dfrac{4}{x^2 - 4}$.

Solution   Multiply each term by the LCD, $(x - 2)(x + 2) = x^2 - 4$.

$$\frac{1}{x - 2} + \frac{1}{x + 2} = \frac{4}{x^2 - 4}$$

Given equation.

$$\frac{(x - 2)(x + 2)}{x - 2} + \frac{(x - 2)(x + 2)}{x + 2} = \frac{4(x^2 - 4)}{x^2 - 4}$$

Multiply by the LCD.

$$(x + 2) + (x - 2) = 4$$

Reduce each rational expression.

$$2x = 4$$

Combine like terms.

$$x = 2$$

Divide by 2.

Check:

$$\frac{1}{2 - 2} + \frac{1}{2 + 2} \overset{?}{=} \frac{4}{2^2 - 4}$$

Substitute $x = 2$.

Note that both sides of the equations are undefined because it is not possible to divide by 0. Therefore 2 is not a solution; rather, it is an *extraneous solution*. That is, there are no solutions.

GRAPHICAL AND NUMERICAL SOLUTIONS

Like other types of equations, rational equations can also be solved graphically and numerically. Graphs of rational expressions are not lines. They are typically curves that can be graphed either by plotting several points and then sketching a graph or by using a graphing calculator.

EXAMPLE 4 Solving a rational equation graphically and numerically

Solve $\frac{2}{x} = x + 1$ graphically and numerically.

Solution *Graphical Solution* To solve $\frac{2}{x} = x + 1$, graph $y_1 = \frac{2}{x}$ and $y_2 = x + 1$. The graph of y_2 is a line with slope 1 and y-intercept 1. To graph y_1 make a table of values, as shown in Table 7.3. Then plot the points and connect them with a smooth curve. Note that y_1 is undefined for $x = 0$. Generally, it is a good idea to plot at least three points on each side of an asymptote. These points were plotted and the curves sketched in Figure 7.3. Note that the graphs of y_1 and y_2 intersect at $(-2, -1)$ and $(1, 2)$. The solutions to the equation are -2 and 1.

TABLE 7.3

x	$\frac{2}{x}$
-3	$-\frac{2}{3}$
-2	-1
-1	-2
0	—
1	2
2	1
3	$\frac{2}{3}$

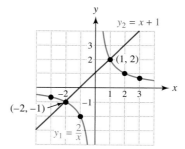

Figure 7.3

Numerical Solution Make a table of values for $y_1 = \frac{2}{x}$ and $y_2 = x + 1$, as shown in Table 7.4. Note that $\frac{2}{x} = x + 1$ when $x = -2$ or $x = 1$.

TABLE 7.4

x	-3	-2	-1	0	1	2	3
$y_1 = \frac{2}{x}$	$-\frac{2}{3}$	-1	-2	—	2	1	$\frac{2}{3}$
$y_2 = x + 1$	-2	-1	0	1	2	3	4

Critical Thinking

If you solve Example 4 symbolically, what will be the result? Verify your answer.

A graphing calculator is a tool that can be used efficiently to create Figure 7.3 and Table 7.4., as illustrated in Figure 7.4.

Calculator Help
To make a table or a graph see
Appendix A (pages AP-2 and
AP-5).

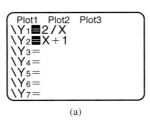
(a)

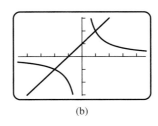
(b)

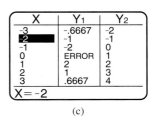
(c)

Figure 7.4

SOLVING A FORMULA FOR A VARIABLE

Formulas in applications often involve both the use of rational expressions and the need to solve an equation for a variable. This idea is illustrated in the next two examples.

EXAMPLE 5 Finding time in a distance problem

If a person travels at a speed, or rate, r for time t, then the distance d traveled is $d = rt$.
(a) How far does a person travel in 2 hours when traveling at 60 miles per hour?
(b) Solve the formula $d = rt$ for t.
(c) How long does it take a person to go 250 miles when traveling at 40 miles per hour?

Solution **(a)** $d = rt = 60 \cdot 2 = 120$ miles
(b) To solve $d = rt$ for t divide both sides by r to obtain

$$\frac{d}{r} = \frac{rt}{r} \quad \text{or} \quad \frac{d}{r} = t.$$

(c) $t = \dfrac{d}{r} = \dfrac{250}{40} = \dfrac{25}{4} = 6.25$ hours, or 6 hours and 15 minutes.

EXAMPLE 6 Solving a formula for a variable

Solve each equation for the specified variable.
(a) $A = \frac{bh}{2}$ for b **(b)** $P = \frac{nRT}{V}$ for V **(c)** $S = 2\pi rh + \pi r^2$ for h

Solution **(a)** First, multiply each side by 2.

$$A = \frac{bh}{2} \qquad \text{Given equation}$$

$$2A = bh \qquad \text{Multiply each side by 2.}$$

$$\frac{2A}{h} = b \qquad \text{Divide each side by } h.$$

(b) Begin by multiplying each side by V.

$$P = \frac{nRT}{V} \qquad \text{Given equation}$$

$$PV = nRT \qquad \text{Multiply each side by } V.$$

$$V = \frac{nRT}{P} \qquad \text{Divide each side by } P.$$

(c) Begin by subtracting πr^2 from each side.

$$S = 2\pi rh + \pi r^2 \qquad \text{Given equation}$$

$$S - \pi r^2 = 2\pi rh \qquad \text{Subtract } \pi r^2 \text{ from each side.}$$

$$\frac{S - \pi r^2}{2\pi r} = h \qquad \text{Divide each side by } 2\pi r.$$

APPLICATIONS

Rational equations sometimes occur in time and rate problems, as demonstrated in the next two examples.

EXAMPLE 7   Mowing a lawn

Two people are mowing a large lawn. One person has a riding mower, and the other person has a push mower. The person with the riding mower can cut the lawn alone in 4 hours, and the person with the push mower can cut the lawn alone in 9 hours. How long does it take for them, working together, to cut the lawn?

Solution   The first person can cut the entire lawn in 4 hours, so this person can cut $\frac{1}{4}$ of the lawn in 1 hour, $\frac{2}{4}$ of the lawn in 2 hours, and in general, $\frac{t}{4}$ of the lawn in t hours. The second person can cut the lawn in 9 hours so (using similar reasoning) this person can cut $\frac{t}{9}$ of the lawn in t hours. Together they can cut

$$\frac{t}{4} + \frac{t}{9}$$

of the lawn in t hours. The job is complete when the fraction of the lawn cut reaches 1. To find out how long this task takes, solve the equation

$$\frac{t}{4} + \frac{t}{9} = 1.$$

Begin by multiplying both sides by the LCD, or 36.

$$\frac{t}{4} + \frac{t}{9} = 1 \qquad \text{Equation to be solved}$$

$$\frac{36t}{4} + \frac{36t}{9} = 36(1) \qquad \text{Multiply each term by 36.}$$

$$9t + 4t = 36 \qquad \text{Reduce.}$$

$$13t = 36 \qquad \text{Combine like terms.}$$

$$t = \frac{36}{13} \qquad \text{Divide each side by 13.}$$

Critical Thinking

If one person can mow a lawn in x hours and another person can mow it in y hours, how long does it take them, working together, to mow the lawn?

Working together they can cut the lawn in $\frac{36}{13} \approx 2.8$ hours.

EXAMPLE 8   Solving a distance problem

Suppose that the winner of a 600-mile car race finishes 12 minutes ahead of the second place finisher. If the winner averages 5 miles per hour faster than the second racer, find the average speed of each racer.

Solution Here we apply the four-step method for solving an application problem.

STEP 1 *Identify any variables.*

x: speed of slower car in miles per hour

$x + 5$: speed of faster car in miles per hour

STEP 2 *Write an equation.* To determine the time required for each car to finish the race we use the equation $t = \frac{d}{r}$. Because the race is 600 miles, the time for the slower car is $\frac{600}{x}$ and the time for the faster car is $\frac{600}{x + 5}$. The difference between these times is 12 minutes, or $\frac{12}{60} = \frac{1}{5}$ hour. Thus to determine x, we solve

$$\frac{600}{x} - \frac{600}{x + 5} = \frac{1}{5}.$$

Note: The cars' speeds are in miles per *hour*, so time must be in hours rather than minutes.

STEP 3 *Solve the equation.* Start by multiplying by the LCD, $5x(x + 5)$.

$\dfrac{600}{x} - \dfrac{600}{x + 5} = \dfrac{1}{5}$	Equation to be solved
$\dfrac{600 \cdot 5x(x + 5)}{x} - \dfrac{600 \cdot 5x(x + 5)}{x + 5} = \dfrac{1 \cdot 5x(x + 5)}{5}$	Multiply each term by the LCD.
$3000(x + 5) - 3000x = x(x + 5)$	Reduce.
$3000x + 15{,}000 - 3000x = x^2 + 5x$	Distributive property
$15{,}000 = x^2 + 5x$	Combine like terms.
$x^2 + 5x - 15{,}000 = 0$	Rewrite the equation.
$(x - 120)(x + 125) = 0$	Factor.
$x - 120 = 0 \quad \text{or} \quad x + 125 = 0$	Zero-product property
$x = 120 \quad \text{or} \quad x = -125$	Solve.

The slower car travels at 120 miles per hour, and the faster car travels 5 miles per hour faster, or 125 miles per hour. (The solution -125 has no meaning in this problem.)

STEP 4 *Check your answer.* The slower car travels 600 miles at 120 miles per hour, which takes $\frac{600}{120} = 5$ hours. The faster car travels 600 miles at 125 miles per hour, which takes $\frac{600}{125} = 4.8$ hours. The difference between their times is $5 - 4.8 = 0.2$ hour, or $0.2 \times 60 = 12$ minutes, so the answer checks.

Technology Note: *Solving Rational Equations Numerically*

Tables sometimes are useful in finding solutions to rational equations. The calculator displays show the positive solution of 120 from Example 8. Note the use of parentheses for entering the formula for Y_1.

Calculator Help
To make a table or enter a formula, see Appendix A (page AP-3 or AP-5).

```
Plot1  Plot2  Plot3
\Y1■600/X−600/(X
+5)
\Y2■1/5
\Y3=
\Y4=
\Y5=
\Y6=
```

X	Y1	Y2
117	.21017	.2
118	.2067	.2
119	.20331	.2
120	.2	.2
121	.19677	.2
122	.19362	.2
123	.19055	.2
X=120		

7.6 PUTTING IT ALL TOGETHER

In this section we discussed two methods of solving rational equations. The following table summarizes these methods.

Concept	Explanation	Examples
Solving the Equation $\dfrac{a}{b} = \dfrac{c}{d}$	Cross multiply to obtain $ad = bc$.	$\dfrac{5}{2x} = \dfrac{3}{6}$ is equivalent to $6x = 30$. $\dfrac{x}{2} = \dfrac{8}{x}$ is equivalent to $x^2 = 16$. (Cross multiplication works only for equations having *one* rational expression on each side.)
Multiplying by the LCD	1. Find the LCD. 2. Multiply each term by the LCD. 3. Reduce each term. 4. Solve the resulting equation.	Solve $\dfrac{1}{x} - \dfrac{2}{3x} = \dfrac{5}{6}$. 1. The LCD is $6x$. 2. $\dfrac{1(6x)}{x} - \dfrac{2(6x)}{3x} = \dfrac{5(6x)}{6}$ 3. $6 - 4 = 5x$ 4. $\dfrac{2}{5} = x$

7.6 EXERCISES

FOR EXTRA HELP

CONCEPTS

1. If an equation contains one or more rational expressions, it is called a _____ equation.

2. Give an example of a rational expression and an example of a rational equation.

3. The equation $\frac{a}{b} = \frac{c}{d}$ is equivalent to _____, provided that _____ and _____ are nonzero.

4. Are the equations $\frac{2}{x-1} = 5$ and $5(x - 1) = 2$ equivalent provided that $x \neq 1$?

5. One way to solve the equation $\frac{5}{3x} + \frac{3}{4x} = 1$ is to multiply each side by the LCD, or _____.

6. To solve the equation $T = \frac{R}{SV}$ for V, multiply each side by the variable _____ and then divide each side by the variable _____.

SOLVING RATIONAL EQUATIONS

Exercises 7–34: Solve and check your answer.

7. $\dfrac{x}{2} = \dfrac{3}{4}$

8. $\dfrac{2x}{3} = \dfrac{2}{5}$

9. $\dfrac{3}{z} = \dfrac{6}{5}$

10. $\dfrac{2}{7} = \dfrac{1}{z}$

11. $\dfrac{12}{7} = \dfrac{2}{t}$

12. $\dfrac{10}{t} = \dfrac{5}{7}$

13. $\dfrac{3y}{4} = \dfrac{7y}{2}$

14. $\dfrac{y}{6} = \dfrac{5y}{3}$

15. $\dfrac{2}{3} = \dfrac{1}{2x + 1}$

16. $\dfrac{1}{x + 4} = \dfrac{3}{5}$

17. $\dfrac{5}{2x} = \dfrac{8}{x + 2}$

18. $\dfrac{1}{x - 1} = \dfrac{5}{3x}$

19. $\dfrac{1}{z - 1} = \dfrac{2}{z + 1}$

20. $\dfrac{4}{z + 3} = \dfrac{2}{z - 2}$

21. $\dfrac{3}{n + 5} = \dfrac{2}{n - 5}$

22. $\dfrac{4}{3n + 2} = \dfrac{1}{n - 1}$

23. $\dfrac{m}{m - 1} = \dfrac{5}{4}$

24. $\dfrac{5m}{2m - 1} = \dfrac{3}{2}$

25. $\dfrac{5x}{5 - x} = \dfrac{1}{3}$

26. $\dfrac{x + 2}{3x} = \dfrac{4}{3}$

27. $\dfrac{6}{5 - 2x} = 2$

28. $\dfrac{x + 1}{x} = 6$

29. $\dfrac{1}{1 - x} = \dfrac{3}{1 + x}$

30. $\dfrac{2x}{1 - 2x} = \dfrac{1}{2}$

31. $\dfrac{1}{z + 2} = -z$

32. $\dfrac{1}{z - 2} = \dfrac{z}{3}$

33. $\dfrac{-1}{2x + 5} = \dfrac{x}{3}$

34. $\dfrac{x}{2} = \dfrac{1}{3x + 5}$

Exercises 35–56: If possible, solve. Check your answer.

35. $\dfrac{x}{2} + \dfrac{x}{4} = 3$

36. $\dfrac{x}{4} - \dfrac{x}{3} = 1$

37. $\dfrac{3x}{4} - \dfrac{x}{2} = 1$

38. $\dfrac{2x}{3} + \dfrac{x}{3} = 6$

39. $\dfrac{4}{t + 1} + \dfrac{1}{t + 1} = -1$

40. $\dfrac{2}{t - 5} - \dfrac{5}{t - 5} = 3$

41. $\dfrac{1}{x} + \dfrac{2}{x} = \dfrac{1}{2}$

42. $\dfrac{1}{2x} - \dfrac{2}{x} = -3$

43. $\dfrac{5}{4z} - \dfrac{2}{3z} = 1$

44. $\dfrac{3}{z + 1} - \dfrac{1}{z + 1} = 2$

45. $\dfrac{4}{y - 1} + \dfrac{1}{y} = \dfrac{6}{5}$

46. $\dfrac{6}{y + 1} + \dfrac{6}{y} = 5$

47. $\dfrac{1}{2x} - \dfrac{1}{x + 3} = 0$

48. $\dfrac{2}{x} - \dfrac{6}{2x - 1} = -1$

49. $\dfrac{1}{x - 2} + \dfrac{1}{x + 2} = \dfrac{6}{x^2 - 4}$

50. $\dfrac{2}{x + 3} - \dfrac{1}{x - 3} = \dfrac{1}{x^2 - 9}$

51. $\dfrac{1}{p + 1} + \dfrac{1}{p + 2} = \dfrac{1}{p^2 + 3p + 2}$

52. $\dfrac{1}{p - 1} - \dfrac{1}{p + 3} = \dfrac{1}{p^2 + 2p - 3}$

53. $\dfrac{1}{x - 2} + \dfrac{3}{2x - 4} = \dfrac{6}{3x - 6}$

54. $\dfrac{4}{x + 1} - \dfrac{4}{2x + 2} = \dfrac{1}{(x + 1)^2}$

55. $\dfrac{1}{r^2 - r - 2} + \dfrac{2}{r^2 - 2r} = \dfrac{1}{r^2 + r}$

56. $\dfrac{3}{r^2 - 1} + \dfrac{1}{r^2 + r} = \dfrac{3}{r^2 - r}$

GRAPHICAL AND NUMERICAL SOLUTIONS

Exercises 57–64: (Refer to Example 4.) Solve the equation
  (a) *graphically and*
  (b) *numerically.*

57. $\dfrac{3}{x} = x + 2$

58. $-\dfrac{2}{x} = 1 - x$

59. $\dfrac{3x}{2} = \dfrac{1}{2}x - 1$

60. $\dfrac{x}{3} = 2 - \dfrac{2}{3}x$

61. $\dfrac{3}{x - 1} = 3$

62. $\dfrac{2}{x + 5} = 1$

63. $\dfrac{4}{x^2} = 1$

64. $\dfrac{-18}{x^2} = -2$

SOLVING AN EQUATION FOR A VARIABLE

Exercises 65–76: (Refer to Examples 5 and 6.) Solve the equation for the specified variable.

65. $m = \dfrac{F}{a}$ for a

66. $m = \dfrac{2K}{v^2}$ for K

67. $I = \dfrac{V}{R + r}$ for r

68. $\dfrac{1}{T} = \dfrac{r}{R - r}$ for R

69. $h = \dfrac{2A}{b}$ for b

70. $h = \dfrac{2A}{b_1 + b_2}$ for b_1

71. $\dfrac{3}{k} = \dfrac{z}{z + 5}$ for z

72. $\dfrac{5}{r} = \dfrac{t + r}{t}$ for t

73. $T = \dfrac{ab}{a + b}$ for b

74. $A = \dfrac{2b}{a - b}$ for b

75. $\dfrac{3}{k} = \dfrac{1}{x} - \dfrac{2}{y}$ for x

76. $\dfrac{1}{R} = \dfrac{1}{R_1} + \dfrac{1}{R_2}$ for R_1

APPLICATIONS

77. *Waiting in Line*   (Refer to the introduction for this section.) Solve the equation

$$\frac{1}{10 - x} = 1$$

to determine the traffic rate x in cars per minute corresponding to an average waiting time of 1 minute.

78. *Waiting in Line*   At a post office customers arrive randomly at an average rate of x people per minute. The clerk can wait on 4 customers per minute. The average time T in minutes spent waiting in line is given by

$$T = \frac{1}{4 - x},$$

where $x < 4$.
   (a) Evaluate T for $x = 3, 3.9$, and 3.99. What happens to the waiting time as the arrival rate nears 4 people per minute?
   (b) Find x when the waiting time is 5 minutes.

79. *Shoveling a Sidewalk*   It takes an older employee 4 hours to shovel the snow from a sidewalk, but a younger employee can shovel the same sidewalk in 3 hours. How long will it take them to clear the walk if they work together?

80. *Pumping Water*   One pump can empty a pool in 5 days, whereas a second pump can empty the pool in 7 days. How long will it take the two pumps, working together, to empty the pool?

81. *Painting a House*   One painter can paint a house in 8 days, yet a more experienced painter can paint the

house in 4 days. How long will it take the two painters, working together, to paint the house?

82. *Working Together*   Suppose that one person can mow a lawn in x hours and that a second person can mow the same lawn in y hours.
   (a) Show that it takes $\dfrac{xy}{x + y}$ hours for the two people, working together, to mow the lawn.
   (b) Use this expression to solve Example 7.

83. *Bicycle Race*   The winner of a 6-mile bicycle race finishes 2 minutes ahead of a teammate and travels, on average, 2 miles per hour faster than the teammate. Find the average speed of each racer.

84. *Freeway Travel*   Two drivers travel 150 miles on a freeway and then stop at a wayside rest area. The first driver travels 5 miles per hour faster and arrives $\frac{1}{7}$ hour ahead of the second. Find the average speed of each car.

85. *Braking Distance*   If a car is traveling *downhill* at 30 miles per hour on wet pavement, then the braking distance B in feet for this car is given by the rational expression

$$B = \frac{30}{0.3 + m},$$

where $m < 0$ is the slope of the hill. (***Source:*** L. Haefner, *Introduction to Transportation Systems.*)
   (a) Find the braking distance for $m = -0.05$ and interpret the result.
   (b) Find the slope of the road if the braking distance is 150 feet.

86. *Slippery Roads*   If a car is traveling at 30 miles per hour on a level road, then its braking distance in feet is $\frac{30}{x}$, where x is the coefficient of friction between the road and the tires. The variable x is positive and satisfies $x \leq 1$. The closer the value of x is to 0, the more slippery is the road.   (***Source:*** L. Haefner.)
   (a) Evaluate the expression for $x = 1, 0.5$, and 0.1. Interpret the results.
   (b) Find x for a braking distance of 150 feet.

87. *River Current*   A boat can travel 36 miles upstream in the same time that it can travel 54 miles downstream. If the speed of the current is 3 miles per hour, find the speed of the boat without a current.

88. *Airplane Speed*   An airplane can travel 380 miles into the wind in the same time that it can travel 420 miles with the wind. If the wind speed is 10 miles per hour, find the speed of the airplane without any wind.

89. *Airplane Speed* An airplane can travel 450 miles into the wind in the same time that it can travel 750 miles with the wind. If the wind speed is 50 miles per hour, find the speed of the airplane without any wind.

90. *River Current* A boat can travel 114 miles upstream in the same time that it can travel 186 miles downstream. If the speed of the current is 6 miles per hour, find the speed of the boat without a current.

91. *Running and Walking* An athlete runs 10 miles and then walks home. The trip home takes 1 hour longer than it took to run that distance. If the athlete runs 5 miles per hour faster than she walks, what are her average running and walking speeds?

92. *Speed Limit* A person drives 390 miles on a stretch of road. Half the distance is driven traveling 5 miles per hour below the speed limit, and half the distance is driven traveling 5 miles per hour above the speed limit. If the time spent traveling at the slower speed exceeds the time spent traveling at the faster speed by 24 minutes, find the speed limit.

93. *Unknown Number* Find a number n that makes $\frac{2 + n}{8 - n}$ equal to 1.

94. *Unknown Number* Find a number n that makes $\frac{2n + 4}{4 - 3n}$ equal to $-\frac{9}{11}$.

95. *Unknown Number* Find a number n that makes $\frac{4 + 6n}{4 - 4n}$ equal to $-\frac{49}{31}$.

96. *Highway Curves* To make a highway curve safe, highway engineers sometimes bank it, as illustrated in the figure. If a curve is designed for a speed of 50 miles per hour and is banked with positive slope m, then a minimum radius R in feet for the curve is computed by

$$R = \frac{2500}{15m + 2}.$$

(*Source:* N. Garber and L. Hoel, *Traffic and Highway Engineering.*)

(a) Find R for $m = 0.1$. Interpret the result.
(b) If $R = 500$, find m. Interpret the result.

WRITING ABOUT MATHEMATICS

97. Do all rational equations have solutions? Explain your answer.

98. Why is it important to check your answer when solving rational equations? Explain your answer.

CHECKING BASIC CONCEPTS SECTIONS 7.5 AND 7.6

1. Simplify each complex fraction.

(a) $\dfrac{\frac{x}{3}}{\frac{2x}{5}}$ **(b)** $\dfrac{\frac{2}{2x} - \frac{1}{3x}}{6x}$

(c) $\dfrac{\frac{1}{a} - \frac{1}{b}}{\frac{1}{a} + \frac{1}{b}}$ **(d)** $\dfrac{\frac{1}{r^2} - \frac{1}{t^2}}{\frac{2}{r} - \frac{2}{t}}$

2. Solve each equation. Check your answer.

(a) $\dfrac{1}{2x} = \dfrac{3}{x + 1}$ **(b)** $\dfrac{x}{2x + 3} = \dfrac{4}{5}$

3. Solve each equation. Check your answer.

(a) $\dfrac{1}{2x} + \dfrac{3}{2x} = 1$ **(b)** $\dfrac{3}{x + 1} - \dfrac{2}{x} = -2$

4. Solve each equation for the specified variable.

(a) $\dfrac{ax}{2} - 3y = b$ for x

(b) $\dfrac{1}{2m - 1} = \dfrac{k}{m}$ for m

5. *Braking Distance* If a car is traveling *uphill* at 60 miles per hour on wet pavement, then the braking distance D in feet for this car is given by the rational expression

$$D = \frac{120}{0.3 + m},$$

where $m > 0$ is the slope of the hill. (*Source:* L. Haefner, *Introduction to Transportation Systems.*)

(a) Find D for $m = 0.1$ and interpret the result.
(b) Find the slope of the road if D is 200 feet.

7.7 PROPORTIONS AND VARIATION

Proportions · Direct Variation · Inverse Variation · Joint Variation

INTRODUCTION

Proportions frequently are used to solve applications. The following are a few examples.

- If someone earns $100 per day, then that person can earn $500 in 5 days.
- If a car goes 210 miles on 10 gallons of gas, then it can go 420 miles on 20 gallons of gas.
- If a person walks 1 mile in 16 minutes, then that person can walk $\frac{1}{2}$ mile in 8 minutes.

Many other applications also involve proportions or variation. In this section we discuss some of them.

PROPORTIONS

A **ratio** is a comparison of two quantities, expressed as a quotient. For example, a math class might have 5 boys for every 6 girls. Thus the boy–girl ratio in this class is *5 to 6*, or $\frac{5}{6}$. In mathematics, ratios are typically expressed as fractions.

Ratios and proportions are sometimes used to determine how much space remains for music on a compact disc (CD). A 650-megabyte CD can store about 74 minutes of music. Suppose that some music has already been recorded on the CD and that 512 megabytes are still available. Using ratios and proportions, we can determine how many more minutes of music could be recorded. A **proportion** is a statement that two *ratios* are equal. (*Source:* Maxell Corporation.)

Let x represent the number of minutes available on a CD. Then **74** minutes are to **650** megabytes as x minutes are to **512** megabytes. By setting the *ratios* $\frac{74}{650}$ and $\frac{x}{512}$ equal to each other, we obtain the *proportion*

$$\frac{74}{650} = \frac{x}{512}.$$

Solving this equation for x gives

$$650x = 74(512)$$

$$x = \frac{74 \cdot 512}{650} \approx 58.3 \text{ minutes.}$$

About 58 minutes are available to record on the CD.

≡ MAKING CONNECTIONS ≡

Proportions and Fractional Parts

We could have solved the preceding problem by noting that the fraction of the CD still available for recording music is $\frac{512}{650}$. So $\frac{512}{650}$ of 74 minutes is

$$\frac{512}{650} \cdot 74 \approx 58.3 \text{ minutes.}$$

EXAMPLE 1 **Calculating the water content in snow**

Six inches of light, fluffy snow are equivalent to about half an inch of rain in terms of water content. If 21 inches of snow fall, estimate the water content.

Solution Let x be the equivalent amount of rain. Then 6 inches of snow is to $\frac{1}{2}$ inch of rain as 21 inches of snow is to x inches of rain, which can be written as the proportion

$$\frac{6}{\frac{1}{2}} = \frac{21}{x}.$$

Solving this equation gives

$$6x = \frac{21}{2} \quad \text{or} \quad x = \frac{21}{12} = 1.75.$$

Thus 21 inches of light, fluffy snow is equivalent to about 1.75 inches of rain. ———

Proportions frequently occur in geometry when we work with similar figures. Two triangles are similar if the measures of their corresponding angles are equal. Corresponding sides of similar triangles are proportional. Figure 7.5 shows two right triangles that are similar because each has angles of 30°, 60°, and 90°.

We can find the length of side x by using proportions. Side x is to 12 as 4.5 is to 6, which can be written as the proportion

$$\frac{x}{12} = \frac{4.5}{6}. \qquad \frac{\text{Hypotenuse}}{\text{Hypotenuse}} = \frac{\text{Shorter leg}}{\text{Shorter leg}}$$

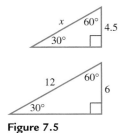

Figure 7.5

Solving yields the equation

$$6x = 4.5(12) \qquad \text{Cross multiply.}$$

$$x = 9. \qquad \text{Divide by 6.}$$

EXAMPLE 2 Calculating the height of a tree

A 6-foot-tall person casts a 4-foot-long shadow. If a nearby tree casts a 44-foot-long shadow, estimate the height of the tree. See Figure 7.6.

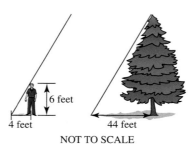

6 feet

4 feet   44 feet

NOT TO SCALE

Figure 7.6

Solution The triangles shown in Figure 7.7 are similar because the measures of the corresponding angles are equal. Therefore their sides are proportional. Let h be the height of the tree.

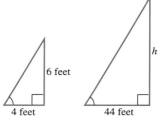

6 feet h

4 feet   44 feet

Figure 7.7

$$\frac{h}{6} = \frac{44}{4} \qquad \frac{\text{Height}}{\text{Height}} = \frac{\text{Shadow length}}{\text{Shadow length}}$$

$$4h = 6(44) \qquad \text{Cross multiply.}$$

$$h = \frac{6(44)}{4} \qquad \text{Divide by 4.}$$

$$h = 66 \qquad \text{Simplify.}$$

The tree is 66 feet tall.

DIRECT VARIATION

If your wage is $9 per hour, the amount you earn is proportional to the number of hours that you work. If you worked H hours, your total pay P satisfies the equation

$$\frac{P}{H} = \frac{9}{1}, \qquad \frac{\text{Pay}}{\text{Hours}}$$

or, equivalently,

$$P = 9H.$$

We say that your pay P is *directly proportional* to the number of hours H worked. The constant of proportionality is 9.

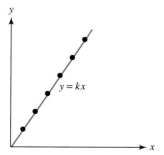

Figure 7.8 Direct Variation, $k > 0$

DIRECT VARIATION

Let x and y denote two quantities. Then y is **directly proportional** to x, or y **varies directly** with x, if there is a nonzero number k such that

$$y = kx.$$

The number k is called the **constant of proportionality**, or the **constant of variation**.

The graph of $y = kx$ is a line passing through the origin, as illustrated in Figure 7.8. Sometimes data in a scatterplot indicate that two quantities are directly proportional. The constant of proportionality k corresponds to the slope of the graph.

EXAMPLE 3   Modeling college tuition

Table 7.5 lists the tuition for taking various numbers of credits.
(a) A scatterplot of the data is shown in Figure 7.9. Could the data be modeled using a line?

TABLE 7.5

Credits	Tuition
3	$189
5	$315
8	$504
11	$693
17	$1071

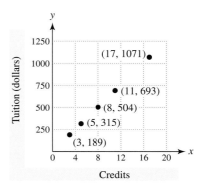

Figure 7.9

(b) Explain why tuition is directly proportional to the number of credits taken.
(c) Find the constant of proportionality. Interpret your result.
(d) Predict the cost of taking 15 credits.

Solution   **(a)** The data are linear and suggest a line passing through the origin.
(b) Because the data can be modeled by a line passing through the origin, tuition is directly proportional to the number of credits taken. Hence doubling the credits will double the tuition and tripling the credits will triple the tuition.
(c) The slope of the line equals the constant of proportionality k. If we use the first and last data points $(3, 189)$ and $(17, 1071)$, the slope is

$$k = \frac{1071 - 189}{17 - 3} = 63.$$

That is, tuition is $63 per credit. If we graph the line $y = 63x$, it models the data, as shown in Figure 7.10. This graph can also be created with a graphing calculator.
(d) If y represents tuition and x represents the credits taken, 15 credits would cost

$$y = 63(15) = \$945.$$

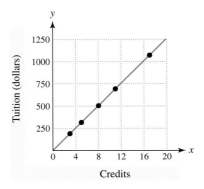

Figure 7.10

====== MAKING CONNECTIONS ======

Ratios and the Constant of Proportionality

The constant of proportionality in Example 3 can also be found by calculating the ratios $\frac{y}{x}$, where y is the tuition and x is the credits taken. Note that each ratio in the table is 63 because the equation $y = 63x$ is equivalent to the equation $\frac{y}{x} = 63$.

x	3	5	8	11	17
y	189	315	504	693	1071
$\frac{y}{x}$	63	63	63	63	63

INVERSE VARIATION

When two quantities vary inversely, an increase in one quantity results in a decrease in the second quantity. For example, at 25 miles per hour a car travels 100 miles in 4 hours, whereas at 50 miles per hour the car travels 100 miles in 2 hours. Doubling the speed (or rate) decreases the travel time by half. Distance equals rate times time, so $d = rt$. Thus

$$100 = rt, \quad \text{or equivalently,} \quad t = \frac{100}{r}.$$

We say that the time t to travel 100 miles is *inversely proportional* to the speed or rate r. The constant of proportionality or constant of variation is 100.

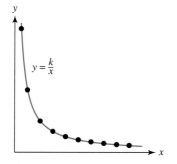

Figure 7.11 Inverse Variation, $k > 0$

███████ **INVERSE VARIATION**

Let x and y denote two quantities. Then y is **inversely proportional** to x, or y **varies inversely** with x, if there is a nonzero number k such that

$$y = \frac{k}{x}.$$

Note: We assume that the constant k is positive.

The data shown in Figure 7.11 represent inverse variation and are modeled by $y = \frac{k}{x}$. Note that, as x increases, y decreases.

A wrench is commonly used to loosen a nut on a bolt. See Figure 7.12. If the nut is difficult to loosen, a wrench with a longer handle is often helpful.

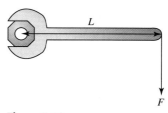

Figure 7.12

EXAMPLE 4   Loosening a nut on a bolt

Table 7.6 lists the force F necessary to loosen a particular nut with wrenches of different lengths L.
(a) Make a scatterplot of the data and discuss the graph. Are the data linear?
(b) Explain why the force F is inversely proportional to the handle length L. Find k so that $F = \frac{k}{L}$ models the data.
(c) Predict the force needed to loosen the nut with an 8-inch wrench.

TABLE 7.6

L (inches)	6	10	12	15	20
F (pounds)	10	6	5	4	3

Solution   **(a)** The scatterplot shown in Figure 7.13 reveals that the data are nonlinear. As the length L of the wrench increases, the force F necessary to loosen the nut decreases.

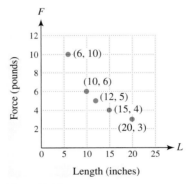

Figure 7.13

(b) If F is inversely proportional to L, then $F = \frac{k}{L}$, or $FL = k$. That is, the product of F and L equals the constant of proportionality k. In Table 7.6, the product of F and L always equals 60 for each data point. Thus F is inversely proportional to L with constant of proportionality $k = 60$.
(c) If $L = 8$, then $F = \frac{60}{8} = 7.5$. A wrench with an 8-inch handle requires a force of 7.5 pounds to loosen the nut.

EXAMPLE 5   Analyzing data

Determine whether the data in each table represent direct variation, inverse variation, or neither.

(a)

x	4	5	10	20
y	50	40	20	10

(b)

x	2	5	9	11
y	14	35	63	77

(c)

x	2	4	6	8
y	10	16	24	48

Solution (a) As x increases, y decreases. Because $xy = 200$ for each data point in the table, the equation $y = \frac{200}{x}$ models the data. The data represent inverse variation.

(b) As $\frac{y}{x} = 7$ for each data point in the table, the equation $y = 7x$ models the data in the table. These data represent direct variation.

(c) Neither the product xy nor the ratio $\frac{y}{x}$ are constant for the data in the table. Therefore these data represent neither direct variation nor inverse variation.

Technology Note: *Scatterplots and Graphs*

A graphing calculator can be used to create scatterplots and graphs. A scatterplot of the data in Table 7.6 is shown in the first figure. In the second figure the data and the equation $y = \frac{60}{x}$ are graphed. Note that each tick mark represents 4 units.

Calculator Help
To make a scatterplot see Appendix A (Page AP-4).

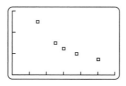
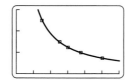

JOINT VARIATION

In many applications a quantity depends on more than one variable. In **joint variation** a quantity varies as the product of more than one variable. For example, the formula for the area A of a rectangle is given by

$$A = LW,$$

where L and W are the length and width, respectively. Thus the area of a rectangle varies jointly with the length and width.

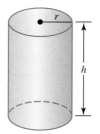

Figure 7.14

> ### JOINT VARIATION
>
> Let x, y, and z denote three quantities. Then z varies jointly as x and y if there is a nonzero number k such that
>
> $$z = kxy.$$

Sometimes joint variation can involve a power of a variable. For example, the volume V of a cylinder is given by $V = \pi r^2 h$, where r is its radius and h is its height, as illustrated in Figure 7.14. In this case we say that the volume varies jointly with the height and the *square* of the radius. The constant of variation is $k = \pi$.

EXAMPLE 6   Strength of a rectangular beam

The strength S of a rectangular beam varies jointly as its width w and the square of its thickness t. See Figure 7.15. If a beam 3 inches wide and 5 inches thick supports 750 pounds, how much can a similar beam 2 inches wide and 6 inches thick support?

Solution The strength of a beam is modeled by $S = kwt^2$, where k is a constant of variation. We can find k by substituting $S = 750$, $w = 3$, and $t = 5$ in the formula.

$$750 = k \cdot 3 \cdot 5^2 \qquad \text{Substitute into } S = kwt^2.$$

$$k = \frac{750}{3 \cdot 5^2} \qquad \text{Solve for } k.$$

$$= 10 \qquad \text{Simplify.}$$

Figure 7.15

Thus $S = 10wt^2$ models the strength of this type of beam. When $w = 2$ and $t = 6$, the beam can support

$$S = 10 \cdot 2 \cdot 6^2 = 720 \text{ pounds.}$$

Critical Thinking

Compare the increased strength of a beam if the width doubles and if the thickness doubles. What happens to the strength of a beam if both the width and thickness triple?

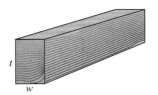

7.7 PUTTING IT ALL TOGETHER

In this section we introduced some basic concepts of proportion and variation. They are summarized in the following table.

Concept	Explanation	Examples
Proportion	A statement that two ratios are equal	$\dfrac{10}{13} = \dfrac{32}{x}$ and $\dfrac{x}{7} = \dfrac{2}{5}$
Direct Variation	Two quantities x and y vary according to the equation $y = kx$, where k is a nonzero constant. The constant of proportionality (or variation) is k.	$y = 3x$ or $\dfrac{y}{x} = 3$ $\begin{array}{c\|ccc} x & 1 & 2 & 4 \\ \hline y & 3 & 6 & 12 \end{array}$ Note that, if x doubles, then y also doubles.
Inverse Variation	Two quantities x and y vary according to the equation $y = \frac{k}{x}$, where k is a nonzero constant. The constant of proportionality (or variation) is k.	$y = \dfrac{2}{x}$ or $xy = 2$ $\begin{array}{c\|ccc} x & 1 & 2 & 4 \\ \hline y & 2 & 1 & \frac{1}{2} \end{array}$ Note that, if x doubles, then y decreases by half.
Joint Variation	Three quantities x, y, and z vary according to the equation $z = kxy$, where k is a constant.	The area A of a triangle varies jointly as b and h according to the equation $A = \frac{1}{2}bh$, where b is its base and h is its height. The constant of variation is $k = \frac{1}{2}$.

7.7 EXERCISES

FOR EXTRA HELP

 Student's Solutions Manual

 InterAct Math

 MathXL

MyMathLab

Math Tutor Center

Digital Video Tutor CD 4 Videotape 10

CONCEPTS

1. What is a proportion?

2. If 5 is to 6 as x is to 7, write a proportion that allows you to find x.

3. Suppose that y is directly proportional to x. If x doubles, what happens to y?

4. Suppose that y is inversely proportional to x. If x doubles, what happens to y?

5. If y varies inversely with x, then xy equals a _____.

6. If y varies directly with x, then $\frac{y}{x}$ equals a _____.

7. If z varies jointly with x and y, then $z =$ _____.

8. If z varies jointly with the square of x and the cube of y, then $z =$ _____.

9. Would the food bill B generally vary directly or inversely with the number of people N being fed? Explain your reasoning.

10. Would the time T needed to paint a building vary directly or inversely with the number of painters N working on the job? Explain your reasoning.

PROPORTIONS

Exercises 11–18: Solve the proportion.

11. $\dfrac{x}{14} = \dfrac{5}{7}$

12. $\dfrac{x}{5} = \dfrac{4}{9}$

13. $\dfrac{8}{x} = \dfrac{2}{3}$

14. $\dfrac{5}{11} = \dfrac{9}{x}$

15. $\dfrac{6}{13} = \dfrac{h}{156}$

16. $\dfrac{25}{a} = \dfrac{15}{8}$

17. $\dfrac{3}{2} = \dfrac{2x}{9}$

18. $\dfrac{7}{4z} = \dfrac{5}{3}$

Exercises 19–26: Do the following.
(a) *Write a proportion that models the situation described.*
(b) *Solve the proportion for x.*

19. 7 is to 9, as 10 is to x

20. x is to 11, as 9 is to 2

21. A triangle has sides of 3, 4, and 6. In a similar triangle the shortest side is 5 and the longest side is x.

22. A rectangle has sides of 9 and 14. In a similar rectangle the longer side is 8 and the shorter side is x.

23. If you earn $78 in 6 hours, then you can earn x dollars in 8 hours.

24. If 12 gallons of gasoline contain 1.2 gallons of ethanol, then 18 gallons of gasoline contain x gallons of ethanol.

25. If 2 cassette tapes can record 90 minutes of music, then 5 cassette tapes can record x minutes.

26. If a gas pump fills a 30-gallon tank in 8 minutes, it can fill a 17-gallon tank in x minutes.

VARIATION

Exercises 27–32: **Direct Variation** *Suppose that y is directly proportional to x.*
(a) *Use the given information to find the constant of proportionality k.*
(b) *Then use $y = kx$ to find y for $x = 7$.*

27. $y = 6$ when $x = 3$

28. $y = 7$ when $x = 14$

29. $y = 5$ when $x = 2$

30. $y = 11$ when $x = 22$

31. $y = -120$ when $x = 16$

32. $y = -34$ when $x = 17$

Exercises 33–38: **Inverse Variation** *Suppose that y is inversely proportional to x.*

(a) *Use the given information to find the constant of proportionality k.*

(b) *Then use $y = \frac{k}{x}$ to find y for $x = 10$.*

33. $y = 5$ when $x = 4$

34. $y = 2$ when $x = 30$

35. $y = 100$ when $x = \frac{1}{2}$

36. $y = \frac{1}{4}$ when $x = 40$

37. $y = 20$ when $x = 20$

38. $y = \frac{45}{4}$ when $x = 8$

Exercises 39–44: **Joint Variation** *Suppose that z varies jointly with x and y.*

(a) *Use the given information to find the constant of variation k.*

(b) *Then use $z = kxy$ to find z for $x = 5$ and $y = 7$.*

39. $z = 6$ when $x = 3$ and $y = 8$

40. $z = 135$ when $x = 2.5$ and $y = 9$

41. $z = 5775$ when $x = 25$ and $y = 21$

42. $z = 1530$ when $x = 22.5$ and $y = 4$

43. $z = 25$ when $x = \frac{1}{2}$ and $y = 5$

44. $z = 12$ when $x = \frac{1}{4}$ and $y = 12$

Exercises 45–50: (Refer to Example 5.)

(a) *Determine whether the data represent direct variation, inverse variation, or neither.*

(b) *If the data represent either direct or inverse variation, find an equation that models the data.*

(c) *Graph the equation and the data when possible.*

45.

x	2	3	4	5
y	3	4.5	6	7.5

46.

x	3	6	9	12
y	12	6	4	3

47.

x	10	20	30	40
y	12	6	5	4

48.

x	2	6	10	14
y	105	35	21	15

49.

x	4	6	12	20
y	10	20	30	40

50.

x	1	5	9	15
y	6	30	54	90

Exercises 51–56: Use the graph to determine whether the data represent direct variation, inverse variation, or neither. Find the constant of variation whenever possible.

51.

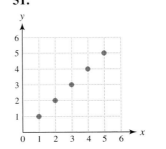

52.

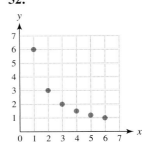

53.

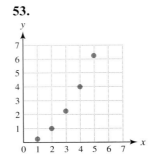

54.

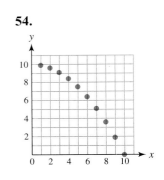

55.

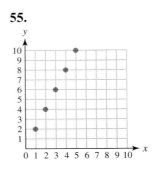

56.

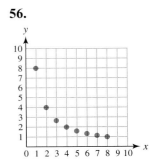

APPLICATIONS

57. *Recording Music*   A 600-megabyte CD can record 68 minutes of music. How many minutes can be recorded on 360 megabytes?

58. *Height of a Tree*   (Refer to Example 2.) A 6-foot person casts a 7-foot shadow, and a nearby tree casts a 27-foot shadow. Estimate the height of the tree.

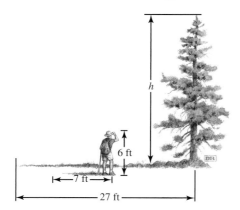

59. *Water Content in Snow*   (Refer to Example 1.) Eight inches of heavy, wet snow are equivalent to an inch of rain. Estimate the water content in 11 inches of heavy, wet snow.

60. *Wages*   If a person working for an hourly wage earns $143 in 11 hours, how much will that person earn in 17 hours?

61. *Rolling Resistance of Cars*   If you were to try and push a car, you would experience *rolling resistance*. This resistance equals the force necessary to keep the car moving slowly in neutral gear. The following table shows the rolling resistance R for passenger cars of different gross weights W.   (*Source:* N. Garber and L. Hoel, *Traffic and Highway Engineering*.)

W (pounds)	2000	2500	3000	3500
R (pounds)	24	30	36	42

(a) Do the data represent direct or inverse variation? Explain.
(b) Find an equation that models the data. Graph the equation with the data.
(c) Estimate the rolling resistance of a 3200-pound car.

62. *Transportation Costs*   The use of a particular toll bridge varies inversely according to the toll. When the toll is $0.75, 6000 vehicles are using the bridge. Esti-

mate the number of users if the toll is $0.40.   (*Source:* N. Garber.)

63. *Flow of Water*   The gallons of water G flowing in 1 minute through a hose with a cross-sectional area A are shown in the table.

A (square inch)	0.2	0.3	0.4	0.5
G (gallons)	5.4	8.1	10.8	13.5

(a) Do the data represent direct or inverse variation? Explain.
(b) Find an equation that models the data. Graph the equation with the data.
(c) Interpret the constant of variation k.

64. *Hooke's Law*   The accompanying table shows the distance D that a spring stretches when a weight W is hung on it.

W (pounds)	2	6	9	15
D (inches)	1.5	4.5	6.75	11.25

(a) Do the data represent direct or inverse variation? Explain.
(b) Find an equation that models the data.
(c) How far will the spring stretch if an 11-pound weight is hung on it, as depicted in the accompanying figure?

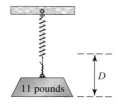

65. *Tightening Lug Nuts*   (Refer to Example 4.) When a tire is mounted on a car, the lug nuts should not be over-tightened. The following table shows the maximum force used with wrenches of different lengths.

L (inches)	8	10	16
F (pounds)	150	120	75

Source: Tires Plus.

(a) Model the data, using the equation $F = \frac{k}{L}$.
(b) How much force should be used with a wrench 20 inches long?

66. *Cost of Tuition*   (Refer to Example 3.) The cost of tuition is directly proportional to the number of credits taken. If 6 credits cost $435, find the cost of 11 credits. What does the constant of proportionality represent?

67. *Air Temperature and Altitude*   In the first 6 miles of Earth's atmosphere, air cools as the altitude increases. The following graph shows the temperature change y in degrees Fahrenheit at an altitude of x miles. (*Source:* A. Miller and R. Anthes, *Meteorology.*)

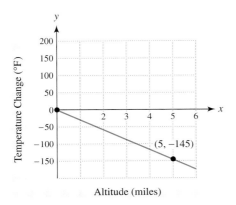

Altitude (miles)

(a) Does this graph represent direct variation or inverse variation?

(b) Find an equation that models the data in the graph.

(c) Is the constant of proportionality k positive or negative? Interpret k.

(d) Estimate the change in air temperature 3.5 miles high.

68. *Ozone and UV Radiation*   Ozone in the upper atmosphere filters out approximately 90% of the harmful ultraviolet (UV) rays from the sun. Depletion of the ozone layer has caused an increase in the amount of UV radiation reaching Earth's surface. An increase in UV radiation is associated with skin cancer. The following graph shows the percentage increase y in UV radiation for a decrease in the ozone layer of x percent. (*Source:* R. Turner, D. Pearce, and I. Bateman, *Environmental Economics.*)

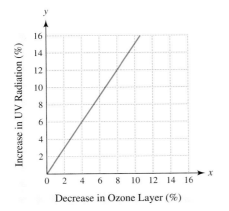

Decrease in Ozone Layer (%)

(a) Does this graph represent direct variation or inverse variation?

(b) Find an equation that models the data in the graph.

(c) Estimate the percentage increase in UV radiation if the ozone layer decreases by 7%.

69. *Electrical Resistance*   The electrical resistance of a wire is directly proportional to its length. If a 35-foot-long wire has a resistance of 2 ohms, find the resistance of a 25-foot-long wire.

70. *Resistance and Current*   The current that flows through an electrical circuit is inversely proportional to the resistance. When the resistance R is 150 ohms, the current I is 0.8 amp. Find the current when the resistance is 40 ohms.

71. *Joint Variation*   The variable z varies jointly as the second power of x and the third power of y. Write a formula for z if $z = 31.9$ when $x = 2$ and $y = 2.5$.

72. *Wind Power*   The electrical power generated by a windmill varies jointly with the square of the diameter of the area swept out by the blades and the cube of the wind velocity. If a windmill with an 8-foot diameter and a 10-mile-per-hour wind generates 2405 watts, how much power would be generated if the blades swept out an area 12 feet in diameter and the wind speed was 15 miles per hour?

73. *Strength of a Beam*   (Refer to Example 6.) If a beam 5 inches wide and 3 inches thick supports 300 pounds, how much can a similar beam 5 inches wide and 2 inches thick support?

74. *Carpeting*   The cost of carpet for a rectangular room varies jointly as its width and length. If a room 11 feet wide and 14 feet long costs $539 to carpet, find the cost to carpet a room 17 feet by 19 feet. Interpret the constant of variation k.

75. Explain in words what it means for a quantity y to be directly proportional to a quantity x.

76. Explain in words what it means for a quantity y to be inversely proportional to a quantity x.

CHECKING BASIC CONCEPTS   SECTION 7.7

1. Solve each proportion.

   (a) $\dfrac{x}{7} = \dfrac{2}{13}$ **(b)** $\dfrac{2}{3} = \dfrac{5}{b}$

2. Write a proportion that models each situation. Then solve it.
   (a) 8 is to 12 as 6 is to x
   (b) If 2 compact discs can record 148 minutes of music, then 7 compact discs can record x minutes of music.

3. Suppose that y is inversely proportional to x. If $y = 6$ when $x = 10$, find the constant of proportionality k. Find y when $x = 15$.

4. Decide whether the data in the table represent direct or inverse variation. Explain your reasoning. Find the constant of variation.

   (a)

x	2	4	6	8
y	3	6	9	12

   (b)

x	2	4	6	8
y	12	6	4	3

5. *Wages*   If a person working for an hourly wage earns \$221 in 17 hours, how much will the person earn in 8 hours?

CHAPTER

7 Summary

Section 7.1   *Introduction to Rational Expressions*

Rational Expression

A rational expression can be written in the form $\dfrac{P}{Q}$, where P and Q are polynomials. A rational expression is defined whenever $Q \neq 0$.

Example: $\dfrac{x^2}{x-5}$ is a rational expression that is defined for all real numbers except $x = 5$.

Simplifying a Rational Expression   To simplify a rational expression factor the numerator and the denominator. Then apply the basic principle of rational expressions,

$$\frac{PR}{QR} = \frac{P}{Q},$$

where P, Q, and R are polynomials.

Example: $\dfrac{x^2 - 4}{x^2 - 3x + 2} = \dfrac{(x+2)(x-2)}{(x-1)(x-2)} = \dfrac{x+2}{x-1}$

CHAPTER 7 Summary **467**

Section 7.2 *Multiplication and Division of Rational Expressions*

Multiplying Rational Expressions To multiply two rational expressions multiply the numerators and multiply the denominators.

$$\frac{A}{B} \cdot \frac{C}{D} = \frac{AC}{BD}$$

Example: $\dfrac{3}{x-1} \cdot \dfrac{4}{x+1} = \dfrac{12}{(x-1)(x+1)}$

Dividing Rational Expressions To divide two rational expressions, multiply by the reciprocal of the divisor.

$$\frac{A}{B} \div \frac{C}{D} = \frac{A}{B} \cdot \frac{D}{C} = \frac{AD}{BC}$$

Example: $\dfrac{x+2}{x^2+3x} \div \dfrac{x+2}{x} = \dfrac{x+2}{x(x+3)} \cdot \dfrac{x}{x+2} = \dfrac{x(x+2)}{x(x+3)(x+2)} = \dfrac{1}{x+3}$

Section 7.3 *Addition and Subtraction with Like Denominators*

Addition of Rational Expressions Having Like Denominators To add two rational expressions having like denominators add their numerators. Keep the same denominator.

$$\frac{A}{C} + \frac{B}{C} = \frac{A+B}{C}$$

Example: $\dfrac{5x}{x+4} + \dfrac{1}{x+4} = \dfrac{5x+1}{x+4}$

Subtraction of Rational Expressions Having Like Denominators To subtract two rational expressions having like denominators subtract their numerators. Keep the same denominator.

$$\frac{A}{C} - \frac{B}{C} = \frac{A-B}{C}$$

Example: $\dfrac{6}{2x+1} - \dfrac{2}{2x+1} = \dfrac{4}{2x+1}$

Section 7.4 *Addition and Subtraction with Unlike Denominators*

Finding Least Common Multiples The least common multiple (LCM) of two or more polynomials can be found as follows.

STEP 1 Factor each polynomial completely.

STEP 2 List each factor the greatest number of times that it occurs in any factorization.

STEP 3 Find the product of this list of factors. It is the LCM.

Example: $4x^2(x + 1) = 2 \cdot 2 \cdot x \cdot x \cdot (x + 1)$

$2x(x^2 - 1) = 2 \cdot x \cdot (x + 1) \cdot (x - 1)$

Listing each factor the greatest number of times and multiplying gives

$$2 \cdot 2 \cdot x \cdot x \cdot (x + 1)(x - 1) = 4x^2(x^2 - 1),$$

which is the LCM of $4x^2(x + 1)$ and $2x(x^2 - 1)$.

Finding the Least Common Denominator   The least common denominator (LCD) is the least common multiple (LCM) of the denominators.

Example: From the preceding example, the LCD for $\dfrac{1}{4x^2(x + 1)}$ and $\dfrac{1}{2x(x^2 - 1)}$ is $4x^2(x^2 - 1)$.

Addition and Subtraction of Rational Expressions Having Unlike Denominators
First write each rational expression by using the LCD. Then add or subtract the resulting rational expressions.

Example: $\dfrac{1}{x - 1} - \dfrac{1}{x} = \dfrac{x}{x(x - 1)} - \dfrac{x - 1}{x(x - 1)} = \dfrac{x - (x - 1)}{x(x - 1)} = \dfrac{1}{x(x - 1)}$

Note that the LCD is $x(x - 1)$.

Section 7.5   *Complex Fractions*

Complex Fractions   A complex fraction is a rational expression that contains fractions in its numerator, denominator, or both. The following equation can be used to simplify a complex fraction.

$$\frac{\dfrac{a}{b}}{\dfrac{c}{d}} = \frac{a}{b} \cdot \frac{d}{c}$$

Example: $\dfrac{\dfrac{x}{3}}{\dfrac{x}{x - 1}} = \dfrac{x}{3} \cdot \dfrac{x - 1}{x} = \dfrac{x(x - 1)}{3x} = \dfrac{x - 1}{3}$

Simplifying Complex Fractions

Method I   Combine terms in the numerator, combine terms in the denominator, and simplify the resulting expression.

Method II   Multiply the numerator and denominator by the LCD for both and simplify the resulting expression.

Example: Method I $\dfrac{\dfrac{1}{a} - \dfrac{1}{b}}{\dfrac{1}{a} + \dfrac{1}{b}} = \dfrac{\dfrac{b - a}{ab}}{\dfrac{b + a}{ab}} = \dfrac{b - a}{ab} \cdot \dfrac{ab}{b + a} = \dfrac{b - a}{b + a}$

Method II   The least common denominator is ab.

$$\frac{\left(\dfrac{1}{a} - \dfrac{1}{b}\right)ab}{\left(\dfrac{1}{a} + \dfrac{1}{b}\right)ab} = \frac{\dfrac{ab}{a} - \dfrac{ab}{b}}{\dfrac{ab}{a} + \dfrac{ab}{b}} = \frac{b - a}{b + a}$$

Section 7.6   *Rational Equations and Formulas*

Solving Rational Equations   One way to solve the equation $\frac{a}{b} = \frac{c}{d}$ is to cross multiply to obtain $ad = bc$. A general way to solve rational equations is to multiply each side by the LCD.

Examples: $\frac{1}{2x} = \frac{2}{x + 3}$ implies that $x + 3 = 4x$, or $x = 1$.

To solve $\frac{5}{x} - \frac{1}{3x} = \frac{7}{3}$ multiply each term by the LCD, $3x$:

$$\frac{5(3x)}{x} - \frac{3x}{3x} = \frac{7(3x)}{3},$$

which simplifies to $15 - 1 = 7x$, or $x = 2$.

Solving for a Variable   Many formulas contain more than one variable. To solve for a particular variable use the rules of algebra to isolate the variable.

Example: To solve $S = \frac{2\pi}{r}$ for r, multiply each side by r to obtain $Sr = 2\pi$ and then divide each side by S to obtain $r = \frac{2\pi}{S}$.

Section 7.7   *Proportions and Variation*

Proportions   A proportion is a statement that two ratios are equal.

Example: $\frac{5}{x} = \frac{4}{7}$

Similar Triangles   Two triangles are similar if the measures of their corresponding angles are equal. Corresponding sides of similar triangles are proportional.

Example: A right triangle has legs of lengths 3 and 4. A similar right triangle has a shorter leg with length 6. Its longer leg can be found by solving the proportion $\frac{3}{6} = \frac{4}{x}$ to obtain $x = 8$.

Direct Variation   A quantity y is *directly proportional* to a quantity x, or y *varies directly* with x, if there is a nonzero constant k such that $y = kx$. The number k is called the *constant of proportionality* or the *constant of variation*.

Example: If y varies directly with x, then the ratios $\frac{y}{x} = k$. The following data satisfy $\frac{y}{x} = 4$, so the constant of variation is 4.

x	1	2	3	4
y	4	8	12	16

Inverse Variation   A quantity y is *inversely proportional* to a quantity x, or y *varies inversely* with x, if there is a nonzero constant k such that $y = \frac{k}{x}$.

Example: If y varies inversely with x, then $xy = k$. The following data satisfy $xy = 12$, so the constant of variation is 12.

x	1	2	4	6
y	12	6	3	2

Joint Variation   The quantity z *varies jointly* as x and y if $z = kxy$, $k \neq 0$.

Example: The area A of a rectangle varies jointly as the length L and width W because $A = LW$.

CHAPTER

7 Review Exercises

SECTION 7.1

Exercises 1–4: If possible, evaluate the expression for the given value of x.

1. $\dfrac{3}{x-3}$ $x = -2$

2. $\dfrac{4x}{5-x^2}$ $x = 3$

3. $\dfrac{-x}{7-x}$ $x = 7$

4. $\dfrac{4x}{x^2-3x+2}$ $x = 2$

5. Complete the table for the rational expression. If a value is undefined, place a dash in the table.

x	-2	-1	0	1	2
$\frac{3x}{x-1}$					

6. Find the x-values that make $\dfrac{8}{x^2-4}$ undefined.

Exercises 7–12: Simplify to lowest terms.

7. $\dfrac{25x^3y^4}{15x^5y}$

8. $\dfrac{x^2-36}{x+6}$

9. $\dfrac{x-9}{9-x}$

10. $\dfrac{x^2-5x}{5x}$

11. $\dfrac{2x^2+5x-3}{2x^2+x-1}$

12. $\dfrac{3x^2+10x-8}{3x^2+x-2}$

SECTION 7.2

Exercises 13–16: Multiply and write in lowest terms.

13. $\dfrac{x-3}{x+1} \cdot \dfrac{2x+2}{x-3}$

14. $\dfrac{2x+5}{(x+5)(x-1)} \cdot \dfrac{x-1}{2x+5}$

15. $\dfrac{(z+3)^2}{(z+3)(z-4)}$

16. $\dfrac{x^2}{x^2-4} \cdot \dfrac{x+2}{x}$

Exercises 17–22: Divide and write in lowest terms.

17. $\dfrac{x+1}{2x} \div \dfrac{3x+3}{5x}$

18. $\dfrac{4}{x^3} \div \dfrac{x+1}{2x^2}$

19. $\dfrac{x-5}{x+2} \div \dfrac{2x-10}{x+2}$

20. $\dfrac{x^2-6x+5}{x^2-25} \div \dfrac{x-1}{x+5}$

21. $\dfrac{x^2-y^2}{x+y} \div \dfrac{x-y}{x+y}$

22. $\dfrac{a^3-b^3}{a+b} \div \dfrac{a-b}{2a+2b}$

SECTION 7.3

Exercises 23–28: Add or subtract and write in lowest terms.

23. $\dfrac{2}{x+10} + \dfrac{8}{x+10}$

24. $\dfrac{9}{x-1} - \dfrac{8}{x-1}$

25. $\dfrac{x+2y}{2x} + \dfrac{x-2y}{2x}$

26. $\dfrac{x}{x+3} + \dfrac{3}{x+3}$

27. $\dfrac{x}{x^2-1} - \dfrac{1}{x^2-1}$

28. $\dfrac{2x}{x^2-25} + \dfrac{10}{x^2-25}$

SECTION 7.4

Exercises 29–34: Find the least common multiple for the expressions. Leave your answer in factored form.

29. $3x, 5x$

30. $5x^2, 10x$

31. $x, x-5$

32. $10x^2, x^2-x$

33. $x^2-1, (x+1)^2$

34. x^2-4x, x^2-16

Exercises 35–40: Rewrite the rational expression by using the given denominator D.

35. $\dfrac{3}{8}, D = 24$

36. $\dfrac{4}{3x}, D = 12x$

37. $\dfrac{3x}{x-2}, D = x^2-4$

38. $\dfrac{2}{x+1}, D = x^2+x$

39. $\dfrac{3}{5x}, D = 5x^2-5x$

40. $\dfrac{2x}{2x-3}, D = 2x^2+x-6$

Exercises 41–50: Simplify the expression.

41. $\dfrac{5}{8} + \dfrac{1}{6}$

42. $\dfrac{3}{4x} + \dfrac{1}{x}$

43. $\dfrac{5}{9x} - \dfrac{2}{3x}$

44. $\dfrac{7}{x-1} - \dfrac{3}{x}$

45. $\dfrac{1}{x+1} + \dfrac{1}{x-1}$

46. $\dfrac{4}{3x^2} - \dfrac{3}{2x}$

47. $\dfrac{1+x}{3x} - \dfrac{3}{2x}$

48. $\dfrac{x}{x^2-1} - \dfrac{1}{x-1}$

49. $\dfrac{2}{x-y} - \dfrac{3}{x+y}$

50. $\dfrac{2}{x} - \dfrac{1}{2x} + \dfrac{2}{3x}$

SECTION 7.5

Exercises 51–60: Simplify the complex fraction.

51. $\dfrac{\frac{3}{4}}{\frac{7}{11}}$

52. $\dfrac{\frac{x}{5}}{\frac{2x}{7}}$

53. $\dfrac{\frac{m}{n}}{\frac{2m}{n^2}}$

54. $\dfrac{\frac{3}{p-1}}{\frac{1}{p+1}}$

55. $\dfrac{\frac{3}{m-1}}{\frac{2m-2}{m+1}}$

56. $\dfrac{\frac{2}{2n+1}}{\frac{8}{2n-1}}$

57. $\dfrac{\frac{1}{2x} - \frac{1}{3x}}{\frac{2}{3x} - \frac{1}{6x}}$

58. $\dfrac{\frac{2}{xy} - \frac{1}{y}}{\frac{2}{xy} + \frac{1}{y}}$

59. $\dfrac{\frac{1}{x} - \frac{1}{x+1}}{\frac{x}{x+1}}$

60. $\dfrac{\frac{2}{x-1} - \frac{1}{x+1}}{\frac{1}{x^2-1}}$

SECTION 7.6

Exercises 61–66: Solve and check your answer.

61. $\dfrac{x}{5} = \dfrac{4}{7}$

62. $\dfrac{4}{x} = \dfrac{3}{2}$

63. $\dfrac{3}{z+1} = \dfrac{1}{2z}$

64. $\dfrac{x+2}{x} = \dfrac{3}{5}$

65. $\dfrac{1}{x+1} = \dfrac{2}{x-2}$

66. $\dfrac{x}{3} = \dfrac{-1}{x+4}$

Exercises 67–78: If possible, solve. Check your answer.

67. $\dfrac{1}{5x} + \dfrac{3}{5x} = \dfrac{1}{5}$

68. $\dfrac{1}{x-1} + \dfrac{2x}{x-1} = 1$

69. $\dfrac{1}{x} + \dfrac{2}{3x} = \dfrac{1}{3}$

70. $\dfrac{1}{x+3} + \dfrac{2x}{x+3} = \dfrac{3}{2}$

71. $\dfrac{5}{x} - \dfrac{3}{x+1} = \dfrac{1}{2}$ **72.** $\dfrac{1}{x-1} - \dfrac{1}{x+1} = \dfrac{1}{4}$

73. $\dfrac{4}{p} - \dfrac{5}{p+2} = 0$

74. $\dfrac{1}{x-3} - \dfrac{1}{x+3} = \dfrac{1}{x^2-9}$

75. $\dfrac{2}{x^2-2x} + \dfrac{1}{x^2-4} = \dfrac{1}{x^2+2x}$

76. $\dfrac{3}{x^2-3x} - \dfrac{1}{x^2-9} = \dfrac{1}{x^2+3x}$

77. $\dfrac{1}{x^2} - \dfrac{5}{x^2+4x} = \dfrac{1}{x^2+4x}$

78. $\dfrac{5}{x^2-1} - \dfrac{1}{x^2+2x+1} = \dfrac{3}{x^2-1}$

Exercises 79 and 80: Solve the equation for the specified variable.

79. $\dfrac{1}{a} + \dfrac{2}{b} = \dfrac{3}{c}$ for b **80.** $y = \dfrac{x}{x-1}$ for x

SECTION 7.7

Exercises 81 and 82: Solve the proportion.

81. $\dfrac{x}{6} = \dfrac{1}{5}$ **82.** $\dfrac{3}{x} = \dfrac{7}{3}$

*Exercises 83 and 84: **Proportions**   Do the following.*
  (a) Write a proportion that models the situation.
  (b) Solve the proportion for x.

83. A rectangle has sides of 5 and 11. In a similar rectangle the longer side is 20 and the shorter side is x.

84. If you earn $117 in 13 hours, then you can earn x dollars in 7 hours.

*Exercises 85 and 86: **Direct Variation**   Suppose that y is directly proportional to x.*
  (a) Use the given information to find the constant of proportionality k.
  (b) Then use y = kx to find y for x = 4.

85. $y = 8$ when $x = 2$ **86.** $y = 21$ when $x = 7$

*Exercises 87 and 88: **Inverse Variation**   Suppose that y is inversely proportional to x.*
  (a) Use the given information to find the constant of proportionality k.
  (b) Then use $y = \frac{k}{x}$ to find y for x = 2.

87. $y = 2.5$ when $x = 4$ **88.** $y = 7$ when $x = 3$

Exercises 89 and 90: Do the following.
  (a) Determine whether the data represent direct or inverse variation.
  (b) Find an equation that models the data.
  (c) Graph the data and your equation.

89.

x	2	3	4	5
y	30	20	15	12

90.

x	2	4	6	8
y	6	12	18	24

Exercises 91 and 92: Use the graph to determine whether the data represent direct or inverse variation. Find the constant of variation whenever possible.

91.

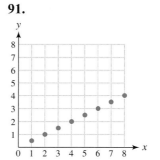

92.

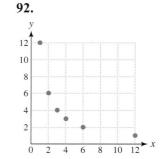

APPLICATIONS

93. *Modeling Traffic Flow*   Fifteen vehicles per minute can pass through an intersection. If vehicles arrive randomly at an average rate of x per minute, the average waiting time T in minutes is given by $T = \dfrac{1}{15-x}$ for $x < 15$.   (*Source:* N. Garber, *Traffic and Highway Engineering.*)
  (a) Evaluate the expression for $x = 10$ and interpret the result.
  (b) Complete the table.

x	5	10	13	14	14.9
T					

(c) What happens to the waiting time as the traffic rate x approaches 15 vehicles per minute?

94. *Distance and Time* A car traveled at 50 miles per hour for 150 miles and then traveled at 75 miles per hour for 150 miles. What was the average speed of the car?

95. *Emptying a Swimming Pool* A large pump can empty a swimming pool in 100 hours, whereas a small pump can empty the pool in 160 hours. How long will it take to empty the pool if both pumps are used?

96. *Jogging* Two athletes jog 10 miles. One of the athletes jogs 2 miles per hour faster and finishes 10 minutes ahead of the other athlete. Find the average speed of each athlete.

97. *River Current* A boat can travel 16 miles upstream in the same time that it can travel 48 miles downstream. If the speed of the current is 4 miles per hour, find the speed of the boat.

98. *Height of a Tree* A 5-foot-tall person has a 6-foot-long shadow, and a nearby tree has a 32-foot-long shadow. Estimate the height of the tree.

99. *Water Content in Snow* Twenty inches of extremely dry, powdery snow is equivalent to an inch of rain. Estimate the water content in 35 inches of this type of snow.

100. *Transportation Costs* Use of a toll road varies inversely with the toll. When the toll is $0.25, 200 vehicles use the road. Estimate the number of users for a toll of $0.50.

101. *Tightening a Bolt* The torque exerted on a nut by a wrench is inversely proportional to the length of the wrench's handle. Suppose that a 10-inch wrench can be used to tighten a nut by using 25 pounds of force. How much force is necessary to tighten the same nut by using a 12-inch wrench?

102. *Cost of Carpet* The cost of carpet is directly proportional to the amount of carpet purchased. If 17 square yards cost $425, find the cost of 13 square yards.

103. *Wind Power* The electric power generated by a windmill varies jointly with the square of the diameter of the area swept out by the blades and the cube of the wind velocity. If a windmill with 6-foot-diameter blades and a 20-mile-per-hour wind generates 10,823 watts, how much power would be generated if the blades were 10 feet in diameter and the wind speed were 12 miles per hour?

104. *Strength of a Beam* The strength of a beam varies jointly as its width w and the square of its thickness t. If a beam 8 inches wide and 5 inches thick supports 650 pounds, how much can a similar beam 6 inches wide and 6 inches thick support?

105. *Cold Water Survival* When a person falls through the ice on a lake, the cold water removes body heat 25 times faster than air at the same temperature. To survive, a person has between 30 and 90 seconds to get out of the water. How long could a person remain in air at the same temperature?

CHAPTER
7 Test

1. Evaluate the expression $\frac{3x}{2x-1}$ for $x = 3$.

2. Find any x-value that makes $\frac{x-1}{x+2}$ undefined.

Exercises 3 and 4: Simplify the expression.

3. $\dfrac{x^2 - 25}{x - 5}$

4. $\dfrac{3x^2 - 15x}{3x}$

Exercises 5–12: Simplify the expression. Write your answer in lowest terms.

5. $\dfrac{x-2}{x+4} \cdot \dfrac{3x+12}{x-2}$

6. $\dfrac{z+1}{z+3} \cdot \dfrac{2z+6}{z+1}$

7. $\dfrac{x+1}{5x} \div \dfrac{2x+2}{x-1}$

8. $\dfrac{2}{x^2} \div \dfrac{x+3}{3x}$

9. $\dfrac{x}{x + 4} + \dfrac{3x + 1}{x + 4}$

10. $\dfrac{4t + 1}{2t - 3} - \dfrac{3t - 6}{2t - 3}$

11. $\dfrac{1}{y^2 + y} - \dfrac{y - 1}{y^2 - y}$

12. $\dfrac{1}{xy} + \dfrac{x}{y} - \dfrac{1}{y^2}$

Exercises 13 and 14: Simplify the complex fraction.

13. $\dfrac{\dfrac{a}{3b}}{\dfrac{5a}{b^2}}$

14. $\dfrac{1 + \dfrac{1}{p - 1}}{1 - \dfrac{1}{p - 1}}$

Exercises 15–20: Solve the equation and check your answer.

15. $\dfrac{2}{7} = \dfrac{5}{x}$

16. $\dfrac{x + 3}{2x} = 1$

17. $\dfrac{1}{2x} + \dfrac{2}{5x} = \dfrac{9}{10}$

18. $\dfrac{1}{x - 1} + \dfrac{2}{x + 2} = \dfrac{3}{2}$

19. $\dfrac{1}{x^2 - 1} - \dfrac{4}{x + 1} = \dfrac{3}{x - 1}$

20. $\dfrac{1}{x^2 - 4x} + \dfrac{2}{x^2 - 16} = \dfrac{2}{x^2 + 4x}$

Exercises 21 and 22: Solve the equation for the specified variable.

21. $y = \dfrac{2}{3x - 5}$ for x

22. $\dfrac{a + b}{ab} = 1$ for b

23. Suppose that y is directly proportional to x.
   (a) If $y = 10$ when $x = 4$, find k so that $y = kx$.
   (b) Then use $y = kx$ to find y for $x = 6$.

24. Use the table to determine whether y varies directly or inversely with x. Find the constant of variation.

x	2	4	8	16
y	32	16	8	4

25. *Emptying a Swimming Pool*   It takes a large pump 40 hours to empty a swimming pool, whereas a small pump can empty the pool in 60 hours. How long will it take to empty the pool if both pumps are used?

26. *Height of a Building*   A 5-foot-tall post has a 4-foot-long shadow, and a nearby building has a 54-foot-long shadow. Estimate the height of the building.

27. *Standing in Line*   A department store clerk can wait on 30 customers per hour. If people arrive randomly at an average rate of x per hour, then the average number of customers N waiting in line is given by $N = \dfrac{x^2}{900 - 30x}$, for $x < 30$. Evaluate the expression for $x = 24$ and interpret your result.

CHAPTER

7 Extended and Discovery Exercises

1. *Graph of a Rational Function*   A car wash can clean 15 cars per hour. If cars arrive randomly at an average rate of x per hour, the average number N of cars waiting in line is given by

$$N = \frac{x^2}{225 - 15x},$$

where $x < 15$.   (*Source*: N. Garber and L. Hoel, *Traffic and Highway Engineering.*)

   (a) Complete the table.

x	3	9	12	13	14
N					

   (b) At what value of x is the expression for N undefined?
   (c) Plot the points from the table in the xy-plane. Then graph $x = 15$ as a vertical dashed line.

(d) Sketch a graph of N that passes through these points. Do not allow your graph to cross the vertical, dashed line, called an asymptote.

(e) Use the graph to explain why a small increase in x can sometimes lead to a long wait.

(f) Explain what happens over a long period of time if the arrival rate x exceeds 15 cars per hour. (*Hint:* The formula for N is not valid when $x \geq 15$.)

Exercises 2–6: Graphing Rational Functions Do the following.

(a) *Use the given equation to complete the table of values for y.*

x	-4	-3	-2	-1	0	1	2	3	4
y									

(b) *Determine any x-value that will make the expression undefined.*

(c) *Sketch a dashed, vertical line (asymptote) in the xy-plane at any undefined values of x.*

(d) *Plot the the points from the table.*

(e) *Sketch a graph of the equation. Do not let your graph cross the vertical dashed line.*

2. $y = \dfrac{1}{x - 1}$

3. $y = \dfrac{1}{x + 1}$

4. $y = \dfrac{4}{x^2 + 1}$

5. $y = \dfrac{x}{x + 1}$

6. $y = \dfrac{x}{x - 1}$

Introduction to Functions

Every day our society creates enormous amounts of data, and mathematics is an important tool for summarizing those data and discovering trends. For example, the table shows the number of Toyota vehicles sold in the United States for selected years.

Year	1998	1999	2000	2001	2002
Vehicles (millions)	1.4	1.5	1.6	1.7	1.8

Source: Autodata.

These data contain an obvious pattern: Sales increased by 0.1 million each year. A scatterplot of these data and a line that models this situation are shown in the figure. In this chapter you will learn how to model these data with a function. (See Section 8.2, Example 6.)

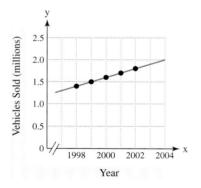

No great thing is created suddenly.
—Epictetus

8.1 FUNCTIONS AND THEIR REPRESENTATIONS

Basic Concepts · Representations of a Function · Definition of a Function · Identifying a Function · Graphing Calculators (Optional)

INTRODUCTION

In Chapter 1 we showed how to use numbers to describe data. For example, instead of simply saying that it is *hot* outside, we might use the number 102°F to describe the temperature. We also showed that data can be modeled with formulas and graphs. Formulas and graphs are sometimes used to represent functions, which are important in mathematics. In this section we introduce functions and their representations.

BASIC CONCEPTS

Input *x*

Function *f*

Output
y = *f*(*x*)

Functions are used to calculate many important quantities. For example, suppose that a person works for $7 per hour. Then we could use a function *f* to calculate the amount of money someone earned after working *x* hours simply by multiplying the *input x* by 7. The result *y* is called the *output*. This concept is shown visually in the following diagram.

For each valid input *x*, a function computes *exactly one* output *y*, which may be represented by the ordered pair (*x*, *y*). If the input is 5 hours, *f* outputs 7 · 5 = $35; if the input is 8 hours, *f* outputs 7 · 8 = $56. These results can be represented by the ordered pairs (5, 35) and (8, 56). Sometimes an input may not be valid. For example, if *x* = −3, there is no reasonable output because a person cannot work −3 hours.

We say that *y is a function of x* because the output *y* is determined by and *depends* on the input *x*. As a result, *y* is called the *dependent variable* and *x* is the *independent variable*. To emphasize that *y* is a function of *x*, we use the notation *y* = *f*(*x*). The symbol *f*(*x*) does not represent multiplication of a variable *f* and a variable *x*. The notation *y* = *f*(*x*) is called *function notation*, is read "*y* equals *f* of *x*," and means that function *f* with input *x* produces output *y*. For example, if *x* = 3 hours, *y* = *f*(3) = $21.

FUNCTION NOTATION

The notation *y* = *f*(*x*) is called **function notation**. The **input** is *x*, the **output** is *y*, and the *name* of the function is *f*.

$$ y = f(x) $$

Name → *f*
Output ↗ *y* =
Input ↖ (*x*)

The variable *y* is called the **dependent variable** and the variable *x* is called the **independent variable**. The expression *f*(4) = **28** is read "*f* of 4 equals 28" and indicates that *f* outputs 28 when the input is 4. A function computes *exactly one* output for each valid input. The letters *f*, *g*, and *h* are often used to denote names of functions.

Functions can be used to compute a variety of quantities. For example, suppose that a boy has a sister that is exactly 5 years older than he is. If the age of the boy is x, then a function g can calculate the age of his sister by adding 5 to x. Thus $g(4) = 4 + 5 = 9$, $g(10) = 10 + 5 = 15$, and in general $g(x) = x + 5$. That is, function g adds 5 to every input x to obtain the output $y = g(x)$.

Functions can be represented by an input–output machine, as illustrated in Figure 8.1. This machine represents function g and receives input $x = 4$, adds 5 to this value, and then outputs $g(4) = 4 + 5 = 9$.

Figure 8.1 Function Machine

REPRESENTATIONS OF A FUNCTION

A function f forms a relation between inputs x and outputs y that can be represented verbally, numerically, symbolically, and graphically. Functions can also be represented with diagrams. We begin by considering a function f that converts yards to feet.

VERBAL REPRESENTATION (WORDS)   To convert x yards to y feet we must multiply x by 3. Therefore, if function f computes the number of feet in x yards, a **verbal representation** of f is "Multiply the input x in yards by 3 to obtain the output y in feet."

NUMERICAL REPRESENTATION (TABLE OF VALUES)   A function f that converts yards to feet is shown in Table 8.1, where $y = f(x)$.

A *table of values* is called a **numerical representation** of a function. Many times it is impossible to list all valid inputs x in a table. On the one hand, if a table does not contain every x-input, it is a *partial* numerical representation. On the other hand, a *complete* numerical representation includes *all* valid inputs. Table 8.1 is a partial numerical representation of f because many valid inputs, such as $x = 10$ or $x = 5.3$, are not shown in it. Note that for each valid input x there is exactly one output y. For a function, inputs are not listed more than once in a table.

TABLE 8.1

x (yards)	y (feet)
1	3
2	6
3	9
4	12
5	15
6	18
7	21

SYMBOLIC REPRESENTATION (FORMULA)   A *formula* provides a **symbolic representation** of a function. The computation performed by f to convert x yards to y feet is expressed by $y = 3x$. A formula for f is $f(x) = 3x$, where $y = f(x)$. We say that function f is *defined by* or *given by* $f(x) = 3x$. Thus $f(2) = 3 \cdot 2 = 6$.

GRAPHICAL REPRESENTATION (GRAPH)   A **graphical representation**, or **graph**, visually associates an x-input with a y-output. The ordered pairs

$$(1, 3), (2, 6), (3, 9), (4, 12), (5, 15), (6, 18), \text{ and } (7, 21)$$

from Table 8.1 are plotted in Figure 8.2(a). This scatterplot suggests a line for the graph f. For each real number x there is exactly one real number y determined by $y = 3x$. If we restrict inputs to $x \geq 0$ and plot all ordered pairs $(x, 3x)$, then a line with no breaks will appear, as shown in Figure 8.2(b).

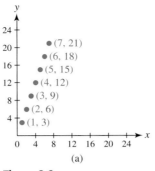

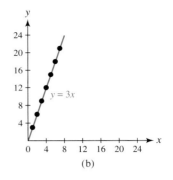

Figure 8.2

(a) Function

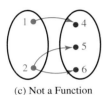

(b) Function

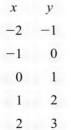

(c) Not a Function

Figure 8.3

$\equiv$ MAKING CONNECTIONS $\equiv$

Functions, Points, and Graphs

If $f(a) = b$, then the point (a, b) lies on the graph of f. Conversely, if the point (a, b) lies on the graph of f, then $f(a) = b$. Thus each point on the graph of f can be written in the form $(a, f(a))$.

DIAGRAMMATIC REPRESENTATION (DIAGRAM) Functions may be represented by **diagrams**. Figure 8.3(a) is a diagram of a function, where an arrow is used to identify the output y associated with input x. For example, input 2 results in output 6, which is written in function notation as $f(2) = 6$. That is, 2 yards are equivalent to 6 feet. Figure 8.3(b) shows a function f even though $f(1) = 4$ and $f(2) = 4$. Although two inputs for f have the same output, each valid input has exactly one output. In contrast, Figure 8.3(c) shows a relation that is not a function because input 2 results in two different outputs, 5 and 6.

$\equiv$ MAKING CONNECTIONS $\equiv$

Four Representations of a Function

Symbolic Representation $f(x) = x + 1$

Numerical Representation   ***Graphical Representation***

x	y
-2	-1
-1	0
0	1
1	2
2	3

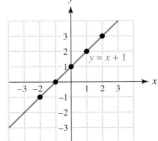

Verbal Representation f adds 1 to an input x to produce an output y.

EXAMPLE 1   Calculating sales tax

Let a function f compute a sales tax of 7% on a purchase of x dollars. Use the given representation to evaluate $f(2)$.
  (a) Verbal Representation   Multiply a purchase of x dollars by 0.07 to obtain a sales tax of y dollars.
  (b) Numerical Representation   Shown in Table 8.2
  (c) Symbolic Representation $f(x) = 0.07x$
  (d) Graphical Representation   Shown in Figure 8.4
  (e) Diagrammatic Representation   Shown in Figure 8.5

TABLE 8.2

x	$f(x)$
$1.00	$0.07
$2.00	$0.14
$3.00	$0.21
$4.00	$0.28

Figure 8.4 Sales Tax of 7%

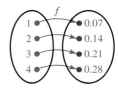

Figure 8.5

Solution **(a)** Multiply the input 2 by 0.07 to obtain 0.14. The sales tax on $2.00 is $0.14.
  (b) From Table 8.2, $f(2) = \$0.14$.
  (c) Because $f(x) = 0.07x, f(2) = 0.07(2) = \mathbf{0.14}$, or $0.14.
  (d) To evaluate $f(2)$ with a graph, first find 2 on the x-axis. Then move vertically upward until you reach the graph of f. The point on the graph may be estimated as $(\mathbf{2}, \mathbf{0.14})$, meaning that $f(\mathbf{2}) = \mathbf{0.14}$, or $0.14 (see Figure 8.6).

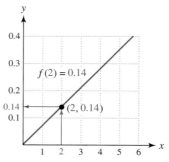

Figure 8.6

  (e) In Figure 8.5, follow the arrow from **2** to **0.14**. Thus $f(2) = 0.14$, or $0.14.

EXAMPLE 2 Evaluating symbolic representations (formulas)

Evaluate each function f at the given value of x.
(a) $f(x) = 3x - 7$ $x = -2$
(b) $f(x) = \dfrac{x}{x + 2}$ $x = 0.5$
(c) $f(x) = \sqrt{x - 1}$ $x = 10$

Solution (a) $f(-2) = 3(-2) - 7 = -6 - 7 = \mathbf{-13}$
(b) $f(0.5) = \dfrac{0.5}{0.5 + 2} = \dfrac{0.5}{2.5} = \mathbf{0.2}$
(c) $f(10) = \sqrt{10 - 1} = \sqrt{9} = \mathbf{3}$

There are many examples of functions. For instance, if we know the radius r of a circle, we can calculate its circumference by using $C(r) = 2\pi r$. The next example illustrates how functions are used in applications.

EXAMPLE 3 Computing crutch length

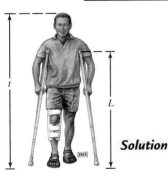

People who sustain leg injuries often require crutches. A proper crutch length can be estimated without using trial and error. The function L, given by $L(t) = 0.72t + 2$, outputs an appropriate crutch length in inches for a person t inches tall. (*Source: Journal of the American Physical Therapy Association.*)
(a) Find $L(60)$ and interpret the result.
(b) If one person is 70 inches tall and another person is 71 inches tall, what should be the difference in their crutch lengths?

Solution (a) $L(60) = 0.72(60) + 2 = \mathbf{45.2}$. Thus a person 60 inches tall needs crutches that are about 45.2 inches long.
(b) From the formula, $L(t) = 0.72t + 2$, we can see that each 1-inch increase in t results in a 0.72-inch increase in L. For example,

$$L(71) - L(70) = 53.12 - 52.4 = 0.72 \text{ inch.}$$

DEFINITION OF A FUNCTION

A function is a fundamental concept in mathematics. Its definition should allow for all representations of a function. *A function receives an input x and produces exactly one output y*, which can be expressed as an ordered pair:

$$(x, y).$$
Input ↗   ↖ Output

A relation is a set of ordered pairs, and a function is a special type of relation.

FUNCTION

A **function** f is a set of ordered pairs (x, y), where each x-value corresponds to exactly one y-value.

The **domain** of f is the set of all x-values (inputs), and the **range** of f is the set of all y-values (outputs). For example, a function f that converts 1, 2, 3, and 4 yards to feet could be expressed as

$$f = \{(1, 3), (2, 6), (3, 9), (4, 12)\}.$$

The domain of f is $D = \{1, 2, 3, 4\}$, and the range of f is $R = \{3, 6, 9, 12\}$.

═══════════════════════ MAKING CONNECTIONS ═══════════════════════

Relations and Functions

A relation can be thought of as a set of input–output pairs. A function is a special type of relation whereby each input results in exactly one output.

EXAMPLE 4 Computing average income

The function f computes the average 1998 individual income in dollars by educational attainment. This function is defined by $f(N) = 16{,}124$, $f(H) = 22{,}895$, $f(B) = 40{,}478$, and $f(M) = 51{,}183$, where N denotes no diploma, H a high school diploma, B a bachelor's degree, and M a master's degree. (**Source:** Bureau of the Census.)
(a) Write f as a set of ordered pairs.
(b) Give the domain and range of f.
(c) Discuss the relationship between education and income.

Solution (a) $f = \{(N, 16124), (H, 22895), (B, 40478), (M, 51183)\}$.
(b) The domain of function f is $D = \{N, H, B, M\}$, and the range of function f is $R = \{16124, 22895, 40478, 51183\}$.
(c) Education pays—the greater the educational attainment the greater are annual earnings.

EXAMPLE 5 Finding the domain and range graphically

Use the graphs of f shown in Figures 8.7 and 8.8 to find each function's domain and range.
(a)

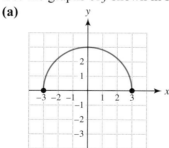

Figure 8.7

(b)

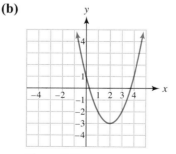

Figure 8.8

Solution (a) The domain is the set of all x-values that correspond to points on the graph of f. Figure 8.9 shows that the domain D includes all x-values satisfying $-3 \leq x \leq 3$. (Recall that the symbol $\leq$ is read "*less than or equal to*.") Because the graph is a semicircle with no breaks, the domain includes all real numbers between -3 and 3. The

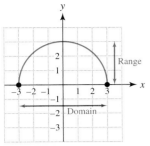

Figure 8.9

range R is the set of y-values that correspond to points on the graph of f. Thus R includes all y-values satisfying $0 \le y \le 3$.

(b) The arrows on the ends of the graph indicate that the graph extends indefinitely left and right, as well as upward. Thus D includes all real numbers. The smallest y-value on the graph is $y = -3$, which occurs when $x = 2$. Thus the range is $y \ge -3$. (Recall that the symbol $\ge$ is read "*greater than or equal to*.")

Critical Thinking

Suppose that a car travels at 50 miles per hour to a city that is 250 miles away. Sketch a graph of a function f that gives the distance y traveled after x hours. Identify the domain and range of f.

Again, the domain of a function is the set of all valid inputs. To determine the domain of a function from a formula, we must determine x-values for which the formula is defined. This concept is demonstrated in the next example.

EXAMPLE 6   Finding the domain of a function

Use $f(x)$ to find the domain of f.

(a) $f(x) = 5x$ **(b)** $f(x) = \dfrac{1}{x - 2}$ **(c)** $f(x) = \sqrt{x}$

Solution   **(a)** Because we can always multiply a real number x by 5, function f is defined for all real numbers. Thus the domain includes all real numbers.

(b) Because we cannot divide by 0, input $x = 2$ is not valid. The expression is defined for all other values of x. Thus the domain includes all real numbers except 2, or $x \ne 2$.

(c) Because square roots of negative numbers are not real numbers, the inputs for f cannot be negative. The domain of f includes all nonnegative numbers, or $x \ge 0$.

IDENTIFYING A FUNCTION

Recall that for a function each valid input x produces exactly one output y. In the next three examples we demonstrate techniques for identifying a function.

EXAMPLE 7   Determining whether a set of ordered pairs is a function

The set S of ordered pairs (x, y) represents the monthly average temperature y in degrees Fahrenheit for the month x in Washington, D.C. Determine whether S is a function.

$S = \{$(January, 33), (February, 37), (March, 45), (April, 53), (May, 66), (June, 73), (July, 77), (August, 77), (September, 70), (October, 51), (November, 48), (December, 37)$\}$

(*Source:* A. Miller and J. Thompson, *Elements of Meteorology*.)

Solution   The input x is the month and the output y is the monthly average temperature. The set S *is a* function because each month x is paired with exactly one monthly average temperature y. Note that, even though an average temperature of 37°F occurs in both February and December, S is nonetheless a function.

EXAMPLE 8 Determining whether a table of values represents a function

Determine whether Table 8.3 represents a function.

TABLE 8.3

x	1	2	3	1	4
y	−4	8	2	5	−6

Solution The table does not represent a function because input $x = 1$ produces two outputs: −4 and 5. That is, the following two ordered pairs both belong to this relation.

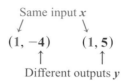

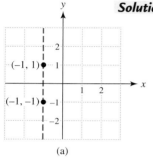

(a)

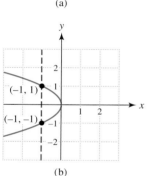

(b)

Figure 8.10

VERTICAL LINE TEST To determine whether a graph represents a function, we must be convinced that it is impossible for an input x to have two or more outputs y. If two distinct points have the same x-coordinate on a graph, then the graph cannot represent a function. For example, the ordered pairs $(-1, 1)$ and $(-1, -1)$ could not lie on the graph of a function because input −1 results in *two* outputs: 1 and −1. When the points $(-1, 1)$ and $(-1, -1)$ are plotted, they lie on the same vertical line, as shown in Figure 8.10(a). A graph passing through these points intersects the vertical line twice, as illustrated in Figure 8.10(b).

To determine whether a graph represents a function, visualize vertical lines moving across the xy-plane. If each vertical line intersects the graph *at most once*, then it is a graph of a function. This test is called the **vertical line test**. Note that the graph in Figure 8.10(b) fails the vertical line test and therefore does not represent a function.

VERTICAL LINE TEST

If every vertical line intersects a graph at no more than one point, then the graph represents a function.

EXAMPLE 9 Determining whether a graph represents a function

Determine whether the graphs shown in Figures 8.11 and 8.12 represent functions.

(a)

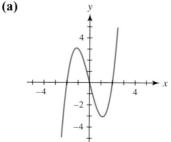

Figure 8.11

(b)

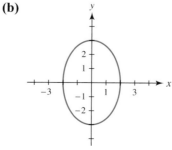

Figure 8.12

Solution (a) Any vertical line will cross the graph at most once, as depicted in Figure 8.13. Therefore the graph *does* represent a function.

(b) The graph *does not* represent a function because some vertical lines can intersect the graph twice, as shown in Figure 8.14.

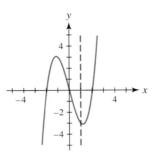

Figure 8.13 Passes Vertical Line Test

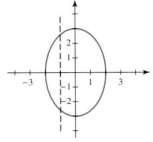

Figure 8.14 Fails Vertical Line Test

GRAPHING CALCULATORS (OPTIONAL)

Graphing calculators provide several features beyond those found on scientific calculators. Graphing calculators have additional keys that can be used to create tables, scatterplots, and graphs.

The **viewing rectangle**, or **window**, on a graphing calculator is similar to the viewfinder in a camera. A camera cannot take a picture of an entire scene. The camera must be centered on some object and can photograph only a portion of the available scenery. A camera can capture different views of the same scene by zooming in and out, as can graphing calculators. The *xy*-plane is infinite, but the calculator screen can show only a finite, rectangular region of the *xy*-plane. The viewing rectangle must be specified by setting minimum and maximum values for both the *x*- and *y*-axes before a graph can be drawn.

We use the following terminology regarding the size of a viewing rectangle. **Xmin** is the minimum *x*-value along the *x*-axis, and **Xmax** is the maximum *x*-value. Similarly, **Ymin** is the minimum *y*-value along the *y*-axis, and **Ymax** is the maximum *y*-value. Most graphs show an *x*-scale and a *y*-scale with tick marks on the respective axes. Sometimes the distance between consecutive tick marks is 1 unit, but at other times it might be 5 or 10 units. The distance represented by consecutive tick marks on the *x*-axis is called **Xscl**, and the distance represented by consecutive tick marks on the y-axis is called **Yscl** (see Figure 8.15). This information about the viewing rectangle can be written as [Xmin, Xmax, Xscl] by [Ymin, Ymax, Yscl]. For example, $[-10, 10, 1]$ by $[-10, 10, 1]$ means that Xmin $= -10$, Xmax $= 10$, Xscl $= 1$, Ymin $= -10$, Ymax $= 10$, and Yscl $= 1$. This setting is referred to as the **standard viewing rectangle**. The window in Figure 8.15 is $[-3, 3, 1]$ by $[-3, 3, 1]$.

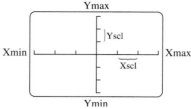

Figure 8.15

EXAMPLE 10 Setting the viewing rectangle

Show the viewing rectangle $[-2, 3, 0.5]$ by $[-100, 200, 50]$ on your calculator.

Solution The window setting and viewing rectangle are displayed in Figure 8.16. Note that in Figure 8.16(b) there are 6 tick marks on the positive x-axis because its length is 3 units and the distance between consecutive tick marks is 0.5 unit.

Calculator Help
To set a viewing rectangle, see
Appendix A (page AP-4).

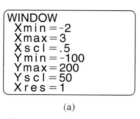

(a)

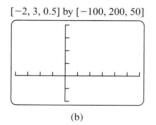

$[-2, 3, 0.5]$ by $[-100, 200, 50]$

(b)

Figure 8.16

We can use graphing calculators to create graphs and tables, usually more efficiently and reliably than pencil-and-paper techniques. However, a graphing calculator uses the same techniques that we might use to sketch a graph. For example, one way to sketch a graph of $y = 2x - 1$ is first to make a table of values, as shown in Table 8.4.

We can plot these points in the xy-plane, as shown in Figure 8.17. Next we might connect the points, as shown in Figure 8.18.

TABLE 8.4

x	y
-1	-3
0	-1
1	1
2	3

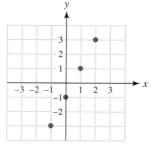

Figure 8.17 Plotting Points

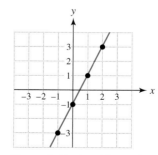

Figure 8.18 Graphing a Line

In a similar manner a graphing calculator plots numerous points and connects them to make a graph. To create a similar graph with a graphing calculator, we enter the formula $Y_1 = 2X - 1$, set an appropriate viewing rectangle, and graph as shown in Figures 8.19(a) and (b). A table of values can also be generated as illustrated in Figure 8.19(c).

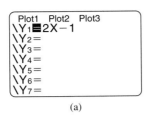
(a)

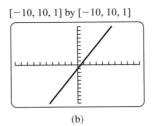
$[-10, 10, 1]$ by $[-10, 10, 1]$
(b)

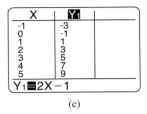
(c)

Calculator Help

To make a graph, see Appendix A (page AP-5). To make a table, see Appendix A (page AP-3).

Figure 8.19

PUTTING IT ALL TOGETHER

One important concept in mathematics is that of a function. A function calculates exactly one output for each valid input and produces input–output ordered pairs in the form (x, y). A function typically computes something, such as area, speed, or sales tax. The following table summarizes some concepts related to functions.

Concept	Explanation	Examples
Function	A set of ordered pairs (x, y), where each x-value corresponds to exactly one y-value	$f = \{(1, 3), (2, 3), (3, 1)\}$ $f(x) = 2x$ A graph of $y = x + 2$ A table of values for $y = 4x$
Independent Variable	The *input* variable for a function	*Function* *Independent Variable* $f(x) = 2x$ x $A(r) = \pi r^2$ r $V(s) = s^3$ s
Dependent Variable	The *output* variable of a function; there is exactly one output for each valid input.	*Function* *Dependent Variable* $y = f(x)$ y $T = F(r)$ T $V = g(r)$ V
Domain and Range of a Function	The domain is the set of all valid inputs. The range is the set of all outputs.	For $S = \{(-1, 0), (3, 4), (5, 0)\}$, $D = \{-1, 3, 5\}$ and $R = \{0, 4\}$. The domain of $f(x) = \frac{1}{x}$ includes all real numbers except 0, or $x \neq 0$.

continued on next page

continued from previous page

Concept	Explanation	Examples
Vertical Line Test	If every vertical line intersects a graph in no more than one point, the graph represents a function.	This graph does *not* pass this test and thus does not represent a function.

A function can be represented verbally, symbolically, numerically, and graphically. The following table summarizes these four representations.

Type of Representation	Explanation	Comments
Verbal	Precise word description of what is computed	May be oral or written Must be stated *precisely*
Symbolic	Mathematical formula	Efficient and concise way of representing a function (e.g., $f(x) = 2x - 3$)
Numerical	List of specific inputs and their outputs	May be in the form of a table or an explicit set of ordered pairs
Graphical, diagrammatic	Shows inputs and outputs visually	No words, formulas, or tables Many types of graphs and diagrams are possible.

8.1   EXERCISES

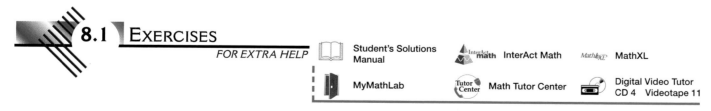

FOR EXTRA HELP

Student's Solutions Manual

MyMathLab

InterAct Math

Math Tutor Center

MathXL

Digital Video Tutor
CD 4   Videotape 11

CONCEPTS

1. The notation $y = f(x)$ is called _____ notation.

2. The notation $y = f(x)$ is read _____.

3. The notation $f(x) = x^2 + 1$ is called a _____ representation of a function.

4. A table of values is a _____ representation of a function.

5. The set of valid inputs for a function is called the _____.

6. The set of outputs for a function is called the _____.

7. A function computes _____ output for each valid input.

8. What is the vertical line test used for?

9. Name four types of representations for a function.

10. If $f(3) = 4$, then the point _____ is on the graph of f. If $(3, 6)$ is on the graph of f, then $f(_) = $ _____.

Exercises 11–14: Determine whether the phrase describes a function.

11. Calculating the square of a number

12. Determining your age to the nearest whole number

13. Listing the students who passed exam x

14. Listing the children of parent x

REPRESENTING AND EVALUATING FUNCTIONS

Exercises 15–24: Evaluate $f(x)$ at the given values of x.

15. $f(x) = 4x - 2 \qquad x = -1, 0$

16. $f(x) = 5 - 3x \qquad x = -4, 2$

17. $f(x) = \sqrt{x} \qquad x = 0, \frac{9}{4}$

18. $f(x) = \sqrt[3]{x} \qquad x = -1, 27$

19. $f(x) = x^2 \qquad x = -5, \frac{3}{2}$

20. $f(x) = x^3 \qquad x = -2, 0.1$

21. $f(x) = 3 \qquad x = -8, \frac{7}{3}$

22. $f(x) = x^2 + 5 \qquad x = -\frac{1}{2}, 6$

23. $f(x) = \dfrac{2}{x + 1} \qquad x = -5, 4$

24. $f(x) = \dfrac{x}{x - 4} \qquad x = -3, 1$

Exercises 25–30: Do the following.

   (a) *Write a formula for the function described.*
   (b) *Evaluate the function for input 10 and interpret the results.*

25. Function I computes the number of inches in x yards.

26. Function M computes the number of miles in x feet.

27. Function A computes the area of a circle with radius r.

28. Function C computes the circumference of a circle with radius r.

29. Function A computes the square feet in x acres. (*Hint:* There are 43,560 square feet in one acre.)

30. Function K computes the number of kilograms in x pounds. (*Hint:* There are about 2.2 pounds in one kilogram.)

Exercises 31–40: Sketch a graph of f.

31. $f(x) = -x + 3$ **32.** $f(x) = -2x + 1$

33. $f(x) = 2x$ **34.** $f(x) = \frac{1}{2}x - 2$

35. $f(x) = 4 - x$ **36.** $f(x) = 6 - 3x$

37. $f(x) = x^2$ **38.** $f(x) = \sqrt{x}$

39. $f(x) = \sqrt{x + 1}$ **40.** $f(x) = \frac{1}{2}x^2 - 1$

Exercises 41–46: Use the graph of f to evaluate the given expressions.

41. $f(0)$ and $f(2)$ **42.** $f(-2)$ and $f(2)$

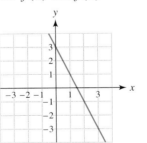 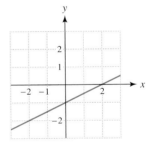

43. $f(-2)$ and $f(1)$ **44.** $f(-1)$ and $f(0)$

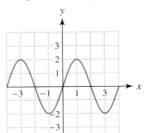 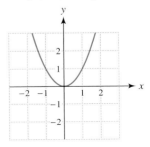

45. $f(1)$ and $f(2)$

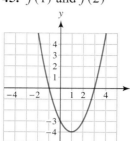

46. $f(-1)$ and $f(4)$

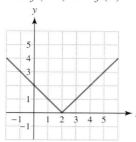

Exercises 47 and 48: Use the table to evaluate the given expression.

47. $f(0)$ and $f(2)$

x	0	1	2	3	4
$f(x)$	5.5	4.3	3.7	2.5	1.9

48. $f(-10)$ and $f(5)$

x	-10	-5	0	5	10
$f(x)$	23	96	-45	-33	23

Exercises 49 and 50: Use the diagram to evaluate $f(1990)$. Interpret your answer.

49. The function f computes average fuel efficiency of new U.S. passenger cars in miles per gallon during year x.   (*Source:* Department of Transportation.)

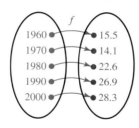

50. The function f computes average cost of tuition at public colleges and universities during academic year x.
(*Source:* The College Board.)

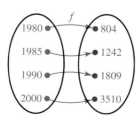

Exercises 51–54: Express the verbal representation for the function f numerically, symbolically, and graphically.

Let $x = -3, -2, -1, \ldots, 3$ for the numerical representation (table), and let $-3 \le x \le 3$ for the graph.

51. Add 5 to the input x to obtain the output y.

52. Square the input x to obtain the output y.

53. Multiply the input x by 5 and then subtract 2 to obtain the output y.

54. Divide the input x by 2 and then add 3 to obtain the output y.

Exercises 55–58: Give a verbal representation for $f(x)$.

55. $f(x) = x - \frac{1}{2}$ **56.** $f(x) = \frac{3}{4}x$

57. $f(x) = \dfrac{x}{3}$ **58.** $f(x) = x^2 + 1$

59. *Cost of Driving*   In 2000, the average cost of driving a new car in the United States was about 50 cents per mile. Symbolically, graphically, and numerically represent a function f that computes the cost in dollars of driving x miles. For the numerical representation (table) let $x = 10, 20, 30, \ldots, 70$.   (*Source:* Associated Press.)

60. *Federal Income Taxes*   In 2001, the lowest U.S. income tax rate was 15 percent. Symbolically, graphically, and numerically represent a function f that computes this tax on a taxable income of x dollars. For the numerical representation (table) let $x = 1000, 2000, 3000, \ldots, 7000$, and for the graphical representation let $0 \le x \le 30{,}000$.   (*Source:* Internal Revenue Service.)

61. *Home Prices*   The median price P of a single-family home in thousands of dollars from 1980 to 1990 can be approximated by $P(x) = 3.421(x - 1980) + 61$, where x is the year. Evaluate $P(1986)$ and interpret the result.

62. *Motorcycles*   The number of Harley-Davidson motorcycles in thousands manufactured between 1985 and 1995 can be approximated by the formula $N(x) = 6.409(x - 1985) + 30.3$, where x is the year. Evaluate $N(1990)$ and interpret the result.

Exercises 63–66: Do the following.

   (a) *Estimate $f(-1)$ graphically.*
   (b) *Use the formula to evaluate $f(-1)$.*

63. $f(x) = 3x + 1$ **64.** $f(x) = 1 - x$

65. $f(x) = 0.5x^2$ **66.** $f(x) = \frac{5}{2}$

IDENTIFYING DOMAINS AND RANGES

Exercises 67–74: Use the graph of f to estimate its domain and range.

67.

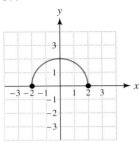

68.

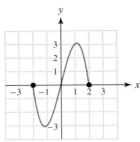

69.

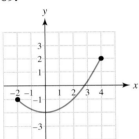

70.

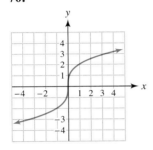

71.

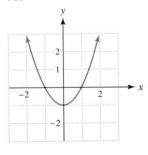

72.

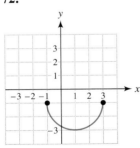

73.

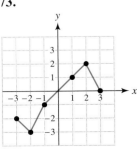

74.

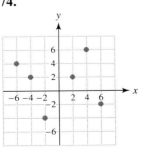

Exercises 75 and 76: Use the diagram to find the domain and range of f.

75.

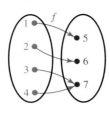

76.

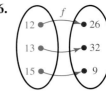

Exercises 77–86: Find the domain.

77. $f(x) = 10x$

78. $f(x) = 5 - x$

79. $f(x) = x^2 - 3$

80. $f(x) = \frac{1}{2}x^2$

81. $f(x) = \dfrac{3}{x - 5}$

82. $f(x) = \dfrac{x}{x + 1}$

83. $f(x) = \dfrac{2x}{x^2 + 1}$

84. $f(x) = \dfrac{6}{1 - x}$

85. $f(x) = \sqrt{x - 1}$

86. $f(x) = |x|$

87. *Accidental Deaths* The function f computes the number of accidental deaths y per 100,000 people during year x. (*Source:* Department of Health and Human Services.)

$$f = \{(1910, 84.4), (1930, 80.5), (1950, 60.3), (1970, 56.2), (1990, 36.9), (2000, 35.5)\}$$

(a) Evaluate $f(1950)$ and interpret the result.
(b) Identify the domain and range of f.
(c) Describe the trend in accidental deaths from 1910 through 2000.

88. *Motor Vehicle Registrations* The following table lists motor vehicle registrations y in millions during year x. Let $y = f(x)$.

x	1920	1940	1960	1980	2000
$f(x)$	9	32	74	156	216

Source: American Automobile Manufacturers Association.

(a) Evaluate $f(1940)$ and interpret the result.
(b) Identify the domain and range of f.
(c) Represent f with a diagram.

IDENTIFYING A FUNCTION

89. *Average Precipitation*   The table lists the monthly average precipitation P in Las Vegas, Nevada, where $x = 1$ is January and $x = 12$ is December.

x (month)	1	2	3	4	5	6
P (inches)	0.5	0.4	0.4	0.2	0.2	0.1

x (month)	7	8	9	10	11	12
P (inches)	0.4	0.5	0.3	0.2	0.4	0.3

Source: J. Williams.

  (a) Determine the value of P during May.
  (b) Is P a function of x? Explain.
  (c) If $P = 0.4$, find x.

90. *Wind Speeds*   The table lists the monthly average wind speed W in Louisville, Kentucky, where $x = 1$ is January and $x = 12$ is December.

x (month)	1	2	3	4	5	6
W (mph)	10.4	12.7	10.4	10.4	8.1	8.1

x (month)	7	8	9	10	11	12
W (mph)	6.9	6.9	6.9	8.1	9.2	9.2

Source: J. Williams.

  (a) Determine the month with the highest average wind speed.
  (b) Is W a function of x? Explain.
  (c) If $W = 6.9$, find x.

Exercises 91–94: Determine whether the diagram could represent a function.

91.
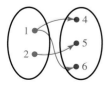

92.
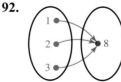

93.
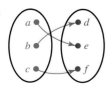

94.
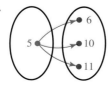

Exercises 95–104: Determine whether the graph represents a function. If it does, identify the domain and range.

95.
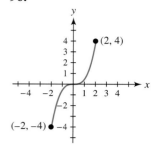

96.
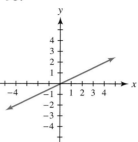

97.
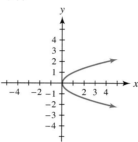

98.
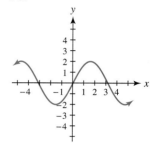

99.
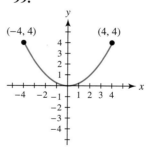

100.
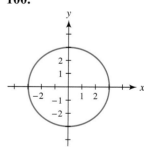

101.
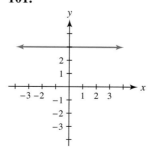

102.
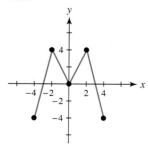

103.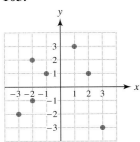

104.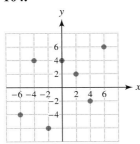

108. *S* is given by the table.

x	−3	−2	−1
y	10	10	10

Exercises 105–108: Determine whether S defines a function.

105. $S = \{(1, 2), (4, 5), (7, 8), (5, 4), (2, 2)\}$

106. $S = \{(4, 7), (-2, 1), (3, 8), (4, 9)\}$

107. *S* is given by the table.

x	5	10	5
y	2	1	0

WRITING ABOUT MATHEMATICS

109. Give an example of a function. Identify the domain and range of your function.

110. Explain in your own words what a function is. How is a function different from a relation?

111. Explain how to evaluate a function by using a graph. Give an example.

112. Give one difficulty that may occur when you use a table of values to evaluate a function.

Group Activity: Working with Real Data

Directions: Form a group of 2 to 4 people. Select someone to record the group's responses for this activity. All members of the group should work cooperatively to answer the questions. If your instructor asks for your results, each member of the group should be prepared to respond.

Large Tires on Vehicles Manufacturers are installing fewer small rims in favor of larger rims. One advantage of bigger rims is that they are better for braking. The following table lists the increase in 16-inch rim sales as a percentage of new vehicle sales for selected years from 1996 to 2002.

Sales of 16-Inch Rims

Year	1996	1998	2000	2002
Percentage	28	36	43	51

Source: USA Today.

1. Plot the data in a scatterplot.

2. Calculate the slope between consecutive points in the table. Interpret each slope.

3. Find the equation of a line that models the data. (*Hint:* You may want to use the point–slope form, $y - y_1 = m(x - x_1)$.)

4. Does your line model the data exactly? Could a line model these data exactly? Explain.

5. Use your equation to write a function *P* that models these data.

6. Evaluate $P(1999)$ and interpret your result.

8.2   LINEAR FUNCTIONS

Basic Concepts · Representations of Linear Functions · Modeling Data with Linear Functions

INTRODUCTION

In applied mathematics, functions are used to model real-world phenomena, such as electricity, weather, and the economy. Because there are so many different applications of mathematics, a wide assortment of functions have been created. In fact, new functions are invented every day for use in business, education, and government. In this section we discuss an important type of function called a *linear function*.

BASIC CONCEPTS

Suppose that the air conditioner is turned on when the temperature inside a house is 80°F. The resulting temperatures are listed in Table 8.5 for various elapsed times. Note that for each 1-hour increase in elapsed time, the temperature decreases by 2°F.

TABLE 8.5   House Temperature

1-hour increase

Elapsed Time (hours)	0	1	2	3	4	5
Temperature (°F)	80	78	76	74	72	70

2°F decrease

A scatterplot is shown in Figure 8.20, which suggests that a line models these data.

Calculator Help

To make a scatterplot, see Appendix A (page AP-4).

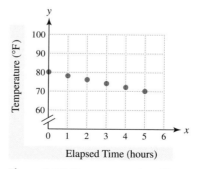

Figure 8.20 Temperature in a Home

Over this 5-hour period, the air conditioner lowers the temperature by 2°F for each hour that it runs. The temperature is found by multiplying the elapsed time x by -2 and adding the initial temperature of 80°F. This situation is modeled by $f(x) = -2x + 80$. For example,

$$f(2.5) = -2(2.5) + 80 = 75$$

means that the temperature is 75°F after the air conditioner has run for 2.5 hours. A graph of $f(x) = -2x + 80$, where $0 \le x \le 5$, is shown in Figure 8.21. We call f a *linear function* because its graph is a *line*. If a function is *not* a linear function, then it is a **nonlinear function**.

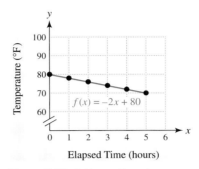

Figure 8.21 A Linear Function

<table>
<tr><td colspan="2">LINEAR FUNCTION</td></tr>
</table>

A function f defined by $f(x) = ax + b$, where a and b are constants, is a **linear function**.

For $f(x) = -2x + 80$, we have $a = -2$ and $b = 80$. The constant a represents the rate at which the air conditioner cools the building, and the constant b represents the initial temperature.

In general, a linear function defined by $f(x) = ax + b$ changes by a units for each unit increase in x. This *rate of change* is an increase if $a > 0$ and a decrease if $a < 0$. For example, if new carpet costs $20 per square yard, then the linear function defined by $C(x) = 20x$ gives the cost of installing x square yards of carpet. The value of $a = 20$ gives the cost (rate of change) for each additional square yard of carpet. For function C, the value of b is 0 because it cost $0 to buy 0 square yards of carpet.

Note: If f is a linear function, then $f(0) = a(0) + b = b$. Thus b can be found by evaluating $f(x)$ at $x = 0$.

EXAMPLE 1   Identifying linear functions

Determine whether f is a linear function. If f is a linear function, find values for a and b so that $f(x) = ax + b$.
(a) $f(x) = 4 - 3x$ **(b)** $f(x) = 8$ **(c)** $f(x) = 2x^2 + 8$

Solution   **(a)** Let $a = -3$ and $b = 4$. Then $f(x) = -3x + 4$, and f is a linear function.
(b) Let $a = 0$ and $b = 8$. Then $f(x) = 0x + 8$, and f is a linear function.
(c) Function f is not linear because its formula contains x^2. The formula for a linear function cannot contain an x with an exponent other than 1.

EXAMPLE 2   Determining linear functions

Use each table of values to determine whether $f(x)$ could represent a linear function. If f could be linear, write a formula for f in the form $f(x) = ax + b$.

(a)

x	0	1	2	3
$f(x)$	10	15	20	25

(b)

x	−2	0	2	4
$f(x)$	4	2	0	−2

(c)

x	0	1	2	3
$f(x)$	1	2	4	7

Solution   (a) For each unit increase in x, $f(x)$ increases by 5 units so $f(x)$ could be linear with $a = 5$. Because $f(0) = 10$, $b = 10$. Thus $f(x) = 5x + 10$.
(b) For each 2-unit increase in x, $f(x)$ decreases by 2 units, or equivalently, each unit increase in x results in a 1-unit decrease in $f(x)$, so $f(x)$ could be linear with $a = -1$. Because $f(0) = 2$, $b = 2$. Thus $f(x) = -x + 2$.
(c) Each unit increase in x does not result in a constant change in $f(x)$. Thus $f(x)$ does not represent a linear function.

REPRESENTATIONS OF LINEAR FUNCTIONS

The graph of a linear function is a line. To graph a linear function f we usually start by making a table of values and then plot two or more points. We can then sketch the graph of f by drawing a line through these points, as demonstrated in the next example.

EXAMPLE 3   Graphing a linear function by hand

Sketch a graph of $f(x) = x - 1$.

Solution   Begin by making a table of values containing at least three points. Pick convenient values of x, such as $x = -1, 0, 1$.

$$f(-1) = -1 - 1 = -2$$
$$f(0) = 0 - 1 = -1$$
$$f(1) = 1 - 1 = 0$$

Display the results, as shown in Table 8.6.

TABLE 8.6
---

x	−1	0	1
y	−2	−1	0

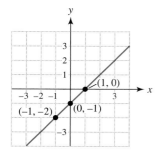

Figure 8.22

Plot the points $(-1, -2)$, $(0, -1)$, and $(1, 0)$. Then sketch a line through the points to obtain the graph of f. A graph of a line results when *infinitely* many points are plotted, as shown in Figure 8.22.

Critical Thinking

Two points determine a line. Why is it a good idea to plot at least three points when graphing a linear function by hand?

EXAMPLE 4 Representing a linear function

A linear function is given by $f(x) = -3x + 2$.
(a) Give a verbal representation of f.
(b) Make a numerical representation (table) of f by letting $x = -1, 0, 1$.
(c) Plot the points in the table from part (b). Then sketch a graph of f.

Solution (a) **Verbal Representation** Multiply the input x by -3 and then add 2 to obtain the output.
(b) **Numerical Representation** Evaluate $f(x) = -3x + 2$ at $x = -1, 0, 1$, which results in Table 8.7.
(c) **Graphical Representation** To make a graph of f by hand, plot the points $(-1, 5)$, $(0, 2)$, and $(1, -1)$ from Table 8.7. Then draw a line passing through these points, as shown in Figure 8.23.

TABLE 8.7

x	$f(x)$
-1	5
0	2
1	-1

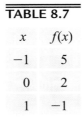

Figure 8.23

In the next example a graphing calculator is used to create a graph and table of a linear function.

EXAMPLE 5 Using a graphing calculator

Give numerical and graphical representations of $f(x) = \frac{1}{2}x - 2$.

Solution **Numerical Representation** To make a numerical representation, construct the table for $Y_1 = .5X - 2$, starting at $x = -3$ and incrementing by 1, as shown in Figure 8.24(a). (Other tables are possible.)

Graphical Representation Graph Y_1 in the standard viewing rectangle, as shown in Figure 8.24(b). (Other viewing rectangles may be used.)

$[-10, 10, 1]$ by $[-10, 10, 1]$

Calculator Help
To make a table, see
Appendix A (page AP-3). To
make a graph, see
Appendix A (page AP-5).

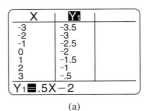

(a)

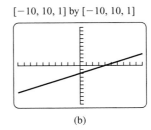

(b)

Figure 8.24

═══ MAKING CONNECTIONS ═══

Mathematics in Newspapers

Think of the mathematics that you see in newspapers. Often percentages are described *verbally*, numbers are displayed in *tables*, and data are shown in *graphs*. Seldom are *formulas* given, which is an important reason not to study only symbolic representations.

MODELING DATA WITH LINEAR FUNCTIONS

A distinguishing feature of a linear function is that each time the input x increases by 1 unit, the output $f(x) = ax + b$ always changes by an amount equal to a. For example, the number of doctors in the private sector from 1970 to 2000 can be modeled by

$$f(x) = 15{,}260x - 29{,}729{,}000,$$

where x is the year. The value $a = 15{,}260$ indicates that the number of doctors has increased, on average, by 15,260 each year.   (*Source:* American Medical Association.)

The following are other examples of quantities that are modeled by linear functions. Try to determine the value of the constant a.

• The wages earned by an individual working x hours at $8 per hour
• The distance traveled by light in x seconds if the speed of light is 186,000 miles per second
• The cost of tuition and fees when registering for x credits if each credit costs $80 and the fees are fixed at $50

When we are modeling data with a linear function defined by $f(x) = ax + b$, the following concepts are helpful to determine a and b.

▌▌▌▌▌▌   **MODELING DATA WITH A LINEAR FUNCTION**

The formula $f(x) = ax + b$ may be interpreted as follows.

$$f(x) = \quad ax \quad + \quad b$$

(New amount) = (Change) + (Fixed amount)

When x represents time, *change* equals (rate of change) × (time).

$$f(x) = \quad a \quad \times \quad x \quad + \quad b$$

(Future amount) = (Rate of change) × (Time) + (Initial amount)

▌▌▌▌▌▌

These concepts are applied in the next three examples.

EXAMPLE 6 Modeling car sales

Table 8.8 shows numbers of Toyota vehicles sold in the United States. (Refer to the introduction to this chapter.)

TABLE 8.8 Toyota Vehicles Sold (millions)

Year	1998	1999	2000	2001	2002
Vehicles	1.4	1.5	1.6	1.7	1.8

Source: Autodata.

(a) What were the sales in 1998?
(b) What was the annual increase in sales?
(c) Find a linear function f that models these data. Let $x = 0$ correspond to 1998, $x = 1$ to 1999, and so on.
(d) Use f to predict sales in 2004.

Solution **(a)** In 1998, 1.4 million vehicles were sold.
(b) Sales have increased by 0.1 million (100,000) vehicles per year. Because this rate of change is the same each year, we can model the data *exactly* with a linear function.
(c) Initial sales in 1998 ($x = 0$) were 1.4 million vehicles, and each year sales increased by 0.1 million vehicles. Thus

$$f(x) \quad = \quad 0.1 \quad \times \quad x \quad + \quad 1.4$$

(Future sales) = (Rate of change in sales) × (Time) + (Initial sales),

or $f(x) = 0.1x + 1.4$.
(d) Because $x = 6$ corresponds to 2004, evaluate $f(6)$.

$$f(6) = 0.1(6) + 1.4 = 2 \text{ million vehicles}$$

In the next example we model tuition and fees.

EXAMPLE 7 Modeling the cost of tuition and fees

Suppose that tuition costs $80 per credit and that student fees are fixed at $50. Give symbolic, numerical, and graphical representations for a linear function that models tuition and fees.

Solution **Symbolic Representation** Total cost is found by multiplying $80 (rate or cost per credit) by the number of credits x and then adding the fixed fees (fixed amount) of $50. Thus $f(x) = 80x + 50$.

Numerical Representation Table 8.9 on the next page lists tuition and fees for 4, 8, 12, and 16 credits. (Other values are also possible.) For example, $f(4) = 80(4) + 50 = 370$, so the cost of 4 credits is $370.

Graphical Representation Start by plotting the points from Table 8.9. Then sketch a line that passes through these points, as shown in Figure 8.25 on the next page.

Calculator Help
To make a scatterplot, see
Appendix A (page AP-4).

TABLE 8.9	Tuition
Credits	Cost
4	$370
8	$690
12	$1010
16	$1330

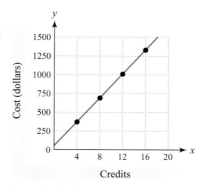

Figure 8.25 Tuition

In the next example we consider a simple linear function that models the speed of a car.

EXAMPLE 8 **Modeling with a constant function**

A car travels on a freeway with its speed recorded at regular intervals, as listed in Table 8.10.

TABLE 8.10	Speed of a Car				
Elapsed Time (hours)	0	1	2	3	4
Speed (miles per hour)	70	70	70	70	70

(a) Discuss the speed of the car during this time interval.
(b) Find a formula for a function f that models these data.
(c) Sketch a graph of f together with the data.

Solution **(a)** The speed of the car appears to be constant at 70 miles per hour.
(b) Because the speed is constant, the rate of change is 0. Thus

$$f(x) = 0x + 70$$
$$\text{(Future speed)} = \text{(Change in speed)} + \text{(Initial speed)}$$

and $f(x) = 70$.
(c) Because $y = f(x)$, graph $y = 70$ with the data points

$$(0, 70), (1, 70), (2, 70), (3, 70), \text{ and } (4, 70)$$

to obtain Figure 8.26.

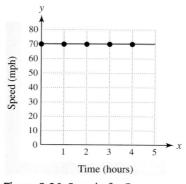

Figure 8.26 Speed of a Car

The function defined by $f(x) = 70$ is an example of a *constant function*. A **constant function** *is a linear function* with $a = 0$ and can be written as $f(x) = b$. Regardless of the input, a constant function always outputs the same value, b. Its graph is a horizontal line. The following are two applications of constant functions.

- A thermostat calculates a constant function regardless of the weather outside by maintaining a set temperature.
- A cruise control in a car calculates a constant function by maintaining a fixed speed, regardless of the type of road or terrain.

PUTTING IT ALL TOGETHER

A linear function is a relatively simple function used to model real data. A clear understanding of linear functions is essential to the understanding of functions in general.

Concept	Explanation	Examples
Linear Function	Can be represented by $f(x) = ax + b$	$f(x) = 2x - 6$, $a = 2$ and $b = -6$ $f(x) = 10$, $a = 0$ and $b = 10$
Constant Function	Can be represented by $f(x) = b$	$f(x) = -7$, $b = -7$ $f(x) = 22$, $b = 22$
Rate of Change for a Linear Function	The output of a linear function changes by a constant amount for each unit increase in the input.	$f(x) = -3x + 8$ decreases 3 units for each unit increase in x. $f(x) = 5$ neither increases nor decreases. The rate of change is 0.

The following table summarizes symbolic, verbal, numerical, and graphical representations of a linear function.

Type of Representation	Comments	Example
Symbolic	Mathematical formula in the form $f(x) = ax + b$	$f(x) = 2x + 1$, where $a = 2$ and $b = 1$
Verbal	Multiply the input x by a and add b.	Multiply the input x by 2 and add 1 to obtain the output.
Numerical (table of values)	For each unit increase in x in the table, the output of $f(x) = ax + b$ changes by an amount equal to a.	1-unit increase x 0 1 2 $f(x)$ 1 3 5 2-unit increase

continued on next page

continued from previous page

Type of Representation	Comments	Example
Graphical	The graph of a linear function is a line. Plot at least 3 points and then sketch the line.	(graph of $y = 2x + 1$)

8.2 EXERCISES

FOR EXTRA HELP

Student's Solutions Manual

InterAct Math   InterAct Math

Math XP   MathXL

MyMathLab

Tutor Center   Math Tutor Center

Digital Video Tutor
CD 4   Videotape 11

CONCEPTS

1. The formula for a linear function is $f(x) =$ _____.

2. The formula for a constant function is $f(x) =$ _____.

3. The graph of a linear function is a _____.

4. The graph of a constant function is a _____.

5. If $f(x) = 7x + 5$, each time x increases by 1 unit, $f(x)$ increases by _____ units.

6. If $f(x) = 5$, each time x increases by 1 unit, $f(x)$ increases by _____ units.

IDENTIFYING LINEAR FUNCTIONS

Exercises 7–14: Determine whether f is a linear function. If f is linear, give values for a and b so that f may be expressed as $f(x) = ax + b$.

7. $f(x) = \frac{1}{2}x - 6$

8. $f(x) = x$

9. $f(x) = \frac{5}{2} - x^2$

10. $f(x) = \sqrt{x} + 3$

11. $f(x) = -9$

12. $f(x) = 1.5 - 7.3x$

13. $f(x) = -9x$

14. $f(x) = \frac{1}{x}$

Exercises 15–18: Determine whether the graph represents a linear function.

15.

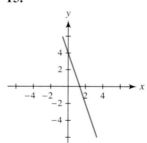

16.

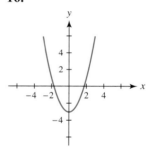

17.

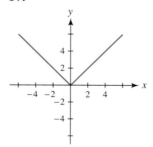

18.

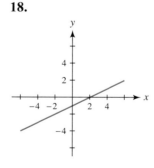

Exercises 19–26: *(Refer to Example 2.) Use the table to determine whether $f(x)$ could represent a linear function. If it could, write $f(x)$ in the form $f(x) = ax + b$.*

19.

x	0	1	2	3
f(x)	−6	−3	0	3

20.

x	0	2	4	6
f(x)	−2	2	6	10

21.

x	−2	0	2	4
f(x)	0	3	6	9

22.

x	0	3	6	9
f(x)	8	4	2	1

23.

x	−2	−1	0	1
f(x)	−5	0	20	40

24.

x	−2	−1	0	1
f(x)	6	3	0	−3

25.

x	1	2	3	4
f(x)	0	2	4	6

26.

x	1	2	3	4
f(x)	0	1	3	7

EVALUATING LINEAR FUNCTIONS

Exercises 27–32: *Evaluate $f(x)$ at the given values of x.*

27. $f(x) = 4x$ $\qquad x = -4, 5$

28. $f(x) = -2x + 1$ $\qquad x = -2, 3$

29. $f(x) = 5 - x$ $\qquad x = -\frac{2}{3}, 3$

30. $f(x) = \frac{1}{2}x - \frac{1}{4}$ $\qquad x = 0, \frac{1}{2}$

31. $f(x) = -22$ $\qquad x = -\frac{3}{4}, 13$

32. $f(x) = 9x - 7$ $\qquad x = -1.2, 2.8$

Exercises 33–38: *Use the graph of f to evaluate the given expressions.*

33. $f(-1)$ and $f(0)$

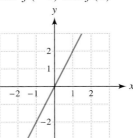

34. $f(-2)$ and $f(2)$

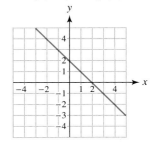

35. $f(-2)$ and $f(4)$

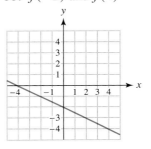

36. $f(0)$ and $f(3)$

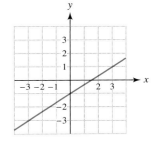

37. $f(-3)$ and $f(1)$

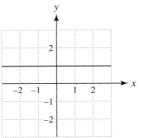

38. $f(1.5)$ and $f(0.5\pi)$

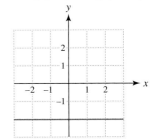

Exercises 39–42: *Use the verbal description to write a formula for $f(x)$. Then evaluate $f(3)$.*

39. Multiply the input by 6.

40. Multiply the input by −3 and add 7.

41. Divide the input by 6 and subtract $\frac{1}{2}$.

42. Output 8.7 for every input.

REPRESENTING LINEAR FUNCTIONS

Exercises 43–46: Match f(x) with its graph (a.–d.).

43. $f(x) = 3x$

44. $f(x) = -2x$

45. $f(x) = x - 2$

46. $f(x) = 2x + 1$

a.

b.

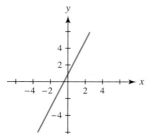
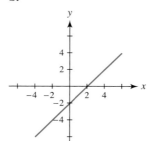

c.

d.

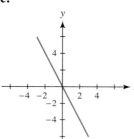
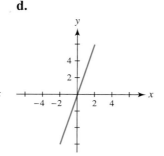

Exercises 47–56: Sketch a graph of y = f(x).

47. $f(x) = 2$

48. $f(x) = -1$

49. $f(x) = 2x$

50. $f(x) = -\frac{1}{2}x$

51. $f(x) = x + 1$

52. $f(x) = x - 2$

53. $f(x) = 3x - 3$

54. $f(x) = -2x + 1$

55. $f(x) = 3 - x$

56. $f(x) = \frac{1}{4}x + 2$

Exercises 57–62: Write a symbolic representation (formula) for a linear function f that calculates the following.

57. The number of pounds in x ounces

58. The number of dimes in x dollars

59. The distance traveled by a car moving at 65 miles per hour for t hours

60. The long-distance phone bill *in dollars* for calling t minutes at 10¢ per minute and a fixed fee of $4.95

61. The total number of hours in a day during day x

62. The cost of skiing x times with a $500 season pass

MODELING

Exercises 63–66: Match the situation with the graph (a.–d.) that models it best, where x-values represent time from 1990 to 2000.

63. The cost of college tuition.

64. The cost of 1 megabyte of computer memory.

65. The distance between Chicago and Denver.

66. The total distance traveled by a satellite orbiting Earth

a.

b.

c.

d.

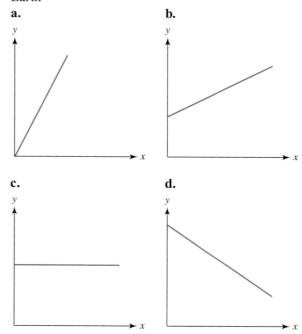

APPLICATIONS

67. *Thermostat*   Let $y = f(x)$ describe the temperature y of a room that is kept at 70°F for x hours.
(a) Represent f symbolically and graphically over a 24-hour period for $0 \le x \le 24$.
(b) Construct a table of f for $x = 0, 4, 8, 12, \ldots, 24$.
(c) What type of function is f?

68. *Cruise Control*   Let $y = f(x)$ describe the speed y of an automobile after x minutes if the cruise control is set at 60 miles per hour.
(a) Represent f symbolically and graphically over a 15-minute period for $0 \le x \le 15$.
(b) Construct a table of f for $x = 0, 1, 2, \ldots, 6$.
(c) What type of function is f?

69. *Distance* A car is initially 50 miles south of the Minnesota–Iowa border traveling south on Interstate 35. Distances D between the car and the border are recorded in the table for various elapsed times t. Find a linear function given by $D(t) = at + b$ that models these data.

t (hours)	0	2	3	5
D (miles)	50	170	230	350

70. *Estimating the Weight of a Bass* Sometimes the weight of a fish can be estimated by measuring its length. The table lists typical weights of bass having various lengths.

Length (inches)	12	14	16	18	20	22
Weight (pounds)	1.0	1.7	2.5	3.6	5.0	6.6

Source: Minnesota Department of Natural Resources.

 (a) Let x be the length and y be the weight. Make a line graph of the data.
(b) Could the data be modeled accurately with a linear function? Explain your answer.

71. *Solid Waste* In 1960, the average American disposed of 2.7 pounds of garbage per day, whereas in 2003 this amount was 4.3 pounds per day. (*Source:* Environmental Protection Agency.)
(a) Find a linear function f that calculates the amount of garbage disposed of by a person in 1960 after x days. Find $f(60)$ and interpret the result.
(b) Repeat part (a) for 2003 with a linear function g.

72. *Rain Forests* Rain forests are defined as forests that grow in regions receiving more than 70 inches of rain per year. The world is losing an estimated 49 million acres of rain forest each year. (*Source:* New York Times Almanac.)
(a) Find a linear function f that calculates the acres of rain forest in millions lost in x years.
(b) Evaluate $f(7)$ and interpret the result.

 73. *Age in the United States* The median age of the population for each year x between 1820 and 1995 can be approximated by

$$f(x) = 0.09x - 147.1$$

(*Source:* Bureau of the Census.)

(a) Graph f in the viewing rectangle

$$[1820, 1995, 20] \text{ by } [0, 40, 10].$$

Discuss any trends shown in the graph.
(b) Construct the table for f starting at $x = 1820$, incrementing by 20. Use the table to evaluate $f(1900)$ and interpret the result.
(c) The value of a in the formula for $f(x)$ is 0.09. Interpret this value.

74. *Temperature and Volume* If a sample of a gas, such as helium, is heated, it will expand. The expression $V(T) = 0.147T + 40$ calculates the volume V in cubic inches of a sample of gas at temperature T in degrees Celsius.
(a) Evaluate $V(0)$ and interpret the result.
(b) If the temperature increases by $10°C$, by how much does the volume increase?
(c) What is the volume of the gas when the temperature is $100°C$?

75. *Temperature and Volume* (Refer to the preceding exercise.) A sample of gas at $0°C$ has a volume V of 137 cubic centimeters, which increases in volume by 0.5 cubic centimeter for every $1°C$ increase in temperature T.
(a) Write a formula $V(T) = aT + b$ that gives the volume of the gas at temperature T.
(b) Find the volume of the gas when $T = 50°C$.

76. *Cost* To make a music video it costs $750 to rent a studio plus $5 for each copy produced.
(a) Write a formula $C(x) = ax + b$ that calculates the cost of producing x videos.
(b) Find the cost of producing 2500 videos.

77. *Baseball* The table shows the average length of a major league baseball game in minutes for various years.

Year	2000	2001	2002
Length (minutes)	180	176	172

Source: Elias Sports Bureau.

(a) What was the average length of a baseball game in 2000?
(b) By how many minutes did the average length change each year?
(c) Find a linear function f that models these data. Let $x = 0$ correspond to 2000.
(d) Use f to predict the average length of a game in 2004.

78. *Wal-Mart Sales* The table shows Wal-Mart's share as a percentage of overall U.S. retail sales for various years. (This percentage excludes restaurants and motor vehicles.)

Year	1998	1999	2000	2001	2002
Share (%)	6	6.5	7	7.5	8

Source: Commerce Department, Wal-Mart.

(a) What was Wal-Mart's share of the market in 1998?
(b) By how much (percent) did Wal-Mart's share increase each year?
(c) Find a linear function f that models these data. Let $x = 0$ correspond to 1998.
(d) Use f to predict Wal-Mart's share in 2005.

79. *Weight Lifting* Lifting weights can increase a person's muscle mass. Each additional pound of muscle burns an extra 40 calories per day. Write a linear function that models the number of calories burned each day by x pounds of muscle. By burning an extra 3500 calories a person can lose 1 pound of fat. How many pounds of muscle are needed to burn the equivalent of 1 pound of fat in 30 days? (*Source: Runner's World.*)

80. *Number of Doctors* The number N of doctors in the private sector from 1970 to 2000 can be approximated by $N(t) = 15{,}260t + 333{,}200$, where t is the year and $t = 0$ corresponds to 1970.
(a) Evaluate $N(0)$ and $N(30)$ and interpret each result.
(b) Explain what the numbers 15,260 and 333,200 represent in the formula.

WRITING ABOUT MATHEMATICS

81. Explain how you can determine whether a function is linear by using its
(a) symbolic representation,
(b) graphical representation, and
(c) numerical representation.

82. Describe one way to determine whether a set of data points can be modeled by a linear function.

CHECKING BASIC CONCEPTS SECTIONS 8.1 AND 8.2

1. Find a formula and sketch a graph for a function that squares the input x and then subtracts 1 from the result.

2. Use the accompanying graph to do the following.
(a) Find the domain and range of f.
(b) Evaluate $f(0)$ and $f(2)$.
(c) Is f a linear function? Explain.

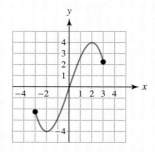

3. Determine whether f is a linear function.
(a) $f(x) = 4x - 2$
(b) $f(x) = 2\sqrt{x} - 5$
(c) $f(x) = -7$
(d) $f(x) = 9 - 2x + 5x$

4. Graph $f(x) = 4 - 3x$. Evaluate $f(-2)$.

5. Find a formula for a linear function that models the data.

x	0	1	2	3	4
$f(x)$	-1	$-\frac{1}{2}$	0	$\frac{1}{2}$	1

6. The median age in the United States from 1970 to 2000 can be approximated by

$$f(x) = 0.264x + 27.7,$$

where $x = 0$ corresponds to 1970, $x = 1$ to 1971, and so on.
(a) Evaluate $f(20)$ and interpret the result.
(b) Explain the meaning of the numbers 0.264 and 27.7.

8.3 COMPOUND INEQUALITIES

Basic Concepts · Symbolic Solutions and Number Lines · Numerical and Graphical Solutions · Interval Notation

INTRODUCTION

A person weighing 143 pounds and needing to purchase a life vest for white-water rafting is not likely to find one designed exactly for this weight. Life vests are manufactured to support a range of body weights. A vest approved for weights between 100 and 160 pounds might be appropriate for this person. In other words, if a person's weight is w, this life vest is safe if $w \geq 100$ *and* $w \leq 160$. This example illustrates the concept of a *compound inequality*.

BASIC CONCEPTS

A **compound inequality** consists of two inequalities joined by the words *and* or *or*. The following are two examples of compound inequalities.

$$2x \geq -3 \quad \text{and} \quad 2x < 5$$
$$x + 2 \geq 3 \quad \text{or} \quad x - 1 < -5$$

If a compound inequality contains the word *and*, a solution must satisfy *both* inequalities. For example, 1 is a solution to the first compound inequality because

$$\underset{\text{True}}{2(1) \geq -3} \quad \text{and} \quad \underset{\text{True}}{2(1) < 5}$$

are both true statements.

If a compound inequality contains the word *or*, a solution must satisfy *at least one* of the two inequalities. Thus 5 is a solution to the second compound inequality, because the first statement is true.

$$\underset{\text{True}}{5 + 2 \geq 3} \quad \text{or} \quad \underset{\text{False}}{5 - 1 < -5}$$

Note that 5 does not need to satisfy the second statement for this compound inequality to be true.

EXAMPLE 1   Determining solutions to compound inequalities

Determine whether the given x-values are solutions to the compound inequalities.
(a) $x + 1 < 9$ and $2x - 1 > 8$ $x = 5, -5$
(b) $5 - 2x \leq -4$ or $5 - 2x \geq 4$ $x = 2, -3$

Solution   **(a)** Substitute $x = 5$ in the given compound inequality.

$$\underset{\text{True}}{5 + 1 < 9} \quad \text{and} \quad \underset{\text{True}}{2(5) - 1 > 8}$$

Both inequalities are true, so 5 is a solution.
Now substitute $x = -5$.

$$\underset{\text{True}}{-5 + 1 < 9} \quad \text{and} \quad \underset{\text{False}}{2(-5) - 1 > 8}$$

To be a solution both inequalities must be true, so -5 is not a solution.

(b) Substitute $x = 2$ into the given compound inequality.

$$5 - 2(2) \leq -4 \quad \text{or} \quad 5 - 2(2) \geq 4$$
$$\text{False} \qquad\qquad\qquad \text{False}$$

Neither inequality is true, so 2 is not a solution.
Now substitute $x = -3$.

$$5 - 2(-3) \leq -4 \quad \text{or} \quad 5 - 2(-3) \geq 4$$
$$\text{False} \qquad\qquad\qquad\quad \text{True}$$

At least one of the two inequalities is true, so -3 is a solution.

SYMBOLIC SOLUTIONS AND NUMBER LINES

We can use a number line to graph solutions to compound inequalities, such as

$$x \leq 6 \quad \text{and} \quad x > -4.$$

The solution set for $x \leq 6$ is shaded to the left of 6, with a bracket placed at $x = 6$, as shown in Figure 8.27. The solution set for $x > -4$ can be shown by shading a different number line to the right of -4 and placing a left parenthesis at -4. Because the inequalities are connected by *and*, the solution set consists of all numbers that are shaded on *both* number lines. The final number line represents the **intersection** of the two solution sets. That is, the solution set includes real numbers where the graphs "overlap."

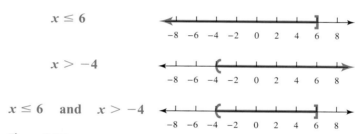

$x \leq 6$

$x > -4$

$x \leq 6$ and $x > -4$

Figure 8.27

Note: A bracket, either [or], is used when an inequality contains $\leq$ or $\geq$. A parenthesis, either (or), is used when an inequality contains $<$ or $>$. This notation makes clear whether an endpoint is included in the inequality.

Critical Thinking

Graph the following inequalities and discuss your results.

1. $x < 2$ and $x > 5$
2. $x > 2$ or $x < 5$

In the next example we use a number line to help solve a compound inequality.

EXAMPLE 2 Solving a compound inequality containing "and"

Solve $2x + 4 > 8$ and $5 - x < 9$. Graph the solution.

Solution   First solve each linear inequality separately.

$$2x + 4 > 8 \quad \text{and} \quad 5 - x < 9$$
$$2x > 4 \quad \text{and} \quad -x < 4$$
$$x > 2 \quad \text{and} \quad x > -4$$

Graph the two inequalities on two different number lines. On a third number line, shade solutions that appear on both of the first two number lines. As shown in Figure 8.28, the solution set is $\{x \mid x > 2\}$.

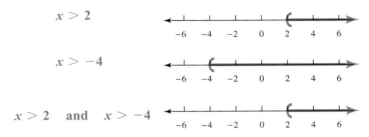

Figure 8.28

Sometimes a compound inequality containing the word *and* can be combined into a three-part inequality. For example, rather than writing

$$x > 5 \quad \text{and} \quad x \le 10,$$

we could write the **three-part inequality**

$$5 < x \le 10.$$

This three-part inequality is represented by the number line shown in Figure 8.29.

$5 < x \le 10$

Figure 8.29

EXAMPLE 3   Solving three-part inequalities

Solve each inequality.

(a) $4 < t + 2 \le 8$ **(b)** $-3 \le 3z \le 6$ **(c)** $-\dfrac{5}{2} < \dfrac{1 - m}{2} < 4$

Solution   **(a)** To solve a three-part inequality, isolate the variable by applying properties of inequalities to each part of the inequality.

$4 < t + 2 \le 8$	Given three-part inequality
$4 - 2 < t + 2 - 2 \le 8 - 2$	Subtract 2 from each part.
$2 < t \le 6$	Simplify each part.

The solution set is $\{t \mid 2 < t \le 6\}$.

(b) To simplify, divide each part by 3.

$$-3 \leq 3z \leq 6 \qquad \text{Given three-part inequality}$$

$$\frac{-3}{3} \leq \frac{3z}{3} \leq \frac{6}{3} \qquad \text{Divide each part by 3.}$$

$$-1 \leq z \leq 2 \qquad \text{Simplify each part.}$$

The solution set is $\{z \mid -1 \leq z \leq 2\}$.

(c) Multiply each part by 2 to clear fractions.

$$-\frac{5}{2} < \frac{1-m}{2} < 4 \qquad \text{Given three-part inequality}$$

$$2 \cdot \left(-\frac{5}{2}\right) < 2 \cdot \left(\frac{1-m}{2}\right) < 2 \cdot 4 \qquad \text{Multiply each part by 2.}$$

$$-5 < 1 - m < 8 \qquad \text{Simplify each part.}$$

$$-5 - 1 < 1 - m - 1 < 8 - 1 \qquad \text{Subtract 1 from each part.}$$

$$-6 < -m < 7 \qquad \text{Simplify each part.}$$

$$-1 \cdot (-6) > -1 \cdot (-m) > -1 \cdot 7 \qquad \text{Multiply each part by } -1; \text{ \textit{reverse} inequality symbols.}$$

$$6 > m > -7 \qquad \text{Simplify each part.}$$

$$-7 < m < 6 \qquad \text{Rewrite inequality.}$$

The solution set is $\{m \mid -7 < m < 6\}$.

Note: Either $6 > m > -7$ or $-7 < m < 6$ is a correct way to write a three-part inequality. *However*, we usually write the smaller number on the left side and the larger number on the right side.

Three-part inequalities occur frequently in applications. In the next example we find altitudes at which the air temperature is within a certain range.

EXAMPLE 4 Solving a three-part inequality

Air temperature is colder at higher altitudes. If the ground-level temperature is $80°F$, the air temperature x miles above Earth's surface can be modeled by $T(x) = 80 - 29x$. Find the altitudes at which the air temperature ranges from $22°F$ to $-7°F$. (*Source:* A. Miller and R. Anthes, *Meteorology.*)

Solution We write and solve the three-part inequality $-7 \leq T(x) \leq 22$.

$$-7 \leq 80 - 29x \leq 22 \qquad \text{Substitute for } T(x).$$

$$-87 \leq -29x \leq -58 \qquad \text{Subtract 80 from each part.}$$

$$\frac{-87}{-29} \geq x \geq \frac{-58}{-29} \qquad \text{Divide by } -29; \text{ \textit{reverse} inequality symbols.}$$

$$3 \geq x \geq 2 \qquad \text{Simplify.}$$

$$2 \leq x \leq 3 \qquad \text{Rewrite inequality.}$$

The air temperature ranges from $22°F$ to $-7°F$ for altitudes between 2 and 3 miles.

We can also solve compound inequalities containing the word *or*. To write the solution to such an inequality we sometimes use *union* notation. For any two sets A and B, the **union** of A and B, denoted $A \cup B$, is defined by

$$A \cup B = \{x \mid x \text{ is an element of } A \text{ or an element of } B\}.$$

If the solution to an inequality is $\{x \mid x < 1\}$ or $\{x \mid x \geq 3\}$, then it can also be written as

$$\{x \mid x < 1\} \cup \{x \mid x \geq 3\}.$$

That is, we can replace the word *or* with the $\cup$ symbol.

EXAMPLE 5 **Solving a compound inequality with "or"**

Solve $x + 2 < -1$ or $x + 2 > 1$.

Solution We first solve each linear inequality.

$$x + 2 < -1 \quad \text{or} \quad x + 2 > 1 \qquad \text{Given compound inequality}$$
$$x < -3 \quad \text{or} \quad x > -1 \qquad \text{Subtract 2.}$$

We can graph the simplified inequalities on different number lines, as shown in Figure 8.30. A solution must satisfy at least one of the two inequalities. Thus the solution set for the compound inequality results from taking the *union* of the first two number lines. We can write the solution, using set-builder notation, as $\{x \mid x < -3\} \cup \{x \mid x > -1\}$ or $\{x \mid x < -3 \text{ or } x > -1\}$.

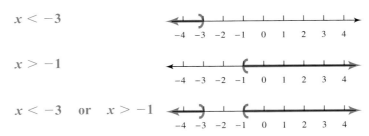

$x < -3$

$x > -1$

$x < -3 \quad \text{or} \quad x > -1$

Figure 8.30

Critical Thinking

Carbon dioxide is emitted when human beings breathe. In one study of college students, the amount of carbon dioxide exhaled in grams per hour was measured during both lectures and exams. The average amount exhaled during lectures L satisfied $25.33 \leq L \leq 28.17$, whereas the average amount exhaled during exams E satisfied $36.58 \leq E \leq 40.92$. What do these results indicate? Explain.
(***Source:*** T. Wang, *ASHRAE Trans.*)

≡ MAKING CONNECTIONS ≡

Writing Three-Part Inequalities

The inequality $-2 < x < 1$ means that $x > -2$ *and* $x < 1$. A three-part inequality should *not* be used when *or* connects a compound inequality. Writing $x < -2$ or $x > 1$ as $1 < x < -2$ is incorrect because it states that x must be both greater than 1 *and* less than -2. It is impossible for any value of x to satisfy this statement.

NUMERICAL AND GRAPHICAL SOLUTIONS

Compound inequalities can also be solved graphically and numerically, as illustrated in the next example.

EXAMPLE 6   Solving a compound inequality numerically and graphically

Tuition at private colleges and universities from 1980 to 1997 can be modeled by $f(x) = 575(x - 1980) + 3600$. Estimate when the average tuition was between $8200 and $10,500.

Solution   **Numerical Solution**   Let $Y_1 = 575(X - 1980) + 3600$. Make a table of values, as shown in Figure 8.31(a). In 1988, the average tuition was $8200 and in 1992 it was $10,500. Therefore from 1988 to 1992 the average tuition ranged from $8200 to $10,500.

Graphical Solution   Let $Y_1 = 575(X - 1980) + 3600$, $Y_2 = 8200$, and $Y_3 = 10,500$. We must find x-values so that $y_2 \le y_1 \le y_3$. Figures 8.31(b) and (c) show that y_1 is between y_2 and y_3 when $1988 \le x \le 1992$.

Calculator Help
To find a point of intersection, see Appendix A (page AP-7).

[1980, 1997, 1] by [3000, 12000, 3000]   [1980, 1997, 1] by [3000, 12000, 3000]

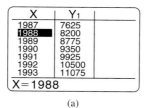

(a)

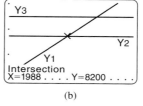

(b)

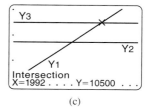

(c)

Figure 8.31

INTERVAL NOTATION

The solution set in Example 4 was $\{x \mid 2 \le x \le 3\}$. This solution set can be graphed on a number line, as shown in Figure 8.32.

Figure 8.32

A convenient notation for number line graphs is called **interval notation**. Instead of drawing the entire number line as in Figure 8.32, the solution set can be expressed as [2, 3] in interval notation. Because the solution set includes the endpoints 2 and 3, brackets are used. A solution set that includes all real numbers satisfying $-2 < x < 3$ can be expressed as $(-2, 3)$. Parentheses indicate that the endpoints are *not* included. The interval $0 \le x < 4$ is represented by $[0, 4)$.

Table 8.11 provides some examples of interval notation. The symbol ∞ refers to infinity, and it does not represent a real number. The interval $(5, \infty)$ represents $x > 5$, which has no maximum x-value, so ∞ is used for the right endpoint. The symbol $-\infty$ may be used similarly and denotes "negative infinity."

TABLE 8.11   Interval Notation

Inequality	Interval Notation	Number Line Graph
$-1 < x < 3$	$(-1, 3)$	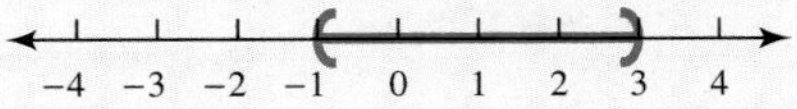
$-3 < x \le 2$	$(-3, 2]$	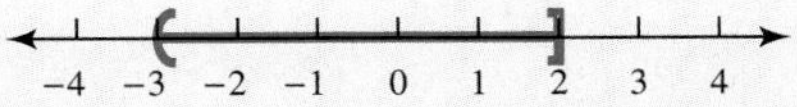
$-2 \le x \le 2$	$[-2, 2]$	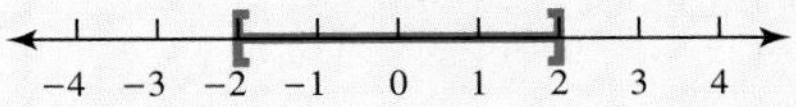
$x < -1$ or $x > 2$	$(-\infty, -1) \cup (2, \infty)$ ($\cup$ is the union symbol.)	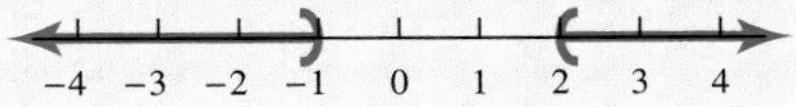
$x > -1$	$(-1, \infty)$	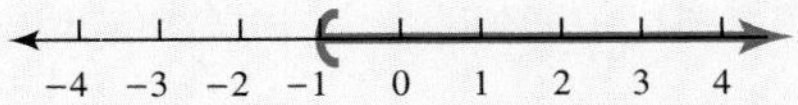
$x \le 2$	$(-\infty, 2]$	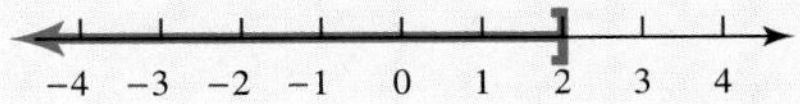

═══ MAKING CONNECTIONS ═══

Points and Intervals

The expression $(1, 2)$ may represent a point in the xy-plane or the interval $1 < x < 2$. To alleviate confusion, phrases such as "the point $(1, 2)$" or "the interval $(1, 2)$" are used.

EXAMPLE 7   Writing inequalities in interval notation

Write each expression in interval notation.
(a) $-2 \le x < 5$ **(b)** $x \ge 3$ **(c)** $x < -5$ or $x \ge 2$
(d) $x > 0$ and $x \le 3$ **(e)** $\{x \mid x \le 1$ or $x \ge 3\}$

Solution   **(a)** $[-2, 5)$ **(b)** $[3, \infty)$ **(c)** $(-\infty, -5) \cup [2, \infty)$
(d) $(0, 3]$ **(e)** $(-\infty, 1] \cup [3, \infty)$

EXAMPLE 8   Solving an inequality

Solve $2x + 1 \le -1$ or $2x + 1 \ge 3$. Write the solution in interval notation.

Solution   First solve each inequality.

$$2x + 1 \le -1 \quad \text{or} \quad 2x + 1 \ge 3 \qquad \text{Given compound inequality}$$
$$2x \le -2 \quad \text{or} \quad 2x \ge 2 \qquad \text{Subtract 1.}$$
$$x \le -1 \quad \text{or} \quad x \ge 1 \qquad \text{Divide by 2.}$$

The solution set may be written as $(-\infty, -1] \cup [1, \infty)$.

8.3   PUTTING IT ALL TOGETHER

The following table summarizes some important topics related to compound inequalities.

Concept	Explanation	Examples
Compound Inequality	Two inequalities joined by *and* or *or*.	$x < -1$ or $x > 2$; $2x \geq 10$ and $x + 2 < 6$
Three-Part Inequality	Can be used to write some types of compound inequalities	$x > -2$ and $x \leq 3$ is equivalent to $-2 < x \leq 3$.
Interval Notation	Notation used to write sets of real numbers rather than using number lines or inequalities	$-2 \leq z \leq 4$ is equivalent to $[-2, 4]$. $x < 4$ is equivalent to $(-\infty, 4)$. $x \leq -2$ or $x > 0$ is equivalent to $(-\infty, -2] \cup (0, \infty)$.

The following table lists basic strategies for solving compound inequalities.

Type of Inequality	Method to Solve Inequality
Solving a Compound Inequality with *and*	**STEP 1:** First solve each inequality individually. **STEP 2:** The solution set includes values that satisfy *both* inequalities from Step 1.
Solving a Compound Inequality with *or*	**STEP 1:** First solve each inequality individually. **STEP 2:** The solution set includes values that satisfy *at least one* of the inequalities from Step 1.
Solving a Three-Part Inequality	Work on all three parts at the same time. Be sure to perform the same step on each part. Continue until the variable is isolated in the middle.

8.3   EXERCISES

FOR EXTRA HELP

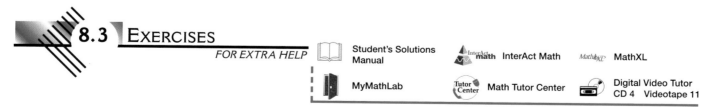

Student's Solutions Manual

MyMathLab

InterAct Math

Math Tutor Center

MathXL

Digital Video Tutor
CD 4   Videotape 11

CONCEPTS

1. Give an example of a compound inequality containing the word *and*.

2. Give an example of a compound inequality containing the word *or*.

3. Is 1 a solution of the compound inequality $x > 3$ and $x \leq 5$?

4. Is 1 a solution of the compound inequality $x < 3$ or $x \geq 5$?

5. Is the compound inequality $x \geq -5$ and $x \leq 5$ equivalent to $-5 \leq x \leq 5$?

6. Name three methods for solving a compound inequality.

Exercises 7–12: Determine whether the given values of x are solutions to the compound inequality.

7. $x - 1 < 5$ and $2x > 3$ $\qquad$ $x = 2, x = 6$

8. $2x + 1 \geq 4$ and $1 - x \leq 3$ $\qquad$ $x = -2, x = 3$

9. $3x < -5$ or $2x \geq 3$ $\qquad$ $x = 0, x = 3$

10. $x + 1 \leq -4$ or $x + 1 \geq 4$ $\qquad$ $x = -5, x = 2$

11. $2 - x > -5$ and $2 - x \leq 4$ $\qquad$ $x = -3, x = 0$

12. $x + 5 \geq 6$ or $3x \leq 3$ $\qquad$ $x = -1, x = 1$

INTERVAL NOTATION

Exercises 13–32: Write the inequality in interval notation.

13. $2 \leq x \leq 10$ $\qquad$ **14.** $-1 < x < 5$

15. $5 < x \leq 8$ $\qquad$ **16.** $-\frac{1}{2} \leq x \leq \frac{5}{6}$

17. $x < 4$ $\qquad$ **18.** $x \leq -3$

19. $x > -2$ $\qquad$ **20.** $x \geq 6$

21. $x \leq -2$ or $x \geq 4$ $\qquad$ **22.** $x \leq -1$ or $x > 6$

23. $x < 1$ or $x \geq 5$ $\qquad$ **24.** $x < -3$ or $x > 3$

25.

26.

27.

28.

29. $\{x \mid x < 4\}$ $\qquad$ **30.** $\{x \mid -1 \leq x < 4\}$

31. $\{x \mid x < 1 \text{ or } x > 2\}$ $\quad$ **32.** $\{x \mid -\infty < x < \infty\}$

SYMBOLIC SOLUTIONS

Exercises 33–40: Solve the compound inequality. Graph the solution set, using a number line.

33. $x \leq 3$ and $x \geq -1$ $\quad$ **34.** $x \geq 5$ and $x > 6$

35. $2x < 5$ and $2x > -4$

36. $2x + 1 < 3$ and $x - 1 \geq -5$

37. $x \leq -1$ or $x \geq 2$

38. $2x \leq -6$ or $x \geq 6$

39. $5 - x > 1$ or $x + 3 \geq -1$

40. $1 - 2x > 3$ or $2x - 4 \geq 4$

Exercises 41–50: Solve the compound inequality. Write your answer in interval notation.

41. $x - 3 \leq 4$ and $x + 5 \geq -1$

42. $2z \geq -10$ and $z < 8$

43. $3t - 1 > -1$ and $2t - \frac{1}{2} > 6$

44. $2(x + 1) < 8$ and $-2(x - 4) > -2$

45. $x - 4 \geq -3$ or $x - 4 \leq 3$

46. $1 - 3n \geq 6$ or $1 - 3n \leq -4$

47. $-x < 1$ or $5x + 1 < -10$

48. $7x - 6 > 0$ or $-\frac{1}{2}x \leq 6$

49. $1 - 7x < -48$ and $3x + 1 \leq -9$

50. $3x - 4 \leq 8$ or $4x - 1 \leq 13$

Exercises 51–72: Solve the three-part inequality. Write your answer in interval notation.

51. $-2 \leq t + 4 < 5$ $\qquad$ **52.** $5 < t - 7 < 10$

53. $-\frac{5}{8} \leq y - \frac{3}{8} < 1$ $\qquad$ **54.** $-\frac{1}{2} < y - \frac{3}{2} < \frac{1}{2}$

55. $-27 \leq 3x \leq 9$ $\qquad$ **56.** $-4 < 2y < 22$

57. $\frac{1}{2} < -2y \leq 8$ $\qquad$ **58.** $-16 \leq -4x \leq 8$

59. $-4 < 5z + 1 \leq 6$ $\qquad$ **60.** $-3 \leq 3z + 6 < 9$

61. $3 \leq 4 - n \leq 6$ $\qquad$ **62.** $-1 < 3 - n \leq 1$

63. $-1 < 2z - 1 < 3$ $\qquad$ **64.** $2 \leq 4z + 5 \leq 6$

65. $-2 \leq 5 - \frac{1}{3}m < 2$ $\quad$ **66.** $-\frac{3}{2} < 4 - 2m < \frac{7}{2}$

67. $100 \leq 10(5x - 2) \leq 200$

68. $-15 < 5(x - 1990) < 30$

69. $-3 < \dfrac{3z + 1}{4} < 1$ $\quad$ **70.** $-3 < \dfrac{z - 1}{2} < 5$

71. $-\dfrac{5}{2} \le \dfrac{2-m}{4} \le \dfrac{1}{2}$ **72.** $\dfrac{4}{5} \le \dfrac{4-2m}{10} \le 2$

NUMERICAL AND GRAPHICAL SOLUTIONS

Exercises 73–76: Use the table to solve the three-part inequality. Write your answer in interval notation.

73. $-3 \le 3x \le 6$

74. $-5 \le 2x - 1 \le 1$

X	Y₁
-2	-6
-1	-3
0	0
1	3
2	6
3	9
4	12

Y₁ ■ 3X

X	Y₁
-4	-9
-3	-7
-2	-5
-1	-3
0	-1
1	1
2	3

Y₁ ■ 2X−1

75. $-1 < 1 - x < 2$

76. $-2 \le -2x < 4$

X	Y₁
-2	3
-1	2
0	1
1	0
2	-1
3	-2
4	-3

Y₁ ■ 1−X

X	Y₁
-3	6
-2	4
-1	2
0	0
1	-2
2	-4
3	-6

Y₁ ■ -2X

Exercises 77–80: Use the graph to solve the compound inequality. Write your answer in interval notation.

77. $-2 \le y_1 \le 2$

78. $1 \le y_1 < 3$

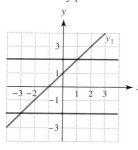

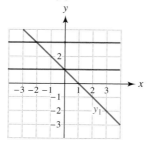

79. $y_1 < -2$ or $y_1 > 2$

80. $y_1 \le -2$ or $y_1 \ge 4$

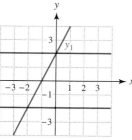

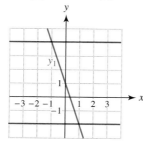

81. *Distance*   The function f gives the distance y in miles between a car and the city of Omaha, Nebraska, after x hours, where $0 \le x \le 6$. The graphs of f and the horizontal lines $y = 100$ and $y = 200$ are shown in the following figure.

(a) Is the car moving toward or away from Omaha? Explain.

(b) Determine the times when the car is 100 miles or 200 miles from Omaha.

(c) When is the car from 100 to 200 miles from Omaha?

(d) When is the car's distance from Omaha greater than or equal to 200 miles?

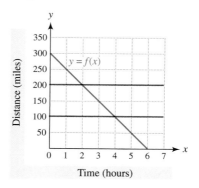

82. Use the following figure to solve each equation or inequality. Assume that the domains of $y_1, y_2,$ and y_3 are $0 \le x \le 5$.

(a) $y_1 = y_2$

(b) $y_2 = y_3$

(c) $y_1 \le y_2 \le y_3$

(d) $y_2 < y_3$

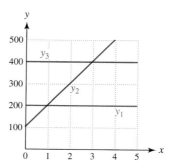

Exercises 83–86: (Refer to Example 6.) Solve the compound inequality numerically or graphically. Write your answer in interval notation.

83. $-2 \le 2x - 4 \le 4$

84. $-1 \le 1 - x \le 3$

85. $x + 1 < -1$ or $x + 1 > 1$

86. $2x - 1 < -3$ or $2x - 1 > 5$

USING MORE THAN ONE METHOD

Exercises 87–90: Solve the compound inequality symbolically, graphically, and numerically. Write the solution set in interval notation.

87. $4 \le 5x - 1 \le 14$

88. $-4 < 2x < 4$

89. $4 - x \ge 1$ or $4 - x < 3$

90. $x + 3 \ge -2$ or $x + 3 \le 1$

APPLICATIONS

91. *Educational Attainment* The line graph shows the percentage of the population that completed 4 years or more of college for selected years. Estimate the years when this percentage ranged from 6% to 17%. (*Source:* Bureau of the Census.)

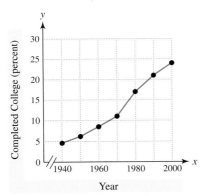

92. *School Enrollment* The line graph shows the school enrollment in millions from kindergarten through university level. Estimate the years when the enrollment was from 57 to 65 million students. (*Source:* Department of Education.)

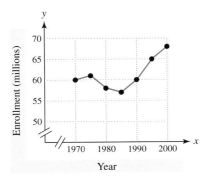

93. *Medicare Costs* Based on current trends, future Medicare costs in billions of dollars may be modeled by $f(x) = 18x - 35,750$, where $1995 \le x \le 2007$. Estimate the years when Medicare costs will be from 250 to 340 billion dollars. (*Source:* Office of Management and Budget.)

94. *Median Home Prices* The median price P of a single-family home from 1980 to 1991 may be modeled by $P(x) = 3400(x - 1980) + 61,000$, where x is the year. Determine the years when the median price ranged from $78,000 to $95,000. (*Source:* Department of Commerce.)

95. *Geometry* For what values of x is the perimeter of the rectangle from 40 to 60 feet?

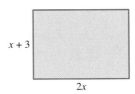

96. *Geometry* A rectangle is three times as long as it is wide. If the rectangle is to have a perimeter from 100 to 160 inches, what values for the width are possible?

97. *Distance and Time* A car's distance in miles from a rest stop is given by $f(x) = 70x + 50$, where x is time in hours.
   (a) Construct a table for f for $x = 4, 5, 6, \ldots, 10$ and use the table to solve the inequality $470 \le f(x) \le 680$. Explain what your result means.
   (b) Solve the inequality in part (a) symbolically.

98. *Altitude and Temperature* If the air temperature at ground level is $70°F$, the air temperature x miles high is given by $T(x) = 70 - 29x$. Determine the altitudes at which the air temperature is from $26.5°F$ to $-2.5°F$. (*Source:* A. Miller and R. Anthes, *Meteorology*.)

99. *Temperature Scales* The formula

$$F = \frac{9}{5}C + 32$$

may be used to convert Celsius temperatures to Fahrenheit temperatures. The greatest temperature ranges on Earth are recorded in Siberia where the temperature has varied from $-90°F$ to $98°F$. Find this temperature range in degrees Celsius.

100. *Temperature Scales*   The formula

$$C = \frac{5}{9}(F - 32)$$

may be used to convert Fahrenheit temperatures to Celsius temperatures. If the Celsius temperature ranged from 5°C to 20°C, use this formula to find the corresponding temperature range in degrees Fahrenheit.

WRITING ABOUT MATHEMATICS

101. Suppose that the solution set for a compound inequality can be written as $x < -3$ or $x > 2$. A student writes it as $2 < x < -3$. Is the student's three-part inequality correct? Explain your answer.

102. How can you determine whether an x-value is a solution to a compound inequality connected by the word *and*? Give an example. Repeat the question for a compound inequality connected by the word *or*.

8.4   OTHER FUNCTIONS AND THEIR PROPERTIES

Expressing Domain and Range in Interval Notation · Absolute Value Function ·
Polynomial Functions · Rational Functions (Optional)

INTRODUCTION

An important application of mathematics is to model and summarize data. Many times when data are plotted they do not lie on a line and, as a result, cannot be modeled with a linear function, as described in Section 8.2. For example, Table 8.12 lists the heart rate in beats per minute (bpm) of an athlete after exercise has stopped. Note that the heart rate does not decrease by the same amount each 2-minute interval. Instead, it decreases faster at first and then begins to level off, indicating that the data are *nonlinear*.

A graph of these data is shown in Figure 8.33. To model such data we can use **nonlinear functions** whose graphs are *not* lines. Because graphs of nonlinear data can have a wide variety of shapes, it has been necessary for mathematicians to create many different types of functions. In this section we discuss some of these functions and use a polynomial function to model the data in Table 8.12 (See Example 7.)

TABLE 8.12 Heart Rate of an Athlete

Time (minutes)	Heart Rate (bpm)
0	175
2	139
4	111
6	91
8	79
10	75

Source: V. Thomas, *Science and Sport.*

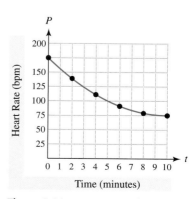

Figure 8.33 Heart Rate of an Athlete

EXPRESSING DOMAIN AND RANGE IN INTERVAL NOTATION

The set of all valid inputs for a function is called the *domain*, and the set of all outputs from a function is called the *range*. For example, all real numbers are valid inputs for $f(x) = x^2$. Rather than writing "the set of all real numbers" for the domain of f, we can use *interval notation* to express the domain as $(-\infty, \infty)$. The symbol ∞ represents infinity and is not a real number. Because $x^2 \geq 0$ for every real number x, the output from $f(x) = x^2$ is never negative. Therefore the range of f is $[0, \infty)$, which denotes all nonnegative real numbers. Note that 0 is in the range of f because $f(0) = 0$, and a bracket "[" is used to indicate that 0 is included in the range of f.

EXAMPLE 1 Writing domains in interval notation

Write the domain for each function in interval notation.

(a) $f(x) = 4x$ **(b)** $g(t) = \sqrt{t - 1}$ **(c)** $h(v) = \dfrac{1}{v + 3}$

Solution **(a)** The expression $4x$ is defined for all real numbers x. Thus the domain of f is $(-\infty, \infty)$.

(b) The square root $\sqrt{t - 1}$ is defined only when $t - 1$ is *not* negative. Thus the domain of g includes all real numbers satisfying $t - 1 \geq 0$ or $t \geq 1$. In interval notation this inequality is written as $[1, \infty)$.

(c) The expression $\frac{1}{v + 3}$ is defined except when $v + 3 = 0$ or $v = -3$. Thus the domain of h includes all real numbers except -3 and can be written as $(-\infty, -3) \cup (-3, \infty)$. Parentheses are used because -3 is not included in the domain of h.

In the next example, we determine the domain and range of a function from its graph. Note that dots placed at each end of a graph indicate that the endpoints are included.

EXAMPLE 2 Writing the domain and range in interval notation

Use the graph of f in Figure 8.34 to write its domain and range in interval notation.

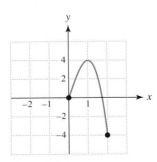

Figure 8.34

Solution Because dots are placed at $(0, 0)$ and $(2, -4)$, the endpoints are included in the graph of f. Thus the graph in Figure 8.34 includes x-values from $x = 0$ to $x = 2$. In interval notation, the domain of f is $[0, 2]$. The range of f includes y-values from -4 to 4 and can be expressed in interval notation as $[-4, 4]$.

ABSOLUTE VALUE FUNCTION

In Chapter 1 we discussed the absolute value of a number. We can define a function called the **absolute value function** as $f(x) = |x|$. We evaluate f as follows.

$$f(11) = |11| = 11, \quad f(-4) = |-4| = 4, \quad \text{and} \quad f(-\pi) = |-\pi| = \pi$$

To graph $y = |x|$ we begin by making a table of values, as shown in Table 8.13. Next we plot these points and sketch the graph, as shown in Figure 8.35. Note that the graph is V-shaped and never lies below the x-axis because the absolute value of a number cannot be negative.

TABLE 8.13

x	y
-2	2
-1	1
0	0
1	1
2	2

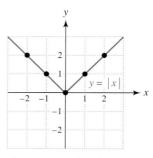

Figure 8.35 Absolute Value

Because valid inputs for $f(x) = |x|$ include all real numbers, the domain of the absolute value function is $(-\infty, \infty)$. The graph of the absolute value function never lies below the x-axis, so the range includes all real numbers greater than or equal to 0 and can be written as $[0, \infty)$.

EXAMPLE 3   Graphing a function

Sketch a graph of $f(x) = |x - 1|$. Write its domain and range in interval notation.

Solution   To graph $f(x) = |x - 1|$ we begin by making a table of values, as shown in Table 8.14. We then plot the points and sketch the V-shaped graph shown in Figure 8.36. Note that the graph of $f(x) = |x - 1|$ has the same shape as the graph of $y = |x|$, except that the graph of $y = f(x)$ is translated 1 unit to the right. The domain of f is $(-\infty, \infty)$, and the range of f is $[0, \infty)$.

TABLE 8.14

x	y
-1	2
0	1
1	0
2	1
3	2

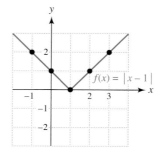

Figure 8.36

POLYNOMIAL FUNCTIONS

In Chapter 5 we introduced polynomials and defined their degrees. The following expressions are examples of polynomials in one variable.

$$1 - 5x, \quad 3t^2 - 5t + 1, \quad \text{and} \quad z^3 + 5$$

(The exponents on variables in polynomials must be nonnegative integers.) Recall that the *degree* of a polynomial in one variable equals the largest exponent on the variable. Thus the degree of $1 - 5x$ is 1, the degree of $3t^2 - 5t + 1$ is 2, and the degree of $z^3 + 5$ is 3.

The equations

$$f(x) = 1 - 5x, \quad g(t) = 3t^2 - 5t + 1, \quad \text{and} \quad h(z) = z^3 + 5$$

define three **polynomial functions in one variable**. Function f is a **linear function** because it has degree 1, function g is a **quadratic function** because it has degree 2, and function h is a **cubic function** because it has degree 3. The domain of any polynomial function is the set of *all* real numbers.

EXAMPLE 4   Identifying polynomial functions

Determine whether $f(x)$ represents a polynomial function. If possible, identify the type of polynomial function and its degree.

(a) $f(x) = 5x^3 - x + 10$ (b) $f(x) = x^{-2.5} + 1$

(c) $f(x) = 1 - 2x$ (d) $f(x) = \dfrac{3}{x - 1}$

Solution   (a) The expression $5x^3 - x + 10$ is a cubic polynomial, so $f(x)$ represents a cubic polynomial function. It has degree 3.

(b) $f(x)$ does not represent a polynomial function because the variables in a polynomial must have *nonnegative integer* exponents.

(c) $f(x) = 1 - 2x$ represents a polynomial function that is linear. It has degree 1.

(d) $f(x)$ does not represent a polynomial function because $\dfrac{3}{x - 1}$ is not a polynomial.

EXAMPLE 5   Evaluating a polynomial function graphically and symbolically

A graph of $f(x) = 4x - x^3$ is shown in Figure 8.37. Evaluate $f(-1)$ graphically and check your result symbolically.

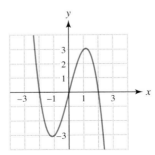

Figure 8.37

Solution ***Graphical Evaluation*** To calculate $f(-1)$ graphically find -1 on the x-axis and move downward until the graph of f is reached. Then move horizontally to the y-axis, as shown in Figure 8.38. Thus, when $x = -1$, $y = -3$ and $f(-1) = -3$.

Symbolic Evaluation Evaluation of $f(x) = 4x - x^3$ for $x = -1$ is performed as follows.

$$f(-1) = 4(-1) - (-1)^3$$
$$= -4 - (-1)$$
$$= -3$$

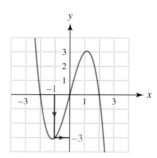

Figure 8.38 $f(-1) = -3$

EXAMPLE 6 Evaluating a polynomial function symbolically

Evaluate $f(x)$ at the given value of x.
(a) $f(x) = -3x^4 - 2$, $x = 2$ **(b)** $f(x) = -2x^3 - 4x^2 + 5$, $x = -3$

Solution **(a)** Be sure to evaluate exponents before multiplying.
$$f(2) = -3(2)^4 - 2 = -3 \cdot 16 - 2 = -50$$
(b) $f(-3) = -2(-3)^3 - 4(-3)^2 + 5 = -2(-27) - 4(9) + 5 = 23$

In the next example we discuss further the application presented in the introduction to this section.

EXAMPLE 7 Modeling the heart rate of an athlete

The quadratic function $P(t) = t^2 - 20t + 175$ models an athlete's pulse, or heart rate, in beats per minute t minutes after exercise has stopped, where $0 \le t \le 10$. See Table 8.12 and Figure 8.33 in the introduction to this section.
(a) Use P to determine the heart rate when $t = 0, 4$, and 10. Do your values agree with those shown in Table 8.12?
(b) Evaluate $P(20)$. Is this value valid?

Solution **(a)** Start by evaluating $P(0)$, $P(4)$, and $P(10)$.
$$P(0) = 0^2 - 20(0) + 175 = \mathbf{175}$$
$$P(4) = 4^2 - 20(4) + 175 = \mathbf{111}$$
$$P(10) = 10^2 - 20(10) + 175 = \mathbf{75}$$
These values agree with those shown in the Table 8.12.

(b) $P(20) = 20^2 - 20(20) + 175 = 175$. Unless the athlete started to exercise again, 175 beats per minute is probably not valid. Note that valid inputs for the formula are limited to $0 \leq t \leq 10$. When t is not in this interval, the results may not be meaningful.

RATIONAL FUNCTIONS (OPTIONAL)

A rational expression is formed when a polynomial is divided by a polynomial. For example, the expressions

$$\frac{2x - 1}{x}, \quad \frac{5}{x^2 + 1}, \quad \text{and} \quad \frac{2x - 5}{x^2 - 9}$$

are rational expressions. Rational expressions can be used to define *rational functions*.

RATIONAL FUNCTION

Let $p(x)$ and $q(x)$ be polynomials. Then a **rational function** is given by

$$f(x) = \frac{p(x)}{q(x)}.$$

The domain of f includes all x-values such that $q(x) \neq 0$.

From this definition, it follows that

$$f(x) = \frac{2x - 1}{x}, \quad g(x) = \frac{5}{x^2 + 1}, \quad \text{and} \quad h(x) = \frac{2x - 5}{x^2 - 9}$$

define rational functions. The domain of f includes all real numbers except 0, the domain of g includes all real numbers because $x^2 + 1 \neq 0$ for any x-value, and the domain of h includes all real numbers except ± 3.

EXAMPLE 8   Identifying the domains of rational functions

Write the domain of each function in interval notation.

(a) $f(x) = \dfrac{1}{x + 2}$ **(b)** $g(x) = \dfrac{2x}{x^2 - 3x + 2}$ **(c)** $h(t) = \dfrac{4}{t^3 - t}$

Solution   **(a)** The domain of f includes all x-values except when the denominator equals 0.

$$\begin{aligned} x + 2 &= 0 \qquad \text{Set the denominator equal to 0.} \\ x &= -2 \qquad \text{Subtract 2.} \end{aligned}$$

Thus $f(-2)$ is undefined and -2 must be excluded from the domain of f. In interval notation the domain of f is $(-\infty, -2) \cup (-2, \infty)$.

(b) The domain of g includes all real numbers except when $x^2 - 3x + 2 = 0$.

$$\begin{aligned} x^2 - 3x + 2 &= 0 \qquad \text{Set the denominator equal to 0.} \\ (x - 1)(x - 2) &= 0 \qquad \text{Factor.} \\ x = 1 \quad \text{or} \quad x &= 2 \qquad \text{Zero-product property} \end{aligned}$$

Because $g(1)$ and $g(2)$ are both undefined, 1 and 2 must be excluded from the domain of g. In interval notation the domain of g is $(-\infty, 1) \cup (1, 2) \cup (2, \infty)$.

(c) The domain of h includes all real numbers except when $t^3 - t = 0$.

$$t^3 - t = 0 \qquad \text{Set the denominator equal to 0.}$$
$$t(t^2 - 1) = 0 \qquad \text{Factor out } t.$$
$$t(t - 1)(t + 1) = 0 \qquad \text{Difference of squares}$$
$$t = 0 \quad \text{or} \quad t = 1 \quad \text{or} \quad t = -1 \qquad \text{Zero-product property}$$

In interval notation the domain of h is $(-\infty, -1) \cup (-1, 0) \cup (0, 1) \cup (1, \infty)$.

Like other functions, rational functions have graphs. To graph a rational function by hand, we usually start by making a table of values, as demonstrated in the next example. Because the graphs of rational functions are typically nonlinear, it is a good idea to plot at least 3 points on each side of an x-value where the function is undefined.

EXAMPLE 9   Graphing a rational function

Graph $f(x) = \frac{1}{x}$.

Solution   Make a table of values for $f(x) = \frac{1}{x}$, as shown in Table 8.15. Notice that $x = 0$ is not in the domain of f, and a dash can be used to denote this undefined value. Start by picking three x-values on each side of 0.

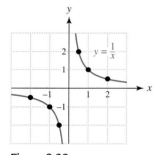

Figure 8.39

TABLE 8.15

x	-2	-1	$-\frac{1}{2}$	0	$\frac{1}{2}$	1	2
$\frac{1}{x}$	$-\frac{1}{2}$	-1	-2	—	2	1	$\frac{1}{2}$

Plot the points shown in Table 8.15 and then connect the points with a smooth curve, as shown in Figure 8.39. Because $f(0)$ is undefined, the graph of $f(x) = \frac{1}{x}$ does not cross the line $x = 0$, the y-axis.

In the next example we evaluate a rational function three different ways.

EXAMPLE 10   Evaluating a rational function

Use Table 8.16, the formula for $f(x)$, and Figure 8.40 to evaluate $f(-1), f(1)$, and $f(2)$.

(a) TABLE 8.16

x	$f(x)$
-3	$\frac{3}{2}$
-2	$\frac{4}{3}$
-1	1
0	0
1	—
2	4
3	3

(b) $f(x) = \dfrac{2x}{x - 1}$

(c)

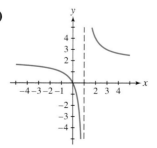

Figure 8.40

Solution (a) *Numerical Evaluation* Table 8.16 shows that

$$f(-1) = 1, \quad f(1) \text{ is undefined,} \quad \text{and} \quad f(2) = 4.$$

(b) *Symbolic Evaluation*

$$f(-1) = \frac{2(-1)}{-1 - 1} = 1$$

$$f(1) = \frac{2(1)}{1 - 1} = \frac{2}{0}, \quad \text{which is undefined. Input 1 is not in the domain of } f.$$

$$f(2) = \frac{2(2)}{2 - 1} = 4$$

(c) *Graphical Evaluation* To evaluate $f(-1)$ graphically, find $x = -1$ on the x-axis and move upward to the graph of f. The y-value is 1 at the point of intersection, so $f(-1) = 1$, as shown in Figure 8.41(a). In Figure 8.41(b) the vertical line $x = 1$ is called a *vertical asymptote*. Because the graph of f does not intersect this line, $f(1)$ is undefined. Figure 8.41(c) reveals that $f(2) = 4$.

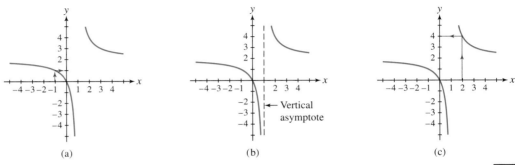

(a) (b) (c)

Figure 8.41

A **vertical asymptote** is a vertical line that is typically drawn in the graph of a rational function when the denominator of the rational expression is 0 but the numerator is not 0. The graph of a rational function *never* crosses a vertical asymptote. In Figure 8.39, the vertical asymptote is the y-axis, or $x = 0$. An asymptote is *not* a part of the graph.

Technology Note: *Asymptotes, Dot Mode, and Decimal Windows*

When rational functions are graphed on graphing calculators, pseudo-asymptotes often occur because the calculator is simply connecting dots to draw a graph. The accompanying figures show the graph of $y = \frac{2}{x - 2}$ in connected mode, dot mode, and with a *decimal*, or *friendly*, *window*. In dot mode, pixels in the calculator screen are not connected. With dot mode (and sometimes with a decimal window) pseudo-asymptotes do not appear. To learn more about these features consult your owner's manual.

Calculator Help

To set a calculator in dot mode or to set a decimal window, see Appendix A (page AP-10).

$[-6, 6, 1]$ by $[-4, 4, 1]$ $[-6, 6, 1]$ by $[-4, 4, 1]$ $[-4.7, 4.7, 1]$ by $[-3.1, 3.1, 1]$

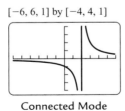 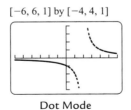 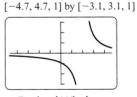

Connected Mode Dot Mode Decimal Window

8.4 PUTTING IT ALL TOGETHER

The following table summarizes some important topics from this section.

Concept	Comments	Examples
Writing Domain and Range in Interval Notation	Interval notation can be used to specify the domain and range of a function.	If $f(x) = x^2 + 1$, the domain of f is $(-\infty, \infty)$, and the range of f is $[1, \infty)$.
Absolute Value Function	Defined by $$f(x) = |x|$$ and has a V-shaped graph	$f(-5) = |-5| = 5$ $f(0) = |0| = 0$ $f(4) = |4| = 4$
Polynomial Function in One Variable	Can be defined by a polynomial; its degree equals the largest exponent of the variable.	Because $x^3 - 4x^2 + 6$ is a polynomial with degree 3, $$f(x) = x^3 - 4x^2 + 6$$ defines a polynomial function of degree 3 and is called a cubic function.
Rational Function	A rational function can be written as $$f(x) = \frac{p(x)}{q(x)},$$ where $p(x)$ and $q(x)$ are polynomials.	Because $2x - 3$ and $x + 1$ are polynomials, $$f(x) = \frac{2x - 3}{x + 1}$$ defines a rational function. Because $f(x)$ is undefined at $x = -1$, the domain of f is $(-\infty, -1) \cup (-1, \infty)$.

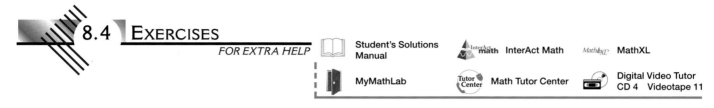

8.4 EXERCISES

FOR EXTRA HELP

Student's Solutions Manual

MyMathLab

InterAct Math

Math Tutor Center

MathXL

Digital Video Tutor
CD 4   Videotape 11

CONCEPTS

1. The set of all valid inputs for a function is called its _____.

2. The set of all outputs for a function is called its _____.

3. The inequality $x \geq 1$ can be written in interval notation as _____.

4. The inequality $x < -1$ can be written in interval notation as _____.

5. The set of all real numbers can be written in interval notation as _____.

6. If the domain of a function includes all real numbers except 5, then its domain can be written in interval notation as ____.

7. The graph of the ____ function is V-shaped.

8. The degree of a polynomial in one variable equals the largest ____ of the variable.

9. A quadratic function has degree ____.

10. If a function is linear, then its degree is ____.

11. If $f(x) = \frac{x}{2x + 1}$, then f is a ____ function.

12. If $f(x) = \frac{x}{2x + 1}$, then the domain of f includes all real numbers except ____.

DOMAIN AND RANGE

Exercises 13–24: Write the domain and the range of the function in interval notation. (Hint: You may want to consider the graph of the function.)

13. $f(x) = -2x$

14. $f(x) = -\frac{1}{4}x + 1$

15. $g(t) = \frac{2}{3}t - 3$

16. $g(t) = 9t$

17. $h(z) = z^2 + 2$

18. $h(z) = z^2 - 1$

19. $f(z) = -z^2$

20. $f(z) = -\frac{1}{4}z^2$

21. $g(x) = \sqrt{x + 1}$

22. $g(x) = \sqrt{x - 2}$

23. $h(x) = |x - 1|$

24. $h(x) = |2x|$

Exercises 25–36: Write the domain of the rational function in interval notation.

25. $f(x) = \dfrac{1}{x - 1}$

26. $f(x) = \dfrac{6}{x}$

27. $f(x) = \dfrac{x}{6 - 3x}$

28. $f(x) = \dfrac{3x}{2x - 4}$

29. $g(t) = \dfrac{2}{t^2 - 4}$

30. $g(t) = \dfrac{5}{1 - t^2}$

31. $g(t) = \dfrac{5t}{t^2 - 2t}$

32. $g(t) = \dfrac{-t}{2t^2 - 3t}$

33. $h(z) = \dfrac{2 - z}{z^3 - 1}$

34. $h(z) = \dfrac{z + 1}{z^3 - z^2}$

35. $f(x) = \dfrac{4}{x^2 - 2x - 3}$

36. $f(x) = \dfrac{1}{x^2 + 4x - 5}$

Exercises 37–42: A graph of a function is shown. Write the domain and range of the function in interval notation.

37.
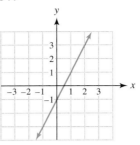

38.

39.
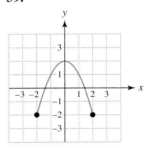

40.

41.
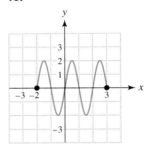

42.

IDENTIFYING POLYNOMIAL FUNCTIONS

Exercises 43–54: Determine whether $f(x)$ represents a polynomial function. If possible, identify the degree and type of polynomial function.

43. $f(x) = 5x - 11$

44. $f(x) = 9 - x$

45. $f(x) = x^3$

46. $f(x) = x^2 + 3$

47. $f(x) = \dfrac{6}{x + 5}$

48. $f(x) = |x|$

49. $f(x) = 1 + 2x - x^2$

50. $\frac{1}{4}x^3 - x$

51. $f(x) = 5x^{-2}$

52. $f(x) = x^2 + x^{-1}$

53. $f(x) = x^4 + 2x^2$

54. $f(x) = x^5 - 3x^3$

EVALUATING FUNCTIONS

Exercises 55–70: If possible, evaluate g(t) for the given values of t.

55. $g(t) = |4t|$ $t = 3, t = 0$

56. $g(t) = |t + 12|$ $t = 18, t = -15$

57. $g(t) = |t - 2|$ $t = 1, t = -\frac{3}{4}$

58. $g(t) = |2t + 1|$ $t = 2, t = -\frac{1}{2}$

59. $g(t) = t^2 - t - 6$ $t = 3, t = -3$

60. $g(t) = 3t^2 - 2t$ $t = -2, t = 4$

61. $g(t) = 2t^3 - t$ $t = 2, t = -2$

62. $g(t) = \frac{1}{3}t^3$ $t = 1, t = -3$

63. $g(t) = t^2 - 2t - 6$ $t = 0, t = -3$

64. $g(t) = 2t^3 - t^2 + 4$ $t = 2, t = -1$

65. $g(t) = \dfrac{1}{t}$ $t = 11, t = -7$

66. $g(t) = \dfrac{2}{3 - t}$ $t = 10, t = 3$

67. $g(t) = -\dfrac{t}{t + 1}$ $t = 5, t = -1$

68. $g(t) = -\dfrac{2 - t}{4t}$ $t = 4, t = -1$

69. $g(t) = \dfrac{t^2}{t^2 - t}$ $t = -5, t = 1$

70. $g(t) = \dfrac{t - 3}{t^2 - 3t + 2}$ $t = -2, t = 1$

Exercises 71–78: If possible, use the graph to evaluate each expression. Then use the formula for f(x) to check your results.

71. $f(0)$ and $f(1)$ 1

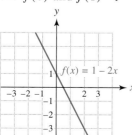

72. $f(-1)$ and $f(2)$

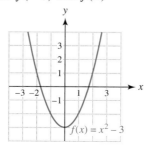

73. $f(-1)$ and $f(2)$

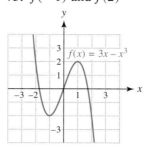

74. $f(0)$ and $f(-2)$

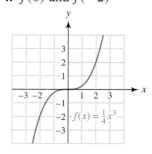

75. $f(-2)$ and $f(2)$

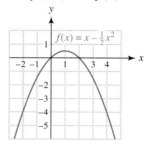

76. $f(-1)$ and $f(0)$

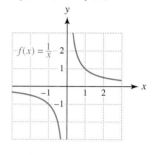

77. $f(-3)$ and $f(-1)$

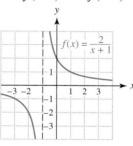

78. $f(0)$ and $f(1)$

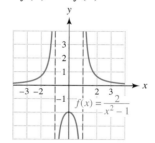

Exercises 79–88: Graph f.

79. $f(x) = |2x|$ **80.** $f(x) = \left|\frac{1}{2}x\right|$

81. $f(x) = |x + 2|$ **82.** $f(x) = |x - 2|$

83. $f(x) = 1 - 2x$ **84.** $f(x) = \frac{1}{2}x + 1$

85. $f(x) = \frac{1}{2}x^2$ **86.** $f(x) = x^2 - 2$

87. $f(x) = \dfrac{1}{x - 1}$ **88.** $f(x) = \dfrac{1}{x + 1}$

APPLICATIONS

89. *Heart Rate of an Athlete*   The following table lists the heart rate of an athlete running a 100-meter race. The race lasts 10 seconds.

Time (seconds)	0	2	4	6	8	10
Heart Rate (bpm)	90	100	113	127	143	160

(a) Does $P(t) = 0.2t^2 + 5t + 90$ model the data in the table exactly? Explain.

(b) Does P provide a reasonable model for the athlete's heart rate?

(c) Does $P(12)$ have significance in this situation? What should be the domain of P?

90. *Heart Rate of an Athlete*   The following table lists an athlete's heart rate after the athlete finishes exercising strenuously.

Time (minutes)	0	2	4	6
Heart Rate (bpm)	180	137	107	90

(a) Does $P(t) = \frac{5}{3}t^2 - 25t + 180$ model the data in the table exactly? Explain.

(b) Does P provide a reasonable model for the athlete's heart rate?

(c) Does $P(12)$ have significance in this situation? What should be the domain of P?

91. *Burning Calories*   The following table lists the calories burned by a 140-pound person walking at 4 miles per hour for various lengths of time.

Time (minutes)	6	12	18	24
Calories	32	64	96	128

Source: Healthstats.com.

(a) Would it be reasonable to assume that the point $(0, 0)$ could be included in the table? Why or why not?

(b) What type of polynomial function would model these data? Explain.

(c) Find a function C that models the calories burned in t minutes.

92. *Burning Calories*   The following table lists the calories burned by a 180-pound person shopping at a mall.

Time (minutes)	10	30	50	80
Calories	30	90	150	240

Source: Healthstats.com.

(a) Would it be reasonable to assume that the point $(0, 0)$ could be included in the table? Why or why not?

(b) What type of polynomial function would model these data? Explain.

(c) Find a function C that models the calories burned in t minutes.

Exercises 93–96: Rational Models   *Match the physical situation with the graph of the rational function that models it best.*

93. A population of fish that increases and then levels off

94. An insect population that dies out

95. The length of a ticket line as the rate at which people arrive in line increases

96. The wind speed during a day that is initially calm, becomes windy, and then is calm again

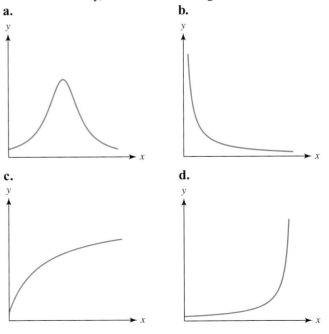

97. *Modeling a Train Track Curve*   When curves are designed for train tracks, sometimes the outer rail is elevated, or banked, so that a locomotive and its cars can safely negotiate the curve at a higher speed than if the tracks were level. Suppose that a circular curve with a radius of r feet is being designed for a train traveling 60 miles per hour. Then $f(r) = \frac{2540}{r}$ calculates the proper elevation y in inches for the outer rail, where $y = f(r)$. See the accompanying figure.
(*Source:* L. Haefner, *Introduction to Transportation Systems.*)

(a) Evaluate $f(300)$ and interpret the result.

 (b) Graph f in the window [0, 600, 100] by [0, 50, 10].

(c) Discuss how the elevation of the outer rail changes as the radius r increases.

98. *Grade of a Highway*   The *grade* x of a hill is a measure of the steepness and corresponds to the slope of the highway. The accompanying figure shows a grade of $x = \frac{10}{100} = 10\%$ because the road rises 10 feet for every 100 feet of horizontal run. The braking distance D in feet for a car traveling 60 miles per hour on a wet uphill grade is given by

$$D(x) = \frac{3600}{30x + 9}.$$

(*Source:* N. Garber and L. Hoel, *Traffic and Highway Engineering.*)

(a) Evaluate $D(0.05)$ and interpret the result.

(b) Find the grade x for a braking distance of 300 feet.

WRITING ABOUT MATHEMATICS

99. Name two functions. Give their formulas, sketch their graphs, and state their domains and ranges.

100. Explain the difference between the domain and the range of a function.

CHECKING BASIC CONCEPTS   SECTIONS 8.3 AND 8.4

1. (a) Is 3 a solution to the compound inequality $x + 2 < 4$ or $2x - 1 \geq 3$?

(b) Is 3 a solution to the compound inequality $x + 2 < 4$ and $2x - 1 \geq 3$?

2. Solve the following compound inequalities. Write your answer in interval notation.

(a) $-5 \leq 2x + 1 \leq 3$

(b) $1 - x \leq -2$ or $1 - x \geq 2$

(c) $-2 < \dfrac{4 - 3x}{2} \leq 6$

3. Write the domain of each function in interval notation.

(a) $f(x) = x^2$

(b) $g(t) = \dfrac{1}{t - 1}$

(c) $h(z) = \sqrt{z}$

4. Use the graph of f to do the following.

(a) Write the domain and range of f in interval notation.

(b) Evaluate $f(0)$ and $f(-2)$.

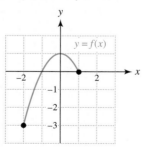

5. Graph $f(x) = |x - 3|$.

8.5 ABSOLUTE VALUE EQUATIONS AND INEQUALITIES

Absolute Value Equations · Absolute Value Inequalities

INTRODUCTION

Monthly average temperatures can vary greatly from one month to another, whereas yearly average temperatures remain fairly constant from one year to the next. In Boston, Massachusetts, the yearly average temperature is $50°F$, but monthly average temperatures can vary from $28°F$ to $72°F$. Because $50°F - 28°F = 22°F$ and $72°F - 50°F = 22°F$, the monthly average temperatures are always within $22°F$ of the yearly average temperature of $50°F$. If T represents a monthly average temperature, we can model this situation by using the absolute value inequality

$$|T - 50| \leq 22.$$

The absolute value is necessary because a monthly average temperature T can be either greater than or less than $50°F$ by as much as $22°F$. In this section we discuss how to solve absolute value equations and inequalities.   (*Source:* A. Miller and J. Thompson, *Elements of Meteorology.*)

ABSOLUTE VALUE EQUATIONS

An equation that contains an absolute value is called an **absolute value equation**. Examples include

$$|x| = 2, \quad |2x - 1| = 5, \quad \text{and} \quad |5 - 3x| - 3 = 1.$$

Consider the absolute value equation $|x| = 2$. This equation has *two* solutions: 2 and -2 because $|2| = 2$ and $|-2| = 2$. We can also demonstrate this result with a table of values or a graph. In Table 8.17, $|x| = 2$ when $x = -2$ or $x = 2$. In Figure 8.42 the graph of $y_1 = |x|$ intersects the graph of $y_2 = 2$ at the points $(-2, 2)$ and $(2, 2)$. The x-values at these points of intersection correspond to the solutions -2 and 2.

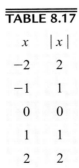

TABLE 8.17

x	$|x|$
-2	2
-1	1
0	0
1	1
2	2

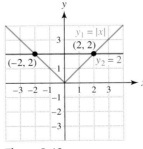

Figure 8.42

We generalize this discussion in the following manner.

SOLVING $|x| = k$

1. If $k > 0$, then $|x| = k$ is equivalent to $x = k$ or $x = -k$.
2. If $k = 0$, then $|x| = k$ is equivalent to $x = 0$.
3. If $k < 0$, then $|x| = k$ has no solutions.

EXAMPLE 1   Solving absolute value equations

Solve each equation.
(a) $|x| = 20$ **(b)** $|x| = -5$

Solution   **(a)** The solutions are -20 and 20.
(b) There are no solutions because $|x|$ is never negative.

We can solve other absolute value equations similarly.

EXAMPLE 2   Solving an absolute value equation

Solve $|2x - 5| = 3$.

Solution   If $|2x - 5| = 3$, then either $2x - 5 = 3$ or $2x - 5 = -3$. Solve each equation separately.

$$2x - 5 = 3 \quad \text{or} \quad 2x - 5 = -3 \qquad \text{Equations to be solved}$$
$$2x = 8 \quad \text{or} \quad 2x = 2 \qquad \text{Add 5.}$$
$$x = 4 \quad \text{or} \quad x = 1 \qquad \text{Divide by 2.}$$

The solutions are 1 and 4.

A table of values can be used to solve the equation $|2x - 5| = 3$ from Example 2. Table 8.18 shows that $|2x - 5| = 3$ when $x = 1$ or $x = 4$.

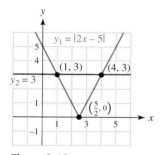

Figure 8.43

TABLE 8.18

x	0	1	2	3	4	5	6
$|2x - 5|$	5	3	1	1	3	5	7

This equation can also be solved by graphing $y_1 = |2x - 5|$ and $y_2 = 3$. To graph y_1, first plot some of the points from Table 8.18. Its graph is V-shaped, as shown in Figure 8.43. Note that the x-coordinate of the "point" or vertex of the V can be found by solving the equation $2x - 5 = 0$ to obtain $\frac{5}{2}$. The graph of y_1 intersects the graph of y_2 at the points $(1, 3)$ and $(4, 3)$, giving the solutions 1 and 4, so the graphical solutions agree with the numerical and symbolic solutions.

This discussion leads to the following result.

ABSOLUTE VALUE EQUATIONS

Let k be a positive number. Then

$$|ax + b| = k$$

is equivalent to

$$ax + b = k \quad \text{or} \quad ax + b = -k.$$

EXAMPLE 3   Solving absolute value equations

Solve.
(a) $|5 - x| - 2 = 8$ **(b)** $\left|\frac{1}{2}(x - 6)\right| = \frac{3}{4}$

Solution   **(a)** Start by adding 2 to each side to obtain

$$|5 - x| = 10.$$

This equation is satisfied by the solution from either of the following equations.

$$
\begin{array}{lcll}
5 - x = 10 & \text{or} & 5 - x = -10 & \text{Equations to be solved} \\
-x = 5 & \text{or} & -x = -15 & \text{Subtract 5.} \\
x = -5 & \text{or} & x = 15 & \text{Multiply by } -1.
\end{array}
$$

The solutions are -5 and 15.

(b) This equation is satisfied by the solution from either of the following equations. To begin solving, multiply by 4 to clear fractions.

$$
\begin{array}{lcll}
\dfrac{1}{2}(x - 6) = \dfrac{3}{4} & \text{or} & \dfrac{1}{2}(x - 6) = -\dfrac{3}{4} & \text{Equations to be solved} \\
2(x - 6) = 3 & \text{or} & 2(x - 6) = -3 & \text{Multiply by 4.} \\
2x - 12 = 3 & \text{or} & 2x - 12 = -3 & \text{Distributive property} \\
2x = 15 & \text{or} & 2x = 9 & \text{Add 12.} \\
x = \dfrac{15}{2} & \text{or} & x = \dfrac{9}{2} & \text{Divide by 2.}
\end{array}
$$

The solutions are $\frac{9}{2}$ and $\frac{15}{2}$.

The next example illustrates absolute value equations that have either 0 or 1 solution.

EXAMPLE 4   Solving absolute value equations

Solve.
(a) $|2x - 1| = -2$ **(b)** $|4 - 2x| = 0$

Solution   **(a)** Because an absolute value can never be negative, the expression $|2x - 1|$ is always greater than or equal to 0. There are no solutions. Figure 8.44 shows that the graph of $y_1 = |2x - 1|$ never intersects the graph of $y_2 = -2$.
(b) If $|y| = 0$, then $y = 0$. Thus the given equation is satisfied when $4 - 2x = 0$ or when $x = 2$. The solution is 2.

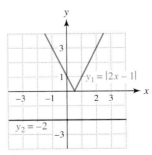

Figure 8.44

Sometimes an absolute value equation can have an absolute value on each side of the equation. An example would be $|2x| = |x - 3|$. In this situation either $2x = x - 3$ (the two expressions are equal), or $2x = -(x - 3)$ (the two expressions are opposites). These concepts are summarized as follows.

SOLVING $|ax + b| = |cx + d|$

Let a, b, c, and d be constants. Then $|ax + b| = |cx + d|$ is equivalent to

$$ax + b = cx + d \quad \text{or} \quad ax + b = -(cx + d).$$

EXAMPLE 5 Solving an absolute value equation

Solve $|2x| = |x - 3|$.

Solution Solve the following compound equality.

$$2x = x - 3 \quad \text{or} \quad 2x = -(x - 3)$$
$$x = -3 \quad \text{or} \quad 2x = -x + 3$$
$$3x = 3$$
$$x = 1$$

The solutions are -3 and 1.

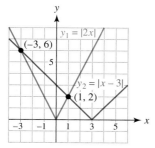

Figure 8.45

The solution set to Example 5 is shown graphically in Figure 8.45. Note that the graphs of $y_1 = |2x|$ and $y_2 = |x - 3|$ are V-shaped and intersect at the points $(-3, 6)$ and $(1, 2)$. Thus the solutions are -3 and 1.

ABSOLUTE VALUE INEQUALITIES

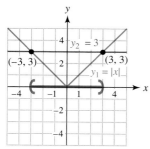

Figure 8.46

As with other inequalities, we can solve absolute value inequalities graphically. For example, to solve $|x| < 3$, let $y_1 = |x|$ and $y_2 = 3$ (see Figure 8.46). Their graphs intersect at the points $(-3, 3)$ and $(3, 3)$. The graph of y_1 is below the graph of y_2 for x-values between, but not including, $x = -3$ and $x = 3$. The solution set is $\{x \mid -3 < x < 3\}$ and is shaded on the x-axis.

Other absolute value inequalities can be solved graphically in a similar way. In Figure 8.47 the solutions to $|2x - 1| = 3$ are -1 and 2. The V-shaped graph of $y_1 = |2x - 1|$ is below the horizontal line $y_2 = 3$ when $-1 < x < 2$. Thus $|2x - 1| < 3$ whenever $-1 < x < 2$. The solution set is shaded on the x-axis.

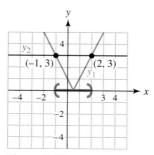

Figure 8.47

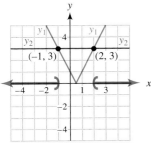

Figure 8.48

In Figure 8.48 the V-shaped graph of $y_1 = |2x - 1|$ is above the horizontal line $y_2 = 3$ both to the left of -1 and to the right of 2. That is, $|2x - 1| > 3$ whenever $x < -1$ or $x > 2$. The solution set is shaded on the x-axis.

This discussion is summarized as follows.

ABSOLUTE VALUE INEQUALITIES

Let the solutions to $|ax + b| = k$ be c and d, where $c < d$ and $k > 0$.

1. $|ax + b| < k$ is equivalent to $c < x < d$.
2. $|ax + b| > k$ is equivalent to $x < c$ or $x > d$.

Similar statements can be made for inequalities involving $\leq$ or $\geq$.

MAKING CONNECTIONS

Graphs of $y = |ax + b|$

The graph of $y = |ax + b|$, $a \neq 0$, is V-shaped and intersects a horizontal line above the x-axis twice. This graph can be used to help visualize the solutions to either $|ax + b| < k$ or $|ax + b| > k$.

EXAMPLE 6 Solving absolute value equations and inequalities

Solve each absolute value equation and inequality.
(a) $|2 - 3x| = 4$ (b) $|2 - 3x| < 4$ (c) $|2 - 3x| > 4$

Solution (a) Solve the given equation.

$$
\begin{array}{lll}
2 - 3x = 4 & \text{or} \quad 2 - 3x = -4 & \text{Equations to be solved} \\
-3x = 2 & \text{or} \qquad -3x = -6 & \text{Subtract 2.} \\
x = -\dfrac{2}{3} & \text{or} \qquad x = 2 & \text{Divide by } -3.
\end{array}
$$

The solutions are $-\frac{2}{3}$ and 2.

(b) Solutions to $|2 - 3x| < 4$ include x-values between, but not including, $-\frac{2}{3}$ and 2. Thus the solution set is $\{x \mid -\frac{2}{3} < x < 2\}$, or in interval notation, $(-\frac{2}{3}, 2)$.

(c) Solutions to $|2 - 3x| > 4$ include x-values to the left of $x = -\frac{2}{3}$ or to the right of $x = 2$. Thus the solution set is $\{x \mid x < -\frac{2}{3} \text{ or } x > 2\}$, or in interval notation, $(-\infty, -\frac{2}{3}) \cup (2, \infty)$.

EXAMPLE 7 Solving an absolute value inequality

Solve $\left|\frac{2x-5}{3}\right| > 3$. Write the solution set in interval notation.

Solution Start by solving $\left|\frac{2x-5}{3}\right| = 3$ as follows.

$$\frac{2x-5}{3} = 3 \quad \text{or} \quad \frac{2x-5}{3} = -3 \qquad \text{Equations to be solved}$$

$$2x - 5 = 9 \quad \text{or} \quad 2x - 5 = -9 \qquad \text{Multiply by 3.}$$

$$2x = 14 \quad \text{or} \quad 2x = -4 \qquad \text{Add 5.}$$

$$x = 7 \quad \text{or} \quad x = -2 \qquad \text{Divide by 2.}$$

Because the inequality symbol is $>$, the solution set is $x < -2$ or $x > 7$, which can be written in interval notation as $(-\infty, -2) \cup (7, \infty)$.

EXAMPLE 8 Modeling temperature in Boston

In the introduction to this section we discussed how the inequality $|T - 50| \leq 22$ models the range for the monthly average temperatures T in Boston, Massachusetts.
(a) Solve this inequality and interpret the result.
(b) Give graphical support for part (a).

Solution **(a)** *Symbolic Solution* Start by solving $|T - 50| = 22$.

$$T - 50 = 22 \quad \text{or} \quad T - 50 = -22 \qquad \text{Equations to be solved}$$

$$T = 72 \quad \text{or} \quad T = 28 \qquad \text{Add 50 to both sides.}$$

Thus the solution set to $|T - 50| \leq 22$ is $\{T \mid 28 \leq T \leq 72\}$. Monthly average temperatures in Boston vary from $28°$F to $72°$F.

(b) *Graphical Solution* The graphs of $y_1 = |x - 50|$ and $y_2 = 22$ intersect at the points $(28, 22)$ and $(72, 22)$, as shown in Figures 8.49(a) and (b), respectively. The V-shaped graph of y_1 intersects the horizontal graph of y_2, or is below it, when $28 \leq x \leq 72$. Thus the solution set is $\{T \mid 28 \leq T \leq 72\}$, which agrees with the symbolic result.

Calculator Help
To graph an absolute value, see Appendix A (page AP-7).

[0, 100, 10] by [0, 70, 10]

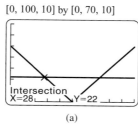

(a)

[0, 100, 10] by [0, 70, 10]

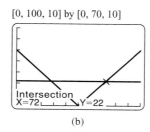

(b)

Figure 8.49

Critical Thinking

Find the solution set for the following inequalities. Discuss your results.

1. $|2x - 5| > -3$ **2.** $|2x - 5| < -3$

PUTTING IT ALL TOGETHER

The absolute value function is given by $f(x) = |x|$, and its graph is V-shaped. Its domain (set of valid inputs) includes all real numbers, and its range (outputs) includes all nonnegative real numbers. The following table summarizes methods for solving absolute value equations and inequalities involving $<$ and $>$ symbols. Inequalities containing $\leq$ and $\geq$ symbols are solved similarly.

Problem	Symbolic Solution	Graphical Solution
$\|ax + b\| = k, k > 0$	Solve the equations $$ax + b = k$$ and $$ax + b = -k.$$	Graph $y_1 = \|ax + b\|$ and $y_2 = k$. Find the x-values of the two points of intersection.
$\|ax + b\| < k, k > 0$	If the solutions to $$\|ax + b\| = k$$ are c and d, $c < d$, then the solutions to $$\|ax + b\| < k$$ satisfy $$c < x < d.$$	Graph $y_1 = \|ax + b\|$ and $y_2 = k$. Find the x-values of the two points of intersection. The solutions are between these x-values on the number line, where the graph of y_1 lies below the graph of y_2.
$\|ax + b\| > k, k > 0$	If the solutions to $$\|ax + b\| = k$$ are c and d, $c < d$, then the solutions to $$\|ax + b\| > k$$ satisfy $$x < c \quad \text{or} \quad x > d.$$	Graph $y_1 = \|ax + b\|$ and $y_2 = k$. Find the x-values of the two points of intersection. The solutions are outside these x-values on the number line, where the graph of y_1 is above the graph of y_2.

8.5 EXERCISES

FOR EXTRA HELP

Student's Solutions Manual

MyMathLab

InterAct Math

Math Tutor Center

MathXL

Digital Video Tutor
CD 4   Videotape 11

CONCEPTS

1. Give an example of an absolute value equation.

2. Give an example of an absolute value inequality.

3. Is -3 a solution to $|x| = 3$?

4. Is -4 a solution to $|x| > 3$?

5. Is $|x| = 5$ equivalent to $x = -5$ or $x = 5$?

6. Is $|x| < 3$ equivalent to $x < -3$ or $x > 3$? Explain.

Exercises 7–12: Determine whether the given values of x are solutions to the absolute value equation or inequality.

7. $|2x - 5| = 1$ $x = -3, x = 3$

8. $|5 - 6x| = 1$ $x = 1, x = 0$

9. $|7 - 4x| \le 5$ $x = -2, x = 2$

10. $|2 + x| < 2$ $x = -4, x = -1$

11. $|7x + 4| > -1$ $x = -\frac{4}{7}, x = 2$

12. $|12x + 3| \ge 3$ $x = -\frac{1}{4}, x = 2$

Exercises 13 and 14: Use the graph of y_1 to solve the equation.

13. $y_1 = 2$

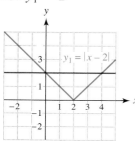

14. $y_1 = 3$

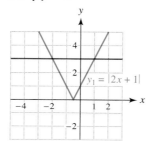

SYMBOLIC SOLUTIONS

Exercises 15–38: Solve the absolute value equation.

15. $|x| = 7$ **16.** $|x| = 4$

17. $|x| = 0$ **18.** $|x| = -6$

19. $|4x| = 9$ **20.** $|-3x| = 7$

21. $|-2x| - 6 = 2$ **22.** $|5x| + 1 = 5$

23. $|2x + 1| = 11$ **24.** $|1 - 3x| = 4$

25. $|-2x + 3| + 3 = 4$

26. $|6x + 2| - 2 = 6$

27. $|\frac{1}{2}x - 1| = 5$ **28.** $|6 - \frac{3}{4}x| = 3$

29. $|2x - 6| = -7$ **30.** $|1 - \frac{2}{3}x| + 2 = 0$

31. $|\frac{2}{3}z - 1| - 3 = 8$ **32.** $|1 - 2z| + 5 = 10$

33. $|z - 1| = |2z|$ **34.** $|2z + 3| = |2 - z|$

35. $|3t + 1| = |2t - 4|$

36. $|\frac{1}{2}t - 1| = |3 - \frac{3}{2}t|$

37. $|\frac{1}{4}x| = |3 + \frac{1}{4}x|$ **38.** $|2x - 1| = |2x + 2|$

Exercises 39–42: Solve the absolute value equation and inequalities.

39. (a) $|2x| = 8$
     (b) $|2x| < 8$
     (c) $|2x| > 8$

40. (a) $|3x - 9| = 6$
     (b) $|3x - 9| \le 6$
     (c) $|3x - 9| \ge 6$

41. (a) $|5 - 4x| = 3$
     (b) $|5 - 4x| \le 3$
     (c) $|5 - 4x| \ge 3$

42. (a) $\left|\dfrac{x - 5}{2}\right| = 2$

     (b) $\left|\dfrac{x - 5}{2}\right| < 2$

     (c) $\left|\dfrac{x - 5}{2}\right| > 2$

Exercises 43–74: Solve the absolute value inequality. Write your answer in interval notation.

43. $|x| \le 3$ **44.** $|x| < 2$

45. $|k| > 4$ **46.** $|k| \ge 5$

47. $|t| \le -3$ **48.** $|t| < -1$

49. $|z| > 0$ **50.** $|2z| \ge 0$

51. $|2x| > 7$ **52.** $|-12x| < 30$

53. $|-4x + 4| < 16$ **54.** $|-5x - 8| > 2$

55. $2|x + 5| \ge 8$ **56.** $-3|x - 1| \ge -9$

57. $|8 - 6x| - 1 \le 2$ **58.** $4 - \left|\dfrac{2x}{3}\right| < -7$

59. $5 + \left|\dfrac{2 - x}{3}\right| \le 9$ **60.** $\left|\dfrac{x + 3}{5}\right| \le 12$

61. $|2x - 1| \le -3$ **62.** $|x + 6| \ge -5$

63. $|x + 1| - 1 > -3$ **64.** $-2|1 - 7x| \ge 2$

65. $|2z - 4| \le -1$ **66.** $|4 - z| \le 0$

67. $|3z - 1| > -3$ **68.** $|2z| \ge -2$

69. $\left|\dfrac{2-t}{3}\right| \geq 5$ **70.** $\left|\dfrac{2t+3}{5}\right| \geq 7$

71. $|t-1| \leq 0.1$ **72.** $|t-2| \leq 0.01$

73. $|b-10| > 0.5$ **74.** $|b-25| \geq 1$

NUMERICAL AND GRAPHICAL SOLUTIONS

Exercises 75 and 76: Use the table of $y = |ax+b|$ to solve each equation or inequality. Write your answers in interval notation for parts (b) and (c).

75. (a) $y = 2$ **(b)** $y < 2$ **(c)** $y > 2$

x	-2	-1	0	1	2	3	4
y	3	2	1	0	1	2	3

76. (a) $y = 6$ **(b)** $y \leq 6$ **(c)** $y \geq 6$

x	-12	-6	0	6	12	18	24
y	9	6	3	0	3	6	9

Exercises 77 and 78: Use the graph of y_1 to solve each equation or inequality. Write your answers in interval notation for parts (b) and (c).

77. (a) $y_1 = 1$ **(b)** $y_1 \leq 1$ **(c)** $y_1 \geq 1$

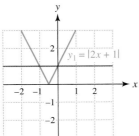

78. (a) $y_1 = 3$ **(b)** $y_1 < 3$ **(c)** $y_1 > 3$

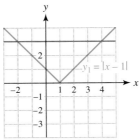

Exercises 79–88: Solve the inequality graphically. Write your answer in interval notation.

79. $|x| \geq 1$ **80.** $|x| < 2$

81. $|x-1| \leq 3$ **82.** $|x+5| \geq 2$

83. $|4-2x| > 2$ **84.** $|1.5x-3| \geq 6$

85. $|10-3x| < 4$ **86.** $|7-4x| \leq 2.5$

87. $|8.1-x| > -2$ **88.** $\left|\dfrac{5x-9}{2}\right| \leq -1$

USING MORE THAN ONE METHOD

Exercises 89–92: Solve the absolute value inequality
  (a) symbolically,
  (b) graphically, and
  (c) numerically.
Write your answer in set-builder notation.

89. $|3x| \leq 9$ **90.** $|5-x| \geq 3$

91. $|2x-5| > 1$ **92.** $|-8-4x| < 6$

APPLICATIONS

Exercises 93–96: Average Temperatures (Refer to Example 8.) The given inequality models the range for the monthly average temperatures T in degrees Fahrenheit at the location specified.
  (a) Solve the inequality.
  (b) Give a possible interpretation of the inequality.

93. $|T-43| \leq 24$, Marquette, Michigan

94. $|T-62| \leq 19$, Memphis, Tennessee

95. $|T-10| \leq 36$, Chesterfield, Canada

96. $|T-61.5| \leq 12.5$, Buenos Aires, Argentina

97. *Highest Elevations* The table lists the highest elevation in each continent.

Continent	Elevation (feet)
Asia	29,028
S. America	22,834
N. America	20,320
Africa	19,340
Europe	18,510
Antarctica	16,066
Australia	7,310

Source: National Geographic.

(a) Calculate the average A of these elevations.

(b) Which continents have their highest elevations within 1000 feet of A?

(c) Which continents have their highest elevations within 5000 feet of A?

98. *Distance*   Suppose that two cars, both traveling at a constant speed of 60 miles per hour, approach each other on a straight highway.

  (a) If they are initially 4 miles apart, sketch a graph of the distance between the cars after x minutes, where $0 \leq x \leq 4$. (*Hint:* 60 miles per hour = 1 mile per minute)

  (b) Write an absolute value equation whose solution gives the times when the cars were 2 miles apart.

  (c) Solve your equation from part (b).

99. *Error in Measurements*   Products are often manufactured to be a given size or shape to within a certain tolerance. For instance, if an aluminum can is supposed to have a diameter of 2.5 inches, either 2.501 inches or 2.499 inches might be acceptable. If the maximum error in the diameter of the can is restricted to 0.002 inch, an acceptable diameter d must satisfy the absolute value inequality

$$|d - 2.5| \leq 0.002.$$

Solve this inequality for d and interpret the result.

100. *Error in Measurements*   (Refer to the preceding exercise.) Suppose that a person can operate a stopwatch accurately to within 0.02 second. If a runner's time in the 400-meter dash is recorded as 51.57 seconds, within what range of values could the true time fall?

101. *Relative Error*   If a quantity is measured to be x and the actual value is t, then the relative error in the measurement is $\left|\frac{x-t}{t}\right|$. If the true measurement is $t = 20$ and you want the relative error to be less than 0.05 (5%), what values for x are possible?

102. *Relative Error*   (Refer to the preceding exercise.) The volume V of a box is 50 cubic inches. How accurately must you measure the volume of the box for the relative error to be less than 3%?

WRITING ABOUT MATHEMATICS

103. If $a \neq 0$, how many solutions are there to the equation $|ax + b| = k$ when
  (a) $k > 0$,   **(b)** $k = 0$, and   **(c)** $k < 0$?
  Explain each answer.

104. Suppose that you know two solutions to the equation $|ax + b| = k$. How can you use these solutions to solve the inequalities $|ax + b| < k$ and $|ax + b| > k$? Give an example.

CHECKING BASIC CONCEPTS   **SECTION 8.5**

Write your answers in interval notation whenever possible.

1. Solve $\left|\frac{3}{4}x - 1\right| - 3 = 5$.

2. Solve the absolute value equation and inequalities.
  (a) $|3x - 6| = 8$ **(b)** $|3x - 6| < 8$
  (c) $|3x - 6| > 8$

3. Solve the inequality $|-2(3 - x)| < 6$. Then solve $|-2(3 - x)| \geq 6$.

4. Use the graph to solve the equation and inequalities.

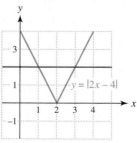

  (a) $|2x - 4| = 2$ **(b)** $|2x - 4| \leq 2$
  (c) $|2x - 4| \geq 2$

CHAPTER

8 Summary

Function   A function is a set of ordered pairs (x, y), where each x-value corresponds to exactly one y-value. A function takes a valid input x and computes exactly one output y, forming the ordered pair (x, y).

Domain and Range   The domain D is the set of all x-values, and the range R is the set of all y-values.

Examples: $f = \{(1, 2), (2, 3), (3, 3)\}$ has $D = \{1, 2, 3\}$ and $R = \{2, 3\}$.

   $f(x) = x^2$ has domain all real numbers and range $y \geq 0$. (See the following graph.)

Function Notation $y = f(x)$ and is read "y equals f of x."

Example: $f(x) = \frac{2x}{x - 1}$ implies that $f(3) = \frac{2 \cdot 3}{3 - 1} = \frac{6}{2} = 3$.

Function Representations   A function can be represented symbolically, numerically, graphically, or verbally.

   Symbolic Representation $f(x) = x^2$

   Numerical Representation   *Graphical Representation*

x	y
-2	4
-1	1
0	0
1	1
2	4

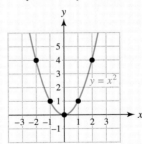

   Verbal Representation f computes the square of the input x.

Vertical Line Test   If every vertical line intersects a graph at most once, then the graph represents a function.

Linear Function   A linear function can be represented by $f(x) = ax + b$. Its graph is a (straight) line. For each unit increase in x, $f(x)$ changes by an amount equal to a.

Example: $f(x) = 2x - 1$ represents a linear function with $a = 2$ and $b = -1$.

Numerical Representation	*Graphical Representation*

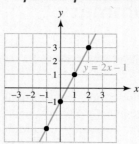

	x	$f(x)$	
1 unit	-1	-3	2 units
1 unit	0	-1	2 units
1 unit	1	1	2 units
	2	3	

Each 1-unit increase in x results in a 2-unit increase in $f(x)$.

Modeling Data with Linear Functions   When data have a constant rate of change, they can be modeled by $f(x) = ax + b$. The constant a represents the rate of change, and the constant b represents the initial amount or the value when $x = 0$. That is,

$$f(x) = (\textbf{constant rate of change})x + (\textbf{initial amount}).$$

Example: In the following table, the y-values decrease by 3 units for each unit increase in x. Also, when $x = 0, y = 4$. Thus the data are modeled by $f(x) = -3x + 4$.

x	-2	-1	0	1	2
y	10	7	4	1	-2

Section 8.3   *Compound Inequalities*

Compound Inequality   Two inequalities connected by *and* or *or*

Examples: For $x + 1 < 3$ *or* $x + 1 > 6$, a solution satisfies *at least* one of the inequalities.

For $2x + 1 < 3$ *and* $1 - x > 6$, a solution satisfies *both* inequalities.

Three-Part Inequality   A compound inequality in the form $x > a$ and $x < b$ can be written as $a < x < b$.

Example: $1 \leq x < 7$ means $x \geq 1$ *and* $x < 7$.

Interval Notation   Can be used to identify intervals on the real number line

Examples: $-2 < x \leq 3$ is equivalent to $(-2, 3]$.

$x < 5$ is equivalent to $(-\infty, 5)$.

Real numbers are denoted $(-\infty, \infty)$.

Section 8.4 *Other Functions and Their Properties*

Domain and Range in Interval Notation   The domain and range of a function can often be expressed in interval notation.

Example: The domain of $f(x) = x^2 - 2$ is all real numbers, or $(-\infty, \infty)$, and its range is real numbers greater than or equal to -2, or $[-2, \infty)$.

Absolute Value Function   The domain of $f(x) = |x|$ is $(-\infty, \infty)$, and the range of $f(x)$ is $[0, \infty)$.

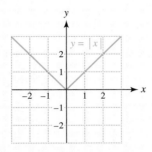

Polynomial Functions   If $f(x)$ is a polynomial, then f is a polynomial function. The degree of a polynomial function (in one variable) equals the largest exponent of a variable. The graphs of polynomial functions with degree greater than 1 are not lines. The domain of a polynomial function is $(-\infty, \infty)$.

Examples: $f(x) = 4x - 1$ defines a linear function with degree 1.
$g(x) = 4x^2 + x - 4$ defines a quadratic function with degree 2.
$h(x) = x^3 + 0.7x - 1$ defines a cubic function with degree 3.

Rational Functions   If $f(x) = \frac{p(x)}{q(x)}$, where $p(x)$ and $q(x)$ are polynomials, f is a rational function. The domain of a rational function includes all real numbers, except x-values that make the denominator 0.

Examples: $f(x) = \frac{1}{x}$ has domain $(-\infty, 0) \cup (0, \infty)$.
$g(x) = \frac{x}{x^2 - 9}$ has domain $(-\infty, -3) \cup (-3, 3) \cup (3, \infty)$.

Section 8.5 *Absolute Value Equations and Inequalities*

Absolute Value Equations   The graph of $y = |ax + b|, a \neq 0$, is V-shaped and intersects the horizontal line $y = k$ twice if $k > 0$. In this case there are two solutions to the equation $|ax + b| = k$ determined by $ax + b = k$ or $ax + b = -k$.

Example: The equation $|2x - 1| = 5$ has two solutions.

Symbolic Solution

$2x - 1 = 5$ or $2x - 1 = -5$
$2x = 6$ or $2x = -4$ Add 1.
$x = 3$ or $x = -2$ Divide by 2.

Graphical Solution   **Numerical Solution**

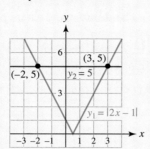

x	-3	-2	-1	0	1	2	3
$\lvert 2x - 1 \rvert$	7	5	3	1	1	3	5

The solutions are -2 and 3.

Absolute Value Inequalities   If the solutions to $\lvert ax + b \rvert = k$ are c and d with $c < d$, then the solution set for $\lvert ax + b \rvert < k$ is $\{x \mid c < x < d\}$, and the solution set for $\lvert ax + b \rvert > k$ is $\{x \mid x < c \text{ or } x > d\}$.

Example: The solutions to the equation $\lvert 2x - 1 \rvert = 5$ are -2 and 3, so the solution set for $\lvert 2x - 1 \rvert < 5$ is $\{x \mid -2 < x < 3\}$, and the solution set for $\lvert 2x - 1 \rvert > 5$ is $\{x \mid x < -2 \text{ or } x > 3\}$.

CHAPTER

8 Review Exercises

SECTION 8.1

Exercises 1–4: Evaluate $f(x)$ for the given values of x.

1. $f(x) = 3x - 1$ $x = -2, \frac{1}{3}$

2. $f(x) = 5 - 3x^2$ $x = -3, 1$

3. $f(x) = \sqrt{x} - 2$ $x = 0, 9$

4. $f(x) = 5$ $x = -5, \frac{7}{5}$

Exercises 5 and 6: Do the following.

  (a) *Write a symbolic representation (formula) for the function described.*
  (b) *Evaluate the function for the input 5 and interpret the result.*

5. Function P computes the number of pints in q quarts.

6. Function f computes 3 less than 4 times a number x.

7. If $f(3) = -2$, then the point _____ lies on the graph of f.

8. If $(4, -6)$ lies on the graph of f, then $f(\text{___}) = $ _____.

Exercises 9–12: Sketch a graph of f.

9. $f(x) = -2x$ **10.** $f(x) = \frac{1}{2}x - \frac{3}{2}$

11. $f(x) = x^2 - 1$ **12.** $f(x) = \sqrt{x + 1}$

Exercises 13 and 14: Use the graph of f to evaluate the given expressions.

13. $f(0)$ and $f(-3)$ **14.** $f(-2)$ and $f(1)$

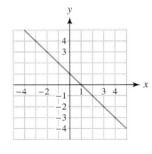

15. Evaluate $f(-1)$ and $f(3)$.

x	-1	1	3	5
$f(x)$	7	3	-1	-5

16. A function f is represented verbally by "Multiply the input x by 3 and then subtract 2." Give numerical, symbolic, and graphical representations for f. Let $x = -3, -2, -1, \ldots, 3$ in the table, and let $-3 \le x \le 3$ for the graph.

Exercises 17 and 18: Use the graph of f to estimate its domain and range.

17. **18.**

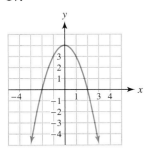

Exercises 19 and 20: Determine whether the graph represents a function.

19. **20.**

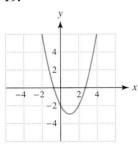

Exercises 21 and 22: Find the domain and range of S. Then state whether S defines a function.

21. $S = \{(-3, 4), (-1, 4), (2, 3), (4, -1)\}$

22. $S = \{(-1, 5), (0, 3), (1, -2), (-1, 2), (2, 4)\}$

Exercises 23–26: Find the domain.

23. $f(x) = -3x + 7$ **24.** $f(x) = \sqrt{x}$

25. $f(x) = \frac{3}{x}$ **26.** $f(x) = x^2 + 2$

SECTION 8.2

Exercises 27 and 28: Determine whether the graph represents a linear function.

27. **28.**

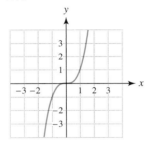
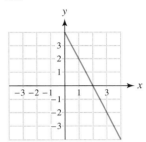

Exercises 29–32: Determine whether f is a linear function. If f is linear, give values for a and b so that f may be expressed as $f(x) = ax + b$.

29. $f(x) = -4x + 5$ **30.** $f(x) = 7 - x$

31. $f(x) = \sqrt{x}$ **32.** $f(x) = 6$

Exercises 33 and 34: Use the table to determine whether $f(x)$ could represent a linear function. If it could, write the formula for f in the form $f(x) = ax + b$.

33.

x	0	2	4	6
$f(x)$	-3	0	3	6

34.

x	-1	0	1	2
$f(x)$	-5	0	10	15

35. Evaluate $f(x) = \frac{1}{2}x + 3$ at $x = -4$.

36. Use the graph to evaluate $f(-2)$ and $f(1)$.

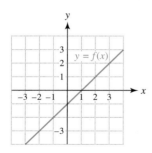

Exercises 37–40: Sketch a graph of $y = f(x)$.

37. $f(x) = x + 1$

38. $f(x) = 1 - 2x$

39. $f(x) = -\frac{1}{3}x$

40. $f(x) = -1$

SECTION 8.3

Use interval notation whenever possible for the remaining exercises.

Exercises 41–44: Solve the compound inequality. Graph the solution set on a number line.

41. $x + 1 \le 3$ and $x + 1 \ge -1$

42. $2x + 7 < 5$ and $-2x \ge 6$

43. $5x - 1 \le 3$ or $1 - x < -1$

44. $3x + 1 > -1$ or $3x + 1 < 10$

45. Use the table to solve $-2 \le 2x + 2 \le 4$.

x	-3	-2	-1	0	1	2	3
$2x + 2$	-4	-2	0	2	4	6	8

46. Use the following figure to solve each equation and inequality.
  (a) $y_1 = y_2$ **(b)** $y_2 = y_3$
  (c) $y_1 \le y_2 \le y_3$
  (d) $y_2 < y_3$

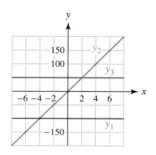

47. The graphs of y_1 and y_2 are shown in the following figure. Solve each equation and inequality.
  (a) $y_1 = y_2$
  (b) $y_1 < y_2$
  (c) $y_1 > y_2$

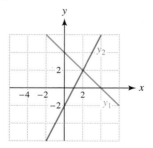

48. The graphs of three linear functions f, g, and h are shown in the following figure. Solve each equation and inequality.
  (a) $f(x) = g(x)$
  (b) $g(x) = h(x)$
  (c) $f(x) < g(x) < h(x)$

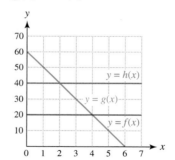

Exercises 49–54: Write the inequality in interval notation.

49. $-3 \le x \le \frac{2}{3}$ **50.** $-6 < x \le 45$

51. $x < \frac{7}{2}$ **52.** $x \ge 1.8$

53. $x > -3$ and $x < 4$ **54.** $x < 4$ or $x > 10$

Exercises 55–60: Solve the three-part inequality. Write the solution set in interval notation.

55. $-4 < x + 1 < 6$ **56.** $20 \le 2x + 4 \le 60$

57. $-3 < 4 - \frac{1}{3}x < 7$ **58.** $2 \le \frac{1}{2}x - 2 \le 12$

59. $-3 \le \dfrac{4 - 5x}{3} - 2 < 3$

60. $30 \le \dfrac{2x - 6}{5} - 4 < 50$

SECTION 8.4

Exercises 61 and 62: Write the domain and the range of the function.

61. $f(t) = \frac{1}{2}t^2$

62. $f(x) = |x + 2|$

63. Write the domain of $f(x) = \frac{x + 1}{2x - 8}$.

64. Write the domain and range of the function shown in the graph.

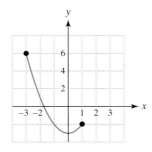

Exercises 65–68: Determine whether $f(x)$ represents a polynomial function. If possible, identify the degree and type of polynomial function.

65. $f(x) = 1 + 2x - 3x^2$

66. $f(x) = 5 + 7x$

67. $f(x) = x^3 + 2x$

68. $f(x) = |2x - 1|$

Exercises 69 and 70: If possible, evaluate $g(t)$ for the given values of t.

69. $g(t) = |1 - 4t|$ $t = 3, t = -\frac{1}{4}$

70. $g(t) = \dfrac{4}{4 - t^2}$ $t = 3, t = -2$

Exercises 71–74: Graph f.

71. $f(x) = |x + 3|$ **72.** $f(x) = x^2 + 1$

73. $f(x) = \dfrac{1}{x}$ **74.** $f(x) = -3x$

SECTION 8.5

75. Use the accompanying table to solve the equation and inequalities.
    (a) $y_1 = 2$ **(b)** $y_1 < 2$ **(c)** $y_1 > 2$

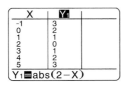

76. Use the graph of $y = |2x + 2|$ to solve the equation and inequalities.
    (a) $|2x + 2| = 4$
    (b) $|2x + 2| \leq 4$
    (c) $|2x + 2| \geq 4$

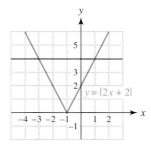

Exercises 77–82: Solve the absolute value equation.

77. $|x| = 22$ **78.** $|2x - 9| = 7$

79. $\left|4 - \frac{1}{2}x\right| = 17$ **80.** $\frac{1}{3}|3x - 1| + 1 = 9$

81. $|2x - 5| = |5 - 3x|$

82. $|-3 + 3x| = |-2x + 6|$

Exercises 83 and 84: Solve each absolute value equation and inequality.

83. **(a)** $|x + 1| = 7$
    (b) $|x + 1| \leq 7$
    (c) $|x + 1| \geq 7$

84. **(a)** $|1 - 2x| = 6$
    (b) $|1 - 2x| \leq 6$
    (c) $|1 - 2x| \geq 6$

Exercises 85–92: Solve the absolute value inequality.

85. $|x| > 3$ **86.** $|-5x| < 20$

87. $|4x - 2| \leq 14$ **88.** $\left|1 - \frac{4}{5}x\right| \geq 3$

89. $|t - 4.5| \leq 0.1$ **90.** $-2|13t - 5| \geq -4$

91. $|5 - 4x| > -5$ **92.** $|2t - 3| \leq 0$

Exercises 93 and 94: Solve the inequality graphically.

93. $|2x| \geq 3$ **94.** $\left|\frac{1}{2}x - 1\right| \leq 2$

APPLICATIONS

95. *Age at First Marriage*   The median age at the first marriage for men from 1890 to 1960 can be modeled by $f(x) = -0.0492x + 119.1$, where x is the year. (*Source:* National Center for Health Statistics.)

(a) Find the median age in 1910.

 (b) Graph f in [1885, 1965, 10] by [22, 26, 1]. What happened to the median age during this time period?

(c) What is the slope of the graph of f? Interpret the slope as a rate of change.

96. *Marriages*   From 1980 to 1997 the number of U.S. marriages in millions could be modeled by the formula $f(x) = 2.4$, where x is the year.

(a) Estimate the number of marriages in 1991.

(b) What information does f give about the number of marriages during this time period?

97. *Fat Grams*   One cup of whole milk contains 8 grams of fat.

(a) Give a formula for $f(x)$ that calculates the number of fat grams in x cups of milk.

(b) What is the slope of the graph of f?

(c) Interpret the slope as a rate of change.

98. *Birth Rate*   The U.S. birth rate per 1000 people from 1990 through 1997 is shown in the table.

Year	1990	1991	1992	1993
Birth Rate	16.7	16.3	16.0	15.7

Year	1994	1995	1996	1997
Birth Rate	15.3	14.8	14.7	14.5

Source: Bureau of the Census.

(a) Make a scatterplot of the data.

(b) Model the data with $f(x) = mx + b$, where x is the year. Answers may vary.

(c) Use f to estimate the birth rate in 2000.

99. *Unhealthy Air Quality*   The Environmental Protection Agency (EPA) monitors air quality in U.S. cities. The function f, represented by the following table, gives the annual number of days with unhealthy air quality in Los Angeles, California, from 1995 through 1999.

x	1995	1996	1997	1998	1999
$f(x)$	113	94	60	56	27

Source: Environmental Protection Agency.

(a) Find $f(1995)$ and interpret your result.

(b) Identify the domain and range of f.

(c) Discuss the trend of air pollution in Los Angeles.

100. *Temperature Scales*   The following table shows equivalent temperatures in degrees Celsius and degrees Fahrenheit.

°C	−40	0	15	35	100
°F	−40	32	59	95	212

(a) Plot the data. Let the x-axis correspond to the Celsius temperature and the y-axis correspond to the Fahrenheit temperature. What type of relation exists between the data?

(b) Find $f(x) = ax + b$ so that f receives the Celsius temperature x as input and outputs the corresponding Fahrenheit temperature. Interpret the slope of the graph of f.

(c) If the temperature is 20°C, what is this temperature in degrees Fahrenheit?

101. *Geometry*   A rectangle is 5 feet longer than twice its width. If the rectangle has a perimeter of 88 feet, what are the dimensions of the rectangle?

102. *Temperature Scales*   The formula

$$F = \frac{9}{5}C + 32$$

may be used to convert Fahrenheit temperature to Celsius temperature. The temperature range at Houghton Lake, Michigan, has varied between −48°F and 107°F. Find this temperature range in Celsius.

103. *Error in Measurements*   A square garden is being fenced along its 160-foot perimeter. If the length L of the fence must be within 1 foot of the garden's perimeter, write an absolute value inequality that gives acceptable values for L. Solve your inequality.

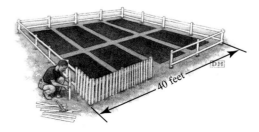

104. *Average Precipitation*   The average rainfall in Houston, Texas, is 3.9 inches per month. Each month's average A is within 1.7 inches of 3.9 inches. (**Source:** J. Williams, *The Weather Almanac 1995.*)

(a) Write an absolute value inequality that models this situation.

(b) Solve the inequality.

105. *Relative Error*   If a quantity is measured to be T and the actual value is A, then the relative error in this measurement is $\left|\frac{T-A}{A}\right|$. If $A = 35$ and the relative error is to be less than 0.08 (8%), what values for T are possible?

CHAPTER 8 Test

1. Evaluate $f(4)$ if $f(x) = 3x^2 - \sqrt{x}$.

2. Sketch a graph of f.

(a) $f(x) = -2x + 1$ (b) $f(x) = x^2 + 1$

3. Use the graph of f to evaluate $f(-3)$ and $f(0)$. Determine the domain and range of f.

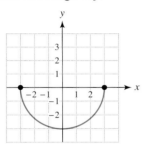

4. A function f is represented verbally by "Square the input x and then subtract 5." Give symbolic, numerical, and graphical representations of f. Let $x = -3, -2, -1, \ldots, 3$ in the numerical representation (table) and let $-3 \le x \le 3$ for the graph.

5. Determine whether the graph represents a function. Explain your reasoning.

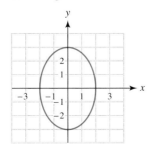

6. Graph the solution set to the compound inequality on a number line.

$$2x + 6 < 2 \text{ and } -3x \ge 3$$

7. Use the following table to solve the compound inequality $-3x < -3$ or $-3x > 6$. Write your answer in interval notation.

x	-3	-2	-1	0	1	2	3
$-3x$	9	6	3	0	-3	-6	-9

8. Use the following figure to solve the equations and inequalities. Write your answers for parts (c) and (d) in interval notation.

(a) $y_1 = y_2$ (b) $y_2 = y_3$

(c) $y_1 \le y_2 \le y_3$ (d) $y_2 < y_3$

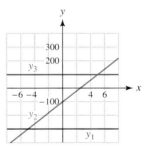

9. Solve the compound inequality

$$-2 < 2 + \frac{1}{2}x < 2$$

and write the solution set in interval notation.

10. Write the domain and range of $f(x) = |2x - 5|$ in interval notation.

11. Determine whether $f(x) = 1 - 2x + x^3$ represents a polynomial function. If possible, identify the degree and type of polynomial function.

12. Evaluate $h(t) = -\frac{4t}{5-t}$ at $t = -2$. Write the domain of h in interval notation.

13. Solve the equation $\left| 2 - \frac{1}{3}x \right| = 6$.

14. Solve the equation and inequalities. Write your answers for parts (b) and (c) in interval notation.
    (a) $|1 - 5x| = 3$ (b) $|1 - 5x| \le 3$
    (c) $|1 - 5x| \ge 3$

15. *Drinking Fluids and Exercise* To determine the number of ounces of fluid that a person should drink in a day, divide his or her weight in pounds by 2 and then add 0.4 ounce for every minute of exercise.
    (a) Write a function that gives the fluid requirements for a person weighing 150 pounds and exercising x minutes a day.

(b) If a 150-pound runner needs 89 ounces of fluid each day, determine the runner's daily minutes of exercise.

16. *Heart Rate of an Athlete* The following table lists the heart rate or pulse of an athlete running a 400-meter race. The race lasts 50 seconds.

Time (seconds)	0	20	30	50
Heart Rate (bpm)	100	134	150	180

(a) Does $P(t) = -\frac{1}{300}t^2 + \frac{53}{30}t + 100$ model the data in the table exactly? Explain.
(b) Does $P(60)$ have significance in this situation? What should be the domain of P?

CHAPTER 8 Extended and Discovery Exercises

1. *Developing a Model* Two identical cylindrical tanks, A and B, each contain 100 gallons of water. Tank A has a pump that begins removing water at a constant rate of 8 gallons per minute. Tank B has a plug removed from its bottom and water begins to flow out—faster at first and then more slowly.
   (a) Assuming that the tanks become empty at the same time, sketch a graph that models the amount of water in each tank. Explain your graphs.
   (b) Which tank is half empty first? Explain.

2. *Modeling Real Data* Per capita personal incomes in the United States from 1990 through 2000 are listed in the following table.

Year	1990	1991	1992	1993
Income	$18,666	$19,091	$20,105	$20,800

Year	1994	1995	1996	2000
Income	$21,809	$23,359	$24,436	$29,676

Source: Department of Commerce.

(a) Make a scatterplot of the data.
(b) Find a function f that models the data. Explain your reasoning.
(c) Use f to estimate per capita personal income in 1998.

3. *Weight of a Small Fish* The figure shows a graph of a function f that models the weight in milligrams of a small fish, *Lebistes reticulatus*, during the first 14 weeks of its life. (*Source:* D. Brown and P. Rothery, *Models in Biology.*)

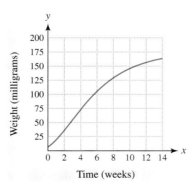

(a) Estimate the weight of the fish when it hatches, at 6 weeks, and at 12 weeks.

(b) If (x_1, y_1) and (x_2, y_2) are points on the graph of a function, the *average rate of change of f from* x_1 *to* x_2 is given by $\frac{y_2 - y_1}{x_2 - x_1}$. Approximate the average rates of change of f from hatching to 6 weeks and from 6 weeks to 12 weeks.

(c) Interpret these rates of change.

(d) During which time period does the fish gain weight the fastest?

4. *Recording Music* A compact disc (CD) can hold approximately 600 million bytes. One *byte* is capable of storing one letter of the alphabet. For example, the word "function" requires 8 bytes to store in computer memory. One million bytes is commonly referred to as a *megabyte* (MB). Recording music requires an enormous amount of memory. The accompanying table lists the megabytes x needed to record y seconds (sec) of music.

x (MB)	0.129	0.231	0.415	0.491
y (sec)	6.010	10.74	19.27	22.83

x (MB)	0.667	1.030	1.160	1.260
y (sec)	31.00	49.00	55.25	60.18

Source: Gateway 2000 System CD.

(a) Make a scatterplot of the data.

(b) What type of relationship seems to exist between x and y? Why does this relationship seem reasonable?

(c) Find the slope–intercept form of a line that models the data. Interpret the slope of this line as a rate of change. Answers may vary.

(d) Check your answer in part (c) by graphing the line and data in the same graph.

(e) Write a linear equation whose solution gives the megabytes needed to record 120 seconds of music.

(f) Solve the equation in part (e) graphically or symbolically.

Systems of Linear Equations

In 1940, a physicist named John Atanasoff at Iowa State University needed to solve 29 equations with 29 variables simultaneously. This task was too difficult to do by hand, so he and a graduate student invented the first fully electronic digital computer. Thus the desire to solve a mathematical problem led to one of the most important inventions of the twentieth century. Today people can solve thousands of equations with thousands of variables. Solutions to such equations have resulted in better airplanes, cars, electronic devices, weather forecasts, and medical equipment.

Equations are also widely used in biology. The following table contains the weight W, neck size N, and chest size C for three black bears. Suppose that park rangers find a bear with a neck size of 22 inches and a chest size of 38 inches. Can they use the data in the table to estimate the bear's weight? Using systems of linear equations, they can answer this question.

W (pounds)	N (inches)	C (inches)
80	16	26
344	28	45
416	31	54
?	22	38

Education is what survives when what has been learned has been forgotten.
—B. F. Skinner

Sources: A. Tucker, *Fundamentals of Computing*; M. Triola, *Elementary Statistics*; Minitab, Inc.

9.1 SYSTEMS OF LINEAR EQUATIONS IN THREE VARIABLES

Basic Concepts · Solving Linear Systems with Substitution and Elimination · Modeling Data

INTRODUCTION

In Chapter 4, we described how to solve systems of linear equations in two variables. In applications linear systems commonly have many variables. Large linear systems are used in the design of electrical circuits, bridges, and ships. They also are used in business, economics, and psychology. Because of the enormous amount of work needed to solve large systems, technology is usually used to find approximate solutions. In this section we discuss symbolic methods for finding solutions of linear systems with three variables. These methods provide the basis for understanding how technology is able to solve large linear systems.

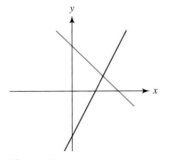

Figure 9.1

BASIC CONCEPTS

When we solve a linear system in two variables, we can express a solution as an ordered pair (x, y). A linear equation in two variables can be represented graphically by a line. A system of two linear equations with a unique solution can be represented graphically by two lines intersecting at a point, as shown in Figure 9.1.

When solving linear systems in three variables, we often use the variables x, y, and z. A solution is expressed as an **ordered triple** (x, y, z), rather than an ordered pair (x, y). For example, if the ordered triple $(1, 2, 3)$ is a solution, $x = 1$, $y = 2$, and $z = 3$ satisfy each equation. A linear equation in three variables can be represented by a flat plane in space. If the solution is unique, we can represent a linear system of three equations in three variables graphically by three planes intersecting at a single point, as illustrated in Figure 9.2.

Critical Thinking

The following figure of three planes in space represents a system of three linear equations in three variables. How many solutions are there? Explain.

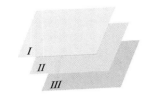

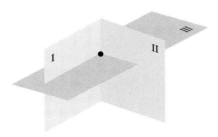

Figure 9.2

The next example shows how to determine whether an ordered triple is a solution.

EXAMPLE 1   Checking for solutions to a system of three equations

Determine whether $(4, 2, -1)$ or $(-1, 0, 3)$ is a solution.

$$2x - 3y + z = 1$$
$$x - 2y + 2z = 5$$
$$2y + z = 3$$

Solution To check $(4, 2, -1)$, substitute $x = 4$, $y = 2$, and $z = -1$ in each equation.

$$2(4) - 3(2) + (-1) \stackrel{?}{=} 1 \qquad \text{True}$$
$$4 - 2(2) + 2(-1) \stackrel{?}{=} 5 \qquad \text{False}$$
$$2(2) + (-1) \stackrel{?}{=} 3 \qquad \text{True}$$

The ordered triple $(4, 2, -1)$ does not satisfy all three equations, so it is not a solution. Next, substitute $x = -1$, $y = 0$, and $z = 3$.

$$2(-1) - 3(0) + 3 \stackrel{?}{=} 1 \qquad \text{True}$$
$$-1 - 2(0) + 2(3) \stackrel{?}{=} 5 \qquad \text{True}$$
$$2(0) + 3 \stackrel{?}{=} 3 \qquad \text{True}$$

The ordered triple $(-1, 0, 3)$ satisfies all three equations, so it is a solution.

In the next example we show how a linear system involving three equations and three variables can be used to model a real-world situation. We solve this system of equations in Example 6.

EXAMPLE 2 Modeling real data with a linear system

The Bureau of Land Management studies antelope populations in Wyoming. It monitors the number of adult antelope, the number of fawns each spring, and the severity of the winter. The first two columns of Table 9.1 contain counts of fawns and adults for three representative winters. The third column shows the severity of each winter. The severity of the winter is measured from 1 to 5, with 1 being mild and 5 being severe.

TABLE 9.1

Fawns (F)	Adults (A)	Winter (W)
405	870	3
414	848	2
272	684	5
?	750	4

We want to use the data in the first three rows of the table to estimate the number of fawns F in the fourth row when the number of adults is 750 and the severity of the winter is 4. To do so, we use the formula

$$F = a + bA + cW,$$

where a, b, and c are constants. Write a system of linear equations whose solution gives appropriate values for a, b, and c.

Solution From the first row in the table, we know that, when $F = 405$, $A = 870$, and $W = 3$, the formula

$$F = a + bA + cW$$

becomes

$$405 = a + b(870) + c(3).$$

Similarly, $F = 414$, $A = 848$, and $W = 2$ gives

$$414 = a + b(848) + c(2),$$

and $F = 272$, $A = 684$, and $W = 5$ yields

$$272 = a + b(684) + c(5).$$

To find values for a, b, and c we can solve the following system of linear equations.

$$405 = a + 870b + 3c$$
$$414 = a + 848b + 2c$$
$$272 = a + 684b + 5c$$

We can also write these equations as a linear system in the following form.

$$a + 870b + 3c = 405$$
$$a + 848b + 2c = 414$$
$$a + 684b + 5c = 272$$

Finding values for a, b, and c will allow us to use the formula $F = a + bA + cW$ to predict the number of fawns F when the number of adults A is 750 and the severity of the winter W is 4 (see Example 6).

Critical Thinking

The model presented in Example 2 considers only two variables that might affect the fawn population. What other factors might be included in the model? If we included these factors, what would happen to the number of variables in the model?

Linear systems of two equations can have no solutions, one solution, or infinitely many solutions. The same is true for larger linear systems. In this section we focus on linear systems having one solution.

SOLVING LINEAR SYSTEMS WITH SUBSTITUTION AND ELIMINATION

When solving systems of linear equations with more than two variables, we usually use both substitution and elimination. However, in the next example we use only substitution to solve a particular type of linear system in three variables.

EXAMPLE 3   Using substitution to solve a linear system of equations

Solve the following system.

$$2x - y + z = 7$$
$$3y - z = 1$$
$$z = 2$$

Solution   Note that the last equation gives us the value of z immediately. We can substitute $z = 2$ into the second equation and determine y.

$$3y - z = 1 \qquad \text{Second equation}$$
$$3y - 2 = 1 \qquad \text{Substitute } z = 2.$$
$$3y = 3 \qquad \text{Add 2 to each side.}$$
$$y = 1 \qquad \text{Divide each side by 3.}$$

Knowing that $y = 1$ and $z = 2$ allows us to find x by using the first equation.

$$2x - y + z = 7 \qquad \text{First equation}$$
$$2x - 1 + 2 = 7 \qquad \text{Let } y = 1 \text{ and } z = 2.$$
$$2x = 6 \qquad \text{Simplify and subtract 1.}$$
$$x = 3 \qquad \text{Divide each side by 2.}$$

Thus $x = 3$, $y = 1$, and $z = 2$ and the solution is $(3, 1, 2)$.

In the next example we use elimination and substitution to solve a system of linear equations. This four-step method is summarized in Putting It All Together at the end of this section.

EXAMPLE 4   **Solving a linear system in three variables**

Solve the following system.

$$x - y + 2z = 6$$
$$2x + y - 2z = -3$$
$$-x - 2y + 3z = 7$$

Solution   **STEP 1**   We begin by eliminating the variable x from the second and third equations. To eliminate x from the second equation we multiply the first equation by -2 and then add it to the second equation. To eliminate x from the third equation we add the first and third equations.

$-2x + 2y - 4z = -12$	First equation times -2		$x - y + 2z = 6$	First equation
$2x + y - 2z = -3$	Second equation		$-x - 2y + 3z = 7$	Third equation
$3y - 6z = -15$	Add.		$-3y + 5z = 13$	Add.

STEP 2   Take the two resulting equations from Step 1 and eliminate either variable. Here we add the two equations to eliminate the variable y.

$$3y - 6z = -15$$
$$-3y + 5z = 13$$
$$-z = -2 \qquad \text{Add the equations.}$$
$$z = 2 \qquad \text{Multiply by } -1.$$

STEP 3   Now we can use substitution to find the values of x and y. We let $z = 2$ in either equation used in Step 2 to find y.

$$3y - 6z = -15$$
$$3y - 6(2) = -15 \qquad \text{Substitute } z = 2.$$
$$3y - 12 = -15 \qquad \text{Multiply.}$$
$$3y = -3 \qquad \text{Add 12.}$$
$$y = -1 \qquad \text{Divide by 3.}$$

STEP 4   Finally, we substitute $y = -1$ and $z = 2$ into any of the given equations to find x.

$$x - y + 2z = 6 \qquad \text{First given equation}$$
$$x - (-1) + 2(2) = 6 \qquad \text{Let } y = -1 \text{ and } z = 2.$$
$$x + 1 + 4 = 6 \qquad \text{Simplify.}$$
$$x = 1 \qquad \text{Subtract 5.}$$

The solution is $(1, -1, 2)$. Check this solution.

In the next example we determine the price of tickets at a play.

EXAMPLE 5 Finding ticket prices

One thousand tickets were sold for a play, which generated $3800 in revenue. The prices of the tickets were $3 for children, $4 for students, and $5 for adults. There were 100 fewer student tickets sold than adult tickets. Find the number of each type of ticket sold.

Solution Let x be the number of tickets sold to children, y be the number of tickets sold to students, and z be the number of tickets sold to adults. The total number of tickets sold was 1000, so

$$x + y + z = 1000.$$

Each child's ticket costs $3, so the revenue generated from selling x tickets would be $3x$. Similarly, the revenue generated from students would be $4y$, and the revenue from adults would be $5z$. Total ticket sales were $3800, so

$$3x + 4y + 5z = 3800.$$

The equation $z - y = 100$ or $y - z = -100$ must also be satisfied, as 100 fewer tickets were sold to students than adults.

To find the price of a ticket we need to solve the following system of linear equations.

$$
\begin{aligned}
x + y + z &= 1000 \\
3x + 4y + 5z &= 3800 \\
y - z &= -100
\end{aligned}
$$

STEP 1 We begin by eliminating the variable x from the second equation. To do so, we multiply the first equation by 3 and subtract the second equation.

$$
\begin{array}{ll}
3x + 3y + 3z = 3000 & \text{First given equation times 3} \\
\underline{3x + 4y + 5z = 3800} & \text{Second equation} \\
-y - 2z = -800 & \text{Subtract.}
\end{array}
$$

STEP 2 We then use the resulting equation from Step 1 and the third equation to eliminate y.

$$
\begin{array}{ll}
-y - 2z = -800 & \text{Equation from Step 1} \\
\underline{y - z = -100} & \text{Third given equation} \\
-3z = -900 & \text{Add the equations.} \\
z = 300 & \text{Divide by } -3.
\end{array}
$$

STEP 3 To find y we can substitute $z = 300$ in the third equation.

$$
\begin{array}{ll}
y - z = -100 & \text{Third given equation} \\
y - 300 = -100 & \text{Let } z = 300. \\
y = 200 & \text{Add 300.}
\end{array}
$$

STEP 4 Finally, we substitute $y = 200$ and $z = 300$ in the first equation.

$$
\begin{array}{ll}
x + y + z = 1000 & \text{First given equation} \\
x + 200 + 300 = 1000 & \text{Let } y = 200 \text{ and } z = 300. \\
x = 500 & \text{Subtract 500.}
\end{array}
$$

Thus 500 tickets were sold to children, 200 to students, and 300 to adults.

MODELING DATA

In the next example we solve the system of equations that we developed to model the data in Example 2.

EXAMPLE 6   Predicting fawns in the spring

Solve the following linear system for a, b, and c. Then use

$$F = a + bA + cW$$

to predict the number of fawns when the number of adults is 750 and the severity of the winter is 4.

$$a + 870b + 3c = 405$$
$$a + 848b + 2c = 414$$
$$a + 684b + 5c = 272$$

Solution **STEP 1** We begin by eliminating the variable a from the second and third equations. To do so, we subtract the second and third equations from the first equation.

$a + 870b + 3c = 405$	First equation	$a + 870b + 3c = 405$	First equation
$a + 848b + 2c = 414$	Second equation	$a + 684b + 5c = 272$	Third equation
$22b + c = -9$	Subtract.	$186b - 2c = 133$	Subtract.

STEP 2 We use the two resulting equations from Step 1 to eliminate c. To do so we multiply $22b + c = -9$ by 2 and add it to the other equation.

$$44b + 2c = -18 \qquad \text{Twice } (22b + c = -9)$$
$$186b - 2c = 133$$
$$\overline{230b = 115} \qquad \text{Add the equations.}$$
$$b = 0.5 \qquad \text{Divide by 230.}$$

STEP 3 To find c we substitute $b = 0.5$ in either equation used in Step 2.

$$44b + 2c = -18$$
$$44(0.5) + 2c = -18 \qquad \text{Let } b = 0.5.$$
$$22 + 2c = -18 \qquad \text{Multiply.}$$
$$2c = -40 \qquad \text{Subtract 22.}$$
$$c = -20 \qquad \text{Divide by 2.}$$

STEP 4 Finally, we substitute $b = 0.5$ and $c = -20$ in any of the given equations to find a.

$$a + 870b + 3c = 405 \qquad \text{First given equation}$$
$$a + 870(0.5) + 3(-20) = 405 \qquad \text{Let } b = 0.5 \text{ and } c = -20.$$
$$a + 435 - 60 = 405 \qquad \text{Multiply.}$$
$$a = 30 \qquad \text{Solve for } a.$$

The solution is $a = 30$, $b = 0.5$, and $c = -20$. Thus we may write

$$F = a + bA + cW$$
$$= 30 + 0.5A - 20W.$$

If there are 750 adults and the winter has a severity of 4, this model predicts

$$F = 30 + 0.5(750) - 20(4)$$
$$= 325 \text{ fawns.}$$

Critical Thinking

Give reasons why the coefficient for A is positive and the coefficient for W is negative in the formula

$$F = 30 + 0.5A - 20W.$$

 9.1 PUTTING IT ALL TOGETHER

In this section we discussed how to solve a system of three linear equations in three variables. Systems of linear equations can have no solutions, one solution, or infinitely many solutions. Our discussion focused on systems that have only one solution. The following table summarizes some of the important concepts presented in this section.

Concept	Explanation
System of Linear Equations in Three Variables	The following is a system of three linear equations in three variables. $$\begin{aligned} x - 2y + z &= 0 \\ -x + y + z &= 4 \\ -y + 4z &= 10 \end{aligned}$$
Solution to a Linear System in Three Variables	The solution to a linear system in three variables is an ordered triple, expressed as (x, y, z). The solution to the preceding system is $(1, 2, 3)$ because substituting $x = 1, y = 2,$ and $z = 3$ in each equation results in a true statement. $$\begin{aligned} (1) - 2(2) + (3) &= 0 \quad \text{True} \\ -(1) + (2) + (3) &= 4 \quad \text{True} \\ -(2) + 4(3) &= 10 \quad \text{True} \end{aligned}$$
Solving a Linear System with Substitution and Elimination	Refer to Example 4. **STEP 1** Eliminate one variable, such as x, from two of the equations. **STEP 2** Use the two resulting equations in two variables to eliminate one of the variables, such as y. Solve for the remaining variable z. **STEP 3** Substitute z in one of the two equations from Step 2. Solve for the unknown variable y. **STEP 4** Substitute values for y and z in one of the given equations and find x. The solution is (x, y, z).

9.1 EXERCISES

FOR EXTRA HELP

 Student's Solutions Manual

 InterAct Math

 MathXL

MyMathLab

 Math Tutor Center

Digital Video Tutor
CD 4   Videotape 12

CONCEPTS

1. Can a system of three linear equations and three variables have two solutions? Explain.

2. Give an example of a system of three linear equations in three variables.

3. Does the ordered triple $(1, 2, 3)$ satisfy the equation $x + y + z = 6$?

4. Does $(3, 4)$ represent a solution to the equation $x + y + z = 7$? Explain.

5. To solve for two variables, how many equations do you usually need?

6. To solve for three variables, how many equations do you usually need?

SOLVING LINEAR SYSTEMS

Exercises 7–10: Determine which ordered triple is a solution to the linear system.

7. $(1, 2, 3), (0, 2, 4)$
$$x + y + z = 6$$
$$x - y - z = -4$$
$$-x - y + z = 0$$

8. $(-1, 0, 2), (0, 4, 4)$
$$2x + y - 3z = -8$$
$$x - 3y + 2z = -4$$
$$3x - 2y + z = -4$$

9. $(1, 0, 3), (-1, 1, 2)$
$$3x - 2y + z = -3$$
$$-x + 3y - 2z = 0$$
$$x + 4y + 2z = 7$$

10. $\left(\frac{1}{2}, \frac{3}{2}, -\frac{1}{2}\right), (-1, 0, -2)$
$$x + 3y - 4z = 7$$
$$-x + 5y + 3z = \frac{11}{2}$$
$$3x - 2y - 7z = 2$$

Exercises 11–16: (Refer to Example 3.) Use substitution to solve the system of linear equations. Check your solution.

11. $x + y - z = 1$
$$2y + z = -1$$
$$z = 1$$

12. $2x + y - 3z = 1$
$$y + 4z = 0$$
$$z = -1$$

13. $-x - 3y + z = -2$
$$2y + 3z = 3$$
$$z = 2$$

14. $3x + 2y - 3z = -4$
$$-y + 2z = 4$$
$$z = 0$$

15. $a - b + 2c = 3$
$$-3b + c = 4$$
$$c = -2$$

16. $5a + 2b - 3c = 10$
$$5b - 2c = -4$$
$$c = 3$$

Exercises 17–34: (Refer to Example 4.) Use elimination and substitution to solve the system of linear equations.

17. $x + y - z = 11$
$$-x + 2y + 3z = -1$$
$$2z = 4$$

18. $x + 2y - 3z = -7$
$$-2x + y + z = -1$$
$$3z = 9$$

19. $x + y - z = -2$
$$-x + z = 1$$
$$y + 2z = 3$$

20. $x + y - 3z = 11$
$$-2x + y + 2z = 1$$
$$-3y + 3z = -21$$

21. $x + y - 2z = -7$
$$y + z = -1$$
$$-y + 3z = 9$$

22. $2x + 3y + z = 5$
$$y + 2z = 4$$
$$-2y + z = 2$$

23. $x + 2y + 2z = 1$
$$x + y + z = 0$$
$$-x - 2y + 3z = -11$$

24. $x + y - z = 0$
$$x - 3y + z = -2$$
$$x - y + 3z = 8$$

25. $x + y + z = 5$
$$y + z = 6$$
$$x + z = 3$$

26. $x + y + z = 0$
$$x - y - z = 6$$
$$-x + y - z = 4$$

27. $x + 2y + 3z = 24$
$$-x + y + 2z = 1$$
$$x + y - 2z = 9$$

28. $5x - 15y + z = 22$
$$-10x + 12y - 2z = -8$$
$$4x - 2y - 3z = 9$$

29. $2x + y + z = 3$
$$2x - y - z = 9$$
$$x + y - z = 0$$

30. $x + 3y + z = -8$
$$x - 2y = 11$$
$$2y - z = -16$$

31. $2x + 6y - 2z = 47$
$$2x + y + 3z = -28$$
$$-x + y + z = -\frac{7}{2}$$

32. $x + y + 2z = 23$
$$3x - y + 3z = 8$$
$$2x + 2y + z = 13$$

33. $x + 3y - 4z = \frac{13}{2}$
$$-2x + 3y - z = \frac{1}{2}$$
$$3x + z = 4$$

34. $x - 2y + z = \frac{9}{2}$
$$4x - y + 3z = 9$$
$$x + 2y = -\frac{3}{2}$$

APPLICATIONS

35. *Finding Costs*   The accompanying table shows the costs of purchasing different combinations of hamburgers, fries, and soft drinks.

Hamburgers	Fries	Soft Drinks	Total Cost
1	2	4	$10
1	4	6	$15
0	3	2	$6

(a) Let x be the cost of a hamburger, y be the cost of fries, and z be the cost of a soft drink. Write a system of three linear equations that represents the data in the table.

(b) Solve the system of linear equations and interpret your answer.

36. *Cost of CDs*   The accompanying table shows the total cost of purchasing various combinations of differently priced CDs. The types of CDs are labeled A, B, and C.

A	B	C	Total Cost
1	1	1	$37
3	2	1	$69
1	1	4	$82

(a) Let x be the cost of a CD of type A, y be the cost of a CD of type B, and z be the cost of a CD of type C. Write a system of three linear equations that represents the data in the table.

(b) Solve this linear system of equations and interpret your answer.

37. *Geometry*   The largest angle in a triangle is $55°$ more than the smallest angle. The sum of the measures of the two smaller angles is $10°$ more than the measure of the largest angle.

(a) Let x, y, and z be the measures of the three angles from largest to smallest. Write a system of three linear equations whose solution gives the measure of each angle. (*Hint:* The sum of the measures of the angles equals $180°$.)

(b) Solve the system by using elimination and substitution.

(c) Check your solution.

38. *Geometry*   The perimeter of a triangle is 90 inches. The longest side is 20 inches longer than the shortest side and 10 inches longer than the remaining side.

(a) Let x, y, and z be the lengths of the three sides from largest to smallest. Write a system of three linear equations whose solution gives the lengths of each side.

(b) Solve the system by using elimination and substitution.

(c) Check your solution.

39. *Mixture Problem*   One type of lawn fertilizer consists of a mixture of nitrogen, N, phosphorus, P, and potassium, K. An 80-pound sample contains 8 more pounds of nitrogen and phosphorus than potassium. There is 9 times as much potassium as phosphorus.

(a) Write a system of three equations whose solution gives the amount of nitrogen, phosphorus, and potassium in this sample.

(b) Solve the system of equations.

40. *Business Production*   A business has three machines that manufacture containers. Together they make 100 containers per day, whereas the two fastest machines can make 80 containers per day. The fastest machine makes 34 more containers per day than the slowest machine.

(a) Let x, y, and z be the number of containers that the machines make from fastest to slowest. Write a system of three equations whose solution gives the number of containers each machine can make.

(b) Solve the system of equations.

41. *Predicting Fawns*   (Refer to Examples 2 and 6.) The accompanying table shows counts for fawns and adult deer and the severity of the winter. These data may be modeled by the equation $F = a + bA + cW$.

Fawns (F)	Adults (A)	Winter (W)
525	600	4
365	400	2
805	900	5
?	500	3

(a) Use the first three rows to write a system of three linear equations in three variables whose solution gives values for a, b, and c.

(b) Solve this system of linear equations.

(c) Predict the number of fawns when there are 500 adults and the winter has severity 3.

42. *Predicting Home Prices* Selling prices of homes can depend on several factors such as size and age. The accompanying table shows the selling price for three homes. In this table, price P is given in thousands of dollars, age A in years, and home size S in thousands of square feet. These data may be modeled by the equation $P = a + bA + cS$.

Price (P)	Age (A)	Size (S)
190	20	2
320	5	3
50	40	1

(a) Write a system of linear equations whose solution gives a, b, and c.

(b) Solve this system of linear equations.

(c) Predict the price of a home that is 10 years old and has 2500 square feet.

43. *Investment Mixture* A sum of $30,000 was invested in three mutual funds. In one year the first fund grew by 8%, the second by 10%, and the third by 15%. Total earnings were $3550. The amount invested in the third fund was $2000 less than the combined amount invested in the other two funds. Use a linear system of equations to determine the amount invested in each fund.

44. *Football Tickets* A total of 2500 tickets were sold at a football game. Prices were $2 for children, $3 for students, and $5 for adults. Twice as many tickets were sold to students as children and ticket revenues were $7250. Use a system of linear equations to determine how many of each type of ticket were sold.

WRITING ABOUT MATHEMATICS

45. In the previous section we solved problems with two variables; to obtain a unique solution we needed two linear equations. In this section we solved problems with three variables; to obtain a unique solution we needed three equations. Try to generalize these results. In the design of aircraft, problems commonly involve 100,000 variables. How many equations are required to solve such problems? Can such problems be solved by hand? If not, how are such problems solved? Explain your answers.

46. In Exercise 42 the price of a home was estimated by its age and size. What other factors might affect the price of a home? Explain how these factors might affect the number of variables and equations in the linear system.

Group Activity: Working with Real Data

Directions: Form a group of 2 to 4 people. Select someone to record the group's responses for this activity. All of the members of the group should work cooperatively to answer the questions. If your instructor asks for your results, each member of the group should be prepared to respond.

CEO Salaries In 2002, hourly wages for the top three CEOs (chief executive officers) of U.S. corporations were calculated, based on working 14 hours per day for 365 days. Together the three CEOs made $44,000 per hour. The top CEO earned $2,000 more per hour than the combined hourly wages of the next two top CEOs, and the top CEO earned $7,000 more per hour than the next top CEO. (*Source: USA Today*, March 31, 2003.)

(a) Let x, y, and z be the hourly wages in *thousands* of dollars for the top CEOs from highest to lowest. Write a system of equations whose solution gives these hourly wages.

(b) Solve the system. Interpret the answer.

(c) The average American worker earned $16.23 per hour in 2002. How many hours does the average American work to earn an amount equal to one hour of work by the top CEO in 2002?

9.2 MATRIX SOLUTIONS OF LINEAR SYSTEMS

Representing Systems of Linear Equations with Matrices · **Gaussian Elimination** · **Using Technology to Solve Systems of Linear Equations (Optional)**

INTRODUCTION

Suppose that the size of a bear's head and its overall length are known. Can its weight be estimated from these variables? Can a bear's weight be estimated if its neck size and chest size are known? In this section we show that systems of linear equations can be used to make such estimates.

In the previous section we solved systems of three linear equations in three variables by using elimination and substitution. In real life, systems of equations often contain thousands of variables. To solve a large system of equations, we need an efficient method. Long before the invention of the computer, Carl Fredrich Gauss (1777–1855) developed a method called *Gaussian elimination* to solve systems of linear equations. Even though it was developed more than 150 years ago, it is still used today in modern computers and calculators. In this section we introduce this method.

REPRESENTING SYSTEMS OF LINEAR EQUATIONS WITH MATRICES

Arrays of numbers are used frequently in many different real-world situations. Spreadsheets often make use of arrays. A **matrix** is a rectangular array of numbers. Each number in a matrix is called an **element**. The following are examples of *matrices* (plural of matrix), with their dimensions written below them.

$$\begin{bmatrix} 2 & 0 \\ 3 & 1 \end{bmatrix} \quad \begin{bmatrix} -1.2 & 5 & 0 \\ 1 & 0 & 1 \\ 4 & -5 & 7 \end{bmatrix} \quad \begin{bmatrix} 3 & -6 & 0 & \sqrt{3} \\ 1 & 4 & 0 & 9 \\ -3 & 1 & 1 & 18 \\ -10 & -4 & 5 & -1 \end{bmatrix} \quad \begin{bmatrix} 4 & 2 \\ 0 & 1 \\ 1 & 0 \end{bmatrix} \quad \begin{bmatrix} 1 & 5 & -1 \\ 3 & 4 & 2 \end{bmatrix}$$

$$\boxed{2 \times 2} \qquad \boxed{3 \times 3} \qquad\qquad \boxed{4 \times 4} \qquad\qquad \boxed{3 \times 2} \qquad \boxed{2 \times 3}$$

$$\boxed{\text{rows} \times \text{columns}}$$

The dimension of a matrix is stated much like the dimensions of a rectangular room. We might say that a room is n feet long and m feet wide. Similarly, the **dimension of a matrix** is $n \times m$ (n by m), if it has n rows and m columns. For example, the last matrix in the preceding group has a dimension of 2×3 because it has 2 rows and 3 columns. If the number of rows and columns are equal, the matrix is a **square matrix**. The first three matrices in that group are square matrices.

Matrices can be used to represent a system of linear equations. For example, if we have the system of equations

$$3x - y + 2z = 7$$
$$x - 2y + z = 0$$
$$2x + 5y - 7z = -9,$$

we can represent the system with the following **augmented matrix**. Note how the coefficients of the variables were placed in the matrix. A vertical line is positioned in the matrix where the equals signs occur in the system of equations. The rows and columns are labeled, and the **main diagonal** of the augmented matrix is circled. The matrix has dimension 3×4.

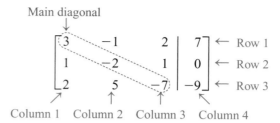

Main diagonal

$$\begin{bmatrix} 3 & -1 & 2 & | & 7 \\ 1 & -2 & 1 & | & 0 \\ 2 & 5 & -7 & | & -9 \end{bmatrix}$$

← Row 1
← Row 2
← Row 3

Column 1   Column 2   Column 3   Column 4

EXAMPLE 1   **Representing a linear system**

Represent each linear system with an augmented matrix. State the dimension of the matrix.

(a) $x - 2y = 9$
$6x + 7y = 16$

(b) $x - 3y + 7z = 4$
$2x + 5y - z = 15$
$2x + y = 8$

Solution   **(a)** This system can be represented by the following 2×3 matrix.

$$\begin{bmatrix} 1 & -2 & | & 9 \\ 6 & 7 & | & 16 \end{bmatrix}$$

(b) This system has three equations in three variables and can be represented by the following 3×4 matrix.

$$\begin{bmatrix} 1 & -3 & 7 & | & 4 \\ 2 & 5 & -1 & | & 15 \\ 2 & 1 & 0 & | & 8 \end{bmatrix}$$

GAUSSIAN ELIMINATION

A convenient matrix form for representing a system of linear equations is **reduced row–echelon form**. The following matrices are examples of reduced row–echelon form. Note that there are 1s on the main diagonal with 0s above and below the 1s.

$$\begin{bmatrix} 1 & 0 & | & 3 \\ 0 & 1 & | & -2 \end{bmatrix} \quad \begin{bmatrix} 1 & 0 & 0 & | & 3 \\ 0 & 1 & 0 & | & 1 \\ 0 & 0 & 1 & | & -1 \end{bmatrix} \quad \begin{bmatrix} 1 & 0 & 0 & | & 8 \\ 0 & 1 & 0 & | & 2 \\ 0 & 0 & 1 & | & 3 \end{bmatrix}$$

If an augmented matrix representing a linear system is in reduced row–echelon form, we can usually determine the solution easily.

EXAMPLE 2   **Determining a solution from reduced row–echelon form**

Each matrix represents a system of linear equations. Find the solution.

(a) $\begin{bmatrix} 1 & 0 & 0 & | & 2 \\ 0 & 1 & 0 & | & -3 \\ 0 & 0 & 1 & | & 5 \end{bmatrix}$ **(b)** $\begin{bmatrix} 1 & 0 & | & 10 \\ 0 & 1 & | & -4 \end{bmatrix}$

Solution (a) The top row represents $1x + 0y + 0z = 2$ or $x = 2$. The second and third rows tell us that $y = -3$ and $z = 5$. The solution is $(2, -3, 5)$.

(b) The system involves two equations in two variables. The solution is $(10, -4)$. _____

We can use a numerical method called **Gaussian elimination** to solve a linear system. It makes use of the following matrix row transformations.

MATRIX ROW TRANSFORMATIONS

For any augmented matrix representing a system of linear equations, the following row transformations result in an equivalent system of linear equations.

1. Any two rows may be interchanged.
2. The elements of any row may be multiplied by a nonzero constant.
3. Any row may be changed by adding to (or subtracting from) its elements a multiple of the corresponding elements of another row.

Gaussian elimination can be used to transform an augmented matrix into reduced row–echelon form. Its objective is to use these matrix row transformations to obtain a matrix that has the following reduced row–echelon form, where (a, b) represents the solution.

$$\begin{bmatrix} 1 & 0 & | & a \\ 0 & 1 & | & b \end{bmatrix}$$

This method is illustrated in the next example.

EXAMPLE 3   Transforming a matrix into reduced row–echelon form

Use Gaussian elimination to transform the augmented matrix of the linear system into reduced row–echelon form. Find the solution.

$$x + y = 5$$
$$-x + y = 1$$

Solution   Both the linear system and the augmented matrix are shown.

$$\begin{array}{cc} \textit{Linear System} & \textit{Augmented Matrix} \end{array}$$

$$\begin{array}{cc} \begin{aligned} x + y &= 5 \\ -x + y &= 1 \end{aligned} & \begin{bmatrix} 1 & 1 & | & 5 \\ -1 & 1 & | & 1 \end{bmatrix} \end{array}$$

First, we want to obtain a 0 in the second row, where the -1 is highlighted. To do so, we add row 1 to row 2 and place the result in row 2. This step is denoted $R_2 + R_1$ and eliminates the x-variable from the second equation.

$$\begin{array}{ccc} \begin{aligned} x + y &= 5 \\ 2y &= 6 \end{aligned} & R_2 + R_1 \rightarrow & \begin{bmatrix} 1 & 1 & | & 5 \\ 0 & 2 & | & 6 \end{bmatrix} \end{array}$$

To obtain a 1 where the 2 in the second row is located, we divide the second row by 2, denoted $\frac{R_2}{2}$.

$$\begin{aligned} x + y &= 5 \\ y &= 3 \end{aligned} \qquad \frac{R_2}{2} \rightarrow \begin{bmatrix} 1 & \boxed{1} & 5 \\ 0 & 1 & 3 \end{bmatrix}$$

Next, we need to obtain a 0 where the 1 is highlighted. We do so by subtracting row 2 from row 1 and placing the result in row 1, denoted $R_1 - R_2$.

$$\begin{aligned} x &= 2 \\ y &= 3 \end{aligned} \qquad R_1 - R_2 \rightarrow \begin{bmatrix} 1 & 0 & 2 \\ 0 & 1 & 3 \end{bmatrix}$$

This matrix is in reduced row–echelon form. The solution is $(2, 3)$.

In the next example, we use Gaussian elimination to solve a system with three linear equations and three variables. To do so we transform the matrix into the following reduced row–echelon form, where (a, b, c) represents the solution.

$$\begin{bmatrix} 1 & 0 & 0 & a \\ 0 & 1 & 0 & b \\ 0 & 0 & 1 & c \end{bmatrix}$$

Several calculations are involved in transforming the system to reduced row–echelon form.

EXAMPLE 4 Transforming a matrix to reduced row–echelon form

Use Gaussian elimination to transform the augmented matrix of the linear system into reduced row–echelon form. Find the solution.

$$\begin{aligned} x + y + 2z &= 1 \\ -x + + z &= -2 \\ 2x + y + 5z &= -1 \end{aligned}$$

Solution The linear system and the augmented matrix are both shown.

Linear System	**Augmented Matrix**

$$\begin{aligned} x + y + 2z &= 1 \\ -x + z &= -2 \\ 2x + y + 5z &= -1 \end{aligned} \qquad \begin{bmatrix} 1 & 1 & 2 & 1 \\ -1 & 0 & 1 & -2 \\ 2 & 1 & 5 & -1 \end{bmatrix}$$

First, we want to put 0s in the second and third rows, where the -1 and 2 are highlighted. To obtain a 0 in the first position of the second row we add row 1 to row 2 and place the result in row 2, denoted $R_2 + R_1$. To obtain a 0 in the first position of the third row we subtract 2 times row 1 from row 3 and place the result in row 3, denoted $R_3 - 2R_1$. Row 1 does not change. These steps eliminate the x-variable from the second and third equations.

$$\begin{aligned} x + y + 2z &= 1 \\ y + 3z &= -1 \\ -y + z &= -3 \end{aligned} \qquad \begin{matrix} \\ R_2 + R_1 \rightarrow \\ R_3 - 2R_1 \rightarrow \end{matrix} \begin{bmatrix} 1 & \boxed{1} & 2 & 1 \\ 0 & 1 & 3 & -1 \\ 0 & \boxed{-1} & 1 & -3 \end{bmatrix}$$

To eliminate the y-variable in row 1, we subtract row 2 from row 1. To eliminate the y-variable from row 3, we add row 2 to row 3.

$$\begin{array}{rl} x \quad - \quad z = & 2 \\ y + 3z = & -1 \\ 4z = & -4 \end{array} \qquad \begin{array}{l} R_1 - R_2 \to \\ \\ R_3 + R_2 \to \end{array} \left[\begin{array}{ccc|c} 1 & 0 & -1 & 2 \\ 0 & 1 & 3 & -1 \\ 0 & 0 & 4 & -4 \end{array}\right]$$

To obtain a 1 in row 3, where the highlighted 4 is located, we divide row 3 by 4.

$$\begin{array}{rl} x \quad - \quad z = & 2 \\ y + 3z = & -1 \\ z = & -1 \end{array} \qquad \begin{array}{l} \\ \\ \dfrac{R_3}{4} \to \end{array} \left[\begin{array}{ccc|c} 1 & 0 & -1 & 2 \\ 0 & 1 & 3 & -1 \\ 0 & 0 & 1 & -1 \end{array}\right]$$

To transform the matrix into reduced row–echelon form, we need to put 0s in the highlighted locations. To do so we first add row 3 to row 1 and then subtract 3 times row 3 from row 2.

$$\begin{array}{rl} x = & 1 \\ y = & 2 \\ z = & -1 \end{array} \qquad \begin{array}{l} R_1 + R_3 \to \\ R_2 - 3R_3 \to \\ \\ \end{array} \left[\begin{array}{ccc|c} 1 & 0 & 0 & 1 \\ 0 & 1 & 0 & 2 \\ 0 & 0 & 1 & -1 \end{array}\right]$$

This matrix is now in reduced row–echelon form. The solution is $(1, 2, -1)$.

Critical Thinking

An *inconsistent* system of linear equations has no solutions, and a *dependent* system of linear equations has infinitely many solutions. Suppose that an augmented matrix row reduces to either of the following matrices. Explain what each matrix indicates about the given system of linear equations.

$$\left[\begin{array}{ccc|c} 1 & 0 & 0 & 2 \\ 0 & 1 & 0 & 3 \\ 0 & 0 & 0 & 1 \end{array}\right] \qquad \left[\begin{array}{ccc|c} 1 & 0 & 0 & 2 \\ 0 & 1 & 2 & 3 \\ 0 & 0 & 0 & 0 \end{array}\right]$$

In the next example we find the amounts invested in three different mutual funds.

EXAMPLE 5   Determining investment amounts

A total of $8000 was invested in three mutual funds that grew at a rate of 5%, 10%, and 20% over 1 year. After 1 year, the combined value of the three funds had grown by $1200. Five times as much money was invested at 20% as at 10%. Find the amount invested in each fund.

Solution   Let x be the amount invested at 5%, y be the amount invested at 10%, and z be the amount invested at 20%. The total amount invested was $8000, so

$$x + y + z = 8000.$$

The growth in the first mutual fund, paying 5% of x, is given by $0.05x$. Similarly, the growths in the other mutual funds are given by $0.10y$ and $0.20z$. As the total growth was $1200, we can write

$$0.05x + 0.10y + 0.20z = 1200.$$

Multiplying this equation by 20 to eliminate decimals results in

$$x + 2y + 4z = 24,000.$$

Five times as much was invested at 20% as at 10%, so $z = 5y$, which can be written as $5y - z = 0$.

These three equations can be written as a system of linear equations and as an augmented matrix.

Linear System | **Augmented Matrix**

$$
\begin{aligned}
x + y + z &= 8{,}000 \\
x + 2y + 4z &= 24{,}000 \\
5y - z &= 0
\end{aligned}
\qquad
\left[\begin{array}{ccc|c}
1 & 1 & 1 & 8{,}000 \\
1 & 2 & 4 & 24{,}000 \\
0 & 5 & -1 & 0
\end{array}\right]
$$

A 0 can be obtained in the highlighted positions by subtracting row 1 from row 2.

$$
\begin{aligned}
x + y + z &= 8{,}000 \\
y + 3z &= 16{,}000 \\
5y - z &= 0
\end{aligned}
\qquad
R_2 - R_1 \rightarrow
\left[\begin{array}{ccc|c}
1 & 1 & 1 & 8{,}000 \\
0 & 1 & 3 & 16{,}000 \\
0 & 5 & -1 & 0
\end{array}\right]
$$

Zeros can be obtained in the highlighted position by subtracting row 2 from row 1 and by subtracting 5 times row 2 from row 3.

$$
\begin{aligned}
x - z &= -8{,}000 \\
y + 3z &= 16{,}000 \\
-16z &= -80{,}000
\end{aligned}
\qquad
\begin{aligned}
R_1 - R_2 \rightarrow \\
\\
R_3 - 5R_2 \rightarrow
\end{aligned}
\left[\begin{array}{ccc|c}
1 & 0 & -2 & -8{,}000 \\
0 & 1 & 3 & 16{,}000 \\
0 & 0 & -16 & -80{,}000
\end{array}\right]
$$

To obtain a 1 in the highlighted position, divide row 3 by -16.

$$
\begin{aligned}
x - 2z &= -8{,}000 \\
y + 3z &= 16{,}000 \\
z &= 5{,}000
\end{aligned}
\qquad
\begin{aligned}
\\
\\
\dfrac{R_3}{-16} \rightarrow
\end{aligned}
\left[\begin{array}{ccc|c}
1 & 0 & -2 & -8{,}000 \\
0 & 1 & 3 & 16{,}000 \\
0 & 0 & 1 & 5{,}000
\end{array}\right]
$$

To obtain a 0 in each of the highlighted positions, add twice row 3 to row 1 and subtract three times row 3 from row 2.

$$
\begin{aligned}
x &= 2{,}000 \\
y &= 1{,}000 \\
z &= 5{,}000
\end{aligned}
\qquad
\begin{aligned}
R_1 + 2R_3 \rightarrow \\
R_2 - 3R_3 \rightarrow \\
\end{aligned}
\left[\begin{array}{ccc|c}
1 & 0 & 0 & 2{,}000 \\
0 & 1 & 0 & 1{,}000 \\
0 & 0 & 1 & 5{,}000
\end{array}\right]
$$

Thus $2000 was invested at 5%, $1000 at 10%, and $5000 at 20%.

USING TECHNOLOGY TO SOLVE SYSTEMS OF LINEAR EQUATIONS (OPTIONAL)

Examples 4 and 5 involve performing a lot of arithmetic. Trying to solve a large system of equations by hand is an enormous—if not impossible—task. In the real world, people use technology to solve large systems. Many types of graphing calculators can solve systems of linear equations.

In the next example we solve the linear systems from Examples 3 and 4 with a graphing calculator.

EXAMPLE 6   Using technology

Use a graphing calculator to solve the following systems of equations.

(a) $x + y = 5$ (b) $x + y + 2z = 1$
      $-x + y = 1$ $-x + z = -2$
                            $2x + y + 5z = -1$

Solution (a) Enter the 2×3 augmented matrix from Example 3 in a graphing calculator, as shown in Figure 9.3(a). Then transform the matrix into reduced row–echelon form (rref), as shown in Figure 9.3(b). The solution is $(2, 3)$.

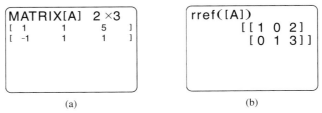

(a)   (b)

Figure 9.3

(b) Enter the 3×4 augmented matrix from Example 4 in a graphing calculator, as shown in Figure 9.4(a). (The fourth column of A can be seen by scrolling right.) Then transform the matrix into reduced row–echelon form (rref), as shown in Figure 9.4(b). The solution is $(1, 2, -1)$.

Calculator Help
To enter a matrix and put it in reduced row–echelon form, see Appendix A (pages AP-8 and AP-9).

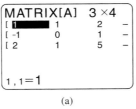

(a)   (b)

Figure 9.4

In the next example we use technology to solve an application.

EXAMPLE 7   Modeling the weight of male bears

The data shown in Table 9.2 give the weight W, head length H, and overall length L of three bears. These data can be modeled with the equation $W = a + bH + cL$, where a, b, and c are constants that we need to determine. (*Sources:* M. Triola, *Elementary Statistics*; Minitab, Inc.)

TABLE 9.2

W (pound)	H (inches)	L (inches)
362	16	72
300	14	68
147	11	52

(a) Set up a system of equations whose solution gives values for constants a, b, and c.

(b) Solve the system.

(c) Predict the weight of a bear with head length $H = 13$ inches and overall length $L = 65$ inches.

Solution **(a)** Substitute each row of the data in the equation $W = a + bH + cL$.

$$362 = a + b(16) + c(72)$$
$$300 = a + b(14) + c(68)$$
$$147 = a + b(11) + c(52)$$

Rewrite this system as

$$a + 16b + 72c = 362$$
$$a + 14b + 68c = 300$$
$$a + 11b + 52c = 147$$

and represent it as the augmented matrix

$$A = \begin{bmatrix} 1 & 16 & 72 & | & 362 \\ 1 & 14 & 68 & | & 300 \\ 1 & 11 & 52 & | & 147 \end{bmatrix}.$$

(b) Enter A in a graphing calculator and put it in reduced row–echelon form, as shown in Figures 9.5(a) and (b), respectively. The solution is $a = -374$, $b = 19$, and $c = 6$.

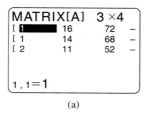

MATRIX[A] 3 ×4
```
[ 1       16      72    –
[ 1       14      68    –
[ 2       11      52    –


1 , 1 = 1
```
(a)

```
rref([A])
   [[1 0 0 -374]
    [0 1 0  19  ]
    [0 0 1  6   ]]
```
(b)

Figure 9.5

(c) For $W = a + bH + cL$, use

$$W = -374 + 19H + 6L$$

to predict the weight of a bear with head length $H = 13$ and overall length $L = 65$. Thus

$$W = -374 + 19(13) + 6(65)$$
$$= 263 \text{ pounds.}$$

9.2 PUTTING IT ALL TOGETHER

A matrix is a rectangular array of numbers. An augmented matrix may be used to represent a system of linear equations. One common method for solving a system of linear equations is Gaussian elimination. Matrix row operations may be used to transform an augmented matrix to reduced row–echelon form. Technology can be used to solve systems of linear equations efficiently. The following table summarizes augmented matrices and reduced row–echelon form.

Concept	Explanation
Augmented Matrix	A linear system can be represented by an augmented matrix. The following matrix has dimension 3×4. $\quad\quad$ **Linear System** $\quad\quad$ **Augmented Matrix** $\begin{aligned} x + 2y - z &= 6 \\ -2x + y - z &= 7 \\ 2x \quad\quad + 3z &= -11 \end{aligned}$ $\quad\quad$ $\begin{bmatrix} 1 & 2 & -1 & 6 \\ -2 & 1 & -1 & 7 \\ 2 & 0 & 3 & -11 \end{bmatrix}$
Reduced Row–Echelon Form	The following augmented matrix is in reduced row–echelon form, which results from transforming the preceding system to reduced row–echelon form. There are 1s along the main diagonal and 0s elsewhere in the first three columns. The solution to the linear system is $(-1, 2, -3)$. $\begin{bmatrix} 1 & 0 & 0 & -1 \\ 0 & 1 & 0 & 2 \\ 0 & 0 & 1 & -3 \end{bmatrix}$

9.2 EXERCISES

FOR EXTRA HELP

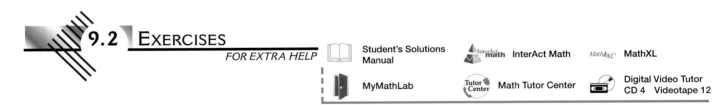

Student's Solutions Manual   InterAct Math   MathXL
MyMathLab   Math Tutor Center   Digital Video Tutor CD 4 Videotape 12

CONCEPTS

1. What is a matrix?

2. Give an example of a matrix and state its dimension.

3. Give an example of an augmented matrix and state its dimension.

4. If an augmented matrix is used to solve a system of three linear equations in three variables, what will be its dimension?

5. Give an example of a matrix in reduced row–echelon form.

6. Identify the elements on the main diagonal in the augmented matrix

$$\begin{bmatrix} 4 & -6 & -1 & 3 \\ 6 & 2 & -2 & 9 \\ 7 & 5 & -3 & 1 \end{bmatrix}.$$

DIMENSIONS OF MATRICES AND AUGMENTED MATRICES

Exercises 7–10: State the dimension of the matrix.

7. $\begin{bmatrix} 3 & -3 & 7 \\ 2 & 6 & -2 \\ 4 & 2 & 5 \end{bmatrix}$

8. $\begin{bmatrix} -2 & 3 & 0 \\ 1 & -8 & 4 \end{bmatrix}$

9. $\begin{bmatrix} 1 & 7 \\ 0 & 2 \\ 2 & -5 \end{bmatrix}$

10. $\begin{bmatrix} 4 & 2 & -3 & -1 \\ 4 & -3 & 2 & -7 \\ 14 & 6 & 4 & 0 \end{bmatrix}$

Exercises 11–14: Represent the linear system as an augmented matrix.

11. $x - 3y = 1$
$-x + 3y = -1$

12. $4x + 2y = -5$
$5x + 8y = 2$

13. $2x - y + 2z = -4$
$x - 2y = 2$
$-x + y - 2z = -6$

14. $3x - 2y + z = 5$
$-x + 2z = -4$
$x - 2y + z = -1$

Exercises 15–20: Write the system of linear equations that the augmented matrix represents. Use the variables x, y, and z.

15. $\left[\begin{array}{cc|c} 1 & 2 & -6 \\ 5 & -1 & 4 \end{array}\right]$

16. $\left[\begin{array}{cc|c} 1 & -5 & 7 \\ 0 & -3 & 6 \end{array}\right]$

17. $\left[\begin{array}{ccc|c} 1 & -1 & 2 & 6 \\ 2 & 1 & -2 & 1 \\ -1 & 2 & -1 & 3 \end{array}\right]$

18. $\left[\begin{array}{ccc|c} 3 & -1 & 2 & -1 \\ 2 & -2 & 2 & 4 \\ 1 & 7 & -2 & 2 \end{array}\right]$

19. $\left[\begin{array}{ccc|c} 1 & 0 & 0 & 4 \\ 0 & 1 & 0 & -2 \\ 0 & 0 & 1 & 7 \end{array}\right]$

20. $\left[\begin{array}{ccc|c} 1 & 0 & 0 & 6 \\ 0 & 1 & 0 & -2 \\ 0 & 0 & 1 & 4 \end{array}\right]$

GAUSSIAN ELIMINATION

Exercises 21–38: Use Gaussian elimination to find the solution. Write the solution as an ordered pair or ordered triple and check the solution.

21. $x + y = 4$
$x + 3y = 10$

22. $x - 3y = -7$
$2x + y = 0$

23. $2x + 3y = 3$
$-2x + 2y = 7$

24. $x + 3y = -14$
$2x + 5y = -24$

25. $x - y = 5$
$x + 3y = -1$

26. $x + 4y = 1$
$3x - 2y = 10$

27. $4x - 8y = -10$
$x + y = 2$

28. $x - 7y = -16$
$4x + 10y = 50$

29. $x + y + z = 6$
$2y - z = 1$
$y + z = 5$

30. $x + y + z = 3$
$x + y - z = 2$
$y + z = 2$

31. $x + 2y + 3z = 6$
$-x + 3y + 4z = 0$
$x + y - 2z = -6$

32. $2x - 4y + 2z = 10$
$-x + 3y - 4z = -19$
$2x - y - 6z = -28$

33. $x + y + z = 0$
$2x + y + 2z = -1$
$x + y = 0$

34. $x + y - 2z = 5$
$x + 2y - 2z = 4$
$-x - y + z = -4$

35. $x + y + z = 3$
$-x - z = -2$
$x + y + 2z = 4$

36. $x + 2y - z = 3$
$-x - y + z = 0$
$x + 2y = 5$

37. $x + 2y + z = 3$
$2x + y - z = -6$
$-x - y + 2z = 5$

38. $x + y + z = -3$
$x - y - z = -1$
$-2x + y + 4z = 4$

*Exercises 39–48: **Technology** Use a graphing calculator to solve the system of linear equations.*

39. $x + 4y = 13$
$5x - 3y = -50$

40. $9x - 11y = 7$
$5x + 6y = 16$

41. $2x - y + 3z = 9$
$-4x + 5y + 2z = 12$
$2x + 7z = 23$

42. $3x - 2y + 4z = 29$
$2x + 3y - 7z = -14$
$5x - y + 11z = 59$

43. $6x + 2y + z = 4$
$-2x + 4y + z = -3$
$2x - 8y = -2$

44. $-x - 9y + 2z = -28.5$
$2x - y + 4z = -17$
$x - y + 8z = -9$

45. $4x + 3y + 12z = -9.25$
$15y + 8z = -4.75 + x$
$7z = -5.5 - 6y$

46. $5x + 4y = 13.3 + z$
$7y + 9z = 16.9 - x$
$x - 3y + 4z = -4.1$

47. $1.2x - 0.9y + 2.7z = 5.37$
$3.1x - 5.1y + 7.2z = 14.81$
$1.8y + 6.38 = 3.6z - 0.2x$

48. $11x + 13y - 17z = 380$
$5x - 14y - 19z = 24$
$-21y + 46z = -676 + 7x$

Exercises 49–54: (Refer to the Critical Thinking box in this section.) Row-reduce the matrix associated with the given system to determine whether the system of linear equations is inconsistent or dependent.

49. $x + 2y = 4$
$-2x - 4y = -8$

50. $x - 5y = 4$
$-2x + 10y = 8$

51. $x + y + z = 3$
$x + y - z = 1$
$x + y = 3$

52. $x + y + z = 5$
$x - y - z = 8$
$2x + 2y + 2z = 6$

53. $x + 2y + 3z = 14$
  $2x - 3y - 2z = -10$
  $3x - y + z = 4$

54. $x + 2y + 3z = 6$
  $-x + 3y + 4z = 6$
  $5y + 7z = 12$

APPLICATIONS

55. *Weight of a Bear*   Use the results of Example 7 to estimate the weight of a bear with a head length of 12 inches and an overall length of 60 inches.

56. *Weight of a Bear*   (Refer to Example 7.) Head length and overall length are not the only variables that can be used to estimate the weight of a bear. The data in the accompanying table list the weight W, neck size N, and chest size C of three bears. These data can be modeled by $W = a + bN + cC$.   (*Sources:* M. Triola, *Elementary Statistics*; Minitab, Inc.)

W (pounds)	N (inches)	C (inches)
80	16	26
344	28	45
416	31	54

(a) Set up a system of equations whose solution gives values for the constants a, b, and c.
(b) Solve this system. Round each value to the nearest tenth.
(c) Predict the weight of a bear with neck size $N = 22$ inches and chest size $C = 38$ inches.

57. *Garbage and Household Size*   A larger household produces more garbage, on average, than a smaller household. If we know the amount of metal M and plastic P waste produced each week, we can estimate the household size H from $H = a + bM + cP$. The table contains representative data for three households.   (*Source:* M. Triola, *Elementary Statistics.*)

H (people)	M (pounds)	P (pounds)
3	2.00	1.40
2	1.50	0.65
6	4.00	3.40

(a) Set up a system of equations whose solution gives values for the constants a, b, and c.
(b) Solve this system with a graphing calculator.

(c) Predict the size of a household that produces 3 pounds of metal waste and 2 pounds of plastic waste each week.

58. *Old Faithful Geyser*   In Yellowstone National Park, Old Faithful Geyser has been a favorite attraction for decades. Although this geyser erupts about every 80 minutes, this time interval varies, as do the duration and height of the eruptions. The accompanying table shows the height H, duration D, and time interval T for three eruptions.   (*Source:* National Park Service.)

H (feet)	D (seconds)	T (minutes)
160	276	94
125	203	84
140	245	79

(a) Assume that these data can be modeled by $H = a + bD + cT$. Set up a system of equations whose solution gives values for the constants a, b, and c.
(b) Solve this system by using a graphing calculator. Round each value to the nearest thousandth.
(c) Use this equation to estimate H when $D = 220$ and $T = 81$.

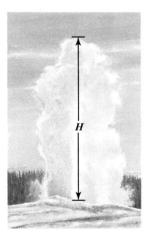

59. *Jogging Speeds*   A runner in preparation for a marathon jogs at 5, 6, and 8 miles per hour. The runner travels a total distance of 12.5 miles in 2 hours and jogs the same length of time at 5 miles per hour and at 8 miles per hour. How long did the runner jog at each speed?

60. *Mixture Problem* Three different types of candy that cost $2, $3, and $4 per pound are to be mixed to produce a 5-pound bag of candy that costs $14.50. If there are to be equal amounts of the $3-per-pound candy and the $4-per-pound candy in the bag, how much of each type of candy should be included in the mixture?

61. *Interest and Investments* (Refer to Example 5.) A total of $3000 is invested at 5%, 8%, and 12% annual interest. The interest earned after 1 year equals $285. The amount invested at 12% is triple the amount invested at 5%. Find the amount invested at each interest rate.

62. *Geometry* The measure of the largest angle in a triangle is twice the measure of the smallest angle. The remaining angle is 10° less than the largest angle. Find the measure of each angle.

WRITING ABOUT MATHEMATICS

63. Explain what the dimension of a matrix means. What is the difference between a matrix that has dimension 3×4 and one that has dimension 4×3?

64. Discuss the advantages of using technology to transform an augmented matrix to reduced row–echelon form. Are there any disadvantages? Explain.

CHECKING BASIC CONCEPTS   SECTIONS 9.1 AND 9.2

1. Determine which ordered triple is the solution to the system of equations. $(5, -4, 0), (1, 3, -1)$

$$\begin{aligned} x - y + 7z &= -9 \\ 2x - 2y + 5z &= -9 \\ -x + 3y - 2z &= 10 \end{aligned}$$

2. Solve the system of equations by using elimination and substitution.

$$\begin{aligned} x - y + z &= 2 \\ 2x - 3y + z &= -1 \\ -x + y + z &= 4 \end{aligned}$$

3. Use an augmented matrix to represent the system of equations. Solve the system using
   (a) Gaussian elimination.
   (b) technology.

$$\begin{aligned} x + 2y + z &= 1 \\ x + y + z &= -1 \\ y + z &= 1 \end{aligned}$$

9.3 DETERMINANTS

Calculation of Determinants · Area of Regions · Cramer's Rule

INTRODUCTION

Surveyors commonly calculate the areas of parcels of land. To do so they frequently divide the land into triangular regions. When the coordinates of the vertices of a triangle are known, determinants may be used to find the area of the triangle. A determinant is a real number that can be calculated for any square matrix. In this section we use determinants to find areas and to solve systems of linear equations.

CALCULATION OF DETERMINANTS

The concept of determinants originated with the Japanese mathematician Seki Kowa (1642–1708), who used them to solve systems of linear equations. Later, Gottfried Leibniz (1646–1716) formally described determinants and also used them to solve systems of linear equations.   (*Source: Historical Topics for the Mathematical Classroom*, NCTM.)

We begin by defining a determinant of a 2×2 matrix.

DETERMINANT OF A 2 × 2 MATRIX

The **determinant** of

$$A = \begin{bmatrix} a & b \\ c & d \end{bmatrix}$$

is a *real number* defined by

$$\det A = ad - cb.$$

EXAMPLE 1   Calculating determinants

Find $\det A$ for each 2×2 matrix.

(a) $A = \begin{bmatrix} 1 & 2 \\ 3 & 4 \end{bmatrix}$ **(b)** $A = \begin{bmatrix} -1 & -3 \\ 2 & -8 \end{bmatrix}$

Solution   **(a)** The determinant is calculated as follows.

$$\det A = \det \begin{bmatrix} 1 & 2 \\ 3 & 4 \end{bmatrix} = (1)(4) - (3)(2) = -2$$

(b) Similarly,

$$\det A = \det \begin{bmatrix} -1 & -3 \\ 2 & -8 \end{bmatrix} = (-1)(-8) - (2)(-3) = 14.$$

We can use determinants of 2×2 matrices to find determinants of 3×3 matrices. This method is called **expansion of a determinant by minors**.

DETERMINANT OF A 3 × 3 MATRIX

$$\det A = \det \begin{bmatrix} a_1 & b_1 & c_1 \\ a_2 & b_2 & c_2 \\ a_3 & b_3 & c_3 \end{bmatrix}$$

$$= a_1 \cdot \det \begin{bmatrix} b_2 & c_2 \\ b_3 & c_3 \end{bmatrix} - a_2 \cdot \det \begin{bmatrix} b_1 & c_1 \\ b_3 & c_3 \end{bmatrix} + a_3 \cdot \det \begin{bmatrix} b_1 & c_1 \\ b_2 & c_2 \end{bmatrix}$$

The 2×2 matrices in this equation are called **minors**.

EXAMPLE 2 Calculating 3 × 3 determinants

Evaluate det A.

(a) $A = \begin{bmatrix} 2 & 1 & -1 \\ -1 & 3 & 2 \\ 4 & -3 & -5 \end{bmatrix}$ (b) $A = \begin{bmatrix} 5 & -2 & 4 \\ 0 & 2 & 1 \\ -1 & 4 & -4 \end{bmatrix}$

Solution (a) We evaluate the determinant as follows.

$$\det \begin{bmatrix} 2 & 1 & -1 \\ -1 & 3 & 2 \\ 4 & -3 & -5 \end{bmatrix} = 2 \cdot \det \begin{bmatrix} 3 & 2 \\ -3 & -5 \end{bmatrix} - (-1) \cdot \det \begin{bmatrix} 1 & -1 \\ -3 & -5 \end{bmatrix}$$

$$+ 4 \cdot \det \begin{bmatrix} 1 & -1 \\ 3 & 2 \end{bmatrix}$$

$$= 2(-9) + 1(-8) + 4(5)$$

$$= -6$$

(b) We evaluate the determinant as follows.

$$\det \begin{bmatrix} 5 & -2 & 4 \\ 0 & 2 & 1 \\ -1 & 4 & -4 \end{bmatrix} = 5 \cdot \det \begin{bmatrix} 2 & 1 \\ 4 & -4 \end{bmatrix} - (0) \cdot \det \begin{bmatrix} -2 & 4 \\ 4 & -4 \end{bmatrix}$$

$$+ (-1) \cdot \det \begin{bmatrix} -2 & 4 \\ 2 & 1 \end{bmatrix}$$

$$= 5(-12) - 0(-8) + (-1)(-10)$$

$$= -50$$

Many graphing calculators can evaluate the determinant of a matrix, as illustrated in the next example, where we evaluate the determinants from Example 2.

EXAMPLE 3 Using technology to find determinants

Find each determinant of A, using a graphing calculator.

(a) $A = \begin{bmatrix} 2 & 1 & -1 \\ -1 & 3 & 2 \\ 4 & -3 & -5 \end{bmatrix}$ (b) $A = \begin{bmatrix} 5 & -2 & 4 \\ 0 & 2 & 1 \\ -1 & 4 & -4 \end{bmatrix}$

Solution (a) Begin by entering the matrix and then evaluate the determinant, as shown in Figure 9.6. The result is det $A = -6$, which agrees with our earlier calculation.

Calculator Help
To find a determinant, see
Appendix A (page AP-9).

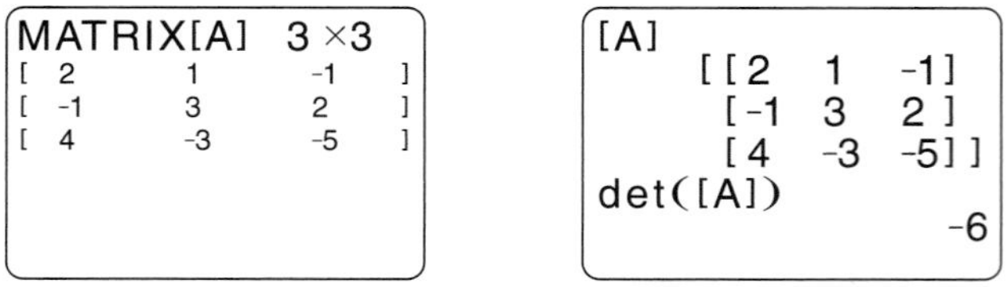

```
MATRIX[A]  3 ×3
[ 2      1      -1    ]
[ -1     3      2     ]
[ 4      -3     -5    ]
```
(a)

```
[A]
    [[2    1    -1]
     [-1   3    2 ]
     [4   -3   -5]]
det([A])
                  -6
```
(b)

Figure 9.6

(b) The determinant of A evaluates to -50 (see Figure 9.7).

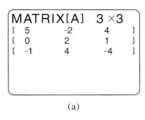

(a)

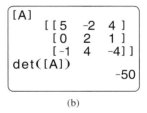

(b)

Figure 9.7

AREA OF REGIONS

A determinant may be used to find the area of a triangle. For example, if a triangle has vertices (a_1, a_2), (b_1, b_2), and (c_1, c_2), its area equals the absolute value of D, where

$$D = \frac{1}{2}\det\begin{bmatrix} a_1 & b_1 & c_1 \\ a_2 & b_2 & c_2 \\ 1 & 1 & 1 \end{bmatrix}.$$

If the vertices are entered in the columns of D counterclockwise, D will be positive.
(**Source:** W. Taylor, *The Geometry of Computer Graphics.*)

EXAMPLE 4   Computing the area of a triangular parcel of land

A triangular parcel of land is shown in Figure 9.8. If all units are miles, find the area of the parcel of land by using a determinant.

Solution   The vertices of the triangular parcel of land are $(2, 2)$, $(5, 4)$, and $(3, 8)$. The area of the triangle is

$$D = \frac{1}{2}\det\begin{bmatrix} 2 & 5 & 3 \\ 2 & 4 & 8 \\ 1 & 1 & 1 \end{bmatrix} = \frac{1}{2} \cdot 16 = 8.$$

The area of the triangle is 8 square miles.

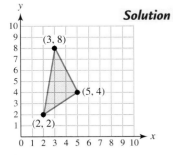

Figure 9.8

Critical Thinking

Suppose that you are given three distinct vertices and $D = 0$. What must be true about the three points?

CRAMER'S RULE

Determinants were developed independently by Gabriel Cramer (1704–1752). His work was published in 1750 and provided a method called **Cramer's rule** for solving systems of linear equations.

CRAMER'S RULE FOR LINEAR SYSTEMS IN TWO VARIABLES

The solution to the system of linear equations

$$a_1x + b_1y = c_1$$

$$a_2x + b_2y = c_2$$

is given by $x = \frac{E}{D}$ and $y = \frac{F}{D}$, where

$$E = \det\begin{bmatrix} c_1 & b_1 \\ c_2 & b_2 \end{bmatrix}, \quad F = \det\begin{bmatrix} a_1 & c_1 \\ a_2 & c_2 \end{bmatrix}, \quad \text{and} \quad D = \det\begin{bmatrix} a_1 & b_1 \\ a_2 & b_2 \end{bmatrix} \neq 0.$$

Note: If $D = 0$, the system has either no solutions or infinitely many solutions.

EXAMPLE 5 Using Cramer's rule

Use Cramer's rule to solve the following linear systems.
(a) $3x - 4y = 18$ **(b)** $-4x + 9y = -24$
      $7x + 5y = -1$ $6x + 17y = -25$

Solution **(a)** $E = \det\begin{bmatrix} c_1 & b_1 \\ c_2 & b_2 \end{bmatrix} = \det\begin{bmatrix} 18 & -4 \\ -1 & 5 \end{bmatrix} = (18)(5) - (-1)(-4) = \mathbf{86}$

$F = \det\begin{bmatrix} a_1 & c_1 \\ a_2 & c_2 \end{bmatrix} = \det\begin{bmatrix} 3 & 18 \\ 7 & -1 \end{bmatrix} = (3)(-1) - (7)(18) = \mathbf{-129}$

$D = \det\begin{bmatrix} a_1 & b_1 \\ a_2 & b_2 \end{bmatrix} = \det\begin{bmatrix} 3 & -4 \\ 7 & 5 \end{bmatrix} = (3)(5) - (7)(-4) = \mathbf{43}$

Because $x = \frac{E}{D} = \frac{86}{43} = 2$ and $y = \frac{F}{D} = \frac{-129}{43} = -3$, the solution is $(2, -3)$.

(b) $E = \det\begin{bmatrix} c_1 & b_1 \\ c_2 & b_2 \end{bmatrix} = \det\begin{bmatrix} -24 & 9 \\ -25 & 17 \end{bmatrix} = (-24)(17) - (-25)(9) = \mathbf{-183}$

$F = \det\begin{bmatrix} a_1 & c_1 \\ a_2 & c_2 \end{bmatrix} = \det\begin{bmatrix} -4 & -24 \\ 6 & -25 \end{bmatrix} = (-4)(-25) - (6)(-24) = \mathbf{244}$

$D = \det\begin{bmatrix} a_1 & b_1 \\ a_2 & b_2 \end{bmatrix} = \det\begin{bmatrix} -4 & 9 \\ 6 & 17 \end{bmatrix} = (-4)(17) - (6)(9) = \mathbf{-122}$

Because $x = \frac{E}{D} = \frac{-183}{-122} = 1.5$ and $y = \frac{F}{D} = \frac{244}{-122} = -2$, the solution is $(1.5, -2)$.

Cramer's rule can be applied to systems that have any number of linear equations. Cramer's rule for three linear equations is discussed in the Extended and Discovery Exercises at the end of this chapter.

USING CRAMER'S RULE IN APPLICATIONS In applications, equations with hundreds of variables are routinely solved. Such systems of equations could be solved with Cramer's rule.

However, using Cramer's rule and the expansion of a determinant with minors to solve a linear system of equations with n variables requires at least

$$1 \cdot 2 \cdot 3 \cdot 4 \cdot \cdots \cdot n \cdot (n + 1)$$

multiplication operations. To solve a linear system involving only 25 variables would require about

$$1 \cdot 2 \cdot 3 \cdot 4 \cdot \cdots \cdot 25 \cdot 26 \approx 4 \times 10^{26}$$

multiplication operations. Supercomputers can perform about 1 trillion (1×10^{12}) multiplication operations per second. This would take about

$$\frac{4 \times 10^{26}}{1 \times 10^{12}} = 4 \times 10^{14} \text{ seconds.}$$

With $60 \times 60 \times 24 \times 365 = 31{,}536{,}000$ seconds in a year, 4×10^{14} seconds equals

$$\frac{4 \times 10^{14}}{31{,}536{,}000} \approx 12{,}700{,}000 \text{ years!}$$

Modern software packages do *not* use Cramer's rule for three or more variables.

9.3 PUTTING IT ALL TOGETHER

The determinant of a square matrix A is a real number, denoted det A. Cramer's rule is a method that uses determinants to solve systems of linear equations. The following table summarizes important topics from this section.

Concept	Explanation
Determinant of a 2×2 Matrix	The determinant of a 2×2 matrix A is given by $$\det A = \det\begin{bmatrix} a & b \\ c & d \end{bmatrix} = ad - cb.$$
Determinant of a 3×3 Matrix	The determinant of a 3×3 matrix A is given by $$\det A = \det\begin{bmatrix} a_1 & b_1 & c_1 \\ a_2 & b_2 & c_2 \\ a_3 & b_3 & c_3 \end{bmatrix}$$ $$= a_1 \cdot \det\begin{bmatrix} b_2 & c_2 \\ b_3 & c_3 \end{bmatrix} - a_2 \cdot \det\begin{bmatrix} b_1 & c_1 \\ b_3 & c_3 \end{bmatrix}$$ $$+ a_3 \cdot \det\begin{bmatrix} b_1 & c_1 \\ b_2 & c_2 \end{bmatrix}$$

continued on next page

continued from previous page

Concept	Explanation
Area of a Triangle	If a triangle has vertices (a_1, a_2), (b_1, b_2), and (c_1, c_2), its area equals the absolute value of D, where $$D = \frac{1}{2}\det\begin{bmatrix} a_1 & b_1 & c_1 \\ a_2 & b_2 & c_2 \\ 1 & 1 & 1 \end{bmatrix}.$$
Cramer's Rule for Linear Systems in Two Variables	The solution to the linear system $$a_1x + b_1y = c_1$$ $$a_2x + b_2y = c_2$$ is given by $x = \frac{E}{D}$ and $y = \frac{F}{D}$, where $$E = \det\begin{bmatrix} c_1 & b_1 \\ c_2 & b_2 \end{bmatrix}, \quad F = \det\begin{bmatrix} a_1 & c_1 \\ a_2 & c_2 \end{bmatrix}, \quad \text{and}$$ $$D = \det\begin{bmatrix} a_1 & b_1 \\ a_2 & b_2 \end{bmatrix} \neq 0.$$ *Note:* If $D = 0$, then the system has either no solutions or infinitely many solutions.

9.3 EXERCISES

FOR EXTRA HELP

Student's Solutions Manual

InterAct Math   InterAct Math

MathXL   MathXL

MyMathLab

Tutor Center   Math Tutor Center

Digital Video Tutor
CD 4   Videotape 12

CALCULATING DETERMINANTS

Exercises 1–16: Evaluate det A by hand, where A is the given matrix.

1. $\begin{bmatrix} 1 & -2 \\ 3 & -8 \end{bmatrix}$

2. $\begin{bmatrix} 5 & -1 \\ 3 & 7 \end{bmatrix}$

3. $\begin{bmatrix} -3 & 7 \\ 8 & -1 \end{bmatrix}$

4. $\begin{bmatrix} 0 & -7 \\ -3 & 1 \end{bmatrix}$

5. $\begin{bmatrix} 23 & 4 \\ 6 & -13 \end{bmatrix}$

6. $\begin{bmatrix} 44 & -51 \\ -9 & 32 \end{bmatrix}$

7. $\begin{bmatrix} 1 & -1 & 2 \\ 0 & 1 & -3 \\ 0 & -4 & 7 \end{bmatrix}$

8. $\begin{bmatrix} 2 & -1 & -5 \\ -1 & 4 & -2 \\ 0 & 1 & 4 \end{bmatrix}$

9. $\begin{bmatrix} 2 & -1 & 0 \\ 1 & -2 & 6 \\ 0 & 1 & 8 \end{bmatrix}$

10. $\begin{bmatrix} 0 & 1 & -4 \\ 3 & -6 & 10 \\ 4 & -2 & 7 \end{bmatrix}$

11. $\begin{bmatrix} -1 & 3 & 5 \\ 3 & -3 & 5 \\ 2 & -3 & 7 \end{bmatrix}$

12. $\begin{bmatrix} 6 & -1 & 9 \\ 7 & 0 & -3 \\ 2 & 5 & -1 \end{bmatrix}$

13. $\begin{bmatrix} 5 & 0 & 0 \\ 0 & -2 & 0 \\ 0 & 0 & 5 \end{bmatrix}$

14. $\begin{bmatrix} 1 & 2 & 3 \\ 2 & 4 & 6 \\ 3 & 6 & 9 \end{bmatrix}$

15. $\begin{bmatrix} 0 & 2 & -3 \\ 0 & 3 & -9 \\ 0 & 5 & 9 \end{bmatrix}$

16. $\begin{bmatrix} 3 & -1 & 2 \\ 0 & 5 & 7 \\ 0 & 0 & -1 \end{bmatrix}$

Exercises 17–20: Use technology to calculate det A, where A is the given matrix.

17. $\begin{bmatrix} 2 & -5 & 13 \\ 10 & 15 & -10 \\ 17 & -19 & 22 \end{bmatrix}$ **18.** $\begin{bmatrix} 1.6 & 3.1 & 5.7 \\ 2.1 & 6.7 & 8.1 \\ -0.4 & -0.8 & -3.1 \end{bmatrix}$

19. $\begin{bmatrix} 17 & 0 & 4 \\ -9 & 14 & 1.5 \\ 13 & 67 & -11 \end{bmatrix}$ **20.** $\begin{bmatrix} 121 & 45 & -56 \\ -45 & 87 & 32 \\ -14 & -34 & 67 \end{bmatrix}$

CALCULATING AREA

Exercises 21–26: Find the area of the figure by using a determinant. Assume that units are feet.

21.
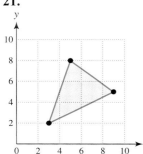

22.
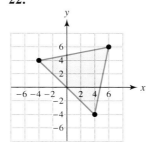

23.
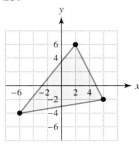

24.
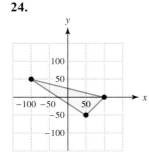

25.
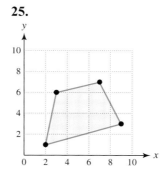

26.
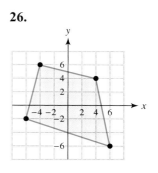

CRAMER'S RULE

Exercises 27–32: Solve the system of equations by using Cramer's rule.

27. $5x + 3y = 4$
$6x - 4y = 20$

28. $-5x + 4y = -5$
$4x + 4y = -32$

29. $7x - 5y = -3$
$-4x + 6y = -8$

30. $-4x - 9y = -17$
$8x + 4y = 9$

31. $8x = 3y - 61$
$-x = 4y - 23$

32. $15y = -188 - 22x$
$23y = -173 - 16x$

WRITING ABOUT MATHEMATICS

33. Suppose that the first column of a 3×3 matrix A is all 0s. What is the value of det A? Give an example and explain your answer.

34. Estimate how long it might take a supercomputer to solve a linear system of equations with 15 variables, using Cramer's rule and expansion of determinant by minors. Explain your calculations. (*Hint:* See the discussion at the end of this section.)

CHECKING BASIC CONCEPTS SECTION 9.3

1. Evaluate det A.

(a) $A = \begin{bmatrix} -3 & 4 \\ -2 & 3 \end{bmatrix}$

(b) $A = \begin{bmatrix} 1 & -2 & 3 \\ 5 & 1 & 1 \\ 0 & 2 & -1 \end{bmatrix}$

2. Use Cramer's rule to solve the system of equations.

$$2x - y = -14$$
$$3x - 4y = -36$$

3. Find the area of a triangle with the following vertices: $(-1, 2)$, $(5, 6)$, and $(2, -3)$.

Summary

Solution to a System of Linear Equations in Three Variables   An ordered triple (x, y, z) that satisfies *every* equation.

Example:
$$x - y + 2z = 3$$
$$2x - y + z = 5$$
$$x + y + z = 6$$

The solution is $(3, 2, 1)$ because these values for (x, y, z) satisfy all three equations.

$$3 - 2 + 2(1) = 3 \quad \text{True}$$
$$2(3) - 2 + 1 = 5 \quad \text{True}$$
$$1 + 2 + 3 = 6 \quad \text{True}$$

Elimination and Substitution   Systems of linear equations in three variables can be solved by elimination and substitution, using the following steps.

STEP 1 Eliminate one variable, such as x, from two of the given equations.

STEP 2 Use the two resulting equations with two variables to eliminate one of the variables, such as y. Solve for the remaining variable z.

STEP 3 Substitute z in one of the two equations from STEP 2. Solve for the unknown variable y.

STEP 4 Substitute values for y and z in one of the given equations. Then find x. The solution is (x, y, z).

Matrix   A rectangular array of numbers is a matrix. If a matrix has n rows and m columns, it has dimension $n \times m$.

Example: Matrix $A = \begin{bmatrix} 3 & -1 & 7 \\ 0 & 6 & -2 \end{bmatrix}$ has dimension 2×3.

Augmented Matrix   Any linear system can be represented with an augmented matrix.

Linear System

$$4x - 3y = 5$$
$$x + 2y = 4$$

Augmented Matrix

$$\begin{bmatrix} 4 & -3 & | & 5 \\ 1 & 2 & | & 4 \end{bmatrix}$$

Gaussian Elimination   A numerical method that uses matrix row transformations to tranform a matrix into reduced row–echelon form from which the solution can be found.

Example: The matrix $\begin{bmatrix} 4 & -3 & | & 5 \\ 1 & 2 & | & 4 \end{bmatrix}$ reduces to $\begin{bmatrix} 1 & 0 & | & 2 \\ 0 & 1 & | & 1 \end{bmatrix}$.

The solution to the system is (2, 1).

Section 9.3   *Determinants*

Determinant for a 2 × 2 Matrix   A determinant is a *real number*. The determinant of a 2 × 2 matrix is

$$\det A = \det \begin{bmatrix} a & b \\ c & d \end{bmatrix} = ad - cb.$$

Example: $\det \begin{bmatrix} 2 & 3 \\ 4 & 5 \end{bmatrix} = (2)(5) - (4)(3) = -2$

Determinant for a 3 × 3 Matrix

$$\det A = \det \begin{bmatrix} a_1 & b_1 & c_1 \\ a_2 & b_2 & c_2 \\ a_3 & b_3 & c_3 \end{bmatrix}$$

$$= a_1 \cdot \det \begin{bmatrix} b_2 & c_2 \\ b_3 & c_3 \end{bmatrix} - a_2 \cdot \det \begin{bmatrix} b_1 & c_1 \\ b_3 & c_3 \end{bmatrix} + a_3 \cdot \det \begin{bmatrix} b_1 & c_1 \\ b_2 & c_2 \end{bmatrix}$$

Example: $\det \begin{bmatrix} 2 & 3 & 2 \\ 3 & 7 & -3 \\ 0 & 0 & -1 \end{bmatrix} = 2 \det \begin{bmatrix} 7 & -3 \\ 0 & -1 \end{bmatrix} - 3 \det \begin{bmatrix} 3 & 2 \\ 0 & -1 \end{bmatrix} + 0$

$$= 2(-7) - 3(-3) = -5$$

Cramer's rule uses determinants to solve linear systems of equations. Determinants can also be used to find areas of triangles. See Putting It All Together for Section 9.3.

CHAPTER
9 Review Exercises

1. Is $(3, -4, 5)$ a solution for $x + y + z = 4$?

2. Decide whether either ordered triple is a solution: $(1, -1, 2)$, $(1, 0, 5)$.
$$2x - 3y + z = 7$$
$$-x - y + 3z = 6$$
$$3x - 2y + z = 7$$

Exercises 3–6: Use elimination and substitution to solve the system of linear equations.

3. $x - y - 2z = -11$
$-x + 2y + 3z = 16$
$3z = 6$

4. $x + y = 4$
$-2x + y + 3z = -2$
$x - 2y + 5z = -26$

5. $2x - y = -5$
$x + 2y + z = 7$
$-2x + y + z = 7$

6. $2x + 3y + z = 6$
$-x + 2y + 2z = 3$
$x + y + 2z = 4$

Exercises 7–10: Write the system of linear equations as an augmented matrix. Then use Gaussian elimination to solve the system, writing the solution as an ordered triple. Check your solution.

7. $x + y + z = -6$
$x + 2y + z = -8$
$y + z = -5$

8. $x + y + z = -3$
$-x + y = 5$
$y + z = -1$

9. $x + 2y - z = 1$
$-x + y - 2z = 5$
$2y + z = 10$

10. $2x + 2y - 2z = -14$
$-2x - 3y + 2z = 12$
$x + y - 4z = -22$

Exercises 11 and 12: Technology   Use a graphing calculator to solve the system of linear equations.

11. $3x - 2y + 6z = -17$
$-2x - y + 5z = 20$
$4y + 7z = 30$

12. $19x - 13y - 7z = 7.4$
$22x + 33y - 8z = 110.5$
$10x - 56y + 9z = 23.7$

Exercises 13–16: Evaluate det A, where A is the given matrix.

13. $\begin{bmatrix} 6 & -5 \\ -4 & 2 \end{bmatrix}$

14. $\begin{bmatrix} 0 & -6 \\ 5 & 9 \end{bmatrix}$

15. $\begin{bmatrix} 3 & -5 & -3 \\ 1 & 4 & 7 \\ 0 & -3 & 1 \end{bmatrix}$

16. $\begin{bmatrix} -2 & -1 & -7 \\ 2 & 1 & -3 \\ 3 & -5 & 8 \end{bmatrix}$

Exercises 17 and 18: Use technology to calculate det A, where A is the given matrix.

17. $\begin{bmatrix} 22 & -45 & 3 \\ 15 & -12 & -93 \\ 5 & 81 & -21 \end{bmatrix}$

18. $\begin{bmatrix} 0.5 & -7.3 & 9.6 \\ 0.1 & 3.1 & 9.2 \\ -0.5 & -1.9 & 5.4 \end{bmatrix}$

Exercises 19 and 20: Find the area of the triangle by using a determinant. Assume that the units are feet.

19.

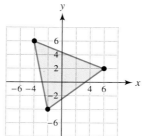

20.

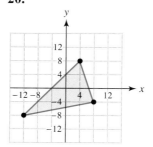

Exercises 21–24: Use Cramer's rule to solve the system.

21. $7x + 6y = 8$
$5x - 8y = 18$

22. $-2x + 5y = 25$
$3x + 4y = -3$

23. $3x - 6y = 1.5$
$7x - 5y = 8$

24. $-5x + 4y = -47$
$6x - 7y = 63$

APPLICATIONS

25. *Pedestrian Fatalities* Forty-seven percent of pedestrian fatalities occur on Friday and Saturday nights. The combined total of pedestrian fatalities in 1988 and 1998 was 12,090. There were 1650 more fatalities in 1988 than in 1998. Find the number of pedestrian fatalities during each year. (*Source:* National Highway Traffic Safety Administration.)

26. *Tickets* Tickets for a football game sold for $8 and $12. If 480 tickets were sold for total receipts of $4620, how many of each type of ticket were sold?

27. *Determining Costs* The accompanying table shows the costs for purchasing different combinations of malts, cones, and ice cream bars.

Malts	Cones	Bars	Total Cost
1	3	5	$14
1	2	4	$11
0	1	3	$5

(a) Let m be the cost of a malt, c the cost of a cone, and b the cost of an ice cream bar. Write a system of three linear equations that represents the data in the table.
(b) Solve this system.

28. *Geometry* The largest angle in a triangle is 20° more than the sum of the two smaller angles. The measure of the largest angle is 85° more than the smallest angle. Find the measure of each angle in the triangle.

29. *Mixture Problem* Three different types of candy that cost $1.50, $2.00, and $2.50 per pound are to be mixed to produce 12 pounds of candy worth $26.00. If there is to be 2 pounds more of the $2.50 candy than the $2.00 candy, how much of each type of candy should be used in the mixture?

30. *Estimating the Chest Size of a Bear* The accompanying table shows the chest size C, weight W, and overall length L of three bears. These data can be modeled with the formula $C = a + bW + cL$. (*Sources:* M. Triola, *Elementary Statistics*; Minitab, Inc.)

C (inches)	W (pounds)	L (inches)
40	202	63
50	365	70
55	446	77

(a) Set up a system of linear equations whose solution gives values for the constants a, b, and c.
 (b) Solve this system. Round each value to the nearest thousandth.
(c) Predict the chest size of a bear weighing 300 pounds and having a length of 68 inches.

CHAPTER
9 Test

1. Can a system of three equations and three variables have exactly three solutions? Explain.

2. Determine which ordered triple is a solution to the linear system.
$$(-4, 3, 3), (1, -2, -2)$$
$$x - 3y + 4z = -1$$
$$2x + y - 3z = 6$$
$$x - y + z = 1$$

3. Use elimination and substitution to solve the system of linear equations.
$$x - y + 2z = 5$$
$$-x + y - 3z = -8$$
$$x - 2y + 2z = 3$$

4. Consider the system of linear equations.
$$2x - 4y = -10$$
$$-3x - 2y = 7$$

   (a) Write the system of linear equations as an augmented matrix.
   (b) Use Gaussian elimination to solve the system, writing the solution as an ordered pair.

5. Consider the system of linear equations.
$$x + y + z = 2$$
$$x - y - z = 3$$
$$2x + 2y + z = 6$$

   (a) Write the system of linear equations as an augmented matrix.
   (b) Use Gaussian elimination to solve the system, writing the solution as an ordered triple.

6. Evaluate $\det A$ if $A = \begin{bmatrix} 3 & 2 & -1 \\ 6 & 2 & -6 \\ 0 & 8 & -3 \end{bmatrix}$.

7. Solve the system of linear equations with Cramer's rule.
$$5x - 3y = 7$$
$$-4x + 2y = 11$$

8. *Jogging Speeds*   A runner in preparation for a race jogs at 6, 7, and 9 miles per hour and travels a total of 7.1 miles in 1 hour. The runner jogs 12 minutes longer at 6 miles per hour than at 9 miles per hour. How many minutes did the runner jog at each speed?

9. *Geometry*   The largest angle in a triangle is 50° more than the smallest angle. The sum of the measures of the smaller two angles is 10° more than the largest angle. Find the measure of each angle in the triangle.

10. *Weight of a Bear*   The data in the accompanying table list the weight W, neck size N, and chest size C of three bears. These data can be modeled by $W = a + bN + cC$.

W (pounds)	N (inches)	C (inches)
168	20	25
270	24	40
405	30	50

   (a) Set up a system of equations whose solution gives values for the constants a, b, and c.
   (b) Solve this system.
   (c) Predict the weight W of a bear with neck size $N = 26$ and chest size $C = 44$.

CHAPTER 9 Extended and Discovery Exercises

CRAMER'S RULE

Exercises 1–6: Cramer's rule can be applied to systems of three equations with three variables. For the system of equations

$$a_1x + b_1y + c_1z = d_1$$
$$a_2x + b_2y + c_2z = d_2$$
$$a_3x + b_3y + c_3z = d_3,$$

the solution can be written as follows.

$$D = \det\begin{bmatrix} a_1 & b_1 & c_1 \\ a_2 & b_2 & c_2 \\ a_3 & b_3 & c_3 \end{bmatrix}, \quad E = \det\begin{bmatrix} d_1 & b_1 & c_1 \\ d_2 & b_2 & c_2 \\ d_3 & b_3 & c_3 \end{bmatrix}$$

$$F = \det\begin{bmatrix} a_1 & d_1 & c_1 \\ a_2 & d_2 & c_2 \\ a_3 & d_3 & c_3 \end{bmatrix}, \quad G = \det\begin{bmatrix} a_1 & b_1 & d_1 \\ a_2 & b_2 & d_2 \\ a_3 & b_3 & d_3 \end{bmatrix}$$

If $D \neq 0$, a unique solution exists and is given by

$$x = \frac{E}{D}, \quad y = \frac{F}{D}, \quad z = \frac{G}{D}.$$

Use Cramer's rule to solve the equations.

1. $\quad x + y + z = 6$
$\quad 2x + y + 2z = 9$
$\qquad\quad y + 3z = 9$

2. $\qquad\quad y + z = 1$
$\quad 2x - y - z = -1$
$\quad x + y - z = 3$

3. $\quad x + \qquad z = 2$
$\quad x + y \qquad = 0$
$\qquad\quad y + 2z = 1$

4. $\quad x + y + 2z = 1$
$\quad -x - 2y - 3z = -2$
$\qquad\quad y - 3z = 5$

5. $\quad x + \qquad 2z = 7$
$\quad -x + y + z = 5$
$\quad 2x - y + 2z = 6$

6. $\quad x + 2y + 3z = -1$
$\quad 2x - 3y - z = 12$
$\quad x + 4y - 2z = -12$

ANALYSIS OF TRAFFIC FLOW

Exercises 7 and 8: Mathematics is frequently used to analyze traffic for timing of traffic lights. The following figure shows three one-way streets with intersections A, B, and C. The average number of cars traveling on certain portions of the streets is given in cars per minute. The variables x, y, and z represent unknown traffic flows that need to be determined. The number of vehicles entering an intersection must equal the number of vehicles exiting an intersection.

(a) *Verify that the accompanying system of linear equations describes the traffic flow at each of the three intersections.*

(b) *Express the system using an augmented matrix.*

(c) *Solve the system and interpret your solution.*

7. A: $8 + 12 = y + x$
B: $z + 4 = 8 + 5$
C: $y + 6 = z + 9$

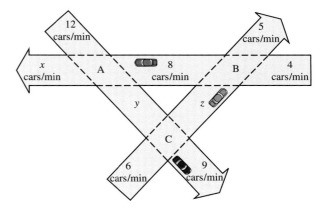

8. A: $6 + 3 = y + x$
B: $z + 10 = 6 + 9$
C: $y + 7 = z + 5$

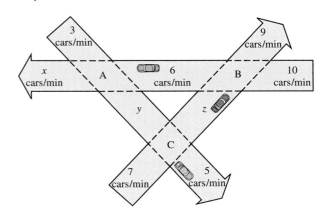

MATRICES AND ROAD MAPS

Exercises 9–12: Adjacency Matrix   A matrix A can be used to represent a map showing distances between cities. Let a_{ij} denote the number in row i and column j of a matrix A. Now consider the following map illustrating freeway distances in miles between four cities. Each city has been assigned a number. For example, there is a direct route from Denver, Colorado (city 1), to Colorado Springs, Colorado (city 2), of approximately 60 miles. Therefore $a_{12} = 60$ in the accompanying matrix A. (Note that a_{12} is the number in row 1 and column 2.) The distance from Colorado Springs to Denver is also 60 miles, so $a_{21} = 60$. As there is no direct freeway connection between Las Vegas, Nevada (city 4), and Colorado Springs (city 2), we let $a_{24} = a_{42} = *$. The matrix A is called an **adjacency matrix.**   (**Source:** S. Baase, *Computer Algorithms: Introduction to Design and Analysis.*)

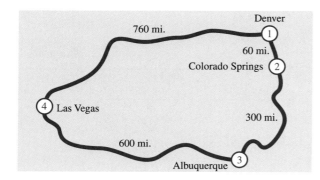

$$A = \begin{bmatrix} 0 & 60 & * & 760 \\ 60 & 0 & 300 & * \\ * & 300 & 0 & 600 \\ 760 & * & 600 & 0 \end{bmatrix}$$

9. Explain how to use A to find the freeway distance from Denver to Las Vegas.

10. Explain how to use A to find the freeway distance from Denver to Albuquerque.

11. If a map shows 20 cities, what would be the dimension of the adjacency matrix? How many elements would there be in this matrix?

12. Why are there only zeros on the main diagonal of A?

Exercises 13 and 14: (Refer to Exercises 9–12.) Determine an adjacency matrix A for the road map.

13.

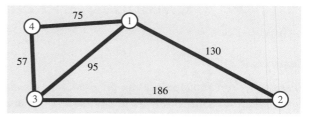

14.

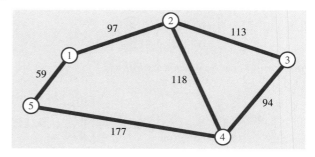

Exercises 15 and 16: (Refer to Exercises 9–12.) Sketch a road map represented by the adjacency matrix A. Is your answer unique?

15. $A = \begin{bmatrix} 0 & 30 & 20 & 5 \\ 30 & 0 & 15 & * \\ 20 & 15 & 0 & 25 \\ 5 & * & 25 & 0 \end{bmatrix}$

16. $A = \begin{bmatrix} 0 & 5 & * & 13 & 20 \\ 5 & 0 & 5 & * & * \\ * & 5 & 0 & 13 & * \\ 13 & * & 13 & 0 & 10 \\ 20 & * & * & 10 & 0 \end{bmatrix}$

SOLVING AN EQUATION IN FOUR VARIABLES

17. *Weight of a Bear*   In Section 9.2 we estimated the weight of a bear by using two variables. We may be able to make more accurate estimates by using three variables. The accompanying table shows the weight W, neck size N, overall length L, and chest size C for four bears.   (**Sources:** M. Triola, *Elementary Statistics;* Minitab, Inc.)

W (pounds)	N (inches)	L (inches)	C (inches)
125	19	57.5	32
316	26	65	42
436	30	72	48
514	30.5	75	54

(b) Solve the system with a graphing calculator. Round each value to the nearest thousandth.

(c) Predict the weight of a bear with $N = 24$, $L = 63$, and $C = 39$. Interpret the result.

(a) We can model these data with the equation $W = a + bN + cL + dC$, where a, b, c, and d are constants. To do so, represent a system of linear equations by a 4×5 augmented matrix whose solution gives values for a, b, c, and d.

CHAPTERS
1–9 Cumulative Review Exercises

1. Write 360 as a product of prime numbers.

2. Translate the sentence "Double a number increased by 7 equals the number decreased by 2" into an equation by using the variable n. Then solve the equation.

Exercises 3 and 4: Evaluate by hand and then reduce to lowest terms.

3. $\frac{2}{3} + \frac{4}{7} \cdot \frac{21}{28}$

4. $\frac{3}{5} \div \frac{6}{5} - \frac{2}{3}$

Exercises 5 and 6: Evaluate by hand.

5. $30 - 4 \div 2 \cdot 6$

6. $\dfrac{3^2 - 2^3}{20 - 5 \cdot 2}$

7. Solve $2(x + 1) - 6x = x - 4$.

8. Solve $4 - 3x = -2$ graphically. Check your answer.

9. Convert 124% to fraction and decimal notation.

10. If $A = 30$ square miles and $h = 10$ miles, use the formula $A = \frac{1}{2}bh$ to find b.

11. Solve $A = \frac{h}{2}(a + b)$ for b.

12. Solve $6t - 1 < 3 - t$. Write the solution set in set-builder notation.

13. Make a scatterplot with the points $(-1, 2)$, $(1, -2)$, $(0, 3)$, $(-2, 0)$, and $(2, 3)$.

Exercises 14–16: Graph the equation and determine any intercepts.

14. $y = -\frac{1}{2}x + 2$

15. $-3x + 4y = 12$ **16.** $x = -2$

17. Use the graph to identify the x-intercept and the y-intercept. Then write the slope–intercept form of the line.

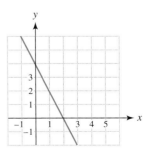

18. Write the slope–intercept form of a line that passes through the point $(-1, 3)$ with slope $m = -2$.

Exercises 19 and 20: Find the slope–intercept form of the line satisfying the given conditions.

19. Passing through the points $(-3, 5)$ and $(2, 8)$

20. Perpendicular to $x + 2y = 5$, passing through the point $(-1, 1)$

21. An insect population P is initially 4000 and increases by 500 insects per day. Write an equation in slope–intercept form that models this population after x days.

22. Solve the system of equations. Write your answer as an ordered pair.

$$-2a + b = -5$$
$$4a - 3b = 0$$

Exercises 23 and 24: Shade the solution set for the inequality.

23. $y \geq -1$ **24.** $3x - 2y \leq 6$

Exercises 25 and 26: Shade the solution set for the system of inequalities.

25. $y \geq x - 1$ **26.** $x - y < 3$
      $y \leq 2x$ $2x + y \geq 6$

Exercises 27–36: Simplify and use positive exponents, when appropriate, to write the expression.

27. $(ab - 6b^2) - (4b^2 + 8ab + 2)$

28. $5 - 3^4$ **29.** $4(4^{-2})(3^{-1})(3^4)$

30. $\dfrac{2^{-4}}{4^{-2}}$ **31.** $(8t^{-3})(3t^2)(t^5)$

32. $2(rt)^4$ **33.** $(2t^3)^{-2}$

34. $(4a^2b^3)^2(2ab)^{-3}$ **35.** $\left(\dfrac{2rt^{-1}}{3r^{-2}t^3}\right)^4$

36. $\left(\dfrac{2a^{-1}}{ab^{-2}}\right)^{-3}$

Exercises 37–42: Multiply the expression.

37. $2a^2(a^2 - 2a + 3)$

38. $(a + b)(a^2 - ab + b^2)$

39. $(5x + 1)(x - 7)$ **40.** $(y - 3)(2y + 3)$

41. $(a + b)(a - b)$ **42.** $(2x + 3y)^2$

43. Write 1.5×10^{-5} in standard form.

44. Write 2,130,000 in scientific notation.

Exercises 45 and 46: Divide.

45. $\dfrac{4x^3 - 8x^2 + 6x}{2x}$

46. $(x^4 - 9x^3 + 23x^2 - 17x + 11) \div (x - 5)$

Exercises 47–54: Factor completely.

47. $10ab^2 - 25a^3b^5$ **48.** $y^3 - 3y^2 + 2y - 6$

49. $6z^2 + 7z - 3$ **50.** $4z^2 - 9$

51. $4y^2 - 20y + 25$ **52.** $a^3 - 27$

53. $4z^4 - 17z^2 + 15$ **54.** $2a^3b + a^2b^2 - ab^3$

Exercises 55–58: Solve the equation.

55. $(x - 1)(x + 2) = 0$ **56.** $x^2 - 9x = 0$

57. $6y^2 - 7y = 3$ **58.** $x^3 = 4x$

Exercises 59 and 60: Simplify to lowest terms.

59. $\dfrac{x^2 - 16}{x + 4}$ **60.** $\dfrac{2x^2 - 11x - 6}{6x^2 - 5x - 4}$

61. Find the x-values that make $\dfrac{x - 3}{16 - x^2}$ undefined.

62. Evaluate $\dfrac{4x + 1}{x - 1}$ for $x = -2$.

Exercises 63–66: Simplify and write in lowest terms.

63. $\dfrac{x^2 - 3x + 2}{x + 7} \div \dfrac{x - 2}{2x + 14}$

64. $\dfrac{x}{2x + 3} + \dfrac{x + 3}{2x + 3}$ **65.** $\dfrac{5x}{x^2 - 1} - \dfrac{3}{x + 1}$

66. $\dfrac{\dfrac{2}{x} - \dfrac{2}{y}}{\dfrac{2}{x} + \dfrac{2}{y}}$

Exercises 67–69: Solve the equation.

67. $\dfrac{x + 2}{5} = \dfrac{x}{4}$ **68.** $\dfrac{1}{3x} + \dfrac{5}{2x} = 2$

69. $\dfrac{1}{x - 2} + \dfrac{2}{x + 2} = \dfrac{1}{x^2 - 4}$

70. Suppose that y is inversely proportional to x and that $y = 25$ when $x = 4$. Find y for $x = 10$.

71. Evaluate $f(x) = 1 - 4x$ for $x = -3$.

72. Graph $f(x) = x^2 - 2$ and identify the domain and range of f. Write your answer in interval notation.

Exercises 73 and 74: Use the graph or table to evaluate $f(0)$ and $f(-2)$.

73.

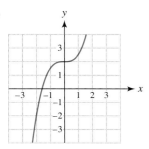

74.

x	-2	-1	0	1	2
$f(x)$	-4	-1	2	5	8

75. Use the table to determine whether $f(x)$ could represent a linear function. If it could, write the formula for f in the form $f(x) = ax + b$.

x	-1	0	1	2	3
$f(x)$	7	3	-1	-5	-9

76. Use the graph to solve the inequality $y_1 \le y_2$. Write the solution set in interval notation.

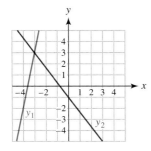

Exercises 77 and 78: Solve the inequality symbolically. Write the solution set in interval notation.

77. $\dfrac{4x - 9}{6} > \dfrac{1}{2}$

78. $\dfrac{2}{3}z - 2 \le \dfrac{1}{4}z - (2z + 2)$

Exercises 79 and 80: Solve the compound inequality. Graph the solution set on a number line.

79. $x + 2 > 1$ and $2x - 1 \le 9$

80. $4x + 7 < 1$ or $3x + 2 \ge 11$

Exercises 81 and 82: Solve the three-part inequality. Write the solution set in interval notation.

81. $-7 \le 2x - 3 \le 5$ **82.** $-8 \le -\frac{1}{2}x - 3 \le 5$

83. Use the graph to solve the equation and inequalities.
   (a) $y_1 = 2$ **(b)** $y_1 \le 2$
   (c) $y_1 \ge 2$

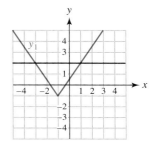

84. Solve the absolute value equation $\left|\frac{2}{3}x - 4\right| = 8$.

Exercises 85 and 86: Solve the absolute value inequality. Write the solution set in interval notation.

85. $|3x + 5| > 13$ **86.** $-3|2t - 11| \ge -9$

87. Use elimination and substitution to solve the system of linear equations.
$$2x + 3y - z = 3$$
$$3x - y + 4z = 10$$
$$2x + y - 2z = -1$$

88. Write the system of linear equations as an augmented matrix. Then use Gaussian elimination to solve the system. Write the solution as an ordered triple.
$$x + y - z = 4$$
$$-x - y - z = 0$$
$$x - 2y + z = -9$$

89. Evaluate $\det A$.
$$A = \begin{bmatrix} 4 & -2 \\ 1 & 3 \end{bmatrix}$$

APPLICATIONS

90. *Modeling Motion* The table lists the distance d in miles traveled by an airplane for various elapsed times t in hours. Find an equation that models these data.

t (hours)	2	3	4	5
d (miles)	650	975	1300	1625

91. *Snowfall* By noon 4 inches of snow had fallen. For the next 6 hours snow fell at $\frac{1}{2}$ inch per hour.
(a) Give a formula for $f(x)$ that calculates the number of inches of snow that fell x hours past noon.
(b) What is the slope of the graph of f?
(c) Interpret the slope as a rate of change.
(d) How much snow had fallen by 4 P.M.?

92. *Graphical Model* An athlete rides a bicycle on a straight road away from home at 20 miles per hour for 1.5 hours. The athlete then turns around and rides home at 15 miles per hour. Sketch a graph that depicts the athlete's distance d from home after t hours.

93. *Ticket Prices* The price of admission to a baseball game is $15 for children and $25 for adults. If a group of 8 people pay $170 to enter the game, find the number of children and the number of adults in the group.

94. *Working Together* One person can shovel the snow from a sidewalk in 2 hours and another person can shovel the same sidewalk in 1.5 hours. How long will it take for them to shovel the snow from the sidewalk if they work together?

95. *Flight of a Golf Ball* If a golf ball is hit upward with a velocity of 88 feet per second (60 miles per hour), then its height h in feet after t seconds can be approximated by
$$h(t) = 88t - 16t^2.$$
(a) What is the height of the golf ball after 2 seconds?
(b) After how long does the golf ball strike the ground?

96. *Height of a Building* A 7-foot-tall stop sign casts a 4-foot-long shadow, while a nearby building casts a 35-foot-long shadow. Find the height of the building.

97. *Traffic Flow* Twenty vehicles per minute can pass through an intersection. If vehicles arrive randomly at an average rate of x per minute, the average waiting time T in minutes is $T = \frac{1}{20 - x}$ for $x < 20$. (*Source:* N. Garber, *Traffic and Highway Engineering.*)
(a) Evaluate the expression for $x = 15$ and interpret the result.
(b) Complete the table.

x	5	10	15	19	19.9
T					

(c) What happens to the waiting time as the traffic rate approaches 20 vehicles per minute?

98. *Error in Measurements* An apprentice carpenter measures the length of a board to be L feet. If the actual measurement of the board is A feet, then the relative error in this measurement is $\left| \frac{L - A}{A} \right|$. If $A = 74$ inches and the relative error is to be less than or equal to 0.005 (0.5%), what values for L are possible?

99. *Determining Costs* The accompanying table shows the costs for purchasing different combinations of burgers, fries, and malts.

Burgers	Fries	Malts	Total Cost
4	3	4	$23
1	2	1	$7
3	1	2	$13

(a) Let b be the cost of a burger, f be the cost of an order of fries, and m be the cost of a malt. Write a system of three linear equations that represents the data in the table.
(b) Solve this system and find the price of each item.

CHAPTER 10

Radical Expressions and Functions

Throughout history, people have created new numbers. Often these new numbers were met with resistance and regarded as being imaginary or unreal. The number 0 was not invented at the same time as the natural numbers. There was no Roman numeral for 0, which is one reason why our calendar started with A.D. 1 and, as a result, the twenty-first century began in 2001. No doubt there were skeptics during the time of the Roman Empire who questioned why anyone needed a number to represent nothing. Negative numbers also met strong resistance. After all, how could anyone possibly have −6 apples?

In this chapter we describe a new number system called *complex numbers*, which involve square roots of negative numbers. The Italian mathematician Cardano (1501–1576) was one of the first mathematicians to work with complex numbers and called them useless. René Descartes (1596–1650) originated the term *imaginary number*, which is associated with complex numbers. However, today complex numbers are used in many applications, such as electricity, fiber optics, and the design of airplanes. We are privileged to study in a period of days what took people centuries to discover.

**Bear in mind that the wonderful things
you learn in schools are the work of many
generations, produced by enthusiastic effort
and infinite labor in every country.
—Albert Einstein**

Source: Historical Topics for the Mathematics Classroom, Thirty-first Yearbook, NCTM.

10.1   RADICAL EXPRESSIONS AND RATIONAL EXPONENTS

Radical Notation · Rational Exponents · Properties of Rational Exponents

INTRODUCTION

Cellular phone technology has become a part of everyday life. In order to have cellular phone coverage, transmission towers, or cellular sites, are spread throughout a region. To estimate the minimum broadcasting distance for each cellular site, radical expressions are needed. (See Example 3.) In this section we discuss radical expressions and rational exponents and show how to manipulate them symbolically.   (*Source:* C. Smith, *Practical Cellular & PCS Design.*)

RADICAL NOTATION

Recall the definition of the square root of a number a.

SQUARE ROOT

The number b is a *square root* of a if $b^2 = a$.

EXAMPLE 1   Finding square roots

Find the square roots of 100.

Solution   The square roots of 100 are 10 *and* -10 because $10^2 = 100$ and $(-10)^2 = 100$. _____

Every positive number a has two square roots, one positive and one negative. Recall that the *positive* square root is called the *principal square root* and is denoted $\sqrt{a}$. The *negative square root* is denoted $-\sqrt{a}$. To identify both square roots we write $\pm\sqrt{a}$. The symbol $\pm$ is read "plus or minus." The symbol $\sqrt{}$ is called the **radical sign**. The expression under the radical sign is called the **radicand**, and an expression containing a radical sign is called a **radical expression**. Examples of radical expressions include

$$\sqrt{6}, \quad 5 + \sqrt{x + 1}, \quad \text{and} \quad \sqrt{\frac{3x}{2x - 1}}.$$

In the next example we show how to find the principal square root of an expression.

EXAMPLE 2   Finding principal square roots

Find the principal square root of each expression.
(a) 25   **(b)** 17   **(c)** 0.49   **(d)** $\frac{4}{9}$ **(e)** $c^2, c > 0$

Solution   **(a)** Because $5 \cdot 5 = 25$, the principal, or positive, square root of 25 is $\sqrt{25} = 5$.
(b) The principal square root of 17 is $\sqrt{17}$. This value is not an integer, but we can approximate it. Figure 10.1 shows that $\sqrt{17} \approx 4.12$, rounded to the nearest hundredth.

Figure 10.1

Note that calculators do not give exact answers when approximating many radical expressions; they give decimal approximations.

(c) Because $(0.7)(0.7) = 0.49$, the principal square root of 0.49 is $\sqrt{0.49} = 0.7$.

(d) Because $\frac{2}{3} \cdot \frac{2}{3} = \frac{4}{9}$, the principal square root of $\frac{4}{9}$ is $\sqrt{\frac{4}{9}} = \frac{2}{3}$.

(e) The principal square root of c^2 is $\sqrt{c^2} = c$, as c is positive.

In the next example we use the principal square root to estimate the minimum transmission distance for a cellular site.

EXAMPLE 3 Estimating cellular phone transmission distance

If the ground is level, a cellular transmission tower will broadcast its signal in roughly a circular pattern, whose radius can be altered by changing the strength of its signal. See Figure 10.2. Suppose that a city has an area of 50 square miles and that there are 10 identical transmission towers spread evenly throughout the city. Estimate a *minimum* transmission radius R for each tower. (Note that a larger distance would probably be necessary to adequately cover the city.)   (*Source:* C. Smith.)

Solution The circular area A covered by one transmission tower is $A = \pi R^2$. The total area covered by 10 towers is $10\pi R^2$, which must equal *at least* 50 square miles.

$$10\pi R^2 = 50$$
$$\pi R^2 = 5 \qquad \text{Divide by 10.}$$
$$R^2 = \frac{5}{\pi} \qquad \text{Divide by } \pi.$$
$$R = \sqrt{\frac{5}{\pi}} \approx 1.26 \qquad \text{Take principal square root.}$$

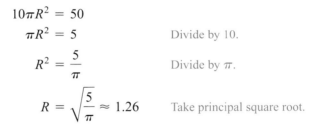

Figure 10.2

Each transmission tower must broadcast with a minimum radius of approximately 1.26 miles.

Another common radical expression is the cube root of a number a, denoted $\sqrt[3]{a}$.

CUBE ROOT

The number b is a *cube root* of a if $b^3 = a$.

Although the square root of a negative number is not a real number, the cube root of a negative number is a negative real number. *Every real number has one real cube root.*

We demonstrate how to find cube roots in the next example.

EXAMPLE 4 Finding cube roots

Find the cube root of each expression.
(a) 8   **(b)** -27 **(c)** 16   **(d)** $\frac{1}{64}$ **(e)** d^6

Solution   **(a)** $\sqrt[3]{8} = 2$ because $2^3 = 2 \cdot 2 \cdot 2 = 8$.

(b) $\sqrt[3]{-27} = -3$ because $(-3)^3 = (-3)(-3)(-3) = -27$.

(c) $\sqrt[3]{16}$ is not an integer. Figure 10.3 shows that $\sqrt[3]{16} \approx 2.52$.

(d) $\sqrt[3]{\frac{1}{64}} = \frac{1}{4}$ because $\left(\frac{1}{4}\right)^3 = \frac{1}{4} \cdot \frac{1}{4} \cdot \frac{1}{4} = \frac{1}{64}$.

(e) $\sqrt[3]{d^6} = d^2$ because $(d^2)^3 = d^2 \cdot d^2 \cdot d^2 = d^{2+2+2} = d^6$.

```
³√(16)
          2.5198421
```

Figure 10.3

Calculator Help

To calculate a cube root, see
Appendix A (page AP-1).

We can generalize square roots and cube roots to include the *n*th root of a number *a*. The number *b* is an **nth root** of *a* if $b^n = a$, where *n* is a positive integer, and the principal *n*th root is denoted $\sqrt[n]{a}$. The number *n* is called the **index**. For the square root the index is 2, although we usually write $\sqrt{a}$ rather than $\sqrt[2]{a}$. When *n* is odd, we are finding an **odd root**, and when *n* is even, we are finding an **even root**. The square root $\sqrt{a}$ is an example of an even root, and the cube root $\sqrt[3]{a}$ is an example of an odd root.

Note: An odd root of a negative number is a negative number, but the even root of a negative number is *not* a real number.

We find *n*th roots in the next example.

EXAMPLE 5   Finding *n*th roots

Find each root, if possible.

(a) $\sqrt[4]{16}$ **(b)** $\sqrt[5]{-32}$ **(c)** $\sqrt[4]{-81}$

Solution   **(a)** $\sqrt[4]{16} = 2$ because $2^4 = 2 \cdot 2 \cdot 2 \cdot 2 = 16$.

(b) $\sqrt[5]{-32} = -2$ because $(-2)^5 = (-2)(-2)(-2)(-2)(-2) = -32$.

(c) The even root of a negative number is not a real number.

Consider the calculations

$$\sqrt{3^2} = \sqrt{9} = 3, \quad \sqrt{(-4)^2} = \sqrt{16} = 4, \quad \text{and} \quad \sqrt{(-6)^2} = \sqrt{36} = 6.$$

In general, the expression $\sqrt{x^2}$ equals $|x|$. Graphical support is shown in Figure 10.4, where the graphs of $Y_1 = \sqrt{(X^2)}$ and $Y_2 = \text{abs}(X)$ appear to be identical.

$[-6, 6, 1]$ by $[-4, 4, 1]$ $[-6, 6, 1]$ by $[-4, 4, 1]$

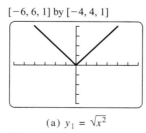

(a) $y_1 = \sqrt{x^2}$

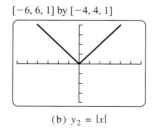

(b) $y_2 = |x|$

Figure 10.4

|||||||| **THE EXPRESSION $\sqrt{x^2}$**

For every real number x, $\sqrt{x^2} = |x|$.

EXAMPLE 6   Simplifying expressions

Write each expression in terms of an absolute value.

(a) $\sqrt{(-3)^2}$ **(b)** $\sqrt{(x+1)^2}$ **(c)** $\sqrt{z^2 - 4z + 4}$

Solution   **(a)** $\sqrt{x^2} = |x|$, so $\sqrt{(-3)^2} = |-3| = 3$

(b) $\sqrt{(x+1)^2} = |x+1|$

(c) $\sqrt{z^2 - 4z + 4} = \sqrt{(z-2)^2} = |z-2|$

RATIONAL EXPONENTS

When m and n are integers, the product rule states that $a^m a^n = a^{m+n}$. This rule can be extended to include exponents that are fractions. For example,

$$4^{1/2} \cdot 4^{1/2} = 4^{1/2 + 1/2} = 4^1 = 4.$$

That is, if we multiply $4^{1/2}$ by itself, the result is 4. Because we also know that $\sqrt{4} \cdot \sqrt{4} = 4$, this discussion suggests that $4^{1/2} = \sqrt{4}$ and motivates the following definition.

THE EXPRESSION $a^{1/n}$

If n is an integer greater than 1, then

$$a^{1/n} = \sqrt[n]{a}.$$

Note: If $a < 0$ and n is an even positive integer, then $a^{1/n}$ is not a real number.

In the next two examples, we show how to interpret rational exponents.

EXAMPLE 7   Interpreting rational exponents

Write each expression in radical notation. Then evaluate the expression to the nearest hundredth when appropriate.

(a) $36^{1/2}$ **(b)** $23^{1/5}$ **(c)** $x^{1/3}$ **(d)** $(5x)^{1/2}$

Solution   **(a)** The exponent $\frac{1}{2}$ indicates a square root. Thus $36^{1/2} = \sqrt{36}$, which evaluates to 6.

(b) The exponent $\frac{1}{5}$ indicates a fifth root. Thus $23^{1/5} = \sqrt[5]{23}$, which is not an integer. Figure 10.5 shows this expression approximated in both exponential and radical notation. In either case $23^{1/5} \approx 1.87$.

(c) The exponent $\frac{1}{3}$ indicates a cube root, so $x^{1/3} = \sqrt[3]{x}$.

(d) The exponent $\frac{1}{2}$ indicates a square root, so $(5x)^{1/2} = \sqrt{5x}$.

```
23^(1/5)
        1.872171231
5^√(23)
        1.872171231
```

Figure 10.5

Calculator Help

To calculate other roots, see Appendix A (page AP-1).

Suppose that we want to define the expression $8^{2/3}$. On the one hand, using properties of exponents we have

$$8^{1/3} \cdot 8^{1/3} = 8^{1/3 + 1/3} = 8^{2/3}.$$

On the other hand, we have

$$8^{1/3} \cdot 8^{1/3} = \sqrt[3]{8} \cdot \sqrt[3]{8} = 2 \cdot 2 = 4.$$

Thus $8^{2/3} = 4$, and that value is obtained whether we interpret $8^{2/3}$ as either

$$8^{2/3} = (8^{1/3})^2 = (\sqrt[3]{8})^2 = 2^2 = 4$$

or

$$8^{2/3} = (8^2)^{1/3} = \sqrt[3]{8^2} = \sqrt[3]{64} = 4.$$

This result suggests the following definition.

THE EXPRESSION $a^{m/n}$

If m and n are positive integers with $\frac{m}{n}$ in lowest terms, then

$$a^{m/n} = \sqrt[n]{a^m} = (\sqrt[n]{a})^m.$$

Note: If $a < 0$ and n is an even integer, then $a^{m/n}$ is not a real number.

EXAMPLE 8   Interpreting rational exponents

Write each expression in radical notation. Then evaluate the expression when the result is an integer.
(a) $(-27)^{2/3}$ **(b)** $12^{3/5}$

Solution **(a)** The exponent $\frac{2}{3}$ indicates that we either take the cube root of -27 and then square it or that we square -27 and then take the cube root. In either case the result will be the same. Thus

$$(-27)^{2/3} = (\sqrt[3]{-27})^2 = (-3)^2 = 9$$

or

$$(-27)^{2/3} = \sqrt[3]{(-27)^2} = \sqrt[3]{729} = 9.$$

(b) The exponent $\frac{3}{5}$ indicates that we either take the fifth root of 12 and then cube it or that we cube 12 and then take the fifth root. Thus

$$12^{3/5} = (\sqrt[5]{12})^3 \quad \text{or} \quad 12^{3/5} = \sqrt[5]{12^3}.$$

This result is not an integer.

Technology Note: *Rational Exponents*

When evaluating expressions with rational (fractional) exponents, be sure to put parentheses around the fraction. For example, most calculators will evaluate 8^(2/3) and 8^2/3 differently. The accompanying figure shows evaluation of $8^{2/3}$ input correctly, 8^(2/3), as 4 but shows evaluation of $8^{2/3}$ input incorrectly, 8^2/3, as $\frac{8^2}{3} = 21.\overline{3}$.

Correct →
Incorrect →

```
8^(2/3)
                    4
8^2/3
       21.33333333
```

From properties of exponents we know that $a^{-n} = \frac{1}{a^n}$, where n is a positive integer. We now define this property for negative rational exponents.

THE EXPRESSION $a^{-m/n}$

If m and n are positive integers with $\frac{m}{n}$ in lowest terms, then

$$a^{-m/n} = \frac{1}{a^{m/n}}, \qquad a \neq 0.$$

EXAMPLE 9 Interpreting negative rational exponents

Write each expression in radical notation and then evaluate.

(a) $(64)^{-1/3}$ **(b)** $(81)^{-3/4}$

Solution **(a)** $(64)^{-1/3} = \dfrac{1}{64^{1/3}} = \dfrac{1}{\sqrt[3]{64}} = \dfrac{1}{4}.$

(b) $(81)^{-3/4} = \dfrac{1}{81^{3/4}} = \dfrac{1}{(\sqrt[4]{81})^3} = \dfrac{1}{3^3} = \dfrac{1}{27}.$

PROPERTIES OF RATIONAL EXPONENTS

Any rational number can be written as a ratio of two integers. That is, if p is a rational number, then $p = \frac{m}{n}$, where m and n are integers. Properties for integer exponents also apply to rational exponents—with one exception. If n is even in the expression $a^{m/n}$ and $\frac{m}{n}$ is written in lowest terms, then a must be nonnegative (not negative) for the result to be a real number.

PROPERTIES OF EXPONENTS

Let p and q be rational numbers written in lowest terms. For all real numbers a and b for which the expressions are real numbers the following properties hold.

1. $a^p \cdot a^q = a^{p+q}$ 　　　　　　　Product rule for exponents

2. $a^{-p} = \dfrac{1}{a^p}, \quad \dfrac{1}{a^{-p}} = a^p$ 　　Negative exponents

3. $\left(\dfrac{a}{b}\right)^{-p} = \left(\dfrac{b}{a}\right)^{p}$ 　　　　　Negative exponents for quotients

4. $\dfrac{a^p}{a^q} = a^{p-q}$ 　　　　　　　Quotient rule for exponents

5. $(a^p)^q = a^{pq}$ 　　　　　　　　Power rule for exponents

6. $(ab)^p = a^p b^p$ 　　　　　　　Power rule for products

7. $\left(\dfrac{a}{b}\right)^{p} = \dfrac{a^p}{b^p}$ 　　　　　　Power rule for quotients

---

In the next two examples, we apply these properties.

EXAMPLE 10 Applying properties of exponents

Write each expression using rational exponents and simplify. Write the answer with a positive exponent. Assume that all variables are positive numbers.

(a) $\sqrt{x} \cdot \sqrt[3]{x}$ **(b)** $\sqrt[3]{27x^2}$ **(c)** $\dfrac{\sqrt[4]{16x}}{\sqrt[3]{x}}$ **(d)** $\left(\dfrac{x^2}{81}\right)^{-1/2}$

Solution **(a)** $\sqrt{x} \cdot \sqrt[3]{x} = x^{1/2} \cdot x^{1/3}$ Use rational exponents.

$\qquad\qquad\qquad = x^{1/2+1/3}$ Product rule for exponents

$\qquad\qquad\qquad = x^{5/6}$ Simplify.

(b) $\sqrt[3]{27x^2} = (27x^2)^{1/3}$ Use rational exponents.

$\qquad\qquad = 27^{1/3}(x^2)^{1/3}$ Power rule for products

$\qquad\qquad = 3x^{2/3}$ Power rule for exponents

(c) $\dfrac{\sqrt[4]{16x}}{\sqrt[3]{x}} = \dfrac{(16x)^{1/4}}{x^{1/3}}$ Use rational exponents.

$\qquad\quad = \dfrac{16^{1/4}x^{1/4}}{x^{1/3}}$ Power rule for products

$\qquad\quad = 16^{1/4}x^{1/4-1/3}$ Quotient rule for exponents

$\qquad\quad = 2x^{-1/12}$ Simplify.

$\qquad\quad = \dfrac{2}{x^{1/12}}$ Negative exponents

(d) $\left(\dfrac{x^2}{81}\right)^{-1/2} = \left(\dfrac{81}{x^2}\right)^{1/2}$ Negative exponents for quotients

$\qquad\qquad = \dfrac{(81)^{1/2}}{(x^2)^{1/2}}$ Power rule for quotients

$\qquad\qquad = \dfrac{9}{x}$ Power rule for exponents; simplify.   ‾‾‾‾

EXAMPLE 11 Applying properties of exponents

Write each expression with positive rational exponents and simplify, if possible.

(a) $\sqrt[3]{4} \cdot \sqrt[6]{4}$ **(b)** $\sqrt[3]{\sqrt{x+1}}$ **(c)** $\sqrt[5]{c^{15}}$ **(d)** $\dfrac{y^{-1/2}}{x^{-1/3}}$ **(e)** $\sqrt{x}(\sqrt{x}-1)$

Solution **(a)** $\sqrt[3]{4} \cdot \sqrt[6]{4} = 4^{1/3} \cdot 4^{1/6} = 4^{1/3+1/6} = 4^{1/2} = \sqrt{4} = 2$

(b) $\sqrt[3]{\sqrt{x+1}} = \left((x+1)^{1/2}\right)^{1/3} = (x+1)^{1/6}$

(c) $\sqrt[5]{c^{15}} = (c^{15})^{1/5} = c^{15/5} = c^3$

(d) $\dfrac{y^{-1/2}}{x^{-1/3}} = \dfrac{x^{1/3}}{y^{1/2}}$

(e) $\sqrt{x}(\sqrt{x}-1) = x^{1/2}(x^{1/2}-1) = x^{1/2}x^{1/2} - x^{1/2} = x - x^{1/2}$ ‾‾‾‾

10.1 PUTTING IT ALL TOGETHER

Properties of radicals and rational exponents are summarized in the following table.

Concept	Explanation	Examples
nth Root of a Real Number	The nth root of a real number a is b if $b^n = a$ and the principal nth root is denoted $\sqrt[n]{a}$. If $a < 0$ and n is even, $\sqrt[n]{a}$ is not a real number.	The square roots of 25 are 5 and -5. The principal square root is $\sqrt{25} = 5$. $\sqrt[3]{-125} = -5$ because $$(-5)^3 = (-5)(-5)(-5) = -125.$$
Rational Exponents	If m and n are positive integers with $\frac{m}{n}$ in lowest terms, $$a^{m/n} = \sqrt[n]{a^m} = \left(\sqrt[n]{a}\right)^m.$$ If $a < 0$ and n is even, $a^{m/n}$ is not a real number.	$8^{4/3} = \left(\sqrt[3]{8}\right)^4 = 2^4 = 16$ and $(-27)^{3/4} = (\sqrt[4]{-27})^3$ is *not* a real number.
Properties of Exponents	Let p and q be rational numbers. For all real numbers a and b for which the expressions are real numbers the following properties hold. 1. $a^p \cdot a^q = a^{p+q}$ 2. $a^{-p} = \dfrac{1}{a^p}, \dfrac{1}{a^{-p}} = a^p$ 3. $\left(\dfrac{a}{b}\right)^{-p} = \left(\dfrac{b}{a}\right)^p$ 4. $\dfrac{a^p}{a^q} = a^{p-q}$ 5. $(a^p)^q = a^{pq}$ 6. $(ab)^p = a^p b^p$ 7. $\left(\dfrac{a}{b}\right)^p = \dfrac{a^p}{b^p}$	$2^{1/3} \cdot 2^{2/3} = 2^{1/3+2/3} = 2^1 = 2$ $2^{-1/2} = \dfrac{1}{2^{1/2}}, \dfrac{1}{3^{-1/4}} = 3^{1/4}$ $\left(\dfrac{3}{4}\right)^{-4/5} = \left(\dfrac{4}{3}\right)^{4/5}$ $\dfrac{7^{2/3}}{7^{1/3}} = 7^{2/3-1/3} = 7^{1/3}$ $(8^{2/3})^{1/2} = 8^{2/6} = 8^{1/3} = 2$ $(2x)^{1/3} = 2^{1/3}x^{1/3}$ $\left(\dfrac{x}{y}\right)^{1/6} = \dfrac{x^{1/6}}{y^{1/6}}$

 **10.1 EXERCISES**

FOR EXTRA HELP

 Student's Solutions Manual

 MyMathLab

 InterAct Math InterAct Math

 Tutor Center Math Tutor Center

 MathXL MathXL

Digital Video Tutor
CD 4 Videotape 13

CONCEPTS

1. What are the square roots of 9?

2. What is the principal square root of 9?

3. What is the cube root of 8?

4. Does every real number have a cube root?

5. If $b^n = a$ and $b > 0$, then $\sqrt[n]{a} = $ _____.

6. Write $a^{1/n}$ in radical notation.

7. Write $a^{m/n}$ in radical notation.

8. Does $(a^{1/3})^2 = a^{2/3}$?

RADICAL EXPRESSIONS

Exercises 9–28: If possible, evaluate the expression by hand. If you cannot, approximate the answer to the nearest hundredth. Variables represent any real number.

9. $\sqrt{9}$

10. $\sqrt{121}$

11. $-\sqrt{5}$

12. $\sqrt{11}$

13. $\sqrt{z^2}$

14. $-\sqrt{(x+2)^2}$

15. $\sqrt[3]{27}$

16. $\sqrt[3]{64}$

17. $\sqrt[3]{-64}$

18. $-\sqrt[3]{-1}$

19. $\sqrt[3]{5}$

20. $\sqrt[3]{-13}$

21. $-\sqrt[3]{x^9}$

22. $\sqrt[3]{(x+1)^6}$

23. $\sqrt[3]{(2x)^6}$

24. $\sqrt[3]{9x^3}$

25. $\sqrt[4]{81}$

26. $\sqrt[5]{-1}$

27. $\sqrt[5]{-7}$

28. $\sqrt[4]{6}$

RATIONAL EXPONENTS

Exercises 29–34: Write the expression in radical notation.

29. $6^{1/2}$

30. $7^{1/3}$

31. $(xy)^{1/2}$

32. $x^{2/3}y^{1/5}$

33. $y^{-1/5}$

34. $\left(\dfrac{x}{y}\right)^{-2/7}$

Exercises 35–56: If possible, evaluate the expression by hand. If you cannot, approximate the answer to the nearest hundredth.

35. $16^{1/2}$

36. $8^{1/3}$

37. $256^{1/4}$

38. $4^{3/2}$

39. $32^{1/5}$

40. $(-32)^{1/5}$

41. $(-8)^{4/3}$

42. $(-1)^{3/5}$

43. $2^{1/2} \cdot 2^{2/3}$

44. $5^{3/5} \cdot 5^{1/10}$

45. $\left(\dfrac{4}{9}\right)^{1/2}$

46. $\left(\dfrac{27}{64}\right)^{1/3}$

47. $\dfrac{4^{2/3}}{4^{1/2}}$

48. $\dfrac{6^{1/5} \cdot 6^{3/5}}{6^{2/5}}$

49. $4^{-1/2}$

50. $9^{-3/2}$

51. $(-8)^{-1/3}$

52. $(49)^{-1/2}$

53. $\left(\dfrac{1}{16}\right)^{-1/4}$

54. $\left(\dfrac{16}{25}\right)^{-3/2}$

55. $(2^{1/2})^3$

56. $(5^{6/5})^{-1/2}$

Exercises 57–86: Simplify the expression. Assume that all variables are positive.

57. $(x^2)^{3/2}$

58. $(y^4)^{1/2}$

59. $(x^2y^8)^{1/2}$

60. $(y^{10}z^4)^{1/4}$

61. $\sqrt[3]{x^3y^6}$

62. $\sqrt{16x^4}$

63. $\sqrt{\dfrac{y^4}{x^2}}$

64. $\sqrt[3]{\dfrac{x^{12}}{z^6}}$

65. $\sqrt{y^3} \cdot \sqrt[3]{y^2}$

66. $\left(\dfrac{x^6}{81}\right)^{1/4}$

67. $\left(\dfrac{x^6}{27}\right)^{2/3}$

68. $\left(\dfrac{1}{x^8}\right)^{-1/4}$

69. $\left(\dfrac{x^2}{y^6}\right)^{-1/2}$

70. $\dfrac{\sqrt{x}}{\sqrt[3]{27x^6}}$

71. $\sqrt{\sqrt{y}}$

72. $\sqrt{\sqrt[3]{(3x)^2}}$

73. $(a^{-1/2})^{4/3}$

74. $(x^{-3/2})^{2/3}$

75. $(a^3b^6)^{1/3}$

76. $(64x^3y^{18})^{1/6}$

77. $\dfrac{(k^{1/2})^{-3}}{(k^2)^{1/4}}$

78. $\dfrac{(b^{3/4})^4}{(b^{4/5})^{-5}}$

79. $\sqrt[3]{b} \cdot \sqrt[4]{b}$

80. $\sqrt[3]{t} \cdot \sqrt[5]{t}$

81. $\sqrt{z} \cdot \sqrt[3]{z^2} \cdot \sqrt[4]{z^3}$

82. $\sqrt{b} \cdot \sqrt[3]{b} \cdot \sqrt[5]{b}$

83. $p^{1/2}(p^{3/2} + p^{1/2})$

84. $d^{3/4}(d^{1/4} - d^{-1/4})$

85. $\sqrt[3]{x}\left(\sqrt{x} - \sqrt[3]{x^2}\right)$

86. $\dfrac{1}{2}\sqrt{x}\left(\sqrt{x} + \sqrt[4]{x^2}\right)$

Exercises 87–102: Simplify the expression. Assume that all variables are real numbers.

87. $\sqrt{(-4)^2}$

88. $\sqrt{9^2}$

89. $\sqrt{y^2}$

90. $\sqrt{z^4}$

91. $\sqrt{(a+3)^2}$

92. $\sqrt{(a-b)^2}$

93. $\sqrt{(x-5)^2}$

94. $\sqrt{(2x-1)^2}$

95. $\sqrt{x^2 - 2x + 1}$ **96.** $\sqrt{4x^2 + 4x + 1}$

97. $\sqrt[4]{y^4}$ **98.** $\sqrt[4]{x^8 z^4}$

99. $\sqrt[4]{x^{12}}$ **100.** $\sqrt[6]{x^6}$

101. $\sqrt[5]{x^5 y^{10}}$ **102.** $\sqrt[5]{32(x + 4)^5}$

APPLICATIONS

103. *Cellular Phone Technology*   (Refer to Example 3.) Suppose that a city has an area of 65 square miles and 15 cellular transmission towers spread evenly throughout it. Estimate a minimum radius R for each tower.

104. *Musical Tones*   One octave on a piano contains 12 keys (including both the black and white keys). The frequency of each successive key increases by a factor of $2^{1/12}$. For example, middle C is two keys below the first D above it. Therefore the frequency of this D is

$$2^{1/12} \cdot 2^{1/12} = 2^{1/6} \approx 1.12$$

times greater than middle C.
(a) If two tones are one octave apart, how do their frequencies compare?
(b) The A tone below middle C has a frequency of 220 cycles per second. Middle C is 3 keys above this A note. Estimate the frequency of middle C.

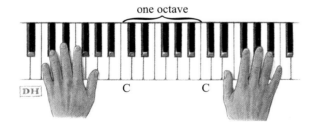

Exercises 105 and 106: Heron's Formula   *Suppose the lengths of the sides of a triangle are a, b, and c as illustrated in the figure.*

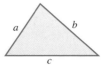

*If the **semiperimeter** of the triangle is $s = \frac{1}{2}(a + b + c)$, then the area of the triangle is*

$$A = \sqrt{s(s - a)(s - b)(s - c)}.$$

Find the area of the triangle with the given sides.

105. $a = 3, b = 4, c = 5$ **106.** $a = 5, b = 9, c = 10$

WRITING ABOUT MATHEMATICS

107. Try to calculate $\sqrt{-7}$, $\sqrt[4]{-56}$, and $\sqrt[6]{-10}$ with a calculator. Describe what happens when you evaluate an even root of a negative number. Does the same difficulty occur when you evaluate an odd root of a negative number? Try to evaluate $\sqrt[3]{-7}$, $\sqrt[5]{-56}$, and $\sqrt[7]{-10}$. Explain.

108. Explain the difference between a root and a power of a number. Give examples.

10.2   SIMPLIFYING RADICAL EXPRESSIONS

Product Rule for Radical Expressions · Quotient Rule for Radical Expressions · Rationalizing Denominators Having Square Roots

INTRODUCTION

In this section we discuss performing arithmetic operations with radical expressions. We demonstrate use of these skills to solve equations that contain radical expressions.

PRODUCT RULE FOR RADICAL EXPRESSIONS

Consider the following examples of multiplying radical expressions.

$$\sqrt{4} \cdot \sqrt{25} = 2 \cdot 5 = 10 \quad \text{and} \quad \sqrt{4 \cdot 25} = \sqrt{100} = 10$$

implies that

$$\sqrt{4} \cdot \sqrt{25} = \sqrt{4 \cdot 25} \quad \text{(see Figure 10.6(a))}.$$

Similarly,

$$\sqrt[3]{8} \cdot \sqrt[3]{27} = 2 \cdot 3 = 6 \quad \text{and} \quad \sqrt[3]{8 \cdot 27} = \sqrt[3]{216} = 6$$

implies that

$$\sqrt[3]{8} \cdot \sqrt[3]{27} = \sqrt[3]{8 \cdot 27} \quad \text{(see Figure 10.6(b))}.$$

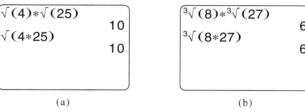

(a)   (b)

Figure 10.6

These examples suggest that the product of two (like) roots is equal to the root of their product.

PRODUCT RULE FOR RADICAL EXPRESSIONS

Let a and b be real numbers, where $\sqrt[n]{a}$ and $\sqrt[n]{b}$ are both defined. Then

$$\sqrt[n]{a} \cdot \sqrt[n]{b} = \sqrt[n]{a \cdot b}.$$

Note: The product rule only works when the radicals have the *same* index. For example, the product $\sqrt{2} \cdot \sqrt[3]{4}$ cannot be simplified because the indexes are 2 and 3. However, by using rational exponents, we can simplify this product. See Example 5.

We apply the product rule in the next two examples.

EXAMPLE 1   Multiplying radical expressions

Multiply each pair of radical expressions.

(a) $\sqrt{5} \cdot \sqrt{20}$ **(b)** $\sqrt[3]{-3} \cdot \sqrt[3]{9}$ **(c)** $\sqrt[3]{-5} \cdot \sqrt[3]{7}$ **(d)** $\sqrt[4]{\frac{1}{3}} \cdot \sqrt[4]{\frac{1}{9}} \cdot \sqrt[4]{\frac{1}{3}}$

Solution **(a)** $\sqrt{5} \cdot \sqrt{20} = \sqrt{5 \cdot 20} = \sqrt{100} = 10$

(b) $\sqrt[3]{-3} \cdot \sqrt[3]{9} = \sqrt[3]{-3 \cdot 9} = \sqrt[3]{-27} = -3$

(c) $\sqrt[3]{-5} \cdot \sqrt[3]{7} = \sqrt[3]{-5 \cdot 7} = \sqrt[3]{-35}$

(d) The product rule can also be applied to three or more factors. Thus

$$\sqrt[4]{\frac{1}{3}} \cdot \sqrt[4]{\frac{1}{9}} \cdot \sqrt[4]{\frac{1}{3}} = \sqrt[4]{\frac{1}{3} \cdot \frac{1}{9} \cdot \frac{1}{3}} = \sqrt[4]{\frac{1}{81}} = \frac{1}{3}$$

because $\frac{1}{3} \cdot \frac{1}{3} \cdot \frac{1}{3} \cdot \frac{1}{3} = \frac{1}{81}$.

EXAMPLE 2   Multiplying radical expressions containing variables

Multiply each pair of radical expressions. Assume that all variables are positive.

(a) $\sqrt{x} \cdot \sqrt{x^3}$ **(b)** $\sqrt[3]{2a} \cdot \sqrt[3]{5a}$ **(c)** $\sqrt{11} \cdot \sqrt{xy}$ **(d)** $\sqrt[5]{\frac{2x}{y}} \cdot \sqrt[5]{\frac{16y}{x}}$

Solution   **(a)** $\sqrt{x} \cdot \sqrt{x^3} = \sqrt{x \cdot x^3} = \sqrt{x^4} = x^2$

(b) $\sqrt[3]{2a} \cdot \sqrt[3]{5a} = \sqrt[3]{2a \cdot 5a} = \sqrt[3]{10a^2}$

(c) $\sqrt{11} \cdot \sqrt{xy} = \sqrt{11xy}$

(d) $\sqrt[5]{\frac{2x}{y}} \cdot \sqrt[5]{\frac{16y}{x}} = \sqrt[5]{\frac{2x}{y} \cdot \frac{16y}{x}}$ Product rule

$$= \sqrt[5]{\frac{32xy}{xy}} \qquad \text{Multiply fractions.}$$

$$= \sqrt[5]{32} \qquad \text{Reduce.}$$

$$= 2 \qquad 2^5 = 32$$

An integer a is a **perfect nth power** if there exists an integer b such that $b^n = a$. Thus 36 is a **perfect square** because $6^2 = 36$, 8 is a **perfect cube** because $2^3 = 8$, and 81 is a *perfect fourth power* because $3^4 = 81$.

The product rule for radicals can be used to simplify radical expressions. For example, because the largest perfect square factor of 50 is 25, the expression $\sqrt{50}$ can be simplified as

$$\sqrt{50} = \sqrt{25} \cdot \sqrt{2} = 5\sqrt{2}.$$

This procedure is generalized as follows.

SIMPLIFYING RADICALS (nth ROOTS)

1. Determine the largest perfect nth power factor of the radicand.
2. Use the product rule to factor out and simplify this perfect nth power.

EXAMPLE 3   Simplifying radical expressions

Simplify each expression.

(a) $\sqrt{300}$ **(b)** $\sqrt[3]{16}$ **(c)** $\sqrt{54}$ **(d)** $\sqrt[4]{512}$

Solution   **(a)** First note that $300 = 100 \cdot 3$ and that 100 is the largest perfect square factor of 300. Thus

$$\sqrt{300} = \sqrt{100} \cdot \sqrt{3} = 10\sqrt{3}.$$

(b) The largest perfect cube factor of 16 is 8. Thus $\sqrt[3]{16} = \sqrt[3]{8} \cdot \sqrt[3]{2} = 2\sqrt[3]{2}$.

(c) $\sqrt{54} = \sqrt{9} \cdot \sqrt{6} = 3\sqrt{6}$

(d) $\sqrt[4]{512} = \sqrt[4]{256} \cdot \sqrt[4]{2} = 4\sqrt[4]{2}$ because $4^4 = 256$.

Note: To simplify a cube root of a negative number we factor out the negative of the largest perfect cube factor. For example, $-16 = -8 \cdot 2$, so $\sqrt[3]{-16} = \sqrt[3]{-8} \cdot \sqrt[3]{2} = -2\sqrt[3]{2}$. This procedure can be used with any negative odd root of a number.

EXAMPLE 4 Simplifying radical expressions

Simplify each expression. Assume that all variables are positive.

(a) $\sqrt{25x^4}$ **(b)** $\sqrt{32n^3}$ **(c)** $\sqrt[3]{-16x^3y^5}$ **(d)** $\sqrt[3]{2a} \cdot \sqrt[3]{4a^2b}$

Solution **(a)** $\sqrt{25x^4} = 5x^2$

(b) $\sqrt{32n^3} = \sqrt{(16n^2)2n}$ $16n^2$ is the largest perfect square factor.

$\qquad\qquad = \sqrt{16n^2} \cdot \sqrt{2n}$ Product rule

$\qquad\qquad = 4n\sqrt{2n}$ $(4n)^2 = 16n^2$

(c) $\sqrt[3]{-16x^3y^5} = \sqrt[3]{(-8x^3y^3)2y^2}$ $8x^3y^3$ is the largest perfect cube factor.

$\qquad\qquad = \sqrt[3]{-8x^3y^3} \cdot \sqrt[3]{2y^2}$ Product rule

$\qquad\qquad = -2xy\sqrt[3]{2y^2}$ $(-2xy)^3 = -8x^3y^3$

(d) $\sqrt[3]{2a} \cdot \sqrt[3]{4a^2b} = \sqrt[3]{(2a)(4a^2b)}$ Product rule

$\qquad\qquad = \sqrt[3]{(8a^3)b}$ $8a^3$ is the largest perfect cube factor.

$\qquad\qquad = \sqrt[3]{8a^3} \cdot \sqrt[3]{b}$ Product rule

$\qquad\qquad = 2a\sqrt[3]{b}$ $(2a)^3 = 8a^3$

The product rule for radical expressions cannot be used if the radicals do not have the same indexes. In this case we use rational exponents, as illustrated in the next example.

EXAMPLE 5 Multiplying radicals with different indexes

Simplify each expression. Write your answer in radical notation.

(a) $\sqrt{5} \cdot \sqrt[4]{5}$ **(b)** $\sqrt{2} \cdot \sqrt[3]{4}$ **(c)** $\sqrt[3]{x} \cdot \sqrt[4]{x}$

Solution **(a)** Because $\sqrt{5} = 5^{1/2}$ and $\sqrt[4]{5} = 5^{1/4}$,

$$\sqrt{5} \cdot \sqrt[4]{5} = 5^{1/2} \cdot 5^{1/4} = 5^{1/2+1/4} = 5^{3/4}.$$

In radical notation, $5^{3/4} = \sqrt[4]{5^3} = \sqrt[4]{125}$.

(b) First note that $\sqrt[3]{4} = \sqrt[3]{2^2} = 2^{2/3}$. Thus

$$\sqrt{2} \cdot \sqrt[3]{4} = 2^{1/2} \cdot 2^{2/3} = 2^{1/2+2/3} = 2^{7/6}.$$

In radical notation, $2^{7/6} = \sqrt[6]{2^7} = \sqrt[6]{2^6 \cdot 2^1} = \sqrt[6]{2^6} \cdot \sqrt[6]{2} = 2\sqrt[6]{2}$.

(c) $\sqrt[3]{x} \cdot \sqrt[4]{x} = x^{1/3} \cdot x^{1/4} = x^{7/12} = \sqrt[12]{x^7}$

QUOTIENT RULE FOR RADICAL EXPRESSIONS

Consider the following examples of dividing radical expressions.

```
√(4/9)▶Frac
              2/3
√(4)/√(9)▶Frac
              2/3
```

Figure 10.7

Calculator Help
To use the Frac feature, see
Appendix A (page AP-2).

$$\sqrt{\frac{4}{9}} = \sqrt{\frac{2}{3} \cdot \frac{2}{3}} = \frac{2}{3} \quad \text{and} \quad \frac{\sqrt{4}}{\sqrt{9}} = \frac{2}{3}$$

implies that

$$\sqrt{\frac{4}{9}} = \frac{\sqrt{4}}{\sqrt{9}}. \quad \text{See Figure 10.7.}$$

These examples suggest that the root of a quotient is equal to the quotient of the roots.

QUOTIENT RULE FOR RADICAL EXPRESSIONS

Let a and b be real numbers, where $\sqrt[n]{a}$ and $\sqrt[n]{b}$ are both defined and $b \neq 0$. Then

$$\sqrt[n]{\frac{a}{b}} = \frac{\sqrt[n]{a}}{\sqrt[n]{b}}.$$

EXAMPLE 6 Simplifying quotients

Simplify each radical expression. Assume that all variables are positive.

(a) $\sqrt[3]{\dfrac{5}{8}}$ **(b)** $\sqrt[4]{\dfrac{x}{16}}$ **(c)** $\sqrt{\dfrac{16}{y^2}}$

Solution **(a)** $\sqrt[3]{\dfrac{5}{8}} = \dfrac{\sqrt[3]{5}}{\sqrt[3]{8}} = \dfrac{\sqrt[3]{5}}{2}$

(b) $\sqrt[4]{\dfrac{x}{16}} = \dfrac{\sqrt[4]{x}}{\sqrt[4]{16}} = \dfrac{\sqrt[4]{x}}{2}$

(c) $\sqrt{\dfrac{16}{y^2}} = \dfrac{\sqrt{16}}{\sqrt{y^2}} = \dfrac{4}{y}$ because $y > 0$.

═══ MAKING CONNECTIONS ═══

Rules for Radical Expressions and Rational Exponents

The rules for radical expressions are a result of the properties of rational exponents.

$$\sqrt[n]{a \cdot b} = \sqrt[n]{a} \cdot \sqrt[n]{b} \quad \text{is equivalent to} \quad (a \cdot b)^{1/n} = a^{1/n} \cdot b^{1/n}.$$

$$\sqrt[n]{\frac{a}{b}} = \frac{\sqrt[n]{a}}{\sqrt[n]{b}} \quad \text{is equivalent to} \quad \left(\frac{a}{b}\right)^{1/n} = \frac{a^{1/n}}{b^{1/n}}.$$

EXAMPLE 7 Simplifying radical expressions

Simplify each radical expression. Assume that all variables are positive.

(a) $\dfrac{\sqrt{40}}{\sqrt{10}}$ **(b)** $\sqrt[4]{\dfrac{16x^3}{y^4}}$ **(c)** $\sqrt{\dfrac{5a^2}{8}} \cdot \sqrt{\dfrac{5a^3}{2}}$

Solution **(a)** $\dfrac{\sqrt{40}}{\sqrt{10}} = \sqrt{\dfrac{40}{10}} = \sqrt{4} = 2$

(b) $\sqrt[4]{\dfrac{16x^3}{y^4}} = \dfrac{\sqrt[4]{16x^3}}{\sqrt[4]{y^4}} = \dfrac{\sqrt[4]{16} \cdot \sqrt[4]{x^3}}{\sqrt[4]{y^4}} = \dfrac{2\sqrt[4]{x^3}}{y}$

(c) To simplify this expression, we use both the product and quotient rules.

$$\sqrt{\dfrac{5a^2}{8}} \cdot \sqrt{\dfrac{5a^3}{2}} = \sqrt{\dfrac{25a^5}{16}} \qquad \text{Product rule}$$

$$= \dfrac{\sqrt{25a^5}}{\sqrt{16}} \qquad \text{Quotient rule}$$

$$= \dfrac{\sqrt{(25a^4)} \cdot \sqrt{a}}{\sqrt{16}} \qquad \text{Factor out largest perfect square.}$$

$$= \dfrac{5a^2\sqrt{a}}{4} \qquad (5a^2)^2 = 25a^4.$$

RATIONALIZING DENOMINATORS HAVING SQUARE ROOTS

Quotients containing radical expressions can appear to be different but actually be equal. For example, $\dfrac{1}{\sqrt{3}}$ and $\dfrac{\sqrt{3}}{3}$ represent the same real number even though they look like they are unequal. To show that they are equal, we multiply the first quotient by 1 in the form $\dfrac{\sqrt{3}}{\sqrt{3}}$:

$$\dfrac{1}{\sqrt{3}} \cdot \dfrac{\sqrt{3}}{\sqrt{3}} = \dfrac{1 \cdot \sqrt{3}}{\sqrt{3} \cdot \sqrt{3}} = \dfrac{\sqrt{3}}{3}.$$

Note: $\sqrt{b} \cdot \sqrt{b} = \sqrt{b^2} = b$ for any *positive* number b.

One way to standardize radical expressions is to remove any radical expressions from the denominator. This process is called **rationalizing the denominator**. Exercise 105 suggests one reason why people rationalized denominators before calculators were invented. The next example demonstrates how to rationalize the denominator of several quotients.

EXAMPLE 8   Rationalizing the denominator

Rationalize each denominator. Assume that all variables are positive.

(a) $\dfrac{1}{\sqrt{2}}$ **(b)** $\dfrac{3}{5\sqrt{3}}$ **(c)** $\sqrt{\dfrac{x}{24}}$ **(d)** $\dfrac{xy}{\sqrt{y^3}}$

Solution **(a)** We start by multiplying this expression by 1 in the form $\dfrac{\sqrt{2}}{\sqrt{2}}$:

$$\dfrac{1}{\sqrt{2}} \cdot \dfrac{\sqrt{2}}{\sqrt{2}} = \dfrac{\sqrt{2}}{\sqrt{4}} = \dfrac{\sqrt{2}}{2}.$$

Note that the expression $\dfrac{\sqrt{2}}{2}$ does not have a radical in the denominator.

(b) We multiply this expression by 1 in the form $\frac{\sqrt{3}}{\sqrt{3}}$:

$$\frac{3}{5\sqrt{3}} \cdot \frac{\sqrt{3}}{\sqrt{3}} = \frac{3\sqrt{3}}{5\sqrt{9}} = \frac{3\sqrt{3}}{5 \cdot 3} = \frac{\sqrt{3}}{5}.$$

(c) Because $\sqrt{24} = \sqrt{4} \cdot \sqrt{6} = 2\sqrt{6}$, we start by simplifying the expression.

$$\sqrt{\frac{x}{24}} = \frac{\sqrt{x}}{\sqrt{24}} = \frac{\sqrt{x}}{2\sqrt{6}}$$

To rationalize the denominator we multiply this expression by 1 in the form $\frac{\sqrt{6}}{\sqrt{6}}$:

$$\frac{\sqrt{x}}{2\sqrt{6}} = \frac{\sqrt{x}}{2\sqrt{6}} \cdot \frac{\sqrt{6}}{\sqrt{6}} = \frac{\sqrt{6x}}{12}.$$

(d) Because $\sqrt{y^3} = \sqrt{y^2} \cdot \sqrt{y} = y\sqrt{y}$ and $y > 0$, we start by simplifying the expression.

$$\frac{xy}{\sqrt{y^3}} = \frac{xy}{y\sqrt{y}} = \frac{x}{\sqrt{y}}$$

To rationalize the denominator we multiply by 1 in the form $\frac{\sqrt{y}}{\sqrt{y}}$:

$$\frac{x}{\sqrt{y}} \cdot \frac{\sqrt{y}}{\sqrt{y}} = \frac{x\sqrt{y}}{y}.$$

 10.2 PUTTING IT ALL TOGETHER

In this section we discussed how to simplify radical expressions. Results are summarized in the following table.

Procedure	Explanation	Examples
Product Rule for Radical Expressions	Let a and b be real numbers, where $\sqrt[n]{a}$ and $\sqrt[n]{b}$ are both defined. Then $$\sqrt[n]{a} \cdot \sqrt[n]{b} = \sqrt[n]{a \cdot b}.$$	$$\sqrt{2} \cdot \sqrt{32} = \sqrt{64} = 8$$
Quotient Rule for Radical Expressions	Let a and b be real numbers, where $\sqrt[n]{a}$ and $\sqrt[n]{b}$ are both defined and $b \neq 0$. Then $$\sqrt[n]{\frac{a}{b}} = \frac{\sqrt[n]{a}}{\sqrt[n]{b}}.$$	$$\frac{\sqrt{60}}{\sqrt{15}} = \sqrt{\frac{60}{15}} = \sqrt{4} = 2 \quad \text{and}$$ $$\sqrt[3]{\frac{x^2}{-27}} = \frac{\sqrt[3]{x^2}}{\sqrt[3]{-27}} = \frac{\sqrt[3]{x^2}}{-3} = \frac{-\sqrt[3]{x^2}}{3}$$
Rationalizing the Denominator	Write the quotient without a radical expression in the denominator.	To rationalize $\frac{5}{\sqrt{7}}$ multiply the expression by 1 in the form $\frac{\sqrt{7}}{\sqrt{7}}$: $$\frac{5}{\sqrt{7}} \cdot \frac{\sqrt{7}}{\sqrt{7}} = \frac{5\sqrt{7}}{\sqrt{49}} = \frac{5\sqrt{7}}{7}.$$

10.2 EXERCISES

CONCEPTS

1. Does $\sqrt{2} \cdot \sqrt{3}$ equal $\sqrt{6}$?

2. Does $\sqrt{5} \cdot \sqrt[3]{5}$ equal 5?

3. $\sqrt[3]{a} \cdot \sqrt[3]{b} = $ _____

4. $\dfrac{\sqrt{a}}{\sqrt{b}} = \sqrt{?}$

5. $\dfrac{\sqrt[n]{a}}{\sqrt[n]{b}} = \sqrt[n]{?}$

6. To rationalize the denominator for $\dfrac{1}{\sqrt{3}}$, multiply by 1 in the form _____.

7. Is $\sqrt[3]{3}$ equal to 1? Explain.

8. Is 64 a perfect cube? Explain.

MULTIPLYING AND DIVIDING

Exercises 9–50: Simplify the expression. Assume that all variables are positive.

9. $\sqrt{3} \cdot \sqrt{3}$

10. $\sqrt{2} \cdot \sqrt{18}$

11. $\sqrt{2} \cdot \sqrt{50}$

12. $\sqrt[3]{-2} \cdot \sqrt[3]{-4}$

13. $\sqrt[3]{4} \cdot \sqrt[3]{16}$

14. $\sqrt[3]{x} \cdot \sqrt[3]{x^2}$

15. $\sqrt{\dfrac{9}{25}}$

16. $\sqrt[3]{\dfrac{x}{8}}$

17. $\sqrt{\dfrac{1}{2}} \cdot \sqrt{\dfrac{1}{8}}$

18. $\sqrt{\dfrac{5}{3}} \cdot \sqrt{\dfrac{1}{3}}$

19. $\sqrt{\dfrac{x}{2}} \cdot \sqrt{\dfrac{x}{8}}$

20. $\sqrt{\dfrac{4}{y}} \cdot \sqrt{\dfrac{y}{5}}$

21. $\dfrac{\sqrt{45}}{\sqrt{5}}$

22. $\dfrac{\sqrt{7}}{\sqrt{28}}$

23. $\sqrt[3]{-4} \cdot \sqrt[3]{-16}$

24. $\sqrt[3]{9} \cdot \sqrt[3]{3}$

25. $\sqrt[4]{9} \cdot \sqrt[4]{9}$

26. $\sqrt[5]{16} \cdot \sqrt[5]{-2}$

27. $\dfrac{\sqrt[5]{64}}{\sqrt[5]{-2}}$

28. $\dfrac{\sqrt[4]{324}}{\sqrt[4]{4}}$

29. $\dfrac{\sqrt{a^2 b}}{\sqrt{b}}$

30. $\dfrac{\sqrt{4xy^2}}{\sqrt{x}}$

31. $\dfrac{\sqrt[3]{54}}{\sqrt[3]{2}}$

32. $\dfrac{\sqrt[3]{x^3 y^7}}{\sqrt[3]{y^4}}$

33. $\sqrt{4x^4}$

34. $\sqrt[3]{-8y^3}$

35. $\sqrt[3]{-5a^6}$

36. $\sqrt{9x^2 y}$

37. $\sqrt[4]{16x^4 y}$

38. $\sqrt[3]{8xy^3}$

39. $\sqrt{3x} \cdot \sqrt{12x}$

40. $\sqrt{6x^5} \cdot \sqrt{6x}$

41. $\sqrt[3]{8x^6 y^3 z^9}$

42. $\sqrt{16x^4 y^6}$

43. $\sqrt[4]{\dfrac{3}{4}} \cdot \sqrt[4]{\dfrac{27}{4}}$

44. $\sqrt[5]{\dfrac{4}{-9}} \cdot \sqrt[5]{\dfrac{8}{-27}}$

45. $\sqrt[3]{12} \cdot \sqrt[3]{ab}$

46. $\sqrt{5x} \cdot \sqrt{5z}$

47. $\sqrt[4]{25z} \cdot \sqrt[4]{25z}$

48. $\sqrt[5]{3z^2} \cdot \sqrt[5]{7z}$

49. $\sqrt[5]{\dfrac{7a}{b^2}} \cdot \sqrt[5]{\dfrac{b^2}{7a^6}}$

50. $\sqrt[3]{\dfrac{8m}{n}} \cdot \sqrt[3]{\dfrac{n^4}{m^2}}$

Exercises 51–56: Use properties of polynomials to simplify the expression. Assume all radicands are positive.

51. $\sqrt{x + 4} \cdot \sqrt{x - 4}$

52. $\sqrt[3]{x - 1} \cdot \sqrt[3]{x^2 + x + 1}$

53. $\sqrt[3]{a + 1} \cdot \sqrt[3]{a^2 - a + 1}$

54. $\sqrt{b - 1} \cdot \sqrt{b + 1}$

55. $\dfrac{\sqrt{x^2 + 2x + 1}}{\sqrt{x + 1}}$

56. $\dfrac{\sqrt{x^2 - 4x + 4}}{\sqrt{x - 2}}$

Exercises 57–62: Complete the equation.

57. $\sqrt{500} = $ _____ $\sqrt{5}$

58. $\sqrt{28} = $ _____ $\sqrt{7}$

59. $\sqrt{8} = $ _____ $\sqrt{2}$

60. $\sqrt{99} = $ _____ $\sqrt{11}$

61. $\sqrt{45} = $ _____ $\sqrt{5}$

62. $\sqrt{243} = $ _____ $\sqrt{3}$

Exercises 63–82: Simplify the radical expression by factoring out the largest perfect nth power. Assume that all variables are positive.

63. $\sqrt{200}$

64. $\sqrt{72}$

65. $\sqrt[3]{81}$

66. $\sqrt[3]{256}$

67. $\sqrt[4]{64}$

68. $\sqrt[5]{27 \cdot 81}$

69. $\sqrt[5]{-64}$

70. $\sqrt[3]{-81}$

71. $\sqrt{b^5}$

72. $\sqrt{t^3}$

73. $\sqrt{8n^3}$

74. $\sqrt{32a^2}$

75. $\sqrt{12a^2b^5}$

76. $\sqrt{20a^3b^2}$

77. $\sqrt[3]{125x^4y^5}$

78. $\sqrt[3]{81a^5b^2}$

79. $\sqrt[3]{5t} \cdot \sqrt[3]{125t}$

80. $\sqrt[4]{4bc^3} \cdot \sqrt[4]{64ab^3c^2}$

81. $\sqrt[4]{\dfrac{9t^5}{r^8}} \cdot \sqrt[4]{\dfrac{9r}{5t}}$

82. $\sqrt[5]{\dfrac{4t^6}{r}} \cdot \sqrt[5]{\dfrac{8t}{r^6}}$

Exercises 83–92: Simplify the expression. Assume that all variables are positive and write your answer in radical notation.

83. $\sqrt{3} \cdot \sqrt[3]{3}$

84. $\sqrt{5} \cdot \sqrt[3]{5}$

85. $\sqrt[4]{8} \cdot \sqrt[3]{4}$

86. $\sqrt[5]{16} \cdot \sqrt{2}$

87. $\sqrt[4]{27} \cdot \sqrt[3]{9} \cdot \sqrt{3}$

88. $\sqrt[5]{16} \cdot \sqrt[3]{16}$

89. $\sqrt[4]{x^3} \cdot \sqrt[3]{x}$

90. $\sqrt[4]{x^3} \cdot \sqrt{x}$

91. $\sqrt[4]{rt} \cdot \sqrt[3]{r^2t}$

92. $\sqrt[3]{a^3b^2} \cdot \sqrt{a^2b}$

Exercises 93–102: Rationalize the denominator.

93. $\dfrac{1}{\sqrt{7}}$

94. $\dfrac{1}{\sqrt{23}}$

95. $\dfrac{4}{\sqrt{3}}$

96. $\dfrac{8}{\sqrt{2}}$

97. $\dfrac{5}{3\sqrt{5}}$

98. $\dfrac{6}{11\sqrt{3}}$

99. $\sqrt{\dfrac{b}{12}}$

100. $\sqrt{\dfrac{5b}{72}}$

101. $\dfrac{rt}{2\sqrt{r^3}}$

102. $\dfrac{m^2n}{2\sqrt{m^5}}$

APPLICATIONS

103. *Bird Wings* Heavier birds tend to have larger wings than lighter birds do. For some birds the relationship between the surface area A of the bird's wings in square inches and its weight W in pounds can be modeled by $A = 100\sqrt[3]{W^2}$. (*Source:* C. Pennycuick, *Newton Rules Biology.*)
 (a) Find the area of the wings when the weight is 8 pounds.
 (b) Write this formula with rational exponents.

104. *Orbits and Distance* Johannes Kepler (1571–1630) discovered a relationship between a planet's distance D from the sun and the time T it takes to orbit the sun. This formula is given by $T = \sqrt{D^3}$, where T is in Earth years and $D = 1$ corresponds to the distance between Earth and the sun, or 93,000,000 miles.
 (a) Neptune is 30 times farther from the sun than Earth ($D = 30$). Estimate the number of years required for Neptune to orbit the sun.
 (b) Write this formula with rational exponents.

WRITING ABOUT MATHEMATICS

105. Suppose that a student knows that $\sqrt{3} \approx 1.73205$ and does not have a calculator. Which of the following expressions would be easier to evaluate by hand? Why?

$$\dfrac{1}{\sqrt{3}} \quad \text{or} \quad \dfrac{\sqrt{3}}{3}$$

106. Explain how the product and quotient rules for radical expressions are the result of properties of rational exponents.

---

1. Find the following.
   (a) The square roots of 49
   (b) The principal square root of 49
   (c) The solutions to $x^2 = 49$

2. Evaluate.
   (a) $\sqrt[3]{-8}$ **(b)** $-\sqrt[4]{81}$

3. Write the expression in radical notation.
   (a) $x^{3/2}$ **(b)** $x^{2/3}$ **(c)** $x^{-2/5}$

4. Simplify $\sqrt{(x-1)^2}$ for any real number x.

5. Simplify each expression. Assume that all variables are positive.
   (a) $(64^{-3/2})^{1/3}$ **(b)** $\sqrt{5} \cdot \sqrt{20}$
   (c) $\sqrt[3]{-8x^4y}$ **(d)** $\sqrt{\dfrac{4b}{5}} \cdot \sqrt{\dfrac{4b^3}{5}}$

6. Simplify $\sqrt[3]{7} \cdot \sqrt{7}$.

7. Rationalize the denominator for $\dfrac{6}{2\sqrt{6}}$.

10.3   OPERATIONS ON RADICAL EXPRESSIONS

Addition and Subtraction · Multiplication · Rationalizing the Denominator

INTRODUCTION

So far we have discussed how to add, subtract, multiply, and divide numbers and variables. In this section we extend these operations to radical expressions. In doing so, we use many of the techniques discussed in Sections 10.1 and 10.2.

ADDITION AND SUBTRACTION

We can add $2x^2$ and $5x^2$ to obtain $7x^2$ because they are *like* terms. That is,

$$2x^2 + 5x^2 = (2 + 5)x^2 = 7x^2.$$

We can add and subtract **like radicals**, which have the same index and the same radicand. For example, we can add $3\sqrt{2}$ and $5\sqrt{2}$ because they are like radicals.

$$3\sqrt{2} + 5\sqrt{2} = (3 + 5)\sqrt{2} = 8\sqrt{2}$$

Sometimes two radical expressions that are not alike can be added by changing them to like radicals. For example, $\sqrt{20}$ and $\sqrt{5}$ are unlike radicals. However,

$$\sqrt{20} = \sqrt{4 \cdot 5} = \sqrt{4} \cdot \sqrt{5} = 2\sqrt{5},$$

so

$$\sqrt{20} + \sqrt{5} = 2\sqrt{5} + \sqrt{5} = 3\sqrt{5}.$$

We cannot combine $x + x^2$ because they are unlike terms. Similarly, we cannot combine $\sqrt{2} + \sqrt{5}$ because they are unlike radicals. When combining radicals, the first step is to see if we can write pairs of terms as like radicals, as demonstrated in the next example.

EXAMPLE 1   Finding like radicals

Write each pair of terms as like radicals, if possible.
(a) $\sqrt{45}, \sqrt{20}$ **(b)** $\sqrt{27}, \sqrt{5}$ **(c)** $5\sqrt[3]{16}, 4\sqrt[3]{54}$

Solution   **(a)** The expressions $\sqrt{45}$ and $\sqrt{20}$ are unlike radicals. However, they can be changed to like radicals as follows.

$$\sqrt{45} = \sqrt{9 \cdot 5} = \sqrt{9} \cdot \sqrt{5} = 3\sqrt{5} \quad \text{and}$$

$$\sqrt{20} = \sqrt{4 \cdot 5} = \sqrt{4} \cdot \sqrt{5} = 2\sqrt{5}$$

The expressions $3\sqrt{5}$ and $2\sqrt{5}$ are like radicals.

(b) The expressions $\sqrt{27} = 3\sqrt{3}$ and $\sqrt{5}$ are unlike radicals and cannot be written as like radicals.

(c) $5\sqrt[3]{16} = 5\sqrt[3]{8 \cdot 2} = 5\sqrt[3]{8} \cdot \sqrt[3]{2} = 5 \cdot 2 \cdot \sqrt[3]{2} = 10\sqrt[3]{2}$ and

$4\sqrt[3]{54} = 4\sqrt[3]{27 \cdot 2} = 4\sqrt[3]{27} \cdot \sqrt[3]{2} = 4 \cdot 3 \cdot \sqrt[3]{2} = 12\sqrt[3]{2}$

The expressions $10\sqrt[3]{2}$ and $12\sqrt[3]{2}$ are like radicals.

We use these techniques to add radical expressions in the next two examples.

EXAMPLE 2   Adding radical expressions

Add the expressions and simplify.
(a) $10\sqrt{11} + 4\sqrt{11}$ **(b)** $5\sqrt[3]{6} + \sqrt[3]{6}$
(c) $\sqrt{12} + 7\sqrt{3}$ **(d)** $3\sqrt{2} + \sqrt{8} + \sqrt{18}$

Solution   **(a)** $10\sqrt{11} + 4\sqrt{11} = (10 + 4)\sqrt{11} = 14\sqrt{11}$

(b) $5\sqrt[3]{6} + \sqrt[3]{6} = (5 + 1)\sqrt[3]{6} = 6\sqrt[3]{6}$

(c) $\sqrt{12} + 7\sqrt{3} = \sqrt{4 \cdot 3} + 7\sqrt{3}$

$$= \sqrt{4} \cdot \sqrt{3} + 7\sqrt{3}$$

$$= 2\sqrt{3} + 7\sqrt{3}$$

$$= 9\sqrt{3}$$

(d) $3\sqrt{2} + \sqrt{8} + \sqrt{18} = 3\sqrt{2} + \sqrt{4 \cdot 2} + \sqrt{9 \cdot 2}$

$$= 3\sqrt{2} + \sqrt{4} \cdot \sqrt{2} + \sqrt{9} \cdot \sqrt{2}$$

$$= 3\sqrt{2} + 2\sqrt{2} + 3\sqrt{2}$$

$$= 8\sqrt{2}$$

Caution: $\sqrt{9 + 4} \neq \sqrt{9} + \sqrt{4} = 3 + 2 = 5$. Rather, $\sqrt{9 + 4} = \sqrt{13} \approx 3.61$.

Critical Thinking

Suppose that $a \neq b$. What must be true about a and b for us to be able to simplify $\sqrt{a} + \sqrt{b}$ using the methods that we have discussed in this section?

EXAMPLE 3   Adding radical expressions

Add the expressions and simplify. Assume that all variables are positive.

(a) $\sqrt[4]{32} + 3\sqrt[4]{2}$ **(b)** $-2\sqrt{4x} + \sqrt{x}$ **(c)** $3\sqrt{3k} + 5\sqrt{12k} + 9\sqrt{48k}$

Solution   **(a)** Because $\sqrt[4]{32} = \sqrt[4]{16 \cdot 2} = \sqrt[4]{16} \cdot \sqrt[4]{2} = 2\sqrt[4]{2}$, we can add and simplify as follows.

$$\sqrt[4]{32} + 3\sqrt[4]{2} = 2\sqrt[4]{2} + 3\sqrt[4]{2} = 5\sqrt[4]{2}$$

(b) Note that $\sqrt{4x} = \sqrt{4} \cdot \sqrt{x} = 2\sqrt{x}$.

$$-2\sqrt{4x} + \sqrt{x} = -2(2\sqrt{x}) + \sqrt{x} = -4\sqrt{x} + \sqrt{x} = -3\sqrt{x}$$

(c) Note that $\sqrt{12k} = \sqrt{4} \cdot \sqrt{3k} = 2\sqrt{3k}$ and that $\sqrt{48k} = \sqrt{16} \cdot \sqrt{3k} = 4\sqrt{3k}$.

$$3\sqrt{3k} + 5\sqrt{12k} + 9\sqrt{48k} = 3\sqrt{3k} + 5(2\sqrt{3k}) + 9(4\sqrt{3k})$$
$$= (3 + 10 + 36)\sqrt{3k}$$
$$= 49\sqrt{3k}$$

Subtraction of radical expressions is similar to addition, as illustrated in the next example.

EXAMPLE 4   Subtracting radical expressions

Subtract and simplify. Assume that all variables are positive.

(a) $5\sqrt{7} - 3\sqrt{7}$ **(b)** $3\sqrt[3]{xy^2} - 2\sqrt[3]{xy^2}$

(c) $\sqrt{16x^3} - \sqrt{x^3}$ **(d)** $\sqrt[3]{\dfrac{5x}{27}} - \dfrac{\sqrt[3]{5x}}{6}$

Solution   **(a)** $5\sqrt{7} - 3\sqrt{7} = (5 - 3)\sqrt{7} = 2\sqrt{7}$

(b) $3\sqrt[3]{xy^2} - 2\sqrt[3]{xy^2} = (3 - 2)\sqrt[3]{xy^2} = \sqrt[3]{xy^2}$

(c) $\sqrt{16x^3} - \sqrt{x^3} = \sqrt{16} \cdot \sqrt{x^3} - \sqrt{x^3}$
$$= 4\sqrt{x^3} - \sqrt{x^3}$$
$$= 3\sqrt{x^3}$$
$$= 3x\sqrt{x}$$

(d) $\sqrt[3]{\dfrac{5x}{27}} - \dfrac{\sqrt[3]{5x}}{6} = \dfrac{\sqrt[3]{5x}}{\sqrt[3]{27}} - \dfrac{\sqrt[3]{5x}}{6}$ Quotient rule for radical expressions

$$= \dfrac{\sqrt[3]{5x}}{3} - \dfrac{\sqrt[3]{5x}}{6}$$ Evaluate $\sqrt[3]{27} = 3$.

$$= \dfrac{2\sqrt[3]{5x}}{6} - \dfrac{\sqrt[3]{5x}}{6}$$ Find a common denominator.

$$= \dfrac{2\sqrt[3]{5x} - \sqrt[3]{5x}}{6}$$ Subtract numerators.

$$= \dfrac{\sqrt[3]{5x}}{6}$$ Simplify.

EXAMPLE 5   Subtracting radical expressions

Subtract and simplify. Assume that all variables are positive.

(a) $\dfrac{5\sqrt{2}}{3} - \dfrac{2\sqrt{2}}{4}$ **(b)** $\sqrt[4]{81a^5b^6} - \sqrt[4]{16ab^2}$ **(c)** $3\sqrt[3]{\dfrac{n^5}{27}} - 2\sqrt[3]{n^2}$

Solution   **(a)** $\dfrac{5\sqrt{2}}{3} - \dfrac{2\sqrt{2}}{4} = \dfrac{5\sqrt{2}}{3}\cdot\dfrac{4}{4} - \dfrac{2\sqrt{2}}{4}\cdot\dfrac{3}{3}$ LCD is 12.

$$= \dfrac{20\sqrt{2}}{12} - \dfrac{6\sqrt{2}}{12}$$ Multiply fractions.

$$= \dfrac{14\sqrt{2}}{12}$$ Subtract numerators.

$$= \dfrac{7\sqrt{2}}{6}$$ Reduce.

(b) $\sqrt[4]{81a^5b^6} - \sqrt[4]{16ab^2} = \sqrt[4]{81a^4b^4}\cdot\sqrt[4]{ab^2} - \sqrt[4]{16}\cdot\sqrt[4]{ab^2}$ Factor out perfect powers.

$$= 3ab\sqrt[4]{ab^2} - 2\sqrt[4]{ab^2}$$ Simplify.

$$= (3ab - 2)\sqrt[4]{ab^2}$$ Distributive property

(c) $3\sqrt[3]{\dfrac{n^5}{27}} - 2\sqrt[3]{n^2} = 3\sqrt[3]{\dfrac{n^3}{27}}\cdot\sqrt[3]{n^2} - 2\sqrt[3]{n^2}$ Factor out perfect cube.

$$= \dfrac{3\sqrt[3]{n^3}}{\sqrt[3]{27}}\cdot\sqrt[3]{n^2} - 2\sqrt[3]{n^2}$$ Quotient rule

$$= n\sqrt[3]{n^2} - 2\sqrt[3]{n^2}$$ Simplify.

$$= (n - 2)\sqrt[3]{n^2}$$ Distributive property

Radicals often occur in geometry. In the next example, we find the perimeter of a triangle by adding radical expressions.

EXAMPLE 6   Finding the perimeter of a triangle

Find the *exact* perimeter of the right triangle shown in Figure 10.8. Then approximate your answer to the nearest hundredth of a foot.

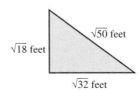

$\sqrt{18}$ feet $\sqrt{50}$ feet

$\sqrt{32}$ feet

Figure 10.8

Solution   The sum of the lengths of the sides of the triangle is

$$\sqrt{18} + \sqrt{32} + \sqrt{50} = 3\sqrt{2} + 4\sqrt{2} + 5\sqrt{2} = 12\sqrt{2}.$$

The perimeter is $12\sqrt{2} \approx 16.97$ feet.

MULTIPLICATION

Some types of radical expressions can be multiplied like binomials. For example, because $(a - b)(a + b) = a^2 - b^2$ we have

$$(\sqrt{x} - 2)(\sqrt{x} + 2) = (\sqrt{x})^2 - (2)^2 = x - 4,$$

provided x is not negative. The next example demonstrates this technique.

EXAMPLE 7   Multiplying radical expressions

Multiply and simplify.

(a) $(4 + \sqrt{3})(4 - \sqrt{3})$ **(b)** $(\sqrt{b} - 4)(\sqrt{b} + 5)$

Solution   **(a)** This expression is in the form $(a + b)(a - b)$, which equals $a^2 - b^2$.

$$(4 + \sqrt{3})(4 - \sqrt{3}) = (4)^2 - (\sqrt{3})^2$$
$$= 16 - 3$$
$$= 13$$

(b) This expression can be multiplied and then simplified.

$$(\sqrt{b} - 4)(\sqrt{b} + 5) = \sqrt{b} \cdot \sqrt{b} + 5\sqrt{b} - 4\sqrt{b} - 4 \cdot 5$$
$$= b + \sqrt{b} - 20$$

Compare this product with $(b - 4)(b + 5) = b^2 + b - 20$.

RATIONALIZING THE DENOMINATOR

In mathematics it is common to write expressions without radicals in the denominator. When the denominator is either a sum or difference containing a square root, we multiply the numerator and denominator by the *conjugate* of the denominator.

$$\frac{1}{1 + \sqrt{2}} = \frac{1}{1 + \sqrt{2}} \cdot \frac{1 - \sqrt{2}}{1 - \sqrt{2}} \qquad \text{Multiply numerator and denominator by the conjugate.}$$

$$= \frac{1 - \sqrt{2}}{(1)^2 - (\sqrt{2})^2} \qquad \text{Sum and difference}$$

$$= \frac{1 - \sqrt{2}}{1 - 2} \qquad \text{Simplify.}$$

$$= \frac{1 - \sqrt{2}}{-1} \qquad \text{Subtract.}$$

$$= -1 + \sqrt{2} \qquad \text{Simplify.}$$

If the denominator consists of two terms, at least one of which contains a radical expression, then the **conjugate** of the denominator is found by changing a $+$ sign to a $-$ sign or vice versa. For example, the conjugate of $\sqrt{2} + \sqrt{3}$ is $\sqrt{2} - \sqrt{3}$ and the conjugate of $\sqrt{3} - 1$ is $\sqrt{3} + 1$. In the next example, we use this method to rationalize the denominator of fractions that contain radicals.

EXAMPLE 8   Rationalizing the denominator

Rationalize the denominator.

(a) $\dfrac{3 + \sqrt{5}}{2 - \sqrt{5}}$ **(b)** $\dfrac{\sqrt{x}}{\sqrt{x} - 2}$

Solution   **(a)** The conjugate of the denominator is $2 + \sqrt{5}$.

$$\dfrac{3 + \sqrt{5}}{2 - \sqrt{5}} = \dfrac{(3 + \sqrt{5})}{(2 - \sqrt{5})} \cdot \dfrac{(2 + \sqrt{5})}{(2 + \sqrt{5})}$$ Multiply by 1.

$$= \dfrac{6 + 3\sqrt{5} + 2\sqrt{5} + (\sqrt{5})^2}{(2)^2 - (\sqrt{5})^2}$$ Multiply.

$$= \dfrac{11 + 5\sqrt{5}}{4 - 5}$$ Combine terms.

$$= -11 - 5\sqrt{5}$$ Simplify.

(b) The conjugate of the denominator is $\sqrt{x} + 2$.

$$\dfrac{\sqrt{x}}{\sqrt{x} - 2} = \dfrac{\sqrt{x}}{(\sqrt{x} - 2)} \cdot \dfrac{(\sqrt{x} + 2)}{(\sqrt{x} + 2)}$$ Multiply by 1.

$$= \dfrac{x + 2\sqrt{x}}{x - 4}$$ Multiply.

PUTTING IT ALL TOGETHER

In this section we discussed how to add, subtract, and multiply radical expressions. Rationalization of the denominator is a technique that can sometimes be helpful when dividing rational expressions. The following table summarizes important topics in this section.

Concept	Explanation	Examples
Adding and Subtracting Radical Expressions	Combine like radicals when adding or subtracting.	$6\sqrt{13} + \sqrt{13} = (6 + 1)\sqrt{13} = 7\sqrt{13}$
	We cannot combine unlike radicals such as $\sqrt{2}$ and $\sqrt{5}$. But sometimes we can rewrite radicals and then combine.	$\sqrt{40} - \sqrt{10} = \sqrt{4} \cdot \sqrt{10} - \sqrt{10}$ $= 2\sqrt{10} - \sqrt{10}$ $= \sqrt{10}$
Like Radicals	Like radicals have the same index and the same radicand.	$7\sqrt{5}$ and $3\sqrt{5}$ are like radicals. $5\sqrt[3]{ab}$ and $\sqrt[3]{ab}$ are like radicals. $\sqrt[3]{5}$ and $\sqrt[3]{4}$ are unlike radicals. $\sqrt[3]{7}$ and $\sqrt{7}$ are unlike radicals.

continued on next page

continued from previous page

Concept	Explanation	Examples
Multiplying Radical Expressions	Radical expressions can sometimes be multiplied like binomials.	$(\sqrt{a} - 5)(\sqrt{a} + 5) = a - 25$ and $(\sqrt{x} - 3)(\sqrt{x} + 1) = x - 2\sqrt{x} - 3$
Conjugate	The conjugate is found by changing a $+$ sign to a $-$ sign or vice versa.	$\begin{array}{cc} \textit{Expression} & \textit{Conjugate} \\ \sqrt{x} + 7 & \sqrt{x} - 7 \\ \sqrt{a} - 2\sqrt{b} & \sqrt{a} + 2\sqrt{b} \end{array}$
Rationalizing the Denominator	Multiply the numerator and denominator by the conjugate of the denominator.	$\dfrac{1}{1 - \sqrt{3}} = \dfrac{1}{1 - \sqrt{3}} \cdot \dfrac{(1 + \sqrt{3})}{(1 + \sqrt{3})}$ $= \dfrac{1 + \sqrt{3}}{1 - 3}$ $= -\dfrac{1}{2} - \dfrac{1}{2}\sqrt{3}$

10.3 EXERCISES

FOR EXTRA HELP

Student's Solutions Manual   InterAct Math   MathXL

MyMathLab   Math Tutor Center   Digital Video Tutor CD 5 Videotape 13

CONCEPTS

1. $\sqrt{a} + \sqrt{a} = $ _____

2. $\sqrt[3]{b} + \sqrt[3]{b} + \sqrt[3]{b} = $ _____

3. You cannot simplify $\sqrt[3]{4} + \sqrt[3]{7}$ because they are not _____ radicals.

4. Can you simplify $4\sqrt{15} - 3\sqrt{15}$? Explain.

5. What is the conjugate of $\sqrt{t} - 5$?

6. To rationalize the denominator of $\dfrac{1}{5 - \sqrt{2}}$, multiply this expression by _____.

LIKE RADICALS

Exercises 7–16: (Refer to Example 1.) Write the pair of terms as like radicals, if possible. Assume that all variables are positive.

7. $\sqrt{12}, \sqrt{24}$ 8. $\sqrt{18}, \sqrt{27}$

9. $\sqrt{7}, \sqrt{28}, \sqrt{63}$ 10. $\sqrt{200}, \sqrt{300}, \sqrt{500}$

11. $\sqrt[3]{16}, \sqrt[3]{-54}$ 12. $\sqrt[3]{80}, \sqrt[3]{10}$

13. $\sqrt{x^2 y}, \sqrt{4y^2}$ 14. $\sqrt{x^5 y^3}, \sqrt{9xy}$

15. $\sqrt[3]{8xy}, \sqrt[3]{x^4 y^4}$ 16. $\sqrt[3]{64x^4}, \sqrt[3]{-8x}$

Exercises 17–52: Simplify the expression. Assume that all variables are positive.

17. $2\sqrt{3} + 7\sqrt{3}$ 18. $8\sqrt{7} + 2\sqrt{7}$

19. $9\sqrt{5} + \sqrt{2} - \sqrt{5}$ 20. $11\sqrt{11} - 5\sqrt{11}$

21. $\sqrt{x} + \sqrt{x} - \sqrt{y}$ 22. $\sqrt{xy^2} - \sqrt{x}$

23. $\sqrt[3]{z} + \sqrt[3]{z}$ 24. $\sqrt[3]{y} - \sqrt[3]{y}$

25. $2\sqrt[3]{6} - 7\sqrt[3]{6}$ 26. $18\sqrt[3]{3} + 3\sqrt[3]{3}$

27. $\sqrt[3]{y^6} - \sqrt[3]{y^3}$ 28. $2\sqrt{20} + 7\sqrt{5} + 3\sqrt{2}$

29. $3\sqrt{28} + 3\sqrt{7}$ 30. $9\sqrt{18} - 2\sqrt{8}$

31. $\sqrt{44} - 4\sqrt{11}$ 32. $\sqrt[4]{5} + 2\sqrt[4]{5}$

33. $2\sqrt[3]{16} + \sqrt[3]{2} - \sqrt{2}$

34. $5\sqrt[3]{x} - 3\sqrt[3]{x}$ **35.** $\sqrt[3]{xy} - 2\sqrt[3]{xy}$

36. $3\sqrt{x^3} - \sqrt{x}$ **37.** $\sqrt{4x + 8} + \sqrt{x + 2}$

38. $\sqrt{2a + 1} + \sqrt{8a + 4}$

39. $\dfrac{4\sqrt{3}}{3} + \dfrac{\sqrt{3}}{6}$ **40.** $\dfrac{8\sqrt{5}}{7} + \dfrac{4\sqrt{5}}{2}$

41. $\dfrac{15\sqrt{8}}{4} - \dfrac{2\sqrt{2}}{5}$ **42.** $\dfrac{23\sqrt{11}}{2} - \dfrac{\sqrt{44}}{8}$

43. $2\sqrt[4]{64} - \sqrt[4]{324} + \sqrt[4]{4}$

44. $2\sqrt[3]{16} - 5\sqrt[3]{54} + 10\sqrt[3]{2}$

45. $5\sqrt[4]{x^5} - \sqrt[4]{x}$ **46.** $20\sqrt[3]{b^4} - 4\sqrt[3]{b}$

47. $\sqrt{64x^3} - \sqrt{x} + 3\sqrt{x}$

48. $2\sqrt{3z} + 3\sqrt{12z} + 3\sqrt{48z}$

49. $\sqrt[4]{81a^5b^5} - \sqrt[4]{ab}$ **50.** $\sqrt[4]{xy^5} - \sqrt[4]{x^5y}$

51. $5\sqrt[3]{\dfrac{n^4}{125}} - 2\sqrt[3]{n}$ **52.** $\sqrt[3]{\dfrac{8x}{27}} - \dfrac{2\sqrt[3]{x}}{3}$

Exercises 53–62: Multiply and simplify.

53. $(3 + \sqrt{7})(3 - \sqrt{7})$ **54.** $(5 - \sqrt{5})(5 + \sqrt{5})$

55. $(11 - \sqrt{2})(11 + \sqrt{2})$

56. $(6 + \sqrt{3})(6 - \sqrt{3})$

57. $(\sqrt{x} + 8)(\sqrt{x} - 8)$

58. $(\sqrt{ab} - 3)(\sqrt{ab} + 3)$

59. $(\sqrt{ab} - \sqrt{c})(\sqrt{ab} + \sqrt{c})$

60. $(\sqrt{2x} + \sqrt{3y})(\sqrt{2x} - \sqrt{3y})$

61. $(\sqrt{x} - 7)(\sqrt{x} + 8)$

62. $(\sqrt{ab} - 1)(\sqrt{ab} - 2)$

Exercises 63–76: Rationalize the denominator.

63. $\dfrac{1}{3 - \sqrt{2}}$ **64.** $\dfrac{1}{\sqrt{3} - 2}$

65. $\dfrac{\sqrt{2}}{\sqrt{5} + 2}$ **66.** $\dfrac{\sqrt{3}}{\sqrt{3} + 2}$

67. $\dfrac{\sqrt{7} - 2}{\sqrt{7} + 2}$ **68.** $\dfrac{\sqrt{3} - 1}{\sqrt{3} + 1}$

69. $\dfrac{1}{\sqrt{7} - \sqrt{6}}$ **70.** $\dfrac{1}{\sqrt{8} - \sqrt{7}}$

71. $\dfrac{\sqrt{z}}{\sqrt{z} - 3}$ **72.** $\dfrac{2\sqrt{z}}{2 - \sqrt{z}}$

73. $\dfrac{\sqrt{a} + \sqrt{b}}{\sqrt{a} - \sqrt{b}}$ **74.** $\dfrac{\sqrt{x} - 2\sqrt{y}}{\sqrt{x} + 2\sqrt{y}}$

75. $\dfrac{1}{\sqrt{x + 1} - \sqrt{x}}$ **76.** $\dfrac{1}{\sqrt{a + 1} + \sqrt{a}}$

GEOMETRY

77. *Perimeter* (Refer to Example 6.) Find the exact perimeter of the right triangle. Then approximate your answer.

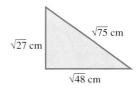

78. *Perimeter* Find the exact perimeter of the rectangle. Then approximate your answer.

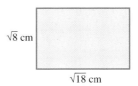

WRITING ABOUT MATHEMATICS

79. What are like radicals? Give examples.

80. A student simplifies an expression *incorrectly:*

$$\sqrt{8} + \sqrt[3]{16} \overset{?}{=} \sqrt{4 \cdot 2} + \sqrt[3]{8 \cdot 2}$$
$$\overset{?}{=} \sqrt{4} \cdot \sqrt{2} + \sqrt[3]{8} \cdot \sqrt[3]{2}$$
$$\overset{?}{=} 2\sqrt{2} + 2\sqrt[3]{2}$$
$$\overset{?}{=} 4\sqrt{4}$$
$$\overset{?}{=} 8.$$

Explain the error that the student made. What would you do differently?

Group Activity: ▪ Working with Real Data

Directions: Form a group of 2 to 4 people. Select someone to record the group's responses for this activity. All members of the group should work cooperatively to answer the questions. If your instructor asks for your results, each member of the group should be prepared to respond.

Designing a Paper Cup A paper drinking cup is being designed in the shape shown in the accompanying figure. The amount of paper needed to manufacture the cup is determined by the surface area S of the cup, which is given by

$$S = \pi r \sqrt{r^2 + h^2},$$

where r is the radius and h is the height.

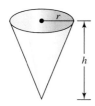

(a) Approximate S to the nearest hundredth when $r = 1.5$ inches and $h = 4$ inches.
(b) Could the formula for S be simplified as follows?

$$\pi r \sqrt{r^2 + h^2} \stackrel{?}{=} \pi r (\sqrt{r^2} + \sqrt{h^2})$$
$$\stackrel{?}{=} \pi r (r + h)$$

Try evaluating this formula with $r = 1.5$ inches and $h = 4$ inches. Is your answer the same as in part (a)? Explain.

(c) Discuss why evaluating real-world formulas correctly is important.
(d) In general, does $\sqrt{a + b}$ equal $\sqrt{a} + \sqrt{b}$? Justify your answer by completing the following table. Approximate answers to the nearest hundredth when appropriate.

a	b	$\sqrt{a + b}$	$\sqrt{a} + \sqrt{b}$
0	4		
4	0		
5	4		
9	7		
4	16		
25	100		

10.4 RADICAL FUNCTIONS

The Square Root Function · **The Square Root Property** · **The Cube Root Function** · **Power Functions** · **Modeling with Power Functions (Optional)**

INTRODUCTION

A good punter can kick a football so that it has a long *hang time*. Hang time is the length of time that the football is in the air. Long hang time gives the kicking team time to run down the field and stop the punt return. Using square roots, we can derive a function that calculates hang time.

THE SQUARE ROOT FUNCTION

50 feet

To derive a function that calculates the hang time of a football we need two facts from physics. First, when a ball is kicked into the air, half the hang time of the ball is spent going up and the other half is spent coming down. Second, the time t in seconds required for a ball to fall from a height of h feet is modeled by the equation

$$16t^2 = h.$$

Solving this equation for t gives half the hang time.

$$16t^2 = h \qquad \text{Given equation}$$

$$t^2 = \frac{h}{16} \qquad \text{Divide by 16.}$$

$$\sqrt{t^2} = \sqrt{\frac{h}{16}} \qquad \text{Take the square root of each side.}$$

$$|t| = \sqrt{\frac{h}{16}} \qquad \sqrt{a^2} = |a|$$

$$t = \frac{\sqrt{h}}{4} \qquad \text{Assume that } t \geq 0 \text{ and simplify.}$$

Half the hang time is $\frac{\sqrt{h}}{4}$, so the total hang time T in seconds is given by

$$T(h) = \frac{\sqrt{h}}{2},$$

where h is the maximum height of the ball.

In the next example, we use this formula to calculate hang time.

EXAMPLE 1   Calculating hang time

If a football is kicked 50 feet into the air, estimate the hang time. Does the hang time double for a football kicked 100 feet into the air?

Solution   A football kicked 50 feet into the air has a hang time of

$$T(50) = \frac{\sqrt{50}}{2} \approx 3.5 \text{ seconds.}$$

If the football is kicked 100 feet into the air, the hang time is

$$T(100) = \frac{\sqrt{100}}{2} = 5 \text{ seconds.}$$

The hang time does not double.

Critical Thinking

How high would a football have to be kicked to have twice the hang time of a football kicked 50 feet into the air?

The square root function is given by $f(x) = \sqrt{x}$. The domain of the square root function is all nonnegative real numbers because we have not defined the square root of a negative number. Table 10.1 lists three points that lie on the graph of $f(x) = \sqrt{x}$. In Figure 10.9 these points are plotted and the graph of $y = \sqrt{x}$ has been sketched. Note that the graph does not appear to the left of the origin because the square root function is undefined for negative inputs.

TABLE 10.1

x	$\sqrt{x}$
0	0
1	1
4	2

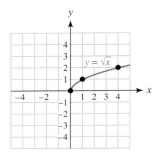

Figure 10.9 Square Root Function

Technology Note: *Square Roots of Negative Numbers*

If a table of values for $y_1 = \sqrt{x}$ includes both negative and positive values for x, then many calculators give error messages when x is negative, as shown in the accompanying figure.

X	Y1
-9	ERROR
-4	ERROR
-1	ERROR
0	0
1	1
4	2
9	3

$Y_1 = \sqrt{(X)}$

EXAMPLE 2 Finding the domain of a function

Let $f(x) = \sqrt{x - 1}$.

(a) Find the domain of f. Write your answer in interval notation.
(b) Graph $y = f(x)$ and compare it to the graph of $y = \sqrt{x}$.

Solution (a) For $f(x)$ to be defined, $x - 1$ cannot be negative. Thus valid inputs for x must satisfy

$$x - 1 \geq 0 \quad \text{or} \quad x \geq 1.$$

The domain is $[1, \infty)$.

(b) Table 10.2 provides points for the graph of $y = \sqrt{x - 1}$. Note in Figure 10.10 that the graph appears only when $x \geq 1$. This graph is similar to $y = \sqrt{x}$ (see Figure 10.9) except that it is shifted one unit to the right.

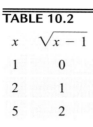

TABLE 10.2

x	$\sqrt{x - 1}$
1	0
2	1
5	2

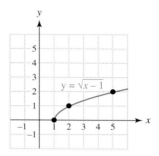

Figure 10.10

EXAMPLE 3 **Finding the domain of a function**

Find the domain of each function. Write your answer in interval notation.
(a) $f(x) = \sqrt{4 - 2x}$ **(b)** $g(x) = \sqrt{x^2 + 1}$

Solution **(a)** To determine when $f(x)$ is defined, we must solve the inequality $4 - 2x \geq 0$.

$$4 - 2x \geq 0 \qquad \text{Inequality to be solved}$$
$$4 \geq 2x \qquad \text{Add } 2x \text{ to each side.}$$
$$2 \geq x \qquad \text{Divide each side by 2.}$$

The domain is $(-\infty, 2]$.

(b) Regardless of the value of x, the expression $x^2 + 1$ is always positive because $x^2 \geq 0$. Thus $g(x)$ is defined for all real numbers, and its domain is $(-\infty, \infty)$.

THE SQUARE ROOT PROPERTY

Square roots can be used to solve the equation $x^2 = k$ for x, whenever $k \geq 0$.

$$x^2 = k \qquad \text{Given equation}$$
$$\sqrt{x^2} = \sqrt{k} \qquad \text{Take the square root of each side.}$$
$$|x| = \sqrt{k} \qquad \sqrt{x^2} = |x| \text{ for all } x.$$
$$x = \pm\sqrt{k} \qquad |x| = b \text{ implies } x = \pm b$$

The fact that $x^2 = k$ is equivalent to $x = \pm\sqrt{k}$ is sometimes called the **square root property**. Note that, if $k < 0$, the equation $x^2 = k$ has no real solutions.

SQUARE ROOT PROPERTY

Let k be a nonnegative number. Then the solutions to the equation

$$x^2 = k$$

are given by $x = \pm\sqrt{k}$.

Note: Recall that the symbol $\pm$ indicates "plus or minus." Thus "± 2" is equivalent to "2 or -2."

EXAMPLE 4 Applying the square root property

Solve each equation.
(a) $x^2 = 4$ **(b)** $n^2 = 7$ **(c)** $(t + 2)^2 = 25$

Solution **(a)** Apply the square root property.

$$x^2 = 4 \qquad \text{Given equation}$$
$$x = \pm\sqrt{4} \qquad \text{Square root property}$$
$$x = \pm 2 \qquad \text{Simplify.}$$

(b) Apply the square root property.

$$n^2 = 7 \qquad \text{Given equation}$$
$$n = \pm\sqrt{7} \qquad \text{Square root property}$$

(c) After applying the square root property, solve for t.

$$(t + 2)^2 = 25 \qquad \text{Given equation}$$
$$t + 2 = \pm\sqrt{25} \qquad \text{Square root property}$$
$$t + 2 = \pm 5 \qquad \text{Simplify.}$$
$$t = -2 \pm 5 \qquad \text{Add } -2 \text{ to each side.}$$

There are two solutions: $-2 + 5 = 3$ and $-2 - 5 = -7$.

═══ MAKING CONNECTIONS ═══

The Square Root Property and Graphical Solutions

Graphically the equation $x^2 = 4$ has two solutions, as shown in the accompanying figures. In the left-hand figure, the graphs of $y_1 = x^2$ and $y_2 = 4$ intersect when $x = \pm 2$. However, the equation $x^2 = -2$ has no solutions because the graphs of $y_1 = x^2$ and $y_2 = -2$ in the right-hand figure have no points of intersection.

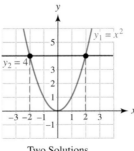

Two Solutions

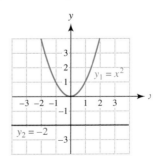

No Solutions

THE CUBE ROOT FUNCTION

The cube root function is given by $f(x) = \sqrt[3]{x}$. Cube roots are defined for both positive and negative numbers, so the domain of the cube root function includes all real numbers.

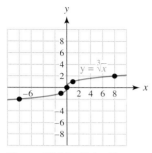

Figure 10.11 Cube Root Function

Table 10.3 lists points that lie on the graph of the cube root function. Figure 10.11 shows a graph of $y = \sqrt[3]{x}$.

TABLE 10.3 Cube Root Function

x	-27	-8	-1	0	1	8	27
$\sqrt[3]{x}$	-3	-2	-1	0	1	2	3

In some parts of the United States, wind power is used to generate electricity. Suppose that the diameter of the circular path created by the blades for a wind-powered generator is 8 feet. Then the wattage W generated by a wind velocity of v miles per hour is modeled by

$$W(v) = 2.4v^3.$$

If the wind blows at 10 miles per hour, the generator can produce about

$$W(10) = 2.4 \cdot 10^3 = 2400 \text{ watts.}$$

(**Source:** *Conquering the Sciences*, Sharp Electronics.)

In the next example, we use this formula to model a windmill.

EXAMPLE 5 Modeling a windmill

The formula $W(v) = 2.4v^3$ is used to calculate the watts generated when there is a wind velocity of v miles per hour.
(a) Find a function f that calculates the wind velocity when W watts are being produced.
(b) If the wattage doubles, has the wind velocity also doubled? Explain.

Solution **(a)** Solve $W = 2.4v^3$ for v.

$$W = 2.4v^3 \qquad \text{Given formula}$$

$$\frac{W}{2.4} = v^3 \qquad \text{Divide by 2.4.}$$

$$\sqrt[3]{\frac{W}{2.4}} = \sqrt[3]{v^3} \qquad \text{Take the cube root of each side.}$$

$$v = \sqrt[3]{\frac{W}{2.4}} \qquad \text{Simplify and rewrite equation.}$$

Thus $f(W) = \sqrt[3]{\frac{W}{2.4}}$.

(b) Table 10.4 shows that, when the power doubles from 1000 watts to 2000 watts, the wind velocity increases only from about 7.5 to 9.4 miles per hour. For the wattage to double, the wind velocity does not need to double.

TABLE 10.4 Wind Speed $f(W) = \sqrt[3]{\dfrac{W}{2.4}}$

W (watts)	0	1000	2000	4000
Wind Speed (mph)	0	7.5	9.4	11.9

Critical Thinking

In Example 5, if the wind speed v doubles, by what factor does the wattage $W = 2.4v^3$ increase?

Conversely, if the wattage W doubles, by what factor does the wind speed $v = \sqrt[3]{\frac{W}{2.4}}$ increase?

In Example 5 we used the fact that $\sqrt[3]{a^3} = a$ to solve the equation for v. Because $\sqrt[3]{a} = a^{1/3}$, from the properties of exponents we have

$$\sqrt[3]{a^3} = (a^3)^{1/3} = a^1 = a$$

for any real number a. We use this fact again in the next example to solve two equations.

EXAMPLE 6 **Using cube roots to solve equations**

Solve each equation.

(a) $x^3 = 64$ **(b)** $2(z - 1)^3 = 16$

Solution **(a)** Take the cube root of each side.

$$x^3 = 64 \qquad \text{Given equation}$$
$$\sqrt[3]{x^3} = \sqrt[3]{64} \qquad \text{Take the cube root of each side.}$$
$$x = 4 \qquad \text{Simplify.}$$

(b) Start by dividing each side by 2.

$$2(z - 1)^3 = 16 \qquad \text{Given equation}$$
$$(z - 1)^3 = 8 \qquad \text{Divide by 2.}$$
$$\sqrt[3]{(z - 1)^3} = \sqrt[3]{8} \qquad \text{Take the cube root of each side.}$$
$$z - 1 = 2 \qquad \text{Simplify.}$$
$$z = 3 \qquad \text{Add 1 to each side.}$$

POWER FUNCTIONS

Power functions are a generalization of root functions. Examples of power functions include

$$f(x) = x^{1/2}, \quad g(x) = x^{2/3}, \quad \text{and} \quad h(x) = x^{-3/5}.$$

The exponents for power functions can be rational numbers. Any rational number can be written in lowest terms as $\frac{m}{n}$, where m and n are integers.

POWER FUNCTION

If a function can be represented by

$$f(x) = x^p,$$

where p is a rational number, then it is a **power function**. If $p = \frac{1}{n}$, where $n \geq 2$ is an integer, then f is also a **root function**, which is given by

$$f(x) = \sqrt[n]{x}.$$

EXAMPLE 7   Evaluating power functions

If possible, evaluate $f(x)$ at the given value of x.
(a) $f(x) = x^{0.75}$ at $x = 16$ **(b)** $f(x) = x^{1/4}$ at $x = -81$

Solution   **(a)** $0.75 = \frac{3}{4}$, so $f(x) = x^{3/4}$. Thus

$$f(16) = 16^{3/4} = (16^{1/4})^3 = 2^3 = 8$$

because $16^{1/4} = \sqrt[4]{16} = 2$.
(b) $f(-81) = (-81)^{1/4} = \sqrt[4]{-81}$, which is undefined. There is no real number a such that $a^4 = -81$ because a^4 is never negative.

In the next example, we investigate the graph of $y = x^p$ for different values of p.

EXAMPLE 8   Graphing power functions

The graphs of three power functions,

$$f(x) = x^{1/3}, \quad g(x) = x^{0.75}, \quad \text{and} \quad h(x) = x^{1.4}$$

are shown in Figure 10.12. Discuss how the value of p affects the graph of $y = x^p$ when $x > 1$ and when $0 < x < 1$.

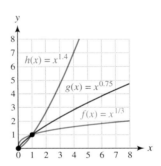

Figure 10.12   Power Functions

Solution   First note that $g(x)$ and $h(x)$ can be written as $g(x) = x^{3/4}$ and $h(x) = x^{7/5}$. When $x > 1$, $h(x) > g(x) > f(x)$ and the graphs increase (rise) faster for larger values of p. When $0 < x < 1$, $h(x) < g(x) < f(x)$. Thus smaller values of p result in larger y-values when $0 < x < 1$. All three graphs appear to intersect at the points $(0, 0)$ and $(1, 1)$.

MODELING WITH POWER FUNCTIONS (OPTIONAL)

Allometry is the study of the relative sizes of different characteristics of an organism. For example, the weight of a bird is related to the surface area of its wings: Heavier birds tend to have larger wings. Allometric relations are often modeled with $f(x) = kx^p$, where k and p are constants.   (***Source:*** C. Pennycuick, *Newton Rules Biology*.)

The next example illustrates the use of modeling with power functions.

EXAMPLE 9 Modeling surface area of wings

The surface area A of a bird's wings with weight w is shown in Table 10.5.

TABLE 10.5

w (kilograms)	0.5	2.0	3.5	5.0
$A(x)$ (square meters)	0.069	0.175	0.254	0.325

Calculator Help
To make a scatterplot, see
Appendix A (page AP-4).

(a) Make a scatterplot of the data. Discuss any trends in the data.
(b) Biologists modeled the data with $A(w) = kw^{2/3}$, where k is a constant. Find k.
(c) Graph A and the data in the same viewing rectangle.
(d) Estimate the area of the wings of a 3-kilogram bird.

Solution (a) A scatterplot of the data is shown in Figure 10.13(a). As the weight of a bird increases so does the surface area of its wings.

[0, 6, 1] by [0, 0.4, 0.1]

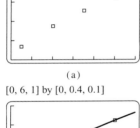

(a)

[0, 6, 1] by [0, 0.4, 0.1]

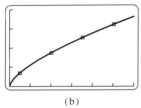

(b)

Figure 10.13

(b) To determine k, substitute one of the data points into $A(w)$.

$$A(w) = kw^{2/3} \qquad \text{Given formula}$$

$$0.175 = k(2)^{2/3} \qquad \begin{array}{l}\text{Let } w = 2 \text{ and } A(w) = 0.175 \\ \text{(any data point could be used).}\end{array}$$

$$k = \frac{0.175}{2^{2/3}} \qquad \text{Solve for } k.$$

$$k \approx 0.11 \qquad \text{Approximate } k.$$

Thus $A(w) = 0.11w^{2/3}$.

(c) The data and graph of $Y_1 = 0.11X^{\wedge}(2/3)$ are shown in Figure 10.13(b). Note that the graph appears to pass through each data point.

(d) $A(3) = 0.11(3)^{2/3} \approx 0.23$ square meter

10.4 **PUTTING IT ALL TOGETHER**

In this section we discussed some new functions involving radicals and rational exponents. Properties of these functions are summarized in the following table.

Function	Explanation
Square Root and Cube Root	The square root and cube root functions are given by $$f(x) = \sqrt{x} \quad \text{and} \quad g(x) = \sqrt[3]{x},$$ respectively. The cube root function is defined for all inputs, whereas the square root function is defined only for nonnegative inputs. Their graphs are shown in the accompanying figures.

Function	Explanation
Square Root and Cube Root (*continued*)	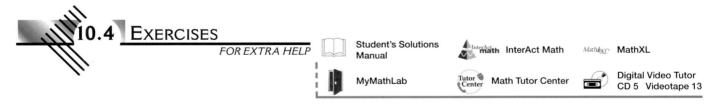 Square Root Function Cube Root Function

Power	If a function f can be defined by $f(x) = x^p$, where p is a rational number, then it is a power function. If $p = \frac{1}{n}$, where $n \geq 2$ is an integer, f is also a root function that can be represented by $f(x) = \sqrt[n]{x}$.
	
	Example: $f(x) = x^{5/3}$ Power function
	            $g(x) = x^{1/4}$ or Both a root and power function
	            $g(x) = \sqrt[4]{x}$

The following table summarizes how to use square roots and cube roots to solve certain types of equations.

Concept	Explanation	Examples
Square Root Property	If $k \geq 0$, the solutions to the equation $x^2 = k$ are given by $x = \pm\sqrt{k}$.	$x^2 = 81$ is equivalent to $x = \pm9$ and $x^2 = 5$ is equivalent to $x = \pm\sqrt{5}$
Solve Equations with Cube Roots	The solution to the equation $x^3 = k$ is given by $x = \sqrt[3]{k}$.	$x^3 = 8$ is equivalent to $x = 2$ and $x^3 = 11$ is equivalent to $x = \sqrt[3]{11}$

10.4 EXERCISES

FOR EXTRA HELP

Student's Solutions Manual   InterAct Math   MathXL

MyMathLab   Math Tutor Center   Digital Video Tutor CD 5   Videotape 13

CONCEPTS

1. Sketch a graph of the square root function.

2. Sketch a graph of the cube root function.

3. What is the domain of the square root function?

4. What is the domain of the cube root function?

5. Give a symbolic representation for a power function.

6. Give a symbolic representation for a root function.

7. What is the domain of $f(x) = \sqrt[4]{x}$?

8. Let $f(x) = x^p$ and $g(x) = x^q$ with $p > q$. When $x > 1$, is $x^p > x^q$ or $x^p < x^q$?

ROOT FUNCTIONS

Exercises 9–12: Evaluate the function at the given value of the variable.

9. $T(h) = \dfrac{\sqrt{h}}{2}$ $h = 100$

10. $L(k) = 2\sqrt{k + 2}$ $k = 23$

11. $f(x) = \sqrt{x + 5} + \sqrt{x}$ $x = 4$

12. $f(x) = \dfrac{\sqrt{x - 5} - \sqrt{x}}{2}$ $x = 9$

Exercises 13–24: (Refer to Example 2.) Find the domain of f. Write your answer in interval notation.

13. $f(x) = \sqrt{x + 1}$ **14.** $f(x) = \sqrt{x - 2}$

15. $f(x) = \sqrt{2x - 4}$ **16.** $f(x) = \sqrt{4x + 2}$

17. $f(x) = \sqrt{1 - x}$ **18.** $f(x) = \sqrt{6 - 3x}$

19. $f(x) = \sqrt{8 - 5x}$ **20.** $f(x) = \sqrt{3 - 2x}$

21. $f(x) = \sqrt{3x^2 + 4}$ **22.** $f(x) = \sqrt{1 + 2x^2}$

23. $f(x) = \dfrac{1}{\sqrt{2x + 1}}$ **24.** $f(x) = \dfrac{1}{\sqrt{x - 1}}$

EQUATIONS AND GRAPHS

Exercises 25–40: Solve the equation.

25. $x^2 = 49$ **26.** $x^2 = 9$

27. $2z^2 = 200$ **28.** $3z^2 = 48$

29. $(t + 1)^2 = 16$ **30.** $(t - 5)^2 = 81$

31. $(4 - 2x)^2 = 100$ **32.** $(3x - 6)^2 = 25$

33. $b^3 = 64$ **34.** $a^3 = 1000$

35. $2t^3 = -128$ **36.** $3t^3 = -81$

37. $(x + 1)^3 = 8$ **38.** $(4 - x)^3 = -1$

39. $(2 - 5z)^3 = -125$ **40.** $(2x + 4)^3 = 125$

Exercises 41–46: Graph the equation. Compare the graph to either $y = \sqrt{x}$ or $y = \sqrt[3]{x}$.

41. $y = \sqrt{x} + 2$ **42.** $y = \sqrt{x} - 1$

43. $y = \sqrt{x + 2}$ **44.** $y = \sqrt[3]{x} + 2$

45. $y = \sqrt[3]{x + 2}$ **46.** $y = \sqrt[3]{x} - 1$

POWER FUNCTIONS

Exercises 47–54: If possible, evaluate f(x) at the given value of x. When appropriate, approximate the answer to the nearest hundredth.

47. $f(x) = x^{5/2}$ $x = 4, x = 5$

48. $f(x) = x^{-3/4}$ $x = 1, x = 3$

49. $f(x) = x^{-7/5}$ $x = -32, x = 10$

50. $f(x) = x^{4/3}$ $x = -8, x = 27$

51. $f(x) = x^{1/4}$ $x = 256, x = -10$

52. $f(x) = x^{3/4}$ $x = 16, x = -1$

53. $f(x) = x^{2/5}$ $x = 32, x = -32$

54. $f(x) = x^{5/6}$ $x = -5, x = 64$

Exercises 55–58: Graph f and g in the window $[0, 6, 1]$ by $[0, 6, 1]$. Which function is greater when $x > 1$?

55. $f(x) = x^{1/5}, g(x) = x^{1/3}$

56. $f(x) = x^{4/5}, g(x) = x^{5/4}$

57. $f(x) = x^{1.2}, g(x) = x^{0.45}$

58. $f(x) = x^{-1.4}, g(x) = x^{1.4}$

APPLICATIONS

59. *Jumping* (Refer to Example 1.) If a person jumps 4 feet off the ground, estimate how long the person is in the air.

60. *Hang Time* (Refer to Example 1.) Find the hang time for a golf ball hit 80 feet into the air.

61. *Aging More Slowly* Albert Einstein in his theory of relativity showed that, if a person travels at nearly the speed of light, then time slows down significantly. Suppose that there are twins; one remains on Earth and the other leaves in a very fast spaceship having velocity v. If the twin on Earth ages T_0 years, then according to Einstein the twin in the spaceship ages T years, where

$$T(v) = T_0 \sqrt{1 - (v/c)^2}.$$

In this formula c represents the speed of light, which is 186,000 miles per second.

(a) Evaluate T when $v = 0.8c$ (eight-tenths the speed of light) and $T_0 = 10$ years. (*Hint:* Simplify $\frac{v}{c}$ without using 186,000 miles per second.)

(b) Interpret your result.

62. *Increasing Your Weight*   (Refer to Exercise 61.) Albert Einstein also showed that the weight (mass) of an object increases when traveling near the speed of light. If a person's weight on Earth is W_0, then the same person's weight W in a spaceship traveling at velocity v is

$$W(v) = \frac{W_0}{\sqrt{1 - (v/c)^2}}.$$

(a) Evaluate W when $v = 0.6c$ (six-tenths the speed of light) and $W_0 = 220$ pounds (100 kilograms).
(b) Interpret your result.

63. *Wind Power*   (Refer to Example 5.) If a wind-powered generator has blades that create a circular path with a diameter of 10 feet, then the wattage W generated by a wind velocity of v miles per hour is modeled by $W(v) = 3.8v^3$.
(a) If the wind velocity doubles, what happens to the wattage generated?
(b) Solve $W = 3.8v^3$ for v.
(c) If the wind generator is producing 30,400 watts, find the wind speed.

 64. *Modeling Wing Span*   (Refer to Example 9.) Biologists have found that the weight W of a bird and the length L of its wing span are related by $L = kW^{1/3}$, where k is a constant. The following table lists L and W for one species of bird.   (*Source:* C. Pennycuick.)

W (kilograms)	0.1	0.4	0.8	1.1
L (meters)	0.422	0.670	0.844	0.938

(a) Use the data to approximate the value of k.
(b) Graph L and the data in the same viewing rectangle. What happens to L as W increases?
(c) Estimate the wing span of a bird weighing 0.7 kilogram.
(d) Find L when $W = 0.65$, and interpret the result.

 65. *Pulse Rate in Animals*   The following table lists typical pulse rates R in beats per minute (bpm) for animals with various weights W in pounds.
(*Source:* C. Pennycuick.)

W (pounds)	20	150	500	1500
R (beats per minute)	198	72	40	23

(a) Describe what happens to the pulse rate as the size of the animal increases.
(b) Plot the data in [0, 1600, 400] by [0, 220, 20].
(c) These data can be modeled by $R = kW^{-1/2}$. Find k.
(d) Find R when $W = 700$, and interpret the result.

66. *Design of Open Channels*   To protect cities from flooding during heavy rains, open channels are sometimes constructed to handle runoff. The rate R at which water flows through the channel is modeled by $R = k\sqrt{m}$, where m is the slope of the channel and k is a constant determined by the shape of the channel. (*Source:* N. Garber and L. Hoel, *Traffic and Highway Design.*)
(a) Suppose that a channel has a slope of $m = 0.01$ (or 1%) and a runoff rate of $R = 340$ cubic feet per second (cfs). Find k.
(b) If the slope of the channel increases to $m = 0.04$ (or 4%), what happens to R? Be specific.

WRITING ABOUT MATHEMATICS

67. Explain why a root function is an example of a power function.

68. Discuss the shape of the graph of $y = x^p$ as p increases. Assume that p is a positive rational number and that x is a positive real number.

1. Simplify each expression.
   (a) $\sqrt{3} \cdot \sqrt{12}$

   (b) $\dfrac{\sqrt[3]{81}}{\sqrt[3]{3}}$

   (c) $\sqrt{36x^6}, x > 0$

2. Simplify each expression.
   (a) $5\sqrt{6} + 2\sqrt{6} + \sqrt{7}$

   (b) $8\sqrt[3]{x} - 3\sqrt[3]{x}$

   (c) $\sqrt{9x} - \sqrt{4x}$

3. Simplify each expression.
   (a) $\sqrt[3]{xy^4} - \sqrt[3]{x^4y}$

   (b) $(4 - \sqrt{2})(4 + \sqrt{2})$

4. Rationalize the denominator of $\dfrac{2}{\sqrt{5} - 1}$.

5. Sketch a graph of each function and then evaluate $f(-1)$, if possible.
   (a) $f(x) = \sqrt{x}$

   (b) $f(x) = \sqrt[3]{x}$

   (c) $f(x) = \sqrt{x^2}$

6. Evaluate $f(x) = 0.2x^{2/3}$ when $x = 64$.

7. Find the domain of $f(x) = \sqrt{x - 4}$. Write your answer in interval notation.

8. Solve the equation $(x + 1)^2 = 16$ for x.

10.5   EQUATIONS INVOLVING RADICAL EXPRESSIONS

Solving Radical Equations · The Distance Formula

INTRODUCTION

In Section 10.4 we showed that for some types of birds there is a relationship between their weight and the size of their wings—heavier birds tend to have larger wings. This relationship can sometimes be modeled by $A = 100\sqrt[3]{W^2}$, where W is weight in pounds and A is area in square inches. Suppose that we want to estimate the weight of a bird whose wings have an area of 600 square inches. To do so we would need to solve the equation

$$600 = 100\sqrt[3]{W^2}$$

for W. This equation contains a radical expression. In this section we explain how to solve this type of equation.   (**Source:** C. Pennycuick, *Newton Rules Biology.*)

SOLVING RADICAL EQUATIONS

Many times, equations contain either radical expressions or rational exponents. Examples include

$$\sqrt{x} = 6, \quad 5x^{1/2} = 1, \quad \text{and} \quad \sqrt[3]{x - 1} = 3.$$

One strategy for solving an equation containing a square root is to isolate the square root and then square each side of the equation. This technique is an example of the *power rule for solving equations*.

POWER RULE FOR SOLVING EQUATIONS

If each side of an equation is raised to the same integer power, then any solutions to the given equation are among the solutions to the new equation. That is, the solutions to the equation $a = b$ are among the solutions to $a^n = b^n$.

We must check our solutions when applying the power rule. For example, consider the equation $2x = 6$. If we square each side, we obtain $4x^2 = 36$. Solving this new equation gives $x^2 = 9$, or $x = \pm 3$. Here, 3 is a solution to both equations, but -3 is an **extraneous solution** that satisfies the second equation but not the given equation.

We illustrate this method in the next example.

EXAMPLE 1   Solving a radical equation symbolically

Solve $\sqrt{2x - 1} = 3$. Check your solution.

Solution   Begin by squaring each side of the equation.

$$\sqrt{2x - 1} = 3 \qquad \text{Given equation}$$
$$(\sqrt{2x - 1})^2 = 3^2 \qquad \text{Square each side.}$$
$$2x - 1 = 9 \qquad \text{Simplify.}$$
$$2x = 10 \qquad \text{Add 1.}$$
$$x = 5 \qquad \text{Divide by 2.}$$

To check our work we substitute $x = 5$ in the given equation.

$$\sqrt{2(5) - 1} \overset{?}{=} 3$$
$$3 = 3 \qquad \text{It checks.}$$

Note: To solve the equation in Example 1, we used the fact that

$$\left(\sqrt{a}\right)^2 = \sqrt{a} \cdot \sqrt{a} = a.$$

The following steps are used to solve a radical equation.

SOLVING A RADICAL EQUATION

STEP 1: Isolate a radical term on one side of the equation.

STEP 2: Apply the power rule by raising each side of the equation to the power equal to the index of the isolated radical term.

STEP 3: Solve the equation. If it still contains a radical, repeat Steps 1 and 2.

STEP 4: Check your answers by substituting each result in the *given* equation.

In the next example, we apply these steps to a radical equation.

EXAMPLE 2 Isolating the radical term

Solve $\sqrt{4 - x} + 5 = 8$.

Solution **STEP 1:** To isolate the radical term, we subtract 5 from each side of the equation.

$$\sqrt{4 - x} + 5 = 8 \qquad \text{Given equation}$$
$$\sqrt{4 - x} = 3 \qquad \text{Subtract 5.}$$

STEP 2: The isolated term involves a square root, so we must square each side.

$$(\sqrt{4 - x})^2 = (3)^2 \qquad \text{Square each side.}$$

STEP 3: Next we solve the resulting equation. (It is not necessary to repeat Steps 1 and 2 because the resulting equation does not contain any radical expressions.)

$$4 - x = 9 \qquad \text{Simplify.}$$
$$-x = 5 \qquad \text{Subtract 4.}$$
$$x = -5 \qquad \text{Multiply by } -1.$$

STEP 4: To check our work we substitute $x = -5$ in the given equation.

$$\sqrt{4 - (-5)} + 5 \stackrel{?}{=} 8$$
$$\sqrt{9} + 5 \stackrel{?}{=} 8$$
$$8 = 8 \qquad \text{It checks.}$$

Example 3 shows why we must check our answers when squaring each side of an equation.

EXAMPLE 3 Solving a radical equation

Solve $\sqrt{3x + 3} = 2x - 1$. Check your results and then solve the equation graphically.

Solution **Symbolic Solution** Begin by squaring each side of the equation.

$$\sqrt{3x + 3} = 2x - 1 \qquad \text{Given equation}$$
$$(\sqrt{3x + 3})^2 = (2x - 1)^2 \qquad \text{Square each side.}$$
$$3x + 3 = 4x^2 - 4x + 1 \qquad \text{Expand.}$$
$$0 = 4x^2 - 7x - 2 \qquad \text{Subtract } 3x + 3.$$
$$0 = (4x + 1)(x - 2) \qquad \text{Factor.}$$
$$x = -\frac{1}{4} \quad \text{or} \quad x = 2 \qquad \text{Solve for } x.$$

To check these values substitute $x = -\frac{1}{4}$ and $x = 2$ in the given equation.

$$\sqrt{3\left(-\frac{1}{4}\right) + 3} \stackrel{?}{=} 2\left(-\frac{1}{4}\right) - 1$$
$$\sqrt{2.25} \stackrel{?}{=} -1.5$$
$$1.5 \neq -1.5 \qquad \text{It does not check.}$$

Thus $-\frac{1}{4}$ is an *extraneous solution.* Next substitute $x = 2$ in the given equation.

$$\sqrt{3 \cdot 2 + 3} \stackrel{?}{=} 2 \cdot 2 - 1$$
$$\sqrt{9} \stackrel{?}{=} 3$$
$$3 = 3 \qquad \text{It checks.}$$

The only solution is 2.

Graphical Solution   The solution 2 is supported graphically in Figure 10.14, where the graphs of $Y_1 = \sqrt{(3X + 3)}$ and $Y_2 = 2X - 1$ intersect at the point $(2, 3)$. Note that the graphical solution does not give an extraneous solution.

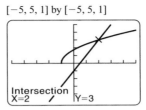

[−5, 5, 1] by [−5, 5, 1]

Intersection
X=2   Y=3

Figure 10.14

Calculator Help
To find a point of intersection, see Appendix A (page AP-7).

Example 3 demonstrates that checking solutions is essential when you are squaring each side of an equation. Squaring may introduce extraneous solutions, which are solutions to the new equation but are not solutions to the given equation.

When an equation contains two or more terms with square roots, it may be necessary to square each side of the equation more than once. In these situations, isolate one of the square roots and then square each side of the equation. If a radical term remains after simplifying, repeat these steps. We apply this technique in the next example.

EXAMPLE 4   **Squaring twice**

Solve $\sqrt{2x} - 1 = \sqrt{x + 1}$.

Solution   Begin by squaring each side of the equation.

$$\sqrt{2x} - 1 = \sqrt{x + 1} \qquad \text{Given equation}$$
$$(\sqrt{2x} - 1)^2 = (\sqrt{x + 1})^2 \qquad \text{Square each side.}$$
$$(\sqrt{2x})^2 - 2(\sqrt{2x})(1) + 1^2 = x + 1 \qquad (a - b)^2 = a^2 - 2ab + b^2$$
$$2x - 2\sqrt{2x} + 1 = x + 1 \qquad \text{Simplify.}$$
$$2x - 2\sqrt{2x} = x \qquad \text{Subtract 1.}$$
$$x = 2\sqrt{2x} \qquad \text{Subtract } x \text{ and add } 2\sqrt{2x}.$$
$$x^2 = 4(2x) \qquad \text{Square each side again.}$$
$$x^2 - 8x = 0 \qquad \text{Subtract } 8x.$$
$$x(x - 8) = 0 \qquad \text{Factor.}$$
$$x = 0 \quad \text{or} \quad x = 8 \qquad \text{Solve.}$$

To check your solutions substitute $x = 0$ in the given equation.

$$\sqrt{2 \cdot 0} - 1 \stackrel{?}{=} \sqrt{0 + 1}$$
$$-1 \neq 1 \qquad \text{It does not check.}$$

Now substitute $x = 8$.

$$\sqrt{2 \cdot 8} - 1 \stackrel{?}{=} \sqrt{8 + 1}$$
$$3 = 3 \qquad \text{It checks.}$$

The only solution is 8.

In the next example we apply the power rule to an equation that contains a cube root.

EXAMPLE 5 Solving an equation containing an isolated cube root

Solve $\sqrt[3]{4x - 7} = 4$.

Solution **STEP 1:** The radical term is already isolated on the left side of the equation, so we proceed to Step 2.

STEP 2: Because the index is 3, we cube each side of the equation.

$$\sqrt[3]{4x - 7} = 4 \qquad \text{Given equation}$$
$$(\sqrt[3]{4x - 7})^3 = (4)^3 \qquad \text{Cube each side.}$$

STEP 3: We solve the resulting equation.

$$4x - 7 = 64 \qquad \text{Simplify.}$$
$$4x = 71 \qquad \text{Add 7 to each side.}$$
$$x = \frac{71}{4} \qquad \text{Divide each side by 4.}$$

STEP 4: To check our work we substitute $x = \frac{71}{4}$ in the given equation.

$$\sqrt[3]{4\left(\frac{71}{4}\right) - 7} \stackrel{?}{=} 4$$
$$\sqrt[3]{64} \stackrel{?}{=} 4$$
$$4 = 4 \qquad \text{It checks.}$$

In the next example we solve the equation presented in the introduction to this section.

EXAMPLE 6 Finding the weight of a bird

Solve the equation $600 = 100\sqrt[3]{W^2}$ to determine the weight in pounds of a bird having wings with an area of 600 square inches.

Solution Begin by dividing each side of the equation by 100.

$$\frac{600}{100} = \sqrt[3]{W^2} \qquad \text{Divide each side by 100.}$$
$$(6)^3 = (\sqrt[3]{W^2})^3 \qquad \text{Cube each side.}$$
$$216 = W^2 \qquad \text{Simplify.}$$
$$W = \sqrt{216} \qquad \text{Take principal square root, } W > 0.$$
$$W \approx 14.7 \qquad \text{Approximate.}$$

The weight of the bird is approximately 14.7 pounds.

Technology Note: *Graphing Radical Expressions*

The equation in Example 6 can be solved graphically. Sometimes it is more convenient to use rational exponents than radical notation. Thus $y = 100 \sqrt[3]{W^2}$ can be entered as $Y_1 = 100X^\wedge(2/3)$. (Be sure to include parentheses around the 2/3.) The accompanying figure shows y_1 intersecting the line $y_2 = 600$ near the point $(14.7, 600)$, which supports our symbolic result.

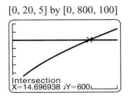

[0, 20, 5] by [0, 800, 100]

Intersection
X=14.696938 ⌐Y=600

In the next example we solve an equation containing different powers of x. This equation would be difficult to solve symbolically, but an *approximate* solution can be found graphically.

EXAMPLE 7   **Solving an equation with a rational exponent**

Solve $x^{2/3} = 3 - x^2$ graphically.

Solution   Graph $Y_1 = X^\wedge(2/3)$ and $Y_2 = 3 - X^\wedge2$. Their graphs intersect near $(-1.34, 1.21)$ and $(1.34, 1.21)$, as shown in Figure 10.15. Thus the solutions are given by $x \approx \pm 1.34$.

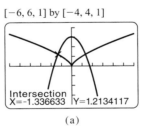

[-6, 6, 1] by [-4, 4, 1]

Intersection
X=-1.336633 Y=1.2134117

(a)

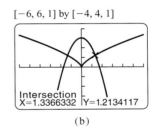

[-6, 6, 1] by [-4, 4, 1]

Intersection
X=1.3366332 Y=1.2134117

(b)

Figure 10.15

THE DISTANCE FORMULA

One of the most famous theorems in mathematics is the **Pythagorean theorem**. It states that, if a right triangle has legs a and b with hypotenuse c (see Figure 10.16), then

$$a^2 + b^2 = c^2.$$

For example, if the legs of a right triangle are $a = 3$ and $b = 4$, the hypotenuse is $c = 5$ because $3^2 + 4^2 = 5^2$.

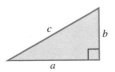

Figure 10.16
$a^2 + b^2 = c^2$

EXAMPLE 8   Applying the Pythagorean theorem

A rectangular television screen has a width of 20 inches and a height of 15 inches. Find the diagonal of the television. Why is it called a 25-inch television?

Solution   In Figure 10.17, let $a = 20$ and $b = 15$. Then the diagonal of the television corresponds to the hypotenuse of a right triangle with legs of 20 inches and 15 inches.

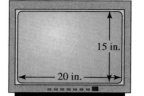

15 in.

20 in.

Figure 10.17

$$c^2 = a^2 + b^2 \qquad \text{Pythagorean theorem}$$
$$c = \sqrt{a^2 + b^2} \qquad \text{Take the principal square root, } c > 0.$$
$$c = \sqrt{20^2 + 15^2} \qquad \text{Substitute } a = 20 \text{ and } b = 15.$$
$$c = 25 \qquad \text{Simplify.}$$

A 25-inch television has a diagonal of 25 inches.

The Pythagorean theorem can be used to determine the distance between two points. Suppose that a line segment has endpoints (x_1, y_1) and (x_2, y_2), as illustrated in Figure 10.18. The lengths of the legs of the right triangle are $x_2 - x_1$ and $y_2 - y_1$. The distance d is the hypotenuse of a right triangle. Applying the Pythagorean theorem, we have

$$d^2 = (x_2 - x_1)^2 + (y_2 - y_1)^2.$$

Distance is nonnegative, so we let d be the principal square root and obtain

$$d = \sqrt{(x_2 - x_1)^2 + (y_2 - y_1)^2}.$$

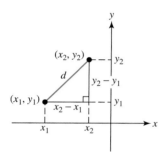

Figure 10.18

DISTANCE FORMULA

The **distance** d between the points (x_1, y_1) and (x_2, y_2) in the xy-plane is

$$d = \sqrt{(x_2 - x_1)^2 + (y_2 - y_1)^2}.$$

EXAMPLE 9   Finding distance between points

Find the distance between the points $(-2, 3)$ and $(1, -4)$.

Solution   Start by letting $(x_1, y_1) = (-2, 3)$ and $(x_2, y_2) = (1, -4)$. Then substitute these values into the distance formula.

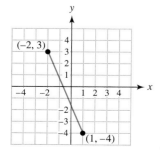

Figure 10.19

$$d = \sqrt{(x_2 - x_1)^2 + (y_2 - y_1)^2} \qquad \text{Distance formula}$$
$$= \sqrt{(1 - (-2))^2 + (-4 - 3)^2} \qquad \text{Substitute.}$$
$$= \sqrt{9 + 49} \qquad \text{Simplify.}$$
$$= \sqrt{58} \qquad \text{Take the square root.}$$
$$\approx 7.62 \qquad \text{Approximate.}$$

The distance between the points, as shown in Figure 10.19, is exactly $\sqrt{58}$ units, or about 7.62 units. Note that we would obtain the same result if we let $(x_1, y_1) = (1, -4)$ and $(x_2, y_2) = (-2, 3)$.

EXAMPLE 10 Designing a highway curve

Figure 10.20 shows a circular highway curve joining a straight section of road. A surveyor is trying to locate the x-coordinate of the *point of curvature PC* where the two sections of the highway meet. The distance between the surveyor and PC should be 400 feet. Estimate the x-coordinate of PC if x is positive.

Solution In Figure 10.20, the distance between the points $(0, 75)$ and $(x, 300)$ is 400 feet. We can apply the distance formula and solve for x.

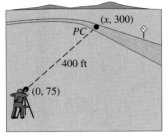

$$d = \sqrt{(x_2 - x_1)^2 + (y_2 - y_1)^2} \quad \text{Distance formula}$$
$$400 = \sqrt{(x - 0)^2 + (300 - 75)^2} \quad \text{Substitute.}$$
$$400^2 = (x - 0)^2 + (300 - 75)^2 \quad \text{Square each side.}$$
$$160{,}000 = x^2 + 50{,}625 \quad \text{Simplify.}$$
$$x^2 = 109{,}375 \quad \text{Solve for } x^2.$$
$$x = \sqrt{109{,}375} \quad \text{Take the principal square root, } x > 0.$$
$$x \approx 330.7 \quad \text{Approximate.}$$

Figure 10.20

The x-coordinate is about 330.7.

10.5 PUTTING IT ALL TOGETHER

In this section we focused on equations that contain either radical expressions or rational exponents. A good strategy for solving an equation containing radical expressions *symbolically* is to raise each side of the equation to the same integer power. However, checking answers is important to eliminate any extraneous solutions. Note that these extraneous solutions do not occur when an equation is being solved graphically.

Concept	Description	Example
Power Rule for Solving Equations	If each side of an equation is raised to the same integer power, any solutions to the given equation are among the solutions to the new equation.	$\sqrt{2x} = x$ $2x = x^2$ Square each side. $x^2 - 2x = 0$ Rewrite equation. $x = 0$ or $x = 2$ Factor and solve. Be sure to check any solutions.
Pythagorean Theorem	If c is the hypotenuse of a right triangle and a and b are its legs, then $a^2 + b^2 = c^2$.	If the sides of the right triangle are $a = 5$, $b = 12$, and $c = 13$, then they satisfy $a^2 + b^2 = c^2$ or $5^2 + 12^2 = 13^2$. (right triangle with hypotenuse 13, sides 5 and 12)

continued on next page

continued from previous page

Concept	Description	Example
Distance Formula	The distance d between the points (x_1, y_1) and (x_2, y_2) is $$d = \sqrt{(x_2 - x_1)^2 + (y_2 - y_1)^2}.$$	The distance between $(2, 3)$ and $(-3, 4)$ is $$d = \sqrt{(-3 - 2)^2 + (4 - 3)^2}$$ $$= \sqrt{(-5)^2 + (1)^2} = \sqrt{26}.$$

10.5 EXERCISES

FOR EXTRA HELP

Student's Solutions Manual

 InterAct Math

 MathXL

MyMathLab

Math Tutor Center

Digital Video Tutor
CD 5   Videotape 13

CONCEPTS

1. What is a good first step for solving $\sqrt{4x - 1} = 5$?

2. What is a good first step for solving $\sqrt[3]{x + 1} = 6$?

3. Can an equation involving rational exponents have more than one solution?

4. When you square each side of an equation to solve for an unknown, what must you do with any answers?

5. What is the Pythagorean theorem used for?

6. If the legs of a right triangle are 3 and 4, what is the length of the hypotenuse?

7. What formula can you use to find the distance d between two points?

8. Write the equation $\sqrt{x} + \sqrt[4]{x^3} = 2$ with rational exponents.

SYMBOLIC SOLUTIONS

Exercises 9–34: Solve the equation symbolically. Check your results.

9. $\sqrt{x} = 8$

10. $\sqrt{3z} = 6$

11. $\sqrt[4]{x} = 3$

12. $\sqrt[3]{x - 4} = 2$

13. $\sqrt{2t + 4} = 4$

14. $\sqrt{y + 4} = 3$

15. $\sqrt{x + 6} = x$

16. $\sqrt{z + 6} = z$

17. $\sqrt[3]{x} = 3$

18. $\sqrt[3]{x + 10} = 4$

19. $\sqrt[3]{2z - 4} = -2$

20. $\sqrt[3]{z - 1} = -3$

21. $\sqrt[4]{t + 1} = 2$

22. $\sqrt[4]{5t} = 5$

23. $\sqrt{5z - 1} = \sqrt{z + 1}$

24. $y = \sqrt{y + 1} + 1$

25. $\sqrt{1 - x} = 1 - x$

26. $\sqrt[3]{4x} = x$

27. $\sqrt{b^2 - 4} = b - 2$

28. $\sqrt{b^2 - 2b + 1} = b$

29. $\sqrt{1 - 2x} = x + 7$

30. $\sqrt{4 - y} = y - 2$

31. $\sqrt{x} = \sqrt{x - 5} + 1$

32. $\sqrt{x - 1} = \sqrt{x + 4} - 1$

33. $\sqrt{2t - 2} + \sqrt{t} = 7$

34. $\sqrt{x + 1} - \sqrt{x - 6} = 1$

GRAPHICAL SOLUTIONS

 Exercises 35–44: Solve the equation graphically. Approximate solutions to the nearest hundredth when appropriate.

35. $\sqrt[3]{x + 5} = 2$ **36.** $\sqrt[3]{x} + \sqrt{x} = 3.43$

37. $\sqrt{2x - 3} = \sqrt{x} - \dfrac{1}{2}$

38. $x^{4/3} - 1 = 2$

39. $x^{5/3} = 2 - 3x^2$ **40.** $x^{3/2} = \sqrt{x + 2} - 2$

41. $z^{1/3} - 1 = 2 - z$ **42.** $z^{3/2} - 2z^{1/2} - 1 = 0$

43. $\sqrt{y + 2} + \sqrt{3y + 2} = 2$

44. $\sqrt{x + 1} - \sqrt{x - 1} = 4$

USING MORE THAN ONE METHOD

Exercises 45–48: Solve the equation
   (a) *symbolically,*
   (b) *graphically, and*
   (c) *numerically.*

45. $2\sqrt{x} = 8$ **46.** $\sqrt[3]{5 - x} = 2$

47. $\sqrt{6z - 2} = 8$ **48.** $\sqrt{y + 4} = \dfrac{y}{3}$

SOLVING AN EQUATION FOR A VARIABLE

Exercises 49–52: Solve the equation for the indicated variable.

49. $T = 2\pi\sqrt{\dfrac{L}{32}}$ for L

50. $Z = \sqrt{L^2 + R^2}$ for R

51. $r = \sqrt{\dfrac{A}{\pi}}$ for A

52. $F = \dfrac{1}{2\pi\sqrt{LC}}$ for C

PYTHAGOREAN THEOREM

Exercises 53–60: If the sides of a triangle are a, b, and c and they satisfy $a^2 + b^2 = c^2$, the triangle is a right triangle. Determine whether the triangle with the given sides is a right triangle.

53. $a = 6$ $b = 8$ $c = 10$

54. $a = 5$ $b = 12$ $c = 13$

55. $a = \sqrt{5}$ $b = \sqrt{9}$ $c = \sqrt{14}$

56. $a = 4$ $b = 5$ $c = 7$

57. $a = 7$ $b = 24$ $c = 25$

58. $a = 1$ $b = \sqrt{3}$ $c = 2$

59. $a = 8$ $b = 8$ $c = 16$

60. $a = 11$ $b = 60$ $c = 61$

Exercises 61–64: Find the length of the missing side in the right triangle.

61.

62.

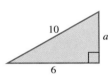

63.

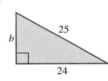

64.

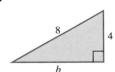

Exercises 65–70: A right triangle has legs a and b with hypotenuse c. Find the length of the missing side.

65. $a = 3, b = 4$ **66.** $a = 4, b = 7$

67. $a = \sqrt{3}, c = 8$ **68.** $a = \sqrt{6}, c = \sqrt{10}$

69. $b = 48, c = 50$ **70.** $b = 10, c = 26$

DISTANCE FORMULA

Exercises 71–74: Find the length of the line segment in the figure.

71.

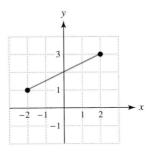

72.

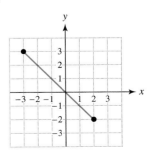

73.

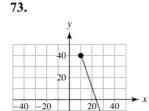

74.

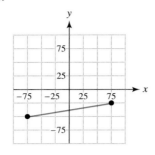

Exercises 75–78: Find the distance between the points.

75. $(-1, 2), (4, 10)$

76. $(5, -40), (-6, 20)$

77. $(0, -3), (4, 0)$

78. $(3, 9), (-4, 2)$

Exercises 79–82: (Refer to Example 10.) Find x if the distance between the given points is d. Assume that x is positive.

79. $(x, 3), (0, 6)$ $d = 5$

80. $(x, -1), (6, 11)$ $d = 13$

81. $(x, -5), (62, 6)$ $d = 61$

82. $(x, 3), (12, -4)$ $d = 25$

APPLICATIONS

Exercises 83 and 84: Weight of a Bird (Refer to Example 6.) Estimate the weight of a bird having wings of area A.

83. $A = 400$ square inches

84. $A = 1000$ square inches

Exercises 85–88: Distance to the Horizon Because of Earth's curvature, a person can see a limited distance to the horizon. The higher the location of the person, the farther that person can see. The distance D in miles to the horizon can be estimated by $D(h) = 1.22\sqrt{h}$, where h is the height of the person above the ground in feet.

85. Find D for a 6-foot-tall person standing on level ground.

86. Find D for a person on top of Mount Everest with a height of 29,028 feet.

87. How high does a person need to be to see 20 miles?

88. How high does a plane need to fly for the pilot to be able to see 100 miles?

89. *Diagonal of a Television* (Refer to Example 8.) A rectangular television screen is 11.4 inches by 15.2 inches. Find the diagonal of the television set.

90. *Dimensions of a Television* The height of a television with a 13-inch diagonal is $\frac{3}{4}$ of its width. Find the width and height of the television set.

91. *DVD and Picture Dimensions* If the picture shown on a television set is h units high and w units wide, the *aspect ratio* of the picture is $\frac{w}{h}$ (see the accompanying figure). Digital video discs support the newer aspect ratio of $\frac{16}{9}$ rather than the older ratio of $\frac{4}{3}$. If the width of a picture with an aspect ratio of $\frac{16}{9}$ is 29 inches, approximate the height and diagonal of the rectangular picture. (*Source: J. Taylor, DVD Demystified.*)

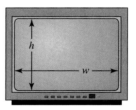

92. *Flood Control* The spillway capacity of a dam is important in flood control. Spillway capacity Q in cubic feet of water per second flowing over the spillway depends on the width W and the depth D of the spillway, as illustrated in the accompanying figure. If W and D are measured in feet, capacity can be modeled by $Q = 3.32WD^{3/2}$. (*Source: D. Callas, Project Director, Snapshots of Applications in Mathematics.*)

(a) Find the capacity of a spillway with $W = 20$ feet and $D = 5$ feet.

(b) A spillway with a width of 30 feet is to have a capacity of $Q = 2690$ cubic feet per second. Estimate the appropriate depth of the spillway.

93. *Sky Diving* When sky divers initially fall from an airplane, their velocity v in miles per hour after free falling d feet can be approximated by $v = \frac{60}{11}\sqrt{d}$. (Because of air resistance, they will eventually reach a terminal velocity and the formula will no longer be valid.) How far do sky divers need to fall to attain the following velocities? (These values for d represent minimum distances.)

(a) 60 miles per hour **(b)** 100 miles per hour

94. *Guy Wire* A guy wire attached to the top of a 30-foot-long pole is anchored 10 feet from the base of the pole, as illustrated in the accompanying figure. Find the length of the guy wire to the nearest tenth of a foot.

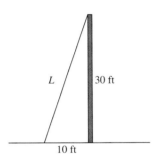

95. *Skid Marks* Vehicles involved in accidents often leave skid marks, which can be used to determine how fast a vehicle was traveling. To determine this speed, officials often use a test vehicle to compare skid marks on the same section of road. Suppose that a vehicle in a crash left skid marks D feet long and that a test vehicle traveling at v miles per hour leaves skid marks d feet long. Then the speed V of the vehicle involved in the crash is given by

$$V = v\sqrt{\frac{D}{d}}.$$

(*Source:* N. Garber and L. Hoel, *Traffic and Highway Engineering.*)

(a) Find V if $v = 30$ mph, $D = 285$ feet, and $d = 178$ feet. Interpret your result.

(b) A test vehicle traveling at 45 mph leaves skid marks 255 feet long. How long would the skid marks be for a vehicle traveling 60 miles per hour?

96. *Highway Curves* If a circular curve without any banking has a radius of R feet, the speed limit L in miles per hour for the curve is $L = 1.5\sqrt{R}$. (*Source:* N. Garber.)

(a) Find the speed limit for a curve having a radius of 400 feet.

(b) If the radius of a curve doubles, what happens to the speed limit?

(c) A curve with a 40-mile-per-hour speed limit is being designed. What should be its radius?

97. *45°–45° Right Triangle* Suppose that the legs of a right triangle with angles of 45° and 45° both have length a, as depicted in the accompanying figure. Find the length of the hypotenuse.

98. *30°–60° Right Triangle* In a right triangle with angles of 30° and 60°, the shortest side is half the length of the hypotenuse (see the accompanying figure). If the hypotenuse has length c, find the length of the other two sides in terms of c.

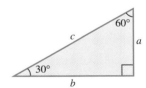

WRITING ABOUT MATHEMATICS

99. A student solves an equation *incorrectly* as follows.

$$\sqrt{3 - x} = \sqrt{x} - 1$$
$$(\sqrt{3 - x})^2 \stackrel{?}{=} (\sqrt{x})^2 - (1)^2$$
$$3 - x \stackrel{?}{=} x + 1$$
$$-2x \stackrel{?}{=} -2$$
$$x \stackrel{?}{=} 1$$

(a) How could you convince the student that the answer is wrong?

(b) Discuss where the error was made.

100. When each side of an equation is squared, you must check your results. Explain why.

Group Activity: Working with Real Data

Directions: Form a group of 2 to 4 people. Select someone to record the group's responses for this activity. All members of the group should work cooperatively to answer the questions. If your instructor asks for your results, each member of the group should be prepared to respond.

Simple Pendulum   Gravity is responsible for an object falling toward Earth. The farther the object falls, the faster it is moving when it hits the ground. For each second that an object falls, its speed increases by a constant amount, called the *acceleration due to gravity*, denoted g. One way to calculate the value of g is to use a simple pendulum. See the accompanying figure.

The time T for a pendulum to swing back and forth once is called its *period* and is given by

$$T = 2\pi\sqrt{\frac{L}{g}},$$

where L equals the length of the pendulum. The accompanying table lists the periods of pendulums with different lengths.

L (feet)	0.5	1.0	1.5
T (seconds)	0.78	1.11	1.36

(a) Solve the formula for g.
(b) Use the table to determine the value of g. (*Note:* The units for g are feet per second per second.)
(c) Interpret your result.

10.6   COMPLEX NUMBERS

Basic Concepts · **Addition, Subtraction, and Multiplication** · **Powers of i** · **Complex Conjugates and Division**

Figure 10.21   A Fractal: *The Cube Roots of Unity*

INTRODUCTION

Mathematics is both applied and theoretical. A common misconception is that abstract or theoretical mathematics is unimportant in today's world. Many new ideas with great practical importance were first developed as abstract concepts with no particular application in mind. For example, complex numbers, which are related to square roots of negative numbers, started as an abstract concept to solve equations. Today complex numbers are used in many sophisticated applications, such as the design of electrical circuits, ships, and airplanes. Even the *fractal image* shown in Figure 10.21 would not have been discovered without complex numbers.   (*Source:* D. Kincaid and W. Cheney, *Numerical Analysis.*)

Basic Concepts

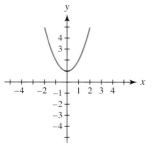

Figure 10.22

A graph of $y = x^2 + 1$ is shown in Figure 10.22. There are no x-intercepts, so the equation $x^2 + 1 = 0$ has no real number solutions.

If we try to solve $x^2 + 1 = 0$ by subtracting 1 from each side, the result is $x^2 = -1$. Because $x^2 \geq 0$ for any real number x, there are no real solutions. However, mathematicians have invented solutions.

$$x^2 = -1$$
$$x = \pm \sqrt{-1} \quad \text{Square root property}$$

We now define a number called the **imaginary unit**, denoted i.

PROPERTIES OF THE IMAGINARY UNIT i

$$i = \sqrt{-1} \quad \text{and} \quad i^2 = -1$$

By creating the number i, the solutions to the equation $x^2 + 1 = 0$ are i and $-i$. Using the real numbers and the imaginary unit i, we can define a new set of numbers called the *complex numbers*. A **complex number** can be written in **standard form**, as $a + bi$, where a and b are real numbers. The **real part** is a and the **imaginary part** is b. Every real number a is also a complex number because it can be written $a + 0i$. A complex number $a + bi$ with $b \neq 0$ is an **imaginary number**. Table 10.6 lists several complex numbers with their real and imaginary parts.

TABLE 10.6

Complex Number: $a + bi$	$-3 + 2i$	5	$-3i$	$-1 + 7i$	$-5 - 2i$	$4 + 6i$
Real Part: a	-3	5	0	-1	-5	4
Imaginary Part: b	2	0	-3	7	-2	6

Figure 10.23 shows how different sets of numbers are related. Note that *the set of complex numbers contains the set of real numbers.*

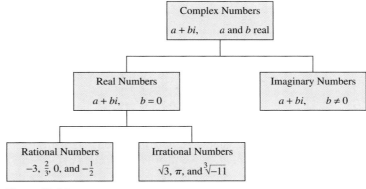

Figure 10.23

Using the imaginary unit i, we may write the square root of a negative number as a complex number. For example, $\sqrt{-2} = i\sqrt{2}$, and $\sqrt{-4} = i\sqrt{4} = 2i$. This method is summarized as follows.

Calculator Help

To set your calculator in $a + bi$ mode, see Appendix A (page AP-11).

THE EXPRESSION $\sqrt{-a}$

If $a > 0$, then $\sqrt{-a} = i\sqrt{a}$.

EXAMPLE 1   Writing the square root of a negative number

Write each square root using the imaginary unit i.

(a) $\sqrt{-25}$ **(b)** $\sqrt{-7}$ **(c)** $\sqrt{-20}$

Solution   **(a)** $\sqrt{-25} = i\sqrt{25} = 5i$ **(b)** $\sqrt{-7} = i\sqrt{7}$
(c) $\sqrt{-20} = i\sqrt{20} = i\sqrt{4}\sqrt{5} = 2i\sqrt{5}$

ADDITION, SUBTRACTION, AND MULTIPLICATION

Arithmetic operations can be defined for complex numbers.

ADDITION AND SUBTRACTION   To add the complex numbers $(-3 + 2i)$ and $(2 - i)$ add the real parts and then add the imaginary parts.

$$(-3 + 2i) + (2 - i) = (-3 + 2) + (2i - i)$$
$$= (-3 + 2) + (2 - 1)i$$
$$= -1 + i$$

This same process works for subtraction.

$$(6 - 3i) - (2 + 5i) = (6 - 2) + (-3i - 5i)$$
$$= (6 - 2) + (-3 - 5)i$$
$$= 4 - 8i$$

This method is summarized as follows.

SUM OR DIFFERENCE OF COMPLEX NUMBERS

Let $a + bi$ and $c + di$ be two complex numbers. Then

$$(a + bi) + (c + di) = (a + c) + (b + d)i \qquad \text{Sum}$$

and

$$(a + bi) - (c + di) = (a - c) + (b - d)i. \qquad \text{Difference}$$

EXAMPLE 2   Adding and subtracting complex numbers

Write each sum or difference in standard form.
(a) $(-7 + 2i) + (3 - 4i)$ **(b)** $3i - (5 - i)$

Solution   **(a)** $(-7 + 2i) + (3 - 4i) = (-7 + 3) + (2 - 4)i = -4 - 2i$
(b) $3i - (5 - i) = 3i - 5 + i = -5 + (3 + 1)i = -5 + 4i$

Technology Note: *Complex Numbers*

Many calculators can perform arithmetic with complex numbers. The figure shows a calculator display for the results in Example 2.

$$\begin{array}{l} (-7+2i)+(3-4i) \\ \quad\quad\quad\quad\quad -4-2i \\ 3i-(5-i) \\ \quad\quad\quad\quad\quad -5+4i \end{array}$$

Calculator Help

To access the imaginary unit i, see Appendix A (page AP-11).

MULTIPLICATION   We multiply two complex numbers in the same way that we multiply binomials and then we apply the property $i^2 = -1$.

EXAMPLE 3   Multiplying complex numbers

Write each product in standard form.
(a) $(2 - 3i)(1 + 4i)$ **(b)** $(5 - 2i)(5 + 2i)$

Solution   **(a)** Multiply the complex numbers like binomials.

$$(2 - 3i)(1 + 4i) = (2)(1) + (2)(4i) - (3i)(1) - (3i)(4i)$$
$$= 2 + 8i - 3i - 12i^2$$
$$= 2 + 5i - 12(-1)$$
$$= 14 + 5i$$

(b) Multiply these complex numbers in the same way.

$$(5 - 2i)(5 + 2i) = (5)(5) + (5)(2i) - (2i)(5) - (2i)(2i)$$
$$= 25 + 10i - 10i - 4i^2$$
$$= 25 - 4(-1)$$
$$= 29$$

$$\begin{array}{l} (2-3i)(1+4i) \\ \quad\quad\quad\quad\quad 14+5i \\ (5-2i)(5+2i) \\ \quad\quad\quad\quad\quad 29 \end{array}$$

Figure 10.24

These results are supported in Figure 10.24.

POWERS OF i

An interesting pattern appears when powers of i are calculated.

$$i^1 = i$$
$$i^2 = -1$$
$$i^3 = i^2 \cdot i = -1 \cdot i = -i$$
$$i^4 = i^2 \cdot i^2 = (-1)(-1) = 1$$
$$i^5 = i^4 \cdot i = (1)i = i$$
$$i^6 = i^4 \cdot i^2 = (1)(-1) = -1$$
$$i^7 = i^4 \cdot i^3 = (1)(-i) = -i$$
$$i^8 = i^4 \cdot i^4 = (1)(1) = 1$$

The powers of i cycle with the pattern i, -1, $-i$, and 1. These examples suggest the following method for calculating powers of i.

▌▌▌▌▌	POWERS OF i

The value of i^n can be found by dividing n by 4. If the remainder is r, then

$$i^n = i^r.$$

Note that $i^0 = 1$, $i^1 = i$, $i^2 = -1$, and $i^3 = -i$.

EXAMPLE 4 **Calculating powers of i**

Evaluate each expression.
(a) i^9 **(b)** i^{19} **(c)** i^{40}

Solution **(a)** When 9 is divided by 4, the result is 2 with remainder 1. Thus $i^9 = i^1 = i$.
(b) When 19 is divided by 4, the result is 4 with remainder 3. Thus $i^{19} = i^3 = -i$.
(c) When 40 is divided by 4, the result is 10 with remainder 0. Thus $i^{40} = i^0 = 1$.

COMPLEX CONJUGATES AND DIVISION

The **complex conjugate** of $a + bi$ is $a - bi$. To find the conjugate, we change the sign of the imaginary part b. Table 10.7 contains examples of complex numbers and their conjugates.

TABLE 10.7 Complex Conjugates

Number	$2 + 5i$	$6 - 3i$	$-2 + 7i$	$-1 - i$	5	$-4i$
Conjugate	$2 - 5i$	$6 + 3i$	$-2 - 7i$	$-1 + i$	5	$4i$

The product of two complex conjugates is a real number, as we demonstrated in Example 3(b). This property is used to divide two complex numbers. To convert the quotient $\frac{2 + 3i}{3 - i}$ into standard form $a + bi$, we multiply the numerator and the denominator by the complex conjugate of the denominator, which is $3 + i$. The next example illustrates this method.

EXAMPLE 5 **Dividing complex numbers**

Write each quotient in standard form.
(a) $\dfrac{2 + 3i}{3 - i}$ **(b)** $\dfrac{4}{2i}$

Solution **(a)** Multiply the numerator and denominator by $3 + i$.

$$\frac{2 + 3i}{3 - i} = \frac{(2 + 3i)(3 + i)}{(3 - i)(3 + i)} \qquad \text{Multiply by 1.}$$

$$= \frac{2(3) + (2)(i) + (3i)(3) + (3i)(i)}{(3)(3) + (3)(i) - (i)(3) - (i)(i)} \qquad \text{Expand.}$$

$$= \frac{6 + 2i + 9i + 3i^2}{9 + 3i - 3i - i^2} \qquad \text{Simplify.}$$

$$= \frac{6 + 11i + 3(-1)}{9 - (-1)} \qquad i^2 = -1$$

$$= \frac{3 + 11i}{10} \qquad \text{Simplify.}$$

$$= \frac{3}{10} + \frac{11}{10}i \qquad \frac{a + bi}{c} = \frac{a}{c} + \frac{b}{c}i$$

(b) Multiply the numerator and denominator by $-2i$.

$$\frac{4}{2i} = \frac{(4)(-2i)}{(2i)(-2i)} \qquad \text{Multiply by 1.}$$

$$= \frac{-8i}{-4i^2} \qquad \text{Simplify.}$$

$$= \frac{-8i}{-4(-1)} \qquad i^2 = -1$$

$$= \frac{-8i}{4} \qquad \text{Simplify.}$$

$$= -2i \qquad \text{Divide.}$$

These results are supported in Figure 10.25.

Figure 10.25

---

continued on next page

10.6 PUTTING IT ALL TOGETHER

In this section we discussed complex numbers and how to perform arithmetic operations with them. Complex numbers allow us to solve equations that could not be solved only with real numbers. The following table summarizes the important concepts in the section.

Concept	Explanation	Examples
Complex Numbers	A complex number can be expressed as $a + bi$, where a and b are real numbers. The imaginary unit i satisfies $i = \sqrt{-1}$ and $i^2 = -1$. As a result, we can write $\sqrt{-a} = i\sqrt{a}$ if $a > 0$.	$\sqrt{-13} = i\sqrt{13}$ and $\sqrt{-9} = 3i$
Addition, Subtraction, and Multiplication	To add (subtract) complex numbers, add (subtract) the real parts and then add (subtract) the imaginary parts. Multiply complex numbers in a similar manner to how *FOIL* is used to multiply binomials. Then apply the property $i^2 = -1$.	$(3 + 6i) + (-1 + 2i)$ Sum $= (3 + -1) + (6 + 2)i$ $= 2 + 8i$ $(2 - 5i) - (1 + 4i)$ Difference $= (2 - 1) + (-5 - 4)i$ $= 1 - 9i$ $(-1 + 2i)(3 + i)$ Product $= (-1)(3) + (-1)(i) + (2i)(3) + (2i)(i)$ $= -3 - i + 6i + 2i^2$ $= -3 + 5i + 2(-1)$ $= -5 + 5i$

continued from previous page

Concept	Explanation	Examples
Complex Conjugates	The conjugate of $a + bi$ is $a - bi$.	The conjugate of $3 - 5i$ is $3 + 5i$.
Division	To simplify a quotient, multiply the numerator and denominator by the complex conjugate of the denominator. Then simplify the expression and write it in standard form as $a + bi$.	$\dfrac{10}{1 + 2i} = \dfrac{10(1 - 2i)}{(1 + 2i)(1 - 2i)}$ $= \dfrac{10 - 20i}{5}$ $= 2 - 4i$

 **10.6** EXERCISES

FOR EXTRA HELP

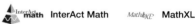

CONCEPTS

1. Give an example of a complex number that is not a real number.

2. Can you give an example of a real number that is not a complex number? Explain.

3. $\sqrt{-1} =$ _____

4. $i^2 =$ _____

5. $\sqrt{-a} =$ _____, if $a > 0$.

6. The complex conjugate of $10 + 7i$ is _____.

7. The standard form for a complex number is _____.

8. Write $\frac{2 + 4i}{2}$ in standard form.

9. The real part of $4 - 5i$ is _____.

10. The imaginary part of $4 - 5i$ is _____.

COMPLEX NUMBERS

Exercises 11–20: Use the imaginary unit to write the expression.

11. $\sqrt{-5}$

12. $\sqrt{-21}$

13. $\sqrt{-100}$

14. $\sqrt{-49}$

15. $\sqrt{-144}$

16. $\sqrt{-64}$

17. $\sqrt{-12}$

18. $\sqrt{-8}$

19. $\sqrt{-18}$

20. $\sqrt{-48}$

Exercises 21–42: Write the expression in standard form.

21. $(5 + 3i) + (-2 - 3i)$

22. $(1 - i) + (5 - 7i)$

23. $(2i) + (-8 + 5i)$ 24. $(-3i) + (5i)$

25. $(2 - 7i) - (1 + 2i)$ 26. $(1 + 8i) - (3 + 9i)$

27. $(5i) - (10 - 2i)$ 28. $(1 + i) - (1 - i)$

29. $4(5 - 3i)$ 30. $(1 + 2i)(-6 - i)$

31. $(-3 - 4i)(5 - 4i)$ 32. $(3 + 5i)(3 - 5i)$

33. $(-4i)(5i)$ 34. $(-6i)(-4i)$

35. $3i + (2 - 3i) - (1 - 5i)$

36. $4 - (5 - 7i) + (3 + 7i)$

37. $(2 + i)^2$ 38. $(-1 + 2i)^2$

39. $2i(-3 + i)$ 40. $5i(1 - 9i)$

41. $i(1 + i)^2$ 42. $2i(1 - i)^2$

Exercises 43–50: (Refer to Example 4.) Simplify the power of i.

43. i^{11} 44. i^{50}

45. i^{21} 46. i^{103}

47. i^{58} 48. i^{61}

49. i^{64} 50. i^{28}

Exercises 51–58: Write the complex conjugate.

51. $3 + 4i$

52. $1 - 4i$

53. $-6i$

54. -10

55. $5 - 4i$

56. $7 + 2i$

57. -1

58. $19i$

Exercises 59–72: Write the expression in standard form.

59. $\dfrac{2}{1 + i}$

60. $\dfrac{-6}{2 - i}$

61. $\dfrac{3i}{5 - 2i}$

62. $\dfrac{-8}{2i}$

63. $\dfrac{8 + 9i}{5 + 2i}$

64. $\dfrac{3 - 2i}{1 + 4i}$

65. $\dfrac{5 + 7i}{1 - i}$

66. $\dfrac{-7 + 4i}{3 - 2i}$

67. $\dfrac{2 - i}{i}$

68. $\dfrac{3 + 2i}{-i}$

69. $\dfrac{1}{i} + \dfrac{1}{2i}$

70. $\dfrac{3}{4i} + \dfrac{2}{i}$

71. $\dfrac{1}{-1 + i} - \dfrac{2}{i}$

72. $-\dfrac{3}{2i} - \dfrac{2}{1 + i}$

APPLICATIONS

Exercises 73 and 74: Corrosion in Airplanes Corrosion in the metal surface of an airplane can be difficult to detect visually. One test used to locate it involves passing an alternating current through a small area on the plane's surface. If the current varies from one region to another, it may indicate that corrosion is occurring. The impedance Z (or opposition to the flow of electricity) of the metal is related to the voltage V and current I by the equation $Z = \frac{V}{I}$, where Z, V, and I are complex numbers. Calculate Z for the given values of V and I. (**Source:** Society for Industrial and Applied Mathematics.)

73. $V = 40 + 70i, I = 2 + 3i$

74. $V = 10 + 20i, I = 3 + 7i$

WRITING ABOUT MATHEMATICS

75. A student multiplies $(2 + 3i)(4 - 5i)$ *incorrectly* to obtain $8 - 15i$. What was the student's mistake?

76. A student divides $\frac{6 - 10i}{3 + 2i}$ *incorrectly* to obtain $2 - 5i$. What was the student's mistake?

CHECKING BASIC CONCEPTS SECTIONS 10.5 AND 10.6

1. Solve each equation. Check your answers.
  (a) $\sqrt{2x - 4} = 2$ **(b)** $\sqrt[3]{x - 1} = 3$
  (c) $\sqrt{3x} = 1 + \sqrt{x + 1}$

2. Find the distance between the points $(-3, 5)$ and $(2, -7)$.

3. A 16-inch diagonal television set has a rectangular picture with a width of 12.8 inches. Find the height of the picture.

4. Use the imaginary unit i to write each expression.
  (a) $\sqrt{-64}$
  (b) $\sqrt{-17}$

5. Simplify each expression.
  (a) $(2 - 3i) + (1 - i)$
  (b) $4i - (2 + i)$
  (c) $(3 - 2i)(1 + i)$
  (d) $\dfrac{3}{2 - 2i}$

Summary

Radicals and Radical Notation

Square Root

b is a square root of a if $b^2 = a$.

Principal Square Root

$\sqrt{a} = b$ if $b^2 = a$ and $b \geq 0$.

Examples: $\sqrt{16} = 4$, $\quad -\sqrt{9} = -3$,

and $\pm\sqrt{36} = \pm6$

Cube Root

b is the cube root of a if $b^3 = a$.

Examples: $\sqrt[3]{27} = 3$ and $\sqrt[3]{-8} = -2$

nth root

b is the nth root of a if $b^n = a$.

Example: $\sqrt[4]{16} = 2$ because $2^4 = 16$.

Note: An *even* root of a *negative* number is not a real number. Also, $\sqrt[n]{a}$ denotes the *principal* nth root. The expressions $|x|$ and $\sqrt{x^2}$ are equivalent.

Absolute Value

Example: $\sqrt{(x+y)^2} = |x+y|$

The Expression $a^{1/n}$

$a^{1/n} = \sqrt[n]{a}$ if n is an integer greater than 1.

Examples: $5^{1/2} = \sqrt{5}$ and $64^{1/3} = \sqrt[3]{64} = 4$

The Expression $a^{m/n}$

$a^{m/n} = \sqrt[n]{a^m}$ or $a^{m/n} = (\sqrt[n]{a})^m$

Examples: $8^{2/3} = \sqrt[3]{8^2} = \sqrt[3]{64} = 4$ and

$8^{2/3} = (\sqrt[3]{8})^2 = (2)^2 = 4$

Properties of Exponents

Product Rule

$a^p a^q = a^{p+q}$

Negative Exponents

$a^{-p} = \dfrac{1}{a^p}, \quad \dfrac{1}{a^{-p}} = a^p$

Negative Exponents for Quotients

$\left(\dfrac{a}{b}\right)^{-p} = \left(\dfrac{b}{a}\right)^p$

Quotient Rule for Exponents

$\dfrac{a^p}{a^q} = a^{p-q}$

Power Rule for Exponents

$(a^p)^q = a^{pq}$

Power Rule for Products

$(ab)^p = a^p b^p$

Power Rule for Quotients

$\left(\dfrac{a}{b}\right)^p = \dfrac{a^p}{b^p}$

Section 10.2   *Simplifying Radical Expressions*

Product Rule for Radical Expressions

$$\sqrt[n]{a} \cdot \sqrt[n]{b} = \sqrt[n]{a \cdot b},$$

provided each expression is defined.

Example: $\sqrt[3]{3} \cdot \sqrt[3]{9} = \sqrt[3]{27} = 3$

Perfect *n*th Power   An integer a is a perfect *n*th power if $b^n = a$ for some integer b.

Examples: 25 is a perfect square, 8 is a perfect cube, and 16 is a perfect fourth power.

Quotient Rule for Radical Expressions

$$\sqrt[n]{\frac{a}{b}} = \frac{\sqrt[n]{a}}{\sqrt[n]{b}},$$

provided each expression is defined.

Example: $\dfrac{\sqrt[3]{24}}{\sqrt[3]{3}} = \sqrt[3]{\dfrac{24}{3}} = \sqrt[3]{8} = 2$

Section 10.3   *Operations on Radical Expressions*

Addition and Subtraction   To add or subtract radical expressions, combine like radicals.

Examples: $2\sqrt[3]{4} + 3\sqrt[3]{4} = 5\sqrt[3]{4}$ and $\sqrt{5} - 2\sqrt{5} = -\sqrt{5}$

Multiplication   Sometimes radical expressions can be multiplied like binomials.

Example: $(5 - \sqrt{3})(5 + \sqrt{3}) = (5)^2 - (\sqrt{3})^2 = 25 - 3 = 22$ because
$(a - b)(a + b) = a^2 - b^2$.

Rationalizing the Denominator   Multiply the numerator and denominator by the conjugate of the denominator.

Example: $\dfrac{1}{4 + \sqrt{2}} = \dfrac{1}{(4 + \sqrt{2})} \cdot \dfrac{(4 - \sqrt{2})}{(4 - \sqrt{2})} = \dfrac{4 - \sqrt{2}}{(4)^2 - (\sqrt{2})^2} = \dfrac{4 - \sqrt{2}}{14}$

Section 10.4   *Radical Functions*

The Square Root Function   The square root function is denoted $f(x) = \sqrt{x}$. Its domain is $\{x \mid x \geq 0\}$ and its graph is shown in the figure.

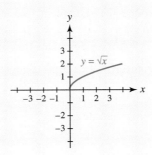

The Square Root Property   Let k be a nonnegative number. Then the solutions to $x^2 = k$ are given by $x = \pm\sqrt{k}$.

Example: $x^2 = 100$ is equivalent to $x = \pm\sqrt{100} = \pm 10$.

The Cube Root Function
The cube root function is denoted $f(x) = \sqrt[3]{x}$. Its domain is all real numbers and its graph is shown in the figure.

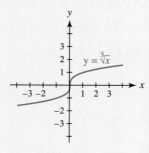

Power Functions   If a function can be defined by $f(x) = x^p$, where p is a rational number, then it is a power function.

Examples: $f(x) = x^{4/5}$ and $g(x) = x^{2.3}$

Section 10.5   *Equations Involving Radical Expressions*

Power Rule for Solving Radical Equations   The solutions to $a = b$ are among the solutions to $a^n = b^n$, where n is an integer.

Example: The solutions to $\sqrt{3x + 3} = 2x - 1$ are among the solutions to $3x + 3 = (2x - 1)^2$.

Solving Radical Equations
  STEP 1: Isolate a radical term on one side of the equation.

  STEP 2: Apply the power rule by raising each side of the equation to the power equal to the index of the isolated radical term.

  STEP 3: Solve the equation. If it still contains a radical, repeat Steps 1 and 2.

  STEP 4: Check your answers by substituting each result in the *given* equation.

Example: To isolate the radical in $\sqrt{x + 1} + 4 = 6$ subtract 4 from each side to obtain $\sqrt{x + 1} = 2$. Next square each side, which gives $x + 1 = 4$ or $x = 3$. Checking verifies that 3 is a solution.

Pythagorean Theorem   If a right triangle has legs a and b with hypotenuse c, then
$$a^2 + b^2 = c^2.$$

Example: If a right triangle has legs 5 and 12, then the hypotenuse equals
$$c = \sqrt{5^2 + 12^2} = \sqrt{169} = 13.$$

The Distance Formula   The distance d between (x_1, y_1) and (x_2, y_2) is

$$d = \sqrt{(x_2 - x_1)^2 + (y_2 - y_1)^2}.$$

Example: The distance between $(-1, 3)$ and $(4, 5)$ is

$$d = \sqrt{\left(4 - (-1)\right)^2 + (5 - 3)^2} = \sqrt{25 + 4} = \sqrt{29}.$$

Section 10.6   *Complex Numbers*

Complex Numbers

Imaginary Unit	$i = \sqrt{-1}$ and $i^2 = -1$
Standard Form	$a + bi$, where a and b are real numbers
	Examples: $4 + 3i,\ 5 - 6i,\ 8,$ and $-2i$
Real Part	The real part of $a + bi$ is a.
	Example: The real part of $3 - 2i$ is 3.
Imaginary Part	The imaginary part of $a + bi$ is b.
	Example: The imaginary part of $2 - i$ is -1.

Arithmetic Operations   Arithmetic operations are similar to arithmetic operations on binomials.

Examples: $(2 + 2i) + (3 - i) = 5 + i,$
$(1 - i) - (1 - 2i) = i,$
$(1 - i)(1 + i) = 1^2 - i^2 = 1 - (-1) = 2,$ and

$$\frac{2}{1 - i} = \frac{2}{1 - i} \cdot \frac{1 + i}{1 + i} = \frac{2 + 2i}{2} = 1 + i$$

Powers of i   The value of i^n equals i^r, where r is the remainder when n is divided by 4.

Example: $i^{21} = i^1 = i$ because when 21 is divided by 4 the remainder is 1.

CHAPTER 10 Review Exercises

SECTION 10.1

Exercises 1–12: Simplify the expression.

1. $\sqrt{4}$

2. $\sqrt{36}$

3. $\sqrt{9x^2}$

4. $\sqrt{(x - 1)^2}$

5. $\sqrt[3]{-64}$

6. $\sqrt[3]{-125}$

7. $\sqrt[3]{x^6}$

8. $\sqrt[3]{27x^3}$

9. $\sqrt[4]{16}$

10. $\sqrt[5]{-1}$

11. $\sqrt[4]{x^8}$

12. $\sqrt[5]{(x+1)^5}$

Exercises 13–16: Write the expression in radical notation.

13. $14^{1/2}$

14. $(-5)^{1/3}$

15. $\left(\dfrac{x}{y}\right)^{3/2}$

16. $(xy)^{-2/3}$

Exercises 17–20: Evaluate the expression.

17. $(-27)^{2/3}$

18. $16^{1/4}$

19. $16^{3/2}$

20. $81^{3/4}$

Exercises 21–24: Simplify the expression. Assume that all variables are positive.

21. $(z^3)^{2/3}$

22. $(x^2y^4)^{1/2}$

23. $\left(\dfrac{x^2}{y^6}\right)^{3/2}$

24. $\left(\dfrac{x^3}{y^6}\right)^{-1/3}$

SECTION 10.2

Exercises 25–40: Simplify the expression. Assume that all variables are positive.

25. $\sqrt{2} \cdot \sqrt{32}$

26. $\sqrt[3]{-4} \cdot \sqrt[3]{2}$

27. $\sqrt[3]{x^4} \cdot \sqrt[3]{x^2}$

28. $\dfrac{\sqrt{80}}{\sqrt{20}}$

29. $\sqrt[3]{-\dfrac{x}{8}}$

30. $\sqrt{\dfrac{1}{3}} \cdot \sqrt{\dfrac{1}{3}}$

31. $\sqrt{48}$

32. $\sqrt{54}$

33. $\sqrt[3]{\dfrac{3}{x}} \cdot \sqrt[3]{\dfrac{9}{x^2}}$

34. $\sqrt{32a^3b^2}$

35. $\sqrt{3xy} \cdot \sqrt{27xy}$

36. $\sqrt[3]{-25z^2} \cdot \sqrt[3]{-5z^2}$

37. $\sqrt{x^2+2x+1}$

38. $\sqrt[4]{\dfrac{2a^2}{b}} \cdot \sqrt[4]{\dfrac{8a^3}{b^3}}$

39. $2\sqrt{x} \cdot \sqrt[3]{x}$

40. $\sqrt[3]{rt} \cdot \sqrt[4]{r^2t^4}$

Exercises 41 and 42: Rationalize the denominator.

41. $\dfrac{4}{\sqrt{5}}$

42. $\dfrac{r}{2\sqrt{t}}$

SECTION 10.3

Exercises 43–52: Simplify the expression. Assume that all variables are positive.

43. $3\sqrt{3} + \sqrt{3}$

44. $\sqrt[3]{x} + 2\sqrt[3]{x}$

45. $3\sqrt[3]{5} - 6\sqrt[3]{5}$

46. $\sqrt[4]{y} - 2\sqrt[4]{y}$

47. $2\sqrt{12} + 7\sqrt{3}$

48. $3\sqrt{18} - 2\sqrt{2}$

49. $7\sqrt[3]{16} - \sqrt[3]{2}$

50. $\sqrt{4x+4} + \sqrt{x+1}$

51. $\sqrt{4x^3} - \sqrt{x}$

52. $\sqrt[3]{ab^4} + 2\sqrt[3]{a^4b}$

Exercises 53–56: Multiply and simplify.

53. $(3 + \sqrt{6})(3 - \sqrt{6})$

54. $(10 - \sqrt{5})(10 + \sqrt{5})$

55. $(\sqrt{a} + \sqrt{2b})(\sqrt{a} - \sqrt{2b})$

56. $(\sqrt{xy} - 1)(\sqrt{xy} + 2)$

Exercises 57–60: Rationalize the denominator.

57. $\dfrac{1}{\sqrt{2}+3}$

58. $\dfrac{2}{5-\sqrt{7}}$

59. $\dfrac{1}{\sqrt{8}-\sqrt{7}}$

60. $\dfrac{\sqrt{a}-\sqrt{b}}{\sqrt{a}+\sqrt{b}}$

SECTION 10.4

Exercises 61 and 62: Graph the equation.

61. $y = \sqrt{x}$

62. $y = \sqrt[3]{x}$

Exercises 63 and 64: Graph the equation. Compare the graph to either $y = \sqrt{x}$ or $y = \sqrt[3]{x}$.

63. $y = \sqrt{x} - 2$

64. $y = \sqrt[3]{x} - 1$

Exercises 65–68: Find the domain of f. Write your answer in interval notation.

65. $f(x) = \sqrt{x-1}$

66. $f(x) = \sqrt{6-2x}$

67. $f(x) = \sqrt{x^2+1}$

68. $f(x) = \dfrac{1}{\sqrt{x+2}}$

Exercises 69–74: Solve.

69. $x^2 = 121$

70. $2z^2 = 32$

71. $(x-1)^2 = 16$

72. $x^3 = 64$

73. $(x-1)^3 = 8$

74. $(2x-1)^3 = 27$

SECTION 10.5

Exercises 75–80: Solve. Check your answer.

75. $\sqrt{x+2} = x$

76. $\sqrt{2x-1} = \sqrt{x+3}$

77. $\sqrt[3]{x-1} = 2$

78. $\sqrt[3]{3x} = 3$

79. $\sqrt{2x} = x - 4$

80. $\sqrt{x} + 1 = \sqrt{x + 2}$

Exercises 81 and 82: Solve the equation graphically. Approximate solutions to the nearest hundredth when appropriate.

81. $\sqrt[3]{2x - 1} = 2$

82. $x^{2/3} = 3 - x$

Exercises 83 and 84: A right triangle has legs a and b with hypotenuse c. Find the length of the missing side.

83. $a = 4, b = 7$

84. $a = 5, c = 8$

Exercises 85 and 86: Find the distance between the points.

85. $(-2, 3), (2, -2)$

86. $(2, -3), (-4, 1)$

Section **10.6**

Exercises 87–92: Write the complex expression in standard form.

87. $(1 - 2i) + (-3 + 2i)$

88. $(1 + 3i) - (3 - i)$

89. $(1 - i)(2 + 3i)$

90. $\dfrac{3 + i}{1 - i}$

91. $\dfrac{i(4 + i)}{2 - 3i}$

92. $(1 - i)^2(1 + i)$

APPLICATIONS

93. *Hang Time* (Refer to Example 1, Section 10.4.) A football is punted and has a hang time of 4.6 seconds. Estimate the height that it was kicked to the nearest foot.

94. *Baseball Diamond* The four bases of a baseball diamond form a square 90 feet on a side. Find the distance from home plate to second base.

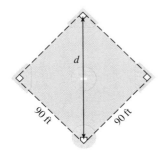

95. *Falling Time* The time T in seconds for an object to fall from a height of h feet is given by $T = \frac{1}{4}\sqrt{h}$. If a person steps off a 10-foot-high board into a swimming pool, how long is the person in the air?

96. *Highway Curves* If a circular highway curve is banked with a slope of $m = \frac{1}{10}$ (see the accompanying figure) and has a radius of R feet, then the speed limit L in miles per hour for the curve is given by

$$L = \sqrt{3.75R}.$$

(*Source:* N. Garber and L. Hoel, *Traffic and Highway Engineering.*)
 (a) Find the speed limit if the curve has a radius of 500 feet.
 (b) With no banking, the speed limit is given by $L = 1.5\sqrt{R}$. Find the speed limit for a curve with no banking and a radius of 500 feet. How does banking affect the speed limit? Does this result agree with your intuition?

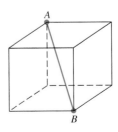

97. *Geometry* Find the length of a side of a square if the square has an area of 7 square feet.

98. *Geometry* A cube has sides of length $\sqrt{5}$.
 (a) Find the area of one side of the cube.
 (b) Find the volume of the cube.
 (c) Find the length of the diagonal of one of the sides.
 (d) Find the distance from A to B in the figure.

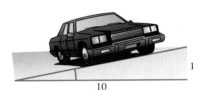

99. *Pendulum* The time required for a pendulum to swing back and forth is called its *period* (see the figure on the next page). The period T of a pendulum does not depend on its weight, only on its length L and gravity. It is given by $T = 2\pi\sqrt{\frac{L}{32.2}}$, where T is in seconds and L is in feet. Estimate the length of a pendulum with a period of 1 second.

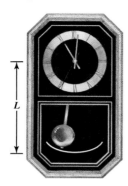

100. *Pendulum*   (Refer to Exercise 99.) If a pendulum were on the moon, its period could be calculated by $T = 2\pi\sqrt{\frac{L}{5.1}}$. Estimate the length of a pendulum with a period of 1 second on the moon. Compare your answer to that for Exercise 99.

101. *Population Growth*   In 1790 the population of the United States was 4 million, and by 2000 it had grown to 281 million. The average annual percentage growth in the population r (expressed as a decimal) can be determined by the polynomial equation

$281 = 4(1 + r)^{210}$. Solve this equation for r and interpret the result.

102. *Surface Area of a Cone*   The surface area of a cone having radius r and height h is given by $S = \pi r\sqrt{r^2 + h^2}$. See the accompanying figure. Estimate the surface area if $r = 11$ inches and $h = 60$ inches.

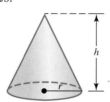

103. *Radioactive Carbon Dating*   Living plants and animals have a constant amount of radioactive carbon in their cells, which comes from the carbon dioxide they breathe. When a plant or animal dies, the exchange of oxygen and carbon dioxide halts and the amount of radioactive carbon starts to decrease. The fraction of radioactive carbon remaining t years after death is given by $2^{-t/5700}$. Find the fraction left after each time period.

(a) 5700 years   (b) 20,000 years

CHAPTER 10 Test

Exercises 1–4: Simplify the expression.

1. $\sqrt{25x^4}$

2. $\sqrt[3]{8z^6}$

3. $\sqrt[4]{16x^4y^5}$

4. $(\sqrt{3} - \sqrt{2})(\sqrt{3} + \sqrt{2})$

Exercises 5 and 6: Write the expression in radical notation.

5. $7^{2/5}$

6. $\left(\dfrac{x}{y}\right)^{-2/3}$

Exercises 7 and 8: Evaluate the expression without a calculator.

7. $(-8)^{4/3}$

8. $36^{-3/2}$

9. Find the domain of $f(x) = \sqrt{4 - x}$. Write your answer in interval notation.

10. Sketch a graph of $y = \sqrt{x + 3}$.

Exercises 11–16: Simplify the expression. Assume that all variables are positive.

11. $(2z^{1/2})^3$

12. $\left(\dfrac{y^2}{z^3}\right)^{-1/3}$

13. $\sqrt{3} \cdot \sqrt{27}$

14. $\dfrac{\sqrt{y^3}}{\sqrt{4y}}$

15. $7\sqrt{7} - 3\sqrt{7} + \sqrt{5}$ 16. $7\sqrt[3]{x} - \sqrt[3]{x}$

17. Solve the equation $\sqrt{2x + 2} = x - 11$.

18. Rationalize the denominator of $\dfrac{1}{\sqrt{14} - \sqrt{13}}$.

19. One leg of a right triangle has length 7 and the hypotenuse has length 13. Find the length of the third side.

20. Find the distance between the points $(-3, 5)$ and $(-1, 7)$.

Exercises 21–24: Write the complex expression in standard form.

21. $(-5 + i) + (7 - 20i)$

22. $(3i) - (6 - 5i)$

23. $\left(\dfrac{1}{2} - i\right)\left(\dfrac{1}{2} + i\right)$ 24. $\dfrac{2i}{5 + 2i}$

25. *Volume of a Sphere*   The volume V of a sphere is given by $V = \frac{4}{3}\pi r^3$, where r is its radius.
    (a) Solve the equation for r.
    (b) Find the radius of a sphere with a volume of 50 cubic inches.

26. *Wing Span of a Bird*   The wing span L of a bird with weight W can sometimes be modeled by $L = 27.4W^{1/3}$, where L is in inches and W is in pounds. Use this formula to estimate the weight of a bird that has a wing span of 30 inches.   (*Source:* C. Pennycuick, *Newton Rules Biology.*)

CHAPTER 10 Extended and Discovery Exercises

   1. *Moons of Jupiter*   The accompanying table lists the orbital distances and periods of several moons of Jupiter. Let x be the distance and y be the period. These data can be modeled by a power function of the form $f(x) = 0.0002x^{m/n}$. Use trial and error to find the value of the fraction $\frac{m}{n}$. Graph f and the data in the same viewing rectangle.

Moon	Distance (10^3 kilometers)	Period (days)
Metis	128	0.29
Almathea	181	0.50
Thebe	222	0.67
Io	422	1.77
Europa	671	3.55
Ganymede	1070	7.16
Callisto	1883	16.69

Source: M. Zeilik, *Introductory Astronomy and Astrophysics.*

2. *Modeling Wood in a Tree*   In forestry the volume of timber in a given area of forest is often estimated. To make such estimates, scientists have developed formulas to find the amount of wood contained in a tree with height h in feet and diameter d in inches. One study concluded that the volume V of wood in a tree could be modeled by the equation $V = kh^{1.12}d^{1.98}$, where k is a constant. Note that the diameter is measured 4.5 feet above the ground.   (*Source:* B. Ryan, B. Joiner, and T. Ryan, *Minitab Handbook.*)
   (a) A tree with an 11-inch diameter and a 47-foot height has a volume of 11.4 cubic feet. Approximate the constant k.
   (b) Estimate the volume of wood in the same type of tree with $d = 20$ inches and $h = 105$ feet.

3. *Area of Skin*   The surface area of the skin covering the human body is a function of more than one variable. Both height and weight influence the surface area of a person's body. Hence a taller person tends to have a larger surface area, as does a heavier person.

A formula to determine the area of a person's skin in square inches is $S = 15.7(w^{0.425})(h^{0.725})$, where w is weight in pounds and h is height in inches. (*Source:* H. Lancaster, *Quantitative Methods in Biological and Medical Sciences.*)

(a) Use S to estimate the area of a person's skin who is 65 inches tall and weighs 154 pounds.
(b) If a person's weight doubles, what happens to the area of the person's skin? Explain.
(c) If a person's height doubles, what happens to the area of the person's skin? Explain.

4. *Minimizing Cost* A natural gas line running along a river is to be connected from point A to a cabin on the other bank located at point D, as illustrated in the accompanying figure. The width of the river is 500 feet, and the distance from point A to point C is 1000 feet. The cost of running the pipe along the shoreline is $30 per foot, and the cost of running it underwater is $50 per foot. The cost of connecting the gas line from A to D is to be minimized.

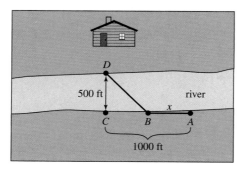

(a) Write an expression that gives the cost of running the line from A to B if the distance between these points is x feet.
(b) Find the distance from B to D in terms of x.
(c) Write an expression that gives the cost of running the line from B to D.
(d) Use your answer from parts (a) and (c) to write an expression that gives the cost of running the line from A to B to D.
 (e) Graph your expression from part (d) in the window [0, 1000, 100] by [40000, 60000, 5000] to determine the value of x that minimizes the cost of the line going from A to D. What is the minimum cost?

Quadratic Functions and Equations

Technology brings with it unforeseen consequences. The powerful insecticide DDT was a miracle of science that promised an adequate food supply for the world. But in a matter of a few decades its environmental impact became a worldwide disaster that had the potential to threaten human beings as a species. Medical technology in the twentieth century increased the average life span from 47 years to 77 years, but it also forced societies to cope with the possibilities created by genetic engineering.

Technology is sometimes used to help teach important concepts in mathematics and to solve applications. Without technology many modern inventions would be impossible. However, technology is not a replacement for mathematical understanding and judgment because mathematical concepts change little over time. On the one hand, the human mind is capable of mathematical insight and decision making, but it is not particularly proficient at performing long, tedious calculations. On the other hand, computers and calculators are incapable of possessing genuine mathematical insight, but they are excellent at performing arithmetic and other routine computation. In this way, technology can complement the human mind in the study and application of mathematics.

People are still the most extraordinary computers of all.
—John F. Kennedy

Source: F. Allen, "Technology at the End of the Century," *Invention and Technology*, Winter 2000.

11.1   QUADRATIC FUNCTIONS AND THEIR GRAPHS

Graphs of Quadratic Functions · Basic Transformations of Graphs ·
Min–Max Applications

INTRODUCTION

Suppose that a hotel is considering giving a group discount on room rates. The regular price is $80, but for each room rented the price decreases by $2. On the one hand, if the hotel rents one room it makes only $78. On the other hand, if the hotel rents 40 rooms, the rooms are all free and the hotel makes nothing. Is there an optimal number of rooms between 1 and 40 that should be rented to maximize the revenue? In this section we use quadratic functions and their graphs to answer this question. Quadratic functions are used not only in business, but they are used also in a wide variety of fields, such as road construction and medicine. In fact, quadratic functions are an important concept throughout mathematics.

GRAPHS OF QUADRATIC FUNCTIONS

In Section 8.4 we discussed how a quadratic function could be represented by a polynomial of degree 2. We now give an alternative definition of a quadratic function.

QUADRATIC FUNCTION

A quadratic function can be written in the form

$$f(x) = ax^2 + bx + c,$$

where a, b, and c are real numbers with $a \neq 0$.

The graph of *any* quadratic function is a *parabola*. Recall that a parabola is a ∪-shaped graph that either opens upward or downward. The graph of the simple quadratic function $y = x^2$ is a parabola that opens upward, with its *vertex* located at the origin, as shown in Figure 11.1(a). The **vertex** is the *lowest* point on the graph of a parabola that opens upward and the *highest* point on the graph of a parabola that opens downward. A parabola opening downward is shown in Figure 11.1(b). Its vertex is the point (0, 2) and is the highest point on a graph. If we were to fold either graph along the y-axis, the left and right sides of the graph would match. That is, both parts of the graph are symmetric with respect to the y-axis. In this case the y-axis is the **axis of symmetry** for the graph. Figure 11.1(c) shows a parabola that opens upward with vertex $(2, -1)$ and axis of symmetry $x = 2$.

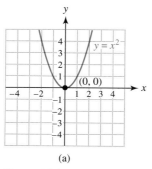

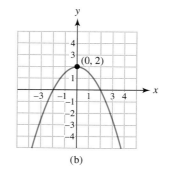

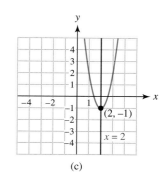

(a)   (b)   (c)

Figure 11.1

Suppose that the graph of $y = x^2$ shown in Figure 11.1(a) represents a valley. If we walk from *left to right*, the valley "goes down" and then "goes up." Mathematically, we say that the graph of $y = x^2$ is *decreasing* when $x \leq 0$ and *increasing* when $x \geq 0$. The vertex represents the point at which the graph switches from decreasing to increasing. In Figure 11.1(b), the graph increases when $x \leq 0$ and decreases when $x \geq 0$, and in Figure 11.1(c) the graph decreases when $x \leq 2$ and increases when $x \geq 2$.

Note: When determining where a graph is increasing and where it is decreasing, we must "walk" along the graph *from left to right*. (We read English from left to right, which might help you remember.)

EXAMPLE 1   Graphing quadratic functions

Graph each quadratic function. Identify the vertex and the axis of symmetry. Then state where the graph is increasing and where it is decreasing.
(a) $f(x) = x^2 - 1$ **(b)** $f(x) = -(x + 1)^2$ **(c)** $x^2 + 4x + 3$

Solution **(a)** Begin by making a convenient table of values (see Table 11.1). Then plot the points and sketch a smooth ∪-shaped curve that opens upward, as shown in Figure 11.2. The lowest point on this graph is $(0, -1)$, which is the vertex. The axis of symmetry is the vertical line $x = 0$, which passes through the vertex and coincides with the y-axis. Note also the symmetry of the y-values in Table 11.1 about the point $(0, -1)$. This graph is decreasing when $x \leq 0$ and increasing when $x \geq 0$.

TABLE 11.1

	x	$y = x^2 - 1$	
	-2	3	
	-1	0	
Vertex →	0	-1	Equal
	1	0	
	2	3	

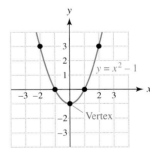

Figure 11.2

(b) Make a table of values (see Table 11.2). Plot the points and sketch a smooth ∩-shaped curve opening downward, as shown in Figure 11.3. The highest point on this graph is $(-1, 0)$, which is the vertex. The axis of symmetry is the vertical line $x = -1$, which passes through the vertex. This graph is increasing when $x \leq -1$ and decreasing when $x \geq -1$.

TABLE 11.2

	x	$y = -(x + 1)^2$	
	-3	-4	
	-2	-1	
Vertex →	-1	0	Equal
	0	-1	
	1	-4	

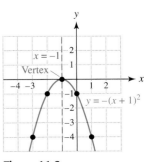

Figure 11.3

(c) Make a table of values (see Table 11.3). Plot the points and sketch a smooth U-shaped graph opening upward, as shown in Figure 11.4. The lowest point on this graph is $(-2, -1)$, which is the vertex. The axis of symmetry is the vertical line $x = -2$, which passes through the vertex. The graph is decreasing when $x \le -2$ and increasing when $x \ge -2$.

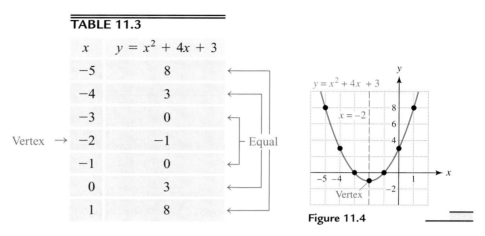

TABLE 11.3

x	$y = x^2 + 4x + 3$
-5	8
-4	3
-3	0
Vertex → -2	-1
-1	0
0	3
1	8

Figure 11.4

Rather than graphing a quadratic function to find its vertex, we can use the following formula. This formula can be derived by completing the square, a technique discussed in the next section.

VERTEX FORMULA

The x-coordinate of the vertex of the graph of $y = ax^2 + bx + c, a \ne 0$, is given by

$$x = -\frac{b}{2a}.$$

To find the y-coordinate of the vertex, substitute this x-value into the equation.

Note: The equation of the axis of symmetry for $f(x) = ax^2 + bx + c$ is $x = -\frac{b}{2a}$, and the vertex is $\left(-\frac{b}{2a}, f\left(-\frac{b}{2a}\right)\right)$.

We apply this formula in the next example.

EXAMPLE 2   Finding the vertex of a parabola

Find the vertex for the graph of $f(x) = 2x^2 - 4x + 1$. Support your answer graphically.

Solution   For $f(x) = 2x^2 - 4x + 1$, $a = 2$ and $b = -4$. The x-value of the vertex is

$$x = -\frac{b}{2a} = -\frac{(-4)}{2(2)} = 1.$$

To find the y-value of the vertex, substitute $x = 1$ in the given formula.

$$f(1) = 2(1)^2 - 4(1) + 1 = -1.$$

Thus the vertex is located at $(1, -1)$, which is supported by Figure 11.5.

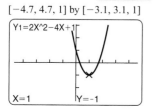

[$-4.7, 4.7, 1$] by [$-3.1, 3.1, 1$]

Figure 11.5

BASIC TRANSFORMATIONS OF GRAPHS

In this subsection we discuss the graph of $y = ax^2$, where $a \neq 0$. First, we consider the case where $a > 0$ by graphing $y_1 = \frac{1}{2}x^2$, $y_2 = x^2$, and $y_3 = 2x^2$, as shown in Figure 11.6(a). Note that $a = \frac{1}{2}$, $a = 1$, and $a = 2$ and that as a increases, the resulting parabola becomes narrower. The graph of $y_1 = \frac{1}{2}x^2$ is wider than the graph of $y_2 = x^2$, and the graph of $y_3 = 2x^2$ is narrower than the graph of $y_2 = x^2$. In general, the graph of $y = ax^2$ is wider than the graph of $y = x^2$ when $0 < a < 1$ and narrower than the graph of $y = x^2$ when $a > 1$. When $a > 0$, the graph of $y = ax^2$ never lies below the x-axis.

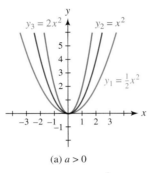

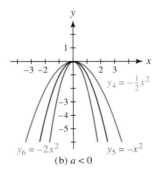

(a) $a > 0$ (b) $a < 0$

Figure 11.6 $y = ax^2$

When $a < 0$, the graph of $y = ax^2$ never lies above the x-axis because, for any input x, the product $ax^2 \leq 0$. The graphs of $y_4 = -\frac{1}{2}x$, $y_5 = -x^2$, and $y_6 = -2x^2$ are shown in Figure 11.6(b). The graph of $y_4 = -\frac{1}{2}x^2$ is wider than the graph of $y_5 = -x^2$ and the graph of $y_6 = -2x^2$ is narrower than the graph of $y_5 = -x^2$.

Note that the graph of $y_1 = \frac{1}{2}x^2$ can be *transformed* into the graph of $y_4 = -\frac{1}{2}x^2$ by *reflecting* it across the x-axis. The graph of y_4 is a **reflection** of the graph of y_1 across the x-axis. That is, if the point (x, y) is on the graph of y_1, then the point $(x, -y)$ is on the graph of y_4. Similarly, the graphs of $y_2 = x^2$ and $y_3 = 2x^2$ can be transformed into the graphs of $y_5 = -x^2$ and $y_6 = -2x^2$, respectively, by reflecting them across the x-axis.

THE GRAPH OF $y = ax^2$

The graph of $y = ax^2$ is a parabola that opens upward when $a > 0$ and opens downward when $a < 0$. As the value of $|a|$ increases, the graph of $y = ax^2$ becomes narrower. The vertex is $(0, 0)$, and the axis of symmetry is the y-axis.

EXAMPLE 3 Graphing $y = ax^2$

Compare the graph of $g(x) = -3x^2$ to the graph of $f(x) = x^2$. Then graph both functions on the same coordinate axes.

Solution Both graphs are parabolas. However, the graph of g opens downward and is narrower than the graph of f. Their graphs are shown in Figure 11.7 (on the following page).

Critical Thinking

Suppose that the graph of y_1 is a reflection of the graph of y_2 across the y-axis. If the point (x, y) lies on the graph of y_1, what point must lie on the graph of y_2?

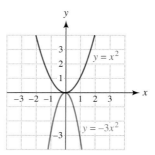

Figure 11.7

MIN–MAX APPLICATIONS

Sometimes when a quadratic function f is used to model real-world data, the vertex provides important information. The reason is that the y-coordinate of the vertex gives either the maximum value of $f(x)$ or the minimum value of $f(x)$. For example, Figure 11.8(a) shows a parabola that opens upward. The minimum y-value on this graph is 1 and occurs at the vertex $(2, 1)$. Similarly, Figure 11.8(b) shows a parabola that opens downward. The maximum y-value on this graph is 3 and occurs at the vertex $(-1, 3)$.

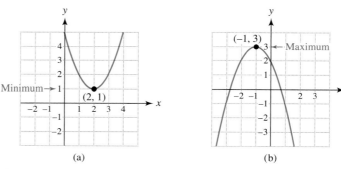

Figure 11.8

In the next example, we demonstrate finding a maximum value.

EXAMPLE 4 Finding maximum height

A baseball is hit into the air and its height h in feet after t seconds can be calculated by $h(t) = -16t^2 + 96t + 3$.
(a) What is the height of the baseball when it is hit?
(b) Determine the maximum height of the baseball.

Solution **(a)** The baseball is hit when $t = 0$, so $h(0) = -16(0)^2 + 96(0) + 3 = 3$ feet.
(b) The graph of h opens downward because $a = -16 < 0$. Thus the maximum height of the baseball occurs at the vertex. To find the vertex, we apply the vertex formula with $a = -16$ and $b = 96$.

$$t = -\frac{b}{2a} = -\frac{96}{2(-16)} = 3 \text{ seconds}$$

The maximum height of the baseball occurs at $t = 3$ seconds and is

$$h(3) = -16(3)^2 + 96(3) + 3 = 147 \text{ feet.}$$

In the next example, we answer the question presented in the introduction to this section.

EXAMPLE 5   Maximizing revenue

A hotel is considering giving the following group discount on room rates. The regular price for a room is $80, but for each room rented the price decreases by $2.

(a) A graph of the revenue received from renting x rooms is shown in Figure 11.9. Interpret the graph.

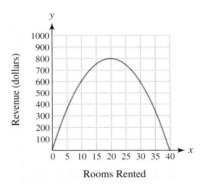

Figure 11.9

(b) What is the maximum revenue? How many rooms should be rented to receive the maximum revenue?
(c) Write a formula for $y = f(x)$ whose graph is shown in Figure 11.9.
(d) Use $f(x)$ to determine symbolically the maximum revenue and the number of rooms that should be rented.

Solution (a) The revenue increases at first, reaches a maximum, which corresponds to the vertex, and then decreases.
(b) In Figure 11.9 the vertex is (20, 800). Thus the maximum revenue of $800 occurs when 20 rooms are rented.
(c) If x rooms are rented, the price for each room is $80 - 2x$. The revenue equals the number of rooms rented times the price of each room. Thus $f(x) = x(80 - 2x)$.
(d) First, multiply $x(80 - 2x)$ to obtain $80x - 2x^2$ and then let $f(x) = -2x^2 + 80x$. The x-coordinate of the vertex is

$$x = -\frac{b}{2a} = -\frac{80}{2(-2)} = 20.$$

The y-coordinate is

$$f(20) = -2(20)^2 + 80(20) = 800.$$

These calculations verify our results in part (b).

Technology Note: *Locating a Vertex*

Calculator Help

To find a minimum or maximum, see Appendix A (page AP-12).

Some graphing calculators can locate a vertex on a parabola with either the MAXIMUM or MINIMUM utility. The maximum for the graph in Example 5 is found in the accompanying figure. This utility is typically more accurate than the TRACE utility.

[0, 50, 10] by [0, 1000, 100]

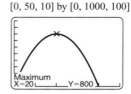

In the next example, we minimize a quadratic function that models percentages of births by cesarean section (C-section).

EXAMPLE 6   Analyzing births by cesarean section

The percentage P of births performed by cesarean section between 1991 and 2001 is modeled by

$$P(t) = 0.105t^2 - 1.08t + 23.6,$$

where t is the year, and $t = 1$ corresponds to 1991, $t = 2$ to 1992, and so on.   (*Source:* The National Center for Health Statistics.)

(a) Estimate the year when the percentage of births by cesarean section was minimum.

(b) What is this minimum percentage?

Solution   **(a)** The graph of P is a parabola that opens upward because $a = 0.105 > 0$. Therefore the t-coordinate of the vertex represents the year when the percentage of births done by cesarean section was minimum, or

$$t = -\frac{b}{2a} = -\frac{-1.08}{2(0.105)} \approx 5.1.$$

Because $t = 5$ corresponds to 1995, the minimum percentage occurred during 1995.

(b) In 1995, this percentage was about $P(5) = 0.105(5)^2 - 1.08(5) + 23.6 \approx 20.8\%$.

11.1   **PUTTING IT ALL TOGETHER**

The following table summarizes some of the important topics in this section.

Concept	Explanation	Examples
Quadratic Function	Can be written as $$f(x) = ax^2 + bx + c, a \neq 0$$	$f(x) = x^2 + x - 2$ and $g(x) = -2x^2 + 4$ $(b = 0)$

Concept	Explanation	Examples
Graph of a Quadratic Function	Its graph is a parabola that opens upward if $a > 0$ and downward if $a < 0$. The value of $|a|$ affects the width of the parabola. The vertex can be used to determine the maximum or minimum output of a quadratic function.	The graph of $y = -\frac{1}{4}x^2$ opens downward, and is wider than the graph of $y = x^2$ as shown in the figure. Each graph has its vertex at $(0, 0)$. 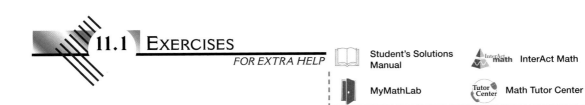
Vertex of a Parabola	The x-coordinate of the vertex for the function $f(x) = ax^2 + bx + c$ with $a \neq 0$ is given by $$x = -\frac{b}{2a}.$$ The y-coordinate of the vertex is found by substituting this x-value in the equation. Hence the vertex is $\left(-\frac{b}{2a}, f\left(-\frac{b}{2a}\right)\right)$.	If $f(x) = -2x^2 + 8x - 7$, then $$x = -\frac{8}{2(-2)} = 2$$ and $$f(2) = -2(2)^2 + 8(2) - 7 = 1.$$ The vertex is $(2, 1)$. The graph of f opens downward because $a < 0$.

11.1 EXERCISES

FOR EXTRA HELP

Student's Solutions Manual   InterAct Math   MathXL

MyMathLab   Math Tutor Center   Digital Video Tutor CD 5 Videotape 14

CONCEPTS

1. The graph of a quadratic function is called a _____.

2. If a parabola opens upward, what is the lowest point on the parabola called?

3. If a parabola is symmetric with respect to the y-axis, the y-axis is called the _____.

4. The vertex of $y = x^2$ is _____.

5. Sketch a parabola that opens downward with a vertex of $(1, 2)$.

6. If $y = ax^2 + bx + c$, the x-coordinate of the vertex is given by $x = $ _____.

7. Compared to the graph of $y = x^2$, the graph of $y = 2x^2$ is _____ (wider/narrower).

8. The graph of $y = -x^2$ is similar to the graph of $y = x^2$ except that it is _____ across the x-axis.

9. Any quadratic function can be written in the form $f(x) = $ _____.

10. If a parabola opens downward, the point with the largest y-value is called the _____.

Exercises 11–14: Use the graph of f to evaluate the expressions.

11. $f(-2)$ and $f(0)$

12. $f(-2)$ and $f(2)$

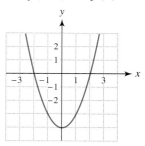

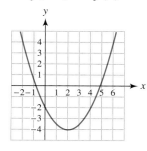

13. $f(-3)$ and $f(1)$

14. $f(-1)$ and $f(2)$

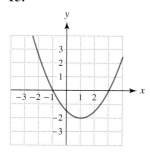

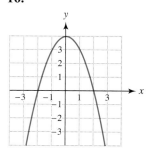

GRAPHS OF QUADRATIC FUNCTIONS

Exercises 15–18: Identify the vertex, axis of symmetry, and whether the parabola opens upward or downward. State where the graph is increasing and where it is decreasing.

15.

16.

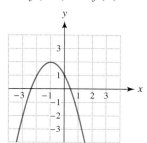

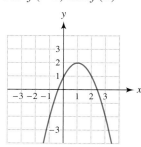

17.

18.

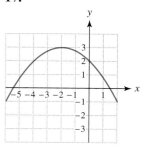

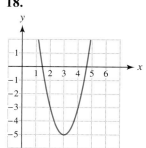

Exercises 19–34: Do the following for the given f(x).
   (a) Graph $y = f(x)$.
   (b) Use the graph to identify the vertex and axis of symmetry.
   (c) Evaluate $f(-2)$ and $f(3)$.

19. $f(x) = x^2 - 2$

20. $f(x) = x^2 - 1$

21. $f(x) = -3x^2 + 1$

22. $f(x) = \frac{1}{2}x^2 + 2$

23. $f(x) = (x - 1)^2$

24. $f(x) = (x + 2)^2$

25. $f(x) = x^2 + x - 2$

26. $f(x) = x^2 - 2x + 2$

27. $f(x) = 2x^2 - 3$

28. $f(x) = 1 - 2x^2$

29. $f(x) = 2x - x^2$

30. $f(x) = x^2 + 2x - 8$

31. $f(x) = -2x^2 + 4x - 1$

32. $f(x) = -\frac{1}{2}x^2 + 2x - 3$

33. $f(x) = \frac{1}{4}x^2 - x + 5$

34. $f(x) = 3 - 6x - 4x^2$

Exercises 35–42: Find the vertex of the parabola.

35. $f(x) = x^2 - 4x - 2$

36. $f(x) = 2x^2 + 6x - 3$

37. $f(x) = -\frac{1}{3}x^2 - 2x + 1$

38. $f(x) = 5 - 4x + x^2$

39. $f(x) = 3 - 2x^2$

40. $f(x) = \frac{1}{4}x^2 - 3x - 2$

41. $f(x) = -0.3x^2 + 0.6x + 1.1$

42. $f(x) = 25 - 10x + 20x^2$

Exercises 43–50: Graph f(x). Compare the graph to $y = x^2$.

43. $f(x) = -x^2$

44. $f(x) = -2x^2$

45. $f(x) = 2x^2$

46. $f(x) = 3x^2$

47. $f(x) = \frac{1}{4}x^2$

48. $f(x) = \frac{1}{2}x^2$

49. $f(x) = -\frac{1}{2}x^2$

50. $f(x) = -\frac{3}{2}x^2$

Exercises 51–56: Find the minimum y-value on the graph of $y = f(x)$. Then state where the graph of f is increasing and where it is decreasing.

51. $f(x) = x^2 + 2x - 1$

52. $f(x) = x^2 + 6x + 2$

53. $f(x) = x^2 - 5x$ **54.** $f(x) = x^2 - 3x$

55. $f(x) = 2x^2 + 2x - 3$ **56.** $f(x) = 3x^2 - 3x + 7$

Exercises 57–62: Find the maximum y-value on the graph of $y = f(x)$. Then state where the graph of f is increasing and where it is decreasing.

57. $f(x) = -x^2 + 2x + 5$

58. $f(x) = -x^2 + 4x - 3$

59. $f(x) = 4x - x^2$ **60.** $f(x) = 6x - x^2$

61. $f(x) = -2x^2 + x - 5$

62. $f(x) = -5x^2 + 15x - 2$

APPLICATIONS

Exercises 63–66: Quadratic Models   Match the physical situation with the graph (a.–d.) that models it best.

63. The height y of a stone thrown from ground level after x seconds.

64. The number of people attending a popular movie x weeks after its opening.

65. The temperature after x hours in a house when the furnace quits and a repair person fixes it.

66. The population of the United States from 1800 to the present.

a.
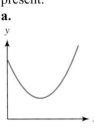

b.
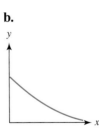

c.
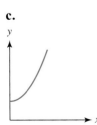

d.
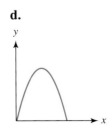

67. *Height Reached by a Baseball*   (Refer to Example 4.) A baseball is hit into the air, and its height h in feet after t seconds is given by $h(t) = -16t^2 + 64t + 2$.
   (a) What is the height of the baseball when it is hit?

(b) After how many seconds did the baseball reach its maximum height?
(c) Determine the maximum height of the baseball.

68. *Height Reached by a Golf Ball*   A golf ball is hit into the air, and its height h in feet after t seconds is given by $h(t) = -16t^2 + 128t$.
   (a) What is the height of the golf ball when it is hit?
   (b) After how many seconds did the golf ball reach its maximum height?
   (c) Determine the maximum height of the golf ball.

69. *Height Reached by a Baseball*   Suppose that a baseball is thrown upward with an initial velocity of 66 feet per second (45 miles per hour) and it is released 6 feet above the ground. Its height h after t seconds is given by

$$h(t) = -16t^2 + 66t + 6.$$

After how many seconds does the baseball reach a maximum height? Estimate this height.

70. *Throwing a Baseball on the Moon*   (Refer to Exercise 69.) If the same baseball were thrown the same way on the moon, its height h above the moon's surface after t seconds would be

$$h(t) = -2.55t^2 + 66t + 6.$$

Does the baseball go higher on the moon or on Earth? What is the difference in these two heights?

71. *Concert Tickets*   (Refer to Example 5.) An agency is promoting concert tickets by offering a group-discount rate. The regular price is $100 and for each ticket bought the price decreases by $1. (One ticket costs $99, two tickets cost $98 each, and so on.)
   (a) A graph of the revenue received from selling x tickets is shown in the figure. When is revenue increasing and when is it decreasing?

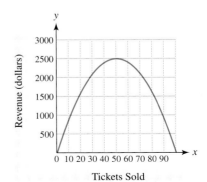

(b) What is the maximum revenue? How many tickets should be sold to a group to maximize revenue?

(c) Write a formula for $y = f(x)$ whose graph is shown in the figure.

(d) Use $f(x)$ to determine symbolically the maximum revenue and the number of tickets that should be sold.

72. *Maximizing Revenue*   The regular price for a round-trip ticket to Las Vegas, Nevada, charged by an airline charter company is $300. For a group rate the company will reduce the price of each ticket by $1.50 for every passenger on the flight.

(a) Write a formula $f(x)$ that gives the revenue from selling x tickets.

(b) Determine how many tickets should be sold to maximize the revenue. What is the maximum revenue?

73. *Maximizing Area*   A rectangular pen being constructed for a pet requires 60 feet of fence.

(a) Write a formula $f(x)$ that gives the area of the pen if one side of the pen has length x.

(b) Find the dimensions of the pen that give the largest area. What is the largest area?

74. *Maximizing Area*   A farmer is fencing a rectangular area for cattle using a straight portion of a river as one side of the rectangle as illustrated in the figure. Note that there is no fence along the river. If the farmer has 1200 feet of fence, find the dimensions for the rectangular area that gives a maximum area for the cattle.

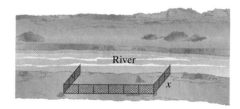

75. *Seedling Growth*   The effect of temperature on the growth of melon seedlings was studied. In this study

the seedlings were grown at different temperatures, and their heights were measured after a fixed period of time. The findings of this study can be modeled by

$$f(x) = -0.095x^2 + 5.4x - 52.2,$$

where x is the temperature in degrees Celsius and the output $f(x)$ gives the resulting average height in centimeters.   (*Source:* R. Pearl, "The growth of *Cucumis melo* seedlings at different temperatures.")

(a) Graph f in $[20, 40, 5]$ by $[0, 30, 5]$.

(b) Estimate graphically the temperature that resulted in the greatest height for the melon seedlings.

(c) Solve part (b) symbolically.

76. *Game Length*   The quadratic function defined by

$$L(x) = -2x^2 + 8000x - 7,999,820$$

approximates the length of the average major league baseball game in minutes during year x, where $1998 \leq x \leq 2002$.   (*Source:* Elias Sports Bureau.)

(a) Determine the year when games were the longest.

(b) How long did the average game last during that year?

WRITING ABOUT MATHEMATICS

77. If a quadratic function is represented by

$$f(x) = ax^2 + bx + c,$$

explain how the values of a and c affect the graph of f.

78. Suppose that a quantity Q is modeled by the formula $Q(x) = ax^2 + bx + c$ with $a < 0$. Explain how to find the x-value that maximizes $Q(x)$. How do you find the maximum value of $Q(x)$?

11.2 PARABOLAS AND MODELING

Vertical and Horizontal Translations · Vertex Form · Modeling with Quadratic Functions (Optional)

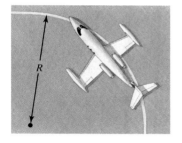

INTRODUCTION

A taxiway used by an airplane to exit a runway often contains curves. A curve that is too sharp for the speed of the plane is a safety hazard. The scatterplot shown in Figure 11.10 gives an appropriate radius R of a curve designed for an airplane taxiing x miles per hour. The data are nonlinear because they do not lie on a line. In this section we explain how a quadratic function may be used to model such data. First, we discuss translations of parabolas.

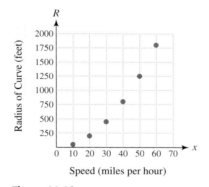

Figure 11.10

VERTICAL AND HORIZONTAL TRANSLATIONS

The graph of $y = x^2$ is a parabola opening upward with vertex $(0, 0)$. Suppose that we graph $y_1 = x^2$, $y_2 = x^2 + 1$, and $y_3 = x^2 - 2$ in the same xy-plane, as calculated for Table 11.4 and shown in Figure 11.11. All three graphs have the same shape. However, compared to the graph of $y_1 = x^2$, the graph of $y_2 = x^2 + 1$ is shifted *upward* 1 unit and the graph of $y_3 = x^2 - 2$ is shifted *downward* 2 units. Such shifts are called **translations** because they do not change the shape of a graph—only its position.

TABLE 11.4

x	$y_1 = x^2$	$y_2 = x^2 + 1$	$y_3 = x^2 - 2$
-2	4	5	2
-1	1	2	-1
0	0	1	-2
1	1	2	-1
2	4	5	2

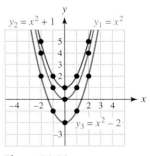

Figure 11.11

Next, suppose that we graph $y_1 = x^2$ and $y_2 = (x - 1)^2$ in the same xy-plane. Compare Tables 11.5 and 11.6. Note that the y-values are equal when the x-value for y_2 is 1 unit *larger* than the x-value for y_1. For example, $y_1 = 4$ when $x = -2$ and $y_2 = 4$ when $x = -1$. Thus the graph of $y_2 = (x - 1)^2$ has the same shape as the graph of $y_1 = x^2$, except that it is translated *horizontally to the right* 1 unit, as illustrated in Figure 11.12.

TABLE 11.5

x	$y_1 = x^2$
-2	4
-1	1
0	0
1	1
2	4

TABLE 11.6

x	$y_2 = (x - 1)^2$
-1	4
0	1
1	0
2	1
3	4

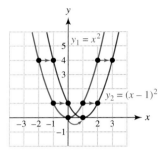

Figure 11.12

The graphs $y_1 = x^2$ and $y_2 = (x + 2)^2$ are shown in Figure 11.13. Note that Tables 11.7 and 11.8 show their y-values to be equal when the x-value for y_2 is 2 units *smaller* than the x-value for y_1. As a result, the graph of $y_2 = (x + 2)^2$ has the same shape as the graph of $y_1 = x^2$ except that it is translated *horizontally to the left* 2 units.

TABLE 11.7

x	$y_1 = x^2$
-2	4
-1	1
0	0
1	1
2	4

TABLE 11.8

x	$y_2 = (x + 2)^2$
-4	4
-3	1
-2	0
-1	1
0	4

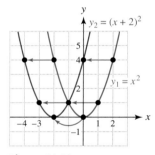

Figure 11.13

These results are summarized as follows.

▌▌▌▌   VERTICAL AND HORIZONTAL TRANSLATIONS OF PARABOLAS

Let h and k be positive numbers.

To graph	*shift the graph of $y = x^2$ by k units*
$y = x^2 + k$	upward.
$y = x^2 - k$	downward.

To graph	*shift the graph of $y = x^2$ by h units*
$y = (x - h)^2$	right.
$y = (x + h)^2$	left.

The next example demonstrates this method.

EXAMPLE 1 Translating the graph $y = x^2$

Sketch the graph of the equation and identify the vertex.
(a) $y = x^2 + 2$ **(b)** $y = (x + 3)^2$ **(c)** $y = (x - 2)^2 - 3$

Solution **(a)** The graph of $y = x^2 + 2$ is similar to the graph of $y = x^2$ except that it has been translated upward 2 units, as shown in Figure 11.14(a). The vertex is $(0, 2)$.

(b) The graph of $y = (x + 3)^2$ is similar to the graph of $y = x^2$ except that it has been translated *left* 3 units, as shown in Figure 11.14(b). The vertex is $(-3, 0)$.

Note: If you are thinking that the graph should be shifted right (instead of left) 3 units, try graphing $y = (x + 3)^2$ on a graphing calculator.

(c) The graph of $y = (x - 2)^2 - 3$ is similar to the graph of $y = x^2$ except that it has been translated downward 3 units *and* right 2 units, as shown in Figure 11.14(c). The vertex is $(2, -3)$.

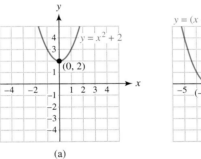

(a)

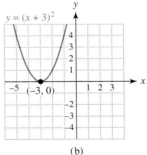

(b)

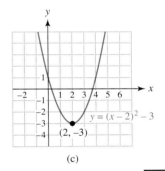

(c)

Figure 11.14

VERTEX FORM

Suppose that a parabola has the equation $y = ax^2 + bx + c$ with vertex (h, k). We can write this equation in a different form called the *vertex form* by transforming the graph of $y = x^2$. The vertex for $y = x^2$ is $(0, 0)$ so we need to translate it h units horizontally and k units vertically. Thus $y = (x - h)^2 + k$ has vertex (h, k). For the graph of our new equation to open correctly and have the same shape as $y = ax^2 + bx + c$, we must be sure that their leading coefficients are identical. That is, the graph of $y = a(x - h)^2 + k$ is identical to $y = ax^2 + bx + c$, provided that the vertex for the second equation is (h, k). This discussion is summarized as follows.

VERTEX FORM OF A PARABOLA

The **vertex form of a parabola** with vertex (h, k) is
$$y = a(x - h)^2 + k,$$
where $a \neq 0$ is a constant. If $a > 0$, the parabola opens upward; if $a < 0$, the parabola opens downward.

In the next three examples, we demonstrate the graphing of parabolas in vertex form, finding their equations, and writing vertex forms of equations.

EXAMPLE 2   Graphing parabolas in vertex form

Compare the graph of $y = f(x)$ to the graph of $y = x^2$. Then sketch a graph of $y = f(x)$ and $y = x^2$ in the same xy-plane.
(a) $f(x) = \frac{1}{2}(x - 5)^2 + 2$ **(b)** $f(x) = -3(x + 5)^2 - 1$

Solution   **(a)** Both graphs are parabolas. However, compared to the graph of $y = x^2$, the graph of $y = f(x)$ is translated 5 units right and 2 units upward. The vertex for $f(x)$ is $(5, 2)$, whereas the vertex of $y = x^2$ is $(0, 0)$. Because $a = \frac{1}{2}$, the graph of $y = f(x)$ opens wider than the graph of $y = x^2$. These graphs are shown in Figure 11.15(a).
(b) Compared to the graph of $y = x^2$, the graph of $y = f(x)$ is translated 5 units left and 1 unit downward. The vertex for $f(x)$ is $(-5, -1)$. Because $a = -3$, the graph of $y = f(x)$ opens downward and is narrower than the graph of $y = x^2$. These graphs are shown in Figure 11.15(b).

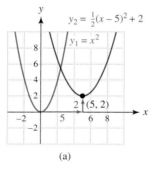

(a)

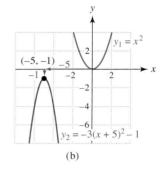

(b)

Figure 11.15

EXAMPLE 3   Finding equations of parabolas

Write the vertex form of a parabola with $a = 2$ and vertex $(-2, 1)$. Then express this equation in the form $y = ax^2 + bx + c$.

Solution   The vertex form of a parabola is

$$y = a(x - h)^2 + k,$$

where the vertex is (h, k). For $a = 2$, $h = -2$, and $k = 1$, the equation becomes

$$y = 2(x - (-2))^2 + 1 \quad \text{or} \quad y = 2(x + 2)^2 + 1.$$

To write this equation in the form $y = ax^2 + bx + c$, do the following.

$$
\begin{aligned}
y &= 2(x^2 + 4x + 4) + 1 && \text{Multiply } (x + 2)^2. \\
&= (2x^2 + 8x + 8) + 1 && \text{Distributive property} \\
&= 2x^2 + 8x + 9 && \text{Add.}
\end{aligned}
$$

The equivalent equation is $y = 2x^2 + 8x + 9$.

If we are given the equation $y = x^2 + 4x + 2$, can we write it in vertex form? The answer is *yes*, and in this process we use the **completing the square method**. Because

$$\left(x + \frac{b}{2}\right)^2 = x^2 + 2x\left(\frac{b}{2}\right) + \left(\frac{b}{2}\right)^2$$
$$= x^2 + bx + \left(\frac{b}{2}\right)^2,$$

we can complete the square for $y = x^2 + bx + c$ by adding and subtracting $\left(\frac{b}{2}\right)^2$. For $y = x^2 + 8x + 2$, we have $b = 8$, and must add *and* subtract $\left(\frac{b}{2}\right)^2 = \left(\frac{8}{2}\right)^2 = 16$ on the right side of the equation to complete the square.

$$y = x^2 + 8x + 2 \qquad \text{Given equation}$$
$$= (x^2 + 8x + 16) - 16 + 2 \qquad \text{Add and subtract 16.}$$
$$= (x + 4)^2 - 14 \qquad \text{Perfect square trinomial}$$

Thus $y = x^2 + 8x + 2$ and $y = (x + 4)^2 - 14$ are equivalent equations. The vertex for this parabola is $(-4, -14)$. Note that we added *and* subtracted 16, so the right side of the equation did not change in value.

EXAMPLE 4   Writing vertex form

Write each equation in vertex form. Identify the vertex.
(a) $y = x^2 - 6x - 1$ **(b)** $y = x^2 + 3x + 4$ **(c)** $y = 2x^2 + 4x - 1$

Solution   **(a)** Because $\left(\frac{b}{2}\right)^2 = \left(\frac{-6}{2}\right)^2 = 9$, add and subtract 9 on the right side.

$$y = x^2 - 6x - 1 \qquad \text{Given equation}$$
$$= (x^2 - 6x + 9) - 9 - 1 \qquad \text{Add and subtract 9.}$$
$$= (x - 3)^2 - 10 \qquad \text{Perfect square trinomial}$$

The vertex is $(3, -10)$.
(b) Because $\left(\frac{b}{2}\right)^2 = \left(\frac{3}{2}\right)^2 = \frac{9}{4}$, add and subtract $\frac{9}{4}$ on the right side.

$$y = x^2 + 3x + 4 \qquad \text{Given equation}$$
$$= \left(x^2 + 3x + \frac{9}{4}\right) - \frac{9}{4} + 4 \qquad \text{Add and subtract } \frac{9}{4}.$$
$$= \left(x + \frac{3}{2}\right)^2 + \frac{7}{4} \qquad \text{Perfect square trinomial}$$

The vertex is $\left(-\frac{3}{2}, \frac{7}{4}\right)$.
(c) This equation is slightly different because the leading coefficient is 2 rather than 1. Start by factoring 2 from the first two terms on the right side.

$$y = 2x^2 + 4x - 1 \qquad \text{Given equation}$$
$$= 2(x^2 + 2x) - 1 \qquad \text{Factor out 2.}$$
$$= 2(x^2 + 2x + 1 - 1) - 1 \qquad \left(\frac{b}{2}\right)^2 = \left(\frac{2}{2}\right)^2 = 1$$
$$= 2(x^2 + 2x + 1) - 2 - 1 \qquad \text{Distributive property}$$
$$= 2(x + 1)^2 - 3 \qquad \text{Perfect square trinomial}$$

The vertex is $(-1, -3)$.

Critical Thinking

What does c represent on the graph of $y = ax^2 + bx + c$?

Note: If $h = -\frac{b}{2a}$ and $k = f\left(-\frac{b}{2a}\right)$, then the formulas $f(x) = ax^2 + bx + c$ and $f(x) = a(x - h)^2 + k$ represent the same quadratic function.

MODELING WITH QUADRATIC FUNCTIONS (OPTIONAL)

In the introduction to this section we discussed airport taxiway curves designed for airplanes. The data previously shown in Figure 11.10 are listed in Table 11.9.

A second scatterplot of the data is shown in Figure 11.16. The data may be modeled by $R(x) = ax^2$ for some value a. To illustrate this relation graph R for different values of a. In Figures 11.17–11.19, R has been graphed for $a = 2$, -1, and $\frac{1}{2}$, respectively. When $a > 0$ the parabola opens upward and when $a < 0$ the parabola opens downward. Larger values of $|a|$ make a parabola narrower, whereas smaller values of $|a|$ make the parabola wider. Through trial and error, $a = \frac{1}{2}$ gives a good fit to the data, so $R(x) = \frac{1}{2}x^2$ models the data.

TABLE 11.9

x (mph)	R (ft)
10	50
20	200
30	450
40	800
50	1250
60	1800

Source: Federal Aviation Administration.

$[-70, 70, 10]$ by $[-2000, 2000, 500]$

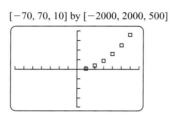

Figure 11.16

$[-70, 70, 10]$ by $[-2000, 2000, 500]$

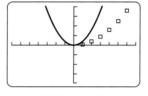

Figure 11.17 $a = 2$

$[-70, 70, 10]$ by $[-2000, 2000, 500]$

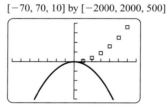

Figure 11.18 $a = -1$

$[-70, 70, 10]$ by $[-2000, 2000, 500]$

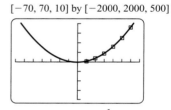

Figure 11.19 $a = \frac{1}{2}$

Calculator Help

To make a scatterplot, see Appendix A (page AP-4).

We can also find this value of a symbolically, as demonstrated in the next example.

EXAMPLE 5 Modeling safe taxiway speed

Find a value for the constant a symbolically so that $R(x) = ax^2$ models the data in Table 11.9. Check your result by making a table of values for $R(x)$.

Solution Table 11.9 shows that, when $x = 10$ miles per hour, the curve radius is $R(x) = 50$ feet. Therefore

$$R(10) = 50 \quad \text{or} \quad a (10)^2 = 50.$$

Solving for a gives

$$a = \frac{50}{10^2} = \frac{1}{2}.$$

X	Y₁
0	0
10	50
20	200
30	450
40	800
50	1250
60	1800

Y₁■.5X²

Figure 11.20

To be sure that $R(x) = \frac{1}{2}x^2$ is correct, construct a table, as shown in Figure 11.20. Its values agree with those in Table 11.9.

Critical Thinking

If the speed of a taxiing airplane doubles, what should happen to the radius of a safe taxiway curve?

In 1981, the first cases of AIDS were reported in the United States. Since then, AIDS has become one of the most devastating diseases of recent times. Table 11.10 lists the *cumulative* number of AIDS cases in the United States for various years. For example, between 1981 and 1990, a total of 199,608 AIDS cases were reported.

A scatterplot of these data is shown in Figure 11.21. To model these nonlinear data, we want to find (the right half of) a parabola with the shape illustrated in Figure 11.22. We do so in the next example.

TABLE 11.10

Years	AIDS Cases
1981	425
1984	11,106
1987	71,414
1990	199,608
1993	417,835
1996	609,933

Source: Department of Health and Human Services.

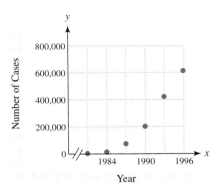

Figure 11.21 AIDS Cases

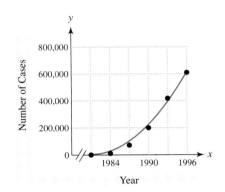

Figure 11.22 Modeling AIDS Cases

EXAMPLE 6 Modeling AIDS cases

Use the data in Table 11.10 to complete the following.
(a) Make a scatterplot of the data in [1980, 1997, 2] by [−10000, 800000, 100000].
(b) The lowest data point in Table 11.10 is (1981, 425). Let this point be the vertex of a parabola that opens upward. Graph $y = a(x - 1981)^2 + 425$ together with the data by first letting $a = 1000$.
(c) Use trial and error to adjust the value of a until the graph models the data.
(d) Use your final equation to estimate the number of AIDS cases in 1992. Compare it to the known value of 338,786.

Solution **(a)** A scatterplot of the data is shown in Figure 11.23.

[1980, 1997, 2] by
[−10000, 800000, 100000]

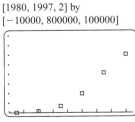

Figure 11.23

[1980, 1997, 2] by
[−10000, 800000, 100000]

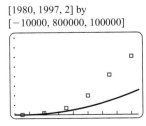

Figure 11.24 $a = 1000$

(b) A graph of $y = 1000(x - 1981)^2 + 425$ is shown in Figure 11.24. To have a better fit of the data, a larger value for a is needed.

[1980, 1997, 2] by
[−10000, 800000, 100000]

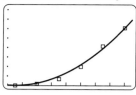

Figure 11.25 $a = 2700$

(c) Figure 11.25 shows the effect of adjusting the value of a to 2700. This value provides a reasonably good fit. (Note that you may decide on a slightly different value for a.)

(d) If $a = 2700$, the modeling equation becomes

$$y = 2700(x - 1981)^2 + 425.$$

To estimate the number of AIDS cases in 1992, substitute $x = 1992$ to obtain

$$y = 2700(1992 - 1981)^2 + 425 = 327{,}125.$$

This number is about 12,000 less than the known value of 338,786.

  PUTTING IT ALL TOGETHER

The following table summarizes some of the important topics in this section.

Concept	Explanation	Examples
Translations of Parabolas	Compared to the graph $y = x^2$, the graph of $y = x^2 + k$ is shifted vertically k units and the graph of $y = (x - h)^2$ is shifted horizontally h units.	Compared to the graph of $y = x^2$, the graph of $y = x^2 - 4$ is shifted downward 4 units. Compared to the graph of $y = x^2$, the graph of $y = (x - 4)^2$ is shifted right 4 units and the graph of $y = (x + 4)^2$ is shifted left 4 units.
Vertex Form of a Parabola	The vertex form of a parabola with vertex (h, k) is $$y = a(x - h)^2 + k,$$ where $a \neq 0$ is a constant. If $a > 0$, the parabola opens upward; if $a < 0$, the parabola opens downward.	The graph of $y = 3(x + 2)^2 - 7$ has a vertex of $(-2, -7)$ and opens upward because $a > 0$.
Completing the Square Method	To complete the square to obtain the vertex form, add and subtract $\left(\frac{b}{2}\right)^2$ on the right side of $y = x^2 + bx + c$. Then factor the perfect square trinomial.	If $y = x^2 + 10x - 3$, then add and subtract $\left(\frac{b}{2}\right)^2 = \left(\frac{10}{2}\right)^2 = 25$ on the right side of this equation. $$y = (x^2 + 10x + 25) - 25 - 3$$ $$= (x + 5)^2 - 28$$ The vertex is $(-5, -28)$.

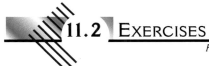 **11.2** EXERCISES

CONCEPTS

1. Compared to the graph of $y = x^2$, the graph of $y =$ _____ is shifted upward 2 units.

2. Compared to the graph of $y = x^2$, the graph of $y =$ _____ is shifted to the right 2 units.

3. The vertex of $y = (x - 1)^2 + 2$ is _____.

4. The vertex of $y = (x + 1)^2 - 2$ is _____.

5. A quadratic function may be written either in the form _____ or _____.

6. The vertex form of a parabola is given by _____ and its vertex is _____.

7. The graph of $y = -x^2$ is a parabola that opens _____.

8. The x-coordinate of the vertex of $y = ax^2 + bx + c$ is $x =$ _____.

GRAPHS OF PARABOLAS

Exercises 9–28: Do the following.
  (a) Sketch a graph of the equation.
  (b) Identify the vertex.
  (c) Compare the graph to the graph of $y = x^2$. (State the transformations used.)

9. $f(x) = x^2 - 4$ **10.** $f(x) = x^2 - 1$

11. $f(x) = 2x^2 + 1$ **12.** $f(x) = \frac{1}{2}x^2 + 1$

13. $f(x) = (x - 3)^2$ **14.** $f(x) = (x + 1)^2$

15. $f(x) = -x^2$ **16.** $f(x) = -(x + 2)^2$

17. $f(x) = 2 - x^2$ **18.** $f(x) = (x - 1)^2$

19. $f(x) = (x + 2)^2$ **20.** $f(x) = (x - 2)^2 - 3$

21. $f(x) = (x + 1)^2 - 2$

22. $f(x) = (x - 3)^2 + 1$

23. $f(x) = (x - 1)^2 + 2$

24. $f(x) = \frac{1}{2}(x + 3)^2 - 3$

25. $f(x) = 2(x - 5)^2 - 4$

26. $f(x) = -3(x + 4)^2 + 5$

27. $f(x) = -\frac{1}{2}(x + 3)^2 + 1$

28. $f(x) = 2(x - 5)^2 + 10$

VERTEX FORM

Exercises 29–32: (Refer to Example 3.) Write the vertex form of a parabola that satisfies the conditions given. Then write the equation in the form $y = ax^2 + bx + c$.

29. Vertex $(3, 4)$ and $a = 3$

30. Vertex $(-1, 3)$ and $a = -5$

31. Vertex $(5, -2)$ and $a = -\frac{1}{2}$

32. Vertex $(-2, -6)$ and $a = \frac{3}{4}$

Exercises 33–36: Write the vertex form of a parabola that satisfies the conditions given. Assume that $a = \pm 1$.

33. Opens upward, vertex $(1, 2)$

34. Opens downward, vertex $(-1, -2)$

35. Opens downward, vertex $(0, -3)$

36. Opens upward, vertex $(5, -4)$

Exercises 37–40: Write the vertex form of the parabola shown in the graph. Assume that $a = \pm 1$.

37.

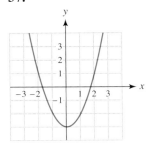

38.

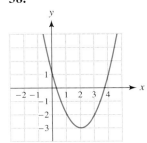

39.

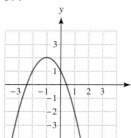

40.

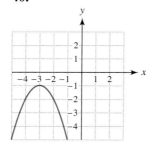

Exercises 41–54: (Refer to Example 4.) Write the equation in vertex form. Identify the vertex.

41. $y = x^2 + 2x - 3$ **42.** $y = x^2 + 4x + 1$

43. $y = x^2 - 4x + 5$ **44.** $y = x^2 - 8x + 10$

45. $y = x^2 + 3x - 2$ **46.** $y = x^2 + 5x - 4$

47. $y = x^2 - 7x + 1$ **48.** $y = x^2 - 3x + 5$

49. $y = 3x^2 + 6x - 1$ **50.** $y = 2x^2 + 4x - 9$

51. $y = 2x^2 - 3x$ **52.** $y = 3x^2 - 7x$

53. $y = -2x^2 - 8x + 5$ **54.** $y = -3x^2 + 6x + 1$

MODELING DATA

Exercises 55–58: Find a value for the constant a so that $f(x) = ax^2$ models the data. If you are uncertain about your value for a, check it by making a table of values.

55.

x	1	2	3
y	2	8	18

56.

x	-2	0	2
y	6	0	6

57.

x	2	4	6	8
y	1.2	4.8	10.8	19.2

58.

x	5	10	15	20
y	17.5	70	157.5	280

 Exercises 59–62: **Modeling Quadratic Data**   *(Refer to Example 6.) Find a quadratic function expressed in vertex form that models the data in the given table.*

59.

x	1	2	3	4
y	-3	-1	5	15

60.

x	-2	-1	0	1	2
y	5	2	-7	-22	-43

61.

x	1980	1990	2000	2010
y	6	55	210	450

62.

x	1990	1995	2000	2005
y	10	60	205	470

63. *Braking Distance*   The table lists approximate stopping distances D in feet for cars traveling at x miles per hour on dry, level pavement.

x	12	24	36	48
D	12	48	108	192

 (a) Make a scatterplot of the data.
(b) Find a function given by $D(x) = ax^2$ that models these data.

64. *Health Care Costs*   The table lists *approximate* percentage increases in the cost of health insurance premiums between 1992 and 2000.

Year	1992	1994	1996	1998	2000
Increase	11%	4%	1%	4%	11%

Source: Kaiser Family Foundation.

(a) Make a scatterplot of the data.
(b) Find a function given by $C(x) = a(x - h)^2 + k$ that models these data. (Answers may vary.)

WRITING ABOUT MATHEMATICS

65. Explain how to find the vertex of $y = x^2 + bx + c$ by completing the square.

66. If $f(x) = a(x - h)^2 + k$, explain how the values of a, h, and k affect the graph of $y = f(x)$.

---

1. Graph each quadratic function. Identify the vertex and axis of symmetry.
   (a) $f(x) = x^2 - 2$
   (b) $f(x) = x^2 - 2x - 2$

2. Compare the graph of $y_1 = 2x^2$ to the graph of $y_2 = -\frac{1}{2}x^2$.

3. Find the maximum y-value on the graph of $y = -3x^2 + 12x - 5$. State where the graph is increasing and where it is decreasing.

4. Sketch a graph of $y = f(x)$. Compare the graph of f to the graph of $y = x^2$.
   (a) $f(x) = (x - 1)^2 + 2$
   (b) $f(x) = -(x + 3)^2$

5. Write the vertex form for each equation.
   (a) $y = x^2 + 14x - 7$
   (b) $y = 4x^2 + 8x - 2$

11.3 QUADRATIC EQUATIONS

Basics of Quadratic Equations · The Square Root Property · Completing the Square · Solving an Equation for a Variable · Applications of Quadratic Equations

INTRODUCTION

In Section 11.2 we modeled curves on airport taxiways by using $R(x) = \frac{1}{2}x^2$. In this formula x represented the airplane's speed in miles per hour, and R represented the radius of the curve in feet. This formula may be used to determine the speed limit for a curve with a radius of 650 feet by solving the *quadratic equation*

$$\frac{1}{2}x^2 = 650.$$

In this section we demonstrate techniques for solving this and other quadratic equations.

BASICS OF QUADRATIC EQUATIONS

Any quadratic function f can be represented by $f(x) = ax^2 + bx + c$ with $a \neq 0$. Examples of quadratic functions include

$$f(x) = 2x^2 - 1, \quad g(x) = -\frac{1}{3}x^2 + 2x, \quad \text{and} \quad h(x) = x^2 + 2x - 1.$$

Quadratic functions can be used to write quadratic equations. Examples of quadratic equations include

$$2x^2 - 1 = 0, \quad -\frac{1}{3}x^2 + 2x = 0, \quad \text{and} \quad x^2 + 2x - 1 = 3.$$

QUADRATIC EQUATION

A **quadratic equation** is an equation that can be written as

$$ax^2 + bx + c = 0,$$

where a, b, and c are real numbers with $a \neq 0$.

Solutions to the quadratic equation $ax^2 + bx + c = 0$ correspond to x-intercepts of the graph of $y = ax^2 + bx + c$. Because the graph of a quadratic function is either ∪-shaped or ∩-shaped, it can intersect the x-axis zero, one, or two times, as illustrated in Figure 11.26. Hence a quadratic equation can have zero, one, or two real solutions.

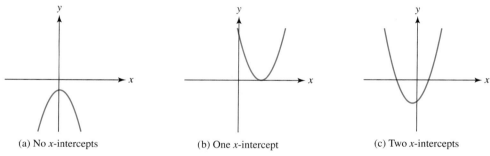

(a) No x-intercepts   (b) One x-intercept   (c) Two x-intercepts

Figure 11.26

We have already solved quadratic equations by factoring, graphing, and constructing tables. In the next example we apply these three techniques to quadratic equations that have zero, one, and two real solutions.

EXAMPLE 1   Solving quadratic equations

Solve each quadratic equation. Support your results numerically and graphically.
(a) $2x^2 + 1 = 0$ (No real solutions)
(b) $x^2 + 4 = 4x$ (One real solution)
(c) $x^2 - 6x + 8 = 0$ (Two real solutions)

Solution   **(a)** *Symbolic Solution*

$$2x^2 + 1 = 0 \qquad \text{Given equation}$$
$$2x^2 = -1 \qquad \text{Subtract 1.}$$
$$x^2 = -\frac{1}{2} \qquad \text{Divide by 2.}$$

This equation has no real-number solutions because $x^2 \geq 0$ for all real numbers x. Note that the graph of $y = 2x^2 + 1$ has no x-intercepts.

Numerical and Graphical Solution   The points in Table 11.11 for $y = 2x^2 + 1$ are plotted in Figure 11.27 and connected with a parabolic graph. The graph of $y = 2x^2 + 1$ has no *x*-intercepts, indicating that there are no real solutions. Note that the *y*-values in Table 11.11 are always positive.

TABLE 11.11

x	y
-2	9
-1	3
0	1
1	3
2	9

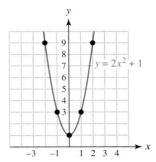

Figure 11.27 No Solutions

(b) Symbolic Solution

$$x^2 + 4 = 4x \qquad \text{Given equation}$$
$$x^2 - 4x + 4 = 0 \qquad \text{Subtract } 4x \text{ from both sides.}$$
$$(x - 2)(x - 2) = 0 \qquad \text{Factor.}$$
$$x - 2 = 0 \quad \text{or} \quad x - 2 = 0 \qquad \text{Zero-product property}$$
$$x = 2 \qquad \text{There is one solution.}$$

Numerical and Graphical Solution   Because the given equation is equivalent to $x^2 - 4x + 4 = 0$, we let $y = x^2 - 4x + 4$. The points in Table 11.12 are plotted in Figure 11.28 and connected with a parabolic graph. The graph of $y = x^2 - 4x + 4$ has one *x*-intercept, 2. Note that in Table 11.12, $y = 0$ when $x = 2$, indicating that the equation has one solution.

TABLE 11.12

x	y
0	4
1	1
2	0
3	1
2	4

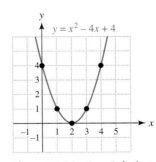

Figure 11.28 One Solution

(c) Symbolic Solution

$$x^2 - 6x + 8 = 0 \qquad \text{Given equation}$$
$$(x - 2)(x - 4) = 0 \qquad \text{Factor.}$$
$$x - 2 = 0 \quad \text{or} \quad x - 4 = 0 \qquad \text{Zero-product property}$$
$$x = 2 \quad \text{or} \quad x = 4 \qquad \text{There are two solutions.}$$

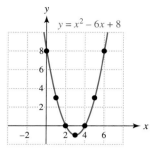

Figure 11.29 Two Solutions

Numerical and Graphical Solution The points in Table 11.13 for $y = x^2 - 6x + 8$ are plotted in Figure 11.29 and connected with a parabolic graph. The graph of $y = x^2 - 6x + 8$ has two x-intercepts, 2 and 4, indicating two solutions. Note that in Table 11.13 $y = 0$ when $x = 2$ or $x = 4$.

TABLE 11.13

x	0	1	2	3	4	5	6
y	8	3	0	-1	0	3	8

THE SQUARE ROOT PROPERTY

The *square root property* is a symbolic method used to solve quadratic equations that are missing x-terms. The following is an example of the square root property.

$$x^2 = 25 \quad \text{is equivalent to} \quad x = \pm 5$$

The expression $x = \pm 5$ (read "x equals plus or minus 5") indicates that either $x = 5$ or $x = -5$. Each value is a solution because $(5)^2 = 25$ and $(-5)^2 = 25$.

SQUARE ROOT PROPERTY

Let k be a nonnegative number. Then the solutions to the equation

$$x^2 = k$$

are given by $x = \pm \sqrt{k}$. If $k < 0$, then this equation has no real solutions.

Before applying the square root property in the next two examples, we review a quotient property of square roots. If a and b are positive numbers, then

$$\sqrt{\frac{a}{b}} = \frac{\sqrt{a}}{\sqrt{b}}.$$

For example,

$$\sqrt{\frac{25}{36}} = \frac{\sqrt{25}}{\sqrt{36}} = \frac{5}{6}.$$

EXAMPLE 2   Using the square root property

Solve each equation.
(a) $x^2 = 7$ **(b)** $16x^2 - 9 = 0$ **(c)** $(x - 4)^2 = 25$

Solution **(a)** $x^2 = 7$ is equivalent to $x = \pm \sqrt{7}$ by the square root property. The solutions are $\sqrt{7}$ and $-\sqrt{7}$.

(b)
$$16x^2 - 9 = 0 \qquad \text{Given equation}$$
$$16x^2 = 9 \qquad \text{Add 9.}$$
$$x^2 = \frac{9}{16} \qquad \text{Divide by 16.}$$
$$x = \pm\sqrt{\frac{9}{16}} \qquad \text{Square root property}$$
$$x = \pm\frac{3}{4} \qquad \text{Simplify.}$$

The solutions are $\frac{3}{4}$ and $-\frac{3}{4}$.

(c)
$$(x - 4)^2 = 25 \qquad \text{Given equation.}$$
$$(x - 4) = \pm\sqrt{25} \qquad \text{Square root property}$$
$$x - 4 = \pm 5 \qquad \text{Simplify.}$$
$$x = 4 \pm 5 \qquad \text{Add 4.}$$
$$x = 9 \quad \text{or} \quad x = -1 \qquad \text{Evaluate } 4 + 5 \text{ and } 4 - 5.$$

The solutions are 9 and -1.

If an object is dropped from a height of h feet, its distance d above the ground after t seconds is given by

$$d(t) = h - 16t^2.$$

This formula can be used to estimate the time it takes for a falling object to hit the ground.

EXAMPLE 3 Modeling a falling object

A toy falls from a window 30 feet above the ground. How long does the toy take to hit the ground?

Solution The height of the window above the ground is 30 feet so let $d(t) = 30 - 16t^2$. The toy strikes the ground when the distance d above the ground equals 0.

$$30 - 16t^2 = 0 \qquad \text{Equation to solve}$$
$$-16t^2 = -30 \qquad \text{Subtract 30.}$$
$$t^2 = \frac{30}{16} \qquad \text{Divide by } -16.$$
$$t = \pm\sqrt{\frac{30}{16}} \qquad \text{Square root property}$$
$$t = \pm\frac{\sqrt{30}}{4} \qquad \text{Simplify.}$$

Time cannot be negative in this problem, so the appropriate solution is $t = \frac{\sqrt{30}}{4} \approx 1.4$. The toy hits the ground after about 1.4 seconds.

COMPLETING THE SQUARE

In Section 11.2, we used the *method of completing the square* to find the vertex of a parabola. This method can also be used to solve quadratic equations. Because

$$x^2 + bx + \left(\frac{b}{2}\right)^2 = \left(x + \frac{b}{2}\right)^2,$$

we can solve a quadratic equation in the form $x^2 + bx = d$, where b and d are constants, by adding $\left(\frac{b}{2}\right)^2$ to each side and then factoring the resulting perfect square trinomial.

In the equation $x^2 + 6x = 7$ we have $b = 6$, so we add $\left(\frac{6}{2}\right)^2 = 9$ to each side.

$$\begin{aligned}
x^2 + 6x &= 7 && \text{Given equation} \\
x^2 + 6x + 9 &= 7 + 9 && \text{Add 9 to each side.} \\
(x + 3)^2 &= 16 && \text{Perfect square trinomial} \\
x + 3 &= \pm 4 && \text{Square root property} \\
x &= -3 \pm 4 && \text{Add } -3 \text{ to each side.} \\
x = 1 \quad \text{or} \quad x &= -7 && \text{Simplify } -3 + 4 \text{ and } -3 - 4.
\end{aligned}$$

The solutions are 1 and -7.

Note that the left side of the equation becomes a perfect square trinomial. We show how to create one in the next example.

EXAMPLE 4 Creating a perfect square trinomial

Find the term that should be added to $x^2 - 10x$ to form a perfect square trinomial.

Solution The coefficient of the x-term is -10, so we let $b = -10$. To complete the square we divide b by 2 and then square the result.

$$\left(\frac{b}{2}\right)^2 = \left(\frac{-10}{2}\right)^2 = 25$$

If we add **25**, a perfect square trinomial is formed.

$$x^2 - 10x + 25 = (x - 5)^2$$

Completing the square can be used to solve quadratic equations when a trinomial does not factor easily, as illustrated in the next two examples.

EXAMPLE 5 Completing the square when the leading coefficient is 1

Solve the equation $x^2 - 4x + 2 = 0$.

Solution Start by writing the equation in the form $x^2 + bx = d$.

$$\begin{aligned}
x^2 - 4x + 2 &= 0 && \text{Given equation} \\
x^2 - 4x &= -2 && \text{Subtract 2.} \\
x^2 - 4x + 4 &= -2 + 4 && \text{Add } \left(\frac{b}{2}\right)^2 = \left(\frac{-4}{2}\right)^2 = 4. \\
(x - 2)^2 &= 2 && \text{Perfect square trinomial} \\
x - 2 &= \pm \sqrt{2} && \text{Square root property} \\
x &= 2 \pm \sqrt{2} && \text{Solve.}
\end{aligned}$$

The solutions are $2 + \sqrt{2} \approx 3.41$ and $2 - \sqrt{2} \approx 0.59$.

EXAMPLE 6 **Completing the square when the leading coefficient is not 1**

Solve the equation $2x^2 + 7x - 5 = 0$.

Solution Start by writing the equation in the form $x^2 + bx = d$. That is, add 5 to each side and then divide the equation by 2 so that the leading coefficient of the x^2-term becomes 1.

$$2x^2 + 7x - 5 = 0 \qquad \text{Given equation}$$

$$2x^2 + 7x = 5 \qquad \text{Add 5.}$$

$$x^2 + \frac{7}{2}x = \frac{5}{2} \qquad \text{Divide by 2.}$$

$$x^2 + \frac{7}{2}x + \frac{49}{16} = \frac{5}{2} + \frac{49}{16} \qquad \text{Add } \left(\frac{b}{2}\right)^2 = \left(\frac{7}{4}\right)^2 = \frac{49}{16}.$$

$$\left(x + \frac{7}{4}\right)^2 = \frac{89}{16} \qquad \text{Perfect square trinomial}$$

$$x + \frac{7}{4} = \pm\frac{\sqrt{89}}{4} \qquad \text{Square root property}$$

$$x = \frac{-7 \pm \sqrt{89}}{4} \qquad \text{Add } -\frac{7}{4}.$$

The solutions are $\frac{-7 + \sqrt{89}}{4} \approx 0.61$ and $\frac{-7 - \sqrt{89}}{4} \approx -4.1$.

Critical Thinking

What happens if you try to solve

$$2x^2 - 13 = 1$$

by completing the square? What method should you use to solve this problem?

SOLVING AN EQUATION FOR A VARIABLE

We often need to solve an equation or formula for a variable. For example, the formula $V = \frac{1}{3}\pi r^2 h$ calculates the volume of the cone shown in Figure 11.30. Let's say that we know the volume V is 120 cubic inches and the height h is 15 inches. We can then find the radius of the cone by solving the equation for r.

$$V = \frac{1}{3}\pi r^2 h \qquad \text{Given equation}$$

$$3V = \pi r^2 h \qquad \text{Multiply by 3.}$$

$$\frac{3V}{\pi} = r^2 h \qquad \text{Divide by } \pi.$$

$$\frac{3V}{\pi h} = r^2 \qquad \text{Divide by } h.$$

$$r = \pm\sqrt{\frac{3V}{\pi h}} \qquad \text{Square root property; rewrite.}$$

Figure 11.30

Because $r \geq 0$, we use the positive or *principal square root*. Thus for $V = 120$ cubic inches and $h = 15$ inches,

$$r = \sqrt{\frac{3(120)}{\pi(15)}} = \sqrt{\frac{24}{\pi}} \approx 2.8 \text{ inches.}$$

EXAMPLE 7 Solving equations for variables

Solve each equation for the specified variable.
(a) $s = -\frac{1}{2}gt^2 + h$, for t **(b)** $d^2 = x^2 + y^2$, for y

Solution **(a)** Begin by subtracting h from each side of the equation.

$s = -\dfrac{1}{2}gt^2 + h$	Given equation
$s - h = -\dfrac{1}{2}gt^2$	Subtract h.
$-2(s - h) = gt^2$	Multiply by -2.
$\dfrac{2h - 2s}{g} = t^2$	Divide by g; simplify.
$t = \pm\sqrt{\dfrac{2h - 2s}{g}}$	Square root property

(b) Begin by subtracting x^2 from each side of the equation.

$d^2 = x^2 + y^2$	Given equation
$d^2 - x^2 = y^2$	Subtract x^2.
$y = \pm\sqrt{d^2 - x^2}$	Square root property

APPLICATIONS OF QUADRATIC EQUATIONS

In the introduction to this section we discussed how the solution to the equation

$$\frac{1}{2}x^2 = 650$$

would give a safe speed limit for a curve with a radius of 650 feet on an airport taxiway. We solve this problem in the next example.

EXAMPLE 8 Finding a safe speed limit

Solve the equation $\frac{1}{2}x^2 = 650$ and interpret any solutions.

Solution Use the square root property to solve this problem.

$\dfrac{1}{2}x^2 = 650$	Given equation
$x^2 = 1300$	Multiply by 2.
$x = \pm\sqrt{1300}$	Square root property

The solutions are $\sqrt{1300} \approx 36$ and $-\sqrt{1300} \approx -36$. The solution of $x \approx 36$ indicates that a safe speed limit for a curve with a radius of 650 feet should be 36 miles per hour. (The negative solution has no physical meaning in this problem.)

In applications, solving a quadratic equation either graphically or numerically is often easier than solving it symbolically. We do so in the next example.

EXAMPLE 9 Modeling Internet users

Use of the Internet in Western Europe has increased dramatically. Figure 11.31 shows a scatterplot of online users in Western Europe, together with a graph of a function f that models the data. The function f is given by

$$f(x) = 0.976x^2 - 4.643x + 0.238,$$

where the output is in millions of users. In this formula $x = 6$ corresponds to 1996, $x = 7$ to 1997, and so on, until $x = 12$ represents 2002. (**Source:** Nortel Networks.)

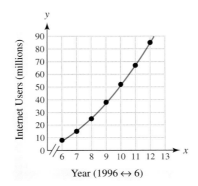

Figure 11.31 Internet Usage in Western Europe

(a) Evaluate $f(10)$ and interpret the result.
(b) Graph f and estimate the year when the number of Internet users reached 85 million. Compare your result with Figure 11.31.
(c) Solve part (b) numerically.

Solution (a) Substituting $x = 10$ into the formula yields

$$f(10) = 0.976(10)^2 - 4.643(10) + 0.238 \approx 51.4.$$

Because $x = 10$ corresponds to 2000, there were about 51.4 million Internet users in Western Europe in 2000.
(b) Graph $Y_1 = .976X^2 - 4.643X + .238$ and $Y_2 = 85$, as shown in Figure 11.32. Their graphs intersect near $x = 12$, which corresponds to 2002 and agrees with Figure 11.31.
(c) Construct the table for y_1, as shown in Figure 11.33, which reveals $y_1 \approx 85$ when $x = 12$.

Calculator Help
To find a point of intersection, see Appendix A (page AP-7).

[5, 13, 1] by [0, 100, 10]

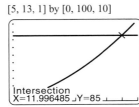

Figure 11.32

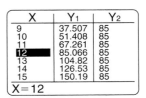

X	Y₁	Y₂
9	37.507	85
10	51.408	85
11	67.261	85
12	85.066	85
13	104.82	85
14	126.53	85
15	150.19	85

X=12

Figure 11.33

11.3   PUTTING IT ALL TOGETHER

Quadratic equations can be expressed in the form $ax^2 + bx + c = 0$, $a \neq 0$. They can have zero, one, or two real solutions and may be solved symbolically, graphically, and numerically. Symbolic techniques for solving quadratic equations include factoring, the square root property, and completing the square. We discussed factoring extensively in Chapter 6, so the following table summarizes only the square root property and the method of completing the square.

Technique	Description	Examples
Square Root Property	If $k \geq 0$, the solutions to the equation $x^2 = k$ are $\pm\sqrt{k}$.	$x^2 = 100$ is equivalent to $x = \pm 10$ and $x^2 = 13$ is equivalent to $x = \pm\sqrt{13}$
Method of Completing the Square	To solve an equation in the form $x^2 + bx = d$, add $\left(\frac{b}{2}\right)^2$ to each side of the equation. Factor the resulting perfect square trinomial and solve for x by applying the square root property.	To solve $x^2 + 8x - 3 = 0$, begin by adding 3 to each side to obtain $x^2 + 8x = 3$. Because $b = 8$, add $\left(\frac{8}{2}\right)^2 = 16$ to each side. $\quad x^2 + 8x + 16 = 3 + 16 \quad$ Add 16 to each side. $\quad (x + 4)^2 = 19 \quad$ Perfect square trinomial $\quad x + 4 = \pm\sqrt{19} \quad$ Square root property $\quad x = -4 \pm \sqrt{19} \quad$ Add -4. $\quad x \approx 0.36, -8.36 \quad$ Approximate.

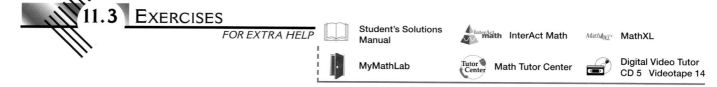

11.3   EXERCISES

FOR EXTRA HELP

Student's Solutions Manual

MyMathLab

InterAct Math

Math Tutor Center

MathXL

Digital Video Tutor
CD 5   Videotape 14

CONCEPTS

1. Give an example of a quadratic equation. How many real solutions can a quadratic equation have?

2. Is a quadratic equation a linear equation or a nonlinear equation?

3. Name three symbolic methods that can be used to solve a quadratic equation.

4. Sketch a graph of a quadratic function that has two x-intercepts and opens downward.

5. Sketch a graph of a quadratic function that has no x-intercepts and opens upward.

6. If the graph of $y = ax^2 + bx + c$ intersects the x-axis twice, how many solutions does the equation $ax^2 + bx + c = 0$ have? Explain.

7. Solve $x^2 = 64$. What property did you use?

8. To solve $x^2 + bx = 6$ by completing the square, what value should be added to each side of the equation?

Exercises 9–16: Determine whether the given equation is quadratic.

9. $x^2 - 3x + 1 = 0$ 10. $2x^2 - 3 = 0$

11. $3x + 1 = 0$

12. $x^3 - 3x^2 + x = 0$

13. $-3x^2 + x = 16$

14. $x^2 - 1 = 4x$

15. $x^2 = \sqrt{x} + 1$

16. $\dfrac{1}{x - 1} = 5$

SOLVING QUADRATIC EQUATIONS

Exercises 17–20: A graph of $y = ax^2 + bx + c$ is given. Use this graph to solve $ax^2 + bx + c = 0$, if possible.

17.

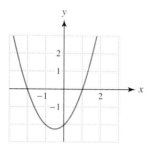

18.

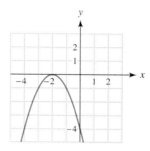

19.

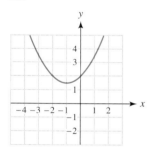

20.

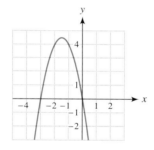

Exercises 21–24: A table of $y = ax^2 + bx + c$ is given. Use this table to solve $ax^2 + bx + c = 0$.

21.

X	Y1
-3	6
-2	0
-1	-4
0	-6
1	-6
2	-4
3	0

Y1≣X^2−X−6

22.

X	Y1
-6	0
-4	-16
-2	-24
0	-24
2	-16
4	0
6	24

Y1≣X^2+2X−24

23.

X	Y1
-2	9
-1.5	4
-1	1
-.5	0
0	1
.5	4
1	9

Y1≣4X^2+4X+1

24.

X	Y1
-4	-2.5
-3	0
-2	1.5
-1	2
0	1.5
1	0
2	-2.5

Y1≣−X^2/2−X+3/2

Exercises 25–32: Solve the quadratic equation. Support your results numerically and graphically.

25. $x^2 - 4x - 5 = 0$

26. $x^2 - x - 6 = 0$

27. $x^2 + 2x = 3$

28. $x^2 + 4x = 5$

29. $x^2 = 9$

30. $x^2 = 4$

31. $4x^2 - 4x - 3 = 0$

32. $2x^2 + x = 1$

Exercises 33–40: Solve by factoring.

33. $x^2 + 2x - 35 = 0$

34. $2x^2 - 7x + 3 = 0$

35. $6x^2 - x - 1 = 0$

36. $x^2 + 4x + 6 = -3x$

37. $2x^2 + x + 3 = 6x$

38. $6x^2 - x = 15$

39. $2(5x^2 + 9) = 27x$

40. $15(3x^2 + x) = 10$

Exercises 41–50: Use the square root property to solve the equation.

41. $x^2 = 144$

42. $4x^2 - 5 = 0$

43. $5x^2 - 64 = 0$

44. $3x^2 = 7$

45. $(x + 1)^2 = 25$

46. $(x + 4)^2 = 9$

47. $(x - 1)^2 = 64$

48. $(x - 3)^2 = 0$

49. $(2x - 1)^2 = 5$

50. $(5x + 3)^2 = 7$

COMPLETING THE SQUARE

Exercises 51–54: To solve the equation by completing the square, what value should you add to each side of the equation?

51. $x^2 + 4x = -3$

52. $x^2 - 6x = 4$

53. $x^2 - 5x = 4$

54. $x^2 + 3x = 1$

Exercises 55–58: (Refer to Example 4.) Find the term that should be added to the expression to form a perfect square trinomial. Write the resulting perfect square trinomial in factored form.

55. $x^2 - 8x$

56. $x^2 - 5x$

57. $x^2 + 9x$

58. $x^2 + x$

Exercises 59–74: Solve the equation by completing the square.

59. $x^2 - 2x = 24$

60. $x^2 - 2x + \frac{1}{2} = 0$

61. $x^2 + 6x - 2 = 0$

62. $x^2 - 16x = 5$

63. $x^2 - 3x = 5$

64. $x^2 + 5x = 2$

65. $x^2 - 5x + 1 = 0$

66. $x^2 - 9x + 7 = 0$

67. $x^2 - 4 = 2x$

68. $x^2 + 1 = 7x$

69. $2x^2 - 3x = 4$

70. $3x^2 + 6x - 5 = 0$

71. $4x^2 - 8x - 7 = 0$

72. $25x^2 - 20x - 1 = 0$

73. $36x^2 + 18x + 1 = 0$

74. $12x^2 + 8x - 2 = 0$

SOLVING EQUATIONS BY MORE THAN ONE METHOD

Exercises 75–80: Solve the quadratic equation
   (a) symbolically,
   (b) graphically, and
   (c) numerically.

75. $x^2 - 3x - 18 = 0$

76. $\frac{1}{2}x^2 + 2x - 6 = 0$

77. $x^2 - 8x + 15 = 0$

78. $2x^2 + 3 = 7x$

79. $4(x^2 + 35) = 48x$

80. $4x(2 - x) = -5$

SOLVING AN EQUATION FOR A VARIABLE

Exercises 81–88: Solve the equation for the specified variable.

81. $x = y^2 - 1$ for y

82. $x = 9y^2$ for y

83. $K = \frac{1}{2}mv^2$ for v

84. $c^2 = a^2 + b^2$ for b

85. $E = \frac{k}{r^2}$ for r

86. $W = I^2R$ for I

87. $LC = \frac{1}{(2\pi f)^2}$ for f

88. $F = \frac{KmM}{r^2}$ for r

APPLICATIONS

89. *Safe Curve Speed* (Refer to Example 8.) Find a safe speed limit x for an airport taxiway curve with the given radius R.
   (a) $R = 450$ feet   **(b)** $R = 800$ feet

90. *Braking Distance* The braking distance y in feet that it takes for a car to stop on wet, level pavement can be estimated by $y = \frac{1}{9}x^2$, where x is the speed of the car in miles per hour. Find the speed associated with each braking distance.   (*Source*: L. Haefner, *Introduction to Transportation Systems.*)
   (a) 25 feet   **(b)** 361 feet   **(c)** 784 feet

91. *Falling Object* (Refer to Example 3.) How long does it take for a toy to hit the ground if it is dropped out of a window 60 feet above the ground? Does it take twice as long as it takes to fall from a window 30 feet above the ground?

92. *Falling Object* If a metal ball is thrown *downward* with an initial velocity of 22 feet per second (15 mph) from a 100-foot water tower, its height h in feet above the ground after t seconds is modeled by
$$h(t) = -16t^2 - 22t + 100.$$
   (a) Determine symbolically when the height of the ball is 62 feet.
   (b) Support your result either graphically or numerically.
   (c) If the ball is thrown *upward* at 22 feet per second, then its height is given by
$$h(t) = -16t^2 + 22t + 100.$$
Determine when the height of the ball is 80 feet.

93. *Distance* Two athletes start jogging at the same time. One jogs north at 6 miles per hour while the second jogs east at 8 miles per hour. After how long are the two athletes 20 miles apart?

94. *Geometry* A triangle has an area of 35 square inches, and its base is 3 inches more than its height. Find the base and height of the triangle.

95. *Construction* A rectangular plot of land has an area of 520 square feet and is 6 feet longer than it is wide.
   (a) Write a quadratic equation in the form $ax^2 + bx + c = 0$, whose solution gives the width of the plot of land.
   (b) Solve the equation.

96. *Modeling Motion* The height h of a tennis ball above the ground after t seconds is shown in the graph. Estimate when the ball was 25 feet above the ground.

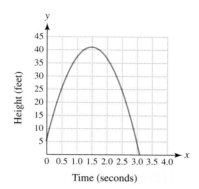

Time (seconds)

97. *Seedling Growth* (Refer to Exercise 75, Section 11.1.) The heights of melon seedlings grown at different temperatures are shown in the following graph. At what temperatures were the heights of the seedlings about 22 centimeters? (*Source:* R. Pearl, "The growth of *Cucumis melo* seedlings at different temperatures.")

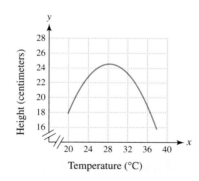

Temperature (°C)

98. *Modeling Internet Users* (Refer to Example 9.) Estimate either graphically or numerically when the number of Internet users in Western Europe is expected to reach 150 million.

99. *U.S. Population* The table shows the population of the United States in millions from 1800 through 2000 at 20-year intervals.

Year	1800	1820	1840	1860
Population	5	10	17	31

Year	1880	1900	1920	1940
Population	50	76	106	132

Year	1960	1980	2000
Population	179	226	269

Source: Bureau of the Census.

(a) Without plotting the data, how do you know that the data are nonlinear?
(b) These data are modeled by
$$f(x) = 0.0066(x - 1800)^2 + 5.$$
Find the vertex of the graph of f and interpret the result.
(c) Estimate when the U.S. population reached 85 million.

100. *Trade Deficit* The U.S. trade deficit in billions of dollars for the years 1997, 1998, and 1999 can be computed by
$$f(x) = 16x^2 - 63{,}861x + 63{,}722{,}378,$$
where x is the year. In which year was the trade deficit \$164 billion? (*Source:* Department of Commerce.)

WRITING ABOUT MATHEMATICS

101. Suppose that you are asked to solve
$$ax^2 + bx + c = 0.$$
Explain how the graph of $y = ax^2 + bx + c$ can be used to find any solutions to the equation.

102. Explain why a quadratic equation could not have more than two solutions. (*Hint:* Consider the graph of $y = ax^2 + bx + c$.)

Group Activity: Working with Real Data

Directions: Form a group of 2 to 4 people. Select someone to record the group's responses for this activity. All members of the group should work cooperatively to answer the questions. If your instructor asks for the results, each member of the group should be prepared to respond.

Minimum Wage   The table shows the minimum wage for three different years.

Year	1940	1968	1997
Wage	$0.25	$1.60	$5.45

Source: Bureau of Labor Statistics.

 (a) Make a scatterplot of the data in the viewing rectangle [1930, 2010, 10] by [0, 6, 1].

(b) Find a quadratic function given by
$$f(x) = a(x - h)^2 + k$$
that models the data.

(c) Estimate the minimum wage in 1976 and compare it to the actual value of $2.30.

(d) Estimate the year when the minimum wage was $1.00.

(e) If current trends continue, predict the minimum wage in 2005.

11.4   THE QUADRATIC FORMULA

Solving Quadratic Equations · The Discriminant · Quadratic Equations Having Complex Solutions

INTRODUCTION

To model the stopping distance of a car, highway engineers compute two quantities. The first quantity is the *reaction distance*, which is the distance a car travels from the time a driver first recognizes a hazard until the brakes are applied. The second quantity is *braking distance*, which is the distance a car travels after a driver applies the brakes. *Stopping distance* equals the sum of the reaction distance and the braking distance. If a car is traveling x miles per hour, highway engineers estimate the reaction distance in feet as $\frac{11}{3}x$ and the braking distance in feet as $\frac{1}{9}x^2$. To estimate the total stopping distance d in feet, they add the two expressions to obtain

$$d(x) = \frac{1}{9}x^2 + \frac{11}{3}x.$$

If a car's headlights don't illuminate the road beyond **500** feet, a safe nighttime speed limit x for the car can be determined by solving the quadratic equation

$$\frac{1}{9}x^2 + \frac{11}{3}x = 500.$$

(**Source:** L. Haefner, *Introduction to Transportation Systems.*)

In this section we learn how to solve this equation with the quadratic formula.

SOLVING QUADRATIC EQUATIONS

Recall that any quadratic equation can be written in the form

$$ax^2 + bx + c = 0.$$

If we solve this equation for x in terms of a, b, and c by completing the square, we obtain the **quadratic formula**. We assume that $a > 0$ and derive it as follows.

$ax^2 + bx + c = 0$	Quadratic equation
$ax^2 + bx = -c$	Subtract c.
$x^2 + \dfrac{b}{a}x = -\dfrac{c}{a}$	Divide by a.
$x^2 + \dfrac{b}{a}x + \dfrac{b^2}{4a^2} = -\dfrac{c}{a} + \dfrac{b^2}{4a^2}$	Add $\left(\dfrac{b/a}{2}\right)^2 = \dfrac{b^2}{4a^2}$.
$\left(x + \dfrac{b}{2a}\right)^2 = -\dfrac{c}{a} + \dfrac{b^2}{4a^2}$	Perfect square trinomial
$\left(x + \dfrac{b}{2a}\right)^2 = -\dfrac{c \cdot 4a}{a \cdot 4a} + \dfrac{b^2}{4a^2}$	Multiply $-\dfrac{c}{a}$ by $\dfrac{4a}{4a}$.
$\left(x + \dfrac{b}{2a}\right)^2 = -\dfrac{4ac}{4a^2} + \dfrac{b^2}{4a^2}$	Simplify.
$\left(x + \dfrac{b}{2a}\right)^2 = \dfrac{-4ac + b^2}{4a^2}$	Add fractions.
$\left(x + \dfrac{b}{2a}\right)^2 = \dfrac{b^2 - 4ac}{4a^2}$	Rewrite.
$x + \dfrac{b}{2a} = \pm\sqrt{\dfrac{b^2 - 4ac}{4a^2}}$	Square root property
$x = -\dfrac{b}{2a} \pm \sqrt{\dfrac{b^2 - 4ac}{4a^2}}$	Add $-\dfrac{b}{2a}$.
$x = -\dfrac{b}{2a} \pm \dfrac{\sqrt{b^2 - 4ac}}{2a}$	Property of square roots
$x = \dfrac{-b \pm \sqrt{b^2 - 4ac}}{2a}$	Combine fractions.

QUADRATIC FORMULA

The solutions to $ax^2 + bx + c = 0$ with $a \neq 0$ are given by

$$x = \frac{-b \pm \sqrt{b^2 - 4ac}}{2a}.$$

Note: The quadratic formula provides the solutions to *any* quadratic equation. It always "works."

≡ MAKING CONNECTIONS ≡

Completing the Square and the Quadratic Formula

The quadratic formula results from completing the square for the equation $ax^2 + bx + c = 0$. When you use the quadratic formula, the work of completing the square has already been done for you.

The next three examples illustrate how to solve quadratic equations symbolically and graphically.

EXAMPLE 1   **Solving a quadratic equation having two solutions**

Solve the equation $2x^2 - 3x - 1 = 0$. Support your results graphically.

Solution   **Symbolic Solution**   Let $a = 2$, $b = -3$, and $c = -1$.

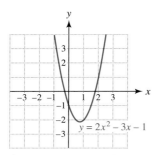

$$x = \frac{-b \pm \sqrt{b^2 - 4ac}}{2a} \qquad \text{Quadratic formula}$$

$$x = \frac{-(-3) \pm \sqrt{(-3)^2 - 4(2)(-1)}}{2(2)} \qquad \text{Substitute for } a, b, \text{ and } c.$$

$$x = \frac{3 \pm \sqrt{17}}{4} \qquad \text{Simplify.}$$

The solutions are $\frac{3 + \sqrt{17}}{4} \approx 1.78$ and $\frac{3 - \sqrt{17}}{4} \approx -0.28$.

Figure 11.34 Two x-intercepts

Graphical Solution   The graph of $y = 2x^2 - 3x - 1$ is shown in Figure 11.34. Note that the two x-intercepts correspond to the two solutions to $2x^2 - 3x - 1 = 0$. Estimating from this graph, we see that the solutions are approximately -0.25 and 1.75, which supports our symbolic solution.

Critical Thinking

Use the results of Example 1 to evaluate each expression mentally.

$$2\left(\frac{3 + \sqrt{17}}{4}\right)^2 - 3\left(\frac{3 + \sqrt{17}}{4}\right) - 1$$

$$2\left(\frac{3 - \sqrt{17}}{4}\right)^2 - 3\left(\frac{3 - \sqrt{17}}{4}\right) - 1$$

EXAMPLE 2   **Solving a quadratic equation having one solution**

Solve the equation $25x^2 + 20x + 4 = 0$. Support your result graphically.

Solution **Symbolic Solution** Let $a = 25$, $b = 20$, and $c = 4$.

$$x = \frac{-b \pm \sqrt{b^2 - 4ac}}{2a} \qquad \text{Quadratic formula}$$

$$= \frac{-20 \pm \sqrt{20^2 - 4(25)(4)}}{2(25)} \qquad \text{Substitute for } a, b, \text{ and } c.$$

$$= \frac{-20 \pm \sqrt{0}}{50} \qquad \text{Simplify.}$$

$$= \frac{-20}{50} = -0.4 \qquad \sqrt{0} = 0$$

There is one solution, -0.4.

Graphical Solution The graph of $y = 25x^2 + 20x + 4$ is shown in Figure 11.35. Note that the one x-intercept, -0.4, corresponds to the solution to $25x^2 + 20x + 4 = 0$.

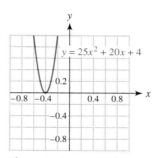

Figure 11.35 One x-intercept

EXAMPLE 3 **Recognizing a quadratic equation having no real solutions**

Solve the equation $5x^2 - x + 3 = 0$. Support your result graphically.

Solution **Symbolic Solution** Let $a = 5$, $b = -1$, and $c = 3$.

$$x = \frac{-b \pm \sqrt{b^2 - 4ac}}{2a} \qquad \text{Quadratic formula}$$

$$= \frac{-(-1) \pm \sqrt{(-1)^2 - 4(5)(3)}}{2(5)} \qquad \text{Substitute for } a, b, \text{ and } c.$$

$$= \frac{1 \pm \sqrt{-59}}{10} \qquad \text{Simplify.}$$

There are no real solutions to this equation because $\sqrt{-59}$ is not a real number. (Later in this section we discuss how to find complex solutions to quadratic equations such as this one.)

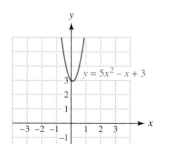

Figure 11.36 No x-intercepts

Graphical Solution The graph of $y = 5x^2 - x + 3$ is shown in Figure 11.36. Note that there are no x-intercepts, indicating that the equation $5x^2 - x + 3 = 0$ has no real solutions.

Earlier in this section we discussed how engineers estimate safe stopping distances for automobiles. In the next example we solve the equation presented in the introduction.

EXAMPLE 4   **Modeling stopping distance**

If a car's headlights do not illuminate the road beyond 500 feet, estimate a safe nighttime speed limit x for the car by solving $\frac{1}{9}x^2 + \frac{11}{3}x = 500$.

Solution   Begin by subtracting 500 from each side of the equation.

$$\frac{1}{9}x^2 + \frac{11}{3}x - 500 = 0.$$

To eliminate fractions multiply each side by the LCD, which is 9. (This step is not necessary, but it makes the problem easier to work.)

$$x^2 + 33x - 4500 = 0.$$

Now let $a = 1$, $b = 33$, and $c = -4500$ in the quadratic formula.

$$x = \frac{-b \pm \sqrt{b^2 - 4ac}}{2a} \qquad \text{Quadratic formula}$$

$$= \frac{-33 \pm \sqrt{(33)^2 - 4(1)(-4500)}}{2(1)} \qquad \text{Substitute for } a, b, \text{ and } c.$$

$$= \frac{-33 \pm \sqrt{19{,}089}}{2} \qquad \text{Simplify.}$$

The solutions are

$$\frac{-33 + \sqrt{19{,}089}}{2} \approx 52.6 \quad \text{and} \quad \frac{-33 - \sqrt{19{,}089}}{2} \approx -85.6.$$

The negative solution has no physical meaning because negative speeds are not possible. The other solution is 52.6, so an appropriate speed limit might be 50 miles per hour.   ———

THE DISCRIMINANT

The expression $b^2 - 4ac$ in the quadratic formula is called the **discriminant**. It provides information about the number of solutions to a quadratic equation.

|||||||| THE DISCRIMINANT AND QUADRATIC EQUATIONS

To determine the number of solutions to the quadratic equation $ax^2 + bx + c = 0$, evaluate the discriminant $b^2 - 4ac$.

1. If $b^2 - 4ac > 0$, there are two real solutions.
2. If $b^2 - 4ac = 0$, there is one real solution.
3. If $b^2 - 4ac < 0$, there are no real solutions; there are two complex solutions.

EXAMPLE 5 Using the discriminant

Use the discriminant to determine the number of solutions to $4x^2 + 25 = 20x$. Then solve the equation, using the quadratic formula.

Solution Write the equation as $4x^2 - 20x + 25 = 0$ so that $a = 4$, $b = -20$, and $c = 25$. The discriminant evaluates to

$$b^2 - 4ac = (-20)^2 - 4(4)(25) = 0.$$

Thus there is one real solution.

$$x = \frac{-b \pm \sqrt{b^2 - 4ac}}{2a} \qquad \text{Quadratic formula}$$

$$= \frac{-(-20) \pm \sqrt{0}}{2(4)} \qquad \text{Substitute.}$$

$$= \frac{20}{8} = 2.5 \qquad \text{Simplify.}$$

The only solution is 2.5.

We also need to be able to analyze graphs of quadratic functions, which we demonstrate in the next example.

EXAMPLE 6 Analyzing graphs of quadratic functions

A graph of $f(x) = ax^2 + bx + c$ is shown in Figure 11.37.
(a) State whether $a > 0$ or $a < 0$.
(b) Solve the equation $ax^2 + bx + c = 0$.
(c) Determine whether the discriminant is positive, negative, or zero.

Solution **(a)** The parabola opens downward, so $a < 0$.
(b) The solutions correspond to the x-intercepts, -3 and 2.
(c) There are two solutions, so the discriminant is positive.

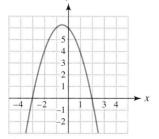

Figure 11.37

QUADRATIC EQUATIONS HAVING COMPLEX SOLUTIONS

The quadratic equation $ax^2 + bx + c = 0$ has no real solutions if the discriminant, $b^2 - 4ac$, is negative. For example, the quadratic equation $x^2 + 4 = 0$ has $a = 1$, $b = 0$, and $c = 4$. Its discriminant is

$$b^2 - 4ac = 0^2 - 4(1)(4) = -16 < 0,$$

so this equation has no real solutions. However, if we use complex numbers, we can solve this equation as follows.

$$x^2 + 4 = 0 \qquad \text{Given equation}$$

$$x^2 = -4 \qquad \text{Subtract 4.}$$

$$x = \pm\sqrt{-4} \qquad \text{Square root property}$$

$$x = \sqrt{-4} \quad \text{or} \quad x = -\sqrt{-4} \qquad \text{Meaning of } \pm$$

$$x = 2i \quad \text{or} \quad x = -2i \qquad \text{The expression } \sqrt{-a}$$

The solutions are $\pm 2i$. We check each solution to $x^2 + 4 = 0$ as follows.

$$(2i)^2 + 4 = (2)^2 i^2 + 4 = 4(-1) + 4 = 0 \qquad \text{It checks.}$$
$$(-2i)^2 + 4 = (-2)^2 i^2 + 4 = 4(-1) + 4 = 0 \qquad \text{It checks.}$$

This result can be generalized as follows.

THE EQUATION $x^2 + k = 0$

If $k > 0$, the solutions to $x^2 + k = 0$ are given by $x = \pm i \sqrt{k}$.

EXAMPLE 7 Solving a quadratic equation having complex solutions

Solve $x^2 + 5 = 0$.

Solution The solutions are $\pm i \sqrt{5}$. That is, $x = i \sqrt{5}$ or $x = -i \sqrt{5}$.

When $b \neq 0$, the preceding method cannot be used. Consider the quadratic equation $2x^2 + x + 3 = 0$, which has $a = 2, b = 1$, and $c = 3$. Its discriminant is

$$b^2 - 4ac = 1^2 - 4(2)(3) = -23 < 0.$$

This equation has two complex solutions as demonstrated in the next example.

EXAMPLE 8 Solving a quadratic equation having complex solutions

Solve $2x^2 + x + 3 = 0$. Write your answer in standard form: $a + bi$.

Solution Let $a = 2, b = 1$, and $c = 3$.

Calculator Help
To set your calculator in $a + bi$ mode or to access the imaginary unit i, see Appendix A (page AP-11).

$$x = \frac{-b \pm \sqrt{b^2 - 4ac}}{2a} \qquad \text{Quadratic formula}$$

$$= \frac{-1 \pm \sqrt{1^2 - 4(2)(3)}}{2(2)} \qquad \text{Substitute for } a, b, \text{ and } c.$$

$$= \frac{-1 \pm \sqrt{-23}}{4} \qquad \text{Simplify.}$$

$$= \frac{-1 \pm i \sqrt{23}}{4} \qquad \sqrt{-23} = i\sqrt{23}$$

$$= -\frac{1}{4} \pm i \frac{\sqrt{23}}{4} \qquad \text{Divide each term by 4.}$$

The solutions are $-\frac{1}{4} + i \frac{\sqrt{23}}{4}$ and $-\frac{1}{4} - i \frac{\sqrt{23}}{4}$.

Critical Thinking

Use the results of Example 8 to evaluate each expression mentally.

$$2\left(-\tfrac{1}{4} + i\tfrac{\sqrt{23}}{4}\right)^2 + \left(-\tfrac{1}{4} + i\tfrac{\sqrt{23}}{4}\right) + 3 \quad \text{and} \quad 2\left(-\tfrac{1}{4} - i\tfrac{\sqrt{23}}{4}\right)^2 + \left(-\tfrac{1}{4} - i\tfrac{\sqrt{23}}{4}\right) + 3$$

Sometimes we can use properties of radicals to simplify a solution to a quadratic equation, as demonstrated in the next example.

EXAMPLE 9   **Solving a quadratic equation having complex solutions**

Solve $\tfrac{3}{4}x^2 + 1 = x$. Write your answer in standard form: $a + bi$.

Solution   Begin by subtracting x from each side of the equation and then multiply by 4 to clear fractions. The resulting equation is $3x^2 - 4x + 4 = 0$. Substitute $a = 3$, $b = -4$, and $c = 4$ in the quadratic formula.

$$x = \frac{-b \pm \sqrt{b^2 - 4ac}}{2a} \qquad \text{Quadratic formula}$$

$$= \frac{-(-4) \pm \sqrt{(-4)^2 - 4(3)(4)}}{2(3)} \qquad \text{Substitute.}$$

$$= \frac{4 \pm \sqrt{-32}}{6} \qquad \text{Simplify.}$$

$$= \frac{4 \pm 4i\sqrt{2}}{6} \qquad \sqrt{-32} = i\sqrt{32} = i\sqrt{16}\sqrt{2} = 4i\sqrt{2}$$

$$= \frac{2}{3} \pm \frac{2}{3}i\sqrt{2} \qquad \text{Divide 6 into each term and reduce.} \quad \underline{}$$

 PUTTING IT ALL TOGETHER

Quadratic equations can be solved symbolically by using factoring, the square root property, completing the square, and the quadratic formula. Graphical and numerical methods can also be used to solve quadratic equations. In this section we discussed the quadratic formula and its discriminant, which we summarize in the following table.

Concept	Explanation	Examples
Quadratic Formula	The quadratic formula can be used to solve *any* quadratic equation written as $ax^2 + bx + c = 0$. The solutions are given by $$x = \frac{-b \pm \sqrt{b^2 - 4ac}}{2a}.$$	For the equation $$2x^2 - 3x + 1 = 0$$ with $a = 2$, $b = -3$, and $c = 1$, the solutions are $$\frac{-(-3) \pm \sqrt{(-3)^2 - 4(2)(1)}}{2(2)} = \frac{3 \pm \sqrt{1}}{4} = 1, \frac{1}{2}.$$

continued on next page

continued from previous page

Concept	Explanation	Examples
The Discriminant	The expression $b^2 - 4ac$ is called the discriminant. The discriminant may be used to determine the number of solutions to $ax^2 + bx + c = 0$. 1. $b^2 - 4ac > 0$ indicates two real solutions. 2. $b^2 - 4ac = 0$ indicates one real solution. 3. $b^2 - 4ac < 0$ indicates no real solutions; rather, there are two complex solutions.	For the equation $$x^2 + 4x - 1 = 0$$ with $a = 1$, $b = 4$, and $c = -1$, the discriminant is $$b^2 - 4ac = 4^2 - 4(1)(-1) = 20 > 0,$$ indicating two real solutions.
Quadratic Formula and Complex Solutions	If the discriminant is negative $(b^2 - 4ac < 0)$, the solutions to a quadratic equation are complex numbers. If $k > 0$, the solutions to $x^2 + k = 0$ are given by $x = \pm i \sqrt{k}$.	$$2x^2 - x + 3 = 0$$ $$x = \frac{1 \pm \sqrt{(-1)^2 - 4(2)(3)}}{2(2)}$$ $$= \frac{1 \pm \sqrt{-23}}{4} = \frac{1}{4} \pm i\frac{\sqrt{23}}{4}$$ $x^2 + 9 = 0$ is equivalent to $x = \pm 3i$ and $x^2 + 7 = 0$ is equivalent to $x = \pm i\sqrt{7}$.

 **11.4**   EXERCISES

FOR EXTRA HELP

📖 Student's Solutions Manual
🚪 MyMathLab
  InterAct Math
 Math Tutor Center
 MathXL
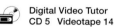 Digital Video Tutor
CD 5   Videotape 14

CONCEPTS

1. What is the quadratic formula used for?

2. What basic algebraic technique is used to derive the quadratic formula?

3. What is the discriminant?

4. If the discriminant evaluates to 0, what does that indicate about the quadratic equation?

5. Name four symbolic techniques for solving a quadratic equation.

6. Does every quadratic equation have at least one real solution? Explain.

THE QUADRATIC FORMULA

Exercises 7–10: Use the quadratic formula to solve the equation. Support your result graphically and numerically. If there are no real solutions, say so.

7. $2x^2 + 11x - 6 = 0$ **8.** $x^2 + 2x - 24 = 0$

9. $-x^2 + 2x - 1 = 0$ **10.** $3x^2 - x + 1 = 0$

Exercises 11–28: Solve by using the quadratic formula. If there are no real solutions, say so.

11. $x^2 - 6x - 16 = 0$ **12.** $2x^2 - 9x + 7 = 0$

13. $4x^2 - x - 1 = 0$ **14.** $-x^2 + 2x + 1 = 0$

15. $-3x^2 + 2x - 1 = 0$ **16.** $x^2 + x + 3 = 0$

17. $36x^2 - 36x + 9 = 0$

18. $4x^2 - 5.6x + 1.96 = 0$

19. $2x(x - 3) = 2$ **20.** $x(x + 1) + x = 5$

21. $(x - 1)(x + 1) + 2 = 4x$

22. $\frac{1}{2}(x - 6) = x^2 + 1$ **23.** $\frac{1}{2}x(x + 1) = 2x^2 - \frac{3}{2}$

24. $\frac{1}{2}x^2 - \frac{1}{4}x + \frac{1}{2} = x$ **25.** $2x(x - 1) = 7$

26. $3x(x - 4) = 4$ **27.** $-3x^2 + 10x - 5 = 0$

28. $-2x^2 + 4x - 1 = 0$

THE DISCRIMINANT

Exercises 29–34: A graph of $y = ax^2 + bx + c$ *is shown.*
  (a) State whether $a > 0$ *or* $a < 0$.
  (b) Solve $ax^2 + bx + c = 0$, *if possible.*
  (c) Determine whether the discriminant is positive, negative, or zero.

29.

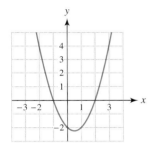

30.

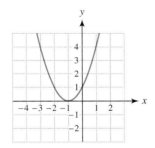

31.

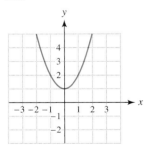

32.

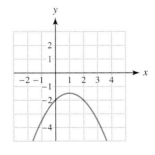

33.

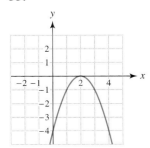

34.

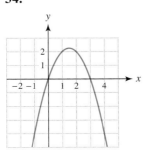

Exercises 35–42: Do the following for the given equation.
  (a) Evaluate the discriminant.
  (b) How many real solutions are there?
  (c) Support your answer for part (b) by using graphing.

35. $3x^2 + x - 2 = 0$ **36.** $5x^2 - 13x + 6 = 0$

37. $x^2 - 4x + 4 = 0$ **38.** $\frac{1}{4}x^2 + 4 = 2x$

39. $\frac{1}{2}x^2 + \frac{3}{2}x + 2 = 0$ **40.** $x - 3 = 2x^2$

41. $x(x + 3) = 3$ **42.** $(4x - 1)(x - 3) = -25$

Exercises 43–52: Use the quadratic formula to find any x-intercepts on the graph of the equation.

43. $y = x^2 - 2x - 1$ **44.** $y = x^2 + 3x + 1$

45. $y = -2x^2 - x + 3$ **46.** $y = -3x^2 - x + 4$

47. $y = x^2 + x + 5$ **48.** $y = 3x^2 - 2x + 5$

49. $y = x^2 + 9$ **50.** $y = x^2 + 11$

51. $y = 3x^2 + 4x - 2$ **52.** $y = 4x^2 - 2x - 3$

COMPLEX SOLUTIONS

Exercises 53–84: Solve the equation. Write complex solutions in standard form.

53. $x^2 + 9 = 0$ **54.** $x^2 + 16 = 0$

55. $x^2 + 80 = 0$ **56.** $x^2 + 20 = 0$

57. $x^2 + \frac{1}{4} = 0$ **58.** $x^2 + \frac{9}{4} = 0$

59. $16x^2 + 9 = 0$ **60.** $25x^2 + 36 = 0$

61. $x^2 = -6$ **62.** $x^2 = -75$

63. $x^2 - 3 = 0$ **64.** $x^2 - 8 = 0$

65. $x^2 + 2 = 0$ **66.** $x^2 + 4 = 0$

67. $x^2 - x + 2 = 0$ **68.** $x^2 + 2x + 3 = 0$

69. $2x^2 + 3x + 4 = 0$ **70.** $3x^2 - x = 1$

71. $x^2 + 1 = 4x$ **72.** $3x^2 + 2 = x$

73. $x^2 + x = -2$ **74.** $x(x - 4) = -8$

75. $5x^2 + 2x + 4 = 0$ **76.** $7x^2 - 2x + 4 = 0$

77. $\frac{1}{2}x^2 + \frac{3}{4}x + 1 = 0$ **78.** $-\frac{1}{3}x^2 + x - 2 = 0$

79. $x(x + 2) = x - 4$ **80.** $x - 5 = 2x(2x + 1)$

81. $x(2x - 1) = 1 + x$ **82.** $2x = x(3 - 4x)$

83. $x^2 = x(1 - x) - 2$ **84.** $2x^2 = 2x(5 - x) - 8$

YOU DECIDE THE METHOD

Exercises 85–92: Find exact solutions to the quadratic equation, using a method of your choice. Explain why you chose the method you did.

85. $x^2 - 3x + 2 = 0$ **86.** $x^2 + 2x + 1 = 0$

87. $0.5x^2 - 1.75x - 1 = 0$ **88.** $\frac{3}{5}x^2 + \frac{9}{10}x - \frac{3}{5} = 0$

89. $x^2 - 5x + 2 = 0$ **90.** $2x^2 - x - 4 = 0$

91. $2x^2 + x = -8$ **92.** $4x^2 = 2x - 3$

APPLICATIONS

Exercises 93–96: Modeling Stopping Distance   (Refer to Example 4.) Use $d = \frac{1}{9}x^2 + \frac{11}{3}x$ to find a safe speed x for the following stopping distances d.

93. 42 feet   **94.** 152 feet

95. 390 feet   **96.** 726 feet

97. *Modeling U.S. AIDS Deaths*   The cumulative numbers in thousands of AIDS deaths from 1984 through 1994 may be modeled by

$$f(x) = 2.39x^2 + 5.04x + 5.1,$$

where $x = 0$ corresponds to 1984, $x = 1$ to 1985, and so on until $x = 10$ corresponds to 1994. See the accompanying graph. Use the formula for $f(x)$ to estimate the year when the total number of AIDS deaths reached 200 thousand. Compare your result with that shown in the graph.

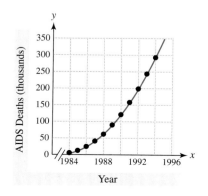

98. *Screen Dimensions*   The width of a rectangular computer screen is 3 inches more than its height. If the area of the screen is 154 square inches, find its dimensions
(a) graphically,
(b) numerically, and
(c) symbolically.

99. *Canoeing*   A camper paddles a canoe 2 miles downstream in a river that has a 2-mile-per-hour current. To return to camp, the canoeist travels upstream on a different branch of the river. It is 4 miles long and has a 1-mile-per-hour current. The total trip (both ways) takes 3 hours. Find the average speed of the canoe in still water. (*Hint:* Time equals distance divided by rate.)

100. *Airplane Speed*   A pilot flies 500 miles against a 20-mile-per-hour wind. On the next day, the pilot flies back home with a 10-mile-per-hour tail wind. The total trip (both ways) takes 4 hours. Find the speed of the airplane without a wind.

101. *Modeling Water Flow*   When water runs out of a hole in a cylindrical container, the height of the water in the container can often be modeled by a quadratic function. The data in the table show the height y in centimeters of water at 30-second intervals in a metal can that had a small hole in it.

Time	0	30	60	90
Height	16	11.9	8.4	5.3

Time	120	150	180
Height	3.1	1.4	0.5

These data are modeled by

$$f(x) = 0.0004x^2 - 0.15x + 16.$$

(a) Explain why a linear function would not be appropriate for modeling these data.
(b) Use the table to estimate the time at which the height was 7 centimeters.
(c) Use $f(x)$ and the quadratic formula to solve part (b).

102. *Hospitals* The general trend in the number of hospitals in the United States from 1945 through 2000 is modeled by

$$f(x) = -1.38x^2 + 84x + 5865,$$

where $x = 5$ corresponds to 1945, $x = 10$ to 1950, and so on until $x = 60$ represents 2000. See the scatterplot and accompanying graph.

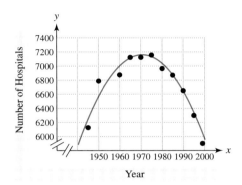

(a) Describe any trends in the numbers of hospitals from 1945 to 2000.

(b) What information does the vertex of the graph of f give?

(c) Use the formula for $f(x)$ to estimate the number of hospitals in 1970 ($x = 30$). Compare your result with that shown in the graph.

(d) Use the formula for $f(x)$ to estimate the year (or years) when there were 6300 hospitals. Compare your result with that shown in the graph.

WRITING ABOUT MATHEMATICS

103. Explain how the discriminant can be used to determine the number of solutions to a quadratic equation.

104. Let $f(x) = ax^2 + bx + c$. If you know the value of $b^2 - 4ac$, what information does this give you about the graph of f? Explain your answer.

CHECKING BASIC CONCEPTS SECTIONS 11.3 AND 11.4

1. Solve the quadratic equation $2x^2 - 7x + 3 = 0$ symbolically and graphically.

2. Use the square root property to solve $x^2 = 5$.

3. Complete the square to solve $x^2 - 4x + 1 = 0$.

4. Solve the equation $x^2 + y^2 = 1$ for y.

5. Use the quadratic formula to solve each equation.
(a) $2x^2 = 3x + 1$

(b) $9x^2 - 24x + 16 = 0$
(c) $x^2 + x + 2 = 0$

6. Calculate the discriminant for each equation and give the number of real solutions.
(a) $x^2 - 5x + 5 = 0$
(b) $2x^2 - 5x + 4 = 0$
(c) $49x^2 - 56x + 16 = 0$

11.5 QUADRATIC INEQUALITIES

Basic Concepts · Graphical and Numerical Solutions · Symbolic Solutions

INTRODUCTION

Quadratic inequalities are nonlinear inequalities. These inequalities are often simple enough to be solved by hand. In this section we discuss how to solve them graphically, numerically, and symbolically.

BASIC CONCEPTS

If the equals sign in a quadratic equation is replaced with $>, \geq, <,$ or $\leq$, a **quadratic inequality** results. Examples of quadratic inequalities include

$$x^2 + 4x - 3 < 0, \quad 5x^2 \geq 5, \quad \text{and} \quad 1 - z \leq z^2.$$

Any quadratic equation can be written as

$$ax^2 + bx + c = 0, \quad a \neq 0,$$

so any quadratic inequality can be written as

$$ax^2 + bx + c > 0, \quad a \neq 0,$$

where $>$ may be replaced with $\geq, <,$ or $\leq$.

The next example demonstrates how to identify a quadratic inequality.

EXAMPLE 1   Identifying a quadratic inequality

Determine whether the inequality is quadratic.
(a) $5x + x^2 - x^3 \leq 0$ (b) $4 + 5x^2 > 4x^2 + x$

Solution   (a) This inequality is not quadratic because x^3 occurs in the inequality.
(b) Write the inequality as follows.

$4 + 5x^2 > 4x^2 + x$	Given inequality
$4 + 5x^2 - 4x^2 - x > 0$	Subtract $4x^2$ and x.
$4 + x^2 - x > 0$	Combine like terms.
$x^2 - x + 4 > 0$	Rewrite the expression.

Because the inequality can be written in the form $ax^2 + bx + c > 0$ with $a = 1$, $b = -1$, and $c = 4$, it is a quadratic inequality.

GRAPHICAL AND NUMERICAL SOLUTIONS

Equality often is the boundary between *greater than* and *less than*, so a first step in solving an inequality is to determine the x-values where equality occurs. We begin by using this concept with graphical and numerical techniques.

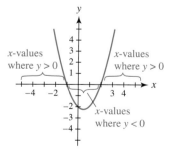

A graph of $y = x^2 - x - 2$ with x-intercepts -1 and 2 is shown in Figure 11.38. The solutions to $x^2 - x - 2 = 0$ are given by $x = -1$ or $x = 2$. Between the x-intercepts the graph dips below the x-axis and the y-values are negative. Thus the solutions to $x^2 - x - 2 < 0$ satisfy $-1 < x < 2$. To check this result we select a **test value**. For example, 0 lies between -1 and 2. If we substitute $x = 0$ in the inequality, it results in a true statement.

$$0^2 - 0 - 2 < 0 \quad \text{True}$$

Figure 11.38

When $x < -1$ or $x > 2$, the graph lies above the x-axis and the y-values are positive. Thus the solutions to $x^2 - x - 2 > 0$ satisfy $x < -1$ or $x > 2$. For example, 3 is greater than 2 and -3 is less than -1. Therefore both 3 and -3 are solutions. We can verify this result by substituting 3 and -3 as test values in the inequality.

$$3^2 - 3 - 2 > 0 \quad \text{True}$$
$$(-3)^2 - (-3) - 2 > 0 \quad \text{True}$$

In the next two examples, we use graphical and numerical methods to solve quadratic inequalities.

EXAMPLE 2   Solving a quadratic inequality

Make a table of values for $y = x^2 - 3x - 4$ and then sketch the graph. Use the table and graph to solve $x^2 - 3x - 4 \leq 0$. Write your answer in interval notation.

Solution   The points calculated for Table 11.14 are plotted in Figure 11.39 and connected with a smooth ∪-shaped graph.

Numerical Solution   Table 11.14 shows that $x^2 - 3x - 4$ equals 0 when $x = -1$ or $x = 4$. Between these values, $x^2 - 3x - 4$ is negative so the solution set to $x^2 - 3x - 4 \leq 0$ is $-1 \leq x \leq 4$ or in interval notation, $[-1, 4]$.

Graphical Solution   In Figure 11.39 the graph of $y = x^2 - 3x - 4$ shows that the x-intercepts are -1 and 4. Between these values, the graph dips *below* the x-axis. Thus the solution set is $[-1, 4]$.

TABLE 11.14

x	$y = x^2 - 3x - 4$
-2	6
-1	0
0	-4
1	-6
2	-6
3	-4
4	0
5	6

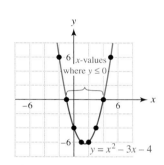

Figure 11.39

EXAMPLE 3   Solving a quadratic inequality

Solve $x^2 > 1$. Write your answer in interval notation.

Solution   First, rewrite $x^2 > 1$ as $x^2 - 1 > 0$. The graph $y = x^2 - 1$ is shown in Figure 11.40 with x-intercepts -1 and 1. The graph lies *above* the x-axis and is shaded red to the left of $x = -1$ and to the right of $x = 1$. Thus the solution set is $x < -1$ or $x > 1$, which can be written in interval notation as $(-\infty, -1) \cup (1, \infty)$.

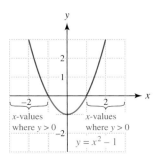

Figure 11.40

Critical Thinking

The graph of $y = -x^2 + x + 12$ is a parabola opening downward with x-intercepts of -3 and 4. Solve each inequality.

i. $-x^2 + x + 12 > 0$ **ii.** $-x^2 + x + 12 < 0$

In the next example we show how quadratic inequalities are used in highway design.

EXAMPLE 4   **Determining elevations on a sag curve**

Parabolas are frequently used in highway design to model hills and sags (valleys) along a proposed route. Suppose that the elevation E in feet of a sag, or *sag curve*, is given by

$$E(x) = 0.00004x^2 - 0.4x + 2000,$$

where x is the horizontal distance in feet along the sag curve and $0 \le x \le 10{,}000$. See Figure 11.41. Estimate graphically the x-values for elevations of 1500 feet or less.   (*Source: F. Mannering and W. Kilareski, Principles of Highway Engineering and Traffic Analysis.*)

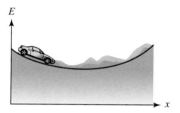

Figure 11.41

Solution   ***Graphical Solution***   We must solve the quadratic inequality

$$0.00004x^2 - 0.4x + 2000 \le 1500.$$

To do so, we let $Y_1 = .00004X^2 - .4X + 2000$ represent the sag or valley in the road and $Y_2 = 1500$ represent a horizontal line with an elevation of 1500 feet. Their graphs intersect at $x \approx 1464$ and $x \approx 8536$, as shown in Figure 11.42. The elevation of the proposed route is less than 1500 feet between these x-values. Therefore the elevation of the road is 1500 feet or less when $1464 \le x \le 8536$ (approximately).

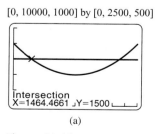

[0, 10000, 1000] by [0, 2500, 500]   [0, 10000, 1000] by [0, 2500, 500]

Intersection
X=1464.4661 Y=1500   Intersection
X=8535.5339 Y=1500

(a)   (b)

Figure 11.42

Symbolic Solutions

To solve a quadratic inequality we first solve the corresponding equality. We can then write the solution to the inequality, using the following method.

SOLUTIONS TO QUADRATIC INEQUALITIES

Let $ax^2 + bx + c = 0$, $a > 0$, have two real solutions p and q, where $p < q$.

$ax^2 + bx + c < 0$ is equivalent to $p < x < q$ (see left-hand figure).

$ax^2 + bx + c > 0$ is equivalent to $x < p$ or $x > q$ (see right-hand figure).

Quadratic inequalities involving $\leq$ or $\geq$ can be solved similarly.

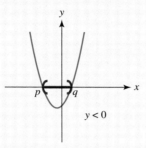

$y < 0$

Solutions lie between p and q.

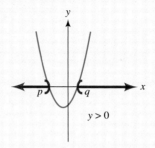

$y > 0$

Solutions lie "outside" p and q.

We demonstrate this method in the next two examples.

EXAMPLE 5   Solving quadratic inequalities

Solve each inequality symbolically. Write your answer in interval notation.
(a) $6x^2 - 7x - 5 \geq 0$ **(b)** $x(3 - x) > -18$

Solution **(a)** Begin by solving $6x^2 - 7x - 5 = 0$.

$$6x^2 - 7x - 5 = 0$$
$$(2x + 1)(3x - 5) = 0 \qquad \text{Factor.}$$
$$2x + 1 = 0 \quad \text{or} \quad 3x - 5 = 0 \qquad \text{Zero-product property}$$
$$x = -\frac{1}{2} \quad \text{or} \quad x = \frac{5}{3} \qquad \text{Solve.}$$

Therefore the solutions to $6x^2 - 7x - 5 \geq 0$ lie "outside" these two values and satisfy $x \leq -\frac{1}{2}$ or $x \geq \frac{5}{3}$. In interval notation the solution set is $\left(-\infty, -\frac{1}{2}\right] \cup \left[\frac{5}{3}, \infty\right)$.

(b) First, rewrite the inequality as follows.

$$x(3 - x) > -18 \qquad \text{Given inequality}$$
$$3x - x^2 > -18 \qquad \text{Distributive property}$$
$$3x - x^2 + 18 > 0 \qquad \text{Add 18.}$$
$$-x^2 + 3x + 18 > 0 \qquad \text{Rewrite.}$$
$$x^2 - 3x - 18 < 0 \qquad \text{Multiply by } -1; \text{ reverse the inequality symbol.}$$

Next, solve $x^2 - 3x - 18 = 0$.

$$(x + 3)(x - 6) = 0 \qquad \text{Factor.}$$
$$x = -3 \quad \text{or} \quad x = 6 \qquad \text{Solve.}$$

Solutions to $x^2 - 3x - 18 < 0$ lie between these two values and satisfy $-3 < x < 6$. In interval notation the solution set is $(-3, 6)$.

Critical Thinking

Graph $y = x^2 + 1$ and solve the following inequalities.

i. $x^2 + 1 > 0$ **ii.** $x^2 + 1 < 0$

Now graph $y = (x - 1)^2$ and solve the following inequalities.

iii. $(x - 1)^2 \geq 0$ **iv.** $(x - 1)^2 \leq 0$

EXAMPLE 6   Finding the dimensions of a building

A rectangular building needs to be 7 feet longer than it is wide, as illustrated in Figure 11.43. The area of the building must be at least 450 square feet. What widths x are possible for this building? Support your results with a table of values.

Figure 11.43

Solution ***Symbolic Solution*** If x is the width of the building, $x + 7$ is the length of the building and its area is $x(x + 7)$. The area must be at least 450 square feet, so the inequality $x(x + 7) \geq 450$ must be satisfied. First solve the following quadratic equation.

$$x(x + 7) = 450 \qquad \text{Quadratic equation}$$

$$x^2 + 7x = 450 \qquad \text{Distributive property}$$

$$x^2 + 7x - 450 = 0 \qquad \text{Subtract 450.}$$

$$x = \frac{-7 \pm \sqrt{7^2 - 4(1)(-450)}}{2(1)} \qquad \begin{array}{l}\text{Quadratic formula; } a = 1,\\ b = 7, \text{ and } c = -450\end{array}$$

$$= \frac{-7 \pm \sqrt{1849}}{2} \qquad \text{Simplify.}$$

$$= \frac{-7 \pm 43}{2} \qquad \sqrt{1849} = 43$$

$$= 18, -25 \qquad \text{Evaluate.}$$

Thus the solutions to $x(x + 7) \geq 450$ are $x \leq -25$ or $x \geq 18$. The width is positive, so the building width must be 18 feet or more.

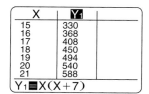

Figure 11.44

Numerical Solution A table of values is shown in Figure 11.44, where $Y_1 = X(X + 7)$ equals 450 when $x = 18$. For $x \geq 18$ the area is *at least* 450 square feet.

11.5 PUTTING IT ALL TOGETHER

The following table summarizes solutions of quadratic inequalities containing the symbols $<$ or $>$. Cases involving $\leq$ or $\geq$ are solved similarly.

Method	Explanation
Solving a Quadratic Inequality Symbolically	Let $ax^2 + bx + c = 0$, $a > 0$, have two real solutions p and q, where $p < q$. $ax^2 + bx + c < 0$ is equivalent to $p < x < q$ $ax^2 + bx + c > 0$ is equivalent to $x < p$ or $x > q$ *Examples:* The solutions to $x^2 - 3x + 2 = 0$ are given by $x = 1$ or $x = 2$. The solutions to $x^2 - 3x + 2 < 0$ are given by $1 < x < 2$. The solutions to $x^2 - 3x + 2 > 0$ are given by $x < 1$ or $x > 2$.
Solving a Quadratic Inequality Graphically	Given $ax^2 + bx + c < 0$ with $a > 0$, graph $y = ax^2 + bx + c$ and locate any x-intercepts. If there are two x-intercepts, then solutions correspond to x-values between the x-intercepts. Solutions to $ax^2 + bx + c > 0$ correspond to x-values "outside" the x-intercepts.
Solving a Quadratic Inequality Numerically	If a quadratic inequality is expressed as $$ax^2 + bx + c < 0 \text{ with } a > 0,$$ solve $$y = ax^2 + bx + c = 0$$ with a table. If there are two solutions, then the solutions to the given inequality lie between these values. Solutions to $ax^2 + bx + c > 0$ lie "outside" these values.

11.5 EXERCISES

FOR EXTRA HELP

Student's Solutions Manual	InterAct Math
MyMathLab	Math Tutor Center
	MathXL
	Digital Video Tutor CD 5 Videotape 14

CONCEPTS

1. How is a quadratic inequality different from a quadratic equation?

2. Do quadratic inequalities typically have two solutions? Explain.

3. Is 3 a solution to $x^2 < 7$?

4. Is 5 a solution to $x^2 \geq 25$?

5. The solutions to $x^2 - 2x - 8 = 0$ are -2 and 4. What are the solutions to $x^2 - 2x - 8 < 0$?

6. The solutions to $x^2 + 2x - 3 = 0$ are -3 and 1. What are the solutions to $x^2 + 2x - 3 > 0$?

Exercises 7–12: Determine whether the inequality is quadratic.

7. $x^2 + 4x + 5 < 0$ 8. $x > x^3 - 5$

9. $x^2 > 19$

10. $x(x - 1) - 2 \geq 0$

11. $4x > 1 - x$

12. $2x(x^2 + 3) < 0$

Exercises 13–18: Determine whether the given value of x is a solution.

13. $2x^2 + x - 1 > 0$ $x = 3$

14. $x^2 - 3x + 2 \leq 0$ $x = 2$

15. $x^2 + 2 \leq 0$ $x = 0$

16. $2x(x - 3) \geq 0$ $x = 1$

17. $x^2 - 3x \leq 1$ $x = -3$

18. $4x^2 - 5x + 1 > 30$ $x = -2$

SOLVING QUADRATIC INEQUALITIES

Exercises 19–24: Use the graph of
$$y = ax^2 + bx + c$$
to solve each quadratic equation or inequality.
   (a) $ax^2 + bx + c = 0$
   (b) $ax^2 + bx + c < 0$
   (c) $ax^2 + bx + c > 0$

19.

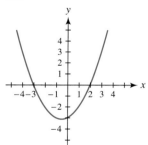

20.

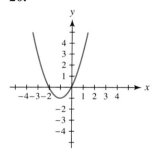

21.

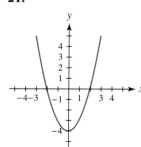

22.

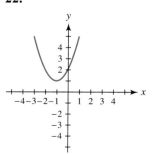

23.

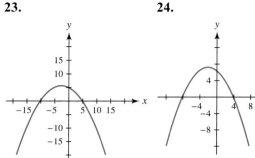

24.

Exercises 25–28: Use the table for
$$y = ax^2 + bx + c$$
to solve each quadratic equation or inequality.
   (a) $ax^2 + bx + c = 0$
   (b) $ax^2 + bx + c < 0$
   (c) $ax^2 + bx + c > 0$

25. $y = x^2 - 4$

x	-3	-2	-1	0	1	2	3
y	5	0	-3	-4	-3	0	5

26. $y = x^2 - x - 2$

x	-3	-2	-1	0	1	2	3
y	10	4	0	-2	-2	0	4

27. $y = x^2 + 4x$

x	-5	-4	-3	-2	-1	0	1
y	5	0	-3	-4	-3	0	5

28. $y = -2x^2 - 2x + 1.5$

x	-2	-1.5	-1	-0.5	0	0.5	1
y	-2.5	0	1.5	2	1.5	0	-2.5

Exercises 29–38: Solve the quadratic inequality by any method. Write your answer in interval notation.

29. $x^2 + 4x + 3 < 0$ **30.** $x^2 + x - 2 \leq 0$

31. $2x^2 - x - 15 \geq 0$ **32.** $3x^2 - 3x - 6 > 0$

33. $2x^2 \leq 8$ **34.** $x^2 < 9$

35. $x^2 > -5$ **36.** $-x^2 \geq 1$

37. $-x^2 + 3x > 0$ **38.** $-8x^2 - 2x + 1 \leq 0$

Exercises 39–42: Solve the quadratic equation in part (a) symbolically. Use the results to solve the inequalities in parts (b) and (c).

39. (a) $x^2 - 4 = 0$
   (b) $x^2 - 4 < 0$
   (c) $x^2 - 4 > 0$

40. (a) $x^2 - 5 = 0$
   (b) $x^2 - 5 \leq 0$
   (c) $x^2 - 5 \geq 0$

41. (a) $x^2 + x - 1 = 0$
   (b) $x^2 + x - 1 < 0$
   (c) $x^2 + x - 1 > 0$

42. (a) $x^2 + 4x - 5 = 0$
   (b) $x^2 + 4x - 5 \leq 0$
   (c) $x^2 + 4x - 5 \geq 0$

Exercises 43–52: Solve the quadratic inequality symbolically. Write your answer in interval notation.

43. $x^2 + 10x + 21 \leq 0$ **44.** $x^2 - 7x - 18 < 0$

45. $3x^2 - 9x + 6 > 0$ **46.** $7x^2 + 34x - 5 \geq 0$

47. $x^2 < 10$ **48.** $x^2 \geq 64$

49. $x(x - 6) > 0$ **50.** $1 - x^2 \leq 0$

51. $x(4 - x) \leq 2$ **52.** $2x(1 - x) \geq 2$

APPLICATIONS

 53. *Highway Design* (Refer to Example 4 and Figure 11.41.) The elevation E of a sag curve in feet is given by

$$E(x) = 0.0000375x^2 - 0.175x + 1000,$$

where $0 \leq x \leq 4000$.
   (a) Estimate graphically the x-values for which the elevation is 850 feet or less. (*Hint:* Use [0, 4000, 1000] by [500, 1200, 100] as a viewing rectangle.)
   (b) For what x-values is the elevation 850 feet or more?

54. *Early Cellular Phone Use* Our society is in transition from an industrial to an informational society. Cellular communication has played an increasingly large role in this transition. The number of cellular subscribers in the United States in thousands from 1985 to 1991 can be modeled by

$$f(x) = 163x^2 - 146x + 205,$$

where x is the year and $x = 0$ corresponds to 1985, $x = 1$ to 1986, and so on. (*Source:* M. Paetsch, *Mobile Communication in the U.S. and Europe.*)

   (a) Write a quadratic inequality whose solution set represents the years when there were 2 million subscribers or more.
   (b) Solve this inequality.

55. *Heart Disease Death Rates* From 1960 to 1995, age-adjusted heart disease rates decreased dramatically. The number of deaths per 100,000 people can be modeled by

$$f(x) = -0.014777x^2 + 54.14x - 49,060,$$

where x is the year, as illustrated in the accompanying figure. (*Source:* Department of Health and Human Services.)

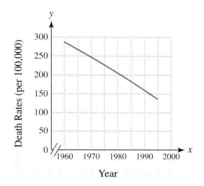

   (a) Evaluate $f(1985)$, using both the formula and the graph. How do your results compare?
   (b) Use the graph to estimate the years when this death rate was 250 or less.
   (c) Solve part (b) by using the quadratic formula.

56. *Accidental Deaths* From 1910 to 1996 the number of accidental deaths per 100,000 people generally decreased and can be modeled by

$$f(x) = -0.001918x^2 + 6.93x - 6156,$$

where x is the year, as shown in the accompanying figure. (*Source:* Department of Health and Human Services.)

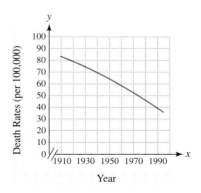

(a) Evaluate $f(1955)$, using both the formula and the graph. How do your results compare?

(b) Use the graph to estimate when this death rate was 60 or more.

(c) Solve part (b) by using the quadratic formula.

57. *Dimensions of a Pen* A rectangular pen for a pet is 5 feet longer than it is wide. Give possible values for the width w of the pen if its area must be between 176 and 500 square feet, inclusively.

58. *Dimensions of a Cylinder* The volume of a cylindrical can is given by $V = \pi r^2 h$, where r is its radius and h is its height. See the accompanying figure. If $h = 6$ inches and the volume of the can must be 50 cubic inches or more, estimate to the nearest tenth of an inch possible values for r.

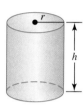

WRITING ABOUT MATHEMATICS

59. Consider the inequality $x^2 < 0$. Discuss the solutions to this inequality and explain your reasoning.

60. Explain how the graph of $y = ax^2 + bx + c$ can be used to solve the inequality

$$ax^2 + bx + c > 0$$

when $a < 0$. Assume that the x-intercepts of the graph are p and q with $p < q$.

11.6   EQUATIONS IN QUADRATIC FORM

Higher Degree Polynomial Equations · Equations Having Rational Exponents

INTRODUCTION

Although many equations are *not* quadratic equations, they can sometimes be put into quadratic form. These equations are *reducible to quadratic form*. To express such an equation in quadratic form we often use substitution. In this section we discuss this process.

HIGHER DEGREE POLYNOMIAL EQUATIONS

Sometimes a fourth degree polynomial can be factored like a quadratic trinomial, provided it does not have an x-term or an x^3-term. Let's consider the equation $x^4 - 5x^2 + 4 = 0$.

$$x^4 - 5x^2 + 4 = 0 \qquad \text{Given equation}$$
$$(x^2)^2 - 5(x^2) + 4 = 0 \qquad \text{Properties of exponents}$$

We use the substitution $u = x^2$.

$$u^2 - 5u + 4 = 0 \qquad \text{Let } u = x^2.$$
$$(u - 4)(u - 1) = 0 \qquad \text{Factor.}$$
$$u - 4 = 0 \quad \text{or} \quad u - 1 = 0 \qquad \text{Zero-product property}$$
$$u = 4 \quad \text{or} \quad u = 1 \qquad \text{Solve each equation.}$$

Because the given equation uses the variable x, we must give the solutions in terms of x. We substitute x^2 for u and then solve to obtain the following four solutions.

$$x^2 = 4 \quad \text{or} \quad x^2 = 1 \qquad \text{Substitute } x^2 \text{ for } u.$$

$$x = \pm 2 \quad \text{or} \quad x = \pm 1 \qquad \text{Square root property}$$

The solutions are -2, -1, 1, and 2.

In the next example we solve a sixth degree polynomial equation.

EXAMPLE 1   Solving equations by substitution

Solve $2x^6 + x^3 = 1$.

Solution   Start by subtracting 1 from each side.

$$2x^6 + x^3 - 1 = 0 \qquad \text{Subtract 1.}$$

$$2(x^3)^2 + (x^3) - 1 = 0 \qquad \text{Properties of exponents}$$

$$2u^2 + u - 1 = 0 \qquad \text{Let } u = x^3.$$

$$(2u - 1)(u + 1) = 0 \qquad \text{Factor.}$$

$$2u - 1 = 0 \quad \text{or} \quad u + 1 = 0 \qquad \text{Zero-product property}$$

$$u = \frac{1}{2} \quad \text{or} \quad u = -1 \qquad \text{Solve.}$$

Now substitute x^3 for u, and solve for x to obtain the following two solutions.

$$x^3 = \frac{1}{2} \quad \text{or} \quad x^3 = -1 \qquad \text{Substitute } x^3 \text{ for } u.$$

$$x = \sqrt[3]{\frac{1}{2}} \quad \text{or} \quad x = -1 \qquad \text{Take cube root of each side.}$$

EQUATIONS HAVING RATIONAL EXPONENTS

Equations that have rational exponents are sometimes reducible to quadratic form. Consider the following example, in which two solutions are presented.

EXAMPLE 2   Solving an equation having negative exponents

Solve $-6m^{-2} + 13m^{-1} + 5 = 0$.

Solution   **Solution I**   Use the substitution $u = m^{-1} = \frac{1}{m}$ and $u^2 = m^{-2} = \frac{1}{m^2}$.

$$-6m^{-2} + 13m^{-1} + 5 = 0 \qquad \text{Given equation}$$

$$-6u^2 + 13u + 5 = 0 \qquad \text{Let } u = m^{-1} \text{ and } u^2 = m^{-2}.$$

$$6u^2 - 13u - 5 = 0 \qquad \text{Multiply by } -1.$$

$$(2u - 5)(3u + 1) = 0 \qquad \text{Factor.}$$

$$2u - 5 = 0 \quad \text{or} \quad 3u + 1 = 0 \qquad \text{Zero-product property}$$

$$u = \frac{5}{2} \quad \text{or} \quad u = -\frac{1}{3} \qquad \text{Solve for } u.$$

Because $u = \frac{1}{m}$, $m = \frac{1}{u}$. Thus $m = \frac{2}{5}$ or $m = -3$.

Solution II Another way to solve this equation is to multiply each side by the LCD, m^2.

$$-6m^{-2} + 13m^{-1} + 5 = 0 \qquad \text{Given equation}$$
$$m^2(-6m^{-2} + 13m^{-1} + 5) = m^2 \cdot 0 \qquad \text{Multiply by } m^2.$$
$$-6m^2m^{-2} + 13m^2m^{-1} + 5m^2 = 0 \qquad \text{Distributive property}$$
$$-6 + 13m + 5m^2 = 0 \qquad \text{Add exponents.}$$
$$5m^2 + 13m - 6 = 0 \qquad \text{Rewrite the equation.}$$
$$(5m - 2)(m + 3) = 0 \qquad \text{Factor.}$$
$$5m - 2 = 0 \quad \text{or} \quad m + 3 = 0 \qquad \text{Zero-product property}$$
$$m = \frac{2}{5} \quad \text{or} \quad m = -3 \qquad \text{Solve.}$$

In the next example we solve an equation having fractional exponents.

EXAMPLE 3   **Solving an equation having rational exponents**

Solve $x^{2/3} - 2x^{1/3} - 8 = 0$.

Solution   Use the substitution $u = x^{1/3}$.

$$x^{2/3} - 2x^{1/3} - 8 = 0 \qquad \text{Given equation}$$
$$(x^{1/3})^2 - 2(x^{1/3}) - 8 = 0 \qquad \text{Rewrite the equation.}$$
$$u^2 - 2u - 8 = 0 \qquad \text{Let } u = x^{1/3}.$$
$$(u - 4)(u + 2) = 0 \qquad \text{Factor.}$$
$$u - 4 = 0 \quad \text{or} \quad u + 2 = 0 \qquad \text{Zero-product property}$$
$$u = 4 \quad \text{or} \quad u = -2 \qquad \text{Solve.}$$

Because $u = x^{1/3}$, $u^3 = (x^{1/3})^3 = x$. Thus $x = 4^3 = 64$ or $x = (-2)^3 = -8$.

11.6   PUTTING IT ALL TOGETHER

The following table demonstrates how to reduce some types of equations to quadratic form.

Equation	Substitution	Examples
Higher Degree Polynomial	Let $u = x^n$ for some integer n.	To solve $x^4 - 3x^2 - 4 = 0$, let $u = x^2$. This equation becomes $$u^2 - 3u - 4 = 0.$$
Rational Exponents	Pick a substitution that reduces the equation to quadratic form.	To solve $n^{-2} + 6n^{-1} + 9 = 0$, let $u = n^{-1}$. This equation becomes $$u^2 + 6u + 9 = 0.$$ To solve $6x^{2/5} - 5x^{1/5} - 4 = 0$, let $u = x^{1/5}$. This equation becomes $$6u^2 - 5u - 4 = 0.$$

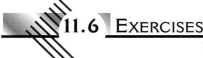

11.6 EXERCISES

FOR EXTRA HELP

Exercises 1–6: Use the given substitution to solve the equation.

1. $x^4 - 7x^2 + 6 = 0$ $u = x^2$

2. $2k^4 - 7k^2 + 6 = 0$ $u = k^2$

3. $3z^6 + z^3 - 10 = 0$ $u = z^3$

4. $2x^6 + 17x^3 + 8 = 0$ $u = x^3$

5. $4n^{-2} + 17n^{-1} + 15 = 0$ $u = n^{-1}$

6. $m^{-2} + 24 = 10m^{-1}$ $u = m^{-1}$

Exercises 7–24: Solve the equation. Find all real solutions.

7. $x^4 = 8x^2 + 9$ **8.** $3x^4 = 10x^2 + 8$

9. $3x^6 - 5x^3 - 2 = 0$ **10.** $6x^6 + 11x^3 + 4 = 0$

11. $2z^{-2} + 11z^{-1} = 40$ **12.** $z^{-2} - 10z^{-1} + 25 = 0$

13. $x^{2/3} - 2x^{1/3} + 1 = 0$ **14.** $3x^{2/3} + 18x^{1/3} = 48$

15. $x^{2/5} - 33x^{1/5} + 32 = 0$

16. $x^{2/5} - 80x^{1/5} - 81 = 0$

17. $x - 13\sqrt{x} + 36 = 0$

18. $x - 17\sqrt{x} + 16 = 0$

19. $z^{1/2} - 2z^{1/4} + 1 = 0$

20. $z^{1/2} - 4z^{1/4} + 4 = 0$

21. $(x + 1)^2 - 5(x + 1) - 14 = 0$

22. $2(x - 5)^2 + 5(x - 5) + 3 = 0$

23. $(x^2 - 1)^2 - 4 = 0$

24. $(x^2 - 9)^2 - 8(x^2 - 9) + 16 = 0$

WRITING ABOUT MATHEMATICS

25. Explain how to solve $ax^4 - bx^2 + c = 0$. Assume that the left side of the equation factors.

26. Explain what it means for an equation to be reducible to quadratic form.

CHECKING BASIC CONCEPTS   SECTIONS 11.5 AND 11.6

1. Solve the quadratic inequality

$$x^2 - x - 6 > 0.$$

Write your answer in interval notation.

2. Solve the quadratic inequality

$$3x^2 + 5x + 2 \le 0.$$

Write your answer in interval notation.

3. Solve $x^6 + 6x^3 - 16 = 0$.

4. Solve $x^{2/3} - 7x^{1/3} - 8 = 0$.

CHAPTER 11 Summary

Quadratic Function   Any quadratic function f can be written as

$$f(x) = ax^2 + bx + c \quad (a \neq 0).$$

Graph of a Quadratic Function   Its graph is ∪-shaped and called a parabola. If $a > 0$, the parabola opens upward; if $a < 0$, the parabola opens downward. The vertex of a parabola is either the lowest point on a parabola that opens upward or the highest point on a parabola that opens downward. The x-coordinate of the vertex is given by $x = -\frac{b}{2a}$.

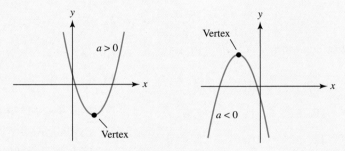

Axis of Symmetry   The axis of symmetry is a vertical line that passes through the vertex of the graph of a quadratic function. If the vertex is (h, k), then the axis of symmetry is $x = h$. The parabola is symmetric with respect to this line.

Vertical and Horizontal Translations   Let h and k be positive numbers.

To graph	*shift the graph of $y = x^2$ by k units*
$y = x^2 + k$	upward.
$y = x^2 - k$	downward.

To graph	*shift the graph of $y = x^2$ by h units*
$y = (x - h)^2$	right.
$y = (x + h)^2$	left.

Example: Compared to $y = x^2$, the graph of $y = (x - 1)^2 + 2$ is translated right 1 unit and upward 2 units.

Vertex Form   Any quadratic function can be expressed in vertex form as

$$f(x) = a(x - h)^2 + k.$$

In this form the point (h, k) is the vertex. A quadratic function can be put in this form by completing the square.

Example: $\quad y = x^2 + 10x - 4 \qquad\qquad$ Given equation

$\qquad\qquad = (x^2 + 10x + 25) - 25 - 4 \qquad \left(\frac{b}{2}\right)^2 = \left(\frac{10}{2}\right)^2 = 25$

$\qquad\qquad = (x + 5)^2 - 29$

The vertex is $(-5, -29)$.

Section 11.3 *Quadratic Equations*

Quadratic Equations   Any quadratic equation can be written as $ax^2 + bx + c = 0$ and can have zero, one, or two real solutions. These solutions correspond to the x-intercepts on the graph of $y = ax^2 + bx + c$. These equations can be solved by factoring or by completing the square.

Example: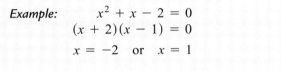

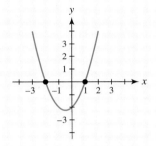

The x-intercepts for $y = x^2 + x - 2$ are -2 and 1.

Completing the Square   To solve $x^2 + bx = d$ by completing the square add $\left(\frac{b}{2}\right)^2$ to each side of the equation.

Section 11.4 *The Quadratic Formula*

The Quadratic Formula   The solutions to $ax^2 + bx + c = 0$ $(a \neq 0)$ are given by

$$x = \frac{-b \pm \sqrt{b^2 - 4ac}}{2a}.$$

Example: Solve $2x^2 + 3x - 1 = 0$ by letting $a = 2, b = 3$, and $c = -1$.

$$x = \frac{-3 \pm \sqrt{3^2 - 4(2)(-1)}}{2(2)} = \frac{-3 \pm \sqrt{17}}{4} \approx 0.28, -1.78$$

The Discriminant   The expression $b^2 - 4ac$ evaluates to a real number and is called the discriminant. If $b^2 - 4ac > 0$, there are two real solutions; if $b^2 - 4ac = 0$, there is one real solution; and if $b^2 - 4ac < 0$, there are no real solutions, rather there are two complex solutions.

Example: For $2x^2 + 3x - 1 = 0$, the discriminant is

$$b^2 - 4ac = 3^2 - 4(2)(-1) = 17 > 0.$$

There are two real solutions to this quadratic equation.

Section 11.5   *Quadratic Inequalities*

Quadratic Inequalities   When the equals sign in a quadratic equation is replaced with $<$, $>$, $\leq$, or $\geq$, a quadratic inequality results. For example,

$$3x^2 - x + 1 = 0$$

is a quadratic equation and

$$3x^2 - x + 1 > 0$$

is a quadratic inequality. Like quadratic equations, quadratic inequalities can be solved symbolically, graphically, and numerically. An important first step in solving a quadratic inequality is to solve the corresponding quadratic equation.

Examples: The solutions to $x^2 - 5x - 6 = 0$ are $x = -1, 6$.

The solutions to $x^2 - 5x - 6 < 0$ satisfy $-1 < x < 6$.

The solutions to $x^2 - 5x - 6 > 0$ satisfy $x < -1$ or $x > 6$.

Section 11.6   *Equations in Quadratic Form*

Equations Reducible to Quadratic Form   An equation that is not quadratic, but can be put into quadratic form by using a substitution is said to be reducible to quadratic form.

Example: To solve $x^{2/3} - 2x^{1/3} - 15 = 0$ let $u = x^{1/3}$. This equation becomes

$$u^2 - 2u - 15 = 0.$$

Factoring results in $(u + 3)(u - 5) = 0$, or $u = -3$ or 5. Because $u = x^{1/3}$, $x = u^3$ and $x = (-3)^3 = -27$ or $x = (5)^3 = 125$.

CHAPTER 11 Review Exercises

SECTION 11.1

Exercises 1 and 2: Identify the vertex, axis of symmetry, and whether the parabola opens upward or downward. State where the graph is increasing and where it is decreasing.

1.

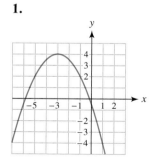

2.

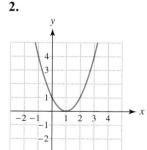

Exercises 3–6: Do the following.
(a) *Graph f.*
(b) *Use the graph to identify the vertex and axis of symmetry.*
(c) *Evaluate f(x) at the given value of x.*

3. $f(x) = x^2 - 2,$ $\qquad x = -1$

4. $f(x) = -x^2 + 4x - 3,$ $\quad x = 3$

5. $f(x) = -\frac{1}{2}x^2 + x + \frac{3}{2},$ $\quad x = -2$

6. $f(x) = 2x^2 + 8x + 5,$ $\qquad x = -3$

7. Find the minimum *y*-value located on the graph of $y = 2x^2 - 6x + 1.$

8. Find the maximum *y*-value located on the graph of $y = -3x^2 + 2x - 5.$

Exercises 9–12: Find the vertex of the parabola.

9. $f(x) = x^2 - 4x - 2$

10. $f(x) = 5 - x^2$

11. $f(x) = -\frac{1}{4}x^2 + x + 1$

12. $f(x) = 2 + 2x + x^2$

SECTION 11.2

Exercises 13–18: Do the following.
(a) *Graph f.*
(b) *Compare the graph of f with the graph of $y = x^2$.*

13. $f(x) = x^2 + 2$ $\qquad$ **14.** $f(x) = 3x^2$

15. $f(x) = (x - 2)^2$ $\qquad$ **16.** $f(x) = (x + 1)^2 - 3$

17. $f(x) = \frac{1}{2}(x + 1)^2 + 2$

18. $f(x) = 2(x - 1)^2 - 3$

19. Write the vertex form of a parabola with $a = -4$ and vertex $(2, -5)$.

20. Write the vertex form of a parabola that opens downward with vertex $(-4, 6)$. Assume that $a = \pm 1$.

Exercises 21–24: Write the equation in vertex form. Identify the vertex.

21. $y = x^2 + 4x - 7$ $\qquad$ **22.** $y = x^2 - 7x + 1$

23. $y = 2x^2 - 3x - 8$ $\qquad$ **24.** $y = 3x^2 + 6x - 2$

Exercises 25 and 26: Find a value for the constant a so that $f(x) = ax^2 - 1$ models the data.

25.

x	1	2	3
f(x)	2	11	26

26.

x	-1	0	1
f(x)	$-\frac{3}{4}$	-1	$-\frac{3}{4}$

SECTION 11.3

Exercises 27–30: Use the graph of
$$y = ax^2 + bx + c$$
to solve $ax^2 + bx + c = 0$.

27. $\qquad\qquad\qquad$ **28.**

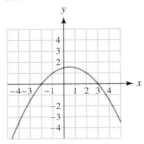 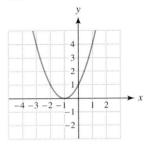

29. $\qquad\qquad\qquad$ **30.**

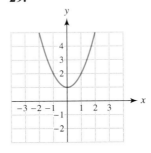 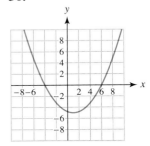

Exercises 31 and 32: Use the table of
$$y = ax^2 + bx + c$$
to solve $ax^2 + bx + c = 0$.

31. $\qquad\qquad\qquad$ **32.**

X	Y1
-20	250
-15	100
-10	0
-5	-50
0	-50
5	0
10	100

Y1■X^2+5X−50

X	Y1
-.75	2
-.5	0
-.25	-1
0	-1
.25	0
.5	2
.75	5

Y1■8X^2+2X−1

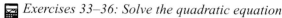 *Exercises 33–36: Solve the quadratic equation*

    (a) *graphically and*
    (b) *numerically.*

33. $x^2 - 5x - 50 = 0$ **34.** $\frac{1}{2}x^2 + x - \frac{3}{2} = 0$

35. $\frac{1}{4}x^2 + \frac{1}{2}x = 2$ **36.** $\frac{1}{2}x + \frac{3}{4} = \frac{1}{4}x^2$

Exercises 37–40: Solve the equation by factoring.

37. $x^2 + x - 20 = 0$ **38.** $x^2 + 11x + 24 = 0$

39. $15x^2 - 4x - 4 = 0$ **40.** $7x^2 - 25x + 12 = 0$

Exercises 41–44: Use the square root property to solve the equation.

41. $x^2 = 100$ **42.** $3x^2 = \frac{1}{3}$

43. $4x^2 - 6 = 0$ **44.** $5x^2 = x^2 - 4$

Exercises 45–48: Solve the equation by completing the square.

45. $x^2 + 6x = -2$ **46.** $x^2 - 4x = 6$

47. $x^2 - 2x - 5 = 0$ **48.** $2x^2 + 6x - 1 = 0$

Exercises 49 and 50: Solve the equation for the specified variable.

49. $F = \dfrac{k}{(R + r)^2}$ for R **50.** $2x^2 + 3y^2 = 12$ for y

SECTION 11.4

Exercises 51–56: Solve the equation, using the quadratic formula.

51. $x^2 - 9x + 18 = 0$ **52.** $x^2 - 24x + 143 = 0$

53. $6x^2 + x = 1$ **54.** $5x^2 + 1 = 5x$

55. $x(x - 8) = 5$ **56.** $2x(2 - x) = 3 - 2x$

Exercises 57–60: A graph of $y = ax^2 + bx + c$ is shown.
    (a) *State whether $a > 0$ or $a < 0$.*
    (b) *Solve $ax^2 + bx + c = 0$.*
    (c) *Determine whether the discriminant is positive, negative, or zero.*

57. 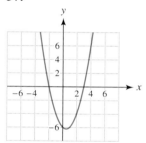   **58.**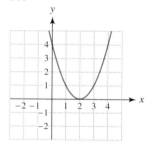

59. 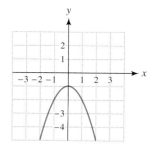   **60.**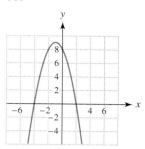

Exercises 61–64: Complete the following for the given equation.
    (a) *Evaluate the discriminant.*
    (b) *How many real solutions are there?*
    (c) *Support your answer for part (b) graphically.*

61. $2x^2 - 3x + 1 = 0$ **62.** $7x^2 + 2x - 5 = 0$

63. $3x^2 + x + 2 = 0$

64. $4.41x^2 - 12.6x + 9 = 0$

Exercises 65–68: Solve the equation. Write complex solutions in standard form.

65. $x^2 + x + 5 = 0$ **66.** $2x^2 + 8 = 0$

67. $2x^2 = x - 1$ **68.** $7x^2 = 2x - 5$

SECTION 11.5

Exercises 69 and 70: Use the graph of
$$y = ax^2 + bx + c$$
to solve each quadratic equation or inequality.

    (a) $ax^2 + bx + c = 0$
    (b) $ax^2 + bx + c < 0$
    (c) $ax^2 + bx + c > 0$

69.

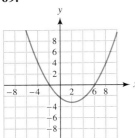

70.

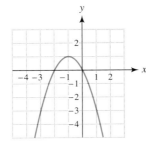

Exercises 71 and 72: Use the table of

$$y = ax^2 + bx + c$$

to solve each quadratic equation or inequality.
(a) $ax^2 + bx + c = 0$
(b) $ax^2 + bx + c < 0$
(c) $ax^2 + bx + c > 0$

71. $y = x^2 - 16$

x	−6	−4	−2	0	2	4	6
y	20	0	−12	−16	−12	0	20

72. $y = x^2 + x - 2$

x	−3	−2	−1	0	1	2	3
y	4	0	−2	−2	0	4	10

Exercises 73 and 74: Solve the quadratic equation in part (a) symbolically. Use the results to solve the inequalities in parts (b) and (c).

73. (a) $x^2 - 2x - 3 = 0$
(b) $x^2 - 2x - 3 < 0$
(c) $x^2 - 2x - 3 > 0$

74. (a) $2x^2 - 7x - 15 = 0$
(b) $2x^2 - 7x - 15 \le 0$
(c) $2x^2 - 7x - 15 \ge 0$

Exercises 75–78: Solve the quadratic inequality. Write your answer in interval notation.

75. $x^2 + 4x + 3 \le 0$

76. $5x^2 - 16x + 3 < 0$

77. $6x^2 - 13x + 2 > 0$

78. $x^2 \ge 5$

SECTION 11.6

Exercises 79–82: Solve the equation.

79. $x^4 - 14x^2 + 45 = 0$

80. $2z^{-2} + z^{-1} - 28 = 0$

81. $x^{2/3} - 9x^{1/3} + 8 = 0$

82. $(x - 1)^2 + 2(x - 1) + 1 = 0$

APPLICATIONS

83. *Construction*   A rain gutter is being fabricated from a flat sheet of metal so that the cross section of the gutter is a rectangle, as shown in the accompanying figure. The width of the metal sheet is 12 inches.

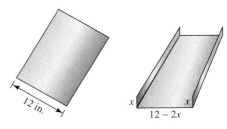

(a) Write a formula $f(x)$ that gives the area of the cross section.
(b) To hold the greatest amount of rainwater, the cross section should have maximum area. Find the dimensions that result in this maximum.

84. *Height of a Stone*   Suppose that a stone is thrown upward with an initial velocity of 44 feet per second (30 miles per hour) and is released 4 feet above the ground. Its height h in feet after t seconds is given by

$$h(t) = -16t^2 + 44t + 4.$$

(a) When does the stone reach a height of 32 feet?
(b) After how many seconds does the stone reach maximum height? Estimate this height.

85. *Maximizing Revenue*   Hotel rooms normally cost $90 per night. However, for a group rate the management is considering reducing the cost of a room by $3 for every room rented.
(a) Write a formula $f(x)$ that gives the revenue from renting x rooms at the group rate.
(b) Graph f in [0, 30, 5] by [0, 800, 100].
(c) How many rooms should be rented to yield revenue of $600?
(d) How many rooms should be rented to maximize revenue?

86. *Airline Complaints* In 2000, major U.S. airlines promised better customer service. From 1997 through 1999, the number of complaints per 100,000 passengers can be modeled by

$$f(x) = 0.4(x - 1997)^2 + 0.8,$$

where x represents the year. (*Source:* Department of Transportation.)
   (a) Evaluate $f(1999)$. Interpret the result.
    (b) Graph f in [1997, 1999, 1] by [0.5, 3, 0.5]. Discuss how complaints changed over this time period.

87. *Braking Distance* On dry pavement a safe braking distance d in feet for a car traveling x miles per hour is $d = \frac{x^2}{12}$. For each distance d, find x. (F. Mannering, *Principles of Highway Engineering and Traffic Control.*)
   (a) $d = 144$ feet   **(b)** $d = 300$ feet

88. *Numbers* The product of two numbers is 143. One number is 2 more than the other.
   (a) Write an equation whose solution gives the smaller number x.
   (b) Solve the equation.

 89. *Educational Attainment* From 1940 through 1991, the percentage of people with a high school diploma increased dramatically, as shown in the table.

Year	1940	1950	1960
H.S. Diploma (%)	25	34	44

Year	1970	1980	1991
H.S. Diploma (%)	55	69	78

Source: Bureau of the Census.

   (a) Plot the data.
   (b) Would it be reasonable to model these data with a linear function rather than a quadratic function? Explain your reasoning.
   (c) Find a function that models the data.

90. *U. S. Energy Consumption* From 1950 to 1970 per capita consumption of energy in millions of Btu can be modeled by

$$f(x) = \frac{1}{4}(x - 1950)^2 + 220,$$

where x is the year. (*Source:* Department of Energy.)

   (a) Find and interpret the vertex.
    (b) Graph f in [1950, 1970, 5] by [200, 350, 25]. What happened to energy consumption during this time period?
   (c) Use f to predict the consumption in 1996. Actual consumption was 354 million Btu. Did f provide a good model for 1996? Explain.

91. *Screen Dimensions* A square computer screen has an area of 123 square inches. Approximate its dimensions.

92. *Flying a Kite* A kite is being flown, as illustrated in the accompanying figure. If 130 feet of string have been let out, find the value of x.

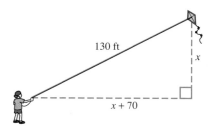

93. *Area* A uniform strip of grass is to be planted around a rectangular swimming pool, as illustrated in the accompanying figure. The swimming pool is 30 feet wide and 50 feet long. If there is only enough grass seed to cover 250 square feet, estimate the width x that the strip of grass should be.

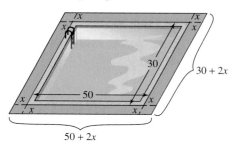

94. *Dimensions of a Cone* The volume V of a cone is given by $V = \frac{1}{3}\pi r^2 h$, where r is its base radius and h is its height. See the accompanying figure. If $h = 20$ inches and the volume of the cone must be between 750 and 1700 cubic inches, inclusively, estimate to the nearest tenth of an inch possible values for r.

1. Find the vertex and axis of symmetry for the graph of $f(x) = -\frac{1}{2}x^2 + x + 1$.

2. Find the minimum y-value located on the graph of $y = x^2 + 3x - 5$.

3. Find the exact value for the constant a so that $f(x) = ax^2 + 2$ models the data in the table.

x	2	0	2	4
$f(x)$	0	2	0	-6

4. Graph $f(x) = \frac{1}{2}(x - 3)^2 + 2$ and compare the graph of f to the graph of $y = x^2$.

5. Write $y = x^2 - 6x + 2$ in vertex form. Identify the vertex.

6. Use the graph of
$$y = ax^2 + bx + c$$
to solve $ax^2 + bx + c = 0$.

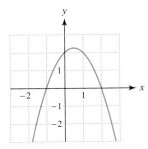

Exercises 7 and 8: Solve the quadratic equation.

7. $3x^2 + 11x - 4 = 0$ 8. $2x^2 = 2 - 6x^2$

9. Solve $x^2 - 8x = 1$ by completing the square.

10. Solve $x(-2x + 3) = -1$, using the quadratic formula.

11. A graph of $y = ax^2 + bx + c$ is shown at the top of the next column.
    (a) State whether $a > 0$ or $a < 0$.
    (b) Solve $ax^2 + bx + c = 0$.
    (c) Determine whether the discriminant is positive, negative, or zero.

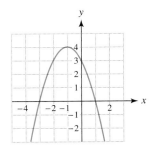

12. Complete the following for $-3x^2 + 4x - 5 = 0$.
    (a) Evaluate the discriminant.
    (b) How many real solutions are there?
    (c) Support your answer for part (b) graphically.

Exercises 13 and 14: Use the graph of
$$y = ax^2 + bx + c$$
to solve each equation or inequality.
    (a) $ax^2 + bx + c = 0$
    (b) $ax^2 + bx + c < 0$
    (c) $ax^2 + bx + c > 0$

13.   14.

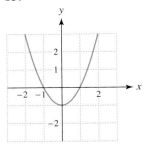

15. Solve the quadratic equation in part (a) symbolically. Use the result to solve the inequalities in parts (b) and (c) and write your answer in interval notation.
    (a) $8x^2 - 2x - 3 = 0$
    (b) $8x^2 - 2x - 3 \leq 0$
    (c) $8x^2 - 2x - 3 \geq 0$

16. Solve $x^6 - 3x^3 + 2 = 0$. Find all real solutions.

17. *Braking Distance*   On wet pavement a safe braking distance d in feet for a car traveling x miles per hour is $d = \frac{x^2}{9}$. What speed corresponds to a braking distance of 250 feet?   (F. Mannering, *Principles of Highway Engineering and Traffic Control.*)

18. *Construction*   A fence is being constructed along a 20-foot building, as shown in the accompanying figure. No fencing is used along the building.
(a) If 200 feet of fence are available, find a formula $f(x)$ that gives the area enclosed by the fence and building.
(b) What value of x gives the greatest area?

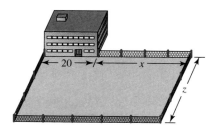

19. *Height of a Stone*   Suppose that a stone is thrown upward with an initial velocity of 88 feet per second (60 miles per hour) and is released 8 feet above the ground. Its height h in feet after t seconds is given by

$$h(t) = -16t^2 + 88t + 8.$$

 (a) Graph h in [0, 6, 1] by [0, 150, 50].
(b) When does the stone strike the ground?
(c) After how many seconds does the stone reach maximum height? Estimate this height.

CHAPTER 11 Extended and Discovery Exercises

MODELING DATA WITH A QUADRATIC FUNCTION

 1. *Survival Rate of Birds*   The survival rate of sparrowhawks varies according to their age. The following table summarizes the results of one study by listing the age in years and the percentage of birds that survived the previous year. For example, 52% of sparrowhawks that reached age 6 lived to be 7 years old. (*Source:* D. Brown and P. Rothery, *Models in Biology: Mathematics, Statistics and Computing.*)

Age	1	2	3	4	5
Percent (%)	45	60	71	67	67

Age	6	7	8	9
Percent (%)	61	52	30	25

(a) Try to explain the relationship between age and the likelihood of surviving the next year.
(b) Make a scatterplot of the data. What type of function might model the data? Explain your reasoning.

(c) Graph each function. Which of the following functions models the data better?

$$f_1(x) = -3.57x + 71.1$$
$$f_2(x) = -2.07x^2 + 17.1x + 33$$

(d) Use one of these functions to estimate the likelihood of a 5.5-year-old sparrowhawk surviving for 1 more year.

 2. *Photosynthesis and Temperature*   Photosynthesis is the process by which plants turn sunlight into energy. At very cold temperatures photosynthesis may halt even though the sun is shining. In one study the efficiency of photosynthesis for an Antarctic species of grass was investigated. The following table lists results for various temperatures. The temperature x is in degrees Celsius, and the efficiency y is given as a percent. The purpose of the research was to determine the temperature at which photosynthesis is most efficient.   (*Source:* D. Brown.)

x (°C)	−1.5	0	2.5	5	7	10	12
y (%)	33	46	55	80	87	93	95

x (°C)	15	17	20	22	25	27	30
y (%)	91	89	77	72	54	46	34

(a) Plot the data.

(b) What type of function might model these data? Explain your reasoning.

(c) Find a function f that models the data.

(d) Use f to estimate the temperature at which photosynthesis is most efficient in this type of grass.

TRANSLATIONS OF PARABOLAS IN COMPUTER GRAPHICS

Exercises 3 and 4: In video games with two-dimensional graphics, the background is often translated to give the illusion that a character in the game is moving. The simple scene on the left shows a mountain and an airplane. To make it appear that the airplane is flying, the mountain can be translated to the left, as shown in the figure on the right. (**Reference:** C. Pokorny and C. Gerald, *Computer Graphics.*)

 3. *Video Games*   Suppose that the mountain in the figure on the left is modeled by $f(x) = -0.4x^2 + 4$ and that the airplane is located at the point $(1, 5)$.

(a) Graph f in $[-4, 4, 1]$ by $[0, 6, 1]$, where the units are kilometers. Plot the point $(1, 5)$ to show the location of the airplane.

(b) Assume that the airplane is moving horizontally to the right at 0.2 kilometer per second. To give a video game player the illusion that the airplane is moving, graph the image of the mountain and the position of the airplane after 10 seconds.

 4. *Video Games*   (Refer to Exercise 3.) Discuss how you could create the illusion of the airplane moving to the left and gaining altitude as it passes over the mountain. Try to perform a translation of this type. Explain your reasoning.

Exercises 5–8:   ***Factoring and the Discriminant***   *If the discriminant of the trinomial $ax^2 + bx + c$ with integer coefficients is a perfect square, then it can be factored. For example, on the one hand, the discriminant of $6x^2 + x - 2$ is*

$$1^2 - 4(6)(-2) = 49,$$

which is a perfect square ($7^2 = 49$), so we can factor the trinomial as

$$6x^2 + x - 2 = (2x - 1)(3x + 2).$$

On the other hand, the discriminant for $x^2 + x - 1$ is

$$1^2 - 4(1)(-1) = 5,$$

which is not a perfect square, so we cannot factor this trinomial by using integers as coefficients. Similarly, if the discriminant is negative, the trinomial cannot be factored by using integer coefficients. Use the discriminant to predict whether the trinomial can be factored. Then test your prediction.

5. $10x^2 - x - 3$ **6.** $4x^2 - 3x - 6$

7. $3x^2 + 2x - 2$ **8.** $2x^2 + x + 3$

Exercises 9–14:   ***Polynomial Inequalities***   *The solution set for a polynomial inequality can be found by first determining the boundary numbers. For example, to solve $f(x) = x^3 - 4x > 0$ begin by solving $x^3 - 4x = 0$. The solutions (boundary numbers) are $-2, 0,$ and 2. The function $f(x) = x^3 - 4x$ is either only positive or only negative on intervals between consecutive zeros. To determine the solution set, we can evaluate test values for each interval as shown on the left below.*

Interval	Test Value	$f(x) = x^3 - 4x$
$(-\infty, -2)$	$x = -3$	$f(-3) = -15 < 0$
$(-2, 0)$	$x = -1$	$f(-1) = 3 > 0$
$(0, 2)$	$x = 1$	$f(1) = -3 < 0$
$(2, \infty)$	$x = 3$	$f(3) = 15 > 0$

$[-6, 6, 1]$ by $[-4, 4, 1]$

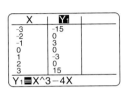 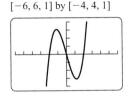

We can see that $f(x) > 0$ for $(-2, 0) \cup (2, \infty)$. These results also are supported graphically in the figure on the right above, where the graph of f is above the x-axis when $-2 < x < 0$ or when $x > 2$.

Use these concepts to solve the polynomial inequality.

9. $x^3 - x^2 - 6x > 0$

10. $x^3 - 3x^2 + 2x < 0$

11. $x^3 - 7x^2 + 14x \le 8$

12. $9x - x^3 \ge 0$

13. $x^4 - 5x^2 + 4 > 0$

14. $1 < x^4$

Exercises 15–20: ***Rational Inequalities*** *Rational inequalities can be solved using many of the same techniques that are used to solve other types of inequalities. However, there is one important difference. For a rational inequality, the boundary between greater than and less than can either be an x-value where equality occurs or an x-value where a rational expression is undefined. For example, consider the inequality $f(x) = \frac{2 - x}{2x} > 0$. The solution to the equation $\frac{2 - x}{2x} = 0$ is 2. The rational expression $\frac{2 - x}{2x}$ is undefined when $x = 0$. Therefore we select test values on the intervals $(-\infty, 0)$, $(0, 2)$, and $(2, \infty)$. The table reveals that $f(x) > 0$ for $(0, 2)$.*

Interval	Test Value	$f(x) = \dfrac{2 - x}{2x}$
$(-\infty, 0)$	$x = -0.5$	$f(-0.5) = -2.5 < 0$
$(0, 2)$	$x = 1$	$f(1) = 0.5 > 0$
$(2, \infty)$	$x = 2.5$	$f(2.5) = -0.1 < 0$

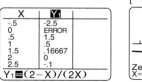

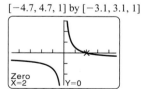

$[-4.7, 4.7, 1]$ by $[-3.1, 3.1, 1]$

Note that $f(x)$ changes from negative to positive at $x = 0$, where $f(x)$ is undefined. These results are supported graphically in the figure on the right above.

Solve the rational inequality.

15. $\dfrac{3 - x}{3x} \ge 0$

16. $\dfrac{x - 2}{x + 2} > 0$

17. $\dfrac{3 - 2x}{1 + x} < 3$

18. $\dfrac{x + 1}{4 - 2x} \ge 1$

19. $\dfrac{5}{x^2 - 4} < 0$

20. $\dfrac{x}{x^2 - 1} \ge 0$

Exponential and Logarithmic Functions

In 1900, the Swedish scientist Svante Arrhenius first predicted a greenhouse effect resulting from emissions of carbon dioxide by industrialized countries. His classic calculation made use of logarithms and predicted that a doubling of the carbon dioxide concentration in the atmosphere would raise the average global temperature by 7°F to 11°F. An increase in world population has resulted in higher emissions of greenhouse gases, such as carbon dioxide, and these emissions have the potential to alter Earth's climate and destroy portions of the ozone layer.

In this chapter we use exponential and logarithmic functions to model a wide variety of phenomena, such as greenhouse gases, acid rain, the decline of the bluefin tuna, the demand for liver transplants, diversity of bird species, hurricanes, and earthquakes. Mathematics plays a key role in understanding, controlling, and predicting natural phenomena and people's effect on them.

Human history becomes more and more a race between education and catastrophe.
—H. G. Wells

Source: M. Kraljic, *The Greenhouse Effect.*

12.1 COMPOSITE AND INVERSE FUNCTIONS

**Composition of Functions · One-to-One Functions · Inverse Functions ·
Tables and Graphs of Inverse Functions**

INTRODUCTION

Suppose that you walk into a classroom, turn on the lights, and sit down at your desk. How could you undo or reverse these actions? You might stand up from the desk, turn off the lights, and walk out of the classroom. Note that you must not only perform the "inverse" of each action, but you also must do them in the *reverse order*. In mathematics, we undo an arithmetic operation by performing its inverse operation. For example, the inverse operation of addition is subtraction, and the inverse operation of multiplication is division. In this section, we explore these concepts further by discussing inverse functions.

COMPOSITION OF FUNCTIONS

Many tasks in life are performed in *sequence*, such as putting on your socks and then your shoes. These types of situations also occur in mathematics. For example, suppose that we want to calculate the number of ounces in 3 tons. Because there are 2000 pounds in one ton, we might first multiply 3 by 2000 to obtain 6000 pounds. There are 16 ounces in a pound, so we could multiply 6000 by 16 to obtain 96,000 ounces. This particular calculation involves a *sequence* of calculations that can be represented by the diagram shown in Figure 12.1.

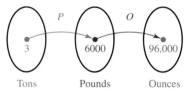

Tons   Pounds   Ounces

Figure 12.1

The results shown in Figure 12.1 can be calculated by using functions. Suppose that we let $P(x) = 2000x$ convert x tons to P pounds and also let $O(x) = 16x$ convert x pounds to O ounces. We can calculate the number of ounces in 3 tons by performing the *composition of O and P*. This method can be expressed symbolically as

$$(O \circ P)(3) = O\big(P(3)\big) \qquad \text{First compute } P(3).$$
$$= O(2000 \cdot 3) \qquad P(x) = 2000x$$
$$= O(6000) \qquad \text{Simplify.}$$
$$= 16(6000) \qquad O(x) = 16x$$
$$= 96{,}000. \qquad \text{Multiply.}$$

Note that the output for function P—namely, $P(3)$—becomes the input for function O. That is, we evaluate $O\big(P(3)\big)$ by calculating $P(3)$ first and then substituting the result of 6000 in function O.

COMPOSITION OF FUNCTIONS

If f and g are functions, then the **composite function** $g \circ f$, or **composition** of g and f, is defined by

$$(g \circ f)(x) = g\big(f(x)\big).$$

Note: We read $g\big(f(x)\big)$ as "g of f of x."

The compositions $g \circ f$ and $f \circ g$ represent evaluating functions f and g in a different order. When evaluating $g \circ f$, function f is performed first followed by function g, whereas for $f \circ g$ function g is performed first followed by function f. Note that in general $(g \circ f)(x) \neq (f \circ g)(x)$. That is, the order in which functions are applied makes a difference, the same way that putting on your socks and then your shoes is quite different from putting on your shoes and then your socks.

EXAMPLE 1   **Finding composite functions**

Evaluate $(g \circ f)(2)$ and then find a formula for $(g \circ f)(x)$.

(a) $f(x) = x^3, g(x) = 3x - 2$ (b) $f(x) = 5x, g(x) = x^2 - 3x + 1$

(c) $f(x) = \sqrt{2x}, g(x) = \dfrac{1}{x - 1}$

Solution   (a) $(g \circ f)(2) = g\big(f(2)\big)$ Composition of functions
$\qquad\qquad = g(\mathbf{8})$ $f(2) = 2^3 = 8$
$\qquad\qquad = 22$ $g(8) = 3(8) - 2 = 22$

Note that the output from f—namely, $f(2)$—becomes the input for g.

$\qquad (g \circ f)(x) = g\big(f(x)\big)$ Composition of functions
$\qquad\qquad = g(x^3)$ $f(x) = x^3$
$\qquad\qquad = 3x^3 - 2$ $g(x) = 3x - 2$

(b) $(g \circ f)(2) = g\big(f(2)\big)$ Composition of functions
$\qquad\qquad = g(\mathbf{10})$ $f(2) = 5(2) = 10$
$\qquad\qquad = 71$ $g(10) = 10^2 - 3(10) + 1 = 71$

$\qquad (g \circ f)(x) = g\big(f(x)\big)$ Composition of functions
$\qquad\qquad = g(\mathbf{5x})$ $f(x) = 5x$
$\qquad\qquad = (\mathbf{5x})^2 - 3(\mathbf{5x}) + 1$ $g(x) = x^2 - 3x + 1$
$\qquad\qquad = 25x^2 - 15x + 1$ Simplify.

(c) $(g \circ f)(2) = g\big(f(2)\big)$ Composition of functions
$\qquad\qquad = g(\mathbf{2})$ $f(2) = \sqrt{2(2)} = 2$
$\qquad\qquad = 1$ $g(2) = \dfrac{1}{2 - 1} = 1$

$\qquad (g \circ f)(x) = g\big(f(x)\big)$ Composition of functions
$\qquad\qquad = g(\sqrt{2x})$ $f(x) = \sqrt{2x}$

$\qquad\qquad = \dfrac{1}{\sqrt{2x} - 1}$ $g(x) = \dfrac{1}{x - 1}$

Composite functions can be evaluated both numerically and graphically, as demonstrated in the next two examples.

EXAMPLE 2   Evaluating composite functions with tables

Use Tables 12.1 and 12.2 to evaluate each expression.
(a) $(f \circ g)(2)$ **(b)** $(g \circ f)(3)$ **(c)** $(f \circ f)(0)$

TABLE 12.1

x	0	1	2	3
$f(x)$	3	2	0	1

TABLE 12.2

x	0	1	2	3
$g(x)$	1	3	2	0

Solution   **(a)** $(f \circ g)(2) = f\big(g(2)\big)$ Composition of functions
$\qquad\qquad\qquad = f(2)$ $g(2) = 2$
$\qquad\qquad\qquad = 0$ $f(2) = 0$

(b) $(g \circ f)(3) = g\big(f(3)\big)$ Composition of functions
$\qquad\qquad\qquad = g(1)$ $f(3) = 1$
$\qquad\qquad\qquad = 3$ $g(1) = 3$

(c) $(f \circ f)(0) = f\big(f(0)\big)$ Composition of functions
$\qquad\qquad\qquad = f(3)$ $f(0) = 3$
$\qquad\qquad\qquad = 1$ $f(3) = 1$

EXAMPLE 3   Evaluating composite functions graphically

Use Figure 12.2 to evaluate $(g \circ f)(2)$.

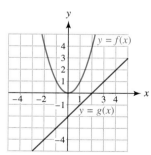

Figure 12.2

Solution   Because $(g \circ f)(2) = g\big(f(2)\big)$, start by using Figure 12.2 to evaluate $f(2)$. Figure 12.3(a) shows that $f(2) = 4$, which becomes the input for g. Figure 12.3(b) reveals that $g(4) = 2$. Thus

$$(g \circ f)(2) = g\big(f(2)\big) = g(4) = 2.$$

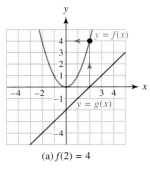

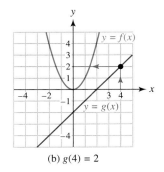

(a) $f(2) = 4$ (b) $g(4) = 2$

Figure 12.3

ONE-TO-ONE FUNCTIONS

If we change the input for a function, does the output also change? Do *different inputs* always result in *different outputs* for every function? The answer is no. For example, if $f(x) = x^2 + 1$, then the inputs –2 and 2 result in the *same* output, 5. That is, $f(-2) = 5$ and $f(2) = 5$. However, for $g(x) = 2x$, *different inputs* always result in *different outputs*. Thus we say that g is a *one-to-one function*, whereas f is not.

ONE-TO-ONE FUNCTION

A function f is **one-to-one** if, for any c and d in the domain of f,

$$c \neq d \quad \text{implies that} \quad f(c) \neq f(d).$$

That is, different inputs always result in different outputs.

One way to determine whether a function f is one-to-one is to look at its graph. Suppose that a function has two different inputs that result in the same output. Then there must be two points on its graph that have the same y-value but different x-values. For example, if $f(x) = 2x^2$, then $f(-1) = 2$ and $f(1) = 2$. Thus the points $(-1, 2)$ and $(1, 2)$ both lie on the graph of f, as shown in Figure 12.4(a). Two points with different x-values and the same y-value determine a horizontal line, as shown in Figure 12.4(b). This horizontal line intersects the graph of f more than once, indicating that different inputs do *not* always have different outputs. Thus $f(x) = 2x^2$ is *not* one-to-one.

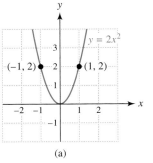

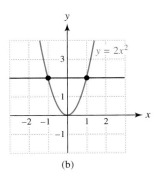

(a)   (b)

Figure 12.4

This discussion motivates the **horizontal line test**.

HORIZONTAL LINE TEST

If every horizontal line intersects the graph of a function f at most once, then f is a one-to-one function.

We apply the horizontal line test in the next example.

EXAMPLE 4   Using the horizontal line test

Determine whether each graph in Figure 12.5 represents a one-to-one function.

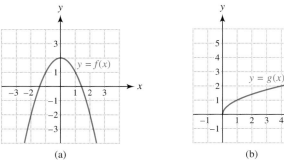

(a)   (b)

Figure 12.5

Solution   Figure 12.6(a) shows one of many horizontal lines that intersect the graph of $y = f(x)$ twice. Therefore function f is *not* one-to-one.

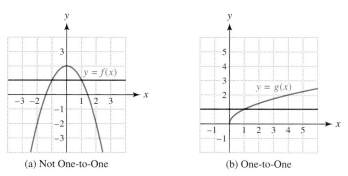

(a) Not One-to-One   (b) One-to-One

Figure 12.6

Figure 12.6(b) suggests that every horizontal line will intersect the graph of $y = g(x)$ *at most* once. Therefore function g is one-to-one.

INVERSE FUNCTIONS

Turning on a light and turning off a light are inverse operations from ordinary life. Inverse operations undo each other. In mathematics, adding **5** to x, and subtracting **5** from x are inverse operations because

$$x + 5 - 5 = x.$$

Similarly, multiplying x by 5 and dividing x by 5 are inverse operations because

$$\frac{5x}{5} = x.$$

In general, addition and subtraction are inverse operations and multiplication and division are inverse operations.

EXAMPLE 5   **Finding inverse operations**

State the inverse operations for each statement. Then write a function f for the given statement and a function g for its inverse operations.
(a) Divide x by 3.
(b) Cube x and then add 1 to the result.

Solution   **(a)** The inverse of dividing x by 3 is to *multiply x by 3*. Thus

$$f(x) = \frac{x}{3} \quad \text{and} \quad g(x) = 3x.$$

(b) When there is more than one operation, we must perform the inverse operations in *reverse order*. The inverse operations in reverse order are to subtract 1 from x and take the cube root of the result. Thus

$$f(x) = x^3 + 1 \quad \text{and} \quad g(x) = \sqrt[3]{x - 1}.$$

Functions f and g in Example 5 are examples of *inverse functions*. Note that, if $f(x) = \frac{x}{3}$ and $g(x) = 3x$, then

$$f(15) = 5 \quad \text{and} \quad g(5) = 15.$$

In general, if f and g are inverse functions, then $f(a) = b$ implies $g(b) = a$. Thus

$$(g \circ f)(a) = g\big(f(a)\big) = g(b) = a$$

for any a in the domain of f, whenever g and f are inverse functions. The composition of a function with its inverse leaves the input unchanged.

INVERSE FUNCTIONS

Let f be a one-to-one function. Then f^{-1} is the **inverse function** of f, if

$$(f^{-1} \circ f)(x) = f^{-1}\big(f(x)\big) = x, \quad \text{for every } x \text{ in the domain of } f, \quad \text{and}$$
$$(f \circ f^{-1})(x) = f\big(f^{-1}(x)\big) = x, \quad \text{for every } x \text{ in the domain of } f^{-1}.$$

Note: In the expression $f^{-1}(x)$, the -1 is *not* an exponent. That is, $f^{-1}(x) \neq \frac{1}{f(x)}$. Rather, if $f(x) = \frac{x}{3}$, then $f^{-1}(x) = 3x$ and, if $f(x) = x^3 + 1$, then $f^{-1}(x) = \sqrt[3]{x - 1}$.

EXAMPLE 6   Verifying inverses

Verify that $f^{-1}(x) = 3x$ if $f(x) = \frac{x}{3}$.

Solution   We must show that $(f^{-1} \circ f)(x) = x$ and that $(f \circ f^{-1})(x) = x$.

$$(f^{-1} \circ f)(x) = f^{-1}\big(f(x)\big) \qquad \text{Composition of functions}$$

$$= f^{-1}\left(\frac{x}{3}\right) \qquad f(x) = \frac{x}{3}$$

$$= 3\left(\frac{x}{3}\right) \qquad f^{-1}(x) = 3x$$

$$= x \qquad \text{Simplify.}$$

$$(f \circ f^{-1})(x) = f\big(f^{-1}(x)\big) \qquad \text{Composition of functions}$$

$$= f(3x) \qquad f^{-1}(x) = 3x$$

$$= \frac{3x}{3} \qquad f(x) = \frac{x}{3}$$

$$= x \qquad \text{Simplify.}$$

The definition of inverse functions states that f must be a one-to-one function. To understand why, consider Figure 12.7. In Figure 12.7(a) a one-to-one function f is represented by a diagram. To find f^{-1} the arrows are reversed. For example, $f(1) = 3$ implies that $f^{-1}(3) = 1$, the arrow from 1 to 3 for f must be redrawn from 3 to 1 for f^{-1}.

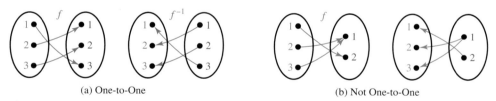

(a) One-to-One   (b) Not One-to-One

Figure 12.7

To be a *function* each input must correspond to exactly one output, which is the case in Figure 12.7(a). However, a different function f that is *not* a one-to-one function because inputs 2 and 3 both result in output 1 is shown in Figure 12.7(b). If the arrows for f are reversed to represent its inverse, then input 1 has two outputs, 2 and 3. Because no inverse *function* can satisfy both $f^{-1}(1) = 2$ and $f^{-1}(1) = 3$ at once, f^{-1} does not exist here.

The following steps can be used to find the inverse of a function symbolically.

FINDING AN INVERSE FUNCTION

To find f^{-1} for a one-to-one function f perform the following steps.

STEP 1: Let $y = f(x)$.

STEP 2: Interchange x and y.

STEP 3: Solve the formula for y. The resulting formula is $y = f^{-1}(x)$.

We apply these steps in the next example.

EXAMPLE 7 Finding an inverse function

Find the inverse of each one-to-one function.
(a) $f(x) = 3x - 7$ (b) $g(x) = (x + 2)^3$

Solution (a) **STEP 1:** Let $y = 3x - 7$.

STEP 2: Write the formula as $x = 3y - 7$.

STEP 3: To solve for y start by adding 7 to each side.

$$x + 7 = 3y \qquad \text{Add 7 to each side.}$$

$$\frac{x + 7}{3} = y \qquad \text{Divide each side by 3.}$$

Thus $f^{-1}(x) = \frac{x + 7}{3}$ or $f^{-1}(x) = \frac{1}{3}x + \frac{7}{3}$.

(b) **STEP 1:** Let $y = (x + 2)^3$.

STEP 2: Write the formula as $x = (y + 2)^3$.

STEP 3: To solve for y start by taking the cube root of each side.

$$\sqrt[3]{x} = y + 2 \qquad \text{Take cube root of each side.}$$
$$\sqrt[3]{x} - 2 = y \qquad \text{Subtract 2 from each side.}$$

Thus $g^{-1}(x) = \sqrt[3]{x} - 2$.

TABLES AND GRAPHS OF INVERSE FUNCTIONS

Inverse functions can be represented with tables and graphs. Table 12.3 shows a table of values for a function f.

TABLE 12.3

x	1	2	3	4	5
$f(x)$	3	6	9	12	15

Because $f(1) = 3$, $f^{-1}(3) = 1$. Similarly, $f(2) = 6$ implies that $f^{-1}(6) = 2$ and so on. Table 12.4 lists values for $f^{-1}(x)$.

TABLE 12.4

x	3	6	9	12	15
$f^{-1}(x)$	1	2	3	4	5

Note that the domain of f is {1, 2, 3, 4, 5} and that the range of f is {3, 6, 9, 12, 15}, whereas the domain of f^{-1} is {3, 6, 9, 12, 15} and the range of f^{-1} is {1, 2, 3, 4, 5}. *The domain of f is the range of f^{-1}, and the range of f is the domain of f^{-1}.* This statement is true in general for a function and its inverse.

If $f(a) = b$, then the point (a, b) lies on the graph of f. This statement also means that $f^{-1}(b) = a$ and that the point (b, a) lies on the graph of f^{-1}. These points are shown in Figure 12.8(a) with a blue line segment connecting them. The line $y = x$ is a perpendicular bisector of this line segment. As a result, the graph of f^{-1} can be obtained from the graph of f by reflecting the graph of f across the line $y = x$. For example, the graphs of a function f and its inverse are shown in Figure 12.8(b).

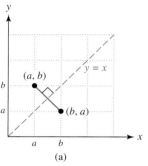 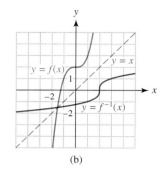

Figure 12.8

GRAPHS OF FUNCTIONS AND THEIR INVERSES

The graph of f^{-1} is a reflection of the graph of f across the line $y = x$.

EXAMPLE 8   Graphing an inverse function

The graph of $y = f(x)$ is shown in Figure 12.9.
(a) Sketch a graph of $y = f^{-1}(x)$.
(b) If $f(x) = 2x - 1$, find $f^{-1}(x)$ symbolically. Does this result agree with your graph?

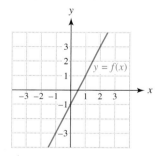

Figure 12.9

Solution   **(a)** The graph of $y = f^{-1}(x)$ is the reflection of the graph of $y = f(x)$ across the line $y = x$ and is shown in Figure 12.10.

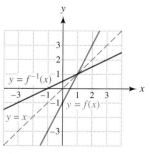

Figure 12.10

(b) Let $y = 2x - 1$ and interchange x and y to obtain $x = 2y - 1$. Solving for y gives $y = \frac{1}{2}x + \frac{1}{2}$. Thus $f^{-1}(x) = \frac{1}{2}x + \frac{1}{2}$, and its graph is a line with slope $\frac{1}{2}$ and y-intercept $\frac{1}{2}$. This result agrees with the graph of $y = f^{-1}(x)$ shown in Figure 12.10. ⎯⎯

12.1 PUTTING IT ALL TOGETHER

The following table summarizes some of the basic concepts about composite and inverse functions.

Concept	Explanation	Examples
Composite Functions	The composite of g and f is given by $$(g \circ f)(x) = g(f(x)),$$ and represents a *new* function whose name is $g \circ f$.	If $f(x) = 1 - 4x$ and $g(x) = x^3$, then $$(g \circ f)(x) = g(f(x))$$ $$= g(1 - 4x)$$ $$= (1 - 4x)^3.$$
One-to-One Functions	Function f is one-to-one if different inputs always give different outputs.	$f(x) = x^2$ is not one-to-one because $f(-4) = f(4) = 16$, whereas $g(x) = x + 1$ is one-to-one because, if two inputs differ, then adding 1 does not change this difference.
Horizontal Line Test	Used to determine whether a function is one-to-one from its graph	$f(x) = x^2$ is not one-to-one because a horizontal line can intersect its graph more than once.
Inverse Functions	f^{-1} will undo the operations performed by f. That is, $$(f^{-1} \circ f)(x) = x \quad \text{and}$$ $$(f \circ f^{-1})(x) = x.$$	If $f(x) = x^3$, then $f^{-1}(x) = \sqrt[3]{x}$ because cubing a number x and then taking its cube root results in the number x.

12.1 EXERCISES

FOR EXTRA HELP

 Student's Solutions Manual

 InterAct Math

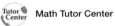 MathXL

MyMathLab

Math Tutor Center

Digital Video Tutor CD 5 Videotape 15

CONCEPTS

1. $(g \circ f)(7) = $ _____

2. $(f \circ g)(x) = $ _____

3. Does $(f \circ g)(x)$ always equal $(g \circ f)(x)$?

4. If a function f is one-to-one, then different _____ always result in different _____.

5. If $f(3) = 5$ and $f(7) = 5$, then could f be one-to-one?

6. If every horizontal line intersects the graph of f at most once, then f is _____.

7. The inverse operation of subtracting 10 is _____.

8. $(f^{-1} \circ f)(7) = $ _____

9. If $f(6) = 8$, then $f^{-1}($_____$) = $ _____.

10. If $f^{-1}(y) = x$, then $f($_____$) = $ _____.

11. For f to have an inverse function, f must be _____.

12. The graph of f^{-1} is a _____ of the graph of f across the line _____.

COMPOSITE FUNCTIONS

Exercises 13–22: For the given $f(x)$ and $g(x)$, find the following.

(a) $(g \circ f)(-2)$ (b) $(f \circ g)(4)$
(c) $(g \circ f)(x)$ (d) $(f \circ g)(x)$

13. $f(x) = x^2$ $g(x) = x + 3$

14. $f(x) = 4x^2$ $g(x) = 5x$

15. $f(x) = 2x$ $g(x) = x^3 - 1$

16. $f(x) = 3x + 1$ $g(x) = x^2 + 4x$

17. $f(x) = \frac{1}{2}x$ $g(x) = |x - 2|$

18. $f(x) = 6x$ $g(x) = \dfrac{2}{x - 5}$

19. $f(x) = \dfrac{1}{x}$ $g(x) = 3 - 5x$

20. $f(x) = \sqrt{x + 3}$ $g(x) = x^3 - 3$

21. $f(x) = 2x$ $g(x) = 4x^2 - 2x + 5$

22. $f(x) = 9x - \dfrac{1}{3x}$ $g(x) = \dfrac{x}{3}$

Exercises 23–28: Use the tables to evaluate the given expression.

x	-2	-1	0	1	2
$f(x)$	2	1	0	-1	-2

x	-2	-1	0	1	2
$g(x)$	0	1	-1	2	-2

23. (a) $(f \circ g)(0)$ (b) $(g \circ f)(-1)$

24. (a) $(f \circ g)(1)$ (b) $(g \circ f)(-2)$

25. (a) $(f \circ f)(-1)$ (b) $(g \circ g)(0)$

26. (a) $(g \circ g)(2)$ (b) $(f \circ f)(1)$

27. (a) $(f^{-1} \circ g)(-2)$ (b) $(g^{-1} \circ f)(2)$

28. (a) $(f \circ g^{-1})(1)$ (b) $(g \circ f^{-1})(-2)$

Exercises 29–30: Use the graph to evaluate each expression.

29. (a) $(f \circ g)(0)$
(b) $(g \circ f)(1)$
(c) $(f \circ f)(-1)$

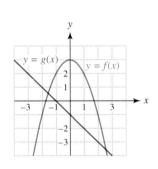

30. (a) $(f \circ g)(1)$
   (b) $(g \circ f)(-2)$
   (c) $(g \circ g)(-2)$

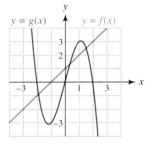

41.

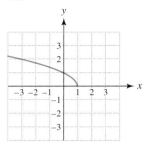

42.

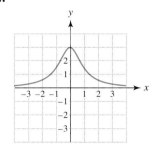

Exercises 31–36: Show that f is not one-to-one by finding two inputs that result in the same output. Answers may vary.

31. $f(x) = 5x^2$

32. $f(x) = 4 - x^2$

33. $f(x) = x^4 + 100$

34. $f(x) = \dfrac{x^2}{x^2 + 1}$

35. $f(x) = x^4 - 3x^2$

36. $f(x) = \sqrt{x^2 - 1}$

Exercises 37–42: Use the horizontal line test to determine whether the graph represents a one-to-one function.

37.

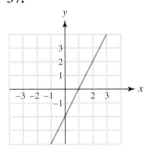

38.

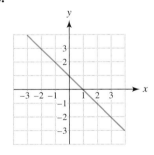

39.

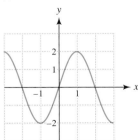

40.

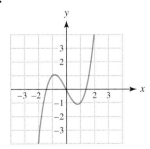

Exercises 43–50: (Refer to Example 5.) Give the inverse operation for the statement. Then write a function f for the given statement and a function g for its inverse.

43. Multiply x by 7.

44. Subtract 10 from x.

45. Add 5 to x and then divide the result by 2.

46. Multiply x by 6 and then add 8 to the result.

47. Multiply x by $\frac{1}{2}$ and then subtract 3 from the result.

48. Divide x by 10 and then add 20 to the result.

49. Cube the sum of x and 5.

50. Take the cube root of x and then subtract 2.

Exercises 51–58: (Refer to Example 6.) Verify that $f(x)$ and $f^{-1}(x)$ are indeed inverse functions.

51. $f(x) = 4x$ $\qquad$ $f^{-1}(x) = \dfrac{x}{4}$

52. $f(x) = \dfrac{2x}{3}$ $\qquad$ $f^{-1}(x) = \dfrac{3x}{2}$

53. $f(x) = 3x + 5$ $\qquad$ $f^{-1}(x) = \dfrac{x - 5}{3}$

54. $f(x) = x + 7$ $\qquad$ $f^{-1}(x) = x - 7$

55. $f(x) = x^3$ $\qquad$ $f^{-1}(x) = \sqrt[3]{x}$

56. $f(x) = \sqrt[3]{x - 4}$ $\qquad$ $f^{-1}(x) = x^3 + 4$

57. $f(x) = \dfrac{1}{x}$ $\qquad$ $f^{-1}(x) = \dfrac{1}{x}$

58. $f(x) = \dfrac{x + 7}{7}$ $\qquad$ $f^{-1}(x) = 7x - 7$

Exercises 59–74: (Refer to Example 7.) Find $f^{-1}(x)$.

59. $f(x) = 12x$

60. $f(x) = \frac{3}{4}x$

61. $f(x) = x + 8$

62. $f(x) = x - 3$

63. $f(x) = 5x - 2$

64. $f(x) = 3x + 4$

65. $f(x) = -\frac{1}{2}x + 1$

66. $f(x) = \frac{3}{4}x - \frac{1}{4}$

67. $f(x) = 8 - x$

68. $f(x) = 5 - x$

69. $f(x) = \dfrac{x + 1}{2}$

70. $f(x) = \dfrac{3 - x}{5}$

71. $f(x) = \sqrt[3]{2x}$

72. $f(x) = \sqrt[3]{x + 4}$

73. $f(x) = x^3 - 8$

74. $f(x) = (x - 5)^3$

Exercises 75–78: Use the table to make a table of values for $f^{-1}(x)$. State the domain and range for f and for f^{-1}.

75.

x	0	1	2	3	4
$f(x)$	0	5	10	15	20

76.

x	−4	−2	0	2	4
$f(x)$	1	2	3	4	5

77.

x	−5	0	5	10	15
$f(x)$	4	2	0	−2	−4

78.

x	0	2	4	6	8
$f(x)$	8	6	4	2	0

Exercises 79–82: (Refer to Example 8.) Use the graph of $y = f(x)$ to sketch a graph of $f^{-1}(x)$. Include the graph of f and the line $y = x$ in your graph.

79.

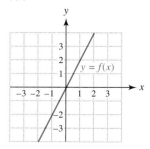

80.

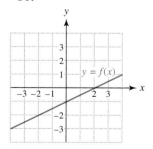

81.

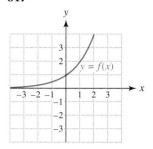

82.

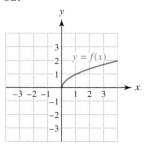

APPLICATIONS

83. *Circular Wave* A stone is dropped in a lake, creating a circular wave. The radius r of the wave in feet after t seconds is $r(t) = 2t$.
 (a) Its circumference C is given by $C(r) = 2\pi r$. Evaluate $(C \circ r)(5)$ and interpret your result.
 (b) Find $(C \circ r)(t)$.

84. *Volume of a Balloon* The volume V of a spherical balloon with radius r is given by $V(r) = \frac{4}{3}\pi r^3$. Suppose that the balloon is being inflated so that the radius in inches after t seconds is $r(t) = \sqrt[3]{t}$.
 (a) Evaluate $(V \circ r)(3)$ and interpret your result.
 (b) Find $(V \circ r)(t)$.

85. *Temperature and Mosquitoes* Temperature can affect the number of mosquitoes observed on a summer night. Graphs of two functions, T and M, are shown. Function T calculates the temperature on a summer evening h hours past midnight, and M calculates the number of mosquitoes observed per 100 square feet when the outside temperature is T.

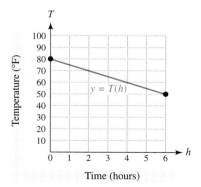

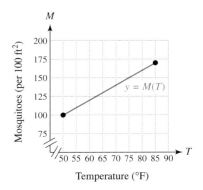

Temperature (°F)

(a) Find $T(1)$ and $M(75)$.
(b) Evaluate $(M \circ T)(1)$ and interpret your result.
(c) What does $(M \circ T)(h)$ calculate?
(d) Find equations for the lines in each graph.
(e) Use your answers from part (d) to write a formula for $(M \circ T)(h)$.

86. *Skin Cancer and Ozone*   Ozone in the stratosphere filters out most of the harmful ultraviolet (UV) rays from the sun. However, depletion of the ozone layer is affecting this protection. The formula $U(x) = 1.5x$ calculates the percent increase in UV radiation for an x percent decrease in the thickness of the ozone layer. The formula $C(x) = 3.5x$ calculates the percent increase in skin cancer cases when the UV radiation increases by x percent.   (*Source:* R. Turner, D. Pierce, and I. Bateman, *Environmental Economics.*)
(a) Evaluate $U(2)$ and $C(3)$ and interpret each result.
(b) Find $(C \circ U)(2)$ and interpret the result.
(c) Find $(C \circ U)(x)$. What does this formula calculate?

87. *College Degree*   The table lists the percentage P of people 25 or older completing 4 or more years of college during year x.

x	1960	1980	2000
$P(x)$	8	16	27

Source: Bureau of the Census.

(a) Evaluate $P(1980)$ and interpret the results.
(b) Make a table for $P^{-1}(x)$.
(c) Evaluate $P^{-1}(16)$.

88. *High School Grades*   The table lists the percentage of college freshmen with a high school grade average of A or A– during year x.

x	1970	1980	1990	2000
$P(x)$	20	26	29	43

Source: Department of Education.

(a) Evaluate $P(1970)$ and interpret the results.
(b) Make a table for $P^{-1}(x)$.
(c) Evaluate $P^{-1}(43)$.

89. *Temperature*   The function given by $f(x) = \frac{5}{9}x + 32$ converts x degrees Celsius to an equivalent temperature in degrees Fahrenheit.
(a) Is f a one-to-one function? Why or why not?
(b) Find a formula for f^{-1} and interpret what it calculates.

90. *Feet and Yards*   The function given by $f(x) = 3x$ converts x yards to feet.
(a) Is f a one-to-one function? Why or why not?
(b) Find a formula for f^{-1} and interpret what it calculates.

91. *Quarts and Gallons*   Write a function f that converts x gallons to quarts. Then find a formula for $f^{-1}(x)$ and interpret what it computes.

92. *One-to-One Function*   The table lists monthly average wind speeds at Hilo, Hawaii, in miles per hour from July through December, where x is the month.

x	July	Aug	Sept	Oct	Nov	Dec
$f(x)$	7	7	7	7	7	7

(a) Is function f one-to-one? Explain.
(b) Does f^{-1} exist?
(c) What happens if you try to make a table for f^{-1}?
(d) Could f be one-to-one if it were computed at a different location? What would have to be true about the monthly average wind speeds?

WRITING ABOUT MATHEMATICS

93. Explain the difference between $(g \circ f)(2)$ and $(f \circ g)(2)$. Are they always equal? If f and g are inverse functions, what are the values of $(g \circ f)(2)$ and $(f \circ g)(2)$?

94. Explain what it means for a function to be one-to-one.

12.2   EXPONENTIAL FUNCTIONS

Basic Concepts · Graphs of Exponential Functions ·
Models Involving Exponential Functions · The Natural Exponential Function

INTRODUCTION

Many times the growth of a quantity depends on the amount or number present. The more money deposited in an account, the more interest the account earns; that is, the interest earned is proportional to the amount of money in an account. For example, if a person has $100 in an account and receives 10% annual interest, the interest accrued the first year will be $10, and the balance in the account will be $110 at the end of 1 year. Similarly, if a person begins with $1000 in a similar account, the balance in the account will be $1100 at the end of 1 year. This type of growth is called *exponential growth* and can be modeled by an *exponential function.*

BASIC CONCEPTS

Suppose that an insect population doubles each week. Table 12.5 shows the populations after x weeks. Note that, as the population of insects becomes larger, the *increase* in population each week becomes greater. The population is increasing by 100%, or doubling numerically, each week. When a quantity increases by a constant percentage (or constant factor) at regular intervals, its growth is exponential.

TABLE 12.5

Week	0	1	2	3	4	5
Population	100	200	400	800	1600	3200

We can model the data in Table 12.5 by using the exponential function

$$f(x) = 100(2)^x.$$

For example,

$$f(0) = 100(2)^0 = 100 \cdot 1 = 100,$$
$$f(1) = 100(2)^1 = 100 \cdot 2 = 200,$$
$$f(2) = 100(2)^2 = 100 \cdot 4 = 400,$$

and so on. Note that the exponential function f has a *variable as an exponent.*

EXPONENTIAL FUNCTION

A function represented by

$$f(x) = Ca^x, \quad a > 0 \quad \text{and} \quad a \neq 1,$$

is an **exponential function with base a and coefficient C.** (Unless stated otherwise, assume that $C > 0$.)

In the formula $f(x) = Ca^x$, a is called the **growth factor** when $a > 1$ and the **decay factor** when $0 < a < 1$. For an exponential function, each time x increases by 1 unit $f(x)$ increases by a *factor* of a when $a > 1$ and decreases by a factor of a when $0 < a < 1$. Moreover, as

$$f(0) = Ca^0 = C(1) = C,$$

the value of C equals the value of $f(x)$ when $x = 0$. If x represents time, C represents the initial value of f when time equals 0. Figure 12.11 illustrates **exponential growth** and **exponential decay** for $x > 0$.

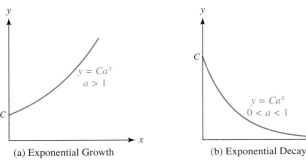

(a) Exponential Growth   (b) Exponential Decay

Figure 12.11

The set of valid inputs (domain) for an exponential function includes all real numbers. The set of corresponding outputs (range) includes all positive real numbers.

In the next example we evaluate some exponential functions. When evaluating an exponential function, we evaluate a^x before multiplying by C. This standard order of precedence is much like doing multiplication before addition. For example, $2(3)^2$ should be evaluated as

$$2(9) = 18 \quad not \ as \quad (6)^2 = 36.$$

EXAMPLE 1   Evaluating exponential functions

Evaluate $f(x)$ for the given value of x.

(a) $f(x) = 10(3)^x \qquad x = 2$ **(b)** $f(x) = 5\left(\frac{1}{2}\right)^x \qquad x = 3$

(c) $f(x) = \frac{1}{3}(2)^x \qquad x = -1$

Solution   **(a)** $f(2) = 10(3)^2 = 10 \cdot 9 = 90$

(b) $f(3) = 5\left(\frac{1}{2}\right)^3 = 5 \cdot \frac{1}{8} = \frac{5}{8}$

(c) $f(-1) = \frac{1}{3}(2)^{-1} = \frac{1}{3} \cdot \frac{1}{2} = \frac{1}{6}$

═══ MAKING CONNECTIONS ═══

The Expressions a^{-x} and $\left(\frac{1}{a}\right)^x$

Using properties of exponents, we can write 2^{-x} as

$$2^{-x} = \frac{1}{2^x} = \left(\frac{1}{2}\right)^x.$$

In general, the expressions a^{-x} and $\left(\frac{1}{a}\right)^x$ are equal.

In the next example, we determine whether a function is linear or exponential.

EXAMPLE 2 **Finding linear and exponential functions**

Use each table to determine whether f is a linear or an exponential function. Find a formula for f.

(a)

x	0	1	2	3	4
$f(x)$	16	8	4	2	1

(b)

x	0	1	2	3	4
$f(x)$	5	7	9	11	13

(c)

x	0	1	2	3	4
$f(x)$	1	3	9	27	81

Solution (a) Each time x increases by 1 unit, $f(x)$ decreases by a factor of $\frac{1}{2}$. Therefore f is an exponential function with a decay factor of $\frac{1}{2}$. Because $f(0) = 16$, $C = 16$ and so $f(x) = 16\left(\frac{1}{2}\right)^x$. This expression can also be written as $f(x) = 16(2)^{-x}$.

(b) Each time x increases by 1 unit, $f(x)$ increases by 2 units. Therefore f is a linear function, and the slope of its graph equals 2. The y-intercept is 5, so $f(x) = 2x + 5$.

(c) Each time x increases by 1 unit, $f(x)$ increases by a factor of 3. Therefore f is an exponential function with a growth factor of 3. Because $f(0) = 1, C = 1$ and so $f(x) = 1(3)^x$, or $f(x) = 3^x$.

MAKING CONNECTIONS

Linear and Exponential Functions

For a *linear function*, given by $f(x) = ax + b$, each time x increases by 1 unit y increases (or decreases) by a units, where a equals the slope of the graph of f.

For an *exponential function*, given by $f(x) = Ca^x$, each time x increases by 1 unit y increases by a factor of a when $a > 1$ and decreases by a factor of a when $0 < a < 1$. The constant a equals either the growth factor or the decay factor.

If $100 are deposited in a savings account paying 10% annual interest, the interest earned after 1 year equals $100 \times 0.10 = \$10$. The total amount of money in the account after 1 year is $100(1 + 0.10) = \$110$. Each year the money in the account increases by a factor of 1.10, so after x years there will be $100(1.10)^x$ dollars in the account. Thus compound interest is an example of exponential growth.

COMPOUND INTEREST

If C dollars are deposited in an account and if interest is paid at the end of each year with an annual rate of interest r, expressed in decimal form, then after x years the account will contain A dollars, where

$$A = C(1 + r)^x.$$

The growth factor is $(1 + r)$.

Note: The compound interest formula takes the form of an exponential function with

$$a = 1 + r.$$

EXAMPLE 3   Calculating compound interest

A 20-year-old worker deposits $2000 in a retirement account that pays 13% annual interest at the end of each year. How much money will be in the account when the worker is 65 years old? What is the growth factor?

Solution   Here, $C = 2000$, $r = 0.13$, and $x = 45$. The amount in the account after 45 years is

$$A = 2000(1 + 0.13)^{45} \approx \$489{,}282.80,$$

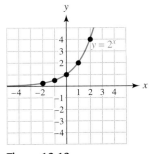

```
2000(1+.13)^45
       489282.8038
```

Figure 12.12

which is supported by Figure 12.12. In this dramatic example of exponential growth, $2000 grows to nearly half a million dollars in 45 years. Each year the amount of money on deposit is multiplied by a factor of $(1 + 0.13)$, so the growth factor is 1.13.

GRAPHS OF EXPONENTIAL FUNCTIONS

We can graph $f(x) = 2^x$ by first plotting some points, as in Table 12.6.

TABLE 12.6

x	−2	−1	0	1	2
2^x	$\frac{1}{4}$	$\frac{1}{2}$	1	2	4

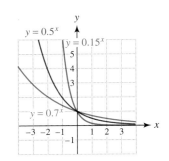

Figure 12.13

If we plot these points and sketch the graph, we obtain Figure 12.13. Note that, for negative values of x, $0 < 2^x < 1$ and that, for positive values of x, $2^x > 1$. The graph of $y = 2^x$ passes through the point $(0, 1)$, never intersects the x-axis, and always lies above the x-axis.

We can investigate the graphs of exponential functions further by graphing $y = 1.3^x$, $y = 1.7^x$, and $y = 2.5^x$ (see Figure 12.14). For $a > 1$ the graph of $y = a^x$ *increases* at a faster rate for larger values of a. We now graph $y = 0.7^x$, $y = 0.5^x$, and $y = 0.15^x$ (see Figure 12.15). Note that, if $0 < a < 1$, the graph of $y = a^x$ *decreases* more rapidly for smaller values of a. The graph of $y = a^x$ is *increasing* when $a > 1$ and *decreasing* when $0 < a < 1$ (from left to right).

Critical Thinking

Every graph of $y = a^x$ passes through what point? Why?

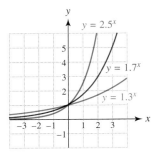

Figure 12.14

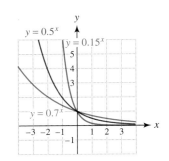

Figure 12.15

In the next example we show the dramatic difference between the outputs of linear and exponential functions.

EXAMPLE 4   Comparing exponential and linear functions

Compare $f(x) = 3^x$ and $g(x) = 3x$ graphically and numerically for $x \geq 0$.

Solution   Graphical Comparison   The graphs of $Y_1 = 3^\wedge X$ and $Y_2 = 3X$ are shown in Figure 12.16. The graph of the exponential function y_1 increases much faster than the graph of the linear function y_2.

Numerical Comparison   The tables of $Y_1 = 3^\wedge X$ and $Y_2 = 3X$ are shown in Figure 12.17. The values for y_1 increase much faster than the values for y_2.

[0, 5, 1] by [0, 120, 20]

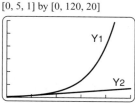

Figure 12.16

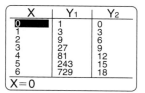

Figure 12.17

The results of Example 4 are true in general: For large enough inputs, exponential functions with $a > 1$ grow far faster than any linear function.

===== MAKING CONNECTIONS =====

Exponential and Polynomial Functions

The function $f(x) = 2^x$ is an exponential function. The base 2 is a constant and the exponent x is a variable, so $f(3) = 2^3 = 8$.

The function $g(x) = x^2$ is a polynomial function. The base x is a variable and the exponent 2 is a constant, so $g(3) = 3^2 = 9$.

The table clearly shows that the exponential function grows much faster than the polynomial function for larger values of x.

x	0	2	4	6	8	10	12
2^x	1	4	16	64	256	1024	4096
x^2	0	4	16	36	64	100	144

MODELS INVOLVING EXPONENTIAL FUNCTIONS

Traffic flow on highways can be modeled by exponential functions whenever traffic patterns occur randomly. In the next example we model traffic at an intersection by using an exponential function.

EXAMPLE 5   Modeling traffic flow

On average, a particular intersection has 360 vehicles arriving randomly each hour. Highway engineers use $f(x) = (0.905)^x$ to estimate the likelihood, or probability, that no vehicle will enter the intersection within a period of x seconds.   (*Source:* F. Mannering and W. Kilareski, *Principles of Highway Engineering and Traffic Analysis.*)

(a) Compute $f(5)$ and interpret the results.

(b) A graph of $y = f(x)$ is shown in Figure 12.18. Discuss this graph.

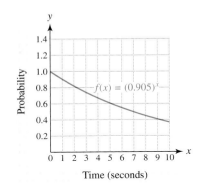

Figure 12.18

(c) Is this function an example of exponential growth or decay?

Solution (a) The result $f(5) = (0.905)^5 \approx 0.61$ indicates that there is a 61% chance that no vehicle will enter the intersection during any particular 5-second interval.

(b) As the number of seconds increases, there is less chance that no car will enter the intersection.

(c) Because the graph is decreasing and $a = 0.905 < 1$, this function is an example of exponential decay.

In the next example, we use an exponential function to model how trees grow in a forest.

EXAMPLE 6 **Modeling tree density in a forest**

Ecologists studied the spacing of individual trees in a British Columbia forest. This pine forest was 40 to 50 years old and contained approximately 1600 randomly spaced trees per acre. The probability or likelihood that no tree is located within a circle of radius x feet can be estimated by $P(x) = (0.892)^x$. For example, $P(4) \approx 0.63$ means that, if a person picks a point at random in the forest, there is a 63% chance that no tree will be located within 4 feet of the person.   (*Source:* E. Pielou, *Populations and Community Ecology.*)

(a) Evaluate $P(8)$ and interpret the result.

(b) Graph P in [0, 20, 5] by [0, 1, 0.1] and discuss the graph.

Solution (a) The probability $P(8) = (0.892)^8 \approx 0.40$ means that there is a 40% chance that no tree is growing within a circle of radius 8 feet.

(b) The graph of $Y_1 = 0.892\char`\^X$, as shown in Figure 12.19, indicates that the larger the circle, the less the likelihood is of no tree being inside the circle.

[0, 20, 5] by [0, 1, 0.1]

Figure 12.19

THE NATURAL EXPONENTIAL FUNCTION

A special type of exponential function is called the *natural exponential function*, $f(x) = e^x$. The base e is a special number in mathematics similar to π. The number π is approximately 3.14, whereas the number e is approximately 2.72. The number e is named for the great Swiss mathematician, Leonhard Euler (1707–1783). Most calculators have a special key that can be used to compute the natural exponential function.

| |||||| NATURAL EXPONENTIAL FUNCTION |

The function represented by

$$f(x) = e^x$$

is the **natural exponential function**, where $e \approx 2.71828$.

This function is frequently used to model **continuous growth**. For example, the fact that births and deaths occur throughout the year, not just at one time during the year, must be recognized when population growth is being modeled. If a population P is growing continuously at r percent per year, expressed as a decimal, we can model this population after x years by

$$P = Ce^{rx},$$

where C is the initial population. To evaluate natural exponential functions, we use a calculator, as in the next example.

EXAMPLE 7   Modeling population

Calculator Help

To evaluate the natural exponential function, see Appendix A (page AP-1).

In 1994 Florida's population was 14 million people and was growing at a continuous rate of 1.53%. This population in millions x years after 1994 could be modeled by

$$f(x) = 14e^{0.0153x}.$$

Estimate the population in 2002.

Solution   As 2002 is 8 years after 1994, we evaluate $f(8)$ and obtain

$$f(8) = 14e^{0.0153(8)} \approx 15.8,$$

```
14e^(.0153*8)
       15.82288531
```

which is supported by Figure 12.20. (Be sure to include parentheses around the exponent of e.) This model estimates that the population of Florida was about 15.8 million in 2002.

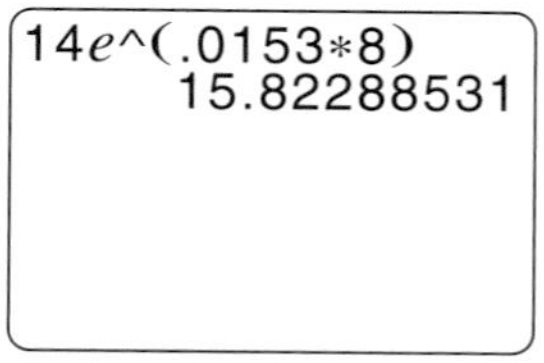

Figure 12.20

Critical Thinking

Sketch a graph of $y = 2^x$ and $y = 3^x$ in the same xy-plane. Then use these two graphs to sketch a graph of $y = e^x$. How do these graphs compare?

12.2 PUTTING IT ALL TOGETHER

The following table summarizes some important concepts of exponential functions and compound interest.

Topic	Explanation	Example
Exponential Function	An exponential function can be written as $f(x) = Ca^x$, where $a > 0$ and $a \neq 1$. If $a > 1$, the function models exponential growth, and if $0 < a < 1$, the function models exponential decay. The natural exponential function has $C = 1$ and $a = e \approx 2.71828$; that is, $f(x) = e^x$.	$f(x) = 3(2)^x$ models exponential growth and $g(x) = 2\left(\frac{1}{3}\right)^x$ models exponential decay.
Compound Interest	If C dollars are deposited in an account and if interest is paid at the end of each year with an annual rate of interest r, expressed as a decimal, then after x years the account will contain A dollars, where $$A = C(1 + r)^x.$$ The growth factor is $(1 + r)$.	If \$1000 are deposited in an account paying 5% annual interest, then after 6 years the amount A in the account is $$A = 1000(1 + 0.05)^6 \approx \$1340.10.$$

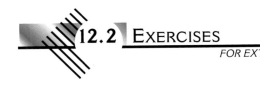 EXERCISES

FOR EXTRA HELP

 Student's Solutions Manual

MyMathLab

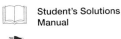 InterAct Math

 Math Tutor Center

 MathXL

Digital Video Tutor
CD 5 Videotape 15

CONCEPTS

1. Give a general formula for an exponential function f.

2. Sketch a graph of an exponential function that illustrates exponential decay.

3. Does the graph of $f(x) = a^x, a > 1$, illustrate exponential growth or decay?

4. Evaluate the expressions 2^x and x^2 for $x = 5$.

5. Give an approximate value for e to the nearest thousandth.

6. Evaluate e^2 and π^2 using your calculator.

7. If a quantity y grows exponentially, then for each unit increase in x, y increases by a constant _____.

8. If $f(x) = 1.5^x$ what is the growth factor?

EVALUATING AND GRAPHING EXPONENTIAL FUNCTIONS

Exercises 9–20: Evaluate the exponential function for the given values of x by hand when possible. Approximate answers to the nearest hundredth when appropriate.

9. $f(x) = 3^x$ $x = -2, x = 2$

10. $f(x) = 5^x$ $x = -1, x = 3$

11. $f(x) = 5(2^x)$ $x = 0, x = 5$

12. $f(x) = 3(7^x)$ $x = -2, x = 0$

13. $f(x) = \left(\frac{1}{2}\right)^x$ $x = -2, x = 3$

14. $f(x) = \left(\frac{1}{4}\right)^x$ $x = 0, x = 2$

15. $f(x) = 5(3)^{-x}$ $x = -1, x = 2$

16. $f(x) = 4\left(\frac{3}{7}\right)^x$ $x = 1, x = 4$

17. $f(x) = 1.8^x$ $x = -3, x = 1.5$

18. $f(x) = 0.91^x$ $x = 5.1, x = 10$

19. $f(x) = 3(0.6)^x$ $x = -1, x = 2$

20. $f(x) = 5(4.5)^{-x}$ $x = -2.1, x = 5.9$

Exercises 21–26: (Refer to Example 2.) A table for a function f is given.
     (a) *Determine whether f represents exponential growth, exponential decay, or linear growth.*
     (b) *Find a formula for f.*

21.

x	0	1	2	3	4
$f(x)$	64	16	4	1	$\frac{1}{4}$

22.

x	0	1	2	3	4
$f(x)$	$\frac{1}{2}$	1	2	4	8

23.

x	0	1	2	3	4
$f(x)$	8	11	14	17	20

24.

x	-2	-1	0	1	2
$f(x)$	4	2	1	$\frac{1}{2}$	$\frac{1}{4}$

25.

x	-2	-1	0	1	2
$f(x)$	2.56	3.2	4	5	6.25

26.

x	-2	-1	0	1	2
$f(x)$	-6	-2	2	6	10

Exercises 27–30: Use the graph of $y = Ca^x$ to determine C and a.

27.

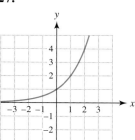

28.

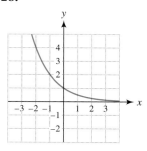

29.

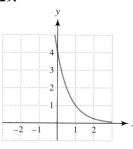

30.

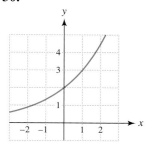

Exercises 31–34: Match the formula with its graph (a.–d.). Do not use a calculator.

31. $f(x) = 1.5^x$ **32.** $f(x) = \frac{1}{4}(2^x)$

33. $f(x) = 4\left(\frac{1}{2}\right)^x$ **34.** $f(x) = \left(\frac{1}{3}\right)^x$

a.

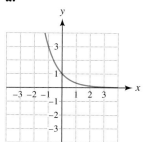

b.

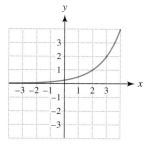

c.

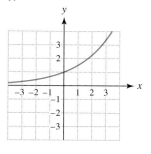

d.

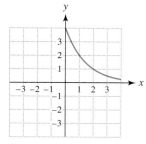

Exercises 35–46: Graph $y = f(x)$. State whether the graph illustrates exponential growth or exponential decay.

35. $f(x) = 2^x$

36. $f(x) = 3^x$

37. $f(x) = \left(\frac{1}{4}\right)^x$

38. $f(x) = \left(\frac{1}{2}\right)^x$

39. $f(x) = 2^{-x}$

40. $f(x) = 3^{-x}$

41. $f(x) = 3^x - 1$

42. $f(x) = 2^x + 1$

43. $f(x) = 2^{x-1}$

44. $f(x) = 2^{x+1}$

45. $f(x) = 4\left(\frac{1}{3}\right)^x$

46. $f(x) = 3\left(\frac{1}{2}\right)^x$

Exercises 47–56: Exponents Use properties of exponents to simplify the expression.

47. $e^2 e^5$

48. $a^2 \cdot a^{-4}$

49. $\dfrac{a^4 a^{-2}}{a^3}$

50. $\dfrac{e^2 e^{-1}}{e}$

51. $e^x e^{y-x}$

52. $a^{1-x} a^{x-1}$

53. $\dfrac{2^{x+3}}{2^x}$

54. $\dfrac{e^{2x}}{e^x}$

55. $5^{-y} \cdot 5^{3y}$

56. $4^{-x} \cdot 4^{2x} \cdot 4^x$

COMPOUND INTEREST

Exercises 57–62: (Refer to Example 3.) If C dollars are deposited in an account paying r percent annual interest, approximate the amount in the account after x years.

57. $C = \$1500$ $r = 9\%$ $x = 10$ years

58. $C = \$1500$ $r = 15\%$ $x = 10$ years

59. $C = \$200$ $r = 20\%$ $x = 50$ years

60. $C = \$5000$ $r = 8.4\%$ $x = 7$ years

61. $C = \$560$ $r = 1.4\%$ $x = 25$ years

62. $C = \$750$ $r = 10\%$ $x = 13$ years

63. *Interest* Suppose that $1000 are deposited in an account paying 8% interest for 10 years. If $2000 had been deposited instead of $1000, would there be twice the money in the account after 10 years? Explain.

64. *Interest* Suppose that $500 are deposited in an account paying 5% interest for 10 years. If the interest rate had been 10% instead of 5%, would the total interest earned after 10 years be twice as much? Explain.

65. *Federal Debt* In 2003 the federal budget deficit was about $300 billion. At the same time, 30-year treasury bonds were paying 4.95% interest. Suppose that U.S. citizens loaned $300 billion to the federal government at 4.95%. If the federal government waited 30 years to pay the entire amount back, including the interest, how much would it be? (*Source:* U.S. Treasury Department.)

66. *Federal Debt* Repeat Exercise 65 but suppose that the interest rate is 2% higher. How much would the federal government owe after 30 years? Is the national debt sensitive to interest rates?

THE NATURAL EXPONENTIAL FUNCTION

Exercises 67–70: Evaluate $f(x)$ for the given value of x. Approximate answers to the nearest hundredth.

67. $f(x) = e^x$ $x = 1.2$

68. $f(x) = 2e^x$ $x = 2$

69. $f(x) = 1 - e^x$ $x = -2$

70. $f(x) = 4e^{-x}$ $x = 1.5$

 Exercises 71–74: Graph $f(x)$ in $[-4, 4, 1]$ by $[0, 8, 1]$. State whether the graph illustrates exponential growth or exponential decay.

71. $f(x) = e^{0.5x}$

72. $f(x) = e^x + 1$

73. $f(x) = 1.5e^{-0.32x}$

74. $f(x) = 2e^{-x} + 1$

APPLICATIONS

75. *Modeling Population* (Refer to Example 7.) In 1997 the population of Arizona was 4.56 million and growing continuously at a rate of 3.1%.

(a) Write a function f that models Arizona's population in millions x years after 1997.

 (b) Graph f in $[0, 10, 1]$ by $[4, 7, 1]$.

(c) Estimate the population of Arizona in 2003.

76. *Dating Artifacts* Radioactive carbon-14 is found in all living things and is used to date objects containing organic material. Suppose that an object initially contains C grams of carbon-14. After x years it will contain A grams, where

$$A = C(0.99988)^x.$$

(a) Let $C = 10$ and graph A over a 20,000-year period. Is this function an example of exponential growth or decay?

(b) How many grams are left after 5700 years? What fraction of the carbon-14 is left?

77. *E. coli Bacteria* A strain of bacteria that inhabits the intestines of animals is named *Escherichia coli* (*E. coli*). These bacteria are capable of rapid growth and can be dangerous to humans—particularly children. The table shows the results of one study of the growth of *E. coli* bacteria, where concentrations are listed in thousands of bacteria per milliliter.

t (minutes)	0	50	100
Concentration	500	1000	2000

t (minutes)	150	200
Concentration	4000	8000

Source: G. S. Stent, *Molecular Biology of Bacterial Viruses.*

(a) Find C and a so that $f(t) = Ca^{t/50}$ models the data.

(b) Use $f(t)$ to estimate the concentration of bacteria after 170 minutes.

(c) Discuss the growth of this strain of bacteria over a 300-minute time period.

78. *Internet Use* Internet use in Western Europe has grown rapidly. The table shows the number of Internet users y in millions during year x, where $x = 0$ corresponds to 2000, $x = 1$ to 2001, and $x = 2$ to 2002.

x (year)	0	1	2
y (millions)	52	67	85

Source: Nortel Networks.

(a) Approximate C and a so that $f(x) = Ca^x$ models the data. (*Hint:* To find a, estimate the factor by which y increases each year.)

(b) Use $f(x)$ to estimate the number of users in 2004 ($x = 4$).

(c) How long is this type of growth likely to continue?

79. *Cellular Phone Use* In 1985, there were about 203,000 cellular phone subscribers in the United States. This number increased to about 84 million users in 2000, as illustrated in the figure at the top of the next column. The rapid growth in cellular phone subscribers in millions can be modeled by $f(x) = 0.0272(1.495)^{x-1980}$, where x is the year. (*Source:* Cellular Telecommunications Industry Association.)

(a) Evaluate $f(1995)$ and interpret the result.

(b) What is the growth factor for $f(x)$? Explain what the growth factor indicates about cellular phone subscribers from 1985 to 2000.

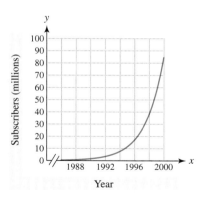

80. *Swimming Pool Maintenance* Chlorine is frequently used to disinfect swimming pools. The chlorine concentration should remain between 1.5 and 2.5 parts per million (ppm). After a warm, sunny day only 80% of the chlorine may remain in the water, with the other 20% dissipating into the air or combining with other chemicals in the water. (*Source:* D. Thomas, *Swimming Pool Operator's Handbook.*)

(a) Let $f(x) = 3(0.8)^x$ model the concentration of chlorine in parts per million after x days. What is the initial concentration of chlorine in the pool?

(b) If no more chlorine is added, estimate when the chlorine level drops below 1.5 parts per million.

81. *Modeling Traffic Flow* (Refer to Example 5.) Construct a table of $f(x) = (0.905)^x$, starting at $x = 0$ and incrementing by 10, until $x = 50$.

(a) Evaluate $f(0)$ and interpret the result.

 (b) After how many seconds is there only a 5% chance that no cars have entered the intersection?

82. *Modeling Tree Density* (Refer to Example 6.)

(a) Evaluate $P(10)$, $P(20)$, and $P(30)$. Interpret the results.

(b) What happens to $P(x)$ as x becomes large? Explain how this probability relates to the spacing of trees in a forest.

WRITING ABOUT MATHEMATICS

83. A student evaluates $f(x) = 4(2)^x$ at $x = 3$ and obtains 512. Did the student evaluate the function correctly? What was the student's error?

84. For a set of data, how can you distinguish between linear growth and exponential growth? Give an example of each type of data.

CHECKING BASIC CONCEPTS ⟨SECTIONS 12.1 AND 12.2⟩

---

1. If $f(x) = 2x^2 + 5x - 1$ and $g(x) = x + 1$, find each expression.
   (a) $(g \circ f)(1)$ (b) $(f \circ g)(x)$

2. Sketch a graph of $f(x) = x^2 - 1$.
   (a) Is f a one-to-one function? Explain.
   (b) Does f have an inverse function?

3. If $f(x) = 4x - 3$, find $f^{-1}(x)$.

4. Evaluate $f(-2)$ if $f(x) = 3(2^x)$.

5. Sketch a graph of $f(x) = \left(\frac{1}{3}\right)^x$.

6. Use the graph of $y = Ca^x$ to determine the constants C and a.

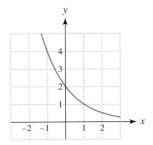

12.3 LOGARITHMIC FUNCTIONS

The Common Logarithmic Function · The Inverse of the Common Logarithmic Function · Logarithms with Other Bases

INTRODUCTION

Logarithmic functions are used in many applications. For example, if one airplane weighs twice as much as another, does the heavier airplane typically need a runway that is twice as long? Using a logarithmic function, we can answer this question. Logarithmic functions are also used to measure the intensity of sound. In this section we discuss logarithmic functions and several of their applications.

THE COMMON LOGARITHMIC FUNCTION

In applications, measurements can vary greatly in size. Table 12.7 lists some examples of objects, with the approximate distances in meters across each.

TABLE 12.7

Object	Distance (meters)
Atom	10^{-9}
Protozoan	10^{-4}
Small Asteroid	10^{2}
Earth	10^{7}
Universe	10^{26}

Source: C. Ronan, *The Natural History of the Universe.*

Each distance is listed in the form 10^k for some k. The value of k distinguishes one measurement from another. The *common logarithmic function* or *base-10 logarithmic function*, denoted *log* or log_{10}, outputs k if the input x can be expressed as 10^k for some real number k. For example, log $10^{-9} = -9$, log $10^2 = 2$, and log $10^{1.43} = 1.43$. For any real number k, log $10^k = k$. Some values for $f(x) = \log x$ are given in Table 12.8.

TABLE 12.8

x	10^{-4}	10^{-3}	10^{-2}	10^{-1}	10^0	10^1	10^2	10^3	10^4
$\log x$	-4	-3	-2	-1	0	1	2	3	4

We use this information to define the common logarithm.

COMMON LOGARITHM

The **common logarithm of a positive number x**, denoted log x, is calculated as follows. If x is written as $x = 10^k$, then

$$\log x = k,$$

where k is a real number. That is, log $10^k = k$.
The function given by

$$f(x) = \log x$$

is called the **common logarithmic function**.

The common logarithmic function outputs an exponent k, which may be positive, negative, or zero. However, a valid input must be positive because 10^k is always positive. *The expression log x equals the exponent k on base* 10 *that gives the number x.* For example, log 1000 $= 3$ because 1000 $= 10^3$.

Note: Previously, we have always used one letter, such as f or g, to *name* a function. The common logarithm is the first function for which we use three letters, *log*, to name it. Thus $f(x)$, $g(x)$, and log (x) all represent functions. Generally, log (x) is written without parentheses as log x. We can also define a function f to be the common logarithmic function by writing $f(x) = \log x$.

EXAMPLE 1   Evaluating common logarithms

Simplify each common logarithm.
(a) log 100   **(b)** log $\frac{1}{10}$ **(c)** log $\sqrt{1000}$ **(d)** log 45

Solution   **(a)** $100 = 10^2$, so log 100 $=$ log $10^2 = 2$
(b) log $\frac{1}{10} =$ log $10^{-1} = -1$
(c) log $\sqrt{1000} =$ log $(10^3)^{1/2} =$ log $10^{3/2} = \frac{3}{2}$

(d) How to write 45 as a power of 10 is not obvious. However, we can use a calculator to determine that log 45 ≈ 1.6532. Thus $10^{\wedge}(1.6532) \approx 45$. Figure 12.21 supports these answers.

Calculator Help
To evaluate the common logarithmic function, see Appendix A (page AP-1).

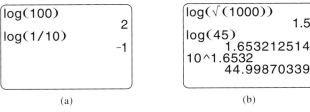

(a)   (b)

Figure 12.21

The points $(10^{-1}, -1)$, $(10^0, 0)$, $(10^{0.5}, 0.5)$, and $(10^1, 1)$ are on the graph of $y = \log x$. Plotting these points, as shown in Figure 12.22(a), and sketching the graph of $y = \log x$ results in Figure 12.22(b). Note some important features of this graph.

- The graph of the common logarithm increases very slowly for large values of x. For example, x must be 100 for log x to reach 2 and x must be 1000 for log x to reach 3.
- The graph passes through the point $(1, 0)$. Thus log $1 = 0$.
- The graph does not exist for negative values of x. The domain of log x includes only positive numbers. The range of log x includes all real numbers.
- When $0 < x < 1$, log x outputs negative values. The y-axis is a vertical asymptote, so as x approaches 0, log x approaches $-\infty$.

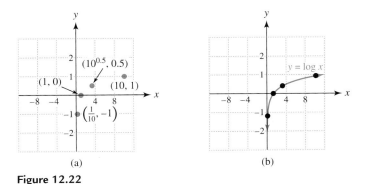

(a)   (b)

Figure 12.22

═══ MAKING CONNECTIONS ═══

The Common Logarithmic Function and the Square Root Function

Much like the square root function, the common logarithmic function does not have an easy-to-evaluate formula. For example, we can calculate $\sqrt{4} = 2$ and $\sqrt{100} = 10$ mentally, but for $\sqrt{2}$ we usually rely on a calculator. Similarly, we can mentally calculate log $1000 = \log 10^3 = 3$, whereas we can use a calculator to approximate log 45. The notation log x is implied to be $\log_{10} x$ much like $\sqrt{x}$ equals $\sqrt[2]{x}$. Another similarity between the square root function and the common logarithmic function is that their domains do not include negative numbers. If only real numbers are allowed as outputs, both $\sqrt{-3}$ and log (-3) are undefined expressions.

The Inverse of the Common Logarithmic Function

The graph of $y = \log x$ just shown in Figure 12.22(b) is a one-to-one function because it passes the horizontal line test. Different inputs always result in different outputs. Thus the common logarithmic function has an inverse function. To determine this inverse function for $\log x$, consider Tables 12.9 and 12.10.

TABLE 12.9

x	-2	-1	0	1	2
10^x	10^{-2}	10^{-1}	10^{0}	10^{1}	10^{2}

TABLE 12.10

x	10^{-2}	10^{-1}	10^{0}	10^{1}	10^{2}
$\log x$	-2	-1	0	1	2

If we start with the number 2, compute 10^2, and then calculate $\log 10^2$, the result is 2. That is,

$$\log(10^2) = 2.$$

In general, $\log 10^x = x$ for any real number x. Now suppose that we perform the calculations in reverse order by taking the common logarithm and then computing a power of 10. For example, suppose that we start with the number 100. The result is

$$10^{\log 100} = 10^2 = 100.$$

In general, $10^{\log x} = x$ for any positive number x.

The *inverse function* of $f(x) = \log x$ is $f^{-1}(x) = 10^x$. That is, if $\log x = y$, then $10^y = x$. Note that composition of these two functions satisfies the definition of an inverse function.

$$(f \circ f^{-1})(x) = f\left(f^{-1}(x)\right) \quad \text{and} \quad (f^{-1} \circ f)(x) = f^{-1}\left(f(x)\right)$$
$$= f(10^x) \qquad\qquad\qquad = f^{-1}(\log x)$$
$$= \log 10^x \qquad\qquad\qquad = 10^{\log x}$$
$$= x \qquad\qquad\qquad\qquad = x$$

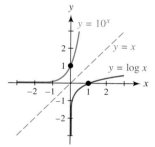

Figure 12.23

In general, the graph of $y = f^{-1}(x)$ is a reflection of the graph of $y = f(x)$ across the line $y = x$. The graphs of $y = \log x$ and $y = 10^x$ are shown in Figure 12.23. Note that the graph of $y = 10^x$ is a reflection of the graph of $y = \log x$ across the line $y = x$.

These inverse properties are summarized as follows.

INVERSE PROPERTIES OF THE COMMON LOGARITHM

The following properties hold for common logarithms.

$$\log 10^x = x, \qquad \text{for any real number } x$$
$$10^{\log x} = x, \qquad \text{for any positive real number } x$$

EXAMPLE 2   Applying inverse properties

Use inverse properties to simplify each expression.
(a) $\log 10^{\pi}$ **(b)** $\log 10^{x^2+1}$ **(c)** $10^{\log 7}$ **(d)** $10^{\log 3x}, x > 0$

Solution   **(a)** Because $\log 10^x = x$ for any real number x, $\log 10^{\pi} = \pi$.
(b) $\log 10^{x^2+1} = x^2 + 1$
(c) Because $10^{\log x} = x$ for any positive real number x, $10^{\log 7} = 7$.
(d) $10^{\log 3x} = 3x$, provided x is a positive number.

We can also graph logarithmic functions, as demonstrated in the next example.

EXAMPLE 3   Graphing the logarithmic functions

Graph f and compare the graph to $y = \log x$.
(a) $f(x) = \log (x - 2)$ **(b)** $f(x) = \log (x) + 1$

Solution   **(a)** We can use our knowledge of translations to sketch the graph of $y = \log (x - 2)$. For example, the graph of $y = (x - 2)^2$ is similar to the graph of $y = x^2$, except that it is translated 2 units to the *right*. Thus the graph of $y = \log (x - 2)$ is similar to the graph of $y = \log x$, except that it is translated 2 units to the right, as shown in Figure 12.24. The graph of $y = \log x$ passes through $(1, 0)$, so the graph of $y = \log (x - 2)$ passes through $(3, 0)$. Also, instead of the y-axis being a vertical asymptote, the line $x = 2$ is the vertical asymptote.

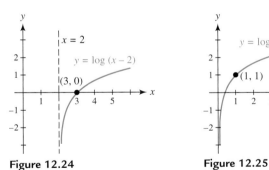

Figure 12.24   **Figure 12.25**

(b) The graph of $y = \log (x) + 1$ is similar to the graph of $y = \log x$, except that it is translated 1 unit *upward*. This graph is shown in Figure 12.25. Note that the graph of $y = \log (x) + 1$ passes through the point $(1, 1)$.

Note: The graph of $y = \log x$ has a vertical asympotote when $x = 0$ and is undefined when $x < 0$. Thus the graph of $\log (x - 2)$ has a vertical asymptote when $x - 2 = 0$, or $x = 2$. It is undefined when $x - 2 < 0$ or $x < 2$.

Logarithms are used to model quantities that vary greatly in intensity. For example, the human ear is extremely sensitive and able to detect intensities on the eardrum ranging from 10^{-16} watts per square centimeter (w/cm^2) to 10^{-4} w/cm^2, which usually is painful. The next example illustrates modeling sound with logarithms.

EXAMPLE 4   Modeling sound levels

Sound levels in decibels (dB) can be computed by $f(x) = 160 + 10 \log x$, where x is the intensity of the sound in watts per square centimeter. Ordinary conversation has an intensity of 10^{-10} w/cm². What decibel level is this?   (*Source:* R. Weidner and R. Sells, *Elementary Classical Physics, Vol. 2.*)

Solution   To find the decibel level for ordinary conversation, evaluate $f(10^{-10})$.

Critical Thinking

If the sound level increases by 10 dB by what factor does the intensity x increase?

$$
\begin{aligned}
f(10^{-10}) &= 160 + 10 \log (10^{-10}) && \text{Substitute } x = 10^{-10}. \\
&= 160 + 10(-10) && \text{Evaluate } \log (10^{-10}). \\
&= 60 && \text{Simplify.}
\end{aligned}
$$

Ordinary conversation corresponds to 60 dB.

LOGARITHMS WITH OTHER BASES

Common logarithms are base-10 logarithms, but we can define logarithms having other bases. For example, base-2 logarithms are frequently used in computer science. Some values for the base-2 logarithmic function, denoted $f(x) = \log_2 x$, are shown in Table 12.11. If x can be expressed as $x = 2^k$ for some real number k, then $\log_2 x = \log_2 2^k = k$.

TABLE 12.11

x	2^{-3}	2^{-2}	2^{-1}	2^0	2^1	2^2	2^3
$\log_2 x$	-3	-2	-1	0	1	2	3

Logarithms with other bases are evaluated in the next three examples.

EXAMPLE 5   Evaluating base-2 logarithms

Simplify each logarithm.
(a) $\log_2 8$ **(b)** $\log_2 \frac{1}{4}$

Solution   **(a)** The expression $\log_2 8$ represents the exponent on base 2 that gives 8. Because $8 = 2^3$, $\log_2 8 = \log_2 2^3 = 3$.

(b) Because $\frac{1}{4} = \frac{1}{2^2} = 2^{-2}$, $\log_2 \frac{1}{4} = \log_2 2^{-2} = -2$.

Some values of base-e logarithms are shown in Table 12.12. A base-e logarithm is referred to as a **natural logarithm** and denoted either $\log_e x$ or $\ln x$. Natural logarithms are used in mathematics, science, economics, electronics, and communications.

Calculator Help

To evaluate the natural logarithmic function, see Appendix A (page AP-1).

TABLE 12.12

x	e^{-3}	e^{-2}	e^{-1}	e^0	e^1	e^2	e^3
$\ln x$	-3	-2	-1	0	1	2	3

To evaluate natural logarithms we usually use a calculator.

EXAMPLE 6   Evaluating natural logarithms

Approximate to the nearest hundredth.
(a) ln 10   **(b)** ln $\frac{1}{2}$

Solution   **(a)** Figure 12.26 shows that ln 10 ≈ 2.30.
(b) Figure 12.26 shows that ln $\frac{1}{2}$ ≈ −0.69.

We now define base-*a* logarithms.

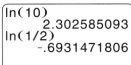
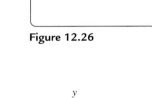

Figure 12.26

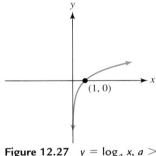

Figure 12.27 $y = \log_a x$, $a > 1$

BASE-*a* LOGARITHMS

The **logarithm with base *a* of a positive number *x*,** denoted $\log_a x$, is calculated as follows. If *x* is written as $x = a^k$, then

$$\log_a x = k,$$

where $a > 0$, $a \neq 1$, and *k* is a real number. That is, $\log_a a^k = k$.
The function given by

$$f(x) = \log_a x$$

is called the **logarithmic function with base *a*.**

Remember that *a logarithm is an exponent.* The expression $\log_a x$ equals the exponent *k* such that $a^k = x$. The graph of $y = \log_a x$ with $a > 1$ is shown in Figure 12.27. Note that the graph passes through the point (1, 0). Thus $\log_a 1 = 0$.

Critical Thinking

Explain why $\log_a 1 = 0$ for any positive base *a*, $a \neq 1$.

Note: The natural logarithm, ln *x*, is a base-*a* logarithm with base *e*. That is, $\ln x = \log_e x$.

EXAMPLE 7   Evaluating base-*a* logarithms

Simplify each logarithm.
(a) $\log_5 25$ **(b)** $\log_4 \frac{1}{64}$ **(c)** $\log_7 1$ **(d)** $\log_3 9^{-1}$

Solution   **(a)** $25 = 5^2$, so $\log_5 25 = \log_5 5^2 = 2$.
(b) $\frac{1}{64} = \frac{1}{4^3} = 4^{-3}$, so $\log_4 \frac{1}{64} = \log_4 4^{-3} = -3$.
(c) $1 = 7^0$, so $\log_7 1 = \log_7 7^0 = 0$. (Note that the logarithm of 1 is always 0, regardless of the base.)
(d) $9^{-1} = (3^2)^{-1} = 3^{-2}$, so $\log_3 9^{-1} = \log_3 3^{-2} = -2$.

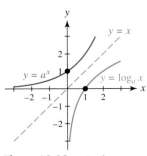

Figure 12.28 $a > 1$

The graph of $y = \log_a x$ in Figure 12.27 passes the horizontal line test, so it is a one-to-one function and it has an inverse. If we let $f(x) = \log_a x$, then $f^{-1}(x) = a^x$. This statement is a generalization of the fact that log *x* and 10^x represent inverse functions. The graphs of $y = \log_a x$ and $y = a^x$ with $a > 1$ are shown in Figure 12.28. Note that the graph of $y = a^x$ is a reflection of the graph of $y = \log_a x$ across the line $y = x$.

These inverse properties for logarithmic and exponential functions are summarized by the following.

INVERSE PROPERTIES

The following properties hold for logarithms with base a.

$$\log_a a^x = x, \qquad \text{for any real number } x$$

$$a^{\log_a x} = x, \qquad \text{for any positive number } x$$

EXAMPLE 8   Applying inverse properties

Simplify each expression.

(a) $\ln e^{0.5x}$ **(b)** $e^{\ln 4}$ **(c)** $2^{\log_2 7x}$ **(d)** $10^{\log (9x-3)}$

Solution   **(a)** $\ln e^{0.5x} = 0.5x$ because $\ln e^k = k$ for all k.
(b) $e^{\ln 4} = 4$ because $e^{\ln k} = k$ for all positive k.
(c) $2^{\log_2 7x} = 7x$ for $x > 0$ because $a^{\log_a k} = k$ for all positive k.
(d) $10^{\log(9x-3)} = 9x - 3$ for $x > \frac{1}{3}$ because $10^{\log k} = k$ for all positive k.

Logarithms occur in many applications. One application is runway length for airplanes, which we discuss in the next example.

EXAMPLE 9   Calculating runway length

There is a mathematical relationship between an airplane's weight x and the runway length required at takeoff. For certain types of airplanes, the minimum runway length L in thousands of feet may be modeled by $L(x) = 1.3 \ln x$, where x is in thousands of pounds.
(**Source:** L. Haefner, *Introduction to Transportation Systems.*)
(a) Estimate the runway length needed for an airplane weighing 10,000 pounds.
(b) Does a 20,000-pound airplane need twice the runway length that a 10,000-pound airplane needs? Explain.

Solution   **(a)** Because $L(10) = 1.3 \ln(10) \approx 3$, an airplane weighing 10,000 pounds requires a runway 3000 feet long.
(b) Because $L(20) = 1.3 \ln(20) \approx 3.9$, a 20,000-pound airplane does not need twice the runway length needed by a 10,000-pound airplane. Rather the heavier airplane needs roughly 3900 feet of runway, or only an extra 900 feet.

12.3   PUTTING IT ALL TOGETHER

Common logarithms are base-10 logarithms. If a positive number x is written as $x = 10^k$, then $\log x = k$. The value of $\log x$ represents the exponent on the base 10 that gives x. We can define logarithms having other bases. For example, the natural logarithm is a base-e logarithm that is usually evaluated using a calculator. The following table summarizes some important concepts related to base-a logarithms.

Concept	Description	Examples
Base-a Logarithms	Logarithms can be defined for any base a, where $a > 0$ and $a \neq 1$. Thus $\log_a x = k$ means $x = a^k$. In other words, if we can write x as $x = a^k$, then $$\log_a x = \log_a a^k = k.$$ When $a = 10$, we write $\log x$, which indicates a common logarithm. When $a = e$, we write $\ln x$, which indicates a natural logarithm. The domain of $f(x) = \log_a x$ is all positive numbers, and its range is all real numbers. The graph of $y = \log_a x$ with $a > 1$ is shown in the figure at the right. Note that the graph passes through the point $(1, 0)$.	$\log 1000 = \log 10^3 = 3$, $\log_2 16 = \log_2 2^4 = 4$, and $\log_3 \dfrac{1}{81} = \log_3 3^{-4} = -4$ *(graph of $y = \log_a x$ passing through $(1, 0)$)*
Inverse Properties	The following properties hold for logarithms with base a. $$\log_a a^x = x, \quad \text{for any real number } x$$ $$a^{\log_a x} = x, \quad \text{for any positive number } x$$	$\log 10^{7.48} = 7.48$ and $2^{\log_2 63} = 63$

12.3 EXERCISES

FOR EXTRA HELP

 Student's Solutions Manual

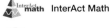 InterAct Math

 MathXL

 MyMathLab

Math Tutor Center

Digital Video Tutor
CD 6 Videotape 15

CONCEPTS

1. What is the base of the common logarithm?

2. What is the base of the natural logarithm?

3. What are the domain and range of $\log x$?

4. What are the domain and range of $\log_a x$?

5. $\log 10^k =$ _____

6. $\ln e^k =$ _____

7. If $\log x = k$, then $10^k =$ _____.

8. $10^{\log x} =$ _____

9. What does k equal if $10^k = 5$?

10. What does k equal if $2^k = 5$?

EVALUATING AND GRAPHING LOGARITHMIC FUNCTIONS

Exercises 11–52: Simplify the expression.

11. $\log 10^5$

12. $\log 10$

13. $\log 10^{-4}$

14. $\log 10^{-1}$

15. $\log 1$

16. $\log \sqrt[3]{100}$

17. $\log \frac{1}{100}$

18. $\log \frac{1}{10}$

19. $\log 10^{4.7}$

20. $\log 10^{2x+4}$

21. $\log_5 5^{6x}$

22. $\log_3 3^3$

23. $\log \sqrt{\frac{1}{1000}}$

24. $\log_2 2^6$

25. $\log_2 4$

26. $\log_2 \frac{1}{32}$

27. $\log_2 \frac{1}{16}$

28. $\log_3 27$

29. $\log_3 \frac{1}{9}$

30. $\log_4 16$

31. $\ln 1$

32. $\ln e^2$

33. $\log 0.001$

34. $\log 0.0001$

35. $\log_5 \frac{1}{25}$

36. $\log_8 64$

37. $10^{\log 2}$

38. $10^{\log 7.5}$

39. $10^{\log x^2}$

40. $10^{\log |x|}$

41. $5^{\log_5 17}$

42. $9^{\log_9 73}$

43. $4^{\log_4 (2x)^2}$

44. $b^{\log_b (x-1)}$

45. $10^{\log 5}$

46. $\ln e^{3/4}$

47. $\ln e^{-5x}$

48. $e^{\ln 2x}$

49. $\log 10^{(2x-7)}$

50. $\log 10^{(8-4x)}$

51. $5^{\log_5 0.6z}$

52. $7^{\log_7 (x-9)}$

Exercises 53–60: Evaluate the logarithm, using a calculator. Round values to the nearest thousandth.

53. $\log 25$

54. $\log 0.501$

55. $\log 1.45$

56. $\log \frac{1}{35}$

57. $\ln 7$

58. $\ln 126$

59. $\ln \frac{4}{7}$

60. $\ln 0.67$

 Exercises 61–64: Graph f in $[-4, 4, 1]$ by $[-4, 4, 1]$. Compare this graph to the graph of $y = \ln x$. Identify the domain of f.

61. $f(x) = \ln |x|$

62. $f(x) = \ln (x) - 2$

63. $f(x) = \ln (x + 2)$

64. $f(x) = 2 \ln x$

Exercises 65–68: Without using a calculator match $f(x)$ with its graph (a.–d.).

65. $f(x) = \log x$

66. $f(x) = \log_3 x$

67. $f(x) = \log_3 (x) + 2$

68. $f(x) = \log (x + 1)$

a.

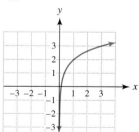

b.

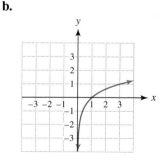

c.

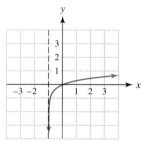

d.

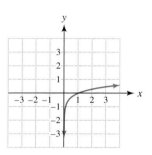

Exercises 69–76: Graph $y = f(x)$. Compare the graph to the graph of $y = \log x$.

69. $f(x) = \log (x) - 1$

70. $f(x) = \log (x) + 2$

71. $f(x) = \log (x + 1)$

72. $f(x) = \log (x + 2)$

73. $f(x) = \log (x - 1)$

74. $f(x) = \log (x - 3)$

75. $f(x) = 2 \log x$

76. $f(x) = -\log x$

Exercises 77 and 78: Complete the table.

77.

x	$\frac{1}{4}$	$\frac{1}{2}$	1	$\sqrt{2}$	64
$\log_2 x$					

78.

x	$\frac{1}{7}$	1	$\sqrt{7}$	7	49
$\log_7 x$					

APPLICATIONS

79. *Modeling Sound*   (Refer to Example 4.) At professional football games in domed stadiums the decibel level may reach 110. The eardrum usually experiences pain when the intensity of the sound reaches 10^{-4} watts per square centimeter. How many decibels does this quantity represent? Is the noise at a football game likely to hurt some people's eardrums?

80. *Runway Length*   (Refer to Example 9.)
   (a) A graph of $L(x) = 1.3 \ln x$ is shown in the accompanying figure. As the weight of the plane increases, what can be said about the length of the runway required?

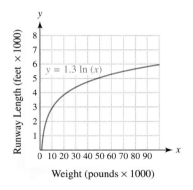

(b) Evaluate $L(50)$ and interpret the result.

81. *Hurricanes* Some of the largest storms on Earth are hurricanes, which have diameters that can exceed 300 miles. The barometric air pressure P in inches of mercury at a distance of d miles from the eye of a severe hurricane can sometimes be modeled by the formula $P(d) = 0.48 \ln(d + 1) + 27$. Average air pressure is about 30 inches of mercury. (*Source:* A. Miller and R. Anthes, *Meteorology.*)

(a) Evaluate $P(0)$ and $P(50)$. Interpret the results.
(b) A graph of $y = P(d)$ is shown in the figure. Describe how the air pressure changes as the distance from the eye of the hurricane increases.

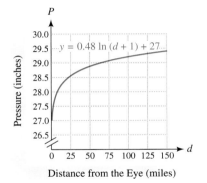

(c) Is the eye of the hurricane a low pressure area or a high pressure area?

82. *Predicting Wind Speed* Wind speed typically varies in the first 20 meters above the ground. Close to the ground, wind speed is often less than it is at 20 meters above the ground. For this reason wind speeds are usually measured at heights from 5 to 10 meters by the U.S. Weather Service. For a particular day, let

$$W(x) = 2.76 \log(h + 1) + 2.3$$

compute the wind speed W in meters per second at a height h meters above the ground. (*Source:* A. Miller.)
(a) Find the wind speed at a height of 10 meters.
(b) A graph of $y = W(h)$ is shown in the accompanying figure. Interpret the graph.

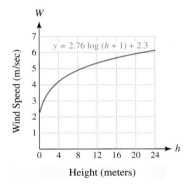

83. *Earthquakes* The Richter scale is used to determine the intensity of earthquakes, which corresponds to the amount of energy released. If an earthquake has an intensity of x, its *magnitude*, as computed by the Richter scale, is given by $R(x) = \log \frac{x}{I_0}$, where I_0 is the intensity of a small, measurable earthquake.
(a) On July 26, 1963, an earthquake in Yugoslavia had a magnitude of 6.0 on the Richter scale, and on August 19, 1977, an earthquake in Indonesia measured 8.0. Find the intensity x for each of these earthquakes if $I_0 = 1$.
(b) How many times more intense was the Indonesian earthquake than the Yugoslavian earthquake?

84. *Growth in Salary* Suppose that a person's salary is initially $40,000 and could be determined by either $f(x)$ or $g(x)$, where x represents the number of years of experience.
  i. $f(x) = 40,000(1.1)^x$
  ii. $g(x) = 40,000 \log(10 + x)$

Would most people prefer that their salaries increase exponentially or logarithmically? Explain your answer.

85. *Calories Consumed and Land Ownership* In developing countries there is a relationship between the amount of land a person owns and the average daily calories consumed. This relationship is modeled by

$$C(x) = 645 \log(x + 1) + 1925,$$

where x is the amount of land owned in acres and $0 \le x \le 4$. (*Source:* D. Grigg, *The World Food Problem.*)

(a) Find the average daily caloric intake for a person who owns 1 acre of land.

(b) A graph of $y = C(x)$ is shown in the figure. How is the number of calories consumed each day affected by the amount of land owned?

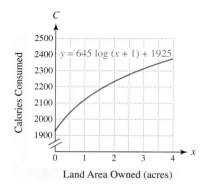

Land Area Owned (acres)

(c) Does a person who owns twice as much land as another person consume twice the calories? Explain.

86. *Magnitude of a Star* The first stellar brightness scale was developed 2000 years ago by two Greek astronomers, Hipparchus and Ptolemy. The brightest star in the sky was given a magnitude of 1, and the faintest star was given a magnitude of 6. In 1856 this scale was described mathematically by the formula

$$M = 6 - 2.5 \log \frac{I}{I_0},$$

where M is the magnitude of a star with an intensity of I and I_0 is the intensity of the faintest star seen in the sky.

Note: The intensity of a star can be measured with instruments in watts per square centimeter. (*Source:* M. Zeilik, *Introductory Astronomy and Astrophysics.*)

(a) Find M if $I = 10$ and $I_0 = 1$.

(b) What is the magnitude of a star that is 100 times more intense than the faintest star?

(c) If the intensity of a star increases by a factor of 10, what happens to its magnitude?

87. *Population of Urban Regions* Although less industrialized urban regions of the world are experiencing exponential population growth, industrialized urban regions are experiencing logarithmic population growth. Population in less industrialized urban regions can be modeled by

$$f(x) = 0.338(1.035)^x,$$

whereas the population in industrialized urban regions can be modeled by

$$g(x) = 0.36 + 0.15 \ln(x + 1).$$

In these formulas the output is in billions of people and x is the year, where $x = 0$ corresponds to 1950, $x = 10$ to 1960, and so on until $x = 80$ corresponds to 2030. (*Source:* D. Meadows, *Beyond The Limits.*)

(a) Evaluate $f(50)$ and $g(50)$. Interpret the results.

(b) Graph f and g in $[0, 80, 10]$ by $[0, 5, 1]$. Compare the two graphs.

88. *Path Loss for Cellular Phones* For cellular phones to work throughout a country, large numbers of cellular towers are necessary. How well the signal is propagated throughout a region depends on the location of these towers. One quick way to estimate the strength of a signal at a distance of x kilometers is to use the formula

$$D(x) = -121 - 36 \log x.$$

This formula computes the decrease in the signal, using decibels, so it is always negative. For example, $D(1) = -121$ means that at a distance of 1 kilometer the signal has decreased in strength by 121 decibels. (*Source:* C. Smith, *Practical Cellular & PCS Design.*)

(a) Evaluate $D(3)$ and interpret the result.

 (b) Graph D in [1, 10, 1] by [−160, −120, 10].

(c) What happens to the signal as the distance increases?

WRITING ABOUT MATHEMATICS

89. Explain in words what $\log_a x$ means and give an example.

90. How would you explain to a student that $\log_a 1 \neq 1$?

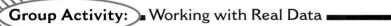

Group Activity: Working with Real Data

Directions: Form a group of 2 to 4 people. Select someone to record the group's responses for this activity. All members of the group should work cooperatively to answer the questions. If your instructor asks for your results, each member of the group should be prepared to respond.

Greenhouse Gases Carbon dioxide (CO_2) is a greenhouse gas in the atmosphere that may raise average temperatures on Earth. The burning of fossil fuels could be responsible for the increased levels of carbon dioxide in the atmosphere. If current trends continue, future concentrations of atmospheric carbon dioxide in parts per million (ppm) are expected to reach the levels shown in the accompanying table. The CO_2 concentration in the year 2000 was greater than it had been at any time in the previous 160,000 years.

Year	2000	2050	2100	2150	2200
CO_2 (ppm)	364	467	600	769	987

Source: R. Turner, Environmental Economics.

(a) Let x represent the year, where $x = 0$ corresponds to 2000, $x = 1$ to 2001, and so on. Find values for C and a so that $f(x) = Ca^x$ models the data.

(b) Graph f and the data together in the same viewing rectangle.

(c) Use $f(x)$ to estimate graphically the year when the carbon dioxide concentration will be double the preindustrial level of 280 ppm.

12.4 PROPERTIES OF LOGARITHMS

Basic Properties · Change of Base Formula

INTRODUCTION

The discovery of logarithms by John Napier (1550–1617) played an important role in the history of science. Logarithms were instrumental in allowing Johannes Kepler (1571–1630) to calculate the positions of the planet Mars, which led to his discovery of the laws of planetary motion. Kepler's laws were used by Isaac Newton (1642–1727) to discover the universal laws of gravity. Although calculators and computers have made tables of logarithms obsolete, applications involving logarithms still play an important role in modern day computation. One reason for their continued importance is that logarithms possess several important properties.

BASIC PROPERTIES

In this subsection we discuss three important properties of logarithms. The first property is the product rule for logarithms.

▌▌▌▌▌ PRODUCT RULE FOR LOGARITHMS

For positive numbers m, n, and $a \neq 1$,

$$\log_a mn = \log_a m + \log_a n.$$

▌▌▌▌▌

This property may be verified by using properties of exponents and the fact that $\log_a a^k = k$ for any real number k. Here, we verify the product property for logarithms. Other properties presented can be verified in a similar manner.

If m and n are positive numbers, we can write $m = a^c$ and $n = a^d$ for some real numbers c and d.

$$\log_a mn = \log_a (a^c a^d) = \log_a (a^{c+d}) = c + d \quad \text{and}$$
$$\log_a m + \log_a n = \log_a a^c + \log_a a^d = c + d$$

Thus $\log_a mn = \log_a m + \log_a n$.

This property is illustrated in Figure 12.29, which shows that

$$\log 10 = \log(2 \cdot 5) = \log 2 + \log 5.$$

In the next two examples, we demonstrate various operations involving logarithms.

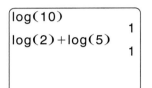

Figure 12.29

EXAMPLE 1 **Writing logarithms as sums**

Write each expression as a sum of logarithms. Assume that x is positive.
(a) $\log 21$ **(b)** $\ln 5x$ **(c)** $\log_2 x^3$

Solution **(a)** $\log 21 = \log(3 \cdot 7) = \log 3 + \log 7$
(b) $\ln 5x = \ln(5 \cdot x) = \ln 5 + \ln x$
(c) $\log_2 x^3 = \log_2 (x \cdot x \cdot x) = \log_2 x + \log_2 x + \log_2 x$

EXAMPLE 2 **Combining logarithms**

Write each expression as one logarithm. Assume that x and y are positive.
(a) $\log 5 + \log 6$ **(b)** $\ln x + \ln xy$ **(c)** $\log_3 2x + \log_3 5x$

Solution **(a)** $\log 5 + \log 6 = \log(5 \cdot 6) = \log 30$
(b) $\ln x + \ln xy = \ln(x \cdot xy) = \ln x^2 y$
(c) $\log_3 2x + \log_3 5x = \log_3 (2x \cdot 5x) = \log_3 10x^2$

The second property is the quotient rule for logarithms.

▌▌▌▌▌ QUOTIENT RULE FOR LOGARITHMS

For positive numbers m, n, and $a \neq 1$,

$$\log_a \frac{m}{n} = \log_a m - \log_a n.$$

▌▌▌▌▌

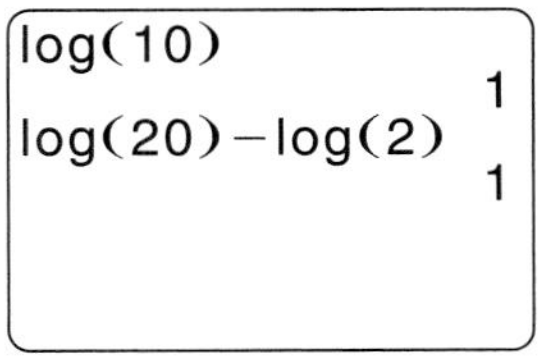

```
log(10)
              1
log(20)−log(2)
              1
```

Figure 12.30

This property is illustrated in Figure 12.30, which shows that

$$\log 10 = \log \frac{20}{2} = \log 20 - \log 2.$$

EXAMPLE 3 Writing logarithms as differences

Write each expression as a difference of two logarithms. Assume that all variables are positive.

(a) $\log \dfrac{3}{2}$ (b) $\ln \dfrac{3x}{y}$ (c) $\log_5 \dfrac{x}{z^4}$

Solution (a) $\log \dfrac{3}{2} = \log 3 - \log 2$ (b) $\ln \dfrac{3x}{y} = \ln 3x - \ln y$

(c) $\log_5 \dfrac{x}{z^4} = \log_5 x - \log_5 z^4$

Note: $\log_a (m + n) \neq \log_a m + \log_a n;$ $\log_a (m - n) \neq \log_a m - \log_a n;$

$$\log_a (mn) \neq \log_a m \cdot \log_a n;   \log_a \left(\frac{m}{n}\right) \neq \frac{\log_a m}{\log_a n}$$

EXAMPLE 4 Combining logarithms

Write each expression as one term. Assume that x is positive.

(a) $\log 50 - \log 25$ (b) $\ln x^3 - \ln x$ (c) $\log_4 15x - \log_4 5x$

Solution (a) $\log 50 - \log 25 = \log \dfrac{50}{25} = \log 2$

(b) $\ln x^3 - \ln x = \ln \dfrac{x^3}{x} = \ln x^2$

(c) $\log_4 15x - \log_4 5x = \log_4 \dfrac{15x}{5x} = \log_4 3$

The third property is the power rule for logarithms. To illustrate this rule we use

$$\log x^3 = \log (x \cdot x \cdot x) = \log x + \log x + \log x = 3 \log x.$$

Thus $\log x^3 = 3 \log x$. This example is generalized in the following rule.

POWER RULE FOR LOGARITHMS

For positive numbers m and $a \neq 1$ and any real number r,

$$\log_a (m^r) = r \log_a m.$$

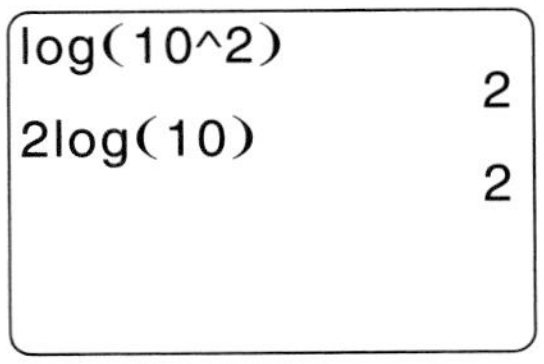

```
log(10^2)
              2
2log(10)
              2
```

Figure 12.31

We use a calculator and the equation

$$\log 10^2 = 2 \log 10$$

to illustrate this property. Figure 12.31 shows the result.

We apply the power rule in the next example.

EXAMPLE 5 Applying the power rule

Rewrite each expression, using the power rule.
(a) $\log 5^6$ **(b)** $\ln (0.55)^{x-1}$ **(c)** $\log_5 8^{kx}$

Solution **(a)** $\log 5^6 = 6 \log 5$
(b) $\ln (0.55)^{x-1} = (x - 1) \ln (0.55)$
(c) $\log_5 8^{kx} = kx \log_5 8$

Sometimes we use more than one property to simplify an expression. We assume that all variables are positive in the next two examples.

EXAMPLE 6 Combining logarithms

Write each expression as one logarithm.
(a) $3 \log x + \log x^2$ **(b)** $2 \ln x - \ln \sqrt{x}$

Solution **(a)** $3 \log x + \log x^2 = \log x^3 + \log x^2$ — Power rule

$= \log (x^3 \cdot x^2)$ — Product rule

$= \log x^5$ — Properties of exponents

(b) $2 \ln x - \ln \sqrt{x} = 2 \ln x - \ln x^{1/2}$ — $\sqrt{x} = x^{1/2}$

$= \ln x^2 - \ln x^{1/2}$ — Power rule

$= \ln \dfrac{x^2}{x^{1/2}}$ — Quotient rule

$= \ln x^{3/2}$ — Properties of exponents

EXAMPLE 7 Expanding logarithms

Write each expression in terms of logarithms of x, y, and z.
(a) $\log \dfrac{x^2 y^3}{\sqrt{z}}$ **(b)** $\ln \sqrt[3]{\dfrac{xy}{z}}$

Solution **(a)** $\log \dfrac{x^2 y^3}{\sqrt{z}} = \log x^2 y^3 - \log \sqrt{z}$ — Quotient rule

$= \log x^2 + \log y^3 - \log z^{1/2}$ — Product rule

$= 2 \log x + 3 \log y - \frac{1}{2} \log z$ — Power rule

(b) $\ln \sqrt[3]{\dfrac{xy}{z}} = \ln \left(\dfrac{xy}{z}\right)^{1/3}$ — $\sqrt[3]{m} = m^{1/3}$

$= \frac{1}{3} \ln \dfrac{xy}{z}$ — Power rule

$= \frac{1}{3} (\ln xy - \ln z)$ — Quotient rule

$= \frac{1}{3} (\ln x + \ln y - \ln z)$ — Product rule

$= \frac{1}{3} \ln x + \frac{1}{3} \ln y - \frac{1}{3} \ln z$ — Distributive property

EXAMPLE 8 Applying properties of logarithms

Using only properties of logarithms and the approximations $\ln 2 \approx 0.7$, $\ln 3 \approx 1.1$, and $\ln 5 \approx 1.6$, find an approximation for each expression.

(a) $\ln 8$ **(b)** $\ln 15$ **(c)** $\ln \dfrac{10}{3}$

Solution **(a)** $\ln 8 = \ln 2^3 = 3 \ln 2 \approx 3(0.7) = 2.1$

(b) $\ln 15 = \ln(3 \cdot 5) = \ln 3 + \ln 5 \approx 1.1 + 1.6 = 2.7$

(c) $\ln \dfrac{10}{3} = \ln\left(\dfrac{2 \cdot 5}{3}\right) = \ln 2 + \ln 5 - \ln 3 \approx 0.7 + 1.6 - 1.1 = 1.2$

CHANGE OF BASE FORMULA

Most calculators only have keys to evaluate common and natural logarithms. Occasionally, it is necessary to evaluate a logarithmic function with a base other than 10 or e. In these situations we use the following change of base formula, which we illustrate in the next example.

> **CHANGE OF BASE FORMULA**
>
> Let x and $a \neq 1$ be positive real numbers. Then
> $$\log_a x = \frac{\log x}{\log a} \quad \text{or} \quad \log_a x = \frac{\ln x}{\ln a}.$$

EXAMPLE 9 Change of base formula

Approximate $\log_2 14$ to the nearest thousandth.

Solution Use the change of base formula,

$$\log_2 14 = \frac{\log 14}{\log 2} \approx 3.807 \quad \text{or} \quad \log_2 14 = \frac{\ln 14}{\ln 2} \approx 3.807.$$

Figure 12.32 supports these results.

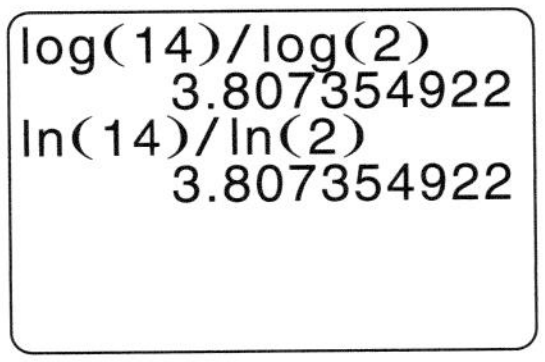

```
log(14)/log(2)
        3.807354922
ln(14)/ln(2)
        3.807354922
```

Figure 12.32

12.4 PUTTING IT ALL TOGETHER

The following table summarizes some important properties for base-a logarithms. Common and natural logarithms satisfy the same properties.

Type of Property	Description	Examples
Logarithmic	The following properties hold for positive numbers m, n, and $a \neq 1$ and for any real number r. **1.** $\log_a mn = \log_a m + \log_a n$ **2.** $\log_a \frac{m}{n} = \log_a m - \log_a n$ **3.** $\log_a (m^r) = r \log_a m$	**1.** $\log 20 = \log 10 + \log 2$ **2.** $\log \frac{45}{6} = \log 45 - \log 6$ **3.** $\ln x^6 = 6 \ln x$
Change of Base	Let x and $a \neq 1$ be positive real numbers. Then $\log_a x = \dfrac{\log x}{\log a}$ and $\log_a x = \dfrac{\ln x}{\ln a}$.	The expression $\log_3 6$ is equivalent to either $\dfrac{\log 6}{\log 3}$ or $\dfrac{\ln 6}{\ln 3}$.

 **12.4 EXERCISES**

FOR EXTRA HELP

Student's Solutions Manual

InterAct Math

MathXL

MyMathLab

Math Tutor Center

Digital Video Tutor
CD 6 Videotape 15

CONCEPTS

1. $\log 12 = \log 3 + \log (\underline{\hspace{1cm}})$

2. $\ln 5 = \ln 20 - \ln (\underline{\hspace{1cm}})$

3. $\log 8 = (\underline{\hspace{1cm}}) \log 2$

4. $\log mn = \underline{\hspace{1cm}}$

5. $\log \dfrac{m}{n} = \underline{\hspace{1cm}}$

6. $\log (m^r) = \underline{\hspace{1cm}}$

7. Does $\log x + \log y$ equal $\log (x + y)$ for all positive numbers x and y?

8. Does $\log x - \log y$ equal $\log \left(\frac{x}{y}\right)$ for all positive numbers x and y?

9. Give the change of base formula.

10. $\log_a 1 = \underline{\hspace{1cm}}$ and $\log_a a = \underline{\hspace{1cm}}$.

BASIC PROPERTIES OF LOGARITHMS

Exercises 11–16: Write the expression as a sum of two or more logarithms.

11. $\ln (3 \cdot 5)$

12. $\log (7 \cdot 11)$

13. $\log_3 xy$

14. $\log_5 y^2$

15. $\ln 10z$

16. $\log x^2 y$

Exercises 17–22: Write the expression as a difference of two logarithms.

17. $\log \frac{7}{3}$

18. $\ln \frac{11}{13}$

19. $\ln \dfrac{x}{y}$

20. $\log \dfrac{2x}{z}$

21. $\log_2 \dfrac{45}{x}$

22. $\log_7 \dfrac{5x}{4z}$

Exercises 23–30: Write the expression as one logarithm.

23. $\log 45 + \log 5$

24. $\log 30 - \log 10$

25. $\ln x + \ln y$

26. $\ln m + \ln n - \ln n$

27. $\ln 7x^2 + \ln 2x$

28. $\ln x + \ln y - \ln z$

29. $\ln x + \ln y^2 - \ln y$

30. $\ln \sqrt{z} - \ln z^3 + \ln y^3$

Exercises 31–40: Rewrite the expression, using the power rule.

31. $\log 3^6$

32. $\log x^8$

33. $\ln 2^x$

34. $\ln (0.77)^{x+1}$

35. $\log_2 5^{1/4}$

36. $\log_3 \sqrt{x}$

37. $\log_4 \sqrt[3]{z}$

38. $\log_7 3^{\pi}$

39. $\log x^{y-1}$

40. $\ln a^{2b}$

Exercises 41–52: Use properties of logarithms to write the expression as one logarithm.

41. $4 \log z - \log z^3$

42. $2 \log_5 y + \log_5 x$

43. $\log x + 2 \log x + 2 \log y$

44. $\log x^2 + 3 \log z - 5 \log y$

45. $\log x - 2 \log \sqrt{x}$

46. $\ln y^2 - 6 \ln \sqrt[3]{y}$

47. $\ln 2^{x+1} - \ln 2$

48. $\ln 8^{1/2} + \ln 2^{1/2}$

49. $2 \log_3 \sqrt{x} - 3 \log_3 x$

50. $\ln \sqrt[3]{x} + \ln \sqrt{x}$

51. $2 \log_a (x + 1) - \log_a (x^2 - 1)$

52. $\log_b (x^2 - 9) - \log_b (x - 3)$

Exercises 53–64: Use properties of logarithms to write the expression in terms of logarithms of x, y, and z.

53. $\log xy^2$

54. $\log \dfrac{x^2}{y^3}$

55. $\ln \dfrac{x^4 y}{z}$

56. $\ln \dfrac{\sqrt{x}}{y}$

57. $\log_4 \dfrac{\sqrt[3]{z}}{\sqrt{y}}$

58. $\log_2 \sqrt{\dfrac{x}{y}}$

59. $\log (x^4 y^3)$

60. $\log (x^2 y^4 z^3)$

61. $\ln \dfrac{1}{y} - \ln \dfrac{1}{x}$

62. $\ln \dfrac{1}{xy}$

63. $\log_4 \sqrt{\dfrac{x^3 y}{z^2}}$

64. $\log_3 \left(\dfrac{x^2 \sqrt{z}}{y^3} \right)$

Exercises 65–68: Graph f and g in the window $[-6, 6, 1]$ by $[-4, 4, 1]$. If the two graphs appear to be identical, prove that they are, using properties of logarithms.

65. $f(x) = \log x^3, g(x) = 3 \log x$

66. $f(x) = \ln x + \ln 3, g(x) = \ln 3x$

67. $f(x) = \ln (x + 5), g(x) = \ln x + \ln 5$

68. $f(x) = \log (x - 2), g(x) = \log x - \log 2$

Exercises 69–78: (Refer to Example 8.) Using only properties of logarithms and the approximations $\log 2 \approx 0.3$, $\log 5 \approx 0.7$, and $\log 13 \approx 1.1$, find an approximation for the expression.

69. $\log 16$

70. $\log 125$

71. $\log 65$

72. $\log 26$

73. $\log 130$

74. $\log 100$

75. $\log \dfrac{5}{2}$

76. $\log \dfrac{26}{5}$

77. $\log \dfrac{1}{13}$

78. $\log \dfrac{1}{65}$

Exercises 79–84: Use the change of base formula to approximate each expression to the nearest hundredth.

79. $\log_3 5$

80. $\log_5 12$

81. $\log_2 25$

82. $\log_7 8$

83. $\log_9 102$

84. $\log_6 293$

APPLICATIONS

85. *Modeling Sound* (Refer to Example 4, Section 12.3.) The formula $f(x) = 10 \log (10^{16} x)$ can be used to calculate the decibel level of a sound with an intensity x. Use properties of logarithms to simplify this formula to

$$f(x) = 160 + 10 \log x.$$

86. *Cellular Phone Technology*   A formula used to calculate the strength of a signal for a cellular phone is

$$L = 110.7 - 19.1 \log h + 55 \log d,$$

where h is the height of the cellular phone tower and d is the distance the phone is from the tower. Use properties of logarithms to write an expression for L that contains only one logarithm.   (*Source:* C. Smith, *Practical Cellular & PCS Design.*)

WRITING ABOUT MATHEMATICS

87. State the three basic properties of logarithms and give an example of each.

88. A student insists that $\log(x - y)$ is equal to $\log x - \log y$. How could you convince the student otherwise?

CHECKING BASIC CONCEPTS   SECTIONS 12.3 AND 12.4

1. Simplify each expression without using a calculator.
   (a) $\log 10^4$ **(b)** $\ln e^x$
   (c) $\log_2 \frac{1}{8}$ **(d)** $\log_5 \sqrt{5}$

2. Sketch a graph of $f(x) = \log x$.
   (a) What are the domain and range of f?
   (b) Evaluate $f(1)$.
   (c) Can the common logarithm of a positive number be negative? Explain.
   (d) Can the common logarithm of a negative number be positive? Explain.

3. Write the expression in terms of logarithms of x, y, and z.
   (a) $\log xy$ **(b)** $\ln \frac{x}{yz}$
   (c) $\ln x^2$ **(d)** $\log \frac{x^2 y^3}{\sqrt{z}}$

4. Write the expression as one logarithm.
   (a) $\log x + \log y$
   (b) $\ln 2x - 3 \ln y$
   (c) $2 \log_2 x + 3 \log_2 y - \log_2 z$

12.5   EXPONENTIAL AND LOGARITHMIC EQUATIONS

Exponential Equations and Models · Logarithmic Equations and Models

INTRODUCTION

Although we have solved many equations throughout this course, one equation that we have not solved *symbolically* is $a^x = k$. This exponential equation occurs frequently in applications and is used to model either exponential growth or decay. Logarithmic equations contain logarithms and are also used in modeling real-world data. To solve exponential equations we use logarithms, and to solve logarithmic equations we use exponential expressions.

EXPONENTIAL EQUATIONS AND MODELS

To solve the equation

$$10 + x = 100$$

we subtract 10 from each side because addition and subtraction are inverse operations.

$$10 + x - 10 = 100 - 10$$
$$x = 90$$

To solve the equation

$$10x = 100$$

we divide each side by 10 because multiplication and division are inverse operations.

$$\frac{10x}{10} = \frac{100}{10}$$
$$x = 10$$

Now suppose that we want to solve the exponential equation

$$10^x = 100.$$

What is new about this type of equation is that the variable x is an *exponent*. The inverse operation of 10^x is log x. Rather than subtracting 10 from each side or dividing each side by 10, we take the base-10 logarithm of each side. Doing so results in

$$\log 10^x = \log 100.$$

Because $\log 10^x = x$ for all real numbers x, the equation becomes

$$x = \log 100, \quad \text{or equivalently,} \quad x = 2.$$

These concepts are applied in the next example.

EXAMPLE 1   Solving exponential equations

Solve.
(a) $10^x = 150$ **(b)** $e^x = 40$ **(c)** $2^x = 50$ **(d)** $0.9^x = 0.5$

Solution   **(a)** $10^x = 150$ Given equation

$\log 10^x = \log 150$ Take the common logarithm of each side.

$x = \log 150 \approx 2.18$ Inverse property: $\log 10^x = x$ for all x

(b) The inverse operation of e^x is ln x, so we take the natural logarithm of each side.

$e^x = 40$ Given equation

$\ln e^x = \ln 40$ Take the natural logarithm of each side.

$x = \ln 40 \approx 3.69$ Inverse property: $\ln e^x = x$ for all x

(c) The inverse operation of 2^x is $\log_2 x$. Calculators do not usually have a base-2 logarithm key, so we take the common logarithm of each side and then apply the power rule.

$2^x = 50$ Given equation

$\log 2^x = \log 50$ Take the common logarithm of each side.

$x \log 2 = \log 50$ Power rule: $\log (m^r) = r \log m$

$x = \dfrac{\log 50}{\log 2} \approx 5.64$ Divide by log 2 and approximate.

(d) This time we begin by taking the natural logarithm of each side.

$$0.9^x = 0.5 \qquad \text{Given equation}$$

$$\ln 0.9^x = \ln 0.5 \qquad \text{Take the natural logarithm of each side.}$$

$$x \ln 0.9 = \ln 0.5 \qquad \text{Power rule: } \ln(m^r) = r \ln m$$

$$x = \frac{\ln 0.5}{\ln 0.9} \approx 6.58 \qquad \text{Divide by } \ln 0.9 \text{ and approximate.}$$

≡ MAKING CONNECTIONS ≡

Logarithms of Quotients and Quotients of Logarithms

The solution in Example 1(c) is $\frac{\log 50}{\log 2}$. Note that

$$\frac{\log 50}{\log 2} \neq \log 50 - \log 2.$$

However, $\log 50 - \log 2 = \log \frac{50}{2} = \log 25$ by the quotient rule for logarithms, as shown in the figure.

```
log(50)/log(2)
         5.64385619
log(50)-log(2)
         1.397940009
log(25)
         1.397940009
```

The next two examples illustrate methods for solving exponential equations.

EXAMPLE 2   Solving exponential equations

Solve each equation.
(a) $2e^x - 1 = 5$ **(b)** $3^{x-5} = 15$ **(c)** $e^{2x} = e^{x+5}$ **(d)** $3^{2x} = 2^{x+3}$

Solution   **(a)** Begin by solving for e^x.

$$2e^x - 1 = 5 \qquad \text{Given equation}$$

$$2e^x = 6 \qquad \text{Add 1 to each side.}$$

$$e^x = 3 \qquad \text{Divide each side by 2.}$$

$$\ln e^x = \ln 3 \qquad \text{Take the natural logarithm.}$$

$$x = \ln 3 \approx 1.10 \qquad \text{Inverse property: } \ln e^k = k$$

(b) Start by taking the common logarithm of each side.

$$3^{x-5} = 15 \qquad \text{Given equation}$$

$$\log 3^{x-5} = \log 15 \qquad \text{Take the common logarithm of each side.}$$

$$(x - 5) \log 3 = \log 15 \qquad \text{Power rule for logarithms}$$

$$x - 5 = \frac{\log 15}{\log 3} \qquad \text{Divide by } \log 3.$$

$$x = \frac{\log 15}{\log 3} + 5 \approx 7.46 \qquad \text{Add 5 to each side and approximate.}$$

(c) Because the bases are equal, the exponents must also be equal. To verify this assertion, take the natural logarithm of each side.

$$e^{2x} = e^{x+5}$$ Given equation

$$\ln e^{2x} = \ln e^{x+5}$$ Take the natural logarithm.

$$2x = x + 5$$ Inverse property: $\ln e^k = k$

$$x = 5$$ Subtract x.

(d) In this equation, the bases are not equal. However, we can still solve the equation by taking a common logarithm of each side. A logarithm of any base could be used.

$$3^{2x} = 2^{x+3}$$ Given equation

$$\log 3^{2x} = \log 2^{x+3}$$ Take the common logarithm.

$$2x \log 3 = (x + 3) \log 2$$ Power rule for logarithms

$$2x \log 3 = x \log 2 + 3 \log 2$$ Distributive property

$$2x \log 3 - x \log 2 = 3 \log 2$$ Subtract $x \log 2$.

$$x(2 \log 3 - \log 2) = 3 \log 2$$ Factor out x.

$$x = \frac{3 \log 2}{2 \log 3 - \log 2}$$ Divide by $2 \log 3 - \log 2$.

$$x \approx 1.38$$ Approximate.

EXAMPLE 3 Solving an exponential equation

Graphs for $f(x) = 0.2e^x$ and $g(x) = 4$ are shown in Figure 12.33.
(a) Use the graphs to estimate the solution to the equation $f(x) = g(x)$.
(b) Check your estimate by solving the equation symbolically.

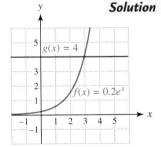

Figure 12.33

Solution **(a)** The graphs intersect near the point $(3, 4)$. Therefore the solution is given by $x \approx 3$.
(b) We must solve the equation $0.2(e^x) = 4$.

$$0.2(e^x) = 4$$ Given equation

$$e^x = 20$$ Divide each side by 0.2.

$$\ln e^x = \ln 20$$ Take the natural logarithm of each side.

$$x = \ln 20$$ Inverse property: $\ln e^k = k$

$$x \approx 2.996$$ Approximate.

Note: The graphical estimate did not give the *exact* solution of $\ln 20$.

In Section 12.2 we showed that, if $1000 are deposited in a savings account paying 10% annual interest at the end of each year, the amount A in the account after x years is given by

$$A(x) = 1000(1.1)^x.$$

After 10 years there will be

$$A(10) = 1000(1.1)^{10} \approx \$2593.74$$

in the account. To calculate how long it will take for $4000 to accrue in the account, we need to solve the exponential equation

$$1000(1.1)^x = 4000.$$

We do so in the next example.

EXAMPLE 4 Solving exponential equations

Solve $1000(1.1)^x = 4000$ symbolically. Give graphical support for your answer.

Solution **Symbolic Solution** Begin by dividing each side of the equation by 1000.

$$1000(1.1)^x = 4000 \qquad \text{Given equation}$$
$$1.1^x = 4 \qquad \text{Divide by 1000.}$$
$$\log 1.1^x = \log 4 \qquad \text{Take the common logarithm of each side.}$$
$$x \log 1.1 = \log 4 \qquad \text{Power rule for logarithms}$$
$$x = \frac{\log 4}{\log 1.1} \approx 14.5 \qquad \text{Divide by } \log 1.1 \text{ and approximate.}$$

Interest is paid at the end of the year, so it will take 15 years for $1000 earning 10% annual interest to grow to $4000.

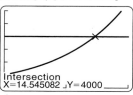

[0, 20, 5] by [0, 6000, 1000]

Intersection
X=14.545082 Y=4000

Figure 12.34

Graphical Solution Graphical support is shown in Figure 12.34, where the graphs of $Y_1 = 1000*1.1\text{^}X$ and $Y_2 = 4000$ intersect when $x \approx 14.5$.

In the next example, we model the life span of a robin with an exponential function.

EXAMPLE 5 Modeling the life span of robins

The life span of 129 robins was monitored over a 4-year period in one study. The formula $f(x) = 10^{-0.42x}$ can be used to calculate the percentage of robins remaining after x years. For example, $f(1) \approx 0.38$ means that after 1 year 38% of the robins were still alive. (**Source:** D. Lack, *The Life Span of a Robin.*)
(a) Evaluate $f(2)$ and interpret the result.
(b) Determine when 5% of the robins remained.

Solution **(a)** $f(2) = 10^{-0.42(2)} \approx 0.145$. After 2 years about 14.5% of the robins were still alive.
(b) Use 5% = 0.05 and solve the following equation.

$$10^{-0.42x} = 0.05 \qquad \text{Equation to solve}$$
$$\log 10^{-0.42x} = \log 0.05 \qquad \text{Take the common logarithm of each side.}$$
$$-0.42x = \log 0.05 \qquad \text{Inverse property: } \log 10^k = k$$
$$x = \frac{\log 0.05}{-0.42} \approx 3.1 \qquad \text{Divide by } -0.42.$$

After about 3 years only 5% of the robins were still alive.

LOGARITHMIC EQUATIONS AND MODELS

To solve an exponential equation we use logarithms. To solve a logarithmic equation we *exponentiate* each side of the equation. To do so we use the fact that if $x = y$, then $a^x = a^y$ for any positive base a. For example, to solve

$$\log x = 3$$

we exponentiate each side of the equation, using base 10.

$$10^{\log x} = 10^3$$

Because $10^{\log x} = x$ for all positive x,

$$x = 10^3 = 1000.$$

To solve logarithmic equations we frequently use the inverse property

$$a^{\log_a x} = x.$$

Examples of this inverse property include

$$e^{\ln 2k} = 2k, \quad 2^{\log_2 x} = x, \quad \text{and} \quad 10^{\log (x+5)} = x + 5.$$

The next two examples show how to solve logarithmic equations, followed by two applications of these methods.

EXAMPLE 6   Solving logarithmic equations

Solve and approximate solutions to the nearest hundredth when appropriate.
(a) $2 \log x = 4$ (b) $\ln 3x = 5.5$ (c) $\log_2 (x + 4) = 7$

Solution   (a) $2 \log x = 4$ Given equation

$\log x = 2$ Divide each side by 2.

$10^{\log x} = 10^2$ Exponentiate each side, using base 10.

$x = 100$ Inverse property: $10^{\log k} = k$

(b) $\ln 3x = 5.5$ Given equation

$e^{\ln 3x} = e^{5.5}$ Exponentiate each side, using base e.

$3x = e^{5.5}$ Inverse property: $e^{\ln k} = k$

$x = \dfrac{e^{5.5}}{3} \approx 81.56$ Divide each side by 3 and approximate.

(c) $\log_2 (x + 4) = 7$ Given equation

$2^{\log_2 (x+4)} = 2^7$ Exponentiate each side, using base 2.

$x + 4 = 2^7$ Inverse property: $2^{\log_2 k} = k$

$x = 2^7 - 4$ Subtract 4 from each side.

$x = 124$ Simplify.

Because the domain of any logarithmic function includes only positive numbers, it is important to check answers, as emphasized in the next example.

EXAMPLE 7   Solving a logarithmic equation

Solve $\log (x + 2) + \log (x - 2) = \log 5$. Check the answers.

Solution   Start by applying the product rule for logarithms.

$\log (x + 2) + \log (x - 2) = \log 5$ Given equation

$\log ((x + 2)(x - 2)) = \log 5$ Product rule

$\log (x^2 - 4) = \log 5$ Multiply.

$10^{\log(x^2-4)} = 10^{\log 5}$ Exponentiate using base 10.

$x^2 - 4 = 5$ Inverse properties

$x^2 = 9$ Add 4.

$x = \pm 3$ Square root property.

Check each answer.

$$\log(3 + 2) + \log(3 - 2) \overset{?}{=} \log 5 \qquad \log(-3 + 2) + \log(-3 - 2) \overset{?}{=} \log 5$$
$$\log 5 + \log 1 \overset{?}{=} \log 5 \qquad\qquad \log(-1) + \log(-5) \neq \log 5$$
$$\log 5 + 0 \overset{?}{=} \log 5$$
$$\log 5 = \log 5$$

Although 3 is a solution, -3 is not, because both $\log(-1)$ and $\log(-5)$ are undefined expressions.

EXAMPLE 8   Modeling runway length

For some types of airplanes with weight x, the minimum runway length L required at takeoff is modeled by

$$L(x) = 3 \log x.$$

In this equation L is measured in thousands of feet and x is measured in thousands of pounds. Estimate the weight of the heaviest airplane that can take off from a runway 5100 feet long.   (*Source:* L. Haefner, *Introduction to Transportation Systems.*)

Solution   Runway length is measured in thousands of feet, so we must solve the equation $L(x) = 5.1$.

$$3 \log x = 5.1 \qquad L(x) = 5.1$$
$$\log x = 1.7 \qquad \text{Divide each side by 3.}$$
$$10^{\log x} = 10^{1.7} \qquad \text{Exponentiate each side, using base 10.}$$
$$x = 10^{1.7} \qquad \text{Inverse property: } 10^{\log k} = k$$
$$x \approx 50.1 \qquad \text{Approximate.}$$

The largest airplane that can take off from this runway weighs about 50,000 pounds.

Critical Thinking

In Example 9, Section 12.3, we used the formula $L(x) = 1.3 \ln x$ to model runway length. Are $L(x) = 1.3 \ln x$ and $L(x) = 3 \log x$ equivalent formulas? Explain.

EXAMPLE 9   Modeling bird populations

Near New Guinea there is a relationship between the number of different species of birds and the size of an island. Larger islands tend to have a greater variety of birds. Table 12.13 lists the number of species of birds y found on islands with an area of x square kilometers.

TABLE 12.13

x (km²)	0.1	1	10	100	1000
y (species)	10	15	20	25	30

Source: B. Freedman, *Environmental Ecology.*

(a) Find values for the constants a and b so that $y = a + b \log x$ models the data.

(b) Predict the number of bird species on an island of 15,000 square kilometers.

Solution **(a)** Because $\log 1 = 0$, substitute $x = 1$ and $y = 15$ in the equation to find a.

$$15 = a + b \log 1$$
$$15 = a + b \cdot 0$$
$$15 = a$$

Thus $y = 15 + b \log x$. To find b substitute $x = 10$ and $y = 20$.

$$20 = 15 + b \log 10$$
$$20 = 15 + b \cdot 1$$
$$5 = b$$

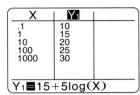

Figure 12.35

The data in Table 12.13 are modeled by $y = 15 + 5 \log x$. This result is supported by Figure 12.35.

(b) To predict the number of species on an island of 15,000 square kilometers, let $x = 15,000$ and find y.

$$y = 15 + 5 \log 15{,}000 \approx 36$$

The model estimates about 36 different species of birds on this island.

PUTTING IT ALL TOGETHER

Exponential and logarithmic equations occur in many applications. When solving exponential equations, we usually take the logarithm of each side. Similarly, when solving logarithmic equations, we usually *exponentiate* each side of the equation. That is, if $x = y$, then $a^x = a^y$ for any positive base a. We do so because a^x and $\log_a x$ represent inverse operations, much like adding and subtracting or multiplying and dividing. Basic steps for solving exponential and logarithmic equations are summarized in the following table.

Type of Equation	Procedure	Example	
Exponential	Begin by solving for the exponential expression a^x. Then take a logarithm of each side.	$4e^x + 1 = 9$	Given equation
		$e^x = 2$	Solve for e^x.
		$\ln e^x = \ln 2$	Take the natural logarithm.
		$x = \ln 2$	Inverse property: $\ln e^k = k$
Logarithmic	Begin by solving for the logarithm in the equation. Then exponentiate each side of the equation, using the same base as the logarithm.	$\dfrac{1}{3} \log 2x = 1$	Given equation
		$\log 2x = 3$	Multiply by 3.
		$10^{\log 2x} = 10^3$	Exponentiate using base 10.
		$2x = 1000$	Inverse property: $10^{\log k} = k$
		$x = 500$	Divide by 2.

12.5 EXERCISES

CONCEPTS

1. To solve $x - 5 = 50$, what should be done?

2. To solve $5x = 50$, what should be done?

3. To solve $10^x = 50$, what should be done?

4. To solve $\log x = 5$, what should be done?

5. $\log 10^x =$ _____

6. $10^{\log x} =$ _____

7. $\ln e^{2x} =$ _____

8. $e^{\ln (x+7)} =$ _____

9. Does $\dfrac{\log 5}{\log 4}$ equal $\log \frac{5}{4}$? Explain.

10. Does $\dfrac{\log 5}{\log 4}$ equal $\log 5 - \log 4$? Explain.

11. How many solutions are there to the equation $\log x = k$, where k is any real number?

12. How many solutions are there to the equation $10^x = k$, where k is a positive number?

EXPONENTIAL EQUATIONS

Exercises 13–40: Solve the equation. Approximate answers to the nearest hundredth when appropriate.

13. $10^x = 1000$

14. $10^x = 0.01$

15. $2^x = 64$

16. $3^x = 27$

17. $2^{x-3} = 8$

18. $3^{2x} = 81$

19. $4^x + 3 = 259$

20. $3(5^{2x}) = 300$

21. $10^{0.4x} = 124$

22. $0.75^x = 0.25$

23. $e^{-x} = 1$

24. $0.5^{-5x} = 5$

25. $e^x - 1 = 6$

26. $2e^{4x} = 15$

27. $2(10)^{x+2} = 35$

28. $10^{3x} + 10 = 1500$

29. $3.1^{2x} - 4 = 16$

30. $5.4^{x-1} = 85$

31. $e^{3x} = e^{2x-1}$

32. $e^{x^2} = e^{3x-2}$

33. $5^{4x} = 5^{x^2-5}$

34. $2^{4x} = 2^{x+3}$

35. $e^{2x} \cdot e^x = 10$

36. $10^{x-2} \cdot 10^x = 1000$

37. $e^x = 2^{x+2}$

38. $2^{2x} = 3^{x-1}$

39. $4^{0.5x} = 5^{x+2}$

40. $3^{2x} = 7^{x+1}$

Exercises 41–44: (Refer to Example 3.) The symbolic and graphical representations of f and g are given.

  (a) Use the graph to solve $f(x) = g(x)$.

  (b) Solve $f(x) = g(x)$ symbolically.

41. $f(x) = 0.2(10^x)$,
    $g(x) = 2$

42. $f(x) = e^x$,
    $g(x) = 7.4$

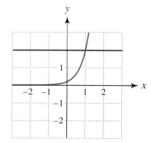

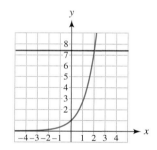

43. $f(x) = 2^{-x}$,
    $g(x) = 4$

44. $f(x) = 0.1(3^x)$,
    $g(x) = 0.9$

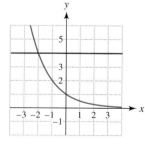

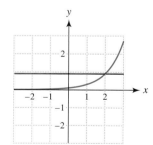

Exercises 45–52: Solve the equation symbolically. Give graphical or numerical support. Approximate answers to the nearest hundredth when appropriate.

45. $10^x = 0.1$

46. $2(10^x) = 2000$

47. $4e^x + 5 = 9$

48. $e^x + 6 = 36$

49. $4^x = 1024$

50. $3^x = 729$

51. $(0.55)^x + 0.55 = 2$ **52.** $5(0.9)^x = 3$

Exercises 53–56: The given equation cannot be solved symbolically. Find any solutions either graphically or numerically to the nearest hundredth.

53. $e^x - x = 2$ **54.** $x \log x = 1$

55. $\ln x = e^{-x}$ **56.** $10^x - 2 = \log(x + 2)$

LOGARITHMIC EQUATIONS

Exercises 57–78: Solve the equation. Approximate answers to the nearest hundredth when appropriate.

57. $\log x = 2$ **58.** $\log x = 0.01$

59. $\ln x = 5$ **60.** $2 \ln x = 4$

61. $\log 2x = 7$ **62.** $6 \ln 4x = 12$

63. $\log_2 x = 4$ **64.** $\log_2 x = 32$

65. $\log_2 5x = 2.3$ **66.** $2 \log_3 4x = 10$

67. $2 \log x + 5 = 7.8$ **68.** $\ln(x - 1) = 3.3$

69. $5 \ln(2x + 1) = 55$ **70.** $5 - \log(x + 3) = 2.6$

71. $\log x^2 = \log x$ **72.** $\ln x^2 = \ln(3x - 2)$

73. $\ln x + \ln(x + 1) = \ln 30$

74. $\log(x - 1) + \log(2x + 1) = \log 14$

75. $\log_3 3x - \log_3(x + 2) = \log_3 2$

76. $\log_4(x^2 - 1) - \log_4(x - 1) = \log_4 6$

77. $\log_2(x - 1) + \log_2(x + 1) = 3$

78. $\log_4(x^2 + 2x + 1) - \log_4(x + 1) = 2$

Exercises 79–82: Two functions, f and g, are given.
(a) *Use the graph to solve $f(x) = g(x)$.*
(b) *Solve $f(x) = g(x)$ symbolically.*

79. $f(x) = \ln x$,
$\quad g(x) = 0.7$

80. $f(x) = \log_2 x$,
$\quad g(x) = 1.6$

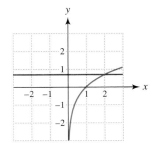

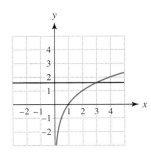

81. $f(x) = 5 \log 2x$,
$\quad g(x) = 3$

82. $f(x) = 2 \ln(x) - 3$,
$\quad g(x) = 0.9$

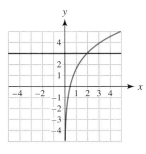

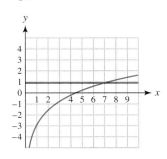

Exercises 83–88: Solve the equation symbolically. Give graphical or numerical support. Approximate answers to the nearest hundredth.

83. $\log x = 1.6$ **84.** $\ln x = 2$

85. $\ln(x + 1) = 1$ **86.** $2 \log(2x + 3) = 8$

87. $17 - 6 \log_3 x = 5$ **88.** $4 \log_2 x + 7 = 12$

APPLICATIONS

89. *Growth of a Mutual Fund*   (Refer to Example 4.) An investor deposits $2000 in a mutual fund that returns 15% at the end of 1 year. Determine the length of time required for the investment to triple its value if the annual rate of return remains the same.

90. *Savings Account*   (Refer to Example 4.) If a savings account pays 6% annual interest at the end of each year, how many years will it take for the account to double in value?

91. *Liver Transplants*   In the United States the gap between available organs for liver transplants and people who need them has widened. The number of individuals waiting for liver transplants can be modeled by

$$f(x) = 2339(1.24)^{x - 1988},$$

where x is the year.   (*Source: United Network for Organ Sharing.*)
(a) Evaluate $f(1994)$ and interpret the result.
(b) Determine when the number of individuals waiting for liver transplants was 20,000.

92. *Life Span of a Robin*   (Refer to Example 5.) Determine when 50% of the robins in the study were still alive.

93. *Runway Length*   (Refer to Example 8.) Determine the weight of the heaviest airplane that can take off from a runway having a length of $\frac{3}{4}$ mile. (*Hint:* 1 mile = 5280 feet.)

94. *Runway Length*   (Refer to Example 8.)
  (a) Suppose that an airplane is 10 times heavier than a second airplane. How much longer should the runway be for the heavier airplane than for the lighter airplane? (*Hint:* Let the heavier airplane have weight $10x$.)
  (b) If the runway length is increased by 3000 feet, by what factor can the weight of an airplane that uses the runway be increased?

95. *The Decline of Bluefin Tuna*   Bluefin tuna are large fish that can weigh 1500 pounds and swim at a speed of 55 miles per hour. They are used for sushi, and a prime fish can be worth more than $30,000. As a result, the number of western Atlantic bluefin tuna has declined dramatically. Their numbers in thousands between 1974 and 1991 can be modeled by $f(x) = 230(10^{-0.055x})$, where x is the number of years after 1974. See the accompanying graph. (*Source:* B. Freedman.)

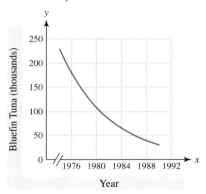

  (a) Evaluate $f(1)$ and interpret the result.
  (b) Use the graph to estimate the year in which bluefin tuna numbered 115 thousand.
  (c) Solve part (b) symbolically.

96. *Insect Populations*   (Refer to Example 9.) The table lists numbers of species of insects y found on islands having areas of x square miles.

x (square miles)	1	2	4
y (species)	1000	1500	2000

x (square miles)	8	16
y (species)	2500	3000

  (a) Find values for constants a and b so that $y = a + b \log_2 x$ models the data.

(b) Construct a table for y and verify that your equation models the data.
(c) Estimate the number of species of insects on an island having an area of 12 square miles.

Exercises 97 and 98: (Refer to Example 9.) Find values for a and b so that $y = a + b \log x$ models the data in the table.

97.

x	0.1	1	10	100
y	22	25	28	31

98.

x	0.01	1	100	1000
y	−10	−2	6	10

99. *Calories Consumed and Land Ownership*   In developing countries there is a relationship between the amount of land a person owns and the average number of calories consumed daily. This relationship is modeled by

$$f(x) = 645 \log (x + 1) + 1925,$$

where x is the amount of land owned in acres and $0 \le x \le 4$.   (*Source:* D. Grigg, *The World Food Problem.*)
   (a) Estimate graphically the number of acres owned by a typical person consuming 2200 calories per day.
  (b) Solve part (a) symbolically.

100. *Population of Industrialized Urban Regions*   The number of people living in industrialized urban regions throughout the world has not grown exponentially. Instead it has grown logarithmically and is modeled by

$$f(x) = 0.36 + 0.15 \ln (x - 1949).$$

In this formula the output is billions of people and the input x is the year, where

$$1950 \le x \le 2030.$$

(*Source:* D. Meadows, *Beyond The Limits.*)
   (a) Determine either graphically or numerically when this population may reach 1 billion.
  (b) Solve part (a) symbolically.

101. *Fertilizer Use*   Between 1950 and 1980, the use of chemical fertilizers increased worldwide. The table lists worldwide average use y in kilograms per acre of cropland during year x. (*Source:* D. Grigg, *The World Food Problem.*)

x	1950	1963	1972	1979
y	5.0	11.3	22.0	31.2

(a) Are the data linear or nonlinear? Explain.
(b) The equation $y = 5(1.06)^{(x-1950)}$ may be used to model the data. The growth factor is 1.06. What does this growth factor indicate about fertilizer use during this time period?
(c) Estimate the year when fertilizer use was 15 kilograms per acre of cropland.

102. *Greenhouse Gases* If current trends continue, future concentrations of atmospheric carbon dioxide (CO_2) in parts per million (ppm) are expected to increase. This increase in concentration of CO_2 has been accelerated by burning fossil fuels and deforestation. The exponential equation $y = 364(1.005)^x$ may be used to model CO_2 in parts per million, where $x = 0$ corresponds to 2000, $x = 1$ to 2001, and so on. (*Source:* R. Turner, *Environmental Economics*.) Estimate the year when the CO_2 concentration could be double the preindustrial level of 280 parts per million.

103. *Modeling Sound* The formula
$$f(x) = 160 + 10 \log x$$
is used to calculate the decibel level of a sound with intensity x measured in watts per square centimeter. The noise level at a basketball game can reach 100 decibels. Find the intensity x of this sound.

104. *Loudness of a Sound* (Refer to Exercise 103.)
(a) Show that, if the intensity of a sound increases by a factor of 10 from x to $10x$, the decibel level increases by 10 decibels. (*Hint:* Show that
$$160 + 10 \log 10x = 170 + 10 \log x.)$$

(b) Find the increase in decibels if the intensity x increases by a factor of 1000.
(c) Find the increase in the intensity x if the decibel level increases by 20.

105. *Hurricanes* (Refer to Exercise 81, Section 12.3.) The barometric air pressure in inches of mercury at a distance of x miles from the eye of a severe hurricane is given by
$$f(x) = 0.48 \ln(x + 1) + 27.$$
(*Source:* A. Miller and R. Anthes, *Meteorology.*)
Determine how far from the eye the pressure is 28 inches of mercury.

106. *Earthquakes* The Richter scale is used to determine the intensity of earthquakes, which corresponds to the amount of energy released. If an earthquake has an intensity of x, its magnitude, as computed by the Richter scale, is given by $R(x) = \log \frac{x}{I_0}$, where I_0 is the intensity of a small, measurable earthquake.
(a) If x is 1000 times greater than I_0, how large is this increase on the Richter scale?
(b) If the Richter scale increases from 5 to 8, by what factor did the intensity x increase?

WRITING ABOUT MATHEMATICS

107. Explain in words the basic steps for solving the equation $a(10^x) - b = c$ and then write the solution.

108. Explain in words the basic steps for solving the equation $a \log 3x = b$ and then write the solution.

CHECKING BASIC CONCEPTS   SECTION 12.5

1. Solve the equation. Approximate answers to the nearest hundredth when appropriate.
   (a) $2(10^x) = 40$ (b) $2^{3x} + 3 = 150$
   (c) $\ln x = 4.1$ (d) $4 \log 2x = 12$

2. Solve $\log(x + 4) + \log(x - 4) = \log 48$. Check the answers.

3. If $500 are deposited in a savings account that pays 3% annual interest at the end of each year, the amount of money A in the account after x years is given by $A = 500(1.03)^x$. Estimate the number of years required for this amount to reach $900.

12 Summary

Composition of Functions If f and g are functions, then the composite function $g \circ f$, or composition of g and f, is defined by $(g \circ f)(x) = g(f(x))$.

Example: If $f(x) = x - 5$ and $g(x) = 2x^2 + 4x - 6$, then $(g \circ f)(x)$ is

$$g(f(x)) = g(x - 5)$$
$$= 2(x - 5)^2 + 4(x - 5) - 6.$$

One-to-One Function A function f is one-to-one if, for any c and d in the domain of f,

$$c \neq d \quad \text{implies that} \quad f(c) \neq f(d).$$

That is, different inputs always result in different outputs.

Example: $f(x) = x^2 + 4$ is *not* one-to-one; $-3 \neq 3$, but $f(-3) = f(3) = 13$.

Horizontal Line Test If every horizontal line intersects the graph of a function f at most once, then f is a one-to-one function.

Examples:

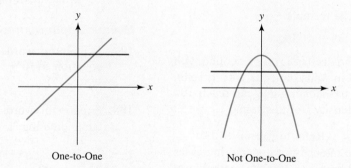

One-to-One   Not One-to-One

Inverse Functions If f is one-to-one, then f has an inverse function, denoted f^{-1}, that satisfies

$$(f^{-1} \circ f)(x) = x \quad \text{and} \quad (f \circ f^{-1})(x) = x.$$

Example: $f(x) = 7x$ and $f^{-1}(x) = \frac{x}{7}$ are inverse functions.

Exponential Function An exponential function is defined by $f(x) = Ca^x$, where $a > 0$, $C > 0$, and $a \neq 1$. Its domain (set of valid inputs) is all real numbers and its range (outputs) is all positive real numbers.

Example: $f(x) = e^x$ is the natural exponential function and $e \approx 2.71828$.

Exponential Growth and Decay When $a > 1$, the graph of $f(x) = Ca^x$ models exponential growth, and when $0 < a < 1$, it models exponential decay. The base a either represents the growth factor or the decay factor. The constant C equals $f(0)$.

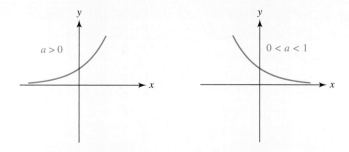

Example: $f(x) = 1.5(2)^x$ is an exponential function with $a = 2$ and $C = 1.5$. It models exponential growth because $a > 1$. The growth factor is 2 because for each unit increase in x, the output from $f(x)$ increases by a factor of 2.

Section 12.3 *Logarithmic Functions*

Base-*a* Logarithms The logarithm with base a of a positive number x is denoted $\log_a x$. If $\log_a x = b$, then $x = a^b$. That is, $\log_a x$ represents the exponent on base a that results in x.

Example: $\log_2 16 = 4$ because $16 = 2^4$.

Domain and Range of Logarithmic Functions The domain (set of valid inputs) of a logarithmic function is the set of all positive real numbers and the range (outputs) is the set of real numbers.

Graph of a Logarithmic Function The graph of a logarithmic function passes through the point $(1, 0)$, as illustrated in the following graph. As x becomes large, $\log_a x$ with $a > 1$ grows very slowly.

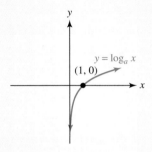

Section 12.4   *Properties of Logarithms*

Basic Properties   Logarithms have several important properties. For positive numbers m, n, and $a \neq 1$ and any real number r,

1. $\log_a mn = \log_a m + \log_a n$.

2. $\log_a \dfrac{m}{n} = \log_a m - \log_a n$.

3. $\log_a (m^r) = r \log_a m$.

Examples: **1.** $\log 5 + \log 20 = \log (5 \cdot 20) = \log 100 = 2$

2. $\log 100 - \log 5 = \log \frac{100}{5} = \log 20 \approx 1.301$

3. $\ln 2^6 = 6 \ln 2 \approx 4.159$

Note: $\log_a 1 = 0$ for any valid base a. Thus $\log 1 = 0$ and $\log_2 1 = 0$.

Inverse Properties   The following inverse properties are important for solving exponential and logarithmic equations.

1. $\log_a a^x = x$,   for any real number x

2. $a^{\log_a x} = x$,   for any positive number x

Examples: **1.** $\log_2 2^\pi = \pi$

2. $10^{\log 2.5} = 2.5$

Section 12.5   *Exponential and Logarithmic Equations*

Solving Equations   The calculations a^x and $\log_a x$ are inverse operations, much like addition and subtraction or multiplication and division. When solving an exponential equation, we usually take a logarithm of each side. When solving a logarithmic equation, we usually exponentiate each side.

Examples:

$2(5)^x = 22$	Exponential equation
$5^x = 11$	Divide by 2.
$\log 5^x = \log 11$	Take the common logarithm.
$x \log 5 = \log 11$	Power rule
$x = \dfrac{\log 11}{\log 5}$	Divide by log 5.

$\log 2x = 2$	Logarithmic equation
$10^{\log 2x} = 10^2$	Exponentiate each side.
$2x = 100$	Inverse properties
$x = 50$	Divide by 2.

CHAPTER

12 Review Exercises

SECTION 12.1

Exercises 1 and 2: For the given f(x) and g(x), find the following.

    (a) $(g \circ f)(-2)$ **(b)** $(f \circ g)(x)$

1. $f(x) = 2x^2 - 4x, \quad g(x) = 5x + 1$

2. $f(x) = \sqrt[3]{x - 6}, \quad g(x) = 4x^3$

3. Use the tables to evaluate each expression.
    (a) $(f \circ g)(2)$ **(b)** $(g \circ f)(1)$

x	0	1	2	3
$f(x)$	3	2	1	0

x	0	1	2	3
$g(x)$	1	2	3	0

4. Use the graph to evaluate each expression.
    (a) $(f \circ g)(-1)$
    (b) $(g \circ f)(2)$
    (c) $(f \circ f)(1)$

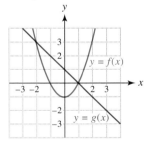

Exercises 5 and 6: Show that f is not one-to-one by finding two inputs that result in the same output.

5. $f(x) = \dfrac{4}{1 + x^2}$ **6.** $f(x) = x^2 - 2x + 1$

Exercises 7 and 8: Use the horizontal line test to determine whether the graph represents a one-to-one function.

7.
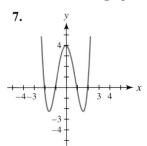

8.
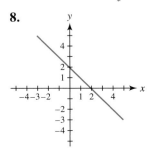

Exercises 9 and 10: Verify that f(x) and $f^{-1}(x)$ are indeed inverse functions.

9. $f(x) = 2x - 9 \qquad f^{-1}(x) = \dfrac{x + 9}{2}$

10. $f(x) = x^3 + 1 \qquad f^{-1}(x) = \sqrt[3]{x - 1}$

Exercises 11–14: Find $f^{-1}(x)$ for the one-to-one function f.

11. $f(x) = 5x$ **12.** $f(x) = x - 11$

13. $f(x) = 2x + 7$ **14.** $f(x) = \dfrac{4}{x}$

15. Use the table to make a table of values for $f^{-1}(x)$. What are the domain and range for f^{-1}?

x	0	1	2	3
$f(x)$	10	8	7	3

16. Use the graph of $y = f(x)$ to sketch a graph of $y = f^{-1}(x)$. Include the graph of f and the line $y = x$ in your graph.

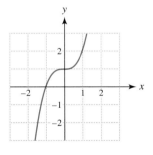

SECTIONS 12.2 AND 12.3

Exercises 17–20: Evaluate the exponential function for the given values of x.

17. $f(x) = 6^x$ $x = -1, \quad x = 2$

18. $f(x) = 5(2^{-x})$ $x = 0, \quad x = 3$

19. $f(x) = \left(\tfrac{1}{3}\right)^x$ $x = -1, \quad x = 4$

20. $f(x) = 3\left(\tfrac{1}{6}\right)^x$ $x = 0, \quad x = 1$

Exercises 21–24: Graph f. State whether the graph illustrates exponential growth, exponential decay, or logarithmic growth.

21. $f(x) = 2^x$ **22.** $f(x) = \left(\frac{1}{2}\right)^x$

23. $f(x) = \ln(x + 1)$ **24.** $f(x) = 3^{-x}$

Exercises 25 and 26: A table for a function f is given.
    (a) *Determine whether f represents linear or exponential growth.*
    (b) *Find a formula for f.*

25.

x	0	1	2	3	4
$f(x)$	5	10	20	40	80

26.

x	0	1	2	3	4
$f(x)$	5	10	15	20	25

27. Use the graph of $y = Ca^x$ to find C and a.

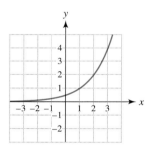

28. Use the graph of $y = k \log_2 x$ to find k.

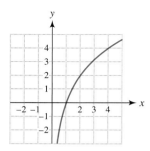

Exercises 29 and 30: If C dollars are deposited in an account that pays r percent annual interest at the end of each year, approximate the amount in the account after x years.

29. $C = \$1200$ $r = 10\%$ $x = 9$ years

30. $C = \$900$ $r = 18\%$ $x = 40$ years

Exercises 31–34: Evaluate f(x) for the given value of x. Approximate answers to the nearest hundredth.

31. $f(x) = 2e^x - 1$ $x = 5.3$

32. $f(x) = 0.85^x$ $x = 2.1$

33. $f(x) = 2 \log x$ $x = 55$

34. $f(x) = \ln(2x + 3)$ $x = 23$

Exercises 35–38: Evaluate the logarithm by hand.

35. $\log 0.001$ **36.** $\log \sqrt{10{,}000}$

37. $\ln e^{-4}$ **38.** $\log_4 16$

Exercises 39–42: Approximate the logarithm to the nearest thousandth.

39. $\log 65$ **40.** $\ln 0.85$

41. $\ln 120$ **42.** $\log_2 \frac{2}{5}$

Exercises 43–46: Simplify, using inverse properties of logarithms.

43. $10^{\log 7}$ **44.** $\log_2 2^{5/9}$

45. $\ln e^{6-x}$ **46.** $e^{2 \ln x}$

SECTION 12.4

Exercises 47–52: Write the expression by using sums and differences of logarithms of x, y, and z.

47. $\ln xy$ **48.** $\log \dfrac{x}{y}$

49. $\ln\left(x^2 y^3\right)$ **50.** $\log \dfrac{\sqrt{x}}{z^3}$

51. $\log_2 \dfrac{x^2 y}{z}$ **52.** $\log_3 \sqrt[3]{\dfrac{x}{y}}$

Exercises 53–56: Write the expression as one logarithm.

53. $\log 45 + \log 5 - \log 3$

54. $\log_4 2x + \log_4 5x$

55. $2 \ln x - 3 \ln y$

56. $\log x^4 - \log x^3 + \log y$

Exercises 57–60: Rewrite the expression, using the power rule.

57. $\log 6^3$

58. $\ln x^2$

59. $\log_2 5^{2x}$

60. $\log_4 (0.6)^{x+1}$

SECTION 12.5

Exercises 61–70: Solve the equation. Approximate answers to the nearest hundredth when appropriate.

61. $10^x = 100$

62. $2^{2x} = 256$

63. $3e^x + 1 = 28$

64. $0.85^x = 0.2$

65. $5 \ln x = 4$

66. $\ln 2x = 5$

67. $2 \log x = 80$

68. $3 \log x - 5 = 1$

69. $2^{x+4} = 3^x$

70. $\ln (2x + 1) + \ln (x - 5) = \ln 13$

Exercises 71 and 72: Graphs and formulas for f and g are given.
(a) *Solve $f(x) = g(x)$ graphically.*
(b) *Solve $f(x) = g(x)$ symbolically.*

71. $f(x) = \frac{1}{2}(2^x)$,
$g(x) = 4$

72. $f(x) = \log_2 2x$,
$g(x) = 3$

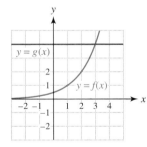

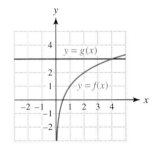

APPLICATIONS

73. *Surface Area of a Balloon* The surface area S of a spherical balloon with radius r is given by $S(r) = 4\pi r^2$. Suppose that the balloon is being inflated so that its radius in inches after t seconds is $r(t) = \sqrt{2t}$.
(a) Evaluate $(S \circ r)(8)$ and interpret your result.
(b) Find $(S \circ r)(t)$.

74. *Sales Tax* Suppose that $f(x) = 0.08x$ calculates the sales tax in dollars on an item that costs x dollars.
(a) Is f a one-to-one function? Why?

(b) Find a formula for f^{-1} and interpret what it calculates.

75. *Growth of a Mutual Fund* An investor deposits $1500 in a mutual fund that returns 12% annually. Determine the time required for the investment to double in value.

76. *Modeling Data* Find values for the constants a and b so that $y = a + b \log x$ models these data.

x	0.1	1	10	100	1000
y	50	100	150	200	250

77. *Modeling Data* Find values for the constants C and a so that $y = Ca^x$ models these data.

x	0	1	2	3	4
y	3	6	12	24	48

78. *Earthquakes* The Richter scale, used to determine the magnitude of earthquakes, is based on the formula $R(x) = \log \frac{x}{I_0}$, where x is the measured intensity. Let $I_0 = 1$. Find the intensity x for an earthquake with $R = 7$.

79. *Modeling Population* In 1997 the population of Nevada was 1.68 million and growing continuously at an annual rate of 4.8%. The population of Nevada in millions x years after 1997 can be modeled by

$$f(x) = 1.68e^{0.048x}.$$

(a) Graph f in $[0, 10, 2]$ by $[0, 4, 1]$. Does this function represent exponential growth or decay?
(b) Predict the population of Nevada in 2003.
(c) Estimate the year when the population might reach 3 million.

80. *Modeling Bacteria* A colony of bacteria can be modeled by $N(t) = 1000e^{0.0014t}$, where N is measured in bacteria per milliliter and t is in minutes.
(a) Evaluate $N(0)$ and interpret the result.
(b) Estimate how long it takes for N to double.

81. *Modeling Wind Speed* Wind speeds are usually measured at heights from 5 to 10 meters above the ground. For a particular day, $f(x) = 1.2 \ln (x) + 5$ computes the wind speed in meters per second x meters above the ground, where $x \geq 1$. (*Source:* A. Miller and R. Anthes, *Meteorology*.)
(a) Find the wind speed at a height of 5 meters.
(b) Estimate the height at which the wind speed is 8 meters per second.

CHAPTER

12 Test

1. If $f(x) = 4x^3 - 5x$ and $g(x) = x + 7$, then evaluate $(g \circ f)(1)$ and $(f \circ g)(x)$.

2. Use the graph to evaluate each expression.
  (a) $(f \circ g)(-1)$
  (b) $(g \circ f)(1)$

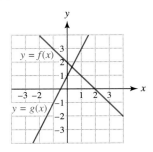

3. Explain why $f(x) = x^2 - 25$ is not a one-to-one function.

4. If $f(x) = 5 - 2x$, find $f^{-1}(x)$.

5. Use the graph of $y = f(x)$ to sketch a graph of $y = f^{-1}(x)$. Include the graph of f and the line $y = x$ in your graph.

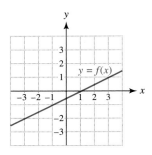

6. Use the table to write a table of values for $f^{-1}(x)$. What are the domain and range of f^{-1}?

x	1	2	3	4
$f(x)$	8	6	4	2

7. Evaluate $f(x) = 3\left(\frac{1}{4}\right)^x$ at $x = 2$.

8. Graph $f(x) = 1.5^{-x}$. State whether the graph illustrates exponential growth, exponential decay, or logarithmic growth.

Exercises 9 and 10: A table for a function f is given.
  (a) *Determine whether f represents linear or exponential growth.*
  (b) *Find a formula for f(x).*

9.

x	-2	-1	0	1	2
$f(x)$	0.75	1.5	3	6	12

10.

x	-2	-1	0	1	2
$f(x)$	-4	-2.5	-1	0.5	2

11. Use the graph of $y = Ca^x$ to find C and a.

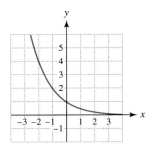

12. If \$750 are deposited in an account paying 7% annual interest at the end of each year, approximate the amount in the account after 5 years.

13. Let $f(x) = 1.5 \ln(x - 5)$. Approximate $f(21)$ to the nearest hundredth.

14. Evaluate $\log \sqrt{10}$ by hand.

15. Approximate $\log_2 43$ to the nearest thousandth.

16. Graph $f(x) = \log(x - 2)$. Compare this graph to the graph of $y = \log x$.

17. Write $\log \dfrac{x^3 y^2}{\sqrt{z}}$, using sums and differences of logarithms of x, y, and z.

18. Write $4 \ln x - 5 \ln y + \ln z$ as one logarithm.

19. Rewrite $\log 7^{2x}$, using the power rule.

20. Simplify $\ln e^{1 - 3x}$, using inverse properties of logarithms.

Exercises 21–24: Solve the equation. Approximate answers to the nearest hundredth when appropriate.

21. $2e^x = 50$ **22.** $3(10)^x - 7 = 143$

23. $5 \log x = 9$ **24.** $3 \ln 5x = 27$

25. *Modeling Data*   Find values for constants a and b so that $y = a + b \log x$ models the data.

x	0.01	0.1	1	10	100
y	−1	2	5	8	11

26. *Modeling Bacteria Growth*   A sample of bacteria is growing continuously at a rate of 9% per hour and can be modeled by

$$f(x) = 4e^{0.09x},$$

where the input x represents elapsed time in hours and the output $f(x)$ is in millions of bacteria.
  (a) What was the initial number of bacteria in the sample?
  (b) Evaluate $f(5)$ and interpret the result.
  (c) Does this function represent exponential growth or decay?
  (d) Determine the elapsed time when there were 6 million bacteria.

CHAPTER 12 Extended and Discovery Exercises

Exercises 1–4: Radioactive Carbon Dating   While an animal is alive, it breathes both carbon dioxide and oxygen. Because a small portion of normal atmospheric carbon dioxide is made up of radioactive carbon-14, a fixed percentage of the animal's body is composed of carbon-14. When the animal dies, it quits breathing and the carbon-14 disintegrates without being replaced. One method used to determine when an animal died is to estimate the percentage of carbon-14 remaining in its bones. The **half-life** of carbon-14 is 5730 years. That is, half the original amount of carbon-14 in bones of a fossil will remain after 5730 years. The percentage P, in decimal form, of carbon-14 remaining after x years is modeled by $P(x) = a^x$.

1. Find the value of a. (*Hint:* $P(5730) = 0.5$.)

2. Calculate the percentage of carbon-14 remaining after 10,000 years.

3. Estimate the age of a fossil with $P = 0.9$.

4. Estimate the age of a fossil with $P = 0.01$.

 *Exercises 5–8: Modeling Blood Flow in Animals*   For medical reasons, dyes may be injected into the bloodstream to determine the health of internal organs. In one study in-volving animals, the dye BSP was injected to assess blood flow in the liver. The results are listed in the accompanying table, where x represents the elapsed time in minutes and y is the concentration of the dye in the bloodstream in milligrams per milliliter (mg/mL). Scientists modeled the data with $f(x) = 0.133(0.878(0.73^x) + 0.122(0.92^x))$.

x (minutes)	1	2	3	4
y (mg/mL)	0.102	0.077	0.057	0.045

x (minutes)	5	7	9	13
y (mg/mL)	0.036	0.023	0.015	0.008

x (minutes)	16	19	22
y (mg/mL)	0.005	0.004	0.003

Source: F. Harrison, "The measurement of liver blood flow in conscious calves."

5. Graph f together with the data. Comment on the fit.

6. Determine the y-intercept and interpret the result.

7. What happens to the concentration of the dye after a long period of time? Explain.

8. Estimate graphically the time at which the concentration of the dye reached 40% of its initial amount. Would you want to solve this problem symbolically? Explain.

Exercises 9 and 10: Acid Rain   *Air pollutants frequently cause acid rain. A measure of acidity is pH, which measures the concentration of the hydrogen ions in a solution, and ranges from 1 to 14. Pure water is neutral and has a pH of 7, acid solutions have a pH less than 7, and alkaline solutions have a pH greater than 7. The pH of a substance can be computed by $f(x) = -\log x$, where x represents the hydrogen ion concentration in moles per liter. Pure water exposed to normal carbon dioxide in the atmosphere has a pH of 5.6. If the pH of a lake drops below this level, it is indicative of an acid lake.* (**Source:** G. Howells, *Acid Rain and Acid Water.*)

9. In rural areas of Europe, rainwater typically has a hydrogen ion concentration of $x = 10^{-4.7}$. Find its pH. What effect might this rain have on a lake with a pH of 5.6?

10. Seawater has a pH of 8.2. Compared to seawater, how many times greater is the hydrogen ion concentration in rainwater from rural Europe?

Exercises 11 and 12: Investment Account   *If x dollars are deposited every 2 weeks (26 times per year) in an account paying an annual interest rate r, expressed in decimal form, the amount A in the account after n years can be approximated by the formula*

$$A = x \left[\frac{(1 + r/26)^{26n} - 1}{(r/26)} \right].$$

11. If $100 are deposited every 2 weeks in an account paying 9% interest, approximate the amount in the account after 10 years.

12. Suppose that your retirement account pays 12% annual interest. Determine how much you should deposit in this account every 2 weeks, in order to have one million dollars at age 65.

Exercises 13–17: Logistic Functions and Modeling Data   *Populations of bacteria, insects, and animals do not continue to grow indefinitely. Initially, population growth may be slow. Then, as the numbers of organisms increase, so does the rate of growth. After a region has become heavily populated or saturated, the growth in population usually levels off because of limited resources. This type of growth may be modeled by a **logistic function**. One of the earliest studies of population growth was done with yeast plants in 1913. A small amount of yeast was placed in a container with a fixed amount of nourishment. The units of yeast were recorded every 2 hours, giving the data shown in the table.*

Time (hours)	0	2	4	6
Yeast (units)	9.6	29.0	71.1	174.6

Time (hours)	8	10	12
Yeast (units)	350.7	513.3	594.8

Time (hours)	14	16	18
Yeast (units)	640.8	655.9	661.8

Source: D. Brown, *Models in Biology.*

13. Make a scatterplot of the data.

14. Describe the growth of yeast and explain the graph.

15. The data are modeled by the logistic function given by

$$Y(t) = \frac{663}{(1 + 71.6(0.579)^t)}.$$

Graph Y and the data in the same viewing rectangle.

16. Determine graphically when the amount of yeast equals 400 units.

17. Solve Exercise 16 symbolically.

CHAPTERS

1–12 Cumulative Review Exercises

1. Classify each real number as one or more of the following: natural number, whole number, integer, rational number, or irrational number.
   (a) $-\frac{11}{7}$ (b) -3π

2. Write 250 as the product of prime numbers.

3. Write an algebraic expression for the phrase "Two cubed decreased by seven."

4. Find the least common denominator for the fractions.
   (a) $\frac{11}{12}, \frac{5}{8}$ (b) $\frac{1}{6}, \frac{2}{15}$

5. Write the expression $4 \cdot 4 \cdot 4 \cdot 4 \cdot 4$ in exponential form.

6. State whether the equation illustrates a commutative, associative, or distributive property.
   $$(5 - y) + 9 = 9 + (5 - y)$$

Exercises 7–10: Solve the equation or inequality. Write the solutions to the inequalities in set-builder notation.

7. $\frac{2}{3}(x - 3) + 8 = -6$

8. $\frac{1}{3}z + 6 < \frac{1}{4}z - (5z - 6)$

9. $\left(\frac{t + 2}{3}\right) - 10 = \frac{1}{3}t - (5t + 8)$

10. $-\frac{3}{5}x - 4 \le -1$

11. Solve the equation $P = \frac{J + 2z}{J}$ for J.

12. Convert 72% to fraction and decimal notation.

13. Use the graph to express the equation of the line in slope–intercept form.

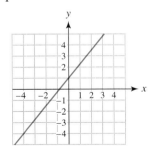

14. Write the equation of the vertical line that passes through the point $(4, 7)$.

15. Calculate the slope of the line passing through the points $(4, -1)$ and $(2, -3)$.

16. Sketch the graph of a line passing through the point $(-1, -2)$ with slope $m = 3$.

Exercises 17 and 18: Write the slope–intercept form for a line satisfying the given conditions.

17. Perpendicular to $y = -\frac{1}{7}x - 8$, passing through $(1, 1)$

18. Parallel to $y = 3x - 1$, passing through $(0, 5)$

Exercises 19–22: Solve the system of equations. Note that these systems may have zero, one, or infinitely many solutions.

19. $4x - 3y = 1$
    $5x + 2y = 7$

20. $2x - 3y = -2$
    $-6x + 9y = 5$

21. $2x - y = 7$
    $2x + 5y = 1$

22. $2x + 8y = -6$
    $-3x - 12y = 9$

23. Shade the solution set in the xy-plane.
    $$x + y > 3$$
    $$2x - y \ge 3$$

24. If the graph of a system of linear equations results in parallel lines, how many solutions does the system have—zero, one, or infinitely many?

25. Write the number 0.000429 using scientific notation.

Exercises 26–29: Simplify the expression. Write the result with positive exponents.

26. $\left(\frac{1}{d^2}\right)^{-2}$

27. $\left(\frac{8a^2}{2b^3}\right)^{-3}$

28. $\frac{(2x^{-2}y^3)^2}{xy^{-2}}$

29. $\frac{x^{-3}y}{4x^2y^{-3}}$

30. Divide.

$$(3x^3 - 2x - 15) \div (x - 2)$$

Exercises 31–34: Factor completely.

31. $2x^3 - 4x^2 + 2x$ **32.** $4a^2 - 25b^2$

33. $8t^3 - 27$

34. $4a^3 - 2a^2 + 10a - 5$

Exercises 35–38: Solve the equation.

35. $6x^2 - 7x - 10 = 0$ **36.** $9x^2 = 4$

37. $x^4 - 2x^3 = 15x^2$ **38.** $5x - 10x^2 = 0$

Exercises 39 and 40: Simplify the expression.

39. $\dfrac{x^2 + 5x + 6}{x^2 - 9} \cdot \dfrac{x - 3}{x + 2}$

40. $\dfrac{x^2 - 2x - 8}{x^2 + x - 12} \div \dfrac{(x - 4)^2}{x^2 - 16}$

Exercises 41 and 42: Solve the rational equation. Check your result.

41. $\dfrac{2}{x + 2} - \dfrac{1}{x - 2} = \dfrac{-3}{x^2 - 4}$

42. $\dfrac{3y}{y^2 + y - 2} = \dfrac{1}{y - 1} - 2$

43. Simplify the complex fraction.

$$\dfrac{\dfrac{3}{x^2} + x}{x - \dfrac{3}{x^2}}$$

44. Suppose that y varies directly as x. If $y = 15$ when $x = 3$, find y when x is 8.

45. Find the domain of $f(x) = \dfrac{10}{x + 3}$.

46. Use the table to write the formula for $f(x) = ax + b$.

x	-2	-1	0	1	2
$f(x)$	-11	-7	-3	1	5

Exercises 47 and 48: Solve the inequality. Write the solution set in interval notation.

47. $-10 \le -\frac{3}{5}x - 4 < -1$

48. $-2|t - 4| \ge -12$

49. Evaluate $\det A$.

$$A = \begin{bmatrix} -1 & -2 \\ 3 & 4 \end{bmatrix}$$

50. Find the area of the triangle by using a determinant. Assume the units are inches.

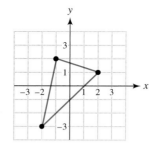

Exercises 51 and 52: Solve the system of equations.

51. $\begin{aligned} 2x - y + 3z &= -2 \\ x + 5y - 2z &= -8 \\ -3x - y - 3z &= 6 \end{aligned}$ **52.** $\begin{aligned} x + y - z &= -1 \\ -x - y - z &= -1 \\ x - 2y + z &= 1 \end{aligned}$

Exercises 53–56: Simplify the expression. Assume that all variables are positive.

53. $\left(\dfrac{x^6}{y^9}\right)^{2/3}$ **54.** $\sqrt[3]{-x^4} \cdot \sqrt[3]{-x^5}$

55. $\sqrt[3]{a^5 b^4} + 3\sqrt[3]{a^5 b}$ **56.** $(5 + \sqrt{5})(5 - \sqrt{5})$

Exercises 57 and 58: Solve. Check your answer.

57. $2(x + 1)^2 = 50$ **58.** $\sqrt{x + 6} = x$

Exercises 59 and 60: Write the complex expression in standard form.

59. $(-2 + 3i) - (-5 - 2i)$ **60.** $\dfrac{3 - i}{1 + 3i}$

61. Find the vertex of the parabola given by the function $f(x) = 3x^2 - 12x + 13$.

62. Find the maximum y-value located on the graph of $y = -2x^2 + 6x - 1$.

Exercises 63 and 64: Solve the quadratic equation using the method of your choice.

63. $x^2 - 13x + 40 = 0$ **64.** $2d^2 - 5 = d$

65. A graph of $y = ax^2 + bx + c$ is shown.
  (a) Solve $ax^2 + bx + c = 0$.
  (b) State whether $a > 0$ or $a < 0$.
  (c) Determine whether the discriminant is positive, negative, or zero.

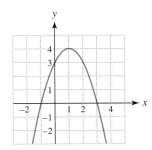

66. Solve the quadratic inequality. Write your answer in interval notation.

$$x^2 + 5x - 14 \geq 0$$

67. For $f(x) = x^2 - 2$ and $g(x) = 2x + 1$, find the following expressions.
  (a) $(f \circ g)(1)$
  (b) $(g \circ f)(x)$

68. If $800 are deposited in an account that pays 7.5% annual interest at the end of each year, approximate the amount in the account after 15 years.

Exercises 69 and 70: Evaluate without a calculator.

69. $\log_3 81$ **70.** $e^{\ln (2x)}$

71. Write the following expression by using sums and differences of logarithms of x and y. Assume x and y are positive.

$$\log \frac{\sqrt{x}}{y^2}$$

72. Write the following expression as one logarithm. Assume x and y are positive.

$$2\ln x + \ln 5x$$

Exercises 73 and 74: Solve the equation. Approximate answers to the nearest hundredth.

73. $6 \log x - 2 = 9$ **74.** $2^{3x} = 17$

APPLICATIONS

75. *Temperature Change* A traveler leaves Bogota, Colombia, where the temperature is 72°F and travels to Minneapolis, Minnesota, where the temperature is $-18°$F. Find the difference between these two temperatures.

76. *Grade Average* A student scores 87, 84, and 93 on three different 100-point tests. If the maximum score on the next test is also 100 points, what score does the student need to maintain an average of at least 90?

77. *Speed of a Truck* At 1:00 P.M. a truck driver traveling on a straight road is 450 miles from Louisville, Kentucky. At 4:00 P.M. the truck is 255 miles from Louisville.
  (a) Find the slope of the line passing through the points $(1, 450)$ and $(4, 255)$.
  (b) Interpret the slope as a rate of change.

78. *Complementary Angles* The larger of two complementary angles is 14° less than three times the smaller angle. Find each angle.

79. *World Population* If current trends continue, world population P in billions may be modeled by $P = 6(1.014)^x$, where x represents the number of years after 2000. Estimate the world population in 2011. (**Source:** United Nations Population Fund.)

80. *Burning Calories* An athlete can burn 12 calories per minute on an elliptical trainer and 10 calories per minute while running. If the athlete burns 810 calories in 75 minutes, how much time is spent on each of these activities?

81. *Ticket Prices* A theater owner discovers that the number of movie-goers varies inversely with the cost of a ticket. When the ticket price is $7.00, there are 400 tickets sold. Estimate the number of tickets that would be sold at a price of $2.00.

82. *Birth Rate* In 1990, the U.S. birth rate per 1000 people was 16.7. By 1994 this number had decreased to 15.3. Find a formula $f(x) = mx + b$, where x is the year, to model this situation. (**Source:** Bureau of the Census.)

83. *Fast Food* Three double cheeseburgers and two orders of jumbo fries cost $6, whereas five double cheeseburgers and three orders of jumbo fries cost $9.50. Find the price of each fast food item.

84. *Population growth*   The population P of a community with an annual percentage growth rate r (expressed as a decimal) after t years is given by $P = P_0(1 + r)^t$, where P_0 represents the initial population of the community. If a community with an initial population of $P_0 = 12,000$ grew to a population of $P = 14,600$ in $t = 5$ years, find the percentage growth rate for this community.

85. *Wing Span of a Bird*   The wing span L of a bird with weight W can sometimes be modeled by $L = 27.4 \sqrt[3]{W}$, where L is in inches and W is in pounds. Use this formula to estimate the weight of a bird that has a wing span of 36 inches.

86. *U.S. Energy Consumption*   From 1950 to 1970 per capita consumption of energy in millions of Btu can be modeled by $f(x) = 0.25x^2 - 975x + 950,845$, where x is the year.   (*Source:* Department of Energy.)
(a) During what year was per capita energy consumption at its lowest?
(b) What was the minimum value for per capita energy consumption?

87. *Braking Distance*   On dry, level pavement a safe braking distance d in feet for a car traveling x miles per hour is $d = \frac{x^2}{12}$. What speed corresponds to a braking distance of 350 feet?   (*Source:* F. Mannering, *Principles of Highway Engineering and Traffic Control.*)

88. *Investing for Retirement*   A college student invests $8000 in an account that pays interest annually. If the student would like this investment to be worth $1,000,000 in 45 years, what annual interest rate would the account need to pay?

89. *Modeling Wind Speed*   Wind speeds are usually measured at heights from 5 to 10 meters above the ground. For a particular day, $f(x) = 1.4 \ln(x) + 7$ computes the wind speed in meters per second x meters above the ground, where $x \geq 1$.   (*Source:* A. Miller and R. Anthes, *Meteorology.*)
(a) Find the wind speed at a height of 8 meters.
(b) Estimate the height at which the wind speed is 10 meters per second.

Conic Sections

Throughout history people have been fascinated with the universe around them and compelled to understand its mysteries. Conic sections, which include parabolas, circles, ellipses, and hyperbolas, have played an important role in gaining this understanding. Although conic sections were described and named by the Greek astronomer Apollonius in 200 B.C., not until much later were they used to model motion in the universe. In the sixteenth century Tycho Brahe, the greatest observational astronomer of the age, recorded precise data on planetary movement in the sky. Using Brahe's data in 1619, Johannes Kepler determined that planets move in elliptical orbits around the sun. In 1686 Newton used Kepler's work to show that elliptical orbits are the result of his famous theory of gravitation. We now know that all celestial objects—including planets, comets, asteroids, and satellites—travel in paths described by conic sections. Today scientists search the sky for information about the universe with enormous radio telescopes in the shape of parabolic dishes.

Conic sections have had a profound influence on people's understanding of their world and the cosmos. In this chapter we introduce you to these age-old curves.

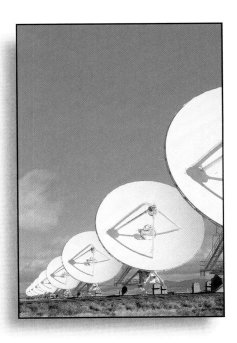

**The art of asking the right questions in
mathematics is more important than
the art of solving them.
—Georg Cantor**

Source: Historical Topics for the Mathematics Classroom, Thirty-first Yearbook, NCTM.

13.1   PARABOLAS AND CIRCLES

Types of Conic Sections · **Graphs of Parabolas with Horizontal Axes of Symmetry** ·
Equations of Circles

INTRODUCTION

In this section we discuss two types of conic sections: parabolas and circles. Recall that we
discussed parabolas with vertical axes of symmetry in Chapter 11. In this section we discuss
parabolas with horizontal axes of symmetry, but first we introduce the three basic types of
conic sections.

TYPES OF CONIC SECTIONS

Conic sections are named after the different ways that a plane can intersect a cone. The three
basic curves are parabolas, ellipses, and hyperbolas. A circle is a special case of an ellipse.
Figure 13.1 shows the three types of conic sections along with an example of the graph associated with each.

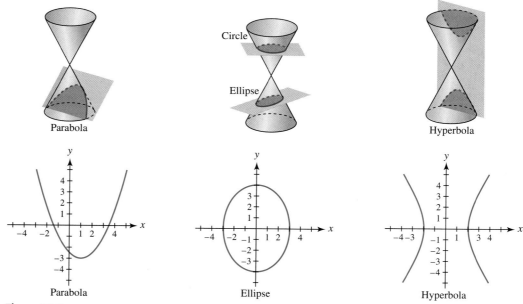

Figure 13.1

GRAPHS OF PARABOLAS WITH HORIZONTAL AXES OF SYMMETRY

Recall that the *vertex form of a parabola* with a vertical axis of symmetry is

$$y = a(x - h)^2 + k,$$

where (h, k) is the vertex. If $a > 0$, the parabola opens upward; if $a < 0$, the parabola opens downward, as shown in Figure 13.2. The preceding equation can also be expressed in the form

$$y = ax^2 + bx + c.$$

In this form the x-coordinate of the vertex is $x = -\dfrac{b}{2a}$.

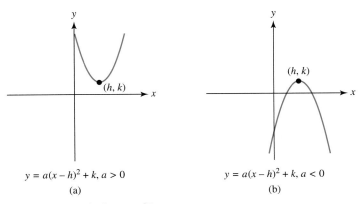

$$y = a(x - h)^2 + k, a > 0$$
(a)

$$y = a(x - h)^2 + k, a < 0$$
(b)

Figure 13.2 Vertical Axes of Symmetry

Interchanging the roles of x and y gives equations for parabolas that open to the right or the left. In this case, their axes of symmetry are horizontal.

PARABOLAS WITH HORIZONTAL AXES OF SYMMETRY

The graph of $x = a(y - k)^2 + h$ is a parabola that opens to the right if $a > 0$ and to the left if $a < 0$. The vertex of the parabola is located at (h, k).

The graph of $x = ay^2 + by + c$ is a parabola opening to the right if $a > 0$ and to the left if $a < 0$. The y-coordinate of its vertex is $y = -\dfrac{b}{2a}$.

These parabolas are illustrated in Figure 13.3.

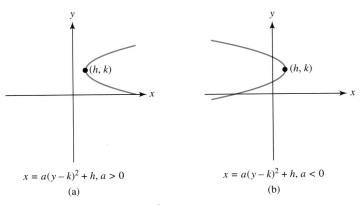

$$x = a(y - k)^2 + h, a > 0$$
(a)

$$x = a(y - k)^2 + h, a < 0$$
(b)

Figure 13.3 Horizontal Axes of Symmetry

EXAMPLE 1   Graphing a parabola

Graph $x = -\frac{1}{2}y^2$. Find its vertex and axis of symmetry.

Solution   The equation can be written in vertex form because $x = -\frac{1}{2}(y - 0)^2 + 0$. The vertex is $(0, 0)$, and because $a = -\frac{1}{2} < 0$, the parabola opens to the left. We can make a table of values, as shown in Table 13.1, and plot a few points to help determine the location and shape of the graph. To obtain Table 13.1, we first choose a y-value and then calculate an x-value. The resulting graph is shown in Figure 13.4. Its axis of symmetry is the x-axis, or $y = 0$.

TABLE 13.1

y	x
-2	-2
-1	$-\frac{1}{2}$
0	0
1	$-\frac{1}{2}$
2	-2

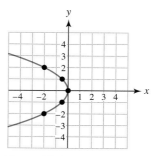

Figure 13.4

EXAMPLE 2   Graphing a parabola

Graph $x = (y - 3)^2 + 2$. Find its vertex and axis of symmetry.

Solution   As $h = 2$ and $k = 3$ in the equation $x = a(y - k)^2 + h$, the vertex is $(2, 3)$, and because $a = 1 > 0$, the parabola opens to the right. This parabola has the same shape as $y = x^2$, except that it opens to the right rather than upward. To graph this parabola we can make a table of values and plot a few points. Table 13.2 can be obtained by first choosing y-values and then calculating corresponding x-values using $x = (y - 3)^2 + 2$.

Sometimes, finding the x- and y-intercepts of the parabola is helpful when you are graphing. To find the x-intercept let $y = 0$ in $x = (y - 3)^2 + 2$. The x-intercept is $x = (0 - 3)^2 + 2 = 11$. To find any y-intercepts let $x = 0$ in $x = (y - 3)^2 + 2$. Here $0 = (y - 3)^2 + 2$ means that $(y - 3)^2 = -2$, which has no real solutions, and that this parabola has no y-intercepts.

Both the graph of the parabola and the points from Table 13.2 are shown in Figure 13.5. Note that there are no y-intercepts and that the x-intercept is correct. The axis of symmetry is $y = 3$ because, if we fold the graph on the horizontal line $y = 3$, the two sides match.

TABLE 13.2

y	x
1	6
2	3
3	2
4	3
5	6

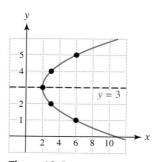

Figure 13.5

EXAMPLE 3   Graphing a parabola and finding its vertex

Identify the vertex and then graph each parabola.
(a) $x = -y^2 + 1$ **(b)** $x = y^2 - 2y - 1$

Solution   **(a)** If we rewrite $x = -y^2 + 1$ as $x = -(y - 0)^2 + 1$, then $h = 1$ and $k = 0$, so the vertex is $(1, 0)$. By letting $y = 0$ in $x = -y^2 + 1$ we find that the x-intercept is $x = -0^2 + 1 = 1$. Similarly, we let $x = 0$ in $x = -y^2 + 1$ to find the y-intercepts. The equation $0 = -y^2 + 1$ has solutions -1 and 1.

      The parabola opens to the left because $a = -1 < 0$. Additional points given in Table 13.3 will help in graphing the parabola shown in Figure 13.6.

TABLE 13.3

y	x
-2	-3
-1	0
0	1
1	0
2	-3

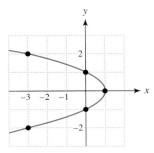

Figure 13.6

(b) The y-coordinate of the vertex is given by

$$y = -\frac{b}{2a} = -\frac{-2}{2(1)} = 1.$$

To find the x-coordinate of the vertex, substitute $y = 1$ into the given equation.

$$x = (1)^2 - 2(1) - 1 = -2$$

The vertex is $(-2, 1)$. The parabola opens to the right because $a = 1 > 0$. The additional points given in Table 13.4 help in graphing the parabola shown in Figure 13.7. Note that the y-intercepts do not have integer values and that the quadratic formula would be needed to find an approximation for these values.

TABLE 13.4

y	x
-1	2
0	-1
1	-2
2	-1
3	2

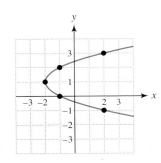

Figure 13.7

EQUATIONS OF CIRCLES

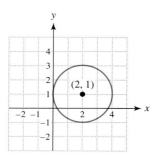

Figure 13.8

A **circle** consists of the set of points in a plane that are the same distance from a fixed point. The fixed distance is called the **radius**, and the fixed point is called the **center.** In Figure 13.8 all points lying on the circle are a distance of 2 units from the center $(2, 1)$. Therefore the radius of the circle equals 2.

We can find the equation of the circle shown in Figure 13.8 by using the distance formula. If a point (x, y) lies on the graph of a circle, its distance from the center $(2, 1)$ is 2 and

$$\sqrt{(x - 2)^2 + (y - 1)^2} = 2.$$

Squaring both sides gives

$$(x - 2)^2 + (y - 1)^2 = 2^2.$$

This equation represents the standard equation for a circle with center $(2, 1)$ and radius 2.

STANDARD EQUATION OF A CIRCLE

The **standard equation of a circle** with center (h, k) and radius r is

$$(x - h)^2 + (y - k)^2 = r^2.$$

EXAMPLE 4   Graphing a circle

Graph $x^2 + y^2 = 9$. Find the radius and center.

Solution   The equation $x^2 + y^2 = 9$ can be written in standard form as

$$(x - 0)^2 + (y - 0)^2 = 3^2.$$

Therefore the center is $(0, 0)$ and the radius is 3. Its graph is shown in Figure 13.9.

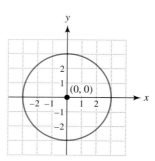

Figure 13.9

EXAMPLE 5   Graphing a circle

Graph $(x + 1)^2 + (y - 3)^2 = 4$. Find the radius and center.

Solution   Write the equation as

$$(x - (-1))^2 + (y - 3)^2 = 2^2.$$

The center is $(-1, 3)$, and the radius is 2. Its graph is shown in Figure 13.10.

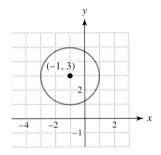

Figure 13.10

In the next example we use the *method of completing the square* to find the center and radius of a circle. (To review completing the square, refer to Sections 11.2 and 11.3.)

EXAMPLE 6   Finding the center of a circle

Find the center and radius of the circle given by $x^2 + 4x + y^2 - 6y = 5$.

Solution   Begin by writing the equation as

$$(x^2 + 4x + \underline{\hspace{0.8cm}}) + (y^2 - 6y + \underline{\hspace{0.8cm}}) = 5.$$

To complete the square, add $\left(\frac{4}{2}\right)^2 = 4$ and $\left(\frac{-6}{2}\right)^2 = 9$ to each side of the equation.

$$(x^2 + 4x + 4) + (y^2 - 6y + 9) = 5 + 4 + 9$$

Factoring the perfect square trinomials yields

$$(x + 2)^2 + (y - 3)^2 = 18.$$

The center is $(-2, 3)$, and because $18 = \left(\sqrt{18}\right)^2$, the radius is $\sqrt{18}$.

Technology Note: *Graphing Circles in a Square Viewing Rectangle*

The graph of a circle does not represent a function. One way to graph a circle with a graphing calculator is to solve the equation for y and obtain two equations. One equation gives the upper half of the circle, and the other equation gives the lower half.

For example, to graph $x^2 + y^2 = 4$ in the viewing rectangle $[-4.7, 4.7, 1]$ by $[-3.1, 3.1, 1]$ begin by solving for y.

$$x^2 + y^2 = 4 \qquad \text{Given equation}$$
$$y^2 = 4 - x^2 \qquad \text{Subtract } x^2$$
$$y = \pm\sqrt{4 - x^2} \qquad \text{Square root property}$$

Then graph $Y_1 = \sqrt{(4 - X^2)}$ and $Y_2 = -\sqrt{(4 - X^2)}$. The graph of y_1 generates the upper half of the circle, and the graph of y_2 generates the lower half of the circle, as shown in Figure 13.11.

Critical Thinking

Does the following equation represent a circle? If so, give its center and radius.

$$x^2 + y^2 + 10y = -32$$

$[-4.7, 4.7, 1]$ by $[-3.1, 3.1, 1]$ $[-4.7, 4.7, 1]$ by $[-3.1, 3.1, 1]$ $[-4.7, 4.7, 1]$ by $[-3.1, 3.1, 1]$

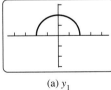

(a) y_1

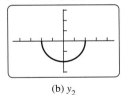

(b) y_2

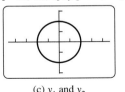

(c) y_1 and y_2

Figure 13.11

(continued)

Calculator Help
To set a square viewing rectangle, see Appendix A (page AP-6).

If a circle is not graphed in a *square viewing rectangle*, it will appear to be an oval rather than a circle. In a square viewing rectangle a circle will appear circular. Figure 13.12 shows the circle graphed in a viewing rectangle that is not square. Consult your owner's manual to learn more about square viewing rectangles for your calculator.

$[-4, 4, 1]$ by $[-5, 5, 1]$

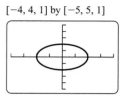

Figure 13.12

  PUTTING IT ALL TOGETHER

The following table summarizes some basic concepts about parabolas and circles.

Concept	Explanation	Example
Parabola with Horizontal Axis	Vertex form: $x = a(y - k)^2 + h$. If $a > 0$, it opens to the right; if $a < 0$, it opens to the left. The vertex is (h, k). These parabolas may also be expressed as $x = ay^2 + by + c$, where the y-coordinate of the vertex is $y = -\frac{b}{2a}$.	$x = 2(y - 1)^2 + 4$ opens to the right and its vertex is $(4, 1)$.
Standard Equation of a Circle	Standard equation: $(x - h)^2 + (y - k)^2 = r^2$. The radius is r and the center is (h, k).	$(x + 2)^2 + (y - 1)^2 = 16$ has center $(-2, 1)$ and radius 4.

13.1 EXERCISES

FOR EXTRA HELP

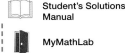
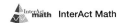

CONCEPTS

1. Name the three general types of conic sections.

2. What is the difference between the graphs of $y = ax^2 + bx + c$ and $x = ay^2 + by + c$?

3. If a parabola has a horizontal axis of symmetry, does it represent a function?

4. Sketch a graph of a parabola with a horizontal axis of symmetry.

5. If a parabola has two y-intercepts, does it represent a function? Why or why not?

6. If $x = a(y - k)^2 + h$, what is the vertex?

7. The graph of $x = -y^2$ opens to the _____.

8. The graph of $x = 2y^2 + y - 1$ opens to the _____.

9. The graph of $(x - h)^2 + (y - k)^2 = r^2$ is a _____ with center _____.

10. The graph of $x^2 + y^2 = r^2$ is a circle with center _____ and radius _____.

PARABOLAS

Exercises 11–30: Graph the parabola. Find the vertex and axis of symmetry.

11. $x = y^2$

12. $x = -y^2$

13. $x = y^2 + 1$

14. $x = y^2 - 1$

15. $x = 2y^2$

16. $x = \frac{1}{4}y^2$

17. $x = (y - 1)^2 + 2$

18. $x = (y - 2)^2 + 1$

19. $y = (x + 2)^2 + 1$

20. $y = (x - 4)^2 + 5$

21. $x = \frac{1}{2}(y + 1)^2 - 3$

22. $x = -2(y + 3)^2 + 1$

23. $x = -3(y - 1)^2$

24. $x = \frac{1}{4}(y + 2)^2 - 3$

25. $y = 2x^2 - x + 1$

26. $y = -x^2 + 2x + 2$

27. $x = \frac{1}{2}y^2 + y - 1$

28. $x = -2y^2 + 3y + 2$

29. $x = 3y^2 + y$

30. $x = -\frac{3}{2}y^2 - 2y + 1$

Exercises 31–34: Use the graph to determine the equation of the parabola. (Hint: Either $a = 1$ or $a = -1$.)

31.

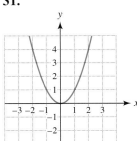

32.

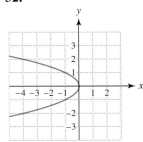

33.

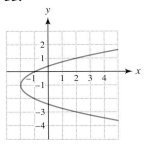

34.

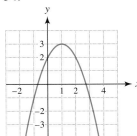

Exercises 35–38: Determine the direction that the parabola opens if it satisfies the given conditions.

35. Passing through $(2, 0)$, $(-2, 0)$, and $(0, -2)$

36. Passing through $(0, -3)$, $(0, 2)$, and $(1, 1)$

37. Vertex $(1, 2)$ passing through $(-1, -2)$ with a vertical axis

38. Vertex $(-1, 3)$ passing through $(0, 0)$ with a horizontal axis

39. What x-values are possible for the graph of $x = 2y^2$?

40. What y-values are possible for the graph of $x = 2y^2$?

41. How many y-intercepts does a parabola
$$x = a(y - k)^2 + h$$
have if $a > 0$ and $h < 0$?

42. Does the graph of $x = ay^2 + by + c$ always have a y-intercept? Explain.

43. What is the *x*-intercept for the graph of
$$x = 3y^2 - y + 1?$$

44. What are the *y*-intercepts for the graph of
$$x = y^2 - 3y + 2?$$

CIRCLES

Exercises 45–50: Write the standard equation of the circle with the given radius r and center C.

45. $r = 1$ $\quad C = (0, 0)$

46. $r = 4$ $\quad C = (2, 3)$

47. $r = 3$ $\quad C = (-1, 5)$

48. $r = 5$ $\quad C = (5, -3)$

49. $r = \sqrt{2}$ $\quad C = (-4, -6)$

50. $r = \sqrt{6}$ $\quad C = (0, 4)$

Exercises 51–54: Use the graph to find the standard equation of the circle.

51.

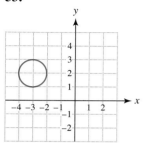

52.

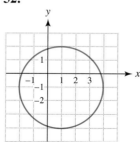

53.

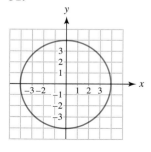

54.

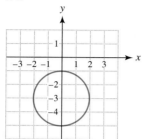

Exercises 55–64: Find the radius and center of the circle. Then graph the circle.

55. $x^2 + y^2 = 9$ $\qquad$ **56.** $x^2 + y^2 = 1$

57. $(x - 1)^2 + (y - 3)^2 = 9$

58. $(x + 2)^2 + (y + 1)^2 = 4$

59. $(x + 5)^2 + (y - 5)^2 = 25$

60. $(x - 4)^2 + (y + 3)^2 = 16$

61. $x^2 + 6x + y^2 - 2y = -1$

62. $x^2 + y^2 + 12y + 32 = 0$

63. $x^2 + 6x + y^2 - 2y + 3 = 0$

64. $x^2 - 4x + y^2 + 4y = -3$

APPLICATIONS

65. *Radio Telescopes* The Parks radio telescope has the shape of a parabolic dish, as depicted in the accompanying figure. A cross section of this telescope can be modeled by $x = \frac{32}{11,025}y^2$, where $-105 \le y \le 105$; the units are feet.

(a) Graph the cross-sectional shape of the dish in $[-40, 40, 10]$ by $[-120, 120, 20]$.

(b) Find the depth *d* of the dish.

66. *Train Tracks* To make a curve safer for trains, parabolic curves are sometimes used instead of circular curves. See the accompanying figure on the next page. (*Source:* F. Mannering and W. Kilareski, *Principles of Highway Engineering and Traffic Analysis.*)

(a) Suppose that a curve must pass through the points $(-1, 0)$, $(0, 2)$, and $(0, -2)$, where the units are kilometers. Find an equation for the train tracks in the form
$$x = a(y - h)^2 + k.$$

(b) Find another point that lies on the train tracks.

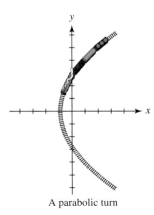

A parabolic turn

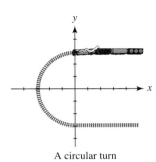

A circular turn

67. *Trajectories of Comets* Under certain circumstances, a comet can pass by the sun once and never return. In this situation the comet may travel in a parabolic path, as illustrated in the accompanying figure. Suppose that a comet's path is given by $x = -2.5y^2$, where the sun is located at $(-0.1, 0)$ and the units are astronomical units (A.U.). One astronomical unit equals 93 million miles. (*Source:* W. Thomson, *Introduction to Space Dynamics.*)

(a) Plot a point for the sun's location and then graph the path of the comet.

(b) Find the distance from the sun to the comet when the comet is located at $(-2.5, 1)$.

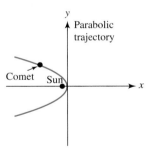

68. *Speed of a Comet* (Continuation of Exercise 67.) The velocity V in meters per second of a comet traveling in a parabolic trajectory about the sun is given by $V = \frac{k}{\sqrt{D}}$, where D is the distance from the sun in meters and $k = 1.15 \times 10^{10}$.

(a) How does the velocity of the comet change as its distance from the sun changes?

(b) Calculate the velocity of the comet when it is closest to the sun. (*Hint:* 1 mile $\approx$ 1609 meters.)

WRITING ABOUT MATHEMATICS

69. Suppose that you are given the equation

$$x = a(y - k)^2 + h.$$

(a) Explain how you can determine the direction that the parabola opens.

(b) Explain how to find the axis of symmetry and the vertex.

(c) If the points $(0, 4)$ and $(0, -2)$ lie on the graph of x, what is the axis of symmetry?

(d) Generalize part (c) if $(0, y_1)$ and $(0, y_2)$ lie on the graph of x.

70. Suppose that you are given the vertex of a parabola. Can you determine the axis of symmetry? Explain.

Group Activity: Working with Real Data

Directions: Form a group of 2 to 4 people. Select someone to record the group's responses for this activity. All members of the group should work cooperatively to answer the questions. If your instructor asks for your results, each member of the group should be prepared to respond.

Radio Telescope   The U.S. Naval Research Laboratory designed a giant radio telescope weighing 3450 tons. Its parabolic dish has a diameter of 300 feet and a depth of 44 feet, as shown in the accompanying figure.   (*Source:* J. Mar, *Structure Technology for Large Radio and Radar Telescope Systems.*)

(a) Determine an equation of the form $x = ay^2, a > 0$, that models a cross section of the dish.

 (b) Graph your equation in an appropriate viewing rectangle.

13.2   ELLIPSES AND HYPERBOLAS

Equations of Ellipses · Equations of Hyperbolas

INTRODUCTION

Celestial objects travel in paths or trajectories determined by conic sections. For this reason, conic sections have been studied for centuries. In modern times physicists have learned that subatomic particles can also travel in trajectories determined by conic sections. Recall that the three main types of conic sections are parabolas, ellipses, and hyperbolas and that circles are a special type of ellipse. In Section 11.2 and Section 13.1, we discussed parabolas and circles. In this section we focus on ellipses and hyperbolas and some of their applications.

EQUATIONS OF ELLIPSES

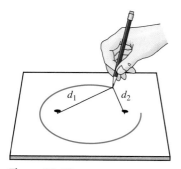

Figure 13.13

One method used to sketch an ellipse is to tie the ends of a string to two nails driven into a flat board. If a pencil is placed against the string anywhere between the nails, as shown in Figure 13.13, and is used to draw a curve, the resulting curve is an ellipse. The sum of the distances d_1 and d_2 between the pencil and each of the nails is always fixed by the length of the string. The location of the nails corresponds to the foci of the ellipse. An **ellipse** is the set of points in a plane, the sum of whose distances from two fixed points is constant. Each fixed point is called a **focus** (plural *foci*) of the ellipse.

Critical Thinking

What happens to the shape of the ellipse shown in Figure 13.13 as the nails are moved farther apart? What happens to its shape as the nails are moved closer together? When would a circle be formed?

In Figure 13.14 the **major axis** and the **minor axis** are labeled for each ellipse. The major axis is the longer of the two axes. Figure 13.14(a) shows an ellipse with a *horizontal* major axis, and Figure 13.14(b) shows an ellipse with a *vertical* major axis. The **vertices**, V_1 and V_2, of each ellipse are located at the endpoints of the major axis, and the **center** of the ellipse is the midpoint of the major axis (or the intersection of the major and minor axes).

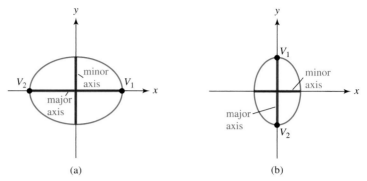

(a)   (b)

Figure 13.14

A vertical line can intersect the graph of an ellipse more than once, so an ellipse cannot be modeled by a function. However, some ellipses can be represented by the following equations.

STANDARD EQUATIONS FOR ELLIPSES CENTERED AT (0, 0)

The ellipse with center at the origin, *horizontal* major axis, and equation

$$\frac{x^2}{a^2} + \frac{y^2}{b^2} = 1, \qquad a > b > 0,$$

has vertices $(\pm a, 0)$ and endpoints of the minor axis $(0, \pm b)$.

The ellipse with center at the origin, *vertical* major axis, and equation

$$\frac{x^2}{b^2} + \frac{y^2}{a^2} = 1, \qquad a > b > 0,$$

has vertices $(0, \pm a)$ and endpoints of the minor axis $(\pm b, 0)$.

Figure 13.15(a) shows an ellipse having a horizontal major axis; Figure 13.15(b) shows one having a vertical major axis. The coordinates of the vertices V_1 and V_2 and endpoints of the minor axis U_1 and U_2 are labeled.

Critical Thinking

Suppose that $a = b$ for an ellipse centered at (0, 0). What can be said about the ellipse? Explain.

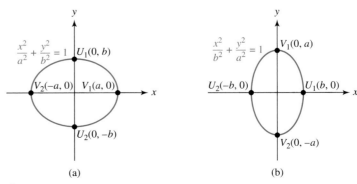

(a)   (b)

Figure 13.15

In the next example we show how to sketch graphs of ellipses.

EXAMPLE 1 Sketching graphs of ellipses

Sketch a graph of each ellipse. Label the vertices and endpoints of the minor axes.

(a) $\dfrac{x^2}{25} + \dfrac{y^2}{4} = 1$ **(b)** $9x^2 + 4y^2 = 36$

Solution **(a)** The equation $\dfrac{x^2}{25} + \dfrac{y^2}{4} = 1$ describes an ellipse with $a^2 = 25$ and $b^2 = 4$. (When you are deciding whether 25 or 4 represents a^2, let a^2 be the larger of the two numbers.) Thus $a = 5$ and $b = 2$, so the ellipse has a horizontal major axis with vertices $(\pm 5, 0)$ and the endpoints of the minor axis are $(0, \pm 2)$. Plot these four points and then sketch the ellipse, as shown in Figure 13.16(a).

(b) The equation $9x^2 + 4y^2 = 36$ can be put into standard form by dividing each side by 36.

$$9x^2 + 4y^2 = 36 \qquad \text{Given equation}$$

$$\frac{9x^2}{36} + \frac{4y^2}{36} = \frac{36}{36} \qquad \text{Divide by 36.}$$

$$\frac{x^2}{4} + \frac{y^2}{9} = 1 \qquad \text{Simplify.}$$

This ellipse has a vertical major axis with $a = 3$ and $b = 2$. The vertices are $(0, \pm 3)$, and the endpoints of the minor axis are $(\pm 2, 0)$, as shown in Figure 13.16(b).

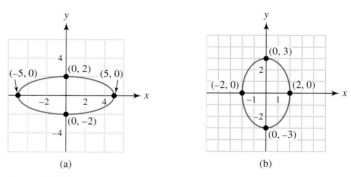

(a)   (b)

Figure 13.16

Planets travel around the sun in elliptical orbits. Astronomers have measured the values of a and b for each planet. Utilizing this information, we can find the equation of a planet's orbit, as illustrated in the next example.

EXAMPLE 2 Modeling the orbit of Mercury

Except for Pluto, the planet Mercury has the least circular orbit of the nine planets. For Mercury $a = 0.387$ and $b = 0.379$. The units are astronomical units (A.U.), where 1 A.U. equals 93 million miles—the distance between Earth and the sun. Graph $\dfrac{x^2}{a^2} + \dfrac{y^2}{b^2} = 1$ to model the orbit of Mercury in $[-0.6, 0.6, 0.1]$ by $[-0.4, 0.4, 0.1]$. Then plot the sun at the point $(0.08, 0)$.   (**Source:** M. Zeilik, *Introductory Astronomy and Astrophysics.*)

Solution   The orbit of Mercury is given by

$$\frac{x^2}{0.387^2} + \frac{y^2}{0.379^2} = 1.$$

To graph an ellipse with some graphing calculators, we must solve the equation for y. Doing so results in two equations.

$$\frac{x^2}{0.387^2} + \frac{y^2}{0.379^2} = 1$$

$$\frac{y^2}{0.379^2} = 1 - \frac{x^2}{0.387^2}$$

$$\frac{y}{0.379} = \pm\sqrt{1 - \frac{x^2}{0.387^2}}$$

$$y = \pm 0.379\sqrt{1 - \frac{x^2}{0.387^2}}$$

The orbit of Mercury results from graphing these two equations. See Figures 13.17(a) and (b). The point $(0.08, 0)$ represents the position of the sun in Figure 13.17(b).

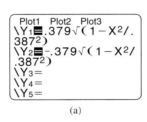

(a)

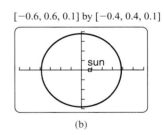

$[-0.6, 0.6, 0.1]$ by $[-0.4, 0.4, 0.1]$

(b)

Figure 13.17

Critical Thinking

Use Figure 13.17 and the information in Example 2 to estimate the minimum and maximum distances that Mercury is from the sun.

EQUATIONS OF HYPERBOLAS

The third type of conic section is the **hyperbola**, which is the set of points in a plane, the difference of whose distances from two fixed points is constant. Each fixed point is called a **focus** of the hyperbola. Figure 13.18 shows a hyperbola whose equation is

$$\frac{x^2}{4} - \frac{y^2}{9} = 1.$$

This hyperbola is centered at the origin and has two **branches**, a *left branch* and a *right branch*. The **vertices** are $(-2, 0)$ and $(2, 0)$, and the line segment connecting the vertices is called the **transverse axis**.

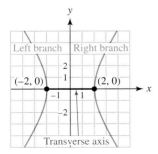

Figure 13.18

By the vertical line test, a hyperbola cannot be represented by a function, but many hyperbolas can be described by the following equations.

║║║║║ STANDARD EQUATIONS FOR HYPERBOLAS CENTERED AT (0, 0)

The hyperbola with center at the origin, *horizontal* transverse axis, and equation

$$\frac{x^2}{a^2} - \frac{y^2}{b^2} = 1$$

has vertices $(\pm a, 0)$.

The hyperbola with center at the origin, *vertical* transverse axis, and equation

$$\frac{y^2}{a^2} - \frac{x^2}{b^2} = 1$$

has vertices $(0, \pm a)$.

Hyperbolas, along with the coordinates of their vertices, are shown in Figure 13.19. The two parts of the hyperbola in Figure 13.19(a) are the *left branch* and *right branch*, whereas in Figure 13.19(b) the hyperbola has an *upper branch* and a *lower branch*. The dashed rectangle in each figure is called the **fundamental rectangle**, and its four vertices are determined by either $(\pm a, \pm b)$ or $(\pm b, \pm a)$. If its diagonals are extended, they correspond to the asymptotes of the hyperbola. The lines $y = \pm\frac{b}{a}x$ and $y = \pm\frac{a}{b}x$ are **asymptotes** for the hyperbolas, respectively, and may be used as an aid to graphing them.

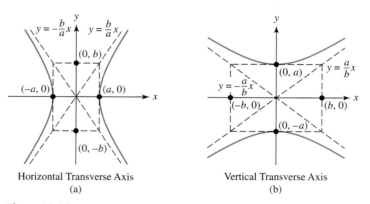

Horizontal Transverse Axis
(a)

Vertical Transverse Axis
(b)

Figure 13.19

Note: A hyperbola consists of two solid curves, or branches. The dashed lines and rectangles are not part of the actual graph but are used as an aid for sketching the graph.

One interpretation of an asymptote of a hyperbola can be based on trajectories of comets as they approach the sun. Comets travel in parabolic, elliptic, or hyperbolic trajectories. If the speed of a comet is too slow, the gravitational pull of the sun captures the comet in an elliptic orbit (see Figure 13.20(a) on the next page). If the speed of the comet is too fast, the sun's gravity is too weak to capture the comet and the comet passes by it in a hyperbolic trajectory. Near the sun the gravitational pull is stronger, and the comet's trajectory is curved. Farther from the sun, the gravitational pull becomes weaker, and the comet eventually returns to a straight-line trajectory determined by the *asymptote* of the hyperbola (see Figure 13.20(b)). Finally, if the speed is neither too slow nor too fast, the comet will travel in a parabolic path (see Figure 13.20(c)).

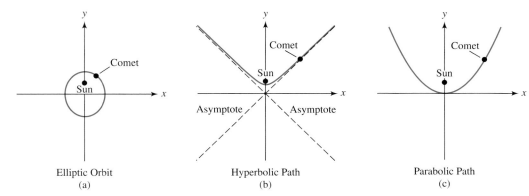

Figure 13.20

Critical Thinking

If a comet is observed at regular intervals, which type of conic section describes its path?

EXAMPLE 3   Sketching the graph of a hyperbola

Sketch a graph of $\frac{y^2}{4} - \frac{x^2}{9} = 1$. Label the vertices and show the asymptotes.

Solution   The equation is in standard form with $a^2 = 4$ and $b^2 = 9$, so $a = 2$ and $b = 3$. It has a vertical transverse axis with vertices $(0, -2)$ and $(0, 2)$. The vertices of the fundamental rectangle are $(\pm 3, \pm 2)$, that is, $(3, 2)$, $(3, -2)$, $(-3, 2)$, and $(-3, -2)$. The asymptotes are the diagonals of this rectangle and are given by $y = \pm\frac{a}{b}x$, or $y = \pm\frac{2}{3}x$. Figure 13.21 shows all these features.

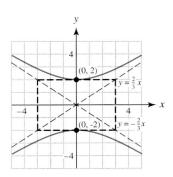

Figure 13.21

EXAMPLE 4   Determining the equation of a hyperbola from its graph

Use the graph shown in Figure 13.22 to determine an equation of the hyperbola.

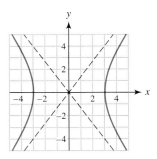

Figure 13.22

Solution The hyperbola has a horizontal transverse axis, so the x^2 term must come first in the equation. The vertices of the hyperbola are $(\pm 3, 0)$, which indicates that $a = 3$ and so $a^2 = 9$. The value of b can be found by noting that one of the asymptotes passes through the point $(3, 4)$. This asymptote has the equation $y = \frac{b}{a}x$ or $y = \frac{4}{3}x$, so let $b = 4$ and $b^2 = 16$. The equation of the hyperbola is $\dfrac{x^2}{9} - \dfrac{y^2}{16} = 1$.

Technology Note: *Graphing a Hyperbola with a Graphing Calculator*

The graph of a hyperbola does not represent a function. One way to graph a hyperbola with a graphing calculator is to solve the equation for y and obtain two equations. One equation gives the upper half of the hyperbola and the other equation gives the lower half. For example, to graph $\dfrac{y^2}{4} - \dfrac{x^2}{8} = 1$ in the viewing rectangle $[-6, 6, 1]$ by $[-4, 4, 1]$ begin by solving for y.

$$\frac{y^2}{4} = 1 + \frac{x^2}{8} \qquad \text{Add } \frac{x^2}{8}.$$

$$y^2 = 4\left(1 + \frac{x^2}{8}\right) \qquad \text{Multiply by 4.}$$

$$y = \pm\, 2\sqrt{1 + \frac{x^2}{8}} \qquad \text{Square root property}$$

Graph $Y_1 = 2\sqrt{(1 + X^2/8)}$ and $Y_2 = -2\sqrt{(1 + X^2/8)}$. The results are shown in Figure 13.23.

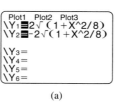

(a)

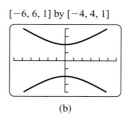

$[-6, 6, 1]$ by $[-4, 4, 1]$

(b)

Figure 13.23

13.2   PUTTING IT ALL TOGETHER

The following table summarizes some basic concepts about ellipses and hyperbolas.

Concept	Description	
Ellipses Centered at $(0, 0)$ with $a > b > 0$	*Horizontal Major Axis* Vertices: $(a, 0)$ and $(-a, 0)$ Endpoints of minor axis: $(0, b)$ and $(0, -b)$ $\dfrac{x^2}{a^2} + \dfrac{y^2}{b^2} = 1$	*Vertical Major Axis* Vertices: $(0, a)$ and $(0, -a)$ Endpoints of minor axis: $(-b, 0)$ and $(b, 0)$ $\dfrac{x^2}{b^2} + \dfrac{y^2}{a^2} = 1$

Concept	Description
Ellipses Centered at $(0, 0)$ with $a > b > 0$ (*continued*)	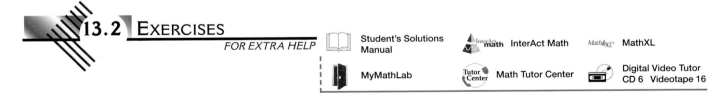
Hyperbolas Centered at $(0, 0)$ with $a > 0$ and $b > 0$	

13.2 EXERCISES

FOR EXTRA HELP

Student's Solutions Manual

MyMathLab

InterAct Math

Math Tutor Center

MathXL

Digital Video Tutor
CD 6   Videotape 16

CONCEPTS

1. Sketch an ellipse with a horizontal major axis.

2. Sketch a hyperbola with a vertical transverse axis.

3. The ellipse whose standard equation is $\frac{x^2}{a^2} + \frac{y^2}{b^2} = 1$, $a > b > 0$, has a _____ major axis.

4. The ellipse whose standard equation is $\frac{y^2}{a^2} + \frac{x^2}{b^2} = 1$, $a > b > 0$, has a _____ major axis.

5. What is the maximum number of times that a line can intersect an ellipse?

6. What is the maximum number of times that a parabola can intersect an ellipse?

7. The hyperbola whose equation is $\frac{x^2}{a^2} - \frac{y^2}{b^2} = 1$ has _____ and _____ branches.

8. The hyperbola whose equation is $\frac{y^2}{a^2} - \frac{x^2}{b^2} = 1$ has _____ and _____ branches.

9. How are the asymptotes of a hyperbola related to the fundamental rectangle?

10. Could an ellipse be centered at the origin and have vertices $(4, 0)$ and $(0, -5)$?

ELLIPSES

Exercises 11–22: Graph the ellipse. Label the vertices and endpoints of the minor axis.

11. $\dfrac{x^2}{9} + \dfrac{y^2}{25} = 1$

12. $\dfrac{y^2}{9} + \dfrac{x^2}{25} = 1$

13. $\dfrac{x^2}{9} + \dfrac{y^2}{4} = 1$

14. $\dfrac{x^2}{3} + \dfrac{y^2}{9} = 1$

15. $x^2 + \dfrac{y^2}{4} = 1$

16. $\dfrac{x^2}{9} + y^2 = 1$

17. $\dfrac{y^2}{5} + \dfrac{x^2}{7} = 1$

18. $\dfrac{y^2}{11} + \dfrac{x^2}{6} = 1$

19. $36x^2 + 4y^2 = 144$

20. $25x^2 + 16y^2 = 400$

21. $6y^2 + 7x^2 = 42$

22. $9x^2 + 5y^2 = 45$

Exercises 23–26: Use the graph to determine the equation of the ellipse.

23.

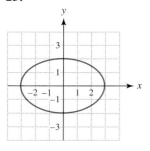

24.

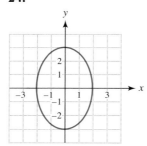

25.

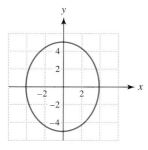

26.

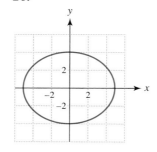

HYPERBOLAS

Exercises 27–38: Graph the hyperbola. Show the asymptotes and vertices.

27. $\dfrac{x^2}{4} - \dfrac{y^2}{9} = 1$

28. $\dfrac{y^2}{4} - \dfrac{x^2}{9} = 1$

29. $\dfrac{x^2}{25} - \dfrac{y^2}{16} = 1$

30. $\dfrac{y^2}{25} - \dfrac{x^2}{16} = 1$

31. $x^2 - y^2 = 1$

32. $y^2 - x^2 = 1$

33. $\dfrac{x^2}{3} - \dfrac{y^2}{4} = 1$

34. $\dfrac{y^2}{5} - \dfrac{x^2}{8} = 1$

35. $9y^2 - 4x^2 = 36$

36. $36x^2 - 25y^2 = 900$

37. $16x^2 - 4y^2 = 64$

38. $y^2 - 9x^2 = 9$

Exercises 39–42: Use the graph to determine an equation of the hyperbola.

39.

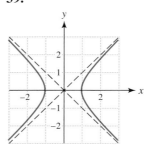

40.

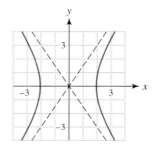

41.

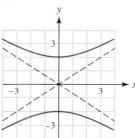

42.

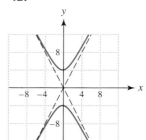

APPLICATIONS

43. *Geometry of an Ellipse* The area inside an ellipse is given by $A = \pi ab$, and its perimeter can be approximated by

$$P = 2\pi\sqrt{\dfrac{a^2 + b^2}{2}}.$$

Approximate A and P to the nearest hundredth for each ellipse.

(a) $\dfrac{x^2}{16} + \dfrac{y^2}{25} = 1$ **(b)** $\dfrac{x^2}{7} + \dfrac{y^2}{2} = 1$

44. *Geometry of an Ellipse* (Refer to Exercise 43.) If $a = b$ in the equation for an ellipse, the ellipse is a circle. Let $a = b$ in the formulas for the area and perimeter of an ellipse. Do the equations simplify to the area and perimeter for a circle? Explain.

45. *Planet Orbit* (Refer to Example 2.) The planet Pluto has the least circular orbit of any planet. For Pluto $a = 39.44$ and $b = 38.20$.

    (a) Graph the elliptic orbit of Pluto in the window $[-60, 60, 10]$ by $[-40, 40, 10]$. Plot the point $(9.82, 0)$ to show the position of the sun. Assume that the major axis is horizontal.

    (b) Use the information in Exercise 43 to determine how far Pluto travels in one orbit around the sun and approximate the area inside its orbit.

46. *Halley's Comet* (Refer to Example 2.) One of the most famous comets is Halley's comet. It travels in an elliptical orbit with $a = 17.95$ and $b = 4.44$ and passes by Earth roughly every 76 years. The most recent pass by Earth was in February 1986. (*Source:* M. Zeilik.)

    (a) Graph the orbit of Halley's comet in $[-21, 21, 5]$ by $[-14, 14, 5]$. Assume that the major axis is horizontal and that all units are in astronomical units. Plot a point at $(17.36, 0)$ to represent the position of the sun.

    (b) Use the formula in Exercise 43 to estimate how many miles Halley's comet travels in one orbit around the sun.

    (c) Estimate the average speed of Halley's comet in miles per hour.

47. *Satellite Orbit* The orbit of Explorer VII and the outline of Earth's surface are shown in the accompanying figure. This orbit is described by

$$\frac{x^2}{4464^2} + \frac{y^2}{4462^2} = 1,$$

and the surface of Earth is described by

$$\frac{(x - 164)^2}{3960^2} + \frac{y^2}{3960^2} = 1.$$

Find the maximum and minimum heights of the satellite above Earth's surface if all units are miles. (*Source:* W. Thomson, *Introduction to Space Dynamics.*)

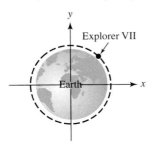

48. *Weight Machines* Elliptic shapes are used rather than circular shapes in modern weight machines. Suppose that the ellipse shown in the accompanying figure is represented by the equation

$$\frac{x^2}{16} + \frac{y^2}{100} = 1,$$

where the units are inches. Find r_1 and r_2.

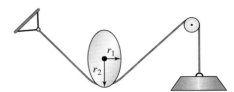

49. *Arch Bridge* The arch under a bridge is designed as the upper half of an ellipse as illustrated in the accompanying figure. Its equation is modeled by

$$400x^2 + 10,000y^2 = 4,000,000,$$

where the units are feet. Find the height and width of the arch.

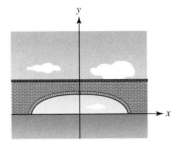

50. *Population Growth* Suppose that the population y of a country can be modeled by the upper right branch of the hyperbola

$$\frac{x^2}{a^2} - \frac{y^2}{b^2} = 1,$$

where x represents time in years. What happens to the population after a long period of time?

WRITING ABOUT MATHEMATICS

51. Explain how the values of a and b affect the graph of $\frac{x^2}{a^2} + \frac{y^2}{b^2} = 1$. Assume that $a > b > 0$.

52. Explain how the values of a and b affect the graph of $\frac{x^2}{a^2} - \frac{y^2}{b^2} = 1$. Assume that a and b are positive.

CHECKING BASIC CONCEPTS   SECTIONS 13.1 AND 13.2

1. Graph the parabola $x = (y - 2)^2 + 1$.   Find the vertex and axis of symmetry.

2. Find the equation of the circle with center $(1, -2)$ and radius 2. Graph the circle.

3. Find the x- and y-intercepts on the graph of
$$\frac{x^2}{4} + \frac{y^2}{9} = 1.$$

4. Graph the following. Label any vertices and state the type of conic section that it represents.

   (a) $x = y^2$

   (b) $\dfrac{x^2}{16} + \dfrac{y^2}{25} = 1$

   (c) $\dfrac{x^2}{4} - \dfrac{y^2}{9} = 1$

   (d) $(x - 1)^2 + (y + 2)^2 = 9$

13.3   NONLINEAR SYSTEMS OF EQUATIONS AND INEQUALITIES

Basic Concepts · Solving Nonlinear Systems of Equations ·
Solving Nonlinear Systems of Inequalities

INTRODUCTION

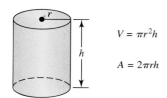

$V = \pi r^2 h$

$A = 2\pi rh$

Figure 13.24 Cylindrical Container

To describe characteristics of curved objects we often need *nonlinear equations*. The equations of the conic sections discussed in this chapter are but a few examples of nonlinear equations. For instance, cylinders have a curved shape, as illustrated in Figure 13.24. If the radius of a cylinder is denoted r and its height h, then its volume V is given by the nonlinear equation

$$V = \pi r^2 h$$

and its side area A is given by the nonlinear equation

$$A = 2\pi rh.$$

If we want to manufacture a cylindrical container that holds 35 cubic inches and whose side area is 50 square inches, we need to solve the following **system of nonlinear equations**. (This system is solved in Example 4.)

$$\pi r^2 h = 35$$
$$2\pi rh = 50$$

In this section we solve systems of nonlinear equations and inequalities.

BASIC CONCEPTS

One way to locate the points at which the line $y = 2x$ intersects the circle $x^2 + y^2 = 5$, is to graph both equations (see Figure 13.25).

The equation describing the circle is nonlinear. Another way to locate the points of intersection is symbolically, by solving the nonlinear system of equations.

$$y = 2x$$
$$x^2 + y^2 = 5$$

Linear systems of equations can have zero, one, or infinitely many solutions. It is possible for a system of nonlinear equations to have *any number* of solutions. Figure 13.25 shows that this nonlinear system of equations has two solutions: $(-1, -2)$ and $(1, 2)$.

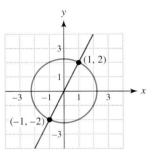

Figure 13.25

SOLVING NONLINEAR SYSTEMS OF EQUATIONS

Nonlinear systems of equations can sometimes be solved graphically, numerically, and symbolically. One symbolic technique is the **method of substitution**, which we demonstrate in the next example.

EXAMPLE 1　Solving a nonlinear system of equations symbolically

Solve

$$y = 2x$$
$$x^2 + y^2 = 5$$

symbolically. Check any solutions.

Solution　Substitute $2x$ for y in the second equation and solve for x.

$$x^2 + (2x)^2 = 5 \qquad \text{Let } y = 2x \text{ in the second equation.}$$
$$x^2 + 4x^2 = 5 \qquad \text{Properties of exponents}$$
$$5x^2 = 5 \qquad \text{Combine like terms.}$$
$$x^2 = 1 \qquad \text{Divide by 5.}$$
$$x = \pm 1 \qquad \text{Square root property}$$

To determine corresponding y-values, substitute $x = \pm 1$ into $y = 2x$; the solutions are $(1, 2)$ and $(-1, -2)$. To check $(1, 2)$, substitute $x = 1$ and $y = 2$ into the given equations.

$$2 \overset{?}{=} 2(1) \qquad \text{True}$$
$$(1)^2 + (2)^2 \overset{?}{=} 5 \qquad \text{True}$$

To check $(-1, -2)$, substitute $x = -1$ and $y = -2$ into the given equations.

$$-2 \overset{?}{=} 2(-1) \qquad \text{True}$$
$$(-1)^2 + (-2)^2 \overset{?}{=} 5 \qquad \text{True}$$

In the next example we solve a nonlinear system of equations graphically and symbolically.

EXAMPLE 2　Solving a nonlinear system of equations

Solve the nonlinear system of equations graphically and symbolically.

$$x^2 - y = 2$$
$$x^2 + y = 4$$

Solution　*Graphical Solution*　Begin by solving each equation for y.

$$y = x^2 - 2$$
$$y = 4 - x^2$$

Graph $Y_1 = X^2 - 2$ and $Y_2 = 4 - X^2$. The solutions are approximately $(-1.73, 1)$ and $(1.73, 1)$, as shown in Figure 13.26. The graphs consist of two parabolas intersecting at two points.

$[-6, 6, 1]$ by $[-4, 4, 1]$

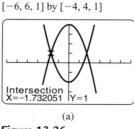

Intersection
X=-1.732051 Y=1

(a)

$[-6, 6, 1]$ by $[-4, 4, 1]$

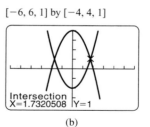

Intersection
X=1.7320508 Y=1

(b)

Figure 13.26

Symbolic Solution Solving the first equation for y gives $y = x^2 - 2$. Substitute this expression for y in the second equation and solve for x.

$$x^2 + y = 4 \qquad \text{Second equation}$$
$$x^2 + (x^2 - 2) = 4 \qquad \text{Substitute } y = x^2 - 2.$$
$$2x^2 = 6 \qquad \text{Combine like terms; add 2.}$$
$$x^2 = 3 \qquad \text{Divide by 2.}$$
$$x = \pm\sqrt{3} \qquad \text{Square root property}$$

To determine y, substitute $x = \pm\sqrt{3}$ in $y = x^2 - 2$.

$$y = (\sqrt{3})^2 - 2 = 3 - 2 = 1$$
$$y = (-\sqrt{3})^2 - 2 = 3 - 2 = 1$$

The *exact* solutions are $(\sqrt{3}, 1)$ and $(-\sqrt{3}, 1)$.

EXAMPLE 3 **Solving a nonlinear system of equations**

Solve the nonlinear system of equations symbolically and graphically.

$$x^2 - y^2 = 3$$
$$x^2 + y^2 = 5$$

Solution ***Symbolic Solution*** Instead of using substitution on this nonlinear system of equations, we use elimination. Note that, if we add the two equations, the y-variable will be eliminated.

$$x^2 - y^2 = 3$$
$$\underline{x^2 + y^2 = 5}$$
$$2x^2 \quad\quad = 8 \qquad \text{Add equations.}$$

Solving gives $x^2 = 4$, or $x = \pm 2$. To determine y, substitute 4 for x^2 in $x^2 + y^2 = 5$.

$$4 + y^2 = 5 \quad \text{or} \quad y^2 = 1$$

Because $y^2 = 1$, $y = \pm 1$. Thus there are four solutions: $(2, 1)$, $(2, -1)$, $(-2, 1)$, and $(-2, -1)$.

Graphical Solution The graph of the first equation is a hyperbola, and the graph of the second is a circle with radius $\sqrt{5}$. The four points of intersection are $(\pm 2, \pm 1)$, as shown in Figure 13.27 on the next page.

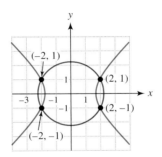

Figure 13.27

In the next example we solve the system of equations presented in the introduction.

EXAMPLE 4 Modeling the dimensions of a can

Find the dimensions of a can having a volume V of 35 cubic inches and a side area A of 50 square inches by solving the following system of nonlinear equations symbolically.

$$\pi r^2 h = 35$$
$$2\pi r h = 50$$

Solution We can find r by solving the following equation.

$$\frac{50}{2\pi r} = \frac{35}{\pi r^2} \qquad h = \frac{50}{2\pi r} \quad \text{and} \quad h = \frac{35}{\pi r^2}$$

$50\pi r^2 = 70\pi r$	Clear fractions.
$50\pi r^2 - 70\pi r = 0$	Subtract $70\pi r$.
$10\pi r(5r - 7) = 0$	Factor out $10\pi r$.
$10\pi r = 0 \quad \text{or} \quad 5r - 7 = 0$	Zero-product property
$r = 0 \quad \text{or} \quad r = \dfrac{7}{5} = 1.4$	Solve.

Because $h = \frac{50}{2\pi r}$, $r = 0$ is not possible, but we can find h by substituting **1.4** for r in the formula.

$$h = \frac{50}{2\pi(1.4)} \approx 5.68$$

A can having a volume of 35 cubic inches and a side area of 50 square inches has a radius of 1.4 inches and a height of about 5.68 inches.

SOLVING NONLINEAR SYSTEMS OF INEQUALITIES

In Section 4.4 we solved systems of linear inequalities. A **system of nonlinear inequalities** can be solved similarly by using graphical techniques. For example, consider the system of nonlinear inequalities

$$y \geq x^2 - 2$$
$$y \leq 4 - x^2.$$

The graph of $y = x^2 - 2$ is a parabola opening upward. The solution set to $y \geq x^2 - 2$ includes all points lying on or above this parabola. See Figure 13.28(a). Similarly, the graph of $y = 4 - x^2$ is a parabola opening downward. The solution set to $y \leq 4 - x^2$ includes all points lying on or below this parabola. See Figure 13.28(b).

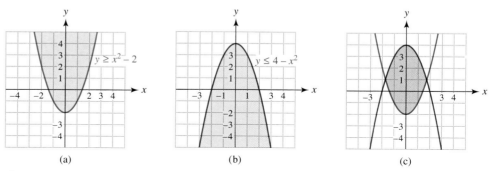

(a)　　　　　(b)　　　　　(c)

Figure 13.28

The solution set for this *system* of nonlinear inequalities includes all points (x, y) in *both* shaded regions. The *intersection* of the shaded regions is shown in Figure 13.28(c).

EXAMPLE 5　Solving a system of nonlinear inequalities graphically

Shade the solution set for the system of inequalities.

$$\frac{x^2}{4} + \frac{y^2}{9} < 1$$

$$y > 1$$

Solution　The solutions to $\frac{x^2}{4} + \frac{y^2}{9} < 1$ lie inside the ellipse $\frac{x^2}{4} + \frac{y^2}{9} = 1$. See Figure 13.29(a). Solutions to $y > 1$ lie above the line $y = 1$, as shown in Figure 13.29(b). The intersection of these two regions is shown in Figure 13.29(c). Note that a dashed curve and line are used when equality is not involved.

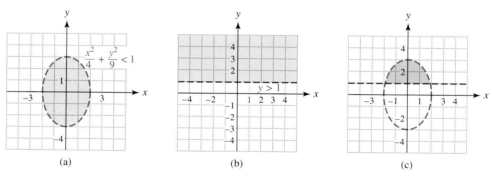

(a)　　　　　(b)　　　　　(c)

Figure 13.29

EXAMPLE 6   Solving a system of nonlinear inequalities graphically

Shade the solution set for the following system of inequalities.

$$x^2 + y \leq 4$$
$$-x + y \geq 2$$

Solution   The solutions to $x^2 + y \leq 4$ lie below the parabola $y = -x^2 + 4$, and the solutions to $-x + y \geq 2$ lie above the line $y = x + 2$. The appropriate shaded region is shown in Figure 13.30. Both the parabola and the line are solid because equality is included in both inequalities.

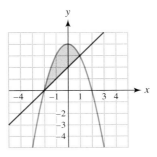

Figure 13.30

In the next example we use a graphing calculator to shade a region that lies above both graphs, using the "Y$_1$=" menu. This feature allows us to shade either above or below the graph of a function.

EXAMPLE 7   Solving a system of inequalities with a graphing calculator

Shade the solution set for the following system of inequalities.

$$y \geq x^2 - 2$$
$$y \geq -1 - x$$

Solution   Enter Y$_1$ = X^2 − 2 and Y$_2$ = −1 − X, as shown in Figure 13.31(a). Note that the option to shade above the graphs of Y$_1$ and Y$_2$ was selected to the left of Y$_1$ and Y$_2$. Then the two equations were graphed in Figure 13.31(b). The solution set corresponds to where there are both vertical and horizontal lines.

$[-4.7, 4.7, 1]$ by $[-3.1, 3.1, 1]$

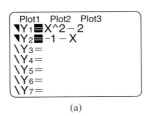

(a)

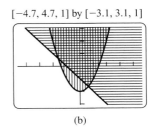

(b)

Calculator Help
To shade the solution set to a
system of inequalities, see
Appendix A (page AP-7).

Figure 13.31

13.3 PUTTING IT ALL TOGETHER

In this section we discussed nonlinear systems of equations in two variables. Unlike a linear system of equations, a nonlinear system of equations can have any number of solutions. Systems of nonlinear equations can be solved symbolically and graphically. Systems of nonlinear inequalities involving two variables usually have infinitely many solutions, which can be represented by a shaded region in the xy-plane. The following table summarizes these concepts.

Concept	Explanation
Nonlinear Systems of Equations	To solve the following system of equations symbolically, using *substitution*, solve the first equation for y. $$x + y = 5 \quad \text{or} \quad y = 5 - x$$ $$x^2 - y = 1$$ Substitute $5 - x$ for y in the second equation and solve the resulting quadratic equation. (*Elimination* can also be used on this system.) $$x^2 - (5 - x) = 1 \quad \text{or} \quad x^2 + x - 6 = 0$$ implies that $$x = -3 \quad \text{or} \quad x = 2.$$ Then $y = 5 - (-3) = 8$ or $y = 5 - 2 = 3$. The solutions are $(-3, 8)$ and $(2, 3)$. Graphical support is shown in the accompanying figure.
Nonlinear Systems of Inequalities	To solve the following system of inequalities graphically, solve each equation for y. $$x + y \le 5 \quad \text{or} \quad y \le 5 - x$$ $$x^2 - y \le 1 \quad \text{or} \quad y \ge x^2 - 1$$ The solutions lie above the parabola and below the line, as shown in the following figure.

13.3 EXERCISES

FOR EXTRA HELP

 Student's Solutions Manual

 MyMathLab

  InterAct Math

  Math Tutor Center

 MathXL

Digital Video Tutor
CD 6   Videotape 16

CONCEPTS

1. How many solutions can a system of nonlinear equations have?

2. If a system of nonlinear equations has two equations, how many equations does a solution have to satisfy?

3. Determine mentally the number of solutions to the following system of equations. Explain your reasoning.

$$y = x$$
$$x^2 + y^2 = 4$$

4. Describe the solution set to $x^2 + y^2 \le 1$.

5. Does $(-2, -1)$ satisfy $5x^2 - 2y^2 > 18$?

6. Does $(3, 4)$ satisfy $x^2 - 2y \ge 4$?

7. Sketch a parabola and ellipse with four points of intersection.

8. Sketch a line and a hyperbola with two points of intersection.

NONLINEAR SYSTEMS OF EQUATIONS

Exercises 9–12: Use the graph to estimate all solutions to the system of equations. Check each solution.

9. $x^2 + y^2 = 10$
$$y = 3x$$

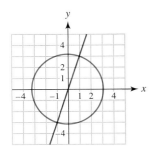

10. $x^2 + 3y^2 = 16$
$$y = -x$$

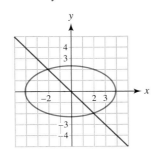

11. $y^2 - x^2 = 1$
$$x^2 + 3y^2 = 3$$

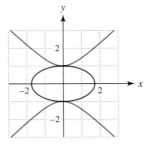

12. $y = 1 - x^2$
$$x^2 + y^2 = 1$$

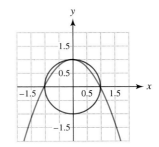

Exercises 13–18: Solve the system of equations symbolically. Check your solutions.

13. $y = 2x$
$$x^2 + y^2 = 45$$

14. $y = x$
$$y^2 = 3 - 2x^2$$

15. $x + y = 1$
$$x^2 - y^2 = 3$$

16. $y - x = -1$
$$y = 2x^2$$

17. $y - x^2 = 0$
$$x^2 + y^2 = 6$$

18. $x^2 - y^2 = 4$
$$x^2 + y^2 = 4$$

Exercises 19–22: Solve the system of equations graphically. Check your solutions.

19. $y = x^2 - 3$
$$2x^2 - y = 1 - 3x$$

20. $x + y = 2$
$$x - y^2 = 3$$

21. $y - x = -4$
$$x - y^2 = -2$$

22. $xy = 1$
$$y = x$$

Exercises 23–26: Solve the system of equations
    (a) symbolically,
    (b) graphically, and
    (c) numerically.

23. $y = -2x$
$$x^2 + y = 3$$

24. $4x - y = 0$
$$x^3 - y = 0$$

25. $xy = 1$
$$x - y = 0$$

26. $x^2 + y^2 = 4$
$$y - x = 2$$

NONLINEAR SYSTEMS OF INEQUALITIES

Exercises 27–30: Shade the solution set in the xy-plane.

27. $y \geq x^2$

28. $y \leq x^2 - 1$

29. $\dfrac{x^2}{4} + \dfrac{y^2}{9} > 1$

30. $x^2 + y^2 \leq 1$

Exercises 31–38: Shade the solution set in the xy-plane. Then use the graph to select one solution.

31. $y > x^2 + 1$
$y < 3$

32. $y > x^2$
$y < x + 2$

33. $x^2 + y^2 \leq 1$
$y < x$

34. $y > x^2 - 2$
$y \leq 2 - x^2$

35. $x^2 + y^2 \leq 1$
$(x - 2)^2 + y^2 \leq 1$

36. $x^2 - y \geq 2$
$(x + 1)^2 + y^2 \leq 4$

37. $x^2 - y^2 \leq 4$
$x^2 + y^2 \leq 9$

38. $3x + 2y < 6$
$x^2 + y^2 \leq 16$

Exercises 39 and 40: Match the inequality or system of inequalities with its graph (a. or b.).

39. $y \leq \dfrac{1}{2}x^2$

40. $y \geq x^2 + 1$
$y \leq 5$

a.

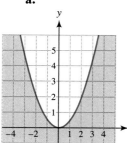

b.

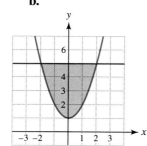

Exercises 41 and 42: Use the graph to write the inequality or system of inequalities.

41.

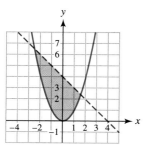

42.

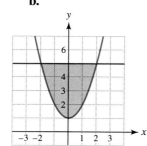

APPLICATIONS

43. *Dimensions of a Can* (Refer to Example 4.) Find the dimensions of a cylindrical container with a volume of 40 cubic inches and a side area of 50 square inches **(a)** graphically and **(b)** symbolically.

44. *Dimensions of a Can* (Refer to Example 4.) Is it possible to design an aluminum can with volume of 60 cubic inches and side area of 60 square inches? If so, find the dimensions of the can.

45. *Dimensions of a Cone* The volume V of a cone is given by $V = \dfrac{1}{3}\pi r^2 h$, and the surface area S of its side is given by $S = \pi r \sqrt{r^2 + h^2}$, where h is the height and r is the radius of the base (see the accompanying figure).

(a) Solve each equation for h.

(b) Estimate r and h graphically for a cone with volume V of 34 cubic feet and surface area S of 52 square feet.

46. *Area and Perimeter* The area of a room is 143 feet, and its perimeter is 48 feet. Let x be the width and y be the length of the room. See the accompanying figure.

(a) Write a system of equations that models this situation.

(b) Solve the system.

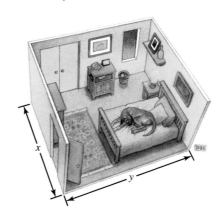

WRITING ABOUT MATHEMATICS

47. A student *incorrectly* changes the system of inequalities

$$x^2 - y \geq 6 \qquad y \geq x^2 - 6$$
$$2x - y \leq -3 \qquad y \leq 2x + 3.$$

to

The student discovers that the point $(1, 2)$ satisfies the second system but not the first. Explain the student's error.

48. Explain graphically how systems of nonlinear equations can have any number of solutions. Sketch graphs of different systems with 0, 1, 2, and 3 solutions.

CHECKING BASIC CONCEPTS   SECTION 13.3

1. Solve the following system of equations symbolically and graphically.

$$x^2 - y = 2x$$
$$2x - y = 3$$

2. Determine mentally the number of solutions to the following system of equations.

$$y = x^2 - 4$$
$$y = x$$

3. The solution set for a system of inequalities is shown in the accompanying figure.
  (a) Find one ordered pair (x, y) that is a solution and one that is not.

(b) Write the system of inequalities represented by the graph.

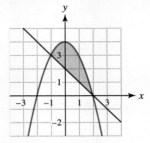

4. Shade the solution set for the following system of inequalities.

$$x^2 + y^2 \leq 4$$
$$y < 1$$

CHAPTER 13 Summary

Section 13.1   *Parabolas and Circles*

There are three basic types of conic sections: parabolas, ellipses, and hyperbolas. A parabola can have a vertical or a horizontal axis. Two forms of an equation for a parabola with a vertical axis are

$$y = ax^2 + bx + c \quad \text{and} \quad y = a(x - h)^2 + k.$$

If $a > 0$ the parabola opens upward, and if $a < 0$ it opens downward (see the figure on the left). The vertex is located at (h, k). Two forms of an equation for a parabola with a horizontal axis are

$$x = ay^2 + by + c \quad \text{and} \quad x = a(y - k)^2 + h.$$

If $a > 0$ the parabola opens to the right, and if $a < 0$ it opens to the left (see the figure on the right). The vertex is located at (h, k).

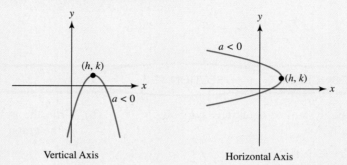

Vertical Axis   Horizontal Axis

The standard equation for a circle with center (h, k) and radius r is

$$(x - h)^2 + (y - k)^2 = r^2.$$

Section 13.2   *Ellipses and Hyperbolas*

Ellipses   The standard equation for an ellipse centered at the origin with a horizontal major axis is $\frac{x^2}{a^2} + \frac{y^2}{b^2} = 1$, $a > b > 0$, and the vertices are $(\pm a, 0)$, as shown in the figure on the left. The standard equation for an ellipse centered at the origin with a vertical major axis is $\frac{x^2}{b^2} + \frac{y^2}{a^2} = 1$, $a > b > 0$, and the vertices are $(0, \pm a)$, as shown in the figure on the right. Circles are a special type of ellipse, with the major and minor axes having equal lengths.

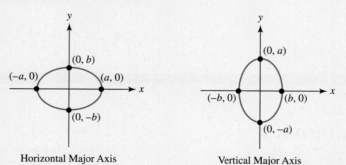

Horizontal Major Axis   Vertical Major Axis

Hyperbolas   The standard equation for a hyperbola centered at the origin with a horizontal transverse axis is $\frac{x^2}{a^2} - \frac{y^2}{b^2} = 1$, the asymptotes are given by $y = \pm\frac{b}{a}x$, and the vertices are $(\pm a, 0)$, as shown in the figure on the left. The standard equation for a hyperbola centered at the origin with a vertical transverse axis is $\frac{y^2}{a^2} - \frac{x^2}{b^2} = 1$, the asymptotes are $y = \pm\frac{a}{b}x$, and the vertices are $(0, \pm a)$, as shown in the figure on the right.

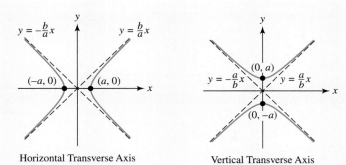

Horizontal Transverse Axis　　Vertical Transverse Axis

Section 13.3 *Nonlinear Systems of Equations and Inequalities*

Systems of nonlinear equations can have any number of solutions. The methods of substitution or elimination can often be used to solve a system of nonlinear equations symbolically. Nonlinear systems can also be solved graphically. The solution set for a system of two nonlinear inequalities with two variables is typically a region in the xy-plane. A solution is an ordered pair (x, y) that satisfies both inequalities. The solution set for $y \geq x^2 - 2$ and $y \leq 4 - \frac{1}{2}x^2$ is shaded in the accompanying figure.

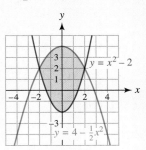

CHAPTER

13 Review Exercises

Exercises 1–6: Graph the parabola. Find the vertex and axis of symmetry.

1. $x = 2y^2$

2. $x = -(y + 1)^2$

3. $x = -2(y - 2)^2$

4. $x = (y + 2)^2 - 1$

5. $x = -3y^2 + 1$

6. $x = \frac{1}{2}y^2 + y - 3$

7. Use the graph to determine the equation of the parabola.

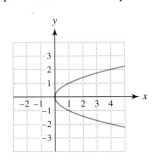

8. Use the graph to find the equation of the circle.

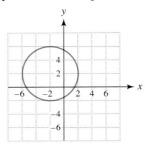

9. Write the equation of the circle with radius 1 and center $(0, 0)$.

10. Write the equation of the circle with radius 4 and center $(2, -3)$.

Exercises 11–14: Find the radius and center of the circle. Then graph the circle.

11. $x^2 + y^2 = 25$

12. $(x - 2)^2 + y^2 = 9$

13. $(x + 3)^2 + (y - 1)^2 = 5$

14. $x^2 - 2x + y^2 + 2y = 7$

SECTION 13.2

Exercises 15–18: Graph the ellipse. Label the vertices and endpoints of the minor axis.

15. $\dfrac{x^2}{4} + \dfrac{y^2}{25} = 1$

16. $x^2 + \dfrac{y^2}{4} = 1$

17. $25x^2 + 20y^2 = 500$

18. $4x^2 + 9y^2 = 36$

19. Use the graph to determine the equation of the ellipse.

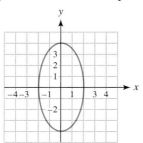

20. Use the graph to determine the equation of the hyperbola.

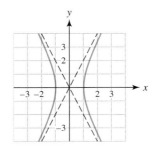

Exercises 21–24: Graph the hyperbola. Show the asymptotes.

21. $\dfrac{x^2}{9} - \dfrac{y^2}{4} = 1$ **22.** $\dfrac{y^2}{25} - \dfrac{x^2}{16} = 1$

23. $y^2 - x^2 = 1$ **24.** $25x^2 - 16y^2 = 400$

SECTION 13.3

Exercises 25–28: Use the graph to estimate all solutions to the system of equations. Check each solution.

25. $x^2 + y^2 = 9$ **26.** $xy = 2$
    $x + y = 3$ $y = 2x$

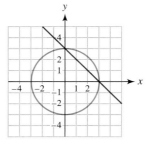   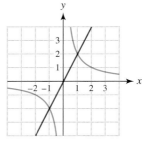

27. $x^2 - y = x$ **28.** $x^2 + y^2 = 5$
    $y = x$ $x^2 - y^2 = 3$

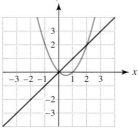   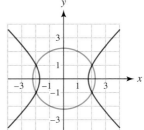

Exercises 29–32: Solve the system of equations. Check your solutions.

29. $\quad y = x$
$\quad x^2 + y^2 = 32$

30. $x - y = 4$
$\quad x^2 + y^2 = 16$

31. $\quad y = x^2$
$\quad 2x^2 + y = 3$

32. $\quad y = x^2 + 1$
$\quad 2x^2 - y = 3x - 3$

Exercises 33 and 34: Solve the system of equations graphically.

33. $2x - y = 4$
$\quad x^2 + y = 4$

34. $x^2 + y = 0$
$\quad x^2 + y^2 = 2$

Exercises 35 and 36: Solve the system of equations
(a) *symbolically,*
(b) *graphically, and*
(c) *numerically.*

35. $\quad y = x$
$\quad x^2 + 2y = 8$

36. $\quad y = x^3$
$\quad x^2 - y = 0$

Exercises 37–44: Shade the solution set in the xy-plane.

37. $y \geq 2x^2$

38. $y < 2x - 3$

39. $y < -x^2$

40. $\dfrac{x^2}{9} + \dfrac{y^2}{16} \leq 1$

41. $y - x^2 \geq 1$
$\quad y \leq 2$

42. $\quad x^2 + y \leq 4$
$\quad 3x + 2y \geq 6$

43. $y > x^2$
$\quad y < 4 - x^2$

44. $\dfrac{x^2}{4} + \dfrac{y^2}{9} > 1$
$\quad x^2 + y^2 < 16$

Exercises 45 and 46: Use the graph to write the system of inequalities.

45.

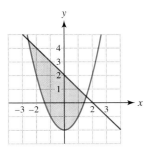

46.

APPLICATIONS

47. *Area and Perimeter*   The area of a desktop is 1000 square inches, and its perimeter is 130 inches. Let x be the width and y be the length of the desktop.
   (a) Write a system of equations that models this situation.
   (b) Solve the system graphically.
   (c) Solve the system symbolically.

48. *Numbers*   The product of two positive numbers is 60, and their difference is 7. Let x be the smaller number and y be the larger number.
   (a) Write a system of equations whose solution gives the two numbers.
   (b) Solve the system graphically.
   (c) Solve the system symbolically.

49. *Dimensions of a Container*   The volume of a cylindrical container is $V = \pi r^2 h$, and its surface area, *excluding* the top and bottom, is $A = 2\pi rh$. Find the dimensions of a container with $A = 100$ square feet and $V = 50$ cubic feet. Is your answer unique?

50. *Dimensions of a Container*   The volume of a cylindrical container is $V = \pi r^2 h$, and its surface area, *including* the top and bottom, is $A = 2\pi rh + 2\pi r^2$. Graphically find the dimensions of a can with $A = 80$ square inches and $V = 35$ cubic inches. Is your answer unique?

51. *Geometry of an Ellipse*   The area inside an ellipse is given by $A = \pi ab$, and its perimeter P can be approximated by

$$P = 2\pi\sqrt{\dfrac{a^2 + b^2}{2}}.$$

   (a) Graph $\dfrac{x^2}{5} + \dfrac{y^2}{12} = 1.$

   (b) Estimate its area and perimeter.

52. *Orbit of Mars*   Mars has an elliptical orbit that is nearly circular, with $a = 1.524$ and $b = 1.517$, where the units are astronomical units (1 A.U. = 93 million miles).   (*Source:* M. Zeilik.)

 (a) Graph the orbit of Mars in $[-3, 3, 1]$ by $[-2, 2, 1]$. Plot the point $(0.14, 0)$ to show the position of the sun. Assume that the major axis is horizontal.

(b) Use the information in Exercise 51 to estimate how far Mars travels in one orbit around the sun. Approximate the area inside its orbit.

CHAPTER

 **13** Test

1. Graph the parabola $x = (y - 4)^2 - 2$. Find the vertex and axis of symmetry.

2. Use the graph to determine the equation of the parabola.

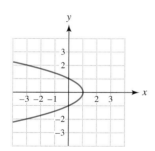

3. Use the graph to find the equation of the circle.

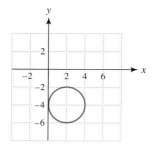

4. Write the equation of the circle with radius 10 and center $(-5, 2)$.

5. Find the radius and center of the circle given by
$$x^2 + 4x + y^2 - 6y = 3.$$
Then graph the circle.

6. Graph the ellipse $\frac{x^2}{16} + \frac{y^2}{49} = 1$. Label the vertices and endpoints of the minor axis.

7. Use the graph to determine the equation of the ellipse.

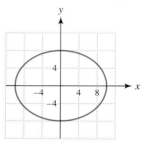

8. Graph the hyperbola $4x^2 - 9y^2 = 36$. Show the asymptotes.

9. Use the graph to estimate all solutions to the system of equations. Check each solution by substitution in the system of equations.
$$x^2 + y^2 = 16 \quad \text{and} \quad x - y = 4$$

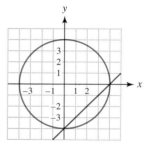

10. Solve the system of equations symbolically.
$$x - y = 3 \quad \text{and} \quad x^2 + y^2 = 17$$

11. Solve the system of equations graphically.

$$2x^2 - y = 4 \quad \text{and} \quad x^2 + y = 8$$

12. Shade the solution set in the *xy*-plane.

$$3x + y > 6 \quad \text{and} \quad x^2 + y^2 < 25$$

13. Use the graph to write the system of inequalities.

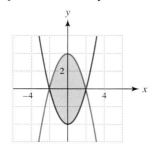

14. *Area and Perimeter*   The area of a rectangular swimming pool is 5000 square feet, and its perimeter is 300 feet. Let *x* be the width and *y* be the length of the pool.
(a) Write a system of equations that models this situation.
(b) Solve the system.

 15. *Dimensions of a Box*   The volume of a rectangular box with a square bottom and open top is $V = x^2 y$, and its surface area is $A = x^2 + 4xy$, where *x* represents its width and length and *y* represents its height. Estimate graphically the dimensions of a box with $V = 1183$ cubic inches and $A = 702$ square inches. Is your answer unique? (*Hint:* Substitute appropriate values for *V* and *A*, and then solve each equation for *y*.)

16. *Orbit of Uranus*   The planet Uranus has an elliptical orbit that is nearly circular, with $a = 19.18$ and $b = 19.16$, where units are astronomical units (1 A.U. = 93 million miles).   (***Source:*** M. Zeilik.)

 (a) Graph the orbit of Uranus in $[-30, 30, 10]$ by $[-20, 20, 10]$. Plot the point $(0.9, 0)$ to show the position of the sun. Assume that the major axis is horizontal.
(b) Find the minimum distance between Uranus and the sun.

CHAPTER
13 Extended and Discovery Exercises

Exercises 1–2: **Foci of Parabolas**   *The focus of a parabola is a point that has special significance. When a parabola is rotated about its axis, it sweeps out a shape called a **paraboloid**, as illustrated in the top figure on the following page. Paraboloids have an important reflective property. When incoming rays of light from the sun or distant stars strike the surface of a paraboloid, each ray is reflected toward the focus, as shown in the figure labeled "Reflective Property." If the rays are sunlight, intense heat is produced, which can be used to generate solar heat. Radio signals from distant space also concentrate at the focus, and scientists can measure these signals by placing a receiver there.*

The same reflective property of a paraboloid can be used in reverse. If a light source is placed at the focus, the light is reflected straight ahead, as depicted in the figure labeled "Headlight." Searchlights, flashlights, and car headlights make use of this reflective property.

The focus is always located inside a parabola, on its axis of symmetry. If the distance between the vertex and the focus is $|p|$, the following equations can be used to locate the focus. Note that the value of p may be either positive or negative.

EQUATION OF A PARABOLA WITH VERTEX (0, 0)

Vertical Axis

The parabola with a focus at $(0, p)$ has the equation $x^2 = 4py$.

Horizontal Axis

The parabola with a focus at $(p, 0)$ has the equation $y^2 = 4px$.

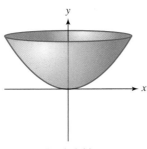

Paraboloid

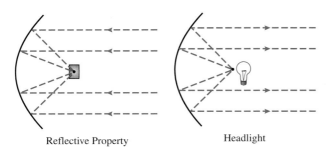

Reflective Property   Headlight

1. Sketch a graph of each parabola. Label the vertex and focus.

(a) $x^2 = 4y$ (b) $y^2 = -8x$ (c) $x = 2y^2$

2. The reflective property of paraboloids is used in satellite dishes and radio telescopes. The U.S. Naval Research Laboratory designed a giant radio telescope weighing 3450 tons. Its parabolic dish has a diameter of 300 feet and a depth of 44 feet, as shown in the accompanying figures.   (*Source:* J. Mar, *Structure Technology for Large Radio and Radar Telescope Systems.*)

(a) Find an equation in the form $y = ax^2$ that describes a cross section of this dish.

(b) If the receiver is located at the focus, how far should it be from the vertex?

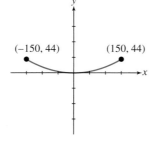

Radio Telescope

*Exercises 3 and 4:   **Translations of Ellipses and Hyperbolas**   Ellipses and hyperbolas can be translated so that they are centered at a point (h, k), rather than at the origin. These techniques are the same as those used for parabolas and circles. To translate a conic section so that it is centered at (h, k) rather than $(0, 0)$, replace x with $(x - h)$ and replace y with $(y - k)$. For example, to center $\frac{x^2}{9} + \frac{y^2}{4} = 1$ at $(-1, 2)$, change its equation to $\frac{(x + 1)^2}{9} + \frac{(y - 2)^2}{4} = 1$. See the accompanying figures.*

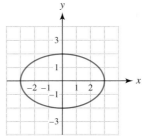

Ellipse Centered at (0, 0)

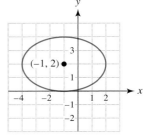

Ellipse Centered at (−1, 2)

3. Graph each conic section. For hyperbolas give the equations of the asymptotes.

(a) $\dfrac{(x - 3)^2}{25} + \dfrac{(y - 1)^2}{9} = 1$

(b) $\dfrac{(x + 1)^2}{4} + \dfrac{(y + 2)^2}{16} = 1$

(c) $\dfrac{(x + 1)^2}{4} - \dfrac{(y - 3)^2}{9} = 1$

(d) $\dfrac{(y - 4)^2}{16} - \dfrac{(x + 1)^2}{4} = 1$

4. Write the equation of an ellipse having the following properties.

(a) Horizontal major axis of length 8, minor axis of length 4, and centered at $(-3, 5)$.

(b) Vertical major axis of length 10, minor axis of length 6, and centered at $(2, -3)$.

5. Determine the center of the conic section by completing the square.

(a) $9x^2 - 18x + 4y^2 + 24y + 9 = 0$

(b) $25x^2 + 150x - 16y^2 + 32y - 191 = 0$

Sequences and Series

In this final chapter we present sequences and series, which are essential topics because they are used to model and approximate important quantities. For example, complicated population growth can be modeled with sequences, and accurate approximations for numbers such as π and e are made with series. Series also are essential to the solution of many modern applied mathematics problems.

Although you may not always recognize the impact of mathematics on everyday life, its influence is nonetheless profound. Mathematics is the *language of technology*—it allows experiences to be quantified. In the preceding chapters we showed numerous examples of mathematics being used to model the real world. Computers, CD players, highway design, weather, hurricanes, electricity, government data, cellular phones, medicine, ecology, business, sports, and psychology represent only some of the applications of mathematics. In fact, if a subject is studied in enough detail, mathematics usually appears in one form or another. Although predicting what the future may bring is difficult, one thing *is* certain—mathematics will continue to play an important role in both theoretical research and the real world.

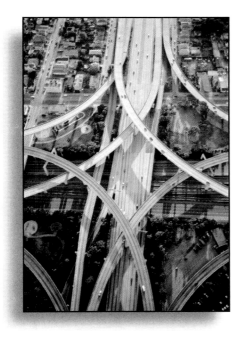

The gains of education are never really lost.
—Franklin Roosevelt

14.1 SEQUENCES

Basic Concepts · Representations of Sequences · Models and Applications

INTRODUCTION

Sequences are ordered lists. For example, names listed alphabetically represent a sequence. Figure 14.1 shows an insect population in thousands per acre over a 6-year period. Listing populations by year is another example of a sequence. In mathematics a sequence is a function, for which valid inputs must be natural numbers. For example, we can use a function f to define this sequence by letting $f(1)$ represent the insect population after 1 year, $f(2)$ represent the insect population after 2 years, and in general let $f(n)$ represent the population after n years. In this section we discuss sequences and how they can be represented.

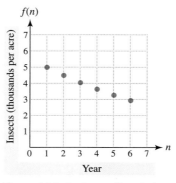

Figure 14.1 An Insect Population

BASIC CONCEPTS

Suppose that an individual's starting salary is $40,000 per year and that the person's salary is increased by 10% each year. This situation is modeled by

$$f(n) = 40{,}000(1.10)^n.$$

We do not allow the input n to be any real number, but rather limit n to a natural number because the individual's salary is constant throughout a particular year. The first 5 terms of the sequence are

$$f(1), f(2), f(3), f(4), f(5).$$

They can be computed as follows.

$$f(1) = 40{,}000(1.10)^1 = 44{,}000$$
$$f(2) = 40{,}000(1.10)^2 = 48{,}400$$
$$f(3) = 40{,}000(1.10)^3 = 53{,}240$$
$$f(4) = 40{,}000(1.10)^4 = 58{,}564$$
$$f(5) = 40{,}000(1.10)^5 \approx 64{,}420$$

This sequence is represented numerically in Table 14.1.

TABLE 14.1

n	1	2	3	4	5
$f(n)$	44,000	48,400	53,240	58,564	64,420

A graphical representation results when each data point in Table 14.1 is plotted in the *xy*-plane, as illustrated in Figure 14.2. Note that graphs of sequences are scatterplots.

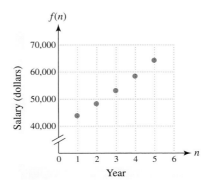

Figure 14.2

The preceding sequence is an example of a *finite sequence* of numbers. The even natural numbers,

$$2, 4, 6, 8, 10, 12, 14, \ldots$$

are an example of an *infinite sequence* represented by $f(n) = 2n$, when n is a natural number. The three dots, or periods (called an *ellipsis*), indicate that the pattern continues indefinitely.

SEQUENCES

A **finite sequence** is a function whose domain is $D = \{1, 2, 3, \ldots, n\}$ for some fixed natural number n.

An **infinite sequence** is a function whose domain is the set of natural numbers.

Because sequences are functions, many of the concepts discussed in previous chapters apply to sequences. Instead of letting y represent the output, however, the convention is to write $a_n = f(n)$, where n is a natural number in the domain of the sequence. The *terms* of a sequence are

$$a_1, a_2, a_3, \ldots, a_n, \ldots.$$

The first term is $a_1 = f(1)$, the second term is $a_2 = f(2)$, and so on. The **nth term**, or **general term**, of a sequence is $a_n = f(n)$.

EXAMPLE 1 Computing terms of a sequence

Write the first four terms of each sequence for $n = 1, 2, 3$, and 4.

(a) $f(n) = 2n - 1$ **(b)** $f(n) = 3(-2)^n$ **(c)** $f(n) = \dfrac{n}{n + 1}$

Solution **(a)** For $a_n = f(n) = 2n - 1$, we write

$$a_1 = f(1) = 2(1) - 1 = 1;$$
$$a_2 = f(2) = 2(2) - 1 = 3;$$
$$a_3 = f(3) = 2(3) - 1 = 5;$$
$$a_4 = f(4) = 2(4) - 1 = 7.$$

The first four terms are 1, 3, 5, and 7.

(b) For $a_n = f(n) = 3(-2)^n$, we write

$$a_1 = f(1) = 3(-2)^1 = -6;$$
$$a_2 = f(2) = 3(-2)^2 = 12;$$
$$a_3 = f(3) = 3(-2)^3 = -24;$$
$$a_4 = f(4) = 3(-2)^4 = 48.$$

The first four terms are $-6, 12, -24$, and 48.

(c) For $a_n = f(n) = \dfrac{n}{n + 1}$, we write

$$a_1 = f(1) = \frac{1}{1 + 1} = \frac{1}{2};$$

$$a_2 = f(2) = \frac{2}{2 + 1} = \frac{2}{3};$$

$$a_3 = f(3) = \frac{3}{3 + 1} = \frac{3}{4};$$

$$a_4 = f(4) = \frac{4}{4 + 1} = \frac{4}{5}.$$

The first four terms are $\frac{1}{2}, \frac{2}{3}, \frac{3}{4}$, and $\frac{4}{5}$. Note that, although the input to a sequence is a natural number, the output need not be a natural number.

Technology Note: *Generating Sequences*

Many graphing calculators can generate sequences if you change the MODE from function (Func) to sequence (Seq). In Figure 14.3 the sequences from Example 1 are generated. On some calculators the sequence utility is found in the LIST OPS menus. The expression

$$\text{seq}(2n - 1, n, 1, 4)$$

represents terms 1 through 4 of the sequence $a_n = 2n - 1$ with the variable n.

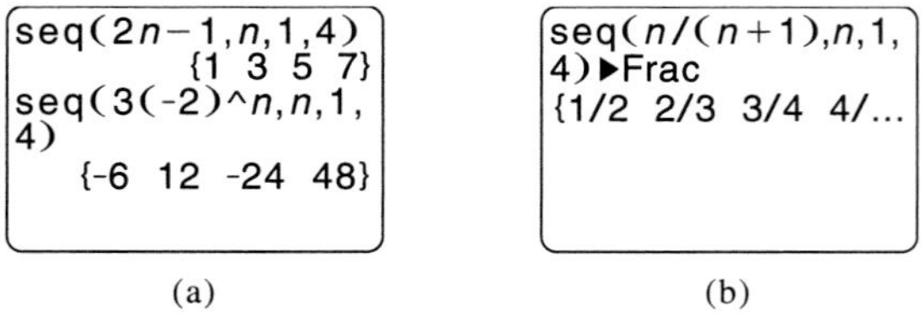

```
seq(2n-1,n,1,4)
          {1 3 5 7}
seq(3(-2)^n,n,1,
4)
    {-6 12 -24 48}
```

```
seq(n/(n+1),n,1,
4)▶Frac
{1/2 2/3 3/4 4/...
```

(a)   (b)

Figure 14.3

REPRESENTATIONS OF SEQUENCES

Because sequences are functions, they can be represented symbolically, graphically, and numerically. The next two examples illustrate such representations.

EXAMPLE 2   Using a graphical representation

Use Figure 14.4 to write the terms of the sequence.

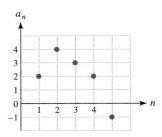

Figure 14.4

Solution   The points $(1, 2), (2, 4), (3, 3), (4, 2), (5, -1)$ are shown in the graph. The terms of the sequence are $a_1 = 2, a_2 = 4, a_3 = 3, a_4 = 2,$ and $a_5 = -1.$

EXAMPLE 3   Representing a sequence

The average person in the United States uses 100 gallons of water each day. Give symbolic, numerical, and graphical representations for a sequence that models the total amount of water used over a 7-day period.

Solution   ***Symbolic Representation***   Let $a_n = 100n$ for $n = 1, 2, 3, \ldots, 7.$

TABLE 14.2   ***Numerical Representation***   Table 14.2 contains the sequence.

Graphical Representation   Plot the points $(1, 100), (2, 200), (3, 300), (4, 400), (5, 500),$ $(6, 600),$ and $(7, 700),$ as shown in Figure 14.5.

n	a_n
1	100
2	200
3	300
4	400
5	500
6	600
7	700

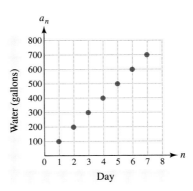

Figure 14.5 Water Usage

MODELS AND APPLICATIONS

A population model for a species of insect with a life span of 1 year can be described with a sequence. Suppose that each adult female insect produces, on average, r female offspring that survive to reproduce the following year. Let a_n represent the female insect population at the beginning of year n. Then the number of female insects is given by

$$a_n = Cr^{n-1},$$

where C is the initial population of female insects.   (**Source:** D. Brown and P. Rothery, *Models in Biology.*)

EXAMPLE 4   **Modeling numbers of insects**

Suppose that the initial population of adult female insects is 500 per acre and that $r = 1.04$. Then the average number of female insects per acre at the beginning of year n is described by

$$a_n = 500(1.04)^{n-1}.$$

Represent the female insect population numerically and graphically. Discuss the results. By what percent is the population increasing each year?

Solution   **Numerical Representation**   Table 14.3 contains approximations for the first 7 terms of the sequence. The insect population increases from 500 to about 633 insects per acre during this time period.

TABLE 14.3

n	1	2	3	4	5	6	7
a_n	500	520	540.8	562.43	584.93	608.33	632.66

Graphical Representation   Plot the points (1, 500), (2, 520), (3, 540.8), (4, 562.43), (5, 584.93), (6, 608.33), and (7, 632.66), as shown in Figure 14.6. These results indicate that the insect population gradually increases. Because the growth factor is 1.04, the population is increasing by 4% each year.

Critical Thinking

Explain how the value of r in Example 4 affects the population of female insects over time. Assume that $r > 0$.

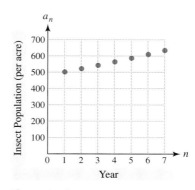

Figure 14.6

Technology Note: *Graphs and Tables of Sequences*

In the sequence mode, many graphing calculators are capable of representing sequences graphically and numerically.

Figure 14.7(a) shows how to enter the sequence from Example 4 to produce the table of values shown in Figure 14.7(b).

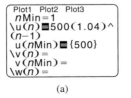

(a)

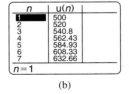

(b)

Figure 14.7

Figures 14.8(a) and (b) show the set-up for graphing the sequence from Example 4 to produce the graph shown in Figure 14.8(c).

[0, 8, 1] by [0, 800, 100]

(a)

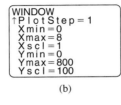

(b)

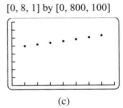

(c)

Figure 14.8

EXAMPLE 5   Interpreting a model

Figure 14.9 shows growth in a sample of bacteria over a 15-hour period when the food supply is limited. Interpret the graph.

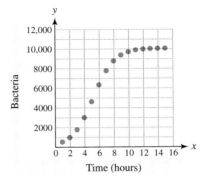

Figure 14.9

Solution   The sample of bacteria grows slowly at first and then more rapidly. Because of the limited food supply, the number of bacteria levels off near 10,000 after several hours.

14.1   PUTTING IT ALL TOGETHER

An infinite sequence is a function whose domain is the set of natural numbers. A finite sequence has the domain $D = \{1, 2, 3, \ldots, n\}$ for some fixed natural number n. Graphs of sequences are scatterplots, *not* continuous lines and curves.

Sequences are functions that may be represented symbolically, numerically, and graphically. Examples of representations of sequences are shown in the following table.

Representation	Example
Symbolic	$a_n = n - 3$ represents a sequence. The first four terms of the sequence are $-2, -1, 0,$ and 1 because $$a_1 = 1 - 3 = -2, a_2 = 2 - 3 = -1,$$ $$a_3 = 3 - 3 = 0, \text{ and } a_4 = 4 - 3 = 1.$$
Numerical	A numerical representation for $a_n = n - 3$ with $n = 1, 2, 3,$ and 4 is shown in the table. $\begin{array}{c\|cccc} n & 1 & 2 & 3 & 4 \\ \hline a_n & -2 & -1 & 0 & 1 \end{array}$
Graphical	For a graphical representation of the first four terms of $a_n = n - 3$, the points $(1, -2), (2, -1), (3, 0),$ and $(4, 1)$ are plotted.

14.1   EXERCISES

CONCEPTS

1. Give an example of a finite sequence.

2. Give an example of an infinite sequence.

3. An infinite sequence is a _____ whose domain is _____.

4. An ordered list is a _____.

5. The third term in the sequence $4, -5, 6, -7, 8$ is _____.

6. The graph of a sequence is not a continuous graph but rather a _____.

7. If $f(n)$ represents a sequence, the second term of the sequence is given by _____.

8. If a_n represents a sequence, the fourth term of the sequence is given by _____.

EVALUATING AND REPRESENTING SEQUENCES

Exercises 9–16: Write the first four terms of the sequence for $n = 1, 2, 3,$ and 4.

9. $f(n) = n^2$

10. $f(n) = 3n + 4$

11. $f(n) = \dfrac{1}{n + 5}$

12. $f(n) = 3^n$

13. $f(n) = 5\left(\frac{1}{2}\right)^n$

14. $f(n) = n^2 + 2n$

15. $f(n) = 9$

16. $f(n) = (-1)^n$

Exercises 17–24: Write the first three terms of the sequence for $n = 1, 2,$ and 3.

17. $a_n = n^3$

18. $a_n = 5 - n$

19. $a_n = \dfrac{4n}{3 + n}$

20. $a_n = 3^{-n}$

21. $a_n = 2n^2 + n - 1$

22. $a_n = n^4 - 1$

23. $a_n = -2$

24. $a_n = n^n$

Exercises 25 and 26: Use the numerical representation to evaluate $\frac{1}{2}(a_1 + a_4)$.

25.

n	1	2	3	4	5
a_n	10	8	6	4	2

26.

n	1	2	3	4	5
a_n	-5	0	10	30	60

Exercises 27–30: Use the graph to write the terms of the sequence.

27.

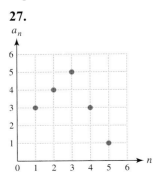

28.

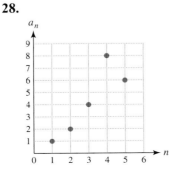

29.

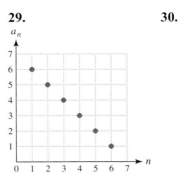

30.

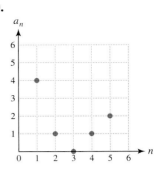

Exercises 31–36: Represent the first seven terms of the sequence numerically and graphically.

31. $a_n = n + 1$

32. $a_n = \frac{1}{2}n - \frac{1}{2}$

33. $a_n = n^2 - n$

34. $a_n = \frac{1}{2}n^2$

35. $a_n = 2^n$

36. $a_n = 2(0.5)^n$

APPLICATIONS

37. *Solid Waste* On average, each U.S. resident in 1995 generated 30 pounds of solid waste per week. Give symbolic, graphical, and numerical representations for a sequence that models the total amount of waste produced over a 7-week period. (*Source:* Environmental Protection Agency.)

38. *Carbon Dioxide Emitters* Because people burn fossil fuels, the United States emits more carbon dioxide than any other country in the world, or about 5.8 billion metric tons per year. (A metric ton is about 2200 pounds.) Give symbolic, graphical, and numerical representations for a sequence that models the total amount of carbon dioxide emitted in the United States in a 5-year period. (*Source:* Energy Information Administration.)

39. *Geometry* The lengths of the sides of a sequence of squares are given by 1, 2, 3, and 4, as shown in the following figure. Write sequences that give
(a) the areas of the squares, and
(b) the perimeters of the squares.

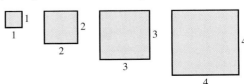

40. *Salaries* An individual's starting salary is $50,000, and the individual receives an increase of 8% per year. Give symbolic, numerical, and graphical representations for this person's salary over 5 years.

41. *Depreciation* Automobiles usually depreciate in value over time. Often, a newer automobile may be worth only 80% of its previous year's value. Suppose that a car is worth $25,000 new.
- **(a)** How much is it worth after 1 year? After 2 years?
- **(b)** Write a formula for a sequence that gives the car's value after *n* years.
- **(c)** Make a table that shows how much the car was worth each year during the first 7 years.

42. *Falling Object* The distance *d* that an object falls during *consecutive* seconds is shown in the table. For example, during the third second (from $n = 2$ to $n = 3$) an object falls a distance of 80 feet.

n (seconds)	1	2	3	4	5
d (feet)	16	48	80	112	144

- **(a)** Find values for *c* and *b* so that $d = cn + b$ models these data.
- **(b)** How far does an object fall during the sixth second?

43. *Auditorium Seating* An auditorium has 50 seats in the first row, 55 seats in the second row, 60 seats in the third row, and so on.
- **(a)** Make a table that shows the number of seats in the first seven rows.
- **(b)** Write a formula that gives the number of seats in row *n*.
- **(c)** How many seats are there in row 23?
- **(d)** Graph the number of seats in each row for $n = 1, 2, 3, \ldots, 10$.

44. *Modeling Insect Populations* (Refer to Example 4.) Suppose that the initial population of insects is 2048 per acre and that $r = 0.5$. Use a sequence to represent the insect population over a 7-year period
- **(a)** symbolically,
- **(b)** numerically, and
- **(c)** graphically.

WRITING ABOUT MATHEMATICS

45. Compare the graph of $f(x) = 2x + 1$, where *x* is a real number, with the graph of the sequence $f(n) = 2n + 1$, where *n* is a natural number.

46. Explain what a sequence is. Describe the difference between a finite and an infinite sequence.

14.2 ARITHMETIC AND GEOMETRIC SEQUENCES

Representations of Arithmetic Sequences · Representations of Geometric Sequences · Applications and Models

INTRODUCTION

Indoor air pollution has become more hazardous as people spend 80% to 90% of their time in tightly sealed, energy-efficient buildings, which often lack proper ventilation. Many contaminants such as tobacco smoke, formaldehyde, radon, lead, and carbon monoxide are often allowed to increase to unsafe levels. One way to alleviate this problem is to use efficient ventilation systems. Mathematics plays an important role in determining the proper amount of ventilation. In this section we use sequences to model ventilation in classrooms. Before implementing this model, however, we discuss the basic concepts relating to two special types of sequences.

REPRESENTATIONS OF ARITHMETIC SEQUENCES

If a sequence is defined by a linear function, it is an *arithmetic sequence*. For example,

$$f(n) = 2n - 3$$

represents an arithmetic sequence because $f(x) = 2x - 3$ defines a linear function. The first five terms of this sequence are shown in Table 14.4.

Each time n increases by 1, the next term is 2 more than the previous term. We say that the *common difference* of this arithmetic sequence is $d = 2$. That is, the difference between successive terms equals 2. When these terms are graphed, they lie on the line $y = 2x - 3$, as illustrated in Figure 14.10. Arithmetic sequences are represented by linear functions and so their graphical representations consist of collinear points (points that lie on a line). The slope m of the line equals the common difference d.

TABLE 14.4

n	$f(n)$
1	-1
2	1
3	3
4	5
5	7

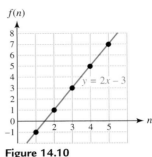

Figure 14.10

ARITHMETIC SEQUENCE

An **arithmetic sequence** is a linear function given by $a_n = dn + c$ whose domain is the set of natural numbers. The value of d is called the **common difference**.

EXAMPLE 1   Recognizing arithmetic sequences

Determine whether f is an arithmetic sequence. If it is, identify the common difference d.

(a) $f(n) = 2 - 3n$

(b)

n	$f(n)$
1	10
2	5
3	0
4	-5
5	-10

(c) A graph of f is shown in Figure 14.11.

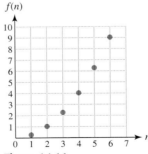

Figure 14.11

Solution   **(a)** This sequence is arithmetic because $f(x) = -3x + 2$ defines a linear function. The common difference is $d = -3$.

(b) The table reveals that each term is found by adding -5 to the previous term. This represents an arithmetic sequence with common difference $d = -5$.

(c) The sequence shown in Figure 14.11 is not an arithmetic sequence because the points are not collinear. That is, there is no common difference.

≣≣≣≣≣ MAKING CONNECTIONS ≣≣≣≣≣

Common Difference and Slope

The common difference d of an arithmetic sequence equals the slope of the line passing through the collinear points. For example, if $a_n = -2n + 4$, the common difference is -2, and the slope of the line passing through the points on the graph of a_n is also -2 (see Figure 14.12).

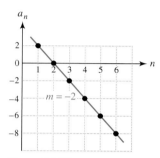

Figure 14.12

EXAMPLE 2 Finding symbolic representations

Find the general term a_n for each arithmetic sequence.
(a) $a_1 = 3$ and $d = 4$ **(b)** $a_1 = 3$ and $a_4 = 12$

Solution **(a)** Let $a_n = dn + c$. For $d = 4$, we write $a_n = 4n + c$, and to find c we use $a_1 = 3$.

$$a_1 = 4(1) + c = 3 \quad \text{or} \quad c = -1$$

Thus $a_n = 4n - 1$.

(b) Because $a_1 = 3$ and $a_4 = 12$, the common difference d equals the slope of the line passing through the points $(1, 3)$ and $(4, 12)$, or

$$d = \frac{12 - 3}{4 - 1} = 3.$$

Therefore $a_n = 3n + c$. To find c we use $a_1 = 3$ and obtain

$$a_1 = 3(1) + c = 3 \quad \text{or} \quad c = 0.$$

Thus $a_n = 3n$.

Consider the arithmetic sequence

$$1, 5, 9, 13, 17, 21, 25, 29, \ldots .$$

The common difference is $d = 4$, and the first term is $a_1 = 1$. To find the second term we add d to the first term. To find the third term we add $2d$ to the first term, and to find the fourth term we add $3d$ to the first term a_1. That is,

$$a_1 = 1,$$
$$a_2 = a_1 + 1d = 1 + 1 \cdot 4 = 5,$$
$$a_3 = a_1 + 2d = 1 + 2 \cdot 4 = 9,$$
$$a_4 = a_1 + 3d = 1 + 3 \cdot 4 = 13,$$

and, in general, a_n is determined by

$$a_n = a_1 + (n-1)d = 1 + (n-1)4.$$

This result suggests the following formula.

GENERAL TERM OF AN ARITHMETIC SEQUENCE

The nth term a_n of an arithmetic sequence is given by

$$a_n = a_1 + (n-1)d,$$

where a_1 is the first term and d is the common difference.

EXAMPLE 3   Finding terms of an arithmetic sequence

If $a_1 = 5$ and $d = 3$, find a_{54}.

Solution   To find a_{54}, apply the formula $a_n = a_1 + (n-1)d$.

$$a_{54} = 5 + (54-1)3 = 164$$

REPRESENTATIONS OF GEOMETRIC SEQUENCES

If a sequence is defined by an exponential function, it is a *geometric sequence*. For example,

$$f(n) = 3(2)^{n-1}$$

represents a geometric sequence because $f(x) = 3(2)^{x-1}$ defines an exponential function. The first five terms of this sequence are shown in Table 14.5.

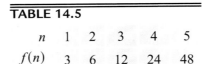

TABLE 14.5

n	1	2	3	4	5
$f(n)$	3	6	12	24	48

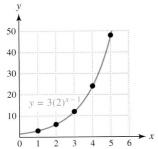

Figure 14.13

Successive terms are found by multiplying the previous term by 2. We say that the *common ratio* of this geometric sequence equals 2. Note that the ratios of successive terms are $\frac{6}{3}, \frac{12}{6}, \frac{24}{12},$ and $\frac{48}{24}$ and that they all equal the common ratio 2. When the terms in Table 14.5 are graphed, they do *not* lie on a line. Rather they lie on the exponential curve $y = 3(2)^{x-1}$, as shown in Figure 14.13. A geometric sequence with a positive common ratio is an exponential function, and its terms reflect either *exponential growth* or *exponential decay*.

GEOMETRIC SEQUENCE

A **geometric sequence** is given by $a_n = a_1(r)^{n-1}$, where n is a natural number and $r \neq 0$ or 1. The value of r is called the **common ratio**, and a_1 is the first term of the sequence.

EXAMPLE 4   Recognizing geometric sequences

Determine whether f is a geometric sequence. If it is, identify the common ratio.
(a) $f(n) = 2(0.9)^{n-1}$

(b)

n	1	2	3	4	5
$f(n)$	8	4	2	1	$\frac{1}{2}$

(c) A graph of f is shown in Figure 14.14.

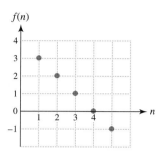

Figure 14.14

Solution **(a)** This sequence is geometric because $f(n) = 2(0.9)^{n-1}$ is an exponential function. The common ratio is $r = 0.9$.
(b) The table shows that each successive term is half the previous term. This progression represents a geometric sequence with a common ratio of $r = \frac{1}{2}$.
(c) The sequence shown in Figure 14.14 is not a geometric sequence because the points are collinear. There is no common ratio.

═══════════ MAKING CONNECTIONS ═══════════

Common Ratios and Growth or Decay Factors

If the common ratio r of a geometric sequence is positive, then r equals either the growth factor or the decay factor for an exponential function.

EXAMPLE 5   Finding symbolic representations

Find a general term a_n for each geometric sequence.
(a) $a_1 = \frac{1}{2}$ and $r = 5$ **(b)** $a_1 = 2$, $a_3 = 18$, and $r < 0$.

Solution **(a)** Let $a_n = a_1(r)^{n-1}$. Because $a_1 = \frac{1}{2}$ and $r = 5$, we can write $a_n = \frac{1}{2}(5)^{n-1}$.
(b) $a_1 = 2$ and $a_3 = 18$, so

$$\frac{a_3}{a_1} = \frac{18}{2} = 9.$$

Now $a_3 = a_1 \cdot r^2$ because a_3 is obtained from multiplying a_1 by r twice. Thus

$$r^2 = 9 \quad \text{or} \quad r = \pm 3.$$

It is specified that $r < 0$, so $r = -3$ and $a_n = 2(-3)^{n-1}$.

Critical Thinking

If we are given a_1 and a_5, can we determine the common ratio of a geometric series? Explain.

EXAMPLE 6 Finding a term of a geometric sequence

If $a_1 = 5$ and $r = 3$, find a_{10}.

Solution To find a_{10}, apply the formula $a_n = a_1(r)^{n-1}$ with $a_1 = 5$, $r = 3$, and $n = 10$.

$$a_{10} = 5(3)^{10-1} = 5(3)^9 = 98{,}415$$

APPLICATIONS AND MODELS

Sequences are frequently used to describe a variety of situations. In the next example, we use a sequence to model classroom ventilation.

EXAMPLE 7 Modeling classroom ventilation

Ventilation is an effective means for removing indoor air pollutants. According to the American Society of Heating, Refrigerating, and Air-Conditioning Engineers (ASHRAE), a classroom should have a ventilation rate of 900 cubic feet per hour per person.
(a) Write a sequence that gives the hourly ventilation necessary for 1, 2, 3, 4, and 5 people in a classroom. Is this sequence arithmetic, geometric, or neither?
(b) Write the general term for this sequence. Why is it reasonable to limit the domain to natural numbers?
(c) Find a_{30} and interpret the result.

Solution **(a)** One person requires 900 cubic feet of air circulated per hour, two people require 1800, three people 2700, and so on. The first five terms of this sequence are

$$900, 1800, 2700, 3600, 4500.$$

This sequence is arithmetic, with a common difference of 900.
(b) The nth term equals $900n$, so we let $a_n = 900n$. Because we cannot have a fraction of a person, limiting the domain to the natural numbers is reasonable.
(c) The result $a_{30} = 900(30) = 27{,}000$ indicates that a classroom with 30 people should have a ventilation rate of 27,000 cubic feet per hour.

Chlorine is frequently added to the water to disinfect swimming pools. The chlorine concentration should remain between 1.5 and 2.5 parts per million (ppm). On a warm, sunny day 30% of the chlorine may dissipate from the water. In the next example we use a sequence to model the amount of chlorine in a pool at the beginning of each day. (*Source:* D. Thomas, *Swimming Pool Operator's Handbook.*)

EXAMPLE 8 Modeling chlorine in a swimming pool

A swimming pool on a warm, sunny day begins with a high chlorine content of 4 parts per million.
(a) Write a sequence that models the amount of chlorine in the pool at the beginning of the first 3 days, assuming that no additional chlorine is added and that the days are warm and sunny. Is this sequence arithmetic, geometric, or neither?
(b) Write the general term for this sequence.
(c) At the beginning of what day does the chlorine first drop below 1.5 parts per million?

Solution **(a)** Because 30% of the chlorine dissipates, 70% remains in the water at the beginning of the next day. If the concentration at the beginning of the first day is 4 parts per million, then at the beginning of the second day it is

$$4 \cdot 0.70 = 2.8 \text{ parts per million,}$$

and at the start of the third day it is

$$2.8 \cdot 0.70 = 1.96 \text{ parts per million.}$$

The first three terms are

$$4, 2.8, 1.96.$$

Successive terms are found by multiplying the previous term by 0.7. Thus the sequence is geometric, with common ratio 0.7.
(b) The initial amount is $a_1 = 4$ and the common ratio is $r = 0.7$, so the sequence can be represented by $a_n = 4(0.7)^{n-1}$.
(c) The table shown in Figure 14.15 reveals that $a_4 = 4(0.7)^{4-1} \approx 1.372 < 1.5$. Thus, at the beginning of the fourth day, the chlorine level in the swimming pool drops below the recommended minimum of 1.5 parts per million.

n	$u(n)$
1	4
2	2.8
3	1.96
4	1.372
5	.9604
6	.67228
7	.4706

$u(n) = 4(.7)^{\wedge}(n-1)$

Figure 14.15

14.2 PUTTING IT ALL TOGETHER

In this section we discussed two types of sequences: arithmetic and geometric. Arithmetic sequences are linear functions, and geometric sequences with positive r are exponential functions. The inputs for both are limited to the natural numbers. Each successive term in an arithmetic sequence is found by adding the common difference d to the previous term. For a geometric sequence each successive term is found by multiplying the previous term by the common ratio r. The graph of an arithmetic sequence consists of points that lie on a line, whereas the graph of a geometric sequence (with a positive r) consists of points that lie on an exponential curve. Examples are shown in the following table.

Sequence	Formula	Example
Arithmetic	$a_n = dn + c$ or $a_n = a_1 + (n - 1)d$, where d is the common difference and a_1 is the first term.	If $a_n = 5n + 2$, then the common difference is $d = 5$ and the terms of the sequence are $$7, 12, 17, 22, 27, 32, 37, \ldots$$ Each term is found by adding 5 to the previous term. The general term can be written as $$a_n = 7 + 5(n - 1).$$
Geometric	$a_n = a_1(r)^{n-1}$, where r is the common ratio and a_1 is the first term.	If $a_n = 4(-2)^{n-1}$, then the common ratio is $r = -2$ and the first term is $a_1 = 4$. The terms of the sequence are $$4, -8, 16, -32, 64, -128, 256, \ldots$$ Each term is found by multiplying the previous term by -2.

14.2 EXERCISES

FOR EXTRA HELP

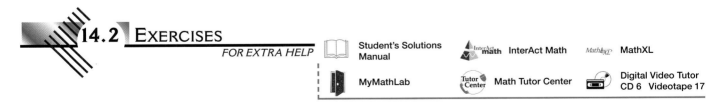

Student's Solutions Manual

InterAct Math

MathXL

MyMathLab

Math Tutor Center

Digital Video Tutor
CD 6 Videotape 17

CONCEPTS

1. An arithmetic sequence is a(n) <u>linear/exponential</u> function.

2. A geometric sequence with $r > 0$ is a(n) <u>linear/exponential</u> function.

3. Give an example of an arithmetic sequence. State the common difference.

4. Give an example of a geometric sequence. State the common ratio.

5. To find successive terms in an arithmetic sequence, _____ the common difference to the _____ term.

6. To find successive terms in a geometric sequence, _____ the previous term by the _____.

7. Find the next term in the arithmetic sequence 3, 7, 11, 15. What is the common difference?

8. Find the next term in the geometric sequence 2, −4, 8, −16. What is the common ratio?

9. Write the general term a_n for a geometric sequence, using a_1 and r.

10. Write the general term a_n for an arithmetic sequence using a_1 and d.

ARITHMETIC SEQUENCES

Exercises 11–28: (Refer to Example 1.) Determine whether f is an arithmetic sequence. Identify the common difference when possible.

11. $f(n) = 10n - 5$ **12.** $f(n) = -3n - 5$

13. $f(n) = 6 - n$ **14.** $f(n) = 6 + \frac{1}{2}n$

15. $f(n) = n^3 + 1$ **16.** $f(n) = 5\left(\frac{1}{3}\right)^{n-1}$

17.

n	1	2	3	4
$f(n)$	3	6	9	12

18.

n	1	2	3	4
$f(n)$	−7	−5	−3	−1

19.

n	1	2	3	4
$f(n)$	10	7	4	1

20.

n	1	2	3	4
$f(n)$	1	2	4	8

21.

n	1	2	3	4
$f(n)$	−4	0	8	12

22.

n	1	2	3	4
$f(n)$	1	2.5	4	5.5

23.   **24.**

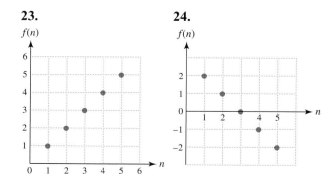

25.   **26.**

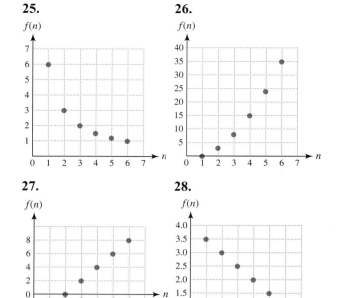

27.   **28.**

Exercises 29–34: (Refer to Example 2.) Find the general term a_n for the arithmetic sequence.

29. $a_1 = 7$ and $d = -2$ **30.** $a_1 = 5$ and $a_2 = 9$

31. $a_1 = -2$ and $a_3 = 6$

32. $a_2 = 7$ and $a_3 = 10$

33. $a_8 = 16$ and $a_{12} = 8$

34. $a_3 = 7$ and $d = -5$

Exercises 35–38: (Refer to Example 3.)

35. If $a_1 = -3$ and $d = 2$, find a_{32}.

36. If $a_1 = 2$ and $d = -3$, find a_{19}.

37. If $a_1 = -3$ and $a_2 = 0$, find a_9.

38. If $a_3 = -3$ and $d = 4$, find a_{62}.

GEOMETRIC SEQUENCES

Exercises 39–54: (Refer to Example 4.) Determine whether f is a geometric sequence. Identify the common ratio when possible.

39. $f(n) = 3^n$ **40.** $f(n) = 2(4)^n$

41. $f(n) = \frac{2}{3}(0.8)^{n-1}$ **42.** $f(n) = 7 - 3n$

43. $f(n) = 2(n-1)^2$ **44.** $f(n) = 2\left(-\frac{3}{4}\right)^{n-1}$

45.

n	1	2	3	4
$f(n)$	2	4	8	16

46.

n	1	2	3	4
$f(n)$	-6	3	-1.5	0.75

47.

n	1	2	3	4
$f(n)$	1	4	9	16

48.

n	1	2	3	4
$f(n)$	7	4	-1	-8

49.

n	1	2	3	4
$f(n)$	2	8	32	128

50.

n	1	2	3	4
$f(n)$	1	$\frac{1}{2}$	$\frac{1}{4}$	$\frac{1}{8}$

51.   **52.**

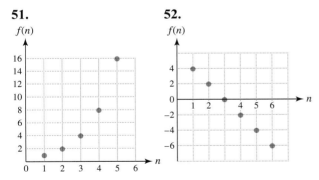

53.   **54.**

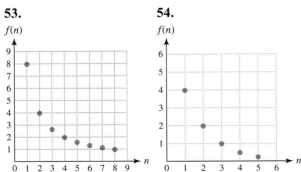

Exercises 55–60: (Refer to Example 5.) Find the general term a_n for the geometric sequence.

55. $a_1 = 1.5$ and $r = 4$ **56.** $a_1 = 3$ and $r = \frac{1}{4}$

57. $a_1 = -3$ and $a_2 = 6$

58. $a_1 = 2$ and $a_4 = 54$

59. $a_1 = 1$, $a_3 = 16$, and $r > 0$

60. $a_2 = 3$, $a_4 = 12$, and $r < 0$

Exercises 61–64: (Refer to Example 6.)

61. If $a_1 = 2$ and $r = 3$, find a_8.

62. If $a_1 = 4$ and $a_2 = 2$, find a_9.

63. If $a_1 = -1$ and $a_2 = 3$, find a_6.

64. If $a_3 = 5$ and $r = -3$, find a_7.

APPLICATIONS

65. *Room Ventilation*   (Refer to Example 7.) In areas such as bars and lounges that allow smoking, the ventilation rate should be 3000 cubic feet per hour per person.   (*Source:* ASHRAE.)
  (a) Write a sequence that gives the hourly ventilation necessary for 1, 2, 3, 4, and 5 people in a barroom. Is this sequence arithmetic, geometric, or neither?
  (b) Write the general term for this sequence.
  (c) Find a_{20} and interpret the result.
  (d) Give a graphical representation for this sequence, using $n = 1, 2, 3, \ldots, 8$. Are the points collinear?

66. *Chlorine in Swimming Pools*   (Refer to Example 8.) Suppose that the water in a swimming pool initially has a chlorine content of 3 parts per million and that 20% of the chlorine dissipates each day.
  (a) If no additional chlorine is added, write the general term for a sequence that gives the chlorine concentration at the beginning of each day.
  (b) Give a graphical representation for this sequence, using $n = 1, 2, 3, \ldots, 8$. Are the points collinear? Is this sequence arithmetic, geometric, or neither?

67. *Salary*   Suppose that an employee receives a $2000 raise each year and that the sequence a_n models the employee's salary after n years. Is this sequence arithmetic, geometric, or neither? Explain.

68. *Salary*   Suppose that an employee receives a 7% increase in salary each year and that the sequence a_n models the employee's salary after n years. Is this sequence arithmetic, geometric, or neither? Explain.

69. *Appreciation of Lake Property* A certain type of lake property in northern Minnesota is increasing in value by 15% per year. Let the sequence a_n give the value of this type of lake property at the beginning of year n.
  (a) Is a_n arithmetic, geometric, or neither? Explain your reasoning.
  (b) Write the general term for this sequence if $a_1 = \$100{,}000$.
  (c) Find a_7 and interpret the result.
  (d) Give a graph for a_n, where $n = 1, 2, 3, \ldots, 10$.

70. *Falling Object* The total distance D_n that an object falls in n seconds is shown in the table. Is the sequence arithmetic, geometric, or neither? Explain your reasoning.

n (seconds)	1	2	3	4	5
D_n (feet)	16	64	144	256	400

71. *Theater Seating* A theater has 40 seats in the first row, 42 seats in the second row, 44 seats in the third row, and so on.
  (a) Can the number of seats in each row be modeled by an arithmetic or geometric sequence? Explain.
  (b) Write the general term for a sequence a_n that gives the number of seats in row n.

(c) How many seats are there in row 20?

72. *Bouncing Ball* A tennis ball bounces back to 85% of the height from which it was dropped and then to 85% of the height of each successive bounce.
  (a) Write the general term for a sequence a_n that gives the maximum height of the ball on the nth bounce. Let $a_1 = 5$ feet.
  (b) Is the sequence arithmetic or geometric? Explain.
  (c) Find a_8 and interpret the result.

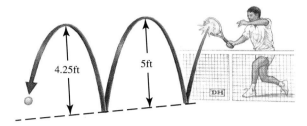
4.25ft   5ft

WRITING ABOUT MATHEMATICS

73. If you have a table of values for a sequence, how can you determine whether it is geometric? Give an example.

74. If you have a graph of a sequence, how can you determine whether it is arithmetic? Give an example.

CHECKING BASIC CONCEPTS SECTIONS 14.1 AND 14.2

1. Write the first four terms of the sequence defined by $a_n = \dfrac{n}{n+4}$

2. Represent the sequence $a_n = n + 1$ graphically and numerically for $n = 1, 2, 3, 4, 5$.

3. Use the table to determine whether the sequence is arithmetic or geometric. Write the general term for the sequence.
  (a)

n	1	2	3	4	5
a_n	−2	1	4	7	10

  (b)

n	1	2	3	4	5
a_n	3	−6	12	−24	48

4. Find the general term a_n for an arithmetic sequence with $a_1 = 5$ and $d = 2$.

5. Find the general term a_n for a geometric sequence with $a_1 = 5$ and $r = 2$.

14.3 SERIES

Basic Concepts · Arithmetic Series · Geometric Series · Summation Notation

INTRODUCTION

Although the terms *sequence* and *series* are sometimes used interchangeably in everyday life, they represent different mathematical concepts. In mathematics a sequence is a function whose domain is the set of natural numbers, whereas a series is a summation of the terms in a sequence. Series have played a central role in the development of modern mathematics. Today series are often used to approximate functions that are too complicated to have formulas. Series are also used to calculate accurate approximations for numbers such as π and e.

BASIC CONCEPTS

Suppose that a person has a starting salary of $30,000 per year and receives a $2000 raise each year. Then the *sequence*

$$30{,}000,\ 32{,}000,\ 34{,}000,\ 36{,}000,\ 38{,}000$$

lists these salaries over a 5-year period. The total amount earned is given by the *series*

$$30{,}000 + 32{,}000 + 34{,}000 + 36{,}000 + 38{,}000,$$

whose sum is $170,000. We now define the concept of a series.

FINITE SERIES

A **finite series** is an expression of the form

$$a_1 + a_2 + a_3 + \cdots + a_n.$$

EXAMPLE 1 Computing total reported AIDS cases

Table 14.6 presents a sequence a_n that computes the number of AIDS cases diagnosed each year from 1991 through 1997, where $n = 1$ corresponds to 1991.

TABLE 14.6

n	1	2	3	4	5	6	7
a_n	60,124	79,054	79,049	71,209	66,233	54,656	31,153

Source: Department of Health and Human Services.

(a) Write a series whose sum represents the total number of AIDS cases diagnosed from 1991 to 1997. Find its sum.
(b) Interpret the sum $a_1 + a_2 + a_3 + \cdots + a_{10}$.

Solution **(a)** A series that represents the total number of AIDS cases diagnosed from 1991 through 1997 is

$$60{,}124 + 79{,}054 + 79{,}049 + 71{,}209 + 66{,}233 + 54{,}656 + 31{,}153 = 441{,}478.$$

Thus 441,478 AIDS cases were diagnosed from 1991 through 1997.

(b) The series $a_1 + a_2 + a_3 + \cdots + a_{10}$ represents the total number of AIDS cases diagnosed from 1991 through 2000.

ARITHMETIC SERIES

Summing the terms of an arithmetic sequence results in an **arithmetic series**. For example, $a_n = 2n - 1$ for $n = 1, 2, 3, \ldots, 7$ defines the arithmetic sequence

$$1, 3, 5, 7, 9, 11, 13.$$

The corresponding arithmetic *series* is

$$1 + 3 + 5 + 7 + 9 + 11 + 13,$$

whose sum is 49. The following formula gives the sum of the first n terms of an arithmetic sequence.

SUM OF THE FIRST n TERMS OF AN ARITHMETIC SEQUENCE

The **sum of the first n terms of an arithmetic sequence**, denoted S_n, is found by averaging the first and nth terms and then multiplying by n. That is,

$$S_n = a_1 + a_2 + a_3 + \cdots + a_n = n\left(\frac{a_1 + a_n}{2}\right).$$

The series $1 + 3 + 5 + 7 + 9 + 11 + 13$ consists of 7 terms, where the first term is **1** and the last term is **13**. Substituting in the formula gives

$$S_7 = 7\left(\frac{1 + 13}{2}\right) = 49,$$

which agrees with the sum obtained by adding the 7 terms.

Because $a_n = a_1 + (n - 1)d$ for an arithmetic sequence, S_n can also be written

$$S_n = n\left(\frac{a_1 + a_n}{2}\right)$$

$$= \frac{n}{2}(a_1 + a_n)$$

$$= \frac{n}{2}(a_1 + a_1 + (n - 1)d)$$

$$= \frac{n}{2}(2a_1 + (n - 1)d).$$

EXAMPLE 2 Finding the sum of a finite arithmetic series

Suppose that a person has a starting annual salary of $30,000 and receives a $1500 raise each year. Calculate the total amount earned after 10 years.

Solution The sequence describing the salaries during year n is given by

$$a_n = 30{,}000 + 1500(n - 1).$$

One way to calculate the sum of the first 10 terms, denoted S_{10}, is to find a_1 and a_{10}, or

$$a_1 = 30{,}000 + 1500(1 - 1) = \mathbf{30{,}000}$$
$$a_{10} = 30{,}000 + 1500(10 - 1) = \mathbf{43{,}500}.$$

Thus the total amount earned during this 10-year period is

$$S_{10} = 10\left(\frac{a_1 + a_{10}}{2}\right)$$
$$= 10\left(\frac{30{,}000 + 43{,}500}{2}\right)$$
$$= \$367{,}500.$$

This sum can also be found with the second formula by letting $d = \mathbf{1500}$.

$$S_n = \frac{n}{2}(2a_1 + (n - 1)d)$$
$$= \frac{10}{2}(2 \cdot 30{,}000 + (10 - 1)1500)$$
$$= 5(60{,}000 + 9 \cdot 1500)$$
$$= \$367{,}500.$$

EXAMPLE 3 Finding the sum of an arithmetic series

Find the sum of the following series.

$$2 + 4 + 6 + \cdots + 100$$

Solution The first term of this series is $a_1 = \mathbf{2}$, and the common difference is $d = \mathbf{2}$. This series represents the even numbers from 2 to 100, so the number of terms is $n = \mathbf{50}$. Using the formula

$$S_n = n\left(\frac{a_1 + a_n}{2}\right)$$

for the sum of an arithmetic series, we obtain

$$S_{50} = 50\left(\frac{2 + 100}{2}\right)$$
$$= 2550.$$

Technology Note: *Sum of a Series*

The "seq(" utility, found on some calculators under the LIST OPS menus, generates a sequence. The "sum(" utility found on some calculators under the LIST MATH menus, calculates the sum of the sequence inside the parentheses. To verify the result in Example 2, let $a_n = 30{,}000 + 1500(n - 1)$. The value 367,500 for S_{10} is shown in Figure 14.16(a). The result found in Example 3 is shown in Figure 14.16(b).

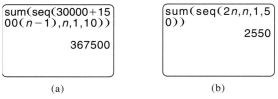

(a)   (b)

Figure 14.16

GEOMETRIC SERIES

A **geometric series** is the sum of the terms of a geometric sequence. For example,

$$1, 2, 4, 8, 16, 32$$

is a geometric sequence with $a_1 = 1$ and $r = 2$. Then

$$1 + 2 + 4 + 8 + 16 + 32$$

is a geometric series. We can use the following formula to sum a finite geometric series.

SUM OF THE FIRST n TERMS OF A GEOMETRIC SEQUENCE

If its first term is a_1 and its common ratio is r, then the **sum of the first n terms of a geometric sequence** is given by

$$S_n = a_1\left(\frac{1 - r^n}{1 - r}\right),$$

provided $r \neq 1$.

EXAMPLE 4   Finding the sum of finite geometric series

Find the sum of each series.
(a) $1 + 2 + 4 + 8 + 16 + 32 + 64 + 128 + 256$
(b) $\frac{1}{2} - \frac{1}{4} + \frac{1}{8} - \frac{1}{16} + \frac{1}{32}$

Solution (a) This series is geometric, with $n = 9$, $a_1 = 1$, and $r = 2$, so

$$S_9 = 1\left(\frac{1 - 2^9}{1 - 2}\right) = 511.$$

(b) This series is geometric, with $n = 5$, $a_1 = \frac{1}{2} = 0.5$, and $r = -\frac{1}{2} = -0.5$, so

$$S_5 = 0.5\left(\frac{1 - (-0.5)^5}{1 - (-0.5)}\right) = \frac{11}{32} = 0.34375.$$

ANNUITIES A sum of money from which regular payments are made is called an **annuity**. An annuity may be purchased with a lump sum deposit or by deposits made at various intervals. Suppose that $1000 are deposited at the end of each year in an annuity account that pays an annual interest rate I expressed as a decimal. At the end of the first year the account contains $1000. At the end of the second year $1000 are deposited again. In addition, the first deposit of $1000 would have received interest during the second year. Therefore the value of the annuity after 2 years is

$$1000 + 1000(1 + I).$$

After 3 years the balance is

$$1000 + 1000(1 + I) + 1000(1 + I)^2,$$

and after n years this amount is given by

$$1000 + 1000(1 + I) + 1000(1 + I)^2 + \cdots + 1000(1 + I)^{n-1}.$$

This series is a geometric series with its first term $a_1 = 1000$ and the common ratio $r = (1 + I)$. The sum of the first n terms is given by

$$S_n = a_1 \left(\frac{1 - (1 + I)^n}{1 - (1 + I)} \right) = a_1 \left(\frac{(1 + I)^n - 1}{I} \right).$$

EXAMPLE 5 Finding the future value of an annuity

Suppose that a 20-year-old worker deposits $1000 into an annuity account at the end of each year. If the interest rate is 12%, find the future value of the annuity when the worker is 65 years old.

Solution Let $a_1 = 1000$, $I = 0.12$, and $n = 45$. The future value of the annuity is

$$S_n = a_1 \left(\frac{(1 + I)^n - 1}{I} \right)$$

$$= 1000 \left(\frac{(1 + 0.12)^{45} - 1}{0.12} \right)$$

$$\approx \$1{,}358{,}230.$$

SUMMATION NOTATION

Summation notation is used to write series efficiently. The symbol Σ, the uppercase Greek letter *sigma*, is used to indicate a sum.

SUMMATION NOTATION

$$\sum_{k=1}^{n} a_k = a_1 + a_2 + a_3 + \cdots + a_n$$

The letter k is called the **index of summation**. The numbers 1 and n represent the subscripts of the first and last terms in the series. They are called the **lower limit** and **upper limit** of the summation, respectively.

EXAMPLE 6   Using summation notation

Evaluate each series.

(a) $\displaystyle\sum_{k=1}^{5} k^2$ (b) $\displaystyle\sum_{k=1}^{4} 5$ (c) $\displaystyle\sum_{k=3}^{6} (2k - 5)$

Solution   (a) $\displaystyle\sum_{k=1}^{5} k^2 = 1^2 + 2^2 + 3^2 + 4^2 + 5^2 = 55$

(b) $\displaystyle\sum_{k=1}^{4} 5 = 5 + 5 + 5 + 5 = 20$

(c) $\displaystyle\sum_{k=3}^{6} (2k - 5) = \underbrace{(2(3) - 5)}_{k=3} + \underbrace{(2(4) - 5)}_{k=4} + \underbrace{(2(5) - 5)}_{k=5} + \underbrace{(2(6) - 5)}_{k=6}$

$$= 1 + 3 + 5 + 7 = 16$$

Summation notation is used frequently in statistics. The next example demonstrates how averages can be expressed in summation notation.

EXAMPLE 7   Applying summation notation

Express the average of the n numbers $x_1, x_2, x_3, \ldots, x_n$ in summation notation.

Solution   The average of n numbers can be written as

$$\frac{x_1 + x_2 + x_3 + \cdots + x_n}{n}.$$

This expression is equivalent to $\frac{1}{n}\left(\sum_{k=1}^{n} x_k\right)$.

Series play an essential role in various applications, as illustrated by the next example.

EXAMPLE 8   Modeling air filtration

Suppose that an air filter removes 90% of the impurities entering it.
(a) Find a series that represents the amount of impurities removed by a sequence of n air filters. Express this answer in summation notation.
(b) How many air filters would be necessary to remove 99.99% of the impurities?

Solution   (a) The first filter removes 90% of the impurities, so 10%, or 0.1, passes through it. Of the 0.1 that passes through the first filter, 90% is removed by the second filter, while 10% of 10%, or 0.01, passes through. Then, 10% of 0.01, or 0.001, passes through the third filter. Figure 14.17 depicts these results, from which we can establish a pattern. If we let 100%, or 1, represent the amount of impurities entering the first air filter, the amount removed by n filters equals

$$(0.9)(1) + (0.9)(0.1) + (0.9)(0.01) + (0.9)(0.001) + \cdots + (0.9)(0.1)^{n-1}.$$

In summation notation we write this series as $\sum_{k=1}^{n} 0.9(0.1)^{k-1}$.

100%   10%   1%   0.1%

Figure 14.17 Impurities
Passing Through Air Filters

(b) To remove 99.99%, or 0.9999, of the impurities requires 4 air filters, because

$$\sum_{k=1}^{4} 0.9(0.1)^{k-1} = (0.9)(1) + (0.9)(0.1) + (0.9)(0.01) + (0.9)(0.001)$$
$$= 0.9 + 0.09 + 0.009 + 0.0009$$
$$= 0.9999.$$

14.3 PUTTING IT ALL TOGETHER

A sequence is an ordered list such as

$$a_1, a_2, a_3, a_4, a_5, \ldots, a_n.$$

A series is the summation of the terms of a sequence and can be expressed as

$$a_1 + a_2 + a_3 + a_4 + a_5 + \cdots + a_n.$$

The following table summarizes concepts related to arithmetic and geometric series.

Series	Description	Example
Finite Arithmetic	$a_1 + a_2 + a_3 + \cdots + a_n,$ where $a_n = dn + c$ or $a_n = a_1 + (n-1)d.$ The sum of the first n terms is $$S_n = n\left(\frac{a_1 + a_n}{2}\right) \quad \text{or}$$ $$S_n = \frac{n}{2}(2a_1 + (n-1)d),$$ where a_1 is the first term and d is the common difference.	The series $$4 + 7 + 10 + 13 + 16 + 19 + 22$$ is defined by $$a_n = 3n + 1 \quad \text{or} \quad a_n = 4 + 3(n-1).$$ Its sum is $$S_7 = 7\left(\frac{4 + 22}{2}\right) = 91 \quad \text{or}$$ $$S_7 = \frac{7}{2}(2 \cdot 4 + (7-1)3) = 91.$$
Finite Geometric	$a_1 + a_2 + a_3 + \cdots + a_n,$ where $a_n = a_1(r)^{n-1}$ for nonzero constants a_1 and $r.$ The sum of the first n terms is $$S_n = a_1\left(\frac{1 - r^n}{1 - r}\right),$$ where a_1 is the first term and r is the common ratio.	The series $$3 + 6 + 12 + 24 + 48 + 96$$ has $a_1 = 3$ and $r = 2.$ Its sum is $$S_6 = 3\left(\frac{1 - 2^6}{1 - 2}\right) = 189.$$

14.3 EXERCISES

CONCEPTS

1. The summation of the terms of a sequence is called a(n) _____.

2. Find the sum of the series $1 + 2 + 3 + 4$.

3. The series $1 + 3 + 5 + 7 + 9$ is an example of a(n) _____ series.

4. The series $1 + 3 + 9 + 27 + 81$ is an example of a(n) _____ series.

5. If $a_1 + a_2 + a_3 + \cdots + a_n$ is an arithmetic series, its sum is $S_n =$ _____.

6. If $a_1 + a_2 + a_3 + \cdots + a_n$ is a geometric series with the common ratio r, its sum is $S_n =$ _____.

7. The symbol $\sum$ is used to indicate a _____.

8. Write $\sum_{k=1}^{4} a_k$ as a sum.

9. $\sum_{n=1}^{5} a_1 + (n-1)d$ is an example of a(n)_____ series.

10. $\sum_{n=1}^{4} a_1 r^{n-1}$ is an example of a(n) _____ series.

SUMS OF SERIES

Exercises 11–16: Find the sum of the arithmetic series by using a formula.

11. $3 + 5 + 7 + 9 + 11 + 13$

12. $7.5 + 6 + 4.5 + 3 + 1.5 + 0 + (-1.5)$

13. $1 + 2 + 3 + 4 + \cdots + 40$

14. $1 + 3 + 5 + 7 + \cdots + 99$

15. $-7 + (-4) + (-1) + 2 + 5$

16. $89 + 84 + 79 + 74 + 69 + 64 + 59 + 54$

Exercises 17–22: Find the sum of the geometric series by using a formula.

17. $3 + 9 + 27 + 81 + 243 + 729 + 2187$

18. $2 - 1 + \frac{1}{2} - \frac{1}{4} + \frac{1}{8} - \frac{1}{16} + \frac{1}{32}$

19. $1 - 2 + 4 - 8 + 16 - 32 + 64 - 128$

20. $2 + \frac{1}{2} + \frac{1}{8} + \frac{1}{32} + \frac{1}{128} + \frac{1}{512}$

21. $0.5 + 1.5 + 4.5 + 13.5 + 40.5 + 121.5$

22. $0.6 + 0.3 + 0.15 + 0.075 + 0.0375$

*Exercises 23–26: **Annuities** (Refer to Example 5.) Find the future value of the annuity.*

23. $a_1 = \$2000$ $I = 0.08$ $n = 20$

24. $a_1 = \$500$ $I = 0.15$ $n = 10$

25. $a_1 = \$10,000$ $I = 0.11$ $n = 5$

26. $a_1 = \$3000$ $I = 0.19$ $n = 45$

SUMMATION NOTATION

Exercises 27–34: Write the terms of the series and find their sum.

27. $\sum_{k=1}^{4} 2k$

28. $\sum_{k=1}^{6} (k - 1)$

29. $\sum_{k=1}^{8} 4$

30. $\sum_{k=2}^{6} (5 - 2k)$

31. $\sum_{k=1}^{7} k^2$

32. $\sum_{k=1}^{4} 5(2)^{k-1}$

33. $\sum_{k=4}^{5} (k^2 - k)$

34. $\sum_{k=1}^{4} \log k$

Exercises 35–38: Write each series in summation notation.

35. $1^4 + 2^4 + 3^4 + 4^4 + 5^4 + 6^4$

36. $1 + \frac{1}{5^1} + \frac{1}{5^2} + \frac{1}{5^3} + \frac{1}{5^4}$

37. $1 + \frac{1}{2^2} + \frac{1}{3^2} + \frac{1}{4^2} + \frac{1}{5^2}$

38. $1 + \frac{1}{10} + \frac{1}{100} + \frac{1}{1000} + \frac{1}{10,000}$

39. Verify that $\sum_{k=1}^{n} k = \frac{n(n+1)}{2}$ by using a formula for the sum of the first n terms of an arithmetic series.

40. Use Exercise 39 to find the sum of the series $\sum_{k=1}^{200} k$.

APPLICATIONS

41. *Prison Escapees*   The table lists the number of escapees from state prisons each year.

Year	1990	1991	1992
Escapees	8518	9921	10,706

Year	1993	1994	1995
Escapees	14,035	14,307	12,249

Source: Bureau of Justice Statistics.

(a) Write a series whose sum is the total number of escapees from 1990 to 1995.
(b) Find its sum.

42. *Captured Prison Escapees*   (Refer to Exercise 41.) The table lists the number of escapees from state prisons who were captured, including inmates who may have escaped during a previous year.

Year	1990	1991	1992
Captured	9324	9586	10,031

Year	1993	1994	1995
Captured	12,872	13,346	12,166

Source: Bureau of Justice Statistics.

(a) Write a series whose sum is the total number of escapees captured from 1990 to 1995.
(b) Find its sum.
(c) Compare the number of escapees to the number captured during this time period.

43. *Area*   A sequence of smaller squares is formed by connecting the midpoints of the sides of a larger square as shown in the figure.

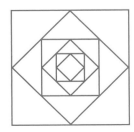

(a) If the area of the largest square is 1 square unit, give the first five terms of a sequence that describes the area of each successive square.
(b) Use a formula to sum the areas of the first 10 squares.

44. *Perimeter*   (Refer to Exercise 43.) Use a formula to find the sum of the perimeters of the first 10 squares.

45. *Stacking Logs*   A stack of logs is made in layers, with one log less in each layer, as shown in the accompanying figure. If the top layer has 6 logs and the bottom row has 14 logs, what is the total number of logs in the pile? Use a formula to find this sum.

46. *Stacking Logs*   (Refer to Exercise 45.) Suppose that a stack of logs has 15 logs in the top layer and a total of 10 layers. How many logs are in the stack?

47. *Salaries*   Suppose that an individual's starting salary is $35,000 per year and that the individual receives a $2000 raise each year. Find the total amount earned over 20 years.

48. *Salaries*   Suppose that an individual's starting salary is $35,000 per year and that the individual receives a 10% raise each year. Find the total amount earned over 20 years.

49. *Bouncing Ball*   A tennis ball first bounces to 75% of the height from which it was dropped and then to 75% of the height of each successive bounce. If it is dropped from a height of 10 feet, find the distance it *falls* between the fourth and fifth bounce.

50. *Bouncing Ball*   A tennis ball first bounces to 75% of the height from which it was dropped and then to 75% of the height of each successive bounce. If it is dropped from a height of 10 feet, find the *total* distance it travels before it reaches its fifth bounce. (*Hint:* Make a sketch.)

51. Discuss the difference between a sequence and a series. Give an example of each.

52. Suppose that an arithmetic series has $a_1 = 1$ and a common difference of $d = 2$, whereas a geometric series has $a_1 = 1$ and a common ratio of $r = 2$. Discuss how their sums compare as the number of terms n becomes large. (*Hint:* Calculate each sum for $n = 10, 20$, and 30.)

Group Activity: Working with Real Data

Directions: Form a group of 2 to 4 people. Select someone to record the group's responses for this activity. All members of the group should work cooperatively to answer the questions. If your instructor asks for your results, each member of the group should be prepared to respond.

Depreciation   For tax purposes, businesses frequently depreciate equipment. Two different methods of depreciation are called *straight-line depreciation* and *sum-of-the-years'-digits*. Suppose that a college student buys a $3000 computer to start a business that provides Internet services. This student estimates the life of the computer at 4 years, after which its value will be $200. The difference between $3000 and $200, or $2800, may be deducted from the student's taxable income over a 4-year period.

In straight-line depreciation, equal portions of $2800 are deducted each year over the 4 years. The sum-of-the-years'-digits method calculates depreciation differently. For a computer having a useful life of 4 years, the sum of the years is computed by

$$1 + 2 + 3 + 4 = 10.$$

With this method, $\frac{4}{10}$ of $2800 is deducted the first year, $\frac{3}{10}$ the second year, and so on, until $\frac{1}{10}$ is deducted the fourth year. Both depreciation methods yield a total deduction of $2800 over the 4 years.   (*Source:* Sharp Electronics Corporation, *Conquering the Sciences.*)

(a) Find an arithmetic sequence that gives the amount depreciated each year by each method.

(b) Write a series whose sum is the amount depreciated over 4 years by each method.

14.4   THE BINOMIAL THEOREM

Pascal's Triangle · Factorial Notation and Binomial Coefficients · Using the Binomial Theorem

INTRODUCTION

In this section we demonstrate how to expand expressions of the form $(a + b)^n$, where n is a natural number. These expressions occur in statistics, finite mathematics, computer science, and calculus. The two methods that we discuss are Pascal's triangle and the binomial theorem.

PASCAL'S TRIANGLE

Expanding $(a + b)^n$ for increasing values of n gives the following results.

$$(a + b)^0 = \qquad\qquad 1$$
$$(a + b)^1 = \qquad\qquad 1a + 1b$$
$$(a + b)^2 = \qquad\qquad 1a^2 + 2ab + 1b^2$$
$$(a + b)^3 = \qquad\qquad 1a^3 + 3a^2b + 3ab^2 + 1b^3$$
$$(a + b)^4 = \qquad\qquad 1a^4 + 4a^3b + 6a^2b^2 + 4ab^3 + 1b^4$$
$$(a + b)^5 = \qquad\qquad 1a^5 + 5a^4b + 10a^3b^2 + 10a^2b^3 + 5ab^4 + 1b^5$$

```
            1
          1   1
        1   2   1
      1   3   3   1
    1   4   6   4   1
  1   5   10  10  5   1
```

Figure 14.18 Pascal's Triangle

Note that $(a + b)^1$ has two terms, starting with a and ending with b; $(a + b)^2$ has three terms, starting with a^2 and ending with b^2; and in general, $(a + b)^n$ has $n + 1$ terms, starting with a^n and ending with b^n. The exponent on a decreases by 1 each successive term, and the exponent on b increases by 1 each successive term.

The triangle formed by the highlighted numbers is called **Pascal's triangle**. This triangle consists of 1s along the sides, and each element inside the triangle is the sum of the two numbers above it, as shown in Figure 14.18. Pascal's triangle is usually written without variables and can be extended to include as many rows as needed.

We can use this triangle to expand $(a + b)^n$, where n is a natural number. For example, the expression $(m + n)^4$ consists of five terms written as

$$(m + n)^4 = \_m^4 + \_m^3n^1 + \_m^2n^2 + \_m^1n^3 + \_n^4.$$

Because there are five terms, the coefficients can be found in the fifth row of Pascal's triangle, which is

$$1 \quad 4 \quad 6 \quad 4 \quad 1.$$

Thus

$$(m + n)^4 = \underline{1}\,m^4 + \underline{4}\,m^3n^1 + \underline{6}\,m^2n^2 + \underline{4}\,m^1n^3 + \underline{1}\,n^4$$
$$= m^4 + 4m^3n + 6m^2n^2 + 4mn^3 + n^4.$$

EXAMPLE 1 Expanding a binomial

Expand each binomial, using Pascal's triangle.
(a) $(x + 2)^5$ **(b)** $(2m - n)^3$

Solution **(a)** To find the coefficients, use the sixth row in Pascal's triangle.

$$(x + 2)^5 = \underline{1}\,x^5 + \underline{5}\,x^4 \cdot 2^1 + \underline{10}\,x^3 \cdot 2^2 + \underline{10}\,x^2 \cdot 2^3 + \underline{5}\,x^1 \cdot 2^4 + \underline{1}\,(2^5)$$
$$= x^5 + 10x^4 + 40x^3 + 80x^2 + 80x + 32$$

(b) To find the coefficients, use the fourth row in Pascal's triangle.

$$(2m - n)^3 = \underline{1}\,(2m)^3 + \underline{3}\,(2m)^2(-n)^1 + \underline{3}\,(2m)^1(-n)^2 + \underline{1}\,(-n)^3$$
$$= 8m^3 - 12m^2n + 6mn^2 - n^3$$

FACTORIAL NOTATION AND BINOMIAL COEFFICIENTS

An alternative to Pascal's triangle is the binomial theorem, which makes use of **factorial notation**.

n FACTORIAL (*n*!)

For any positive integer,

$$n! = 1 \cdot 2 \cdot 3 \cdots \cdot n.$$

We also define $0! = 1$.

Note: Because multiplication is commutative, *n* factorial can also be defined as

$$n! = n \cdot (n - 1) \cdot (n - 2) \cdots \cdot 2 \cdot 1.$$

Examples include the following.

$$0! = 1$$
$$1! = 1$$
$$2! = 1 \cdot 2 = 2$$
$$3! = 1 \cdot 2 \cdot 3 = 6$$
$$4! = 1 \cdot 2 \cdot 3 \cdot 4 = 24$$
$$5! = 1 \cdot 2 \cdot 3 \cdot 4 \cdot 5 = 120$$

Figure 14.19 supports these results. On some calculators, factorial (!) can be accessed in the MATH PRB menus.

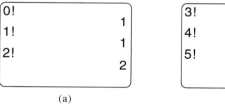

0!	
1!	1
2!	1
	2

3!	6
4!	24
5!	120

(a)   (b)

Figure 14.19

EXAMPLE 2   Evaluating factorial expressions

Simplify the expression.

(a) $\frac{5!}{3!2!}$ **(b)** $\frac{4!}{4!0!}$

Solution   **(a)** $\frac{5!}{3!2!} = \frac{1 \cdot 2 \cdot 3 \cdot 4 \cdot 5}{(1 \cdot 2 \cdot 3)(1 \cdot 2)} = \frac{120}{6 \cdot 2} = 10$

(b) $0! = 1$, so $\frac{4!}{4!0!} = \frac{4!}{4!(1)} = \frac{4!}{4!} = 1$

The expression $_nC_r$ represents a *binomial coefficient* that can be used to calculate the numbers in Pascal's triangle.

BINOMIAL COEFFICIENT $_nC_r$

For *n* and *r* nonnegative integers, $n \geq r$,

$$_nC_r = \frac{n!}{(n - r)! \, r!}$$

is a **binomial coefficient.**

Values of $_nC_r$ for $r = 0, 1, 2, \ldots, n$ correspond to the $n + 1$ numbers in row $n + 1$ of Pascal's triangle.

EXAMPLE 3 Calculating $_nC_r$

Calculate $_3C_r$ for $r = 0, 1, 2, 3$ by hand. Check your results on a calculator. Compare these numbers with the fourth row in Pascal's triangle.

Solution

$$_3C_0 = \frac{3!}{(3-0)!0!} = \frac{6}{6 \cdot 1} = 1$$

$$_3C_1 = \frac{3!}{(3-1)!1!} = \frac{6}{2 \cdot 1} = 3$$

$$_3C_2 = \frac{3!}{(3-2)!2!} = \frac{6}{1 \cdot 2} = 3$$

$$_3C_3 = \frac{3!}{(3-3)!3!} = \frac{6}{1 \cdot 6} = 1$$

These results are supported in Figure 14.20. The fourth row of Pascal's triangle is

$$1 \quad 3 \quad 3 \quad 1,$$

which agrees with the calculated values for $_3C_r$. On some calculators, the MATH PRB menus are used to calculate $_nC_r$.

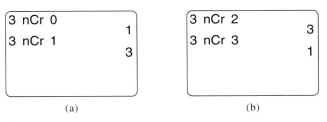

(a)   (b)

Figure 14.20

USING THE BINOMIAL THEOREM

The binomial coefficients can be used to expand expressions of the form $(a + b)^n$. To do so, we use the **binomial theorem**.

BINOMIAL THEOREM

For any positive integer n and any numbers a and b,

$$(a + b)^n = {_nC_0}a^n + {_nC_1}a^{n-1}b^1 + \cdots + {_nC_{n-1}}a^1b^{n-1} + {_nC_n}b^n.$$

Using the results of Example 3, we write

$$(a + b)^3 = {_3C_0}a^3 + {_3C_1}a^2b^1 + {_3C_2}a^1b^2 + {_3C_3}b^3$$

$$= 1a^3 + 3a^2b + 3ab^2 + 1b^3$$

$$= a^3 + 3a^2b + 3ab^2 + b^3.$$

EXAMPLE 4 Expanding a binomial

Use the binomial theorem to expand each expression.
(a) $(x + y)^5$ **(b)** $(3 - 2x)^4$

Solution **(a)** The coefficients are calculated as follows.

$$_5C_0 = \frac{5!}{(5-0)!0!} = 1, \qquad _5C_1 = \frac{5!}{(5-1)!1!} = 5, \qquad _5C_2 = \frac{5!}{(5-2)!2!} = 10$$

$$_5C_3 = \frac{5!}{(5-3)!3!} = 10, \qquad _5C_4 = \frac{5!}{(5-4)!4!} = 5, \qquad _5C_5 = \frac{5!}{(5-5)!5!} = 1$$

Using the binomial theorem, we arrive at the following result.

$$(x + y)^5 = {}_5C_0x^5 + {}_5C_1x^4y^1 + {}_5C_2x^3y^2 + {}_5C_3x^2y^3 + {}_5C_4x^1y^4 + {}_5C_5y^5$$
$$= 1x^5 + 5x^4y + 10x^3y^2 + 10x^2y^3 + 5xy^4 + 1y^5$$
$$= x^5 + 5x^4y + 10x^3y^2 + 10x^2y^3 + 5xy^4 + y^5$$

(b) The coefficients are calculated as follows.

$$_4C_0 = \frac{4!}{(4-0)!0!} = 1, \qquad _4C_1 = \frac{4!}{(4-1)!1!} = 4, \qquad _4C_2 = \frac{4!}{(4-2)!2!} = 6,$$

$$_4C_3 = \frac{4!}{(4-3)!3!} = 4, \qquad _4C_4 = \frac{4!}{(4-4)!4!} = 1$$

Using the binomial theorem with $a = 3$ and $b = (-2x)$, we arrive at the following result.

$$(3 - 2x)^4 = {}_4C_0(3)^4 + {}_4C_1(3)^3(-2x) + {}_4C_2(3)^2(-2x)^2$$
$$+ {}_4C_3(3)(-2x)^3 + {}_4C_4(-2x)^4$$
$$= 1(81) + 4(27)(-2x) + 6(9)(4x^2) + 4(3)(-8x^3) + 1(16x^4)$$
$$= 81 - 216x + 216x^2 - 96x^3 + 16x^4$$

The binomial theorem gives *all* of the terms of $(a + b)^n$. However, we can find any individual term by noting that the $(r + 1)$st term in the binomial expansion for $(a + b)^n$ is given by the formula $_nC_ra^{n-r}b^r$, for $0 \le r \le n$. The next example shows how to use this formula to find the $(r + 1)$st term of $(a + b)^n$.

EXAMPLE 5 Finding the *k*th term in a binomial expansion

Find the third term of $(x - y)^5$.

Solution In this example the $(r + 1)$st term is the *third* term in the expansion of $(x - y)^5$. That is, $r + 1 = 3$, or $r = 2$. Also, the exponent in the expression is $n = 5$. To get this binomial into the form $(a + b)^n$, we note that the first term in the binomial is $a = x$ and that the second term in the binomial is $b = -y$. Substituting the values for r, n, a, and b in the formula for the $(r + 1)$st term yields

$$_5C_2(x)^{5-2}(-y)^2 = 10x^3y^2.$$

The third term in the binomial expansion of $(x - y)^5$ is $10x^3y^2$.

14.4 PUTTING IT ALL TOGETHER

In this section we showed how to expand the expression $(a + b)^n$ by using Pascal's triangle and the binomial theorem. The following table outlines important topics from this section.

Topic	Explanation	Example
Pascal's Triangle	$$\begin{array}{ccccccccccc} & & & & & 1 & & & & & \\ & & & & 1 & & 1 & & & & \\ & & & 1 & & 2 & & 1 & & & \\ & & 1 & & 3 & & 3 & & 1 & & \\ & 1 & & 4 & & 6 & & 4 & & 1 & \\ 1 & & 5 & & 10 & & 10 & & 5 & & 1 \end{array}$$ To expand $(a + b)^n$, use row $n + 1$ in the triangle.	$(a + b)^3 = 1a^3 + 3a^2b + 3ab^2 + 1b^3$ (Row 4)
Factorial Notation	The expression $n!$ equals $1 \cdot 2 \cdot 3 \cdot \cdots \cdot n.$	$5! = 1 \cdot 2 \cdot 3 \cdot 4 \cdot 5 = 120$
Binomial Coefficient $_nC_r$	$_nC_r = \dfrac{n!}{(n-r)!\,r!}$	$_6C_4 = \dfrac{6!}{(6-4)!\,4!} = \dfrac{6!}{2!\,4!} = \dfrac{720}{2 \cdot 24} = 15$
Binomial Theorem	$(a + b)^n = {}_nC_0 a^n + {}_nC_1 a^{n-1}b^1 + \cdots$ $+ {}_nC_{n-1} a^1 b^{n-1} + {}_nC_n b^n$	$(a + b)^4 = {}_4C_0 a^4 + {}_4C_1 a^3 b + {}_4C_2 a^2 b^2$ $+ {}_4C_3 ab^3 + {}_4C_4 b^4$ $= 1a^4 + 4a^3b + 6a^2b^2 + 4ab^3 + 1b^4$ $= a^4 + 4a^3b + 6a^2b^2 + 4ab^3 + b^4$

14.4 EXERCISES

FOR EXTRA HELP

CONCEPTS

1. How many terms result from expanding $(a + b)^4$?

2. How many terms result from expanding $(a + b)^n$?

3. To find the coefficients for the expansion of $(a + b)^3$, what row of Pascal's triangle do you use?

4. Write down the first 5 rows of Pascal's triangle.

5. $4! =$ _____

6. $1 \cdot 2 \cdot 3 \cdot 4 \cdot 5 \cdot 6 =$ _____

7. $_nC_r =$ _____   **8.** $(a + b)^2 =$ _____

USING PASCAL'S TRIANGLE

Exercises 9–16: Use Pascal's triangle to expand the expression.

9. $(x + y)^3$ **10.** $(x + y)^4$

11. $(2x + 1)^4$

12. $(2x - 1)^4$

13. $(a - b)^5$

14. $(3x + 2y)^3$

15. $(x^2 + 1)^3$

16. $\left(\frac{1}{2} - x^2\right)^5$

FACTORIALS AND BINOMIAL COEFFICIENTS

Exercises 17–30: Evaluate the expression.

17. $3!$

18. $6!$

19. $\frac{4!}{3!}$

20. $\frac{6!}{3!}$

21. $\frac{2!}{0!}$

22. $\frac{5!}{1!}$

23. $\frac{5!}{2!3!}$

24. $\frac{6!}{4!2!}$

25. $_5C_4$

26. $_3C_1$

27. $_6C_5$

28. $_2C_2$

29. $_4C_0$

30. $_4C_3$

Exercises 31–36: Evaluate the binomial coefficient with a calculator.

31. $_{12}C_7$

32. $_{13}C_8$

33. $_9C_5$

34. $_{25}C_{14}$

35. $_{19}C_{11}$

36. $_{10}C_6$

THE BINOMIAL THEOREM

Exercises 37–48: Use the binomial theorem to expand the expression.

37. $(m + n)^3$

38. $(m + n)^5$

39. $(x - y)^4$

40. $(1 - 3x)^4$

41. $(2a + 1)^3$

42. $(x^2 - 1)^3$

43. $(x + 2)^5$

44. $(a - 3)^5$

45. $(3 + 2m)^4$

46. $(m - 3n)^3$

47. $(2x - y)^3$

48. $(2a + 3b)^4$

Exercises 49–54: The $(r + 1)$st term of the expression $(a + b)^n$, $0 \le r \le n$, is given by $_nC_r a^{n-r} b^r$. Find the specified term. Refer to Example 5.

49. The first term of $(a + b)^8$

50. The second term of $(a - b)^{10}$

51. The fourth term of $(x + y)^7$

52. The sixth term of $(a + b)^9$

53. The first term of $(2m + n)^9$

54. The eighth term of $(2a - b)^8$

WRITING ABOUT MATHEMATICS

55. Explain how to find the numbers in Pascal's triangle.

56. Compare the expansion of $(a + b)^n$ to the expansion of $(a - b)^n$. Give an example.

CHECKING BASIC CONCEPTS   SECTIONS 14.3 AND 14.4

1. Determine whether the series is arithmetic or geometric.
   (a) $\frac{1}{2} + \frac{1}{4} + \frac{1}{8} + \cdots + \frac{1}{256}$
   (b) $\frac{1}{2} + \frac{5}{2} + \frac{9}{2} + \frac{13}{2} + \frac{17}{2}$

2. Use a formula to find the sum of the arithmetic series

$$4 + 8 + 12 + \cdots + 48.$$

3. Use a formula to find the sum of the geometric series

$$1 - 2 + 4 - 8 + 16 - 32 + 64 - 128$$
$$+ 256 - 512.$$

4. Use Pascal's triangle to expand $(x - y)^4$.

5. Use the binomial theorem to expand $(x + 2)^3$.

CHAPTER
14 Summary

An infinite sequence is a function whose domain is the natural numbers. A finite sequence is a function whose domain is $D = \{1, 2, 3, \ldots, n\}$ for some natural number n. Because sequences are functions, they have symbolic, graphical, and numerical representations.

Example: $a_n = 2n$ is a symbolic representation of the even natural numbers. The first six terms of this sequence are represented numerically and graphically in the table and figure.

n	1	2	3	4	5	6
a_n	2	4	6	8	10	12

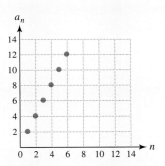

Two common types of sequences are arithmetic and geometric.

Arithmetic Sequence   An arithmetic sequence is determined by a linear function of the form $f(n) = dn + c$ or $f(n) = a_1 + (n - 1)d$. Successive terms in an arithmetic sequence are found by adding the common difference d to the previous term. The sequence $1, 3, 5, 7, 9, 11, \ldots$ is an arithmetic sequence with its first term $a_1 = 1$, common difference $d = 2$, and general term $a_n = 2n - 1$.

Geometric Sequence   The general term for a geometric sequence is given by $f(n) = a_1 r^{n-1}$. Successive terms in a geometric sequence are found by multiplying the previous term by the common ratio r. The sequence $3, 6, 12, 24, 48, \ldots$ is a geometric sequence with its first term $a_1 = 3$, common ratio $r = 2$, and general term $a_n = 3(2)^{n-1}$.

Section 14.3 Series

Series   A series results when the terms of a sequence are summed. The series associated with the sequence 2, 4, 6, 8, 10 is

$$2 + 4 + 6 + 8 + 10,$$

and its sum equals 30. An arithmetic series results when the terms of an arithmetic sequence are summed, and a geometric series results when the terms of a geometric sequence are summed. In this chapter, we discussed formulas for finding sums of arithmetic and geometric series. See Putting It All Together for Section 14.3.

Summation Notation   Summation notation can be used to write series efficiently. For example,

$$1^2 + 2^2 + 3^2 + 4^2 + 5^2 = \sum_{k=1}^{5} k^2.$$

Section 14.4 The Binomial Theorem

Pascal's triangle may be used to find the coefficients for the expansion of $(a + b)^n$, where n is a natural number.

$$
\begin{array}{ccccccccccc}
 & & & & & 1 & & & & & \\
 & & & & 1 & & 1 & & & & \\
 & & & 1 & & 2 & & 1 & & & \\
 & & 1 & & 3 & & 3 & & 1 & & \\
 & 1 & & 4 & & 6 & & 4 & & 1 & \\
1 & & 5 & & 10 & & 10 & & 5 & & 1 \\
\end{array}
$$

The binomial theorem can also be used to expand powers of binomials.

Example: To expand $(x + y)^4$, use the fifth row of Pascal's triangle.

$$(x + y)^4 = 1x^4 + 4x^3y + 6x^2y^2 + 4xy^3 + 1y^4$$
$$= x^4 + 4x^3y + 6x^2y^2 + 4xy^3 + y^4$$

CHAPTER 14 Review Exercises

SECTION 14.1

Exercises 1–4: Write the first four terms of the sequence for $n = 1, 2, 3,$ and 4.

1. $f(n) = n^3$

2. $f(n) = 5 - 2n$

3. $f(n) = \dfrac{2n}{n^2 + 1}$

4. $f(n) = (-2)^n$

Exercises 5–6: Use the graph to write the terms of the sequence.

5.

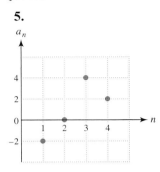

6.

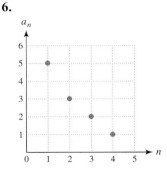

Exercises 7–10: Represent the first seven terms of the sequence with $n = 1, 2, \ldots, 7$ numerically and graphically.

7. $a_n = 2n$

8. $a_n = n^2 - 4$

9. $a_n = 4\left(\frac{1}{2}\right)^n$

10. $a_n = \sqrt{n}$

SECTION 14.2

Exercises 11–18: Determine whether f is an arithmetic sequence. Identify the common difference when possible.

11. $f(n) = 5n - 1$

12. $f(n) = 4 - n^2$

13. $f(n) = 2^n$

14. $f(n) = 4 - \frac{1}{3}n$

15.

n	1	2	3	4
$f(n)$	20	17	14	11

16.

n	1	2	3	4
$f(n)$	-3	0	6	12

17.

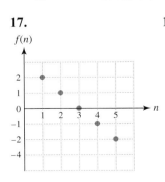

18.

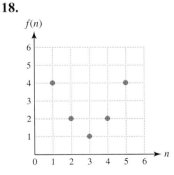

Exercises 19 and 20: Find the general term a_n for the arithmetic sequence.

19. $a_1 = -3$ and $d = 4$ **20.** $a_1 = 2$ and $a_2 = -3$

Exercises 21–28: Determine whether f is a geometric sequence. Identify the common ratio when possible.

21. $f(n) = 2(4)^n$ **22.** $f(n) = 2n^4$

23. $f(n) = 1 - 2n$ **24.** $f(n) = 5(0.7)^n$

25.

n	1	2	3	4
$f(n)$	5	4	3	1

26.

n	1	2	3	4
$f(n)$	27	-9	3	-1

27.   **28.**

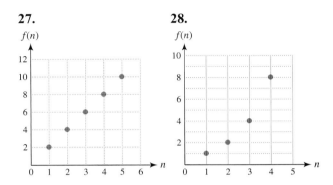

Exercises 29 and 30: Find the general term a_n for the geometric sequence.

29. $a_1 = 5$ and $r = 0.9$ **30.** $a_1 = 2$ and $a_2 = 8$

SECTION 14.3

Exercises 31–34: Find the sum, using a formula.

31. $4 + 9 + 14 + 19 + 24 + 29 + 34 + 39 + 44$

32. $4.5 + 3.0 + 1.5 + 0 - 1.5$

33. $1 - 4 + 16 - 64 + \cdots + 4096$

34. $1 + \frac{1}{2} + \frac{1}{4} + \frac{1}{8} + \frac{1}{16} + \cdots + \frac{1}{256}$

Exercises 35–38: Write the terms of the series.

35. $\sum\limits_{k=1}^{5} 2k + 1$ **36.** $\sum\limits_{k=1}^{4} \frac{1}{k+1}$

37. $\sum\limits_{k=1}^{4} k^3$ **38.** $\sum\limits_{k=2}^{7} (1 - k)$

Exercises 39–42: Write the series in summation notation.

39. $1 + 2 + 3 + \cdots + 20$

40. $1 + \frac{1}{2} + \frac{1}{3} + \cdots + \frac{1}{20}$

41. $\frac{1}{2} + \frac{2}{3} + \frac{3}{4} + \cdots + \frac{9}{10}$

42. $1^2 + 2^2 + 3^2 + 4^2 + 5^2 + 6^2 + 7^2$

SECTION 14.4

Exercises 43–46: Use Pascal's triangle to expand the expression.

43. $(x + 4)^3$ **44.** $(2x + 1)^4$

45. $(x - y)^5$

46. $(a - 1)^6$

Exercises 47–50: Evaluate the expression.

47. $3!$ **48.** $\frac{5!}{3!2!}$

49. $_6C_3$ **50.** $_4C_3$

Exercises 51–54: Use the binomial theorem to expand the expression.

51. $(m + 2)^4$ **52.** $(a + b)^5$

53. $(x - 3y)^4$ **54.** $(3x - 2)^3$

APPLICATIONS

55. *Salaries* An individual's starting salary is $45,000, and the individual receives a 10% raise each year. Give symbolic, numerical, and graphical representa-

tions for this person's salary over 7 years. What type of sequence is it?

56. *Salaries* An individual's starting salary is $45,000, and the individual receives an increase of $5000 each year. Give symbolic, numerical, and graphical representations for this person's salary over 7 years. What type of sequence is it?

57. *Rain Forests* Rain forests are defined as forests that grow in regions that receive more than 70 inches of rain each year. The world is losing an estimated 49 million acres of rain forests annually. Give symbolic, graphical, and numerical representations for a sequence that models the total number of acres (in millions) lost over a 7-year period. (*Source: New York Times Almanac, 1999.*)

58. *Home Mortgage Payments* The average home mortgage payment in 1996 was $1087 per month. Since then, mortgage payments have risen, on average, by 2.5% per year.
   (a) Write a sequence a_n that models the average mortgage payment in year n, where $n = 1$ corresponds to 1996, $n = 2$ to 1997, and so on.
   (b) Is a_n arithmetic, geometric, or neither? Explain your reasoning.
   (c) Find a_5 and interpret the result.
   (d) Give a graphical representation for a_n, where $n = 1, 2, 3, \ldots, 10$.

CHAPTER
14 Test

1. Write the first four terms of the sequence for $n = 1, 2, 3,$ and 4.

$$f(n) = \frac{n^2}{n + 1}.$$

2. Use the graph to write the terms of the sequence.

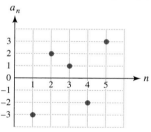

3. List the first seven terms of $a_n = n^2 - n$ in a table. Let $n = 1, 2, \ldots, 7$.

4. Expand the expression $(2x - 1)^4$.

Exercises 5 and 6: Determine whether the sequence is arithmetic or geometric. Identify either the common difference or the common ratio.

5. $f(n) = 7 - 3n$

6.
n	1	2	3	4
$f(n)$	-2	4	-8	16

7. Find the general term a_n for the arithmetic sequence if $a_1 = 2$ and $d = -3$.

8. Find the general term a_n for the geometric sequence if $a_1 = 2$ and $a_3 = 4.5$.

Exercises 9 and 10: Determine whether f is a geometric sequence. Identify the common ratio when possible.

9. $f(n) = -3(2.5)^n$

10.

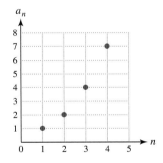

11. $-1 + 2 + 5 + 8 + 11 + 14 + 17 + 20 + 23$

12. $1 - \frac{2}{3} + \frac{4}{9} - \frac{8}{27} + \frac{16}{81} - \frac{32}{243} + \frac{64}{729}$

13. Write the terms of the series $\sum_{k=2}^{7} 3k$.

14. Write the series $1^3 + 2^3 + 3^3 + \cdots + 60^3$ in summation notation.

15. Evaluate $\frac{7!}{4!\,3!}$.   16. Evaluate $_5C_3$.

17. *Auditorium Seating*   An auditorium has 50 seats in the first row, 57 seats in the second row, 64 seats in the third row, and so on. Use a formula to find the total number of seats in the first 45 rows.

18. *Median Home Price*   In 1997 the median price of a single-family home was \$159,700 and was increasing at a rate of 4% per year. Give symbolic, numerical, and graphical representations for the median home price over a 7-year period, starting in 1997. What type of sequence is it?

19. *Tent Worms*   Large numbers of tent worms can defoliate trees and ruin crops. After they mature, they spin a cocoon and develop into moths that lay eggs. Suppose that an initial population of 2000 tent worms doubles every 5 days.
    (a) Write a formula for a_n that models the number of tent worms after $n - 1$ five-day time periods. (*Hint:* $a_1 = 2000$, $a_2 = 4000$, and $a_3 = 8000$.)
    (b) Is a_n arithmetic, geometric, or neither? Explain your reasoning.
    (c) Find a_6 and interpret the result.
    (d) Give a graph for a_n, where $n = 1, 2, 3, 4, 5, 6$.

CHAPTER 14 Extended and Discovery Exercises

SEQUENCES AND SERIES

Exercises 1 and 2: Recursive Sequences   Some sequences are not defined by a formula for a_n. Instead they are defined recursively. With a **recursive formula** you must find terms a_1 through a_{n-1} before you can find a_n. For example, let

$$a_1 = 2$$
$$a_n = a_{n-1} + 3, \qquad \text{for } n \geq 2.$$

To find a_2, a_3, and a_4, we let $n = 2, 3, 4$.

$$a_2 = a_1 + 3 = 2 + 3 = 5$$
$$a_3 = a_2 + 3 = 5 + 3 = 8$$
$$a_4 = a_3 + 3 = 8 + 3 = 11$$

The first four terms of the sequence are 2, 5, 8, 11.

1. **Fibonacci Sequence**   The Fibonacci sequence dates back to 1202 and is one of the most famous sequences in mathematics. It can be defined recursively as follows.

$$a_1 = 1, \qquad a_2 = 1$$
$$a_n = a_{n-1} + a_{n-2}, \qquad \text{for } n \geq 3$$

Find the first 12 terms of this sequence.

2. **Insect Populations**   Frequently the population of a particular insect does not continue to grow indefinitely. Instead, its population grows rapidly at first and then levels off because of competition for limited resources. In one study, the behavior of the winter moth was modeled with a sequence similar to the following, where a_n gives the population density in thousands per acre during year n.

$$a_1 = 1$$
$$a_n = 2.85a_{n-1} - 0.19a_{n-1}^2, \qquad n \geq 2$$

(*Source:* G. Varley and G. Gradwell, "Population models for the winter moth.")

   (a) Make a table for $n = 1, 2, 3, \ldots, 7$. Describe what happens to the population density of the winter moth.

   Note: Many graphing calculators are capable of generating tables and graphs for a recursive sequence.

   (b) Graph the sequence for $n = 1, 2, 3, \ldots, 20$. Discuss the graph.

3. **Bode's Law**   The average distances of the planets from the sun display a pattern first described by Johann Bode in 1772. This relationship is called Bode's law and was proposed before Uranus, Neptune, and Pluto were discovered. It is a sequence defined by

$$a_1 = 0.4$$
$$a_n = 0.3(2)^{n-2} + 0.4, \qquad \text{for } n = 2, 3, 4, \ldots, 10.$$

In this sequence, a distance of 1 unit corresponds to the average Earth–sun distance of 93 million miles. The number n represents the nth planet. The actual distances of the planets, including an average dis-

tance for the asteroids, are listed in the accompanying table.   (*Source:* M. Zeilik, *Introductory Astronomy and Astrophysics.*)
   (a) Find a_4 and interpret the result.
   (b) Calculate the terms of Bode's sequence. Compare them with the values in the table.
   (c) If there is another planet beyond Pluto, use Bode's law to predict its distance from the sun.

Planet	Distance
Mercury	0.39
Venus	0.72
Earth	1.00
Mars	1.52
Asteroids	2.8
Jupiter	5.20
Saturn	9.54
Uranus	19.2
Neptune	30.1
Pluto	39.5

4. **Calculating π**   The quest for an accurate estimation for π is a fascinating story covering thousands of years. Because π is an irrational number, it cannot be represented exactly by a fraction. Its decimal expansion neither repeats nor has a pattern. The ability to compute π was essential to the development of societies because π appears in formulas used in construction, surveying, and geometry. In early historical records, π was given the value of 3. Later the Egyptians used a value of

$$\frac{256}{81} \approx 3.1605.$$

Not until the discovery of series was an exceedingly accurate decimal approximation of π possible. In 1989, π was computed to 1,073,740,000 digits, which required 100 hours of supercomputer time. Why would anyone want to compute π to so many decimal places? One practical reason is to test electrical circuits in new computers. If a computer has a small defect in its hardware, there is a good chance that an error will appear after it has performed trillions of arithmetic calculations during the computation of π. (*Source:* P. Beckmann, *A History of Pi.*) The series given by

$$\frac{\pi^4}{90} \approx \frac{1}{1^4} + \frac{1}{2^4} + \frac{1}{3^4} + \frac{1}{4^4} + \frac{1}{5^4} + \cdots + \frac{1}{n^4}$$

can be used to estimate π, where larger values of n give better approximations.

(a) Approximate π by finding the sum of the first four terms.

 (b) Use a calculator to approximate π by summing the first 50 terms. Compare the result to the actual value of π.

5. *Infinite Series*   The sum S of an infinite geometric series can be found if its common ratio r satisfies $|r| < 1$. It is given by

$$S = \frac{a_1}{1 - r}.$$

If $|r| \geq 1$, this sum does not exist. For example, the infinite geometric series

$$1 + \frac{1}{2} + \frac{1}{4} + \frac{1}{8} + \frac{1}{16} + \cdots$$

has $a_1 = 1$ and $r = \frac{1}{2}$. Therefore its sum S equals

$$S = \frac{1}{1 - \frac{1}{2}} = 2.$$

You might want to add terms of this series to see how increasing the number of terms results in a number closer to 2. Find the sum of each infinite geometric series.

(a) $2 - 1 + \frac{1}{2} - \frac{1}{4} + \frac{1}{8} - \frac{1}{16} + \cdots$

(b) $1 + \frac{1}{3} + \frac{1}{9} + \frac{1}{27} + \frac{1}{81} + \cdots$

(c) $0.1 + 0.01 + 0.001 + 0.0001 + \cdots$

(d) $0.12 + 0.0012 + 0.000012$
$$+ 0.00000012 + \cdots$$

CHAPTERS
1–14 Cumulative Review Exercises

1. State whether the equation illustrates an identity, commutative, associative, or distributive property.

$$29(102) = 29(100) + 29(2)$$

2. Translate the sentence, "Three times a number decreased by seven equals the number increased by 3," to an equation using the variable x. Then solve the equation.

Exercises 3–8: Solve the equation or inequality. Write the solution set to the inequalities in interval notation.

3. $\frac{2}{5}(x - 4) = -12$

4. $\frac{2}{5}z + \frac{1}{4}z > 2 - (z - 1)$

5. $-3|t - 5| \leq -18$ **6.** $\left|4 + \frac{2}{3}x\right| = 6$

7. $\frac{1}{4}t - (2t + 5) + 6 = \frac{t + 3}{4}$

8. $-3 \leq \frac{2}{3}x + 5 < 11$

9. Write the equation of the horizontal line that passes through the point $(2, 3)$.

10. Find the slope and the y-intercept of the graph of $f(x) = -3x + 5$.

Exercises 11 and 12: Write the slope–intercept form for a line satisfying the given conditions.

11. Perpendicular to $y = -\frac{2}{3}x - 4$, passing through $(1, 4)$

12. Parallel to $y = 2x - 7$, passing through $(5, 2)$

13. Use the table to write the formula for $f(x) = ax + b$.

x	-2	-1	0	1	2
$f(x)$	5	3	1	-1	-3

14. Determine which of the following is a solution to the given system of equations.

$$(3, -2), (-1, 3)$$
$$3x + y = 7$$
$$-2x - 3y = 0$$

15. Solve the linear system of equations.

$$x - 2y = 1$$
$$-2x + 7y = 4$$

16. Shade the solution set in the xy-plane.

$$x - y < 4$$
$$x + 2y \geq 7$$

Exercises 17 and 18: Multiply the expression.

17. $2x^3(4x^4 - 3x^3 + 5)$ **18.** $(2z - 7)(3z + 4)$

Exercises 19–22: Simplify the expression. Write the result using positive exponents.

19. $\dfrac{x^{-2}y^3}{(3xy^{-2})^3}$ **20.** $\left(\dfrac{3b}{6a^2}\right)^{-4}$

21. $\left(\dfrac{1}{z^2}\right)^{-5}$ **22.** $\dfrac{8x^{-3}y^2}{4x^3y^{-1}}$

Exercises 23 and 24: Factor completely.

23. $4x^2 - 9y^2$ **24.** $2a^3 - a^2 + 8a - 4$

Exercises 25 and 26: Solve the equation.

25. $4x^2 - x - 3 = 0$ **26.** $x^4 - 10x^3 = -24x^2$

Exercises 27 and 28: Simplify the expression.

27. $\dfrac{x^2 - 7x + 10}{x^2 - 25} \cdot \dfrac{x + 5}{x + 1}$

28. $\dfrac{x^2 + 7x + 12}{x^2 - 9} \div \dfrac{x^2 - 5x + 6}{(x - 3)^2}$

Exercises 29 and 30: Solve the rational equation. Check your result.

29. $\dfrac{2}{x + 5} = \dfrac{-3}{x^2 - 25} + \dfrac{1}{x - 5}$

30. $\dfrac{2y}{y^2 - 3y + 2} = \dfrac{1}{y - 2} + 2$

31. Solve the equation for W.

$$R = \dfrac{3C - 2W}{5}$$

32. Simplify the complex fraction.

$$\dfrac{\dfrac{1}{x^2} + \dfrac{2}{x}}{\dfrac{1}{x^2} - \dfrac{4}{x}}$$

33. Identify the domain and range of the relation $S = \{(-6, 5), (-2, 1), (0, 3), (2, 0)\}$.

34. Find the domain of $f(x) = \dfrac{-5}{x - 8}$.

35. Solve the linear system of equations.

$$x + y + z = 5$$
$$-2x - y + z = -10$$
$$x + 2y + 8z = 1$$

36. Evaluate $\det A$.

$$A = \begin{bmatrix} 4 & -3 \\ 3 & 2 \end{bmatrix}$$

Exercises 37 and 38: Simplify the expression. Assume that all variables are positive.

37. $\sqrt[3]{x^4y^4} - 2\sqrt[3]{xy}$ **38.** $(4 + \sqrt{2})(4 - \sqrt{2})$

Exercises 39 and 40: Solve. Check your answer.

39. $8(x - 3)^2 = 200$ **40.** $3\sqrt{2x + 6} = 6x$

Exercises 41 and 42: Write the complex expression in standard form.

41. $(-3 + i)(-4 - 2i)$ **42.** $\dfrac{2 - 6i}{1 + 2i}$

43. Find the minimum y-value located on the graph of $y = 3x^2 + 8x + 5$.

44. Write the equation $y = 2x^2 + 8x + 17$ in vertex form and identify the vertex.

Exercises 45 and 46: Solve the quadratic equation using the method of your choice. Write any complex solutions in standard form.

45. $x^2 - 4x + 13 = 0$ **46.** $z^2 - 4z = 32$

47. A graph of $y = ax^2 + bx + c$ is shown.
   (a) Solve $ax^2 + bx + c = 0$.
   (b) State whether $a > 0$ or $a < 0$.

(c) Determine whether the discriminant is positive, negative, or zero.

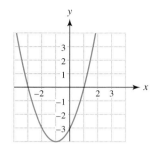

48. Solve the quadratic inequality. Write your answer in interval notation.

$$x^2 + 2x + 3 > 0$$

49. For $f(x) = x^2 + 1$ and $g(x) = 3x - 2$, find the following expression.
  (a) $(f \circ g)(-2)$ **(b)** $(g \circ f)(x)$

50. Find $f^{-1}(x)$ for the one-to-one function given by $f(x) = \frac{3x + 1}{2}$.

51. Write the following expression by using sums and differences of logarithms of x and y. Assume x and y are positive.

$$\ln(x^3 \sqrt{y})$$

52. Write the following expression as one logarithm. Assume x and y are positive.

$$2\log x - \log 4xy$$

Exercises 53 and 54: Solve the equation. Approximate answers to the nearest hundredth.

53. $8\log x + 3 = 17$ **54.** $4^{2x} = 5$

55. Graph the parabola $x = (y - 3)^2 + 1$. Find the vertex and axis of symmetry.

56. Find the center and the radius of the circle given by $x^2 - 6x + y^2 + 2y = -6$.

Exercises 57 and 58: Graph the ellipse or hyperbola. Label the vertices and the endpoints of the minor axis on the ellipse. Show the asymptotes on the hyperbola.

57. $\dfrac{x^2}{4} + \dfrac{y^2}{9} = 1$ **58.** $\dfrac{x^2}{16} - \dfrac{y^2}{4} = 1$

Exercises 59 and 60: Use the graph to determine the equation of the ellipse or hyperbola.

59.

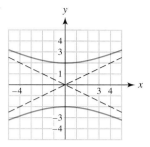

60.

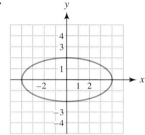

61. Solve the system of nonlinear equations. Check your solutions.

$$y = x^2 + 1$$
$$x^2 + 2y = 5$$

62. Shade the solution set in the xy-plane.

$$y \geq x^2 - 2$$
$$y \leq -x$$

Exercises 63–66: Determine whether f is an arithmetic or a geometric sequence. If it is arithmetic, find the common difference. If it is geometric, find the common ratio.

63. $f(n) = 5 - 2n$ **64.** $f(n) = 3(0.2)^n$

65. $f(n) = 7(4)^n$ **66.** $f(n) = 6n + 1$

67. Find the general term a_n for the arithmetic sequence where $a_1 = 2$ and $a_2 = 5$.

68. Find the general term a_n for the geometric sequence where $a_1 = 4$ and $a_2 = 12$.

Exercises 69 and 70: Find the sum using a formula.

69. $3 + 7 + 11 + 15 + 19 + \cdots + 35$

70. $1 - 2 + 4 - 8 + 16 - \cdots + 1024$

Exercises 71 and 72: Expand the binomial expression.

71. $(2x + 3)^4$ **72.** $(2a - 5b)^3$

APPLICATIONS

73. *Radius of a Circle* If a circle has an area of A square units, its radius r is given by $r = \sqrt{\frac{A}{\pi}}$. Find the radius of a circle with an area of 14 square inches. Approximate this radius to the nearest hundredth of an inch.

74. *Distance from Work* Starting at a warehouse, a delivery truck driver travels down a straight highway for 3 hours at 40 miles per hour, stops and unloads the truck for 1 hour, and then returns to the warehouse at 60 miles per hour. Sketch a graph that shows the distance between the truck and the warehouse over this period of time.

75. *Exercise and Fluid Consumption* When a person exercises, the total amount of fluid needed for the day increases depending on the person's weight and the duration of the exercise. To determine the total number of ounces of fluid needed, divide the person's weight by 2 and then add 0.4 ounces for every minute of exercise. (*Source: Runner's World.*)
(a) Write a function that gives the total fluid requirements for a person weighing 170 pounds who exercises for x minutes a day.

(b) If an athlete who exercises for 90 minutes requires a total of 130 ounces of fluid, determine the athlete's weight.

76. *Airplane Speed* An airplane travels 1080 miles into the wind in 3 hours. The return trip with the wind takes 2.7 hours. Find the average ground speed of the airplane and the average wind speed.

77. *Size of a Tent* The length of a rectangular tent floor is 6 feet shorter than twice the width. If the area of the tent floor is 108 square feet, what are the dimensions of the tent?

78. *Working Together* Suppose that one person can weed a garden in 60 minutes and a second person can weed the same garden in 90 minutes. How long would it take these two people to weed the garden if they worked together?

79. *Numbers* The product of two positive numbers is 96. If the larger number is subtracted from 3 times the smaller number, the result is 12. Let x be the smaller number and let y be the larger number.
(a) Write a system of equations for this situation.
(b) What are the two numbers?

80. *Marching Band* A band is marching in a triangular formation so that 1 person is in the first row, 3 people are in the second row, 5 people are in the third row, and so on. Use a formula to find the total number of musicians in the marching band if the last row contains 23 people.

Appendix A   Using the Graphing Calculator

OVERVIEW OF THIS APPENDIX

The intent of this appendix is to provide instruction for the TI-83, TI-83 Plus, and TI-84 Plus graphing calculators that may be used in conjunction with this textbook. It includes specific keystrokes needed to work several examples from the text. Students are also advised to consult the *Graphing Calculator Guidebook* provided by the manufacturer.

ENTERING MATHEMATICAL EXPRESSIONS

```
π
            3.141592654
√(200)
            14.14213562
10^4
                   10000
```

Figure A.1

EVALUATING π: To evaluate π use the following keystrokes, as shown in the first and second lines of Figure A.1. (Do *not* use 3.14 or $\frac{22}{7}$ for π.)

[2nd] [^[π]] [ENTER]

EVALUATING A SQUARE ROOT: To evaluate a square root, such as $\sqrt{200}$, use the following keystrokes, as shown in the third and fourth lines of Figure A.1.

[2nd] [x²[√]] [2] [0] [0] [)] [ENTER]

EVALUATING AN EXPONENTIAL EXPRESSION: To evaluate an exponential expression, such as 10^4, use the following keystrokes, as shown in the last two lines of Figure A.1.

[1] [0] [^] [4] [ENTER]

```
³√(64)
                       4
5ˣ√23
             1.872171231
```

Figure A.2

EVALUATING A CUBE ROOT: To evaluate a cube root, such as $\sqrt[3]{64}$, use the following keystrokes, as shown in the first and second lines of Figure A.2.

[MATH] [4] [6] [4] [)] [ENTER]

EVALUATING OTHER ROOTS: To evaluate a fifth root, such as $\sqrt[5]{23}$, use the following keystrokes, as shown in the third and fourth lines of Figure A.2.

[5] [MATH] [5] [2] [3] [ENTER]

```
14e^(.0153*8)
             15.82288531
log(100)
                       2
ln(10)
             2.302585093
```

Figure A.3

EVALUATING THE NATURAL EXPONENTIAL FUNCTION: To evaluate $14e^{0.0153(8)}$, use the following keystrokes, as shown in the first and second lines of Figure A.3.

[1] [4] [2nd] [LN [eˣ]] [.] [0] [1] [5] [3] [×] [8] [)] [ENTER]

EVALUATING THE COMMON LOGARITHMIC FUNCTION: To evaluate $\log(100)$, use the following keystrokes, as shown in the third and fourth lines of Figure A.3.

[LOG] [1] [0] [0] [)] [ENTER]

EVALUATING THE NATURAL LOGARITHMIC FUNCTION: To evaluate $\ln(10)$, use the following keystrokes, as shown in the last two lines of Figure A.3.

[LN] [1] [0] [)] [ENTER]

	SUMMARY: ENTERING MATHEMATICAL EXPRESSIONS

To access the *number* π, use (2nd)(^[π]).

To evaluate a *square root*, use (2nd)(x^2[$\sqrt{\ }$]).

To evaluate an *exponential expression*, use the (^) key. To square a number, the (x^2) key can also be used.

To evaluate a *cube root*, use (MATH)(4).

To evaluate a *kth root*, use (k)(MATH)(5).

To access the *natural exponential function*, use (2nd)(LN [e^x]).

To access the *common logarithmic function*, use (LOG).

To access the *natural logarithmic function*, use (LN).

EXPRESSING ANSWERS AS FRACTIONS

To evaluate $\frac{1}{3} + \frac{2}{5} - \frac{4}{9}$ in fraction form, use the following keystrokes, as shown in Figure 1.24 on page 51.

(()(1)(÷)(3)())(+)(()(2)(÷)(5)())(−)(()(4)(÷)(9)())(MATH)(1)(ENTER)

	SUMMARY: EXPRESSING ANSWERS AS FRACTIONS

Enter the arithmetic expression. To access the "Frac" feature, use the keystrokes (MATH)(1). Then press (ENTER).

DISPLAYING NUMBERS IN SCIENTIFIC NOTATION

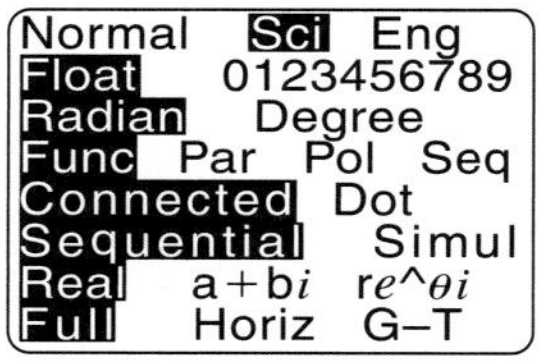

Figure A.4

To display numbers in scientific notation, set the graphing calculator in scientific mode (Sci), by using the following keystrokes. See Figure A.4. (These keystrokes assume that the calculator is in normal mode.)

(MODE)(▷)(ENTER)(2nd)(MODE [QUIT])

In scientific mode we can display the numbers 5432 and 0.00001234 in scientific notation, as shown in Figure A.5.

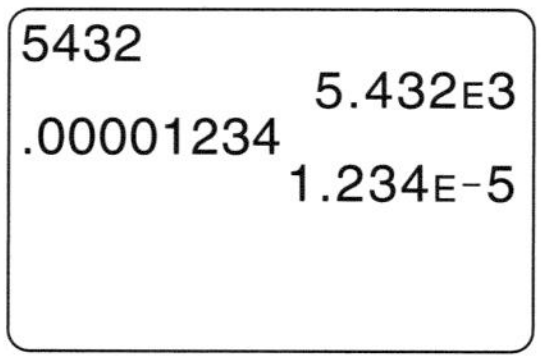

Figure A.5

	SUMMARY: SETTING SCIENTIFIC MODE

If your calculator is in normal mode, it can be set in scientific mode by pressing

(MODE)(▷)(ENTER)(2nd)(MODE [QUIT]).

These keystrokes return the graphing calculator to the home screen.

ENTERING NUMBERS IN SCIENTIFIC NOTATION

Numbers can be entered in scientific notation. For example, to enter 4.2×10^{-3} in scientific notation, use the following keystrokes. (Be sure to use the negation key $(-)$ rather than the subtraction key.)

$$\boxed{4}\ \boxed{.}\ \boxed{2}\ \boxed{\text{2nd}}\ \boxed{,[\text{EE}]}\ \boxed{(-)}\ \boxed{3}$$

This number can also be entered using the following keystrokes. See Figure A.6.

$$\boxed{4}\ \boxed{.}\ \boxed{2}\ \boxed{\times}\ \boxed{1}\ \boxed{0}\ \boxed{\wedge}\ \boxed{(}\ \boxed{(-)}\ \boxed{3}\ \boxed{)}$$

```
4.2E-3
              .0042
4.2*10^(-3)
              .0042
```

Figure A.6

SUMMARY: ENTERING NUMBERS IN SCIENTIFIC NOTATION

One way to enter a number in scientific notation is to use the keystrokes

$$\boxed{\text{2nd}}\ \boxed{,[\text{EE}]}$$

to access an exponent (EE) of 10.

MAKING A TABLE

To make a table of values for $y = 3x + 1$ starting at $x = 4$ and incrementing by 2, begin by pressing $\boxed{\text{Y}=}$ and then entering the formula $Y_1 = 3X + 1$, as shown in Figure A.7. To set the table parameters, press the following keys. See Figure A.8.

$$\boxed{\text{2nd}}\ \boxed{\text{WINDOW [TBLSET]}}\ \boxed{4}\ \boxed{\text{ENTER}}\ \boxed{2}$$

```
Plot1  Plot2  Plot3
\Y1■3X+1
\Y2=
\Y3=
\Y4=
\Y5=
\Y6=
\Y7=
```

Figure A.7

These keystrokes specify a table that starts at $x = 4$ and increments the x-values by 2. Therefore, the values of Y_1 at $x = 4, 6, 8, \ldots$ appear in the table. To create this table, press the following keys.

$$\boxed{\text{2nd}}\ \boxed{\text{GRAPH [TABLE]}}$$

We can scroll through x- and y-values by using the arrow keys. See Figure A.9. Note that there is no first or last x-value in the table.

```
TABLE SETUP
 TblStart=4
 △Tbl=2
Indpnt:  Auto  Ask
Depend:  Auto  Ask
```

Figure A.8

```
  X    | Y1
  4    | 13
  6    | 19
  8    | 25
 10    | 31
 12    | 37
 14    | 43
 16    | 49
Y1■3X+1
```

Figure A.9

SUMMARY: MAKING A TABLE

1. Enter the formula for the equation using $\boxed{\text{Y}=}$.
2. Press $\boxed{\text{2nd}}\ \boxed{\text{WINDOW [TBLSET]}}$ to set the starting x-value and the increment between x-values appearing in the table.
3. Create the table by pressing $\boxed{\text{2nd}}\ \boxed{\text{GRAPH [TABLE]}}$.

Setting the Viewing Rectangle (Window)

```
ZOOM MEMORY
1:ZBox
2:Zoom In
3:Zoom Out
4:ZDecimal
5:ZSquare
6:ZStandard
7↓ZTrig
```

Figure A.10

```
WINDOW
 Xmin=-10
 Xmax=10
 Xscl=1
 Ymin=-10
 Ymax=10
 Yscl=1
 Xres=1
```

Figure A.11

There are at least two ways to set the standard viewing rectangle of $[-10, 10, 1]$ by $[-10, 10, 1]$. The first involves pressing (ZOOM) followed by (6). See Figure A.10. The second method for setting the standard viewing rectangle is to press (WINDOW), and enter the following keystrokes. See Figure A.11.

(−) (1) (0) (ENTER) (1) (0) (ENTER) (1) (ENTER)

(−) (1) (0) (ENTER) (1) (0) (ENTER) (1) (ENTER)

(Be sure to use the negation key (−) rather than the subtraction key.) Other viewing rectangles can be set in a similar manner by pressing (WINDOW) and entering the appropriate values. To see the viewing rectangle, press (GRAPH). An example is shown in Figure 8.16 on page 486.

|||||| **SUMMARY: SETTING THE VIEWING RECTANGLE**

To set the standard viewing rectangle, press (ZOOM)(6). To set any viewing rectangle, press (WINDOW) and enter the necessary values. To see the viewing rectangle, press (GRAPH).

Note: You do not need to change "Xres" from 1. ||||||||

Making a Scatterplot or a Line Graph

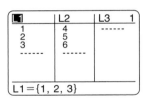

Figure A.12

To make a scatterplot with the points $(-5, -5)$, $(-2, 3)$, $(1, -7)$, and $(4, 8)$ begin by following these steps.

1. Press (STAT) followed by (1).
2. If list L1 is not empty, use the arrow keys to place the cursor on L1, as shown in Figure A.12. Then press (CLEAR) followed by (ENTER). This deletes all elements in the list. Similarly, if L2 is not empty, clear the list.
3. Input each x-value into list L1 followed by (ENTER). Input each y-value into list L2 followed by (ENTER). See Figure A.13.

```
L1      L2      L3     1
 -5      -5     ------
 -2       3
  1      -7
  4       8
------  ------
L1(5) =
```

Figure A.13

It is essential that both lists have the same number of values—otherwise an error message appears when a scatterplot is attempted. Before these four points can be plotted, "STAT-PLOT" must be turned on. It is accessed by pressing

(2nd)(Y = [STAT PLOT]),

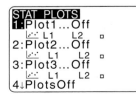

Figure A.14

as shown in Figure A.14.

There are three possible "STATPLOTS", numbered 1, 2, and 3. Any one of the three can be selected. The first plot can be selected by pressing (1). Next, place the cursor over "On" and press (ENTER) to turn "Plot1" on. There are six types of plots that can be selected. The first type is a *scatterplot* and the second type is a *line graph*, so place the cursor over the first type of plot and press (ENTER) to select a scatterplot. (To make the line graph, place the cursor over the second type of plot and press (ENTER).) The x-values are stored in list L1, so select L1 for "Xlist" by pressing (2nd)(1). Similarly, press (2nd)(2) for the "Ylist," since the y-values are stored in list L2. Finally, there are three styles of marks that can be used to show data points in the graph. We will usually use the first, because it is largest and shows

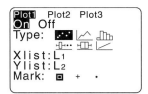

Figure A.15

$[-10, 10, 1]$ by $[-10, 10, 1]$

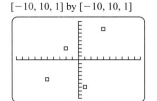

Figure A.16

up the best. Make the screen appear as in Figure A.15. Before plotting the four data points, be sure to set an appropriate viewing rectangle. Then press (GRAPH). The data points appear as in Figure A.16.

Remark 1: A fast way to set the viewing rectangle for any scatterplot is to select the "ZOOMSTAT" feature by pressing (ZOOM)(9). This feature automatically scales the viewing rectangle so that all data points are shown.

Remark 2: If an equation has been entered into the (Y =) menu and selected, it will be graphed with the data. This feature is used frequently to model data.

▍▍▍   SUMMARY: MAKING A SCATTERPLOT OR A LINE GRAPH

The following are basic steps necessary to make either a scatterplot or a line graph.

1. Use (STAT)(1) to access lists L1 and L2.
2. If list L1 is not empty, place the cursor on L1 and press (CLEAR)(ENTER). Repeat for list L2, if it is not empty.
3. Enter the x-values into list L1 and the y-values into list L2.
4. Use (2nd)(Y = [STAT PLOT]) to select appropriate parameters for the scatterplot or line graph.
5. Set an appropriate viewing rectangle. Press (GRAPH). Otherwise, press (ZOOM)(9). This feature automatically sets the viewing rectangle and plots the data.

Note: (ZOOM)(9) *cannot* be used to set a viewing rectangle for the graph of a function. ▕▏▏▏▏▏▏▏▏

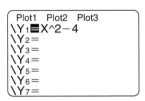

Figure A.17

ENTERING A FORMULA

To enter a formula, press (Y =). For example, use the following keystrokes after "$Y_1 = $ " to enter $y = x^2 - 4$. See Figure A.17.

(Y =)(CLEAR)(X, T, θ, n)(^)(2)(−)(4)

Note that there is a built-in key to enter the variable X. If "$Y_1 = $ " does not appear after pressing (Y =), press (MODE) and make sure the calculator is set in *function mode,* denoted "Func". See Figure A.18.

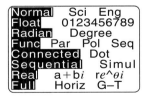

Figure A.18

▍▍▍   SUMMARY: ENTERING A FORMULA

To enter a formula, press (Y =). To delete an existing formula, press (CLEAR). ▕▏▏▏▏▏▏▏

$[-10, 10, 1]$ by $[-10, 10, 1]$

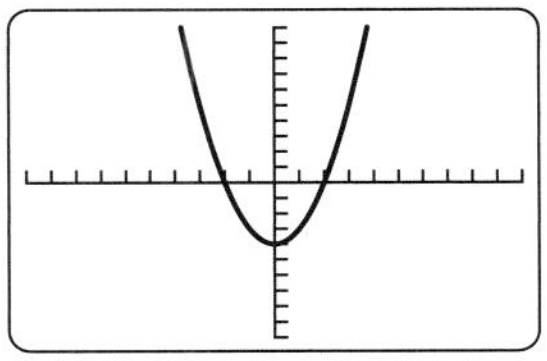

GRAPHING AN EQUATION OR A FUNCTION

To graph an equation, such as $y = x^2 - 4$, start by pressing (Y =) and enter $Y_1 = X^\wedge 2 - 4$. If there is an equation already entered, remove it by pressing (CLEAR). The equals signs in "$Y_1 = $ " should be in reverse video (a dark rectangle surrounding a white equals sign), which indicates that the equation will be graphed. If the equals sign is not in reverse video, place the cursor over it and press (ENTER). Set an appropriate viewing rectangle and then press (GRAPH). The graph of f will appear in the specified viewing rectangle. See Figures A.17 and A.19.

Figure A.19

SUMMARY: GRAPHING AN EQUATION OR A FUNCTION

1. Use the Y= menu to enter the formula for the function.
2. Use the WINDOW menu to set an appropriate viewing rectangle.
3. Press GRAPH.

GRAPHING A VERTICAL LINE

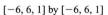

Figure A.20

$[-6, 6, 1]$ by $[-6, 6, 1]$

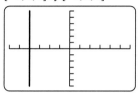

Figure A.21

Set an appropriate window (or viewing rectangle). Then return to the home screen by pressing

2nd MODE [QUIT].

To graph a vertical line, such as $x = -4$, press

2nd PRGM [DRAW] 4 (−) 4.

See Figure A.20. Pressing ENTER will make the vertical line appear, as shown in Figure A.21.

SUMMARY: GRAPHING THE VERTICAL LINE $x = h$

1. Set an appropriate window by pressing WINDOW.
2. Return to the home screen by pressing 2nd MODE [QUIT].
3. Draw a vertical line by pressing 2nd PRGM [DRAW] 4 h ENTER.

SQUARING A VIEWING RECTANGLE

Figure A.22

In a square viewing rectangle the graph of $y = x$ is a line that makes a 45° angle with the positive x-axis, a circle appears circular, and all sides of a square have the same length. An approximate square viewing rectangle can be set if the distance along the x-axis is 1.5 times the distance along the y-axis. Examples of viewing rectangles that are (approximately) square include

$$[-6, 6, 1] \text{ by } [-4, 4, 1] \quad \text{and} \quad [-9, 9, 1] \text{ by } [-6, 6, 1].$$

Square viewing rectangles can be set automatically by pressing either

ZOOM 4   or   ZOOM 5.

ZOOM 4 provides a *decimal window*, which is discussed later. See Figure A.22.

SUMMARY: SQUARING A VIEWING RECTANGLE

Either ZOOM 4 or ZOOM 5 may be used to produce a square viewing rectangle. An (approximately) square viewing rectangle has the form

$$[-1.5k, 1.5k, 1] \text{ by } [-k, k, 1],$$

where k is a positive number.

LOCATING A POINT OF INTERSECTION

Figure A.23

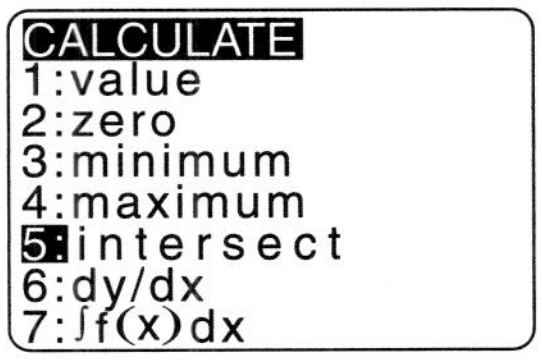

To find the point of intersection for the graphs of

$$y_1 = 3(1 - x) \quad \text{and} \quad y_2 = 2,$$

start by entering Y_1 and Y_2, as shown in Figure A.23. Set the window, and graph both equations. Then press the following keys to find the intersection point.

$$\boxed{\text{2nd}}\;\boxed{\text{TRACE [CALC]}}\;\boxed{5}$$

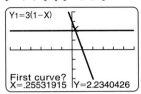

Figure A.24

See Figure A.24, where the "intersect" utility is being selected. The calculator prompts for the first curve, as shown in Figure A.25. Use the arrow keys to locate the cursor near the point of intersection and press $\boxed{\text{ENTER}}$. Repeat these steps for the second curve. Finally we are prompted for a guess. For each of the three prompts, place the free-moving cursor near the point of intersection and press $\boxed{\text{ENTER}}$. The approximate coordinates of the point of intersection will be shown.

$[-6, 6, 1]$ by $[-4, 4, 1]$

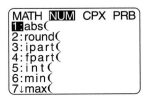

Figure A.25

SUMMARY: FINDING A POINT OF INTERSECTION

1.   Graph the two equations in an appropriate viewing rectangle.
2.   Press $\boxed{\text{2nd}}\;\boxed{\text{TRACE [CALC]}}\;\boxed{5}$.
3.   Use the arrow keys to select an approximate location for the point of intersection. Press $\boxed{\text{ENTER}}$ to make the three selections for "First curve?", "Second curve?", and "Guess?". (Note that if the cursor is near the point of intersection, you usually do not need to move the cursor for each selection. Just press $\boxed{\text{ENTER}}$ three times.)

ACCESSING THE ABSOLUTE VALUE

Figure A.26

To graph $y_1 = |x - 50|$, begin by entering $Y_1 = \text{abs}(X - 50)$. The absolute value (abs) is accessed by pressing

$$\boxed{\text{MATH}}\;\boxed{\triangleright}\;\boxed{1}.$$

See Figure A.26.

SUMMARY: ACCESSING THE ABSOLUTE VALUE

1.   Press $\boxed{\text{MATH}}$.
2.   Position the cursor over "NUM".
3.   Press $\boxed{1}$ to select the absolute value.

SHADING A SYSTEM OF INEQUALITIES

To shade the solution set to the system of linear inequalities $2x + y \leq 5$, $-2x + y \geq 1$, begin by solving each inequality for y to obtain $y \leq 5 - 2x$ and $y \geq 2x + 1$. Then let $Y_1 = 5 - 2X$ and $Y_2 = 2X + 1$, as shown in Figure A.27 on the next page. Position the

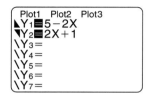

Figure A.27

$[-15, 15, 5]$ by $[-10, 10, 5]$

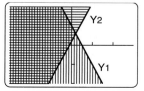

Figure A.28

cursor to the left of Y_1, and press (ENTER) three times. The triangle that appears indicates that the calculator will shade the region below the graph of Y_1. Next locate the cursor to the left of Y_2 and press (ENTER) twice. This triangle indicates that the calculator will shade the region above the graph of Y_2. After setting the viewing rectangle to $[-15, 15, 5]$ by $[-10, 10, 5]$ press (GRAPH). The result is shown in Figure A.28.

SUMMARY: SHADING A SYSTEM OF INEQUALITIES

1. Solve each inequality for y.
2. Enter each formula as Y_1 and Y_2 in the (Y =) menu.
3. Locate the cursor to the left of Y_1 and press (ENTER) two or three times, to shade either above or below the graph of Y_1. Repeat for Y_2.
4. Set an appropriate viewing rectangle.
5. Press (GRAPH).

Note: The "Shade" utility in the DRAW menu can also be used to shade the region *between* two graphs.

ENTERING THE ELEMENTS OF A MATRIX

In Example 6(b), Section 9.2 on page 569, the elements of a matrix are entered. The augmented matrix A is given by

$$A = \begin{bmatrix} 1 & 1 & 2 & | & 1 \\ -1 & 0 & 1 & | & -2 \\ 2 & 1 & 5 & | & -1 \end{bmatrix}.$$

Use the following keystrokes on the TI-83 Plus or TI-84 Plus to define a matrix A with dimension 3×4. (*Note:* On the TI-83 the matrix menu is found by pressing (MATRIX).)

(2nd) (x^{-1} [MATRIX]) (▷) (▷) (1) (3) (ENTER) (4) (ENTER)

See Figure 9.4(a).

Then input the 12 elements of the matrix A, row by row. Finish each entry by pressing (ENTER). After these elements have been entered, press

(2nd) (MODE [QUIT])

to return to the home screen. To display the matrix A, press

(2nd) (x^{-1} [MATRIX]) (1) (ENTER).

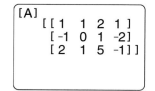

Figure A.29

See Figure A.29.

SUMMARY: ENTERING THE ELEMENTS OF A MATRIX A

1. Begin by accessing the matrix A by pressing (2nd) (x^{-1} [MATRIX]) (▷) (▷) (1).
2. Enter the dimension of A by pressing (m) (ENTER) (n) (ENTER), where the dimension of the matrix is $m \times n$.
3. Input each element of the matrix, row by row. Finish each entry by pressing (ENTER). Use (2nd) (MODE [QUIT]) to return to the home screen.

Note: On the TI-83, replace the keystrokes (2nd) (x^{-1} [MATRIX]) with (MATRIX).

REDUCED ROW-ECHELON FORM

In Example 6(b), Section 9.2 on page 569, the reduced row-echelon form of a matrix is found. To find this reduced row-echelon form, use the following keystrokes from the home screen on the TI-83 Plus or TI-84 Plus.

[2nd] [x^{-1} [MATRIX]] [▷] [ALPHA] [APPS [B]] [2nd] [x^{-1} [MATRIX]] [1] [)] [ENTER]

The resulting matrix is shown in Figure 9.4(b). On the TI-83 graphing calculator use the following keystrokes to find the reduced row-echelon form.

[MATRIX] [▷] [ALPHA] [MATRIX [B]] [MATRIX] [1] [)] [ENTER]

▥ SUMMARY: FINDING REDUCED ROW-ECHELON FORM OF A MATRIX

1. To make rref([A]) appear on the home screen, use the following keystrokes for the TI-83 Plus or TI-84 Plus graphing calculator.

   [2nd] [x^{-1} [MATRIX]] [▷] [ALPHA] [APPS [B]] [2nd] [x^{-1} [MATRIX]] [1] [)]

2. Press [ENTER] to calculate the reduced row-echelon form.
3. Use arrow keys to access elements that do not appear on the screen.

Note: On the TI-83, replace the keystrokes [2nd] [x^{-1} [MATRIX]] with [MATRIX] and [APPS [B]] with [MATRIX [B]].

EVALUATING A DETERMINANT

In Example 3(a), Section 9.3 on page 576, a graphing calculator is used to evaluate a determinant of a matrix. Start by entering the 9 elements of the 3×3 matrix A, as shown in Figure 9.6(a). To compute det A, perform the following keystrokes from the home screen.

[2nd] [x^{-1} [MATRIX]] [▷] [1] [2nd] [x^{-1} [MATRIX]] [1] [)] [ENTER]

The results are shown in the last two lines of Figure 9.6(b).

▥ SUMMARY: EVALUATING A DETERMINANT OF A MATRIX

1. Enter the dimension and elements of the matrix A.
2. Return to the home screen by pressing

   [2nd] [MODE [QUIT]].

3. On the TI-83 Plus or TI-84 Plus, perform the following keystrokes.

   [2nd] [x^{-1} [MATRIX]] [▷] [1] [2nd] [x^{-1} [MATRIX]] [1] [)] [ENTER]

Note: On the TI-83, replace the keystrokes [2nd] [x^{-1} [MATRIX]] with [MATRIX].

LOCATING AN *X*-INTERCEPT OR ZERO

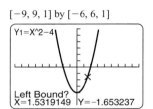

```
CALCULATE
1:value
2:zero
3:minimum
4:maximum
5:intersect
6:dy/dx
7:∫f(x)dx
```

Figure A.30

To locate an *x*-intercept or *zero* of $f(x) = x^2 - 4$, start by entering $Y_1 = X^2 - 4$ into the $\boxed{Y =}$ menu. Set the viewing rectangle to $[-9, 9, 1]$ by $[-6, 6, 1]$ and graph Y_1. Afterwards, press the following keys to invoke the zero finder. See Figure A.30.

$$\boxed{\text{2nd}}\ \boxed{\text{TRACE [CALC]}}\ \boxed{2}$$

The graphing calculator prompts for a left bound. Use the arrow keys to set the cursor to the left of the *x*-intercept and press $\boxed{\text{ENTER}}$. The graphing calculator then prompts for a right bound. Set the cursor to the right of the *x*-intercept and press $\boxed{\text{ENTER}}$. Finally the graphing calculator prompts for a guess. Set the cursor roughly at the *x*-intercept and press $\boxed{\text{ENTER}}$. See Figures A.31–A.33. The calculator then approximates the *x*-intercept or zero automatically, as shown in Figure A.34. The zero of -2 can be found similarly.

$[-9, 9, 1]$ by $[-6, 6, 1]$

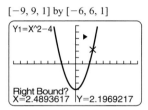

Figure A.31

$[-9, 9, 1]$ by $[-6, 6, 1]$

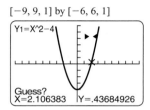

Figure A.32

$[-9, 9, 1]$ by $[-6, 6, 1]$

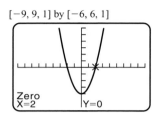

Figure A.33

$[-9, 9, 1]$ by $[-6, 6, 1]$

Figure A.34

▌▌▌▌▌ SUMMARY: LOCATING AN *x*-INTERCEPT OR ZERO

1. Graph the function in an appropriate viewing rectangle.
2. Press $\boxed{\text{2nd}}\ \boxed{\text{TRACE [CALC]}}\ \boxed{2}$.
3. Select the left and right bounds, followed by a guess. Press $\boxed{\text{ENTER}}$ after each selection. The calculator then approximates the *x*-intercept or zero.

SETTING CONNECTED AND DOT MODE

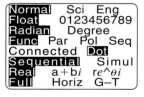

```
Normal   Sci  Eng
Float    0123456789
Radian    Degree
Func  Par  Pol  Seq
Connected   Dot
Sequential   Simul
Real   a+bi  re^θi
Full    Horiz  G-T
```

Figure A.35

To set your graphing calculator in dot mode, press $\boxed{\text{MODE}}$, position the cursor over "Dot", and press $\boxed{\text{ENTER}}$. See Figure A.35. Graphs will now appear in dot mode rather than connected mode.

▌▌▌▌▌ SUMMARY: SETTING CONNECTED OR DOT MODE

1. Press $\boxed{\text{MODE}}$.
2. Position the cursor over "Connected" or "Dot". Press $\boxed{\text{ENTER}}$.

SETTING A DECIMAL WINDOW

With a decimal window, the cursor stops on convenient *x*-values. In the decimal window $[-9.4, 9.4, 1]$ by $[-6.2, 6.2, 1]$ the cursor stops on *x*-values that are multiples of 0.2. If we reduce the viewing rectangle to $[-4.7, 4.7, 1]$ by $[-3.1, 3.1, 1]$, the cursor stops on *x*-values

Figure A.36

that are multiples of 0.1. To set this smaller window automatically, press (ZOOM)(4). See Figure A.36. Decimal windows are also useful when graphing rational functions with asymptotes in connected mode.

SUMMARY: SETTING A DECIMAL WINDOW

1. Press (ZOOM)(4) to set the viewing rectangle $[-4.7, 4.7, 1]$ by $[-3.1, 3.1, 1]$.
2. A larger decimal window is $[-9.4, 9.4, 1]$ by $[-6.2, 6.2, 1]$.

SETTING $a + bi$ MODE

To evaluate expressions containing square roots of negative numbers, such as $\sqrt{-25}$, set your calculator in $a + bi$ mode by using the following keystrokes.

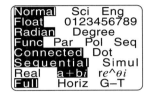

Figure A.37

(MODE)(▽)(▽)(▽)(▽)(▽)(▽)(▷)(ENTER)(2nd)(MODE [QUIT])

See Figures A.37 and A.38.

SUMMARY: SETTING $a + bi$ MODE

1. Press (MODE).
2. Move the cursor to the seventh line and highlight $a + bi$.
3. Press (2nd)(MODE [QUIT]) and return to the home screen.

Figure A.38

EVALUATING COMPLEX ARITHMETIC

Complex arithmetic can be performed much like other arithmetic expressions. This is done by entering

(2nd)(. [i])

to obtain the imaginary unit i from the home screen. For example, to find the sum $(-2 + 3i) + (4 - 6i)$, perform the following keystrokes on the home screen.

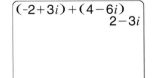

Figure A.39

(()((-))(2)(+)(3)(2nd)(. [i])())(+)(()(4)(-)(6)(2nd)(. [i])())(ENTER)

The result is shown in Figure A.39. Other complex arithmetic operations are done similarly.

SUMMARY: EVALUATING COMPLEX ARITHMETIC

Enter a complex expression in the same way as you would any arithmetic expression. To obtain the complex number i, use (2nd)(. [i]).

FINDING MAXIMUM AND MINIMUM VALUES

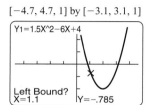

Figure A.40

To find a minimum y-value y (or vertex) on the graph of $f(x) = 1.5x^2 - 6x + 4$, start by entering $Y_1 = 1.5X^2 - 6X + 4$ from the ⟨Y =⟩ menu. Set the viewing rectangle and then perform the following keystrokes to find the minimum y-value.

⟨2nd⟩ ⟨TRACE [CALC]⟩ ⟨3⟩

See Figure A.40.

The calculator prompts for a left bound. Use the arrow keys to position the cursor left of the vertex and press ⟨ENTER⟩. Similarly, position the cursor to the right of the vertex for the right bound and press ⟨ENTER⟩. Finally, the graphing calculator asks for a guess between the left and right bounds. Place the cursor near the vertex and press ⟨ENTER⟩. See Figures A.41–A.43. The minimum value is shown in Figure A.44.

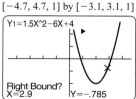
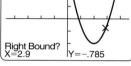

Figure A.41

Figure A.42

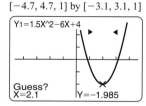

Figure A.43

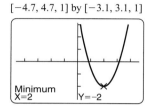

Figure A.44

A maximum of the function f on an interval can be found in a similar manner, except enter

⟨2nd⟩ ⟨TRACE [CALC]⟩ ⟨4⟩.

The calculator prompts for left and right bounds, followed by a guess. Press ⟨ENTER⟩ after the cursor has been located appropriately for each prompt. The graphing calculator will display the maximum y-value. For example, see the Technology Note on page 668.

SUMMARY: FINDING MAXIMUM AND MINIMUM VALUES

1. Graph the function in an appropriate viewing rectangle.
2. Press ⟨2nd⟩ ⟨TRACE [CALC]⟩ ⟨3⟩ to find a minimum y-value.
3. Press ⟨2nd⟩ ⟨TRACE [CALC]⟩ ⟨4⟩ to find a maximum y-value.
4. Use the arrow keys to locate the left and right x-bounds, followed by a guess. Press ⟨ENTER⟩ to select each position of the cursor.

Appendix B   Sets

BASIC TERMINOLOGY

A **set** is a collection of things, and the members of a set are called **elements**. A set can be described by listing its elements between braces. For example, the set W containing the *weekdays* is

$$W = \{\text{Monday, Tuesday, Wednesday, Thursday, Friday}\}.$$

This set has 5 elements. For example, Monday *is an element of* W, which is denoted

$$\text{Monday} \in W.$$

However Sunday *is not an element of* W, which is denoted

$$\text{Sunday} \notin W.$$

If a set contains no elements, then it is called the **empty set** or **null set**. The empty set is denoted $\varnothing$, or $\{\ \ \}$. For example, the set Z that contains the names of U.S. states starting with the letter Z is the empty set. That is, $Z = \varnothing$ or $Z = \{\ \ \}$.

Note: Do *not* write the empty set as $\{\varnothing\}$.

EXAMPLE 1   Listing the elements of a set

Use set notation to list the elements of each set S described.
(a) The natural numbers from 1 to 12 that are odd
(b) The days of the week that start with the letter T
(c) The last names of U.S. presidents in office during the 1990s

Solution   **(a)** The list of the natural numbers from 1 to 12 is

$$1, 2, 3, 4, 5, 6, 7, 8, 9, 10, 11, \text{ and } 12.$$

The set of odd natural numbers from this list is

$$S = \{1, 3, 5, 7, 9, 11\}.$$

(b) $S = \{\text{Tuesday, Thursday}\}$
(c) $S = \{\text{Bush, Clinton}\}$

EXAMPLE 2   Determining the elements of sets

Use $\in$ or $\notin$ to make each statement true.
(a) 5 _____ $\{1, 2, 3, 4, 5, 6\}$
(b) -2 _____ $\{-4, 0, 2, 4, 6\}$
(c) $\frac{1}{2}$ _____ $\{0, 0.5, 1.0, 1.5, 2.0\}$

Solution   **(a)** Because 5 is an element of $\{1, 2, 3, 4, 5, 6\}$, we write

$$5 \in \{1, 2, 3, 4, 5, 6\}.$$

(b) Because -2 is not an element of $\{-4, 0, 2, 4, 6\}$, we write

$$-2 \notin \{-4, 0, 2, 4, 6\}.$$

(c) Because $\frac{1}{2} = 0.5$, we write

$$\frac{1}{2} \in \{0, 0.5, 1.0, 1.5, 2.0\}.$$

Universal Set

When discussing sets, we assume that there is a *universal set*. The **universal set** contains all elements under consideration. For example, if the universal set U is all days of the week, then the set S containing the days that start with the letter S is

$$S = \{\text{Sunday, Saturday}\}.$$

However if the universal set U is only the weekdays, then the set S containing the days starting with the letter S is the *empty set*, or $S = \{\ \}$.

EXAMPLE 3   Using different universal sets

Determine the set O of odd integers that belong to each universal set U.
(a) $U = \{1, 2, 3, 4, 5, 6, 7, 8, 9, 10\}$
(b) $U = \{1, 6, 11, 16, 21, 26\}$
(c) $U = \{1, 2, 3, 4, \ldots\}$

Solution **(a)** The odd integers in U are 1, 3, 5, 7, and 9, so

$$O = \{1, 3, 5, 7, 9\}.$$

(b) The odd integers in $U = \{1, 6, 11, 16, 21, 26\}$ are 1, 11, and 21. Thus

$$O = \{1, 11, 21\}.$$

(c) The three dots in $\{1, 2, 3, 4, \ldots\}$ indicate that U contains all natural numbers. Thus the set O contains all odd natural numbers, or

$$O = \{1, 3, 5, 7, 9, \ldots\}.$$

Subsets

If every element in a set B is contained in a set A, then we say that B is a **subset** of A, denoted $B \subseteq A$. For example, if $A = \{1, 2, 3, 4\}$ and $B = \{2, 4\}$, then $B \subseteq A$ because every element in B belongs to A. However, A is not a subset of B, denoted $A \nsubseteq B$, because the elements 1 and 3 are in A but *not* in B. The symbol $\nsubseteq$ is read "is not a subset of."

If every element in set A is in set B and every element in set B is in set A, then A and B are **equal sets**, denoted $A = B$. Note that if $A \subseteq B$ and $B \subseteq A$, then $A = B$. Why?

EXAMPLE 4   Determining subsets

Let $A = \{a, b, c, d, e\}$, $B = \{b, c, d\}$, $C = \{b, e\}$, and $D = \{e, b\}$. Determine whether each statement is true or false.
(a) $A \subseteq B$ **(b)** $B \subseteq A$ **(c)** $C = D$ **(d)** $C \nsubseteq A$ **(e)** $\varnothing \subseteq B$ **(f)** $A \subseteq A$

Solution   **(a)** False; the elements a and e in A are not in B, so A is *not* a subset of B.

        (b) True; every element in B is in A, so B is a subset of A.

        (c) True; although the elements in C and D are listed in a different order, they contain exactly the same elements, so C and D are equal.

        (d) False; every element in C is in A, so C *is* a subset of A.

        (e) True; the empty set, or null set, is a subset of *every* set.

        (f) True; every element in A is in A, so A is a subset of itself.

Note: Every set is a subset of itself.

VENN DIAGRAMS

Venn diagrams are often used to depict relationships among sets. A large rectangle typically represents the universal set, and subsets of the universal set are represented by regions within the universal set. In Figure B.1 the universal set U is represented by everything inside the large rectangle. The set A is represented by the red circular region within this rectangle because A is a subset of U.

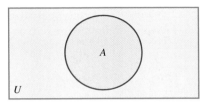

Figure B.1 $A \subseteq U$

    The **complement** of a set A, denoted A', is the set containing all elements in the universal set that are *not* in A. That is, if $a \notin A$, then $a \in A'$. For example, if

$$U = \{1, 2, 3, 4, 5, 6\} \quad \text{and} \quad A = \{1, 2, 3\},$$

then

$$A' = \{4, 5, 6\}$$

because the elements 4, 5, and 6 are found in U but not in A. This situation is illustrated by the Venn diagram in Figure B.2. The red region is A, and the blue region is A'. Together, the red and blue regions comprise the universal set U.

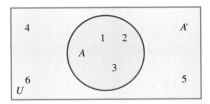

Figure B.2 A and A'

Note: Every element in U must be in either A or A', but not both.

EXAMPLE 5   Determining complements

Let the universal set U be $U = \{$red, blue, yellow, green, black, white$\}$, and let two subsets of U be $A = \{$red, blue, yellow$\}$ and $B = \{$black, white$\}$. Find each of the following.

(a) A'
(b) B'
(c) U'

Solution **(a)** The elements in U that are not in A are in $A' = \{$green, black, white$\}$.
(b) $B' = \{$red, blue, yellow, green$\}$
(c) Because the universal set U contains every element under consideration, the complement of U, or U', is empty. That is, $U' = \{\ \}$.

UNION AND INTERSECTION

Although we do not perform arithmetic operations, such as multiplication or division, on sets, we can find the *union* or *intersection* of two or more sets. The **union** of two sets A and B, denoted $A \cup B$ and read "A union B," is the set containing every element of A *and* every element of B. If an element is in both A and B, then this element is listed only once in $A \cup B$. For example, if

$$A = \{1, 2, 3, 4\} \quad \text{and} \quad B = \{3, 4, 5, 6\},$$

then

$$A \cup B = \{1, 2, 3, 4, 5, 6\}.$$

Note that elements 3 and 4 are in both A and B but are listed only once in $A \cup B$. A Venn diagram of this situation is shown in Figure B.3. The region that represents the union is shaded blue.

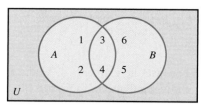

Figure B.3

The **intersection** of two sets A and B, denoted $A \cap B$ and read "A intersect B," is the set containing elements that *belong to both A and B*. For example, if

$$A = \{1, 2, 3, 4\} \quad \text{and} \quad B = \{3, 4, 5, 6\},$$

then

$$A \cap B = \{3, 4\}.$$

Note that elements 3 and 4 belong to *both* A and B. This situation is illustrated by the Venn diagram in Figure B.4. The purple region containing elements 3 and 4 represents the intersection of A and B.

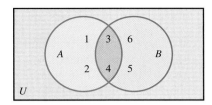

Figure B.4

EXAMPLE 6 Finding unions and intersections of sets

Let $A = \{a, b, c, x, z\}$, $B = \{a, x, y\}$, and $C = \{c, x, y, z\}$. Find each set.
(a) $A \cup B$ **(b)** $B \cap C$ **(c)** $A \cup C$ **(d)** $A \cap B \cap C$

Solution **(a)** The union contains the elements belonging to either A or B or both. Thus

$$A \cup B = \{a, b, c, x, y, z\}.$$

(b) The intersection contains the elements belonging to both B and C. Thus

$$B \cap C = \{x, y\}.$$

(c) $A \cup C = \{a, b, c, x, y, z\}$

(d) For an element to be in the intersection of three sets, the element must belong to each set. The only element belonging to A and B and C is x. Thus $A \cap B \cap C = \{x\}$.

EXAMPLE 7 Using set operations

Let $U = \{1, 2, 3, 4, 5, 6, 7, 8\}$, $A = \{2, 4, 6, 8\}$, $B = \{1, 3, 5, 6\}$, and $C = \{5, 6, 7, 8\}$. Find each set.
(a) $A \cap B$ **(b)** $B \cap B'$ **(c)** $B \cup C$ **(d)** $A \cup C'$

Solution **(a)** $A \cap B = \{6\}$
(b) Because $B = \{1, 3, 5, 6\}$ and $B' = \{2, 4, 7, 8\}$, $B \cap B' = \{\ \}$.

Note: The intersection of a set and its complement is always the empty set.

(c) $B \cup C = \{1, 3, 5, 6, 7, 8\}$
(d) Because $A = \{2, 4, 6, 8\}$ and $C' = \{1, 2, 3, 4\}$, $A \cup C' = \{1, 2, 3, 4, 6, 8\}$.

FINITE AND INFINITE SETS

Sets can either have a finite number of elements or infinitely many elements. The elements of a **finite set** can be listed explicitly, whereas the elements of an **infinite set** cannot be listed because there are infinitely many elements. For example, the set of integers S from -3 to 3 is a finite set with 7 elements because

$$S = \{-3, -2, -1, 0, 1, 2, 3\}.$$

In contrast, the set of natural numbers

$$N = \{1, 2, 3, 4, \ldots\},$$

is an infinite set because the list continues without end.

EXAMPLE 8 Recognizing finite and infinite sets

Identify each set as finite or infinite.
(a) The set of integers
(b) The set of natural numbers between 1 and 5, inclusive
(c) The set of rational numbers between 1 and 5, inclusive

Solution **(a)** There are infinitely many integers, so the set of integers is infinite.
(b) The set of natural numbers between 1 and 5, inclusive, is $\{1, 2, 3, 4, 5\}$. Because we can list each element of this set, it is a finite set.
(c) Because there are infinitely many rational numbers (fractions) between 1 and 5, this set is an infinite set.

 B EXERCISES

BASIC CONCEPTS

Exercises 1–10: Use set notation to list the elements of the set described.

1. The natural numbers less than 8

2. The natural numbers greater than 4 and less than or equal to 10

3. The days of the week starting with the letter S

4. U.S. states whose names start with the letter A

5. The letters of the alphabet from A to G

6. The letters of the alphabet from D to M

7. Dogs that can fly under their own power

8. People having an income of $10 trillion in 2004

9. The even integers greater than −2 and less than 11

10. The odd integers less than 21 and greater than 12

Exercises 11–20: Insert $\in$ or $\notin$ to make the statement true.

11. 10 _____ $\{5, 10, 15, 20\}$

12. 7 _____ $\{1, 3, 5, 7, 9, 11\}$

13. 5 _____ $\{2, 4, 6, 8, 10\}$

14. −1 _____ $\{0, 1, 2, 3, 4, 5\}$

15. $\frac{1}{4}$ _____ $\{0, 0.25, 0.5, 0.75, 1\}$

16. $\frac{2}{5}$ _____ $\{0, 0.2, 0.4, 0.6, 0.8, 1\}$

17. $\frac{1}{3}$ _____ $\{0, 0.33, 0.67, 1\}$

18. $\frac{4}{2}$ _____ $\{1, 3, 4, 5, 7\}$

19. red _____ $\{$blue, green, red, yellow$\}$

20. M _____ $\{$M, T, W, R, F$\}$

Exercises 21–24: Determine the set E of even integers that belong to the given universal set U.

21. $U = \{1, 2, 3, 4, 5, 6, 7, 8, 9, 10\}$

22. $U = \{-3, -2, -1, 0, 1, 2, 3\}$

23. $U = \{1, 2, 3, 4, \ldots\}$

24. $U = \{1, 3, 5, 7, 9, 11, \ldots\}$

Exercises 25–28: Determine the set A of words starting with the letter A that belong to the given universal set U.

25. $U = \{$Apple, Orange, Pear, Apricot$\}$

26. $U = \{$Apple, Orange$\}$

27. $U = \{\text{Calculus, Algebra, Geometry}\}$

28. $U = \{\text{Calculus, Algebra, Geometry, Arithmetic}\}$

SUBSETS

Exercises 29–36: Let $A = \{1, 2, 3, 4\}$, $B = \{3, 4, 5, 6\}$, and $C = \{3, 6\}$. Determine whether the statement is true or false.

29. $A \subseteq B$ **30.** $B \subseteq A$

31. $C \subseteq B$ **32.** $\varnothing \subseteq C$

33. $C \subseteq C$ **34.** $A \nsubseteq B$

35. $B \nsubseteq C$ **36.** $C \subseteq A$

Exercises 37–44: Let $A = \{a, b, c, d, e\}$, $B = \{b, d\}$, and $C = \{a, c, e\}$. Determine whether the statement is true or false.

37. $B \subseteq A$ **38.** $C \subseteq A$

39. $B \nsubseteq C$ **40.** $C = B$

41. $C \subseteq \varnothing$ **42.** $C \nsubseteq B$

43. $A \subseteq A$ **44.** $C \nsubseteq C$

VENN DIAGRAMS

Exercises 45–52: Let A and B be two sets and U be the universal set. Sketch a Venn diagram and shade the region that illustrates the given set.

45. $A \cup B$ **46.** $A \cap B$

47. A' **48.** B'

49. $A \cap U$ **50.** $B \cup U$

51. $(A \cap B)'$ **52.** $(A \cup B)'$

UNIONS, INTERSECTIONS, AND COMPLEMENTS

Exercises 53–62: Write the expression in terms of one set.

53. $\{1, 2, 3\} \cup \{3, 4\}$ **54.** $\{1, 2, 3\} \cup \{6\}$

55. $\{a, b, c\} \cap \{a, b, d\}$ **56.** $\{x, y, z\} \cap \{a, b, c\}$

57. $\{a, b, c\} \cup \{\ \}$ **58.** $\{a, b, c\} \cap \varnothing$

59. $\{a, b\} \cup \{c, b\} \cup \{d, a\}$

60. $\{1, 3, 5\} \cup \{4, 5\} \cup \{4, 5, 6\}$

61. $\{4, 5, 8, 9\} \cap \{3, 5, 8, 9\} \cap \{4, 5, 8\}$

62. $\{1, 2, 5\} \cap \{2, 1\} \cap \{1, 2, 3\}$

Exercises 63–80: Let

$$U = \{1, 2, 3, 4, 5, 6, 7, 8, 9, 10\},$$
$$A = \{1, 3, 5, 7, 9\}, B = \{2, 4, 6, 8, 10\},$$
$$C = \{3, 4, 5, 6, 7\}, \text{ and } D = \{4, 5, 6, 9\}.$$

Use set notation to list the elements in the given set.

63. $A \cup B$ **64.** $A \cap B$

65. $C \cap D$ **66.** $A \cup D$

67. $U \cap \varnothing$ **68.** $\varnothing \cup U$

69. $A \cup \varnothing$ **70.** $\varnothing \cap B$

71. $D \cap D$ **72.** $A \cup A$

73. U' **74.** A'

75. B' **76.** D'

77. $A \cup D'$ **78.** $A' \cap B'$

79. $A \cap B \cap C$ **80.** $B \cup C \cup D$

Exercises 81–86: Determine whether the given set is finite or infinite.

81. The set of whole numbers

82. The set of real numbers

83. The set of natural numbers less than 1000

84. The set of natural numbers greater than 4 and less than 20

85. The days of the week

86. The names of the students in your class

Linear Programming

BASIC CONCEPTS

Suppose that a small business sells candy for $3 per pound and fresh ground coffee for $5 per pound. All inventory is sold by the end of the day. The revenue R collected in dollars is given by

$$R = 3x + 5y,$$

where x is the pounds of candy sold and y is the pounds of coffee sold. For example, if the business sells **80** pounds of candy and **40** pounds of coffee during a day, then its revenue is

$$R = 3(80) + 5(40) = \$440.$$

The function $R = 3x + 5y$ is called an **objective function**.

Suppose also that the company cannot package more than 150 pounds of candy and coffee per day. Then the inequality

$$x + y \leq 150$$

represents a **constraint** on the objective function, which limits the company's revenue for any one day.

A goal of this business might be to maximize

$$R = 3x + 5y,$$

subject to the constraints

$$x + y \leq 150$$
$$x \geq 0, \ y \geq 0.$$

Note that the constraints $x \geq 0$ and $y \geq 0$ are included because the number of pounds of candy or coffee cannot be negative. The problem that we have described is called a *linear programming problem*. Before learning how to solve a linear programming problem, we need to discuss the set of *feasible solutions*.

REGION OF FEASIBLE SOLUTIONS

The constraints for a linear programming problem consist of linear inequalities. These inequalities are satisfied by some points in the xy-plane but not by others. The set of solutions to these constraints is called the **feasible solutions**. For example, the region of feasible solutions to the constraints for the business just described is shaded in Figure C.1.

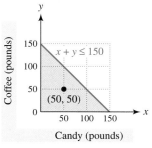

Figure C.1 Constraints on Sales

The point (50, 50) lies in the shaded region and represents the small business selling 50 pounds of candy and 50 pounds of coffee. In the next example, we shade the region of feasible solutions to a set of constraints.

EXAMPLE 1   Finding the region of feasible solutions

Shade the region of feasible solutions to the following constraints.

$$x + 2y \leq 30$$
$$2x + y \leq 30$$
$$x \geq 0, \; y \geq 0$$

Solution   The feasible solutions are the ordered pairs (x, y) that satisfy all four inequalities. They lie below the lines $x + 2y = 30$ and $2x + y = 30$ and above the line $y = 0$ and to the right of $x = 0$, as shown in Figure C.2. Note that the inequalities $x \geq 0$ and $y \geq 0$ restrict the feasible solutions to quadrant I.

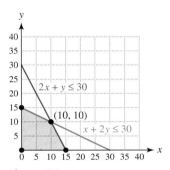

Figure C.2

SOLVING LINEAR PROGRAMMING PROBLEMS

A **linear programming problem** consists of an *objective function* and a system of linear inequalities called *constraints*. The solution set for the system of linear inequalities is called the *region of feasible solutions*. The objective function describes a quantity that is to be optimized. The **optimal value** to a linear programming problem often results in maximum revenue or minimum cost.

When the system of constraints has only two variables, the boundary of the region of feasible solutions often consists of line segments intersecting at points called *vertices* (plural of **vertex**.) To solve a linear programming problem we use the *fundamental theorem of linear programming*.

||||||| FUNDAMENTAL THEOREM OF LINEAR PROGRAMMING

If the optimal value for a linear programming problem exists, then it occurs at a vertex of the region of feasible solutions.

The fundamental theorem of linear programming is used to solve the following linear programming problem.

EXAMPLE 2   Maximizing an objective function

Maximize the objective function $R = 2x + 3y$ subject to

$$x + 2y \leq 30$$
$$2x + y \leq 30$$
$$x \geq 0, \; y \geq 0.$$

Solution   The region of feasible solutions is shaded in Figure C.2, previously shown. Note that the vertices on the boundary of feasible solutions are $(0, 0)$, $(15, 0)$, $(10, 10)$, and $(0, 15)$. To find the maximum value of R, substitute each vertex in the formula for R, as shown in Table C.1. The maximum value of R is 50 when $x = 10$ and $y = 10$.

TABLE C.1

Vertex	$R = 2x + 3y$	
$(0, 0)$	$2(0) + 3(0) = 0$	
$(15, 0)$	$2(15) + 3(0) = 30$	
$(10, 10)$	$2(10) + 3(10) = 50$	$\leftarrow$ Maximum R
$(0, 15)$	$2(0) + 3(15) = 45$	

The following steps are helpful in solving linear programming word problems.

STEPS FOR SOLVING A LINEAR PROGRAMMING WORD PROBLEM

STEP 1: Read the problem carefully. Consider making a table to display the information given.

STEP 2: Write the objective function and all the constraints.

STEP 3: Sketch a graph of the region of feasible solutions. Identify all vertices or corner points.

STEP 4: Evaluate the objective function at each vertex. A maximum (or a minimum) occurs at a vertex.

Note: If the region is unbounded, a maximum (or minimum) may not exist.

EXAMPLE 3   Minimizing the cost of vitamins

A breeder is mixing two different vitamins, Brand X and Brand Y, into pet food. Each serving of pet food should contain at least 60 units of vitamin A and 30 units of vitamin C. Brand X costs 80 cents per ounce and Brand Y costs 50 cents per ounce. Each ounce of Brand X contains 15 units of vitamin A and 10 units of vitamin C, whereas each ounce of Brand Y contains 20 units of vitamin A and 5 units of vitamin C. Determine how much of each brand of vitamin should be mixed to produce a minimum cost per serving.

Solution   **STEP 1:** Begin by listing the information, as illustrated in Table C.2.

Table C.2

Brand	Amount	Vitamin A	Vitamin C	Cost
X	x	15	10	80 cents
Y	y	20	5	50 cents
Minimum		60	30	

STEP 2: If x ounces of Brand X are purchased at 80 cents per ounce and if y ounces of Brand Y are purchased at 50 cents per ounce, then the total cost C is given by $C = 80x + 50y$. Because each ounce of Brand X contains 15 units of vitamin A and each ounce of Brand Y contains 20 units of vitamin A, the total number of units of vitamin A is $15x + 20y$. If each serving of pet food must contain at least 60 units of vitamin A, the constraint is $15x + 20y \geq 60$. Similarly, because each serving requires at least 30 units of vitamin C, $10x + 5y \geq 30$. The linear programming problem then becomes the following.

Minimize: $C = 80x + 50y$ Cost (in cents)

Subject to: $15x + 20y \geq 60$ Vitamin A

$10x + 5y \geq 30$ Vitamin C

$x \geq 0,\ y \geq 0$

STEP 3: The region containing the feasible solutions is shown in Figure C.3.

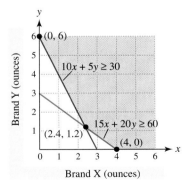

Figure C.3

Note: To determine the vertex (2.4, 1.2) solve the system of equations

$$15x + 20y = 60$$
$$10x + 5y = 30$$

by using elimination.

STEP 4: The vertices for this region are $(0, 6)$, $(2.4, 1.2)$, and $(4, 0)$. Evaluate the objective function C at each vertex, as shown in Table C.3.

TABLE C.3

Vertex	$C = 80x + 50y$	
$(0, 6)$	$80(0) + 50(6) = 300$	
$(2.4, 1.2)$	$80(2.4) + 50(1.2) = 252$	$\leftarrow$ Minimum cost (cents)
$(4, 0)$	$80(4) + 50(0) = 320$	

The minimum cost occurs when 2.4 ounces of Brand X and 1.2 ounces of Brand Y are mixed, at a cost of $2.52 per serving.

C EXERCISES

CONCEPTS

1. A procedure used in business to optimize quantities such as cost and profit is called _____.

2. In linear programming, the function to be optimized is called the _____ function.

3. The region in the xy-plane that satisfies the constraints is called the region of _____.

4. In linear programming, the constraints typically consist of a system of _____.

5. If the optimal value for a linear programming problem exists, then it occurs at a _____ of the region of feasible solutions.

6. To find the optimal value in a linear programming problem, substitute each vertex in the _____ function.

REGIONS OF FEASIBLE SOLUTIONS

Exercises 7–20: Shade the region of feasible solutions for the following constraints.

7. $x + y \leq 3$
$x \geq 0, y \geq 0$

8. $2x + y \leq 4$
$x \geq 0, y \geq 0$

9. $4x + 3y \leq 12$
$x \geq 0, y \geq 0$

10. $5x + 3y \leq 15$
$x \geq 0, y \geq 0$

11. $x \leq 5$
$y \leq 2$
$x \geq 0, y \geq 0$

12. $x \leq 3$
$y \leq 4$
$x \geq 1, y \geq 1$

13. $x + y \leq 5$
$x + y \geq 2$
$x \geq 0, y \geq 0$

14. $2x + y \leq 6$
$x + y \geq 3$
$x \geq 0, y \geq 0$

15. $3x + 2y \leq 6$
$2x + 3y \leq 6$
$x \geq 0, y \geq 0$

16. $5x + 3y \leq 30$
$3x + 5y \leq 30$
$x \geq 0, y \geq 0$

17. $x + y \leq 3$
$x + 3y \geq 3$
$x \geq 0, y \geq 0$

18. $3x + y \geq 6$
$x + 2y \geq 6$
$x \geq 0, y \geq 0$

19. $x + 2y \geq 4$
$3x + 2y \geq 6$
$x \geq 0, y \geq 0$

20. $4x + 3y \geq 12$
$3x + 4y \geq 12$
$x \geq 0, y \geq 0$

Exercises 21–24: Find the maximum of R on the region of feasible solutions.

21. $R = 4x + 5y$

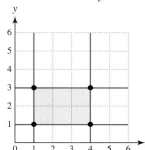

22. $R = 2x + 3y$

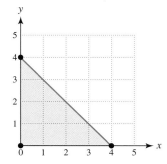

23. $R = x + 3y$

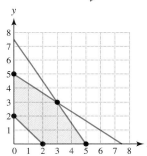

24. $R = 12x + 9y$

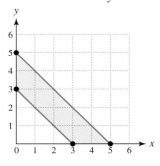

Exercises 25–28: Use the figures in Exercises 21–24 to complete Exercises 25–28, respectively, to minimize C.

25. $C = 2x + 3y$

26. $C = 3x + y$

27. $C = 5x + y$

28. $C = 2x + 7y$

Exercises 29–36: Maximize the objective function R, subject to the given constraints.

29. $R = 3x + 5y$
$x + y \le 150$
$x \ge 0, y \ge 0$

30. $R = 6x + 5y$
$x + y \le 8$
$x \ge 0, y \ge 0$

31. $R = 3x + 2y$
$2x + y \le 6$
$x \ge 0, y \ge 0$

32. $R = x + 3y$
$x + 2y \le 4$
$x \ge 0, y \ge 0$

33. $R = 12x + 9y$
$3x + y \le 6$
$x + 3y \le 6$
$x \ge 0, y \ge 0$

34. $R = 10x + 30y$
$x + 3y \le 12$
$3x + y \le 12$
$x \ge 0, y \ge 0$

35. $R = 4x + 5y$
$x + y \ge 2$
$x + 2y \le 4$
$x \ge 0, y \ge 0$

36. $R = 3x + 7y$
$x + y \ge 1$
$3x + y \le 3$
$x \ge 0, y \ge 0$

Exercises 37–42: Minimize the objective function C, subject to the given constraints.

37. $C = x + 2y$
$x \le 3, y \le 2$
$x \ge 0, y \ge 0$

38. $C = 3x + y$
$x \le 5, y \le 3$
$x \ge 1, y \ge 1$

39. $C = 8x + 15y$
$x + y \ge 4$
$x \ge 0, y \ge 0$

40. $C = x + 2y$
$3x + 4y \ge 12$
$x \ge 0, y \ge 0$

41. $C = 30x + 40y$
$2x + y \le 6$
$x + y \ge 2$
$x \ge 0, y \ge 0$

42. $C = 50x + 70y$
$2x + 3y \le 6$
$x + y \ge 1$
$x \ge 0, y \ge 0$

APPLICATIONS

Exercises 43–48: Use linear programming to solve the problem.

43. *Maximizing Revenue*   A small business sells candy for $4 per pound and coffee for $6 per pound. The business can package and sell at most a total of 100 pounds of candy and coffee per day, but at least 20 pounds of candy must be sold each day. Determine how many pounds of candy and coffee need to be sold each day to maximize revenue.

44. *Minimizing Cost*   It costs a business $20 to make one compact disc player and $10 to make one radio. Each week the company must make a combined total of at least 50 compact disc players and radios. At least as many compact disc players as radios must be manufactured. Determine how many compact disc players and radios should be made to minimize weekly costs.

45. *Vitamin Cost*   A pet owner is mixing two different vitamins, Brand X and Brand Y, into pet food. Brand X costs 90 cents per ounce and Brand Y costs 60 cents per ounce. Each serving is a mixture of the two brands and should contain at least 40 units of vitamin A and 30 units of vitamin C. Each ounce of Brand X contains 20 units of vitamin A and 10 units of vitamin C, whereas each ounce of Brand Y contains 10 units of vitamin A and 10 units of vitamin C. Determine how much of each brand of vitamin should be mixed to produce a minimum cost per serving.

46. *Pet Food Cost*   A pet owner is buying two brands of food, X and Y, for his animals. Each serving of the mixture of the two foods should contain at least 60 grams of protein and 40 grams of fat. Brand X costs 75 cents per unit and Brand Y costs 50 cents per unit. Each unit of Brand X contains 20 grams of protein and 10 grams of fat, whereas each unit of Brand Y contains 10 grams of protein and 10 grams of fat. Determine how much of each brand should be bought to obtain a minimum cost per serving.

47. *Raising Animals*   A breeder can raise no more than 50 hamsters and mice but no more than 20 hamsters. If she sells the hamsters for $15 each and the mice for $10 each, find the maximum revenue produced.

48. *Maximizing Profit*   A business manufactures two parts, X and Y. Machines A and B are needed to make each part. To make part X, machine A is needed 3 hours and machine B is needed 1 hour. To make part Y, machine A is needed 1 hour and machine B is needed 2 hours. Machine A is available 60 hours per week and machine B is available 50 hours per week. The profit from part X is $300 and the profit from part Y is $250. How many parts of each type should be made to maximize weekly profit?

Appendix D   Synthetic Division

BASIC CONCEPTS

A shortcut called **synthetic division** can be used to divide $x - k$, where k is a number, into a polynomial. For example, to divide $x - 2$ into $3x^3 - 8x^2 + 7x - 6$, we do the following (with the equivalent long division shown at the right).

Synthetic Division

$$\begin{array}{r|rrrr} 2 & 3 & -8 & 7 & -6 \\ & & 6 & -4 & 6 \\ \hline & 3 & -2 & 3 & 0 \end{array}$$

Long Division of Polynomials

$$\begin{array}{r} 3x^2 - 2x + 3 \\ x - 2 \overline{\smash{)}3x^3 - 8x^2 + 7x - 6} \\ \underline{3x^3 - 6x^2} \\ -2x^2 + 7x \\ \underline{-2x^2 + 4x} \\ 3x - 6 \\ \underline{3x - 6} \\ 0 \end{array}$$

Note how the highlighted numbers in the expression for long division correspond to the third row in synthetic division. The remainder is 0, which is the last number in the third row. The quotient is $3x^2 - 2x + 3$. Its coefficients are 3, -2, and 3 and are located in the third row. To divide $x - 2$ into $3x^3 - 8x^2 + 7x - 6$ with synthetic division use the following steps.

STEP 1. In the top row write 2 (the value of k) on the left and then write the coefficients of the dividend $3x^3 - 8x^2 + 7x - 6$.

STEP 2. **(a)** Copy the leading coefficient 3 of $3x^3 - 8x^2 + 7x - 6$ in the third row and multiply it by 2 (the value of k). Write the result 6 in the second row below -8. Add -8 and 6 in the second column to obtain the -2 in the third row.

   (b) Repeat the process by multiplying -2 by 2 and place the result -4 below 7. Then add 7 and -4 to obtain 3.

   (c) Multiply 3 by 2 and place the result 6 below the -6. Adding 6 and -6 gives 0.

STEP 3. The last number in the third row is 0, which is the remainder. The other numbers in the third row are the coefficients of the quotient, which is $3x^2 - 2x + 3$.

EXAMPLE 1   Performing synthetic division

Use synthetic division to divide $x^4 - 5x^3 + 9x^2 - 10x + 3$ by $x - 3$.

Solution   Because the divisor is $x - 3$, the value of k is 3.

$$\begin{array}{r|rrrrr} 3 & 1 & -5 & 9 & -10 & 3 \\ & & 3 & -6 & 9 & -3 \\ \hline & 1 & -2 & 3 & -1 & 0 \end{array}$$

The quotient is $x^3 - 2x^2 + 3x - 1$ and the remainder is 0. This result is also expressed by

$$\frac{x^4 - 5x^3 + 9x^2 - 10x + 3}{x - 3} = x^3 - 2x^2 + 3x - 1 + \frac{0}{x - 3}, \quad \text{or}$$

$$\frac{x^4 - 5x^3 + 9x^2 - 10x + 3}{x - 3} = x^3 - 2x^2 + 3x - 1.$$

EXAMPLE 2   **Performing synthetic division**

Use synthetic division to divide $2x^3 - x + 5$ by $x + 1$.

Solution   Write $2x^3 - x + 5$ as $2x^3 + 0x^2 - x + 5$. The divisor $x + 1$ can be written as

$$x + 1 = x - (-1),$$

so we let $k = -1$.

$$
\begin{array}{r|rrrr}
-1 & 2 & 0 & -1 & 5 \\
   &   & -2 & 2 & -1 \\
\hline
   & 2 & -2 & 1 & 4
\end{array}
$$

The remainder is 4, and the quotient is $2x^2 - 2x + 1$. This result can also be expressed as

$$\frac{2x^3 - x + 5}{x + 1} = 2x^2 - 2x + 1 + \frac{4}{x + 1}.$$

D | EXERCISES

Exercises 1–12: Use synthetic division to divide.

1. $\dfrac{x^2 + 3x - 1}{x - 1}$

2. $\dfrac{2x^2 + x - 1}{x - 3}$

3. $(3x^2 - 22x + 7) \div (x - 7)$

4. $(5x^2 + 29x - 6) \div (x + 6)$

5. $\dfrac{x^3 + 7x^2 + 14x + 8}{x + 4}$

6. $\dfrac{2x^3 + 3x^2 + 2x + 4}{x + 1}$

7. $\dfrac{2x^3 + x^2 - 1}{x - 2}$

8. $\dfrac{x^3 + x - 2}{x + 3}$

9. $(x^3 - 2x^2 - 2x + 4) \div (x - 4)$

10. $(2x^4 + 3x^2 - 4) \div (x + 2)$

11. $(b^4 - 1) \div (b - 1)$

12. $(a^2 + a) \div (a + 2.5)$

Answers to Selected Exercises

CHAPTER 1: INTRODUCTION TO ALGEBRA

SECTION 1.1 (PP. 8–10)

1. counting **3.** 1 **5.** prime **7.** factors **9.** variable
11. Composite; $4 = 2 \times 2$ **13.** Neither **15.** Prime
17. Composite; $92 = 2 \times 2 \times 23$
19. Composite; $225 = 3 \times 3 \times 5 \times 5$ **21.** Prime
23. $6 = 2 \times 3$ **25.** $12 = 2 \times 2 \times 3$
27. $32 = 2 \times 2 \times 2 \times 2 \times 2$
29. $294 = 2 \times 3 \times 7 \times 7$
31. $300 = 2 \times 2 \times 3 \times 5 \times 5$ **33.** Yes **35.** No
37. Yes **39.** Yes **41.** 10 **43.** 7 **45.** 4 **47.** 18
49. 4 **51.** 9 **53.** 8 **55.** 14 **57.** $y = 1$
59. $y = 28$ **61.** $F = 7$ **63.** $F = 5$ **65.** $y = 18$
67. $y = 3$ **69.** $3s$; s is the cost of soda.
71. $x + 5$; x is the number. **73.** $3n$; n is the number.
75. $p - 200$; p is the population. **77.** $\frac{z}{6}$; z is the number.
79. $\frac{x + 7}{y}$; x is one number, y is the other number.

81.

Yards (y)	1	2	3	4	5	6	7
Feet (F)	3	6	9	12	15	18	21

$F = 3y$

83. $P = 100D$ **85.** $M = 50H$; 150 miles
87. $B = 70x$; 4200 beats **89.** $C = 12x$
91. 198 square feet

SECTION 1.2 (PP. 21–24)

1. $\frac{3}{4}$; $\frac{1}{4}$ **3.** 0 **5.** numerators **7.** $\frac{ac}{bd}$ **9.** $\frac{1}{4}$ **11.** $\frac{ad}{bc}$
13. $\frac{a - c}{b}$ **15.** 2 **17.** 4 **19.** 25 **21.** 10 **23.** $\frac{3}{5}$
25. $\frac{3}{5}$ **27.** $\frac{1}{2}$ **29.** $\frac{1}{3}$ **31.** $\frac{2}{5}$ **33.** $\frac{1}{3}$ **35.** $\frac{2}{5}$ **37.** $\frac{1}{4}$
39. $\frac{1}{6}$ **41.** $\frac{3}{20}$ **43.** 1 **45.** $\frac{3}{5}$ **47.** $\frac{12}{5}$ **49.** $\frac{3}{4}$ **51.** 1
53. $\frac{3a}{2b}$ **55.** $\frac{3}{16}$ **57.** 4 **59.** $\frac{1}{3}$
61. (a) $\frac{1}{5}$ (b) $\frac{1}{7}$ (c) $\frac{7}{4}$ (d) $\frac{8}{9}$
63. (a) 2 (b) 9 (c) $\frac{101}{12}$ (d) $\frac{17}{31}$
65. $\frac{3}{2}$ **67.** $\frac{3}{2}$ **69.** 8 **71.** $\frac{4}{3}$ **73.** $\frac{63}{8}$ **75.** 12 **77.** $\frac{3}{10}$
79. $\frac{a}{2}$ **81.** 1 **83.** (a) 1 (b) $\frac{1}{3}$ **85.** (a) 2 (b) 1
87. (a) $\frac{7}{33}$ (b) $\frac{1}{11}$ **89.** 10 **91.** 45 **93.** 15 **95.** 24
97. 12 **99.** 24 **101.** $\frac{5}{6}$ **103.** $\frac{13}{16}$ **105.** $\frac{1}{4}$ **107.** $\frac{1}{6}$

109. $\frac{59}{70}$ **111.** $\frac{13}{36}$ **113.** $\frac{17}{600}$ **115.** $\frac{9}{8}$ **117.** $\frac{961}{1260}$
119. $32\frac{5}{16}$ inches **121.** $5\frac{1}{4}$ cups **123.** $\frac{5}{8}$ yd^2
125. $8\frac{1}{4}$ miles **127.** $\frac{7}{100}$

CHECKING BASIC CONCEPTS 1.1 & 1.2 (P. 24)

1. (a) Prime (b) Composite; $28 = 2 \times 2 \times 7$
(c) Neither (d) Composite; $180 = 2 \times 2 \times 3 \times 3 \times 5$
2. 2 **3.** 30 **4.** $x + 5$ **5.** $127\frac{1}{2}$ ft^2
6. (a) $\frac{5}{7}$ (b) $\frac{2}{3}$ **7.** $\frac{3}{4}$
8. (a) $\frac{1}{2}$ (b) $\frac{1}{4}$ (c) $\frac{2}{5}$ (d) $\frac{7}{12}$ **9.** $3\frac{1}{3}$ cups

SECTION 1.3 (PP. 29–31)

1. add **3.** a^6 **5.** 6^2 **7.** 17; multiplication; addition
9. 4; left; right **11.** No; $2^3 = 8$, but $3^2 = 9$. **13.** 2^5
15. 3^4 **17.** $\left(\frac{1}{2}\right)^4$ **19.** a^5 **21.** (a) 16 (b) 16
23. (a) 6 (b) 1 **25.** (a) 32 (b) 1000
27. (a) $\frac{4}{9}$ (b) $\frac{1}{32}$ **29.** (a) $\frac{8}{125}$ (b) $\frac{81}{49}$ **31.** 2^3
33. 5^2 **35.** 7^2 **37.** 10^3 **39.** $\left(\frac{1}{2}\right)^4$ **41.** 29 **43.** 4
45. 90 **47.** 3 **49.** 2 **51.** 19 **53.** 32 **55.** 125
57. 3 **59.** 4 **61.** 80 **63.** $\frac{49}{16}$ **65.** $2^3 - 8$; 0
67. $30 - 4 \cdot 3$; 18 **69.** $\frac{4^2}{2^3}$; 2 **71.** $\frac{40}{10} + 2$; 6
73. $100(2 + 3)$; 500 **75.** About 1,509,949 bytes
77. (a) $k = 10$ (b) About 20 years

SECTION 1.4 (PP. 39–40)

1. $-b$ **3.** b **5.** rational **7.** irrational
9. $\sqrt{2}$ (answers may vary) **11.** approximately equal
13. 0 **15.** origin **17.** $>$ **19.** $=$
21. (a) -9 (b) 9 **23.** (a) $-\frac{2}{3}$ (b) $\frac{2}{3}$
25. (a) -8 (b) 8 **27.** (a) $-a$ (b) a
29. 6 **31.** $\frac{1}{2}$ **33.** 0.25 **35.** 0.875 **37.** 1.5
39. 0.05 **41.** $0.\overline{6}$ **43.** $0.\overline{7}$
45. Natural, whole, integer, and rational
47. Natural, whole, integer, and rational
49. Whole, integer, and rational **51.** Rational
53. Irrational **55.** Natural, integer, and rational
57. Irrational
59.

61.

63.

65.

67.

69. 5.23   **71.** 7   **73.** $\frac{1}{2}$ **75.** $\pi - 3$ **77.** $-b$

79. $<$ **81.** $>$ **83.** $>$ **85.** $<$

87. $-9, -2^3, -3, 0, 1$ **89.** $-2, -\frac{3}{2}, \frac{1}{3}, \sqrt{5}, \pi$

91. $-4^2, -\frac{17}{28}, -\frac{4}{7}, \sqrt{2}, \sqrt{7}$

93. (a) 14.6 million   **(b)** (*answers may vary*)
(c) 14.775 million

Checking Basic Concepts 1.3 & 1.4 (p. 41)

1. (a) 8   **(b)** 10,000   **(c)** $\frac{8}{27}$ **(d)** -81
2. (a) 26   **(b)** 9   **(c)** 2   **(d)** $\frac{1}{2}$ **(e)** 4   **(f)** 0
3. $5^3 \div 3$, or $\frac{53}{3}$ **4. (a)** 17   **(b)** $-a$
5. (a) Natural, integer, and rational   **(b)** Integer and rational   **(c)** Irrational   **(d)** Rational
6.

7. (a) 12   **(b)** a **8.** $-7, -1.6, 0, \frac{1}{3}, \sqrt{3}, 3^2$

Section 1.5 (pp. 45–47)

1. 0   **3.** addends   **5.** difference   **7.** negative
9. addition   **11.** $-25; 0$ **13.** $\sqrt{21}; 0$ **15.** $-5.63; 0$
17. 4   **19.** 2   **21.** -3 **23.** 2   **25.** 10   **27.** -150
29. 1   **31.** -7 **33.** $\frac{1}{4}$ **35.** $-\frac{9}{14}$ **37.** $-\frac{5}{4}$ **39.** -1.1
41. 34   **43.** -1 **45.** 0   **47.** -3 **49.** 7   **51.** 1
53. $-\frac{1}{14}$ **55.** $\frac{1}{2}$ **57.** 2.9   **59.** -164 **61.** -11
63. -9 **65.** 42   **67.** -5 **69.** 50   **71.** 8.8   **73.** $\frac{1}{2}$
75. -1 **77.** $2 + (-5); -3$ **79.** $-5 + 7; 2$
81. $-(2^3); -8$ **83.** $-6 - 7; -13$
85. $6 + (-10) - 5; -9$ **87.** \$230   **89.** 64,868 feet

Section 1.6 (pp. 54–56)

1. factors   **3.** negative   **5.** quotient   **7.** $\frac{1}{a}$
9. reciprocal or multiplicative inverse   **11.** positive
13. 2   **15.** -12 **17.** -18 **19.** 0   **21.** 60   **23.** $\frac{1}{4}$
25. -1 **27.** 200   **29.** -5000 **31.** 120   **33.** $-\frac{3}{2}$
35. 1   **37.** -2 **39.** 10   **41.** -4 **43.** -3 **45.** -32
47. $-\frac{1}{22}$ **49.** $\frac{4}{15}$ **51.** $-\frac{15}{16}$ **53.** Undefined   **55.** -1
57. $-\frac{4}{3}$ **59.** 0.5   **61.** 0.1875   **63.** 3.5   **65.** $5.\overline{6}$
67. 1.4375   **69.** 0.875   **71.** $\frac{1}{4}$ **73.** $\frac{4}{25}$ **75.** $\frac{5}{8}$ **77.** $\frac{11}{16}$
79. (a) 49.5 million   **(b)** 48.56 million
81. \$10,500; (*answers may vary*)
83. (a) The number of subscribers is increasing.
(b) 5.3 to 5.5 million; (*answers may vary*)

Checking Basic Concepts 1.5 & 1.6 (p. 56)

1. (a) 0   **(b)** -19 **2. (a)** $\frac{8}{9}$ **(b)** -3.2
3. $144°$ F   **4. (a)** 35   **(b)** $\frac{4}{15}$ **5. (a)** $-\frac{15}{2}$ **(b)** $\frac{15}{32}$
6. $-\frac{6}{7}$ **7. (a)** -5 **(b)** -5 **(c)** -5 **(d)** 5
8. (a) 0.6   **(b)** 3.875
9. (a) The incarceration rate is increasing.   **(b)** 539 to 596; (*answers may vary*)

Section 1.7 (pp. 64–66)

1. commutative; addition   **3.** associative; addition
5. distributive   **7.** identity; addition   **9.** $-a$
11. $10 + (-6)$ **13.** $6 \cdot (-5)$ **15.** $10 + a$ **17.** $7b$
19. $1 + (2 + 3)$ **21.** $(2 \cdot 3) \cdot 4$ **23.** $a + (5 + c)$
25. $x \cdot (3 \cdot 4)$ **27.** 20   **29.** $ab - 8a$ **31.** $-t - z$
33. $-5 + a$ **35.** $3a + 15$ **37.** $3z - 18$
39. Commutative (multiplication)
41. Associative (addition)   **43.** Distributive
45. Distributive, commutative (multiplication)
47. Distributive   **49.** Associative (multiplication)
51. Distributive   **53.** Identity (addition)
55. Identity (multiplication)   **57.** Identity (multiplication)
59. Inverse (multiplication)   **61.** Inverse (addition)
63. 30   **65.** 100   **67.** 178   **69.** 477   **71.** 79   **73.** 90
75. 816   **77.** 1   **79.** $\frac{1}{6}$
81. (a) 410   **(b)** 9970   **(c)** -6300 **(d)** $-140,000$
83. (a) 19,000   **(b)** $-45,100$ **(c)** 60,000
(d) $-7,900,000$
85. (a) 1.256   **(b)** 0.96   **(c)** 0.0987   **(d)** -0.0056
(e) 120   **(f)** 457.8
87. Commutative (addition)   **89.** 19.8 miles
91. (a) 200,000 pixels   **(b)** Commutative (multiplication)

Section 1.8 (pp. 71–73)

1. term   **3.** coefficient   **5.** like; unlike   **7.** Yes; 91
9. Yes; -6 **11.** No   **13.** Yes; 1   **15.** No
17. Yes; -9 **19.** Like   **21.** Like   **23.** Unlike
25. Like   **27.** Unlike   **29.** $8x$ **31.** $14y$
33. Not possible   **35.** Not possible   **37.** $3x^2$ **39.** $-xy$
41. $3x + 2$ **43.** $-2z + \frac{1}{2}$ **45.** $11y$ **47.** $4z - 1$
49. $12y - 7z$ **51.** $-3x - 4$ **53.** $-6x - 1$
55. $-\frac{1}{3}x + \frac{2}{3}$ **57.** $\frac{2}{5}x + \frac{3}{5}y + \frac{1}{5}$ **59.** $0.4x^2$
61. $7x^2 - 7x$ **63.** x **65.** 3   **67.** z **69.** $5x + 6x; 11x$
71. $x^2 + 2x^2; 3x^2$ **73.** $6x - 4x; 2x$
75. (a) $1570w$ **(b)** 65,940 square feet
77. (a) $50x$ **(b)** 2400 cubic feet   **(c)** 24 minutes

Checking Basic Concepts 1.7 & 1.8 (p. 73)

1. (a) $18y$ **(b)** $x + 10$ **2.** $20y$
3. (a) $5 - x$ **(b)** $5x - 35$ **4.** Distributive   **5.** 0

6. (a) 60 **(b)** 7 **(c)** 368
7. (a) $14z$ **(b)** $-3y + 3$
8. (a) $-3y - 3$ **(b)** $-14y$ **(c)** x **(d)** 35

CHAPTER 1 REVIEW EXERCISES (PP. 78–81)

1. Prime **2.** Composite; $27 = 3 \times 3 \times 3$
3. Composite; $108 = 2 \times 2 \times 3 \times 3 \times 3$
4. Composite; $91 = 7 \times 13$ **5.** 3 **6.** 5 **7.** 12
8. 6 **9.** 7 **10.** 7
11. $5c$, where c is the cost of the CD
12. $x - 5$, where x is the number **13.** $3^2 + 5$
14. $2^3 \div (3 + 1)$
15. (a) $\frac{5}{8}$ **(b)** $\frac{3}{4}$ **16. (a)** $\frac{3}{4}$ **(b)** $\frac{3}{5}$ **17.** $\frac{5}{8}$ **18.** $\frac{2}{9}$
19. $\frac{3}{11}$ **20.** $\frac{12}{23}$ **21.** $\frac{3}{35}$
22. (a) $\frac{1}{8}$ **(b)** 1 **(c)** $\frac{19}{5}$ **(d)** $\frac{2}{3}$ **23.** 9 **24.** $\frac{9}{14}$
25. 12 **26.** $\frac{1}{8}$ **27.** 24 **28.** 42 **29.** $\frac{1}{3}$ **30.** $\frac{1}{2}$
31. $\frac{19}{24}$ **32.** $\frac{9}{22}$ **33.** $\frac{5}{12}$ **34.** $\frac{13}{18}$ **35.** 5^6 **36.** $\left(\frac{7}{6}\right)^3$
37. 3^4 **38.** x^5 **39. (a)** 64 **(b)** 49 **(c)** 8 **40.** 5
41. 25 **42.** 7 **43.** 3 **44.** 25 **45.** 1 **46.** 2
47. 0 **48.** 0 **49.** 10 **50.** 1 **51.** 5 **52.** 19
53. Whole, integer, and rational **54.** Rational
55. Integer and rational **56.** Irrational **57.** Irrational
58. Rational
59.

60. (a) 5 **(b)** π **(c)** $\sqrt{2} - 1$
61. (a) $<$ **(b)** $>$
62. $-3, -\frac{2}{3}, \sqrt{3}, \pi - 1, 3$ **63.** 1 **64.** -5 **65.** 1
66. -2 **67.** -44 **68.** 40 **69.** $-\frac{11}{4}$ **70.** -7
71. $-\frac{5}{4}$ **72.** $-\frac{2}{9}$ **73.** $\frac{2}{7}$ **74.** $-\frac{4}{35}$ **75.** 4 **76.** $-\frac{3}{4}$
77. $3 + (-5); -2$ **78.** $2 - (-4); 6$ **79.** $0.\overline{7}$
80. 2.2 **81.** $\frac{3}{5}$ **82.** $\frac{3}{8}$
83. Commutative (multiplication)
84. Associative (addition) **85.** Distributive
86. Commutative (addition)
87. Identity (multiplication)
88. Associative (multiplication) **89.** Distributive
90. Identity (addition) **91.** Inverse (addition)
92. Inverse (multiplication) **93.** 40 **94.** 301
95. 2475 **96.** 6580 **97.** 549.8 **98.** 43.56
99. Yes; 55 **100.** Yes; -1 **101.** No **102.** No
103. $-6x$ **104.** $15z$ **105.** $4x^2$ **106.** $3x + 1$
107. $\frac{1}{2}z + 2$ **108.** $x - 18$ **109.** 5 **110.** c
111. (a) $7x$ **(b)** 420 square feet **(c)** 24 minutes
112. 16 square feet
113.

Gallons (G)	1	2	3	4	5	6
Pints (P)	8	16	24	32	40	48

$P = 8G$

114. $C = 0.25x$ **115.** $\frac{3}{20}$ **116.** $1\frac{3}{20}$ feet
117. $15\frac{3}{8}$ miles
118. (a) 1,042,000 **(b)** 1,725,000; (*answers may vary*)
119. $1101 **120.** $124°$F
121. (a) 3.2 inches **(b)** 48.1 inches (*answers may vary*)
122. $9000 (*answers may vary*)

CHAPTER 1 TEST (PP. 81–82)

1. Composite; $56 = 2 \times 2 \times 2 \times 7$ **2.** $\frac{15}{7}$
3. $4^2 - 3$; 13 **4.** $\frac{3}{4}$
5. (a) $\frac{3}{4}$ **(b)** $\frac{16}{45}$ **(c)** $\frac{2}{7}$ **(d)** $\frac{15}{4}$
6. (a) 8 **(b)** 71 **(c)** -40
7. (a) Distributive **(b)** Associative (multiplication)
(c) Commutative (addition)
8. (a) $12 - 4z$ **(b)** $15x - 6$ **(c)** $x - 19$
9. (a) $\frac{19}{12}x$ **(b)** $12\frac{2}{3}$ acres
10. (a) $C = 13x$ **(b)** $221 **11.** $2\frac{3}{5}$ feet **12.** $640

CHAPTER 2: LINEAR EQUATIONS AND INEQUALITIES

SECTION 2.1 (PP. 91–92)

1. equals **3.** solution set **5.** solutions **7.** $b + c$
9. 1 **11.** 22 **13.** -5 **15.** 8 **17.** 17 **19.** 2
21. 5 **23.** -15 **25.** 1963 **27.** 5 **29.** 3 **31.** 0
33. 3 **35.** $\frac{5}{4}$ **37.** 5 **39.** 7 **41.** $\frac{3}{2}$
43. (a)

Hours (x)	0	1	2	3	4	5	6
Rainfall (R)	3	3.5	4	4.5	5	5.5	6

(b) $R = 0.5x + 3$ **(c)** 4.5 inches; yes
(d) 4.125 inches
45. (a) $L = 300x$ **(b)** $870 = 300x$ **(c)** 2.9
47. 17.5 years **49.** $25,000

SECTION 2.2 (PP. 100–102)

1. $ax + b = 0$ **3.** Addition, multiplication **5.** LCD
7. Infinitely many **9.** Yes; $a = 3, b = -7$
11. Yes; $a = \frac{1}{2}, b = 0$ **13.** No
15. Yes; $a = 1.1, b = -0.9$ **17.** Yes; $a = 2, b = -6$
19. No
21.

x	1	2	3	4	5
$-4x + 8$	4	0	-4	-8	-12

2

23.

x	-2	-1	0	1	2
$4 - 2x$	8	6	4	2	0

-1

25. $\frac{3}{11}$ **27.** 23 **29.** 7 **31.** $-\frac{11}{5}$ **33.** $-\frac{7}{2}$ **35.** 3
37. $\frac{9}{4}$ **39.** $-\frac{5}{2}$ **41.** $\frac{4}{7}$ **43.** $\frac{9}{8}$ **45.** 1 **47.** -0.055
49. 8 **51.** $\frac{1}{7}$ **53.** 1 **55.** Zero **57.** One **59.** Zero
61. Zero

63. Infinitely many

65. (a)

Hours (x)	0	1	2	3	4
Distance (D)	4	12	20	28	36

(b) $D = 8x + 4$ **(c)** 28 miles; yes   **(d)** 2.25 hours; the bicyclist is 22 miles from home after 2 hours and 15 minutes.

67. 1997   **69.** 1992

CHECKING BASIC CONCEPTS 2.1 & 2.2 (P. 102)

1.

x	3	3.5	4	4.5	5
4x − 3	9	11	13	15	17

4

2. (a) 18   **(b)** $\frac{1}{6}$ **(c)** $2.\overline{6}$ **(d)** 3

3. (a) One   **(b)** Infinitely many   **(c)** Zero

4. (a) $D = 300 - 75x$ **(b)** $0 = 300 - 75x$

(c) 4 hours

SECTION 2.3 (PP. 110–112)

1. Check your solution.   **3.** $\frac{x}{100}$ **5.** 50

7. $\frac{P_2 - P_1}{P_1} \times 100$ **9.** $2 + x = 12$; 10

11. $\frac{x}{5} = x - 24$; 30   **13.** $\frac{x+5}{2} = 7$; 9   **15.** $\frac{x}{2} = 17$; 34

17. 31, 32, 33   **19.** 34   **21.** 8   **23.** 21   **25.** 11

27. $\frac{37}{100}$; 0.37   **29.** $\frac{37}{25}$; 1.48   **31.** $\frac{69}{1000}$; 0.069

33. $\frac{1}{2000}$; 0.0005   **35.** 45%   **37.** 180%   **39.** 40%

41. 75%   **43.** $83.\overline{3}\%$ **45.** About 21.8%

47. $988 per credit   **49.** About 123 million

51. About 49,412   **53.** 1500   **55.** $d = 8$ miles

57. $r = 20$ feet per second   **59.** $t = 5$ hours

61. 60 mph   **63.** $\frac{3}{8}$ hours

65. 0.5 hr at 6 mph; 0.8 hr at 5 mph   **67.** 3.75 hours

69. 30 ounces   **71.** $1500 at 6%; $2500 at 5%

73. 6 gallons

SECTION 2.4 (PP. 121–123)

1. formula   **3.** $\frac{1}{2}bh$ **5.** 360   **7.** $2\pi r$ **9.** LWH

11. 18 ft^2 **13.** 9 in^2 **15.** $16\pi \approx 50.3$ ft^2 **17.** 11 ft^2

19. 91 in^2 **21.** 36 in^2 **23.** $8\pi \approx 25.1$ in.

25. 4602 ft^2 **27.** 65°   **29.** 81°   **31.** 30°

33. 36°, 36°, 108°

35. $C = 12\pi \approx 37.7$ in.; $A = 36\pi \approx 113.1$ in^2

37. $r = 1$ in.; $A = \pi \approx 3.1$ in^2

39. $V = 2640$ in^3; $S = 1208$ in^2

41. $V = 2$ ft^3; $S = \frac{32}{3}$ ft^2 **43.** 20π in^3 **45.** 600π in^3

47. $\frac{147}{64}\pi \approx 7.2$ ft^3 **49.** $W = \frac{A}{L}$ **51.** $h = \frac{V}{\pi r^2}$

53. $a = \frac{2A}{h} - b$ **55.** $W = \frac{V}{LH}$ **57.** $b = 2s - a - c$

59. $b = a - c$ **61.** $a = \frac{cd}{b-d}$ **63.** 15 in.   **65.** 2.98

67. 2.24   **69.** 77°F   **71.** −40°F   **73.** −5°C

75. −20°C   **77.** 2.4 mi

CHECKING BASIC CONCEPTS 2.3 & 2.4 (P. 124)

1. (a) $3x = 36$; 12   **(b)** $35 - x = 43$; −8

2. −30, −31, −32   **3.** 0.095   **4.** 125%

5. About 4185   **6.** 6.5 hr

7. $5000 at 6%; $3000 at 7%   **8.** 12 in.

9. $A = 9\pi \approx 28.3$ ft^2; $C = 6\pi \approx 18.8$ ft   **10.** 30°

11. $l = \dfrac{A - \pi r^2}{\pi r}$

SECTION 2.5 (PP. 133–137)

1. equals sign; $<$, $\leq$, $>$, $\geq$ **3.** one

5. number line   **7.** $>$ **9.** $>$

11.

13.

15.

17.

19. $x < 0$ **21.** $x \leq 3$ **23.** $x \geq 10$ **25.** Yes

27. Yes   **29.** Yes   **31.** Yes   **33.** No   **35.** $x > -2$

37. $x < 1$

39.

x	1	2	3	4	5
−2x + 6	4	2	0	−2	−4

$x \geq 3$

41.

x	−3	−2	−1	0	1
5 − x	8	7	6	5	4
x + 7	4	5	6	7	8

$x < -1$

43. $x > 3$;

45. $y \geq -2$;

47. $z > 8$;

49. $t \leq -5$;

51. $x < 5$;

53. $t \leq -2$;

55. $y > -\frac{3}{20}$;

57. $z \geq -\frac{14}{3}$;

59. $\{x \mid x < 7\}$ **61.** $\{x \mid x \leq -\frac{4}{3}\}$ **63.** $\{x \mid x > -\frac{39}{2}\}$
65. $\{x \mid x \leq \frac{3}{2}\}$ **67.** $\{x \mid x > \frac{1}{2}\}$ **69.** $\{x \mid x \leq \frac{5}{4}\}$
71. $\{x \mid x > -\frac{8}{3}\}$ **73.** $\{x \mid x \leq -\frac{3}{10}\}$
75. $\{x \mid x \leq -\frac{1}{3}\}$ **77.** $\{x \mid x < 5\}$ **79.** $\{x \mid x \leq -\frac{3}{2}\}$
81. $\{x \mid x > -\frac{9}{8}\}$ **83.** $\{x \mid x \leq \frac{4}{7}\}$ **85.** $\{x \mid x < \frac{21}{11}\}$
87. $x > 60$ **89.** $x \geq 21$ **91.** $x > 40,000$
93. $x \leq 70$ **95.** Less than 10 feet
97. It is less than 20 inches.   **99.** 86 or more
101. 4.5 hours   **103.** 4 days
105. (a) $C = 1.5x + 2000$ (b) $R = 12x$
(c) $P = 10.5x - 2000$ (d) 191 or more compact discs
107. (a) After 3.5 hours   (b) After more than 3.5 hours
109. Altitudes less than 2 miles   **111.** 1999 and earlier

CHECKING BASIC CONCEPTS 2.5 (P. 137)

1.

2. $x < 1$
3.

x	-2	-1	0	1	2
$5 - 2x$	9	7	5	3	1

$; x \geq -1$

4. (a) $\{x \mid x > 3\}$ (b) $\{x \mid x \geq -35\}$ (c) $\{x \mid x \leq -1\}$
5. More than 13 inches

CHAPTER 2 REVIEW EXERCISES (PP. 141–143)

1. -6 **2.** 2   **3.** $\frac{9}{4}$ **4.** $-\frac{1}{2}$ **5.** 3   **6.** $-\frac{7}{3}$ **7.** $-\frac{5}{2}$
8. $-\frac{7}{2}$ **9.** Yes; $a = 5, b = -3$
10. Yes; $a = -4, b = 1$ **11.** Yes; $a = 0.55, b = -0.05$
12. No   **13.** 2   **14.** 22   **15.** $\frac{27}{5}$ **16.** $-\frac{2}{3}$ **17.** $\frac{9}{4}$
18. 0   **19.** $\frac{7}{8}$ **20.** $\frac{11}{4}$ **21.** Zero   **22.** Infinitely many
23. Infinitely many   **24.** One
25.

x	0.5	1.0	1.5	2.0	2.5
$-2x + 3$	2	1	0	-1	-2

1.5

26.

x	-2	-1	0	1	2
$-(x+1)+3$	4	3	2	1	0

0

27. $6x = 72; 12$ **28.** $x + 18 = -23; -41$
29. $2x - 5 = x + 4; 9$ **30.** $x + 4 = 3x; 2$
31. 16, 17, 18, 19   **32.** $-52, -51, -50$ **33.** $\frac{17}{20}; 0.85$
34. $\frac{7}{125}; 0.056$ **35.** $\frac{3}{10,000}; 0.0003$ **36.** $\frac{171}{50}; 3.42$
37. 89%   **38.** 0.5%   **39.** 230%   **40.** 100%
41. $d = 24$ mi   **42.** $d = 3850$ ft
43. $r = 25$ yards per second   **44.** $t = \frac{25}{3}$ hr

45. 7.5 m^2 **46.** $36\pi \approx 113.1$ ft^2 **47.** 864 in^2, or 6 ft^2
48. 45.5 in^2 **49.** $18\pi \approx 56.5$ ft   **50.** $25\pi \approx 78.5$ in^2
51. $50°$ **52.** $22.5°$ **53.** $625\pi \approx 1963.5$ in^3
54. 1620 in^2, or 11.25 ft^2 **55.** $x = a - y$
56. $x = \frac{P - 2y}{2}$ **57.** $y = \frac{z}{2x}$ **58.** $b = 3S - a - c$
59. $b = \frac{12T - 4a}{3}$ **60.** $c = \frac{ab}{d - b}$

61.

62.

63.

64.

65. $x < 3$ **66.** $x \geq -1$ **67.** Yes   **68.** Yes   **69.** Yes
70. No
71.

x	0	1	2	3	4
$5 - x$	5	4	3	2	1

$x < 2$

72.

x	1	1.5	2	2.5	3
$2x - 5$	-3	-2	-1	0	1

$x \leq 2.5$

73. $\{x \mid x > 3\}$ **74.** $\{x \mid x \geq -5\}$ **75.** $\{x \mid x \leq -1\}$
76. $\{x \mid x < \frac{23}{3}\}$ **77.** $\{x \mid x \leq \frac{1}{9}\}$ **78.** $\{x \mid x \geq \frac{9}{4}\}$
79. $x < 50$ **80.** $x \leq 45,000$ **81.** $x \geq 16$
82. $x < 1995$
83. (a)

Time	12:00	1:00	2:00	3:00	4:00	5:00
Rainfall (R)	2	2.75	3.5	4.25	5	5.75

(b) $R = \frac{3}{4}x + 2$ (c) $\frac{23}{4} = 5\frac{3}{4}$ in.; yes
(d) $\frac{77}{16} = 4\frac{13}{16}$ in.
84. $2125
85. (a)

Hours (x)	1	2	3	4	5
Distance (D)	40	30	20	10	0

(b) $D = 50 - 10x$ (c) 20 miles; yes
(d) 3 hours or less or from noon to 3:00 p.m.
86. 1998   **87.** About 82%   **88.** $\frac{1}{6}$ hr, or 10 min
89. 50 mL   **90.** $500 at 3%; $800 at 5%
91. 33 in. by 23 in.   **92.** 25 inches or less
93. 74 or more   **94.** 6 hr
95. (a) $C = 85x + 150,000$ (b) $R = 225x$
(c) $P = 140x - 150,000$ (d) 1071 or fewer

CHAPTER 2 TEST (P. 144)

1. -6 **2.** $\frac{5}{2}$ **3.** $-\frac{16}{11}$ **4.** $\frac{25}{6}$ **5.** Zero
6.

x	0	1	2	3	4
$6 - 2x$	6	4	2	0	-2

3

7. $x + (-7) = 6; 13$ **8.** $2x + 6 = x - 7; -13$

9. 111, 112, 113   **10.** $\frac{7}{125}$; 0.056   **11.** 34.5%
12. $37.50   **13.** 1056 ft/sec   **14.** 7.5 in^2
15. $C = 30\pi \approx 94.2$ in.; $A = 225\pi \approx 706.9$ in^2
16. 30°   **17.** $x = \frac{y - z}{3y}$ **18.** $\left\{x \mid x > -\frac{1}{7}\right\}$
19. (a) $S = 2x + 5$ **(b)** 21 in.   **(c)** 17.5 in.
20. 2000 mL   **21.** 300%

CHAPTER 3: GRAPHING EQUATIONS

SECTION 3.1 (PP. 153–155)

1. xy-plane   **3.** 4   **5.** III   **7.** scatterplot
9. $(-2, -2), (-2, 2), (0, 0), (2, 2)$
11. $(-1, 0), (0, -3), (0, 2), (2, 0)$
13. (a) I   **(b)** III   **15. (a)** None   **(b)** I
17. (a) II   **(b)** IV
19.

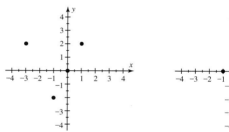

21.

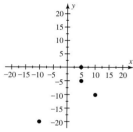

23.

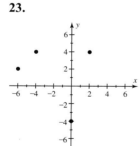

25.

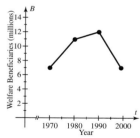

27.

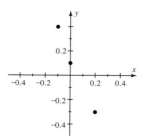

29.

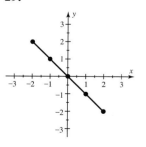

31.

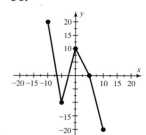

33.

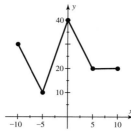

35. (1970, 29), (1980, 41), (1990, 79), (2000, 62); in 1970 the United States spent $29 billion on military personnel. (Answers may vary slightly.)

37. (a)

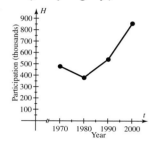

(b) Head Start participation decreased and then increased.

39. (a)

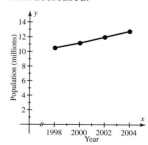

(b) The number of welfare beneficiaries increased and then decreased.

41. (a)

(b) The Asian-American population increased.

SECTION 3.2 (PP. 162–163)

1. two   **3.** linear   **5.** graph   **7.** Yes   **9.** No
11. No   **13.** Yes   **15.** No

17.

x	−2	−1	0	1	2
y	−8	−4	0	4	8

19.

x	−8	−4	0	4	8
y	−4	0	4	8	12

21.

x	6	3	0	−3	−9
y	−2	0	2	4	8

23.

x	−3	0	3	6
y	−9	0	9	18

25.

x	−8	−4	0	4
y	−2	0	2	4

27.

x	8	6	4	2
y	−2	0	2	4

29.

x	$-\frac{1}{2}$	$-\frac{1}{4}$	0	$\frac{1}{4}$
y	−2	−1	0	1

31.

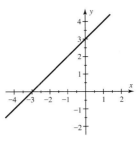

33.

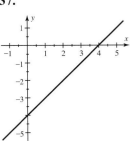

35.

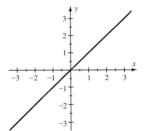

37.

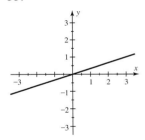

39.

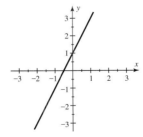

41.

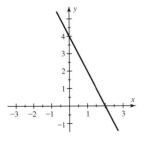

43.

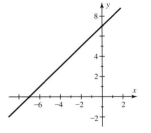

45.

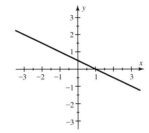

47.

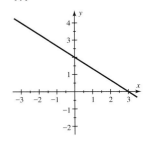

49.

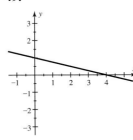

51.

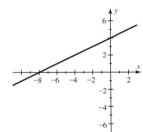

53.

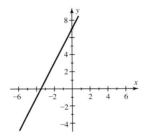

55.

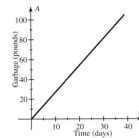

57.

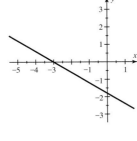

59. (a)

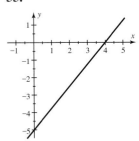

(b) About 37 days
61. (a) 67; 89

(b)

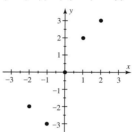

(c) 1999

(b)

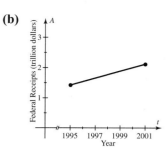

(c) 1996

CHECKING BASIC CONCEPTS 3.1 & 3.2 (P. 164)

1. $(-2, 2)$, II; $(-1, -2)$, III; $(1, 3)$, I; $(3, 0)$, none

2.

3.

x	-2	-1	0	1	2
y	5	3	1	-1	-3

4. **(a)** 5   **(b)** 4   **(c)** -2

5. **(a)**

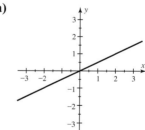

(b)

6. $(1800, 1)$, $(1955, 3)$, $(2000, 6)$; in 2000, the world population was 6 billion.

7. **(a)** In 1995, receipts were $1.425 trillion; in 2001, receipts were $2.115 trillion.

SECTION 3.3 (PP. 172–175)

1. Two   **3.** x-intercept   **5.** y-intercept
7. horizontal; 3   **9.** vertical; 3   **11.** 3; -2 **13.** 0; 0
15. -2 and 2; 4   **17.** 1; 1

19.

x	-2	-1	0	1	2
y	0	1	2	3	4

-2; 2

21.

x	-4	-2	0	2	4
y	-6	-4	-2	0	2

1; -2

23. x-intercept: 3; y-intercept: -2

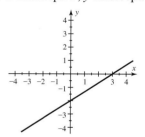

25. x-intercept: 5; y-intercept: -3

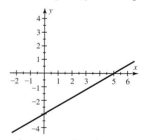

27. x-intercept: 6; y-intercept: -2

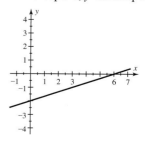

29. x-intercept: -1; y-intercept: 6

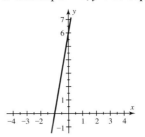

31. x-intercept: 7; y-intercept: 3

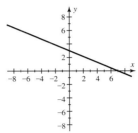

33. x-intercept: 4; y-intercept: -3

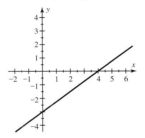

35. x-intercept: 4; y-intercept: -2

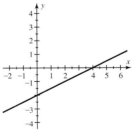

37. x-intercept: -4; y-intercept: 3

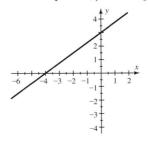

39. x-intercept: 3; y-intercept: 2

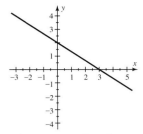

41. x-intercept: -2; y-intercept: 5

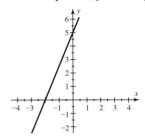

43. (a)

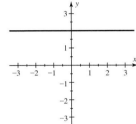

(b)

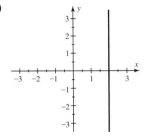

45. (a)

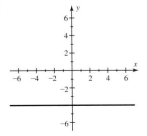

(b)

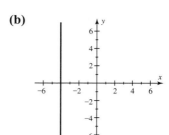

47. (a)

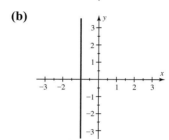

(b)

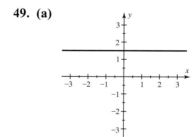

49. (a)

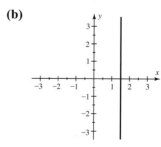

(b)

51. $y = 4$ **53.** $x = -1$ **55.** $y = -6$ **57.** $x = 5$
59. $y = 1$ **61.** $x = -6$ **63.** $y = 2; x = 1$
65. $y = -45; x = 20$ **67.** $y = 5; x = 0$ **69.** $x = -1$
71. $y = -\frac{5}{6}$ **73.** $x = 4$ **75.** $x = -\frac{2}{3}$

77. (a) y-intercept: 200; x-intercept: 4   **(b)** The driver was initially 200 miles from home; the driver arrived home after 4 hours.
79. (a) y-intercept: 2000; x-intercept: 4   **(b)** The pool initially contained 2000 gallons; the pool was empty after 4 hours.
81. (a) v-intercept: 128; t-intercept: 4

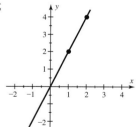

(b) The initial velocity was 128 ft/sec; the velocity after 4 seconds was 0.

SECTION 3.4 (PP. 185–190)

1. horizontal   **3.** rise; run   **5.** vertical   **7.** Negative
9. Zero   **11.** Negative
13. 0; the rise always equals 0.
15. 1; the graph rises 1 unit for each unit of run.
17. 2; the graph rises 2 units for each unit of run.
19. Undefined; the run always equals 0.
21. $\frac{1}{2}$; the graph rises 1 unit for each 2 units of run.
23. -1; the graph falls 1 unit for each unit of run.
25. 2;

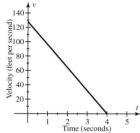

27. $-\frac{2}{3}$;

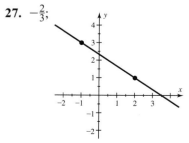

29. $\frac{5}{7}$ **31.** $-\frac{6}{7}$ **33.** 0   **35.** Undefined   **37.** $\frac{13}{20}$
39. $\frac{9}{100}$ **41.** $-\frac{5}{7}$ **43.** 49

45.

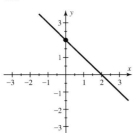

47.

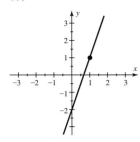

67. (a)

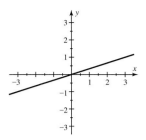

(b) $\frac{1}{3}$

49.

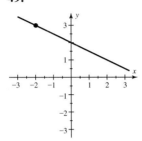

51.

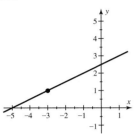

69. (a)

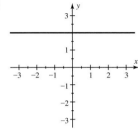

(b) 0

53. 2; 1; −2   **55.** −9; −1; −9

57.

x	0	1	2	3
y	−4	−2	0	2

59.

x	1	2	3	4
y	4	1	−2	−5

61.

x	−4	−2	0	2
y	0	3	6	9

63. (a)

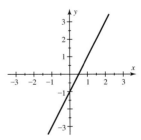

(b) 2

65. (a)

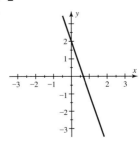

(b) −3

71. c.   **73.** b.

75. (a) $m_1 = 1000$; $m_2 = -1000$ **(b)** $m_1 = 1000$: Water is being added to the pool at a rate of 1000 gallons per hour. $m_2 = -1000$: Water is being removed from the pool at a rate of 1000 gallons per hour.   **(c)** Initially the pool contained 2000 gallons of water. Over the first 3 hours, water was pumped into the pool at a rate of 1000 gallons per hour. For the next 2 hours, water was pumped out of the pool at a rate of 1000 gallons per hour.

77. (a) $m_1 = 50$; $m_2 = 0$; $m_3 = -50$ **(b)** $m_1 = 50$: The car is moving away from home at a rate of 50 mph. $m_2 = 0$: The car is not moving. $m_3 = -50$: The car is moving toward home at a rate of 50 mph.   **(c)** Initially the car is at home. Over the first 2 hours, the car travels away from home at a rate of 50 mph. Then the car is parked for 1 hour. Finally, the car travels toward home at a rate of 50 mph.

79.

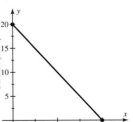

81.

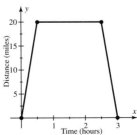

83. (a) 2   **(b)** The birth rate increased on average by 2000 children per year.

85. (a) 2.5   **(b)** The revenue is $2.50 per screwdriver.

87. (a) $1; 2$ **(b)** $\frac{1}{50}$ **(c)** Oil should be added at a rate of 1 gallon of oil per 50 gallons of gasoline.

89. (a)

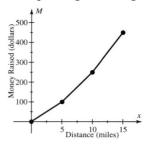

(b) $m_1 = 20; m_2 = 30; m_3 = 40$ **(c)** $m_1 = 20$: Each mile between 0 and 5 miles is worth $20 per mile. $m_2 = 30$: Each mile between 5 and 10 miles is worth $30 per mile. $m_3 = 40$: Each mile between 10 and 15 miles is worth $40 per mile.

91. (a) 1200 **(b)** Median family income increased on average by $1200 per year over this time period. **(c)** $48,000

CHECKING BASIC CONCEPTS 3.3 & 3.4 (PP. 190–191)

1. $-2; 3$

2.

x	-2	-1	0	1	2
y	-6	-4	-2	0	2

$1; -2$

3. (a) x-intercept: 6; y-intercept: -3

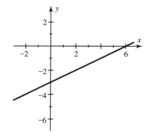

(b) y-intercept: 2

(c) x-intercept: -1

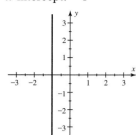

4. $y = 4; x = -2$ **5. (a)** $\frac{3}{4}$ **(b)** 0 **(c)** Undefined

6. $-\frac{1}{2}$

7.

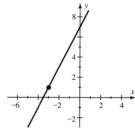

8. (a) $m_1 = 0; m_2 = 2; m_3 = -\frac{2}{3}$ **(b)** $m_1 = 0$: the depth is not changing. $m_2 = 2$: the depth increased at a rate of 2 feet per hour. $m_3 = -\frac{2}{3}$: the depth decreased at a rate of $\frac{2}{3}$ foot per hour. **(c)** Initially the pond had a depth of 5 feet. For the first hour, there was no change in the depth of the pond. For the next hour, the depth of the pond increased at a rate of 2 feet per hour to a depth of 7 feet. Finally, the depth of the pond decreased for 3 hours at a rate of $\frac{2}{3}$ foot per hour until it was 5 feet deep.

9. (a) 28 **(b)** For counties between 450 and 600 square miles, the population increases at an average rate of 28 people per square mile. **(c)** No. We do not know if this trend continues.

SECTION 3.5 (PP. 197–200)

1. $y = mx + b$ **3.** y-intercept **5.** origin **7.** f.

9. a. **11.** e. **13.** $y = x - 1$ **15.** $y = -2x + 1$

17. $y = \frac{1}{2}x - 2$ **19.** $y = -2x$ **21.** $y = \frac{3}{4}x + 2$

23. $y = x + 2$

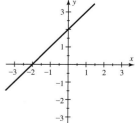

25. $y = 2x - 1$

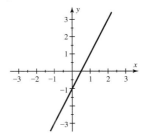

27. $y = -\frac{1}{2}x - 2$

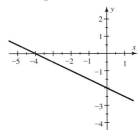

29. $y = \frac{1}{3}x$

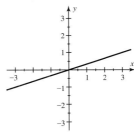

31. $y = -2x + 1$

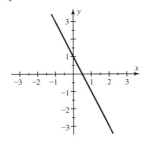

33. (a) $y = -x + 4$ **(b)** $-1; 4$
35. (a) $y = -2x + 4$ **(b)** $-2; 4$
37. (a) $y = \frac{1}{2}x + 2$ **(b)** $\frac{1}{2}; 2$
39. (a) $y = \frac{2}{3}x - 2$ **(b)** $\frac{2}{3}; -2$
41. (a) $y = \frac{1}{4}x + \frac{3}{2}$ **(b)** $\frac{1}{4}; \frac{3}{2}$
43. (a) $y = -\frac{1}{3}x + \frac{2}{3}$ **(b)** $-\frac{1}{3}; \frac{2}{3}$

45.

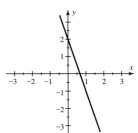

47.

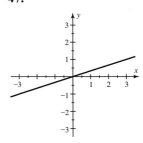

49.

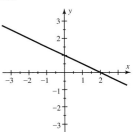

51.

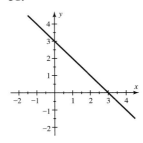

53.

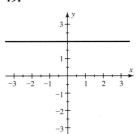

55.

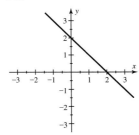

57. $y = 2x + 2$ **59.** $y = x - 2$ **61.** $y = \frac{4}{7}x + 3$
63. $y = 3x$ **65.** $y = -\frac{1}{2}x + \frac{5}{2}$ **67.** $y = 2x - 4$
69. $y = \frac{1}{3}x + \frac{1}{3}$
71. (a) \$25   **(b)** 25 cents   **(c)** 25; the fixed cost of renting the car   **(d)** 0.25; the cost per mile of driving the car
73. (a) \$7.45   **(b)** $C = 0.07x + 3.95$ **(c)** 67 min
75. (a) $m = 0.3; b = 189.2$ **(b)** The fixed cost of owning the car

SECTION 3.6 (PP. 207–210)

1. One   **3.** $y = mx + b$
5. Yes; every nonvertical line has exactly one slope and one y-intercept.
7. Yes   **9.** No   **11.** Yes   **13.** $y - 2 = \frac{3}{4}(x - 1)$
15. $y + 1 = -\frac{1}{2}(x - 3)$ **17.** $y + 2 = -3(x - 1)$
19. $y - 30 = 1.5(x - 2000)$ **21.** $y - 4 = \frac{7}{3}(x - 2)$
23. $y = \frac{3}{5}(x - 5)$ **25.** $y - 15 = 5(x - 1990)$

27. $y = 3x - 2$ **29.** $y = \frac{1}{3}x$ **31.** $y = -2x + 9$
33. $y = -16x - 19$ **35.** $y = -2x + 5$
37. $y = -x + 1$ **39.** $y = -\frac{1}{9}x + \frac{1}{3}$ **41.** $y = 2x - 7$
43. $y = 2x - 15$ **45.** $y = -2x - 1$ **47.** $y = \frac{1}{2}x - \frac{5}{2}$
49. (a) Toward   **(b)** After 1 hour the person is 250 miles from home. After 4 hours the person is 100 miles from home.   **(c)** 50 mph   **(d)** $y = -50x + 300$; the car is traveling toward home at 50 mph.
51. (a) Entering; 300 gallons   **(b)** 100; initially the tank contains 100 gallons   **(c)** $y = 50x + 100$; the amount of water is increasing at a rate of 50 gallons per minute.
(d) 8 minutes
53. (a) $-5°$F per hour   **(b)** $T = -5x + 45$; the temperature is decreasing at a rate of $5°$F per hour.   **(c)** 9; at 9 A.M. the temperature was $0°$F.
(d)

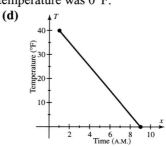

(e) At 4 A.M. the temperature was $25°$F.
55. (a) 628,000; 1,396,000   **(b)** 64; the number of inmates increased on average by 64,000 per year.
57. (a) In 1984, tuition and fees were \$1225; in 1987, tuition and fees were \$1621.   **(b)** $y - 1225 = 132(x - 1984)$ or $y - 1621 = 132(x - 1987)$; tuition and fees increased on average by \$132 per year.   **(c)** About \$2000; \$2017
59. (a) $y = 3.1x - 6165.6$ **(b)** 43.7 million

CHECKING BASIC CONCEPTS 3.5 & 3.6 (P. 210)

1. $y = -3x + 1$ **2.** $y = \frac{4}{5}x - 4; \frac{4}{5}; -4$
3.

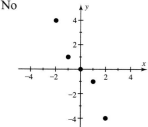

4. (a) $y = 3x - 2$ **(b)** $y = -\frac{3}{2}x$ **(c)** $y = -\frac{7}{3}x - \frac{5}{3}$
5. $y = 2x + 3$
6. (a) $y = -12x + 48$ **(b)** 12 mph   **(c)** 4 P.M.
(d) 48 miles

7. (a) 5 inches   **(b)** 2 inches per hour   **(c)** 5; total inches of snow that fell before noon   **(d)** 2; the rate of snowfall was 2 inches per hour.

SECTION 3.7 (PP. 216–218)

1. abstraction   **3.** linear   **5.** approximate   **7.** f.   **9.** a.
11. e.   **13.** Yes   **15.** No   **17.** No
19. Exact; $y = 2x - 2$ **21.** Approximate; $y = 2x + 2$
23. Approximate; $y = 2$ **25.** $y = 5x + 40$
27. $y = -20x - 5$ **29.** $y = 8$
31. (a) Yes

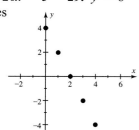

(b)

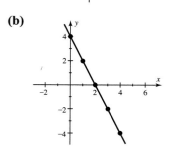

(c) $y = -2x + 4$
33. (a) No

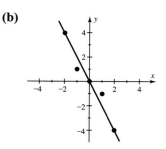

(b)

(c) $y = -2x$

35. (a) Yes

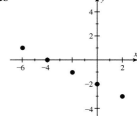

(b)

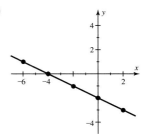

(c) $y = -\frac{1}{2}x - 2$

37. (a) No

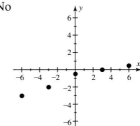

(b)

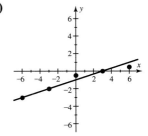

(c) $y = \frac{1}{3}x - 1$

39. $g = 5t + 200$, where g represents gallons of water and t represents time in minutes.

41. $d = 6t + 5$, where d represents distance in miles and t represents time in hours.

43. $p = 8t + 200$, where p represents total pay in dollars and t represents time in hours.

45. $r = t + 5$, where r represents total number of roofs shingled and t represents time in days.

47. (a) $-\frac{2}{45}$ acre per year   **(b)** $A = -\frac{2}{45}t + 5$

49. (a) $y = \frac{32}{3}x - 21{,}188$ (approx.)   **(b)** About 177,000

51. (a)

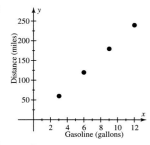

(b)

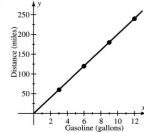

(c) 20; the mileage is 20 miles per gallon

(d) $y = 20x$ **(e)** 140 miles

CHECKING BASIC CONCEPTS 3.7 (P. 219)

1. Yes   **2.** Approximate; $y = x - 1$

3. (a) $y = 10x + 50$ **(b)** $y = -2x + 200$

4. (a) Yes

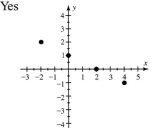

(b)

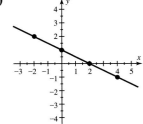

(c) $y = -\frac{1}{2}x + 1$

5. (a) $T = 0.05x$ **(b)** $0.8°F$

CHAPTER 3 REVIEW EXERCISES (PP. 225–229)

1. $(-2, 0)$: none; $(-1, 2)$: II; $(0, 0)$: none; $(1, -2)$: IV; $(1, 3)$: I

2.
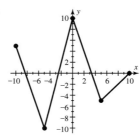

3.
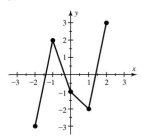

17.
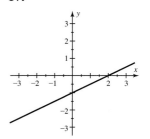

18.
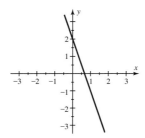

4.
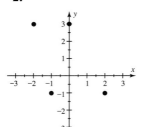

19.
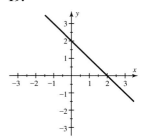

20.
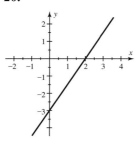

5. Yes   **6.** No   **7.** No   **8.** Yes

9.

x	−2	−1	0	1	2
y	6	3	0	−3	−6

10.

x	4	3	2.5	2	1
y	−3	−1	0	1	3

11.

x	−2	0	2	4
y	−4	2	8	14

12.

x	1	2	3	4
y	6	5	4	3

13.

x	−0.5	0	0.5	1
y	−1	0	1	2

14.

x	−1	−3	−5	−7
y	1	2	3	4

21.
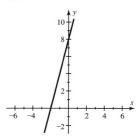

22.
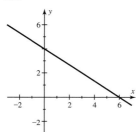

23. 3; −2
24. −2 and 2; −4
25.

x	−2	−1	0	1	2
y	4	3	2	1	0

2; 2

26.

x	−4	−2	0	2	4
y	−4	−3	−2	−1	0

4; −2

27. x-intercept: 3; y-intercept: −2

15.
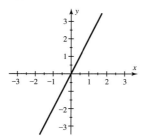

16.
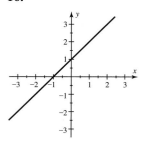

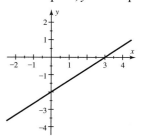

28. *x*-intercept: 1; *y*-intercept: -5

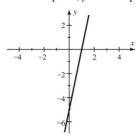

29. *x*-intercept: 4; *y*-intercept: -2

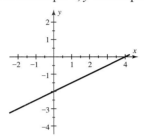

30. *x*-intercept: 2; *y*-intercept: 3

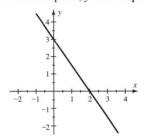

31. (a)

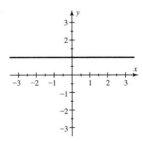

(b)

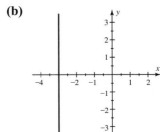

32. $x = -1; y = 1$ **33.** $y = 1$ **34.** $x = 3$
35. $y = 3; x = -2$
36. (a) *y*-intercept: 90; *x*-intercept: 3   **(b)** The driver is initially 90 miles from home; the driver arrives home after 3 hours.
37. 2   **38.** $-\frac{2}{5}$ **39.** 0   **40.** Undefined   **41.** 3
42. $-\frac{1}{2}$
43. (a)

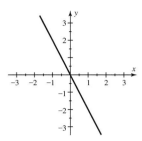

(b) $-2; 0$
44. (a)

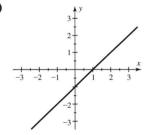

(b) $1; -1$
45. (a)

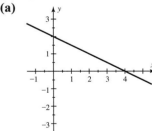

(b) $-\frac{1}{2}; 2$
46. (a)

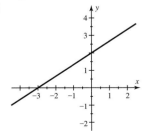

(b) $\frac{2}{3}; 2$

47.
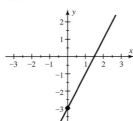

48.
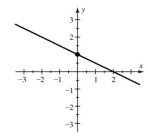

61.
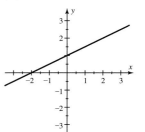

62.
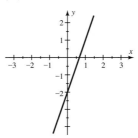

49.
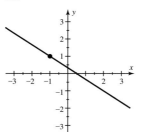

50.
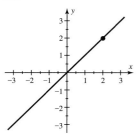

63.
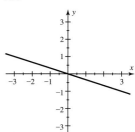

64.
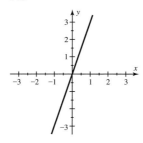

51. slope: 2; x-intercept: 1; y-intercept: -2

52.

x	0	1	2	3
y	1	$\frac{3}{2}$	2	$\frac{5}{2}$

53. $y = x + 1$ **54.** $y = -2x + 2$

55. $y = 2x - 2$
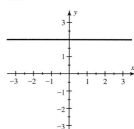

65.
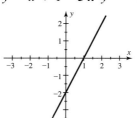

66.
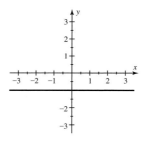

56. $y = -\frac{3}{4}x + 3$

67.
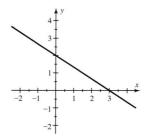

68.
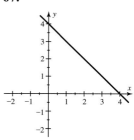

57. (a) $y = -x + 3$ **(b)** $-1; 3$
58. (a) $y = \frac{3}{2}x - 3$ **(b)** $\frac{3}{2}; -3$
59. (a) $y = 2x - 20$ **(b)** $2; -20$
60. (a) $y = \frac{5}{6}x - 5$ **(b)** $\frac{5}{6}; -5$

69. $y = 5x - 5$ **70.** $y = -2x$ **71.** $y = -\frac{5}{6}x + 2$
72. $y = -2x + 1$ **73.** $y = \frac{2}{3}x - 2$ **74.** $y = -\frac{1}{5}x - 2$
75. Yes   **76.** No   **77.** $y = 5x - 3$ **78.** $y = 20x - 65$
79. $y = -\frac{2}{3}x - \frac{1}{3}$ **80.** $y = 3x - 90$ **81.** $y = \frac{4}{3}x - 4$
82. $y = 2x - 1$ **83.** $y = 2x - 3$ **84.** $y = -\frac{2}{3}x - \frac{2}{3}$
85. No   **86.** Approximate; $y = -x + 5$

87. $y = -2x + 40$ **88.** $y = 20x + 200$ **89.** $y = 50$
90. $y = 5x - 20$
91. (a) Yes

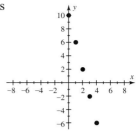

(b)

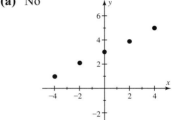

(c) $y = -4x + 10$
92. (a) No

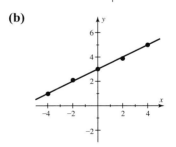

(b)

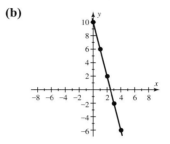

(c) $y = \frac{1}{2}x + 3$
93. (a)

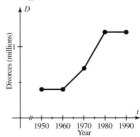

(b) The number of divorces remained unchanged from 1950 to 1960, increased significantly between 1960 and 1980, and then remained unchanged from 1980 to 1990.

94. (a) $G = 100t$
(b)

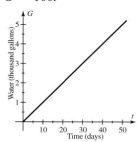

(c) 50 days
95. (a)

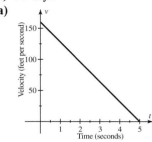

(b) v-intercept: 160; t-intercept 5; the initial velocity was 160 ft/sec, and the velocity after 5 seconds was 0.
96. (a) $m_1 = 0$; $m_2 = -1500$; $m_3 = 500$ **(b)** $m_1 = 0$: The population remained unchanged. $m_2 = -1500$: The population decreased at a rate of 1500 insects per week. $m_3 = 500$: The population increased at a rate of 500 insects per week.   **(c)** For the first week the population did not change from its initial value of 4000. Over the next two weeks the population decreased at a rate of 1500 insects per week until it reached 1000. Finally, the population increased at a rate of 500 per week for two weeks, reaching 2000.
97.

98. (a) -175 **(b)** The number of nursing homes decreased at an average rate of 175 per year.
99. (a) 0.5   **(b)** School enrollment is increasing at an average rate of 0.5 million students per year.
(c) 80 million students
100. (a) \$35   **(b)** 20¢   **(c)** 35; the fixed cost of renting the car   **(d)** 0.2; the cost for each mile driven
101. (a) Toward; the slope is negative.   **(b)** After 1 hour the car is 200 miles from home; after 3 hours the car is 100 miles from home.   **(c)** $y = -50x + 250$; the car is moving toward home at 50 mph.   **(d)** 150 miles; 150 miles
102. (a) 77,400; 90,120   **(b)** 1.59; deaths from pneumonia increased at an average rate of 1590 per year.

103. (a)

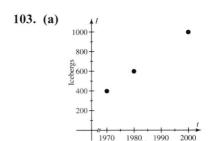

(b) $I = 20t - 39{,}000$; the number of icebergs increased at an average rate of 20 per year.   **(c)** Yes
(d) 1100 icebergs

104. (a)

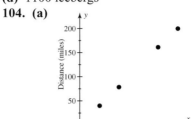

(b)

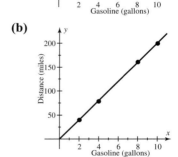

(c) 20; the mileage is 20 miles per gallon.
(d) $y = 20x$; no   **(e)** About 180 miles

CHAPTER 3 TEST (PP. 230–231)

1. $(-2, -2)$: III; $(-2, 1)$: II; $(0, -2)$: none; $(1, 0)$: none; $(2, 3)$: I; $(3, -1)$: IV; $(3, 1)$: I

2.

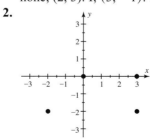

3.

x	-2	-1	0	1	2
y	-8	-6	-4	-2	0

$2; -4$

4.

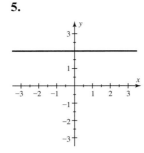

5.

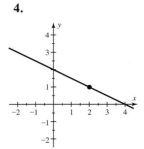

6.

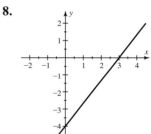

7.

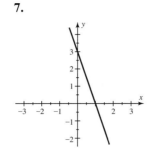

8.

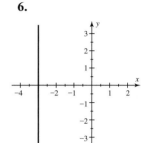

9. $y = -x + 2$ **10.** $y = 2x + \frac{1}{2}$; 2; $\frac{1}{2}$
11. $y = -\frac{4}{3}x - 5$ **12.** $y = 3x - 11$
13. $y = -3x + 5$ **14.** $y = -\frac{1}{2}x$
15. Approximate; $y = x - 1$
16. (a) $m_1 = 2$; $m_2 = -9$; $m_3 = 2$; $m_4 = 5$
(b) $m_1 = 2$: The population increased at a rate of 2000 fish per year. $m_2 = -9$: The population decreased at a rate of 9000 fish per year. $m_3 = 2$: The population increased at a rate of 2000 fish per year. $m_4 = 5$: The population increased at a rate of 5000 fish per year.   **(c)** For the first year the population increased from an initial value of 8000 to 10,000 at a rate of 2000 fish per year. During the second year the population dropped dramatically to 1000 at a rate of 9000 fish per year. Over the third year the population grew to 3000 at a rate of 2000 fish per year. Finally, over the fourth year, the population grew at a rate of 5000 fish per year to reach 8000.

17.

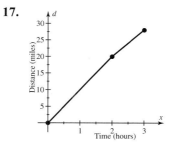

18. (a) 1.3 **(b)** Delinquency cases increased at a rate of 1.3 cases per 1000 youths per year, on average. **(c)** 67.8 cases per 1000 youths
19. $N = 100x + 2000$

CHAPTERS 1–3 CUMULATIVE REVIEW (PP. 232–235)

1. Composite; $40 = 2 \times 2 \times 2 \times 5$ **2.** Prime   **3.** 7
4. 2   **5.** $n + 10$ **6.** $n^2 - 2$ **7.** 8   **8.** $\frac{4}{5}$ **9.** $\frac{1}{10}$
10. $\frac{3}{7}$ **11.** $\frac{7}{6}$ **12.** $\frac{2}{3}$ **13.** 2   **14.** $\frac{17}{30}$ **15.** 4^3
16. $2 \times 2 \times 2 \times 2$; 16   **17.** 14   **18.** 7   **19.** $\frac{5}{3}$
20. -9 **21.** 25   **22.** 2   **23.** Rational   **24.** Irrational
25.

26. 2   **27.** $-3.1, -\frac{1}{2}, \sqrt{2}, 3, \pi$ **28.** $2^2 - (-4)$; 8
29. -6 **30.** 15   **31.** 18   **32.** -4
33. Commutative and associative properties
34. 50   **35.** $7x + 1$ **36.** $x - 4$ **37.** -3 **38.** 21
39. 2   **40.** 0
41.

x	-2	-1	0	1	2
y	10	8	6	4	2

1

42. $2n + 2 = n - 5$; -7 **43.** $-26, -25, -24, -23$
44. $\frac{49}{500}$; 0.098   **45.** 23.4%   **46.** 8 hours   **47.** 720 ft²
48. 20 ft²   **49.** $4\pi \approx 12.6$ ft²   **50.** 62°   **51.** $b = \frac{2A}{h}$
52. $L = \frac{P - 2W}{2}$ **53.** $x < 2$
54.

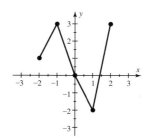

55. Yes
56.

x	0	1	2	3	4
$2x - 3$	-3	-1	1	3	5

$x \leq 2$

57. $\{x \mid x > 0\}$ **58.** $\{x \mid x \leq \frac{1}{2}\}$
59. $(-2, -3)$: III; $(0, 3)$: none; $(2, -2)$: IV; $(2, 1)$: I

60.

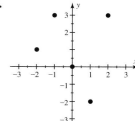

61. Yes
62.

x	-2	-1	0	1	2
y	3	2.5	2	1.5	1

63.

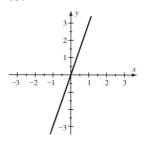

64.

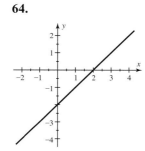

65.

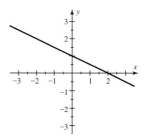

66.

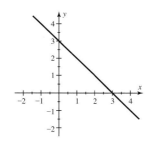

67.

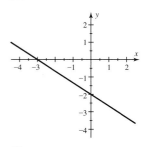

68.

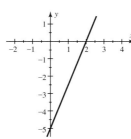

69.

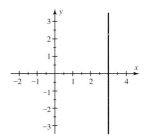

70.

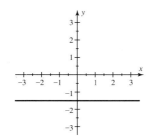

71. $\frac{3}{2}$; -3; $y = 2x - 3$ **72.** 2; 1; $y = -\frac{1}{2}x + 1$

73. x-intercept: -10; y-intercept: 8 **74.** $x = \frac{3}{2}$; $y = -2$

75.

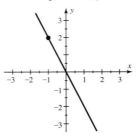

76.

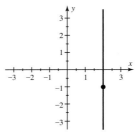

77. $y = 3x - 3$

78. $y = \frac{3}{5}x - 3$;

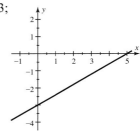

79. $y = -\frac{1}{3}x + 7$ **80.** $y = -\frac{2}{3}x - 3$

81. $y = -2x + 1$ **82.** $y = \frac{1}{4}x + \frac{1}{2}$

83. Approximate; $y = \frac{1}{2}x + 1$ **84.** $y = 5x + 100$

85. $I = 5000x + 20,000$

86. (a) Yes

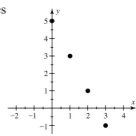

(b)

(c) $y = -2x + 5$

87. (a) $50x + 30x$; $80x$ **(b)** 250 min   **88.** $C = 8x$

89. $\frac{99}{200}$ **90.** $10\frac{1}{2}$ mi   **91.** $\$5500$

92. (a) $\$807.40$ **(b)** 1990 **93.** About 13.4%

94. $\$1000$ at 3%; $\$2000$ at 4% **95.** $w > 1135$ in.

96. (a) $C = 0.25x + 200$ **(b)** $R = 10x$

(c) $P = 9.75x - 200$ **(d)** 21 CDs

97. (a) $m_1 = 1.5$; $m_2 = 1$; $m_3 = 0$ **(b)** $m_1 = 1.5$: The fish grew 1.5 inches for each additional pound of food. $m_2 = 1$: The fish grew 1 inch for each additional pound of food. $m_3 = 0$: The fish stopped growing.   **(c)** There was more food than the fish could eat.

98. (a) 0.16 **(b)** Participation increased at an average rate of 0.16 million students per year.   **(c)** 28 million

99. (a) $\$85$ **(b)** $\$25$ **(c)** $30¢$

100. (a) Selling 20 hats generates $\$240$ in revenue; selling 50 hats generates $\$600$ in revenue.   **(b)** $y = 12x$; the hats cost $\$12$ each.

CHAPTER 4: SYSTEMS OF LINEAR EQUATIONS IN TWO VARIABLES

SECTION 4.1 (PP. 244–246)

1. ordered   **3.** $(11, 9)$ **5.** table   **7.** 1 **9.** 2 **11.** -2

13. 2 **15.** $(1, 1)$ **17.** $(2, -3)$ **19.** $(2, 0)$ **21.** $(2, 1)$

23. $(3, 2)$ **25.** $(-1, 1)$ **27.** $(2, 4)$ **29.** $(3, 1)$

31. $(1, 3)$;

x	0	1	2	3
$y = x + 2$	2	3	4	5
$y = 4 - x$	4	3	2	1

33. (a) $(-1, 1)$ **(b)** $(-1, 1)$ **35. (a)** $(3, 1)$ **(b)** $(3, 1)$

37. (a) $(1, 3)$ **(b)** $(1, 3)$ **39.** $(-1, -2)$ **41.** $(1, -1)$

43. $(1, 2)$ **45.** $(3, 2)$ **47.** $(3, 1)$ **49.** $(1, 2)$

51. (a) $C = 0.5x + 50$ **(b)** 60 mi   **(c)** 60 mi

53. (a) $x + y = 35$; $x = 2.5y$ **(b)** $(25, 10)$

55. (a) $x - y = 4$; $2x + 2y = 28$ where x is length, y is width   **(b)** $(9, 5)$; the rectangle is 9×5 in.

SECTION 4.2 (PP. 253–256)

1. substitution   **3.** It has infinitely many solutions.

5. consistent   **7.** independent   **9.** $(3, 6)$ **11.** $(2, 1)$

13. $(-1, 0)$ **15.** $(3, 0)$ **17.** $\left(\frac{1}{2}, 0\right)$ **19.** $(0, 4)$

21. $(1, 2)$ **23.** $(6, -4)$ **25.** $(4, -3)$ **27.** $\left(1, -\frac{1}{2}\right)$

29. $(1, 3)$ **31.** $(0, -2)$ **33.** $(2, 2)$ **35.** $(-3, 5)$

37. One; consistent; independent

39. Infinitely many; consistent; dependent

41. Zero; inconsistent   **43.** No solutions; inconsistent

45. $(2, 1)$; consistent; independent

47. Infinitely many; consistent; dependent

49. No solutions; inconsistent   **51.** No solutions

53. Infinitely many   **55.** $(3, 1)$ **57.** Infinitely many

59. No solutions   **61.** $\left(-\frac{4}{3}, \frac{11}{9}\right)$ **63.** $(0, 0)$
65. No solutions   **67.** No solutions
69. (a) $L - W = 10$
               $2L + 2W = 72$
      (b) $(23, 13)$
71. (a) $x = \frac{1}{2}y$; $x + y = 90$ (b) $(30, 60)$ (c) $(30, 60)$
73. (a) $x - y = 1.68$; $y = 0.98x$ (b) $(84, 82.32)$
75. 94×50 ft   **77.** $17, 53$
79. $3.\overline{3}$ L of 20% solution, $6.\overline{6}$ L of 50% solution
81. 10 mph; 2 mph
83. Superior: $32,000$ mi^2; Michigan: $22,000$ mi^2

CHECKING BASIC CONCEPTS 4.1 & 4.2 (P. 256)

1. (a) -2 (b) 6   **2.** $(4, 2)$ **3.** $(2, 1)$
4. (a) No solutions   (b) $(1, -1)$; one   (c) Infinitely many
5. (a) $x + y = 300$
               $50x + 60y = 17,000$
      (b) $(100, 200)$

SECTION 4.3 (PP. 265–267)

1. Substitution; elimination   **3.** $=$
5. Add the equations.   **7.** $(1, 1)$ **9.** $(-2, 1)$
11. No solutions   **13.** Infinitely many   **15.** $(6, 1)$
17. $(-1, 4)$ **19.** $(2, 4)$ **21.** $(2, 1)$ **23.** $(-4, 1)$
25. $(5, 8)$ **27.** $(-2, -5)$ **29.** $(-4, 2)$ **31.** $(2, 2)$
33. $(4, -1)$ **35.** $\left(\frac{4}{5}, \frac{9}{5}\right)$ **37.** $(3, 2)$ **39.** $(0, 1)$
41. $(2, 1)$ **43.** $(4, -3)$ **45.** $(1, 0)$
47. Infinitely many;

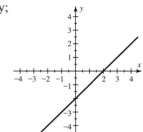

49. One;

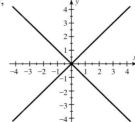

51. Zero;

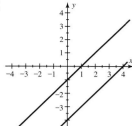

53. One;

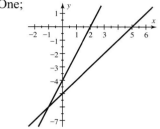

55. Infinitely many;

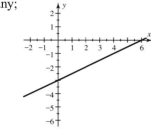

57. Men: 31,700; women: 24,700
59. Bicycle:18 min; stair climber: 12 min
61. Current: 2 mph; boat: 6 mph
63. $2000 at 3%; $3000 at 5%   **65.** $-43, 26$
67. (a) 20×40 in.   (b) 20×40 in.

SECTION 4.4 (PP. 276–279)

1. All points below and including the line $y = k$
3. All points above and including the line $y = x$
5. dashed   **7.** $Ax + By = C$ **9.** Yes   **11.** No
13. No   **15.** Yes   **17.** Yes   **19.** No
21. $(3, 1)$;

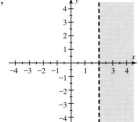

23. $(-2, -1)$;

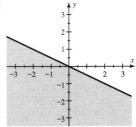

25. Yes   **27.** No   **29.** Yes
31. The region containing $(1, 2)$
33. The region containing $(1, 0)$
35. $x > 1$ **37.** $y \geq 2$ **39.** $y < x$ **41.** $-x + y \leq 1$
43.   **45.**

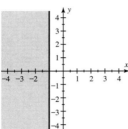   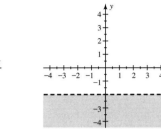

47.   **49.**

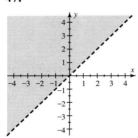   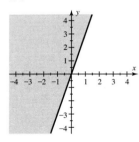

51.   **53.**

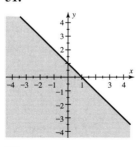   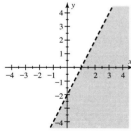

55.   **57.**

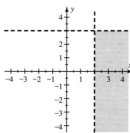   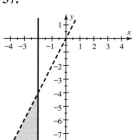

59.   **61.**

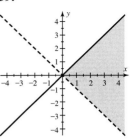   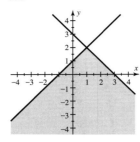

63.   **65.**

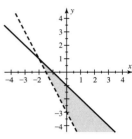   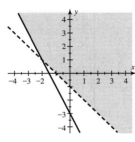

67.   **69.**

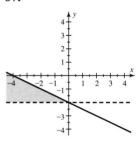   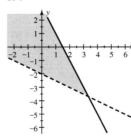

71.   **73.**

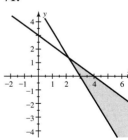   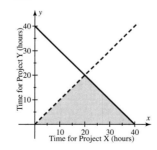

75. (a) 200 bpm; 150 bpm
    (b)

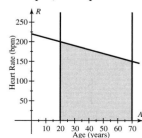

    (c) The shaded region represents possible heart rates
for ages 20 to 70.
77. 150 to 200 lb

CHECKING BASIC CONCEPTS 4.3 & 4.4 (P. 279)

1. $(1, 1)$ **2. (a)** $(0, -1)$; one   **(b)** No solutions
(c) Infinitely many
3. Infinitely many
4. (a)

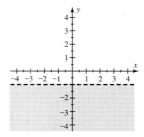

(b)

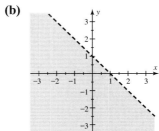

5.

6. (a) $x + y = 11$
        $x - y = 5$
(b) $(8, 3)$

CHAPTER 4 REVIEW EXERCISES (PP. 283–286)

1. 3   **2.** 2   **3.** $(1, 2)$ **4.** $(5, 2)$ **5.** $(4, 3)$
6. $(2, -4)$ **7.** $(2, 2)$ **8.** $(1, 2)$ **9.** $(2, 6)$ **10.** $(1, 1)$
11. $(4, -3)$ **12.** $(1, 2)$ **13.** $(1, 1)$ **14.** $(1, 2)$
15. $(1, 1)$ **16.** $(-2, -1)$ **17.** $(2, 6)$ **18.** $(2, -10)$
19. $(1, 3)$ **20.** $(5, 10)$ **21.** $(-2, 1)$ **22.** $(-4, -4)$
23. (a) Zero   **(b)** Inconsistent
24. (a) One   **(b)** Consistent; independent
25. (a) Infinitely many   **(b)** Consistent; dependent
26. (a) Zero   **(b)** Inconsistent   **27.** No solutions
28. No solutions   **29.** Infinitely many   **30.** $(1, 1)$
31. $(2, 1)$ **32.** $(-1, 2)$ **33.** $(11, -1)$ **34.** $(1, 0)$
35. $\left(\frac{9}{4}, \frac{7}{4}\right)$ **36.** $\left(\frac{5}{2}, 1\right)$ **37.** $(5, -7)$ **38.** $(8, 2)$
39. $(-2, 3)$ **40.** $(1, 0)$ **41.** $(1, 3)$ **42.** $(2, -1)$
43. Infinitely many   **44.** Infinitely many   **45.** Zero
46. One   **47.** Yes   **48.** No   **49.** No   **50.** Yes

51. $(1, -2)$;

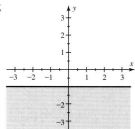

52. $(1, 1)$;

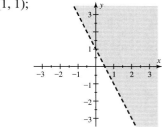

53. Yes   **54.** No   **55.** The region containing $(2, -2)$
56. The region containing $(1, 3)$ **57.** $y > 1$
58. $y \le 2x + 1$

59.

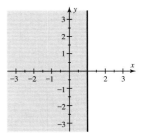

60.

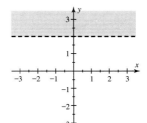

61.

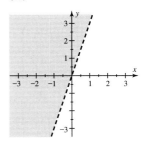

62.

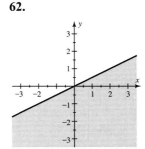

63.

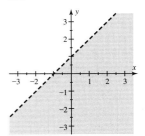

64.

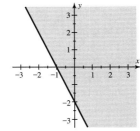

65.
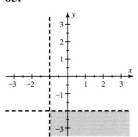

66.
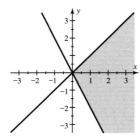

67.
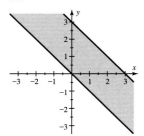

68.
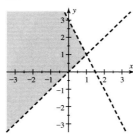

69.
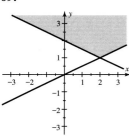

70.
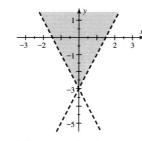

71. 3100 deaths in 1912; 41,230 deaths in 1999
72. Men: 102,500 cases; women: 82,500 cases
73. **(a)** $C = 0.2x + 40$ **(b)** 250 mi   **(c)** 250 mi
74. 20×24 ft
75. **(a)** $2x + y = 180$
$\qquad 2x - y = 40$
(b) $(55, 70)$ **(c)** $(55, 70)$
76. $75°, 105°$
77. **(a)** $\quad x + \quad y = \quad 10$
$\qquad 80x + 120y = 920$; x is \$80 rooms, y is \$120 rooms.
(b) $(7, 3)$
78. 7 lb of \$2 candy; 11 lb of \$3 candy
79. Bicycle: 35 min; stair climber: 25 min   **80.** 2 mph
81. **(a)** 16×24 ft; answers may vary slightly.
(b) 16×24 ft

82.
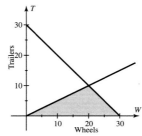

83. **(a)** 136 bpm; 108 bpm
(b)
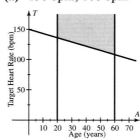

(c) The shaded region represents target heart rates above 70% of the maximum heart rate for ages 20 to 60.

CHAPTER 4 TEST (P. 287)

1. $(1, 2)$ **2.** $(-2, -1)$ **3.** $(-1, -2)$ **4.** $(-2, 3)$
5. $(1, 3)$
6. **(a)** $(2, 1)$; one; consistent   **(b)** No solutions; zero; inconsistent
7. $(-3, 4)$ **8.** No solutions   **9.** $y \le -\frac{1}{2}x$
10.
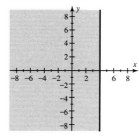

11.

12.
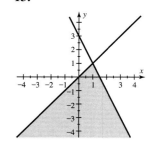

13.

14. \$1.8 trillion in 1998; \$1.9 trillion in 1999
15. $\frac{2}{3}$ hr at 6 mph; $\frac{1}{3}$ hr at 9 mph

CHAPTER 5: POLYNOMIALS AND EXPONENTS

SECTION 5.1 (PP. 296–298)

1. base; exponent   **3.** $\left(\frac{1}{2}\right)^3$ **5.** a^{m+n} **7.** $a^n b^n$ **9.** 64
11. -8 **13.** -8 **15.** 1   **17.** 32   **19.** 26   **21.** 8
23. $\frac{1}{2}$ **25.** $4^8 = 65{,}536$ **27.** $2^5 = 32$ **29.** x^9 **31.** z^4
33. x^6 **35.** $20x^7$ **37.** $-3x^3 y^4$ **39.** 2^6 **41.** n^{12}
43. x^7 **45.** $49b^2$ **47.** $a^3 b^3$ **49.** $4x^4$ **51.** $-64b^6$
53. $x^{14}y^{21}$ **55.** $x^{12}y^9$ **57.** $\frac{8}{27}$ **59.** $\frac{a^5}{b^5}$ **61.** $\frac{8x^3}{125}$
63. $\frac{27x^6}{125y^{12}}$ **65.** $(a+b)^5$ **67.** 6   **69.** $a^3 + 2ab^2$
71. $r^2 t + rt^2$ **73.** $10x^4$ square units   **75.** $8x^3$
77. $9\pi x^4$ **79.** \$1157.63
81.

83.

SECTION 5.2 (PP. 306–308)

1. monomial   **3.** polynomial   **5.** binomial   **7.** 2; 3
9. like   **11.** opposite   **13.** 2; 3   **15.** 2; -1
17. 2; -5 **19.** 0; -6 **21.** Yes; 1; 1; 1   **23.** Yes; 3; 1; 2
25. No   **27.** Yes; 2; 2; 4   **29.** Yes; x **31.** Yes; $-5x^3$
33. No   **35.** Yes; $2ab$ **37.** Yes; $4xy^2$ **39.** $-x + 9$
41. $4x^2 + 8x - 5$ **43.** $5a^3 + 1$
45. $4y^3 + 3y^2 + 4y - 9$ **47.** $4xy + 1$ **49.** $2a^2 b^3$
51. $9x^2 + x - 6$ **53.** $x^2 - 7x - 1$ **55.** $-5x^2$
57. $-3a^2 + a - 4$ **59.** $2t^2 + 3t - 4$
61. $3x^3 + x - 5$ **63.** $-9xy + x^2$ **65.** $4x - 2$
67. $-3x^2 + 5x + 10$ **69.** $-3a^2 - 5a$
71. $z^3 - 6z^2 - 6z - 1$ **73.** $3xy + 2x^2 y^2$ **75.** $-a^3 b$
77. $-x^2 - 5x - 4$ **79.** $-2x^3 - 6x - 2$ **81.** 27 in^3
83. (a) 200 bpm   **(b)** 100 bpm   **(c)** It decreases quickly at first, then more slowly.
85. $2z^2$; 200 in^2 **87.** $2x^2 + x^2$ or $3x^2$; 108 ft^2
89. $\pi x^2 + \pi y^2$; 13π ft^2
91. (a) $m_1 \approx 0.077$; $m_2 \approx 0.083$; $m_3 \approx 0.077$. A line is a reasonable estimate. It is not exact.   **(b)** Substituting different years for t shows that the polynomial gives a reasonable estimate.

CHECKING BASIC CONCEPTS 5.1 & 5.2 (P. 309)

1. (a) -25 **(b)** 1
2. (a) 10^8 **(b)** $-12x^7$ **(c)** $a^6 b^2$ **(d)** $\frac{x^4}{z^{12}}$
3. 3; 2; 4   **4. (a)** $2W^2 H$ **(b)** 2880 in^3

5. (a) $3a^2 + 6$ **(b)** $2z^3 + 7z - 8$ **(c)** $4xy$ **6.** 34

SECTION 5.3 (PP. 314–316)

1. The product rule   **3.** $x^2 - 7$; answers may vary.
5. term; term   **7.** x^7 **9.** $-12a^2$ **11.** $20x^5$
13. $4x^2 y^3$ **15.** $-12x^3 y^3$ **17.** $3x + 12$ **19.** $-45x - 5$
21. $4z - z^2$ **23.** $-5y - 3y^2$ **25.** $15x^3 - 12x$
27. $6x^3 - 6x^2$ **29.** $-32t^2 - 8t - 8$
31. $-8m^3 + 28m^2 - 24m$ **33.** $-5n^4 + n^3 - 2n^2$
35. $x^2 y + xy^2$ **37.** $x^4 y - x^3 y^2$ **39.** $-a^4 b + 2ab^4$
41. $x^2 + 3x$ **43.** $x^2 + 4x + 4$ **45.** $x^2 + 9x + 18$
47. $x^2 + 8x + 15$ **49.** $x^2 - 17x + 72$
51. $6z^2 - 19z + 10$ **53.** $64b^2 - 1$
55. $10y^2 - 3y - 7$ **57.** $5 - 13a + 6a^2$ **59.** $1 - 9x^2$
61. $x^3 - x^2 + x - 1$ **63.** $4x^3 - 3x^2 + 16x - 12$
65. $2n^3 + n^2 + 6n + 3$ **67.** $m^3 + 4m^2 + 4m + 1$
69. $6x^3 - 7x^2 + 14x - 8$ **71.** $x^3 + 1$
73. $4b^4 + 3b^3 + 19b^2 + 9b + 21$
75. (a) $h^3 - 2h^2 - 8h$ **(b)** 14,175 in^3
77. $6x^2 + 12x + 6$
79. (a) $64t - 16t^2$ **(b)** 64; 64   **(c)** Yes; yes
81. (a) $50x - x^2$ **(b)** 625

SECTION 5.4 (PP. 322–324)

1. $a^2 - b^2$ **3.** $a^2 + 2ab + b^2$
5. No; $x = 1, y = 1$; answers may vary.
7. No; $z = 1$; answers may vary.   **9.** $x^2 - 1$
11. $16x^2 - 1$ **13.** $1 - 4a^2$ **15.** $4x^2 - 9y^2$
17. $a^2 b^2 - 25$ **19.** $a^2 b^2 - 16$ **21.** $a^4 - b^4$
23. $x^6 - y^6$ **25.** 9999   **27.** 391   **29.** 9900
31. $x^2 + 2x + 1$ **33.** $a^2 - 4a + 4$
35. $4x^2 + 12x + 9$ **37.** $9b^2 + 30b + 25$
39. $\frac{9}{16}a^2 - 6a + 16$ **41.** $1 - 2b + b^2$
43. $25 + 10y^3 + y^6$ **45.** $a^4 + 2a^2 b + b^2$
47. $a^3 + 3a^2 + 3a + 1$ **49.** $x^3 - 6x^2 + 12x - 8$
51. $8x^3 + 12x^2 + 6x + 1$
53. $216u^3 - 108u^2 + 18u - 1$
55. $t^4 + 6t^3 + 12t^2 + 8t$ **57.** $20x + 36$
59. $x^2 + 2x - 35$ **61.** $9x^2 - 30x + 25$
63. $25x^2 + 35x + 12$ **65.** $16b^2 - 25$
67. $-20x^3 + 35x^2 - 10x$ **69.** $64 - 48a + 12a^2 - a^3$
71. (a) $x^2 + 4x + 4$ **(b)** $x^2 + 4x + 4$
73. (a) $4x^2 + 12x + 9$ **(b)** $4x^2 + 12x + 9$
75. (a) $6x^2 + 60x + 150$ **(b)** $x^3 + 15x^2 + 75x + 125$
77. (a) $1 + 2x + x^2$ **(b)** 1.21; the money increases by 1.21 times in 2 years if the interest rate is 10%.
79. (a) $1 - 2x + x^2$ **(b)** 0.25; if the chance of rain on each day is 50%, then there is a 25% chance that it will not rain on either day.
81. (a) $32z + 256$ **(b)** 2176; the area of an 8-foot-wide sidewalk around a 60 $\times$ 60 foot pool is 2176 ft^2.

Checking Basic Concepts 5.3 & 5.4 (p. 324)

1. (a) $-15x^3y^5$ **(b)** $-6x + 4x^2$
(c) $3a^3b - 6a^2b^2 + 3ab^3$
2. (a) $4x^2 + 9x - 9$ **(b)** $2x^4 - 2$ **(c)** $x^3 + y^3$
3. (a) $25x^2 - 4$ **(b)** $x^2 + 6x + 9$
(c) $4 - 28x + 49x^2$ **(d)** $t^3 + 6t^2 + 12t + 8$
4. (a) $4\pi a^2 + 56\pi a + 196\pi$ **(b)** 900π
5. (a) $m^2 + 10m + 25$ **(b)** $m^2 + 10m + 25$

Section 5.5 (pp. 332–334)

1. $\dfrac{1}{a^n}$ **3.** a^n **5.** a^{m-n} **7.** $\left(\dfrac{b}{a}\right)^n$ **9.** $1 \le |b| < 10$
11. $\dfrac{1}{4}$ **13.** 9 **15.** 2 **17.** 100 **19.** $\dfrac{1}{81}$ **21.** $\dfrac{1}{8}$
23. $\dfrac{1}{576}$ **25.** 64 **27.** 1 **29.** 64 **31.** 25 **33.** $\dfrac{49}{4}$
35. $\dfrac{1}{x}$ **37.** $\dfrac{1}{a^4}$ **39.** $\dfrac{1}{x^2}$ **41.** $\dfrac{1}{a^8}$ **43.** $\dfrac{y^3}{x^3}$ **45.** $\dfrac{1}{x^3y^3}$
47. $\dfrac{1}{16t^4}$ **49.** $\dfrac{1}{(x+1)^7}$ **51.** a^8 **53.** $\dfrac{1}{r^2t^6}$ **55.** $\dfrac{b^2}{a^4}$
57. x^2 **59.** a^{13} **61.** $\dfrac{2}{z^3}$ **63.** $-\dfrac{2y^3}{3x^2}$ **65.** $\dfrac{1}{x^3}$
67. $2b$ **69.** $\dfrac{a^3}{b^3}$ **71.** y^5 **73.** $2t^3$ **75.** x^2y^2
77. a^6b^3 **79.** $\dfrac{b^2}{a^2}$ **81.** $\dfrac{4v}{u}$ **83.** Thousand
85. Billion **87.** Hundredth **89.** 2000 **91.** 45,000
93. 0.008 **95.** 0.000456 **97.** 39,000,000
99. $-500,000$ **101.** 2×10^3 **103.** 5.67×10^2
105. 1.2×10^7 **107.** 4×10^{-3} **109.** 8.95×10^{-4}
111. -5×10^{-2} **113.** 1,500,000 **115.** -1.5
117. 2000 **119.** 0.002
121. (a) About 5.859×10^{12} mi **(b)** About 2.5×10^{13} mi
123. About 9.2×10^9 mi
125. (a) 9.963×10^{12} **(b)** About \$35,456

Section 5.6 (pp. 340–341)

1. $\dfrac{a}{d} + \dfrac{b}{d}$ **3.** No **5.** 9; 4; 1 **7.** $2x$ **9.** $z^3 + z^2$
11. $\dfrac{a^2}{2} - 3$ **13.** $\dfrac{4}{x} - 7x^2$ **15.** $3x^3 - 1 + \dfrac{2}{x}$
17. $4y^2 - 1 + \dfrac{2}{y}$ **19.** $3m^2 - 2m + 4$
21. $2x + 1 + \dfrac{3}{x-2}$ **23.** $x + 1$ **25.** $x^2 + 1 + \dfrac{-1}{x-1}$
27. $x^2 - x + 2 + \dfrac{1}{4x+1}$ **29.** $x^2 + 2x + 3 + \dfrac{8}{x-2}$
31. $3x^2 + 3x + 3 + \dfrac{5}{x-1}$ **33.** $x + 3 + \dfrac{-x-2}{x^2+1}$
35. $x + 1$ **37.** $x^2 - 2x + 4$ **39.** $4x$
41. $x + 2$;

Checking Basic Concepts 5.5 & 5.6 (p. 342)

1. (a) $\dfrac{1}{81}$ **(b)** $\dfrac{1}{2x^7}$ **(c)** $\dfrac{b^8}{16a^2}$
2. (a) z^5 **(b)** $\dfrac{y^6}{x^3}$ **(c)** $\dfrac{x^6}{27}$
3. (a) 4.5×10^4 **(b)** 2.34×10^{-4} **(c)** 1×10^{-2}
4. (a) 47,100 **(b)** 0.006 **5.** $5a - 3$
6. $3x + 2$; R: -2
7. (a) 9.3×10^7 **(b)** 500 sec (about 8 min 20 sec)

Chapter 5 Review Exercises (pp. 346–349)

1. 125 **2.** -81 **3.** 4 **4.** 11 **5.** -5 **6.** 1 **7.** 6^5
8. 10^{12} **9.** z^9 **10.** y^6 **11.** $30x^9$ **12.** a^4b^4
13. 2^{10} **14.** m^{20} **15.** a^3b^3 **16.** x^8y^{12} **17.** x^7y^{11}
18. 1 **19.** $(r-t)^9$ **20.** $(a+b)^6$ **21.** 7; 6
22. 5; -1 **23.** Yes; 1; 1; 1 **24.** Yes; 4; 1; 3
25. Yes; 3; 2; 2 **26.** No **27.** $5x^2 - x + 3$
28. $-6x^2 + 3x + 7$ **29.** $3x + 4$ **30.** $-2x^2 - 13$
31. $-2x^2 + 9x + 5$ **32.** $2a^3 - a^2 + 7a$
33. $5y^2 - 3xy$ **34.** $4xy$ **35.** $-x^5$ **36.** $-r^3t^4$
37. $-6t + 15$ **38.** $2y - 12y^2$ **39.** $18x^5 + 30x^4$
40. $-x^3 + 2x^2 - 9x$ **41.** $-a^3b + 2a^2b^2 - ab^3$
42. $a^2 + 3a - 10$ **43.** $8x^2 + 13x - 6$
44. $-2x^2 + 3x - 1$ **45.** $2y^3 + y^2 + 2y + 1$
46. $2y^4 - y^2 - 1$ **47.** $z^3 + 1$
48. $4z^3 - 15z^2 + 13z - 3$ **49.** $z^2 + z$ **50.** $2x^2 + 4x$
51. $z^2 - 4$ **52.** $25z^2 - 81$ **53.** $1 - 9y^2$
54. $25x^2 - 16y^2$ **55.** $r^2t^2 - 1$ **56.** $4m^4 - n^4$
57. $x^2 + 2x + 1$ **58.** $16x^2 + 24x + 9$
59. $y^2 - 6y + 9$ **60.** $4y^2 - 20y + 25$
61. $16 + 8a + a^2$ **62.** $16 - 8a + a^2$
63. $x^4 + 2x^2y^2 + y^4$ **64.** $x^2y^2 - 4xy + 4$
65. $z^3 + 15z^2 + 75z + 125$ **66.** $8z^3 - 12z^2 + 6z - 1$
67. $\dfrac{1}{9}$ **68.** $\dfrac{1}{9}$ **69.** 4 **70.** $\dfrac{1}{1000}$ **71.** 36 **72.** $\dfrac{1}{25}$
73. $\dfrac{1}{z^2}$ **74.** $\dfrac{1}{y^4}$ **75.** $\dfrac{1}{a^2}$ **76.** $\dfrac{1}{x^2}$ **77.** $\dfrac{1}{4t^2}$ **78.** $\dfrac{1}{a^3b^6}$
79. $\dfrac{1}{y^3}$ **80.** x^4 **81.** $\dfrac{2}{x^3}$ **82.** $\dfrac{2x^4}{3y^3}$ **83.** $\dfrac{a^5}{b^5}$ **84.** $4t^4$
85. $\dfrac{27}{x^3}$ **86.** $2ab$ **87.** $\dfrac{y^2}{x^2}$ **88.** $\dfrac{2v}{3u}$ **89.** 600
90. 52,400 **91.** 0.0037 **92.** 0.06234 **93.** 1×10^4
94. 5.61×10^7 **95.** 5.4×10^{-5} **96.** 1×10^{-3}
97. 24,000,000 **98.** 0.2 **99.** $\dfrac{5}{3}x + 1$
100. $3b^2 - 2 + \dfrac{1}{b^2}$ **101.** $3x + 2 + \dfrac{4}{x-1}$
102. $3x - 4 + \dfrac{6}{3x+2}$ **103.** $x^2 - 3x - 1$
104. $x^2 + \dfrac{-1}{2x-1}$ **105.** $x - 1 + \dfrac{-2x+2}{x^2+1}$
106. $x^2 + 2x + 5$
107. (a) 60 bpm **(b)** 160 bpm **(c)** It increases.

108. $6xy$; 72 ft^2 **109.** $10z^2 + 25z$; 510 in^2 **110.** x^4y^2
111. $\$833.71$
112. (a) $2x^2 + 10x$ **(b)** $x^2 + 7x + 10$ **(c)** $2x^2 + 4x$
(d) $10x^2 + 42x + 20$
113. $\frac{4}{3}\pi x^3 + 8\pi x^2 + 16\pi x + \frac{32}{3}\pi$
114. (a) $96t - 16t^2$ **(b)** 128; after 2 sec the ball is 128 ft
high.
115. (a) $600L - L^2$ **(b)** $27{,}500$; a rectangular building
with a perimeter of 1200 ft and a side of length 50 ft has an
area of 27,500 ft^2.
116. (a) $x^2 + 10x + 25$ **(b)** $x^2 + 10x + 25$
117. (a) $16x$ **(b)** 1600
118. About 6.43 billion   **119.** About $\$8795$ per person
120. About $466{,}310{,}000$ gal or 4.6631×10^8 gal

Chapter 5 Test (pp. 349–350)

1. Yes; 3; 2; 3   **2.** $x^3 - 4x + 8$ **3.** $4x + 6$
4. $x^2 + x - 7$ **5.** $4a^3 + 2ab$
6. (a) -6 **(b)** $\frac{1}{64}$ **(c)** 8 **(d)** -3
7. x^4 **8.** $\dfrac{a^3}{b^6}$ **9.** $a^3b - ab^3$ **10.** $\dfrac{4}{9a^4b^6}$ **11.** $\dfrac{2y^3}{x}$
12. $12x^5 - 18x^3 + 3x^2$ **13.** $2z^2 - 2z - 12$
14. $49y^4 - 9$ **15.** $9x^2 - 12x + 4$
16. $m^3 + 9m^2 + 27m + 27$ **17.** 0.0061
18. 5.41×10^3 **19.** $3x - 2 + \frac{1}{x}$
20. $x^2 - x + 1 + \frac{-1}{x + 2}$
21. (a) $20t$ **(b)** $2t + 2000$ **(c)** $18t - 2000$; profit
from selling t tickets
22. $12x^2$; 1200 ft^2 **23.** $3x^3 + 27x^2 + 54x$
24. (a) $88t - 16t^2$ **(b)** 120; after 3 sec the ball is 120 ft
high.

CHAPTER 6: FACTORING POLYNOMIALS AND SOLVING EQUATIONS

Section 6.1 (pp. 357–358)

1. factoring   **3.** greatest common factor (GCF)
5. $1, 2, x, 2x$
7. $2(x + 2)$;

9. $z(z + 4)$;

11. $6x$; $6x(1 - 3x)$ **13.** $4y^2$; $4y^2(2y - 3)$
15. $3z$; $3z(2z^2 + z + 3)$ **17.** x^2; $x^2(x^2 - 5x - 4)$
19. $5y^2$; $5y^2(y^3 + 2y^2 - 3y + 2)$ **21.** x; $x(y + z)$
23. ab; $ab(b - a)$ **25.** $5x^2y^3$; $5x^2y^3(y + 2x)$
27. ab; $ab(a + b + 1)$ **29.** $(x - 2)(x + 1)$
31. $(z + 4)(z + 5)$ **33.** $y(y - 4)(y + 7)$

35. $(4x^3 + 1)(x - 5)$ **37.** $(x^2 + 3)(x + 2)$
39. $(y^2 + 1)(2y + 1)$ **41.** $(2z^2 + 5)(z - 3)$
43. $(4t^2 + 3)(t - 5)$ **45.** $(3r^2 - 2)(3r + 2)$
47. $(7x^2 - 2)(x + 3)$ **49.** $(y^2 - 2)(2y - 7)$
51. $(z^2 - 7)(z - 4)$ **53.** $(x^3 + 2)(2x - 3)$
55. $(x + y)(a + b)$ **57. (a)** $16t$ **(b)** $16t(5 - t)$
59. (a) 168 in^3 **(b)** $4x(x^2 - 15x + 50)$

Section 6.2 (pp. 363–364)

1. F: multiply the *first* terms.
   O: multiply the *outside* terms.
   I: multiply the *inside* terms.
   L: multiply the *last* terms.
3. mn; $m + n$ **5.** 1, 12; 2, 6; 3, 4   **7.** 4, 7   **9.** 3, 10
11. $-5, 10$ **13.** $-7, -4$ **15.** $(x + 1)(x + 2)$
17. $(y + 2)(y + 2)$ **19.** $(z + 3)(z + 3)$
21. $(x + 3)(x + 5)$ **23.** $(m + 4)(m + 9)$
25. $(n + 10)(n + 10)$ **27.** $(x - 1)(x - 5)$
29. $(y - 3)(y - 4)$ **31.** $(z - 5)(z - 8)$
33. $(a - 7)(a - 9)$ **35.** $(b - 5)(b - 25)$
37. $(x - 5)(x + 18)$ **39.** $(m - 5)(m + 9)$
41. $(n - 10)(n + 20)$ **43.** $(x - 1)(x + 23)$
45. $(a - 4)(a + 8)$ **47.** $(b + 4)(b - 5)$
49. $(x + 8)(x - 9)$ **51.** $(y + 2)(y - 17)$
53. $(z + 6)(z - 11)$ **55.** $(x + 1)(x + 5)$
57. $(x - 1)(x - 3)$ **59.** $(6 - x)(2 + x)$
61. $(8 + x)(4 - x)$
63. $x + 1$;

65. $x + 1, x + 2$;

67. $x + 1$ **69.** $(x + 2)(x + 6)$

Checking Basic Concepts 6.1 & 6.2 (p. 364)

1. $4x$ **2.** $6z^2(2z - 3)$
3. (a) $(6y + 5)(y - 2)$ **(b)** $(x^2 + 5)(2x + 1)$
4. (a) $(x + 2)(x + 4)$ **(b)** $(x - 7)(x + 6)$
5. $(x + 5)(x + 5)$

Section 6.3 (pp. 369–370)

1. ac; b **3.** $+$; $+$ **5.** $-$; $-$ **7.** 3; x **9.** 1; $2x$
11. $(x + 3)(2x + 1)$ **13.** $(x + 1)(3x + 1)$
15. $(2x + 3)(3x + 1)$ **17.** $(x - 2)(5x - 1)$
19. $(y - 1)(2y - 5)$ **21.** $(z - 5)(7z - 2)$

23. $(t - 3)(3t + 2)$ **25.** $(3r + 2)(5r - 3)$
27. $(3m - 4)(8m + 3)$ **29.** $(5x - 1)(5x + 2)$
31. $(x + 2)(6x - 1)$ **33.** $(3n + 1)(7n - 1)$
35. $(2y + 3)(7y + 1)$ **37.** $(4z - 3)(7z - 1)$
39. $(3x - 2)(10x - 3)$ **41.** $(2t + 3)(9t - 2)$
43. $(7x + 1)(x + 2)$ **45.** $(2x - 1)(x - 2)$
47. $-(4x + 3)(2x - 1)$ **49.** $-(x + 5)(2x - 3)$
51. $-(x - 3)(5x + 1)$
53. $3x + 2$ by $2x + 1$;

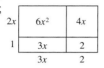

55. $(2x + 1)(x + 3)$

Section 6.4 (pp. 375–376)

1. $(a - b)(a + b)$ **3.** $6x$; $7y$ **5.** $(a - b)^2$ **7.** $20rt$
9. $(a - b)(a^2 + ab + b^2)$ **11.** $-$; $+$
13. $(x - 1)(x + 1)$ **15.** $(z - 10)(z + 10)$
17. $(2y - 1)(2y + 1)$ **19.** $(6z - 5)(6z + 5)$
21. $(3 - x)(3 + x)$ **23.** $(1 - 3y)(1 + 3y)$
25. $(2a - 3b)(2a + 3b)$ **27.** $(6m - 5n)(6m + 5n)$
29. $(9r - 7t)(9r + 7t)$ **31.** $(x + 4)^2$
33. Not possible   **35.** $(x - 3)^2$ **37.** $(3y + 1)^2$
39. $(2z - 1)^2$ **41.** Not possible   **43.** $(3x + 5)^2$
45. $(2a - 9)^2$ **47.** $(x + y)^2$ **49.** $(r - 5t)^2$
51. Not possible   **53.** $(z + 1)(z^2 - z + 1)$
55. $(x + 4)(x^2 - 4x + 16)$ **57.** $(y - 2)(y^2 + 2y + 4)$
59. $(n - 1)(n^2 + n + 1)$
61. $(2x + 1)(4x^2 - 2x + 1)$
63. $(m - 4n)(m^2 + 4mn + 16n^2)$
65. $2(2a + b)(4a^2 - 2ab + b^2)$
67. $(2x + 5y)(4x^2 - 10xy + 25y^2)$
69. $4(5r - 2t)(25r^2 + 10rt + 4t^2)$ **71.** $3x^3(3x - 1)$
73. $(2y - 3)(2y + 3)$ **75.** $(5z - 8)^2$
77. $(x - 7)(2x - 1)$ **79.** $(n + 9)(2n + 3)$
81. $(2x - 3)(4x^2 + 6x + 9)$ **83.** $(x^2 - 5)(x + 2)$
85. $(a + 9b)^2$
87. $2x + 3$;

Checking Basic Concepts 6.3 & 6.4 (p. 376)

1. (a) $(x - 4)(2x + 3)$ **(b)** $(2x + 7)(3x - 2)$
2. $(3x + 2)(x + 3)$
3. (a) $(z - 8)(z + 8)$ **(b)** $(3r - 2t)(3r + 2t)$
(c) $(x + 6)^2$ **(d)** $(3a - 2b)^2$
4. (a) $(m - 3)(m^2 + 3m + 9)$
(b) $(5n + 3)(25n^2 - 15n + 9)$

Section 6.5 (pp. 382–384)

1. 0; 0 **3.** $2x = 0$; $x + 6 = 0$
5. Apply the zero-product property.
7. $ax^2 + bx + c = 0$ with $a \neq 0$ **9.** $x = 0$ or $y = 0$
11. $-8, 0$ **13.** $1, 2$ **15.** $\frac{1}{2}, \frac{3}{4}$ **17.** $\frac{1}{3}, \frac{3}{7}$ **19.** $0, 5, 8$
21. $0, 1$ **23.** $0, 5$ **25.** $-\frac{3}{2}, 0$ **27.** $-1, 1$ **29.** $-\frac{1}{2}, \frac{1}{2}$
31. $-2, -1$ **33.** $5, 7$ **35.** $-2, \frac{1}{2}$ **37.** $-\frac{5}{2}, -\frac{2}{3}$
39. $-5, 5$ **41.** $0, 5$ **43.** $-3, 0$ **45.** $-1, 6$ **47.** $-\frac{5}{3}, \frac{3}{4}$
49. $-2, 1$ **51.** $-3, \frac{1}{2}$ **53.** 12 ft   **55.** 4 in.   **57.** 4
59. (a) 6 sec
(b)

Time (t)	0	1	2	3	4	5	6
Height (h)	0	80	128	144	128	80	0

; 3 sec

61. (a) 81.8 ft; 327.3 ft; when the speed doubles, the braking distance quadruples.   **(b)** About 19 mph   **(c)** About 19 mph; yes
63. (a) About 11 million; about 66 million   **(b)** About 1981
65. 40 by 50 pixels

Section 6.6 (pp. 389–390)

1. GCF   **3.** $(z^2 + 1)(z^2 + 2)$
5. Subtract x from each side.   **7.** $5(x - 3)(x + 2)$
9. $-4(y + 2)(y + 6)$ **11.** $-10(z + 5)(2z + 1)$
13. $-4(t + 3)(7t - 5)$ **15.** $r(r - 1)(r + 1)$
17. $3x(x - 2)(x + 3)$ **19.** $12z(2z - 1)(3z + 2)$
21. $x^2(x - 2)(x + 2)$ **23.** $t^2(t - 1)(t + 2)$
25. $(x^2 - 3)(x^2 - 2)$ **27.** $(x^2 + 3)(2x^2 + 1)$
29. $(y^2 + 3)^2$ **31.** $(x^2 - 3)(x^2 + 3)$
33. $(x - 3)(x + 3)(x^2 + 9)$ **35.** $z^3(z + 1)^2$
37. $(x + y)(2x - y)$ **39.** $(a + b)^2(a - b)^2$
41. $x(x + y)(x - y)$ **43.** $x(2x + y)^2$
45. (a) $x(x - 2)(x + 2)$ **(b)** $-2, 0, 2$
47. (a) $2y(y - 6)(y + 3)$ **(b)** $-3, 0, 6$
49. $-8, -3$ **51.** $-1, 3$ **53.** $-1, 0, 4$ **55.** $-6, 0, 4$
57. $-6, 0, 6$ **59.** $-7, 0, 1$ **61.** $-3, -2, 2, 3$
63. $-1, 1$ **65.** $-3, 3$ **67.** $-1, 1, 2$ **69.** 5
71. (a) $x < 7.5$ in. because the width is 15 in.
(b) $300 - 4x^2$ **(c)** 2.5 in.
73. 18.6 trillion ft^3
75. Factoring is very difficult. (*answers may vary*)

Checking Basic Concepts 6.5 & 6.6 (p. 390)

1. (a) $0, \frac{3}{2}$ **(b)** $-1, \frac{3}{5}$ **2.** $-3, 1$ **3.** After $\frac{11}{2}$ sec
4. (a) $(x - 2)^2(x + 2)^2$ **(b)** $y(y + 10)(2y - 3)$
(c) $(x - 2y)(x + 2y)(x^2 + 4y^2)$
5. $-4, 0, 3$

CHAPTER 6 REVIEW EXERCISES (PP. 394–395)

1. $4z^2$; $4z^2(2z - 1)$ 2. $3x^2$; $3x^2(2x^2 + x - 4)$
3. $3y$; $3y(3x + 5z^2)$ 4. a^2b^2; $a^2b^2(b + a)$
5. $(x - 3)(x + 2)$ 6. $y(y + 3)(x - 5)$
7. $(z^2 + 5)(z - 2)$ 8. $(t^2 + 8)(t + 1)$
9. $(x^2 + 6)(x - 3)$ 10. $(x - y)(a + b)$ 11. 4, 5
12. $-3, 7$ 13. $-9, -4$ 14. $-25, 4$
15. $(x - 4)(x + 3)$ 16. $(x + 4)(x + 6)$
17. $(x - 2)(x + 8)$ 18. $(x - 7)(x + 6)$
19. $(x - 1)(x + 3)$ 20. $(x + 10)(x + 12)$
21. $(2 - x)(5 - x)$ 22. $(6 - x)(4 + x)$
23. $(3x - 1)(3x + 2)$ 24. $(x - 1)(2x + 5)$
25. $(x + 3)(3x + 5)$ 26. $(5x - 1)(7x + 1)$
27. $(3x + 1)(8x - 5)$ 28. $(x + 9)(4x - 3)$
29. $(3 - 2x)(4 + x)$ 30. $(1 - 2x)(1 + 5x)$
31. $(z - 2)(z + 2)$ 32. $(3z - 8)(3z + 8)$
33. $(6 - y)(6 + y)$ 34. $(10a - 9b)(10a + 9b)$
35. $(x + 7)^2$ 36. $(x - 5)^2$ 37. $(2x - 3)^2$
38. $(3x + 8)^2$ 39. $(2t - 1)(4t^2 + 2t + 1)$
40. $(3r + 2t)(9r^2 - 6rt + 4t^2)$ 41. $m = 0$ or $n = 0$
42. 0   43. $-9, \frac{3}{4}$ 44. $-\frac{6}{5}, \frac{1}{4}$ 45. 0, 1, 2   46. 0, 7
47. $-8, 8$ 48. $-7, -2$ 49. $-2, 3$ 50. $-\frac{3}{2}, \frac{2}{5}$
51. $-4, 18$ 52. $-2, \frac{5}{2}$ 53. $5(x - 5)(x + 2)$
54. $-3(x - 3)(x + 5)$ 55. $y(y - 2)(y + 2)$
56. $3y(y - 1)(y + 3)$ 57. $2z^2(z + 2)(z + 5)$
58. $8z^2(z - 2)(z + 2)$ 59. $(x^2 - 3)^2$
60. $(x - 3)(x + 3)(2x^2 + 3)$ 61. $(a + 5b)^2$
62. $x(x + y)(x - y)$ 63. $-\frac{1}{2}, 5$ 64. $0, \frac{5}{2}, 3$
65. $-5, 0, 5$ 66. 0, 3, 4   67. $-2, 2$ 68. $-4, 4$
69. -4 70. $-1, 1$
71. $3x + 7$;

	$3x$	7
$3x$	$9x^2$	$21x$
7	$21x$	49

72. $x + 1$ by $x + 5$;

	x	5
x	x^2	$5x$
1	x	5

73. $x + 1$ 74. $(x + 1)(x + 3)$ 75. $(2x + 3)(x + 6)$
76. 2   77. 8 by 15 ft   78. 2 sec and 3 sec
79. (a) 390 ft   (b) 15 mph   (c) 15 mph; yes
80. (a) $10,000   (b) $50 or $150   (c) $50 or $150; yes
81. (a) 372 million   (b) 1975   82. 50 by 80 pixels
83. (a) $2000 - 4x^2$ (b) 5 in.

CHAPTER 6 TEST (P. 396)

1. $4xy$; $4xy(x - 5y + 3)$ 2. $3a^2b^2$; $3a^2b^2(3a + 1)$
3. $(y + z)(a + b)$ 4. $(x^2 - 5)(3x + 1)$
5. $(y - 2)(y + 6)$ 6. $(2x + 5)^2$

7. $(z - 4)(4z - 3)$ 8. $(t - 7)(2t - 3)$ 9. $-4, 4$
10. $-4, 5$ 11. $\frac{4}{3}$ 12. $-6, 11$ 13. $3x(x + 1)(2x - 1)$
14. $2(z - 3)(z + 3)(z^2 + 3)$ 15. $-3, 0, 3$
16. $-2, -1, 1, 2$ 17. $3x + 5$ 18. $(x + 2)(x + 3)$
19. (a) 275 ft   (b) 33 mph   20. 1 sec and 2 sec

CHAPTERS 1–6 CUMULATIVE REVIEW (PP. 397–401)

1. $2 \times 2 \times 2 \times 2 \times 3 \times 3$ 2. 16   3. $2n + 5$
4. $-\frac{5}{8}$ 5. $\frac{3}{7}$ 6. $\frac{2}{5}$ 7. $\frac{3}{4}$ 8. $\frac{7}{10}$ 9. 17   10. $-\frac{7}{2}$
11.

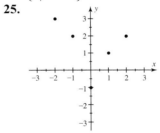

12. $\frac{2}{9}$ 13. $2x + 2$ 14. $y - 11$ 15. 8
16. 1;

x	-2	-1	0	1	2
$2x + 3$	-1	1	3	5	7

17. $3n - 5 = n - 7$; -1 18. $\frac{57}{1000}$; 0.057   19. 12.3%
20. 2.5 hr   21. $\frac{25}{4}\pi$ ft^2 22. $W = \frac{P - 2L}{2}$
23.

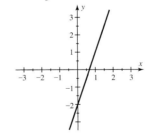

24. $\{z \mid z > 2\}$
25.

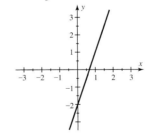

26. x-int: $\frac{2}{3}$; y-int: -2

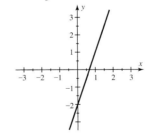

27. x-int: 3; y-int: -2

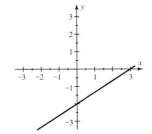

28. x-int: 1

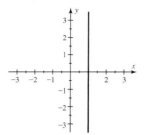

29. y-int: -2

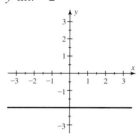

30. $-1; -2; y = -2x - 2$
31. $y = 3x + 5$

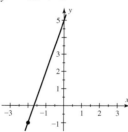

32. $y = 2x - 1$ **33.** $y = -\frac{2}{3}x + \frac{1}{3}$ **34.** $y = -\frac{3}{2}x + \frac{7}{2}$
35. $y = \frac{4}{3}x + \frac{11}{3}$ **36.** $N = 200x + 2000$ **37.** $(1, -2)$
38. $(1, 2)$ **39.** $(1, 2)$ **40.** $(1, -1)$ **41.** $(-2, 5)$
42. $(2, -8)$ **43. (a)** One   **(b)** Consistent; independent
44. (a) Zero   **(b)** Inconsistent
45. Infinitely many solutions
46. No solutions   **47.** $(1, 2)$
48. $(0, 0)$; (*answers may vary*)

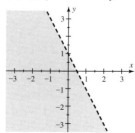

49.

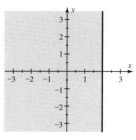

50.

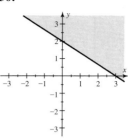

51.

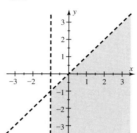

52.

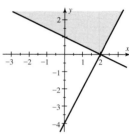

53. $4x^2 + 1$ **54.** $7xy - x^2$ **55.** -16 **56.** 1
57. $15z^8$ **58.** a^3b^6 **59.** a^3b^3 **60.** $\dfrac{x^{10}}{y^4}$
61. $-14x^5 + 21x^4$ **62.** $a^3 - b^3$
63. $21x^2 + 29x - 10$ **64.** $10y^4 - 11y^2 - 6$
65. $4r^2t^2 - 9$ **66.** $4t^2 + 20t + 25$
67. $x^2 - 14x + 49$ **68.** $x^4 - 2x^2y^2 + y^4$ **69.** 2
70. 9   **71.** $\dfrac{1}{a^2}$ **72.** $\dfrac{1}{4t^6}$ **73.** $\dfrac{1}{x^2y^5}$ **74.** $\dfrac{2}{x^2}$
75. $32x^5y^{10}$ **76.** $83{,}000$ **77.** 0.00623
78. 5.43×10^5 **79.** 1.23×10^{-3} **80.** $2x^2 + 4x$
81. $2x + 1 + \dfrac{4}{x - 1}$ **82.** $3x + \dfrac{-4x + 1}{x^2 + 1}$
83. $5z^2$; $5z^2(4z - 3)$ **84.** $3xy$; $3xy(4x + 5y)$
85. $(2y^2 - 5)(x + 2)$ **86.** $(t^2 + 1)(t + 6)$
87. $(x - 4)(x + 7)$ **88.** $(2y + 3)(3y - 4)$
89. $(3 - x)(2 + 5x)$ **90.** $(3z - 2)(3z + 2)$
91. $(5x - 2y)(5x + 2y)$ **92.** $(t + 8)^2$ **93.** $(8x - 1)^2$
94. $(3t - 2)(9t^2 + 6t + 4)$ **95.** $-4(x - 3)(x + 2)$
96. $x(x - 4)(x + 4)$ **97.** $x(x - 9)(x + 11)$
98. $(x - 3)(x + 3)(x^2 - 3)$ **99.** $(a + 3b)^2$
100. $x^2y(x - y)$ **101.** $-5, \frac{7}{2}$ **102.** $0, 2$ **103.** $-7, 2$
104. $-5, 0, 5$ **105.** $-2, 1$ **106.** $-\frac{7}{2}, 0, \frac{7}{2}$ **107.** $-3, 3$
108. (a) $10x + 8x = 18x$ **(b)** 50 min
109. $6\frac{11}{12}$ mi   **110.** $\$21{,}000$
111. 25 min skiing; 35 min running
112. (a) 24.1   **(b)** Participation in Head Start increased by 24.1 thousand children per year, on average.   **(c)** 1099 thousand or about 1.1 million
113. (a) $C = 0.25x + 20$ **(b)** 320 mi

114. **(a)** $2x + y = 180$; $2x - y = 20$ **(b)** $(50, 80)$, the angles are $50°$, $50°$, and $80°$.

115.

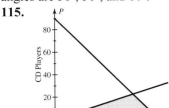

116. **(a)** 21.6; in 1995 natural gas consumption was about 21.6 trillion ft^3.   **(b)** 2000

117. $18xy$; 108 yd^2 **118.** $8x^3y^6$

119. **(a)** $x^2 + 4x + 4$
(b) $x^2 + 2x + 2x + 4 = x^2 + 4x + 4$

120. $x + 6$ **121.** **(a)** 4 sec   **(b)** 1 sec and 3 sec

CHAPTER 7: RATIONAL EXPRESSIONS

SECTION 7.1 (PP. 409–411)

1. $\frac{P}{Q}$; polynomials   **3.** the denominator equals zero

5. $\frac{a}{b}$ **7.** -1 **9.** $-\frac{3}{7}$ **11.** $-\frac{4}{9}$ **13.** $-\frac{1}{4}$

15. Undefined   **17.** Undefined   **19.** 1

21.

x	-2	-1	0	1	2
$\frac{x}{x+1}$	2	—	0	$\frac{1}{2}$	$\frac{2}{3}$

23.

x	-2	-1	0	1	2
$\frac{3x}{2x^2+1}$	$-\frac{2}{3}$	-1	0	1	$\frac{2}{3}$

25. 0 **27.** 3 **29.** $-\frac{4}{5}$ **31.** None   **33.** $-5, 5$

35. $-3, -2$ **37.** $1, \frac{5}{2}$ **39.** $\frac{2}{3}$ **41.** $\frac{1}{2}$ **43.** $-\frac{2}{5}$

45. $-\frac{1}{3}$ **47.** $\frac{1}{2x^2}$ **49.** $\frac{4y}{3x}$ **51.** $\frac{1}{2}$ **53.** $\frac{3}{5}$ **55.** $\frac{x+1}{x+6}$

57. $\frac{5y+3}{y+2}$ **59.** -1 **61.** -1 **63.** $-\frac{3x+5}{3x-5}$

65. $\frac{n-1}{n-5}$ **67.** $\frac{x}{6}$ **69.** $\frac{z-2}{z-3}$ **71.** $\frac{x+4}{3x+2}$

73. $\frac{1}{3x-2}$ **75.** 1 **77.** $-\frac{1}{2}$

79. **(a)** $\frac{1}{2}$; when traffic arrives at a rate of 3 vehicles per minute, the average wait is $\frac{1}{2}$ minute.

(b)

x	2	4	4.5	4.9	4.99
T	$\frac{1}{3}$	1	2	10	100

As x nears 5 vehicles per minute, a small increase in x increases the wait dramatically.

81. $\frac{1}{2}$

83. **(a)** $\frac{3}{n}$ **(b)** $\frac{n-3}{n}$; $\frac{97}{100}$; there is a 97% chance that a winning ball will not be drawn.

85. **(a)** 6 hr   **(b)** $\frac{M}{60}$

87. **(a)** $x = 5$ **(b)** As the average rate nears 5 cars per minute, a small increase in x increases the wait dramatically.

SECTION 7.2 (PP. 416–417)

1. $\frac{AC}{BD}$ **3.** $\frac{x+7}{x+1}$ **5.** No; it is equal to $1 + \frac{2}{x}$.

7. $\frac{2}{5}$ **9.** $\frac{12}{7}$ **11.** $\frac{2}{11}$ **13.** 4 **15.** $\frac{8}{15}$ **17.** 10 **19.** 1

21. $\frac{z+1}{z+4}$ **23.** $\frac{2}{3}$ **25.** $\frac{x+2}{x-2}$ **27.** $\frac{8(x+1)}{x^2}$

29. $\frac{x-3}{x}$ **31.** $\frac{z+3}{z-7}$ **33.** 1 **35.** $(t+1)(t+2)$

37. $\frac{x(x+4)}{x^2+4}$ **39.** $\frac{z-1}{z+2}$ **41.** $\frac{y}{y-1}$ **43.** $\frac{x+1}{x-3}$

45. $\frac{x-3}{x-1}$ **47.** $\frac{2}{2x+3}$ **49.** -2 **51.** $\frac{z-1}{z+1}$

53. $\frac{y+2}{2y+1}$ **55.** $\frac{4(t-1)}{t^2+1}$ **57.** $\frac{y-3}{y-5}$ **59.** $2x$

61. $\frac{2z+1}{z+5}$ **63.** $\frac{1}{t+6}$ **65.** $\frac{2a+3b}{a+b}$ **67.** $x - y$

69. **(a)** $\frac{1}{n+1}$ **(b)** $\frac{1}{100}$

CHECKING BASIC CONCEPTS 7.1 & 7.2 (P. 417)

1. Undefined; $\frac{3}{8}$ **2.** **(a)** $\frac{2x}{5y}$ **(b)** 5 **(c)** $\frac{x+2}{x+4}$

3. **(a)** $\frac{4}{9}$ **(b)** $\frac{2}{x-1}$ **4.** **(a)** $\frac{5z}{6}$ **(b)** $x + 1$

5. **(a)**

x	0.5	1.0	1.5	1.9
T	$\frac{2}{3}$	1	2	10

(b) As x nears 2 persons per minute, a small increase in x increases the wait dramatically.

SECTION 7.3 (PP. 423–424)

1. add; numerators; denominators   **3.** $\frac{A+B}{C}$ **5.** $4 - x$

7. $x + 5$ **9.** 1 **11.** $\frac{6}{5}$ **13.** 1 **15.** $\frac{3}{7}$ **17.** $\frac{1}{2}$

19. $\frac{1}{2}$ **21.** $\frac{2}{3}$ **23.** $\frac{3}{x}$ **25.** $\frac{1}{2}$ **27.** 3 **29.** $\frac{4z}{4z+3}$

31. 2 **33.** 1 **35.** 2 **37.** $y - 1$ **39.** $\frac{x}{2}$ **41.** $z - 2$

43. $x - 3$ **45.** $\frac{7n}{2n^2-n+5}$ **47.** $\frac{6}{x+3}$

49. 1 **51.** $a - b$ **53.** $2x - 3y$

55. **(a)** $\frac{14}{n+1}$ **(b)** $\frac{7}{50}$; there are 7 chances out of 50 chances that a defective battery is chosen.

SECTION 7.4 (PP. 431–433)

1. Examples include 36 and 54; answers may vary.

3. $\frac{3}{3}$ **5.** 12 **7.** 12 **9.** 6 **11.** 30 **13.** 72 **15.** $12x$

17. $10x^2$ **19.** $x(x+1)$ **21.** $(2x+1)(x+3)$

23. $36x^3$ **25.** $x(x-1)(x+1)$ **27.** $(x+1)^2$

29. $(2x-1)^3(x+3)$ **31.** $(2x-1)(2x+1)$

33. $(x-1)(x+1)$ **35.** $4x(x^2+4)$

37. $(2x+3)(x+2)(x+3)$ **39.** $\frac{3}{9}$ **41.** $\frac{15}{21}$ **43.** $\frac{2x^2}{8x^3}$

45. $\frac{x-2}{x^2-4}$ **47.** $\frac{x}{x^2+x}$ **49.** $\frac{2x^2+2x}{x^2+2x+1}$ **51.** $\frac{13}{10}$

53. $\frac{2}{9}$ **55.** $\frac{13}{12x}$ **57.** $\frac{5z-7}{z^3}$ **59.** $\frac{y-x}{xy}$ **61.** $\frac{a^2+b^2}{ab}$

63. $\frac{7}{2(x+2)}$ **65.** $\frac{t+2}{t(t-2)}$ **67.** $\frac{n^2+4n+5}{(n-1)(n+1)}$

69. $-\frac{3}{x-3}$ **71.** 0   **73.** $\frac{6x-4}{(x-1)^2}$ **75.** $\frac{3}{2y-1}$

77. $\frac{x-1}{x(x+2)}$ **79.** $\frac{2x+2}{(x+2)^2}$

81. $\frac{x^2+2x-1}{(x+1)(x+2)(x+3)}$ **83.** $-\frac{2b}{(a+b)(a-b)}$

85. 0   **87.** $\frac{13}{12a}$ **89.** 0   **91.** $\frac{3}{x}$ **93.** $\frac{1}{75}; \frac{1}{75}$; yes

95. $\frac{D-F}{FD}$

CHECKING BASIC CONCEPTS 7.3 & 7.4 (P. 433)

1. (a) 1   (b) $\frac{2-x}{3x}$ (c) z
2. (a) $15x$ (b) $4x(x+1)$ (c) $(x+1)(x-1)$
3. (a) $\frac{6x+5}{x(x+1)}$ (b) $\frac{4}{x-3}$ (c) $-\frac{x+4}{2x+1}$

4. $\frac{a^2+b^2}{(a-b)(a+b)}$

SECTION 7.5 (PP. 440–442)

1. $\frac{1}{2}\cdot\frac{4}{3}=\frac{2}{3}$ **3.** fractions   **5.** Division   **7.** $\frac{\frac{a}{b}}{\frac{c}{d}}$

9. 30   **11.** $(x-1)(x+1)$ **13.** $x(2x-1)(2x+1)$

15. $\frac{4}{5}$ **17.** $\frac{1}{2}$ **19.** $\frac{3y}{x}$ **21.** 3   **23.** $\frac{p+1}{p+2}$ **25.** $\frac{5}{z}$

27. $\frac{y}{y-3}$ **29.** 1   **31.** $\frac{n+m}{n-m}$ **33.** $\frac{2x+y}{2x-y}$ **35.** $\frac{1+b}{1-a}$

37. $\frac{q+2}{q}$ **39.** $2x+3$ **41.** $\frac{2x}{(x+1)(2x-1)(2x+1)}$

43. $\frac{1}{ab}$ **45.** $\frac{ab}{a+b}$ **47.** $\frac{P(1+\frac{r}{26})^{52}-P}{\frac{r}{26}}$

49. $R=\frac{ST}{S+T}$

SECTION 7.6 (PP. 450–453)

1. rational   **3.** $ad=bc$; b; d **5.** $12x$ **7.** $\frac{3}{2}$ **9.** $\frac{5}{2}$

11. $\frac{7}{6}$ **13.** 0   **15.** $\frac{1}{4}$ **17.** $\frac{10}{11}$ **19.** 3   **21.** 25   **23.** 5

25. $\frac{5}{16}$ **27.** 1   **29.** $\frac{1}{2}$ **31.** -1 **33.** $-\frac{3}{2}, -1$ **35.** 4

37. 4   **39.** -6 **41.** 6   **43.** $\frac{7}{12}$ **45.** $\frac{1}{6}, 5$ **47.** 3

49. 3   **51.** No solutions   **53.** No solutions   **55.** -2

57. (a) $-3, 1$ (b) $-3, 1$ **59.** (a) -1 (b) -1

61. (a) 2   (b) 2   **63.** (a) $-2, 2$ (b) $-2, 2$

65. $a=\frac{F}{m}$ **67.** $r=\frac{V}{I}-R$ **69.** $b=\frac{2A}{h}$

71. $z=\frac{15}{k-3}$ **73.** $b=\frac{aT}{a-T}$ **75.** $x=\frac{ky}{3y+2k}$

77. 9 cars per minute   **79.** $\frac{12}{7}\approx 1.7$ hours

81. $\frac{8}{3}\approx 2.7$ days   **83.** 18 mph; 20 mph

85. (a) 120; the braking distance is 120 feet when the slope of the road is -0.05.   (b) -0.1

87. 15 mph   **89.** 200 mph

91. 10 mph running; 5 mph walking   **93.** 3   **95.** 32

CHECKING BASIC CONCEPTS 7.5 & 7.6 (PP. 453–454)

1. (a) $\frac{5}{6}$ (b) $\frac{1}{9x^2}$ (c) $\frac{b-a}{b+a}$ (d) $\frac{r+t}{2rt}$

2. (a) $\frac{1}{5}$ (b) -4 **3.** (a) 2   (b) $-2, \frac{1}{2}$

4. (a) $x=\frac{2(b+3y)}{a}$ (b) $m=\frac{k}{2k-1}$

5. (a) 300; when the slope of the hill is 0.1, the braking distance is 300 feet.   (b) 0.3; the braking distance is 200 feet when the slope of the road is 0.3.

SECTION 7.7 (PP. 462–466)

1. A statement that two ratios are equal   **3.** It doubles.
5. constant   **7.** kxy
9. Directly; if the number being fed doubles, the bill will double.

11. 10   **13.** 12   **15.** 72   **17.** $\frac{27}{4}$

19. (a) $\frac{7}{9}=\frac{10}{x}$ (b) $\frac{90}{7}$ **21.** (a) $\frac{3}{5}=\frac{6}{x}$ (b) 10

23. (a) $\frac{78}{6}=\frac{x}{8}$ (b) \$104

25. (a) $\frac{2}{90}=\frac{5}{x}$ (b) 225 minutes   **27.** (a) 2   (b) 14

29. (a) $\frac{5}{2}$ (b) $\frac{35}{2}$ **31.** (a) $-\frac{15}{2}$ (b) $-\frac{105}{2}$

33. (a) 20   (b) 2   **35.** (a) 50   (b) 5

37. (a) 400   (b) 40   **39.** (a) $\frac{1}{4}$ (b) $\frac{35}{4}$

41. (a) 11   (b) 385   **43.** (a) 10   (b) 350

45. (a) Direct   (b) $y=\frac{3}{2}x$
(c)

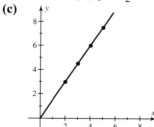

47. (a) Neither   (b) NA   (c) NA
49. (a) Neither   (b) NA   (c) NA
51. Direct; 1   **53.** Neither   **55.** Direct; 2
57. 40.8 minutes   **59.** 1.375 inches

61. (a) Direct; the ratios $\frac{R}{W}$ always equal 0.012.
(b) $R=0.012W$

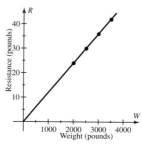

(c) 38.4 pounds

63. (a) Direct; the ratios $\frac{G}{A}$ always equal 27.

(b) $G = 27A$

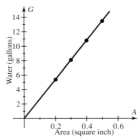

(c) For each square-inch increase in the cross-sectional area of the hose, the flow increases by 27 gallons per minute.

65. (a) $F = \frac{1200}{L}$ **(b)** 60 pounds

67. (a) Direct   **(b)** $y = -29x$ **(c)** Negative; for each 1-mile increase in altitude the temperature decreases by 29°F.   **(d)** 101.5°F decrease

69. About 1.43 ohms   **71.** $z = 0.5104x^2y^3$

73. About 133 pounds

Checking Basic Concepts 7.7 (p. 466)

1. (a) $\frac{14}{13}$ **(b)** $\frac{15}{2}$

2. (a) $\frac{8}{12} = \frac{6}{x}$; 9   **(b)** $\frac{2}{148} = \frac{7}{x}$; 518 minutes   **3.** 60; 4

4. (a) Direct; the ratios $\frac{y}{x}$ always equal $\frac{3}{2}$; $\frac{3}{2}$.

(b) Inverse; the products xy always equal 24; 24.

5. $104

Chapter 7 Review Exercises (pp. 470–473)

1. $-\frac{3}{5}$ **2.** -3 **3.** Undefined   **4.** Undefined

5.

x	-2	-1	0	1	2
$\frac{3x}{x-1}$	2	$\frac{3}{2}$	0	—	6

6. $-2, 2$ **7.** $\frac{5y^3}{3x^2}$ **8.** $x - 6$ **9.** -1 **10.** $\frac{x-5}{5}$

11. $\frac{x+3}{x+1}$ **12.** $\frac{x+4}{x+1}$ **13.** 2   **14.** $\frac{1}{x+5}$ **15.** $\frac{z+3}{z-4}$

16. $\frac{x}{x-2}$ **17.** $\frac{5}{6}$ **18.** $\frac{8}{x(x+1)}$ **19.** $\frac{1}{2}$ **20.** 1

21. $x + y$ **22.** $2(a^2 + ab + b^2)$ **23.** $\frac{10}{x+10}$

24. $\frac{1}{x-1}$ **25.** 1   **26.** 1   **27.** $\frac{1}{x+1}$ **28.** $\frac{2}{x-5}$

29. $15x$ **30.** $10x^2$ **31.** $x(x-5)$ **32.** $10x^2(x-1)$

33. $(x-1)(x+1)^2$ **34.** $x(x-4)(x+4)$

35. $\frac{9}{24}$ **36.** $\frac{16}{12x}$ **37.** $\frac{3x^2+6x}{x^2-4}$ **38.** $\frac{2x}{x^2+x}$

39. $\frac{3x-3}{5x^2-5x}$ **40.** $\frac{2x^2+4x}{2x^2+x-6}$ **41.** $\frac{19}{24}$ **42.** $\frac{7}{4x}$

43. $-\frac{1}{9x}$ **44.** $\frac{4x+3}{x(x-1)}$ **45.** $\frac{2x}{(x-1)(x+1)}$

46. $\frac{8-9x}{6x^2}$ **47.** $\frac{2x-7}{6x}$ **48.** $-\frac{1}{(x-1)(x+1)}$

49. $\frac{5y-x}{(x-y)(x+y)}$ **50.** $\frac{13}{6x}$ **51.** $\frac{33}{28}$ **52.** $\frac{7}{10}$

53. $\frac{n}{2}$ **54.** $\frac{3(p+1)}{p-1}$ **55.** $\frac{3(m+1)}{2(m-1)^2}$ **56.** $\frac{2n-1}{4(2n+1)}$

57. $\frac{1}{3}$ **58.** $\frac{2-x}{2+x}$ **59.** $\frac{1}{x^2}$ **60.** $x + 3$ **61.** $\frac{20}{7}$

62. $\frac{8}{3}$ **63.** $\frac{1}{5}$ **64.** -5 **65.** -4 **66.** $-3, -1$

67. 4   **68.** -2 **69.** 5   **70.** 7   **71.** $-2, 5$

72. $-3, 3$ **73.** 8   **74.** No solutions   **75.** -3

76. -12 **77.** $\frac{4}{5}$ **78.** -3 **79.** $b = \frac{2ac}{3a-c}$

80. $x = \frac{y}{y-1}$ **81.** $\frac{6}{5}$ **82.** $\frac{9}{7}$

83. (a) $\frac{5}{x} = \frac{11}{20}$ **(b)** $\frac{100}{11}$

84. (a) $\frac{117}{13} = \frac{x}{7}$ **(b)** $63   **85. (a)** 4   **(b)** 16

86. (a) 3   **(b)** 12   **87. (a)** 10   **(b)** 5

88. (a) 21   **(b)** $\frac{21}{2}$

89. (a) Inverse   **(b)** $y = \frac{60}{x}$

(c)

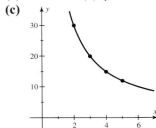

90. (a) Direct   **(b)** $y = 3x$

(c)

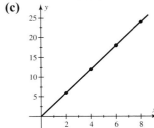

91. Neither; not possible   **92.** Inverse; 12

93. (a) $\frac{1}{5} = 0.2$; when the rate of arrival is 10 cars per minute, the wait is 0.2 minutes or 12 seconds.

(b)

x	5	10	13	14	14.9
T	$\frac{1}{10}$	$\frac{1}{5}$	$\frac{1}{2}$	1	10

(c) It increases dramatically.

94. 60 mph   **95.** $\frac{800}{13} \approx 61.5$ hours

96. 10 mph and 12 mph   **97.** 8 mph

98. About 26.7 feet   **99.** 1.75 inches   **100.** 100 vehicles

101. About 20.8 pounds   **102.** $325   **103.** 6493.8 watts

104. 702 pounds

105. 750 to 2250 seconds or 12.5 to 37.5 minutes

CHAPTER 7 TEST (PP. 473–474)

1. $\frac{9}{5}$ **2.** -2 **3.** $x + 5$ **4.** $x - 5$ **5.** 3   **6.** 2
7. $\frac{x-1}{10x}$ **8.** $\frac{6}{x(x+3)}$ **9.** $\frac{4x+1}{x+4}$ **10.** $\frac{t+7}{2t-3}$
11. $-\frac{1}{y+1}$ **12.** $\frac{x^2y - x + y}{xy^2}$ **13.** $\frac{b}{15}$
14. $\frac{p}{p-2}$ **15.** $\frac{35}{2}$ **16.** 3   **17.** 1   **18.** $-1, 2$
19. $\frac{2}{7}$ **20.** -12 **21.** $x = \frac{2+5y}{3y}$ **22.** $b = \frac{a}{a-1}$
23. (a) $\frac{5}{2}$ (b) 15   **24.** Inversely; 64
25. 24 hours   **26.** 67.5 feet
27. $\frac{16}{5} = 3.2$; when the arrival rate is 24 people per hour, there are about 3 people in line.

CHAPTER 8: INTRODUCTION TO FUNCTIONS

SECTION 8.1 (PP. 488–493)

1. function   **3.** symbolic   **5.** domain   **7.** one
9. Verbal, numerical, symbolic, and graphical or diagrammatic
11. Yes   **13.** No   **15.** $-6; -2$ **17.** $0; \frac{3}{2}$ **19.** $25; \frac{9}{4}$
21. $3; 3$ **23.** $-\frac{1}{2}; \frac{2}{5}$
25. (a) $I(x) = 36x$ (b) $I(10) = 360$
27. (a) $A(r) = \pi r^2$ (b) $A(10) = 100\pi \approx 314.2$
29. (a) $A(x) = 43,560x$ (b) $A(10) = 435,600$
31.   **33.**

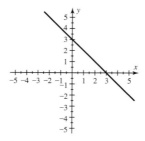   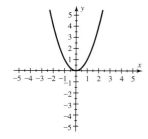

35.   **37.**

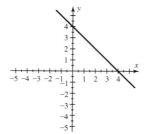

39.

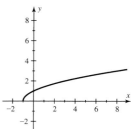

41. $3; -1$ **43.** $0; 2$ **45.** $-4; -3$ **47.** $5.5; 3.7$
49. In 1990 the average fuel efficiency was 26.9 mpg.
51.

x	-3	-2	-1	0	1	2	3
$y = f(x)$	2	3	4	5	6	7	8

$y = x + 5$

53.

x	-3	-2	-1	0	1	2	3
$y = f(x)$	-17	-12	-7	-2	3	8	13

$y = 5x - 2$

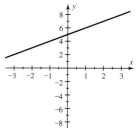

55. Subtract $\frac{1}{2}$ from the input x to obtain the output y.
57. Divide the input x by 3 to obtain the output y.
59. $f(x) = 0.50x$

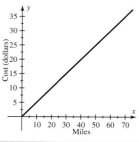

Miles	10	20	30	40	50	60	70
Cost	$5	$10	$15	$20	$25	$30	$35

61. 81.526; in 1986 the median price was $81,526.

63. -2 **65.** 0.5 **67.** $D: -2 \le x \le 2; R: 0 \le y \le 2$
69. $D: -2 \le x \le 4; R: -2 \le y \le 2$
71. D: All real numbers; $R: y \ge -1$
73. $D: -3 \le x \le 3; R: -3 \le y \le 2$
75. $D = \{1, 2, 3, 4\}; R = \{5, 6, 7\}$
77. All real numbers **79.** All real numbers **81.** $x \ne 5$
83. All real numbers **85.** $x \ge 1$
87. **(a)** 60.3
(b) $D = \{1910, 1930, 1950, 1970, 1990, 2000\}$
$R = \{35.5, 36.9, 56.2, 60.3, 80.5, 84.4\}$
(c) Decreased
89. **(a)** 0.2 **(b)** Yes. Each month has one average
amount of precipitation. **(c)** 2, 3, 7, 11
91. No **93.** Yes
95. Yes
D: All real numbers
R: All real numbers
97. No
99. Yes
$D: -4 \le x \le 4$
$R: 0 \le y \le 4$
101. Yes
D: All real numbers
$R: y = 3$
103. No **105.** Yes **107.** No

Section 8.2 (pp. 502–506)

1. $ax + b$ **3.** line **5.** 7 **7.** Linear: $a = \frac{1}{2}, b = -6$
9. Nonlinear **11.** Linear: $a = 0, b = -9$
13. Linear: $a = -9, b = 0$ **15.** Yes **17.** No
19. Yes; $f(x) = 3x - 6$ **21.** Yes; $f(x) = \frac{3}{2}x + 3$
23. No **25.** Yes; $f(x) = 2x - 2$ **27.** $-16; 20$
29. $\frac{17}{3}; 2$ **31.** $-22; -22$ **33.** $-2; 0$ **35.** $-1; -4$
37. $1; 1$ **39.** $f(x) = 6x; 18$
41. $f(x) = \frac{x}{6} - \frac{1}{2}; 0$ **43.** d **45.** b
47. **49.**

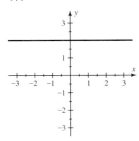

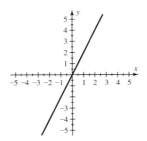

51.

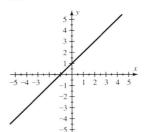

53.

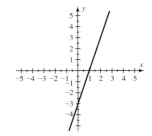

55.

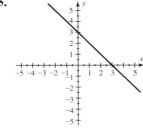

57. $f(x) = \frac{1}{16}x$ **59.** $f(t) = 65t$ **61.** $f(x) = 24$
63. b **65.** c
67. **(a)** $f(x) = 70$

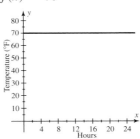

(b)

Hours	0	4	8	12	16	20	24
Temp. (°F)	70	70	70	70	70	70	70

(c) Constant
69. $D(t) = 60t + 50$
71. **(a)** $f(x) = 2.7x; 162$
(b) $g(x) = 4.3x; 258$
73. **(a)** $[1820, 1995, 20]$ by $[0, 40, 10]$

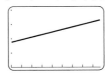

Median age is increasing.
(b)

Year (x)	1820	1840	1860	1880	1900	1920	1940
Median Age	16.7	18.5	20.3	22.1	23.9	25.7	27.5

$f(1900) = 23.9$; in 1900 the median age was
about 23.9 years.
(c) Each year the median age increased by 0.09 year,
on average.

75. (a) $V(T) = 0.5T + 137$ **(b)** 162 cm^3
77. (a) 180 minutes   **(b)** Decreased by 4 minutes
(c) $f(x) = -4x + 180$ **(d)** 164 minutes
79. $f(x) = 40x$; about 2.92 lb

CHECKING BASIC CONCEPTS 8.1 & 8.2 (P. 506)

1. $f(x) = x^2 - 1$

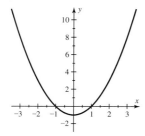

2. (a) $D: -3 \le x \le 3$; $R: -4 \le y \le 4$ **(b)** 0; 4
(c) No. The graph is not a line.
3. (a) Yes   **(b)** No   **(c)** Yes   **(d)** Yes
4. 10

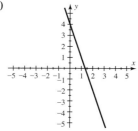

5. $f(x) = \frac{1}{2}x - 1$
6. (a) 32.98: In 1990 the median age was about 33 years.
(b) 0.264: The median age is increasing by 0.264 year each year. 27.7: In 1970 the median age was 27.7 years.

SECTION 8.3 (PP. 514–518)

1. $x > 1$ and $x \le 7$ (*answers may vary*)   **3.** No   **5.** Yes
7. Yes, no   **9.** No, yes   **11.** No, yes   **13.** [2, 10]
15. (5, 8]   **17.** $(-\infty, 4)$ **19.** $(-2, \infty)$
21. $(-\infty, -2] \cup [4, \infty)$ **23.** $(-\infty, 1) \cup [5, \infty)$
25. (-3, 5]   **27.** $(-\infty, -2)$ **29.** $(-\infty, 4)$
31. $(-\infty, 1) \cup (2, \infty)$
33. $\{x \mid -1 \le x \le 3\}$

35. $\{x \mid -2 < x < 2.5\}$

37. $\{x \mid x \le -1 \text{ or } x \ge 2\}$

39. All real numbers

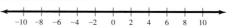

41. [-6, 7]   **43.** $\left(\frac{13}{4}, \infty\right)$ **45.** $(-\infty, \infty)$
47. $\left(-\infty, -\frac{11}{5}\right) \cup (-1, \infty)$ **49.** No solutions
51. [-6, 1)   **53.** $\left[-\frac{1}{4}, \frac{11}{8}\right)$ **55.** [-9, 3]
57. $\left[-4, -\frac{1}{4}\right)$ **59.** (-1, 1]   **61.** [-2, 1]   **63.** (0, 2)
65. (9, 21]   **67.** $\left[\frac{12}{5}, \frac{22}{5}\right]$ **69.** $\left(-\frac{13}{3}, 1\right)$ **71.** [0, 12]
73. [-1, 2]   **75.** (-1, 2)   **77.** [-3, 1]
79. $(-\infty, -2) \cup (0, \infty)$
81. (a) Toward, because distance is decreasing
(b) 4 hours, 2 hours   **(c)** From 2 to 4 hours   **(d)** During the first 2 hours
83. [1, 4]   **85.** $(-\infty, -2) \cup (0, \infty)$ **87.** [1, 3]
89. $(-\infty, \infty)$ **91.** From 1950 to 1980
93. From 2000 to 2005   **95.** From $5.\overline{6}$ to 9 feet
97. (a) [6, 9]; the car is from 470 to 680 miles from the rest stop for times between 6 and 9 hours.

x	4	5	6	7	8	9	10
$f(x) = 70x + 50$	330	400	470	540	610	680	750

(b) [6, 9]
99. $-67.\overline{7}°C$ to $36.\overline{6}°C$

SECTION 8.4 (PP. 526–530)

1. domain   **3.** $[1, \infty)$ **5.** $(-\infty, \infty)$ **7.** absolute value
9. 2   **11.** rational   **13.** $D: (-\infty, \infty)$; $R: (-\infty, \infty)$
15. $D: (-\infty, \infty)$; $R: (-\infty, \infty)$ **17.** $D: (-\infty, \infty)$; $R: [2, \infty)$
19. $D: (-\infty, \infty)$; $R: (-\infty, 0]$ **21.** $D: [-1, \infty)$; $R: [0, \infty)$
23. $D: (-\infty, \infty)$; $R: [0, \infty)$ **25.** $(-\infty, 1) \cup (1, \infty)$
27. $(-\infty, 2) \cup (2, \infty)$
29. $(-\infty, -2) \cup (-2, 2) \cup (2, \infty)$
31. $(-\infty, 0) \cup (0, 2) \cup (2, \infty)$ **33.** $(-\infty, 1) \cup (1, \infty)$
35. $(-\infty, -1) \cup (-1, 3) \cup (3, \infty)$
37. $D: (-\infty, \infty)$; $R: (-\infty, \infty)$ **39.** $D: [-2, 2]$; $R: [-2, 2]$
41. $D: [-2, 3]$; $R: [-2, 2]$ **43.** Yes; 1; linear
45. Yes; 3; cubic   **47.** No   **49.** Yes; 2; quadratic
51. No   **53.** Yes; 4; fourth degree   **55.** 12; 0   **57.** 1; $\frac{11}{4}$
59. 0; 6   **61.** 14; -14   **63.** -6; 9   **65.** $\frac{1}{11}$; $-\frac{1}{7}$
67. $-\frac{5}{6}$; undefined   **69.** $\frac{5}{6}$; undefined   **71.** 1; -1
73. -2; -2   **75.** -4; 0   **77.** -1; undefined
79.   **81.**

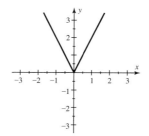   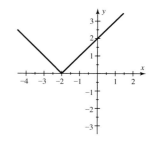

83.

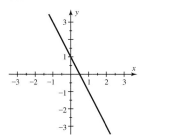

85.

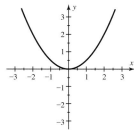

87.

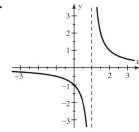

89. (a) No (*answers may vary*)   **(b)** Yes   **(c)** No;
$0 \leq t \leq 10$
91. (a) Yes. Walking for 0 minutes burns 0 calories.
(b) Linear. The data lie on a straight line.   **(c)** $C(t) = \frac{16}{3}t$
93. c   **95.** d
97. (a) $8.4\overline{6}$; the elevation of the outer rail should be about
8.5 inches when the radius of the curve is 300 feet.
   (b) [0, 600, 100] by [0, 50, 10]

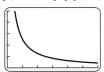

   (c) It decreases.

CHECKING BASIC CONCEPTS 8.3 & 8.4 (P. 530)

1. (a) Yes   **(b)** No
2. (a) $[-3, 1]$ **(b)** $(-\infty, -1] \cup [3, \infty)$ **(c)** $\left[-\frac{8}{3}, \frac{8}{3}\right]$
3. (a) $(-\infty, \infty)$ **(b)** $(-\infty, 1) \cup (1, \infty)$ **(c)** $[0, \infty)$
4. (a) $D: [-2, 1]$; $R: [-3, 1]$ **(b)** 1; -3
5.

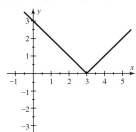

SECTION 8.5 (PP. 537–540)

1. $|3x + 2| = 6$ (*answers may vary*)   **3.** Yes   **5.** Yes
7. No, yes   **9.** No, yes   **11.** Yes, yes   **13.** 0, 4
15. $-7, 7$ **17.** 0   **19.** $-\frac{9}{4}, \frac{9}{4}$ **21.** $-4, 4$ **23.** $-6, 5$
25. 1, 2   **27.** $-8, 12$ **29.** No solutions   **31.** $-15, 18$
33. $-1, \frac{1}{3}$ **35.** $-5, \frac{3}{5}$ **37.** -6
39. (a) $-4, 4$ **(b)** $\{x \mid -4 < x < 4\}$ **(c)** $\{x \mid x < -4$
or $x > 4\}$
41. (a) $\frac{1}{2}, 2$ **(b)** $\{x \mid \frac{1}{2} \leq x \leq 2\}$
(c) $\{x \mid x \leq \frac{1}{2}$ or $x \geq 2\}$
43. $[-3, 3]$ **45.** $(-\infty, -4) \cup (4, \infty)$ **47.** No solutions
49. $(-\infty, 0) \cup (0, \infty)$ **51.** $\left(-\infty, -\frac{7}{2}\right) \cup \left(\frac{7}{2}, \infty\right)$
53. $(-3, 5)$ **55.** $(-\infty, -9] \cup [-1, \infty)$ **57.** $\left[\frac{5}{6}, \frac{11}{6}\right]$
59. $[-10, 14]$ **61.** No solutions   **63.** $(-\infty, \infty)$
65. No solutions   **67.** $(-\infty, \infty)$
69. $(-\infty, -13] \cup [17, \infty)$ **71.** $[0.9, 1.1]$
73. $(-\infty, 9.5) \cup (10.5, \infty)$
75. (a) $-1, 3$ **(b)** $(-1, 3)$ **(c)** $(-\infty, -1) \cup (3, \infty)$
77. (a) $-1, 0$ **(b)** $[-1, 0]$ **(c)** $(-\infty, -1] \cup [0, \infty)$
79. $(-\infty, -1] \cup [1, \infty)$ **81.** $[-2, 4]$
83. $(-\infty, 1) \cup (3, \infty)$ **85.** $(2, 4.\overline{6})$
87. $(-\infty, \infty)$ **89.** $\{x \mid -3 \leq x \leq 3\}$
91. $\{x \mid x < 2$ or $x > 3\}$
93. (a) $\{T \mid 19 \leq T \leq 67\}$ **(b)** Monthly average temperatures vary from 19°F to 67°F.
95. (a) $\{T \mid -26 \leq T \leq 46\}$ **(b)** Monthly average temperatures vary from −26°F to 46°F.
97. (a) About 19,058 feet   **(b)** Africa and Europe
(c) South America, North America, Africa, Europe, and
Antarctica
99. $\{d \mid 2.498 \leq d \leq 2.502\}$; the diameter can vary from
2.498 to 2.502 inches.
101. Values between 19 and 21, exclusively

CHECKING BASIC CONCEPTS 8.5 (P. 540)

1. $-\frac{28}{3}, 12$ **2. (a)** $-\frac{2}{3}, \frac{14}{3}$ **(b)** $\left(-\frac{2}{3}, \frac{14}{3}\right)$
(c) $\left(-\infty, -\frac{2}{3}\right) \cup \left(\frac{14}{3}, \infty\right)$
3. $(0, 6)$ and $(-\infty, 0] \cup [6, \infty)$
4. (a) $1, 3$ **(b)** $[1, 3]$ **(c)** $(-\infty, 1] \cup [3, \infty)$

CHAPTER 8 REVIEW EXERCISES (PP. 544–549)

1. $-7; 0$ **2.** $-22; 2$ **3.** $-2; 1$ **4.** $5; 5$
5. (a) $P(q) = 2q$ **(b)** $P(5) = 10$, there are 10 pints in
5 quarts.
6. (a) $f(x) = 4x - 3$ **(b)** $f(5) = 17$, three less than
four times 5 is 17.
7. $(3, -2)$ **8.** $4; -6$

9.
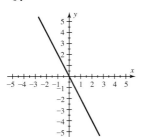

10.
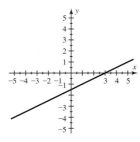

37.
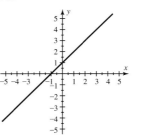

38.
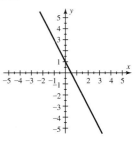

11.
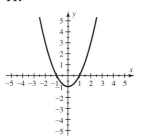

12.
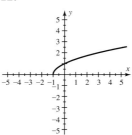

39.
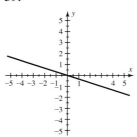

40.
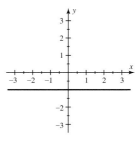

13. 1; 4   **14.** 1; −2   **15.** 7; −1
16. Numerical:

x	−3	−2	−1	0	1	2	3
$y = f(x)$	−11	−8	−5	−2	1	4	7

Symbolic: $f(x) = 3x - 2$
Graphical:

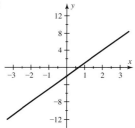

17. D: All real numbers; R: $y \le 4$
18. D: $-4 \le x \le 4$; R: $-4 \le y \le 0$ **19.** Yes   **20.** No
21. $D = \{-3, -1, 2, 4\}$; $R = \{-1, 3, 4\}$; yes
22. $D = \{-1, 0, 1, 2\}$; $R = \{-2, 2, 3, 4, 5\}$; no
23. All real numbers   **24.** $x \ge 0$ **25.** $x \ne 0$
26. All real numbers   **27.** No   **28.** Yes
29. Yes; $a = -4, b = 5$ **30.** Yes; $a = -1, b = 7$
31. No   **32.** Yes; $a = 0, b = 6$
33. Yes; $f(x) = \frac{3}{2}x - 3$ **34.** No   **35.** 1
36. $f(-2) = -3$; $f(1) = 0$

41. $[-2, 2]$

42. $(-\infty, -3]$

43. $\left(-\infty, \frac{4}{5}\right] \cup (2, \infty)$

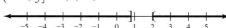

44. $(-\infty, \infty)$

45. $[-2, 1]$
46. (a) −4 (b) 2 (c) $[-4, 2]$ (d) $(-\infty, 2)$
47. (a) 2 (b) $(2, \infty)$ (c) $(-\infty, 2)$
48. (a) 4 (b) 2 (c) $(2, 4)$ **49.** $\left[-3, \frac{2}{3}\right]$ **50.** $(-6, 45]$
51. $\left(-\infty, \frac{7}{2}\right)$ **52.** $[1.8, \infty)$ **53.** $(-3, 4)$
54. $(-\infty, 4) \cup (10, \infty)$ **55.** $(-5, 5)$ **56.** $[8, 28]$
57. $(-9, 21)$ **58.** $[8, 28]$ **59.** $\left(-\frac{11}{5}, \frac{7}{5}\right]$ **60.** $[88, 138]$
61. $D = (-\infty, \infty)$; $R = [0, \infty)$
62. $D = (-\infty, \infty)$; $R = [0, \infty)$
63. $(-\infty, 4) \cup (4, \infty)$ **64.** $D = [-3, 1]$; $R = [-3, 6]$
65. Yes; 2; quadratic **66.** Yes; 1; linear

67. Yes; 3; cubic **68.** No **69.** 11; 2 **70.** $-\frac{4}{5}$; undefined
71. **72.**

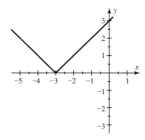 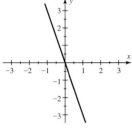

73. **74.**

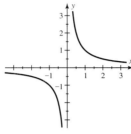

75. (a) 0, 4 **(b)** (0, 4) **(c)** $(-\infty, 0) \cup (4, \infty)$
76. (a) $-3, 1$ **(b)** $[-3, 1]$ **(c)** $(-\infty, -3] \cup [1, \infty)$
77. $-22, 22$ **78.** 1, 8 **79.** $-26, 42$ **80.** $-\frac{23}{3}, \frac{25}{3}$
81. 0, 2 **82.** $-3, \frac{9}{5}$
83. (a) $-8, 6$ **(b)** $[-8, 6]$ **(c)** $(-\infty, -8] \cup [6, \infty)$
84. (a) $-\frac{5}{2}, \frac{7}{2}$ **(b)** $\left[-\frac{5}{2}, \frac{7}{2}\right]$ **(c)** $\left(-\infty, -\frac{5}{2}\right] \cup \left[\frac{7}{2}, \infty\right)$
85. $(-\infty, -3) \cup (3, \infty)$ **86.** $(-4, 4)$ **87.** $[-3, 4]$
88. $\left(-\infty, -\frac{5}{2}\right] \cup [5, \infty)$ **89.** $[4.4, 4.6]$ **90.** $\left[\frac{3}{13}, \frac{7}{13}\right]$
91. $(-\infty, \infty)$ **92.** $\frac{3}{2}$ **93.** $(-\infty, -1.5] \cup [1.5, \infty)$
94. $[-2, 6]$
95. (a) About 25.1
(b) Decreased

[1885, 1965, 10] by [22, 26, 1]

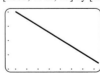

(c) -0.0492; the median age decreased, on average, by about 0.0492 year per year.

96. (a) 2.4 million **(b)** The number of marriages each year did not change.
97. (a) $f(x) = 8x$ **(b)** 8 **(c)** The total fat increases at the rate of 8 grams per cup.
98. (a)

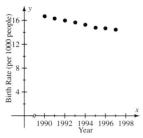

(b) $f(x) = -0.31x + 633.6$
(c) About 13.6 per 1000 people (*answers may vary*)
99. (a) 113; in 1995 there were 113 unhealthy days.
(b) $D = \{1995, 1996, 1997, 1998, 1999\}$
$R = \{27, 56, 60, 94, 113\}$
(c) The number of unhealthy days is decreasing.
100. (a) Linear

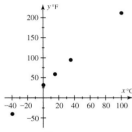

(b) $f(x) = \frac{9}{5}x + 32$; a 1°C change equals a $\frac{9}{5}$°F change. **(c)** 68°F
101. 13 feet by 31 feet **102.** $-44.\overline{4}$°C to $41.\overline{6}$°C
103. $|L - 160| \le 1$; $159 \le L \le 161$
104. (a) $|A - 3.9| \le 1.7$ **(b)** $2.2 \le A \le 5.6$
105. Values between 32.2 and 37.8

CHAPTER 8 TEST (PP. 549–550)

1. 46
2. (a)

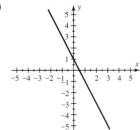

(b)

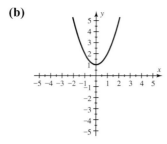

3. $0, -3$; D: $-3 \le x \le 3$, R: $-3 \le y \le 0$
4. Symbolic: $f(x) = x^2 - 5$
Numerical:

x	-3	-2	-1	0	1	2	3
$y = f(x)$	4	-1	-4	-5	-4	-1	4

Graphical:

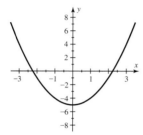

5. No, it fails the vertical line test.
6.

7. $(-\infty, -2) \cup (1, \infty)$
8. (a) -5 **(b)** 5 **(c)** $[-5, 5]$ **(d)** $(-\infty, 5)$
9. $(-8, 0)$
10. D: $(-\infty, \infty)$; R: $[0, \infty)$ **11.** Yes; 3; cubic
12. $\frac{8}{7}$; $(-\infty, 5) \cup (5, \infty)$ **13.** $-12, 24$
14. (a) $-\frac{2}{5}, \frac{4}{5}$ **(b)** $\left[-\frac{2}{5}, \frac{4}{5}\right]$ **(c)** $\left(-\infty, -\frac{2}{5}\right] \cup \left[\frac{4}{5}, \infty\right)$
15. (a) $f(x) = 0.4x + 75$ **(b)** 35 minutes
16. (a) Yes; $P(0) = 100$, $P(20) = 134$, $P(30) = 150$,
$P(50) = 180$ **(b)** Probably not because the race is over
after 50 seconds; $[0, 50]$.

CHAPTER 9: SYSTEMS OF LINEAR EQUATIONS

SECTION 9.1 (PP. 560–562)

1. No; three planes cannot intersect at exactly two points.
3. Yes **5.** Two **7.** $(1, 2, 3)$ **9.** $(-1, 1, 2)$
11. $(3, -1, 1)$ **13.** $\left(\frac{17}{2}, -\frac{3}{2}, 2\right)$ **15.** $(5, -2, -2)$
17. $(11, 2, 2)$ **19.** $(1, -1, 2)$ **21.** $(0, -3, 2)$
23. $(-1, 3, -2)$ **25.** $(-1, 2, 4)$ **27.** $(8, 5, 2)$
29. $(3, -3, 0)$ **31.** $\left(-\frac{3}{2}, 5, -10\right)$ **33.** $\left(\frac{3}{2}, 1, -\frac{1}{2}\right)$

35. (a) $x + 2y + 4z = 10$; $x + 4y + 6z = 15$;
$3y + 2z = 6$ **(b)** $(2, 1, 1.5)$; a hamburger costs \$2, fries
\$1, and a soft drink \$1.50.
37. (a) $x + y + z = 180$; $x - z = 55$;
$x - y - z = -10$ **(b)** $x = 85°, y = 65°$, and $z = 30°$
(c) These values check.
39. (a) $N + P + K = 80$
$N + P - K = 8$
$9P - K = 0$
(b) $(40, 4, 36)$; 40 pounds nitrogen, 4 pounds phosphorus, 36 pounds potassium
41. (a) $a + 600b + 4c = 525$; $a + 400b + 2c = 365$;
$a + 900b + 5c = 805$ **(b)** $a = 5, b = 1, c = -20$;
$F = 5 + A - 20W$ **(c)** 445 fawns
43. \$7500 at 8\%, \$8500 at 10\%, and \$14,000 at 15\%

SECTION 9.2 (PP. 571–574)

1. A rectangular array of numbers
3. $\begin{bmatrix} 1 & 3 & | & 10 \\ 2 & -6 & | & 4 \end{bmatrix}$; 2×3 (*answers may vary*)

5. $\begin{bmatrix} 1 & 0 & | & -3 \\ 0 & 1 & | & 5 \end{bmatrix}$ (*answers may vary*)

7. 3×3 **9.** 3×2 **11.** $\begin{bmatrix} 1 & -3 & | & 1 \\ -1 & 3 & | & -1 \end{bmatrix}$

13. $\begin{bmatrix} 2 & -1 & 2 & | & -4 \\ 1 & -2 & 0 & | & 2 \\ -1 & 1 & -2 & | & -6 \end{bmatrix}$

15. $x + 2y = -6$; $5x - y = 4$
17. $x - y + 2z = 6$; $2x + y - 2z = 1$;
$-x + 2y - z = 3$
19. $x = 4$; $y = -2$; $z = 7$ **21.** $(1, 3)$ **23.** $\left(-\frac{3}{2}, 2\right)$
25. $\left(\frac{7}{2}, -\frac{3}{2}\right)$ **27.** $\left(\frac{1}{2}, \frac{3}{2}\right)$ **29.** $(1, 2, 3)$ **31.** $(3, -3, 3)$
33. $(-1, 1, 0)$ **35.** $(1, 1, 1)$ **37.** $(-3, 2, 2)$
39. $(-7, 5)$ **41.** $(1, 2, 3)$ **43.** $(1, 0.5, -3)$
45. $(0.5, 0.25, -1)$ **47.** $(0.5, -0.2, 1.7)$
49. Dependent **51.** Inconsistent **53.** Dependent
55. 214 pounds
57. (a) $a + 2b + 1.4c = 3$; $a + 1.5b + 0.65c = 2$;
$a + 4b + 3.4c = 6$ **(b)** $a = 0.6, b = 0.5$, and $c = 1$
(c) $4.1 \approx 4$ people
59. $\frac{1}{2}$ hour at 5 miles per hour, 1 hour at 6 miles per hour,
and $\frac{1}{2}$ hour at 8 miles per hour
61. \$500 at 5\%, \$1000 at 8\%, and \$1500 at 12\%

CHECKING BASIC CONCEPTS 9.1 & 9.2 (P. 574)

1. $(1, 3, -1)$ **2.** $(1, 2, 3)$ **3.** $(-2, 2, -1)$

SECTION 9.3 (PP. 580–581)

1. -2 **3.** -53 **5.** -323 **7.** -5 **9.** -36
11. -42 **13.** -50 **15.** 0 **17.** -3555 **19.** -7466.5
21. 15 square feet   **23.** 52 square feet
25. 25.5 square feet   **27.** $(2, -2)$ **29.** $\left(-\frac{29}{11}, -\frac{34}{11}\right)$
31. $(-5, 7)$

CHECKING BASIC CONCEPTS 9.3 (P. 581)

1. (a) -1 **(b)** 17 **2.** $(-4, 6)$ **3.** 21 square units

CHAPTER 9 REVIEW EXERCISES (PP. 584–585)

1. Yes   **2.** $(1, -1, 2)$ **3.** $(-4, 3, 2)$ **4.** $(-1, 5, -3)$
5. $(-1, 3, 2)$ **6.** $(1, 1, 1)$

7. $\begin{bmatrix} 1 & 1 & 1 & | & -6 \\ 1 & 2 & 1 & | & -8 \\ 0 & 1 & 1 & | & -5 \end{bmatrix}; (-1, -2, -3)$

8. $\begin{bmatrix} 1 & 1 & 1 & | & -3 \\ -1 & 1 & 0 & | & 5 \\ 0 & 1 & 1 & | & -1 \end{bmatrix}; (-2, 3, -4)$

9. $\begin{bmatrix} 1 & 2 & -1 & | & 1 \\ -1 & 1 & -2 & | & 5 \\ 0 & 2 & 1 & | & 10 \end{bmatrix}; (-5, 4, 2)$

10. $\begin{bmatrix} 2 & 2 & -2 & | & -14 \\ -2 & -3 & 2 & | & 12 \\ 1 & 1 & -4 & | & -22 \end{bmatrix}; (-4, 2, 5)$

11. $(-7, 4, 2)$ **12.** $(5.4, 2.1, 9.7)$ **13.** -8 **14.** 30
15. 89 **16.** 130 **17.** $181{,}845$ **18.** 67.688
19. 46 square feet   **20.** 128 square feet   **21.** $(2, -1)$
22. $(-5, 3)$ **23.** $\left(\frac{3}{2}, \frac{1}{2}\right)$ **24.** $(7, -3)$
25. 6870 in 1988; 5220 in 1998
26. \$8 tickets: 285; \$12 tickets: 195
27. (a) $m + 3c + 5b = 14; m + 2c + 4b = 11;$
$c + 3b = 5$ **(b)** Malts: \$3; cones: \$2; bars: \$1
28. $100°, 65°,$ and $15°$
29. 2 pounds of \$1.50 candy, 4 pounds of \$2 candy,
6 pounds of \$2.50 candy
30. (a) $a + 202b + 63c = 40; a + 365b + 70c = 50;$
$a + 446b + 77c = 55$ **(b)** $a \approx 27.134; b \approx 0.061;$
$c \approx 0.009$ **(c)** About 46 inches

CHAPTER 9 TEST (P. 586)

1. No. Three planes cannot intersect at exactly three
points.
2. $(1, -2, -2)$ **3.** $(1, 2, 3)$
4. (a) $\begin{bmatrix} 2 & -4 & | & -10 \\ -3 & -2 & | & 7 \end{bmatrix}$ **(b)** $(-3, 1)$

5. (a) $\begin{bmatrix} 1 & 1 & 1 & | & 2 \\ 1 & -1 & -1 & | & 3 \\ 2 & 2 & 1 & | & 6 \end{bmatrix}$ **(b)** $\left(\frac{5}{2}, \frac{3}{2}, -2\right)$

6. 114 **7.** $\left(-\frac{47}{2}, -\frac{83}{2}\right)$
8. 6 mph: 30 min; 7 mph: 12 min; 9 mph: 18 min
9. $85°, 60°,$ and $35°$
10. (a) $a + 20b + 25c = 168; a + 24b + 40c = 270;$
$a + 30b + 50c = 405$ **(b)** $a = -270, b = 20.1,$
$c = 1.44$ **(c)** About 316 lb

CHAPTERS 1–9 CUMULATIVE REVIEW (PP. 589–592)

1. $2^3 \cdot 3^2 \cdot 5$ **2.** $2n + 7 = n - 2; -9$ **3.** $\frac{23}{21}$
4. $-\frac{1}{6}$ **5.** 18 **6.** $\frac{1}{10}$ **7.** $\frac{6}{5}$ **8.** 2 **9.** $\frac{31}{25}; 1.24$
10. 6 miles   **11.** $b = \frac{2A}{h} - a$ **12.** $\left\{ t \mid t < \frac{4}{7} \right\}$
13.

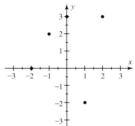

14. x-intercept: 4; y-intercept: 2

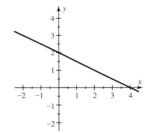

15. x-intercept: -4; y-intercept: 3

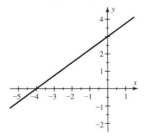

16. x-intercept: -2; y-intercept: none

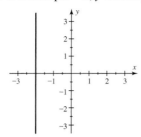

17. x-intercept: 2; y-intercept: 4; $y = -2x + 4$
18. $y = -2x + 1$ **19.** $y = \frac{3}{5}x + \frac{34}{5}$ **20.** $y = 2x + 3$
21. $P = 500x + 4000$ **22.** $(7.5, 10)$
23.   **24.**

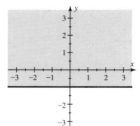

25.   **26.**

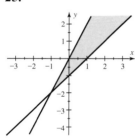

27. $-7ab - 10b^2 - 2$ **28.** -76 **29.** $\frac{27}{4}$ **30.** 1
31. $24t^4$ **32.** $2r^4t^4$ **33.** $\frac{1}{4t^6}$ **34.** $2ab^3$ **35.** $\frac{16r^{12}}{81t^{16}}$
36. $\frac{a^6}{8b^6}$ **37.** $2a^4 - 4a^3 + 6a^2$ **38.** $a^3 + b^3$
39. $5x^2 - 34x - 7$ **40.** $2y^2 - 3y - 9$ **41.** $a^2 - b^2$
42. $4x^2 + 12xy + 9y^2$ **43.** 0.000015 **44.** 2.13×10^6
45. $2x^2 - 4x + 3$ **46.** $x^3 - 4x^2 + 3x - 2 + \frac{1}{x - 5}$
47. $5ab^2(2 - 5a^2b^3)$ **48.** $(y - 3)(y^2 + 2)$
49. $(2z + 3)(3z - 1)$ **50.** $(2z - 3)(2z + 3)$
51. $(2y - 5)^2$ **52.** $(a - 3)(a^2 + 3a + 9)$
53. $(z^2 - 3)(4z^2 - 5)$ **54.** $ab(a + b)(2a - b)$
55. $-2, 1$ **56.** $0, 9$ **57.** $-\frac{1}{3}, \frac{3}{2}$ **58.** $-2, 0, 2$
59. $x - 4$ **60.** $\frac{x - 6}{3x - 4}$ **61.** $-4, 4$ **62.** $\frac{7}{3}$
63. $2(x - 1)$ **64.** 1   **65.** $\frac{2x + 3}{(x - 1)(x + 1)}$ **66.** $\frac{y - x}{y + x}$
67. 8   **68.** $\frac{17}{12}$ **69.** 1   **70.** 10   **71.** 13

72. $D = (-\infty, \infty); R = [-2, \infty)$

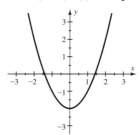

73. $2, -2$ **74.** $2, -4$ **75.** Yes; $f(x) = -4x + 3$
76. $(-\infty, -3]$ **77.** $(3, \infty)$ **78.** $(-\infty, 0]$
79. $(-1, 5]$

80. $\left(-\infty, -\frac{3}{2}\right) \cup [3, \infty)$

81. $[-2, 4]$ **82.** $[-16, 10]$
83. (a) $-3, 1$ **(b)** $[-3, 1]$ **(c)** $(-\infty, -3] \cup [1, \infty)$
84. $-6, 18$ **85.** $(-\infty, -6) \cup \left(\frac{8}{3}, \infty\right)$ **86.** $[4, 7]$
87. $(1, 1, 2)$ **88.** $(-1, 3, -2)$ **89.** 14   **90.** $d = 325t$
91. (a) $f(x) = \frac{1}{2}x + 4$ **(b)** $\frac{1}{2}$ **(c)** Snow is falling at a rate of $\frac{1}{2}$ inch per hour.   **(d)** 6 inches
92.

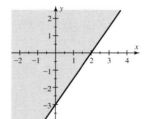

93. 3 children and 5 adults   **94.** $\frac{6}{7}$ hour
95. (a) 112 feet   **(b)** 5.5 seconds
96. 61.25 feet
97. (a) $\frac{1}{5}$ minute; the average wait is $\frac{1}{5}$ minute or 12 seconds when vehicles arrive at 15/min.
(b)

x	5	10	15	19	19.9
T	$\frac{1}{15}$	$\frac{1}{10}$	$\frac{1}{5}$	1	10

(c) The waiting time increases dramatically.
98. $73.63 \le L \le 74.37$
99. (a) $4b + 3f + 4m = 23; b + 2f + m = 7$; $3b + f + 2m = 13$ **(b)** Burgers: \$2; fries: \$1; malt: \$3

CHAPTER 10: RADICAL EXPRESSIONS AND FUNCTIONS

SECTION 10.1 (PP. 601–603)

1. ± 3 **3.** 2   **5.** b **7.** $(\sqrt[n]{a})^m$ or $\sqrt[n]{a^m}$ **9.** 3
11. -2.24 **13.** $|z|$ **15.** 3   **17.** -4 **19.** 1.71
21. $-x^3$ **23.** $4x^2$ **25.** 3   **27.** -1.48 **29.** $\sqrt{6}$
31. $\sqrt{xy}$ **33.** $\dfrac{1}{\sqrt[5]{y}}$ **35.** 4   **37.** 4   **39.** 2   **41.** 16
43. 2.24   **45.** $\frac{2}{3}$ **47.** 1.26   **49.** $\frac{1}{2}$ **51.** $-\frac{1}{2}$ **53.** 2
55. 2.83   **57.** x^3 **59.** xy^4 **61.** xy^2 **63.** $\dfrac{y^2}{x}$
65. $y^{13/6}$ **67.** $\dfrac{x^4}{9}$ **69.** $\dfrac{y^3}{x}$ **71.** $y^{1/4}$ **73.** $\dfrac{1}{a^{2/3}}$
75. ab^2 **77.** $\dfrac{1}{k^2}$ **79.** $b^{3/4}$ **81.** $z^{23/12}$ **83.** $p^2 + p$
85. $x^{5/6} - x$ **87.** 4   **89.** $|y|$ **91.** $|a + 3|$
93. $|x - 5|$ **95.** $|x - 1|$ **97.** $|y|$ **99.** $|x^3|$
101. xy^2 **103.** $R \approx 1.17$ miles   **105.** $A = 6$

SECTION 10.2 (PP. 610–611)

1. Yes   **3.** $\sqrt[3]{ab}$ **5.** $\frac{a}{b}$ **7.** No, since $1^3 \neq 3$ **9.** 3
11. 10   **13.** 4   **15.** $\frac{3}{5}$ **17.** $\frac{1}{4}$ **19.** $\frac{x}{4}$ **21.** 3   **23.** 4
25. 3   **27.** -2 **29.** a **31.** 3   **33.** $2x^2$
35. $-a^2\sqrt[3]{5}$ **37.** $2x\sqrt[4]{y}$ **39.** $6x$ **41.** $2x^2yz^3$ **43.** $\frac{3}{2}$
45. $\sqrt[3]{12ab}$ **47.** $5\sqrt{z}$ **49.** $\frac{1}{a}$ **51.** $\sqrt{x^2 - 16}$
53. $\sqrt[3]{a^3 + 1}$ **55.** $\sqrt{x + 1}$ **57.** 10   **59.** 2   **61.** 3
63. $10\sqrt{2}$ **65.** $3\sqrt[3]{3}$ **67.** $2\sqrt{2}$ **69.** $-2\sqrt[5]{2}$
71. $b^2\sqrt{b}$ **73.** $2n\sqrt{2n}$ **75.** $2ab^2\sqrt{3b}$
77. $5xy\sqrt[3]{xy^2}$ **79.** $5\sqrt[3]{5t^2}$ **81.** $\dfrac{3t}{r\sqrt[4]{5r^3}}$ **83.** $\sqrt[6]{3^5}$
85. $2\sqrt[12]{2^5}$ **87.** $3\sqrt[12]{3^{11}}$ **89.** $x\sqrt[12]{x}$ **91.** $\sqrt[12]{r^{11}t^7}$
93. $\dfrac{\sqrt{7}}{7}$ **95.** $\dfrac{4\sqrt{3}}{3}$ **97.** $\dfrac{\sqrt{5}}{3}$ **99.** $\dfrac{\sqrt{3b}}{6}$ **101.** $\dfrac{t\sqrt{r}}{2r}$
103. (a) 400 in^2 (b) $A = 100W^{2/3}$

CHECKING BASIC CONCEPTS 10.1 & 10.2 (P. 612)

1. (a) ± 7 (b) 7   (c) ± 7 **2.** (a) -2 (b) -3
3. (a) $\sqrt{x^3}$ or $(\sqrt{x})^3$ (b) $\sqrt[3]{x^2}$ or $(\sqrt[3]{x})^2$
(c) $\dfrac{1}{\sqrt[5]{x^2}}$ or $\dfrac{1}{(\sqrt[5]{x})^2}$
4. $|x - 1|$
5. (a) $\frac{1}{8}$ (b) 10   (c) $-2x\sqrt[3]{xy}$ (d) $\dfrac{4b^2}{5}$
6. $\sqrt[6]{7^5}$ **7.** $\dfrac{\sqrt{6}}{2}$

SECTION 10.3 (PP. 618–619)

1. $2\sqrt{a}$ **3.** like   **5.** $\sqrt{t + 5}$ **7.** Not possible
9. $\sqrt{7}, 2\sqrt{7}, 3\sqrt{7}$ **11.** $2\sqrt[3]{2}, -3\sqrt[3]{2}$
13. Not possible   **15.** $2\sqrt[3]{xy}, xy\sqrt[3]{xy}$ **17.** $9\sqrt{3}$
19. $8\sqrt{5} + \sqrt{2}$ **21.** $2\sqrt{x} - \sqrt{y}$ **23.** $2\sqrt[3]{z}$
25. $-5\sqrt[3]{6}$ **27.** $y^2 - y$ **29.** $9\sqrt{7}$ **31.** $-2\sqrt{11}$
33. $5\sqrt[3]{2} - \sqrt{2}$ **35.** $-\sqrt[3]{xy}$ **37.** $3\sqrt{x + 2}$
39. $\dfrac{3\sqrt{3}}{2}$ **41.** $\dfrac{71\sqrt{2}}{10}$ **43.** $2\sqrt{2}$ **45.** $(5x - 1)\sqrt[4]{x}$
47. $(8x + 2)\sqrt{x}$ or $2\sqrt{x}(4x + 1)$ **49.** $(3ab - 1)\sqrt[4]{ab}$
51. $(n - 2)\sqrt[3]{n}$ **53.** 2   **55.** 119   **57.** $x - 64$
59. $ab - c$ **61.** $x + \sqrt{x} - 56$ **63.** $\dfrac{3 + \sqrt{2}}{7}$
65. $\sqrt{10} - 2\sqrt{2}$ **67.** $\dfrac{11 - 4\sqrt{7}}{3}$ **69.** $\sqrt{7} + \sqrt{6}$
71. $\dfrac{z + 3\sqrt{z}}{z - 9}$ **73.** $\dfrac{a + 2\sqrt{ab} + b}{a - b}$ **75.** $\sqrt{x + 1} + \sqrt{x}$
77. $12\sqrt{3} \approx 20.8$ cm

SECTION 10.4 (PP. 629–631)

1.

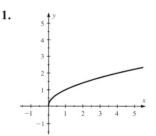

3. $\{x \mid x \geq 0\}$ **5.** $f(x) = x^p$, p is rational.
7. $\{x \mid x \geq 0\}$ **9.** 5   **11.** 5   **13.** $[-1, \infty)$
15. $[2, \infty)$ **17.** $(-\infty, 1]$ **19.** $\left(-\infty, \frac{8}{5}\right]$ **21.** $(-\infty, \infty)$
23. $\left(-\frac{1}{2}, \infty\right)$ **25.** ± 7 **27.** ± 10 **29.** $-5, 3$
31. $-3, 7$ **33.** 4   **35.** -4 **37.** 1   **39.** $\frac{7}{5}$
41.   **43.**

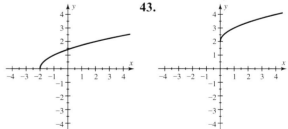

Shifted 2 units left   Shifted 2 units upward

45.

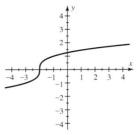

Shifted 2 units left

47. 32; 55.90   **49.** $-\frac{1}{128} \approx -0.01; 0.04$

51. 4; Not possible   **53.** 4; 4

55. [0, 6, 1] by [0, 6, 1]

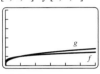

$g(x)$

57. [0, 6, 1] by [0, 6, 1]

$f(x)$

59. 1 second

61. (a) 6 years   **(b)** The twin in the spaceship will be 4 years younger than the twin on Earth.

63. (a) It increases by a factor of 8.   **(b)** $v = \sqrt[3]{\frac{W}{3.8}}$

(c) 20 miles per hour

65. (a) It decreases.

(b) [0, 1600, 400] by [0, 220, 20]

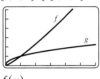

(c) $k \approx 885$ **(d)** $R \approx 33$; a 700-pound animal will have a pulse rate of about 33 beats per minute.

CHECKING BASIC CONCEPTS 10.3 & 10.4 (P. 632)

1. (a) 6   **(b)** 3   **(c)** $6x^3$

2. (a) $7\sqrt{6} + \sqrt{7}$ **(b)** $5\sqrt[3]{x}$ **(c)** $\sqrt{x}$

3. (a) $(y - x)\sqrt[3]{xy}$ **(b)** 14

4. $\frac{\sqrt{5} + 1}{2}$

5. (a)

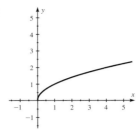

$f(-1)$ is undefined.

(b)

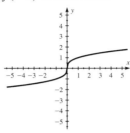

$f(-1) = -1$

(c)

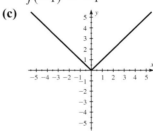

$f(-1) = 1$

6. 3.2   **7.** $[4, \infty)$ **8.** $-5, 3$

SECTION 10.5 (PP. 640–643)

1. Square each side.   **3.** Yes

5. Finding an unknown side of a right triangle

7. $d = \sqrt{(x_2 - x_1)^2 + (y_2 - y_1)^2}$ **9.** 64   **11.** 81

13. 6   **15.** 3   **17.** 27   **19.** -2 **21.** 15   **23.** $\frac{1}{2}$

25. 0, 1   **27.** 2   **29.** -4 **31.** 9   **33.** 9   **35.** 3

37. 1.88   **39.** $-1, 0.70$ **41.** 1.79   **43.** -0.47 **45.** 16

47. 11   **49.** $L = \frac{8T^2}{\pi^2}$ **51.** $A = \pi r^2$ **53.** Yes

55. Yes   **57.** Yes   **59.** No   **61.** $\sqrt{32} = 4\sqrt{2}$ **63.** 7

65. $c = 5$ **67.** $b = \sqrt{61}$ **69.** $a = 14$

71. $\sqrt{20} = 2\sqrt{5}$ **73.** $\sqrt{4000} = 20\sqrt{10}$ **75.** $\sqrt{89}$

77. 5   **79.** 4   **81.** 2, 122   **83.** $W = 8$ pounds

85. About 3 miles   **87.** About 269 feet

89. 19 inches   **91.** $h \approx 16.3$ inches, $d \approx 33.3$ inches

93. (a) 121 feet   **(b)** About 336 feet

95. (a) About 38 miles per hour; the vehicle involved in the accident was traveling about 38 miles per hour.

(b) About 453 feet

97. $a\sqrt{2}$

SECTION 10.6 (PP. 650–651)

1. $2 + 3i$ (*answers may vary*)   **3.** i **5.** $i\sqrt{a}$
7. $a + bi$ **9.** 4   **11.** $i\sqrt{5}$ **13.** $10i$ **15.** $12i$
17. $2i\sqrt{3}$ **19.** $3i\sqrt{2}$ **21.** 3   **23.** $-8 + 7i$
25. $1 - 9i$ **27.** $-10 + 7i$ **29.** $20 - 12i$
31. $-31 - 8i$ **33.** 20   **35.** $1 + 5i$ **37.** $3 + 4i$
39. $-2 - 6i$ **41.** -2 **43.** $-i$ **45.** i **47.** -1
49. 1   **51.** $3 - 4i$ **53.** $6i$ **55.** $5 + 4i$ **57.** -1
59. $1 - i$ **61.** $-\frac{6}{29} + \frac{15}{29}i$ **63.** $2 + i$ **65.** $-1 + 6i$
67. $-1 - 2i$ **69.** $-\frac{3}{2}i$ **71.** $-\frac{1}{2} + \frac{3}{2}i$ **73.** $\frac{290}{13} + \frac{20}{13}i$

CHECKING BASIC CONCEPTS 10.5 & 10.6 (P. 651)

1. (a) 4   (b) 28   (c) 3   **2.** 13   **3.** 9.6 inches
4. (a) $8i$ (b) $i\sqrt{17}$
5. (a) $3 - 4i$ (b) $-2 + 3i$ (c) $5 + i$ (d) $\frac{3}{4} + \frac{3}{4}i$

CHAPTER 10 REVIEW EXERCISES (PP. 655–658)

1. 2   **2.** 6   **3.** $3|x|$ **4.** $|x - 1|$ **5.** -4
6. -5 **7.** x^2 **8.** $3x$ **9.** 2   **10.** -1 **11.** x^2
12. $x + 1$ **13.** $\sqrt{14}$ **14.** $\sqrt[3]{-5}$
15. $\left(\sqrt{\frac{x}{y}}\right)^3$ or $\sqrt{\left(\frac{x}{y}\right)^3}$ **16.** $\frac{1}{\sqrt[3]{(xy)^2}}$ or $\frac{1}{(\sqrt[3]{xy})^2}$ **17.** 9
18. 2   **19.** 64   **20.** 27   **21.** z^2 **22.** xy^2 **23.** $\frac{x^3}{y^9}$
24. $\frac{y^2}{x}$ **25.** 8   **26.** -2 **27.** x^2 **28.** 2   **29.** $-\frac{\sqrt[3]{x}}{2}$
30. $\frac{1}{3}$ **31.** $4\sqrt{3}$ **32.** $3\sqrt{6}$ **33.** $\frac{3}{x}$ **34.** $4ab\sqrt{2a}$
35. $9xy$ **36.** $5z\sqrt[3]{z}$ **37.** $x + 1$ **38.** $\frac{2a\sqrt[4]{a}}{b}$
39. $2\sqrt[6]{x^5}$ **40.** $\sqrt[6]{r^5 t^8}$ or $t\sqrt[6]{r^5 t^2}$ **41.** $\frac{4\sqrt{5}}{5}$ **42.** $\frac{r\sqrt{t}}{2t}$
43. $4\sqrt{3}$ **44.** $3\sqrt[3]{x}$ **45.** $-3\sqrt[3]{5}$ **46.** $-\sqrt[4]{y}$
47. $11\sqrt{3}$ **48.** $7\sqrt{2}$ **49.** $13\sqrt[3]{2}$ **50.** $3\sqrt{x + 1}$
51. $(2x - 1)\sqrt{x}$ **52.** $(b + 2a)\sqrt[3]{ab}$ **53.** 3   **54.** 95
55. $a - 2b$ **56.** $xy + \sqrt{xy} - 2$ **57.** $\frac{3 - \sqrt{2}}{7}$
58. $\frac{5 + \sqrt{7}}{9}$ **59.** $\sqrt{8} + \sqrt{7}$ **60.** $\frac{a - 2\sqrt{ab} + b}{a - b}$
61.
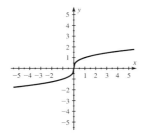
62.

63.
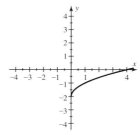
Shifted 2 units downward

64.
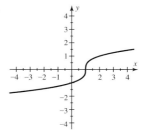
Shifted 1 unit to the right

65. $[1, \infty)$ **66.** $(-\infty, 3]$ **67.** $(-\infty, \infty)$ **68.** $(-2, \infty)$
69. ± 11 **70.** ± 4 **71.** $-3, 5$ **72.** 4   **73.** 3
74. 2   **75.** 2   **76.** 4   **77.** 9   **78.** 9   **79.** 8   **80.** $\frac{1}{4}$
81. 4.5   **82.** 1.62   **83.** $c = \sqrt{65}$ **84.** $b = \sqrt{39}$
85. $\sqrt{41}$ **86.** $\sqrt{52} = 2\sqrt{13}$ **87.** -2 **88.** $-2 + 4i$
89. $5 + i$ **90.** $1 + 2i$ **91.** $-\frac{14}{13} + \frac{5}{13}i$ **92.** $2 - 2i$
93. About 85 feet   **94.** $\sqrt{16,200} \approx 127.3$ feet
95. About 0.79 second
96. (a) About 43 miles per hour   (b) About 34 miles per hour; a steeper bank allows for a higher speed limit; yes
97. $\sqrt{7} \approx 2.65$ feet
98. (a) 5 square units   (b) $5\sqrt{5}$ cubic units
(c) $\sqrt{10}$ units   (d) $\sqrt{15}$ units
99. About 0.82 foot   **100.** About 0.13 foot; it is shorter.
101. $r = \sqrt[210]{\frac{281}{4}} - 1 \approx 0.02$; from 1790 through 2000 the average annual percentage growth rate was about 2%.
102. About 2108 square inches
103. (a) $0.5 = \frac{1}{2}$ (b) About 0.09, or $\frac{9}{100}$

CHAPTER 10 TEST (PP. 658–659)

1. $5x^2$ **2.** $2z^2$ **3.** $2xy\sqrt[4]{y}$ **4.** 1   **5.** $\sqrt[5]{7^2}$ or $(\sqrt[5]{7})^2$
6. $\sqrt[3]{\left(\frac{y}{x}\right)^2}$ or $\left(\sqrt[3]{\frac{y}{x}}\right)^2$ **7.** 16   **8.** $\frac{1}{216}$ **9.** $(-\infty, 4]$
10.
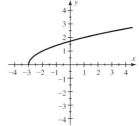

11. $8z^{3/2}$ **12.** $\dfrac{z}{y^{2/3}}$ **13.** 9   **14.** $\dfrac{y}{2}$ **15.** $4\sqrt{7} + \sqrt{5}$

16. $6\sqrt[3]{x}$ **17.** 17   **18.** $\sqrt{14} + \sqrt{13}$

19. $\sqrt{120} \approx 10.95$ **20.** $\sqrt{8} = 2\sqrt{2}$ **21.** $2 - 19i$

22. $-6 + 8i$ **23.** $\dfrac{5}{4}$ **24.** $\dfrac{4}{29} + \dfrac{10}{29}i$

25. (a) $r = \sqrt[3]{\dfrac{3V}{4\pi}}$ **(b)** About 2.29 inches

26. 1.31 pounds

CHAPTER 11: QUADRATIC FUNCTIONS AND EQUATIONS

SECTION 11.1 (PP. 669–672)

1. parabola   **3.** axis of symmetry

5.

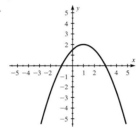

7. narrower   **9.** $ax^2 + bx + c$ with $a \neq 0$ **11.** 0, -4

13. $-2, -2$

15. $(1, -2)$; $x = 1$; upward; incr.: $x \geq 1$; decr.: $x \leq 1$

17. $(-2, 3)$; $x = -2$; downward; incr.: $x \leq -2$; decr.: $x \geq -2$

19. (a)

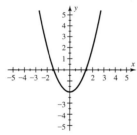

(b) $(0, -2)$; $x = 0$ **(c)** 2; 7

21. (a)

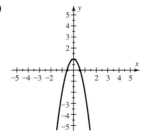

(b) $(0, 1)$; $x = 0$ **(c)** $-11; -26$

23. (a)

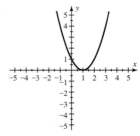

(b) $(1, 0)$; $x = 1$ **(c)** 9; 4

25. (a)

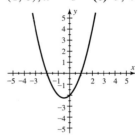

(b) $(-0.5, -2.25)$; $x = -0.5$ **(c)** 0; 10

27. (a)

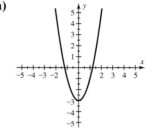

(b) $(0, -3)$; $x = 0$ **(c)** 5; 15

29. (a)

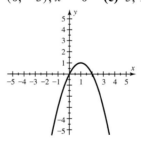

(b) $(1, 1)$; $x = 1$ **(c)** $-8; -3$

31. (a)

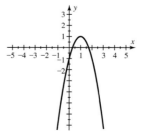

(b) $(1, 1)$; $x = 1$ **(c)** $-17; -7$

33. (a)

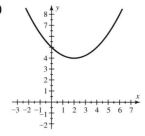

(b) $(2, 4)$; $x = 2$ **(c)** 8; 4.25

35. $(2, -6)$ **37.** $(-3, 4)$ **39.** $(0, 3)$ **41.** $(1, 1.4)$

43.

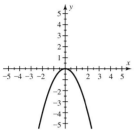

Reflected across the x-axis

45.

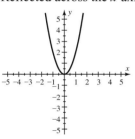

Narrower

47.

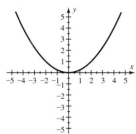

Wider

49.

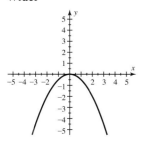

Reflected across the x-axis and wider

51. -2; incr.: $x \geq -1$; decr.: $x \leq -1$

53. $-\frac{25}{4}$; incr.: $x \geq \frac{5}{2}$; decr.: $x \leq \frac{5}{2}$

55. $-\frac{7}{2}$; incr.: $x \geq -\frac{1}{2}$; decr.: $x \leq -\frac{1}{2}$

57. 6; incr.: $x \leq 1$; decr.: $x \geq 1$

59. 4; incr.: $x \leq 2$; decr.: $x \geq 2$

61. $-\frac{39}{8}$; incr.: $x \leq \frac{1}{4}$; decr.: $x \geq \frac{1}{4}$ **63.** d. **65.** a.

67. (a) 2 feet **(b)** 2 seconds **(c)** 66 feet

69. $\frac{66}{32} \approx 2$ seconds; about 74 feet

71. (a) The revenue increases when $x \leq 50$ and it decreases when $x \geq 50$. **(b)** $2500; 50
(c) $f(x) = x(100 - x)$ **(d)** $2500; 50

73. (a) $f(x) = x(30 - x)$ **(b)** 15 feet by 15 feet; 225 square feet

75. (a) [20, 40, 5] by [0, 30, 5]

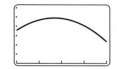

(b) About 28.4°C **(c)** About 28.4°C

SECTION 11.2 (PP. 681–682)

1. $x^2 + 2$ **3.** $(1, 2)$

5. $f(x) = ax^2 + bx + c$; $f(x) = a(x - h)^2 + k$

7. downward

9. (a)

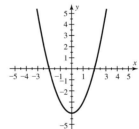

(b) $(0, -4)$ **(c)** Down 4 units

11. (a)

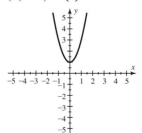

(b) $(0, 1)$ **(c)** Narrower and up 1 unit

13. (a)

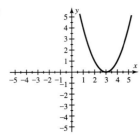

(b) $(3, 0)$ **(c)** Right 3 units

15. (a)

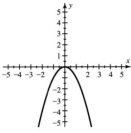

(b) $(0, 0)$ **(c)** Reflected across the x-axis

17. (a)

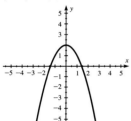

(b) $(0, 2)$ **(c)** Reflected across the x-axis and up 2 units

19. (a)

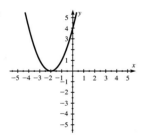

(b) $(-2, 0)$ **(c)** Left 2 units

21. (a)

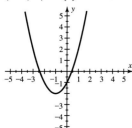

(b) $(-1, -2)$ **(c)** Left 1 unit and down 2 units

23. (a)

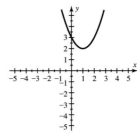

(b) $(1, 2)$ **(c)** Right 1 unit and up 2 units

25. (a)

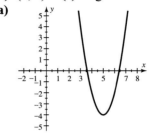

(b) $(5, -4)$ **(c)** Narrower, right 5 units and down 4 units

27. (a)

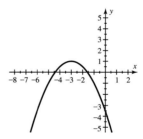

(b) $(-3, 1)$ **(c)** Wider, reflected across the x-axis, left 3 units and up 1 unit

29. $y = 3(x - 3)^2 + 4; y = 3x^2 - 18x + 31$

31. $y = -\frac{1}{2}(x - 5)^2 - 2; y = -\frac{1}{2}x^2 + 5x - \frac{29}{2}$

33. $y = (x - 1)^2 + 2$ **35.** $y = -(x - 0)^2 - 3$

37. $y = (x - 0)^2 - 3$ **39.** $y = -(x + 1)^2 + 2$

41. $y = (x + 1)^2 - 4; (-1, -4)$

43. $y = (x - 2)^2 + 1; (2, 1)$

45. $y = \left(x + \frac{3}{2}\right)^2 - \frac{17}{4}; \left(-\frac{3}{2}, -\frac{17}{4}\right)$

47. $y = \left(x - \frac{7}{2}\right)^2 - \frac{45}{4}; \left(\frac{7}{2}, -\frac{45}{4}\right)$

49. $y = 3(x + 1)^2 - 4; (-1, -4)$

51. $y = 2\left(x - \frac{3}{4}\right)^2 - \frac{9}{8}; \left(\frac{3}{4}, -\frac{9}{8}\right)$

53. $y = -2(x + 2)^2 + 13; (-2, 13)$ **55.** $a = 2$

57. $a = 0.3$ **59.** $y = 2(x - 1)^2 - 3$

61. $y = 0.5(x - 1980)^2 + 6$

63. (a)

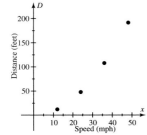

(b) $D(x) = \frac{1}{12}x^2$

CHECKING BASIC CONCEPTS 11.1 & 11.2 (P. 683)

1. (a)

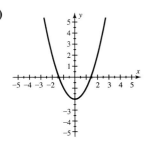

$(0, -2); x = 0$

(b)

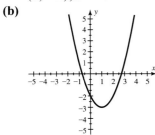

$(1, -3); x = 1$

2. y_1 opens upward, whereas y_2 opens downward, y_1 is narrower than y_2.

3. 7; incr.: $x \leq 2$; decr.: $x \geq 2$

4. (a)

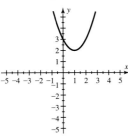

1 unit right, 2 units up

(b)

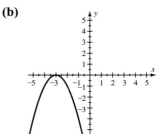

Reflected across the x-axis, 3 units left

5. (a) $y = (x + 7)^2 - 56$ **(b)** $y = 4(x + 1)^2 - 6$

SECTION 11.3 (PP. 692–695)

1. $x^2 + 3x - 2 = 0$ (*answers may vary*); 0, 1, or 2 solutions

3. Factoring, square root property, completing the square

5.

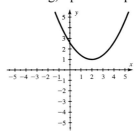

(*answers may vary*)

7. ± 8; the square root property   **9.** Yes   **11.** No

13. Yes   **15.** No   **17.** $-2, 1$ **19.** No real solutions

21. $-2, 3$ **23.** -0.5 **25.** $-1, 5$ **27.** $-3, 1$

29. $-3, 3$ **31.** $-\frac{1}{2}, \frac{3}{2}$ **33.** $-7, 5$ **35.** $-\frac{1}{3}, \frac{1}{2}$ **37.** $1, \frac{3}{2}$

39. $\frac{6}{5}, \frac{3}{2}$ **41.** ± 12 **43.** $\pm\frac{8}{\sqrt{5}}$ **45.** $-6, 4$ **47.** $-7, 9$

49. $\frac{1 \pm \sqrt{5}}{2}$ **51.** 4   **53.** $\frac{25}{4}$ **55.** 16; $(x - 4)^2$

57. $\frac{81}{4}$; $\left(x + \frac{9}{2}\right)^2$ **59.** $-4, 6$ **61.** $-3 \pm \sqrt{11}$

63. $\frac{3 \pm \sqrt{29}}{2}$ **65.** $\frac{5 \pm \sqrt{21}}{2}$ **67.** $1 \pm \sqrt{5}$

69. $\frac{3 \pm \sqrt{41}}{4}$ **71.** $\frac{2 \pm \sqrt{11}}{2}$ **73.** $\frac{-3 \pm \sqrt{5}}{12}$ **75.** $-3, 6$

77. $3, 5$ **79.** $5, 7$ **81.** $y = \pm\sqrt{x + 1}$

83. $v = \pm\sqrt{\frac{2K}{m}}$ **85.** $r = \pm\sqrt{\frac{k}{E}}$ **87.** $f = \pm\frac{1}{2\pi\sqrt{LC}}$

89. (a) 30 miles per hour   **(b)** 40 miles per hour

91. About 1.9 seconds; no   **93.** 2 hours

95. (a) $x^2 + 6x - 520 = 0$ **(b)** -26 or 20; 20 feet

97. About 23°C and 34°C

99. (a) The rate of increase for each 20-year period is not constant.   **(b)** $(1800, 5)$; in 1800 the population was 5 million.   **(c)** 1910

SECTION 11.4 (PP. 704–707)

1. To solve quadratic equations of the form $ax^2 + bx + c = 0$
3. $b^2 - 4ac$
5. Factoring, square root property, completing the square, and the quadratic formula
7. $-6, \frac{1}{2}$ **9.** 1   **11.** $-2, 8$ **13.** $\frac{1 \pm \sqrt{17}}{8}$
15. No real solutions   **17.** $\frac{1}{2}$ **19.** $\frac{3 \pm \sqrt{13}}{2}$
21. $2 \pm \sqrt{3}$ **23.** $\frac{1 \pm \sqrt{37}}{6}$ **25.** $\frac{1 \pm \sqrt{15}}{2}$
27. $\frac{5 \pm \sqrt{10}}{3}$ **29.** (a) $a > 0$ (b) $-1, 2$ (c) Positive
31. (a) $a > 0$ (b) No real solutions   (c) Negative
33. (a) $a < 0$ (b) 2   (c) Zero
35. (a) 25   (b) 2   **37.** (a) 0   (b) 1
39. (a) $-\frac{7}{4}$ (b) 0   **41.** (a) 21   (b) 2
43. $1 \pm \sqrt{2}$ **45.** $-\frac{3}{2}, 1$ **47.** None   **49.** None
51. $\frac{-2 \pm \sqrt{10}}{3}$ **53.** $\pm 3i$ **55.** $\pm 4i\sqrt{5}$ **57.** $\pm \frac{1}{2}i$
59. $\pm \frac{3}{4}i$ **61.** $\pm i\sqrt{6}$ **63.** $\pm\sqrt{3}$ **65.** $\pm i\sqrt{2}$
67. $\frac{1}{2} \pm i\frac{\sqrt{7}}{2}$ **69.** $-\frac{3}{4} \pm i\frac{\sqrt{23}}{4}$ **71.** $2 \pm \sqrt{3}$
73. $-\frac{1}{2} \pm i\frac{\sqrt{7}}{2}$ **75.** $-\frac{1}{5} \pm i\frac{\sqrt{19}}{5}$ **77.** $-\frac{3}{4} \pm i\frac{\sqrt{23}}{4}$
79. $-\frac{1}{2} \pm i\frac{\sqrt{15}}{2}$ **81.** $\frac{1 \pm \sqrt{3}}{2}$ **83.** $\frac{1}{4} \pm i\frac{\sqrt{15}}{4}$
85. 1, 2; (*answers may vary*)
87. $-\frac{1}{2}, 4$; (*answers may vary*)
89. $\frac{5 \pm \sqrt{17}}{2}$; quadratic formula or completing the square
91. $-\frac{1}{4} \pm \frac{3}{4}i\sqrt{7}$; quadratic formula or completing the square
93. 9 miles per hour   **95.** 45 miles per hour
97. $x \approx 8.04$, or about 1992; this agrees with the graph.
99. $\frac{1 \pm \sqrt{17}}{2} \approx 2.6$ mph
101. (a) The rate of change is not constant   (b) 75 seconds; (*answers may vary*)   (c) 75 seconds

CHECKING BASIC CONCEPTS 11.3 & 11.4 (P. 707)

1. $\frac{1}{2}, 3$ **2.** $\pm\sqrt{5}$ **3.** $2 \pm \sqrt{3}$ **4.** $y = \pm\sqrt{1 - x^2}$
5. (a) $\frac{3 \pm \sqrt{17}}{4}$ (b) $\frac{4}{3}$ (c) $-\frac{1}{2} \pm i\frac{\sqrt{7}}{2}$
6. (a) 5; two real solutions   (b) -7; no real solutions   (c) 0; one real solution

SECTION 11.5 (PP. 713–716)

1. It has an inequality symbol rather than an equals sign.
3. No   **5.** $-2 < x < 4$ **7.** Yes   **9.** Yes   **11.** No
13. Yes   **15.** No   **17.** No
19. (a) $-3, 2$ (b) $-3 < x < 2$ (c) $x < -3$ or $x > 2$

21. (a) $-2, 2$ (b) $-2 < x < 2$ (c) $x < -2$ or $x > 2$
23. (a) $-10, 5$ (b) $x < -10$ or $x > 5$ (c) $-10 < x < 5$
25. (a) $-2, 2$ (b) $-2 < x < 2$ (c) $x < -2$ or $x > 2$
27. (a) $-4, 0$ (b) $-4 < x < 0$ (c) $x < -4$ or $x > 0$
29. $(-3, -1)$ **31.** $(-\infty, -2.5] \cup [3, \infty)$ **33.** $[-2, 2]$
35. $(-\infty, \infty)$ **37.** $(0, 3)$
39. (a) $-2, 2$ (b) $-2 < x < 2$ (c) $x < -2$ or $x > 2$
41. (a) $\frac{-1 \pm \sqrt{5}}{2}$ (b) $\frac{-1 - \sqrt{5}}{2} < x < \frac{-1 + \sqrt{5}}{2}$
(c) $x < \frac{-1 - \sqrt{5}}{2}$ or $x > \frac{-1 + \sqrt{5}}{2}$
43. $[-7, -3]$ **45.** $(-\infty, 1) \cup (2, \infty)$
47. $(-\sqrt{10}, \sqrt{10})$ **49.** $(-\infty, 0) \cup (6, \infty)$
51. $(-\infty, 2 - \sqrt{2}] \cup [2 + \sqrt{2}, \infty)$
53. (a) From 1131 feet to 3535 feet (approximately)
(b) Before 1131 feet or after 3535 feet (approximately)
55. (a) About 183; they agree.   (b) About 1970 or after
(c) About 1969 or after
57. From 11 feet to 20 feet

SECTION 11.6 (P. 719)

1. $\pm 1, \pm\sqrt{6}$ **3.** $-\sqrt[3]{2}, \sqrt[3]{\frac{5}{3}}$ **5.** $-\frac{4}{5}, -\frac{1}{3}$ **7.** $-3, 3$
9. $-\sqrt[3]{\frac{1}{3}}, \sqrt[3]{2}$ **11.** $-\frac{1}{8}, \frac{2}{5}$ **13.** 1
15. 1, $32^5 = 33{,}554{,}432$
17. 16, 81   **19.** 1   **21.** $-3, 6$ **23.** $-\sqrt{3}, \sqrt{3}$

CHECKING BASIC CONCEPTS 11.5 & 11.6 (P. 719)

1. $(-\infty, -2) \cup (3, \infty)$ **2.** $\left[-1, -\frac{2}{3}\right]$ **3.** $-2, \sqrt[3]{2}$
4. $-1, 8^3 = 512$

CHAPTER 11 REVIEW EXERCISES (PP. 722–726)

1. $(-3, 4)$; $x = -3$; downward; incr.: $x \leq -3$; decr.: $x \geq -3$
2. $(1, 0)$; $x = 1$; upward; incr.: $x \geq 1$; decr.: $x \leq 1$
3. (a)

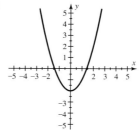

(b) $(0, -2)$; $x = 0$ (c) -1

4. (a)

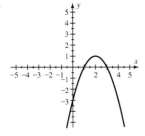

(b) $(2, 1)$; $x = 2$ **(c)** 0

5. (a)

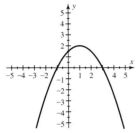

(b) $(1, 2)$; $x = 1$ **(c)** -2.5

6. (a)

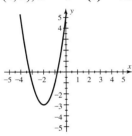

(b) $(-2, -3)$; $x = -2$ **(c)** -1

7. $-\frac{7}{2}$ **8.** $-\frac{14}{3}$ **9.** $(2, -6)$ **10.** $(0, 5)$ **11.** $(2, 2)$
12. $(-1, 1)$
13. (a)

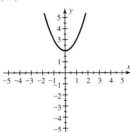

(b) Shifted up 2 units
14. (a)

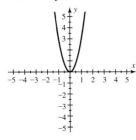

(b) Narrower

15. (a)

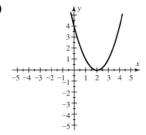

(b) Shifted right 2 units
16. (a)

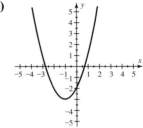

(b) Shifted left 1 unit and down 3 units
17. (a)

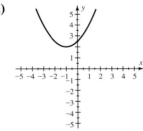

(b) Wider, shifted left 1 unit and up 2 units
18. (a)

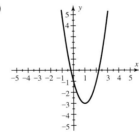

(b) Narrower, shifted right 1 unit and down 3 units
19. $y = -4(x - 2)^2 - 5$ **20.** $y = -(x + 4)^2 + 6$
21. $y = (x + 2)^2 - 11$; $(-2, -11)$
22. $y = \left(x - \frac{7}{2}\right)^2 - \frac{45}{4}$; $\left(\frac{7}{2}, -\frac{45}{4}\right)$
23. $y = 2\left(x - \frac{3}{4}\right)^2 - \frac{73}{8}$; $\left(\frac{3}{4}, -\frac{73}{8}\right)$
24. $y = 3(x + 1)^2 - 5$; $(-1, -5)$ **25.** $a = 3$
26. $a = \frac{1}{4}$ **27.** $-2, 3$ **28.** -1 **29.** No real solutions
30. $-4, 6$ **31.** $-10, 5$ **32.** $-0.5, 0.25$ **33.** $-5, 10$
34. $-3, 1$ **35.** $-4, 2$ **36.** $-1, 3$ **37.** $-5, 4$
38. $-8, -3$ **39.** $-\frac{2}{5}, \frac{2}{3}$ **40.** $\frac{4}{7}, 3$ **41.** ± 10 **42.** $\pm\frac{1}{3}$
43. $\pm\frac{\sqrt{6}}{2}$ **44.** No real solutions   **45.** $-3 \pm \sqrt{7}$
46. $2 \pm \sqrt{10}$ **47.** $1 \pm \sqrt{6}$ **48.** $\frac{-3 \pm \sqrt{11}}{2}$

49. $R = -r \pm \sqrt{\frac{k}{F}}$ **50.** $y = \pm\sqrt{\frac{12 - 2x^2}{3}}$ **51.** 3, 6

52. 11, 13   **53.** $-\frac{1}{2}, \frac{1}{3}$ **54.** $\frac{5 \pm \sqrt{5}}{10}$ **55.** $4 \pm \sqrt{21}$

56. $\frac{3 \pm \sqrt{3}}{2}$ **57.** (a) $a > 0$ (b) $-2, 3$ (c) Positive

58. (a) $a > 0$ (b) 2   (c) Zero

59. (a) $a < 0$ (b) No real solutions   (c) Negative

60. (a) $a < 0$ (b) $-4, 2$ (c) Positive

61. (a) 1   (b) 2   **62.** (a) 144   (b) 2

63. (a) -23 (b) 0   **64.** (a) 0   (b) 1

65. $-\frac{1}{2} \pm i\frac{\sqrt{19}}{2}$ **66.** $\pm 2i$ **67.** $\frac{1}{4} \pm i\frac{\sqrt{7}}{4}$ **68.** $\frac{1}{7} \pm i\frac{\sqrt{34}}{7}$

69. (a) $-2, 6$ (b) $-2 < x < 6$ (c) $x < -2$ or $x > 6$

70. (a) $-2, 0$ (b) $x < -2$ or $x > 0$ (c) $-2 < x < 0$

71. (a) $-4, 4$ (b) $-4 < x < 4$ (c) $x < -4$ or $x > 4$

72. (a) $-2, 1$ (b) $-2 < x < 1$ (c) $x < -2$ or $x > 1$

73. (a) $-1, 3$ (b) $-1 < x < 3$ (c) $x < -1$ or $x > 3$

74. (a) $-\frac{3}{2}, 5$ (b) $-\frac{3}{2} \le x \le 5$ (c) $x \le -\frac{3}{2}$ or $x \ge 5$

75. $[-3, -1]$ **76.** $\left(\frac{1}{5}, 3\right)$ **77.** $\left(-\infty, \frac{1}{6}\right) \cup (2, \infty)$

78. $(-\infty, -\sqrt{5}] \cup [\sqrt{5}, \infty)$ **79.** $\pm\sqrt{5}, \pm 3$

80. $-\frac{1}{4}, \frac{2}{7}$ **81.** 1, 512   **82.** 0

83. (a) $f(x) = x(12 - 2x)$ (b) 6 inches by 3 inches

84. (a) 1 second and 1.75 seconds   (b) 1.375 seconds; 34.25 feet

85. (a) $f(x) = x(90 - 3x)$
    (b) [0, 30, 5] by [0, 800, 100]

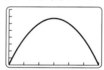

    (c) 10 or 20 rooms   (d) 15 rooms

86. (a) 2.4; in 1999, there were 2.4 complaints per 100,000 passengers.
    (b) [1997, 1999, 1] by [0.5, 3, 0.5]

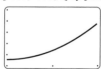

    Complaints have increased.

87. (a) $\sqrt{1728} \approx 41.6$ miles per hour   (b) 60 miles per hour

88. (a) $x(x + 2) = 143$ (b) $x = -13$ or $x = 11$; the numbers are -13 and -11 or 11 and 13.

89. (a) [1935, 1995, 10] by [0, 100, 10]

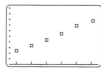

    (b) Yes, the data are nearly linear.
    (c) $f(x) = 1.04(x - 1940) + 25$ (*answers may vary*)

90. (a) (1950, 220); in 1950, the per capita consumption was at a low of 220 million Btu.
    (b) [1950, 1970, 5] by [200, 350, 25]

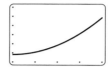

    It increased.
    (c) $f(1996) = 749$; no; the trend represented by this model did not continue after 1970.

91. About 11.1 inches by 11.1 inches   **92.** 50 feet

93. About 1.5 feet   **94.** About 6.0 to 9.0 inches

CHAPTER 11 TEST (PP. 727–728)

1. $\left(1, \frac{3}{2}\right); x = 1$ **2.** $-\frac{29}{4}$ **3.** $a = -\frac{1}{2}$

4.

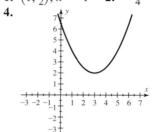

It is wider, translated right 3 units and translated upward 2 units.

5. $y = (x - 3)^2 - 7; (3, -7)$ **6.** $-1, 2$ **7.** $-4, \frac{1}{3}$

8. $-\frac{1}{2}, \frac{1}{2}$ **9.** $4 \pm \sqrt{17}$ **10.** $\frac{3 \pm \sqrt{17}}{4}$

11. (a) $a < 0$ (b) $-3, 1$ (c) Positive

12. (a) -44 (b) No real solutions   (c) The graph of $y = -3x^2 + 4x - 5$ does not intersect the x-axis.

13. (a) $-1, 1$ (b) $-1 < x < 1$ (c) $x < -1$ or $x > 1$

14. (a) $-10, 20$ (b) $x < -10$ or $x > 20$
(c) $-10 < x < 20$

15. (a) $-\frac{1}{2}, \frac{3}{4}$ (b) $\left[-\frac{1}{2}, \frac{3}{4}\right]$ (c) $\left(-\infty, -\frac{1}{2}\right] \cup \left[\frac{3}{4}, \infty\right)$

16. 1, $\sqrt[3]{2}$ **17.** $\sqrt{2250} \approx 47.4$ miles per hour

18. (a) $f(x) = (x + 20)(90 - x)$ (b) 35

19. (a) [0, 6, 1] by [0, 150, 50]

    (b) After about 5.6 seconds   (c) 2.75 seconds; 129 feet

CHAPTER 12: EXPONENTIAL AND LOGARITHMIC FUNCTIONS

Section 12.1 (pp. 742–745)

1. $g(f(7))$ **3.** No   **5.** No   **7.** adding 10   **9.** 8; 6
11. one-to-one
13. (a) 7   **(b)** 49   **(c)** $(g \circ f)(x) = x^2 + 3$
(d) $(f \circ g)(x) = (x + 3)^2$
15. (a) -65 **(b)** 126   **(c)** $(g \circ f)(x) = 8x^3 - 1$
(d) $(f \circ g)(x) = 2x^3 - 2$
17. (a) 3   **(b)** 1   **(c)** $(g \circ f)(x) = \left|\frac{1}{2}x - 2\right|$
(d) $(f \circ g)(x) = \frac{1}{2}|x - 2|$
19. (a) $\frac{11}{2}$ **(b)** $-\frac{1}{17}$ **(c)** $(g \circ f)(x) = 3 - \frac{5}{x}$
(d) $(f \circ g)(x) = \frac{1}{3 - 5x}$
21. (a) 77   **(b)** 122   **(c)** $(g \circ f)(x) = 16x^2 - 4x + 5$
(d) $(f \circ g)(x) = 8x^2 - 4x + 10$
23. (a) 1   **(b)** 2   **25. (a)** -1 **(b)** 1
27. (a) 0   **(b)** 2   **29. (a)** 2   **(b)** -3 **(c)** -1
31. $f(1) = f(-1) = 5$ **33.** $f(1) = f(-1) = 101$
35. $f(2) = f(-2) = 4$ **37.** Yes   **39.** No   **41.** Yes
43. Divide x by 7; $f(x) = 7x$; $g(x) = \frac{x}{7}$
45. Multiply x by 2 then subtract 5;
$f(x) = \frac{x + 5}{2}$; $g(x) = 2x - 5$
47. Add 3 to x and multiply the result by 2;
$f(x) = \frac{1}{2}x - 3$; $g(x) = 2(x + 3)$
49. Take the cube root of x and subtract 5;
$f(x) = (x + 5)^3$; $g(x) = \sqrt[3]{x} - 5$
51. $(f \circ f^{-1})(x) = 4\left(\frac{x}{4}\right) = x$; $(f^{-1} \circ f)(x) = \frac{4x}{4} = x$
53. Show $(f \circ f^{-1})(x) = (f^{-1} \circ f)(x) = x$. See Exercise 51 answer.
55. Show $(f \circ f^{-1})(x) = (f^{-1} \circ f)(x) = x$. See Exercise 51 answer.
57. Show $(f \circ f^{-1})(x) = (f^{-1} \circ f)(x) = x$. See Exercise 51 answer.
59. $f^{-1}(x) = \frac{x}{12}$ **61.** $f^{-1}(x) = x - 8$
63. $f^{-1}(x) = \frac{x + 2}{5}$ **65.** $f^{-1}(x) = -2(x - 1)$
67. $f^{-1}(x) = 8 - x$ **69.** $f^{-1}(x) = 2x - 1$
71. $f^{-1}(x) = \frac{x^3}{2}$ **73.** $f^{-1}(x) = \sqrt[3]{x + 8}$
75.

x	0	5	10	15	20
$f^{-1}(x)$	0	1	2	3	4

Domain of f = range of f^{-1} = $\{0, 1, 2, 3, 4\}$
Range of f = domain of f^{-1} = $\{0, 5, 10, 15, 20\}$

77.

x	4	2	0	-2	-4
$f^{-1}(x)$	-5	0	5	10	15

Domain of f = range of f^{-1} = $\{-5, 0, 5, 10, 15\}$
Range of f = domain of f^{-1} = $\{-4, -2, 0, 2, 4\}$

79.

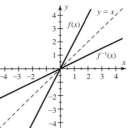

81.

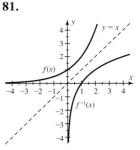

83. (a) 20π; after 5 seconds, the wave has a circumference of $20\pi \approx 62.8$ feet.   **(b)** $(C \circ r)(t) = 4\pi t$
85. (a) $75°$; 150   **(b)** 150; One hour after midnight there are 150 mosquitoes per 100 square feet.   **(c)** The number of mosquitoes per 100 square feet, h hours after midnight
(d) $T(h) = -5h + 80$; $M(T) = 2T$
(e) $(M \circ T)(h) = -10h + 160$
87. (a) 16; in 1980, 16% of people 25 or older completed four or more years of college.

(b)

x	8	16	27
$P^{-1}(x)$	1960	1980	2000

(c) 1980

89. (a) Yes, different inputs result in different outputs.
(b) $f^{-1}(x) = \frac{9}{5}(x - 32)$ converts x degrees Fahrenheit to an equivalent temperature in degrees Celsius.
91. $f(x) = 4x$; $f^{-1}(x) = \frac{x}{4}$ converts x quarts to gallons.

Section 12.2 (pp. 753–756)

1. $f(x) = Ca^x$ **3.** Growth   **5.** 2.718   **7.** factor
9. $\frac{1}{9}$; 9   **11.** 5; 160   **13.** 4; $\frac{1}{8}$ **15.** 15; $\frac{5}{9}$
17. 0.17; 2.41   **19.** 5; 1.08
21. (a) Exponential decay   **(b)** $f(x) = 64\left(\frac{1}{4}\right)^x$
23. (a) Linear growth   **(b)** $f(x) = 3x + 8$
25. (a) Exponential growth   **(b)** $f(x) = 4(1.25)^x$
27. $C = 1, a = 2$ **29.** $C = 4, a = \frac{1}{4}$ **31.** c.   **33.** d.
35.

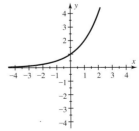

Growth

37.

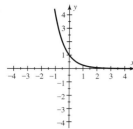

Decay

39.

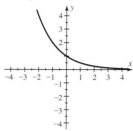

Decay

41.

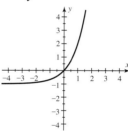

Growth

43.

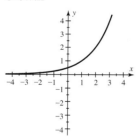

Growth

45.

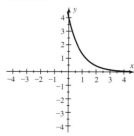

Decay

47. e^7 **49.** $\frac{1}{a}$ **51.** e^y **53.** $2^3 = 8$ **55.** $5^{2y} = 25^y$
57. \$3551.05   **59.** \$1,820,087.63   **61.** \$792.75

63. Yes; this is equivalent to having two accounts, each containing \$1000 initially.   **65.** About \$1.28 trillion
67. 3.32   **69.** 0.86
71. $[-4, 4, 1]$ by $[0, 8, 1]$

Growth

73. $[-4, 4, 1]$ by $[0, 8, 1]$

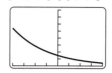

Decay
75. (a) $f(x) = 4.56e^{0.031x}$
(b) $[0, 10, 1]$ by $[4, 7, 1]$

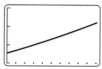

(c) About 5.49 million
77. (a) $C = 500, a = 2$ **(b)** About 5278 thousand per
milliliter   **(c)** The growth is exponential.
79. (a) About 11.3; in 1995 there were about 11.3 million
cellular phone subscribers.   **(b)** 1.495; each year from
1985 to 2000 the number of subscribers increased by a factor of 1.495, or by 49.5%.
81.

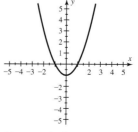

(a) 1; the probability that no vehicle will enter the intersection during a period of 0 seconds is 1 or 100%.
(b) About 30 seconds

CHECKING BASIC CONCEPTS 12.1 & 12.2 (P. 757)

1. (a) 7   **(b)** $(f \circ g)(x) = 2x^2 + 9x + 6$
2.

(a) No, it does not pass the horizontal line test.
(b) No

3. $f^{-1}(x) = \frac{x+3}{4}$ **4.** $\frac{3}{4}$
5.

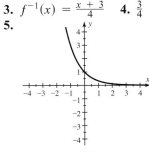

6. $C = 2; a = \frac{1}{2}$

SECTION 12.3 (PP. 765–769)

1. 10   **3.** $D = \{x \mid x > 0\}$; R: all real numbers   **5.** k
7. x **9.** $\log 5$ **11.** 5   **13.** -4 **15.** 0   **17.** -2
19. 4.7   **21.** $6x$ **23.** $-\frac{3}{2}$ **25.** 2   **27.** -4 **29.** -2
31. 0   **33.** -3 **35.** -2 **37.** 2   **39.** x^2 **41.** 17
43. $(2x)^2$ **45.** 5   **47.** $-5x$ **49.** $2x - 7$ **51.** $0.6z$
53. 1.398   **55.** 0.161   **57.** 1.946   **59.** -0.560
61. $[-4, 4, 1]$ by $[-4, 4, 1]$

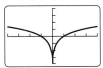

This is a reflection across the y-axis together with the graph of $y = \ln x$; $D = \{x \mid x \neq 0\}$
63. $[-4, 4, 1]$ by $[-4, 4, 1]$

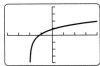

Shifted 2 units to the left; $D = \{x \mid x > -2\}$
65. d.   **67.** a.
69.

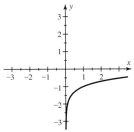

1 unit downward

71.

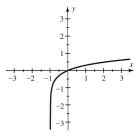

1 unit to the left
73.

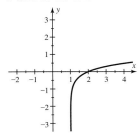

1 unit to the right
75.

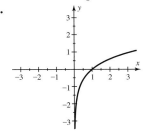

Increases faster
77.

x	1/4	1/2	1	$\sqrt{2}$	64
$\log_2 x$	-2	-1	0	1/2	6

79. 120 dB; yes
81. **(a)** $P(0) = 27$; the air pressure at the eye of the hurricane is 27 inches of mercury. $P(50) \approx 28.9$; the air pressure 50 miles from the eye is about 28.9 inches of mercury.
(b) It increases rapidly at first and then more slowly.
(c) Low
83. **(a)** 10^6; 10^8 **(b)** 100 times
85. **(a)** About 2119 calories   **(b)** Caloric intake increases as the amount of land increases.   **(c)** No; the growth levels off and is not linear.

87. (a) $f(50) \approx 1.89$; in 2000 the population of less industrialized regions was about 1.89 billion. $g(50) \approx 0.95$; in 2000 the population of industrialized regions was about 0.95 billion.

(b) $[0, 80, 10]$ by $[0, 5, 1]$

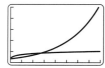

The population of less industrialized regions grew faster than industrialized regions.

SECTION 12.4 (PP. 774–776)

1. 4   **3.** 3   **5.** $\log m - \log n$ **7.** No

9. $\log_a x = \frac{\log x}{\log a}$ or $\log_a x = \frac{\ln x}{\ln a}$ **11.** $\ln 3 + \ln 5$

13. $\log_3 x + \log_3 y$ **15.** $\ln 2 + \ln 5 + \ln z$

17. $\log 7 - \log 3$ **19.** $\ln x - \ln y$

21. $\log_2 45 - \log_2 x$ **23.** $\log 225$ **25.** $\ln xy$

27. $\ln 14x^3$ **29.** $\ln xy$ **31.** $6 \log 3$ **33.** $x \ln 2$

35. $\frac{1}{4} \log_2 5$ **37.** $\frac{1}{3} \log_4 z$ **39.** $(y - 1)\log x$

41. $\log z$ **43.** $\log x^3 y^2$ **45.** 0   **47.** $\ln 2^x$ **49.** $\log_3 \frac{1}{x^2}$

51. $\log_a \frac{x + 1}{x - 1}$ **53.** $\log x + 2 \log y$

55. $4 \ln x + \ln y - \ln z$ **57.** $\frac{1}{3} \log_4 z - \frac{1}{2} \log_4 y$

59. $4 \log x + 3 \log y$ **61.** $\ln x - \ln y$

63. $\frac{3}{2} \log_4 x + \frac{1}{2} \log_4 y - \log_4 z$

65. $[-6, 6, 1]$ by $[-4, 4, 1]$

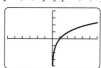

$[-6, 6, 1]$ by $[-4, 4, 1]$

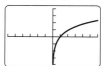

By the power rule, $\log x^3 = 3 \log x$

67. $[-6, 6, 1]$ by $[-4, 4, 1]$

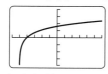

$[-6, 6, 1]$ by $[-4, 4, 1]$

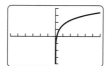

Not the same

69. 1.2   **71.** 1.8   **73.** 2.1   **75.** 0.4   **77.** -1.1

79. 1.46   **81.** 4.64   **83.** 2.10

85. $10 \log (10^{16}x) = 10(\log 10^{16} + \log x) =$
$10(16 + \log x) = 160 + 10 \log x$

CHECKING BASIC CONCEPTS 12.3 & 12.4 (P. 776)

1. (a) 4   **(b)** x **(c)** -3 **(d)** $\frac{1}{2}$

2.

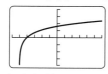

(a) $D = \{x \mid x > 0\}$; R: all real numbers   **(b)** 0
(c) Yes; for example, $\log \frac{1}{10} = -1$.   **(d)** No; negative numbers are not in the domain of $\log x$.

3. (a) $\log x + \log y$ **(b)** $\ln x - \ln y - \ln z$
(c) $2 \ln x$ **(d)** $2 \log x + 3 \log y - \frac{1}{2} \log z$

4. (a) $\log xy$ **(b)** $\ln \frac{2x}{y^3}$ **(c)** $\log_2 \frac{x^2 y^3}{z}$

SECTION 12.5 (PP. 784–787)

1. Add 5 to each side.
3. Take the common logarithm of each side.
5. x **7.** $2x$ **9.** No; $\log \frac{5}{4} = \log 5 - \log 4$ **11.** 1

13. 3   **15.** 6   **17.** 6   **19.** 4   **21.** $\frac{\log 124}{0.4} \approx 5.23$

23. 0   **25.** $\ln 7 \approx 1.95$ **27.** $\log \frac{35}{2} - 2 \approx -0.76$

29. $\frac{\log 20}{2 \log 3.1} \approx 1.32$ **31.** -1 **33.** $-1, 5$ **35.** $\frac{\ln 10}{3} \approx 0.77$

37. $\frac{2 \ln 2}{1 - \ln 2} \approx 4.52$ **39.** $\frac{2 \log 5}{0.5 \log 4 - \log 5} \approx -3.51$

41. (a) 1   **(b)** 1   **43. (a)** -2 **(b)** -2 **45.** -1

47. 0   **49.** 5   **51.** $\frac{\log 1.45}{\log 0.55} \approx -0.62$ **53.** $-1.84, 1.15$

55. 1.31   **57.** 100   **59.** $e^5 \approx 148.41$ **61.** 5,000,000

63. 16   **65.** $\frac{2^{2.3}}{5} \approx 0.98$ **67.** $10^{1.4} \approx 25.12$

69. $\frac{e^{11} - 1}{2} \approx 29,936.57$ **71.** 1   **73.** 5   **75.** 4

77. 3   **79. (a)** 2   **(b)** $e^{0.7} \approx 2.01$

81. (a) 2   **(b)** $\frac{1}{2}(10^{0.6}) \approx 1.99$

83. $10^{1.6} \approx 39.81$ **85.** $e - 1 \approx 1.72$ **87.** 9

89. 8 years

91. (a) About 8503; in 1994 about 8503 people were waiting for liver transplants.   **(b)** In about 1998

93. About 20,893 pounds

95. (a) About 203; in 1975 there were about 203 thousand bluefin tuna.   **(b)** In 1979   **(c)** In 1979

97. $a = 25, b = 3$
99. (a) About 1.67 acres   **(b)** About 1.67 acres
101. (a) Nonlinear; they do not increase at a constant rate.
(b) Each year the amount of fertilizer increases by a factor of 1.06, or by 6%.   **(c)** In 1968
103. 10^{-6} watts/square centimeter   **105.** About 7 miles

CHECKING BASIC CONCEPTS 12.5 (P. 787)

1. (a) $\log 20 \approx 1.30$ **(b)** $\frac{\log 147}{3 \log 2} \approx 2.40$
(c) $e^{4.1} \approx 60.34$ **(d)** 500
2. 8   **3.** 20 years

CHAPTER 12 REVIEW EXERCISES (PP. 791–793)

1. (a) 81   **(b)** $(f \circ g)(x) = 50x^2 - 2$
2. (a) -32 **(b)** $(f \circ g)(x) = \sqrt[3]{4x^3 - 6}$
3. (a) 0   **(b)** 3   **4. (a)** 3   **(b)** -2 **(c)** -1
5. $f(1) = f(-1) = 2$ **6.** $f(0) = f(2) = 1$ **7.** No
8. Yes
9. $(f \circ f^{-1})(x) = 2\left(\frac{x+9}{2}\right) - 9 = x$
$(f^{-1} \circ f)(x) = \frac{(2x-9)+9}{2} = x$
10. $(f \circ f^{-1}) = (\sqrt[3]{x-1})^3 + 1 = x$
$(f^{-1} \circ f)(x) = \sqrt[3]{(x^3+1)-1} = x$
11. $f^{-1}(x) = \frac{x}{5}$ **12.** $f^{-1}(x) = x + 11$
13. $f^{-1}(x) = \frac{x-7}{2}$ **14.** $f^{-1}(x) = \frac{4}{x}$
15.

x	10	8	7	3
$f^{-1}(x)$	0	1	2	3

$D = \{3, 7, 8, 10\}; R = \{0, 1, 2, 3\}$
16.
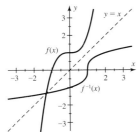
17. $\frac{1}{6}$; 36   **18.** 5; $\frac{5}{8}$ **19.** 3; $\frac{1}{81}$ **20.** 3; $\frac{1}{2}$
21.
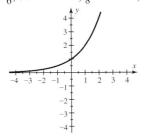
Exponential growth

22.
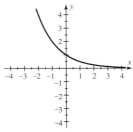
Exponential decay
23.
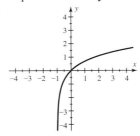
Logarithmic growth
24.
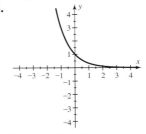
Exponential decay
25. (a) Exponential growth   **(b)** $f(x) = 5(2)^x$
26. (a) Linear growth   **(b)** $f(x) = 5x + 5$
27. $C = \frac{1}{2}, a = 2$ **28.** $k = 2$ **29.** \$2829.54
30. \$675,340.51   **31.** 399.67   **32.** 0.71   **33.** 3.48
34. 3.89   **35.** -3 **36.** 2   **37.** -4 **38.** 2   **39.** 1.813
40. -0.163 **41.** 4.787   **42.** -1.322 **43.** 7   **44.** $\frac{5}{9}$
45. $6 - x$ **46.** x^2 **47.** $\ln x + \ln y$ **48.** $\log x - \log y$
49. $2 \ln x + 3 \ln y$ **50.** $\frac{1}{2}\log x - 3 \log z$
51. $2 \log_2 x + \log_2 y - \log_2 z$ **52.** $\frac{1}{3}\log_3 x - \frac{1}{3}\log_3 y$
53. $\log 75$ **54.** $\log_4 (10x^2)$ **55.** $\ln \frac{x^2}{y^3}$ **56.** $\log xy$
57. $3 \log 6$ **58.** $2 \ln x$ **59.** $(2x) \log_2 5$
60. $(x + 1) \log_4 0.6$ **61.** 2   **62.** 4   **63.** $\ln 9 \approx 2.20$
64. $\frac{\log (0.2)}{\log (0.85)} \approx 9.90$ **65.** $e^{0.8} \approx 2.23$ **66.** $\frac{1}{2}e^5 \approx 74.21$
67. 10^{40} **68.** 100   **69.** $\frac{4 \log 2}{\log 3 - \log 2} \approx 6.84$ **70.** 6
71. (a) 3   **(b)** 3   **72. (a)** 4   **(b)** 4
73. (a) 64π; after 8 seconds, the balloon has a surface area of $64\pi \approx 201$ in².   **(b)** $(S \circ r)(t) = 8\pi t$
74. (a) Yes, different inputs result in different outputs.
(b) $f^{-1}(x) = \frac{x}{0.08}$ calculates the cost of an item whose sales tax is x dollars.

75. 7 years   **76.** $a = 100, b = 50$ **77.** $C = 3, a = 2$
78. 10^7
79. (a) $[0, 10, 2]$ by $[0, 4, 1]$

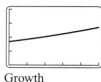

Growth
(b) About 2.24 million   **(c)** 2009
80. (a) 1000; there were 1000 bacteria/mL initially.
(b) About 495.11 minutes
81. (a) About 6.93 meters/second   **(b)** 12.18 meters

CHAPTER 12 TEST (PP. 794–795)

1. 6; $(f \circ g)(x) = 4(x + 7)^3 - 5(x + 7)$
2. (a) 3   **(b)** 3   **3.** $-5 \neq 5$, but $f(-5) = f(5) = 0$
4. $f^{-1}(x) = \frac{5 - x}{2}$
5.

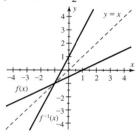

6.

x	8	6	4	2
$f^{-1}(x)$	1	2	3	4

$D = \{2, 4, 6, 8\}; R = \{1, 2, 3, 4\}$
7. $\frac{3}{16}$
8.

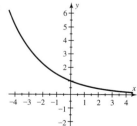

Exponential decay
9. (a) Exponential growth   **(b)** $f(x) = 3(2)^x$
10. (a) Linear growth   **(b)** $f(x) = 1.5x - 1$
11. $C = 1, a = \frac{1}{2}$ **12.** $1051.91   **13.** 4.16
14. $\frac{1}{2}$ **15.** 5.426

16.

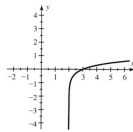

Shifted to the right 2 units
17. $3 \log x + 2 \log y - \frac{1}{2} \log z$ **18.** $\ln \frac{x^4 z}{y^5}$
19. $2x \log 7$ **20.** $1 - 3x$ **21.** $\ln 25 \approx 3.22$
22. $\log 50 \approx 1.70$ **23.** $10^{1.8} \approx 63.10$
24. $\frac{1}{5}e^9 \approx 1620.62$ **25.** $a = 5, b = 3$
26. (a) 4 million   **(b)** $4e^{0.45} \approx 6.27$; after 5 hours there
were about 6.27 million bacteria.   **(c)** Growth   **(d)** After
4.51 hours

CHAPTERS 1–12 CUMULATIVE REVIEW (PP. 797–800)

1. (a) Rational   **(b)** Irrational
2. $2 \times 5 \times 5 \times 5$ **3.** $2^3 - 7$
4. (a) 24   **(b)** 30
5. 4^5 **6.** Commutative   **7.** -18 **8.** $\{z \mid z < 0\}$
9. $\frac{4}{15}$ **10.** $\{x \mid x \geq -5\}$ **11.** $J = \frac{2z}{P - 1}$ **12.** $\frac{18}{25}, 0.72$
13. $y = \frac{5}{4}x + 1$ **14.** $x = 4$ **15.** 1
16.

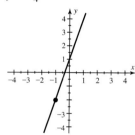

17. $y = 7x - 6$ **18.** $y = 3x + 5$ **19.** $(1, 1)$
20. No solutions   **21.** $(3, -1)$ **22.** Infinitely many
23.

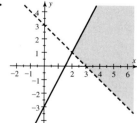

24. Zero   **25.** 4.29×10^{-4} **26.** d^4 **27.** $\dfrac{b^9}{64a^6}$

28. $\dfrac{4y^8}{x^5}$ **29.** $\dfrac{y^4}{4x^5}$ **30.** $3x^2 + 6x + 10 + \dfrac{5}{x-2}$

31. $2x(x-1)^2$ **32.** $(2a - 5b)(2a + 5b)$

33. $(2t - 3)(4t^2 + 6t + 9)$ **34.** $(2a^2 + 5)(2a - 1)$

35. $-\dfrac{5}{6}, 2$ **36.** $-\dfrac{2}{3}, \dfrac{2}{3}$ **37.** $-3, 0, 5$ **38.** $0, \dfrac{1}{2}$ **39.** 1

40. $\dfrac{x+2}{x-3}$ **41.** 3   **42.** -3 **43.** $\dfrac{x^3 + 3}{x^3 - 3}$ **44.** 40

45. $\{x \mid x \neq -3\}$ **46.** $f(x) = 4x - 3$ **47.** $(-5, 10]$

48. $[-2, 10]$ **49.** 2   **50.** 8 in^2 **51.** $(-4, 0, 2)$

52. $(0, 0, 1)$ **53.** $\dfrac{x^4}{y^6}$ **54.** x^3 **55.** $(b + 3)a\sqrt[3]{a^2 b}$

56. 20   **57.** $-6, 4$ **58.** 3   **59.** $3 + 5i$ **60.** $-i$

61. $(2, 1)$ **62.** $\dfrac{7}{2}$ **63.** $5, 8$ **64.** $\dfrac{1 \pm \sqrt{41}}{4}$

65. (a) $-1, 3$ **(b)** $a < 0$ **(c)** Positive

66. $(-\infty, -7] \cup [2, \infty)$

67. (a) 7   **(b)** $(g \circ f)(x) = 2x^2 - 3$

68. \$2367.10   **69.** 4   **70.** $2x$ **71.** $\dfrac{1}{2} \log x - 2 \log y$

72. $\ln(5x^3)$ **73.** $10^{11/6} \approx 68.13$ **74.** $\dfrac{\log 17}{3 \log 2} \approx 1.36$

75. 90°F   **76.** 96 or more

77. (a) -65 **(b)** The truck is moving toward Louisville at a rate of 65 mph.

78. 26°, 64°   **79.** About 7 billion

80. 30 min on elliptical trainer; 45 min running   **81.** 1400

82. $f(x) = -0.35x + 713.2$

83. Double cheeseburger: \$1; jumbo fries: \$1.50   **84.** 4%

85. 2.27 lb

86. (a) 1950   **(b)** 220 million Btu

87. $\sqrt{4200} \approx 64.8$ mph   **88.** About 11.3%

89. (a) 9.91 m/sec   **(b)** 8.52 m

CHAPTER 13: CONIC SECTIONS

SECTION 13.1 (PP. 809–811)

1. Parabola, ellipse, hyperbola   **3.** No

5. No; it does not pass the vertical line test.   **7.** left

9. circle; (h, k)

11.
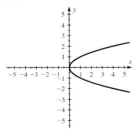
$(0, 0); y = 0$

13.
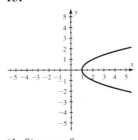
$(1, 0); y = 0$

15.
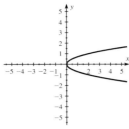
$(0, 0); y = 0$

17.
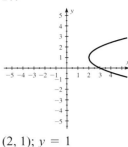
$(2, 1); y = 1$

19.
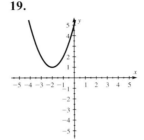
$(-2, 1); x = -2$

21.

$(-3, -1); y = -1$

23.
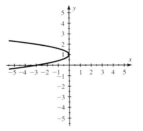
$(0, 1); y = 1$

25.
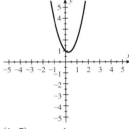
$\left(\dfrac{1}{4}, \dfrac{7}{8}\right); x = \dfrac{1}{4}$

27.
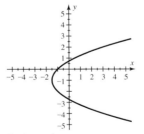
$\left(-\dfrac{3}{2}, -1\right); y = -1$

29.
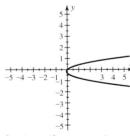
$\left(-\dfrac{1}{12}, -\dfrac{1}{6}\right); y = -\dfrac{1}{6}$

31. $y = x^2$ **33.** $x = (y + 1)^2 - 2$ **35.** Upward

37. Downward   **39.** $x \geq 0$ **41.** Two   **43.** 1

45. $x^2 + y^2 = 1$ **47.** $(x + 1)^2 + (y - 5)^2 = 9$

49. $(x + 4)^2 + (y + 6)^2 = 2$ **51.** $x^2 + y^2 = 16$

53. $(x + 3)^2 + (y - 2)^2 = 1$

55. 3; (0, 0)

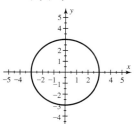

57. 3; (1, 3)

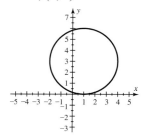

59. 5; (−5, 5)

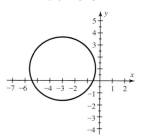

61. 3; (−3, 1)

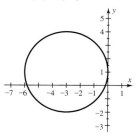

63. $\sqrt{7}$; (−3, 1)

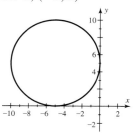

65. (a) [−40, 40, 10] by [−120, 120, 20]

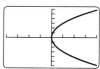

(b) 32 ft

67. (a) [−1.5, 1.5, 0.5] by [−1, 1, 0.5]

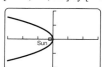

(b) 2.6 A.U., or 241,800,000 miles

SECTION 13.2 (PP. 819–821)

1.

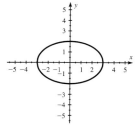

(*Answers may vary*)

3. horizontal   **5.** 2   **7.** left; right
9. They are the diagonals extended.

11.

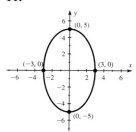

13.

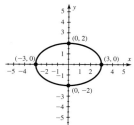

15.

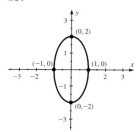

17.

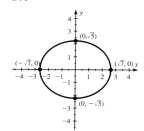

19.

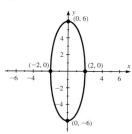

21.

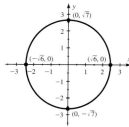

23. $\dfrac{x^2}{9} + \dfrac{y^2}{4} = 1$ **25.** $\dfrac{y^2}{25} + \dfrac{x^2}{16} = 1$

27.

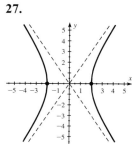

29.

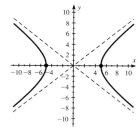

31.

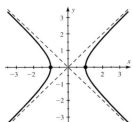

33.

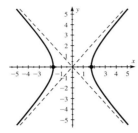

2. $(x - 1)^2 + (y + 2)^2 = 4$

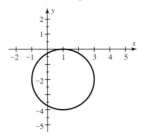

35.

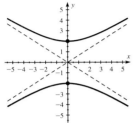

37.

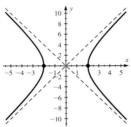

3. x-intercepts: ± 2; y-intercepts: ± 3

4. (a)

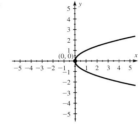

Parabola

39. $x^2 - y^2 = 1$ **41.** $\dfrac{y^2}{4} - \dfrac{x^2}{9} = 1$

43. (a) $A \approx 62.83$; $P \approx 28.45$
(b) $A \approx 11.75$; $P \approx 13.33$
45. (a) $[-60, 60, 10]$ by $[-40, 40, 10]$

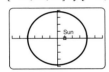

(b) $P \approx 243.9$ A.U., or about 2.27×10^{10} miles;
$A \approx 4733$ square A.U., or about 4.09×10^{19} square miles
47. Maximum: 668 miles; minimum: 340 miles
49. Height: 20 feet; width: 200 feet

(b)

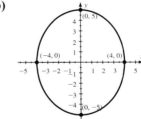

Ellipse

(c)

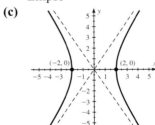

Hyperbola

CHECKING BASIC CONCEPTS 13.1 & 13.2 (P. 822)

1.

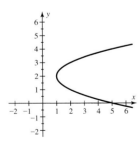

$(1, 2)$; $y = 2$

(d)

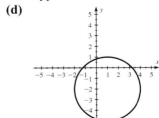

Circle (and ellipse)

SECTION 13.3 (PP. 829–831)

1. Any number
3. Two; the line intersects the circle twice.   **5.** No
7.

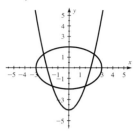

(*Answers may vary*)
9. $(1, 3), (-1, -3)$ **11.** $(0, -1), (0, 1)$
13. $(3, 6), (-3, -6)$ **15.** $(2, -1)$
17. $(-\sqrt{2}, 2), (\sqrt{2}, 2)$ **19.** $(-1, -2), (-2, 1)$
21. $(7, 3), (2, -2)$ **23.** $(-1, 2), (3, -6)$
25. $(-1, -1), (1, 1)$
27.   **29.**

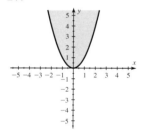   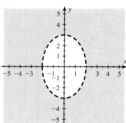

31.   **33.**

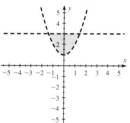   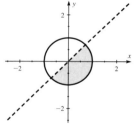

$(0, 2)$, (*answers may vary*) $\left(\frac{1}{2}, -\frac{1}{2}\right)$, (*answers may vary*)

35.   **37.**

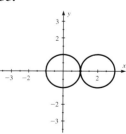   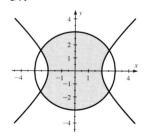

$(1, 0)$ $(0, 0)$, (*answers may vary*)

39. a.   **41.** $y \geq x^2$; $y < 4 - x$
43. $r = 1.6$ inches; $h \approx 4.97$ inches
45. (a) $h = \dfrac{3V}{\pi r^2}$; $h = \sqrt{\left(\frac{S}{\pi r}\right)^2 - r^2}$ **(b)** $r \approx 2.02$ feet,
$h \approx 7.92$ feet.; $r \approx 3.76$ feet, $h \approx 2.30$ feet

CHECKING BASIC CONCEPTS 13.3 (P. 831)

1. $(1, -1), (3, 3)$ **2.** 2
3. (a) $(0, 3), (4, 4)$ (*answers may vary*)
(b) $y \geq 2 - x$ and $y \leq 4 - x^2$
4.

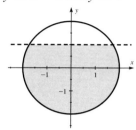

CHAPTER 13 REVIEW EXERCISES (PP. 833–836)

1.   **2.**

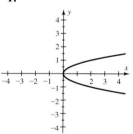   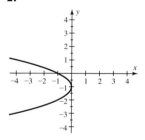

$(0, 0); y = 0$ $(0, -1); y = -1$
3.   **4.**

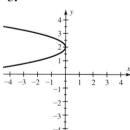   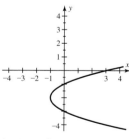

$(0, 2); y = 2$ $(-1, -2); y = -2$

5.

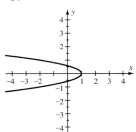

$(1, 0); y = 0$

6.

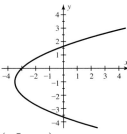

$\left(-\frac{7}{2}, -1\right); y = -1$

7. $x = y^2$ **8.** $(x + 2)^2 + (y - 2)^2 = 16$

9. $x^2 + y^2 = 1$ **10.** $(x - 2)^2 + (y + 3)^2 = 16$

11. $5; (0, 0)$

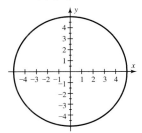

12. $3; (2, 0)$

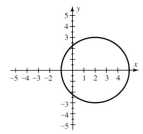

13. $\sqrt{5}; (-3, 1)$

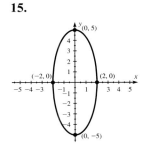

14. $3; (1, -1)$

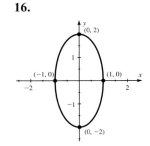

15.

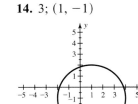

16.

17.

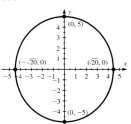

18.

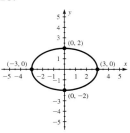

19. $\dfrac{y^2}{16} + \dfrac{x^2}{4} = 1$ **20.** $x^2 - \dfrac{y^2}{4} = 1$

21.

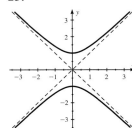

22.

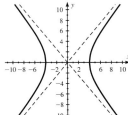

23.

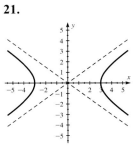

24.

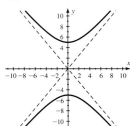

25. $(0, 3), (3, 0)$ **26.** $(-1, -2), (1, 2)$ **27.** $(0, 0), (2, 2)$

28. $(-2, -1), (-2, 1), (2, -1), (2, 1)$

29. $(-4, -4), (4, 4)$ **30.** $(0, -4), (4, 0)$

31. $(-1, 1), (1, 1)$ **32.** $(1, 2), (2, 5)$

33. $(-4, -12), (2, 0)$ **34.** $(-1, -1), (1, -1)$

35. $(2, 2), (-4, -4)$ **36.** $(1, 1), (0, 0)$

37.

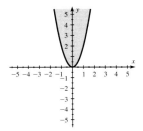

38.

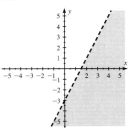

39.

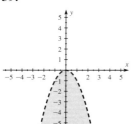

40.

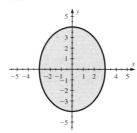

41.

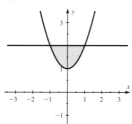

42.

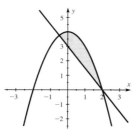

43.

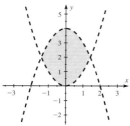

44.

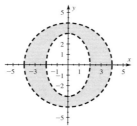

45. $y \geq x^2 - 2$; $y \leq 2 - x$ **46.** $y \geq x$; $x^2 + y^2 \leq 4$
47. (a) $xy = 1000$; $2x + 2y = 130$ **(b)** $x = 25$ inches;
$y = 40$ inches **(c)** $x = 25$ inches; $y = 40$ inches
48. (a) $xy = 60$; $y - x = 7$ **(b)** $x = 5$; $y = 12$
(c) $x = 5$; $y = 12$
49. $r = 1$ foot, $h \approx 15.92$ feet; yes
50. Either $r \approx 0.94$ inches, $h \approx 12.60$ inches or $r \approx 3.00$
inches, $h \approx 1.23$ inches; no
51. (a) $[-7.5, 7.5, 1]$ by $[-5, 5, 1]$

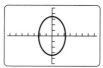

(b) $A \approx 24.33$, $P \approx 18.32$
52. (a) $[-3, 3, 1]$ by $[-2, 2, 1]$

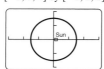

(b) $P \approx 9.55$ A.U., or about 8.9×10^8 miles;
$A \approx 7.26$ square A.U., or about 6.3×10^{16} square miles

CHAPTER 13 TEST (PP. 836–837)

1.

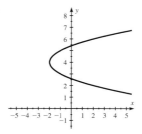

$(-2, 4)$; $y = 4$
2. $x = -y^2 + 1$ **3.** $(x - 2)^2 + (y + 4)^2 = 4$
4. $(x + 5)^2 + (y - 2)^2 = 100$
5. $r = 4$, center $= (-2, 3)$

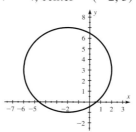

6.

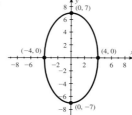

7. $\dfrac{x^2}{100} + \dfrac{y^2}{64} = 1$

8.

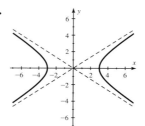

9. $(0, -4)$, $(4, 0)$ **10.** $(-1, -4)$, $(4, 1)$
11. $(-2, 4)$, $(2, 4)$
12.

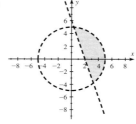

13. $y \leq 4 - x^2$; $y \geq x^2 - 4$

14. (a) $xy = 5000$; $2x + 2y = 300$ **(b)** 50 feet by 100 feet

15. Either $x \approx 22.08$ in., $y \approx 2.43$ in. or $x \approx 7.29$ in., $y \approx 22.24$ in.; no

16. (a) $[-30, 30, 10]$ by $[-20, 20, 10]$

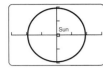

(b) 18.28 A.U., or about 1,700,040,000 miles

CHAPTER 14: SEQUENCES AND SERIES

SECTION 14.1 (PP. 846–848)

1. 1, 2, 3, 4; answers may vary.

3. function; the set of natural numbers   **5.** 6   **7.** $f(2)$

9. 1, 4, 9, 16   **11.** $\frac{1}{6}, \frac{1}{7}, \frac{1}{8}, \frac{1}{9}$ **13.** $\frac{5}{2}, \frac{5}{4}, \frac{5}{8}, \frac{5}{16}$

15. 9, 9, 9, 9   **17.** 1, 8, 27   **19.** $1, \frac{8}{5}, 2$ **21.** 2, 9, 20

23. $-2, -2, -2$ **25.** 7   **27.** 3, 4, 5, 3, 1

29. 6, 5, 4, 3, 2, 1

31. Numerical

n	1	2	3	4	5	6	7
a_n	2	3	4	5	6	7	8

Graphical

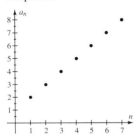

33. Numerical

n	1	2	3	4	5	6	7
a_n	0	2	6	12	20	30	42

Graphical

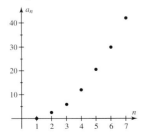

35. Numerical

n	1	2	3	4	5	6	7
a_n	2	4	8	16	32	64	128

Graphical

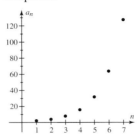

37. $a_n = 30n$ for $n = 1, 2, 3, \ldots, 7$

Graphical

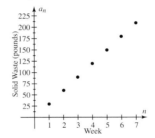

Numerical

n	1	2	3	4	5	6	7
a_n	30	60	90	120	150	180	210

39. (a) 1, 4, 9, 16   **(b)** 4, 8, 12, 16

41. (a) $20,000; $16,000

(b) $a_n = 25,000(0.8)^n$

(c)

n	1	2	3	4	5	6	7
a_n	20,000	16,000	12,800	10,240	8192	6553.6	5242.9

43. (a)

n	1	2	3	4	5	6	7
a_n	50	55	60	65	70	75	80

(b) $a_n = 50 + 5(n - 1)$ or $a_n = 45 + 5n$

(c) 160 seats

(d)

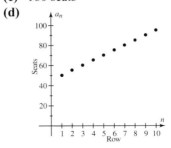

SECTION 14.2 (PP. 855–858)

1. linear   **3.** $a_n = 3n + 1$; 3   **5.** add; previous

7. 19; 4   **9.** $a_n = a_1(r)^{n-1}$ **11.** Yes; 10   **13.** Yes; -1

15. No   **17.** Yes; 3   **19.** Yes; -3 **21.** No
23. Yes; 1   **25.** No   **27.** Yes; 2   **29.** $a_n = -2n + 9$
31. $a_n = 4n - 6$ **33.** $a_n = -2n + 32$ **35.** 59
37. 21   **39.** Yes; 3   **41.** Yes; 0.8   **43.** No   **45.** Yes; 2
47. No   **49.** Yes; 4   **51.** Yes; 2   **53.** No
55. $a_n = 1.5(4)^{n-1}$ **57.** $a_n = -3(-2)^{n-1}$
59. $a_n = 1(4)^{n-1}$ **61.** 4374   **63.** 243
65. (a) 3000, 6000, 9000, 12,000, 15,000; arithmetic
(b) $a_n = 3000n$ **(c)** 60,000; when there are 20 people, the ventilation rate should be 60,000 cubic feet per hour.
(d)

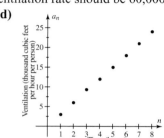

Yes
67. Arithmetic; the common difference is 2000.
69. (a) Geometric; the common ratio is 1.15.
(b) $a_n = 100,000(1.15)^{n-1}$ **(c)** $a_7 \approx 231,306$; at the beginning of the seventh year, it will be worth about $231,306.
(d)

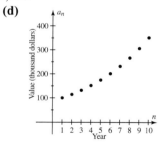

71. (a) Arithmetic; the common difference is 2.
(b) $a_n = 40 + 2(n - 1)$ or $a_n = 38 + 2n$ **(c)** 78

CHECKING BASIC CONCEPTS 14.1 & 14.2 (P. 858)

1. $\frac{1}{5}, \frac{1}{3}, \frac{3}{7}, \frac{1}{2}$
2. Graphical

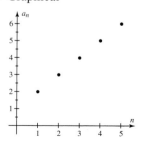

Numerical

n	1	2	3	4	5
a_n	2	3	4	5	6

3. (a) Arithmetic; $a_n = 3n - 5$ **(b)** Geometric; $a_n = 3(-2)^{n-1}$
4. $a_n = 2n + 3$ **5.** $a_n = 5(2)^{n-1}$

SECTION 14.3 (PP. 866–868)

1. series   **3.** arithmetic
5. $n\left(\frac{a_1 + a_n}{2}\right)$ or $\frac{n}{2}(2a_1 + (n-1)d)$ **7.** sum
9. arithmetic   **11.** 48   **13.** 820   **15.** -5 **17.** 3279
19. -85 **21.** 182   **23.** $91,523.93   **25.** $62,278.01
27. $2 + 4 + 6 + 8$; 20
29. $4 + 4 + 4 + 4 + 4 + 4 + 4 + 4$; 32
31. $1 + 4 + 9 + 16 + 25 + 36 + 49$; 140
33. $12 + 20$; 32   **35.** $\sum_{k=1}^{6} k^4$ **37.** $\sum_{k=1}^{5} \frac{1}{k^2}$
39. $\sum_{k=1}^{n} k = n\left(\frac{a_1 + a_n}{2}\right) = n\left(\frac{1 + n}{2}\right) = \frac{n(n + 1)}{2}$
41. (a) $8518 + 9921 + 10,706 + 14,035 + 14,307 + 12,249$ **(b)** 69,736
43. (a) $1, \frac{1}{2}, \frac{1}{4}, \frac{1}{8}, \frac{1}{16}$ **(b)** $\frac{1023}{512}$
45. 90 logs   **47.** $1,080,000   **49.** About 3.16 feet

SECTION 14.4 (PP. 873–874)

1. 5   **3.** 4   **5.** 24   **7.** $\frac{n!}{(n-r)! \, r!}$
9. $x^3 + 3x^2y + 3xy^2 + y^3$
11. $16x^4 + 32x^3 + 24x^2 + 8x + 1$
13. $a^5 - 5a^4b + 10a^3b^2 - 10a^2b^3 + 5ab^4 - b^5$
15. $x^6 + 3x^4 + 3x^2 + 1$ **17.** 6   **19.** 4   **21.** 2
23. 10   **25.** 5   **27.** 6   **29.** 1   **31.** 792   **33.** 126
35. 75,582   **37.** $m^3 + 3m^2n + 3mn^2 + n^3$
39. $x^4 - 4x^3y + 6x^2y^2 - 4xy^3 + y^4$
41. $8a^3 + 12a^2 + 6a + 1$
43. $x^5 + 10x^4 + 40x^3 + 80x^2 + 80x + 32$
45. $81 + 216m + 216m^2 + 96m^3 + 16m^4$
47. $8x^3 - 12x^2y + 6xy^2 - y^3$ **49.** a^8 **51.** $35x^4y^3$
53. $512m^9$

CHECKING BASIC CONCEPTS 14.3 & 14.4 (P. 874)

1. (a) Geometric   **(b)** Arithmetic
2. 312   **3.** -341 **4.** $x^4 - 4x^3y + 6x^2y^2 - 4xy^3 + y^4$
5. $x^3 + 6x^2 + 12x + 8$

CHAPTER 14 REVIEW EXERCISES (PP. 876–878)

1. 1, 8, 27, 64   **2.** 3, 1, -1, -3 **3.** $1, \frac{4}{5}, \frac{3}{5}, \frac{8}{17}$
4. $-2, 4, -8, 16$ **5.** $-2, 0, 4, 2$ **6.** 5, 3, 2, 1

7. Numerical

n	1	2	3	4	5	6	7
a_n	2	4	6	8	10	12	14

Graphical

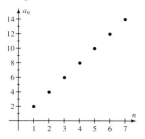

8. Numerical

n	1	2	3	4	5	6	7
a_n	−3	0	5	12	21	32	45

Graphical

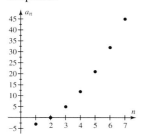

9. Numerical

n	1	2	3	4	5	6	7
a_n	2	1	0.5	0.25	0.125	0.0625	0.0313

Graphical

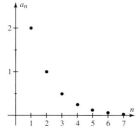

10. Numerical

n	1	2	3	4	5	6	7
a_n	1	1.4142	1.7321	2	2.2361	2.4495	2.6458

Graphical

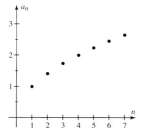

11. Yes; 5   **12.** No   **13.** No   **14.** Yes; $-\frac{1}{3}$
15. Yes; -3 **16.** No   **17.** Yes; -1 **18.** No
19. $a_n = 4n - 7$ **20.** $a_n = -5n + 7$ **21.** Yes; 4
22. No   **23.** No   **24.** Yes; 0.7   **25.** No   **26.** Yes; $-\frac{1}{3}$
27. No   **28.** Yes; 2   **29.** $a_n = 5(0.9)^{n-1}$
30. $a_n = 2(4)^{n-1}$ **31.** 216   **32.** 7.5   **33.** 3277
34. $\frac{511}{256}$ **35.** $3 + 5 + 7 + 9 + 11$ **36.** $\frac{1}{2} + \frac{1}{3} + \frac{1}{4} + \frac{1}{5}$
37. $1 + 8 + 27 + 64$
38. $-1 + (-2) + (-3) + (-4) + (-5) + (-6)$
39. $\sum_{k=1}^{20} k$ **40.** $\sum_{k=1}^{20} \frac{1}{k}$ **41.** $\sum_{k=1}^{9} \frac{k}{k+1}$ **42.** $\sum_{k=1}^{7} k^2$

43. $x^3 + 12x^2 + 48x + 64$
44. $16x^4 + 32x^3 + 24x^2 + 8x + 1$
45. $x^5 - 5x^4y + 10x^3y^2 - 10x^2y^3 + 5xy^4 - y^5$
46. $a^6 - 6a^5 + 15a^4 - 20a^3 + 15a^2 - 6a + 1$
47. 6   **48.** 10   **49.** 20   **50.** 4
51. $m^4 + 8m^3 + 24m^2 + 32m + 16$
52. $a^5 + 5a^4b + 10a^3b^2 + 10a^2b^3 + 5ab^4 + b^5$
53. $x^4 - 12x^3y + 54x^2y^2 - 108xy^3 + 81y^4$
54. $27x^3 - 54x^2 + 36x - 8$
55. $a_n = 45,000(1.1)^{n-1}$ for $n = 1, 2, 3, \ldots, 7$;
geometric
Numerical

n	1	2	3	4	5	6	7
a_n	45,000	49,500	54,450	59,895	65,885	72,473	79,720

Graphical

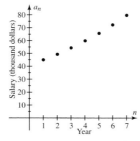

56. $a_n = 45,000 + 5000(n - 1)$ for
$n = 1, 2, 3, \ldots, 7$; arithmetic
Numerical

n	1	2	3	4	5	6	7
a_n	45,000	50,000	55,000	60,000	65,000	70,000	75,000

Graphical

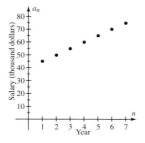

57. $a_n = 49n$ for $n = 1, 2, 3, \ldots, 7$
Graphical

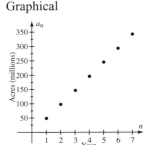

Numerical

n	1	2	3	4	5	6	7
a_n	49	98	147	196	245	294	343

58. (a) $a_n = 1087(1.025)^{n-1}$ **(b)** Geometric; the common ratio is 1.025. **(c)** About 1200; the average mortgage payment in 2000 was about $1200 per month.
(d)

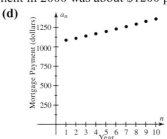

CHAPTER 14 TEST (PP. 878–879)

1. $\frac{1}{2}, \frac{4}{3}, \frac{9}{4}, \frac{16}{5}$ **2.** $-3, 2, 1, -2, 3$
3.

n	1	2	3	4	5	6	7
a_n	0	2	6	12	20	30	42

4. $16x^4 - 32x^3 + 24x^2 - 8x + 1$ **5.** Arithmetic; -3
6. Geometric; -2
7. $a_n = 2 - 3(n - 1)$ or $a_n = 5 - 3n$
8. $a_n = 2(1.5)^{n-1}$ **9.** Yes; 2.5 **10.** No **11.** 99
12. $\frac{463}{729}$ **13.** $6 + 9 + 12 + 15 + 18 + 21$ **14.** $\sum_{k=1}^{60} k^3$
15. 35 **16.** 10 **17.** 9180

18. $a_n = 159{,}700(1.04)^{n-1}$ for $n = 1, 2, 3, \ldots, 7$; geometric
Numerical

n	1	2	3	4	5	6	7
a_n	159,700	166,088	172,732	179,641	186,826	194,299	202,071

Graphical

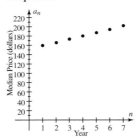

19. (a) $a_n = 2000(2)^{n-1}$ **(b)** Geometric; the common ratio is 2. **(c)** 64,000; after 30 days there are 64,000 worms.
(d)

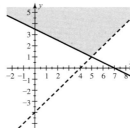

CHAPTERS 1–14 CUMULATIVE REVIEW (PP. 881–884)

1. Distributive **2.** $3x - 7 = x + 3$; 5 **3.** -26
4. $\left(\frac{20}{11}, \infty\right)$ **5.** $(-\infty, -1] \cup [11, \infty)$ **6.** $-15, 3$ **7.** $\frac{1}{8}$
8. $[-12, 9)$ **9.** $y = 3$ **10.** -3; 5 **11.** $y = \frac{3}{2}x + \frac{5}{2}$
12. $y = 2x - 8$ **13.** $f(x) = -2x + 1$ **14.** $(3, -2)$
15. $(5, 2)$
16.

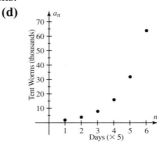

17. $8x^7 - 6x^6 + 10x^3$ **18.** $6z^2 - 13z - 28$ **19.** $\dfrac{y^9}{27x^5}$
20. $\dfrac{16a^8}{b^4}$ **21.** z^{10} **22.** $\dfrac{2y^3}{x^6}$ **23.** $(2x - 3y)(2x + 3y)$
24. $(a^2 + 4)(2a - 1)$ **25.** $-\frac{3}{4}, 1$ **26.** 0, 4, 6
27. $\dfrac{x - 2}{x + 1}$ **28.** $\dfrac{x + 4}{x - 2}$ **29.** 12 **30.** $\frac{1}{2}, 3$

31. $W = \frac{3C - 5R}{2}$ **32.** $\frac{1 + 2x}{1 - 4x}$

33. $D = \{-6, -2, 0, 2\}$
$R = \{0, 1, 3, 5\}$

34. $D = \{x \mid x \neq 8\}$ **35.** $(3, 3, -1)$ **36.** 17

37. $(xy - 2)\sqrt[3]{xy}$ **38.** 14   **39.** $-2, 8$ **40.** $\frac{3}{2}$

41. $14 + 2i$ **42.** $-2 - 2i$ **43.** $-\frac{1}{3}$

44. $y = 2(x + 2)^2 + 9; (-2, 9)$ **45.** $2 \pm 3i$

46. $-4, 8$

47. **(a)** $-3, 1$ **(b)** $a > 0$ **(c)** Positive

48. $(-\infty, \infty)$

49. **(a)** 65   **(b)** $(g \circ f)(x) = 3x^2 + 1$

50. $f^{-1}(x) = \frac{2x - 1}{3}$ **51.** $3\ln x + \frac{1}{2}\ln y$ **52.** $\log \frac{x}{4y}$

53. $10^{7/4} \approx 56.23$ **54.** $\frac{\log 5}{2 \log 4} \approx 0.58$

55.

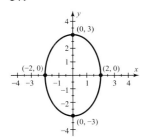

$(1, 3); y = 3$

56. $(3, -1); 2$

57.   **58.**

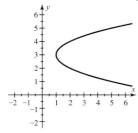   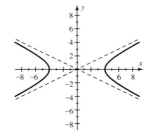

59. $\frac{y^2}{4} - \frac{x^2}{16} = 1$ **60.** $\frac{x^2}{16} + \frac{y^2}{4} = 1$

61. $(1, 2), (-1, 2)$

62.

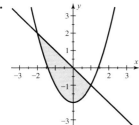

63. Arithmetic; -2 **64.** Geometric; 0.2

65. Geometric; 4   **66.** Arithmetic; 6   **67.** $a_n = 3n - 1$

68. $a_n = 4(3)^{n-1}$ **69.** 171   **70.** 683

71. $16x^4 + 96x^3 + 216x^2 + 216x + 81$

72. $8a^3 - 60a^2b + 150ab^2 - 125b^3$

73. $\sqrt{\dfrac{14}{\pi}} \approx 2.11$ in.

74.

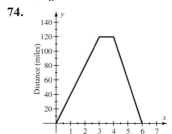

75. **(a)** $f(x) = 0.4x + 85$ **(b)** 188 lb

76. Airplane: 380 mph; wind: 20 mph

77. 9 by 12 ft   **78.** 36 min

79. **(a)** $xy = 96; 3x - y = 12$ **(b)** 8 and 12

80. 144

APPENDICES

APPENDIX B (PP. AP-18–AP-19)

1. $\{1, 2, 3, 4, 5, 6, 7\}$ **3.** $\{\text{Sunday, Saturday}\}$

5. $\{A, B, C, D, E, F, G\}$ **7.** $\varnothing$ **9.** $\{0, 2, 4, 6, 8, 10\}$

11. $\in$ **13.** $\notin$ **15.** $\in$ **17.** $\notin$ **19.** $\in$

21. $E = \{2, 4, 6, 8, 10\}$ **23.** $E = \{2, 4, 6, 8, 10, \dots\}$

25. $A = \{\text{Apple, Apricot}\}$ **27.** $A = \{\text{Algebra}\}$

29. False   **31.** True   **33.** True   **35.** True   **37.** True

39. True   **41.** False   **43.** True

45.   **47.**

49.   **51.**

53. $\{1, 2, 3, 4\}$ **55.** $\{a, b\}$ **57.** $\{a, b, c\}$

59. $\{a, b, c, d\}$ **61.** $\{5, 8\}$

63. $\{1, 2, 3, 4, 5, 6, 7, 8, 9, 10\}$ **65.** $\{4, 5, 6\}$ **67.** $\varnothing$

69. $\{1, 3, 5, 7, 9\}$ **71.** $\{4, 5, 6, 9\}$ **73.** $\varnothing$

75. $\{1, 3, 5, 7, 9\}$ **77.** $\{1, 2, 3, 5, 7, 8, 9, 10\}$ **79.** $\varnothing$

81. Infinite   **83.** Finite   **85.** Finite

APPENDIX C (PP. AP-24–AP-26)

1. linear programming **3.** feasible solutions **5.** vertex

7.
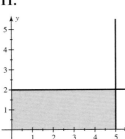

9.
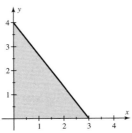

11.
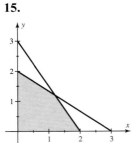

13.
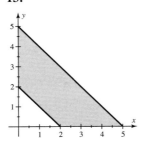

15.
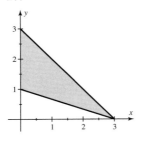

17.

19.
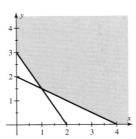

21. $R = 31$ **23.** $R = 15$ **25.** $C = 5$ **27.** $C = 2$
29. $R = 750$ **31.** $R = 12$ **33.** $R = 31.5$
35. $R = 16$ **37.** $C = 0$ **39.** $C = 32$ **41.** $C = 60$
43. 20 pounds of candy, 80 pounds of coffee
45. 1 ounce Brand X, 2 ounces Brand Y **47.** \$600

APPENDIX D (P. AP-28)

1. $x + 4 + \dfrac{3}{x-1}$ **3.** $3x - 1$ **5.** $x^2 + 3x + 2$
7. $2x^2 + 5x + 10 + \dfrac{19}{x-2}$ **9.** $x^2 + 2x + 6 + \dfrac{28}{x-4}$
11. $b^3 + b^2 + b + 1$

Glossary

absolute value A real number a, written $|a|$, is equal to its distance from the origin on the number line.

absolute value equation An equation that contains an absolute value.

absolute value function The function defined by $f(x) = |x|$.

absolute value inequality An inequality that contains an absolute value.

addends In an addition problem, the two numbers that are added.

addition property of equality If a, b, and c are real numbers, then $a = b$ is equivalent to $a + c = b + c$.

additive identity The number 0.

additive inverse (opposite) The additive inverse or opposite of a number a is $-a$.

adjacency matrix A matrix used to represent a map showing distances between cities.

algebraic expression An expression consisting of numbers, variables, operation symbols, such as $+$, $-$, $\times$, and $\div$, and grouping symbols, such as parentheses.

annuity A sum of money from which regular payments are made.

approximately equal The symbol $\approx$ indicates that two quantities are nearly equal.

arithmetic sequence A linear function given by $a_n = dn + c$ whose domain is the set of natural numbers.

arithmetic series The sum of the terms of an arithmetic sequence.

associative property for addition For any real numbers a, b, and c, $(a + b) + c = a + (b + c)$.

associative property for multiplication For any real numbers a, b, and c, $(a \cdot b) \cdot c = a \cdot (b \cdot c)$.

asymptotes of a hyperbola The two lines determined by the diagonals of the hyperbola's fundamental rectangle.

augmented matrix A matrix used to represent a system of linear equations; a vertical line is positioned in the matrix where the equal signs occur in the system of equations.

average The result of adding up the numbers of a set and then dividing the sum by the number of elements in the set.

axis of symmetry The line passing through the vertex of the parabola that divides the parabola into two symmetric parts.

base The value of b in the expression b^n.

basic principle of fractions When simplifying fractions, the principle that states $\frac{a \cdot c}{b \cdot c} = \frac{a}{b}$.

binary operation An operation that requires two numbers to calculate an answer.

binomial A polynomial with two terms.

binomial coefficient The expression $_nC_r = n!/((n - r)!r!)$, where n and r are nonnegative integers, $n \geq r$, that can be used to calculate the numbers in Pascal's triangle.

binomial theorem A theorem that provides a formula to expand expressions of the form $(a + b)^n$.

braces { }, used to enclose the elements of a set.

branches A hyperbola has two branches, a left branch and a right branch, or an upper branch and a lower branch.

byte A unit of computer memory, capable of storing one letter of the alphabet.

center of a circle The fixed point in the center of a circle.

center of an ellipse The midpoint of the major axis.

circle The set of points in a plane that are the same distance from a fixed point.

circumference The perimeter of a circle.

coefficient The numeric constant of a term.

coefficient of a monomial The number in a monomial.

common difference The value of d in an arithmetic sequence, $a_n = dn + c$.

common logarithmic function The function given by $f(x) = \log x$.

common logarithm of a positive number x Denoted $\log x$, it may be calculated as follows: if x is expressed as $x = 10^k$, then $\log x = k$, where k is a real number. That is, $\log 10^k = k$.

common ratio The value of r in a geometric sequence, $a_n = a_1(r)^{n-1}$.

commutative property for addition For any real numbers a and b, $a + b = b + a$.

commutative property for multiplication For any real numbers a and b, $a \cdot b = b \cdot a$.

complement The set containing all elements in the universal set that are not in A, denoted A'.

completing the square method An important technique in mathematics that involves adding a constant to a binomial so that a perfect square trinomial results.

complex conjugate The complex conjugate of $a + bi$ is $a - bi$.

complex fraction A rational expression that contains fractions in its numerator, denominator, or both.

complex number A complex number can be written in standard form as $a + bi$, where a and b are real numbers and i is the imaginary unit.

composite function If f and g are functions, then g of f, or composition of g and f, is defined by $(g \circ f)(x) = g(f(x))$ and is read "g of f of x."

composite number A natural number greater than 1 that is not a prime number.

compound inequality Two inequalities joined by the words *and* or *or*.

compound interest A type of interest paid at the end of each year using the formula $A = C(1 + r)^x$, where C is the amount deposited, r is the annual interest rate in decimal form, and x is the number of years the account will contain A dollars.

conic section The curve formed by the intersection of a plane and a cone.

conjugate The conjugate of $a + b$ is $a - b$.

consistent system A system of linear equations with at least one solution.

constant function A linear function with $a = 0$ and can be written as $f(x) = b$.

constant of proportionality (constant of variation) In the equation $y = kx$, the number k.

constraint In linear programming, an inequality which limits the objective function.

continuous growth Growth in a quantity that is directly proportional to the amount present.

Cramer's rule A method that uses determinants to solve linear systems of equations.

cube root The number b is a cube root of a if $b^3 = a$.

cube root function The function defined by $f(x) = \sqrt[3]{x}$.

cubic function A function f represented by $f(x) = ax^3 + bx^2 + cx + d$, where $a, b, c,$ and d are real numbers and $a \neq 0$.

decay factor The value of a in an exponential function, $f(x) = Ca^x$, when $0 < a < 1$.

degree A degree, (°), is 1/360 of a revolution.

degree of a monomial The sum of the exponents of the variables.

degree of a polynomial The degree of the term (or monomial) with highest degree.

dependent equations Equations in a linear system that have infinitely many solutions.

dependent variable The variable that represents the output of a function.

determinant A real number associated with a square matrix.

diagrammatic representation A function represented by a diagram.

difference The answer to a subtraction problem.

difference of two cubes Expression in the form $a^3 - b^3$, which can be factored as $(a - b)(a^2 + ab + b^2)$.

difference of two squares Expression in the form $a^2 - b^2$, which can be factored as $(a - b)(a + b)$.

dimension of a matrix The size expressed in number of rows and columns. For example, if a matrix has n rows and m columns, its dimension is $n \times m$ (n by m).

directly proportional A quantity y is directly proportional to x if there is a nonzero number k such that $y = kx$.

discriminant The expression $b^2 - 4ac$ in the quadratic formula.

distance formula The distance d between the points (x_1, y_1) and (x_2, y_2) in the xy-plane is $d = \sqrt{(x_2 - x_1)^2 + (y_2 - y_1)^2}$.

distributive properties For any real numbers $a, b,$ and c, $a(b + c) = ab + ac$ and $a(b - c) = ab - ac$.

dividend In a division problem, the number being divided.

divisor In a division problem, the number being divided *into* the dividend.

domain The set of all x-values of the ordered pairs in a function.

element of a matrix Each number in a matrix.

elements of a set The members of a set.

elimination method A symbolic method used to solve a system of equations that is based on the property that if "equals are added to equals the results are equal."

ellipse The set of points in a plane, the sum of whose distances from two fixed points is constant.

empty set (null set) A set that contains no elements.

equal sets If every element in a set A is in set B and every element in set B is in set A, then A and B are equal sets, denoted $A = B$.

equation A mathematical statement that two algebraic expressions are equal.

equivalent equations Equations that have the same solution set.

even root The nth root, $\sqrt[n]{a}$, where n is even.

expansion of a determinant by minors A method of finding a 3×3 determinant by using determinants of 2×2 matrices.

exponent The value of n in the expression b^n.

exponential decay When $0 < a < 1$, the graph of $f(x) = Ca^x$ models exponential decay.

exponential equation An equation that has a variable as an exponent.

exponential expression An expression that has an exponent.

exponential function with base a and coefficient C A function represented by $f(x) = Ca^x$, where $a > 0, C > 0,$ and $a \neq 1$.

exponential growth When $a > 1$, the graph of $f(x) = Ca^x$ models exponential growth.

extraneous solution A solution that does not satisfy the given equation.

factors In a multiplication problem, the two numbers multiplied.

factorial notation $n! = 1 \cdot 2 \cdot 3 \cdot \cdots \cdot n$ for any positive integer.

factoring The process of writing a polynomial as a *product* of lower degree polynomials.

factoring by grouping A technique that uses the associative and distributive properties by grouping four terms of a polynomial in such a way that the polynomial can be factored even though its greatest common factor is 1.

feasible solutions In linear programming, the set of solutions that satisfy the constraints.

finite sequence A function with domain $D = \{1, 2, 3, \ldots, n\}$ for some fixed natural number n.

finite series A series that contains a finite number of terms, and can be expressed in the form $a_1 + a_2 + a_3 + \cdots + a_n$ for some n.

finite set A set where the elements can be listed explicitly.

focus (plural foci) A fixed point used to determine the points that form a parabola, an ellipse, or a hyperbola.

FOIL A method for multiplying two binomials $(A + B)$ and $(C + D)$. Multiply First terms AC, Outside terms AD, Inside terms BC, and Last terms BD; then combine like terms.

formula A special type of equation used to calculate one quantity from given values of other quantities.

function A set of ordered pairs (x, y), where each x-value corresponds to exactly one y-value.

function notation The notation $y = f(x)$, where the input x produces output y.

fundamental rectangle The rectangle of a hyperbola whose four vertices are determined by either $(\pm a, \pm b)$ or $(\pm b, \pm a)$, where $\frac{x^2}{a^2} - \frac{y^2}{b^2} = 1$ or $\frac{y^2}{a^2} - \frac{x^2}{b^2} = 1$.

Gaussian elimination A method used to solve a linear system in which matrix row transformations are applied to an augmented matrix.

general term (nth term) of a sequence a_n, where n is a natural number in the domain of a sequence $a_n = f(n)$.

geometric sequence An exponential function given by $a_n = a_1(r)^{n-1}$, where n is a natural number and $r \neq 0$ or 1.

geometric series The sum of the terms of a geometric sequence.

graphical representation A graph of a function.

graphical solution A solution to an equation obtained by graphing.

greater than If a real number b is located to the right of a real number a on the number line, we say that b is greater than a, and write $b > a$.

greater than or equal to If a real number a is greater than or equal to b, denoted $a \geq b$, then either $a > b$ or $a = b$ is true.

greatest common factor (GCF) The term with the highest degree and greatest coefficient that is a factor of all terms in the polynomial.

growth factor The value of a in the exponential function, $f(x) = Ca^x$, when $a > 1$.

half-life The time it takes for a radioactive sample to decay to half its original amount.

horizontal line test If every horizontal line intersects the graph of a function f at most once, then f is a one-to-one function.

hyperbola The set of points in a plane, the difference of whose distances from two fixed points is constant.

identity property of 1 If any number a is multiplied by 1, the result is a, that is, $a \cdot 1 = 1 \cdot a = a$.

identity property of 0 If 0 is added to any real number, a, the result is a, that is, $a + 0 = 0 + a = a$.

imaginary number A complex number $a + bi$ with $b \neq 0$.

imaginary part The value of b in the complex number $a + bi$.

imaginary unit A number denoted i whose properties are $i = \sqrt{-1}$ and $i^2 = -1$.

improper fraction A fraction whose numerator is greater than its denominator.

inconsistent system A system of linear equations that has no solution.

independent equations Equations in a linear system that have one solution.

independent variable The variable that represents the input of a function.

index The value of n in the expression $\sqrt[n]{a}$.

index of summation The variable k in the expression $\sum_{k=1}^{n}$.

inequality When the equals sign in an equation is replaced with any one of the symbols $<$, $\leq$, $>$, or $\geq$.

infinite sequence A function whose domain is the set of natural numbers.

infinite set A set with infinitely many elements.

input An element of the domain of a function.

integers A set of numbers including natural numbers, their opposites, and 0, or $\ldots, -3, -2, -1, 0, 1, 2, 3, \ldots$.

intercept form A linear equation in the form $x/a + y/b = 1$.

intersection Denoted $A \cap B$ and read "A intersect B" is the set containing elements that belong to *both* A and B.

intersection-of-graphs method A graphical technique for solving two equations.

interval notation A notation for number line graphs that eliminates the need to draw the entire line.

inverse function If f is a one-to-one function, then f^{-1} is the inverse function of f, if $(f^{-1} \circ f)(x) = f^{-1}(f(x)) = x$, for every x in the domain of f, and $(f \circ f^{-1})(x) = f(f^{-1}(x)) = x$, for every x in the domain of f^{-1}.

inversely proportional A quantity y is inversely proportional to x if there is a nonzero number k such that $y = k/x$.

irrational numbers Real numbers that cannot be expressed as fractions, such as π or $\sqrt{2}$.

joint variation A quantity z varies jointly as x and y if there is a nonzero number k such that $z = kxy$.

leading coefficient In a polynomial of one variable, the coefficient of the monomial with highest degree.

least common denominator (LCD) The common denominator with the fewest factors.

least common multiple (LCM) The smallest number that two or more numbers will divide into evenly.

less than If a real number a is located to the left of a real number b on the number line, we say that a is less than b and write $a < b$.

less than or equal to If a real number a is less than or equal to b, denoted $a \leq b$, then either $a < b$ or $a = b$ is true.

like radicals Radicals that have the same index and the same radicand.

like terms Two terms, or monomials, that contain the same variables raised to the same powers.

linear equation An equation that can be written in the form $ax + b = 0$, where $a \neq 0$.

linear equation in two variables An equation that can be written in the form $Ax + By = C$, where A, B, and C are fixed numbers and A and B are not both equal to 0.

linear function A function f represented by $f(x) = ax + b$, where a and b are constants.

linear inequality A linear inequality results whenever the equals sign in a linear equation is replaced with any one of the symbols $<$, $\leq$, $>$, or $\geq$.

linear inequality in two variables When the equals sign in a linear equation of two variables is replaced with $<$, $\leq$, $>$, or $\geq$, a linear inequality in two variables results.

linear polynomial A polynomial of degree 1 that can be written as $ax + b$, where $a \neq 0$.

line graph The resulting graph when the data points in a scatterplot are connected with straight line segments.

linear programming problem A problem consisting of an objective function and a system of linear inequalities called constraints.

logistic function A function used to model growth of a population.

logarithm with base a of a positive number x Denoted by $\log_a x$, it may be calculated as follows: If x can be expressed as $x = a^k$, then $\log_a x = k$, where $a > 0$, $a \neq 1$, and k is a real number. That is, $\log_a a^k = k$.

logarithmic function with base a The function denoted $f(x) = \log_a x$.

lower limit In summation notation, the number representing the subscript of the first term of the series.

lowest terms A fraction is in lowest terms if its numerator and denominator have no factors in common.

main diagonal In an augmented matrix, the diagonal set of numbers from the upper left of the matrix to the lower right.

major axis The longest axis of an ellipse, which connects the vertices.

matrix A rectangular array of numbers.

method of substitution A symbolic method for solving a system of equations in which one equation is solved for one of the variables and then the result is substituted into the other equation.

minor axis The shortest axis of an ellipse.

minors The 2×2 matrices that are used to find a determinant of a 3×3 matrix.

monomial A number, a variable, or a product of numbers and variables raised to natural number powers.

multiplication property of equality If a, b, and c are real numbers with $c \neq 0$, then $a = b$ is equivalent to $ac = bc$.

multiplicative identity The number 1.

multiplicative inverse (reciprocal) The multiplicative inverse of a nonzero number a is $1/a$.

natural exponential function The function represented by $f(x) = e^x$, where $e \approx 2.71828$.

natural logarithm The base-e logarithm, denoted either $\log_e x$ or $\ln x$.

natural numbers The set of (counting) numbers expressed as $1, 2, 3, 4, 5, 6, \ldots$.

negative slope On a graph, the slope of a line that falls from left to right.

negative square root The negative square root is denoted $-\sqrt{a}$.

nonlinear data If data points do not lie on a (straight) line, the data are nonlinear.

nonlinear function A function that is *not* a linear function; its graph is not a line.

nth root The number b is an nth root of a if $b^n = a$, where n is a positive integer, and is denoted $\sqrt[n]{a} = b$.

nth term (general term) of a sequence Denoted $a_n = f(n)$.

null set (empty set) A set that contains no elements.

numerical representation A table of values for a function.

numerical solution A solution often obtained by using a table of values.

objective function The given function in a linear programming problem.

odd root The nth root, $\sqrt[n]{a}$, where n is odd.

one-to-one function A function f in which for any c and d in the domain of f, $c \neq d$ implies that $f(c) \neq f(d)$. That is, different inputs always result in different outputs.

opposite (additive inverse) The opposite, or additive inverse, of a number a is $-a$.

opposite of a polynomial The polynomial obtained by negating each term in a given polynomial.

optimal value In linear programming, the value that often results in maximum revenue or minimum cost.

ordered pair A pair of numbers written in parentheses (x, y), in which the order of the numbers is important.

ordered triple Can be expressed as (x, y, z), where x, y, and z are numbers and represent a solution to a linear system in three variables.

origin On the number line, the point associated with the real number 0; in the xy-plane, the point where the axes intersect, $(0, 0)$.

output An element of the range of a function.

parabola The U-shaped graph of a quadratic function that either opens upward, downward, to the right, or to the left.

parallel lines Two or more lines in the same plane that never intersect; they have the same slope.

Pascal's triangle A triangle made up of numbers in which there are 1s along the sides and each element inside the triangle is the sum of the two numbers above it.

percent change If the quantity changes from x to y, then the percent change is $[(y - x)/x] \times 100$.

perfect cube The value of a if there exists an integer b such that $b^3 = a$.

perfect nth power The value of a if there exists an integer b such that $b^n = a$.

perfect square The value of a if there exists an integer b such that $b^2 = a$.

perfect square trinomial A trinomial that can be factored as the square of a binomial, for example, $a^2 + 2ab + b^2 = (a + b)^2$ and $a^2 - 2ab + b^2 = (a - b)^2$.

perpendicular lines Two lines in a plane that intersect to form a right (90°) angle.

point–slope form The line with slope m passing through the point (x_1, y_1), given by $y - y_1 = m(x - x_1)$, or equivalently, $y = m(x - x_1) + y_1$.

polynomial The sum of one or more monomials.

polynomial functions in one variable Functions that contain a polynomial in one variable.

polynomials of one variable Polynomials that contain one variable.

positive slope On a graph, the slope of a line that rises from left to right.

power function A function that can be represented by $f(x) = x^p$, where p is a rational number.

prime number A natural number greater than 1 that has *only* itself and 1 as natural number factors.

principal square root The square root of a that is nonnegative, denoted $\sqrt{a}$.

probability A real number between 0 and 1. A probability of 0 indicates that an event is impossible, whereas a probability of 1 indicates that an event is certain.

product The answer to a multiplication problem.

proportion A statement that two ratios are equal.

Pythagorean theorem If a right triangle has legs a and b with hypotenuse c, then $a^2 + b^2 = c^2$.

quadrants The four regions determined by the xy-plane.

quadratic equation An equation that can be written as $ax^2 + bx + c = 0$, where a, b, and c are real numbers, with $a \neq 0$.

quadratic formula The solutions of the quadratic equation, $ax^2 + bx + c = 0$, $a \neq 0$, are given by $x = (-b \pm \sqrt{b^2 - 4ac})/(2a)$.

quadratic function A function f represented by $f(x) = ax^2 + bx + c$, where a, b, and c are real numbers, with $a \neq 0$.

quadratic inequality If the equals sign in a quadratic equation is replaced with $>$, $\geq$, $<$, or $\leq$, a quadratic inequality results.

quadratic polynomial A polynomial of degree 2 that can be written as $ax^2 + bx + c$, with $a \neq 0$.

quotient The answer to a division problem.

radical expression An expression that contains a radical sign.

radical sign The symbol $\sqrt{}$ or $\sqrt[n]{}$ for some n.

radicand The expression under the radical sign.

radius The fixed distance between the center and any point on the circle.

range The set of all y-values of the ordered pairs in a relation.

rate of change Slope can be interpreted as a rate of change. It indicates how fast the graph of a line is changing.

ratio A comparison of two quantities, expressed as a quotient.

rational equation An equation that contains one or more rational expressions.

rational expression A polynomial divided by a nonzero polynomial.

rational function A function defined by $f(x) = p(x)/q(x)$, where $p(x)$ and $q(x)$ are polynomials and the domain of f includes all x-values such that $q(x) \neq 0$.

rational number Any number that can be expressed as the ratio of two integers p/q, where $q \neq 0$; a fraction.

rationalizing the denominator The process of removing radicals from a denominator so that the denominator contains only rational numbers.

real numbers All rational and irrational numbers; any number that can be represented by decimal numbers.

real part The value of a in a complex number $a + bi$.

reciprocal (multiplicative inverse) The reciprocal of a nonzero number a is $1/a$.

rectangular coordinate system (xy-plane) The xy-plane used to plot points and graph data.

reduced row–echelon form A matrix form for representing a system of linear equations in which there are 1s on the main diagonal with 0s above and below each 1.

reflection If the point (x, y) is on the graph of a function, then $(x, -y)$ is on the graph of its reflection across the x-axis.

relation A set of ordered pairs.

rise The change in y between two points on a line, that is, $y_2 - y_1$.

root function In the power function, $f(x) = x^p$, if $p = 1/n$, where $n \geq 2$ is an integer, then f is also a root function, which is given by $f(x) = \sqrt[n]{x}$.

run The change in x between two points on a line, that is, $x_2 - x_1$.

scatterplot A graph of distinct points plotted in the xy-plane.

scientific notation A real number a written as $b \times 10^n$, where $1 \leq |b| < 10$ and n is an integer.

set A collection of things.

set-builder notation Notation to describe a set of numbers without having to list all of the elements. For example, $\{x \mid x > 5\}$ is read as "the set of all real numbers x such that x is greater than 5."

slope The ratio of the change in y (rise) to the change in x (run) along a line. The slope m of a line passing through the points (x_1, y_1) and (x_2, y_2) is $m = (y_2 - y_1)/(x_2 - x_1)$, where $x_1 \neq x_2$.

slope–intercept form The line with slope m and y-intercept b is given by $y = mx + b$.

solution Each value of the variable that makes the equation true.

solution set The set of all solutions to an equation.

solution to a system In a system of equations, an ordered pair, (x, y), that makes *both* equations true.

square matrix A matrix in which the number of rows and columns is equal.

square root The number b is a square root of a number a if $b^2 = a$.

square root function The function given by $f(x) = \sqrt{x}$, where $x \geq 0$.

square root property If k is a nonnegative number, then the solutions to the equation $x^2 = k$ are $x = \pm\sqrt{k}$. If $k < 0$, then this equation has no real solutions.

standard equation of a circle The standard equation of a circle with center (h, k) and radius r is $(x - h)^2 + (y - k)^2 = r^2$.

standard form of a complex number $a + bi$, where a and b are real numbers.

standard form of an equation for a line The equation $Ax + By = C$, where A, B, and C are fixed numbers, with A and B not both 0.

standard form of a quadratic equation The equation $ax^2 + bx + c = 0$, where $a \neq 0$.

standard viewing rectangle of a graphing calculator Xmin $= -10$, Xmax $= 10$, Xscl $= 1$, Ymin $= -10$, Ymax $= 10$, and Yscl $= 1$, denoted $[-10, 10, 1]$ by $[-10, 10, 1]$.

subscript The symbol x_1 has a subscript of 1 and is read "x sub one" or "x one".

subset If every element in a set B is contained in a set A, then we say that B is a subset of A, denoted $B \subseteq A$.

sum The answer to an addition problem.

summation notation Notation in which the uppercase Greek letter *sigma* represents the sum, for example,

$$\sum_{k=1}^{n} a_k = a_1 + a_2 + a_3 + \cdots + a_n.$$

sum of the first n terms of an arithmetic sequence Denoted S_n, is found by averaging the first and nth terms and then multiplying by n.

sum of the first n terms of a geometric sequence Given by $S_n = a_1(1 - r^n)/(1 - r)$, if its first term is a_1 and its common ratio is r, provided $r \neq 1$.

sum of two cubes Expression in the form $a^3 + b^3$, which can be factored as $(a + b)(a^2 - ab + b^2)$.

symbolic representation Representing a function with a formula; for example, $f(x) = x^2 - 2x$.

symbolic solution A solution to an equation obtained by using properties of equations; the resulting solution set is exact.

synthetic division A shortcut that can be used to divide $x - k$, where k is a number, into a polynomial.

system of linear equations in two variables A system of equations in which each equation can be written in the form $Ax + By = C$.

system of linear inequalities in two variables Two or more inequalities to be solved at the same time, the solution to which must satisfy both inequalities.

system of nonlinear equations Two or more equations, at least one of which is nonlinear.

system of nonlinear inequalities Two or more inequalities, at least one of which is nonlinear.

table of values An organized way to display the inputs and outputs of a function; a numerical representation.

term A number, a variable, or a product of numbers and variables raised to powers.

terms of a sequence $a_1, a_2, a_3, \ldots a_n, \ldots$ where the first term is $a_1 = f(1)$, the second term is $a_2 = f(2)$, and so on.

test value When graphing the solution set of an inequality, a point chosen to determine which region of the xy-plane to include in the solution set.

three-part inequality A compound inequality written in the form $a < x < b$, where $\leq$ may replace $<$.

translations The shifting of a graph upward, downward, to the right, or to the left in such a way that the shape of the graph stays the same.

transverse axis In a hyperbola, the line segment that connects the vertices.

trinomial A polynomial with three terms.

unary operation An operation that requires only one number.

union Denoted $A \cup B$ and read "A union B," it is the set containing every element of A *and* every element of B.

universal set A set that contains all elements under consideration.

upper limit In summation notation, the number representing the subscript of the last term of the series.

variable A symbol, such as x, y, or z, used to represent any unknown quantity.

varies directly A quantity y varies directly with x if there is a nonzero number k such that $y = kx$.

varies inversely A quantity y varies inversely with x if there is a nonzero number k such that $y = k/x$.

varies jointly A quantity z varies jointly as x and y if there is a nonzero number k such that $z = kxy$.

Venn diagrams Diagrams used to depict relationships between sets.

verbal representation A description of what a function computes in words.

vertex The lowest point on the graph of a parabola that opens upward or the highest point on the graph of a parabola that opens downward.

vertex form of a parabola The vertex form of a parabola with vertex (h, k) is $y = a(x - h)^2 + k$, where $a \neq 0$ is a constant.

vertical asymptote A vertical asymptote typically occurs in the graph of a rational function when the denominator of the rational expression is 0, but the numerator is not 0; it can be represented by a vertical line in the graph of a rational function.

vertical line test If every vertical line intersects a graph in no more than one point, then the graph represents a function.

vertices of an ellipse The endpoints of the major axis.

vertices of a hyperbola The endpoints of the transverse axis.

viewing rectangle (window) On a graphing calculator, the window that determines the x- and y-values shown in the graph.

whole numbers The set of numbers given 0, 1, 2, 3, 4, 5, ...

x-axis The horizontal axis in the xy-plane.

x-intercept The x-coordinate of a point where a graph intersects the x-axis.

Xmax Regarding the viewing rectangle of a graphing calculator, Xmax is the maximum x-value along the x-axis.

Xmin Regarding the viewing rectangle of a graphing calculator, Xmin is the minimum x-value along the x-axis.

Xscl The distance represented by consecutive tick marks on the x-axis.

y-axis The vertical axis in the xy-plane.

xy-plane (rectangular coordinate system) The system used to plot points and graph data.

y-intercept The y-coordinate of a point where a graph intersects the y-axis.

Ymax Regarding the viewing rectangle of a graphing calculator, Ymax is the maximum y-value along the y-axis.

Ymin Regarding the viewing rectangle of a graphing calculator, Ymin is the minimum y-value along the y-axis.

Yscl The distance represented by consecutive tick marks on the y-axis.

zero-product property If the product of two numbers is 0, then at least one of the numbers must be 0, that is, $ab = 0$ implies $a = 0$ or $b = 0$ (or both).

zero of a polynomial An x-value that results in an output of 0 when it is substituted into a polynomial; for example, the zeros of $f(x) = x^2 - 4$ are 2 and -2.

Bibliography

Baase, S. *Computer Algorithms: Introduction to Design and Analysis.* 2nd ed. Reading, Mass.: Addison-Wesley Publishing Company, 1988.

Beckmann, P. *A History of PI.* New York: Barnes and Noble, Inc., 1993.

Brown, D., and P. Rothery. *Models in Biology: Mathematics, Statistics and Computing.* West Sussex, England: John Wiley and Sons Ltd, 1993.

Burden, R., and J. Faires. *Numerical Analaysis.* 5th ed. Boston: PWS-KENT Publishing Company, 1993.

Callas, D. *Snapshots of Applications in Mathematics.* Deli, New York: State University College of Technology, 1994.

Conquering the Sciences. Sharp Electronics Corporation, 1986.

Eves, H. *An Introduction to the History of Mathematics.* 5th ed. Philadelphia: Saunders College Publishing, 1983.

Freedman, B. *Environmental Ecology: The Ecological Effects of Pollution, Disturbance, and Other Stresses.* 2nd ed. San Diego: Academic Press, 1995.

Garber, N., and L. Hoel. *Traffic and Highway Engineering.* Boston, Mass.: PWS Publishing Co., 1997.

Grigg, D. *The World Food Problem.* Oxford: Blackwell Publishers, 1993.

Haefner, L. *Introduction to Transportation Systems.* New York: Holt, Rinehart and Winston, 1986.

Harrison, F., F. Hills, J. Paterson, and R. Saunders. "The measurement of liver blood flow in conscious calves." *Quarterly Journal of Experimental Physiology* 71: 235–247.

Historical Topics for the Mathematics Classroom, Thirty-first Yearbook. National Council of Teachers of Mathematics, 1969.

Horn, D. *Basic Electronics Theory.* Blue Ridge Summit, Penn.: TAB Books, 1989.

Howells, G. *Acid Rain and Acid Waters.* 2nd ed. New York: Ellis Horwood, 1995.

Huffman, R. *Atmospheric Ultraviolet Remote Sensing.* San Diego: Academic Press, 1992.

Kincaid, D., and W. Cheney. *Numerical Analysis.* Pacific Grove, Calif.: Brooks/Cole Publishing Company, 1991.

Kraljic, M. *The Greenhouse Effect.* New York: The H. W. Wilson Company, 1992.

Lack, D. *The Life of a Robin.* London: Collins, 1965.

Lancaster, H. *Quantitative Methods in Biological and Medical Sciences: A Historical Essay.* New York: Springer-Verlag, 1994.

Mannering, F., and W. Kilareski. *Principles of Highway Engineering and Traffic Analysis.* New York: John Wiley and Sons, 1990.

Mar, J., and H. Liebowitz. *Structure Technology for Large Radio and Radar Telescope Systems.* Cambridge, Mass.: The MIT Press, 1969.

Mason, C. *Biology of Freshwater Pollution.* New York: Longman Scientific and Technical, John Wiley and Sons, 1991.

Meadows, D. *Beyond the Limits.* Post Mills, Vermont: Chelsea Green Publishing Co., 1992.

Miller, A., and R. Anthes. *Meteorology.* 5th ed. Columbus, Ohio: Charles E. Merrill Publishing Company, 1985.

Miller, A., and J. Thompson. *Elements of Meteorology.* 2nd ed. Columbus, Ohio: Charles E. Merrill Publishing Company, 1975.

Paetsch, M. *Mobile Communications in the U.S. and Europe: Regulation, Technology, and Markets.* Norwood, Mass.: Artech House, Inc., 1993.

Pearl, R., T. Edwards, and J. Miner. "The growth of *Cucumis melo* seedlings at different temperatures." *J. Gen. Physiol.* 17: 687–700.

Pennycuick, C. *Newton Rules Biology.* New York: Oxford University Press, 1992.

Pielou, E. *Population and Community Ecology: Principles and Methods.* New York: Gordon and Breach Science Publishers, 1974.

Pokorny, C., and C. Gerald. *Computer Graphics: The Principles behind the Art and Science.* Irvine, Calif.: Franklin, Beedle, and Associates, 1989.

Ronan, C. *The Natural History of the Universe.* New York: MacMillan Publishing Company, 1991.

Ryan, B., B. Joiner, and T. Ryan. *Minitab Handbook.* Boston: Duxbury Press, 1985.

Smith, C. *Practical Cellular and PCS Design.* New York: McGraw-Hill, 1998.

Stent, G. S. *Molecular Biology of Bacterial Viruses.* San Francisco: W. H. Freeman, 1963.

Taylor, J. *DVD Demystified.* New York: McGraw-Hill, 1998.

Taylor, W. *The Geometry of Computer Graphics.* Pacific Grove, Calif.: Wadsworth and Brooks/Cole, 1992.

Thomas, D. *Swimming Pool Operators Handbook.* National Swimming Pool Foundation of Washington, D.C., 1972.

Thomas, V. *Science and Sport.* London: Faber and Faber, 1970.

Thomson, W. *Introduction to Space Dynamics.* New York: John Wiley and Sons, 1961.

Toffler, A., and H. Toffler. *Creating a New Civilization: The Politics of the Third Wave.* Kansas City, Mo.: Turner Publications, 1995.

Triola, M. *Elementary Statistics.* 7th ed. Reading, Mass.: Addison-Wesley Publishing Company, 1998.

Tucker, A., A. Bernat, W. Bradley, R. Cupper, and G. Scragg. *Fundamentals of Computing Logic: Problem Solving, Programs, and Computers.* New York: McGraw-Hill, 1995.

Turner, R. K., D. Pierce, and I. Bateman. *Environmental Economics, An Elementary Approach.* Baltimore: The Johns Hopkins University Press, 1993.

Van Sickle, J. *GPS for Land Surveyors.* Chelsey, Mich.: Ann Arbor Press, 1996.

Varley, G., and G. Gradwell. "Population models for the winter moth." *Symposium of the Royal Entomological Society of London* 4: 132–142.

Wang, T. *ASHRAE Trans.* 81, Part 1 (1975): 32.

Weidner, R., and R. Sells. *Elementary Classical Physics,* Vol. 2. Boston: Allyn and Bacon, Inc., 1965.

Williams, J. *The Weather Almanac 1995.* New York: Vintage Books, 1994.

Wright, J. *The New York Times Almanac 1999.* New York: Penguin Group, 1998.

Zeilik, M., S. Gregory, and D. Smith. *Introductory Astronomy and Astrophysics.* 3rd ed. Philadelphia: Saunders College Publishers, 1992.

Index

Videotape and CD Index

	Examples	Exercises
1.1	1, 4, 7	43, 47, 51, 53, 63, 69, 71
1.2	3, 11ab, 13	19, 41, 45, 49, 67, 71, 83, 91
1.3	2, 4	17, 19, 21, 29, 61, 65, 67, 73
1.4	1, 5, 7, 9	69, 71, 73, 89
1.5	2, 4, 6	15, 19, 47, 49, 57, 79, 83
1.6	4, 6	15, 19, 25, 31, 37, 43, 47, 53, 61, 67
1.7	4, 6, 9	13, 15, 21, 23, 29, 33, 53, 55, 59
1.8	3, 5	7, 9, 13, 15, 21, 23, 41, 43, 49, 55, 59
2.1	2, 4	13, 17, 21, 31, 35, 39
2.2	1, 8	29, 33, 37, 41, 47, 51
2.3	3, 8, 10	11, 15, 17, 21, 27, 29, 31
2.4	3, 6, 9, 10	11, 13, 15, 31, 45, 53
2.5	5, 6, 11	13, 15, 17, 25, 29, 33, 49, 71, 77, 83
3.1	1, 5	19, 21, 29, 31
3.2	1, 3, 5	41, 49, 61
3.3	2, 6	11, 15, 23, 31, 43, 45, 81
3.4	2, 4, 10	25, 31, 35, 53, 85
3.5	5, 6, 8	15, 27, 35, 39, 53
3.6	4, 5	13, 17, 21, 23, 29
3.7	2, 5	13, 17, 21, 33
4.1	1, 4	11, 17, 19, 35
4.2	3, 4	9, 15, 31, 75
4.3	4, 5	19, 21, 27, 31, 57
4.4	2, 4	17, 19, 33, 69, 73
5.1	1, 2, 8	25, 29, 37, 43, 51, 59, 65
5.2	5, 7, 9	23, 25, 31, 33, 39, 45, 65, 67
5.3	2, 4, 8	9, 13, 25, 39, 47, 51, 63, 69
5.4	4, 6	11, 15, 23, 35, 39, 41, 49
5.5	3, 8	11, 25, 37, 57, 63, 67, 73, 79, 89, 93, 97, 105, 109
5.6	2, 4	11, 19, 29, 33
6.1	1, 4	17, 19, 25, 41, 49, 55
6.2	3	15, 23, 27, 31, 37
6.3	1, 3	11, 15, 21, 25, 27, 39
6.4	3	15, 19, 25, 31, 37, 45, 49
6.5	3, 6	11, 15, 25, 29, 35, 45
6.6	2, 5	11, 17, 21, 27, 33, 43, 63

	Examples	Exercises
7.1	2, 4, 6	11, 13, 17, 51, 57, 67, 69
7.2	1, 4	15, 17, 27, 31, 35, 37, 41, 43
7.3	1, 5	23, 27, 29, 37, 41, 45, 53
7.4	1, 2	55, 59, 63, 65, 69, 73, 77, 83
7.5	1	27, 31, 35, 39, 43
7.6	3, 6, 8	9, 15, 21, 27, 57
7.7	1, 4	27, 29, 71
8.1	1, 2ac, 8	19
8.2	—	7, 9, 11, 25, 33, 39, 53, 55, 73
8.3	1	37, 39
8.4	1, 4, 8	61, 63, 73, 75, 81
8.5	—	7, 15, 21, 29, 39, 53, 91
9.1	2, 3, 4, 6	23
9.2	1b, 2a, 4	13
9.3	—	3, 9, 11, 19, 23, 31
10.1	2, 4, 5, 7ab, 8, 9, 10	65
10.2	—	11, 25, 35, 39, 77, 81, 97, 101
10.3	—	1, 9, 31, 47, 49, 57, 61, 65, 75
10.4	5ab, 9	15, 47
10.5	1, 4, 6, 8, 10	31
10.6	1, 4, 5	21, 25, 29, 31, 33
11.1	2, 3, 6	19, 25, 47, 69
11.2	1, 3	21, 43, 49, 57, 59
11.3	1c, 2a, 5	43, 47, 61
11.4	1	9, 15, 33, 55, 67
11.5	1, 4, 6	53
11.6	1	11, 15, 19, 21
12.1	1, 4, 7, 8	23, 45, 69, 77
12.2	—	9, 11, 13, 17, 21, 23, 57, 75
12.3	1, 3, 5, 6, 7	29
12.4	7, 8, 9	11, 13, 17, 19, 31, 33, 41, 43
12.5	—	15, 23, 47, 57, 67, 87, 95, 99
13.1	—	13, 17, 47, 59, 61
13.2	1, 2, 3	—
13.3	1, 2, 7	15
14.1	1, 3, 4	33
14.2	2, 5	35, 57, 71
14.3	—	11, 15, 19, 23, 27, 33, 37, 47
14.4	1, 2, 3, 4a	47

Library of Functions
Basic Functions

Several important functions are used in algebra. The following provides symbolic, numerical, and graphical representations for several of these basic functions.

Absolute Value Function: $f(x) = |x|$

x	-2	-1	0	1	2
$|x|$	2	1	0	1	2

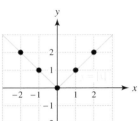

Domain: $(-\infty, \infty)$
Range: $[0, \infty)$

Square Function: $f(x) = x^2$

x	-2	-1	0	1	2
x^2	4	1	0	1	4

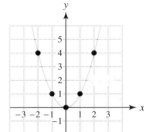

Domain: $(-\infty, \infty)$
Range: $[0, \infty)$

Cube Function: $f(x) = x^3$

x	-2	-1	0	1	2
x^3	-8	-1	0	1	8

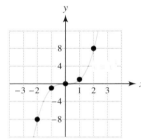

Domain: $(-\infty, \infty)$
Range: $(-\infty, \infty)$

Square Root Function: $f(x) = \sqrt{x}$

x	0	1	4	9
$\sqrt{x}$	0	1	2	3

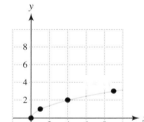

Domain: $[0, \infty)$
Range: $[0, \infty)$

Cube Root Function: $f(x) = \sqrt[3]{x}$

x	-8	-1	0	1	8
$\sqrt[3]{x}$	-2	-1	0	1	2

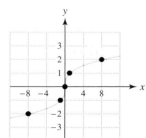

Domain: $(-\infty, \infty)$
Range: $(-\infty, \infty)$

Reciprocal Function: $f(x) = \dfrac{1}{x}$

x	-2	-1	0	1	2
$\dfrac{1}{x}$	$-\dfrac{1}{2}$	-1	$-$	1	$\dfrac{1}{2}$

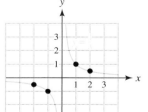

Domain: $(-\infty, 0) \cup (0, \infty)$
Range: $(-\infty, 0) \cup (0, \infty)$